Handbook of POROUS MEDIA

Handbook of POROUS MEDIA

edited by

Kambiz Vafai

The Ohio State University
Columbus, Ohio

assistant editor

Hamid A. Hadim

Stevens Institute of Technology
Hoboken, New Jersey

Marcel Dekker, Inc. New York • Basel

Library of Congress Cataloging-in-Publication Data

Handbook of porous media / edited by Kambiz Vafai.
p. cm.
ISBN 0-8427-8886-9 (alk. paper)
1. Porous materials–Handbooks, manuals, etc. I. Vafai, K. (Kambiz)
TA418.9.P6 H36 2000
620.1′16–dc21 00-031591

ISBN: 0-8247-8886-9

This book is printed on acid-free paper.

Headquarters
Marcel Dekker, Inc.
270 Madison Avenue, New York, NY 10016
tel: 212-696-9000; fax: 212-685-4540

Eastern Hemisphere Distribution
Marcel Dekker AG
Hutgasse 4, Postfach 812, CH-4001 Basel, Switzerland
tel: 41-61-261-8482; fax: 41-61-261-8896

World Wide Web
http://www.dekker.com

The publisher offers discounts on this book when ordered in bulk quantities. For more information, write to Special Sales/Professional Marketing at the headquarters address above.

Current printing (last digit):
10 9 8 7 6 5 4 3 2 1

PRINTED IN THE UNITED STATES OF AMERICA

Preface

Fundamental and applied research in heat and mass transfer in porous media has generated increasing interest over the past three decades because of the importance of porous media in many engineering applications. Consequently, a large amount of literature has been generated on this subject. Significant advances have been made in modeling fluid flow and heat transfer in porous media including several important physical phenomena. The non-Darcy effects of inertia and solid boundaries on momentum and energy transport in porous media have been studied extensively for various geometrical configurations and boundary conditions, and most of the studies used what is now commonly known as the Brinkman-Forchheimer-extended Darcy or the generalized model.

Other important topics that have received significant attention include porosity variation, thermal dispersion, the effects of local thermal equilibrium between the fluid phase and the solid phase, partially porous configurations, and anisotropic porous media. Advanced measurement techniques have been developed, for example, more efficient measurement of effective thermal conductivity, flow and heat transfer measurement, and flow visualization. The main objective of this handbook is to compile and present all the important up-to-date research information related to heat transfer in porous media, including practical applications for analysis and design of engineering devices and systems involving porous media. It also describes recent studies related to current and future challenges for further advances in fundamental as well as applied research in porous media including random porous media (e.g., fractured media), multiphase flow and heat transfer, turbulent flow and heat transfer, improved measurement and flow visualization techniques, and improved design correlations.

It is important to recognize that different models can be found in the literature and in the present handbook in the area of fluid flow, heat and mass transfer in porous media. An in-depth analysis of these models is essential in resolving any confusion in utilizing them (see Tien, C. L., and Vafai, K. 1989. Convective and radiative heat transfer in porous media. *Advances in Applied Mechanics*, **27**: 225–282). In a recent study, analysis of variants within the transport models for fluid flow and heat transfer in porous media is presented in an article by Alazmi and Vafai (see Alazmi, B. and Vafai, K. 2000. Analysis of variants within the porous media transport models. *Journal of Heat Transfer*, **122**, In Press). In that work, the pertinent models for fluid flow and heat transfer in porous media for four major categories were analyzed. Another important aspect of modeling in porous media relates to interface conditions between a porous medium and a fluid layer. As such, analysis of fluid flow and heat transfer in the neighborhood of an interface region for the pertinent interfacial models is presented in another article by Alazmi and Vafai (see Alazmi, B. and Vafai, K. 2000. Analysis of fluid flow and heat transfer interfacial conditions between a porous medium and a fluid layer, to appear in *International Journal of Heat and Mass Transfer*). Finally, competing models for multiphase transport models in porous media were analyzed in detail in Vafai and Sozen (Vafai, K. and Sozen, M. 1990. A comparative analysis of multiphase transport models in porous media. *Annual Review of Heat Transfer*, **3**, 145–162). In that work, a critical analysis of various heat and mass transfer models including the phase change process was presented. These three studies provide some clarification and insight for understanding several pertinent aspects of modeling of transport phenomena in porous media utilized in the literature and this handbook.

This handbook is targeted at researchers and practicing engineers, as well as beginners in this field. A leading expert in the related subject area presents each topic. An attempt has been made to present the topics in a cohesive, homogeneous yet complementary way with common format. Nomenclature common to various sections was used as much as possible.

The handbook is arranged into eight sections with a total of 19 chapters. The material in Part I covers fundamental topics of transport in porous media, including theoretical models of fluid flow and the local volume averaging technique, capillary and viscous effects in porous media, and application of fractal and percolation concepts in characterizing porous materials. Part II covers basic aspects of conduction in porous media. In Part III, various aspects of forced convection in porous media including numerical modeling are explored. Natural convection, thermal stability, and double diffusive convection in porous media are reviewed in Part IV. Part V presents mixed convection in porous media. Part VI discusses radiative transfer in porous media, and Part VII covers turbulence in porous media. The final part covers several important applications of transport in porous media, including packed bed chemical reactors, environmental applications (e.g., soil remediation), and drying and liquid composite molding (e.g., RTM and SRIM), which has received significant recent interest. Other applications, such as forced convection heat transfer enhancement, are reviewed along with other material presented in earlier chapters.

Chapter 1 deals with the basic governing equations describing conductive and convective heat transfer for flow in rigid porous media. Both homogeneous and heterogeneous thermal sources are considered in the analysis, which leads to Darcy-scale, or volume-averaged, transport equations for the temperature and velocity. Closure problems are developed that can be used to predict the values of the effective transport coefficients. The problem of local thermal equilibrium is considered in detail, and the constraints associated with this condition are developed. The one-equation model that is associated with local thermal equilibrium is derived, and predicted values of the longitudinal thermal dispersion coefficient are compared with experimental values. When the constraints associated with local thermal equilibrium are not valid, a two-equation model is required to accurately describe the heat transfer process. This two-equation model is also developed in this chapter.

In Chapter 2, displacement of a fluid by another immiscible fluid under quasi-static conditions and in low Reynolds number flow is considered. The microscopic aspects are analyzed in terms of capillary models, and, whenever feasible, the macroscopic phenomenological displacement relationships are either derived from or interpreted in terms of the capillary models. Particular importance is attached to waterflooding of oil and to mobilization of trapped oil blobs by the mechanism of spreading oil on water in the presence of an inert gas. Steady, concurrent two-phase flow is discussed, with particular emphasis on the effects of contact angle, interfacial tension, and viscosities on the relative permeabilities and on the widely debated issue of viscous coupling.

Finishing the first part is Chapter 3. It begins with a consideration of field-scale porous media that are highly heterogeneous at many length scales, ranging from microscopic to macroscopic and megascopic. Modeling flow and transport in such porous media depend critically on their correct characterization and modeling, which have been hampered, however, by an insufficient amount of data, variations in the data, and their great uncertainty. Since about a decade ago, it has become increasingly clear that the distributions of the heterogeneities of field-scale porous media follow fractal stochastic processes. An important consequence of this is that the heterogeneities contain long-range correlations whose extent may be as large as the linear size of the system. In addition, it has been firmly established that for highly heterogeneous porous media, percolation theory provides a realistic description of the media and of the flow and transport processes that take place there. Percolation theory quantifies the effect of the connectivity of various zones or regions of a system on its large-scale properties. This chapter attempts to summarize the progress that has been made in applying such concepts and ideas to characterization and modeling of field-scale porous media and flow and transport processes.

In Chapter 4, conduction in porous media is considered. Heat conduction in materials consisting of substances with different thermal properties is encountered frequently in many engineering applications. However, the traditional treatment of heat conduction is still based largely on the lumped mixture model under the local thermal equilibrium assumption. With these assumptions, the issue becomes the determination of effective thermal conductivity. This chapter extends the scope of validity in treating the transient heat conduction in porous materials saturated with fluids to the domain of local thermal non-equilibrium with a two-equation model. The conditions for the validity of the local thermal equilibrium assumption are assessed by exercising the two-equation model for one-dimensional heat conduction in a porous slab subjected to a sudden change of boundary condition. The deviation from local thermal equilibrium in the two phases is discussed based on the parameters such as interfacial transfer, thermal conduction ratio, heat capacity ratio and porosity. It is noted that the treatment can be extended easily to a multicomponent porous medium.

The section on forced convection begins with Chapter 5. This chapter presents a review of forced convective heat transfer in porous media for boundary-layer flows over a flat plate embedded in a fluid-saturated, porous medium and internal forced convection through parallel-plate channels, circular pipes, and annular ducts filled or partially filled with porous media. Models for the momentum and energy transport that have recently been applied are examined. Variable porosity and thermal dispersion effects, as well as departure from local thermal equilibrium, are discussed. Forced convection in porous-filled ducts saturated with non-Newtonian fluids is also documented at the end of the chapter.

Chapter 6 concentrates on the analytical study of fully developed steady laminar forced convection flow in a number of classical composite configurations. Both Poiseuille flow and Couette flow in composite parallel-plate channels are considered. Poiseuille flow is investigated in four composite configurations: (1) a channel with a fluid core, (2) a channel with a porous core, (3) a channel heated from the fluid side, and (4) a channel heated from the porous side. Couette flow is investigated in two composite configurations. It is assumed that momentum flow in the porous region can be described by the Brinkman-Forchheimer-extended Darcy equation. Utilizing the boundary-layer technique, analytical solutions are obtained for the velocity and temperature distributions, as well as for the Nusselt number. These new analytical solutions make it possible to investigate possibilities for heat transfer enhancement in composite channels. These new solutions are also valuable for gaining deeper insight into and understanding of the transport processes at the porous/fluid interface and of ways for testing numerical codes.

Chapter 7 provides the basic concepts and fundamentals of convective boundary layers in porous media for external flows. The governing equations, along with their important simplifications, are presented so as to include the dimensionless parameters that arise and the basic nature of the transport processes. Free and mixed convection over vertical and horizontal surfaces embedded in a porous medium are discussed in detail, and the resulting heat transfer expressions are presented. The case of free convection boundary layer near the forward stagnation point of a cylindrical surface in a porous medium has also been considered. Emphasis is placed on the modern developments in this field, including numerical and analytical techniques, and a considerable number of recent references are also included.

A post-1993 review of publications related to the application of porous media for enhancing incompressible forced-convection heat transfer is presented in Chapter 8. A brief introduction to the fundamentals of convection heat transfer through porous media is followed by the description of enhancing designs for particular applications grouped into five main categories: (1) porous inserts and cavities, (2) microsintering, porous coatings, and porous fins, (3) microchannels, (4) permeable fences and perforated baffles, and (5) cylinder arrays.

The flow and thermal convection in rotating porous media is developed and a systematic classification and identification of the relevant problems in such a configuration is introduced in Chapter 9. An initial distinction between rotating flows in isothermal heterogeneous porous systems and free convection in homogeneous non-isothermal porous systems provides the two major classes of problems to be considered in this chapter. Examples of solutions to selected problems are presented, highlighting the significant impact of rotation on the flow in porous media.

Chapter 10 presents the recent development of numerical analyses of thermofluid behavior in fluid-saturated porous media based on a structural model. A regular array of obstacles is used to describe microscopic porous structures. The idea of microscopic numerical simulations at pore scale to determine macroscopic flow and heat transfer characteristics is elucidated, and a series of numerical results are examined. The permeability and thermal dispersion coefficients, determined purely from the theoretical basis without any empiricism, agree quite well with available experimental data.

Starting the fourth part, on natural and double diffusive convection in porous media, is Chapter 11. This chapter is concerned with buoyancy-driven flows in saturated, porous, media-filled (or partially filled) enclosures. The discussion begins with a consideration of vertical, inclined, and horizontal rectangular enclosures. The governing equations are presented and various solutions are discussed. Equations for the Nusselt number are given. The flow in non-rectangular, porous, media-filled enclosures is then considered, emphasis being placed on flow in annular enclosures. Attention is then given to enclosures that are partly filled with a saturated porous medium. The flow in enclosures filled with anisotropic porous media, maximum density effects, double-diffusive flows, and non-Darcy effects are discussed briefly. The chapter is intended to provide a broad introduction to the subject.

In Chapter 12, an attempt has been made to give a comprehensive account of the state of the art of research into the Darcy–Bénard problem. Some attention has been paid to analytical techniques used in determining the stability of convective flows, and detailed descriptions of how flow changes occur are presented. The manner in which the Darcy–Bénard problem is altered by the inclusion of extra terms is discussed. The discussion includes, but is not limited to, various effects such as form drag, internal heating, local thermal nonequilibrium, and modified boundary conditions. A detailed list of references is provided to enable the new researcher to gain a good overview of the subject.

The objective of Chapter 13 is to investigate natural convection driven by two buoyancy sources, such as heat and mass, in a porous medium saturated with Newtonian fluids. A comprehensive review of the literature concerning double-diffusive convection in fluid-saturated, porous media, the Soret effect and the thermogravitational diffusion in multicomponent systems is presented. This chapter mainly reports analytical, numerical, and scale analysis studies of the onset of a double-diffusive convective regime in a tilted rectangular cavity filled with a porous medium saturated with a binary fluid. The instabilities that can occur when two walls are maintained at different uniform temperature and concentration while the two other walls are impervious and adiabatic are investigated.

Research on mixed convection in saturated porous media conducted over the last 20 years is reviewed in Chapter 14. Although emphasis has been placed on the Darcy flow regime, non-Darcy effects are included whenever the need for a complete understanding of the subject is called for. The results are classified into general categories: external and internal flows. In each category, works are reviewed according to their geometry and thermal boundary condition. The effects of variable properties are included for a complete discussion. Heat transfer correlations are provided for engineering applications. Finally, the direction and challenge of future research are suggested.

Chapter 15, on radiative transfer in porous media, provides recent information on the contemporary understanding of the importance of dependent scattering in porous media. It is pointed out that even very high porosity beds are subject to dependent scattering effects, which were not included in earlier analyses of radiative transfer in these systems. Thus, early methods for predicting radiative properties and for treating radiation in packed and porous materials are suspect. Guidance is provided on determining the importance of dependent scattering.

Chapter 16 reviews the weak turbulence regime associated with porous media non-steady and non-periodic convection in models allowing temporal irregular (i.e., chaotic) solutions, and the conditions for the regime's validity are specified. The rich dynamics linked to the transition from steady convection to chaos is demonstrated and explained analytically as well as computationally.

The final part, on applications, begins with Chapter 17. Over the past 25 years, modeling of the drying process has evolved from the solution of the mass diffusion equation to the solution of the balance equations for the gas and liquid phases, the various constituents of the gas and liquid phases, and the energy equation. This chapter summarizes the development of a drying model based on three dependent variables—moisture content, temperature, and gas phase pressure—from the general balance equations. For many practical drying problems this complex set of nonlinear partial differential equations can be greatly simplified using scale analysis. Scale analysis is utilized to illustrate the relative importance of the various transport phenomena for the thermal transient, constant drying rate, and low- and high-intensity drying regimes.

Chapter 18 briefly describes several technologies for remediation of soils contaminated with hydrocarbons. Mathematical models for hydrodynamic, thermal, and mass transfer processes that take place during removal of multispecies contaminants by soil venting and steam injection are discussed in detail. The models have been shown to describe, in general, the behavior observed in laboratory experiments. A detailed validation of the models in a typical field site is lacking. However, soil venting with preheating of the incoming air and steam injection have been shown to be very effective technologies for remediation of soil contaminated with volatile and semi-volatile hydrocarbons.

Finally, Chapter 19 focuses on non-isothermal mold filling during the liquid composite manufacturing (LCM) process. The LCM processes such as resin transfer molding (RTM), vacuum resin transfer molding (VARTM), and structural reaction injection molding (SRIM) are modeled using Darcy's law. In such processes, a viscous resin (about 100 to 1000 times more viscous than water) is injected to saturate the fiber preform (which is considered as a stationary porous media) placed inside a hot mold cavity. To predict the flow front, one needs only the average Darcian velocity. To predict the temperature distribution, however, one must account for micro-convection due to the interstitial local microvelocity. Important issues related to composite processing are highlighted and future outlook is discussed.

In each of these chapters, whenever applicable, pertinent aspects of experimental work and techniques are discussed. Experts in the field rigorously reviewed each chapter. Overall, many reviewers were involved. The authors and I are very thankful for the valuable and constructive comments that were received.

Kambiz Vafai

Contents

PART III: FORCED CONVECTION IN POROUS MEDIA

PART IV: NATURAL AND DOUBLE-DIFFUSIVE CONVECTION IN POROUS MEDIA

PART V: MIXED CONVECTION IN POROUS MEDIA

PART VI: RADIATIVE TRANSFER IN POROUS MEDIA

Contributors

Suresh G. Advani Professor, Department of Mechanical Engineering, University of Delaware, Newark, Delaware

Marie-Catherine Charrier-Mojtabi Professor, Mechanical Engineering Department, Institut de Mécanique des Fluides, Université Paul Sabatier, Toulouse, France

V. K. Dhir Professor, Mechanical and Aerospace Engineering Department, University of California, Los Angeles, California

F. A. L. Dullien Professor Emeritus, Department of Chemical Engineering, University of Waterloo, Waterloo, Ontario, Canada

Riad Ghafir Ph.D. Student, Material Science Department, Université de Marne-la-Vallée, Marne-la-Vallée, France

John R. Howell Baker-Hughes Centennial Professor, Department of Mechanical Engineering, University of Texas at Austin, Austin, Texas

Kuang-Ting Hsiao Postdoctoral Fellow, Department of Mechanical Engineering, University of Delaware, Newark, Delaware

Chin-Tsau Hsu Senior Lecturer, Mechanical Engineering Department, The Hong Kong University of Science and Technology, Kowloon, Hong Kong

D. B. Ingham Professor, Applied Mathematics Department, University of Leeds, Leeds, West Yorkshire, England

F. Kuwahara Associate Professor, Department of Mechanical Engineering, Shizuoka University, Hamamatsu, Japan

A. V. Kuznetsov Assistant Professor, Mechanical and Aerospace Engineering Department, North Carolina State University, Raleigh, North Carolina

José L. Lage Associate Professor, Mechanical Engineering Department, Southern Methodist University, Dallas, Texas

F. C. Lai Associate Professor, Aerospace and Mechanical Engineering Department, University of Oklahoma, Norman, Oklahoma

Guy Lauriat Professor, Material Science Department, Université de Marne-la-Vallée, Marne-la-Vallée, France

Abdelkader Mojtabi Professor, Mechanical Engineering Department, Institut de Mécanique des Fluides, Université Paul Sabatier, Toulouse, France

A. Nakayama Professor, Department of Mechanical Engineering, Shizuoka University, Hamamatsu, Japan

Arunn Narasimhan Research Scholar, Mechanical Engineering Department, Southern Methodist University, Dallas, Texas

P. H. Oosthuizen Professor, Department of Mechanical Engineering, Queen's University, Kingston, Ontario, Canada

O. A. Plumb Professor, School of Mechanical and Materials Engineering, Washington State University, Pullman, Washington

I. Pop Professor, Faculty of Mathematics, University of Cluj, Cluj, Romania

Michel Quintard Director of Research, Institut de Mécanique des Fluides, C.N.R.S., Toulouse, France

D. A. S. Rees Reader, Department of Mechanical Engineering, University of Bath, Bath, England

Muhammad Sahimi Professor, Chemical Engineering Department, University of Southern California, Los Angeles, California

P. Vadasz Professor, Department of Mechanical Engineering, University of Durban-Westville, Durban, South Africa

Stephen Whitaker Professor, Department of Chemical Engineering and Material Science, University of California, Davis, California

1

Theoretical Analysis of Transport in Porous Media

Michel Quintard
C.N.R.S., Toulouse, France

Stephen Whitaker
University of California, Davis, California

1. INTRODUCTION

In this study of transport in porous media, we consider the system illustrated in Figure 1 where the σ-phase represents a rigid, impermeable solid phase and the β-phase represents a Newtonian fluid. The nomenclature used in this presentation differs from that employed in other chapters in this volume where the *solid phase* is denoted by a subscript s and the *fluid phase* by a subscript f. Petroleum engineers prefer to designate fluids using either w or n depending on whether they are *wetting* or *non-wetting*, and chemical engineers favor g or ℓ depending on whether the fluid is a *gas* or a *liquid*. In our case, we prefer a *discipline-free* nomenclature in which Greek subscripts are used to identify *distinct regions in space*. In this chapter, we are concerned with only two distinct regions in space and we refer to them as the σ-phase and the β-phase, as indicated in Figure 1. This type of nomenclature carries over quite conveniently to hierarchical porous media (Cushman, 1990, 1997) where one must identify *various regions* in addition to the *several phases* that exist in those regions. One often thinks of hierarchical porous media as geological in origin and therefore a special case; however, the typical packed bed catalytic reactor (Whitaker, 1989) is hierarchical in nature and transport processes must be examined in both *phases* and *regions*.

In this treatment of heat transfer in porous media, *coupling* between the transport processes under consideration will be ignored. There are many

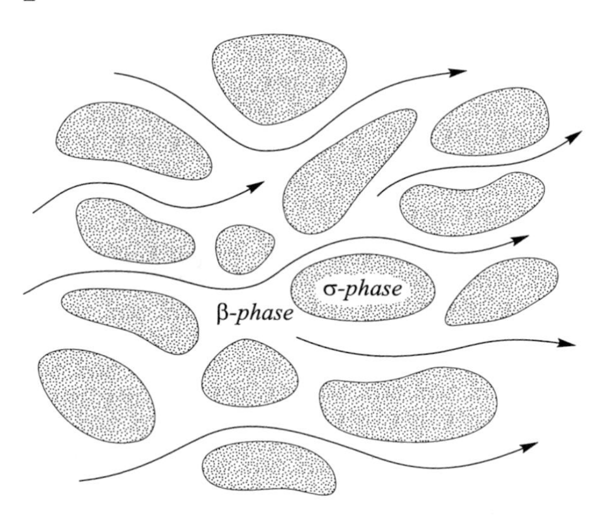

Figure 1. Flow in a rigid porous medium.

important transport processes in porous media in which coupling occurs both at the *macroscopic level* and at the *microscopic level*, and the effects of macroscopic coupling are well documented. Coupling at the microscopic level, or the level of closure, is less well understood, but it is clearly important in the process of drying, as indicated by the work of Azizi et al. (1988) and Moyne et al. (1988) and by the more recent study of Whitaker (1998).

In the absence of coupling the governing equations and boundary conditions under consideration in this review can be expressed as

$$\nabla \cdot \mathbf{v}_\beta = 0, \text{ in the } \beta\text{-phase} \tag{1}$$

$$(\rho c_p)_\beta \frac{\partial T_\beta}{\partial t} + (\rho c_p)_\beta \mathbf{v}_\beta \cdot \nabla T_\beta = \nabla \cdot (k_\beta \nabla T_\beta), \qquad \text{in the } \beta\text{-phase} \tag{2}$$

$$\text{B.C.1} \qquad T_\beta = T_\sigma, \text{ at } \mathcal{A}_{\beta\sigma} \tag{3}$$

$$\text{B.C.2} \qquad \mathbf{n}_{\beta\sigma} \cdot k_\beta \nabla T_\beta = \mathbf{n}_{\beta\sigma} \cdot k_\sigma \nabla T_\sigma + \Omega, \qquad \text{at } \mathcal{A}_{\beta\sigma} \tag{4}$$

$$(\rho c_p)_\sigma \frac{\partial T_\sigma}{\partial t} = \nabla \cdot (k_\sigma \nabla T_\sigma) + \Phi_\sigma, \qquad \text{in the } \sigma\text{-phase} \tag{5}$$

$$\rho_\beta \frac{\partial \mathbf{v}_\beta}{\partial t} + \rho_\beta \mathbf{v}_\beta \cdot \nabla \mathbf{v}_\beta = -\nabla p_\beta + \rho_\beta \mathbf{g} + \mu_\beta \nabla^2 \mathbf{v}_\beta, \qquad \text{in the } \beta\text{-phase} \tag{6}$$

$$\text{B.C.3} \qquad \mathbf{v}_\beta = 0, \qquad \text{at } \mathcal{A}_{\beta\sigma} \tag{7}$$

Here we have used $\mathcal{A}_{\beta\sigma}$ to represent the *entire area* of the β–σ interface contained in the macroscopic region illustrated in Figure 2. Under normal circumstances, the heterogeneous thermal source in Eq. (4) would be *coupled* to the appropriate mass transfer problem involving chemical reaction, adsorption, or phase change (Carberry, 1976; Gates, 1992; Whitaker, 1998) at the β–σ interface. In addition, the homogeneous thermal source would normally be *coupled* to a diffusion and reaction process (Aris, 1975) or to an electromagnetic heating process (Constant and Moyne, 1996); however, in this study we will simply treat Ω and Φ_σ as specified functions in order to determine how these functions influence the structure of the volume-averaged transport equations.

In addition to the interfacial boundary conditions indicated by Eqs. (3), (4), and (7), we also need boundary conditions at the entrances and exits of the macroscopic system shown in Figure 2. In general, these conditions will not be available to us in terms of the point values of the temperature and velocity, and we need to construct jump conditions in terms of volume-averaged values of temperature and velocity. This has been done for the case of passive heat transport (Ochoa-Tapia and Whitaker, 1997, 1998); how-

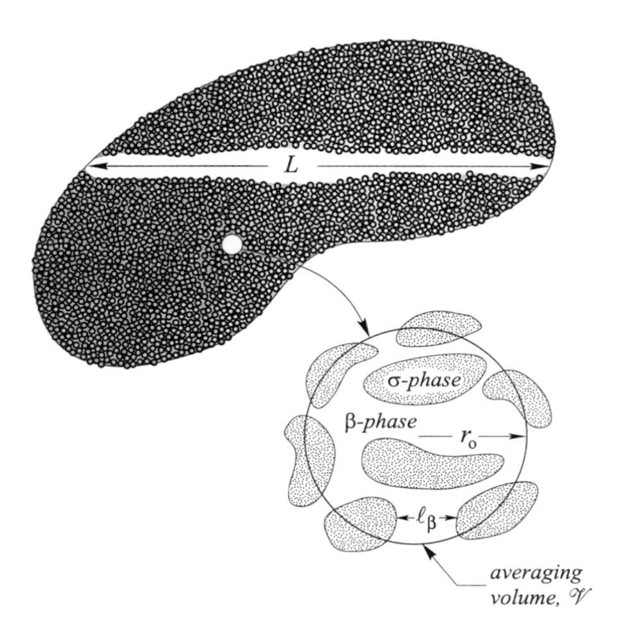

Figure 2. Macroscopic region and averaging volume.

ever, the influence of Ω and Φ_σ on the jump conditions remains to be clarified.

II. VOLUME AVERAGING

In order to develop the volume-averaged forms of the transport equations for mass, energy, and momentum, one associates an average volume, $\mathcal{V}$, with every point in space and defines average values in terms of that volume. Such an averaging volume is illustrated in Figure 2, and can be represented in terms of the volumes of the individual phases according to

$$\mathcal{V} = V_\beta(\mathbf{x}) + V_\sigma(\mathbf{x}) \tag{8}$$

In Figure 3 we have indicated that the centroid of an averaging volume is located by the position vector $\mathbf{x}$, and that points in the β-phase are located by the *relative position* vector $\mathbf{y}_\beta$.

We will use the averaging volume $\mathcal{V}$ to define three averages: the *superficial average*, the *intrinsic average*, and the *spatial average*. Each of these averages is routinely used in the description of multiphase transport processes, and it is important to clearly define each one. In terms of the averaging volume, we define the *superficial average* of some function ψ_β according to

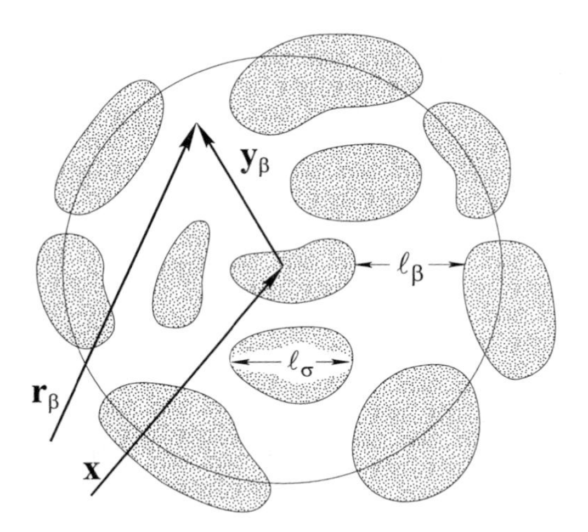

Figure 3. Position vectors associated with an averaging volume.

$$\langle \psi_\beta \rangle|_{\mathbf{x}} = \frac{1}{\mathcal{V}} \int_{V_\beta(\mathbf{x})} \psi_\beta(\mathbf{x} + \mathbf{y}_\beta) \mathrm{d}V_y \tag{9}$$

This clearly indicates that $\langle \psi_\beta \rangle$ is associated with the centroid and that integration is carried out with respect to the components of the relative position vector $\mathbf{y}_\beta$. In addition to the superficial average, we will make use of the *intrinsic average* defined by

$$\langle \psi_\beta \rangle^\beta|_{\mathbf{x}} = \frac{1}{V_\beta(\mathbf{x})} \int_{V_\beta(\mathbf{x})} \psi_\beta(\mathbf{x} + \mathbf{y}_\beta) \mathrm{d}V_y \tag{10}$$

It is important to remember the precise notation used in Eqs. (9) and (10); however, in our analysis we will make use of the simpler representations given by

$$\langle \psi_\beta \rangle = \frac{1}{\mathcal{V}} \int_{V_\beta} \psi_\beta \mathrm{d}V \tag{11}$$

$$\langle \psi_\beta \rangle^\beta = \frac{1}{V_\beta} \int_{V_\beta} \psi_\beta \mathrm{d}V \tag{12}$$

These two averages are related according to

$$\langle \psi_\beta \rangle = \varepsilon_\beta \langle \psi_\beta \rangle^\beta \tag{13}$$

in which ε_β is the volume fraction of the β-phase, defined explicitly as

$$\varepsilon_\beta = V_\beta / \mathcal{V} \tag{14}$$

In this notation for the volume averages, a *Greek subscript* is used to identify the particular phase under consideration and a *Greek superscript* is used to identify an intrinsic average. Since the intrinsic and superficial averages differ by a factor of ε_β, it is essential to make use of a notation that clearly distinguishes between the two.

When we form the volume average of any transport equation, we are immediately confronted with the average of a gradient or the average of a divergence and it is the gradient or divergence of the average that we are seeking. In order to interchange integration and differentiation, we will make use of the spatial averaging theorem (Howes and Whitaker, 1985; Gray et al., 1993; Whitaker, 1999). For the two-phase system illustrated in Figure 3 this theorem can be expressed as

$$\langle \nabla \psi_\beta \rangle = \nabla \langle \psi_\beta \rangle + \frac{1}{\mathcal{V}} \int_{A_{\beta\sigma}} \mathbf{n}_{\beta\sigma} \psi_\beta \mathrm{d}A \tag{15}$$

in which ψ_β is any function associated with the β-phase. Here $A_{\beta\sigma}$ represents the interfacial area contained within the averaging volume, and we have used $\mathbf{n}_{\beta\sigma}$ to represent the unit normal vector pointing *from* the β-phase *toward* the σ-phase.

A. Continuity Equation

Use of the spatial averaging theorem with Eq. 1-1 yields

$$\langle \nabla \cdot \mathbf{v}_\beta \rangle = \nabla \cdot \langle \mathbf{v}_\beta \rangle + \frac{1}{\mathcal{V}} \int_{A_{\beta\sigma}} \mathbf{n}_{\beta\sigma} \cdot \mathbf{v}_\beta \mathrm{d}A = 0 \tag{16}$$

and the fact that the σ phase is assumed to be rigid and impermeable allows us to express the superficial average form of the continuity equation as

$$\nabla \cdot \langle \mathbf{v}_\beta \rangle = 0 \tag{17}$$

The fact that the *superficial average* velocity $\langle \mathbf{v}_\beta \rangle$ is solenoidal encourages its use as the *preferred representation* of the macroscopic or volume-averaged velocity field. We will also have occasion to use the continuity equation in terms of the *intrinsic average velocity* $\langle \mathbf{v}_\beta \rangle^\beta$, thus making use of the relation between the superficial velocity and the intrinsic velocity

$$\langle \mathbf{v}_\beta \rangle = \varepsilon_\beta \langle \mathbf{v}_\beta \rangle^\beta \tag{18}$$

to obtain an alternate form of the continuity equation given by

$$\nabla \cdot \langle \mathbf{v}_\beta \rangle^\beta = -\varepsilon_\beta^{-1} \nabla \varepsilon_\beta \cdot \langle \mathbf{v}_\beta \rangle^\beta \tag{19}$$

This form will be used in the development of the closure problem that is discussed in a subsequent section.

B. Thermal Energy Equation

Our analysis of the energy transport process begins with the β-phase. We form the superficial average of Eq. 1-2 to obtain

$$\frac{1}{\mathcal{V}} \int_{V_\beta} (\rho c_p)_\beta \frac{\partial T_\beta}{\partial t} \mathrm{d}V + \frac{1}{\mathcal{V}} \int_{V_\beta} (\rho c_p)_\beta \mathbf{v}_\beta \cdot \nabla T_\beta \mathrm{d}V = \frac{1}{\mathcal{V}} \int_{V_\beta} \nabla \cdot (k_\beta \nabla T_\beta) \mathrm{d}V \tag{20}$$

It is possible that the physical properties of the β-phase will vary significantly over the macroscopic region shown in Figure 2; however, variations over the averaging volume can generally be neglected (Whitaker, 1999) and this allows us to express Eq. (20) as

$$(\rho c_p)_\beta \frac{1}{\mathcal{V}} \int_{V_\beta} \frac{\partial T_\beta}{\partial t} \mathrm{d}V + (\rho c_p)_\beta \frac{1}{\mathcal{V}} \int_{V_\beta} \nabla \cdot (\mathbf{v}_\beta T_\beta) \, \mathrm{d}V = \frac{1}{\mathcal{V}} \int_{V_\beta} \nabla \cdot (k_\beta \nabla T_\beta) \mathrm{d}V \tag{21}$$

Here we have made use of the continuity equation to express the convective transport in a form that will be convenient for use with the averaging theorem. Since the limits of integration are independent of time, we can interchange integration and differentiation in the first term in (21) to obtain

$$(\rho c_p)_\beta \frac{1}{\mathcal{V}} \int_{V_\beta} \frac{\partial T_\beta}{\partial t} \mathrm{d}V = (\rho c_p)_\beta \frac{\partial \langle T_\beta \rangle}{\partial t} = \varepsilon_\beta (\rho c_p)_\beta \frac{\partial \langle T_\beta \rangle^\beta}{\partial t} \tag{22}$$

Here we have chosen to represent the average temperature in terms of the *intrinsic average*, with the idea that this temperature best represents the thermal energy transport process taking place in the β-phase. Moving on to the convective transport term, we apply the averaging theorem to obtain

$$(\rho c_p)_\beta \frac{1}{\mathcal{V}} \int_{V_\beta} \nabla \cdot (\mathbf{v}_\beta T_\beta) \mathrm{d}V = (\rho c_p)_\beta \nabla \cdot \langle \mathbf{v}_\beta T_\beta \rangle + (\rho c_p)_\beta \frac{1}{\mathcal{V}} \int_{A_{\beta\sigma}} \mathbf{n}_{\beta\sigma} \cdot \mathbf{v}_\beta T_\beta \mathrm{d}A \tag{23}$$

and we can again make use of the fact that the σ-phase is impermeable and rigid in order to express the volume-averaged convective transport in the form

$$(\rho c_p)_\beta \frac{1}{\mathcal{V}} \int_{V_\beta} \nabla \cdot (\mathbf{v}_\beta T_\beta) \mathrm{d}V = (\rho c_p)_\beta \nabla \cdot \langle \mathbf{v}_\beta T_\beta \rangle \tag{24}$$

Use of this result, along with Eq. (22), in the volume-averaged convective transport equation given by Eq. (21) leads to

$$\varepsilon_\beta (\rho c_p)_\beta \frac{\partial \langle T_\beta \rangle^\beta}{\partial t} + (\rho c_p)_\beta \nabla \cdot \langle \mathbf{v}_\beta T_\beta \rangle = \frac{1}{\mathcal{V}} \int_{V_\beta} \nabla \cdot (k_\beta \nabla T_\beta) \mathrm{d}V \tag{25}$$

Turning our attention to the conductive transport term, we apply the averaging theorem once to obtain

$$\begin{aligned} &\varepsilon_\beta (\rho c_p)_\beta \frac{\partial \langle T_\beta \rangle^\beta}{\partial t} + (\rho c_p)_\beta \nabla \cdot \langle \mathbf{v}_\beta T_\beta \rangle \\ &= \nabla \cdot \langle k_\beta \nabla T_\beta \rangle + \frac{1}{\mathcal{V}} \int_{A_{\beta\sigma}} \mathbf{n}_{\beta\sigma} \cdot k_\beta \nabla T_\beta \mathrm{d}A \end{aligned} \tag{26}$$

The last term in this result will *connect* the β-phase equation to the σ-phase equation (5), and this connection will take into account the heterogeneous thermal source that appears in boundary condition (4). Once again we ignore the variation of a physical property *within the averaging volume* and apply the averaging theorem a second time to express the first term on the right-hand side of (26) as

$$\nabla \cdot \langle k_\beta \nabla T_\beta \rangle = \nabla \cdot [k_\beta \langle \nabla T_\beta \rangle] = \nabla \cdot \left[k_\beta \left(\nabla \langle T_\beta \rangle + \frac{1}{\mathcal{V}} \int_{A_{\beta\sigma}} \mathbf{n}_{\beta\sigma} T_\beta \mathrm{d}A \right) \right] \tag{27}$$

Here we are confronted with the *superficial average* temperature that we would like to replace with the *intrinsic average* temperature, and we are also confronted with the *point* temperature that we are attempting to avoid. The first problem can be resolved by the use of Eq. (13) in the form

$$\langle T_\beta \rangle = \varepsilon_\beta \langle T_\beta \rangle^\beta \tag{28}$$

whereas the second problem requires that we introduce a *decomposition* in terms of the intrinsic average temperature and a spatial deviation temperature. We must also introduce this decomposition for the velocity, and we list these two decompositions as

$$T_\beta = \langle T_\beta \rangle^\beta + \tilde{T}_\beta, \qquad \mathbf{v}_\beta = \langle \mathbf{v}_\beta \rangle^\beta + \tilde{\mathbf{v}}_\beta \tag{29}$$

Use of Eq. (28) and the first of Eqs. (29) in (27) leads to

$$\nabla \cdot \langle k_\beta \nabla T_\beta \rangle = \nabla \cdot \left[k_\beta \left(\varepsilon_\beta \nabla \langle T_\beta \rangle^\beta + \langle T_\beta \rangle^\beta \nabla \varepsilon_\beta + \frac{1}{\mathcal{V}} \int_{A_{\beta\sigma}} \mathbf{n}_{\beta\sigma} \langle T_\beta \rangle^\beta \mathrm{d}A \right.\right.$$
$$\left.\left. + \frac{1}{\mathcal{V}} \int_{A_{\beta\sigma}} \mathbf{n}_{\beta\sigma} \tilde{T}_\beta \mathrm{d}A \right) \right] \tag{30}$$

This represents a *non-local form* in terms of the average temperature $\langle T_\beta \rangle^\beta$ since this quantity is evaluated over the β–σ interface at points other than the centroid of the averaging volume illustrated in Figure 3. If we can remove $\langle T_\beta \rangle^\beta$ from the area integral over $A_{\beta\sigma}$, the averaging theorem (15) leads to

$$\frac{1}{\mathcal{V}} \int_{A_{\beta\sigma}} \mathbf{n}_{\beta\sigma} \langle T_\beta \rangle^\beta \mathrm{d}A = \langle T_\beta \rangle^\beta \left\{ \frac{1}{\mathcal{V}} \int_{A_{\beta\sigma}} \mathbf{n}_{\beta\sigma} \mathrm{d}A \right\} = -\langle T_\beta \rangle^\beta \nabla \varepsilon_\beta \tag{31}$$

Under these circumstances, Eq. (30) can be simplified to

$$\nabla \cdot \langle k_\beta \nabla T_\beta \rangle = \nabla \cdot \left[k_\beta \left(\varepsilon_\beta \nabla \langle T_\beta \rangle^\beta + \frac{1}{\mathcal{V}} \int_{A_{\beta\sigma}} \mathbf{n}_{\beta\sigma} \tilde{T}_\beta \mathrm{d}A \right) \right] \tag{32}$$

The length-scale constraints associated with this simplification have been carefully discussed by Quintard and Whitaker (1994) and by Whitaker (1999). Substitution of (32) into (26) provides the following form of the volume-averaged energy transport equation

$$\varepsilon_\beta(\rho c_p)_\beta \frac{\partial \langle T_\beta \rangle^\beta}{\partial t} + (\rho c_p)_\beta \nabla \cdot \langle \mathbf{v}_\beta T_\beta \rangle$$

$$= \nabla \cdot \left[k_\beta \left(\varepsilon_\beta \nabla \langle T_\beta \rangle^\beta + \frac{1}{\mathcal{V}} \int_{A_{\beta\sigma}} \mathbf{n}_{\beta\sigma} \tilde{T}_\beta \mathrm{d}A \right) \right] + \frac{1}{\mathcal{V}} \int_{A_{\beta\sigma}} \mathbf{n}_{\beta\sigma} \cdot k_\beta \nabla T_\beta \mathrm{d}A \quad (33)$$

Our treatment of the convective transport term makes use of the decompositions given by (29) to obtain

$$\langle \mathbf{v}_\beta T_\beta \rangle = \langle \langle \mathbf{v}_\beta \rangle^\beta \langle T_\beta \rangle^\beta + \tilde{\mathbf{v}}_\beta \langle T_\beta \rangle^\beta + \langle \mathbf{v}_\beta \rangle^\beta \tilde{T}_\beta + \tilde{\mathbf{v}}_\beta \tilde{T}_\beta \rangle \quad (34)$$

The analysis of the right-hand side of this result is rather complex. The details are given by Whitaker (1999) who suggests that the first and last terms dominate so that the volume-averaged convective transport takes the form

$$\langle \mathbf{v}_\beta T_\beta \rangle = \varepsilon_\beta \langle \mathbf{v}_\beta \rangle^\beta \langle T_\beta \rangle^\beta + \langle \tilde{\mathbf{v}}_\beta \tilde{T}_\beta \rangle \quad (35)$$

Here we have been careful to use

$$\langle 1 \rangle = \varepsilon_\beta \quad (36)$$

on the basis of the Eqs. (11)–(14).

At this point we can use Eq. (35) in (33) to obtain an important form of the β-phase convective transport equation given by

$$\underbrace{\varepsilon_\beta(\rho c_p)_\beta \frac{\partial \langle T_\beta \rangle^\beta}{\partial t}}_{\text{accumulation}} + \underbrace{\varepsilon_\beta(\rho c_p)_\beta \langle \mathbf{v}_\beta \rangle^\beta \cdot \nabla \langle T_\beta \rangle^\beta}_{\text{convection}}$$

$$= \underbrace{\nabla \cdot \left[k_\beta \left(\varepsilon_\beta \nabla \langle T_\beta \rangle^\beta + \frac{1}{\mathcal{V}} \int_{A_{\beta\sigma}} \mathbf{n}_{\beta\sigma} \tilde{T}_\beta \mathrm{d}A \right) \right]}_{\text{conduction}}$$

$$- \underbrace{(\rho c_p)_\beta \nabla \cdot \langle \tilde{\mathbf{v}}_\beta \tilde{T}_\beta \rangle}_{\text{dispersion}} + \underbrace{\frac{1}{\mathcal{V}} \int_{A_{\beta\sigma}} \mathbf{n}_{\beta\sigma} \cdot k_\beta \nabla T_\beta \mathrm{d}A}_{\text{interfacial flux}} \quad (37)$$

The development that led to this result can be repeated to obtain the analogous result for the σ-phase, that takes the form

$$\underbrace{\varepsilon_\sigma(\rho c_p)_\sigma \frac{\partial \langle T_\sigma \rangle^\sigma}{\partial t}}_{\text{accumulation}} = \underbrace{\nabla \cdot \left[k_\sigma \left(\varepsilon_\sigma \nabla \langle T_\sigma \rangle^\sigma + \frac{1}{\mathcal{V}} \int_{A_{\sigma\beta}} \mathbf{n}_{\sigma\beta} \tilde{T}_{\sigma \mathrm{d}A} \right) \right]}_{\text{conduction}}$$

$$+\underbrace{\frac{1}{\mathcal{V}}\int_{A_{\sigma\beta}} \mathbf{n}_{\sigma\beta}\cdot k_\sigma \nabla T_\sigma \mathrm{d}A}_{\text{interfacial flux}} + \underbrace{\varepsilon_\sigma \langle \Phi_\sigma \rangle^\sigma}_{\substack{\text{homogeneous}\\ \text{thermal source}}} \tag{38}$$

The presence of the convective transport term in Eq. (37) indicates that we need to be able to determine the β-phase velocity $\langle \mathbf{v}_\beta \rangle^\beta$. This leads to an analysis of the momentum transport process.

C. Momentum Equation

The momentum equation (6) is similar in form to the convective heat transport equation (2), except for the fact that it contains the pressure gradient and the gravitational force and that it is subject to the no-slip condition (7). The details of the averaging of (6) are given elsewhere (Whitaker, 1996, 1997) and the form that is analogous to (37) is given by

$$\underbrace{\varepsilon_\beta \rho_\beta \frac{\partial \langle \mathbf{v}_\beta \rangle^\beta}{\partial t}}_{\text{accumulation}} + \underbrace{\varepsilon_\beta \rho_\beta \langle \mathbf{v}_\beta \rangle^\beta \cdot \nabla \langle \mathbf{v}_\beta \rangle^\beta}_{\text{convection}}$$

$$= \underbrace{-\varepsilon_\beta \nabla \langle p_\beta \rangle^\beta + \frac{1}{\mathcal{V}}\int_{A_{\beta\sigma}} \mathbf{n}_{\beta\sigma} \tilde{p}_\beta \mathrm{d}A}_{\text{pressure force}} + \underbrace{\varepsilon_\beta \rho_\beta \mathbf{g}}_{\substack{\text{gravity}\\ \text{force}}}$$

$$\underbrace{\mu_\beta \nabla \cdot \nabla (\varepsilon_\beta \langle \mathbf{v}_\beta \rangle^\beta)}_{\text{viscous force}} - \underbrace{\rho_\beta \nabla \cdot \langle \tilde{\mathbf{v}}_\beta \tilde{\mathbf{v}}_\beta \rangle}_{\text{dispersion}} + \underbrace{\frac{1}{\mathcal{V}}\int_{A_{\beta\sigma}} \mathbf{n}_{\beta\sigma} \cdot \mu_\beta \nabla \mathbf{v}_\beta \mathrm{d}A}_{\text{interfacial viscous force}} \tag{39}$$

It is of some interest to note that Eqs. (17) and (37)–(39) are all superficial transport equations in the sense that each one represents some quantity *per unit volume of the porous medium* and not per unit volume of either the β-phase or the σ-phase. Most multiphase transport equations are used in the superficial form; however, the momentum equation is an exception and the traditional applications make use of an *intrinsic form*, Darcy's law being the classical example.

The next step in the analysis of Eq. (39) is the development of a *closure problem* that will provide a representation for the velocity deviation $\tilde{\mathbf{v}}_\beta$. To develop the closure problem, we first need to express (39) entirely in terms of intrinsic average quantities, $\langle \mathbf{v}_\beta \rangle^\beta$ and $\langle p_\beta \rangle^\beta$, and spatial deviation quantities, $\tilde{\mathbf{v}}_\beta$ and $\tilde{p}_\beta$, and this means that the last area integral in (39) must be expressed in terms of $\langle \mathbf{v}_\beta \rangle^\beta$ and $\tilde{\mathbf{v}}_\beta$. One can use the point- and volume-averaged equations to develop a boundary value problem for $\tilde{\mathbf{v}}_\beta$ and $\tilde{p}_\beta$, and this leads to representations that can be used to produce the closed form of Eq. (39). This situation is also true of the two thermal energy equations, and this means that the interfacial flux terms in Eqs. (37) and (38) must be expressed in terms of intrinsic average temperatures and spatial deviation temperatures. However, if the two intrinsic average temperatures $\langle T_\beta\rangle^\beta$ and $\langle T_\sigma\rangle^\sigma$ are *close enough* so that we can make use of the *approximation*

$$\langle T_\beta\rangle^\beta = \langle T_\sigma\rangle^\sigma \tag{40}$$

we can add (37) and (38) to obtain a one-equation model for the thermal energy transport process. What is meant by *close enough* is that the condition of local thermal equilibrium is valid, and we will identify that condition in the next section.

III. LOCAL THERMAL EQUILIBRIUM

Rather than accept Eq. (40) as a plausible assumption and proceed to construct the one-equation model on the basis of (37) and (38), it is more prudent to represent the temperatures in the β- and σ-phases in terms of a single temperature and Darcy-scale deviations. These decompositions are given by

$$\langle T_\beta\rangle^\beta = \langle T\rangle + \hat{T}_\beta \tag{41a}$$

$$\langle T_\sigma\rangle^\sigma = \langle T\rangle + \hat{T}_\sigma \tag{41b}$$

in which $\langle T\rangle$ represents the *spatial average* temperature defined explicitly by

$$\langle T\rangle = \frac{1}{\mathcal{V}}\int_{\mathcal{V}} T\,\mathrm{d}V = \varepsilon_\beta\langle T_\beta\rangle^\beta + \varepsilon_\sigma\langle T_\sigma\rangle^\sigma \tag{42}$$

Use of Eqs. (41) in (37) and (38) allows us to add those two volume-averaged transport equations to obtain

$$[\varepsilon_\beta(\rho c_p)_\beta + \varepsilon_\sigma(\rho c_p)_\sigma]\frac{\partial\langle T\rangle}{\partial t} + \varepsilon_\beta(\rho c_p)_\beta\langle \mathbf{v}_\beta\rangle^\beta\cdot\nabla\langle T\rangle$$

$$= \nabla\cdot\left[(\varepsilon_\beta k_\beta + \varepsilon_\sigma k_\sigma)\nabla\langle T\rangle + \frac{k_\beta}{\mathcal{V}}\int_{A_{\beta\sigma}}\mathbf{n}_{\beta\sigma}\tilde{T}_\beta\mathrm{d}A + \frac{k_\sigma}{\mathcal{V}}\int_{A_\sigma}\mathbf{n}_{\sigma\beta}\tilde{T}_\sigma\mathrm{d}A\right]$$

$$-(\rho c_p)_\beta\nabla\cdot\langle\tilde{\mathbf{v}}_\beta\tilde{T}_\beta\rangle + a_\mathrm{v}\langle\Omega\rangle_{\beta\sigma} + \varepsilon_\sigma\langle\Phi_\sigma\rangle^\sigma$$

$$-\left[\varepsilon_\beta(\rho c_p)_\beta\frac{\partial\hat{T}_\beta}{\partial t}+\varepsilon_\sigma(\rho c_p)_\sigma\frac{\partial\hat{T}_\sigma}{\partial t}+\varepsilon_\beta(\rho c_p)_\beta\langle\mathbf{v}_\beta\rangle^\beta\cdot\nabla\hat{T}_\beta-(\varepsilon_\beta k_\beta\nabla\hat{T}_\beta+\varepsilon_\sigma k_\sigma\nabla\hat{T}_\sigma)\right] \tag{43}$$

Here we have made use of the flux boundary condition (4) in order to obtain

$$\frac{1}{\mathcal{V}}\int_{A_{\beta\sigma}}\mathbf{n}_{\beta\sigma}\cdot k_\beta\nabla T_\beta \mathrm{d}A=\frac{1}{\mathcal{V}}\int_{A_{\beta\sigma}}\mathbf{n}_{\beta\sigma}\cdot k_\sigma\nabla T_\sigma \mathrm{d}A+a_\mathrm{v}\langle\Omega\rangle_{\beta\sigma} \tag{44}$$

where a_v represents the interfacial area per unit volume and $\langle\Omega\rangle_{\beta\sigma}$ is the area-averaged value of the heterogeneous thermal source. When the last four terms in Eq. (43) are *small enough* to be neglected, the condition of local thermal equilibrium is valid and we can express the spatial deviation temperatures in terms of the *closure variables* according to

$$\tilde{T}_\beta=\mathbf{b}_\beta\cdot\nabla\langle T\rangle,\qquad \tilde{T}_\sigma=\mathbf{b}_\sigma\cdot\nabla\langle T\rangle \tag{45}$$

These vector fields have the characteristic that

$$\mathbf{b}_\beta=\mathbf{b}_\sigma,\quad \text{at}\quad A_{\beta\sigma} \tag{46}$$

and this allows us to express the conductive terms in (43) as

$$(\varepsilon_\beta k_\beta+\varepsilon_\sigma k_\sigma)\nabla\langle T\rangle+\frac{k_\beta}{\mathcal{V}}\int_{A_{\beta\sigma}}\mathbf{n}_{\beta\sigma}\tilde{T}_\beta \mathrm{d}A+\frac{k_\sigma}{\mathcal{V}}\int_{A\sigma\beta}\mathbf{n}_{\sigma\beta}\tilde{T}_\sigma \mathrm{d}A$$

$$=\left[(\varepsilon_\beta k_\beta+\varepsilon_\sigma k_\sigma)\mathbf{I}+\frac{(k_\beta-k_\sigma)}{\mathcal{V}}\int_{A_{\beta\sigma}}\mathbf{n}_{\beta\sigma}\mathbf{b}_\beta \mathrm{d}A\right]\cdot\nabla\langle T\rangle=\mathbf{K}_\mathrm{eff}\cdot\nabla\langle T\rangle \tag{47}$$

In addition, the dispersive transport takes the form

$$(\rho c_p)_\beta\nabla\cdot\langle\tilde{\mathbf{v}}_\beta\tilde{T}_\beta\rangle=-\nabla\cdot(\varepsilon_\beta\mathbf{K}_\mathrm{D}\cdot\nabla\langle T\rangle) \tag{48}$$

in which $\mathbf{K}_\mathrm{D}$ is the *hydrodynamic* thermal dispersion tensor for the β-phase. Use of Eqs. (47) and (48) in the local thermal equilibrium form of (43) leads to the one-equation model given by

$$\langle\rho\rangle C_p\frac{\partial\langle T\rangle}{\partial t}+\varepsilon_\beta(\rho c_p)_\beta\langle\mathbf{v}_\beta\rangle^\beta\cdot\nabla\langle T\rangle=\nabla\cdot(\mathbf{K}^*\cdot\nabla\langle T\rangle)+a_\mathrm{v}\langle\Omega\rangle_{\beta\sigma}+\varepsilon_\sigma\langle\Phi_\sigma\rangle^\sigma \tag{49}$$

Here the heat capacity per unit volume is given by

$$\langle\rho\rangle C_p=\varepsilon_\beta(\rho c_p)_\beta+\varepsilon_\sigma(\rho c_p)_\sigma \tag{50}$$

and the total effective thermal conductivity tensor includes the contribution due to thermal dispersion and takes the form

$$\mathbf{K}^* = \mathbf{K}_{\text{eff}} + \varepsilon_\beta \mathbf{K}_{\text{D}} \tag{51}$$

The simplicity of the one-equation model, with the single undetermined coefficient, provides strong motivation for its use and thus strong motivation for knowing exactly when the condition of local thermal equilibrium occurs.

At this point we are in a position to determine what we mean by the statement that $\langle T_\beta \rangle^\beta$ and $\langle T_\sigma \rangle^\sigma$ are *close enough* so that Eq. (40) is an acceptable approximation. What we mean by that statement is that the last four terms in (43) are *negligible*. If we recognize that the Darcy-scale spatial deviations take the form

$$\hat{T}_\beta = \varepsilon_\sigma(\langle T_\beta \rangle^\beta - \langle T_\sigma \rangle^\sigma \qquad \text{and} \qquad \hat{T}_\sigma = \varepsilon_\beta(\langle T_\sigma \rangle^\sigma - \langle T_\beta \rangle^\beta) \tag{52}$$

we can express the last four terms in (43) as

$$\varepsilon_\beta (\rho c_p)_\beta \frac{\partial \hat{T}_\beta}{\partial t} + \varepsilon_\sigma (\rho c_p)_\sigma \frac{\partial \hat{T}_\sigma}{\partial t} = \varepsilon_\beta \varepsilon_\sigma [(\rho c_p)_\beta - (\rho c_p)_\sigma] \frac{\partial}{\partial t} (\langle T_\beta \rangle^\beta - \langle T_\sigma \rangle^\sigma) \tag{53a}$$

$$\varepsilon_\beta (\rho c_p)_\beta \langle \mathbf{v}_\beta \rangle^\beta \cdot \nabla \hat{T}_\beta = \varepsilon_\beta \varepsilon_\sigma (\rho c_p)_\beta \langle \mathbf{v}_\beta \rangle^\beta \cdot \nabla (\langle T_\beta \rangle^\beta - \langle T_\sigma \rangle^\sigma) \tag{53b}$$

$$\nabla \cdot (\varepsilon_\beta k_\beta \nabla \hat{T}_\beta + \varepsilon_\sigma k_\sigma \nabla \langle T_\sigma \rangle^\sigma) = \nabla \cdot [\varepsilon_\beta \varepsilon_\sigma (k_\beta - k_\sigma) \nabla (\langle T_\beta \rangle^\beta - \langle T_\sigma \rangle^\sigma)] \tag{53c}$$

If these terms are small compared to the conductive transport term in (43), and if *small causes give rise to small effects*, we can neglect the terms containing the temperature difference, $\langle T_\beta \rangle^\beta - \langle T_\sigma \rangle^\sigma$, in order to obtain the one-equation model given by (49). On the basis of Eqs. (53) and the representation of the conductive transport term given by (47), we can express the *restrictions* associated with local thermal equilibrium as

$$\varepsilon_\beta \varepsilon_\sigma [(\rho c_p)_\beta - (\rho c_p)_\sigma] \frac{\partial}{\partial t} (\langle T_\beta \rangle^\beta - \langle T_\sigma \rangle^\sigma) \ll \nabla \cdot \ (\mathbf{K}_{\text{eff}} \cdot \nabla \langle T \rangle) \tag{54a}$$

$$\varepsilon_\beta \varepsilon_\sigma (\rho c_p)_\beta \langle \mathbf{v}_\beta \rangle^\beta \cdot \nabla (\langle T_\beta \rangle^\beta - \langle T_\sigma \rangle^\sigma) \ll \nabla \cdot (\mathbf{K}_{\text{eff}} \cdot \nabla \langle T \rangle) \tag{54b}$$

$$\nabla \cdot [\varepsilon_\beta \varepsilon_\sigma (k_\beta - k_\sigma) \nabla (\langle T_\beta \rangle^\beta - \langle T_\sigma \rangle^\sigma)] \ll \nabla \cdot (\mathbf{K}_{\text{eff}} \cdot \nabla \langle T \rangle) \tag{54c}$$

In order to develop the *constraints* associated with these three restrictions, we need an estimate of the temperature difference. This estimate can be obtained from the governing differential equation, the initial condition, and the boundary conditions for $\langle T_\beta \rangle^\beta - \langle T_\sigma \rangle^\sigma$.

A. Governing Equation for $\langle T_\beta\rangle^\beta - \langle T_\sigma\rangle^\sigma$

The governing equation for $\langle T_\beta\rangle^\beta - \langle T_\sigma\rangle^\sigma$ can be developed only in an approximate sense. One reasonable approximation is to ignore variations in ε_β and divide Eq. (37) by ε_β to obtain the intrinsic form given by

$$(\rho c_p)_\beta \frac{\partial \langle T_\beta\rangle^\beta}{\partial t} + (\rho c_p)_\beta \langle \mathbf{v}_\beta\rangle^\beta \cdot \nabla \langle T_\beta\rangle^\beta$$

$$= \nabla \cdot \left[k_\beta \nabla \langle T_\beta\rangle^\beta + \frac{k_\beta \varepsilon_\beta^{-1}}{\mathcal{V}} \int_{A_{\beta\sigma}} \mathbf{n}_{\beta\sigma} \tilde{T}_\beta \mathrm{d}A \right]$$

$$-(\rho c_p)_\beta \nabla \cdot \langle \tilde{\mathbf{v}}_\beta \tilde{T}_\beta \rangle^\beta + \frac{\varepsilon_\beta^{-1}}{\mathcal{V}} \int_{A_{\beta\sigma}} \mathbf{n}_{\beta\sigma} \cdot k_\beta \nabla T_\beta \mathrm{d}A \tag{55}$$

We can repeat this process for the σ-phase transport equation to obtain

$$(\rho c_p)_\sigma \frac{\partial \langle T_\sigma\rangle^\sigma}{\partial t} = \nabla \cdot \left(k_\sigma \nabla \langle T_\sigma\rangle^\sigma + \frac{k_\sigma \varepsilon_\sigma^{-1}}{\mathcal{V}} \int_{A_{\sigma\beta}} \mathbf{n}_{\sigma\beta} \tilde{T}_\sigma \mathrm{d}A \right)$$

$$+ \frac{\varepsilon_\sigma^{-1}}{\mathcal{V}} \int_{A_{\sigma\beta}} \mathbf{n}_{\sigma\beta} \cdot k_\sigma \nabla T_\sigma \mathrm{d}A + \langle \Phi_\sigma \rangle^\sigma \tag{56}$$

Subtracting (56) from (55) and following the algebraic manipulation indicated by Whitaker (1991) leads to

$$(\rho c_p)_{\beta\sigma} \frac{\partial}{\partial t} (\langle T_\beta\rangle^\beta - \langle T_\sigma\rangle^\sigma) + \varepsilon_\sigma (\rho c_p)_\beta \langle \mathbf{v}_\beta\rangle^\beta \cdot \nabla (\langle T_\beta\rangle^\beta - \langle T_\sigma\rangle^\sigma)$$

$$-\nabla \cdot [k_{\beta\sigma} \nabla (\langle T_\beta\rangle^\beta - \langle T_\sigma\rangle^\sigma)] = -[(\rho c_p)_\beta - (\rho c_p)_\sigma] \frac{\partial \langle T\rangle}{\partial t} - (\rho c_p)_\beta \langle \mathbf{v}_\beta\rangle^\beta \cdot \nabla \langle T\rangle$$

$$+ \nabla \cdot \left[(k_\beta - k_\sigma) \nabla \langle T\rangle + \frac{k_\beta}{V_\beta} \int_{A_{\beta\sigma}} \mathbf{n}_{\beta\sigma} \tilde{T}_\beta \mathrm{d}A - \frac{k_\sigma}{V_\sigma} \int_{A_{\beta\sigma}} \mathbf{n}_{\sigma\beta} \tilde{T}_\sigma \mathrm{d}A \right]$$

$$- (\rho c_p)_\beta \nabla \cdot \langle \tilde{\mathbf{v}}_\beta \tilde{T}_\beta \rangle^\beta + \frac{(\varepsilon_\beta \varepsilon_\sigma)^{-1}}{\mathcal{V}} \int_{A_{\beta\sigma}} \mathbf{n}_{\beta\sigma} \cdot k_\beta \nabla T_\beta \mathrm{d}A - \varepsilon_\sigma^{-1} a_{\mathrm{v}} \langle \Omega \rangle_{\beta\sigma} - \langle \Phi_\sigma \rangle^\sigma \tag{57}$$

Here we have made use of the *mixed-mode parameters* defined by

$$(\rho c_p)_{\beta\sigma} = \varepsilon_\sigma (\rho c_p)_\beta + \varepsilon_\beta (\rho c_p)_\sigma, \quad k_{\beta\sigma} = \varepsilon_\sigma k_\beta + \varepsilon_\beta k_\sigma \tag{58}$$

that always appear in this type of analysis, and we have used the interfacial flux condition given by Eq. (4) in order to incorporate the heterogeneous thermal source into the equation for the temperature difference.

In our study of the two-equation model in Section V, we show that the interfacial heat fluxes can be expressed as

$$\frac{1}{\mathcal{V}}\int_{A_{\beta\sigma}} \mathbf{n}_{\beta\sigma} \cdot k_\beta \nabla T_\beta \mathrm{d}A = -a_\mathrm{v}h(\langle T_\beta\rangle^\beta - \langle T_\sigma\rangle^\sigma) + a_\mathrm{v}\xi\langle\Omega\rangle_{\beta\sigma} \tag{59a}$$

$$\frac{1}{\mathcal{V}}\int_{A_{\sigma\beta}} \mathbf{n}_{\sigma\beta} \cdot k_\sigma \nabla T_\sigma \mathrm{d}A = -a_\mathrm{v}h(\langle T_\sigma\rangle^\sigma - \langle T_\beta\rangle^\beta) + a_\mathrm{v}(1-\xi)\langle\Omega\rangle_{\beta\sigma} \tag{59b}$$

Here $a_\mathrm{v}h$ is the volumetric interfacial heat transfer coefficient and ξ is a fraction, bounded by zero and one, that determines how the heterogeneous thermal source is distributed between the two phases. Both $a_\mathrm{v}h$ and ξ are determined by the solution of the two-equation model closure problem discussed in Section VI, and the expressions for the flux given by Eqs. (59) neglect the contribution of the non-traditional convective transport terms.

The results given by Eqs. (45) and (46) can be used to construct the *phase geometry tensor* that takes the form

$$\mathbf{C}_{\beta\sigma} = \frac{(\varepsilon_\beta\varepsilon_\sigma)^{-1}k_{\beta\sigma}}{\mathcal{V}}\int_{A_{\beta\sigma}} \mathbf{n}_{\beta\sigma}\mathbf{b}_\beta \mathrm{d}A \tag{60}$$

and this can be used, along with (59a), to express Eq. (57) as

$$\begin{aligned}
&(\rho c_p)_{\beta\sigma}\frac{\partial}{\partial t}(\langle T_\beta\rangle^\beta - \langle T_\sigma\rangle^\sigma) + \varepsilon_\sigma(\rho c_p)_\beta\langle\mathbf{v}_\beta\rangle^\beta\cdot\nabla(\langle T_\beta\rangle^\beta - \langle T_\sigma\rangle^\sigma)\\
&\quad + \frac{a_\mathrm{v}h}{\varepsilon_\beta\varepsilon_\sigma}(\langle T_\beta\rangle^\beta - \langle T_\sigma\rangle^\sigma) - \nabla\cdot[k_{\beta\sigma}\nabla(\langle T_\beta\rangle^\beta - \langle T_\sigma\rangle^\sigma)]\\
&\quad = -[(\rho c_p)_\beta - (\rho c_p)_\sigma]\frac{\partial\langle T\rangle}{\partial t} - (\rho c_p)_\beta\langle\mathbf{v}_\beta\rangle^\beta\cdot\nabla\langle T\rangle\\
&\quad + \nabla\cdot[(k_\beta - k_\sigma)(\mathbf{I} + \mathbf{C}_{\beta\sigma})\cdot\nabla\langle T\rangle] - (\rho c_p)_\beta\nabla\cdot\langle\tilde{\mathbf{v}}_\beta\tilde{T}_\beta\rangle^\beta\\
&\quad + \left(\frac{\xi - \varepsilon_\beta}{\varepsilon_\beta\varepsilon_\sigma}\right)a_\mathrm{v}\langle\Omega\rangle_{\beta\sigma} - \langle\Phi_\sigma\rangle^\sigma
\end{aligned} \tag{61}$$

As a final simplification to this governing differential equation for the temperature difference, we again use a result from local thermal equilibrium to express the intrinsic dispersive thermal flux according to

$$(\rho c_p)_\beta\nabla\cdot\langle\tilde{\mathbf{v}}_\beta\tilde{T}_\beta\rangle^\beta = -\nabla\cdot(\mathbf{K}_\mathrm{D}\cdot\nabla\langle T\rangle) \tag{62}$$

Substitution of this result into Eq. (61) leads to the final form of our governing differential equation for $\langle T_\beta\rangle^\beta - \langle T_\sigma\rangle^\sigma$

$$(\rho c_p)_{\beta\sigma}\frac{\partial}{\partial t}(\langle T_\beta\rangle^\beta - \langle T_\sigma\rangle^\sigma) + \varepsilon_\sigma(\rho c_p)_\beta\langle \mathbf{v}_\beta\rangle\cdot\nabla(\langle T_\beta\rangle^\beta - \langle T_\sigma\rangle^\sigma)$$
$$+\frac{a_v h}{\varepsilon_\beta\varepsilon_\sigma}(\langle T_\beta\rangle^\beta - \langle T_\sigma\rangle^\sigma) - \nabla\cdot[k_{\beta\sigma}\nabla(\langle T_\beta\rangle^\beta - \langle T_\sigma\rangle^\sigma)]$$
$$= -[(\rho c_p)_\beta - (\rho c_p)_\sigma]\frac{\partial\langle T\rangle}{\partial t} - (\rho c_p)_\beta\langle \mathbf{v}_\beta\rangle^\beta\cdot\nabla\langle T\rangle$$
$$+\nabla\cdot[(k_\beta - k_\sigma)(\mathbf{I}+\mathbf{C}_{\beta\sigma}) + \mathbf{K}_\mathrm{D}]\cdot\nabla\langle T\rangle + \left(\frac{\xi-\varepsilon_\beta}{\varepsilon_\beta\varepsilon_\sigma}\right)a_\mathrm{v}\langle\Omega\rangle_{\beta\sigma} - \langle\Phi_\sigma\rangle^\sigma \quad (63)$$

All the terms on the left-hand side of this result contain the dependent variable $\langle T_\beta\rangle^\beta - \langle T_\sigma\rangle^\sigma$, while all the terms on the right-hand side represent the *sources* or *generators* of the temperature difference. If all the terms on the right-hand side of (63) were zero, we would expect $\langle T_\beta\rangle^\beta$ to be equal to $\langle T_\sigma\rangle^\sigma$ and local thermal equilibrium would exist. Equation (63) has the interesting quality that the influence of the heterogeneous thermal source appears to be zero when the fraction ξ is just equal to the volume fraction of the β-phase. This is somewhat misleading since $a_\mathrm{v}\langle\Omega\rangle_{\beta\sigma}$ can influence the time and space derivatives of $\langle T\rangle$ and therefore have an effect on the right-hand side of (63) even when $\xi = \varepsilon_\beta$.

It is important to keep in mind that the temperature difference will be determined by the approximate governing differential equation represented by Eq. (63) *in addition to* the appropriate boundary and initial conditions. Since the boundary and initial conditions will depend on the specific applications, we are forced to base any general estimate on the governing equation (63).

B. Order-of-Magnitude Estimates

We begin with the time and space derivatives of the spatial average temperature, and make use of the convention followed in a previous development (Whitaker, 1991). This leads to the estimates given by

$$\frac{\partial\langle T\rangle}{\partial t} = \mathbf{O}\left(\frac{\Delta\langle T\rangle}{t^*}\right) \quad (64a)$$

$$\langle \mathbf{v}_\beta\rangle^\beta\cdot\nabla\langle T\rangle = \mathbf{O}\left(\frac{\langle v_\beta\rangle^\beta\Delta\langle T\rangle}{L_{\rho c_p}}\right) \quad (64b)$$

$$\nabla\nabla\langle T\rangle = \mathbf{O}\left(\frac{\Delta\langle T\rangle}{L_\mathrm{c}^2}\right) \quad (64c)$$

Here $L_{\rho c_p}$ represents a *convective* length-scale, i.e., it represents the distance along a volume-averaged streamline over which significant changes in $\langle T\rangle$

occur. On the other hand, L_c represents a conductive length-scale, i.e., it represents the smallest distance in any direction over which significant changes in $\langle T \rangle$ occur. Typically, the direction associated with L_c is orthogonal to the direction associated with $L_{\rho c_p}$. For the special case in which $\nabla \langle T \rangle$ is a constant, one must be careful to remember that L_c is infinite and according to the definition given by Eq. (64c). It is important to recognize that it is up to the individual to make judgments about the characteristic time- and length-scales associated with $\langle T \rangle$ since every process will have its own special characteristics.

Our estimates of the time and space derivatives of the temperature difference take the same form as those given by Eqs. (64), however, one must remember that the temperature difference is a Darcy-scale spatial *deviation* (see Eqs. (41) and (52)) and the change associated with a deviation is on the order of the deviation itself. This leads to

$$\frac{\partial}{\partial t}(\langle T_\beta \rangle^\beta - \langle T_\sigma \rangle^\sigma) = \mathbf{O}\left(\frac{(\langle T_\beta \rangle^\beta - \langle T_\sigma \rangle^\sigma)}{t^*}\right) \tag{65a}$$

$$\langle \mathbf{v}_\beta \rangle^\beta \cdot \nabla(\langle T_\beta \rangle^\beta - \langle T_\sigma \rangle^\sigma) = \mathbf{O}\left(\frac{\langle \mathrm{v}_\beta \rangle^\beta (\langle T_\beta \rangle^\beta - \langle T_\sigma \rangle^\sigma)}{L_{\rho c_p}}\right) \tag{65b}$$

$$\nabla\nabla(\langle T_\beta \rangle^\beta - \langle T_\sigma \rangle^\sigma) = \mathbf{O}\left(\frac{\Delta(\langle T_\beta \rangle^\beta - \langle T_\sigma \rangle^\sigma)}{L_c^2}\right) \tag{65c}$$

With the aid of Eqs. (64) and (65), the order-of-magnitude form of (63) is given by

$$\left[\frac{(\rho c_p)_{\beta\sigma}}{t^*} + \mathbf{O}\left(\frac{\varepsilon_\sigma (\rho c_p)_\beta \langle \mathrm{v}_\beta \rangle^\beta}{L_{\rho c_p}}\right) + \mathbf{O}\left(\frac{a_\mathrm{v} h}{\varepsilon_\beta \varepsilon_\sigma}\right) + \mathbf{O}\left(\frac{k_{\beta\sigma}}{L_c^2}\right)\right](\langle T_\beta \rangle^\beta - \langle T_\sigma \rangle^\sigma)$$

$$= \left[\frac{(\rho c_p)_\beta - (\rho c_p)_\sigma}{t^*} + \mathbf{O}\left(\frac{(\rho c_p)_\beta \langle \mathrm{v}_\beta \rangle^\beta}{L_{\rho c_p}}\right) + \mathbf{O}\left(\frac{(k_\beta - k_\sigma)(\mathbf{I} + \mathbf{C}_{\beta\sigma})}{L_c^2}\right) + \mathbf{O}\left(\frac{\mathbf{K}_D}{L_c^2}\right)\right]\Delta\langle T \rangle$$

$$+\mathbf{O}\left[\left(\frac{\xi - \varepsilon_\beta}{\varepsilon_\beta \varepsilon_\sigma}\right) a_\mathrm{v} \langle \Omega \rangle_{\beta\sigma}\right] + \mathbf{O}(\langle \Phi_\sigma \rangle^\sigma) \tag{66}$$

At this point we follow the work of Quintard and Whitaker (1995), who concluded that the term $a_\mathrm{v} h$ was a key parameter in the determination of the condition of local thermal equilibrium. Thus we divide Eq. (66) by $a_\mathrm{v} h / \varepsilon_\beta \varepsilon_\sigma$ and arrange the result in the following form

$$\frac{\langle T_\beta\rangle^\beta - \langle T_\sigma\rangle^\sigma}{\Delta\langle T\rangle} = \mathbf{O}\left(\frac{\ell_{\beta\sigma}}{L_c}\right)^2 \frac{[A + B + C + D + \Sigma]}{\left[1 + \mathbf{O}(\ell_{\beta\sigma}/L_c)^2 + \mathbf{O}(\ell_{\beta\sigma}^2/\alpha_{\beta\sigma}t^*) + \mathbf{O}\left(\frac{\varepsilon_\sigma \ell_{\beta\sigma}^2}{\ell_\beta L_{\rho c_p}}\right)\frac{k_\beta}{k_{\beta\sigma}}\mathrm{Pe}\right]} \tag{67}$$

Here the mixed-mode quantities are defined by

$$\ell_{\beta\sigma}^2 = \frac{\varepsilon_\beta \varepsilon_\sigma k_{\beta\sigma}}{a_v h}, \qquad \alpha_{\beta\sigma} = \frac{k_{\beta\sigma}}{(\rho c_p)_{\beta\sigma}} \tag{68}$$

while the heat transfer Péclet number is loosely defined according to

$$\mathrm{Pe} = \frac{(\rho c_p)_\beta \langle \mathrm{v}_\beta\rangle^\beta \ell_\beta}{k_\beta} \tag{69}$$

On the right-hand side of Eq. (67) we have used the following definitions:

$$A = \frac{(k_\beta - k_\sigma)(\mathrm{I} + \mathrm{C}_{\beta\sigma})}{k_{\beta\sigma}} \tag{70a}$$

$$B = \frac{L_c^2}{\alpha_{\beta\sigma}t^*}\left[\frac{(\rho c_p)_\beta - (\rho c_p)_\sigma}{(\rho c_p)_{\beta\sigma}}\right] \tag{70b}$$

$$C = \frac{\mathrm{K_D}}{k_{\beta\sigma}} \tag{70c}$$

$$D = \left(\frac{L_c^2}{\ell_\beta L_{\rho c_p}}\right)\frac{k_\beta}{k_{\beta\sigma}}\mathrm{Pe} \tag{70d}$$

The upper-case Greek letter on the right-hand side of (67) has been reserved for terms containing the estimated *temperature difference* and is given by

$$\Sigma = \frac{(\xi - \varepsilon_\beta)a_v\langle\Omega\rangle_{\beta\sigma}L_c^2}{\varepsilon_\beta\varepsilon_\sigma k_{\beta\sigma}\Delta\langle T\rangle} + \frac{\langle\Phi_\sigma\rangle^\sigma L_c^2}{k_{\beta\sigma}\Delta\langle T\rangle} \tag{71}$$

For transient heat conduction processes at short times (Quintard and Whitaker, 1995), one can encounter situations for which

$$\mathbf{O}(\ell_{\beta\sigma}/L_c)^2 \approx \mathbf{O}(\ell_{\beta\sigma}^2/\alpha_{\beta\sigma}t_\sigma^* \approx \mathbf{O}(1) \tag{72}$$

and the estimation of $\langle T_\beta\rangle^\beta - \langle T_\sigma\rangle^\sigma$ relative to $\Delta\langle T\rangle$ becomes quite difficult. On the other hand, for most heat transfer processes these quantities are more likely to be related by

$$\mathbf{O}(\ell_{\beta\sigma}/L_c)^2 \ll \mathbf{O}(1), \quad \mathbf{O}(\ell_{\beta\sigma}^2/\alpha_{\beta\sigma}t^*) \ll \mathbf{O}(1) \tag{73}$$

and Eq. (67) can be simplified accordingly. In addition, many practical processes will be constrained by

$$A = \mathbf{O}(1), \quad B = \mathbf{O}(1), \quad C = \mathbf{O}(1) \tag{74}$$

and when these results, along with those given by Eq. (73), are used in (67) we obtain

$$\frac{\langle T_\beta\rangle^\beta - \langle T_\sigma\rangle^\sigma}{\Delta\langle T\rangle} = \mathbf{O}\left(\frac{\ell_{\beta\sigma}}{L_c}\right)^2 \frac{\left[1 + \mathbf{O}\left(\frac{L_c^2}{\ell_\beta L_{\rho c_p}}\right)\frac{k_\beta}{k_{\beta\sigma}}\mathrm{Pe} + \mathbf{O}(\Sigma)\right]}{\left[1 + \mathbf{O}\left(\frac{\varepsilon_\sigma \ell_{\beta\sigma}^2}{\ell_\beta L_{\rho c_p}}\right)\frac{k_\beta}{k_{\beta\sigma}}\mathrm{Pe}\right]} \tag{75}$$

When the Péclet number increases, the convective length also increases as does the heat transfer coefficient. This leads to a situation for which

$$\mathrm{Pe} \to \text{large}, \qquad L_{\mathrm{pc}_p} \to \text{large}, \qquad \ell_{\beta\sigma} \to \text{small} \tag{76}$$

and suggests that for many situations Eq. (75) can be simplified to

$$\frac{\langle T_\beta\rangle^\beta - \langle T_\sigma\rangle^\sigma}{\Delta\langle T\rangle} = \mathbf{O}\left(\frac{\ell_{\beta\sigma}}{L_c}\right)^2 \left[1 + \mathbf{O}\left(\frac{L_c^2}{\ell_\beta L_{\rho c_p}}\frac{k_\beta}{k_{\beta\sigma}}\mathrm{Pe}\right) + \mathbf{O}(\Sigma)\right] \tag{77}$$

This form of our estimate for $\langle T_\beta\rangle^\beta - \langle T_\sigma\rangle^\sigma$ implies that convective effects, and thermal sources as indicated by Σ, may be primarily responsible for the failure of local thermal equilibrium and the necessity for a two-equation model. It is difficult to estimate Σ; however, if we make use of Eq. (49) we can develop the crude estimate given by

$$\Sigma \approx \mathbf{O}\left[\frac{\langle\rho\rangle C_p L_c^2}{k_{\beta\sigma}t^*} + \mathbf{O}\left(\frac{L_c^2}{\ell_\beta L_{\rho c_p}}\frac{k_\beta}{k_{\beta\sigma}}\mathrm{Pe}\right) + \mathbf{O}\left(\frac{\mathrm{K}^*}{k_{\beta\sigma}}\right)\right] \tag{78}$$

This provides the somewhat counter-intuitive result that the heterogeneous and homogeneous thermal sources do not play a particularly important role in breaking the condition of local thermal equilibrium, and that a reasonable estimate of the temperature difference is given by

$$\frac{\langle T_\beta\rangle^\beta - \langle T_\sigma\rangle^\sigma}{\Delta\langle T\rangle} = \mathbf{O}\left(\frac{\ell_{\beta\sigma}}{L_c}\right)^2 \left[1 + \mathbf{O}\left(\frac{L_c^2}{\ell_\beta L_{\rho c_p}}\frac{k_\beta}{k_{\beta\sigma}}\rho\mathrm{e}\right)\right] \tag{79}$$

This result certainly needs to be explored more thoroughly than we have done using order-of-magnitude analysis. The numerical experiments of Quintard and Whitaker (1995) indicate how such a study could be carried out.

C. Constraints for Local Thermal Equilibrium

Given the estimate indicated by Eq. (79), we now are in a position to return to the restrictions given by (54) and develop the following three constraints associated with the condition of local thermal equilibrium

$$\varepsilon_\beta \varepsilon_\sigma \frac{[(\rho c_p)_\beta - (\rho c_p)_\sigma]\ell_{\beta\sigma}^2}{\mathrm{K}_{\mathrm{eff}} t^*} \left[1 + \mathbf{O}\left(\frac{L_c^2}{\ell_\beta L_{\rho c_p}} \frac{k_\beta}{k_{\beta\sigma}} \mathrm{Pe} \right) \right] \ll 1 \tag{80a}$$

$$\varepsilon_\beta \varepsilon_\sigma \frac{\ell_{\beta\sigma}^2}{\ell_\beta L_{\rho c_p}} \frac{k_\beta}{\mathrm{K}_{\mathrm{eff}}} \mathrm{Pe} \left[1 + \mathbf{O}\left(\frac{L_c^2}{\ell_\beta L_{\rho c_p}} \frac{k_\beta}{k_{\beta\sigma}} \mathrm{Pe} \right) \right] \ll 1 \tag{80b}$$

$$\varepsilon_\beta \varepsilon_\sigma \frac{(k_\beta - k_\sigma)}{\mathrm{K}_{\mathrm{eff}}} \frac{\ell_{\beta\sigma}^2}{L_c^2} \left[1 + \mathbf{O}\left(\frac{L_c^2}{\ell_\beta L_{\rho c_p}} \frac{k_\beta}{k_{\beta\sigma}} \mathrm{Pe} \right) \right] \ll 1 \tag{80c}$$

When these three constraints are satisfied, the condition of local thermal equilibrium should be valid *provided that* the boundary and/or initial conditions do not create values of $\langle T_\beta \rangle^\beta - \langle T_\sigma \rangle^\sigma$ that are significantly different from the estimate given by (79). In general, the initial and boundary conditions are responsible for the creation of non-zero values of $\partial \langle T \rangle / \partial t$ and $\nabla \langle T \rangle$ that appear as *sources* on the right-hand side of (61). Our *estimates* of these sources are just that, and they should be examined carefully for every particular problem. As a reminder of the importance of boundary and/or initial conditions, we list another estimate of $\langle T_\beta \rangle^\beta - \langle T_\sigma \rangle^\sigma$ that always needs to be considered:

$$\frac{\langle T_\beta \rangle^\beta - \langle T_\sigma \rangle^\sigma}{\Delta \langle T \rangle} = \mathbf{O}\begin{pmatrix} \text{boundary conditions} \\ \text{and} \\ \text{initial conditions} \end{pmatrix} \tag{81}$$

An example of the importance of the boundary conditions is given by the work of Yagi et al. (1960), who measured longitudinal thermal dispersion coefficients using a steady-state process in which the departure from local thermal equilibrium was created by an imposed thermal flux at the outlet of a packed bed. For that process, the restrictions associated with local thermal equilibrium are still correctly given by Eqs. (54); however, the estimate of $\langle T_\beta \rangle^\beta - \langle T_\sigma \rangle^\sigma$ should be considered in terms of the specific system used by Yagi et al. (1960).

IV. CLOSED FORMS FOR MOMENTUM AND ENERGY

When local thermal equilibrium is valid, one can use Eq. (49) to determine the spatial average temperature, provided that the velocity is known and the thermal sources can be determined. Traditionally, the thermal energy transport equation is represented in terms of the *superficial velocity* according to

$$\langle\rho\rangle C_p \frac{\partial\langle T\rangle}{\partial t} + (\rho c_p)_\beta \langle\mathbf{v}_\beta\rangle \cdot \nabla\langle T\rangle = \nabla\cdot(\mathbf{K}^* \cdot \nabla\langle T\rangle) + a_v\langle\Omega\rangle_{\beta\sigma} + \varepsilon_\sigma\langle\Phi_\sigma\rangle^\sigma \tag{82}$$

and this means that a closed form of Eq. (39) must be developed in order to determine $\langle\mathbf{v}_\beta\rangle$. The analysis is given elsewhere (Whitaker, 1996), and here we will only list the result known as the Forchheimer equation

$$\langle\mathbf{v}_\beta\rangle = -\frac{\mathbf{K}_\beta}{\mu_\beta}\cdot\left[\nabla\langle p_\beta\rangle^\beta - \rho_\beta\mathbf{g} - \underbrace{(\mu_\beta/\varepsilon_\beta)\nabla^2\langle\mathbf{v}_\beta\rangle}_{\substack{\text{Brinkman}\\ \text{correction}}}\right] - \underbrace{\mathbf{F}_\beta\cdot\langle\mathbf{v}_\beta\rangle}_{\substack{\text{Forchheimer}\\ \text{correction}}} \tag{83}$$

in which $\mathbf{K}_\beta$ represents the Darcy's law permeability tensor and $\mathbf{F}_\beta$ represents the Forchheimer correction tensor. The latter is essentially a linear function of the velocity and the Forchheimer correction will be negligible when the Reynolds number is less than one. Even though the governing differential equations for energy and momentum are very similar, as are the volume-averaged forms given by Eqs. (37) and (39), the closed forms given by Eqs. (82) and (83) are dramatically different. This difference is caused by the form of the interfacial boundary conditions and their influence on the closure problem. This difference has been examined in detail by Whitaker (1997).

The Brinkman (1947) correction in Eq. (83) is often associated with a *Brinkman viscosity*; however, the theoretical analysis leading to the Forchheimer equation provides no indication that μ_β is anything other than the fluid velocity. In addition to the energy and momentum equations, the solution for $\langle T\rangle$ and $\langle\mathbf{v}_\beta\rangle$ requires the use of the continuity equation

$$\nabla\cdot\langle\mathbf{v}_\beta\rangle = 0 \tag{84}$$

since this allows us to determine $\langle p_\beta\rangle^\beta$ as part of the solution of Eqs. (82) and (83).

A. Boundary Conditions

In order to solve the energy, momentum, and continuity equations in the macroscopic region illustrated in Figure 2, we need to impose conditions on $\langle T\rangle$ and $\langle\mathbf{v}_\beta\rangle$ at the boundary between the porous medium and the surrounding homogeneous fluid. A portion of this boundary is shown in Figure 4 where we have identified the porous medium as the ω-region and the surrounding homogeneous fluid as the η-region. The conditions that apply to the volume-averaged velocity and temperature at the boundary between the ω- and η-regions has been the subject of a series of studies by Ochoa-Tapia and Whitaker (1995, 1997, 1998), which lead to rather complex representations involving surface excess mass, momentum, and the surface transport of these quantities. In this review we will present the jump conditions in the form that they take when the tangential transport of surface excess quantities can be neglected.

In the ω-region, we represent the three transport equations under consideration according to

$$\nabla \cdot \langle \mathbf{v}_\beta \rangle_\omega = 0 \tag{85}$$

$$\langle \mathbf{v}_\beta \rangle_\omega = -\frac{\mathbf{K}_{\beta\omega}}{\mu_\beta} \cdot [\nabla \langle p_\beta \rangle_\omega^\beta - \rho_\beta \mathbf{g} - (\mu_\beta / \varepsilon_{\beta\omega}) \nabla^2 \langle \mathbf{v}_\beta \rangle_\omega] - \mathbf{F}_{\beta\omega} \cdot \langle \mathbf{v}_\beta \rangle_\omega \tag{86}$$

$$\begin{aligned} &(\langle \rho \rangle C_p)_\omega \frac{\partial \langle T \rangle_\omega}{\partial t} + (\rho c_p)_\beta \langle \mathbf{v}_\beta \rangle_\omega \cdot \nabla \langle T \rangle_\omega \\ &= \nabla \cdot (\mathbf{K}_\omega^* \cdot \nabla \langle T \rangle_\omega) + (a_\mathrm{v} \langle \Omega \rangle_{\beta\sigma})_\omega + (\varepsilon_\sigma \langle \Phi \rangle^\sigma)_\omega \end{aligned} \tag{87}$$

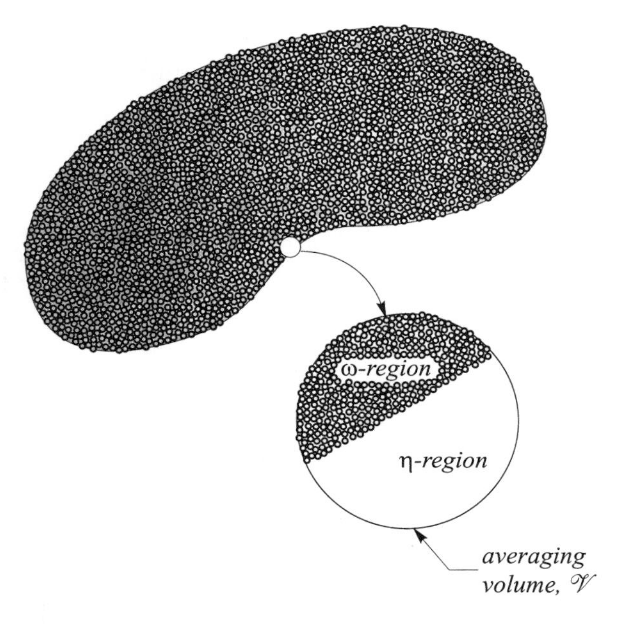

Figure 4. Boundary between a porous medium and a homogeneous fliud.

The motivation for the subscript ω is to indicate clearly that these results were derived on the basis of certain length-scale constraints and that we expect these length-scale constraints to fail in the neighborhood of the ω–η boundary.

The jump conditions at the ω–η boundary associated with the continuity equation simply requires that the normal component of the velocity be continuous. We express this idea as

$$\text{B.C.1} \qquad \langle \mathbf{v}_\beta \rangle_\omega \cdot \mathbf{n}_{\omega\eta} = \langle \mathbf{v}_\beta \rangle_\eta \cdot \mathbf{n}_{\omega\eta}, \qquad \text{at the } \omega\text{–}\eta \text{ boundary} \tag{88}$$

in which $\mathbf{n}_{\omega\eta}$ represents the unit normal vector directed from the ω-region toward the η-region. For the present, we will assume that some appropriate governing equation exists that allows us to determine $\langle \mathbf{v}_\beta \rangle_\eta$ and we will return to this matter in the following paragraphs.

Even neglecting the surface transport of surface excess quantities, the momentum jump condition is quite complex and takes the form

$$\text{B.C.2} \qquad \mathbf{n}_{\omega\eta} \cdot [-\mathbf{I}(\langle p_\beta \rangle_\omega^\beta - \langle p_\beta \rangle_\eta^\beta) + \mu_\beta(\varepsilon_{\beta\omega}^{-1} \nabla \langle \mathbf{v}_\beta \rangle_\omega - \nabla \langle \mathbf{v}_\beta \rangle_\eta)]$$

$$= \mu_\beta [\delta^{-1} \mathbf{A} (\varepsilon_{\beta\omega} - 1)^2 (\varepsilon_{\beta\omega}^{-3} + 1) - \delta \mathbf{D} \cdot \mathbf{H}_{\beta\omega}^{-1}] \cdot \langle \mathbf{v}_\beta \rangle_\omega$$

$$+ \mathbf{n}_{\omega\eta} \cdot [\rho_\beta \langle \mathbf{v}_\beta \rangle_\omega \cdot \mathbf{E}^{(4)} \cdot \langle \mathbf{v}_\beta \rangle_\omega] \tag{89}$$

Here $\mathbf{A}$ and $\mathbf{D}$ are dimensionless second-order tensors of order one, δ is a measure of the thickness of the boundary region, and $\mathbf{E}^{(4)}$ is a fourth-order tensor of order one. These quantities must be determined experimentally and it would appear to be impossible to determine $\mathbf{A}$, $\mathbf{D}$, and δ independently. However, $\mathbf{E}^{(4)}$ could be determined by experiments in which the Reynolds number was varied, and the tensor $\mathbf{H}_{\beta\omega}$ is available to us in terms of the closure problem (Whitaker, 1996) since it is related to $\mathbf{K}_{\beta\omega}$ and $\mathbf{F}_{\beta\omega}$ by

$$\mathbf{H}_{\beta\omega}^{-1} = \mathbf{K}_{\beta\omega}^{-1} + \mathbf{K}_{\beta\omega}^{-1} \cdot \mathbf{F}_{\beta\omega} \tag{90}$$

Ochoa-Tapia and Whitaker (1998) have presented arguments indicating that $\mathbf{E}^{(4)}$ is related to surface roughness whereas $\mathbf{A}$ and $\mathbf{D}$ are connected to the basic structure of the porous medium. For smooth surfaces, such as those studied by Beavers and Joseph (1967), the inertial term in (89) can be neglected and the equation simplifies to

$$\text{B.C.2} \qquad \mathbf{n}_{\omega\eta} \cdot [-\mathbf{I}(\langle p_\beta \rangle_\omega^\beta - \langle p_\beta \rangle_\eta^\beta) + \mu_\beta(\varepsilon_{\beta\omega}^{-1} \nabla \langle \mathbf{v}_\beta \rangle_\omega - \nabla \langle \mathbf{v}_\beta \rangle_\eta)]$$

$$= \mu_\beta \mathbf{C} \cdot \langle \mathbf{v}_\beta \rangle_\omega \tag{91}$$

in which **C** is a second-order tensor on the order of δ^{-1} that must be determined experimentally. In a comparison with the experimental data of Beavers and Joseph (1967), Ochoa-Tapia and Whitaker (1995) were able to *adjust* the values of a single component of **C** to obtain good agreement with experiment. However, at this point in time there is not much justification for the values that were used, other than the fact that they were of the proper order of magnitude.

The thermal energy jump condition developed by Ochoa-Tapia and Whitaker (1998) is based on the assumption that $\langle T\rangle_\omega$ is equal to $\langle T\rangle_\eta$ at the ω–η boundary. This led to the following condition for the thermal flux:

$$\text{B.C.3} \qquad \mathbf{n}_{\omega\eta}\cdot\mathbf{K}^*_\omega\cdot\nabla\langle T\rangle_\omega = \mathbf{n}_{\omega\eta}\cdot k_\beta\nabla\langle T\rangle_\eta + \Phi_s \tag{92}$$

Here Φ_s represents a thermal source caused by the departure from local thermal equilibrium in the boundary region. The idea here is that the constraints given by Eqs. (80) might be valid in the homogeneous region of the porous media but are likely to fail in the boundary region, and this gives rise to the parameter Φ_s that might be determined experimentally.

In the η-region, the continuity equation, the thermal energy, and the momentum equation are given by Eqs. (1), (2), and (6), and at the ω–η boundary illustrated in Figure 4 there is a mismatch of length scales between those *point equations* and the volume-averaged equations given by Eqs. (85)–(87). This problem can be avoided by forming the volume average of the point equations in the homogeneous η-region in order to develop transport equations for $\langle\mathbf{v}_\beta\rangle_\eta$, $\langle T\rangle_\eta$, and $\langle p_\beta\rangle^\beta_\eta$. These can then be used in conjunction with the transport equations for the homogeneous ω-region to construct the jump conditions. In order to use the point form of the appropriate transport equations in the η-region, one must demonstrate under what circumstances the point forms are equivalent to the volume-average forms. Given the length scales based on the following estimates (Whitaker, 1999)

$$\nabla\mathbf{v}_\beta = \mathbf{O}\left(\frac{\Delta\mathbf{v}_\beta}{L_\mathrm{v}}\right), \quad \nabla(\nabla\mathbf{v}_\beta) = \mathbf{O}\left(\frac{\nabla\mathbf{v}_\beta}{L_{\mathrm{v}1}}\right) \tag{93a}$$

$$\nabla \mathrm{T}_\beta = \mathbf{O}\left(\frac{\Delta T_\beta}{L_T}\right), \quad \nabla(\nabla T_\beta) = \mathbf{O}\left(\frac{\nabla T_\beta}{L_{T1}}\right) \tag{93b}$$

$$\nabla p_\beta = \mathbf{O}\left(\frac{\Delta p_\beta}{L_p}\right), \quad \nabla(\nabla p_\beta) = \mathbf{O}\left(\frac{\nabla p_\beta}{L_{p1}}\right) \tag{93c}$$

Ochoa-Tapia and Whitaker (1995, 1997) showed that the η-region volume-averaged equations were essentially equal to the η-region point equations when the following constraints were satisfied in the homogeneous η-region

$$\frac{r_o^2}{L_{v1}L_v} \ll 1, \quad \frac{r_o^2}{L_{T1}L_T} \ll 1, \quad \frac{r_o^2}{L_{p1}L_p} \ll 1 \tag{94}$$

Under these circumstances, the point equations in the homogeneous η-region can be used with the volume-averaged equations for the homogeneous ω-region and the jump conditions described above in order to develop solutions for the velocity, pressure, and temperature in the porous medium illustrated in Figure 2 and in the surrounding homogeneous fluid.

B. Comparison with Experiment

The comparison between theory and experiment for the one-equation model represented by Eq. (82) requires only the theoretical and experimental determination of the total thermal dispersion tensor $\mathbf{K}^*$. The theoretical prediction of $\mathbf{K}^*$ requires the closure variables $\mathbf{b}_\beta$ and $\mathbf{b}_\sigma$. The general closure problem is discussed in detail in Section 5 and from that presentation one can extract the closure problem that is applicable to the case of local thermal equilibrium. This is given by

$$(\rho c_p)_\beta \tilde{\mathbf{v}}_\beta + (\rho c_p)_\beta \mathbf{v}_\beta \cdot \nabla \mathbf{b}_\beta = k_\beta \nabla^2 \mathbf{b}_\beta - \varepsilon_\beta^{-1} \mathbf{c}, \qquad \text{in the } \beta\text{–phase} \tag{95a}$$

$$\text{B.C.1} \qquad \mathbf{b}_\beta = \mathbf{b}_\sigma, \qquad \text{at } A_{\beta\sigma} \tag{95b}$$

$$\text{B.C.2} \qquad \mathbf{n}_{\beta\sigma} \cdot k_\beta \nabla \mathbf{b}_\beta = \mathbf{n}_{\beta\sigma} \cdot k_\sigma \nabla \mathbf{b}_\sigma - \mathbf{n}_{\beta\sigma}(k_\beta – k_\sigma), \qquad \text{at } A_{\beta\sigma} \tag{95c}$$

$$0 = k_\sigma \nabla^2 \mathbf{b}_\sigma + \varepsilon_\sigma^{-1} \mathbf{c}, \qquad \text{in the } \sigma\text{-phase} \tag{95d}$$

$$\text{Periodicity:} \quad \mathbf{b}_\beta(\mathbf{r} + \ell_i) = \mathbf{b}_\beta(\mathbf{r}), \;\; \mathbf{b}_\sigma(\mathbf{r} + \ell_i) = \mathbf{b}_\sigma(\mathbf{r}), \qquad i = 1, 2, 3 \tag{95e}$$

$$\text{Average:} \quad \langle \mathbf{b}_\beta \rangle^\beta = 0, \qquad \langle \mathbf{b}_\sigma \rangle^\sigma = 0 \tag{95f}$$

Here $\mathbf{c}$ represents an undetermined constant that is evaluated by the application of (95f). The details of this computation are given by Quintard and Whitaker (1993) and are discussed in Section 5.

Given values for the closure variable, $\mathbf{b}_\beta$, one can determine $\mathbf{K}^*$ by the expression

$$\mathbf{K}^* = (\varepsilon_\beta k_\beta + \varepsilon_\sigma k_\sigma)\mathbf{I} + \frac{(k_\beta - k_\sigma)}{\mathcal{V}} \int_{A_{\beta\sigma}} \mathbf{n}_{\beta\sigma} \mathbf{b}_\beta \mathrm{d}A - (\rho c_p)_\beta \langle \tilde{\mathbf{v}}_\beta \mathbf{b}_\beta \rangle \tag{96}$$

In Figure 5 we have compared theoretical values of k_{zz}^*/k_β with the experimental results obtained by Yagi et al. (1960) and by Gunn and De Souza (1974) as a function of the *particle* Péclet number defined by

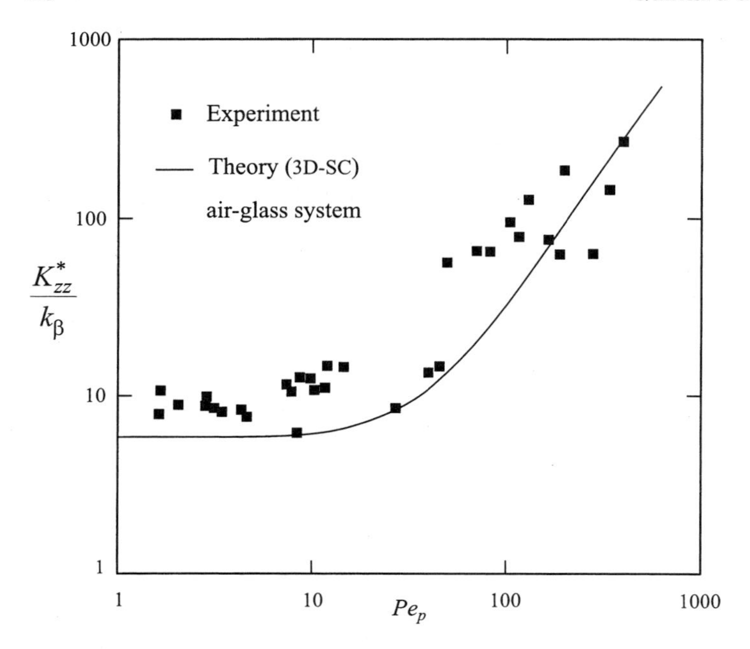

Figure 5. Theoretical and experimental values for the longitudinal thermal dispersion coefficient.

$$\mathrm{Pe_p} = \frac{(\rho c_p)_\beta \langle \mathrm{v}_\beta \rangle^\beta d_\mathrm{p}}{k_\beta} \left(\frac{\varepsilon_\beta}{1 - \varepsilon_\beta} \right) \tag{97}$$

Here d_p represents the diameter of the glass spheres used in these studies, $\langle \mathrm{v}_\beta \rangle^\beta$ represents the intrinsic average velocity, and the term involving the porosity results in the use of the *hydraulic diameter* as a characteristic length (Eidsath et al., 1983). The theoretical calculations were carried out for a simple cubic (SC), three-dimensional lattice of spheres. The dimensionless parameters for the system are

$$k_\sigma / k_\beta = 23, \qquad \varepsilon_\beta = 0.40 \tag{98}$$

and the values of k_β and $(\rho c_p)_\beta$ were taken to be that for air at 40°C and one atmosphere. The values for the physical properties are consistent with those of Yagi et al. (1960) and we assume that they are representative of the glass–air system studied by Gunn and De Souza (1974). The work of Yagi et al. was based on a steady-state experiment which was interpreted in terms of a one-equation model. Gunn and De Souza used a frequency response technique that was interpreted in terms of a two-equation model; however, the solid-phase transport was treated as spherically symmetric conduction in isolated particles. This means that any longitudinal transport in the solid phase was, by default, attributed to the fluid phase. At this point, we are not in a position to comment on whether the experiments of Gunn and De Souza and those of Yagi et al. satisfied the conditions of local thermal equilibrium; however, both sets of data are in reasonably good agreement and their measured values of K^*_{zz} are consistent with the theoretical calculations. It should be obvious that the simple cubic lattice used in the theoretical calculations is not the best model of a packed bed of spheres; nevertheless, the agreement between theory and experiment in the absence of adjustable parameters is very attractive.

The theory has also been compared with the experiments of Grangeot et al. (1990, 1994) in which layers of spheres were arranged in hexagonal packing. These experiments are particularly attractive since the packed beds represent spatially periodic porous media and there is no ambiguity in terms of the choice of a representative unit cell. The experiments were carried out by creating an abrupt change in the inlet temperature and monitoring the temperature as a function of time at several locations in the bed. Temperatures were measured in both the solid and fluid phases for brass–water and nylon–water systems that had the following properties

$$\begin{aligned} k_\sigma/k_\beta &= 180, \qquad \text{brass–water system} \\ k_\sigma/k_\beta &= 0.45, \qquad \text{nylon–water system} \end{aligned} \tag{99}$$

In order to determine the longitudinal thermal dispersion coefficient, the experiments were analyzed using the one-equation model. The results are shown in Figure 6 as a function of a cell Péclet number defined by

$$\mathrm{Pe}_{\mathrm{cell}} = \frac{(\rho c_p)_\beta \langle \mathrm{v}_\beta \rangle^\beta d_\mathrm{p}}{k_\beta} \tag{100}$$

The experiment values of K^*_{zz}/k_β for the brass–water system are higher than those predicted by the theory. This may be due to the particle–particle contact that occurs in the laboratory system but was not accounted for in the theoretical calculations. It is possible to account for particle–particle contact using an *adjustable parameter* to fit the data and this has been done by Nozad et al. (1985). The theoretical studies of Nozad et al. (1985) and the experimental studies of Shonnard and Whitaker (1989) indicate that particle–particle contact becomes important for values of k_σ/k_β larger than 100; thus the difference between theory and experiment for the brass–water system may be associated with the additional conductive transport in the solid phase. On the other hand, the experimental values of K^*_{zz}/k_β

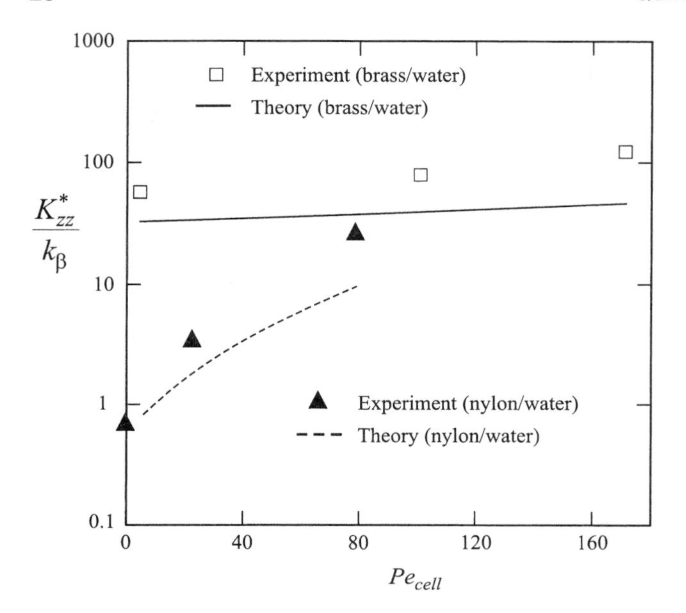

Figure 6. Theoretical and experimental values of the longitudinal thermal dispersion coefficient for spatially periodic systems.

for the nylon–water system are also higher than the theoretical values and particle–particle contact is not important for this system. If one recognizes the difficulty associated with measuring longitudinal thermal dispersion coefficients, one concludes that the agreement between theory and experiment is quite reasonable. In addition, the general trends for the two different systems are predicted with reasonable accuracy in the absence of any adjustable parameters.

V. TWO-EQUATION MODEL FOR HEAT TRANSFER

When the condition of local thermal equilibrium is not valid, we must return to Eqs. (37) and (38) and develop separate equations for $\langle T_\beta \rangle^\beta$ and $\langle T_\sigma \rangle^\sigma$. We begin with the interfacial flux term and decompose the temperature to obtain

$$\frac{1}{\mathcal{V}}\int_{A_{\beta\sigma}} \mathbf{n}_{\beta\sigma} \cdot k_\beta \nabla T_\beta \mathrm{d}A = \frac{1}{\mathcal{V}}\int_{A_{\beta\sigma}} \mathbf{n}_{\beta\sigma} \cdot k_\beta \nabla \langle T_\beta \rangle^\beta \mathrm{d}A + \frac{1}{\mathcal{V}}\int_{A_{\beta\sigma}} \mathbf{n}_{\beta\sigma} \cdot k_\beta \nabla \tilde{T}_\beta \mathrm{d}A \quad (101)$$

When the standard length-scale constraints are valid (Carbonell and Whitaker, 1984), the gradient of the average temperature can be removed from inside the integral, leading to

$$\frac{1}{\mathcal{V}}\int_{A_{\beta\sigma}} \mathbf{n}_{\beta\sigma}\cdot k_\beta \nabla T_\beta \mathrm{d}A = \left\{\frac{1}{\mathcal{V}}\int_{A_{\beta\sigma}} \mathbf{n}_{\beta\sigma}\mathrm{d}A\right\}\cdot k_\beta \nabla\langle T_\beta\rangle^\beta + \frac{1}{\mathcal{V}}\int_{A_{\beta\sigma}} \mathbf{n}_{\beta\sigma}\cdot k_\beta\nabla\tilde{T}_\beta \mathrm{d}A \tag{102}$$

From the averaging theorem we have the geometrical relation

$$\frac{1}{\mathcal{V}}\int_{A_{\beta\sigma}} \mathbf{n}_{\beta\sigma}\mathrm{d}A = \nabla\varepsilon_\beta \tag{103}$$

which allows us to express (102) in the form

$$\frac{1}{\mathcal{V}}\int_{A_{\beta\sigma}} \mathbf{n}_{\beta\sigma}\cdot k_\beta \nabla T_\beta \mathrm{d}A = -\nabla\varepsilon_\beta\cdot k_\beta\nabla\langle T_\beta\rangle^\beta + \frac{1}{\mathcal{V}}\int_{A_{\beta\sigma}} \mathbf{n}_{\beta\sigma}\cdot k_\beta\nabla\tilde{T}_\beta \mathrm{d}A \tag{104}$$

Substitution of this expression for the interfacial flux into the β-phase transport equation (37) leads to

$$\begin{aligned}
&\varepsilon_\beta(\rho c_p)_\beta \frac{\partial\langle T_\beta\rangle^\beta}{\partial t} + \varepsilon_\beta(\rho c_p)_\beta\langle\mathbf{v}_\beta\rangle^\beta\cdot\nabla\langle T_\beta\rangle^\beta \\
&= \nabla\cdot\left[k_\beta\left(\varepsilon_\beta\nabla\langle T_\beta\rangle^\beta + \frac{1}{\mathcal{V}}\int_{A_{\beta\sigma}}\mathbf{n}_{\beta\sigma}\tilde{T}_\beta\mathrm{d}A\right)\right] - (\rho c_p)_\beta\nabla\cdot\langle\tilde{\mathbf{v}}_\beta\tilde{T}_\beta\rangle \\
&-\nabla\varepsilon_\beta\cdot k_\beta\nabla\langle T_\beta\rangle^\beta + \frac{1}{\mathcal{V}}\int_{A_{\beta\sigma}}\mathbf{n}_{\beta\sigma}\cdot k_\beta\cdot\nabla\tilde{T}_\beta\mathrm{d}A
\end{aligned} \tag{105}$$

and we see that part of the interfacial flux has a *convective-like* characteristic in that it is proportional to the gradient of the temperature. The analogous equation for the σ-phase is given by

$$\begin{aligned}
&\varepsilon_\sigma(\rho c_p)_\sigma \frac{\partial\langle T_\sigma\rangle^\sigma}{\partial t} = \nabla\cdot\left[k_\sigma\left(\varepsilon_\sigma\nabla\langle T_\sigma\rangle^\sigma + \frac{1}{\mathcal{V}}\int_{A_{\sigma\beta}}\mathbf{n}_{\sigma\beta}\tilde{T}_\sigma\mathrm{d}A\right)\right] \\
&-\nabla\varepsilon_\sigma\cdot k_\sigma\nabla\langle T_\sigma\rangle^\sigma + \frac{1}{\mathcal{V}}\int_{A_{\sigma\beta}}\mathbf{n}_{\sigma\beta}\cdot k_\sigma\nabla\tilde{T}_\sigma\mathrm{d}A + \varepsilon_\sigma\langle\Phi_\sigma\rangle^\sigma
\end{aligned} \tag{106}$$

and we need only develop the closure problem for $\tilde{T}_\beta$ and $\tilde{T}_\sigma$ to complete the theoretical analysis for the two-equation model.

A. Closure Problem

In order to develop a governing differential equation for $\tilde{T}_\beta$, we need the point equation given by

$$(\rho c_p)_\beta \frac{\partial T_\beta}{\partial t} + (\rho c_p)_\beta \mathbf{v}_\beta \cdot \nabla T_\beta = \nabla \cdot (k_\beta \nabla T_\beta) \tag{107}$$

and the intrinsic form of the governing differential equation for $\langle T_\beta \rangle^\beta$. This can be obtained by dividing (105) by ε_β and arranging the result in the form

$$\begin{aligned}(\rho c_p)_\beta \frac{\partial \langle T_\beta \rangle^\beta}{\partial t} &+ (\rho c_p)_\beta \langle \mathbf{v}_\beta \rangle^\beta \cdot \nabla \langle T_\beta \rangle^\beta \\ &= \nabla \cdot (k_\beta \nabla \langle T_\beta \rangle^\beta) + \varepsilon_\beta^{-1} \nabla \cdot \left(\frac{k_\beta}{\mathcal{V}} \int_{A_{\beta\sigma}} \mathbf{n}_{\beta\sigma} \tilde{T}_\beta \mathrm{d}A \right) \\ &- \varepsilon_\beta^{-1} (\rho c_p)_\beta \nabla \cdot \langle \tilde{\mathbf{v}}_\beta \tilde{T}_\beta \rangle + \frac{\varepsilon_\beta^{-1}}{\mathcal{V}} \int_{A_{\beta\sigma}} \mathbf{n}_{\beta\sigma} \cdot k_\beta \nabla \tilde{T}_\beta \mathrm{d}A \end{aligned} \tag{108}$$

At this point we recall the decompositions given by Eqs. (29)

$$T_\beta = \langle T_\beta \rangle^\beta + \tilde{T}_\beta, \qquad \mathbf{v}_\beta = \langle \mathbf{v}_\beta \rangle^\beta + \tilde{\mathbf{v}}_\beta \tag{109}$$

so that it becomes clear that subtracting (108) from (107) will lead to a governing differential equation for $\tilde{T}_\beta$ given by

$$\begin{aligned}(\rho c_p)_\beta \frac{\partial \tilde{T}_\beta}{\partial t} &+ (\rho c_p)_\beta \mathbf{v}_\beta \cdot \nabla \tilde{T}_\beta + (\rho c_p)_\beta \tilde{\mathbf{v}}_\beta \cdot \nabla \langle T_\beta \rangle^\beta \\ &= \nabla \cdot (k_\beta \nabla \tilde{T}_\beta) - \underbrace{\varepsilon_\beta^{-1} \nabla \cdot \left(\frac{k_\beta}{\mathcal{V}} \int_{A_{\beta\sigma}} \mathbf{n}_{\beta\sigma} \tilde{T}_\beta \mathrm{d}A \right)}_{\text{non-local conduction}} \\ &+ \underbrace{\varepsilon_\beta^{-1} (\rho c_p)_\beta \nabla \cdot \langle \tilde{\mathbf{v}}_\beta \tilde{T}_\beta \rangle}_{\text{non-local convection}} - \frac{\varepsilon_\beta^{-1}}{\mathcal{V}} \int_{A_{\beta\sigma}} \mathbf{n}_{\beta\sigma} \cdot k_\beta \nabla \tilde{T}_\beta \mathrm{d}A \end{aligned} \tag{110}$$

Here we have used the word *non-local* to describe terms in which $\tilde{T}_\beta$ is evaluated at points different from the centroid of the averaging volume. We can simplify this result on the basis of the restrictions

$$\nabla \cdot (k_\beta \nabla \tilde{T}_\beta) \gg \underbrace{\varepsilon_\beta^{-1} \nabla \cdot \left(\frac{k_\beta}{\mathcal{V}} \int_{A_{\beta\sigma}} \mathbf{n}_{\beta\sigma} \tilde{T}_\beta \mathrm{d}A \right)}_{\text{non-local conduction}} \tag{111a}$$

$$(\rho c_p)_\beta \mathbf{v}_\beta \cdot \nabla \tilde{T}_\beta \gg \underbrace{\varepsilon_\beta^{-1} (\rho c_p)_\beta \nabla \cdot \langle \tilde{\mathbf{v}}_\beta \tilde{T}_\beta \rangle}_{\text{non-local convection}} \tag{111b}$$

Both these restrictions will be satisfied whenever $\ell_\beta \ll L$ (Quintard and Whitaker, 1994; Whitaker, 1999); thus the non-local terms will generally be negligible in the *spatial deviation transport equation* and the imposition of these restrictions allows us to simplify Eq. (110) in the obvious manner. The governing equation for $\tilde{T}_\sigma$ can be derived following the same procedure, and the boundary conditions for the spatial deviation temperatures are obtained directly from Eqs. (3) and (4), leading to the following closure problem

$$(\rho c_p)_\beta \frac{\partial \tilde{T}_\beta}{\partial t} + (\rho c_p)_\beta \mathbf{v}_\beta \cdot \nabla \tilde{T}_\beta + \underbrace{(\rho c_p)_\beta \tilde{\mathbf{v}}_\beta \cdot \nabla \langle T_\beta \rangle^\beta}_{\text{source}}$$

$$= \nabla \cdot (k_\beta \nabla \tilde{T}_\beta) - \frac{\varepsilon_\beta^{-1}}{\mathcal{V}} \int_{A_{\beta\sigma}} \mathbf{n}_{\beta\sigma} \cdot k_\beta \nabla \tilde{T}_\beta \mathrm{d}A \tag{112}$$

B.C.1 $$\tilde{T}_\beta = \tilde{T}_\sigma - \underbrace{(\langle T_\beta \rangle^\beta - \langle T_\sigma \rangle^\sigma)}_{\text{source}}, \qquad \text{at } \mathcal{A}_{\beta\sigma} \tag{113}$$

B.C.2 $$\mathbf{n}_{\beta\sigma} \cdot k_\beta \nabla \tilde{T}_\beta = \mathbf{n}_{\beta\sigma} \cdot k_\sigma \nabla \tilde{T}_\sigma - \underbrace{\mathbf{n}_{\beta\sigma} \cdot k_\beta \nabla \langle T_\beta \rangle^\beta}_{\text{source}}$$

$$+ \underbrace{\mathbf{n}_{\beta\sigma} \cdot k_\sigma \nabla \langle T_\sigma \rangle^\sigma}_{\text{source}} + \underbrace{\langle \Omega \rangle_{\beta\sigma}}_{\text{source}} + \tilde{\Omega}, \qquad \text{at } \mathcal{A}_{\beta\sigma} \tag{114}$$

$$(\rho c_p)_\sigma \frac{\partial \tilde{T}_\sigma}{\partial t} = \nabla \cdot (k_\sigma \nabla \tilde{T}_\sigma) - \frac{\varepsilon_\sigma^{-1}}{\mathcal{V}} \int_{A_{\sigma\beta}} \mathbf{n}_{\sigma\beta} \cdot k_\sigma \nabla \tilde{T}_\sigma \mathrm{d}A + \tilde{\Phi}_\sigma \tag{115}$$

Here we have decomposed the heterogeneous thermal source and the homogeneous thermal source in the usual manner, i.e.

$$\Omega = \langle \Omega \rangle_{\beta\sigma} + \tilde{\Omega}, \qquad \Phi_\sigma = \langle \Phi_\sigma \rangle^\sigma + \tilde{\Phi}_\sigma \tag{116}$$

and we can see that these two thermal sources play different roles in the closure problem. The heterogeneous thermal source gives rise to both a source (in the closure sense) and a coupling term whereas the homogeneous thermal source gives rise to only a coupling term. In a detailed study of the drying process (Whitaker, 1998), one sees that terms such as $\tilde{\Omega}$ and $\tilde{\Phi}_\sigma$ can give rise to coupling at the *microscopic level*, while $\langle \Omega \rangle_{\beta\sigma}$ and $\langle \Phi_\sigma \rangle^\sigma$ can cause coupling at the *macroscopic level*. In this particular study, we will set $\tilde{\Omega}$ and $\tilde{\Phi}_\sigma$ equal to zero and thus neglect coupling at the microscopic level. In addition to ignoring the spatial deviations in the thermal sources, we will also simplify the closure problem by imposing the quasi-steady condition. This requires the constraints

$$\frac{\alpha_\beta t^*}{\ell_\beta^2} \gg 1, \qquad \frac{\alpha_\sigma t^*}{\ell_\sigma^2} \gg 1 \tag{117}$$

in which α_β and α_σ are the thermal diffusivities for the β- and σ-phases. The closure problem can now be simplified to

$$(\rho c_p)_\beta \mathbf{v}_\beta \cdot \nabla \tilde{T}_\beta + \underbrace{(\rho c_p)_\beta \tilde{\mathbf{v}}_\beta \cdot \nabla \langle T_\beta \rangle^\beta}_{\text{source}} = \nabla \cdot (k_\beta \nabla \tilde{T}_\beta) - \frac{\varepsilon_\beta^{-1}}{\mathcal{V}} \int_{A_{\beta\sigma}} \mathbf{n}_{\beta\sigma} \cdot k_\beta \nabla \tilde{T}_\beta \mathrm{d}A \tag{118}$$

$$\text{B.C.1} \qquad \tilde{T}_\beta = \tilde{T}_\sigma - \underbrace{(\langle T_\beta \rangle^\beta - \langle T_\sigma \rangle^\sigma)}_{\text{source}}, \qquad \text{at} \qquad \mathcal{A}_{\beta\sigma} \tag{119}$$

$$\text{B.C.2} \qquad \mathbf{n}_{\beta\sigma} \cdot k_\beta \nabla \tilde{T}_\beta = \mathbf{n}_{\beta\sigma} \cdot k_\sigma \nabla \tilde{T}_\sigma - \underbrace{\mathbf{n}_{\beta\sigma} \cdot k_\beta \nabla \langle T_\beta \rangle^\beta}_{\text{source}}$$

$$+ \underbrace{\mathbf{n}_{\beta\sigma} \cdot k_\sigma \nabla \langle T_\sigma \rangle^\sigma}_{\text{source}} + \underbrace{\Omega}_{\text{source}}, \qquad \text{at} \qquad \mathcal{A}_{\beta\sigma} \tag{120}$$

$$0 = \nabla \cdot (k_\sigma \nabla \tilde{T}_\sigma) - \frac{\varepsilon_\sigma^{-1}}{\mathcal{V}} \int_{A_{\sigma\beta}} \mathbf{n}_{\sigma\beta} \cdot k_\sigma \nabla \tilde{T}_\sigma \mathrm{d}A \tag{121}$$

$$\text{Average:} \qquad \langle \tilde{T}_\beta \rangle^\beta = 0, \qquad \langle \tilde{T}_\sigma \rangle^\sigma = 0 \tag{122}$$

$$\text{Periodicity:} \quad \tilde{T}_\beta(\mathbf{r} + \ell_i) = \tilde{T}_\beta(\mathbf{r}), \quad \tilde{T}_\sigma(\mathbf{r} + \ell_i) = \tilde{T}_\sigma(\mathbf{r}), \quad i = 1, 2, 3 \tag{123}$$

Here we have imposed a *periodicity condition* on the spatial deviation temperatures with the idea that we only need to solve the closure problem in some *representative region* that can be treated as a unit cell in a spatially periodic model of a porous medium (Bensoussan et al., 1978, Sanchez-Palencia, 1980; Brenner, 1980). We have shown such a representative region in Figure 7. In order to be effective, that region must contain the elements of the porous media that will pass through the filters (Whitaker, 1999) represented by the area and volume integrals in Eqs. (105) and (106). In addition to the periodicity condition, we also require that the average of the spatial deviation temperatures be zero; this is necessary in order to determine the value of the area integrals in Eqs. (118) and (121). We can make use of the

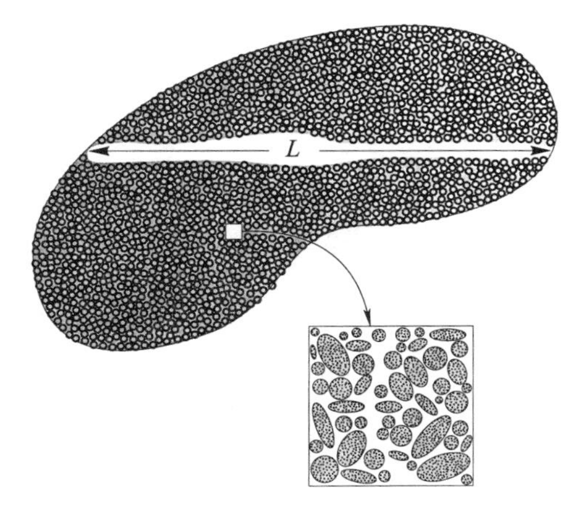

Figure 7. Representative region.

flux boundary condition (12) to show that these two area integrals are related by

$$\frac{1}{\mathcal{V}}\int_{A_{\beta\sigma}} \mathbf{n}_{\beta\sigma} \cdot k_\beta \nabla \tilde{T}_\beta \mathrm{d}A = \frac{1}{\mathcal{V}}\int_{A_{\beta\sigma}} \mathbf{n}_{\beta\sigma} \cdot k_\sigma \nabla \tilde{T}_\sigma \mathrm{d}A + a_{\mathrm{v}} \langle \Omega \rangle_{\beta\sigma} \tag{124}$$

The evaluation of these integrals is described by Quintard and Whitaker (1993) and by Quintard et al. (1997).

Given the four sources in the closure problem, we can follow previous studies (Ryan et al., 1981; Carbonell and Whitaker, 1984; Zanotti and Carbonell, 1984; Quintard and Whitaker, 1993) and express the spatial deviation temperatures according to

$$\tilde{T}_\beta = \mathbf{b}_{\beta\beta} \cdot \nabla \langle T_\beta \rangle^\beta + \mathbf{b}_{\beta\sigma} \cdot \nabla \langle T_\sigma \rangle^\sigma - s_\beta(\langle T_\beta \rangle^\beta - \langle T_\sigma \rangle^\sigma) + r_\beta \langle \Omega \rangle_{\beta\sigma} \tag{125a}$$

$$\tilde{T}_\sigma = \mathbf{b}_{\sigma\beta} \cdot \nabla \langle T_\beta \rangle^\beta + b_{\sigma\sigma} \cdot \nabla \langle T_\sigma \rangle^\sigma + s_\sigma(\langle T_\sigma \rangle^\sigma - \langle T_\beta \rangle^\beta) + r_\sigma \langle \Omega \rangle_{\beta\sigma} \tag{125b}$$

The new variables, $\mathbf{b}_{\beta\sigma}$, s_β, r_σ, etc., are the *closure variables* or the *mapping variables*, and we only need to determine these variables in some *representative region* in order to evaluate the terms in Eqs. (105) and (106) that contain the spatial deviation variables. The first closure problem is associated with $\nabla \langle T_\beta \rangle^\beta$ and takes the form

Problem I

$$(\rho c_p)_\beta \tilde{\mathbf{v}}_\beta + (\rho c_p)_\beta \mathbf{v}_\beta \cdot \nabla \mathbf{b}_{\beta\beta} = k_\beta \nabla^2 \mathbf{b}_{\beta\beta} - \varepsilon_\beta^{-1} \mathbf{c}_{\beta\beta}, \quad \text{in the } \beta\text{-phase} \tag{126a}$$

$$\text{B.C.1} \quad \mathbf{b}_{\beta\beta} = \mathbf{b}_{\sigma\beta}, \quad \text{at } A_{\beta\sigma} \tag{126b}$$

$$\text{B.C.2} \quad \mathbf{n}_{\beta\sigma} \cdot k_\beta \nabla \mathbf{b}_{\beta\beta} = \mathbf{n}_{\beta\sigma} \cdot k_\sigma \nabla \mathbf{b}_{\sigma\beta} - \mathbf{n}_{\beta\sigma} k_\beta, \quad \text{at } A_{\beta\sigma} \tag{126c}$$

$$0 = k_\sigma \nabla^2 \mathbf{b}_{\sigma\beta} + \varepsilon_\sigma^{-1} \mathbf{c}_{\beta\beta}, \quad \text{in the } \sigma\text{-phase} \tag{126d}$$

$$\text{Periodicity:} \quad \mathbf{b}_{\beta\beta}(\mathbf{r} + \ell_i) = \mathbf{b}_{\beta\beta}(\mathbf{r}), \quad \mathbf{b}_{\sigma\beta}(\mathbf{r} + \ell_i) = \mathbf{b}_{\sigma\beta}(\mathbf{r}), \quad i = 1, 2, 3 \tag{126e}$$

$$\text{Average:} \quad \langle \mathbf{b}_{\beta\beta} \rangle^\beta = 0, \quad \langle \boldsymbol{\beta}_{\sigma\beta} \rangle^\sigma = 0 \tag{126f}$$

Here $\mathbf{c}_{\beta\beta}$ is the unknown integral represented by

$$\mathbf{c}_{\beta\beta} = \frac{1}{\mathcal{V}} \int_{A_{\beta\sigma}} \mathbf{n}_{\beta\sigma} \cdot k_\beta \nabla \mathbf{b}_{\beta\beta} \mathrm{d}A \tag{126g}$$

and it is the constraints on the averages of $\mathbf{b}_{\beta\beta}$ and $\mathbf{b}_{\sigma\beta}$ that allows us to determine the constant vector $\mathbf{c}_{\beta\beta}$. A detailed description of the evaluation of this unknown integral is given by Quintard and Whitaker (1993), Quintard *et al.* (1997), and Whitaker (1999).

The term $\nabla \langle T_\sigma \rangle^\sigma$ is also a source in the closure problem for $\tilde{T}_\beta$ and $\tilde{T}_\sigma$. The boundary value problem associated with the mapping variable for $\nabla \langle T_\sigma \rangle^\sigma$ is given by

Problem II

$$(\rho c_p)_\beta \mathbf{v}_\beta \cdot \nabla \mathbf{b}_{\beta\sigma} = k_\beta \nabla^2 \mathbf{b}_{\beta\sigma} - \varepsilon_\beta^{-1} \mathbf{c}_{\beta\sigma}, \quad \text{in the } \beta\text{-phase} \tag{127a}$$

$$\text{B.C.1} \quad \mathbf{b}_{\beta\sigma} = \mathbf{b}_{\sigma\sigma}, \quad \text{at } A_{\beta\sigma} \tag{127b}$$

$$\text{B.C.2} \quad \mathbf{n}_{\beta\sigma} \cdot k_\beta \nabla \mathbf{b}_{\beta\sigma} = \mathbf{n}_{\beta\sigma} \cdot k_\sigma \nabla \mathbf{b}_{\sigma\sigma} + \mathbf{n}_{\beta\sigma} k_\sigma, \quad \text{at } A_{\beta\sigma} \tag{127c}$$

$$0 = k_\sigma \nabla^2 \mathbf{b}_{\sigma\sigma} + \varepsilon_\sigma^{-1} \mathbf{c}_{\beta\sigma}, \quad \text{in the } \sigma\text{-phase} \tag{127d}$$

$$\text{Periodicity:} \quad \mathbf{b}_{\beta\sigma}(\mathbf{r} + \ell_i) = \mathbf{b}_{\beta\sigma}(\mathbf{r}), \quad \mathbf{b}_{\sigma\sigma}(\mathbf{r} + \ell_i) = \mathbf{b}_{\sigma\sigma}(\mathbf{r}), \quad i = 1, 2, 3 \tag{127e}$$

$$\text{Average:} \quad \langle \mathbf{b}_{\beta\sigma} \rangle^\beta = 0, \quad \langle \mathbf{b}_{\sigma\sigma} \rangle^\sigma = 0 \tag{127f}$$

Here the single undetermined constant is given by

$$\mathbf{c}_{\beta\sigma} = \frac{1}{\mathcal{V}} \int_{A_{\beta\sigma}} \mathbf{n}_{\beta\sigma} \cdot k_\beta \nabla \mathbf{b}_{\beta\sigma} \mathrm{d}A \tag{127g}$$

and can be evaluated by means of Eqs. (127f).

Moving on to the source represented by $\langle T_\beta\rangle^\beta - \langle T_\sigma\rangle^\sigma$ in Eq. (113), we construct the following boundary value problem for the *mapping scalars* s_β and s_σ:

Problem III

$$(\rho c_p)_\beta \mathbf{v}_\beta \cdot \nabla s_\beta = k_\beta \nabla^2 s_\beta - \varepsilon_\beta^{-1}(a_\mathrm{v} h), \quad \text{in the } \beta\text{-phase} \tag{128a}$$

$$\text{B.C.1} \qquad \mathbf{n}_{\beta\sigma} \cdot k_\beta \nabla s_\beta = \mathbf{n}_{\beta\sigma} \cdot k_\sigma \nabla s_\sigma, \qquad \text{at } A_{\beta\sigma} \tag{128b}$$

$$\text{B.C.2} \qquad s_\beta = s_\sigma + 1, \qquad \text{at } A_{\beta\sigma} \tag{128c}$$

$$0 = k_\sigma \nabla^2 s_\sigma + \varepsilon_\sigma^{-1}(a_\mathrm{v} h), \qquad \text{in the } \sigma\text{-phase} \tag{128d}$$

$$\text{Periodicity:} \quad s_\beta(\mathbf{r} + \ell_i) = s_\beta(\mathbf{r}), \qquad s_\sigma(\mathbf{r} + \ell_i) = s_\sigma(\mathbf{r}), \qquad i = 1, 2, 3 \tag{128e}$$

$$\text{Average:} \quad \langle s_\beta \rangle^\beta = 0, \qquad \langle s_\sigma \rangle^\sigma = 0 \tag{128f}$$

In this closure problem, the undetermined constant is represented by

$$a_\mathrm{v} h = \frac{1}{\mathcal{V}} \int_{A_{\beta\sigma}} \mathbf{n}_{\beta\sigma} \cdot k_\beta \nabla s_\beta \mathrm{d}A \tag{128g}$$

and this volumetric heat transfer coefficient is determined by means of (128f). Analytic expressions for the purely conductive case are given by Quintard and Whitaker (1993, 1995), and numerical results for the general case are given by Quintard et al. (1997).

The final closure problem is used to determine how the heterogeneous thermal source is distributed between the two phases. This closure problem takes the form

Problem IV

$$(\rho c_p)_\beta \mathbf{v}_\beta \cdot \nabla r_\beta = k_\beta \nabla^2 r_\beta - a_\mathrm{v} \varepsilon_\beta^{-1} \xi_\beta, \quad \text{in the } \beta\text{-phase} \tag{129a}$$

$$\text{B.C.1} \qquad \mathbf{n}_{\beta\sigma} \cdot k_\beta \nabla r_\beta = \mathbf{n}_{\beta\sigma} \cdot k_\sigma \nabla r_\sigma + 1, \qquad \text{at } A_{\beta\sigma} \tag{129b}$$

$$\text{B.C.2} \qquad r_\beta = r_\sigma \qquad \text{at } A_{\beta\sigma} \tag{129c}$$

$$0 = k_\sigma \nabla^2 r_\sigma - a_\mathrm{v} \varepsilon_\sigma^{-1} \xi_\sigma, \qquad \text{in the } \sigma\text{-phase} \tag{129d}$$

$$\text{Periodicity:} \quad r_\beta(\mathbf{r} + \ell_i) = r_\beta(\mathbf{r}), \qquad r_\sigma(\mathbf{r} + \ell_i) = r_\sigma(\mathbf{r}), \qquad i = 1, 2, 3 \tag{129e}$$

Average: $\langle r_\beta \rangle^\beta = 0, \qquad \langle r_\sigma \rangle^\sigma = 0$ (129f)

Here we have defined the fraction ξ_β and ξ_σ explicitly by

$$\xi_\beta = \frac{1}{A_{\beta\sigma}} \int_{A_{\beta\sigma}} \mathbf{n}_{\beta\sigma} \cdot k_\beta \nabla r_\beta \mathrm{d}A, \qquad \xi_\sigma = \frac{1}{A_{\sigma\beta}} \int_{A_{\sigma\beta}} \mathbf{n}_{\sigma\beta} \cdot k_\sigma \nabla r_\sigma \mathrm{d}A \tag{130}$$

but we have not yet imposed the flux condition given by (124). When we impose that condition, we can quickly conclude that the fractions are constrained by

$$\xi_\beta + \xi_\sigma = 1 \tag{131}$$

We designate the single independent fraction by

$$\xi = \xi_\beta \tag{132}$$

and we refer to ξ as the *distribution coefficient.*

1. Distribution Coefficient

To illustrate how ξ is determined, we follow the ideas of Quintard et al. (1997) and introduce the decompositions given by

$$r_\beta = r_\beta^{\mathrm{o}} + \xi R_\beta, \qquad r_\sigma = r_\sigma^{\mathrm{o}} + \xi R_\sigma \tag{133}$$

This leads to two independent closure problems that are given by

Problem IV′

$$(\rho c_p)_\beta \mathbf{v}_\beta \cdot \nabla r_\beta^{\mathrm{o}} = k_\beta \nabla^2 r_\beta^{\mathrm{o}}, \qquad \text{in the } \beta\text{-phase} \tag{134a}$$

B.C.1 $\quad \mathbf{n}_{\beta\sigma} \cdot k_\beta \nabla r_\beta^{\mathrm{o}} = \mathbf{n}_{\beta\sigma} \cdot k_\sigma \nabla r_\sigma^{\mathrm{o}} + 1, \qquad \text{at } A_{\beta\sigma}$ (134b)

B.C.2 $\quad r_\beta^{\mathrm{o}} = r_\sigma^{\mathrm{o}}, \qquad \text{at } A_{\beta\sigma}$ (134c)

$$0 = k_\sigma \nabla^2 r_\sigma^{\mathrm{o}} - a_{\mathrm{v}} \varepsilon_\sigma^{-1}, \qquad \text{in the } \sigma\text{-phase} \tag{134d}$$

Periodicity: $r_\beta^{\mathrm{o}}(\mathbf{r} + \ell_i) = r_\beta^{\mathrm{o}}(\mathbf{r}), \qquad r_\sigma^{\mathrm{o}}(\mathbf{r} + \ell_i) = r_\sigma^{\mathrm{o}}(\mathbf{r}), \qquad i = 1, 2, 3$ (134e)

Average: $\varepsilon_\beta \langle r_\beta^{\mathrm{o}} \rangle^\beta + \varepsilon_\sigma \langle r_\sigma^{\mathrm{o}} \rangle^\sigma = 0$ (134f)

Problem IV″

$$(\rho c_p)_\beta \mathbf{v}_\beta \cdot \nabla R_\beta = k_\beta \nabla^2 R_\beta - a_{\mathrm{v}} \varepsilon_\beta^{-1}, \qquad \text{in the } \beta\text{-phase} \tag{135a}$$

B.C.1 $\quad \mathbf{n}_{\beta\sigma} \cdot k_\beta \nabla R_\beta = \mathbf{n}_{\beta\sigma} \cdot k_\sigma \nabla R_\sigma, \qquad \text{at } A_{\beta\sigma}$ (135b)

$$\text{B.C.2} \qquad R_\beta = R_\sigma, \qquad \text{at } A_{\beta\sigma} \tag{135c}$$

$$0 = k_\sigma \nabla^2 R_\sigma - a_v \varepsilon_\sigma^{-1}, \qquad \text{in the } \sigma\text{-phase} \tag{135d}$$

$$\text{Periodicity:} \quad R_\beta(\mathbf{r} + \ell_i) = R_\beta(\mathbf{r}), \quad R_\sigma(\mathbf{r} + \ell_i) = R_\sigma(\mathbf{r}), \quad i = 1, 2, 3 \tag{135e}$$

$$\text{Average:} \quad \varepsilon_\beta \langle R_\beta \rangle^\beta + \varepsilon_\sigma \langle R_\sigma \rangle^\sigma = 0 \tag{135f}$$

The conditions imposed on the averages by Eqs. (129f) are not automatically satisfied by these two closure problems, and we can use (129f), (134f), and (135f) to determine ξ according to

$$\xi = -\langle r_\beta^{\mathrm{o}} \rangle^\beta / \langle R_\beta \rangle^\beta \tag{136}$$

The two closure problems that provide the distribution coefficient can be solved numerically following the methodology outlined in Quintard et al. (1997), and a three-dimensional numerical algorithm has been developed. Analytical solutions are also available in the case of simple unit cells.

(a) Stratified Systems. The closure problem was solved analytically for the stratified system shown in Figure 8. For stratified systems, there is no effect of the velocity field; thus ξ is independent of the Péclet number and is given by

$$\xi = \frac{\varepsilon_\sigma k_\beta}{\varepsilon_\beta k_\sigma + \varepsilon_\sigma k_\beta} \tag{137}$$

The evolution of ξ versus the thermal conductivity ratio k_σ / k_β is plotted in Figure 9, as well as some results obtained numerically for the same unit cell. There is excellent agreement between the analytical solution and the numer-

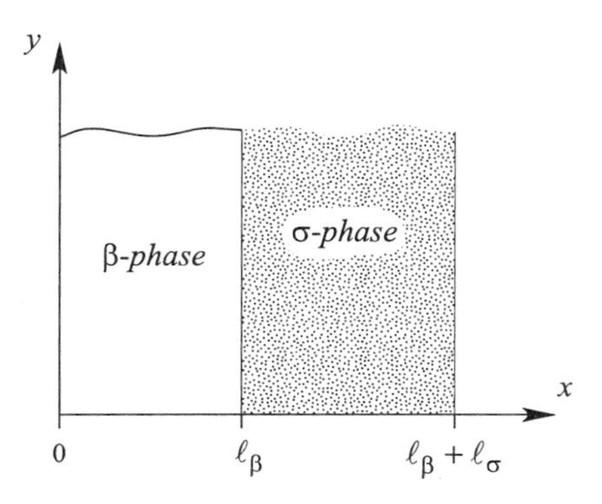

Figure 8. Unit cell for a stratified system.

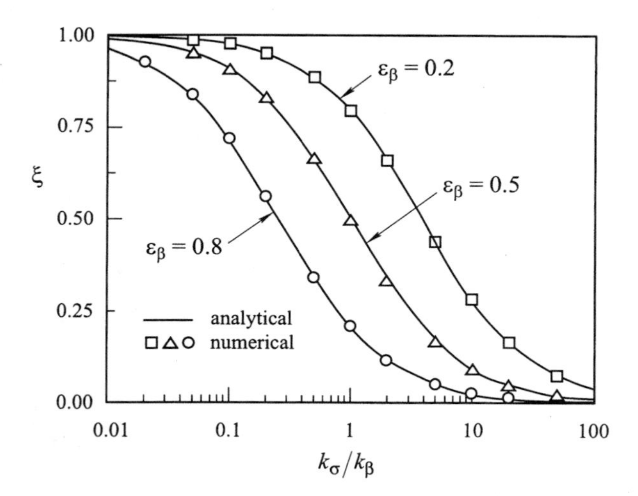

Figure 9. Comparison between analytical and numerical results for stratified systems.

ical results, and this provides verification of the numerical method which we will use for the treatment of more complex unit cells.

(b) Chang's Unit Cell. Analytical results have been obtained for the Chang's unit cell (Chang 1982, 1983; Ochoa-Tapia et al., 1994) represented in Figure 10. In the three-dimensional case, the distribution coefficient is given by

$$\xi = \frac{\varepsilon_\beta^2}{\varepsilon_\beta^2 + (k_\beta/k_\sigma)[(\alpha^3 + 3\alpha^2 + 6\alpha + 5)(1-\alpha)^3]} \tag{138}$$

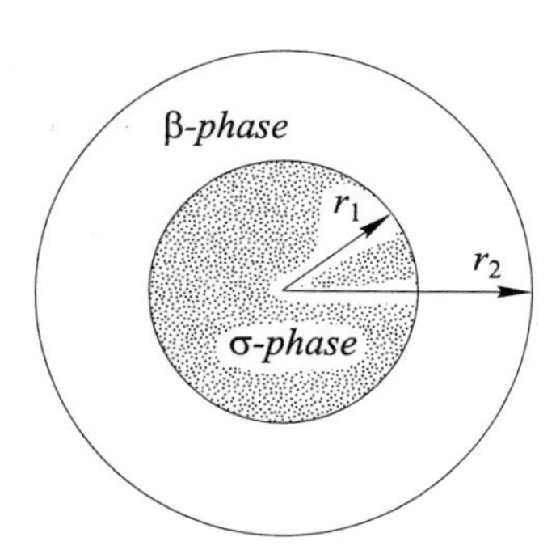

Figure 10. Chang's unit cell.

where the coefficient α is related to the volume fraction of the solid phase by

$$\alpha = \varepsilon_\sigma^{1/3} \tag{139}$$

The results for a three-dimensional version of Chang's unit cell are compared with numerical results for simple cubic packing in Figure 11. For high values of the porosity there is a very good agreement between both predictions. For low values of the porosity, where the assumptions associated with Chang's unit cell are not valid (Ochoa-Tapia et al., 1994), the agreement is less attractive. When convective effects are important, the complete version of the closure problem is required. The influence of the Péclet number is studied in the next section.

(c) Numerical Results for Convection. Numerical results can be obtained for two- and three-dimensional periodic unit cells. The numerical model is based on the methodology presented in Quintard et al. (1997). The domain and phase geometry is discretized over a Cartesian grid as illustrated in Figure 12. The finite difference scheme is based on an upstream scheme corrected for numerical diffusion as explained in Quintard and Whitaker (1994). The diffusive part is discretized as indicated in Quintard (1993) to account for the phase repartition. Examples of numerical results are given in

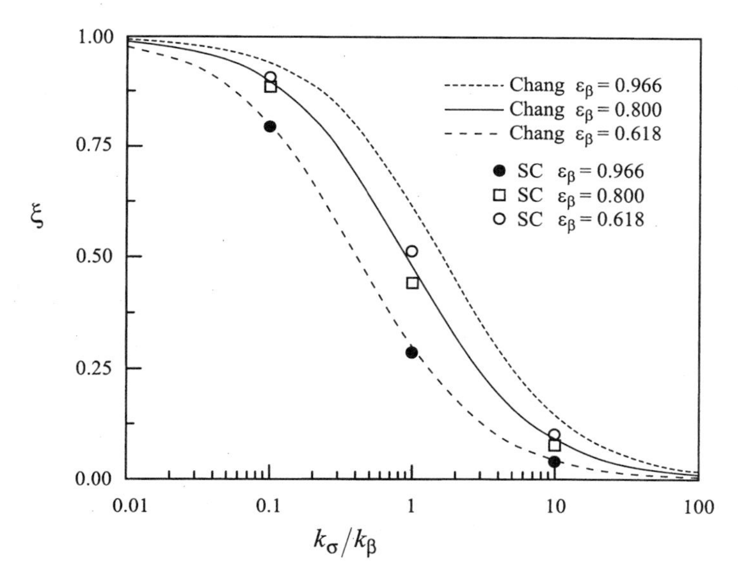

Figure 11. Comparison between three-dimensional numerical computation for arrays of spheres and results obtained with Chang's unit cell.

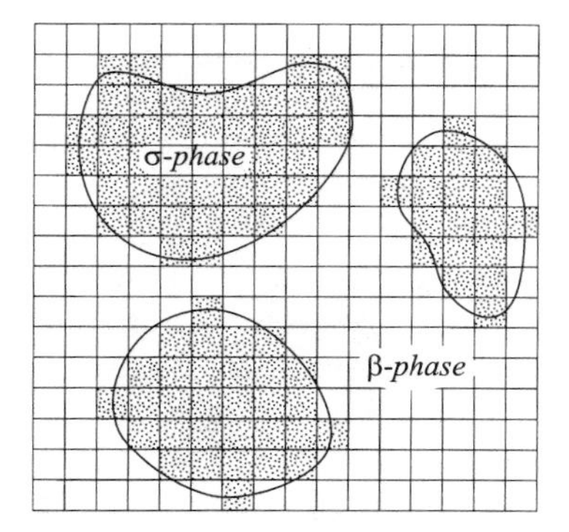

Figure 12. Example of discretized unit cell.

Figure 13 for the inline and staggered two-dimensional unit cells illustrated in Figure 14, and in Figure 15 for a three-dimensional, simple cubic packing of spheres. The results show a behavior similar to that of the case for pure conduction illustrated in Figures 9 and 11. The dependence of ξ on the Péclet number is seen to be quite weak for values of the Péclet number less than 10. It is of some interest to note that the Péclet number dependence is stronger for the two-dimensional arrays than for the three-dimensional

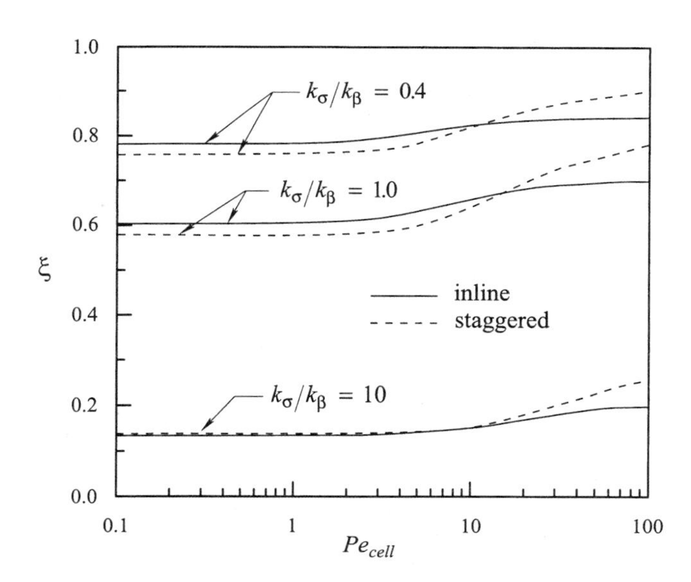

Figure 13. Distribution coefficient for two-dimensional arrays of cylinders ($\varepsilon_\beta = 0.38$).

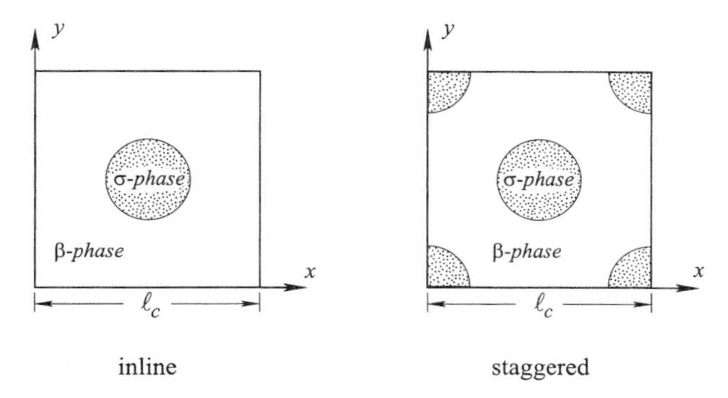

Figure 14. Inline and staggered arrays of cylinders.

simple cubic packing. This is caused by the fact that two-dimensional systems generate a more intense lateral convective transport at the same porosity as a three-dimensional system. In addition, the porosity is 0.38 for the two-dimensional results shown in Figure 13, and this is significantly lower than the value of 0.47 used for the three-dimensional calculations presented in Figure 15. Since the heterogeneous thermal source represents a source *at the fluid–solid interface* where the fluid velocity is zero, it is not surprising that the *distribution coefficient* is dominated by the conductivities of the two phases rather than the convective transport. In both Figure 13 and Figure 15, we have used a *cell Péclet number* defined by

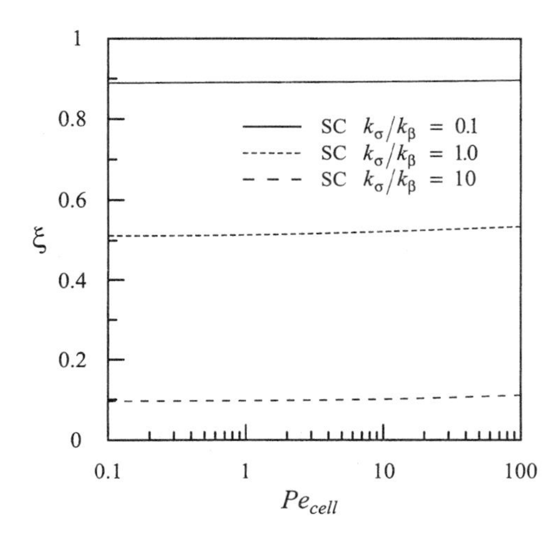

Figure 15. Distribution coefficient for a simple cubic packing of spheres ($\varepsilon_\beta = 0.47$).

$$\text{Pe}_{\text{cell}} = \frac{(\rho c_p)_\beta \langle \text{v}_\beta \rangle^\beta \ell_\text{c}}{k_\beta} \tag{140}$$

where the distance ℓ_c is illustrated in Figure 14.

At this point in time, no experimental studies have been carried out with the intention of measuring the distribution coefficient ξ; however, this quantity is required whenever a two-equation model is needed to describe a heat transfer process in which there is a heterogeneous thermal source.

B. Closed Forms

In order to develop the closed forms of the β- and σ-phase transport equations represented by (105) and (106), we need only make use of the representations for the spatial deviation temperatures given by Eqs. (125). For the β-phase we obtain a closed form given by

$$\begin{aligned}\varepsilon_\beta(\rho c_p)_\beta \frac{\partial \langle T_\beta \rangle^\beta}{\partial t} &+ \varepsilon_\beta(\rho c_p)_\beta \langle \mathbf{v}_\beta \rangle^\beta \cdot \nabla \langle T_\beta \rangle^\beta - \mathbf{u}_{\beta\beta} \cdot \nabla \langle T_\beta \rangle^\beta - \mathbf{u}_{\beta\sigma} \cdot \nabla \langle T_\sigma \rangle^\sigma \\ &= \nabla \cdot (\mathbf{K}^*_{\beta\beta} \cdot \nabla \langle T_\beta \rangle^\beta + \mathbf{K}^*_{\beta\sigma} \cdot \nabla \langle T_\sigma \rangle^\sigma) - a_\text{v} h(\langle T_\beta \rangle^\beta - \langle T_\sigma \rangle^\sigma) + a_\text{v}\xi \langle \Omega \rangle_{\beta\sigma}\end{aligned} \tag{141}$$

in which $\mathbf{K}^*_{\beta\beta}$ and $\mathbf{K}^*_{\beta\sigma}$ represent both *conductive* and *convective* transport and are defined by

$$\mathbf{K}^*_{\beta\beta} = \varepsilon_\beta k_\beta \mathbf{I} + \frac{k_\beta}{\mathcal{V}} \int_{A_{\beta\sigma}} \mathbf{n}_{\beta\sigma} \mathbf{b}_{\beta\beta} \mathrm{d}A - (\rho c_p)_\beta \langle \tilde{\mathbf{v}}_\beta \mathbf{b}_{\beta\beta} \rangle \tag{142}$$

$$\mathbf{K}^*_{\beta\sigma} = \frac{k_\beta}{\mathcal{V}} \int_{A_{\beta\sigma}} \mathbf{n}_{\beta\sigma} \mathbf{b}_{\beta\sigma} \mathrm{d}A - (\rho c_p)_\beta \langle \tilde{\mathbf{v}}_\beta \mathbf{b}_{\beta\sigma} \rangle \tag{143}$$

We refer to $\mathbf{K}^*_{\beta\beta}$ as the *dominant* thermal dispersion tensor and to $\mathbf{K}^*_{\beta\sigma}$ as the *coupling* thermal dispersion tensor. The heat transfer coefficient is determined by the solution of Problem III, and the result is given by

$$a_\text{v} h = \frac{1}{\mathcal{V}} \int_{A_{\beta\sigma}} \mathbf{n}_{\beta\sigma} \cdot k_\beta \nabla s_\beta \mathrm{d}A \tag{144}$$

while the manner in which the heterogeneous thermal source is distributed can be extracted from Problem IV, and the distribution coefficient is represented by

$$a_\text{v} \xi = \frac{1}{\mathcal{V}} \int_{A_{\beta\sigma}} \mathbf{n}_{\beta\sigma} \cdot k_\beta \nabla r_\beta \mathrm{d}A \tag{145}$$

Because we have assumed that Ω is a constant, two terms involving the heterogeneous thermal source are zero. We list these as

$$\nabla \cdot \left[\left(\frac{1}{\mathcal{V}} \int_{A_{\beta\sigma}} \mathbf{n}_{\beta\sigma} r_\beta \mathrm{d}A \right) \Omega \right] = 0 \tag{146}$$

$$(\rho c_p)_\beta \nabla \cdot (\langle \tilde{\mathbf{v}}_\beta r_\beta \rangle \Omega) = 0 \tag{147}$$

The two non-traditional convective transport terms in Eq. (141) depend on the coefficients $\mathbf{u}_{\beta\beta}$ and $\mathbf{u}_{\beta\sigma}$ that are determined by

$$\mathbf{u}_{\beta\beta} = \frac{1}{\mathcal{V}} \int_{A_{\beta\sigma}} \mathbf{n}_{\beta\sigma} \cdot k_\beta \nabla \mathbf{b}_{\beta\beta} \mathrm{d}A - \frac{k_\beta}{\mathcal{V}} \int_{A_{\beta\sigma}} \mathbf{n}_{\beta\sigma} s_\beta \mathrm{d}A + (\rho c_p)_\beta \langle \tilde{\mathbf{v}}_\beta s_\beta \rangle \tag{148}$$

$$\mathbf{u}_{\beta\sigma} = \frac{1}{\mathcal{V}} \int_{A_{\beta\sigma}} \mathbf{n}_{\beta\sigma} \cdot k_\beta \nabla \mathbf{b}_{\beta\sigma} \mathrm{d}A + \frac{k_\beta}{\mathcal{V}} \int_{A_{\beta\sigma}} \mathbf{n}_{\beta\sigma} s_\beta \mathrm{d}A - (\rho c_p)_\beta \langle \tilde{\mathbf{v}}_\beta s_\beta \rangle \tag{149}$$

It is of some interest to note that in the closed form of the β-phase transport equation we have discarded the term $k_\beta \nabla \varepsilon_\beta \cdot \nabla \langle T_\beta \rangle^\beta$ which appears in Eq. (105) as part of the interfacial flux. If one is confronted with a process in which this term is *not negligible*, it is likely that the errors occurring in the four closure problems are also *not negligible* and one needs to think very carefully about the validity of Eq. (141).

The closed form of the σ-phase transport equation is given by

$$\begin{aligned} \varepsilon_\sigma (\rho c_p)_\sigma \frac{\partial \langle T_\sigma \rangle^\sigma}{\partial t} - \mathbf{u}_{\sigma\beta} \cdot \nabla \langle T_\beta \rangle^\beta - \mathbf{u}_{\sigma\sigma} \cdot \nabla \langle T_\sigma \rangle^\sigma = \nabla \cdot (\mathbf{K}_{\sigma\beta} \cdot \nabla \langle T_\beta \rangle^\beta + \mathbf{K}_{\sigma\sigma} \cdot \nabla \langle T_\sigma \rangle^\sigma) \\ - a_\mathrm{v} h (\langle T_\sigma \rangle^\sigma - \langle T_\beta \rangle^\beta) + a_\mathrm{v} (1 - \xi) \langle \Omega \rangle_{\beta\sigma} + \varepsilon_\sigma \langle \Phi_\sigma \rangle^\sigma \end{aligned} \tag{150}$$

in which $\mathbf{K}_{\sigma\beta}$ and $\mathbf{K}_{\sigma\sigma}$ represent *only conductive transport*. These effective coefficients are defined by

$$\mathbf{K}_{\sigma\beta} = \frac{k_\sigma}{\mathcal{V}} \int_{A_{\beta\sigma}} \mathbf{n}_{\sigma\beta} \mathbf{b}_{\sigma\beta} \mathrm{d}A \tag{151}$$

$$\mathbf{K}_{\sigma\sigma} = \varepsilon_\sigma k_\sigma \mathbf{I} + \frac{k_\sigma}{\mathcal{V}} \int_{A_{\beta\sigma}} \mathbf{n}_{\sigma\beta} \mathbf{b}_{\sigma\sigma} \mathrm{d}A \tag{152}$$

We refer to $\mathbf{K}_{\sigma\beta}$ as the *coupling* thermal conductivity tensor and to $\mathbf{K}_{\sigma\sigma}$ as the *effective* thermal conductivity tensor. In the σ-phase, the non-traditional convective transport is entirely conductive in nature and the velocity-like coefficients are given by

$$\mathbf{u}_{\sigma\beta} = \frac{1}{\mathcal{V}} \int_{A_{\sigma\beta}} \mathbf{n}_{\sigma\beta} \cdot k_\sigma \nabla \mathbf{b}_{\sigma\beta} \mathrm{d}A - \frac{k_\sigma}{\mathcal{V}} \int_{A_{\sigma\beta}} \mathbf{n}_{\sigma\beta} s_\sigma \mathrm{d}A \tag{153}$$

$$\mathbf{u}_{\sigma\sigma} = \frac{1}{\mathcal{V}} \int_{A_{\sigma\beta}} \mathbf{n}_{\sigma\beta} \cdot k_\sigma \nabla \mathbf{b}_{\sigma\sigma} \mathrm{d}A + \frac{k_\sigma}{\mathcal{V}} \int_{A_{\sigma\beta}} \mathbf{n}_{\sigma\beta} s_\sigma \mathrm{d}A \tag{154}$$

It is important to remember that the closure problem has been developed for the case in which both Ω and Φ_σ have been treated as constants. Under these circumstances, we have imposed the following simplifcations

$$\langle \Omega \rangle_{\beta\sigma} = \Omega, \qquad \langle \Phi_\sigma \rangle^\sigma = \Phi_\sigma \tag{155}$$

This means that there is no coupling at the *microscopic level* associated with these thermal sources; however, coupling can occur at the *macroscopic level.*

When the condition of local thermal equilibrium is valid, one can add Eqs. (141) and (142) and show that the non-traditional convective transport terms sum to zero to obtain the one-equation model, given earlier by Eq. (49), that we list here as

$$\langle \rho \rangle C_p \frac{\partial \langle T \rangle}{\partial t} + \varepsilon_\beta (\rho c_p)_\beta \langle \mathbf{v}_\beta \rangle^\beta \cdot \nabla \langle T \rangle = \nabla \cdot (\mathbf{K}^* \cdot \nabla \langle T \rangle) + a_\mathrm{v} \langle \Omega \rangle_{\beta\sigma} + \varepsilon_\sigma \langle \Phi_\sigma \rangle^\sigma \tag{156}$$

Here the total effective thermal conductivity tensor is given by

$$\mathbf{K}^* = \mathbf{K}^*_{\beta\beta} + \mathbf{K}^*_{\beta\sigma} + \mathbf{K}_{\sigma\beta} + \mathbf{K}_{\sigma\sigma} \tag{157}$$

and the single closure problem that is used to determine $\mathbf{K}^*$ is given by Eqs. (95). The simplicity of the one-equation model and the associated closure problem certainly motivates the use of this form of the volume-averaged thermal energy equation; however, the one-equation model is valid only when the constraints associated with local thermal equilibrium are satisfied.

C. Comparison with Experiment

Because of the complexity of Eqs. (141) and (150), the comparison between theory and experiment is probably best done in terms of theoretical and experimental temperature fields as opposed to attempting to extract experimental values of the various coefficients. Quintard and Whitaker (1993) have used numerical experiments to carry out this type of comparison for the conductive transport problem; that comparison certainly supports the theoretical development of the two-equation model in the absence of convective transport. In the studies of Grangeot et al. (1990, 1994), the experimental determination of the volumetric heat transfer coefficient was carried out by approximating the various effective thermal coefficients as follows:

$$\mathbf{K}^*_{\beta\beta} = \varepsilon_\beta k_\beta \mathbf{I}, \qquad \mathbf{K}^*_{\beta\sigma} = \mathbf{K}_{\sigma\beta} = 0, \qquad \mathbf{K}_{\sigma\sigma} = \varepsilon_\sigma k_\sigma \mathbf{I} \tag{158}$$

whereas the non-traditional convective terms were neglected according to

$$\mathbf{u}_{\beta\beta} = \mathbf{u}_{\beta\sigma} = \mathbf{u}_{\sigma\beta} = \mathbf{u}_{\sigma\sigma} = 0 \tag{159}$$

On the basis of the calculations of Quintard et al. (1997), we know that first of Eqs. (158) significantly underestimates $\mathbf{K}^*_{\beta\beta}$ for Péclet numbers greater than 10. In addition, we know that the coupling tensors $\mathbf{K}^*_{\beta\sigma}$ and $\mathbf{K}_{\sigma\beta}$ are non-negligible for Péclet numbers greater than 100 when k_σ/k_β is large compared to one. Furthermore, calculations (Quintard et al., 1997) clearly indicate that the third of Eqs. (158) overestimates $\mathbf{K}_{\sigma\sigma}$ for non-touching particles, and from the studies of Nozad et al. (1985) and Shonnard and Whitaker (1989) one can deduce that it also overestimates $\mathbf{K}_{\sigma\sigma}$ for touching particles.

Given the simplifications indicated by Eqs. (158) and (159), Grangeot et al. (1990, 1994) estimated the volumetric heat transfer coefficient by comparing solutions of the two-equation model with the measured fluid and solid temperatures. The comparison between theory and experiment is illustrated in Figure 16 where the dimensionless heat transfer coefficient is shown as a function of the *cell Péclet number* for the brass–water system and the nylon–water system. The quantity $a_v h \ell_\beta^2 / k_\beta$ is essentially a Nusselt number; thus Figure 16 represents a classic plot of the Nusselt number as a function of the Péclet number. The values of k_σ / k_β for these two systems are given by Eqs. (99) and the theory certainly provides the proper trend with respect to this parameter. For the brass–water system, the theoretical values are too small by about a factor of four and it seems that this results from an underestimation of $\mathbf{K}^*_{\beta\beta}$ coupled with an overestimation of $\mathbf{K}_{\sigma\sigma}$. For the nylon–water system, the theoretical values are too large but the agreement between theory and experiment must be considered very good.

One experimental characteristic that is captured by the theory is the relatively weak dependence on the Péclet number. Many experimental studies of heat transfer in packed beds have been interpreted with some success using boundary layer models (Wakao and Kaguei, 1982). The empirical analysis of Whitaker (1972) makes use of a Péclet number dependence between one-half and two-thirds, and a reasonably good correlation is obtained, whereas the results shown in Figure 16 indicate a much weaker dependence. It is quite possible that the different dependence on the Péclet number results from the fact that the theory is only valid when the length-scale constraints (Whitaker, 1999) associated with periodicity in the closure problem are satisfied. By *periodicity* we mean

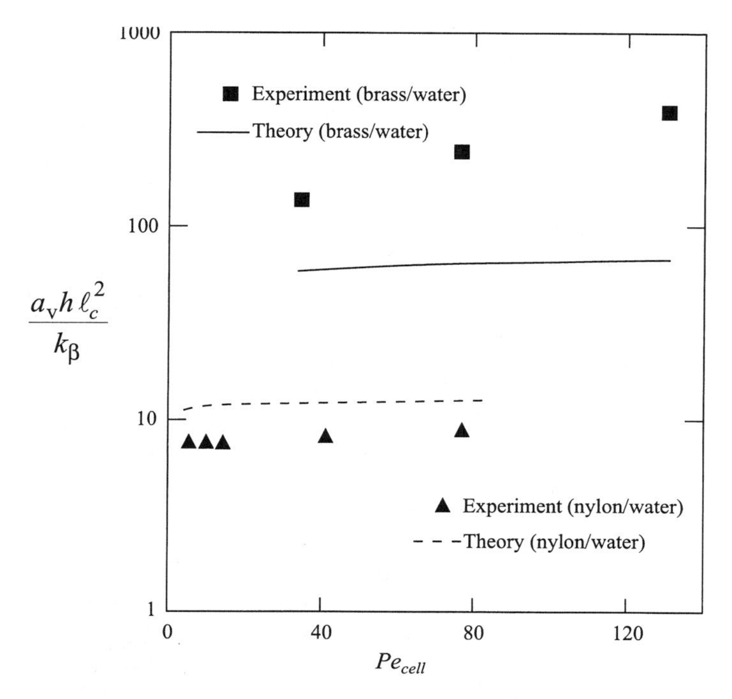

Figure 16. Theoretical and experimental volumetric heat transfer coefficients.

not only that the geometry is spatially periodic, but also that the spatial deviation temperatures $\tilde{T}_\beta$ and $\tilde{T}_\sigma$ are periodic. Periodicity in $\tilde{T}_\beta$ and $\tilde{T}_\sigma$ requires *small temperature gradients* whereas the typical experimental study of heat transfer in a packed bed is set up to produce *large temperature gradients* that lead to more easily measured heat transfer coefficients. For heat transfer to laminar flow in a tube, the classic Graetz solution (Jakob, 1949) indicates that the asymptotic Nusselt number is *independent* of the Péclet number; thus it should not be surprising that a theory constrained by periodicity in the closure problem gives rise to a relatively weak dependence on the Péclet number. While laboratory experiments are generally constructed to facilitate the measurement of a particular coefficient, practical applications are often associated with a parameter space that is different from the one encountered in the laboratory. In order to examine the complete spectrum of practical cases, the method of volume averaging should be extended to include cases that are not restricted to periodic boundary conditions in the closure problem. That problem remains as a challenge.

VI. CONCLUSIONS

In this chapter we have examined the simple process of convective heat transfer in a porous medium when there is a homogeneous thermal source in the solid phase and a heterogeneous thermal source at the fluid–solid interface. Heat transfer processes of this type are often coupled to mass transfer processes; however, the coupling has not been explored in this study. The condition of local thermal equilibrium is investigated and constraints associated with this condition are given. When local thermal equilibrium exists, the energy transport process can be described by a one-equation model containing a single parameter, the total thermal dispersion tensor. When local thermal equilibrium does not exist, the transport process must be described by a two-equation model. The details of this model have been presented. In addition to the four effective conduction and dispersion tensors and the volumetric heat transfer coefficient, one also requires information about the distribution of the heterogeneous thermal source between the two phases; this is determined by a closure problem that allows one to calculate the distribution coefficient. Both the one-equation model and the two-equation model have been compared with experimental results, and reasonably good agreement is obtained in both cases.

ACKNOWLEDGMENT

The authors would like to acknowledge the helpful conversations with Alberto Ochoa-Tapia.

NOMENCLATURE

Roman Letters

a_v — $A_{\beta\sigma}/\mathcal{V}$, interfacial area per unit volume, m^{-1}

$\mathcal{A}_{\beta\sigma}$ — area of the β–σ interface contained within the entire macroscopic region, m^2

$A_{\beta\sigma}$ — $A_{\sigma\beta}$, area of the β–σ interface contained within the averaging volume, m^2

$\mathbf{b}_{\beta\beta}$ — vector field that maps $\nabla\langle T_\beta\rangle^\beta$ onto $\tilde{T}_\beta$, m

$\mathbf{b}_{\beta\sigma}$ — vector field that maps $\nabla\langle T_\sigma\rangle^\sigma$ onto $\tilde{T}_\beta$, m

$\mathbf{b}_{\sigma\beta}$ — vector field that maps $\nabla\langle T_\beta\rangle^\beta$ onto $\tilde{T}_\sigma$, m

$\mathbf{b}_{\sigma\sigma}$ — vector field that maps $\nabla\langle T_\sigma\rangle^\sigma$ onto $\tilde{T}_\sigma$, m

$\mathbf{b}_\sigma$ — vector field that maps $\nabla\langle T\rangle$ onto $\tilde{T}_\sigma$, m

$\mathbf{b}_\beta$	vector field that maps $\nabla\langle T\rangle$ onto $\tilde{T}_\beta$, m
$\mathbf{C}_{\beta\sigma}$	phase geometry tensor
$(c_p)_\alpha$	constant pressure heat capacity in the α-phase ($\alpha = \beta, \sigma$), J/kg K
C_p	$(\varepsilon_\beta(\rho c_p)_\beta + \varepsilon_\sigma(\rho c_p)_\sigma)/\langle\rho\rangle$, mass fraction weighted constant pressure heat capacity, J/kg K
$\mathbf{F}_\beta$	Forchheimer correction tensor
$\mathbf{g}$	gravitational acceleration, m/s^2
h	heat transfer coefficient, J/m^2 s K
$\mathbf{I}$	unit tensor
k_α	thermal conductivity in the α phase ($\alpha = \beta, \sigma$), J/m s K
$k_{\beta\sigma}$	$\varepsilon_\sigma k_\beta + \varepsilon_\beta k_\sigma$, mixed-mode thermal conductivity, J/m s K
$\mathbf{K}_\beta$	Darcy's law permeability tensor, m^2
$\mathbf{K}_\mathrm{D}$	hydrodynamic thermal dispersion tensor for the β-phase, J/m s K
$\mathbf{K}_\mathrm{eff}$	effective thermal conductivity tensor for the one-equation model, J/m s K
$\mathbf{K}^*$	$\mathbf{K}_\mathrm{eff} + \varepsilon_\beta \mathbf{K}_\mathrm{D}$, total effective thermal conductivity tensor, J/m s K
$\mathbf{K}^*_{\beta\beta}$	dominant thermal dispersion tensor for the β-phase, J/m s K
$\mathbf{K}^*_{\beta\sigma}$	coupling thermal dispersion tensor for the β-phase, J/m s K
$\mathbf{K}_{\sigma\beta}$	coupling thermal conductivity tensor for the σ-phase, J/m s K
$\mathbf{K}_{\sigma\sigma}$	effective thermal conductivity tensor for the σ-phase, J/m s K
$L_{\rho c_p}$	convective length scale, m
L_c	conductive length scale, m
ℓ_α	characteristic length scale for the α-phase ($\alpha = \beta, \sigma$), m
$\ell_{\beta\sigma}$	$\sqrt{\varepsilon_\beta \varepsilon_\sigma k_{\beta\sigma}/a_\mathrm{v} h}$, mixed-mode small length scale, m
ℓ_i	$i = 1, 2, 3$, lattice vectors, m
$\mathbf{n}_{\beta\sigma}$	$-\mathbf{n}_{\sigma\beta}$, unit normal vector directed from the β-phase toward the σ-phase
$\mathbf{n}_{\omega\eta}$	$-\mathbf{n}_{\eta\omega}$, unit normal vector directed from the ω-region toward the η-region
p_β	pressure in the β-phase, N/m^2
$\langle p_\beta\rangle^\beta$	intrinsic average pressure in the β-phase, N/m^2
$\langle p_\beta\rangle$	$\varepsilon_\beta\langle p_\beta\rangle^\beta$, superficial average pressure in the β-phase, N/m^2
$\tilde{p}_\beta$	$p_\beta - \langle p_\beta\rangle^\beta$, spatial deviation pressure, N/m^2
Pe	$(\rho c_p)_\beta \langle \mathrm{v}_\beta\rangle^\beta \ell_\beta / k_\beta$, Péclet number
$\mathrm{Pe}_\mathrm{cell}$	$(\rho c_p)_\beta \langle \mathrm{v}_\beta\rangle^\beta \ell_c / k_\beta$ or $(\rho c_p)_\beta \langle \mathrm{v}_\beta\rangle^\beta d_p / k_\beta$, cell Péclet number
$\mathbf{r}$	position vector, m
r_α	scalar mapping variable that maps Ω onto $\tilde{T}_\alpha$ ($\alpha = \beta, \sigma$), K m^2 s/J
s_α	scalar mapping variable that maps $\langle T_\sigma\rangle^\sigma - \langle T_\beta\rangle^\beta$ onto $\tilde{T}_\alpha$ ($\alpha = \beta, \sigma$)
t	time, s
t^*	characteristic process time, s
T_α	temperature of the α-phase ($\alpha = \beta, \sigma$), K
$\langle T_\alpha\rangle^\alpha$	intrinsic average temperature in the α-phase ($\alpha = \beta, \sigma$), K
$\langle T_\alpha\rangle$	$\varepsilon_\alpha\langle T_\alpha\rangle^\alpha$, superficial average temperature in the α-phase ($\alpha = \beta, \sigma$), K
$\tilde{T}_\alpha$	$T_\alpha - \langle T_\alpha\rangle^\alpha$, spatial deviation temperature in the α-phase ($\alpha = \beta, \sigma$), K
$\langle T\rangle$	$\varepsilon_\beta\langle T_\beta\rangle^\beta + \varepsilon_\sigma\langle T_\sigma\rangle^\sigma$, spatial average temperature, K
$\hat{T}_\alpha$	$\langle T_\alpha\rangle^\alpha - \langle T\rangle$, large-scale spatial deviation temperature in the α-phase ($\alpha = \beta, \sigma$), K
$\mathbf{u}_{\beta\beta}$	non-traditional, convective transport coefficient for the β-phase, J/m^2 s K

$\mathbf{u}_{\beta\sigma}$ non-traditional, coupling convective transport coefficient for the β-phase, J/m^2 s K
$\mathbf{u}_{\sigma\beta}$ non-traditional, coupling convective transport coefficient for the σ-phase, J/m^2 s K
$\mathbf{u}_{\sigma\sigma}$ non-traditional, convective transport coefficient for the σ-phase, J/m^2 s K
$\mathbf{v}_\beta$ velocity vector in the β-phase, m/s
$\langle \mathbf{v}_\beta \rangle^\beta$ intrinsic average velocity in the β-phase, m/s
$\langle \mathbf{v}_\beta \rangle$ $\varepsilon_\beta \langle \mathbf{v}_\beta \rangle^\beta$, superficial average velocity in the β-phase, m/s
$\tilde{\mathbf{v}}_\beta$ $\mathbf{v}_\beta - \langle \mathbf{v}_\beta \rangle^\beta$, spatial deviation velocity in the β-phase, m/s
$\mathcal{V}$ averaging volume, m^3
V_α volume of the α-phase contained in the averaging volume ($\alpha = \beta, \sigma$), m^3
$\mathbf{x}$ position vector locating the centroid of the averaging volume, m
$\mathbf{y}_\alpha$ relative position vector locating points in the α-phase relative to the centroid of the averaging volume, m

Greek Letters

α_β $k_\beta/(\rho c_p)_\beta$, thermal diffusivity for the β-phase, m^2/s
α_σ $k_\sigma/(\rho c_p)_\sigma$, thermal diffusivity for the σ-phase, m^2/s
$\alpha_{\beta\sigma}$ $k_{\beta\sigma}/(\rho c_p)_{\beta\sigma}$, mixed-mode thermal diffusivity, m^2/s
ε_α $V_\alpha/\mathcal{V}$, volume fraction of the β-phase
ρ_α density of the α-phase, kg/m^3
$\langle \rho \rangle$ $\varepsilon_\beta \rho_\beta + \varepsilon_\sigma \rho_\sigma$, spatial average density, kg/m^3
$(\rho c_p)_{\beta\sigma}$ $\varepsilon_\sigma(\rho c_p)_\beta + \varepsilon_\beta(\rho c_p)_\sigma$, mixed-mode volumetric heat capacity, J/m^3 K
Ω heterogeneous thermal source, J/m^2 s
Φ_σ homogeneous thermal source, J/m^3 s
μ_β viscosity of the β-phase, N/s m^2
ξ_β ξ, fraction of the energy from the homogeneous thermal source that flows to the β-phase
ξ_σ $1 - \xi_\beta$, fraction of the energy from the homogeneous thermal source that flows to the σ-phase.

REFERENCES

Aris R. The Mathematical Theory of Diffusion and Reaction in Permeable Catalysts, London: Clarendon Oxford, 1975.
Azizi S, Moyne C, Degiovanni A. Approche expérimentale et théorique de la conductivité thermique des milieux poreux humides I: Expérimentation. Int J Heat Mass Transfer 31:2305–2317, 1988.
Beavers GS, Joseph DD. Boundary conditions at a naturally permeable wall. J Fluid Mech 30:197–207, 1967.
Bensoussan A, Lions JL, Papanicolaou G. Asymptotic Analysis for Periodic Structures. Amsterdam: North-Holland, 1978.

Brenner H. Dispersion resulting from flow through spatially periodic porous media. Trans Roy Soc (Lond) 297:81–133, 1980.

Brinkman HC. A calculation of the viscous force exerted by a flowing fluid on a dense swarm of particles. App Sci Res A1:27–34, 1947.

Carberry JJ, Chemical and Catalytic Reaction Engineering. New York: McGraw-Hill, 1976.

Carbonell RG, Whitaker S. Heat and mass transfer in porous media. In: Bear J, Corapcioglu MY, eds. Fundamentals of Transport Phenomena in Porous Media. Dordrecht: Martinus Nijhoff, 1984, pp 123–198.

Chang H-C. Multiscale analysis of effective transport in periodic heterogeneous media. Chem Eng Commun 15:83–91, 1982.

Chang H-C. Effective diffusion and conduction in two-phase media: A unified approach. AIChE J 29:846–853, 1983.

Constant T, Moyne C. Drying with internal heat generation: Theoretical aspects and application to microwave heating. AIChE J 42:359–368, 1996.

Cushman JH. Dynamics of Fluids in Hierarchical Porous Media. London: Academic Press, 1990.

Cushman JH. The Physics of Fluids in Hierarchical Porous Media: Angstroms to Miles. Dordrecht: Kluwer, 1997.

Eidsath AB, Carbonell RG, Whitaker S, Hermann LR. Dispersion in pulsed systems III:Comparison between theory and experiments for packed beds. Chem Eng Sci 38:1803–1816, 1983.

Gates BC. Catalytic Chemistry. New York:John Wiley, 1992.

Grangeot G, Quintard M, Whitaker S. Heat transfer in packed beds: Interpretation of experiments in terms of one and two-equation models. 10th International Heat Transfer Conference, Brighton, UK, 1994, vol 5, pp 291–296.

Grangeot G, Quintard M, Combarnous M. Heat transfer in packed beds at both pore scale and macroscopic level. 9th International Heat Transfer Conference, Jerusalem, 1990, vol 5, pp 237–242.

Gray WG, Leijnse A, Kolar RL, Blain CA. Mathematical Tools for Changing Spatial Scales in the Analysis of Physical Systems, Boca Raton, FL:CRC Press, 1993.

Gunn DJ, De Souza JFC. Heat transfer and axial dispersion in packed beds. Chem Eng Sci 29:291–296, 1974.

Howes, FA, Whitaker S. The spatial averaging theorem revisited. Chem Eng Sci 40:1387–1392, 1985.

Jakob M. Heat Transfer. New York:John Wiley, 1949.

Moyne C, Batsale J-C, Degiovanni A. Approche expérimentale et théorique de la conductivité thermique des milieux poreux humides II:Théorie. Int J Heat Mass Transfer 31:2319–2330, 1988.

Nozad I, Carbonell RG, Whitaker S. Heat conduction in multiphase systems I:Theory and experiment for two-phase systems. Chem Eng Sci 40:843–855, 1985.

Ochoa-Tapia JA, Whitaker S. Momentum transfer at the boundary between a porous medium and a homogeneous fluid I:Theoretical development. Int J Heat Mass Transfer 38:2635–2646, 1995a.

Ochoa-Tapia JA, Whitaker S. Momentum transfer at the boundary between a porous medium and a homogeneous fluid II: Comparison with experiment. Int J Heat Mass Transfer 38:2647–2655, 1995b.

Ochoa-Tapia JA, Whitaker S. Heat transfer at the boundary between a porous medium and a homogeneous fluid. Int J Heat Mass Transfer 40:2691–2707, 1997.

Ochoa-Tapia JA, Whitaker S. Heat transfer at the boundary between a porous medium and a homogeneous fluid: The one-equation model. J Porous Media 1:31–46, 1998a.

Ochoa-Tapia, JA, Whitaker S. Momentum transfer at the boundary between a porous medium and a homogeneous fluid: Inertial effects. J Porous Media 1:201–217, 1998b.

Ochoa-Tapia JA, Stroeve P, Whitaker S. Diffusive transport in two-phase media: Spatially periodic models and Maxwell's theory for isotropic and anisotropic systems. Chem Eng Sci 49:709–726, 1994

Quintard M. Diffusion in isotropic and anisotropic porous systems:Three-dimensional calculations. Transp Porous Media 11:187–199, 1993.

Quintard M, Whitaker S. One and two-equation models for transient diffusion processes in two-phase systems. In: Hartnett, JP, ed. Advances in Heat Transfer, Vol 23. New York:Academic Press, 1993, pp 369–465.

Quintard M, Whitaker S. Convective and diffusive heat transfer in porous media:3D calculations of macroscopic transport properties. In:EUROTHERM Seminar 36 on Advanced Concepts and Techniques in Thermal Modelling Paris: Elsevier, 1994a, pp 301–307.

Quintard M, Whitaker S. Transport in ordered and disordered porous media I: The cellular average and the use of weighting functions. Transp Porous Media 14:163–177, 1994b.

Quintard M, Whitaker S. Transport in ordered and disordered porous media II: Generalized volume averaging. Transp Porous Media 14:179–206, 1994c.

Quintard M, Whitaker S. Transport in ordered and disordered porous media III: Closure and comparison between theory and experiment. Transp Porous Media 15:31–49, 1994d.

Quintard M, Whitaker S. Transport in ordered and disordered porous media IV: Computer generated porous media. Transp Porous Media 15:51–70, 1994e.

Quintard M, Whitaker S. Transport in ordered and disordered porous media V: Geometrical results for two-dimensional systems. Transp Porous Media 15:183–196, 1994f.

Quintard M, Whitaker S. Local thermal equilibrium for transient heat conduction: Theory and comparison with numerical experiments. Int J Heat Mass Transfer 38:2779–2796, 1995.

Quintard M, Kaviany M, Whitaker S. Two-medium treatment of heat transfer in porous media: Numerical results for effective parameters. Adv Water Resour 20:77–94, 1997.

Ryan D, Carbonell RG, and Whitaker S. A theory of diffusion and reaction in porous media. AIChE Symposium Series #202, Vol 71, pp 46–62, 1981.

Sanchez-Palencia, E. Non-homogeneous media and vibration theory. Lecture Notes in Physics 127. New York: Springer-Verlag. 1980.

Shonnard DR, Whitaker S. The effective thermal conductivity for a point-contact porous medium: An experimental study. Int J Heat and Mass Trasnfer 32: 503–512, 1989.

Wakao N, Kaguei S. Heat and Mass Transfer in Packed Beds. New York:Gordon and Breach, 1982.

Whitaker S. Forced convection heat transfer correlations for flow in pipes, past flat plates, single cylinders, single spheres, and for flow in packed beds and tube bundles. AIChE J 18:361–371, 1972.

Whitaker S. Heat transfer in catalytic packed bed reactors. In:Handbook of Heat and Mass Transfer, Vol 3, Chapter 10, Catalysis, Kinetics & Reactor Engineering. Cheremisinoff NP, ed. Matawan, NJ:Gulf, 1989.

Whitaker S. Improved constraints for the principle of local thermal equilibrium. Ind & Eng Chem 30:983–997, 1991.

Whitaker S. The Forchheimer equation:A theoretical development. Transp Porous Media 25:27–61, 1996.

Whitaker S. Volume averaging of transport equations. In:Flow in Porous Media, Chapter 1. Du Plessis, JP, ed. Southampton, UK:Computational Mechanics Publications, 1997.

Whitaker S. Coupled transport in multiphase systems:A theory of drying In:Advances in Heat Transfer, Vol 31, Chapter 1. New York:Academic Press, 1998.

Whitaker S. The Method of Volume Averaging. Dordrecht:Kluwer, 1999.

Yagi S, Kunii D, Wakao N. Studies on axial effective thermal conductivities in packed beds. AIChE J 6:543–547, 1960.

Zanotti F, Carbonell RG. Development of transport equations for multiphase systems I:General development for two-phase systems. Chem Eng Sci 39:263–278, 1984a.

Zanotti F, Carbonell RG. Development of transport equations for multiphase systems II:Application to one-dimensional axisymmetric flows of two-phases. Chem Eng Sci 39:279–297, 1984b.

Zanotti F, Carbonell RG. Development of transport equations for multiphase systems III:Application to heat transfer in packed beds. Chem Eng Sci 39:299–311, 1984c.

2

Capillary and Viscous Effects in Porous Media

F. A. L. Dullien
University of Waterloo, Waterloo, Ontario, Canada

I. INTRODUCTION

In this paper the role played by capillary and viscous forces in statistically homogeneous porous media containing two or more fluids which are in low Reynolds number flow is discussed in the light of some recent experimental results and some new mathematical models. The special case of quasistatic displacement of one fluid by another fluid is also included. The effects of preferential wettability of the pore surface by the fluids and of the microscopic pore morphology on the type of fluid motion observed are emphasized. Macroscopic properties represent average behavior of a sample containing many pores. Therefore, the pore scale properties are essential for an understanding and an explanation of the macroscopic properties. Throughout this chapter pore scale and macroscopic scale are treated side-by-side in an attempt to emphasize the relationship between the two scales. Flow phenomena of the type discussed in this paper are of importance in many fields of technology, e.g., petroleum production, groundwater hydrology, fuel cells, nuclear reactors, to name just a few.

II. BACKGROUND: CONFIGURATION OF TWO IMMISCIBLE FLUIDS IN CAPILLARY EQUILIBRIUM

A. Capillary Pressure and Wettability

Immiscible fluids in porous media are separated from each other by curved interfaces across which there exists a step change in pressure, called "capillary pressure" P_c. The immiscible fluids that may be present simultaneously in the porous medium exert forces of adhesion or attraction of different intensities toward the pore surface, resulting in a competition by these fluids for the occupancy of the pore surface. For any pair of immiscible fluids, the relative magnitudes of forces of attraction toward a smooth solid surface may be characterized by means of the "contact angle" θ (Figure 1). The contact angle is defined for the range 0--180°. Although it is an oversimplification of facts, it is a widespread custom to call the fluid through which $0° \leq \theta < 90°$ the (preferentially) "wetting fluid," whereas the other fluid through which $90° < \theta \leq 180°$ is called the "nonwetting fluid." The actual value of θ in a pore is also determined (in addition to the magnitudes of the forces of adhesion between the two immiscible fluids and solid surface) by the rugosity of the surface (Figure 2) and by the surface contaminations, which may consist also of a continuous or a discontinuous film of either of the two fluids. If the pore walls are not parallel but converge to form a throat, the effective contact angle is $\theta + \phi$, where ϕ is the angle shown in Figure 3a.

The effective value of contact angle is an important parameter for determining the curvature of the interface between the two fluids in a pore. In most porous media gravitational forces have only negligible effects on the interface. The value of P_c is determined by the mean radius of curvature r_m of the fluid/fluid interface and the interfacial tension σ between the two fluids, via Laplace's equation

$$P_c \equiv P'' - P' = 2\sigma/r_m \tag{1}$$

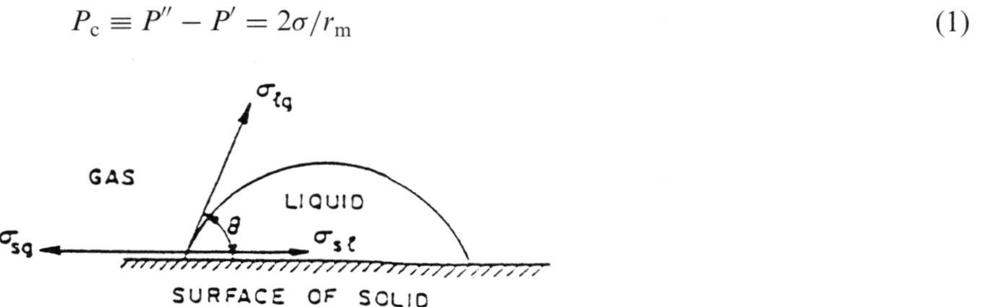

Figure 1. Definition of contact angle θ: $\sigma_{lg}\cos\theta = \sigma_{sg} - \sigma_{sl}(\sigma_{lg}, \sigma_{sg}$, and σ_{sl} are the liquid/gas, solid/gas, and solid/liquid interfacial tensions, respectively) (Dullien 1992a).

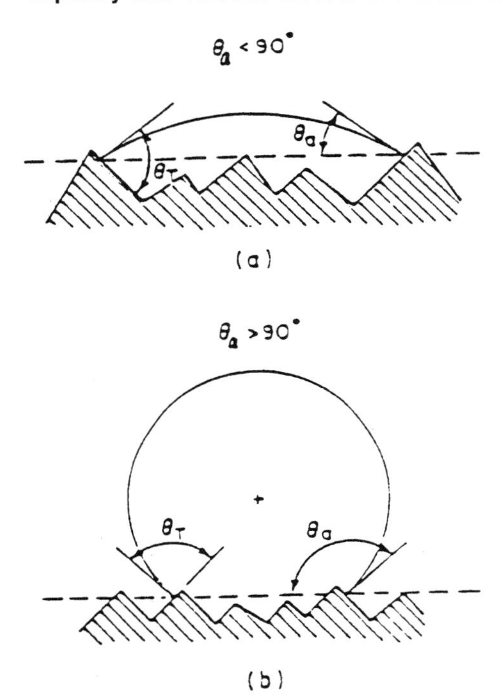

Figure 2. Example of possible effects of surface roughness on apparent contact angle θ_a. θ_T is the contact angle measured on a smooth, flat surface. The droplet is (a) the preferentially wetting fluid, $\theta_a < 90°$, and (b) the nonwetting fluid, $\theta_a > 90°$ (Popiel, 1978).

where P'' is the hydrostatic pressure on the concave side, P' is the hydrostatic pressure on the convex side of the fluid/fluid interface, and $1/r_m = [(1/r_1) + (1/r_2)]$, r_1 and r_2 being the two principal radii of curvature of the interface. (In the case of selloid interfaces no distinction between concave and convex sides is possible.) For an interface in a circular capillary, $r_1 = r_2 = r$ is the radius of a sphere. According to Eq. (1) the capillary pressure is, by definition, a positive quantity.

B. Imbibition and Drainage

The pattern of occupancy of pores by two or more immiscible fluids, which are present in the porous medium simultaneously, is very complicated. The complexity due to the pore structure will be introduced below step-by-step.

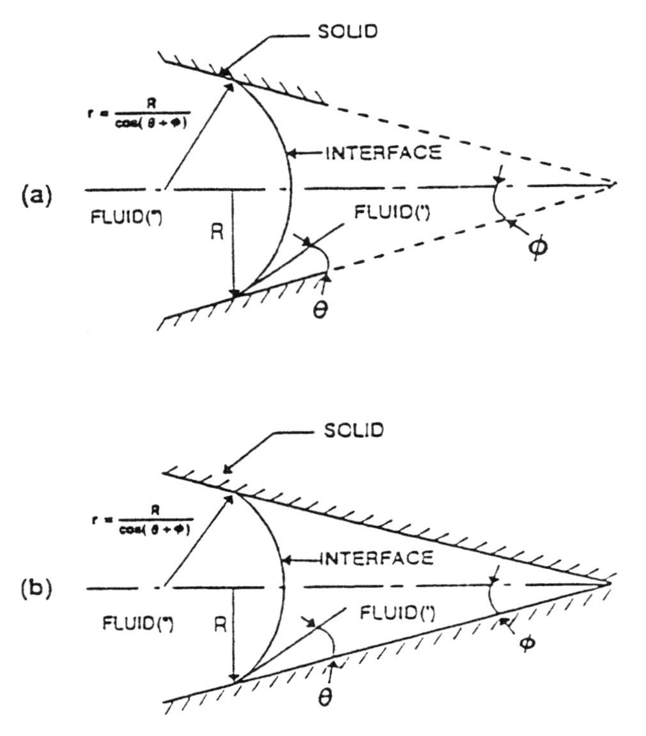

Figure 3. (a) Schematic showing effective contact angle $\theta + \phi$ in a circular capillary with convergent walls (Dullien, 1992a). (b) Effective contact angle $\theta + \phi$ of a thick film in an edge of an angular capillary (Dong et al., 1995).

1. Circular Capillary

The simplest case is represented by a straight circular capillary of uniform diameter with smooth and clean walls. If such a capillary is filled initially with the nonwetting fluid and is then exposed at one end to the wetting fluid (Figure 4), while there is zero pressure differential between the two extremities of the capillary, thereby ensuring that the only driving force is $P_c = 2\sigma \cos\theta / R$ (where R is the capillary radius), then the nonwetting fluid will be displaced from the capillary by the wetting fluid in a process called "free spontaneous imbibition". If, at the end of this displacement, a positive pressure differential $P_2 - P_1 > P_c$ is imposed between the two extremities of the capillary, then the nonwetting fluid will displace the wetting fluid from the capillary in a "drainage"-type displacement, leaving a very thin film of the latter on the capillary walls. The nonwetting fluid

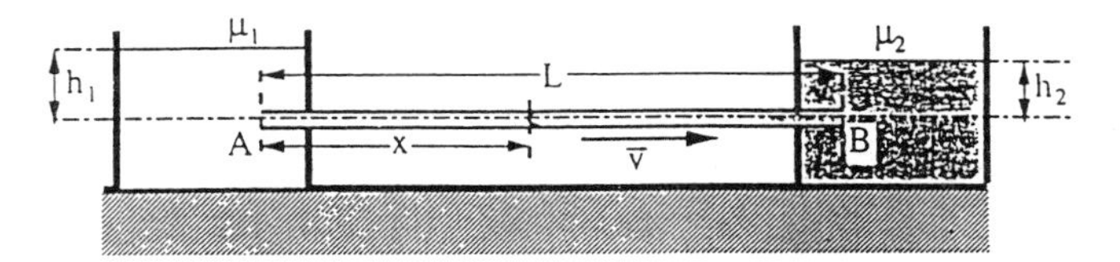

Figure 4. Schematic of imbibition/drainage experiment in a capillary tube (Dullien, 1992b).

occupying the capillary at the end of drainage is in contact with a solid surface of modified physical properties because of the presence of the thin film of the preferentially wetting fluid. This explains why the contact angle in imbibition, in which the solid surface is "dry," is greater (advancing contact angle) than in drainage, in which the solid surface is "wet" (receding contact angle).

2. Angular Capillary

Straight circular capillaries of uniform diameter and a smooth surface differ from pores in porous media which have angular cross-sections, rough walls, and nonuniform diameters.

Both the effects of an angular cross-section and those of rough walls can be illustrated by the example of a square capillary. If a free spontaneous imbibition experiment, described in the preceding paragraph, is carried out with a square capillary, the wetting fluid will imbibe faster along the edges than in the central part of the capillary (Figure 5). At the end of the process, all of the nonwetting fluid will be displaced from the square capillary. In a drainage-type displacement, following imbibition, however, there will be a "thick film" of the wetting fluid left in each edge of the square capillary if $\theta + \phi/2 < 90°$, where ϕ is the angle of the corner (Figure 3b) (for the square capillary $\phi = 90°$). The equilibrium "displacement pressure" P_d is, for $\theta = 0°$ (Lenormand, 1981; Lenormand et al., 1983; Legait, 1983)

$$P_d = 0.943 \frac{4\sigma}{x} \tag{2}$$

where x is the edge length of the square. The radius of curvature of the thick film's meniscus in the corners of the square in Figure 5a is approximately $x/4 = R/2$, where R is the radius of the inscribed circle. (The other principal radius of curvature is infinite.) The thick films occupy 6% of the cross-section of the square capillary. Therefore, the displacement pressure in a square capillary of edge length $x = 2R$ is approximately the same as in a circular capillary of radius R; however, the radius of curvature of the menis-

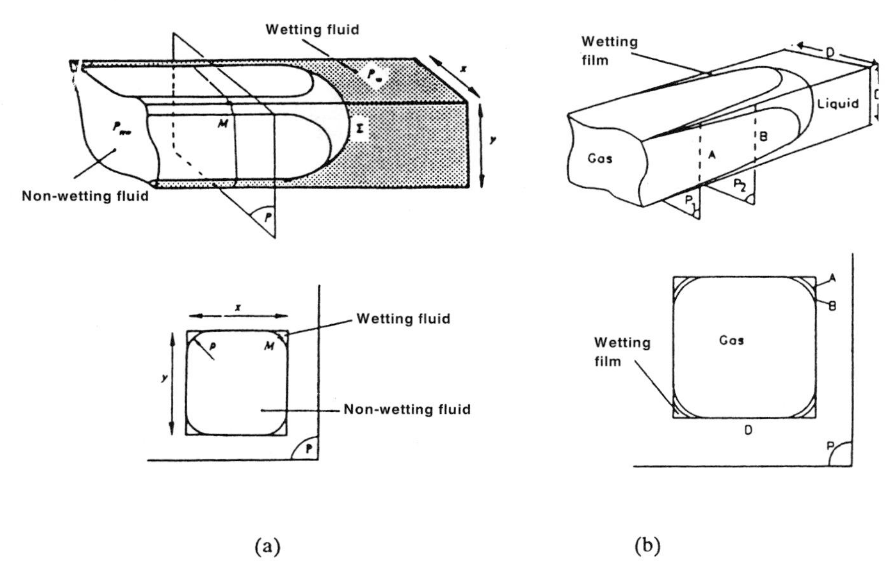

Figure 5. (a) Condition of capillary equilibrium in a square capillary tube (Lenormand et al., 1983). (b) Imbibition into a square capillary tube (Dong et al., 1995).

cus is $R/2$ instead of R. If the pressure difference between the two extremities of the square capillary is increased over the value given by Eq. (2), this will result in a decrease of the thickness of the films in the edges.

The important consequence of the presence of both edges and rugosity in the pores is that an interconnected network of grooves may be formed on the pore surface and, therefore, the wetting phase may remain continuous and mobile in porous media even at the lowest wetting phase saturations attainable in practice. (A notable exception is presented by packs of smooth glass beads that do not form an interconnected network of edges.)

If, after completion of the drainage displacement, the pressures at the two extremities of the square capillary are adjusted to be equal, the wetting fluid will displace the nonwetting one in a free spontaneous "secondary imbibition" process in which the wetting fluid will penetrate both the thick films present in the edges, if they have a radius of curvature less than approximately $x/4$, and the central part of the square capillary. Penetration into the thick films makes the thickness of the films increase to a maximum value which is defined by a radius of curvature of the meniscus approximately equal to $x/4$.

3. Branching Capillaries of Irregular Cross-section and Nonuniform Diameter

The next step in the direction of increased complexity of pore structure is represented by a capillary of irregular cross-section and nonuniform diameter, containing branches (Figure 6) (Chatzis and Dullien, 1983). It is apparent from this figure that in such a capillary much nonwetting fluid is trapped and bypassed by the imbibing wetting fluid. Trapping takes place either in bulges, so-called "pore bodies," or in ganglia consisting of two or more pore bodies separated by "throats." The wetting fluid is never trapped because it can advance from one end of the capillary to the other, sometimes filling the entire capillary cross-section. The wetting fluid occasionally flows in edges past trapped nonwetting fluid. The mechanism of trapping in imbibition is "snap-off" of the nonwetting fluid, a result of "choke-off" by the wetting fluid. There is the possibility of completely filling a pore throat with the wetting fluid, whereas in the adjacent pore bodies the wetting fluid may occupy only the edges. The conditions resulting in snap-off, or choke-off, arise because of the following:

1. For the same fractional filling by the wetting fluid, there is a smaller radius of curvature of the meniscus in the edges of a throat than in those of a pore body. This results in a lower pressure (i.e., suction) in the wetting fluid in a throat and in the transport of wetting fluid from the edges of the pore body into the edges of the throat, with a tendency toward equalizing the capillary pressures.
2. The concomitant increase of the radius of curvature of the thick films of wetting fluid in the throat edges may lead to the nonwetting fluid in the throat losing contact with the wall (Figure 7). At

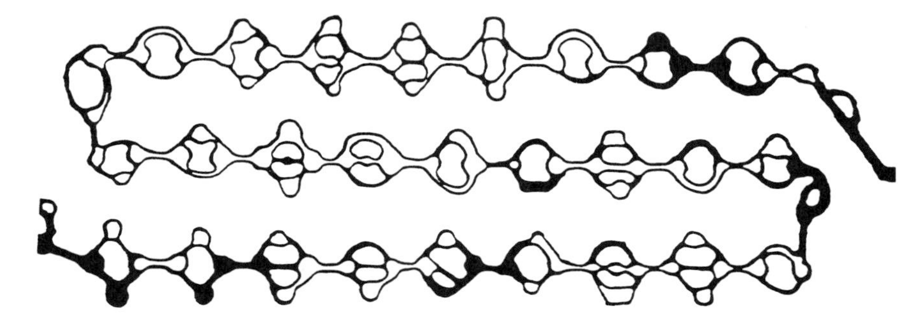

Figure 6. Imbibition into a capillary with rejoining branches (Chatzis and Dullien, 1983).

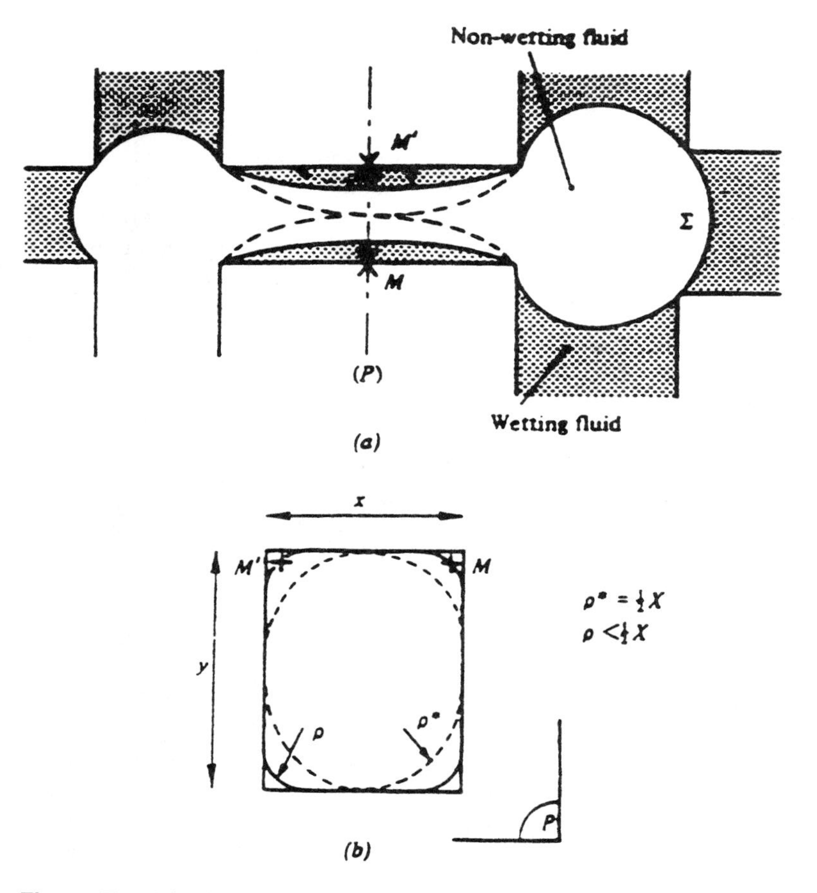

Figure 7. Criterion of capillary instability in a rectangular capillary. The solid curve shows the equilibrium position. The dashed curve shows the critical position of instability (Lenormand et al., 1983).

this point the thread of nonwetting fluid becomes unstable and snap-off occurs. (Note that the same radius of curvature of the thick film's meniscus in an edge of a large-diameter square capillary, corresponding to a saturation equal to or less than 6%, may result in 21.5% saturation in a smaller diameter square capillary; in which case the menisci in the edges are defined by an inscribed circle in the square, corresponding to the thick film's stability limit.)

In a drainage-type displacement the presence of an alternating sequence of pore throats and bodies does not result in trapping. In branching structures (so-called "pore doublets," when a capillary splits into two branches that join further down the path to form again a single capillary) there may be trapping of the wetting phase, given the right pore structure and displacement conditions that are controlled by capillary rather than viscous forces. Under these conditions (Figure 8) the wetting fluid is trapped in the branch of the smaller diameter.

4. Intersecting Capillaries

The next step toward the topological complexity of pore structure that exists in porous media is represented by intersecting capillaries. A great deal of attention has been paid to the problem of how an imbibing wetting fluid

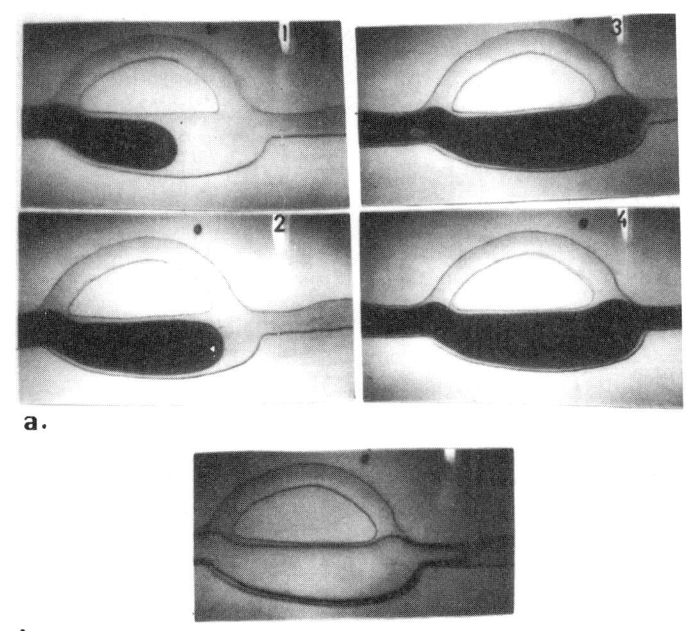

Figure 8. Experimental results illustrating the mechanism of displacement under drainage conditions in a pore doublet: (a) n-decane (black) displacing water, and (b) air displacing water (Chatzis and Dullien, 1983).

traverses an intersection between capillaries, also called a "node" (Lenormand and Zarcone, 1983; Li and Wardlaw, 1986a, b). The mechanism depends, among other things, on whether the wetting phase reaches the node in three branches, two adjacent branches, two opposite branches, or only a single branch. The topological details of the intersection and the contact angle are also of great importance for the outcome. In the presence of substantial surface roughness the intersection can be crossed by the wetting fluid by imbibition into the surface capillaries of the rugosity, with the formation of a network of thick films on the pore surface. If this mechanism prevails, then the presence of an intersection presents no obstacle to the imbibing wetting fluid.

5. Pore Networks

The final step to reaching the complexity of porous media topology is represented by interconnected three-dimensional (3D) networks. A drainage type of displacement in such networks, generated by the computer, has been studied extensively (e.g., Chatzis and Dullien, 1982, 1985; Diaz et al., 1987), largely by application of percolation theory. Trapping of the wetting phase was assumed to occur if all of the pores in contact with a cluster of wetting phase were occupied by the nonwetting phase. The fact that the wetting phase may "leak" past the surrounding nonwetting phase in the form of thick films through the surface capillaries of the rugosity, or through pore edges, was not considered in the simulations.

In typical network modeling the pore structure is assumed to consist of a 3D network of throats (bonds) and pore bodies (nodes) (stick-and-ball model). Different diameters are assigned to both the throats and the pore bodies by using certain rules. One such plausible rule that gave realistic results consists of assigning different diameters, distributed according to some probability density distribution, randomly to the nodes, whereas the bond diameters, distributed according to a different probability density distribution, are subject to the constraint that no bond can have a larger diameter than the smaller one of the two nodes that it connects. The diameter of the bond is correlated with the diameter of the smaller one of the two nodes. This type of percolation is called "bond-correlated site percolation." In simulated drainage-type displacements the nonwetting phase is assumed to advance in a piston-like manner in the pore throats. The penetration is started at one of the faces of the network. The bonds (throats) are penetrated by the invading nonwetting fluid step-by-step, starting with the largest throat diameter and continuing in the direction of decreasing throat diameter. Evidently, the bond diameters control the drainage, whereas the volume penetrated is contributed mostly by the nodes. The predictions closely matched the drainage capillary pressure curves of actual porous media samples (mostly sandstones), consisting of plots of equilibrium nonwetting fluid saturation of the sample vs. the capillary pressure (the pressure difference between the nonwetting and wetting phases) applied externally (see Figure 9, where "secondary drainage" is drainage-type displacement following secondary imbibition).

In simulated secondary imbibition-type displacement the nonwetting phase was assumed to advance in a piston-like manner in the pore bodies. The penetration is started at one of the network faces. The nodes are penetrated step-by-step by the invading wetting fluid, starting with the smallest diameter and continuing in the direction of increasing diameter. This displacement process is controlled by the node diameters. As seen in Figure 9, the simulation resulted in zero imbibition until a low capillary pressure was reached, at which point there was abrupt displacement. This behavior is in contradiction with experiment. Subsequently, imbibition was simulated by

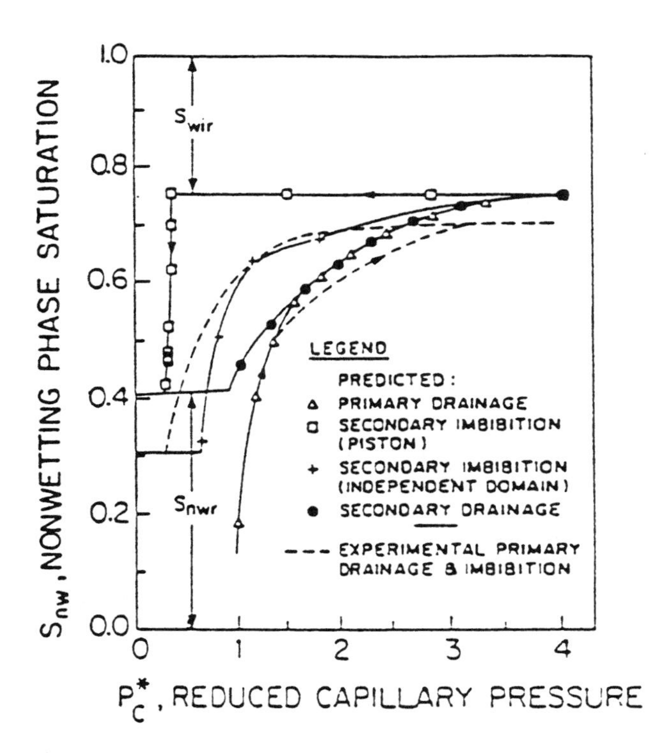

Figure 9. Comparison of simulated and measured capillary pressure curves of sandstone sample (Dullien, 1992).

assuming that the sites anywhere in the network were directly accessible to the wetting phase because of the existence of an interconnected network of thick films of wetting phase on the pore walls. In this simulation each and every node was invaded by the wetting phase at the capillary pressure corresponding to its size. As seen in Figure 9, this simulation gave results in much better agreement with the experiment.

In the course of imbibition, portions of the nonwetting phase become trapped by the choke-off (snap-off) mechanism, while the rest is displaced from the medium. The end of imbibition corresponds to the conditions when all of the nonwetting phase remaining in the medium is trapped, ie., is discontinuous. The final nonwetting phase saturation is usually called "residual" saturation, but the term "irreducible saturation" would be entirely appropriate as long as the displacement is limited to no-flow conditions. Under static conditions displacement of the trapped nonwetting phase is no longer possible, even if the wetting phase is placed under a higher pressure than the nonwetting phase.

In agreement with experiment, percolation theory-based network simulation predicts that in 3D both (incompressible) fluid phases are continuous only in a range of saturations centered in the neighborhood of 50%. By "continuous phase" is meant that there is a continuous path between any two points in the phase that never goes outside that phase.

In network simulation of a drainage-type displacement, at low capillary pressures applied in the early stages, displacement is limited to a thin layer of the porous medium. At a critical or threshold value of the capillary pressure the depth of penetration is without limit and at that point the nonwetting phase breaks through the sample. Therefore, this value is referred to also as "breakthrough" capillary pressure, which is more descriptive than the somewhat ambiguous term "displacement pressure." At this point the nonwetting phase becomes continuous and, therefore, conductive. Breakthrough conditions in a two-dimensional (2D) square lattice are shown in Figure 10. It is evident that in 2D both phases cannot be continuous simultaneously. An important feature is that a portion of the nonwetting phase, although continuous, is not conductive because it forms dead-end branches, also called "dendritic structures." In 3D networks, however, the wetting phase remains continuous and conductive at breakthrough. The wetting phase saturation at breakthrough is typically 70–80%. With continued increase of the applied capillary pressure the nonwetting phase displaces additional wetting phase by invading the smaller throats, while portions of the wetting phase are cut off and become discontinuous. Eventually a capillary pressure is reached where all of the wetting fluid remaining in the network is discontinuous and is therefore "trapped." The wetting phase saturation corresponding to this point is usually called "irre-

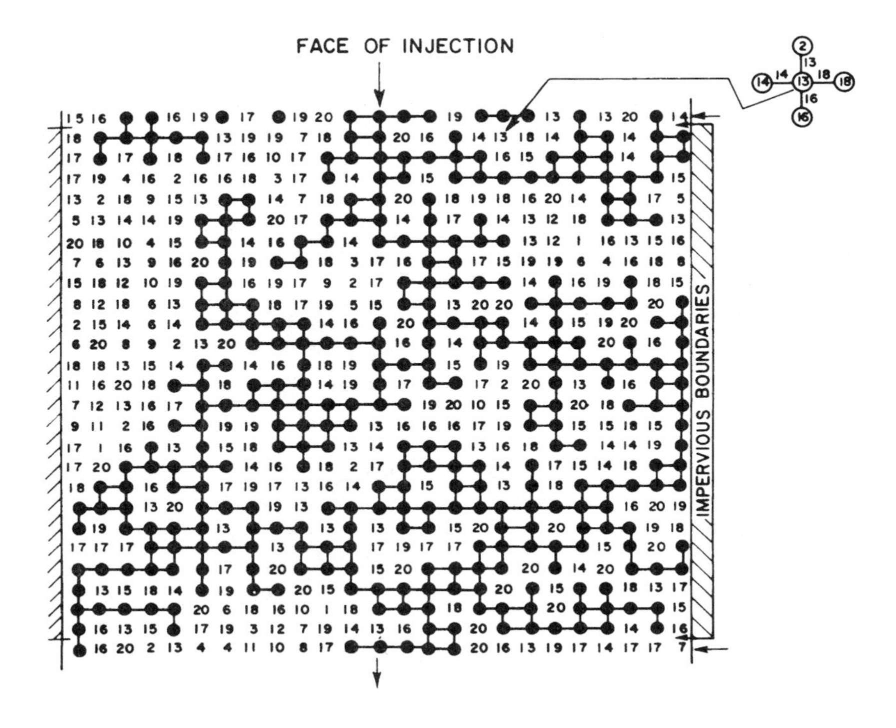

Figure 10. Breakthrough in a 25 × 25 square network (Chatzis and Dullien, 1985).

ducible saturation" and is typically 20–50%. It is important to point out, however, that in the network simulation the role played by the thick films of the wetting phase remaining in the grooves on the pore walls after the pore has been penetrated by the nonwetting fluid has not been considered. As these grooves form an interconnected network, the thick films can be squeezed and made thinner by increasing the applied capillary pressure and allowing enough time for further displacement to take place (Dullien et al., 1989). Therefore, in the presence of surface grooves there is no irreducible wetting phase saturation, but only a residual saturation that decreases at a very slow rate as the capillary pressure is increased.

At intermediate saturations, where both phases are continuous in the network of pore bodies and pore throats, there exist two subnetworks of pores: one subnetwork that is completely filled with the wetting fluid, and the other in which only the main (central) parts of the pores are filled with the nonwetting fluid, which is surrounded by the wetting fluid present in the form of thick films on the pore walls. The hydraulic conductivity of the wetting fluid present in the form of thick films is negligibly small compared with that of the wetting fluid present in its own pore subnetwork.

Throughout this entire section it has been assumed that one of the fluids is preferentially wetting the pore surface, i.e., it will spontaneously imbibe into the porous medium, and that the pore surface is "uniformly wetted," i.e., it does not have portions that are wetted preferentially by one of the two fluids, mixed with portions that are preferentially wetted by the other fluid. However, when the contact angle is in the vicinity of 90°, approximately ($60^\circ < \theta < 120^\circ$), neither fluid can be called preferentially wetting, and the term "neutral wettability" is often used in such situations. If a spontaneous imbibition test, described earlier, is attempted in a porous sample involving two fluids that are neutrally wetting the medium, there will be no spontaneous displacement of either fluid by the other because the capillary pressure is zero at certain controlling points in the pore network (Dullien and Fleury, 1994). The only way either fluid can be displaced by the other under conditions of neutral wettability is if an excess pressure is imposed on the displacing fluid, i.e., by a drainage mechanism. Capillary pressure curves for such conditions are shown in Figure 11. In Figures 11 and 12 the capillary pressure curves have been plotted in conformity with the definitions used by the author, according to which the capillary pressure is a positive quantity and, in a drainage type of displacement, the displacing phase is under a higher pressure than the displaced phase (conversely, in imbibition, the displacing phase is under a lower pressure than the displaced phase). When phase B displaces phase A, and phase B is under a higher pressure than phase A, the capillary pressure is $P_c = P_B - P_A > 0$, and this also is a drainage-type displacement.

The conditions may exist in some porous media, particularly in petroleum-bearing sands or rocks, when portions of the pore surface are wetted preferentially by one of two fluids, i.e., water, whereas other portions are wetted preferentially by the other fluid, i.e., oil. Such media are best referred to as "fractionally wetted." Sometimes the term "mixed wettability" is also used, but this may have some special connotations (Salathiel, 1973). It has been observed that in fractionally wetted systems either of the two fluids (i.e., oil *or* water) may occasionally imbibe spontaneously and displace the other fluid to a certain extent (Dullien et al., 1990). This behavior can be explained by assuming that both the oil-wet and the water-wet regions in the pore network are continuous (Dullien and Fleury, 1994).

If such a medium is initially saturated with fluid A and then brought into contact with fluid B, conditions of capillary equilibrium require that fluid A be at a higher pressure than fluid B, otherwise fluid B would imbibe spontaneously into the sample (Figure 12). Lowering the pressure difference $P_A - P_B > 0$ step-by-step results in the displacement of fluid A

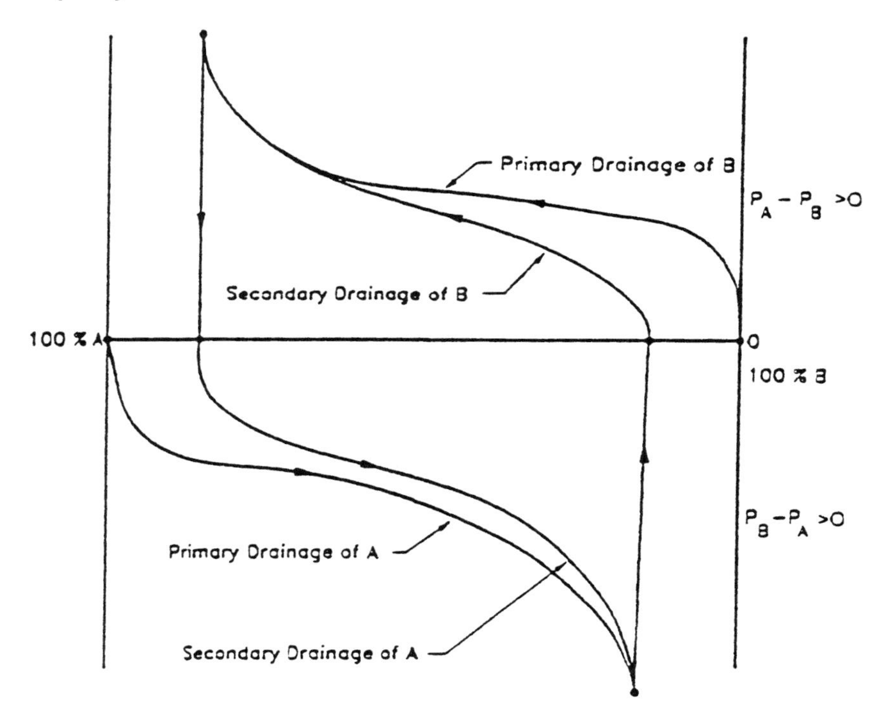

Figure 11. Schematic capillary pressure curves in a neutrally wetted system (Dullien and Fleury, 1994).

from pores that are preferentially wetted by fluid B, starting with the smallest pore diameter and continuing in the direction of increasing diameter. No displacement of fluid A from pores preferentially wetted by fluid A takes place as long as $P_A - P_B \geq 0$. The displacement in this range of pressures is of the imbibition type and $P_c \equiv P_A - P_B > 0$. If, after the point $P_A - P_B = 0$ has been reached, fluid B is placed under increasingly higher pressure, i.e., now $P_B - P_A > 0$, then further displacement of fluid A takes place from those pores that are preferentially wetted by fluid A. The displacement mechanism in this range of pressures is drainage and $P_c \equiv P_B - P_A > 0$. There is also trapping of fluid A in the entire displacement process. If, at some point beyond which there is only an incremental additional displacement of A, the pressure difference $P_B - P_A > 0$ is gradually decreased, then fluid B is displaced by fluid A in secondary imbibition of A from pores preferentially wetted by fluid A in the pressure range $P_B - P_A \geq 0$. If, after reaching the point $P_B - P_A = 0$,

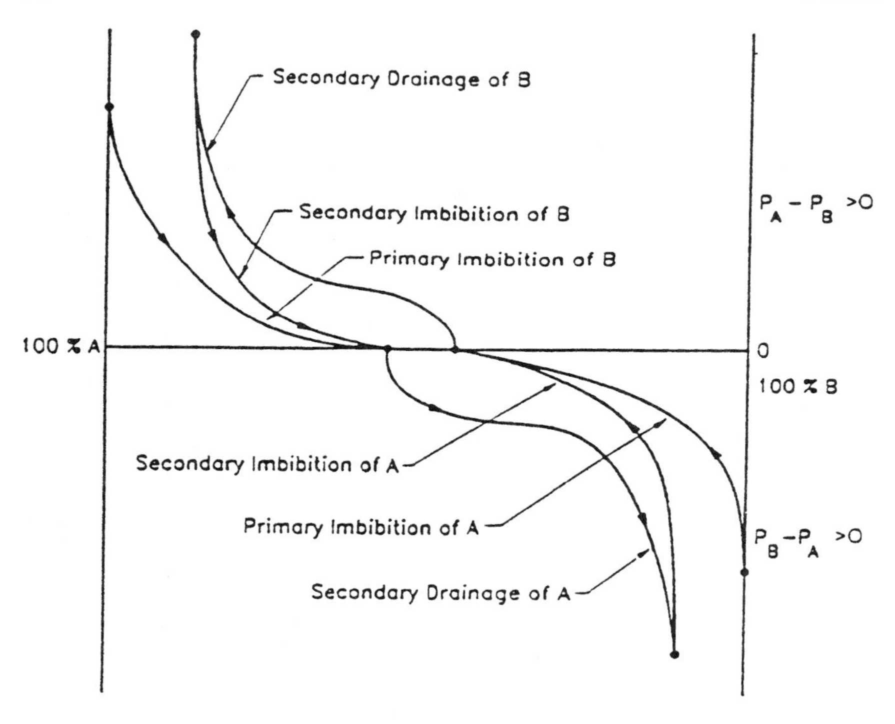

Figure 12. Schematic capillary pressure curves in a fractionally wetted system (50% A-wet and 50% B-wet) (Dullien and Fleury, 1994).

fluid A is placed again under the higher pressure, then fluid B is displaced by A by a secondary drainage mechanism in the pressure range $P_A - P_B > 0$ from those pores preferentially wetted by fluid B.

III. STEADY COCURRENT TWO-PHASE FLOW

By definition, "steady state" means the lack of change with time of the parameter values characterizing the system that are, in this case, flow rates, capillary pressure, and saturation. It is not implied by steady state, however, that the values of all of these quantities must be the same at all points in the porous medium. (In a porous medium the meaning of a "point" is actually a small volume, sometimes called "representative elementary volume" or REV, and the value of a property at a point is taken to be equal to the value of that property suitably averaged over that small volume.) In practice, conditions of steady cocurrent flow are usually established by injecting both fluids at constant rates and allowing time for the discharge rates to become equal to the injection rates. Under these conditions it has been found in many experiments that the saturation and the capillary pressure are approximately independent of position. As shown in a later section of this paper, however, it is also possible to maintain steady-state conditions with a saturation gradient across the sample.

The distribution of the two fluids in cocurrent steady flow depends on a number of factors, including the saturation, the wettability conditions of the pore surface, the interfacial tension σ, the fluid viscosities μ, and the pore velocity ν. The case that has been of great interest corresponds to the conditions when one of the two fluids wets the pore surface preferentially, the interfacial tension is great, and the viscosities and the velocities are low. These parameters are often combined in the form of a dimensionless group called the capillary number Ca:

$$Ca = \frac{\mu\nu}{\sigma} \tag{3}$$

The conditions stipulated in Eq. (3) are shown schematicaly in Figures 13a and 13b. In Figure 13a both fluids flow in separate channels (Dullien, 1988). Fluid I wets the solid surface preferentially and therefore tends to occupy the finer pores in the network. The contribution by the thick films to the flow of the wetting fluid has been assumed to be negligible in this case. These conditions exist at intermediate saturations. In Figure 13b, however, both fluids flow in the same pores; the wetting fluid is in the form of thick films (this is a greatly exaggerated thickness for the purpose of easier visualization). Such conditions exist at low wetting fluid saturations, where the wetting fluid is no longer continuous in the main (central) parts of the pores. Distributions of both fluids at various saturations have been presented photomicrographically in sandstone by Yadav et al. (1987).

Under conditions of neutral wettability there are no thick films of either fluid, and either fluid may fill small as well as large pores. It is even conceivable that the two fluids may flow side-by-side in some pores, as shown in Figure 13c. The same situation may exist also in the case of fractional wettability if one "side" of a pore is preferentially wetted by fluid A whereas the other "side" is preferentially wetted by fluid B.

Finally, in cocurrent two-phase flow at a very high capillary number, which in practice usually corresponds to very low values of interfacial tension, one of the two fluids may become discontinuous, forming an emulsion, as shown in Figure 13d.

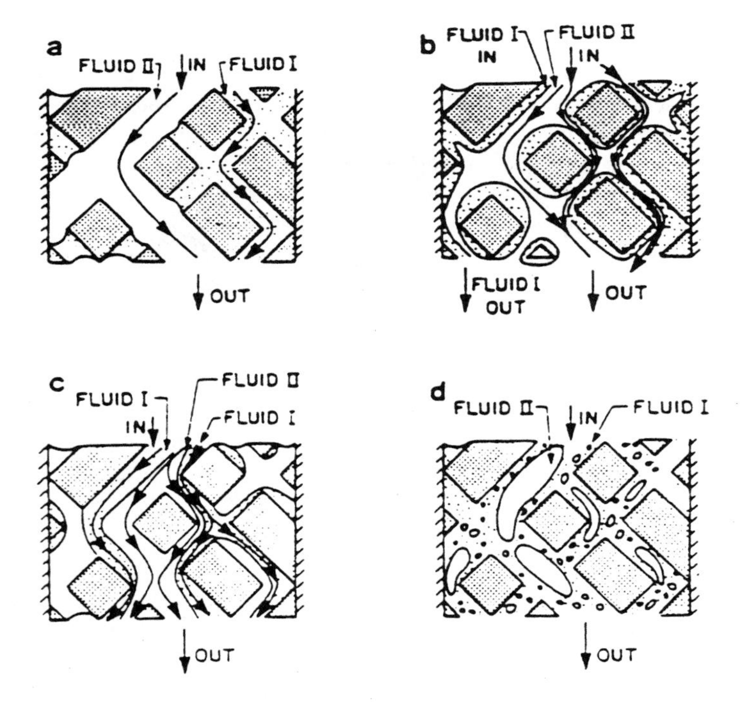

Figure 13. Two-dimensional representation of different patterns of cocurrent steady two-phase flow in porous media (Dullien, 1992b).

A. Mathematical Description of Steady Cocurrent Two-phase Flow

Darcy's law has been generalized (e.g., Dullien, 1992a) for both steady and unsteady multiphase flow in porous media:

$$\mathbf{v}_i = -\left(\frac{k_i}{\mu_i}\right)(\nabla P_i - \rho_i \mathbf{g}) \tag{4}$$

where $\mathbf{v}_i = (\delta q_i / \delta A)\mathbf{n}$. $\mathbf{n}$ is the unit normal vector of the surface δA, δq_i is the volume flow of fluid i across this surface, $\mathbf{g}$ is the gravitational acceleration vector, ρ_i is the density of fluid i, k_i is the effective permeability to fluid i of the medium at the saturation S_i at the point considered, and ∇P_i is the pressure gradient in fluid i.

It has been widely assumed in the literature that the effective permeabilities k_i (or the relative permeabilities $k_{ri} = k_i/k$, where k is the absolute permeability of the sample or, sometimes, the effective oil permeability at the lowest water saturation) in a sample of the porous medium do not depend on the viscosities μ of the fluids. The idea behind this assumption is that the two fluids flow in separate channels, corresponding to case of Figure 13a, and therefore the viscosity of fluid I has no effect on the flow of fluid II, and vice versa. It is shown later in this section that the validity of this assumption is limited to certain conditions.

Steady-state relative permeabilities are calculated from measured values of flow rates q_i and pressure drops ΔP_i, using the integrated form of Darcy's law for steady unidirectional flow:

$$q_i = \left(\frac{kk_{ri}A}{\mu_i}\right)\frac{\Delta P_i}{L} \tag{5}$$

where A is the cross-sectional area and L is the length of the sample, k_{ri} are functions of the saturations S_i in the sample. It is customary to distinguish between drainage and (secondary) "imbibition" relative permeabilities (Figure 14).

B. Effect of Contact Angle

The effect of the contact angle θ on relative permeabilities has been studied in considerable detail (Owens and Archer, 1971; McCaffery and Bennion, 1974). Various interpretations of the results are possible, but probably it is safe to conclude that in the range of contact angles $0 \leq \theta \leq 60°$ (the upper limit is approximate), measured through the wetting fluid, there is no change in the relative permeabilities, whereas in the region of neutral wettability the relative permeabilities are markedly different from the values measured in

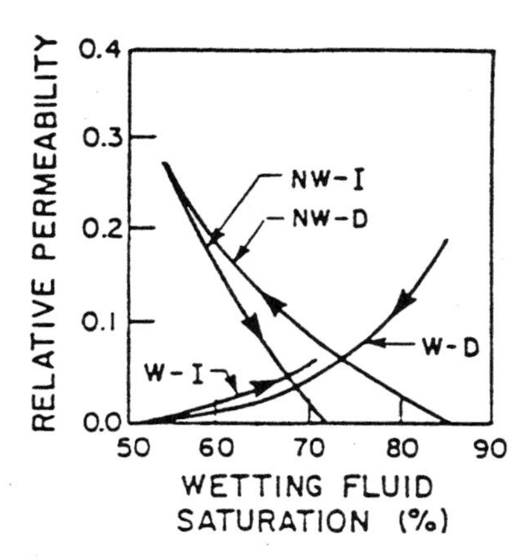

Figure 14. Relative permeability curves (Dullien, 1992a).

systems such that one of the two fluids preferentially wets the pore surface (Figure 15). This behavior is easily explained by the different topologies of the fluids existing under these different conditions.

C. Effect of Interfacial Tension

In the range of interfacial tensions greater than approximately 1 mN/m, the interfacial tension has no effect on the conventional relative permeabilities. At very low interfacial tensions, however, the networks of the two fluids begin to break up, first in the relatively large pores, and then, at even lower interfacial tensions, also in the finer pores. Ultimately, an emulsion is formed in the entire pore network (Figure 13d) and, at that point, each fluid flows everywhere in proportion to its saturation in the sample, and therefore the relative permeability curves become straight diagonal lines (Figure 16) (Bardon and Longeron, 1978).

D. Effect of Viscosity

Probably the most striking example of the effect of viscosity is represented by the case of emulsion formation, illustrated in Figure 13d. Evidently, in

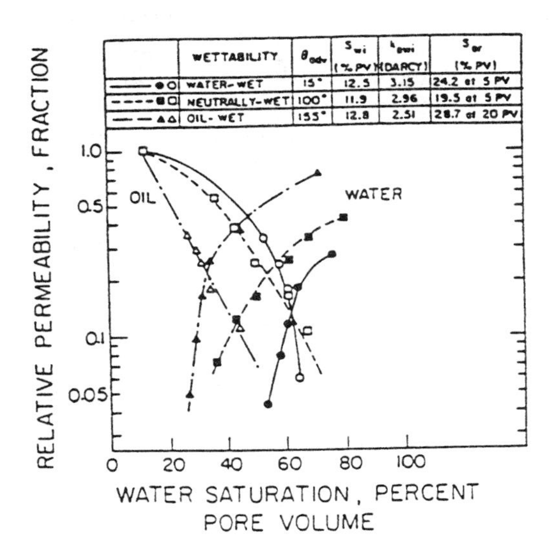

Figure 15. Effect of wettability on relative permeability in dolomite pack with water and oil treated with octanoic acid. Relative permeabilities are based on the effective oil permeability at the initial water saturation (Morrow et al., 1973).

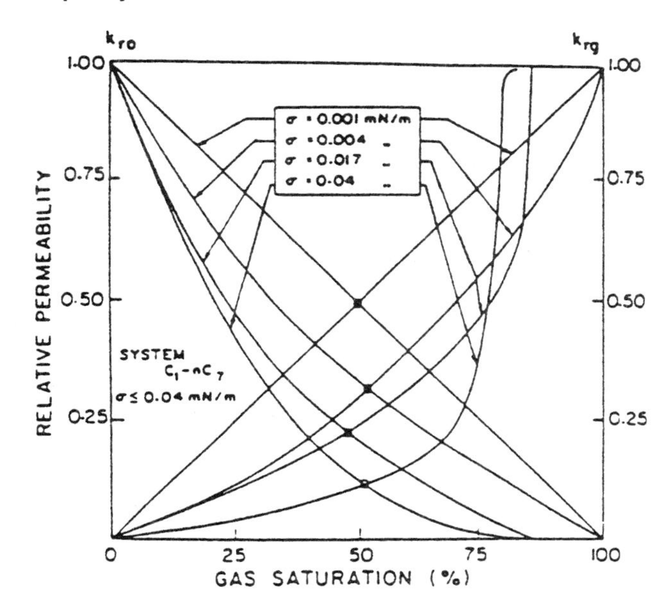

Figure 16. Gas–oil permeability of Fontainebleau sandstone for very low interfacial tension values (Bardon and Longeron, 1978).

this case the flow rate of phase I is $Q_I = Q_t S_I$, Q_t being the total flow rate and S_I being the saturation (i.e., volume fraction) of this phase, the controlling viscosity is the viscosity of the emulsion μ, and the driving force is the pressure drop ΔP measured in the emulsion. Hence, in this case, application of Darcy's law gives

$$Q_I = Q_t S_I = \left(\frac{k k_{ri}^* A}{\mu}\right) \frac{\Delta P}{L} \tag{6}$$

and

$$Q_t = \left(\frac{kA}{\mu}\right) \frac{\Delta P}{L} \tag{7}$$

whence

$$Q_I = \left(\frac{k S_I A}{\mu}\right) \frac{\Delta P}{L} \tag{8}$$

Comparison of Eqs. (6) and (8) gives the relative permeability $k_{ri}^* = S_I$, i.e., the volume fraction of phase I at all saturations. This is different from the value calculated from the conventional definition of k_{rI} given by Eq. (5), i.e.

$$Q_{\rm I} = \left(\frac{kk_{\rm rI}A}{\mu_{\rm I}}\right)\frac{\Delta P}{L} \tag{9}$$

because $k_{\rm rI} = k^*_{\rm rI}(\mu_{\rm I}/\mu)$. Similar considerations apply also to fluid II.

The role played by the viscosity in the conventional case (Figures 13a and 13b) is controversial. There is experimental evidence that at low water saturations, when the wetting phase is no longer continuous in the main (i.e., central) parts of the pores, in low permeability media the lower viscosity of the water may exert a "lubricating effect" on the oil flow under the conditions of steady-state relative permeability measurements where both the oil and the water are pumped through the sample at constant rates. The essence of the lubricating effect is that, as a result of the presence of the lower viscosity fluid (water) near the pore walls, the oil flows at a higher rate than it would if oil occupied also the regions near the wall, instead of water. In such cases, using the oil viscosity in Eq. (5) gives anomalously high oil relative permeabilities. (This could be corrected by using a (lower) effective viscosity that takes into account the viscosity of water.) Under these conditions, the oil relative permeability shows an increasing trend with the oil-to-water viscosity ratio $\mu_{\rm o}/\mu_{\rm w}$, as shown in Figures 17 and 18 (Odeh, 1959; Danis and Jacquin, 1983). It can be easily understood that the water flowing on the pore walls in the form of thick films provides the oil

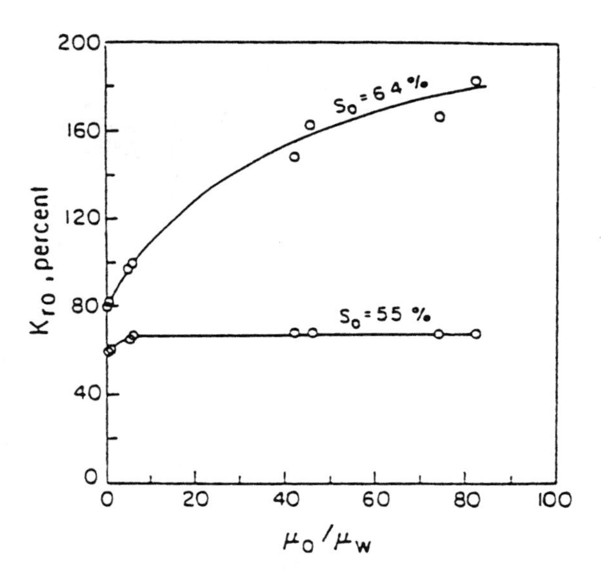

Figure 17. Relative permeability to oil vs. viscosity ratio at two low water saturations in low permeability sandstone ($S_{\rm o}$ = oil saturation) (Odeh, 1959).

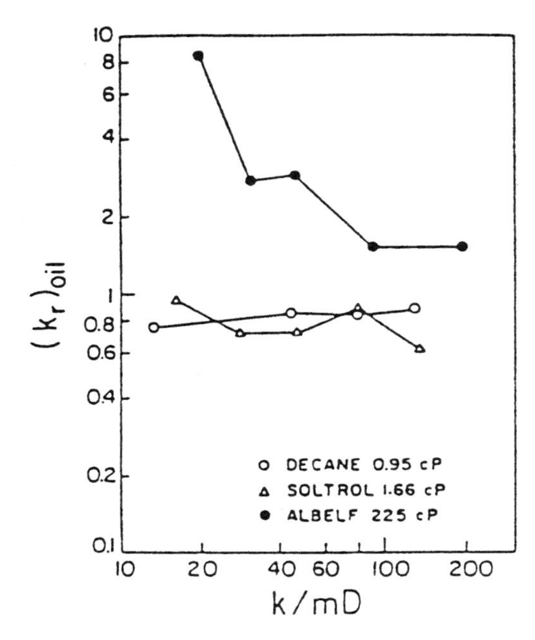

Figure 18. Relative permeabilities to oil of Rouffach limestone samples of different permeabilities k, in millidarcy units, at connate water saturation (Danis and Jacquin, 1983).

with a moving boundary, as shown schematically on the right-hand side of Figure 19 (Dullien, 1993). The likely reason for the strong dependence of this effect on the permeability is that the pores in low-permeability media are finer than in high-permeability ones whereas the thickness of the thick films of water is not necessarily less in the low-permeability media. Therefore, in low-permeability media the position of the moving boundary, expressed as a fraction of the pore "radius", is farther away from the pore walls than in high-permeability media, and this results in a greater lubricating effect. This physical picture also explains why the water relative permeability has never been found to vary with the viscosity ratio. The probable explanation of the absence of a lubricating effect at higher water saturations, as illustrated on the left-hand side of Figure 19, is that the moving boundary of flowing oil is interrupted at frequent intervals at such points in the pore network where there is a nonflow (dendritic) branch of oil. Under conditions existing in the case of either "fractional" or "neutral" wettability, to the best of this author's knowledge there are no experimental data on the effects exerted by viscosity on steady-state relative permeabilities. Provided

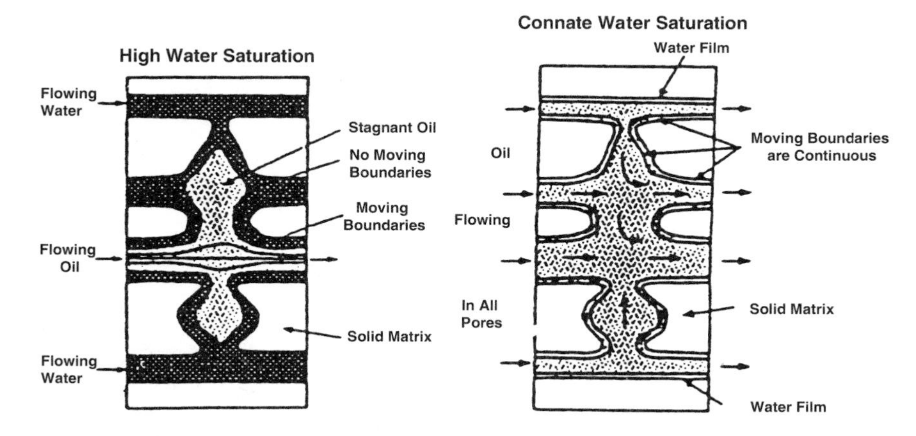

Figure 19. Schematic distribution of oil and water in the pores at high (left-hand side) and low (right-hand side) water saturations (Dullien, 1993).

that both fluids flow side-by-side at least in some pores, the flow rates of both fluids are expected to be influenced by both viscosities and therefore Eq. (5) is expected to be not applicable.

IV. IMMISCIBLE DISPLACEMENT

In the introductory section of this paper immiscible displacement was considered mainly from the point of view of the role played by capillary forces; i.e., rate effects involving viscous forces were left outside of consideration. In the present section the roles played by both capillary and viscous forces are considered from a purely macroscopic, phenomenological point of view. Pore scale dynamics, a subject that has been addressed by Buyevich (1995), lies outside the scope of this paper.

A. Rate of Displacement in a Circular Capillary

Consider the displacement experiment in Figure 4, where the pressure difference $\Delta P \equiv P_1 - P_2$ between the two extremities of the capillary can be zero, positive, or negative, and, if negative, then in absolute value it can be either greater or less than or equal to the capillary pressure P_c. The net driving force of the displacement is $\Delta P + P_c$ and the velocity of the interface dx/dt is given by the following form of the Poiseuille equation, also called the Washburn equation

$$v = \frac{dx}{dt} = \frac{D^2}{32} \frac{\Delta P + P_c}{\mu_1 x + \mu_2 (L - x)} \tag{10}$$

In the domains $\Delta P \geq 0$ and $-P_c < \Delta P < 0$ there is imbibition, $\Delta P = -P_c$ corresponds to capillary equilibrium, and for $\Delta P < -P_c$ there is drainage. The special case $\Delta P = 0$ corresponds to free spontaneous imbibition, driven by the capillary pressure P_c, $\Delta P > 0$ represents forced imbibition, and $-P_c < \Delta P < 0$ is the region of controlled spontaneous imbibition.

Using the system cyclohexane–water, for which $\mu_1 = \mu_2 = 1$ cp, primary imbibition, followed by primary drainage, and finally secondary imbibition displacement tests have been carried out over a wide range of capillary numbers ($Ca \equiv v\mu/\sigma$) with the results shown in Figure 20 (Calvo et al., 1991). It is apparent that the capillary pressure under dynamic, i.e., displacement conditions can vary over a wide range of values (the meaning of the negative values is that at the meniscus the pressure in the water exceeded the pressure in the oil). Discussions of these controversial results can be found in Dullien (1992b).

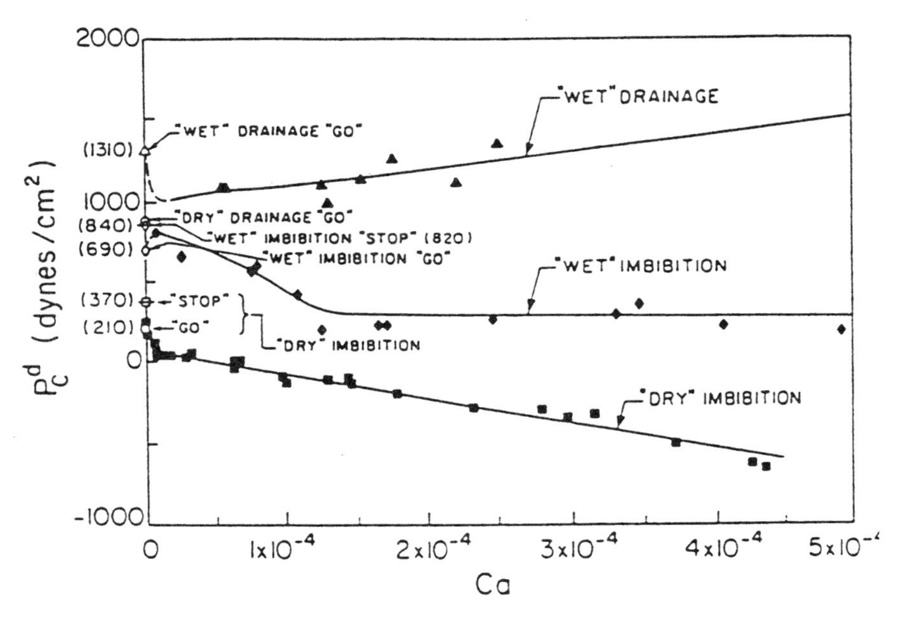

Figure 20. Summary of dynamic capillary pressure results (Calvo et al., 1991).

B. Complete Capillary Number *CA*

Although the capillary number *Ca* is usually referred to as a measure of the ratio of viscous to capillary forces, it does not really deserve this description. For example, in Figure 20 almost all of the values of *Ca* are less than unity, which may be interpreted (erroneously) that capillary forces were in complete control throughout these experiments.

The true value of the ratio of viscous to capillary forces in immiscible displacement in a circular capillary tube is readily obtained as follows (Dullien, 1992a). The viscous force F_v is equal to the wall shear stress τ_w multiplied by the surface area of the tube, that is

$$F_v = \tau_w(\pi DL) \tag{11}$$

where D is tube diameter, L is tube length and

$$\tau_w = 8\mu v/D \tag{12}$$

with v the average flow velocity, whence

$$F_v = 8\pi\mu vL \tag{13}$$

The capillary force F_c is equal to the capillary pressure P_c times the cross-sectional area of the tube, i.e. $\pi D^2/4$. As

$$P_c = 4\sigma\cos\theta/D \tag{14}$$

there follows

$$F_c = \pi D\sigma\cos\theta \tag{15}$$

Hence

$$CA \equiv \frac{F_v}{F_c} = \frac{8\mu vL}{\sigma\cos\theta D} = \frac{8Ca}{\cos\theta}\frac{L}{D} \tag{16}$$

Since, in the experiments resulting in the plots of Figure 20, $L = 85$ cm, $D = 0.11$ cm, and $\theta \approx 75°$, there follows that $CA = 23{,}885\, Ca$. Therefore, for the most part of these tests, viscous forces controlled the displacement.

C. Evaluation of *CA* in Imbibition-type Displacements in Porous Media

Displacement of fluids in porous media is often of great practical importance, even if nothing else besides petroleum production by waterflooding or the cyclic movements of ground water is considered. The variation of saturation profiles with time is of particular interest. It has been known for a long time (Rapoport and Leas, 1953) that the saturation profiles are functions of the "scaling coefficient" $Lv\mu_w$, the oil-to-water viscosity ratio, and the preferential wettability, i.e., whether the medium is "water-wet" or "oil-wet". There has not been proposed, however, a representation of displacement histories in a given porous medium in terms of dimensionless groups.

It seems likely on physical grounds that, at least in statistically homogeneous media, such a representation may be possible in terms of the two dimensionless groups, the ratio of viscous to capillary forces and the viscosity ratio. The displacement histories are evidently different in imbibition and in drainage. Presently there are data available for the imbibition case only, and for a viscosity ratio equal to one (Dong et al., 1998). A fluid pair of the same viscosity was used and the ratio of viscous to capillary forces was varied to ascertain its effect on the evolution of saturation profiles during an imbibition-type displacement. The ratio of viscous to capillary forces is represented by an expression formally identical to the one derived rigorously for the case of a circular capillary tube, i.e.

$$CA = (8\mu \nu L)/(D_{\mathrm{eq}}\sigma \cos\theta) \tag{17}$$

where ν is mean pore velocity and D_{eq} is a pore size scale. The value of CA in a waterflood conducted at a constant pumping rate Q_{t} of water in a sample of uniform cross-section and of length L, with an oil/water system of viscosity ratio 1, is obtained with reference to the initial rate of free spontaneous imbibition Q of water into an identical sample of the same porous medium/water/oil system. In free spontaneous imbibition $\Delta P = 0$ and, because under these conditions the viscous forces in the porous medium are equal to the capillary forces, therefore $CA = 1$. As also $Q_{\mathrm{t}}/Q = \nu/\nu_0$ (ν_0 = initial mean pore velocity in free spontaneous imbibition), applications of Eq. (17) to a waterflood and to a free spontaneous imbibition experiment yield two equations which, after division, result in the following value of CA in a waterflood

$$CA = Q_{\mathrm{t}}/Q \tag{18}$$

because the expression $8\mu L/D_{\mathrm{eq}}\sigma \cos\theta$ is the same in both equations and therefore it cancels out on division.

In a free spontaneous imbibition experiment $CA = 1$ and therefore Eq. (17) becomes

$$(D_{\mathrm{eq}}\sigma \cos\theta)/(8\mu) = \nu_0 L = QL/A\phi \tag{19}$$

where A is the cross-section and ϕ the porosity of the sample.

In Eq. (19) every term on the left-hand side is a constant for a given fluid pair and a given type of porous medium and therefore, with A kept at the same value, Q is expected to be inversely proportional to L. A test of the validity of this relationship was carried out by performing free spontaneous imbibition experiments, using the same fluid pair and two packs of the same sand (sand A) of $L_1 = 35\,\text{cm}$ and $L_2 = 50\,\text{cm}$, using Soltrol 100 oil, $\mu_o = 1\,\text{cp}$, and water, $\mu_w = 1\,\text{cp}$ (see Figure 21).

By performing constant rate waterfloods at different values of Q_t, a wide range of values of CA, from much less than 1 to much greater than 1, has been covered, and for each value of CA the evolution of saturation profiles has been determined experimentally in sand packs. It has been found that for $CA \ll 1$ the saturation profile remains practically horizontal and keeps rising uniformly throughout the sample; for $CA \gg 1$ the saturation profile approaches a step function, i.e., the displacement is almost piston-like, and for intermediate values of CA the saturation profiles are S-shaped curves (Figures 22a, b, and c).

As illustrated by the example of Figure 23, the evolutions of saturation profiles in waterfloods carried out with samples 1 and 2 of the same sand A of lengths L_1 and L_2 have been found by Dong et al. (1998) to be identical, within experimental error, at the same value of CA if plotted against a normalized distance coordinate. It follows from Eqs. (18) and (19) that the condition of $CA_1 = CA_2$ can also be expressed as $Q_{t1}/Q_{t2} = L_2/L_1$,

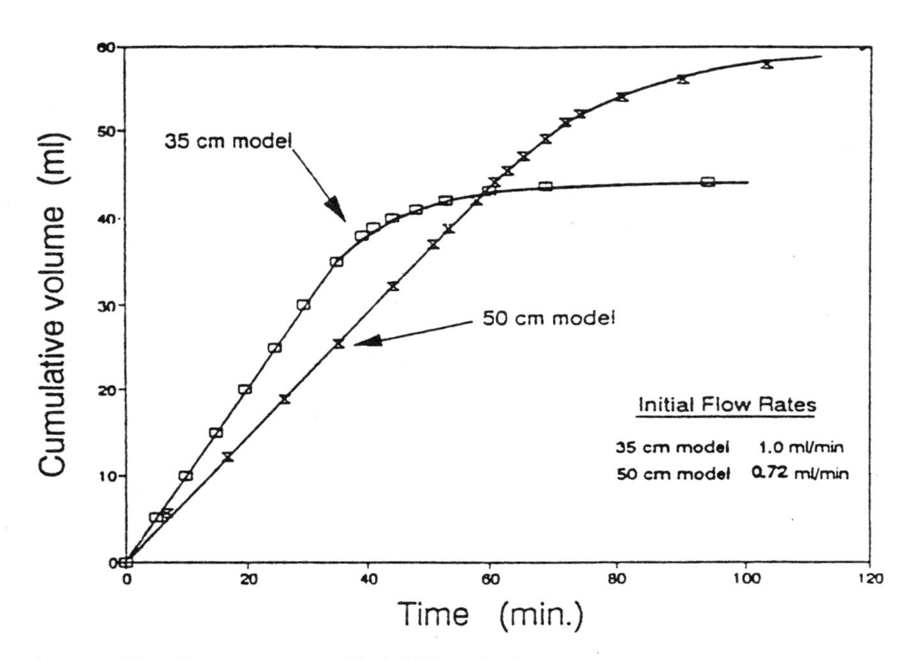

Figure 21. Free spontaneous imbibition displacement rates in two packs of sand A of different lengths (Dong et al., 1998).

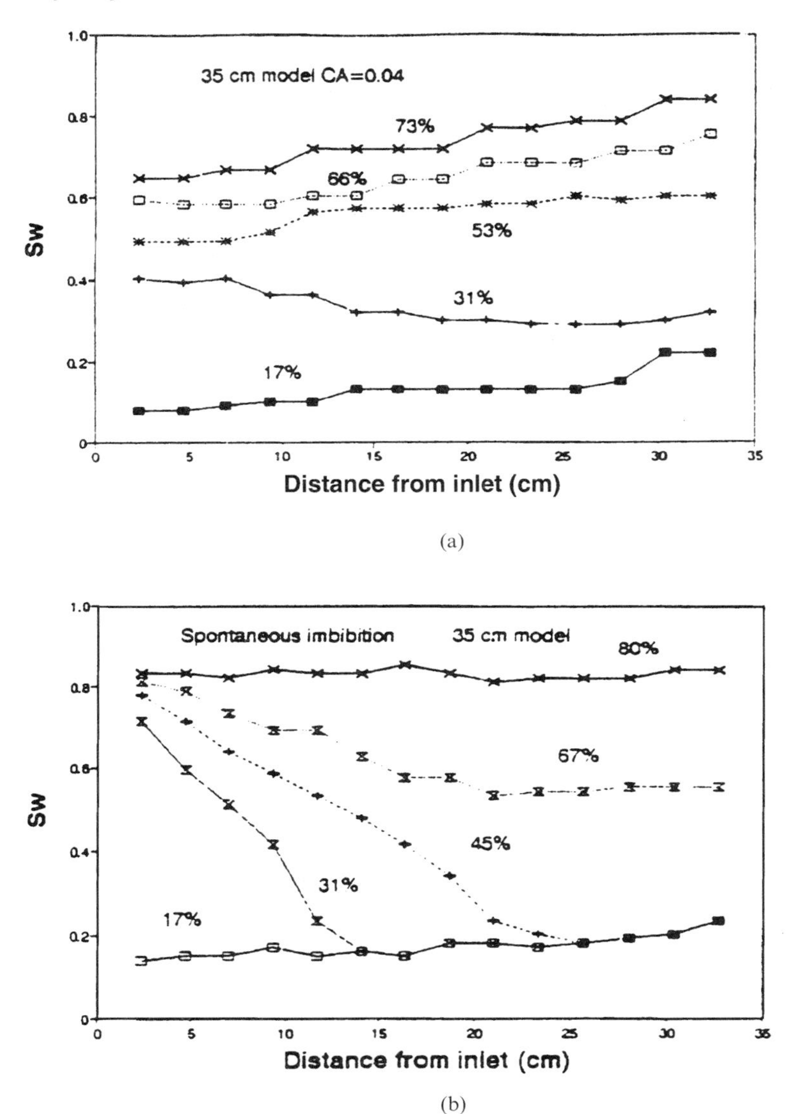

(a)

(b)

Figure 22. Evolution of saturation profiles in a pack of sand A of 35 cm length at (a) $CA = 0.04$, (b) $CA = 1$, and (c) $CA = 20$ (Dong et al., 1998).

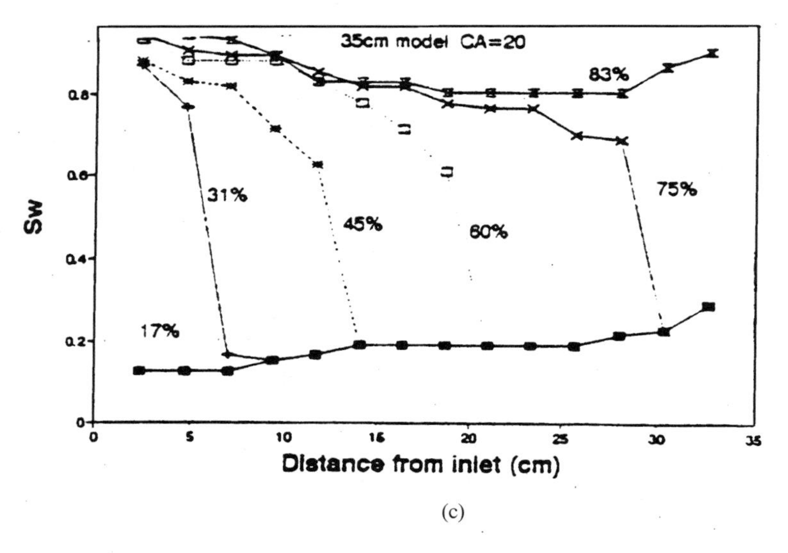

(c)

Figure 22 (*continued*)

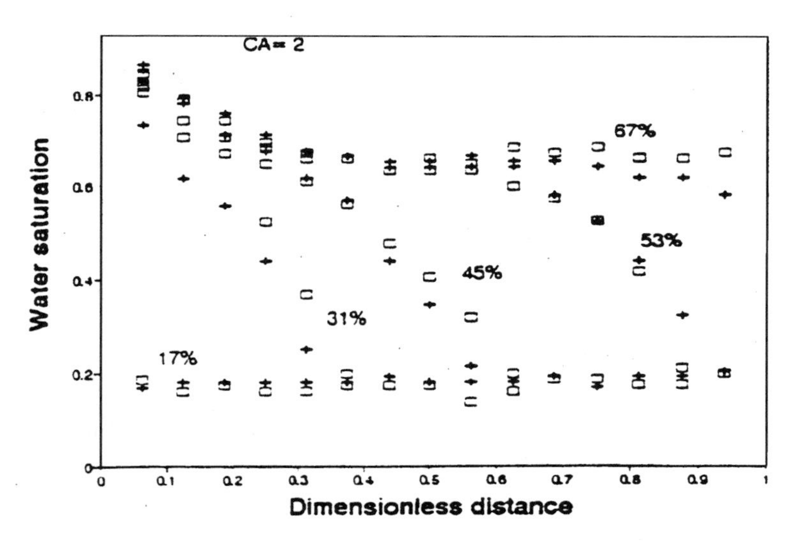

Figure 23. Normalized saturation profile histories of sand A at $CA = 2$ (Dong et al., 1998).

provided that the two waterfloods are carried out in samples of the same uniform cross-sectional area.

It has been found by Dong et al. (1998), that saturation profile histories in waterfloods at 1:1 viscosity ratio in sand B, of about 1/10 of the permeability of sand A, could be matched with the saturation profile histories measured in sand A. The matching occurred not at the same value of CA in the two different sands but for $CA(\mathrm{B}) \approx 4CA(\mathrm{A})$, $CA(\mathrm{B})$ and $CA(\mathrm{A})$ being the values of CA for the waterfloods in sand B and sand A, respectively, as illustrated by the example shown in Figure 24. This result is a consequence of different pore structures of the two sands. It is interesting to note that matching occurred at a ratio of about 4 of the two sets of values of CA which is, for the purpose of such a comparison, the same as the square root of the ratio of the two permeabilities, i.e., 3.2.

D. Mathematical Model of Countercurrent Imbibition

Dong and Dullien (1997) studied countercurrent imbibition in thin horizontal sandpacks. The experiments were started by creating a step-function saturation profile in the sandpack initially at "connate" water saturation and then isolating the system by closing the inlet and outlet valves to the sandpack (Figure 25a). The saturation profile changed with time, as shown schematically in Figure 25b, as a result of countercurrent imbibition which was driven entirely by the gradient of the capillary pressure $\partial P_c/\partial x$.

Figure 26 illustrates the capillary model of countercurrent imbibition of Dong and Dullien (1997). The following relations apply

$$\Delta P_c = \Delta P_w + \Delta P_o \tag{20}$$

where $\Delta P_c = P_{c2} - P_{c1}$, and ΔP_w and ΔP_o are the viscous pressure drops in water and oil, respectively, and

$$q_w = \frac{k_a A_a}{\mu_w} \frac{\Delta P_w}{L} \tag{21}$$

$$q_o = \frac{k_b A_b}{\mu_o} \frac{\Delta P_o}{L} \tag{22}$$

where k_a and k_b are the permeabilitis of tube a and tube b, respectively, L is the length of each tube, and A_a and A_b are the cross-sectional areas of tubes a and b, respectively. Combining Eqs. (20)–(22) results in

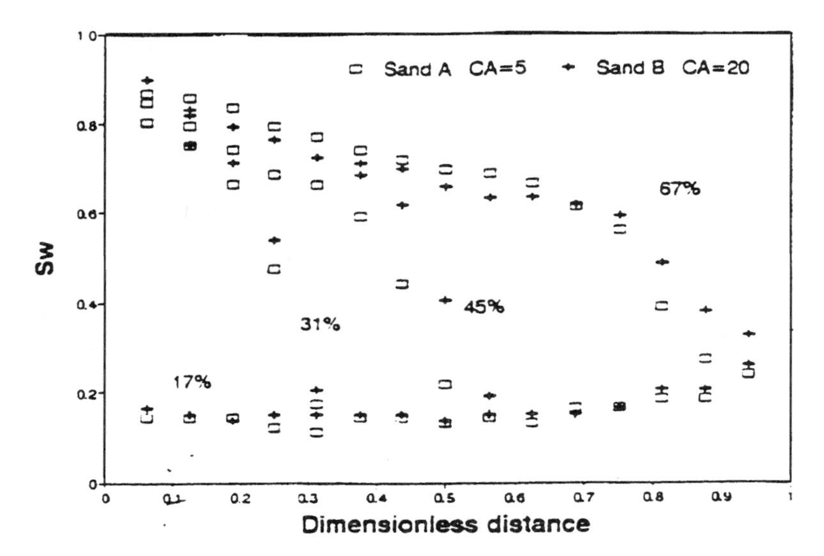

Figure 24. Matching of saturation profile histories of packs of sand A and B; $CA(\mathrm{A}) = 5$, $CA(\mathrm{B}) = 20$ (Dong et al., 1998).

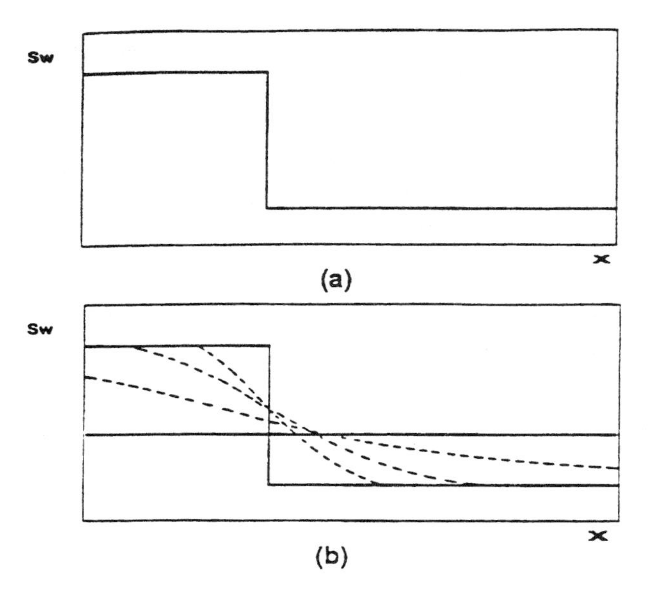

Figure 25. Schematic diagram of saturation profiles during countercurrent imbibition inside a porous medium: (A) step-function displacement front; (b) change of saturation profile from a step function to uniform saturation (Dong and Dullien, 1997).

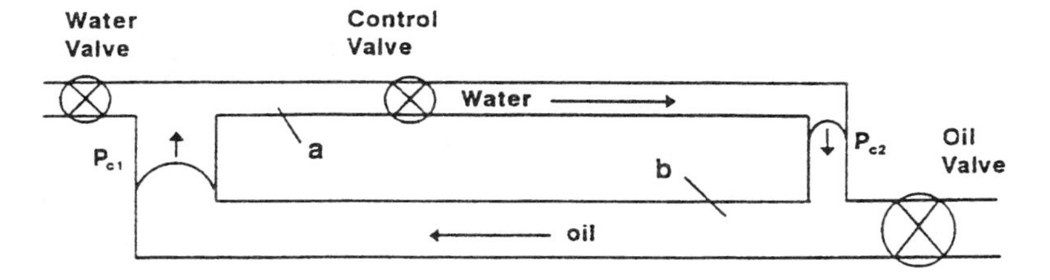

Figure 26. Capillary model showing the countercurrent flow caused by capillary pressure difference (Dong and Dullien, 1997).

$$q_{\mathrm{w}} = -q_{\mathrm{o}} = \frac{\lambda_{\mathrm{a}}\lambda_{\mathrm{b}}}{\lambda_{\mathrm{a}} + \lambda_{\mathrm{b}}} \frac{\Delta P_{\mathrm{c}}}{L} \tag{23}$$

where

$$\lambda_{\mathrm{a}} = \frac{k_{\mathrm{a}} A_{\mathrm{a}}}{\mu_{\mathrm{w}}} \tag{24}$$

and

$$\lambda_{\mathrm{b}} = \frac{k_{\mathrm{b}} A_{\mathrm{b}}}{\mu_{\mathrm{o}}} \tag{25}$$

For a porous medium Eqs. (23)--(25) become

$$q_{\mathrm{w}} = -q_{\mathrm{o}} = \frac{\lambda_{\mathrm{w}}\lambda_{\mathrm{o}}}{\lambda_{\mathrm{w}} + \lambda_{\mathrm{o}}} \frac{\partial P_{\mathrm{c}}}{\partial x} \tag{26}$$

with

$$\lambda_{\mathrm{w}} = \frac{k k_{\mathrm{rw}}}{\mu_{\mathrm{w}}} A \tag{27}$$

and

$$\lambda_{\mathrm{o}} = \frac{k k_{\mathrm{ro}}}{\mu_{\mathrm{o}}} A \tag{28}$$

where k is the permeability and A the cross-sectional area of the sample, and k_{rw} and k_{ro} are the relative permeabilities to water and to oil, respectively.

E. Mathematical Model of Waterflooding Based on the Model of Countercurrent Imbibition

It has been pointed out by Dong and Dullien (1997) that, in a waterflood, pumping of water makes both the water and the oil flow in the forward direction, whereas the capillary pressure gradient, which is present due to a saturation gradient, produces, on the one hand, additional forward flow of water but, on the other hand, it also produces oil flow in the direction opposite to the water flow. Hence the net flows of both water and oil may be regarded as superpositions of the flows due to pumping and the flows due to countercurrent imbibition, i.e.

$$q_{\mathrm{w}} = -\lambda_{\mathrm{w}} \frac{\partial p}{\partial x} + \frac{\lambda_{\mathrm{w}}\lambda_{\mathrm{o}}}{\lambda_{\mathrm{w}} + \lambda_{\mathrm{o}}} \frac{\partial P_{\mathrm{c}}}{\partial x} \tag{29}$$

and

$$q_{\mathrm{o}} = -\lambda_{\mathrm{o}} \frac{\partial p}{\partial x} - \frac{\lambda_{\mathrm{w}}\lambda_{\mathrm{o}}}{\lambda_{\mathrm{w}} + \lambda_{\mathrm{o}}} \frac{\partial P_{\mathrm{c}}}{\partial x} \tag{30}$$

where $\partial p/\partial x$ is the pressure gradient associated with the flow of water and oil, according to Darcy's law, due to pumping of water at a rate Q_{t}. Q_{t} is equal to the total flow rate at any location of the sample, i.e.

$$Q_{\mathrm{t}} = q_{\mathrm{w}} + q_{\mathrm{o}} = -(\lambda_{\mathrm{w}} + \lambda_{\mathrm{o}})\partial p/\partial x \tag{31}$$

Combination of Eq. (31) with (29) and (30) yields the following expressions for q_{w} and q_{o}

$$q_{\mathrm{w}} = \frac{\lambda_{\mathrm{w}}}{\lambda_{w} + \lambda_{\mathrm{o}}} Q_{\mathrm{t}} + \frac{\lambda_{\mathrm{o}}\lambda_{\mathrm{w}}}{\lambda_{\mathrm{w}} + \lambda_{\mathrm{o}}} \frac{\partial P_{\mathrm{c}}}{\partial x} \tag{32}$$

and

$$q_{\mathrm{o}} = \frac{\lambda_{\mathrm{o}}}{\lambda_{w} + \lambda_{\mathrm{o}}} Q_{\mathrm{t}} - \frac{\lambda_{\mathrm{o}}\lambda_{\mathrm{w}}}{\lambda_{\mathrm{w}} + \lambda_{\mathrm{o}}} \frac{\partial P_{\mathrm{c}}}{\partial x} \tag{33}$$

Eq. (32) is equivalent to the complete fractional equation of water which was first derived by Leverett (1941).

Waterfloods have been simulated for a mathematical model of a porous medium defined by values of permeability k, porosity ϕ, and relative permeabilities k_{rw} and k_{ro} and the capillary pressure P_{c} as functions of saturation S_{w}. Specifying an initial saturation profile in the model and also the values of Q_{t}, μ_{w}, μ_{o}, $A = 1\,\mathrm{cm}^2$, the "connate" and "irreducible" water saturations, and the "residual" oil saturation, Eq. (32) has been solved numerically for $q_{\mathrm{w}} \equiv v_{\mathrm{w}}$ vs. distance x. The conservation of mass equation

$$\phi \frac{\partial S_{\mathrm{w}}}{\partial t} + \frac{\partial \mathrm{v}_{\mathrm{w}}}{\partial x} = 0 \tag{34}$$

has been used to calculate new saturation profiles for different values of time t (Dong and Dullien, 1997). These calculations yield the saturation profiles in a straightforward manner. Unlike in the case of the widely used Buckley–

Leverett method, there is no need to resort to "shock solution", for triple-valued saturation profiles never arise in these simulations. The assumption of negligible capillary pressure gradient, made in the Buckley–Leverett model, is incorrect in the case of interfacial tensions encountered in petroleum reservoirs and is the cause of obtaining triple-valued saturation profiles.

The predicted saturation profile evolutions for different values of Q_t (for $\mu_w = \mu_o$) are similar to the measured evolutions of saturation profiles of Dong et al. (1998) which were characterized by the "complete" capillary number CA as parameter. In order to assign a value of CA to each calculated evolution with the help of Eq. (18) it is necessary also to be able to predict the initial value of free spontaneous imbibition Q in the model system used in the simulations of Dong and Dullien (1997). The mathematical model used for predicting the rate of free spontaneous imbibition Q is described in the next section.

F. Mathematical Model of Cocurrent Imbibition

A simple capillary model has been introduced by Dong et al. (1998) which is endowed with the physical properties that control cocurrent imbibition. This model, illustrated in Figure 27a, consists of two parallel capillaries of radii R_1 and R_2, separated by a membrane permitting equilibration of pressures between the two tubes at every distance x measured along the axes of the tubes, but preventing flow from one tube into the other. Equilibration of pressures has the result that, in agreement with experiments using physical capillary models of porous media (Chatzis and Dullien, 1983), the water imbibes first into the narrow tube and also advances at a faster rate there than in the wide capillary. This is the opposite behavior to that predicted by the well-known Washburn equation where the rate of advance is faster in the wide capillary. The reason for this difference is that the Washburn equation applies to individual capillaries and does not consider any interaction between two or more parallel capillaries. In most porous media, however, the pores are interconnected and therefore there is pressure equilibration between adjacent pores.

The pressure profiles in the two capillary tubes of Figure 27a, for the case of free spontaneous imbibition of water where $\Delta P = 0$, are shown in Figure 27b. Elementary analysis, using the Hagen–Poiseuille equation, yields the following equations for the flows q_1 and q_2 in tubes 1 and 2, respectively

$$q_1 = \frac{\lambda_2 Q - \lambda_2 \lambda_3 (P_{c2} - P_{c1})}{\lambda_2 + \lambda_3} \tag{35}$$

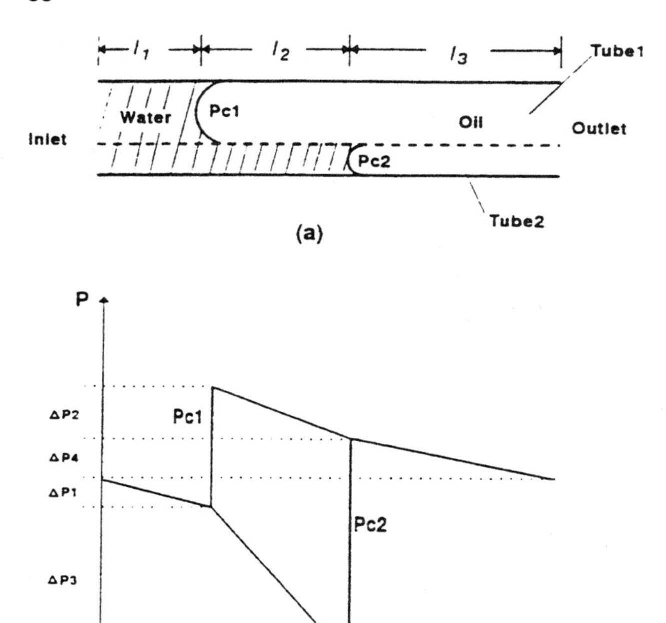

Figure 27. Free spontaneous imbibition in the simple capillary model: (a) schematic of model; (b) pressure profiles in the model (Dong et al., 1998).

$$q_2 = \frac{\lambda_3 Q - \lambda_2 \lambda_3 (P_{c2} - P_{c1})}{\lambda_2 + \lambda_3} \tag{36}$$

and for $Q = q_1 + q_2$

$$Q = \frac{\lambda_2 P_{c1} + \lambda_3 P_{c2}}{1 + \dfrac{(\lambda_2 + \lambda_3)(\lambda_1 + \lambda_4)}{\lambda_1 \lambda_4}} \tag{37}$$

with $\lambda_1 = \pi(R_1^4 + R_2^4)/8\mu_w \ell_1$, $\lambda_2 = \pi R_1^4/8\mu_o \ell_2$, $\lambda_3 = \pi R_2{}^4/8\mu_w \ell_2$, and $\lambda_4 = \pi(R_1{}^4 + R_2{}^4)/8\mu_o \ell_3$. Substituting these expressions into (37) and letting $\mu_w = \mu_o = \mu$, the following expression is obtained for the rate of free spontaneous imbibition

$$Q = \frac{\pi(R_1^4 P_{c1} + R_2^4 P_{c2})}{8\mu L} \tag{38}$$

As $P_{c2}/P_{c1} = R_1/R_2$, Eq. (38) can be written as follows (Dullien and Dong, 1998)

$$Q = AP_{c1}k_{\text{eff}\,1}/\mu L \tag{39}$$

where

$$k_{\text{eff}\,1} = \frac{R_1({R_1}^3 + {R_2}^3)}{8({R_1}^2 + {R_2}^2)} \tag{40}$$

and

$$A = \pi({R_1}^2 + {R_2}^2) \tag{41}$$

It is evident from Eq. (39) that the rate of free spontaneous imbibition in the model is proportional to P_{c1} and an effective permeability $k_{\text{eff}\,1}$ of the model and does not involve relative permeabilities of water and oil.

Equation (38) can also be written as follows

$$Q = AP_{c2}k_{\text{eff}\,2}/\mu L \tag{42}$$

where now

$$k_{\text{eff}\,2} = \frac{R_2({R_1}^3 + {R_2}^3)}{8({R_1}^2 + {R_2}^2)} \tag{43}$$

Evidently, either P_{c1} or P_{c2} can be chosen as the driving force of imbibition because the correspondingly different values of the effective permeability of the system will guarantee that the same value is obtained for Q.

The Darcy permeability k of the model is

$$k = \frac{{R_1}^4 + {R_2}^4}{8({R_1}^2 + {R_2}^2)} \tag{44}$$

which has a similar magnitude to either $k_{\text{eff}\,1}$ or $k_{\text{eff}\,2}$.

By the analogy of Eq. (39) or (42), the following expression has been written by Dullien and Dong (1998) for Q in the case of a porous medium

$$Q = A\frac{P_c(S_w)}{\mu L}k_{\text{eff}}(S_w) \tag{45}$$

where $P_c(S_w)$ is the capillary pressure at any saturation S_w and $k_{\text{eff}}(S_w)$ is the corresponding effective permeability of the medium.

Equation (45) is a special form of Darcy's law. $P_c(S_w)$ and $k_{\text{eff}}(S_w)$ are constant values. The prediction by Eq. (45) of constant Q agrees with the experimental results of Dong et al. (1998), illustrated in Figure 21, where Q

remained constant in the case of free spontaneous imbibition of water into sandpacks, displacing oil, until the imbibition front reached the exit end of the pack. Substituting the measured value Q, along with the rest of the known values of A, μ, L, and P_c (0.6) into Eq. (45), Dullien and Dong (1998) calculated k_{eff} (0.6) and obtained a value which was within 10% of k, the measured Darcy permeability of the sandpack. This result supports the soundness of the physical basis of the capillary model of cocurrent imbibition.

The capillary model of cocurrent imbibition has been applied by Dong et al. (1998) also to "forced imbibition", i.e. a waterflood carried out at a constant pumping rate Q_t, resulting in a pressure drop ΔP across the model. In this case the pressure profiles of Figure 27b become

$$\Delta P_1 + \Delta P_2 + \Delta P_4 = P_{c1} + \Delta P \tag{46}$$

and

$$\Delta P_1 + \Delta P_3 + \Delta P_4 = P_{c2} + \Delta P \tag{47}$$

These equations have been used by Dong et al. (1998) to solve the Hagen-Poiseuille equation for q_1 and q_2, with the following results

$$q_1 = \frac{\lambda_2}{\lambda_2 + \lambda_3} Q_t - \frac{\lambda_2 \lambda_3}{\lambda_2 + \lambda_3} \Delta P_c \tag{48}$$

and

$$q_2 = \frac{\lambda_3}{\lambda_2 + \lambda_3} Q_t + \frac{\lambda_2 \lambda_3}{\lambda_2 + \lambda_3} \Delta P_c \tag{49}$$

where

$$Q_t = q_1 + q_2 \tag{50}$$

and

$$\Delta P_c = P_{c2} - P_{c1} \tag{51}$$

Equations (48) and (49), written for the transition region ℓ_2 of the capillary model, where $q_1 = q_o$ and $q_2 = q_w$, after introduction of the permeability k (Eq. 44), the cross-section $A = \pi(R_1{}^2 + R_2{}^2)$, and the relative permeabilities

$$k_{r1} = k_{ro} = \frac{R_1{}^4}{R_1{}^4 + R_2{}^4} \tag{52}$$

and

$$k_{r2} = k_{rw} = \frac{R_2{}^4}{R_1{}^4 + R_2{}^4} \tag{53}$$

of the model, result in expressions which are identical with Eqs. (32) and (33).

The conclusion can be reached that the flows of oil and water in immiscible displacement can be construed *either* as countercurrent imbibition on which there are superimposed flows of water and oil due to pumping water at a rate Q_t, *or* as "forced" cocurrent imbibition due to a pressure difference ΔP imposed across the medium. The resulting expressions are identical in both cases.

G. Comparison of Saturation Profile Histories Measured in Sandpacks with those Calculated for a Model of a Porous Medium

The parameter used by Dullien and Dong (1998) to characterize each saturation profile history is the "complete" capillary number CA. For the measured profiles, CA was calculated by Eq. (18) from the experimental water injection rates Q_t and the measured value of the initial rate of free spontaneous imbibition Q into the sandpack. In the case of the calculated profiles, the value of Q was estimated by Dullien and Dong (1998), using Eq. (45), by substituting the values of A, μ, L, $k \approx k_{\text{eff}}$, and P_c (0.6) used in the simulations. The best matches between the saturation profile histories measured in sandpacks and those calculated in the case of a mathematical model of a porous medium of less than 1/10 of the permeability of the sandpacks were found for a ratio $CA(\text{cac.})/CA(\text{exp.}) \approx 3$. An example of the matching pairs is shown in Figures 28a and b, where $CA(\text{calc.}) = 2.62$ and $CA(\text{exp.}) = 1$. The measured saturation profile histories of sand B (Dong et al. 1998) match the simulated histories of the model for about the same value of CA in both systems. As both of these media, i.e., sand B and the model medium, have about the same permeability, this result suggests that porous media of the same permeability may have matching saturation histories for the same value of CA.

H. Drainage-type Displacement Patterns as a Function of Capillary Number *Ca* and Viscosity Ratio

The evolution of saturation profiles in drainage as a function of CA has not yet been studied. On the other hand, there has been a 2D network simulation study of drainage displacement patterns as a function of $Ca = \mu v/\sigma$ and the viscosity ratio $\kappa = \mu_2/\mu_1$($2 =$ displacing fluid), also including unfavorable

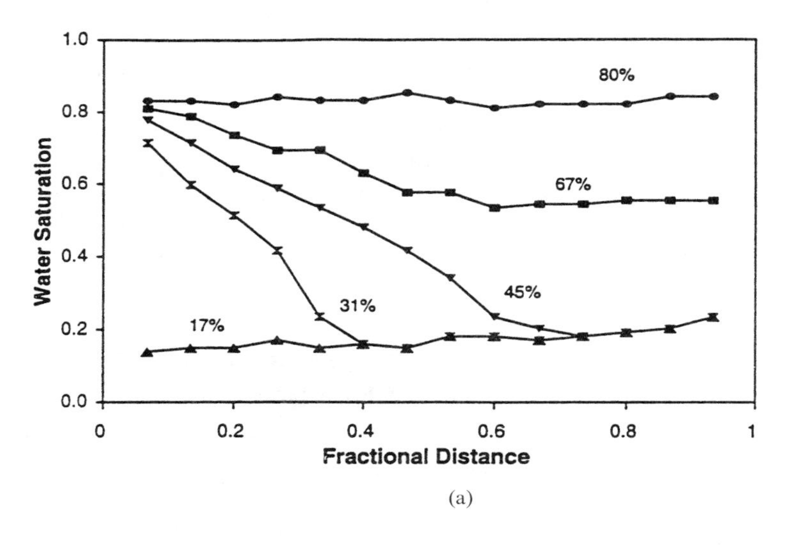

(a)

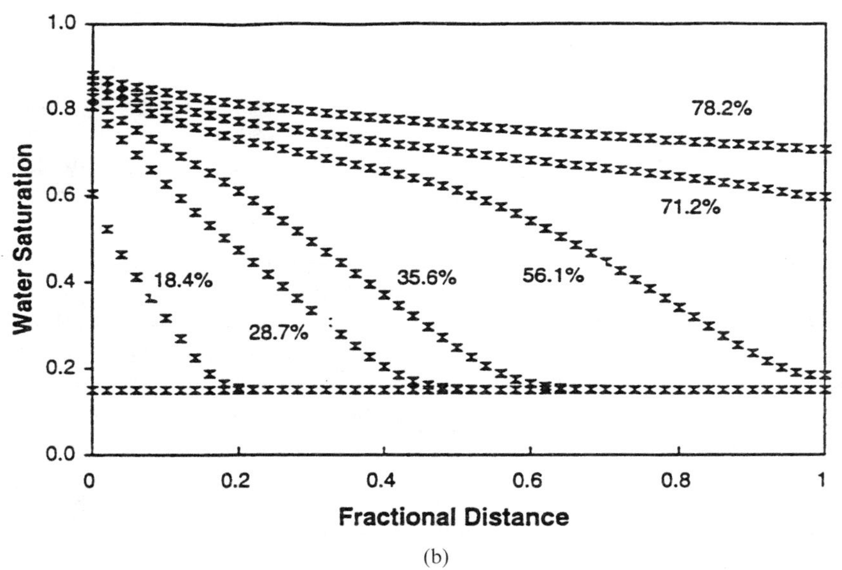

(b)

Figure 28. (a) Measured saturation profiles at $CA = 1$ (Dullien and Dong, 1998). (b) Calculated saturation profiles at $CA = 2.62$ (Dullien and Dong, 1998).

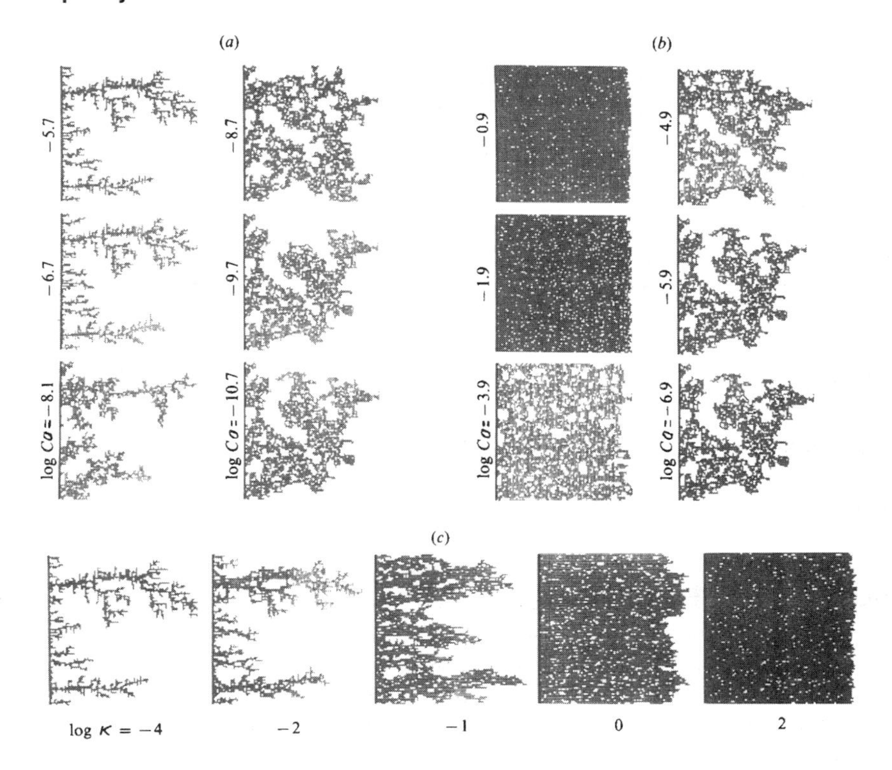

Figure 29. Network (100 × 100) simulations of drainage-type displacement at various viscosity ratios κ and capillary numbers Ca: (a) $\log \kappa = -4.7$ (from viscous to capillary fingering); (b) $\log \kappa = 1.0$ (from stable displacements to capillary fingering); (c) $\log Ca = 0$ (from viscous fingering to stable displacement) (Lenormand et al., 1988).

viscosity ratios in other words $\kappa < 1$ (Lenormand et al., 1988), the results of which are shown in Figure 29. It is evident that, for unfavorable viscosity ratios and values of Ca corresponding to $CA \gg 1$, i.e., viscous forces controlling, that there is viscous fingering, whereas for values of Ca corresponding to $CA \ll 1$, i.e., capillary forces controlling, there is "capillary fingering". For favorable viscosity ratios, the saturation profile is a step function when viscous forces control, whereas when capillary forces control there is capillary fingering. Displacement experiments, carried out in etched capillary "micromodels" with the fluid pairs consisting of air/very viscous oil, mercury/hexane, mercury/air, and glucose solution/oil, yielded results that were in good qualitative agreement with the simulations. Inspection of Figure 29 shows that the values of Ca do not represent the ratio of viscous-to-capillary forces.

V. VISCOUS COUPLING OF TWO IMMISCIBLE FLUIDS IN COCURRENT FLOW IN THE PRESENCE OF A SATURATION GRADIENT

The discussion of this section is limited to conditions such that one of the two fluids is strongly preferentially wetting, the interfacial tension is sufficiently high to prevent emulsification, and the viscosity of neither fluid is high enough to take control of the pore level phenomena.

In the generalized form of Darcy's law for multiphase flow, expressed by Eq. (5), the flows as well as the pressure gradients of the various immiscible fluids present in the porous medium have been assumed to be independent of each other. In steady cocurrent flow, where both fluids are pumped through the sample at constant rates, near-uniform saturation was found experimentally, and as the value of the capillary pressure is approximately uniform the pressure gradients of the two immiscible fluids are approximately equal. The special type of what might be called "kinematic coupling" that can arise under such conditions, in which the wetting fluid of a lower viscosity may enhance the flow of a nonwetting fluid of a higher viscosity by means of the so-called lubricating effect, has already been discussed in an earlier section of this paper. In this section the general case of cocurrent two-phase flow is considered where a saturation gradient and, therefore, also a capillary pressure gradient exist. Under these conditions the following relationship between the pressure gradients of the two flows has been assumed to exist (Leverett, 1941)

$$\nabla P_{\mathrm{c}} = \nabla P_{\mathrm{nw}} - \nabla P_{\mathrm{w}} \tag{54}$$

Here, P_{c} is not necessarily the same value as under static conditions. As a result of this relationship the pressure gradients of the two fluids cannot be either specified or varied independently. It is possible, however, to measure both pressure gradients simultaneously. The problem at hand is whether or not the correct flows of the two fluids are predicted by Eq. (5) if the appropriate measured (i.e., known) pressure gradients and viscosities of the two fluids are substituted, using effective permeabilities that are assumed to be independent of the pressure gradients and the viscosities.

Many authors (e.g., Rose, 1971; deGennes, 1983; de la Cruz and Spanos, 1983) have proposed that there exists viscous coupling between the flows of two immiscible fluids; i.e., the flow of either fluid is affected by the pressure gradient of the other fluid and, consequently, the generalized form of Darcy's law given by Eq. (5) is incorrect. One form of the coupled equations that has been proposed is, for the case of one-dimensional flow

$$v_1 = \frac{k_{11}}{\mu_1}\frac{dP_1}{dz} - \frac{k_{12}}{\mu_2}\frac{dP_2}{dz} \tag{55}$$

and

$$v_2 = \frac{k_{21}}{\mu_1}\frac{dP_1}{dz} - \frac{k_{22}}{\mu_2}\frac{dP_2}{dz} \tag{56}$$

Experimental determination of the four effective permeability coefficients k_{11}, k_{12}, k_{21}, and k_{22} in Eqs. (55) and (56) has presented some problems. The terms containing the coefficients k_{11} and k_{22}, considered in isolation, represent Darcy's law; whereas the terms with the cross-coefficients k_{12} and k_{21} represent the effect of coupling. To determine this effect on the flow of fluid 1, the contribution to the value of v_1 resulting from dP_2/dz must be measured. In general, however, according to Eq. (54), imposing a pressure gradient dP_2/dz on fluid 2 results in a pressure gradient of dP_1/dz also in fluid 1 and, therefore, the value of v_1 that is measured in the experiment results from the pressure gradients in both fluids. However, under experimental conditions where $dP_2/dz \neq 0$ but $dP_1/dz = 0$, i.e., when

$$v_1 = -(k_{12}/\mu_2)(dP_2/dz) \tag{57}$$

and

$$v_2 = -(k_{22}/\mu_2)(dP_2/dz) \tag{58}$$

k_{12} and k_{22} can be calculated from measured values of v_1 and v_2. The coefficients k_{21} and k_{11} are determined in separate experiments, where $dP_1/dz \neq 0$ and $dP_2/dz = 0$, from the following equations

$$v_1 = -(k_{11}/\mu_1)(dP_1/dz) \tag{59}$$

and

$$v_2 = -(k_{21}/\mu_2)(dP_1/dz) \tag{60}$$

These conditions have been realized in a horizontal sandpack, using Soltrol 100 oil of 1 cp viscosity and water of the same viscosity (Dullien and Dong, 1996). First the conventional steady-state conditions were established, corresponding to some saturation of the pack (Figure 30). Then, after both the inlet and the outlet of water had been shut, oil was pumped at a constant rate until the water pressure gradient became zero in the entire pack (Figure 31). Simultaneously, a saturation and capillary pressure gradient was built up in the sandpack. At this point the water inlet and outlet were opened, and as a result of viscous coupling water was sucked into the pack from a reservoir and produced into another reservoir via a discharge tube (Figure 32, piezometers not shown), while steps were taken to ensure that the zero pressure gradient in the water in the sandpack was maintained.

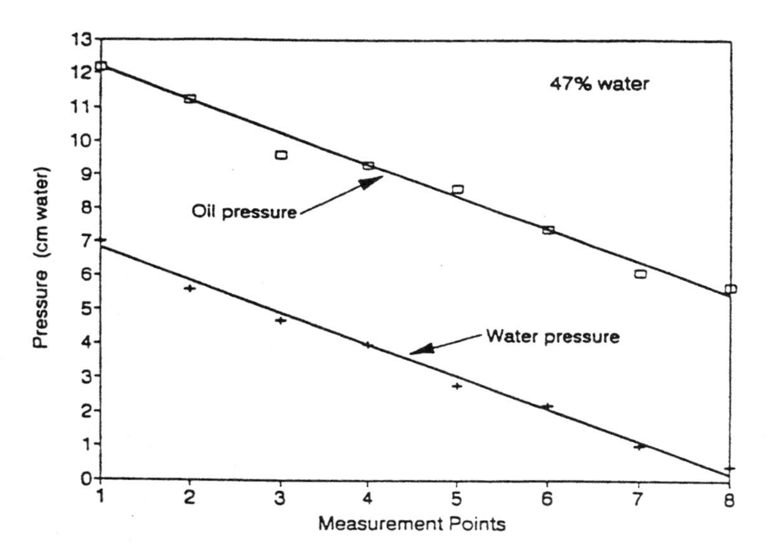

Figure 30. Conventional steady-state pressure profiles in sandpack (Dullien and Dong, 1996).

Steady-state conditions could be maintained, in this case in the presence of gradients of capillary pressure and saturation. The shape of the oil pressure gradient was the same in all of the tests, carried out at different average saturations. This shape can be understood if it is realized that, whereas the

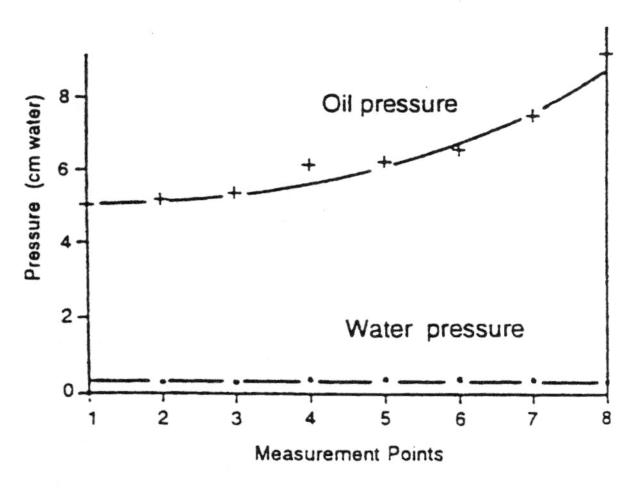

Figure 31. Steady-state pressure profiles with a saturation gradient in sandpack; zero pressure gradient in water (Dullien and Dong, 1996).

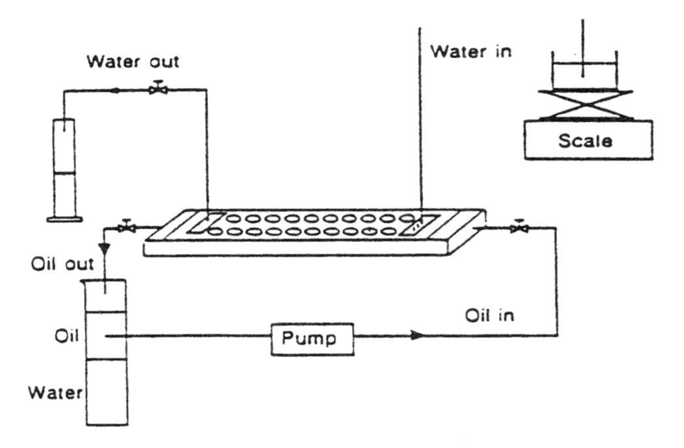

Figure 32. Experimental conditions for measuring viscous coupling (Dullien and Dong, 1996).

water flow rate was the same value throughout the sandpack, the water saturation near the inlet was much lower than near the outflow. At low water saturations, however, the water permeability is much less than at high water saturations. Therefore, it required a much greater oil pressure gradient to make the water flow by viscous coupling at the same rate at low water saturations than at high values of the water saturation.

Corresponding experiments were carried out in which the roles played by the water and the oil were reversed, i.e., the water made the oil flow by viscous coupling. The measured flow rates are shown in Tables 1 and 2, and the values of the four effective permeability coefficients are plotted vs. the average saturations in the sandpack in Figure 33 along with the Darcy's law

Table 1. Oil and water flow rates measured at the conditions $dP_2/dz = 0$ and $dP_1/dz \neq 0$

Order of runs	Water saturation (%)	Q_o (ml/min)	Q_w (ml/min)	Q_o/Q_w
4	64.5	0.20	0.05	4.0
5	61.5	0.26	0.15	1.7
6	60.5	0.30	0.18	1.7
1	51.5	0.33	0.15	2.2
2	43.0	1.15	0.06	19
3	27.0	1.50	0.01	150

Table 2. Water and oil flow rates measured at the conditions $dP_1/dz = 0$ and $dP_2/dz \neq 0$

Order of runs	Water saturation (%)	Q_w (ml/min)	Q_o (ml/min)	Q_w/Q_o
4	47	0.27	0.04	6.8
3	54	0.44	0.04	11
2	57	0.53	0.05	10.6
1	65	0.80	0.05	16

effective permeabilities. It is evident that the coupling coefficients measured in this work are significant and also that $\langle k_{12} \rangle \neq \langle k_{21} \rangle$. The latter effect can be explained by different distributions of the two fluids at the same saturation in the two experiments leading to the determination of the two coupling coefficients (Lasseux et al., 1996).

It is important to note that the values of the coupling coefficients at a given average saturation can be expected to be smaller when the pressure gradient in neither fluid is zero. As this situation exists in all conventional two-phase flows, under normal conditions of two-phase flow the value of the coupling coefficients can be expected to be a very small fraction of the

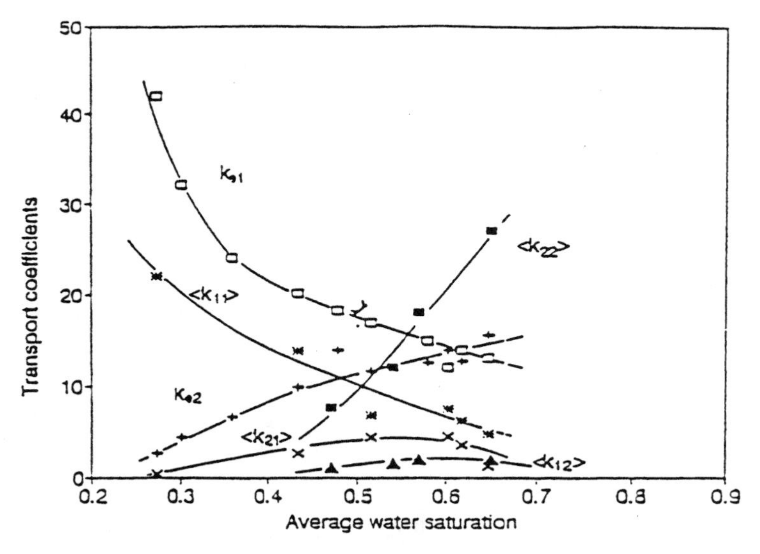

Figure 33. Effective permeability coefficients $\langle k_{11} \rangle$, $\langle k_{12} \rangle$, $\langle k_{21} \rangle$, and $\langle k_{22} \rangle$ vs. average water saturation (Dullien and Dong, 1996).

Darcy's law effective permeabilities and besides, as pointed out earlier, their values are not amenable to experimental determination. Only the effect of viscosity ratio, i.e., the "lubricating effect" in low permeability media and at low wetting phase saturations, can be expected to result in significant coupling (Dullien, 1993).

Zarcone and Lenormand (1994) published experimental results obtained on viscous coupling between mercury and water in a sandpack, with mercury the driving and water the driven fluid. The induced water flow rate at all the saturations tested was less than 1% of the mercury flow. These authors assumed that the water pressure could be rendered uniform everywhere in their model by connecting the point of water entry into the model with the point of water exit from the model by a tube. A critical discussion of this work and that by Rakotomalala et al. (1995) is presented by Dullien and Dong (1996).

VI. SPREADING OF OIL OVER WATER IN PORES

In previous sections of this paper, discussion was limited to conditions involving only two fluid phases. There are situations of considerable practical importance, however, where in addition to two liquid phases there is also a gas phase present in immiscible displacement in porous media. One such instance involves the problem of mobilization of oil (petroleum or LNAPL) blobs that are surrounded by water and trapped in the pores by capillary forces. Trapping of oil blobs happens regularly in water-wet oil reservoirs after water flooding and it can also happen in the soil in the course of attempted clean-up operations.

It was demonstrated by numerous researchers (e.g., Chatzis et al., 1988; Kalaydjian et al., 1993) that, if the water is drained from a porous medium containing trapped residual oil, the oil blobs start spreading spontaneously immediately after they have been contacted by the air or other inert gas, such as nitrogen, used to displace the water from the main (central) parts of the pores (Figure 34). The explanation of this phenomenon is that oil spreads on the surface of the thick water films remaining on the pore walls after the bulk of the water has drained out if its spreading coefficient S, defined as

$$S = \sigma_{gw} - (\sigma_{go} + \sigma_{ow}) \tag{61}$$

has a positive value (σ_{gw}, σ_{go}, σ_{ow} is the gas-water, gas-oil, and oil-water interfacial tension, respectively).

The phenomenon of spreading oil blobs has resulted in the recovery of a very high percentage of the residual oil, because the oil films thus formed

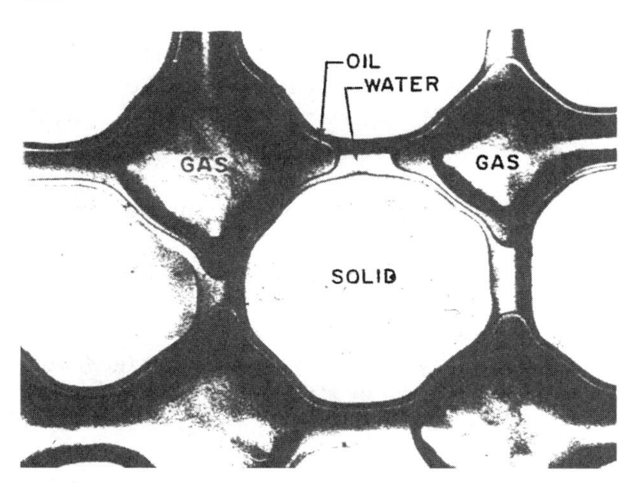

Figure 34. Spontaneous spreading of oil blobs on the surface of connate water in a capillary micromodel after being contacted by air (Chatzis et al., 1988).

drain under the influence of gravity on the surface of the thick water films, and oil is produced at the low end of the medium (Catalan et al., 1994).

The problem with the explanation offered for the spreading of trapped oil blobs under the circumstances described above is that there have been instances reported in the literature where an unexpectedly large percentage of a nonspreading type of oil, i.e., when the spreading coefficient S has a negative value, was recovered by the same process. In light of these reports the question of spreading of oil drops on the surface of thick water films present in pores has been re-examined (Dong et al., 1995), and it was noted that the conditions of spreading of an oil drop over a thick water film in a pore edge are different than on a water surface because of the presence of a solid, represented by the pore walls (Figure 34), which is also contacted by the oil. As a result, even an oil drop with a negative spreading coefficient may spread over the thick water film under certain conditions and form a film of definite thickness; i.e., the spreading does not result in a film of molecular thickness. The driving force of this kind of spreading is the capillary pressure and, therefore, it can be regarded as a special case of imbibition.

It has been observed that when a small quantity of nonspreading ($S < 0$) oil is put in contact with a thick water film in an edge, a small lens of oil is formed. If one adds a little more oil to the lens, the thickness of the center of the lens and the length of the lens along the edge increase. After the addition of a certain amount of oil, the thickness of the lens does not continue to increase, but its length keeps increasing in direct proportion to the amount of oil added. The lens has become a thick film. The constant thickness of the oil film (Figure 35) is called the "critical" film thickness for imbibition into an edge. The critical film thickness has been calculated by combined application of Laplace's equation of capillary pressure and the condition of minimum surface energy as a function of the water film thickness in the edge, the half-angle ϕ of the edge (Figure 3b), the oil–water interfacial tension, and the (negative) spreading coefficient (Dong et al., 1995). The results of these calculations are shown in Figure 36.

These results indicate that the critical oil film thickness for imbibition over a thick water film along an open edge strongly depends on the water saturation, the geometry of the edge, and the negativity of the spreading coefficient of the oil on water. There is, however, a difference between imbibition into an open edge, where there is no limit to the total thickness, i.e., water plus oil film thickness, and imbibition into the edges of a pore, modeled by a square or a triangular capillary, because in the latter the edges are not open but are closed in the form of a polygon. Consequently, in a pore the total film thickness can never exceed the equilibrium value, pointed out earlier in this paper, which corresponds to 6% of the cross-section of a square capillary (12.5% of the cross-section of an equilateral triangular capillary).

Under the conditions when, in porous media, initially there are oil blobs trapped and surrounded by water in the pores, after drainage of the bulk of the water the oil drops come in contact with an inert gas (e.g., air), the pore walls, and the thick water films present in the pore edges or grooves. Using a square capillary as the pore model, the oil blob can imbibe into all four edges of the square capillary if there is enough oil present in the

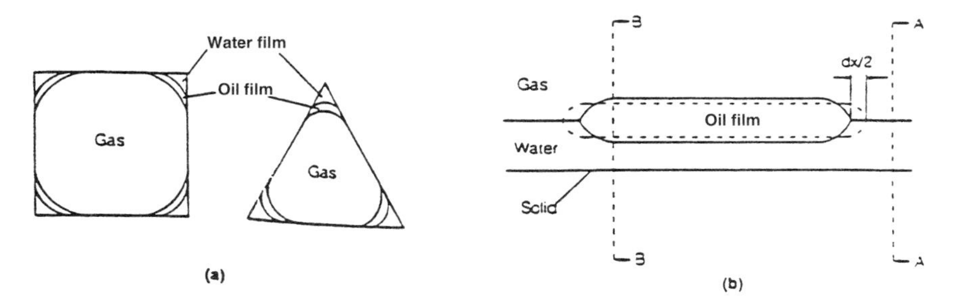

Figure 35. Three phases in equilibrium in an edge: (a) three phases in the edges of a square and a triangular capillary, and (b) oil film as seen in a diagonal slice through an edge (Dong et al., 1995).

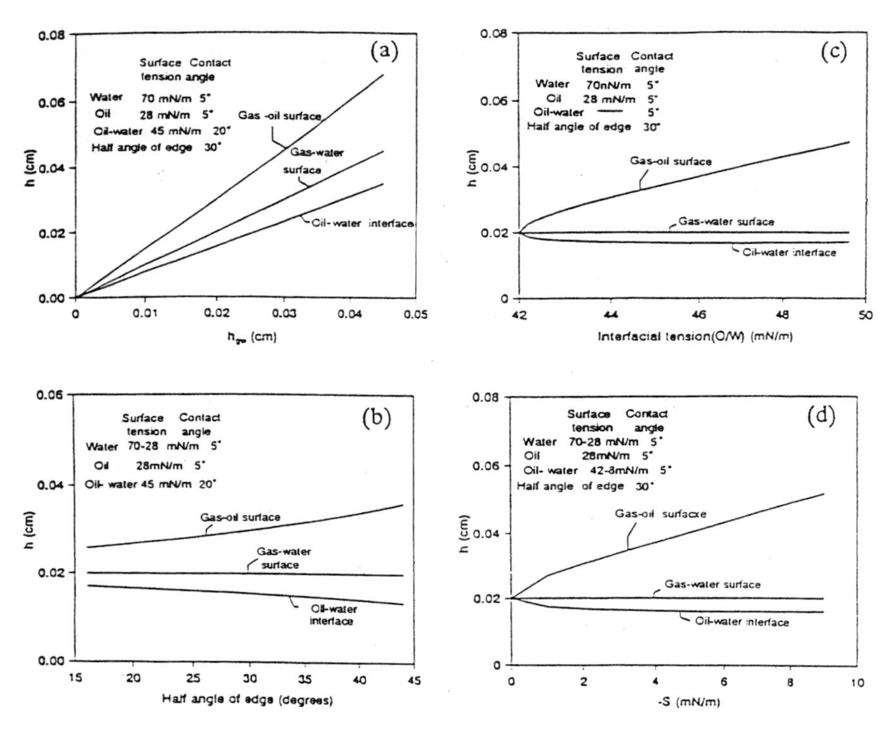

Figure 36. Critical thickness of oil film as a function of (a) the water film thickness in an edge, (b) the half-angle of an edge, (c) oil–water interfacial tension, and (d) the spreading coefficient (< 0) (Dong et al., 1995).

blob to form a film of the critical thickness and if the sum of the thick water film thickness and the critical oil film thicknes is not greater than the value corresponding to 6% occupancy of the cross-section of the capillary. The range of possible total film thicknesses corresponds to the portion of the plots in Figure 37 lying under the horizontal line. (H is the maximum total equilibrium film thickness, corresponding in the case of a triangular capillary to 12.5% occupancy of the cross-section, h_{go} is the total critical film thickness, and h_{gw} is the water film thickness.) System 1 is characterized by a more negative spreading coefficient than system 2 and, therefore, in it the maximum possible total film thickness corresponds to a lower water saturation (i.e., a smaller water film thickness). This means, of course, that in this example the oil blob cannot imbibe into the edges in system 1 if the dimensionless water film thickness is greater than approximately 0.45; whereas in

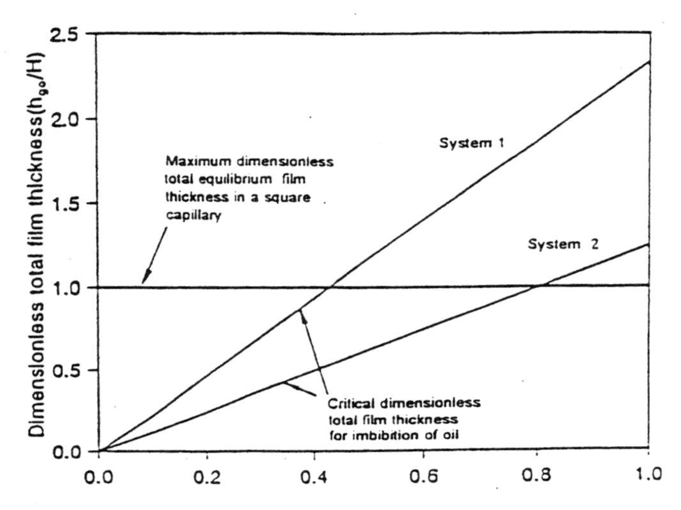

Figure 37. Dimensionless total critical film thickness as a function of the dimensionless water film thickness in a square capillary (Dong et al., 1995).

system 2 the oil blob can imbibe as long as the dimensionless film thickness is less than approximately 0.8.

The previous predictions have been confirmed by experiments (Dong et al., 1995) in a 0.05 cm square capillary tube with 0.6% water saturation, using benzene (spreading coefficient, equilibrated with water: $S = -1.4\,\text{mN/m}$) and a light paraffin oil ($S = -7.9\,\text{mN/m}$), respectively, as the oil phase. For these conditions, the calculated critical total film thicknesses were, for the benzene–water system, $h_{go} = 4.22 \times 10^{-3}$ cm; whereas, for the light paraffin oil–water system, $h_{go} = 6.17 \times 10^{-3}$ cm. Comparing these values with the maximum total equilibrium film thickness $H = 5.47 \times 10^{-3}$ cm, it follows that, although a slug of benzene can be expected to imbibe into edges over the water there, a slug of light paraffin oil is not. These predictions were confirmed by experiments in which a slug of benzene did imbibe, whereas a slug of the light paraffin oil did not.

VII. SUMMARY AND CONCLUSIONS

1. Interfacial tension, wettability, pore morphology, and displacement history are the fundamental parameters that determine the topologies of immiscible fluids at a given saturation in capillary equilibrium in porous media.

2. The preferentially wetting fluid remains continuous and conductive at all saturations in most porous media (notable exception: packs of smooth glass beads) because of the presence of thick films (of the wetting fluid) in pore edges and grooves, whereas the nonwetting fluid can become discontinuous as a result of being trapped by the wetting fluid.
3. The trapping of nonwetting fluid is by snap-off of threads of nonwetting fluid, choked off by annular wetting fluid formed from the thick films.
4. Special cases of wettability are neutral wettability, when neither fluid is preferentially wetting, and fractional wettability, when different portions of the pore surface are preferentially wetted by different fluids.
5. In equilibrium displacement different saturations are reached at the same capillary pressure, depending on whether the wetting fluid (imbibition) or the nonwetting fluid (drainage) is the displacing phase. This phenomenon is called capillary hysteresis.
6. In steady cocurrent two-phase flow the fluid topologies at a given saturation depend on the viscosities and fluid velocities, in addition to the parameters listed under 1. The validity of Darcy's law is limited to certain ranges of the values of these parameters.
7. Relative permeabilities become different under conditions of neutral wettability or very low interfacial tensions.
8. The oil relative permeability may become anomalously great in water-wet systems of low permeability at low water saturations, due to the "lubricating" effect of the water of lower viscosity than oil.
9. In the imbibition type of immiscible displacement (waterflood) in horizontal sandpacks the evolution of saturation profiles has been respresented in terms of a "complete" capillary number $CA = Q_t/Q$. Q_t is the constant injection rate of water in the waterflood and Q is the initial rate of free spontaneous imbibition of water into the same system at 1:1 viscosity ratio. For $CA \ll 1$ the saturation profiles are approximately horizontal, whereas for $CA \gg 1$ they are approximately piston-like.
10. The evolutions of saturation profiles in waterfloods in two samples of the same sandpack, of lengths L_1 and L_2 at 1:1 viscosity ratio, have been found to be approximately identical at the same value of CA in the two waterfloods if plotted against normalized distance coordinates.
11. The evolutions of saturation profiles in waterfloods at 1:1 viscosity ratio in sandpacks of very different permeabilities were

found to be approximately identical for different values of CA in the two sands.

12. A mathematical model of countercurrent imbibition has been developed, based on a capillary doublet model.
13. The (fractional) flow equations of water and of oil have been obtained by treating the flows of water and oil as superposition of flows due to pumping of water and flows due to countercurrent imbibition. This model permits straightforward calculation of saturation profiles. Calculated saturation profile evolutions for 1:1 viscosity ratio in a mathematical model of a porous medium are similar to the evolutions measured in sandpacks.
14. The cause of triple-valued saturation profiles in the Buckley–Leverett model is the incorrect assumption that the capillary pressure gradient can be neglected in systems characterized by oil/water interfacial tensions encountered in petroleum reservoirs.
15. A mathematical model of cocurrent imbibition based on a simple capillary model has been developed which, when applied to a waterflood in a porous medium, results in identical flow equations to those of the model based on the superposition of flows due to pumping and countercurrent imbibition.
16. The simple capillary model of cocurrent imbibition has been used also to obtain a mathematical model of the initial rate of free spontaneous imbibition Q. This is a special form of Darcy's law with a capillary pressure as the driving force and an effective permeability which is approximately equal to the Darcy permeability of the medium.
17. Calculated saturation profile evolutions for a mathematical model of a porous medium have also been characterized by values of the "complete" capillary number CA and have been matched with saturation profile evolutions measured in sandpacks. The best matches were obtained for a ratio of approximately 3 of the values of CA for the two systems.
18. There is significant viscous coupling between two immiscible fluids of the same viscosity in the presence of a saturation gradient in one of the fluids which drives the other fluid in which zero pressure gradient is maintained. Under conditions where there are nonzero unequal pressure gradients in both fluids of the same viscosity, viscous coupling cannot be measured and may be presumed to be negligible.
19. Oil blobs trapped in the pores by water spread on the surface of thick water films left in the edges and grooves of pores after the

bulk of the water has been drained from the pores. If the spreading coefficient is positive, the oil spreads on the water according to the known laws of spreading. For negative spreading coefficients, however, the oil may still imbibe on the surface of thick water films in the form of films of a definite critical thickness, which is a function of the thickness of the water film, the edge angle, and the (negative) spreading coefficient. Imbibition does not take place if the sum of the critical thickness of the oil film and the thickness of the water film is greater than the equilibrium film thickness characteristic of the particular pore of an angular cross-section.

ACKNOWLEDGMENT

The author gratefully acknowledges a subsidy grant by the Porous Media Research Institute of the University of Waterloo, from which the cost of production of the manuscript of this paper was covered.

NOMENCLATURE

Roman Letters

A	cross-sectional area
A_a, A_b	cross-sectional areas of tubes of model of countercurrent imbibition
Ca	capillary number
CA	"complete" capillary number
D	diameter of capillary
D_{eq}	equivalent pore size
F_c	capillary force
F_v	viscous force
$\mathbf{g}$	gravitational acceleration vector
h	film thickness
k	Darcy permeability
k_a, k_b	permeabilities of tubes of model of countercurrent imbibition
k_{eff}	effective permeability
k_{eff1}, k_{eff2}	effective permeabilities in free spontaneous imbibition into capillary model
k_{ri}	relative permeability to fluid i
k_{11}, k_{12}, k_{21}, k_{22}	permeability coefficients in coupled flow equations

ℓ_1, ℓ_3	lengths in capillary model of cocurrent imbibition, occupied by water and oil, respectively
ℓ_2	length of transition region in capillary model of cocurrent imbibition
L	length of sample
$\mathbf{n}$	normal vector of surface
P	hydrostatic pressure
P_c	capillary pressure
P_{c1}, P_{c2}	capillary pressures in capillaries of models of counter-current and cocurrent imbibition
P''	hydrostatic pressure on concave side of fluid/fluid interface
P'	hydrostatic pressure on convex side of fluid/fluid interface
$\partial p/\partial x$	pressure gradient associated with flow of both water and oil, according to Darcy's law, due to pumping of water at rate Q_t
$\partial P_c/\partial x$	capillary pressure gradient
q_i	flow rate of fluid i
q_w, q_o	water and oil flow rates, respectively
q_1, q_2	flow rates in tubes 1 and 2, respectively, of model of cocurrent imbibition
Q	initial flow rate in free spontaneous imbibition
Q_I, Q_{II}	flow rates of fluid I and fluid II, respectively, in an emulsion
r_m	mean radius of curvature of fluid/fluid interface
r_1, r_2	principal radii of curvature of fluid/fluid interface
R	radius of capillary
R_1, R_2	radii of capillary 1 and 2, respectively, of model of cocurrent imbibition
S	spreading coefficient
S_i	saturation of fluid i
v_i	superficial velocity of fluid i
$\mathbf{v}_i$	Darcy velocity vector of fluid i
v_w	superficial velocity of water
x, z	distance coordinates

Greek Letters

θ	contact angle
κ	visocity ratio
λ_a, λ_b	coefficients defined by Eqs. (24) and (25)
λ_w, λ_o	coefficients defined by Eqs. (27) and (28)
λ_1, λ_2, λ_3	coefficients in Eqs. (35), (36), (37), (48), and (49)
λ_4	coefficient in Eq. (37)

μ_i	viscosity of fluid i
ν	mean pore velocity
ν_0	initial mean pore velocity in free spontaneous imbibition
ρ_i	density of fluid i
σ	interfacial tension
τ_w	wall sheer stress
ϕ	angle of a corner, porosity

Subscripts

a	refers to tube a
b	refers to tube b
eff	effective
i	refers to fluid i
m	mean
nw	nonwetting phase
o	refers to oil
r	relative
w	refers to water, wetting phase, wall

Special Symbols

Δ	difference operator
∇	"del" or "nabla" operator
δ	denotes a small quantity
∂	partial differential operator

REFERENCES

Bardon C, Longeron D. Influence of very low interfacial tensions on relative permeability. Preprint for 53rd Annual Fall Technical Conference and Exhibition, SPE of AIME, Houston, Texas, 1978.

Buyevich YA. Towards a theory of nonequilibrium multiphase filtration flow. Transp Porous Media 21:145–162, 1995.

Calvo A, Paterson I, Chertcoff R, Rosen M, Hulin JP. Dynamic capillary pressure variations in diphasic flows through glass capillaries. J Colloid Interface Sci 141:384, 1991.

Chatzis I, Dullien FAL. Mise en oeuvre de la théorie de la percolation pour modeliser le drainage des milieux poreux et la perméabilité relative au liquide non muillant injecté. Revue de l'IFP 183:1982.

Chatzis I, Dullien FAL. Application of the percolation theory to modeling of drainage in porous media and of the relative permeability of the injected

nonwetting liquid. International Chemical Engineering 25:47–66, 1985. (English translation of Chatzis and Dullien (1982).)
Chatzis I, Dullien FAL. Dynamic immiscible displacement mechanisms in pore doublets: Theory versus experiment. J Colloid Interface Sci 91:199–222, 1983.
Chatzis I, Kantzas A, Dullien FAL. On the investigation of gravity assisted inert gas injection, using micromodels, long Berea cores and computer assisted tomography. SPE 63rd Meeting, SPE 18284, Houston, Texas, 1988.
Catalan L, Dullien FAL, Chatzis I. The effect of wettability and heterogeneities on the recovery of waterflood residual oil with low pressure inert gas injection assisted by gravity drainage. Soc Pet Eng Adv Technol 2:140–149, 1994.
Danis M, Jacquin C. Influence du contraste de viscosité sur les perméabilités relatives lors du drainage:Experimentation et modélisation. Revue de l'IFP 38:1983.
deGennes PG. Theory of slow biphasic flows in porous media. Phys Chem Hydr 4:175–185, 1983.
de la Cruz V, Spanos TJT. Modelization of oil ganglia. AIChE J 29:854–858, 1983.
Diaz CE, Chatzis I, Dullien FAL. Simulation of capillary pressure curves using bond correlated site percolation on a simple cubic network. Transp Porous Media 2:215–240, 1987.
Dong M, Dullien FAL. A new model for immiscible displacement in porous media. Transp Porous Media 27:185–204, 1997.
Dong M, Dullien FAL, Chatzis I. The imbibition of oil in film form over water present in edges of capillaries with an angular cross section. J Colloid Interface Sci 172:21–36, 1995.
Dong M, Dullien FAL, Zhou J. Characterization of waterflood saturation profile histories by the "complete" capillary numbers. Transp Porous Media 31:213–237, 1998.
Dullien FAL. Two-phase flow in porous media. Chem Eng Technol 11:407–429, 1988.
Dullien FAL. Porous Media–Fluid Transport and Pore Structure. 2nd ed. Academic Press, San Diego, California, 1992a, p 339.
Dullien FAL. Comments on the differences observed between static and dynamic capillary pressures. In: Toulhoat H, Lecourtier J, eds, Physical Chemistry of Colloids and Interfaces in Oil Production. Paris: Editions Techniques, 1992b, pp 115–123.
Dullien FAL. Physical interpretation of hydrodynamic coupling in steady two-phase flow, Proceedings of American Geophysical Union 13th Annual Hydrology Days, Fort Collins, Colorado, 1993. Atherton, CA: Hydrology Days Publications, pp 363–377.
Dullien FAL, Dong M. Experimental determination of the flow transport coefficients in the coupled equations of two-phase flow in porous media. Transp Porous Media 25:97–120, 1996.
Dullien FAL, Dong M. Mathematical model of imbibition. *Proceedings of the 6th Symposium on Mining Chemistry*, Siófok, Hungary, 1998. Budapest: Dunaprint KFT, 1998, pp 3–8.

Dullien FAL, Fleury M. Analysis of the USBM wettability test. Transp Porous Media 16:175–188, 1994.

Dullien FAL, Zarcone C, Macdonald IF, Collins A, Bochard DE. The effects of surface roughness on the capillary pressure curves and the heights of capillary rise in glass bead packs. J Colloid Interface Sci 127:362–372, 1989.

Dullien FAL, Allsop, HA, Macdonald IF, Chatzis I. Wettability and inmiscible displacement in Pembina Cardium sandstone. J Can Pet Technol 29:63, 1990.

Kalaydjian F, Moulu J-C, Vizika O, Munkenrud PK. Three-phase flow in waterwet porous media: Determination of gas/oil relative permeabilities under various spreading conditions. SPE 68th Meeting. SPE 26671, Houston, Texas, 1993.

Lasseux D, Quintard M, Whitaker S. Determination of permeability tensors for two-phase flow in homogeneous porous media:Theory. Transp Porous Media 24:107–137, 1996.

Legait B. Laminar flow of two phases through a capillary tube with variable square cross-section. J Colloid Interface Sci 96:28–38, 1983.

Lenormand R. PhD dissertation, University of Toulouse, France, 1981.

Lenormand R, Zarcone C. Description des mecanismes d'imbibition dans un reseau de capillaires, CR Acad Sci Paris 297, Serie II:393–396, 1983.

Lenormand R, Zarcone C, Sarr A. Mechanisms of the displacement of one fluid by another in a network of capillary ducts. J Fluid Mech 135:337–353, 1983.

Lenormand R, Touboul E, Zarcone C. Numerical models and experiments on immiscible displacements in porous media. J Fluid Mech 189:165–187, 1988.

Leverett MC. Capillary behaviour in porous solids. Pet Trans AIME 142:152–169, 1941.

Li Y, Wardlaw NC. The influence of wettability and critical pore throat size ratio on snap-off. J Colloid Interface Sci 109:461–472, 1986a.

Li Y, Wardlaw NC. Mechanism of nonwetting phase trapping during imbibition at slow rates. J Colloid Interface Sci 109:473–486, 1986b.

McCaffery FG, Bennion DW. The effect of wettability on two-phase relative permeabilities. J Can Pet Technol 13:42–53, 1974.

Morrow NR, Cram PJ, McCaffery FG. Displacement studies in dolomite with wettability control by octanoic acid. Soc Pet Eng J Trans AIME 255:221–232, 1973.

Odeh AS. Effect of viscosity ratio on relative permeability. Pet Trans AIME 216:346–353, 1959.

Owens WW, Archer DL. The effect of rock wettability on oil–water relative permeability relationships. J Pet Tech 23:873, 1971.

Popiel WJ. Introduction to Colloid Science. Hicksville, NY: Exposition Press, 1978.

Rakotomalala N, Salin D, Yortsos YC. Viscous coupling in a model porous medium geometry: Effect of fluid contact area. *Appl Sci Res* 55:155–169, 1995.

Rapoport LA, Leas WJ. Properties of linear waterfloods. Trans Am Inst Min Eng 198:139–148, 1953.

Rose W. Petroleum reservoir engineering at the crossroads (ways of thinking). Iran Pet Inst Bull 46:23–27, 1972.

Salathiel RA. Oil recovery by surface film drainage in mixed-wettability rocks. J Pet Tech 25:2216–2224, 1973.
Yadav GD, Dullien FAL, Chatzis I, Macdonald IF. Microscopic distribution of wetting and non-wetting phases in sandstones during immiscible displacements. SPE–Reservoir Eng 2:137–147, 1987.
Zarcone C, Lenormand R. Determination expérimentale du couplage visqueux dans les écoulements diphasiques en milieu poreux. C R Acad Sci Paris 318, Série II:1429–1435, 1994.

3

Characterization of Geology of, and Flow and Transport in, Field-scale Porous Media

Application of Fractal and Percolation Concepts

Muhammad Sahimi
University of Southern California, Los Angeles, California

I. INTRODUCTION

Field-scale porous media (FSPM), such as oil and gas reservoirs and groundwater aquifers, are highly heterogeneous at many length scales. Their heterogeneities manifest themselves at three different scales which are: (1) microscopic, which is at the level of pores and grains; (2) macroscopic, which is at length scales comparable with core plugs; (3) megascopic which includes the entire reservoir or aquifer. Modeling flow and transport in such porous media depends critically on accurate characterization of their structure, and in particular on the distribution of their heterogeneities. However, although characterization of laboratory-scale (macroscopic) porous media has been studied in great detail and reasonable understanding of such porous media has emerged (for recent reviews see, for example, Sahimi, 1993b, 1995b), the same is not true of FSPM, whose characterization is plagued by lack of sufficient data and hampered by the wide variations in the data that are collected at various locations throughout the system, e.g., along wells in oil or gas reservoirs.

Three important characteristics of FSPM are their porosity logs, often measured by in-situ methods, their permeability distributions, which are usually obtained by collecting a number of core plugs at various depths along the wells and measuring their permeabilities (or by in-situ methods, such as nuclear magnetic resonance), and the structure of their fracture network, if they are indeed fractured. Fractures are crucial to flow of oil in reservoir rock, development of groundwater resources, and generation of heat and vapor from geothermal reservoirs for use in power plants. They provide high permeability paths for fluid flow in FSPM that are otherwise of very low permeability and porosity, e.g., many fractured carbonate oil reservoirs that are in the Middle East.

In a pioneering work, Hewett (1986) proposed that the porosity logs and permeability distributions of many FSPM follow fractal statistics. More specifically, he provided evidence that the porosity logs in the direction perpendicular to the bedding may obey the statistics of fractional Gaussian noise, while those parallel to the bedding follow a fractional Brownian motion. Barton and Larsen (1985) appear to be the first to point out the possibility of characterizing the fracture network of heterogeneous rock by fractal concepts. One goal of this paper is to review the progress that has been made over the past decade toward characterization of FSPM using fractal statistics.

Even before application of fractal concepts to various phenomena became popular, percolation theory (for a simple introduction to percolation theory see Stauffer and Aharony, 1992) had already become a powerful, much-used tool for investigating transport processes in disordered systems, and in particular porous media. The popularity of percolation theory stems from its relevance to a wide variety of phenomena (Sahimi, 1994), and from the fact that, despite the simplicity of its underlying concepts, it leads to non-trivial predictions for the morphology and transport properties of disordered media. In particular, one of the most successful applications of percolation theory has been the modeling of flow and transport in heterogeneous porous media, the focus of this paper. At certain length scales (see below) a percolation system possesses fractal properties, so that there is a close connection between percolation processes and the more general fractal structures and phenomena.

Characterization of the morphology of FSPM is not the only fruitful application of fractal and percolation concepts to problems involving porous media. Percolation and fractal concepts have also provided us with a much deeper understanding of reaction and precipitation in porous media (for reviews see, for example, Avnir, 1990; Sahimi et al., 1990; Sahimi, 1992a, 1994) which result in dynamical evolution and restructuring of porous media, multiphase flow in porous media (for recent reviews see, for example, Sahimi 1993b, 1995b), and nucleation and propagation of fractures in rock (Sahimi, 1992b, 1998; Sahimi et al., 1993; Sahimi and Arbabi,

1992, 1996). It is impossible to discuss all the applications of fractal and percolation concepts to porous media, as the task would be daunting and could be the subject of a book by itself. Instead, we focus on just two such applications. The first is characterization of morphological properties of FSPM, i.e., their porosity logs, permeability distributions, and the structure of their fracture network, while the second application is modeling of single-phase flow and transport in FSPM.

This review is organized as follows. In Section II we discuss the basic concepts of fractal geometry that will be employed in the rest of this paper. Fractality of a system implies the existence of long-range correlations in that system, and therefore we discuss in Section III three important stochastic processes that generate such correlations. Section IV outlines the basic concepts of percolation theory. In Section V we discuss fractal and percolation characterization of FSPM. We then discuss in Section VI the application of fractal and percolation concepts to modeling single-phase flow in porous media, while Section VII describes the phenomenon of fractal diffusion in porous media. In the last section, application of fractal and percolation concepts to modeling hydrodynamic dispersion in FSPM is described.

II. FRACTAL CONCEPTS

In general, we divide fractal systems into two classes. One class contains those whose morphology exhibits fractal properties. We call such systems *geometrical fractals*. In the second class are those systems whose dynamics possesses fractal properties. We refer to these as *dynamical fractals*. In this section we explain and summarize a few key concepts of fractal geometry that we will use in the rest of this paper.

A. Self-similar Fractals

While it is possible to give a formal mathematical definition of a fractal system or set (Mandelbrot, 1982), an intuitive definition is probably more useful: in a geometrical fractal *the part is reminiscent of the whole*. This implies that the system is self-similar and scale-invariant, i.e., its morphology repeats itself at various length scales. That is, there exist pieces of the system above a certain length scale—the lower cutoff scale for the fractality of the system—that can be magnified to recover the structure of the system up to another length scale—the upper cutoff for its fractality. Some systems are fractal at *any* length scale, while natural systems that exhibit fractal properties typically lose their fractality at sufficiently small or large length scales. Moreover, it should be pointed out that natural systems are generally not, in the strict mathematical sense, fractal; rather, their behavior *approaches* what is envisioned in fractal geometry (see, for example, Avnir et al., 1998, for a discussion of this), which is why it is useful to use fractals for describing them.

The simplest characteristic of a geometrical fractal is its fractal dimension D_f, defined as follows. The fractal system is covered by non-overlapping d-dimensional hyperspheres of Euclidean radius r, and the number $N(r)$ of such spheres which is required for the coverage is counted. For a fractal system one has

$$N(r) \sim r^{-D_f} \tag{1}$$

where $\sim$ implies an asymptotic proportionality. Equation (1) can be rewritten as

$$D_f = \frac{\ln N}{\ln(1/r)} \tag{2}$$

Thus, for Euclidean objects such as straight lines, squares, or spheres one has $N(r) \sim r^{-1}$, r^{-2}, and r^{-3} respectively, and their effective fractal dimension coincides with the Euclidean dimension. Another way of defining the fractal dimension is by considering a segment of a fractal system of linear dimension L and studying its volume $V(L)$ as L is varied. If $V(L)$ is calculated by covering the system by spheres of radius unity, then $V(L) = N(L)$, where N is the number of such spheres required to cover the system. For a fractal system one finds that

$$N(L) \sim L^{D_f} \tag{3}$$

where D_f is the same as in Eq. (1). Note that, with this definition, we implicitly assume the existence of lower and upper cutoffs for the fractal behavior of the system, namely, the radius of hyperspheres (the lower cutoff) and the system's linear size L (the upper cutoff). Calculating D_f using Eq. (3) is called the *box-counting method*.

In Figure 1 we show the construction of a classical self-similar fractal system called the *Sierpinski gasket*. In each generation every triangle is replaced by $N = 3$ smaller triangles that have been scaled down by a factor $r = 1/2$, and therefore, using Eq. (2), we obtain $D_f = \ln 3/\ln 2 \simeq 1.58$, while for the three-dimensional (3D) version of the gasket we have $D_f = \ln 4/\ln 2 = 2$. Although the Sierpinski gasket appears rather artificial, it has been used in many applications as a model of disordered media.

The Sierpinski gasket is a well-known example of what we call *exact* fractals, because its self-similarity is repeated exactly at every stage of its construction. Another type of geometrical fractal is what we call *statistically self-similar* fractals, because their self-similarity is only in an average sense.

Figure 1. The two-dimensional Sierpinski gasket.

One of the most important examples of such fractals is one that is generated by the diffusion-limited aggregation model (Witten and Sander, 1981). In this model the site at the center of a lattice is occupied by a stationary particle. A new particle is then injected into the lattice, far from the center, which diffuses on the lattice until it reaches a surface site, i.e., an empty site which is a nearest neighbor of the stationary particle, at which time the particle sticks to it and remains there permanently. Another diffusing particle is then injected into the lattice to reach another surface (empty) site and stick to it, and so on. If this process is continued for a long time, a large aggregate is formed. The most important property of diffusion-limited aggregates is that they are self-similar and fractal. This means that if $N(L)$ is the number of the elementary particles in an aggregate of radius L (clearly, N is proportional to the mass of the aggregate), then $N(L)$ follows Eq. (3). Extensive computer simulations (for a review see, for example, Meakin, 1988) indicate that $D_f \simeq 1.7$ and 2.45 for 2D and 3D aggregates, respectively. A 2D example of such aggregates is shown in Figure 2. A fractal dimension less than the Euclidean dimension of the space implies that the fractal object cannot fill the space and has a sparse structure, as is evident in Figures 1 and 2. Diffusion-limited aggregates have found wide applications, ranging from colloidal systems to miscible displacement processes in porous media.

Because diffusion-limited aggregates and many other disordered media, such as percolation networks discussed below, are only statistically

Figure 2. An example of a two-dimensional diffusion-limited aggregate.

self-similar, it may be more appropriate to use the term scale-invariant for describing them. As these systems are disordered, visual inspection of their self-similarity may be very difficult. A practical and powerful method for testing their self-similarity is based on constructing a *correlation function* $\mathcal{C}(\mathbf{r})$ defined by

$$\mathcal{C}(\mathbf{r}) = \frac{1}{V}\sum_{\mathbf{r}'} s(\mathbf{r}')s(\mathbf{r}+\mathbf{r}') \tag{4}$$

where V is the volume of the system, $s(\mathbf{r}) = 1$ if a point at $\mathbf{r}$ belongs to the system, and $s(\mathbf{r}) = 0$ otherwise. For example, for a diffusion-limited aggregate, $s(\mathbf{r}) = 1$ if a point at $\mathbf{r}$ belongs to the aggregate, and $s(\mathbf{r}) = 0$ otherwise. If a disordered medium is scale-invariant, then its correlation function defined by Eq. (4) should remain the same, up to a constant factor, if all of its length scales are rescaled by a constant factor b. Thus, one must have $\mathcal{C}(br) \sim b^{-x}\mathcal{C}(r)$, where $r = |\mathbf{r}|$. It is not difficult to see that only a power law, $\mathcal{C}(r) \sim r^{-x}$, has this property, and that one must have $x = d - D_f$. Therefore, for a fractal medium the correlation function $\mathcal{C}(r)$ decays as

$$\mathcal{C}(r) \sim r^{D_f - d} \tag{5}$$

Equation (5) not only provides a test of self-similarity or fractality of a disordered medium, it also gives us a means of estimating its fractal dimension since, if one prepares a logarithmic plot of $\mathcal{C}(r)$ versus r, then for a fractal medium one should obtain a straight line with a slope $D_f - d$.

Equation (5) also indicates that in a fractal system *there are long-range correlations*, since $\mathcal{C}(r) \to 0$ only if $r \to \infty$.

B. Self-affine Fractals

The self-similarity of a fractal system implies that its structure is invariant under an isotropic rescaling of lengths; i.e., the invariance is preserved if all the lengths in all the directions are rescaled by the same scale factor. However, there are many fractal systems that preserve their scale-invariance only if the length scales in different directions are rescaled by different direction-dependent scale factors. In other words, the scale-invariance of such systems is preserved only if the length scales in the x, y and z directions are rescaled by scale factors b_x, b_y, and b_z which, in general, are not equal. This scale-invariance under a direction-dependent rescaling implies that the fractal system is *anisotropic*. Such fractal systems are called *self-affine fractals* (Mandelbrot, 1985). A self-affine fractal can no longer be described by a single fractal dimension. Instead, its local or small-scale properties are described by an *effective* fractal dimension, whereas its large-scale behavior is characterized by another *integer* dimension which is, however, *less* than d. An important example of a self-affine fractal is the surfaces that are generated by a fractional Brownian motion (see below). A well-known example of a process that gives rise to a self-affine fractal is the marginally stable growth of an interface. For example, if a wetting fluid displaces a non-wetting fluid in a porous medium, the interface between the two fluids is a self-affine fractal.

Before closing this section, we should point out that another way of generating porous media that resemble fractal systems is by constructal theory (see, for example, Bejan and Errera, 1997; Errera and Bejan, 1998; Bejan et al., 1998; Bejan and Tondeur, 1998). In this theory, the channels, locations, tributaries, and cutoffs are results of a geometric optimization process, subject to local and global constraints. Examples include volume-point flows in fractured porous media, lung-like systems, and river drainage networks. In the past, there has been speculation as to why fractal structures seem to be abundant in nature; constructal theory may explain this abundance as nature's way of optimizing the structures of such systems.

III. FRACTAL STOCHASTIC PROCESSES WITH LONG-RANGE CORRELATIONS

As emphasized above, fractality of a system implies the existence of long-range correlations in that system. On the other hand, long-range correlations are ubiquitous in nature and, as we see below, are important to characterization of FSPM. Thus, let us discuss three key stochastic processes that generate such long-range correlations and have found wide applications in characterization of FSPM.

A. Fractional Brownian Motion and Fractional Gaussian Noise

A fractional Brownian motion (fBm) is a stochastic process $B_H(\mathbf{r})$ (Mandelbrot and van Ness, 1968) with the properties that $\langle B_H(\mathbf{r}) - B_H(\mathbf{r}_0)\rangle = 0$, and

$$\langle [B_H(\mathbf{r}) - B_H(\mathbf{r}_0)]^2\rangle \sim |\mathbf{r} - \mathbf{r}_0|^{2H} \tag{6}$$

where $\mathbf{r} = (x, y, z)$ and $\mathbf{r}_0 = (x_0, y_0, z_0)$ are two arbitrary points, and H is called the Hurst exponent (Hurst, 1951). A remarkable property of the fBm is that it generates correlations whose extent is *infinite* (i.e. their extent is as large as the linear size of the system). Moreover, the type of the correlations can be tuned by varying H. If $H > 1/2$, then the fBm displays *persistence*; i.e., a trend (for example, a high or low value) at x is likely to be followed by a similar trend at $x + \Delta x$. If $H < 1/2$, then the fBm generates *antipersistence*; i.e., a trend at x is not likely to be followed by a similar trend at $x + \Delta x$. For $H = 1/2$ the trace of the fBm is similar to that of a random walk, and the *increments* in $B_H(\mathbf{r})$ are uncorrelated. Figure 3 shows examples of 1D and 2D fBm.

A convenient way of representing a stochastic function is through its spectral density $\mathcal{S}(\boldsymbol{\omega})$, the Fourier transform of its variance. For a d-dimensional fBm it can be shown that

$$\mathcal{S}(\boldsymbol{\omega}) = \frac{a_d}{(\sum_{i=1}^{d} \omega_i^2)^{H+d/2}} \tag{7}$$

where $\boldsymbol{\omega} = (\omega_1, \ldots, \omega_d)$ and a_d is a d-dependent constant. As mentioned above, many fractal systems lose their fractality above a cutoff length scale. The spectral representation of a stochastic process allows a convenient way of introducing this cutoff length scale. Thus, we introduce a cutoff length scale $\ell_{co} = 1/\sqrt{\omega_{co}}$ such that

$$\mathcal{S}(\boldsymbol{\omega}) = \frac{a_d}{(\omega_{co} + \sum_{i=1}^{d} \omega_i^2)^{H+d/2}} \tag{8}$$

The cutoff length scale ℓ_{co} allows us to control the scale over which the spatial properties of the system are correlated. Hence, for length scales $\ell < \ell_{co}$ they preserve their correlations (for $H > 0.5$) or anticorrelations (for $H < 0.5$), whereas for $\ell > \ell_{co}$ the correlations (anticorrelations) are lost.

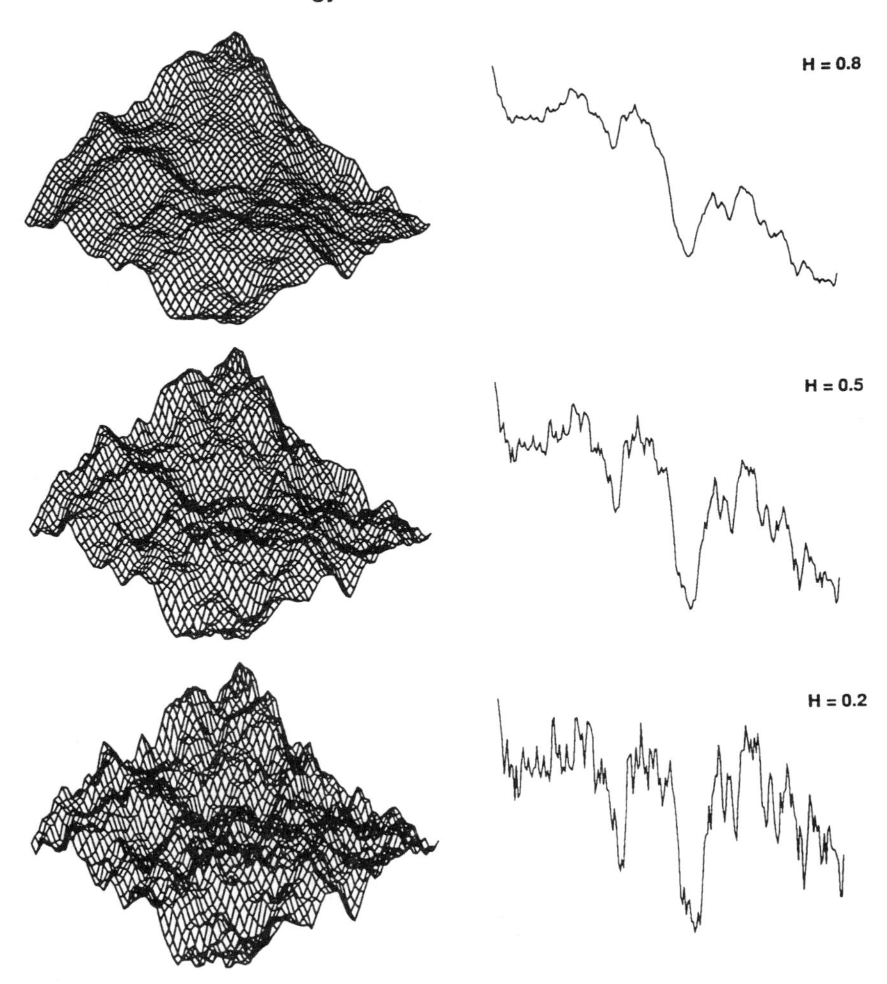

Figure 3. Examples of 1D and 2D fractional Brownian motion.

Fractional Brownian motion increments are stationary but not ergodic. The variance of fBm for a large enough array is divergent. Its trace in d dimensions is a self-affine fractal with a local fractal dimension $D_f = d + 1 - H$. It is not differentiable, but by smoothing it over an interval one can obtain its numerical "derivative" which is the fractional Gaussian noise (fGn) whose spectral density in, e.g., 1D is given by

$$\mathcal{S}(\omega) = \frac{b_1}{\omega^{2H-1}} \tag{9}$$

where b_1 is another constant. Figure 4 shows an example of a 1D fGn, which should be compared with its fBm counterpart in Figure 3. We should point out that, if v is a stochastic variable that obeys the statistics of the fBm, then the increments $v(\mathbf{r} + \Delta\mathbf{r}) - v(\mathbf{r})$ are Gaussian variables.

Given the Hurst exponent H, which determines the nature of the correlations, one can generate synthetic porosity logs or permeability distributions for use in modeling of FSPM and computer simulation of flow and transport in them, if such properties follow the statistics of the fBm or fGn. To do this, one must have an efficient and accurate method of generating a fBm or fGn. There are several methods of generating a fBm (or a fGn), two of which are discussed here. Since the data that one analyzes for FSPM are usually one-dimensional (collected along a well), we restrict our discussion to 1D fBm, or fGn, but their generalization to higher dimensions is straightforward.

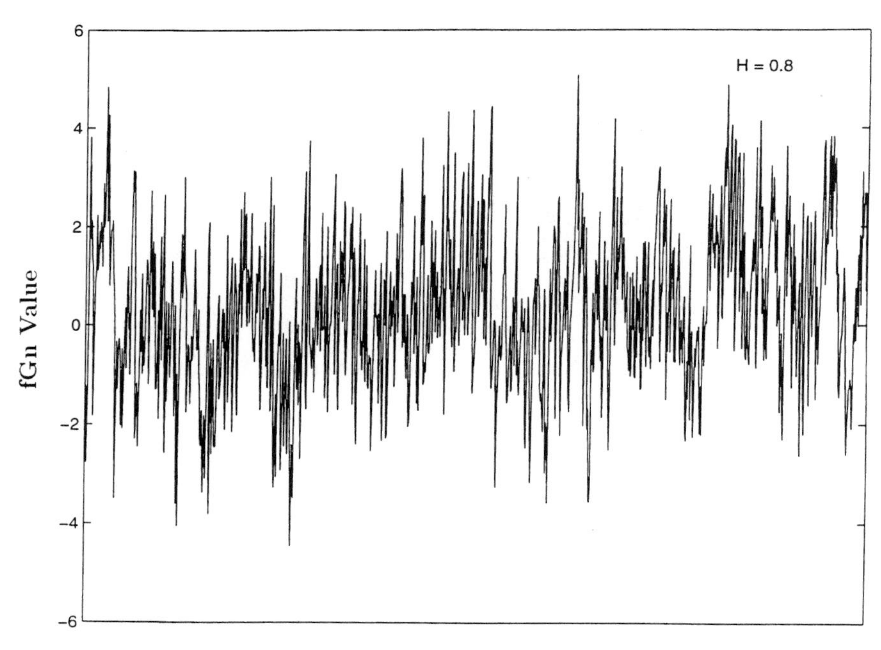

Figure 4. One-dimensional fractional Gaussian noise with $H = 0.8$.

1. Fast Fourier Transformation Method

Equations (7)–(9) provide a convenient method of generating an array of numbers that obey the fBm or fGn statistics using a fast Fourier transformation technique. In this method, one first generates random numbers, uniformly distributed in [0,1), and assigns them to the sites of a d-dimensional lattice (for example, a linear chain for 1D data). The Fourier transform of the resulting d-dimensional array of the numbers is then calculated numerically. The Fourier-transformed numbers are multiplied by $\sqrt{S(\omega)}$, and the results are inverse-Fourier-transformed back into the real space. The array so obtained obeys the fBm or fGn statistics. To avoid the problem associated with the periodicity of the numbers arising as a result of their Fourier transformation, one has to generate the array with a much larger lattice size than the actual size which is to be used in the analysis, and use the central part of the array.

2. Successive Random Additions

In this method (Voss, 1985) one starts with the two end-points on [0,1) and assigns a zero value to them. Then Gaussian random numbers Δ_0 are added to these values. In the next stage, new points are added at a fraction r of the previous stage by interpolating between the old points (by either linear or spline interpolation), and Gaussian random numbers Δ_1 are added to the new points. Thus, given a sample of N_i points at stage i with resolution λ, stage $i+1$ with resolution $r\lambda$ is determined by first interpolating the $N_{i+1} = N_i/r$ new points from the old points, and then Gaussian random numbers Δ_i are added to all of the new points. At stage i with $r < 1$, the Gaussian random numbers have a variance (see Eq. (6))

$$\sigma_i^2 \sim r^{2iH} \tag{10}$$

The process is continued until the desired length of the data array is reached.

B. Lévy Flights and Fractional Lévy Motion

Consider a d-dimensional lattice on which a particle executes a random walk; i.e., at each site of the lattice it selects a direction randomly and makes a jump of length ℓ to a new site in that direction. Suppose that $P(\ell)$ is the probability for a random walk jump of vector displacement ℓ. For simplicity, we consider a 1D walk and choose

$$P(\ell) = \frac{n_j - 1}{2n_j} \sum_{j=0}^{\infty} n_j^{-j} (\delta_{\ell,b^j} + \delta_{\ell,-b^j}) \tag{11}$$

where $n_j > 1$ and $b > 1$. During this random walk, jumps of all orders of magnitude can occur in base b, but each successive order of magnitude displacement occurs with an order of magnitude less probability in base n_j. The random walker makes n_j jumps of unit length before a jump of length b occurs. Eventually, the set of the sites visited by the random walker forms a fractal set. To understand this stochastic process better, consider the Fourier transform of $P(\ell)$

$$\tilde{P}(\omega) = \frac{n_j - 1}{n_j} \sum_{j=0}^{\infty} n_j^{-j} \cos(b^j \omega) \tag{12}$$

This function was first used by Weierstrass as an example of a continuous but *everywhere* non-differentiable function for $b > n_j$. If ω is small and $\langle \ell^2 \rangle = \sum_j \ell^2 P(\ell)$ is finite, then

$$\tilde{P}(\omega) \sim 1 - \frac{1}{2} \omega^2 \langle \ell^2 \rangle \tag{13}$$

On the other hand, if $b^2 > n_j$, then $\langle \ell^2 \rangle \sim \sum_{j=0}^{\infty} (b^2/n_j)^j$ is divergent. In this case

$$\tilde{P}(b\omega) = n_j \tilde{P}(\omega) - (n_j - 1) \cos \omega \tag{14}$$

and the random walk is called a Lévy flight. For small ω, Eq. (14) yields

$$\tilde{P}(\omega) \sim 1 - (C_{\mathrm{L}} \omega)^{\alpha} \sim \exp[-(C_{\mathrm{L}} \omega)^{\alpha}] \tag{15}$$

where

$$\alpha = \frac{\ln n_j}{\ln b} \tag{16}$$

and $C_{\mathrm{L}} > 0$ is a constant. α is the fractal dimension of the set of the sites visited by the Lévy walker. If $\alpha = 2$ (as is the case when $\langle \ell^2 \rangle$ is finite), then we have Gaussian behavior. If $\alpha < 2$, then $\langle \ell^2 \rangle$ is divergent, the random walk is the Lévy flight, and the trail of the walk has a fractal dimension of $D_f = \alpha$. An example of a Lévy flight is given in Figure 5 for $n_j = 4$ and $b = 8$. From Eq. (15) we find that

$$P(x) = \frac{1}{\pi} \int_0^{\infty} \exp[-(C_{\mathrm{L}} \omega)^{\alpha}] \cos(\omega x) \mathrm{d}\omega \tag{17}$$

We shall use Eq. (17) in characterizing FSPM. Note that, when $\alpha < 2$, the second and higher moments of $P_j(x)$ are all divergent. This divergence is caused by *rare events*, i.e., those whose probability of occurring is very small, but when they do occur their values are very large, giving rise to long tails in the distribution and also to divergent moments. The presence of long tails in the distribution of the *increments* of some natural data has

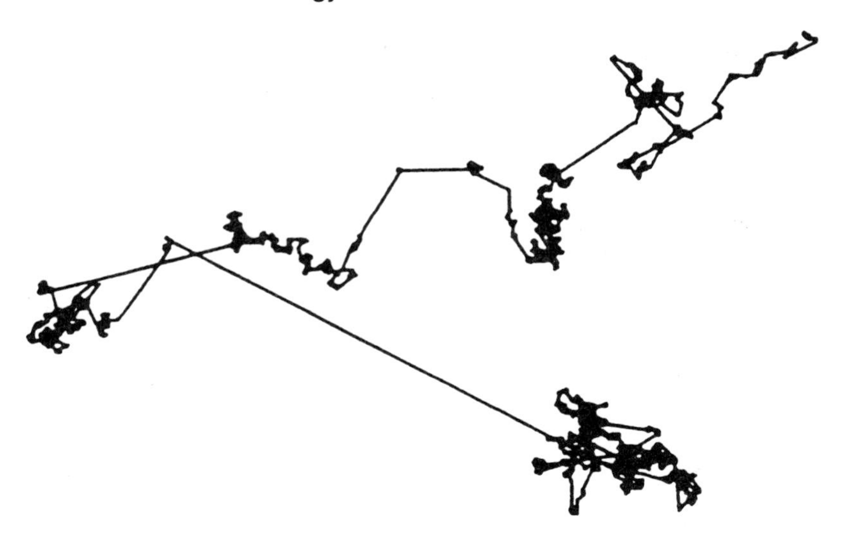

Figure 5. The trail of a 2D Lèvy flight with $\alpha = 5/4$.

led to the consideration of a general form of fractal distributions, called the fractional Lévy motions (fLm) (Taqqu, 1987), which has stationary increments whose distribution is the Lévy-stable distribution given by Eq. (17). Since the increments $v(x + \Delta x) - v(x)$ of a variable v, that obeys the statistics of the fBm, are Gaussian variables, the limit $\alpha = 2$ corresponds to the fBm case. Shlesinger (1988) and Hughes (1995) have given extensive discussions of the applications of Lévy processes.

IV. PERCOLATION PROCESSES

To describe applications of percolation theory to flow and transport in a porous medium, one must have a realistic model of the medium itself. Any porous medium can, in principle, be mapped onto an equivalent network of pore throats and pore bodies (Mohanty, 1981; Lin and Cohen, 1982). For brevity, we refer to pore bodies and pore throats as pores and throats, respectively. In this section we describe the basic properties of percolation processes in random network models of porous media. We start with the classical percolation in which the disorder is random with no correlation in the system, but, as the nature of the disorder in FSPM is not random, we will also discuss correlated percolation.

A. Random Percolation

Consider a 2D or 3D network of coordination number Z, where Z is the number of the bonds that are connected to the same site. The bonds of the network represent the flow passages or the throats in a porous medium, while the sites are the pores where the flow passages meet. Suppose that a randomly selected fraction p of the bonds are open to flow (i.e., their permeability or hydraulic conductance is nonzero). For convenience we call these open bonds or throats. If p is small enough, no sample-spanning cluster of the open throats or bonds is formed and macroscopic flow and transport do not exist, whereas for large values of p macroscopie flow and transport do occur. The transition between a system with no macroscopic transport and one in which macroscopic flow and transport do occur is characterized by a well-defined value of p, called the *bond percolation threshold* p_{cb}, whose value depends on Z and the dimensionality d of the system. For example, for the square network $(Z = 4), p_{cb} = 1/2$, for the cubic network $(Z = 6$, $p_{cb} \simeq 0.2488$, and in general $p_{cb} \simeq d/[Z(d-1)]$ for d-dimensional network. Figure 6 shows a square network in the bond percolation process. In the context of modeling flow through porous media, bond percolation is relevant to single-phase flow through a strongly disordered porous medium (see

Figure 6. Random bond percolation on the square network with $p = 0.6$.

Section 6), flow through a fracture network, and also to drainage, i.e., displacement of a wetting fluid by a non-wetting fluid in a porous medium.

We can also consider a *site percolation* process which is very similar to bond percolation. Consider, for example, a square network and suppose that we color a randomly selected fraction p of the sites black; see Figure 7. Similar to bond percolation, if $p \simeq 0$, then the size of any cluster of black sites is small, and thus there is no sample-spanning cluster of such sites. On the other hand, if $p \simeq 1$, almost all the sites are black, and there exists a sample-spanning cluster of such sites. Thus, one has a well-defined *site percolation threshold* p_{cs}, the analog of p_{cb}, at which a transition takes place from a disconnected system with no sample-spanning cluster of the black sites to one with a sampling-spanning cluster of such sites. For the square network of Figure 7 one has $p_{cs} \simeq 0.592\,77$, and for the simple-cubic network, $p_{cs} \simeq 0.3116$. In general, one has $p_{cs} \geq p_{cb}$. Aside from the numerical values of p_{cs} and p_{cb}, all the properties of bond and site percolation are qualitatively similar, and thus hereafter we do not distinguish between them and denote by p_c the (site or bond) percolation threshold of the system. In the context of modeling multiphase flow through a porous medium, site percolation is relevant to imbibition, i.e., displacement of a non-wetting fluid by a wetting fluid.

The typical radius of percolation clusters below p_c, and the typical radius of the holes in between the percolation clusters (i.e., clusters of uncolored sites or closed bonds) above p_c, is called the *correlation length* ξ_p of the

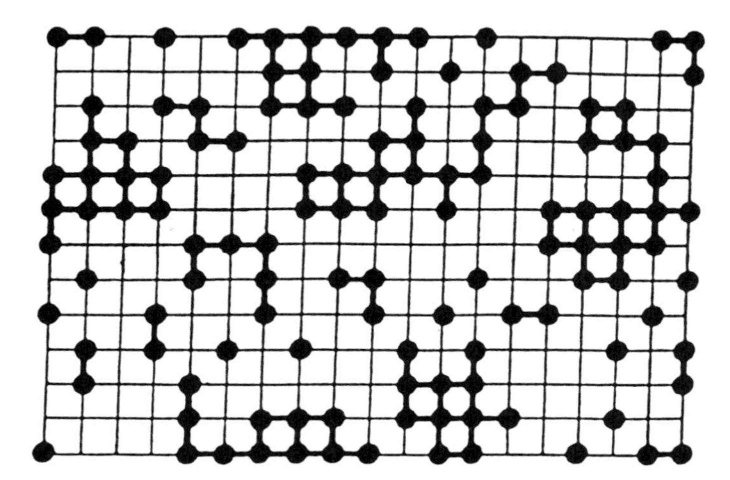

Figure 7. Random site percolation on the square network with $p = 0$. Solid circles are the occupied sites.

network. Below p_c, as p increases, the radii of percolation clusters also increase, whereas above p_c, as p decreases, it is the radii of the holes that increase. Thus, in both cases ξ_p increases as p_c is approached, and in fact at p_c the correlation length ξ_p is divergent. Near p_c the correlation length obeys the following *power law*

$$\xi_p \sim |p - p_c|^{-\nu} \tag{18}$$

This power law is *universal* in the sense that ν, the *critical exponent* of ξ_p, does not depend on the microscopic structure of the network (e.g., its Z) or on whether one has a site or bond percolation process, and depends only on its dimensionality d, so long as there are at most short-range correlations in the system.

At *any* length scale $L \gg \xi_p$ the percolation system is macroscopically homogeneous. However, for $\ell_b \ll L \ll \xi_p$, where ℓ_b is the length of a bond, the system is macroscopically heterogeneous, and in fact at such length scales the sample-spanning cluster is a fractal and statistically self-similar object. Obviously, since a percolation system has a random structure, it may not look self-similar at all. However, if we look at many different *realizations* of the sample-spanning cluster and study them at length scales $L \ll \xi_p$, then we find that *on average* these clusters are self-similar and fractal. That is, the *average* number $N(L)$ of the black sites or occupied bonds within a cluster of linear size $L \ll \xi_p$ obeys Eq. (3), where the averaging is taken over *all* the possible realizations (configurations) or the percolation clusters that one can generate for a fixed value of p. Thus, the significance of ξ_p is that it is the upper cutoff length scale for fractality of a percolation system. Since, according to Eq. (18), the correlation length is divergent at p_c, at this particular point the sample-spanning cluster is a fractal object at *any* length scale. Figure 8 shows the sample-spanning cluster at $p_{cs} = 1/2$, the site percolation threshold of a triangular network. The size of the network is $10^4 \times 10^4$.

An important percolation quantity is the *accessible fraction* $X_A(p)$ of the open bonds (or throats), which is the fraction of the open bonds that are in the sample-spanning cluster (some of the open bonds are isolated and not connected to this cluster). The sample-spanning cluster can be divided into two parts: the dead-end bonds that carry no flow, and the *backbone* which is the multiply-connected part of the cluster through which a fluid flows. Consequently, we define the *backbone fraction* $X_B(p)$ as the fraction of the open throats or bonds that are in the backbone. It should be obvious that $X_B(p) < X_A(p)$. Figure 9 presents the sample-spanning cluster and its backbone in a square network just above p_c.

So far, we have discussed the *connectivity* characteristics of percolation systems. Their effective flow and transport properties are also defined in a straightforward manner. Suppose that the hydraulic conductance or the

Figure 8. A 2D sample-spanning percolation cluster on a $10^4 \times 10^4$ triangular network at $p_{cs} = 1/2$.

permeability of the open bonds or throats has been assigned according to a given distribution with or without correlations. If a unit potential or pressure gradient is applied to two opposite faces of the network, and if the network is above its percolation threshold, then we will have macroscopic flow. We denote by G and K the overall conductance and permeability of the system, respectively, which obviously depend on p, the fraction of the open bonds or throats, with $K(p = p_c) = 0$ and $G(p = p_c) = 0$. In a similar manner, we define the overall diffusivity D of the system, which is proportional to G.

Conceptually, numerical calculation of flow and transport properties of percolation networks is simple. Consider, as an example, calculation of the permeability K. One applies a unit pressure gradient to the network. The bonds' permeabilities are selected from a given distribution, such as a fBm, or a fGn, or a fLm. Assuming a uniform cross-sectional area for the bonds, one writes a mass balance for a node i of the networks, $\sum_j Q_{ij} = 0$ (the constant fluid density has been ignored). Here, $Q_{ij} = k_{ij} \Delta P_{ij}$, where k_{ij} is

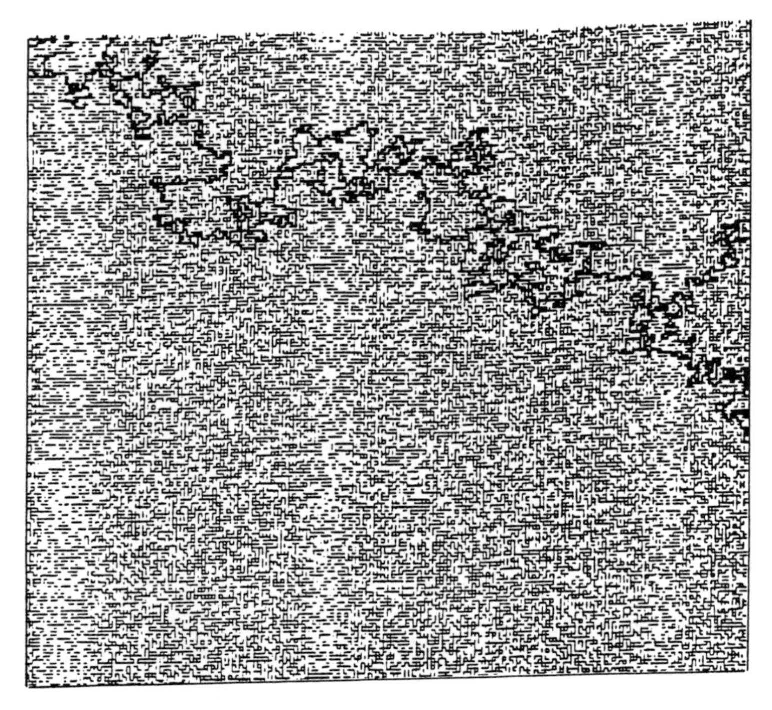

Figure 9. The sample-spanning percolation cluster and its backbone (dark area) on the square network just above the bond percolation threshold $p_{cb} = 1/2$.

the permeability of the bond ij (which is proportional to r_{ij}^2, where r_{ij} is the effective radius of the throat ij), and ΔP_{ij} the pressure difference between nodes i and j. Hence, $\sum_j k_{ij}\Delta P_{ij} = 0$, yielding a set of simultaneous linear equations for the nodal pressures P_i when we write the mass balance for every node of the network. One usually uses periodic boundary conditions in the direction(s) perpendicular to the direction of the applied pressure gradient, which removes sample-size effects in those directions. The resulting set of equations is usually solved by the conjugate-gradient method, from the solution of which the effective permeability of the network is calculated. There are also several approximate analytical methods of estimating K and G, which are discussed extensively by Sahimi (1995b).

One of the most important features of percolation systems is their *universal* properties near p_c. As p_c is approached, the correlation length diverges according to the power law (18). Moreover, near p_c one has

$$X_A(p) \sim (p - p_c)^{\beta} \sim \xi_p^{-\beta/\nu} \tag{19}$$

$$X_B(p) \sim (p - p_c)^{\beta_b} \sim \xi_p^{-\beta_b/\nu} \tag{20}$$

$$G(p) \sim (p - p_c)^{\chi} \sim \xi_p^{-\chi/\nu} \tag{21}$$

$$K(p) \sim (p - p_c)^{e} \sim \xi_p^{-e/\nu} \tag{22}$$

$$D(p) \sim (p - p_c)^{\chi-\beta} \sim \xi_p^{-\theta} \tag{23}$$

where $\theta = (\chi - \beta)/\nu$. Similar to ν, the critical exponents β and β_b are also universal and depend only on the dimensionality of the system if there are at most short-range correlations in the system. The flow and transport coefficients e and χ are equal and also universal if there are at most short-range correlations in the system, and if (Kogut and Straley, 1979; Sahimi et al., 1983b)

$$f_{-1} = \int_0^{\infty} \frac{f(x)}{x} \mathrm{d}x < \infty \tag{24}$$

where $f(x)$ is the conductance or permeability distribution of the bonds. With long-range correlations, the universality of the percolation critical exponents is lost (see below), in which case $e \neq \chi$. Note that although scaling laws (21)–(24) are supposed to be valid in the critical region close to p_c, a region whose extent is roughly $p - p_c \leq 1/Z$, in many disordered media they are actually valid over a much broader region, and thus they provide useful equations for estimating the effective flow and transport properties.

As mentioned above, at any length scale $L \ll \xi_p$ the sample-spanning cluster is fractal and statistically self-similar. The fractal dimension D_f of the cluster can be shown to be given by

$$D_f = d - \beta/\nu \tag{25}$$

At length scales $L \ll \xi_p$, the backbone is also a fractal and statistically self-similar object with a fractal dimension

$$D_b = d - \beta_b/\nu \tag{26}$$

and it is obvious that $D_b < D_f < d$. Table 1 summarizes the most accurate estimate of various percolation exponents and fractal dimensions.

When the sample-spanning percolation cluster is fractal, then flow and transport properties of the percolation networks are *scale-dependent*. Since for $L \ll \xi_p$ the only relevant length scale of the system is L, we should replace ξ_p in Eqs. (21)–(23) by L, which yields

$$G \sim L^{-\chi/\nu} \tag{27}$$

Table 1. Currently accepted values of critical exponents and fractal dimensions in d dimensions*

d	β	ν	β_b	D_f	D_b	θ	χ	e
2	5/36	4/3	0.48	91/48	1.64	0.87	1.3	1.3
3	0.41	0.88	0.99	2.52	1.87	1.80	2.0	2.0

*Rational numbers represent exact values.

$$K \sim L^{-e/\nu} \tag{28}$$

$$D \sim L^{-\theta} \tag{29}$$

Therefore, *scale-dependent flow and transport properties are a signature of a fractal system*, implying the existence of long-range correlations. More importantly, because of such correlations, classical equations *cannot* describe flow and transport in fractal systems. For example, diffusion in a fractal medium cannot be described by the classical diffusion equation with a constant diffusivity. We will return to this in Section VII.

Let us point out that, instead of regular networks such as the square and simple-cubic networks, one can study percolation and flow and transport processes in topologically random networks, i.e., those in which the coordination number of the network varies from site to site. Jerauld et al. (1984) and Sahimi and Tsotsis (1997) have carried out such a study, and have shown that percolation, flow, and transport in such networks are completely similar to those in regular networks.

B. Percolation with Long-range Correlations

As mentioned in the Introduction and also discussed below, the porosity logs and permeability distributions of many FSPM are described by either a fBm, a fGn, or a fLm. Flow and transport in such porous media cannot be described by a random percolation model of the type discussed above. Instead, one must consider flow and transport in percolation systems with long-range correlations, with the correlations being generated by a fBm, a fGn, or a fLm. This type of correlated percolation was first studied by Sahimi (1995a) and Sahimi and Mukhopadhyay (1996). We describe a percolation process in which the correlations are generated by a fBm. Its generalization to those based on the fGn or the fLm will then be clear.

To study such a correlated percolation model, we first generate a correlated permeability field by assigning a permeability to each bond of a network which is selected from a fBm. To construct a percolation network and to preserve the correlations between the bonds, we remove those bonds that have been assigned the *smallest* permeabilities (i.e., we change their permeabilities to zero). The idea is that, in a porous medium with a broad distribution of the permeabilities, a finite (volume) fraction of the medium should have such small permeabilities that they make negligible contribution to the overall permeability. Figure 10 shows a 2D system in which the permeabilities have been selected according to a fBm with $H = 0.8$, and 30% of the smallest permeabilities have been removed. Because of the positive correlations induced by the fBm with $H > 0.5$, most of the large or small permeabilities are clustered together. Moreover, as Figure 10 indicates, the sample-spanning cluster generated by this model appears to be compact (non-fractal).

There are several major differences between random percolation and the correlated percolation discussed here. Typically, the correlations significantly decrease the percolation threshold of the system from its value for random percolation, which results from the clustering of the low or high

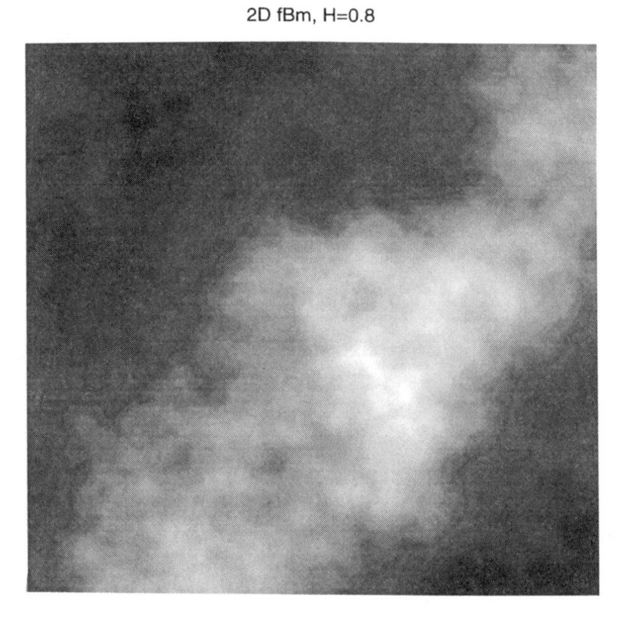

Figure 10. A 2D permeability distribution generated by a fBm with $H = 0.8$. Darkest and lightest areas correspond to the lowest and highest permeability regions, respectively, and 30% of the lowest permeabilities have been set to zero.

permeability regions. This is consistent with many experimental observations that indicate that many porous media are permeable down to very low porosities, i.e., down to very low percolation thresholds. Moreover, unlike random percolation, in the present correlated percolation there are large variations among different realizations of the system with the same H, and as a result, although for a given realization p_c is well defined, its value varies widely among all the realizations. This implies that in this correlated percolation model p_c is not *self-averaging*. The practical implication of this is that, although the porosity logs and permeability distributions of many FSPM may be described by a fBm, or a fGn, or a fLm with the *same* value of H or α, their percolation thresholds, i.e., the minimum porosity at which a fluid can flow through them, can be very different from each other.

The second difference between random and correlated percolation is in the structure of the sample-spanning cluster. It has been shown (Sahimi and Mukhopadhyay, 1996) that in this correlated percolation model the sample-spanning cluster is compact and non-fractal and hence $D_f = d$, in contrast with random percolation discussed above. This compactness is also evident in Figure 10. However, unlike the sample-spanning cluster, the backbone is a fractal object with a fractal dimension D_b that depends on H (for random percolation D_b is universal), such that D_b increases with H and $D_b \to d$ as $H \to 1$. Moreover, unlike random percolation for which the critical exponents are universal, for the present correlated percolation most of the exponents are non-universal and depend on H. The implications of the non-universality of the critical exponent of correlated percolation for modeling flow and transport in FSPM are discussed in Sections VI and VIII.

V. FRACTAL ANALYSIS OF GEOLOGICAL DATA

To give the reader some idea about the complexities that are involved in the characterization of FSPM, we show in Figure 11 a vertical porosity log that was collected along an oil well in Iran. The well's depth was about 600 m and the porosity ϕ was measured every 20 cm, so that over 3000 data points were collected. Figure 12 presents the permeability distribution along the same well. The question then is: given a porosity log or a permeability distribution of FSPM, how can one accurately analyze it? In particular, if such data follow the statistics of a fBm, or a fGn, or a fLm, how can one estimate H and/or α? In the case of fBm- or fGn-type data, we have shown (Mehrabi et al., 1997) that the most efficient and accurate analysis is done by either the orthonormal wavelet decomposition method or the maximum entropy method, both of which we now briefly discuss.

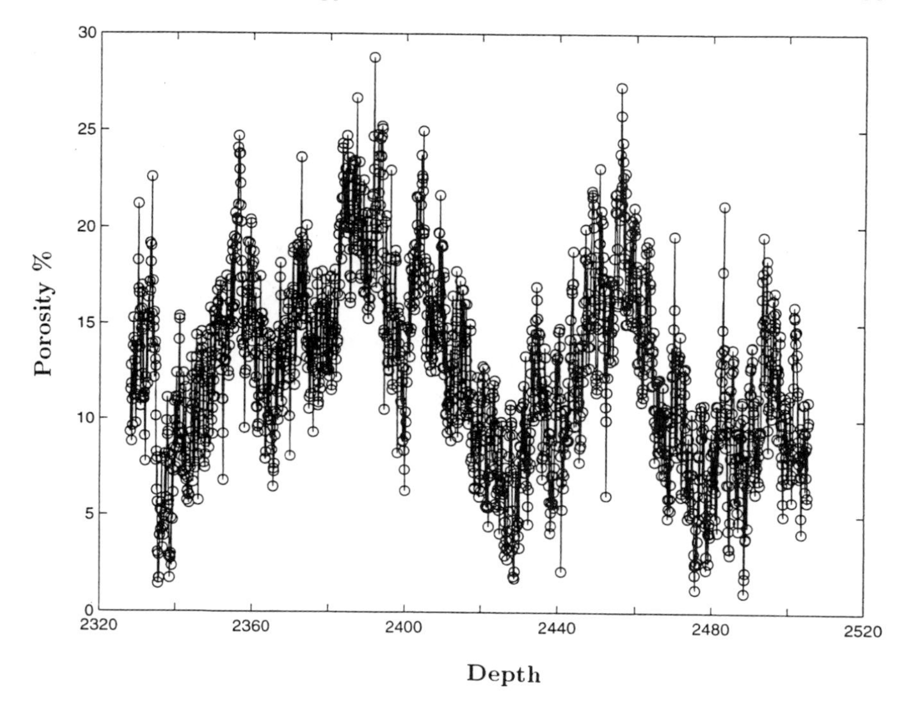

Figure 11. The porosity log along a well in a fractured oil reservoir in Iran. The depth is in meters.

A. Wavelet Decomposition Method

This is a suitable tool for analyzing the fBm-type data (Flandrin, 1992). For such data one calculates the following quantity

$$d_j(k) = 2^{-j/2} \int_{-\infty}^{\infty} B_H(x)\psi(2^{-j}x - k)\mathrm{d}x \tag{30}$$

where $d_j(k)$ are called the wavelet-detail coefficients of the fBm, ψ is the wavelet function, $k = 1, 2, \ldots, n$, where n is the size of the data array, and the js are integers. Thus, in this method one fixes j and varies k to calculate $d_j(k)$. For each j one determines n such numbers and calculates their variance $\sigma^2(j)$. Then it can be shown that, regardless of the wavelet function ψ, one has

$$\log_2[\sigma^2(j)] = (2H + 1)j + \text{const.} \tag{31}$$

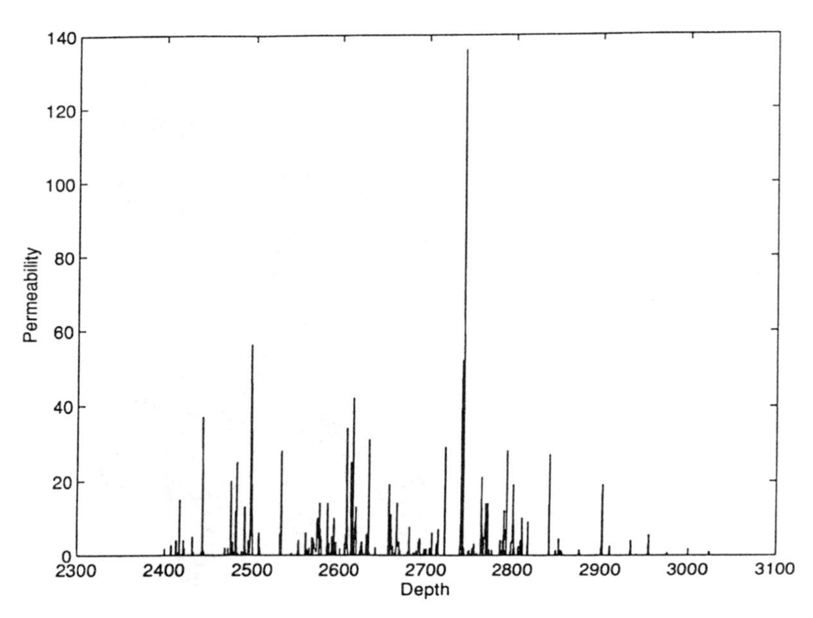

Figure 12. The permeability distribution corresponding to the porosity log of Figure 11.

Thus, plotting $\log_2[\sigma^2(j)]$ versus j yields H. One can use a variety of wavelet functions; a typical one is the Daubechies function shown in Figure 13.

B. Maximum Entropy Method

This is a method of estimating the spectral density of the data *without* using a fast Fourier transformation. In this method the spectral density is approximated by

$$\mathcal{S}(\omega) \simeq \frac{a_0}{|1 + \sum_{k=1}^{M} a_k z^k|^2} \tag{32}$$

where the coefficients a_k are calculated such that Eq. (32) matches the Laurent series, $\mathcal{S}(\omega) = \sum_{-M}^{M} b_i z^i$. Here z is the frequency in the z-transform plane, $z \equiv e^{2\pi i \omega \Delta}$, and Δ is the sampling interval in the real space. In practice, to calculate the coefficients a_i one first computes the correlations functions

$$\mathcal{C}_j = \langle v_i v_{i+j} \rangle \simeq \frac{1}{n-j} \sum_{i=1}^{n-j} v_i v_{i+j} \tag{33}$$

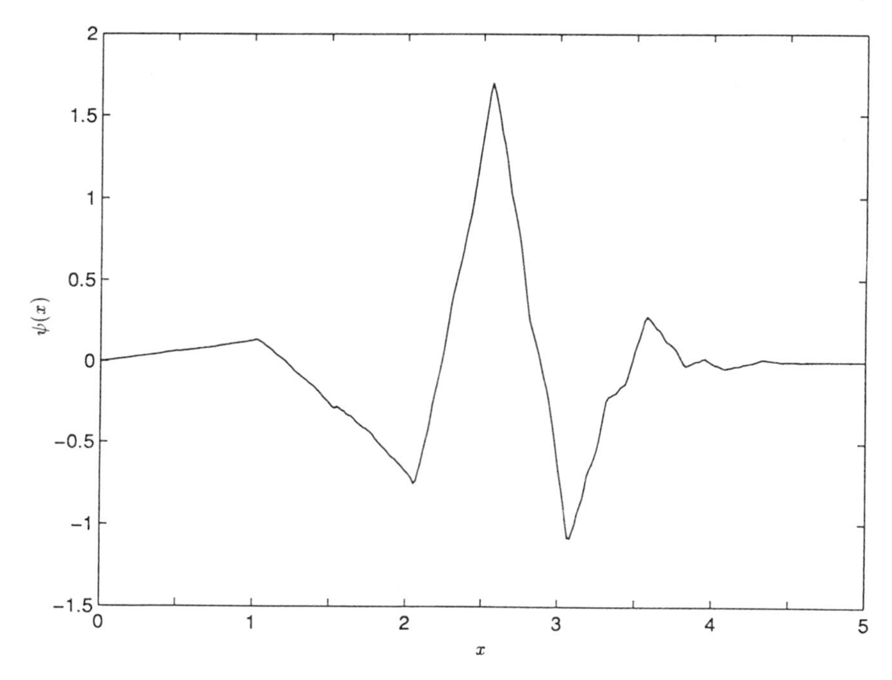

Figure 13. The Daubechies function $\psi(x)$.

where n is the number of data points and v_i is the datum at point i. The coefficients a_i are then calculated from

$$\sum_{j=1}^{M} a_j \mathcal{C}_{|j-k|} = \mathcal{C}_k, \quad k = 1, 2, \ldots, M \tag{34}$$

The advantage of Eq. (32) over the Laurent series is that, if $\mathcal{S}(\omega)$ contains sharp peaks, then (32) detects them easily since the peaks manifest themselves as the poles of the equation, whereas one may have to use a very large number of terms in the Laurent series to detect the same peaks.

While these two methods are both highly accurate, in terms of the required size of the data array for reliable characterization of long-range correlations the maximum entropy method offers the most accurate tool for analyzing a given data array (Mehrabi et al., 1997). Using this method, one extracts valuable information even from a small data array, which is a great advantage for characterization of FSPM since collecting a large data base for such porous media is costly and time consuming.

C. Fractal Analysis of Porosity Logs and Permeability Distributions

Extensive studies by several research groups have provided compelling evidence that the porosity of logs often obey the statistics of fGn (Hewett, 1986; Crane and Tubman, 1990; Sahimi and Yortsos, 1990; Taggart and Salisch, 1991; Aasum et al., 1991; Hardy, 1992; Sahimi and Mehrabi, 1995; Sahimi et al., 1995). For example, Crane and Tubman (1990) analyzed three horizontal wells and four vertical wells in a carbonate reservoir, and found their porosity logs to be described well by a fGn with $H \simeq 0.88$. Likewise, Hardy (1992) analyzed 240 porosity logs from a carbonate formation, and found them to be described well by a fGn with $H \simeq 0.82$. The analysis of a single horizontal well in the same formation also indicated fGn statistics with a slightly higher H. The analysis of porosity logs of several Iranian oil reservoirs (Sahimi and Mohradi, 1995), which are also mostly carbonate formations, revealed that at least some of the logs are described reasonably well by a fGn. In all of these studies the lower limit of fractal behavior was *below* the resolution of the instruments used for measuring the data. A surprising discovery of most of these studies was that the same value of H was found for the porosity and resistivity logs, which, in general, is not expected to be the case.

Similarly, permeability measurements on outcrop surfaces have provided additional evidence that fractal behavior in oil reservoirs and other natural formations is not the exception but the rule. For example, the analysis of Goggin et al. (1992) on sandstone outcrop data indicated that the logarithm of the permeabilities along the lateral trace obeyed fGn statistics with $H \simeq 0.85$. Figure 14 shows their data, along with their power spectrum analysis.

There is also evidence that some reservoir properties follow the statistics of a fBm. In the original analysis of Hewett (1986), a fBm was used for generating horizontal properties of the reservoir. This was justified on the basis of data obtained from groundwater plumes (see, for example, Pickens and Grisak, 1981). Emanuel et al. (1989), Mathews et al. (1989), and Hewett and Behrens (1990) all obtained accurate reservoir descriptions using a fBm. Neuman (1994) provided further evidence that the permeability distributions of many aquifers obey the statistics of a fBm with $H < 0.5$. In general, the description of a reservoir by a fBm does not differ much from that by a fGn, if the wells are closely spaced. Moreover, both fBm and fGn descriptions approach a stratified medium as the variations in the reservoir properties between the wells decrease with decreasing well spacing. We should, however, point out that, under practical circumstances that one encounters with FSPM, fGn with

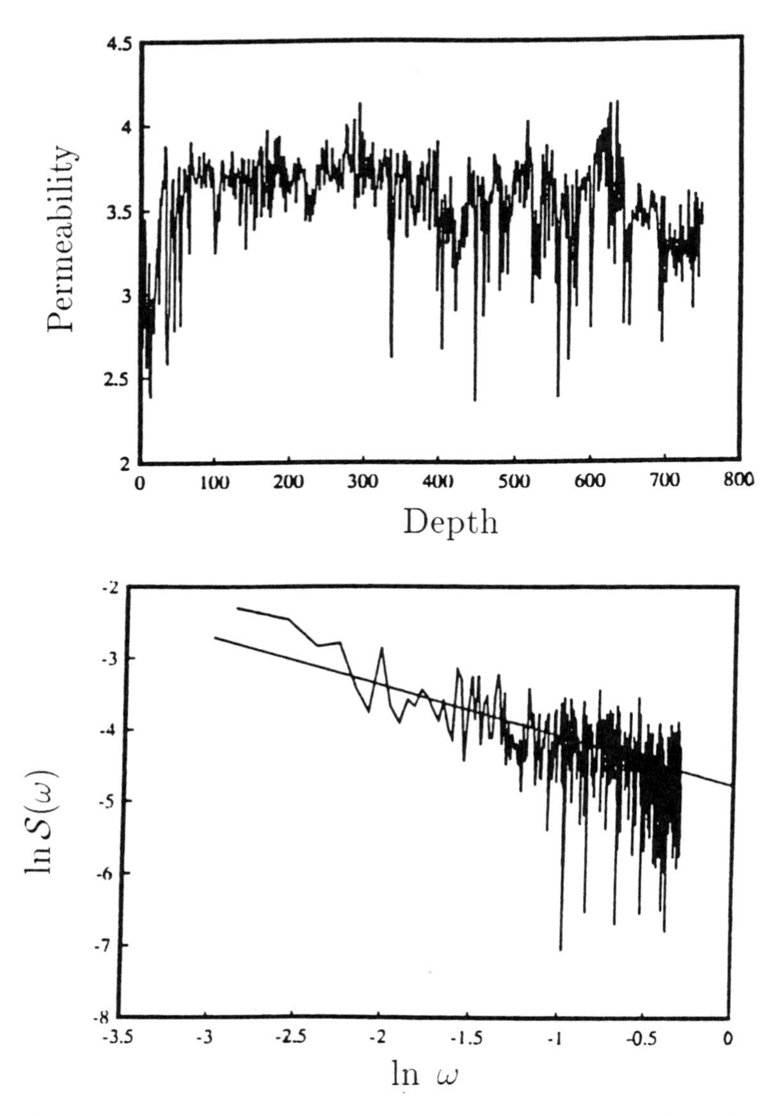

Figure 14. The permeability distribution measured by Goggin et al. (1992) (top) and its spectral density (bottom).

$0 < H < 1$ generates synthetic data that are very similar to fBm with $1 < H < 2$.

However, fBm and fGn do not provide accurate characterization of the porosity logs and the permeability distributions if such data contain long tails. For example, if an oil reservoir or aquifer is fractured, then one must detect large jumps in the permeability distribution, indicating the presence of the fractures, since, as one moves along a well and passes from the reservoir's or the aquifer's porous matrix to its fractures, the permeability should increase significantly, as the permeabilities of the fractures are much larger than that of the porous matrix. Because, compared with the pores, the fractures are relatively rare, their presence represents rare events (jumps) in the permeability distribution, giving rise to long tails in the distribution. In this case a fLm may provide a more accurate and complete characterization of the data (Mehrabi et al., 1997). Moreover, the Lévy stable distribution and the associated fLm have recently been used even in the analysis of the porosity logs (Painter and Paterson, 1984; Sahimi et al., 1995; Painter, 1996; Mehrabi et al., 1997), seismic data (Painter et al., 1995), and the permeability distribution (Mehrabi et al., 1997) of non-fractured FSPM.

To use a fLm for describing the permeability and porosity data of fractured porous media, one first estimates the parameters H and α by studying the dependence of C_L on the lag ℓ. For each lag or separation distance ℓ between two data measured at two locations one calculates the increments in the data and fits the results to Eq. (17) to obtain estimates of C_L and α. Then it can be shown that $C_L(\ell)$ is related to the lag ℓ by

$$C_L(\ell) \sim \ell^H \tag{35}$$

The limit $H = 1/\alpha$ corresponds to the case of independent increments. For positive (negative) correlations (anticorrelations) in the increments we have $H > 1/\alpha (H < 1/\alpha)$. Since $P(x)$ cannot be obtained in closed form, C_L and α are estimated numerically.

To illustrate the method, consider the porosity log of Figure 11. We first construct the experimental histogram of the increments $\varepsilon(x + \ell) - \varepsilon(x)$, where we fix the separation length ℓ. The results are shown in Figure 15 for $\ell = 20\,\text{cm}$, together with the fit of the incremental data to the Lévy distribution, Eq. (17). They show clear deviations from a Gaussian distribution and indicate the existence of long tails in the distribution, presumably caused by the fractures in the reservoir. Indeed, the oil reservoir, one of whose porosity logs is shown in Figure 11, is known to be highly fractured. A Gaussian distribution of the increments would have indicated that the data either are completely random or follow a fBm. From Figure 15 we obtain $\alpha \simeq 1.49$.

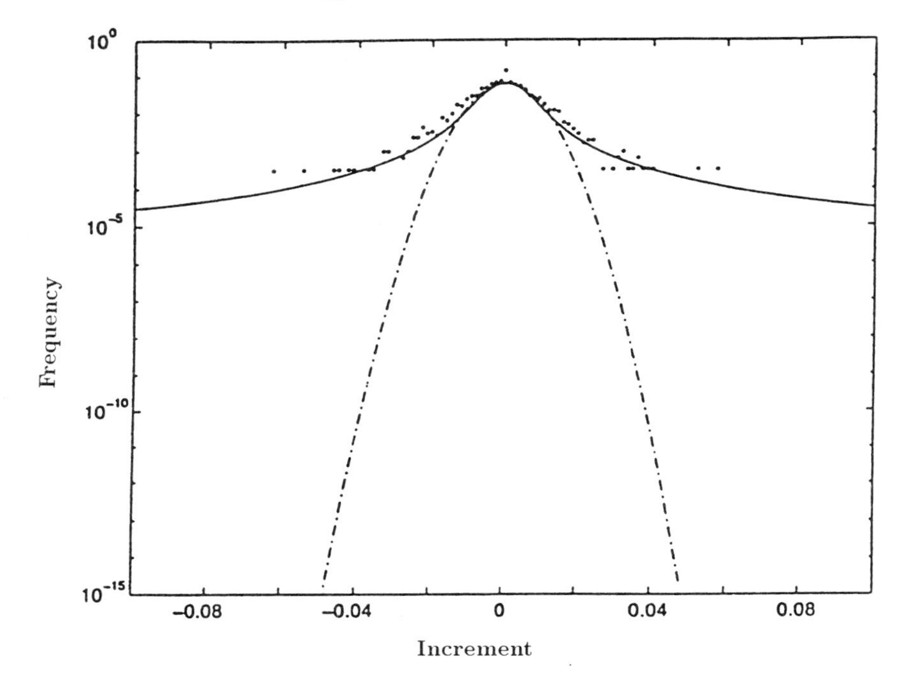

Figure 15. The frequency distribution of the porosity increments of Figure 11. Solid and dashed curves are, respectively, the fit of the data to the Lévy and Gaussian distributions.

We also used Eq. (35) to estimate H. The results are shown in Figure 16. For short separation distances we obtain $H \simeq 0.68 \simeq \alpha^{-1} \simeq 0.67$, indicating that at such length scales the porosities are distributed essentially at random, presumably because at such short length scales one samples the reservoir's properties within a stratum where the morphology is more or less random. However, at large separation distances that are of interest to reservoir simulation (which span many strata), the reservoir is highly heterogeneous with the strata's properties differing greatly; thus we expect negative correlations, and indeed from Figure 16 we obtain $H \simeq 0.13 \ll \alpha^{-1} \simeq 0.67$, i.e., the data exhibit negative correlations, completely consistent with the porosity log shown in Figure 11 (where large jumps are followed by much smaller fluctuations and vice versa). If we now use $\alpha \simeq 1.49$ that we obtained from the analysis of the porosity log and generate a permeability distribution using a fLm, we obtain the results shown in Figure 17, whose similarity with Figure 12 is striking.

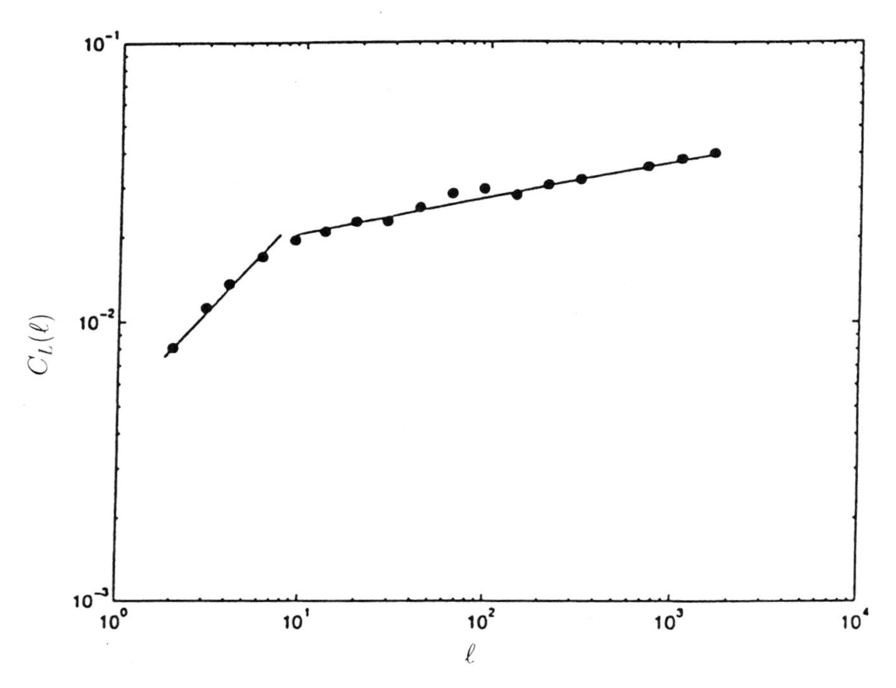

Figure 16. Dependence of the coefficient $C_L(\ell)$ on the separation distance ℓ for the porosity log of Figure 11.

D. Fractal Characterization of Fractures and Faults

Prior to the first application of fractal geometry to characterization of fracture network of FSPM, collection of data was limited mostly to 1D sampling of spacing between fractures intersected along a traverse. This was done without any regard for, or knowledge of, orientation or size of the fractures, and the data analysis was limited to calculation of arithmetic averages of the properties of interest, using such distributions as the log-normal or exponential, which did not provide deep insight into the properties of fracture networks. In the early 1980s a step in the right direction was taken by several research groups by exploring the use of semivariograms as a method of analyzing fracture spacing along a scanline. Semivariograms plot the second moment of the distribution of the number of fractures per unit length of the scanline as a function of the length of sampling increment over some range of increment size. This is similar to the box-counting method discussed in Section II for measuring the fractal dimension of a set. In fact, such semivariograms can be recast into fractal plots if we plot

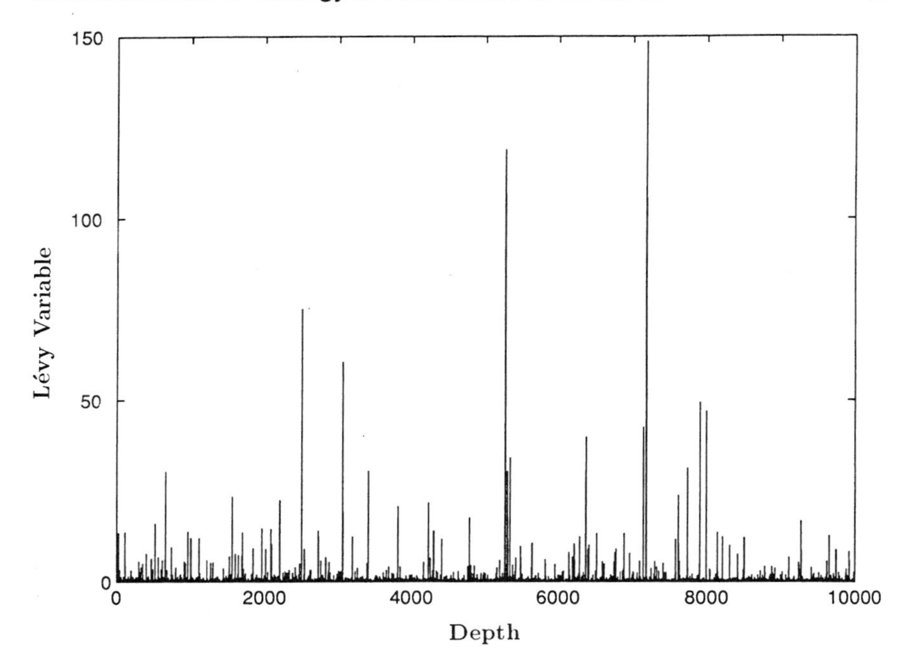

Figure 17. The permeability distribution corresponding to Figure 12, generated by the fractional Lévy motion.

logarithm of semivariance versus logarithm of sample increment size. The slope of the resulting diagram provides an estimate of the fractal dimension of the set.

Application of fractal concepts to characterization of fracture network of FSPM was first explored in 1985 in a study carried out by the U.S. Geological Survey as part of the program to characterize the geologic and hydrologic framework at Yucca Mountain in Nevada (Barton and Larsen, 1985; Barton et al., 1987; summarized in Barton and Hsieh, 1989). This site is being evaluated by the U.S. Department of Energy as a potential underground repository for high-level radioactive waste. Barton and Larsen (1985) developed the *pavement method* of clearing a subplanar surface and mapping the fracture surface in order to measure its connectivity, trace length, density, and fractal scaling in addition to the orientation, surface roughness, and aperture. The most significant observation of the Yucca Mountain study was that the fractured pavements had a fractal geometry and were scale-invariant, and that it was possible to represent the distribution of fractures ranging from 20 cm to 20 m by a single parameter, the fractal dimension D_f, calculated by the box-counting method, Eq. (3), in which $N(L)$ was taken to be the number of fractures of length L; for the fractured surfaces analyzed, $D_f \simeq 1.6$–1.7. Note that a fractal dimension $D_f \simeq 1.6$–1.7 for a fractured surface implies $D_f \simeq 2.6$–2.7 for the corresponding 3D fracture network, if the network is essentially isotropic. LaPointe (1988) carried out a careful reanalysis of three fracture-trace maps of Barton and Larsen (1985) and found that for the corresponding 3D fracture networks $D_f \simeq 2.37$, 2.52, and 2.68, with an average of about $D_f \simeq 2.52$. Two-dimensional maps of fracture traces spanning nearly *ten* orders of magnitude, ranging from microfractures in Archean Albites to large fractures in South Atlantic seafloors, were analyzed by Barton (1992), who reported that $D_f \simeq 1.3$–1.7, with an average of about 1.5, implying an average of $D_f \simeq 2.5$ for the corresponding 3D fracture networks.

A similar result was obtained for the Geysers geothermal field in northeast California (Sahimi et al., 1993), one of the most significant geothermal fields in the world. An example of its fracture is shown in Figure 18. The reservoir's fractures are detected during drilling since they produce a sudden and measurable increase in the steam pressure. Using a discrete model of fracture of solids developed by Sahimi and Goddard (1986), Sahimi et al. (1993) proposed that at *large* length scales (of the order of a few hundred meters or more) the fracture network of the field must have the structure of the sample-spanning percolation cluster at p_c and thus $D_f \simeq 2.5$ for the 3D fracture network of the reservoir. Watanabe and Takahashi (1995) also used fractal concepts to characterize the structure of geothermal reservoirs.

Nolen-Hoeksema and Gordon (1987) and Chelidze and Gueguen (1990) studied the fracture patterns in Stockbridge (near Canamn, Connecticut) dolomite marble. The fracture pattern in this rock is very branched and appears to be a highly interconnected network. Analysis of Chelidze and Gueguen (1990) indicated that the 3D fracture network is a fractal object with $D_f \simeq 2.5$, the same as that of the sample-spanning percolation cluster at p_c. There are also many papers in which percolation networks have been used as models of fracture or fault network of rock. These include Englman et al. (1983), Robinson (1984), Charlaix et al. (1987), Hestir and Long (1990), Balberg et al. (1991), Mukhopadhyay and Sahimi (1992), Berkowitz (1995), Mourzenko et al. (1996, 1999), Huseby et al. (1997), and Bour and Davy (1997, 1998). These modeling efforts are based on the fact that fractures are the main pathways for fluid flow in a fractured rock, as their permeabilities are much larger than those of the matrix in which they are embedded, and thus fractured rock may be envisioned as a mixture of impermeable or closed bonds (the pores of the matrix) and the highly permeable or open bonds (the fractures), i.e., a percolation

Figure 18. A sample of fracture surface of the Geysers geothermal field.

network. These models have provided considerable insight into the connectivity properties of fracture networks, and provide further evidence for the applicability of percolation models to the modeling of porous and fractured rock. Finally, Sahimi (1991) showed how fractal patterns can arise as a result of the consumption of a porous medium by a reactant.

The applicability of fractal geometry to characterization of fractured rock is not restricted to describing the connectivity of the fracture networks. For example, the internal surface of a fracture is typically very rough, unlike the traditional smooth surfaces that have been used in modeling flow through fractures. It has been argued (see, for example, Brown and Scholz, 1985; Brown, 1987) that this roughness can be described as a self-affine fractal surface. For example, if we consider the 2D roughness profile of the fracture surface, then the average height $\langle h \rangle$ of the profile is related to its length L by

$$\langle h \rangle \sim L^H \tag{36}$$

where H is the roughness or Hurst exponent defined above. A value of $H \simeq 0.85$ was found by Schmittbuhl et al. (1993) for granitic faults, implying positive and long-range correlations. Odling (1994) obtained more scattered data for rock joints in the range $0.46 \leq H \leq 0.85$. Cox and Wang (1993) provide a review of this subject. The roughness of the internal surface of a fracture also has two implications. One is that, as Eq. (36) indicates, a fBm can be used to generate very realistic rough surfaces for rock fractures (see Figure 3). The second implication is that the task of simulation of fluid flow in realistic models of fractures is very complex. However, new methods of simulating fluid flow based on lattice-gas automata and lattice Boltzmann models (for reviews see, for example, Sahimi, 1995b; Rothman and Zaleski, 1997) provide potentially powerful methods for this problem (Gutfraind and Hansen, 1995; Zhang et al., 1996; Di Pietro, 1996).

In addition to fractures, faults also possess fractal properties. Faults are usually created when two strata or layers move with respect to each other. The interface between the two displaced layers is what constitutes a fault. Therefore, in some sense faults are similar to fractures and one often finds large faults in almost any kind of reservoir. However, unlike fractures which are created by a variety of processes, ranging from diagenetic to mechanical, faults are usually manifestations of tectonic processes that reservoirs experienced in the past. Moreover, unlike fractures which provide large permeability zones and facilitate transport of fluids in reservoir rock, faults may or may not do so. Sometimes they hinder fluid transport in the reservoir by compartmentalizing reservoirs and isolating large portions of them.

Tchalenko (1970) observed that over many orders of magnitude in length scale, ranging from millimeters to hundreds of meters, shear deformation zones in rock are similar, thus suggesting strongly that the fault patterns are fractal (although Tchalenko was not aware of fractal geometry). The work of others (Andrews, 1980; Aki, 1981; King, 1984) has confirmed this suggestion. In particular, Okubo and Aki (1987) and Aviles et al. (1987) analyzed maps of the San Andreas fault system in California and obtained fractal dimensions for fault surfaces varying from 1.1 to 1.4.

Summarizing this section, ample evidence indicates that the geology of oil reservoirs, aquifers, and other natural rock formations can be described accurately by fractal methods, and in particular by stochastic fractal processes such as the fBm, fGn, and the fLm. Percolation, on the other hand, gives us information about the connectivity of the fracture network of rock at large length scales. Thus, together the provide the tools for realistic modeling of geology of heterogeneous rock, which is perhaps the most crucial element of accurate reservoir simulations and modeling of groundwater flow.

VI. PERCOLATION MODELS OF FLOW IN POROUS MEDIA

We now discuss applications of fractal and percolation concepts to modeling of flow and transport in porous media. The literature on this subject is vast, and thus we restrict ourselves to single-phase flow, diffusion, and hydrodynamic dispersion in porous media. We first consider modeling of flow in macroscopic (laboratory-scale) and megascopic (field-scale) porous media. We then discuss diffusion and dispersion in porous media.

A. Single-phase Flow in Macroscopic Porous Media

Katz and Thompson (1986, 1987) used percolation theory successfully to predict the effective permeability K and electrical conductivity G of (fluid-saturated) macroscopic porous media. Their method was based on a concept called the *critical path method*, first proposed by Ambegaokar, Halperin, and Langer (AHL) in 1971, who argued that transport in a disordered medium with a *broad* distribution of conductances is dominated by those conductances whose magnitudes are larger than some characteristic g_c, which is the smallest conductance such that the set of conductances $\{g|g > g_c\}$ forms a sample-spanning cluster called the critical path. Therefore, transport in a disordered medium with a broad conductance distribution reduces to a percolation problem with a threshold conductance g_c. Shante (1977) and Kirkpatrick (1979) modified this idea by assigning g_c to all the local conductances with $g \geq g_c$ and setting all conductances with $g < g_c$ to zero (since the contribution of such bonds to the overall conductivity is negligible), and obtained a trial solution for G which has the same form as Eq. (21):

$$G = ag_c[p(g_c) - p_c]^{\chi} \tag{37}$$

where $p(g_c)$ is the probability that a given conductance $g \geq g_c$, and a is a constant. Equation (37) was then maximized with respect to g_c to obtain an estimate of g_c and thus G.

Katz and Thompson (1986, 1987) extended the ideas of Ambegaokar et al. (1971) to estimate the permeability and electrical conductivity of porous media. In a porous medium the local hydraulic conductance is a function of a length scale ℓ. Thus, there is a characteristic length scale ℓ_c corresponding to g_c. The length that signals the percolation threshold in the flow problem also defines the threshold in the conduction problem. Thus we rewrite Eq. (37) as

$$G = \varepsilon g_c(\ell)[p(\ell) - p_c]^{\chi} \tag{38}$$

where the porosity ε ensures a proper normalization of the fluid or the electric-charge density. $g_c(\ell)$ is equal to $c_f\ell^3$ and $c_c\ell$ for the flow and electrical conduction problems, respectively. For appropriate choices of $p(\ell)$, $G(\ell)$ achieves a maximum for some $\ell_{\max} \leq \ell_c$. In general, $\ell^f_{\max}$ for the flow problem is different from $\ell^c_{\max}$ for the conduction problem, because the transport paths have different weights for the two problems. If $p(\ell)$ allows for a maximum in the conductance which occurs for $\ell_{\max} \leq \ell_c$, then for $\ell^f_{\max} = \ell_c - \Delta\ell_f$ and $\ell^c_{\max} = \ell_c - \Delta\ell_c$, and $\ell_c\chi p''(\ell_c)/p'(\ell_c) \ll 1$ (which is valid if the pore size distribution of the porous medium is broad), we can write

$$\ell^f_{\max} = \ell_c\left[1 - \frac{\chi}{1 + \chi + \ell_c\chi p''(\ell_c)/p'(\ell_c)}\right] \simeq \ell_c\left(1 - \frac{\chi}{1+\chi}\right) \simeq \frac{1}{3}\ell_c \tag{39}$$

$$\ell^c_{\max} = \ell_c\left[1 - \frac{\chi}{3 + \chi + \ell_c\chi p''(\ell_c)/p'(\ell_c)}\right] \simeq \ell_c\left(1 - \frac{\chi}{3+\chi}\right) \simeq \frac{3}{5}\ell_c \tag{40}$$

where we used $\chi \simeq 2$ for 3D percolation (see Table 1). Thus, writing

$$G = a_1\varepsilon[p(\ell^c_{\max}) - p_c]^{\chi} \tag{41}$$

and

$$K = a_2\varepsilon(\ell^f_{\max})^2[p(\ell^f_{\max}) - p_c]^{\chi} \tag{42}$$

we obtain, to first order in $\Delta\ell_c$ or in $\Delta\ell_f$, $p(\ell^{f,c}_{\max}) - p_c = -\Delta\ell_{f,c}p'(\ell_c)$. To obtain a_1 and a_2, Katz and Thompson (1986) assumed that at the local level the conductivity of the porous medium is the same as the conductivity g_f of the fluid (usually brine) that saturates the pore space, and that the local throat geometry is cylindrical, implying that $a_1 = g_f$ and $a_2 = 1/32$. Therefore, one obtains

$$K = \frac{1}{226}\frac{G}{g_f}\ell_c^2 \tag{43}$$

and (Katz and Thompson, 1987)

$$\frac{G}{g_f} = \frac{\ell^c_{\max}}{\ell_c}\varepsilon S(\ell^c_{\max}) \tag{44}$$

where $S(\ell^c_{\max})$ is the volume fraction of connected pore space involving pore widths of size $\ell^c_{\max}$ and larger. Note that it has been assumed that the permeability and conductivity exponents are equal, $\chi = e$, which is true if inequality (24) is satisfied.

Equations (43) and (44) involve no adjustable parameters. To obtain the characteristic length ℓ_c, Katz and Thompson (1986, 1987) used mercury porosimetry (a common method of measuring the pore size distribution), and proposed that $\ell_c = -(4\sigma \cos\theta)/P_c$, where σ is the surface tension between the mercury and the vacuum, θ the contact angle between the mercury and the pore surface, and P_c the capillary pressure at the point when a sample-spanning cluster of mercury is formed in the pore space for the first time. (Mercury porosimetry itself is a percolation process; see, for example, Knackstedt et al., 1998). Thus, a single set of porosimetry measurement yields ℓ_c and hence K and G. Figure 19 compares the logarithm of the calculated K for a set of various sandstones with the measured values. The dashed lines mark a factor of two; the agreement between the predictions and the data is very good. Equally impressive predictions were obtained for the electrical conductivity of the same porous media. A somewhat similar percolation model was developed for estimating the effective permeability of a porous medium saturated by a *non-Newtonian* fluid (Sahimi, 1993a) such as a polymer solution.

B. Single-phase Flow in Megascopic Porous Media

As discussed above, the permeability distribution and porosity logs of FSPM contain long-range correlations and may follow stochastic processes such as the fGn, or the fBm, or the fLm. Since such distributions are typically broad, by the critical path method discussed in the last section one can

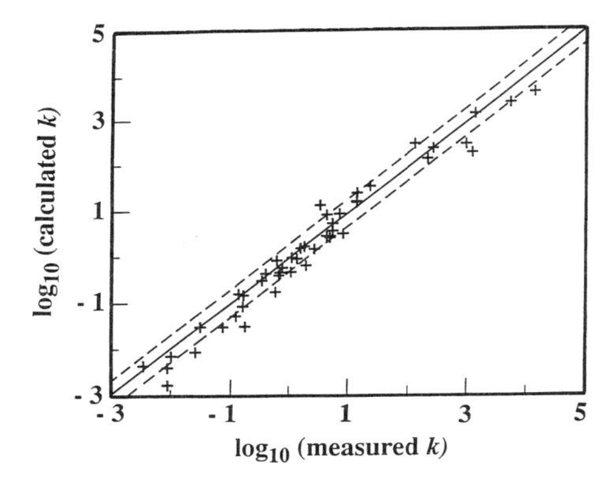

Figure 19. Percolation predictions of the permeability K (in millidarcy) of various porous media versus the measured values. Dashed lines denote a factor 2 of possible difference (after Katz and Thompson, 1986).

reduce flow in such FSPM to flow through the sample-spanning cluster of an appropriate percolation model (Sahimi and Mukhopadhyay, 1996). However, unlike flow through macroscopic porous media discussed above, the application to FSPM which contain large-scale correlated heterogeneities reduces the problem to a percolation problem with long-range correlations of the type discussed in Section IV.B. Computer simulations of Moreno and Tsang (1994) and Heweijer and Dubrule (1995) have confirmed the applicability of percolation to flow in FSPM. These authors found that, in a 3D model porous medium with a broad permeability distribution $F(K)$, the flow paths are restricted to only those regions that have the *largest* permeabilities, precisely the essence of the critical path method, although these authors did not mention percolation.

This idea has been exploited (Sahimi and Mukhopadhyay, 1996) for developing an equation for the permeability K of FSPM with a permeability distribution $F(K)$. Suppose that K_c is the critical permeability, the analog of g_c, so that all the permeabilities less than K_c are set to zero and the permeability of the rest of the pore space is assigned the same value K_c. Equation (22) tells us that

$$K \simeq K_c[p(K) - p(K_c)]^e \tag{45}$$

where $p(K) = \int_K^\infty F(K)\mathrm{d}K$ is the fraction of the regions of the pore space having a permeability larger than K. We must eliminate $p(K) - p(K_c)$ from Eq. (45), as it cannot be measured directly, and replace it with a measurable quantity. To do this, we maximize (45) with respect to K_c, which yields $p(K) - p(K_c) = eK_cF(K_c)$, implying that

$$K \simeq e^e K_c^{1+e}[F(K_c)]^e \tag{46}$$

Therefore, given a broad permeability distribution $F(K)$, we first estimate the exponent e and the percolation threshold p_c, from which the critical permeability K_c is estimated. Then, Eq. (46) provides us with an estimate of the overall permeability K of the FSPM. In particular, if $F(K)$ contains long-range correlations, then, as discussed in Section IV.B, one will have non-universal values for the exponent e which will depend on the properties of $F(K)$. For example, if $F(K)$ follows the statistics of a fBm or a fGn, then e will depend on H.

VII. FRACTAL DIFFUSION IN POROUS MEDIA

As mentioned in Section IV.A, diffusion in a percolating porous medium at length scales $L \ll \xi_p$ cannot be described by the classical Fick's law with a constant diffusivity. To gain insight into such a transport process, we consider a useful characteristic of diffusion, namely the mean square displacement $\langle R^2(t)\rangle$ of the diffusing particles at time t, which in d dimensions is related to the effective diffusivity D by

$$\langle R^2(t)\rangle = 2dDt \tag{47}$$

and therefore, depending on how $\langle R^2(t)\rangle$ grows with t, one can have three distinct diffusion regimes. Let us write quite generally

$$\langle R^2(t)\rangle \sim t^{2/D_w} \tag{48}$$

where D_w is called the fractal dimension of the random walk or diffusion. Equation (48) is a manifestation of dynamical fractals mentioned in Section II. In general, there are three possibilities.

1. $D_w = 2$, which means that $\langle R^2(t)\rangle$ grows linearly with t, and D is a constant. This happens when $R_d = \sqrt{\langle R^2(t)\rangle}$ is larger than the dominant length scale of the medium, which for percolation disorder is the correlation length ξ_p. This is the classical diffusion described by the Fick's law, since in this case D is constant.
2. If $D_w > 2$, $\langle R^2(t)\rangle$ grows with t slower than linearly. In this case diffusion is called anomalous (Gefen et al., 1983) or fractal (Sahimi et al., 1983b). In this regime, $D \sim t^{2/D_w - 1}$ vanishes as $t \to \infty$, i.e., diffusion is very slow and inefficient. This happens if R_d is smaller than the dominant length scale of the medium, which for percolation disorder is ξ_p. In general, diffusion in *all* fractal media is anomalous. Note that ξ_p is divergent at p_c, and therefore one always has fractal diffusion at p_c.
3. If $D_w < 2$, then $\langle R^2(t)\rangle$ grows with t faster than linearly. This type of transport process is called *superdiffusion* (Sahimi, 1987). Turbulent flow is an example of a superdiffusive transport process, as is hydrodynamic dispersion in FSPM, discussed in the next section. Neither fractal diffusion nor superdiffusion is described by the classical diffusion equation, and the exact form of the governing equation is unknown.

In case 2, for $L \ll \xi_p$ (or equivalently at p_c), we can relate D_w to the critical exponents of percolation defined in Section IV. Suppose that diffusion is only in the sample-spanning cluster. Equation (23) tells us that, near p_c, $D \sim \xi_p^{-\theta}$, where $\theta = (\chi - \beta)/\nu$. Because fractal diffusion occurs only if $R_d < \xi_p$, we can replace ξ_p with $R_d = \langle R^2(t)\rangle^{1/2}$ and write $D \sim \langle R^2(t)\rangle^{-\theta/2} \sim \mathrm{d}\langle R^2(t)\rangle/\mathrm{d}t$, which after integration yields $\langle R^2(t)\rangle \sim t^{2/(2+\theta)}$, implying that (Gefen et al., 1983)

$$D_w = 2 + \frac{\chi - \beta}{\nu} = 2 + \theta \tag{49}$$

Thus, we find that (see Table 1) $D_w(d=2) \simeq 2.87$ and $D_w(d=3) \simeq 3.8$. The crossover between fractal and Fickian diffusion occurs at a crossover time t_{co} such that

$$t_{co} \sim \xi_p^{D_w} \sim (p - p_c)^{-2\nu+\beta-\chi} \tag{50}$$

so that for $t \ll t_{co}$ diffusion is fractal, whereas for $t \gg t_{co}$ it is Fickian. Equation (50) indicates that t_{co} diverges as $p \to p_c$, which is due to the fact that as $p \to p_c$ the correlation length ξ_p diverges, and thus the percolation cluster is a fractal object on a rapidly increasing length scale. If diffusion takes place on all the percolation clusters, then it can be shown that (Gefen et al., 1983)

$$D_w = 2 + \frac{2\chi}{2\nu - \beta} \tag{51}$$

Equation (49) was verified experimentally by Knackstedt et al. (1995), who obtained $D_w \simeq 3.7 \pm 0.2$, consistent with $D_w \simeq 3.8$ predicted by Eq. (49).

VIII. HYDRODYNAMIC DISPERSION IN POROUS MEDIA

The last application of fractal and percolation concepts that we discuss is modeling of hydrodynamic dispersion in FSPM. If one fluid is injected into another miscible fluid which is flowing in a porous medium with a non-uniform velocity field, then, in addition to mixing of the two fluids by molecular diffusion, additional mixing is caused by the non-uniform velocity field which itself is caused by the morphology of the medium, the nature of fluid flow (laminar, turbulent, etc.), and chemical or physical interactions with the solid surface of the medium. This mixing process, called *hydrodynamic dispersion*, is important to a wide variety of processes. For example, dispersion is important to miscible displacements which are used in enhanced recovery of oil. When oil production from a reservoir declines, a fluid is injected into the reservoir to mobilize the remaining oil in the reservoir and increase its production. The injected fluid can be immiscible or miscible with the oil in place. If the injected fluid and the oil are miscible, and if the viscosity of the displacing miscible fluid is less than that of oil, large fingers of the displacing fluid are formed that advance throughout the medium, leaving behind a large amount of oil. Dispersive mixing of the displacing fluid and the oil can help the fingers join and displace more oil, thus increasing the efficiency of the process. Dispersion is also important to salt water intrusion in coastal aquifers, where fresh and salt water mix by a dispersion process, to in-situ study of the characteristics of an aquifer, where a classical method of determining such characteristics is to inject fluid tracers to mix with the water and to measure their travel times, to the pollution of subsurface water because of industrial and nuclear wastes, and to flow and reaction in packed-bed chemical reactors.

Two basic mechanisms drive dispersion in disordered porous media. The first mechanism is *kinematic*: streamlines along which the fluid particles travel diverge and converge repeatedly at the junctions of flow passages in the pore space. This convergence and divergence of streamlines is accentuated by the widely varying orientations of flow passages and coordination number of the pore space. The result is a wide variation in the lengths of the streamlines and the downstream transverse separations of the streamlines. The second mechanism is *dynamic*: the speed with which a given flow passage is traversed depends on the flow resistance or hydraulic conductance of the passage, its orientation, and the local pressure field.

These two fundamental mechanisms of dispersion do *not* depend on molecular diffusion. Diffusion modifies convective mixing not only by transferring the particles into and out of the stagnant or very slow regions of the pore space, but also by moving them out of the boundary layers that are formed near the solid surfaces. Thus, the modification of dispersion by diffusion depends on pore space morphology and how it in turn affects local flow and concentration fields. However, diffusion does not play an important role in dispersion in FSPM, since in this case dispersion is caused principally by large-scale fluctuations in the velocity field, which are caused by the spatial variations of the porosity and permeability fields.

A most important task in modeling dispersion is to develop the appropriate equation for the evolution of the solute concentration. Traditional modeling of dispersion processes in microscopically disordered and macroscopically homogeneous porous media is usually based on the convective-diffusion equation (CDE)

$$\frac{\partial C}{\partial t} + \mathbf{v} \cdot \nabla C = D_L \frac{\partial^2 C}{\partial x^2} + D_T \nabla_T^2 C \tag{52}$$

where $\mathbf{v}$ is the macroscopic mean velocity, and ∇_T^2 the Laplacian in transverse directions. Thus, dispersion is modeled as anisotropic diffusion of the solute or the injected fluid, with the diffusivities being the longitudinal dispersion coefficient D_L (in the direction of macroscopic flow) and the transverse dispersion coefficient D_T, augmented by the flow field which is represented by $\mathbf{v} \cdot \nabla C$. The anisotropy is dynamic and is caused by the flow field since the solute mixes with and spreads in the flowing fluid better in the longitudinal direction than the transverse direction. Dispersion is said to be *diffusive* or Gaussian if it obeys a CDE. However, dispersion in FSPM does not follow a CDE, and what follows is a brief description of the past modeling efforts for describing dispersion in such porous media.

A. Non-local Dispersion in Field-scale Porous Media

Field measurements of dispersion coefficients are often costly and time consuming. One needs to drill many observation wells to monitor the spread of the solute, and the spreading itself is often very time consuming and slow; a few *years* may be needed for completing the investigations. The level of uncertainties in all the operations and measurements is quite high. Despite such difficulties, a considerable amount of data has been collected which indicate unequivocally that D_L and D_T measured in FSPM are larger by several orders of magnitude than those measured in a laboratory, and an entirely deterministic approach to this problem, based on the classical transport equations, cannot provide a satisfactory explanation for such data. Moreover, field experiments of Sudicky and Cherry (1981), Pickens and Grisak (1981), Molz et al. (1983), and Sudicky et al. (1983, 1985) (for a review see Gelhar et al., 1992) indicate that D_L and D_T are scale dependent, and that the apparent dispersivities $\alpha_L = v/D_L$ and $\alpha_T = v/D_T$ seem to increase with the transit times of the solute particles. For example, over 130 dispersivities were collected from the literature by Arya et al. (1988) and others. They vary anywhere from less than 1 mm to over 1 km, collected on length scales that vary from less than 10 cm to more than 100 km. The data show large scatter, but from a logarithmic plot of α_L versus the length scale of the measurements a straight line fit of the data can be obtained by regression analysis, which indicates that at least 75% of the data do follow the regression line with 95% confidence, implying that

$$\alpha_L \sim L_m^{\delta} \tag{53}$$

where L_m is the length scale of measurements or distance from the source (where the tracer particles are injected into the porous medium). The analysis of Arya et al. (1988) yielded $\delta \simeq 0.88$. Neuman (1990) presented another regression analysis of these and other data, and proposed that there are in fact two regimes, one for $L_m \ll 100\,\mathrm{m}$ for which $\delta \simeq 1.5$, and another one for $L_m \gg 100\,\mathrm{m}$ for which $\delta \simeq 0.92$, in agreement with the results of Arya et a., (1988). A considerable amount of field data was also collected and analyzed by Gelhar et al. (1992), some of which are shown in Figure 20. They exhibit trends similar to the data of Arya et al. (1988). However, in general, the exponent δ is not expected to be universal. As pointed out by Gelhar et al. (1992), it is inappropriate to represent all the field data by a single universal scaling law; this is similar to the fact that,

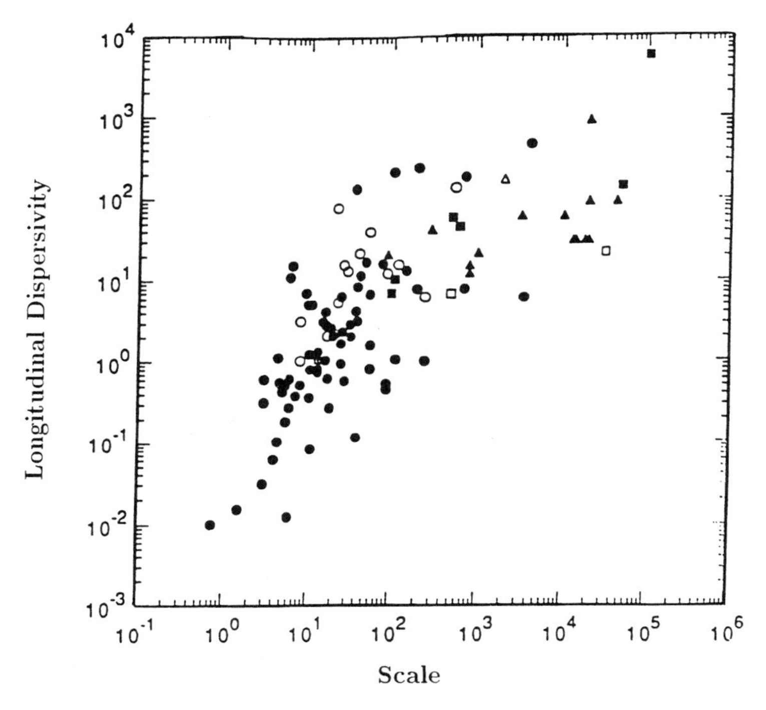

Figure 20. Field-scale longitudinal dispersivity α_L versus the measurement length scale L_m. Solid symbols represent data for unfractured media, whereas the open symbols are for fractured porous media (after Gelhar et al., 1992).

when the porosity logs and permeability distributions of FSPM follow a fBm or a fGn, the Hurst exponent H is not the same for all FSPM. Moreover, as percolation with long-range correlations has taught us (see Section IV.B), flow and transport properties of a heterogeneous medium with long-range correlations, i.e., with fractal distributions, follow non-universal scaling laws.

Gelhar et al. (1992) also analyzed a considerable amount of field data on transverse dispersivities. Their analysis indicated that the vertical transverse dispersivities are typically smaller than the horizontal ones by about one order of magnitude. An important issue raised by Gelhar et al. (1992) was the reliability of the collected data. Although very large dispersivities have been reported in the literature, varying over many orders of magnitude, Gelhar et al. (1992) showed that dispersivities that vary by only about 2–3 orders of magnitude are more reliable. Another important finding of Gelhar et al. (1992) was that, when the field data are classified according to porous versus fractured media, there does not appear to be any significant difference between the two types of media; see Figure 20.

As mentioned in Section IV.A, scale-dependent properties of a heterogeneous medium are the signature of fractal properties and/or structure, and therefore scale-dependent dispersion coefficients and dispersivities are also indicative of fractal morphology for FSPM. Scale-dependence of the dispersivities also implies that they depend anomalously on the time. Let us then write

$$\alpha_L \sim t^{\gamma} \tag{54}$$

where $\gamma = 1/2$ for diffusive dispersion. Generally speaking, there have been three theoretical approaches for investigating non-local dispersion in FSPM and deriving exact or approximate relations for the exponent γ, and what follows is a brief discussion of each approach.

B. Stochastic Continuum Models of Non-local Dispersion

These models were developed to take into account the effect of the uncertainties in the measured dispersivities and other important quantities, and also the effect of the spatial variations of the permeabilities. A typical example of this class of models is the stochastic-spectral model of Gelhar et al. (1979) and Gelhar and Axness (1983), which we discuss below. In their original form, these models were not intended for investigating dispersion in fractal FSPM, since at the time of their development the notion of fractal geometry had not even been introduced into the science of porous media. The assumption of ergodicity is implicit in all such stochastic approaches. That is, one assumes that dispersion of a solute in an ensemble of porous media with the assigned statistical properties mimics the phenomenon in a real field, which is a single realization of FSPM with large-scale variations of permeability, porosity, and other properties. This assumption is valid if the length scale of the flow system is large compared with the correlation length of the system.

Let us briefly discuss the work of Gelhar et al. (1979) for a 2D porous medium (for its extension to 3D see Gelhar and Axness, 1983). The starting point is a CDE at the local level

$$\frac{\partial C}{\partial t} + \frac{\partial}{\partial x}(vC) = \frac{\partial}{\partial x}\left(D_L \frac{\partial C}{\partial x}\right) + \frac{\partial}{\partial z}\left(D_T \frac{\partial C}{\partial z}\right) \tag{55}$$

where it is assumed that the *local* C, D_L, D_T, and K are random processes with

$$C(x, z, t) = C_m(x, t) + c(x, z, t) \tag{56}$$

$$v = v_m + u \tag{57}$$

$$D_L = D_{Lm} + d_L \tag{58}$$

$$D_T = D_{Tm} + d_T \tag{59}$$

$$K = K_m + k \tag{60}$$

Here subscript m denotes a mean value, e.g., $C_m(x, t) = \langle C(x, z, t)\rangle$, with the averaging being taken with respect to the vertical depth $z.c(x, z, t)$, d_L, d_T, and k are *fluctuations* such that their mean values are zero, and are assumed to be random processes which, similar to the fBm and fGn, are represented by their spectral densities. One substitutes Eqs. (56)–(59) into (55) and averages (55) with respect to z. The resulting equation is then subtracted from (55), and a new coordinate, $\zeta = x - v_m t$, is introduced. It is then assumed that the perturbation u is small, so that the second-order terms, such as $\partial(uc - \langle uc\rangle)/\partial\zeta$, can be neglected in the resulting equation. Moreover, Gelhar et al. (1979) also assumed that the dispersivity α_L is proportional to $\sqrt{K}$ (which is intuitively clear, as $\sqrt{K}$ and α_L both represent some sort of a length scale), and that $d_L/D_{Lm} = 3k/(2K_m)$, which the experimental data indicate to be roughly correct. Then, with $\alpha_L = D_{Lm}/v_m$, $a_T = D_{Tm}/v_m$, $\beta_T = a_T v_m \omega^2$, and

$$A = \int_{-\infty}^{\infty} \frac{S_k(\omega)}{k_m^2} \frac{1 - e^{-\beta_T t}}{a_T \omega^2} d\omega, \quad B = \int_{-\infty}^{\infty} \frac{S_k(\omega)}{k_m^2} \frac{1 - e^{-\beta_T t}}{a_T^2 \omega^4} d\omega \tag{61}$$

one finally obtains the following equation

$$\frac{\partial C_m}{\partial t} = (A + a_L)v_m \frac{\partial^2 C_m}{\partial\zeta^2} - B\frac{\partial^3 C_m}{\partial\zeta^2 \partial t} - 3a_L A v_m \frac{\partial^3 C_m}{\partial\zeta^3} + 3a_L B \frac{\partial^4 C_m}{\partial\zeta^3 \partial t} + \left(a_L B v_m + \frac{9}{4}a_L^2 A v_m\right)\frac{\partial^4 C_m}{\partial\zeta^4} + \ldots \tag{62}$$

where $S_k(\omega)$ is the spectral density of the permeability perturbations k. Equation (62) indicates that the average concentration C_m is not governed by a CDE. One now specifies a spectral density $S_k(\omega)$ and calculates A and B. Having determined A and B, one proceeds to analyze Eq. (62).

The quantity A is essentially a large scale or megadispersivity. If the permeabilities obey fractal statistics with a given spectral density $S_k(\omega)$, one can determine the megadispersivity A and study its properties. For example, suppose that the permeability perturbations k are represented by a fBm so that $S_k(\omega)/k_m^2 = a_0/\omega^{2H+1}$. Then

$$A = \frac{2a_0}{a_T} \int_0^\infty \frac{1 - e^{-a_T v \omega^2 t}}{\omega^{2H+3}} d\omega \tag{63}$$

and hence Eq. (63) may be used to investigate the time-dependence of A. However, note that (63) predicts that A becomes independent of t as $t \to \infty$, which is possible only if there is a cutoff length scale $\ell_{co} = 1/\sqrt{\omega_{co}}$ [see Eq. (8)], such that beyond ℓ_{co} the fractality of the permeability distribution is lost. Otherwise, for FSPM which are fractal at *any* length scale, A will never be independent of t. Despite this contradiction, Lemblowski and Wen (1993) and Zhan and Wheatcraft (1996) have carried out extensive analyses of A using Eqs. (7) and (63). The contradiction lies in the assumption that the length scale of the flow system is large compared with the correlation length of the system, whereas with a fBm or a fGn the correlation length is infinite.

Koch and Brady (1988) also studied dispersion in FSPM using more sophisticated averaging techniques. They were able to show that if the correlation length ξ_k of the permeability distribution is finite, then dispersion is diffusive; otherwise dispersion is non-local. They also showed that if the correlation function for $\log K$ varies as $C(r) \sim r^{-\Omega}$, then

$$\gamma = \max\left\{\frac{1}{2}, 1 + \frac{\Omega}{2}, 1 + \frac{\zeta\Omega}{2}\right\} \tag{64}$$

where ζ is the exponent that characterizes the long distance cutoff length scale which one needs if $\Omega < 0$ (otherwise the correlation function diverges as $r \to \infty$). Equation (64) was also derived by Glimm and Sharp (1991) and Zhang and Glimm (1992).

C. Fractal Models of Non-local Dispersion

A more sophisticated stochastic continuum approach to dispersion was developed by Hewett (1986), Arya et al. (1988), Philip (1986), Ababou and Gelhar (1990), and Neuman (1990), using a fractal representation of the permeability and/or porosity distribution. Hewett (1986) analyzed dispersion in FSPM assuming, on the basis of his analysis of field data, that the porosity logs can be represented by a fGn. In his work, the dispersivity α_L is predicted (implicitly) to vary with the time as

$$\alpha_L \sim t^{2H-1} \tag{65}$$

so that one must have $H > 0.5$. A similar result was obtained by Philip (1986). However, Philip's work also predicted that at short times, $\alpha_L \sim t$. Moreover, the constraint $0.5 < H < 1$ was proposed, which is consistent with Hewett's analysis of porosity logs. Arya et al. (1988) developed a model of dispersion in FSPM in which

$$\alpha_L \sim \langle x \rangle^{H/(2-H)} \tag{66}$$

where $\langle x \rangle$ is the average displacement of the center of mass of the solutes. Ababou and Gelhar (1990) developed stochastic continuum equations of transport based on the earlier work of Gelhar et al. (1979) discussed above, except that they assumed that $\mathcal{S}_k(\omega)$ is a fGn, and also proposed that $H \simeq 0$. Ababou and Gelhar (1990) did propose that the dispersivity α_L varies according to Eq. (65). Neuman (1990) assumed that the permeabilities of FSPM are distributed according to a fBm, and coupled it to a quasi-linear theory of dispersion. He then proposed that Eq. (65) should describe the variation of the dispersivity with time.

Zhang (1992) developed a multi-scale or multi-fractal theory for dispersion in FSPM. Starting from a linear transport equation

$$\frac{\partial S}{\partial t} + \mathbf{v} \cdot \nabla S = 0 \tag{67}$$

where S is the saturation of the fluid, he developed an equation for the ensemble average $\langle S \rangle$ of the saturation

$$\frac{\partial \langle S \rangle}{\partial t} + \mathbf{v}_0 \cdot \nabla \langle S \rangle = \nabla \cdot \mathbf{D} \cdot \nabla \langle S \rangle + \cdots \tag{68}$$

where $\mathbf{D}$ is a diffusion matrix and $\mathbf{v}_0$ is the ensemble-average flow velocity. Assuming that the velocity field is a stochastic process with a velocity correlation function $\mathcal{C}_v(r)$, Zhang derived the following equation for the exponent γ

$$\gamma(\ell) \frac{1}{2} \left[1 - \frac{\int_0^\ell \xi \mathcal{C}_v(\xi) \mathrm{d}\xi}{\ell \int_0^\ell \mathcal{C}_v(\xi) \mathrm{d}\xi} \right]^{-1} \tag{69}$$

so that (i) γ depends on a length scale ℓ, and therefore (ii) γ depends in some sense on the complete history of the dispersion process up to the length scale ℓ. However, there is no experimental evidence that γ actually changes *continuously* with ℓ. Moreover, if $\varpi = \mathrm{d} \ln \mathcal{C}_v(r) / \mathrm{d} \ln r$, and $\varpi_\infty = \lim_{r \to \infty} \varpi(r)$ exists, then Zhang proposed that

$$\gamma_\infty = \max\left\{ \frac{1}{2}, 1 + \frac{\varpi_\infty}{2} \right\} \tag{70}$$

provided that $\varpi_\infty \geq -1$. Finally, Zhang (1992) derived the lower bound, $\gamma \geq 1/2$. Similar results were derived by Glimm et al. (1993).

D. Percolation Model of Non-local Dispersion

We have already discussed in Sections VI.A and VI.B the concept of the critical path method according to which FSPM with broadly distributed heterogeneities (permeability and porosity) are reduced to equivalent percolation systems. This suggests the third approach to modeling dispersion in FSPM, namely the notion that the flow-carrying part of FSPM is essentially the backbone of a percolation cluster which, because of the long-range correlations in the permeabilities and porosities, will also contain long-range correlations. The properties of this percolation model have already been discussed in Section IV.B. In this section we discuss its application to predicting the scale- and time-dependence of the dispersivity α_L which are described by Eqs. (53) and (54). Dispersion in percolation porous media was first studied by Sahimi et al. (1983a, 1986) and Sahimi and Imdakm (1988); see also Koplik et al. (1988). Experimental investigations of dispersion in percolation porous media were carried out by Charlaix et al. (1988), Hulin et al. (1988), and Gist et al. (1990). The percolation model of dispersion in FSPM that we discuss below was first proposed by Sahimi (1993c, 1995a).

We have already discussed the scaling behavior of the permeability K and diffusivity D of a porous medium near its p_c. Similar to fractal diffusion in porous media, discussed in Section VII, let us define a random walk fractal dimension D_w for the dispersion process. Assuming that the megascopic flow is in the x direction, we define D_w by

$$\langle \Delta x^2 \rangle \sim t^{2/D_w} \tag{71}$$

where $\langle \Delta x^2 \rangle = \langle (x - \langle x \rangle)^2 \rangle = \langle x^2 \rangle - \langle x \rangle^2$. The longitudinal dispersion coefficient D_L is then given by

$$D_L = \frac{1}{2}\frac{\mathrm{d}\langle \Delta x^2 \rangle}{\mathrm{d}t} \tag{72}$$

If dispersion is diffusive, then $D_w = 2$ and the description of dispersion by a CDE is adequate. However, as discussed in Section VII, we may also have $D_w > 2$, in which $\langle \Delta x^2 \rangle$ grows with time more slowly than linearly, and $D_L \to 0$ as $t \to \infty$, or $D_w < 2$, in which case $\langle \Delta x^2 \rangle$ grows with time faster than linearly, and $D_L \to \infty$ as $t \to \infty$, so that in this case dispersion is superdiffusive. In both cases dispersion is not described by a CDE.

For dispersion in a porous medium two average velocities may be defined. If the solute spends a considerable amount of time in the dead-end pores, then the average solute velocity should be defined in terms of the total travel time in the pore space (dead-end pores plus the backbone). This can happen only if the solute diffuses into the dead-end pores and spends a considerable amount of time in there. This, however, is unlikely (unless the porous medium contains a large number of fractures, many of which are dead-ended) since, as we discussed above, molecular diffusion plays no significant role in dispersion in FSPM. Instead, the main contributing factor to dispersion in FSPM is the fluctuations in the flow velocity field, arising as the result of large-scale variations in the permeability and porosity. Thus, the average velocity of the solute tracers should be defined only in terms of the travel times in the backbone (since the flow velocity field is non-vanishing only in the backbone). Since $K \sim vX_B$, we obtain, using Eqs. (20) and (23)

$$v \sim (p - p_c)^{e-\beta_b} \sim \xi_p^{-\theta_b} \tag{73}$$

where $\theta_b = (e - \beta_b)/\nu$.

Because dispersion in FSPM is caused by the variations in the flow velocities, D_L and v are linearly related, $D_L \sim \alpha_L v$. It is not unreasonable to replace α_L by ξ_p, the percolation correlation lengths, as ξ_p is the dominant length scale in the system. Therefore, we obtain

$$D_L \sim \xi_p v \sim \xi_p^{1-\theta_b} \sim (p - p_c)^{e-\beta_b-\nu} \tag{74}$$

For $L \ll \xi_p$, we replace ξ_p with L to obtain $D_L \sim L^{1-\theta_b}$, and since $\alpha_L = D_L/v$, we obtain

$$\alpha_L \sim L \tag{75}$$

which, when compared with Eq. (53), implies that $\delta = 1$, in reasonable agreement with the empirical result of Arya et al. (1988), $\delta \simeq 0.88$, and of Neuman (1990), $\delta \simeq 0.92$. On the other hand, Eq. (72) tells that $D_L \sim \mathrm{d}\langle \Delta x^2 \rangle/\mathrm{d}t \sim L^{1-\theta_b} \sim \langle \Delta x^2 \rangle^{(1-\theta_b)/2}$. A simple integration then yields $\langle \Delta x^2 \rangle \sim t^{2/(1+\theta_b)}$, implying that, for dispersion in FSPM,

$$D_w = 1 + \theta_b \tag{76}$$

Equation (72) also implies that

$$D_L \sim t^{(1-\theta_b)/(1+\theta_b)} \tag{77}$$

Since $\theta_b < 1$, (77) tells us that D_L increases with t, and thus dispersion in FSPM is superdiffusive. Equation (73) tells us that for $L \ll \xi_p$, i.e., the fractal regime, we may write $v \sim L^{-\theta_b} \sim \langle \Delta x \rangle^{-\theta_b}$. Since $v = \mathrm{d}\langle \Delta x \rangle/\mathrm{d}t$, we obtain $\mathrm{d}\langle \Delta x \rangle/\mathrm{d}t \sim \langle \Delta x \rangle^{-\theta_b}$, implying that $\langle \Delta x \rangle \sim t^{1/(1+\theta_b)}$, and therefore

$$v \sim t^{-\theta_b/(1+\theta_b)} \tag{78}$$

Equation (78) is interesting in that it tells us that the mean flow velocity in fractal FSPM *decays* with the time. Most previous investigations have assumed that v is independent of time, whereas flow and transport properties in fractal media are scale- and time-dependent quantities. While the background flow may be steady, and its average velocity a constant, the *solute* velocity is a scale- and time-dependent quantity. Having obtained the scaling relations for D_L and v, we finally obtain the time-dependence of the dispersivity:

$$\alpha_L \sim t^{1/(1+\theta_b)} \tag{79}$$

which, when compared with Eq. (54), implies that

$$\gamma = \frac{1}{1+\theta_b} \tag{80}$$

Equation (80) indicates that, since $\theta_b < 1$, for superdiffusive dispersion we have $\gamma > 1/2$, consistent with Zhang's (1992) prediction that $\gamma \geq 1/2$.

Thus, one advantage of the percolation model is that it expresses γ in terms of the well-defined percolation exponents e, β_b and ν. These exponents are non-universal when long-range correlations are present in the system (see Section IV.B), and therefore we also obtain non-universal values for γ for dispersion in FSPM, consistent with the field observations.

IX. CONCLUSIONS

There is now considerable evidence that the porosity logs and permeability distributions of many field-scale porous media follow fractal distributions and, in particular, fractional Brownian motion, fractional Gaussian noise, and fractional Lévy motion. The fracture network of rock at large length scales may be similar to the sample-spanning percolation cluster at the percolation threshold, a well defined and well studied fractal structure. Thus, fractal and percolation concepts provide quantitative tools for characterizing FSPM. Moreover, they help us gain a much deeper understanding of flow and dispersion in FSPM, and provide rational explanations for much of the field data.

In this paper, we have only discussed application of fractals and percolation to single-phase flow and dispersion in porous media. For extensive discussions of application of these concepts to miscible displacements and multiphase flow in porous media, the interested reader is referred to Sahimi (1993b, 1995b).

ACKNOWLEDGMENTS

My own work discussed here is the result of collaboration with many of my past students and colleagues. I would like to thank them all. Various portions of my work discussed here have been supported by the National Science Foundation and the Department of Energy. The preparation of this paper was partially supported by the Petroleum Research Fund, administered by the American Chemical Society.

NOMENCLATURE

Roman Letters

B_H fractional Brownian motion
b jump base in Lévy flights
C solute concentration
$\mathcal{C}(r)$ correlation function
C_L scale factor for Lévy flights
d Euclidean dimensionality
d_j wavelet-detail coefficient
D diffusivity
D_b fractal dimension of the percolation backbone
D_f fractal dimension
D_L longitudinal dispersion coefficient
D_T transverse dispersion coefficient
D_w fractal dimension of random walk
e permeability exponent
f conductance or permeability distribution
G conductivity of porous media
g_c critical conductance
g_f fluid (brine) conductivity
H Hurst exponent
h height of a rough surface
K effective permeability of porous media
L length scale
ℓ_c critical length scale
N number of spheres for covering a system
n_j number of jumps during a Lévy flight
P probability of jumps during a Lévy flight
p fraction of open bonds or sites
p_c percolation threshold
R random walk displacement
S fluid saturation
$\mathcal{S}$ spectral density
t time
v flow velocity
X_A accessible fraction
X_B backbone fraction
Z coordination number

Greek Letters

α fractal dimension of Lévy flights
α_L longitudinal dispersivity
α_T transverse dispersivity
β critical exponent of the accessible fraction
β_b critical exponent of the backbone fraction
γ longitudinal dispersivity exponent (scaling with the time)
δ longitudinal dispersivity exponent (scaling with the length scale)
ε porosity
θ diffusivity exponent
ν percolation correlation length exponent
ξ_p percolation correlation length
χ conductivity critical exponent
ψ wavelet function
ω Fourier component

REFERENCES

Aasum, Y, Kelkar MG, Gupta SP. An application of geostatistics and fractal geometry for reservoir characterization. SPE Form Eval 6:11–19, 1991.

Ababou R, Gelhar LW. Self-similar randomness and spectral conditioning: Analysis of scale effects in subsurface hydrology. In: Cushman JH, ed. Dynamics of Fluids in Hierarchical Porous Media. San Diego: Academic Press, 1990, pp 393–428.

Aki K. A probabilistic synthesis of precursory phenomena. In: Simpson DW, Richards PG, eds. Earthquake Prediction: An Introduction Review. Washington: American Geophysical Union, 1981, pp 566–574.

Ambegaokar V, Halperin BI, Langer S. Hopping conductivity in disordered systems. Phys Rev B 4:2612–2618, 1971.

Andrews DJ. A stochastic fault model. 1. Static case. J Geophys Res 85:3867–3877, 1980.

Arya A, Hewett TA, Larson RG, Lake LW. Dispersion and reservoir heterogeneity. SPE Reservoir Eng 3:139–148, 1988.

Aviles CA, Scholz CH, Boatwright J. Fractal analysis applied to characteristic segments of the San Andreas fault. J Geophys Res 92:331–344, 1987.

Avnir D, ed. The Fractal Approach to Heterogeneous Chemistry. New York: Wiley, 1990.

Avnir D, Biham O, Lidar D, Malcai O. Is the geometry of nature fractal? Science 279:39–40, 1998.

Balberg I, Berkowitz B, Drachsler GE. Application of a percolation model to flow in fractured hard rocks. J Geophys Res 96:10015–10021, 1991.

Barton CC. Fractal analysis of the spatial clustering of fractures. In: Barton CC, LaPointe PR, eds. Fractals and their Use in the Earth Sciences. Las Vegas: Geological Society of America, 1992, pp 126–154.

Barton CC, Hsieh PA. Physical and Hydrological-flow Properties of Fractures, Guidebook T385. Las Vegas: American Geophysical Union, 1989.

Barton CC, Larsen E. Fractal geometry of two-dimensional fracture networks at Yucca Mountain, southwest Nevada. In: Stephenson O, ed. Proceedings of the International Symposium on Fundamentals of Rock Joints, Bjorkliden, Sweden. Lulea, Sweden: Centek Publishers, 1985, pp 74–84.

Barton C, Schutter TA, Page WR, Samuel JK. Computer generation of fracture networks for hydrologic-flow modeling. Trans Am Geophys Union 68:1295, 1987.

Bejan A, Errera MR. Geometry of minimal flow resistance between a volume and one point. Fractals 5:685–695, 1997.

Bejan A, Tondeur D. Equipartion, optimal allocation, and the constructal approach to predicting organization in nature. Rev Gén Therm 37:165–180, 1998.

Bejan A, Ikegami Y, Ledezma GA. Constructural theory of natural crack pattern formation for fastest cooling. Int J Heat Mass Transfer 41:1945–1954, 1998.

Berkowitz B. Analysis of fracture network connectivity using percolation theory. Math Geol 27:467–483, 1995.

Bour O, Davy P. Connectivity of random fault networks following a power-law fault length distribution. Water Resour Res 33:1567–1583, 1997.

Bour O, Davy P. On the connectivity of three-dimensional fault networks. Water Resour Res 34:2611–2622, 1998.

Brown SR. A note on the description of surface roughness using fractal dimensions. Geophys Res Lett 14:1095–1098, 1987.

Brown SR, Scholz CH. Broad bandwidth study of the topography of natural rock surfaces. J Geophys Res 90B:12575–12583, 1985.

Charlaix E, Guyon E, Roux S. Permeability of a random array of fractures of widely varying apertures. Transp Porous Media 2:31–42, 1987.

Charlaix E,. Hulin JP, Plona TJ. Experimental study of tracer dispersion in flow through two-dimensional networks of etched capillaries. J Phys D 21:1727–1732, 1988.

Cheldize T, Gueguen Y. Evidence of fractal fracture. Int J Rock Mech Min Sci Geomech Abstr 27:223–225, 1990.

Cox BL, Wang JSY. Fractals 1:87–110, 1993.

Crane SE, Tubman KM. Reservoir variability and modeling with fractals. SPE Paper 20606, New Orleans, LA, 1990.

Di Pietro L. Application of a lattice-gas numerical algorithm to modelling water transport in fractured porous media. Transp Porous Media 22:307–325, 1996.

Emanuel AS, Alameda GD, Behrens RA, Hewett TA. Reservoir performance prediction methods based on fractal geostatistics. SPE Reservoir Eng 4:311–318, 1989.

Englman R, Gur Y, Jaeger Z. Fluid flow through a crack network in rocks. J Appl Mech 50:707–711, 1983.

Errera MR, Bejan A. Deterministic tree networks for river drainage basins. Fractals 6:245–261, 1998.

Flandrin P. Wavelet analysis and synthesis of fractional Brownian motion. IEEE Trans Inf Theory 38:910–917, 1992.

Gefen Y, Aharony A, Alexander S. Anomalous diffusion on percolation clusters. Phys Rev Lett 50:77–80, 1983.

Gelhar LW, Axness CL. Three-dimensional stochastic analysis of macrodispersion in aquifers. Water Resour Res 19:161–180, 1983.

Gelhar LW, Gutjahr AL, Naff RL. Stochastic analysis of macrodispersion in a stratified aquifer. Water Resour Res. 15:1387–1397, 1979.

Gelhar LW, Welty C, Rehfeldt KR. A critical review of data on field-scale dispersion in aquifers. Water Resour Res 28:1955–1974, 1992.

Gist GA, Thompson AH, Katz AJ, Higgins RL. Hydrodynamic dispersion and pore geometry in consolidated rock. Phys Fluids A 2:1533–1544, 1990.

Glimm J, Sharp DH. A random field model for anomalous diffusion in heterogeneous porous media. J Stat Phys 62:415–424, 1991.

Glimm J, Lindquist WB, Pereira F, Zhang Q. A theory of macrodispersion for the scale-up problem. Transp Porous Media 13:97–122, 1993.

Goggin DJ, Chandler MA, Kocurek G, Lake LW. Permeability transects of Eolian sands and their use in generating random permeability fields. SPE Form Eval 7:7–16, 1992.

Gutfraind R, Hansen A. Study of fracture permeability using lattice gas automata. Transp Porous Media 18:131–149, 1995.

Hardy HH. The generation of reservoir property distributions in cross section for reservoir simulation based on core and outcrop photos. SPE Paper 23968, Midland, TX, 1992.

Herweijer JC, Dubrule ORF. Screening of geostatistical reservoir models with pressure transients. J Pet Tech 47:973–979, 1995.

Hestir K, Long JCS. Analytical expressions for the permeability of random two-dimensional Poisson fracture networks based on regular lattice percolation and equivalent media theories. J. Geophys Res B95:21565–21581, 1990.

Hewett TA. Fractal distributions of reservoir heterogeneity and their influence on fluid transport. SPE Paper 15386, New Orleans, LA, 1986.

Hewett TA, Behrens RA. Conditional simulation of reservoir heterogeneity with fractals. SPE Form Eval 5:217–225, 1990.

Hughes BD. Random Walks and Random Environments. London: Oxford University Press, 1995, vol 1.

Hulin JP, Charlaix E. Plona TJ, Oger L. Guyon E. Tracer dispersion in sintered glass beads with a bidisperse size distribution. AIChE J 34:610–617, 1988.

Hurst HE. Long-term storage capacity of reservoirs. Trans Am Soc Civil Eng 116:770–808, 1951.

Huseby O, Thovert JF, Adler PM. Geometry and topology of fracture systems. J Phys A 30:1415–1444, 1997.

Jerauld GR, Scriven LE, Davis HT. Percolation and conduction on the 3D Voronoi and regular networks: a second case study in topological disorder. J Phys C 17:3429–3439, 1984.

Katz AJ, Thompson AH. Quantitative prediction of permeability in porous rock. Phys Rev B 34:8179–8181, 1986.

Katz AJ, Thompson AH. Prediction of rock electrical conductivity from mercury injection measurements. J Geophys Res B92:599–607, 1987.

Kemblowski MW, Wen JC. Contaminant spreading in stratified soils with fractal permeability distribution. Water Resour Res 29:419–425, 1993.

King GCP. The accommodation of large strains in the upper lithosphere of the earth and other solids by self-similar fault systems: The geometrical origin of *b*-value. Pure Appl Geophys 121:761–815, 1984.

Kirkpatrick S. Models of disorder. In: Balian R, Maynard R, Toulouse G, eds. Ill-Condensed Matter. Amsterdam: North Holland, 1979, pp 321–403.

Knackstedt MA, Ninham BW, Monduzzi M. Diffusion in model disordered media. Phys Rev Lett 75:653–656, 1995.

Knackstedt MA, Sheppard AP, Pinczewski WV. Simulation of mercury porosimetry on correlated grids: Evidence for extended correlated heterogeneity at the pore scale in rocks. Phys Rev E 58:R6923–R6926, 1998.

Koch DL, Brady JF. Anomalous diffusion in heterogeneous porous media. Phys Fluids 31: 965–973, 1988.

Kogut PM, Straley JP. Distribution induced non-universality of the percolation conductivity exponents. J Phys C 12:2151–2159, 1979.

Koplik J, Redner S, Wilkinson D. Transport and dispersion in random networks with percolation disorder. Phys Rev A 37:2619–2636, 1988.

LaPointe PR. A method to characterize fracture density and connectivity through fractal geometry. Int J Rock Mech Sci Geomech Abstr 25:421–429, 1988.

Lin C, Cohen MH. Quantitative methods for microgeometric modeling. J Appl Phys 53:4152–4165, 1982.

Mandelbrot BB. The Fractal Geometry of Nature. San Francisco: Freeman, 1982.

Mandelbrot BB. Self-affine fractals and fractal dimensions. Phys Script 32:257–260, 1985.

Mandelbrot BB, Van Ness JW. Fractional Brownian motion, fractional Gaussian noise, and their applications. SIAM Rev 10:422–437, 1968.

Mathews JL, Emanuel AS, Edwards KA. Fractal methods improve Mitsue miscible predictions. J Pet Tech 41:1136–1142, 1989.

Meakin P. In: Domb C, Lebowitz JL, eds. Phase Transitions and Critical Phenomena. London: Academic Press, 1988, pp 335–415.

Mehrabi AR, Rassamdana H, Sahimi M. Characterization of long-range correlations in complex distributions and profiles. Phys Rev E 56:712–722, 1997.

Mohanty KK, PhD dissertation, University of Minnesota, Minneapolis, MN, 1981.

Molz FJ, Güven, O, Melville JG. An examination of scale-dependent dispersion coefficients. Ground Water 21:715–725, 1983.

Moreno L, Tsang CF. Flow channeling in strongly heterogeneous porous media: A numerical study. Water Resour Res 30:1421–1430, 1994.

Mourzenko VV, Thovert JF, Adler PM. Geometry of simulated fractures. Phys Rev E 53:5606–5626, 1996.

Mourzenko VV, Thovert JF, Adler PM. Percolation and conductivity of self-affine fractures. Phys Rev E 59:4265–4284, 1999.
Mukhopadhyay S, Sahimi M. Heat transfer and two-phase flow in fractured reservoirs. SPE Paper 24043, Bakersfield, CA, 1992.
Neuman SP. Universal scaling of hydraulic conductivities and dispersivities in geologic media. Water Resour Res 26:1749–1758, 1990.
Neuman SP. Generalized scaling of permeabilities: validation and effect of support scale. Geophys Res Lett 21:349–352, 1994.
Nolen-Hoeksema RC, Gordon RB. Optical detection of crack patterns in the opening-mode fracture of marble. Int J Rock Mech Min Sci Geomech Abstr 24:135–144, 1987.
Odling NE. Natural fracture profiles, fractal dimension and joint roughness coefficients. Rock Mech Rock Eng 27:135–153, 1994.
Okubo PG, Aki K. Fractal geometry in the San Andreas fault system. J Geophys Res B92:345–355, 1987.
Painter S. Evidence for non-gaussian scaling behavior in heterogeneous sedimentary formations. Water Resour Res 32:1183–1195, 1996.
Painter S, Paterson L. Fractional Lévy motion as a model for spatial variability in sedimentary rock. Geophys Res Lett 21:2857–2860, 1994.
Painter S, Beresford G, Paterson L. On the distribution of seismic amplitudes and seismic reflection coefficients. Geophysics 60:1187–1194, 1995.
Philip JR. Issues in flow and transport in heterogeneous porous media. Transp Porous Media 1:319–338, 1986.
Pickens JF, Grisak GE. Scale-dependent dispersion in a stratified granular aquifer. Water Resour Res 17:1191–1211, 1981.
Robinson PC. Numerical calculations of critical densities for lines and planes. J Phys A 17:2823–2830, 1984.
Rothman DH, Zaleski S. Lattice-gas Cellular Automata. London: Cambridge University Press, 1997.
Sahimi M. Hydrodynamic dispersion near the percolation threshold: scaling and probability densities. J Phys A 20:L1293–L1298, 1987.
Sahimi M. Transport, reaction and fragmentation in evolving porous media. Phys Rev A 43:5367–5376, 1991.
Sahimi M. Fractal concepts in chemistry: Chemtech 22:687–693, 1992a.
Sahimi M. Brittle fracture in disordered media: from reservoir rocks to composite solids. Physica A 186:160–182, 1992b.
Sahimi M. Nonlinear transport processes in disordered media. AIChE J 39:369–386, 1993a.
Sahimi M. Flow phenomena in rocks: from continuum models to fractals, percolation, cellular automata and simulated annealing. Rev Mod Phys 65:1393–1534, 1993b.
Sahimi M. Fractal and superdiffusive transport and hydrodynamic dispersion in heterogeneous porous media. Transp Porous Media 13:3–40, 1993c.
Sahimi M. Applications of Percolation Theory. London: Taylor & Francis, 1994.

Sahimi M. Effect of long-range correlations on transport phenomena in disordered media. AIChE J 41:229–240, 1995a.
Sahimi M. Flow and Transport in Porous Media and Fractured Rock. Weinheim: VCH, 1995b.
Sahimi M. Non-linear and non-local transport processes in heterogeneous media: from long-range correlated percolation to fracture and materials breakdown. Phys Rep 306:213–395, 1998.
Sahimi M, Arbabi S. Percolation and fracture in disordered solids and granular media: approach to a fixed point. Phys Rev Lett 68:608–611, 1992.
Sahimi M, Arbabi S. Scaling laws for fracture of heterogeneous materials and rock. Phys Rev Lett 77: 3689–3692, 1996.
Sahimi M, Goddard JD. Elastic percolation models for cohesive mechanical failure in heterogeneous systems. Phys Rev B 33:7848–7851, 1986.
Sahimi M, Imdakm AO. The effect of morphological disorder on hydrodynamic dispersion in flow through porous media. J Phys A 21:3833–3870, 1988.
Sahimi M, Mehrabi AR. Report to NIOC, unpublished, 1995.
Sahimi M, Mukhopadhyay S. Scaling properties of a percolation model with long-range correlations. Phys Rev E 54:3870–3880, 1996.
Sahimi M, Tsotsis TT. Transient diffusion and conduction in heterogeneous media: beyond the classical effective-medium approximation. Ind Eng Chem Res 36: 3043–3052, 1997.
Sahimi M, Yortsos Y. Application of fractal geometry to porous media: a review. SPE Paper 20476, New Orleans, LA, 1990.
Sahimi M, Davis HT, Scriven LE. Dispersion in disordered porous media. Chem Eng Commun 23:329–341, 1983a.
Sahimi M, Hughes BD, Scriven LE, Davis HT. Stochastic transport in disordered systems. J Chem Phys 78:6849–6864, 1983b.
Sahimi M, Hughes BD, Scriven LE, Davis HT. Dispersion in flow through porous media. I. One-phase flow. Chem Eng Sci 41:2103–2122, 1986.
Sahimi M, Gavalas GR, Tsotsis TT. Statistical and continuum models of fluid–solid reactions in porous media. Chem Eng Sci 45:1443–1502, 1990.
Sahimi M, Robertson MC, Sammis CG. Fractal distribution of earthquake hypocenters and its relation with fault patterns and percolation. Phys Rev Lett 70:2186–2189, 1993.
Sahimi M, Rassamdana H, Mehrabi AR. Fractals in porous media: from pore to field scale. MRS Proc 367:203–214, 1995.
Schmittbuhl J, Gentier S, Roux S. Field measurements of the roughness of fault surfaces. Geophys Res Lett 20:639–641, 1993.
Shante VKS. Hopping conduction in quasi-one-dimensional disordered compounds. Phys Rev B 16:2597–2612, 1977.
Shlesinger MF. Fractal time in condensed matter. Annu Rev Phys Chem 39:269–290, 1988.
Stauffer D, Aharony A. Introduction to Percolation Theory, 2nd ed. London: Taylor & Francis, 1992.

Sudicky EA, Cherry JA. Field observations of tracer dispersion under natural flow conditions in an unconfined sandy aquifer. Water Pollution Res Can 14:1–17, 1981.

Sudicky EA, Cherry JA, Frind EO. Migration of contaminants in groundwater at a landfill: A cast study. 4. A natural-gradient dispersion test. J Hydrol 63:81–108, 1983.

Sudicky EA, Gilham RW, Frind EO. Experimental investigation of solute transport in stratified porous media. 1. The non-reactive case. Water Resour Res 21:1035–1042, 1985.

Taggart IJ, Salisch HA. Fractal geometry, reservoir characterization, and oil recovery. APEA J 31:377–385, 1991.

Taqqu MS. Random processes with long-range dependence and high variability. J Geophys Res D92:9683–9689, 1987.

Tchalenko JS. Similarities between shear zones of different magnitudes. Geol Soc Am Bull 81:1625–1640, 1970.

Voss RF. Random fractal forgeries. In: Earnshaw RA, ed. Fundamental Algorithms for Computer Graphics. NATO ASI Series, vol. 17. Heidelberg: Springer-Verlag, 1985, pp 805–835.

Watanabe K, Takahashi H. Fractal geometry characterization of geothermal reservoir fracture network. J Geophys Res 100:521–528, 1995.

Witten TA, Sander LM. Diffusion-limited aggregation, a kinetic critical phenomenon. Phys Rev Lett 47:1400–1403, 1981.

Zhan H, Wheatcraft S. Macrodispersivity tensor for nonreactive solute transport in isotropic and anisotropic fractal porous media: Analytical solutions. Water Resour Res 32:3461–3474, 1996.

Zhang Q. A multi-length-scale theory of the anomalous mixing-length growth for tracer flow in heterogeneous porous media. J Stat Phys 66:485–501, 1992.

Zhang Q, Glimm J. Inertial range scaling of laminar shear flow as a model of turbulent transport. Commun Math Phys 146:217–229, 1992.

Zhang X, Knackstedt MA, Sahimi M. Fluid flow across fractal volumes and self-affine surfaces. Physica A 233:835–847, 1996.

4
Heat Conduction in Porous Media

Chin-Tsau Hsu
The Hong Kong University of Science and Technology, Kowloon, Hong Kong

I. INTRODUCTION

Heat conduction in materials consisting of substances of different thermal properties has been encountered frequently in many engineering applications. One such example is the manufacture of composite materials where one substance with a distinct material property is added to a matrix substance to enhance the overall property performance. Other examples, involving substances of solid and fluid phases, commonly referred to as porous materials, are the extraction of oil from reservoirs, energy production in geothermal engineering, the disposal of nuclear waste, and the confinement of heat in thermal insulation, to name a few. In this chapter, we will restrict our discussion to heat conduction in porous materials saturated with fluids. It is noted that the results of the present discussion are generally applicable to two-component composite materials and can be extended easily to multi-component materials.

Heat conduction in porous materials is usually described macroscopically by averaging the microscopic heat transfer processes over a representative elementary volume (REV). Traditional treatment of heat conduction in porous materials is based largely on the mixture theory, assuming local thermal equilibrium within the solid and fluid phases, so that the heat transfer processes in the two phases can be lumped into a process described by a single heat conduction equation. The problem then becomes the construction of an appropriate composite model for the effective stagnant thermal conductivity of the mixture. This type of approach can be traced back to the works of Maxwell (1873) and Rayleigh (1892). Continuing efforts in the past forty years include those of Deissler and Boegli (1958), Kunii and Smith (1960), Zehner and Schlunder (1970), Nozad et al. (1985), Sahraoui and Kaviany (1993), and others. Most recently, Hsu et al. (1994, 1995, 1996) provided a more systematic account of the modeling of effective stagnant thermal conductivity. Summaries of the existing models for effective stagnant thermal conductivity were given in the book by Kaviany (1991) and in the review article by Cheng and Hsu (1998).

It has been generally realized that the assumption of local thermal equilibrium may not be applicable to transient heat transfer processes in porous media, especially when the differences in thermal properties of fluid and solid are large. This requires the separation of the macroscopic heat conduction equations for fluid and solid. One then encounters the so-called closure problem in which schemes are required to construct closure relations for new unknowns. Furthermore, the final equations are usually coupled to each other, and their solutions cannot be obtained easily without employing a complicated numerical procedure. For these reasons, there exists little literature on transient heat conduction in porous media. Amiri and Vafai (1994) studied transient convective heat transfer in porous media on the basis of a two-equation model. Quintard and Whitaker (1993) gave a comprehensive review of closure modeling for transient heat conduction. They evaluated the closure coefficients by numerically solving the microscopic temperature fluctuation equations for periodic arrays of particles. More recently, Quintard and Whitaker (1995) discussed the validity of assuming local thermal equilibrium for transient heat conduction in solid–fluid mixtures. They also provided a closure model for interfacial heat transfer based on Chang's (1983) unit cell. Most recently, Hsu (1999) proposed a closure scheme for transient heat transfer in porous media. However, unlike Quintard and Whitaker (1993), Hsu analytically evaluated the closure coefficients of thermal tortuosity and interfacial heat transfer. Hsu and Wu (2000) executed the model of Hsu (1999) for the problem of one-dimensional transient heat conduction to assess the validity of the assumption of local thermal equilibrium.

In this chapter, we shall discuss heat conduction in porous media in a general way by first using a volumetric averaging procedure to obtain the macroscopic heat conduction equations for the solid and fluid phases, respectively. This leads to the closure problem, which requires the construction of closure models for interfacial thermal tortuosity and interfacial heat transfer. Closure relations for thermal tortuosity and interfacial heat transfer are then constructed under a microscopic quasi-steady assumption. The macroscopic transient heat conduction equations are then lumped into a single equation for the solid–fluid mixture under the local thermal equilibrium condition to recapture the concept of effective stagnant thermal conductivity of the mixture. Several models of effective stagnant thermal conductivity are then reviewed. The scheme of Hsu (1999) to determine the closure coefficients for thermal tortuosity and interfacial heat transfer is reviewed. The dependencies of these coefficients on the porosity and the thermal properties of fluids and solids are illustrated. We then follow the approach of Hsu and Wu (2000) to discuss the applicability of the local thermal equilibrium assumption. The limitation of the current macroscopic model of transient heat conduction in porous media and the possible extension of the model to nonisotropic media are discussed.

II. MACROSCOPIC TRANSIENT HEAT CONDUCTION EQUATIONS

Let us assume that porous materials consist of packed solid particles surrounded by fluids, as depicted in Figure 1. For simplicity, but without loss of generality, we should consider the spherical particles to be of uniform size. The diameter d_p of the spheres then represents the microscopic scale of the media. We also assume that d_p is much larger than the typical size of molecules, so that fluid and solid are regarded microscopically as continuous. Hence, the microscopic transient heat conduction equations for the fluid and solid are described by

$$\rho_f C p_f \frac{\partial T_f}{\partial t} = \nabla \cdot (k_f \nabla T_f) \tag{1}$$

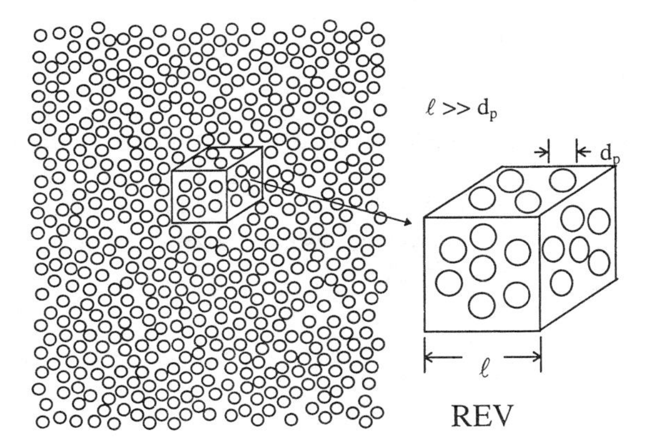

Figure 1. Schematic of porous media and the representative elementary volume (REV)

and

$$\rho_s Cp_s \frac{\partial T_s}{\partial t} = \nabla \cdot (k_s \nabla T_s) \tag{2}$$

where the subscripts f and s refer to fluid and solid, respectively. In (1) and (2), the material properties such as thermal conductivity k_f and k_s, density ρ_f and ρ_s, and thermal capacitance Cp_f and Cp_s, are assumed constant. The proper boundary conditions on the fluid–solid interface A_{fs} are

$$T_f = T_s \qquad \text{on} \qquad A_{fs} \tag{3}$$

and

$$\mathbf{n}_{fs} \cdot k_f \nabla T_f = \mathbf{n}_{fs} \cdot k_s \nabla T_s \qquad \text{on} \qquad A_{fs} \tag{4}$$

where $\mathbf{n}_{fs}$ is the unit vector out normal from fluid to solid.

It is impractical to solve the equation system (1–4) in detail, especially when the number of solid particles is large. In any case, we are more interested in the global characteristics of heat conduction in the porous materials. To this end, we now introduce a representative elementary volume (REV) of size V, as depicted in Figure 1. The volumetric phase-average procedure is then defined as

$$\bar{F}_i = \frac{1}{V_i} \iiint_{V_i} F_i \mathrm{d}V \tag{5}$$

where F_i is a physical quantity under consideration and V_i is the volume of the i-phase in REV, with i representing f or s for fluid or solid, respectively. Hence, $V = V_f + V_s$ and the porosity of the media is defined by $\varepsilon = V_f / V$. The length scale l of REV is presumed to be much larger than d_p but much smaller than the global scale of the domain under consideration. Hence, the phase-averaged quantity $\bar{F}_i$ is regarded as continuous in the global domain.

If (1) and (2) are averaged over REV with respect to fluid and solid phases, the volumetric phase-averaged equations, after invoking the divergence theorem, become

$$\rho_f Cp_f \frac{\partial(\varepsilon \bar{T}_f)}{\partial t} = \bar{\nabla} \cdot \left[k_f \bar{\nabla}(\varepsilon \bar{T}_f)\right] + \bar{\nabla} \cdot (k_f \bar{\Lambda}_{fs}) + Q_{fs} \tag{6}$$

$$\rho_s Cp_s \frac{\partial[(1-\varepsilon)\bar{T}_s]}{\partial t} = \bar{\nabla} \cdot \left[k_s \bar{\nabla}((1-\varepsilon)\bar{T}_s)\right] - \bar{\nabla} \cdot (k_s \bar{\mathbf{\Lambda}}_{fs}) - Q_{fs} \tag{7}$$

where $\bar{\nabla}$ is the gradient operator in the macroscopic coordinate system. In (6) and (7),

$$\bar{\mathbf{\Lambda}}_{fs} = \frac{1}{V}\iint_{A_{fs}} T_f \mathbf{ds} = \frac{1}{V}\iint_{A_{fs}} T_s \mathbf{ds} \tag{8}$$

represents the contribution due to thermal tortuosity, and

$$Q_{fs} = \frac{1}{V}\iint_{A_{fs}} k_f \nabla T_f \cdot \mathbf{ds} = \frac{1}{V}\iint_{A_{fs}} k_s \nabla T_s \cdot \mathbf{ds} \tag{9}$$

that due to interfacial heat transfer. The last equalities in (8) and (9) are evident from the interfacial boundary conditions (3) and (4), since $\mathbf{ds} = \mathbf{n}_{fs} \mathrm{d}A_{fs}$.

In (6) and (7), the time scales for thermal diffusion through fluid and solid in REV of length scale l are l^2/α_f and l^2/α_s, respectively, where $\alpha_f = k_f/\rho_f Cp_f$ is the thermal diffusivity of fluid and $\alpha_s = k_s/\rho_s Cp_s$ that of solid. These time scales for the two phases will be quite different if the thermal diffusivity ratio $\beta(= \alpha_s/\alpha_f)$ is either very small or large. As a result, $\bar{T}_f \neq \bar{T}_s$ locally and the transient heat conduction process in the two-phase system cannot be regarded locally as being in thermal equilibrium. This renders the problem more complicated since (6) and (7) have to be solved simultaneously. Equations (6) and (7) are the macroscopic equations which describe the local thermal non-equilibrium transient heat conduction in porous materials.

The term on the left-hand side of (6) represents the rate of thermal energy stored in REV. On the right-hand side of (6), the first term represents the rate of heat entering the fluid phase through the REV boundary by conduction. The second term is associated with the tortuosity effect—an elongation in thermal path due to the existence of solid particles. The last term represents the interfacial heat transfer to fluid from solid. Similar interpretations can be given to the respective terms in (7) for solid phase. The negative sign in the last two terms of (7) reflects the fact that the source terms in (6) have to become the sink terms in (7), or vice versa. Note that the magnitudes of the thermal tortuosity terms in solid and fluid phases are different because of the difference in thermal conductivity. On the other hand, the magnitudes of the interfacial heat transfer terms are the same. These two integral terms representing the effects due to thermal tortuosity and interfacial heat transfer are the new unknowns in the equation system (6) and (7). Closure modeling for these integral terms then becomes inevitable.

III. CLOSURE MODELING

To close the equation system (6) and (7), we need to construct the constitutive equations which relate the integral terms to the macroscopically averaged temperatures $\bar{T}_f$ and $\bar{T}_s$. Quintard and Whitaker (1993) provided a detailed account of the closure scheme. A treatment similar to that of Quintard and Whitaker (1993), but mathematically more concise, was provided by Hsu (1999); therefore we repeat it here. To this end, T_f and T_s are first decomposed into

$$T_f = \bar{T}_f + T_f' \tag{10}$$

and

$$T_s = \bar{T}_s + T_s' \tag{11}$$

where T_f' and T_s' represent the microscopic temperature variations from the volumetric averaged values in REV. For the REV of length scale l, the time scale for macroscopic conduction in the fluid phase, l^2/α_f, is much longer than the time scale for microscopic conduction, d_p^2/α_f, since $l \gg d_p$; similarly for the solid phase. Note that the macroscopic time scales in solid and fluid phases will be different if $\alpha_f \neq \alpha_s$. With respect to the macroscopic time scale, it is plausible to assume that the local microscopic heat conduction process is quasi-steady. After the substitution of (10) and (11), Eqs. (1) and (2) under this quasi-steady assumption become

$$\nabla \cdot \left(k_f \nabla T_f'\right) = 0 \tag{12}$$

and

$$\nabla \cdot \left(k_s \nabla T_s'\right) = 0 \tag{13}$$

The interfacial boundary conditions now read

$$T_f' = T_s' + \left(\bar{T}_s - \bar{T}_f\right) \qquad \text{on} \qquad A_{fs} \tag{14}$$

and

$$\mathbf{n}_{fs} \cdot \nabla T_f' = \mathbf{n}_{fs} \cdot \sigma \nabla T_s' + \mathbf{n}_{fs} \cdot \left(\sigma \bar{\nabla} \bar{T}_s - \bar{\nabla} \bar{T}_f\right) \qquad \text{on} \qquad A_{fs} \tag{15}$$

where $\sigma = k_s/k_f$ is the conductivity ratio of solid to fluid.

Since (12) and (13) are linear, the interfacial boundary conditions (14) and (15) suggest that the solutions to T_f' and T_s' will take the forms of

$$T_f' = f_0'\left(\bar{T}_s - \bar{T}_f\right) + \mathbf{f}_1' \cdot d_p\left(\bar{\nabla} \bar{T}_f - \sigma \bar{\nabla} \bar{T}_s\right) \tag{16}$$

and

$$T_s' = g_0'\left(\bar{T}_s - \bar{T}_f\right) + \mathbf{g}_1' \cdot d_p\left(\bar{\nabla} \bar{T}_f - \sigma \bar{\nabla} \bar{T}_s\right) \tag{17}$$

The substitution of (16) and (17) into (12)–(15) will lead to equations and boundary conditions for f_0', $\mathbf{f}_1'$, g_0' and $\mathbf{g}_1'$. The details of their solutions depend on the local geometry of the solid particles and require elaborate work. However, for the aim of constructing the closure relations, the solution forms appearing in (16) and (17) will suffice. The substitution of (16) and (17) into (10) and (11), and then into (8) and (9), results in

$$\Lambda_{fs} = \mathbf{G}_0(\bar{T}_s - \bar{T}_f) + \boldsymbol{G}_1 \cdot (\bar{\nabla}\bar{T}_f - \sigma\bar{\nabla}\bar{T}_s) \tag{18}$$

where

$$\mathbf{G}_0 = \frac{1}{V}\iint_{A_{fs}} f_0' \mathbf{ds} \quad \text{and} \quad \boldsymbol{G}_1 = \frac{d_p}{V}\iint_{A_{fs}} \mathbf{f}_1' \mathbf{ds} \tag{19a, b}$$

and

$$Q_{fs} = M_0(\bar{T}_s - \bar{T}_f) + \mathbf{M}_1 \cdot (\bar{\nabla}\bar{T}_f - \sigma\bar{\nabla}\bar{T}_s) \tag{20}$$

where

$$M_0 = \frac{1}{V}\iint_{A_{fs}} k_f \nabla f_0' \cdot \mathbf{ds} \quad \text{and} \quad \mathbf{M}_1 = \frac{d_p}{V}\iint_{A_{fs}} k_f \nabla \mathbf{f}_1' \cdot \mathbf{ds} \tag{21a, b}$$

The above closure relations are basically similar to those of Quintard and Whitaker (1993), except for the slight difference in forms for thermal tortuosity. It is important to note that the closure coefficients, $\mathbf{G}_0$, $\boldsymbol{G}_1$, M_0, and $\mathbf{M}_1$, depend only on the local geometry of particles and on the material thermal properties, not on the macroscopic quantities. For randomly distributed spherical particles, the volumetric averaged tensor $\boldsymbol{G}_1$ has to be axially symmetric to exhibit an isotropic property, i.e., $\boldsymbol{G}_1 = G\boldsymbol{I}$ where $\boldsymbol{I}$ is the unit matrix and G is a scalar quantity. According to Newton's law, the interfacial heat transfer is proportional to the total fluid–solid interfacial area A_{fs} in REV. Therefore, we have $M_0 = h_{fs}a_{fs}$ where $a_{fs}(= A_{fs}/V)$ is the specific interfacial area and h_{fs} is the interfacial heat transfer coefficient. From the physical point of view, a non-zero value of $\mathbf{G}_0$ or $\mathbf{M}_1$ will lead to a convective behavior associated with the mean temperature gradients, where the vectors $\mathbf{G}_0$ and $\mathbf{M}_1$ resemble the convection velocities. This convective behavior does not have a physical ground since all these transfer processes arise essentially from conduction. Therefore, we have $\mathbf{G}_0 = \mathbf{M}_1 = 0$, as is also demonstrated mathematically by Quintard and Whitaker (1993). Equations (18) and (20) then reduce to

$$\Lambda_{fs} = G(\bar{\nabla}\bar{T}_f - \sigma\bar{\nabla}\bar{T}_s) \tag{22}$$

and

$$Q_{fs} = h_{fs} a_{fs} (\bar{T}_s - \bar{T}_f) \tag{23}$$

Equations (6) and (7), with the thermal tortuosity and interfacial heat transfer terms given by the closure relations (22) and (23), are the governing equations for transient heat conduction in porous media. The determination of the closure coefficients G and h_{fs} then becomes the critical task. We want to re-emphasize that the values of G and h_{fs} depend only on the microscopic interfacial geometry and the thermal properties of the solid and fluid.

IV. LUMPED MIXTURE MODEL UNDER LOCAL THERMAL EQUILIBRIUM CONDITION

We now examine heat conduction in porous media under the local thermal equilibrium condition. The constraints for invoking the assumption of local thermal equilibrium have been discussed by Quintard and Whitaker (1995) and recently by Hsu and Wu (2000). This assumption implies

$$\bar{T}_f = \bar{T}_s = \bar{T} \tag{24}$$

Equations (6) and (7), after invoking (24), are added to produce the following heat conduction equation for the solid–fluid mixture:

$$(\rho Cp)_m \frac{\partial \bar{T}}{\partial t} = \bar{\nabla} \cdot [k_e \bar{\nabla} \bar{T}] \tag{25}$$

where

$$(\rho Cp)_m = \varepsilon \rho_f Cp_f + (1 - \varepsilon)\rho_s Cp_s \tag{26}$$

is the heat capacitance and

$$k_e = \varepsilon k_f + (1 - \varepsilon)k_s + k_f(1 - \sigma)^2 G \tag{27}$$

is the effective stagnant thermal conductivity of the mixture. Note that the value of k_e depends only on the interfacial geometry and the solid–fluid thermal properties. Equation (25) resembles the transient heat conduction of a pure substance, solutions of which (subject to different initial and boundary conditions) have been obtained extensively in open literature. Therefore, the main task becomes the determination of the stagnant effective thermal conductivity.

V. EFFECTIVE STAGNANT THERMAL CONDUCTIVITY

The determination of effective stagnant thermal conductivity has been a subject of great effort for more than a century (Maxwell, 1873). A large number of experiments have been carried out to measure the effective stagnant thermal conductivity. Kunii and Smith (1960), Krupiczka (1967), and Crane and Vachon (1977) have compiled these early experimental data. The experimental methods for determining k_e were also reviewed by Tsotsas and Martin (1987). Most of these measurements were carried out for materials in the range of $1 < \sigma < 10^3$. Effective thermal conductivities of porous materials with higher values of σ were obtained experimentally by Swift (1966) and Nozad et al. (1985), while those with lower σ were found by Prasad et al. (1989). With advances in computer technology, the effective stagnant thermal conductivities were determined numerically. Deissler and Boegli (1958) were the first to obtain k_e for cubic-packing spheres on the basis of a finite-difference scheme, followed by Wakao and Kato (1969) and by Wakao and Vortmeyer (1971) for a periodic orthorhombic structure. More recent numerical results were obtained by Nozad et al. (1985) and Sahraoui and Kaviany (1993). It should be noted that all the numerical investigations assumed periodic porous structures so that the computation domain could be confined to a unit cell. Since Maxwell (1873), several analytical composite-layer models have been proposed for k_e (Kunii and Smith, 1960; Zehner and Schlunder, 1970). Recently, Hsu et al. (1994) extended the model of Zehner and Schlunder (1970) by introducing a particle touching parameter. The model of Kunii and Smith (1960) was improved by Hsu et al. (1995), using the touching and non-touching geometry of Nozad et al. (1985); they found the predicted results of k_e agree remarkably well with the experimental data of Nozad et al. (1985). Kaviany (1991) and Cheng and Hsu (1998) have reviewed the existing models of effective thermal conductivity in detail. To provide a clear picture of the behavior of effective stagnant thermal conductivity, some of the analytical composite-layer models are briefly described below.

A. Phase-segregated-layer Model

The simplest model for effective stagnant thermal conductivity is to assume that the solid and fluid phases can be segregated into a solid layer and a fluid layer, either in parallel or in series with respect to the direction of thermal gradient. For layers in parallel, the simple conservation of total heat flux under the same temperature gradient leads to

$$k_e/k_f = \varepsilon + (1 - \varepsilon)\sigma \tag{28}$$

On the other hand, for layers in series the balance of total temperature change under the same heat flux condition leads to

$$k_e/k_f = \frac{\sigma}{\varepsilon\sigma + (1-\varepsilon)} \tag{29}$$

The predictions of k_e/k_f based on Eqs. (28) and (29) are plotted in Figure 2. Clearly the layer-in-parallel results are much larger than those of the layer-in-series model, especially when σ is large. As pointed out by Deissler and Boegli (1958), Eqs. (28) and (29) represent the ideal cases of perfect contact and no contact between solid particles, so that k_e/k_f attains maximum and minimum values, respectively.

B. Kunii and Smith's Model and its Extension

Using the concept of unit cell, Kunii and Smith (1960) simplified the media of packed spheres at point-contact into a fluid layer in parallel with a second layer consisting of fluid and solid layers in series. Therefore, the effective thermal conductivity is given by $k_e/k_f = \varepsilon + (1-\varepsilon)k^*/k_f$, where k^* is the effective thermal conductivity of the second layer. This was given by Kunii and Smith (1960) as $k^*/k_f = \sigma/(\alpha_K\sigma + 2/3)$, with α_K being an empirical

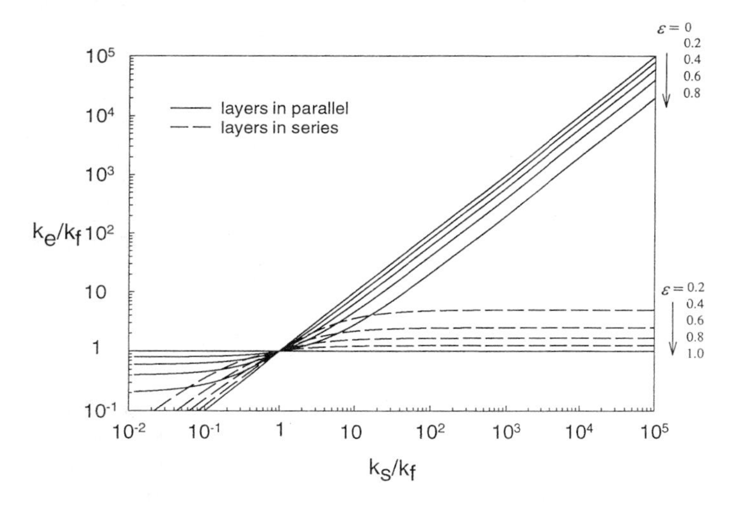

Figure 2. Prediction of effective thermal conductivity based on phase-segregated layer models: solid curves for the layer-in-parallel model and dashed curves for layer-in-series.

parameter depending on the porosity. As a result, the expression for k_e/k_f becomes

$$k_e/k_f = \varepsilon + \frac{(1-\varepsilon)\sigma}{(\alpha_K \sigma + 2/3)} \tag{30}$$

Evidently, the effective thermal conductivity given by (30) is a more advanced model since it combines the results of (28) and (29). However, the applicability of (30) remains limited because of the large uncertainty associated with the determination of α_K.

Hsu et al. (1995) adopted Kunii and Smith's (1960) concept of unit cell to periodic in-line arrays of square cylinders, used by Nozad et al. (1985) for 2D direct numerical simulation, as shown in Figure 3. The unit cell shown in the dashed square of Figure 3 is composed of a solid layer in parallel with two other layers consisting of solid and fluid layers in series. By applying the rules of layer-in-series and layer-in-parallel, Hsu et al. (1995) obtained the following expression for k_e/k_f:

$$k_e/k_f = \gamma_a \gamma_c \sigma + \frac{\gamma_a(1-\gamma_c)\sigma}{(1-\gamma_a)\sigma + \gamma_a} + \frac{(1-\gamma_a)\sigma}{(1-\gamma_a\gamma_c)\sigma + \gamma_a\gamma_c} \tag{31}$$

In (31), $\gamma_a (= a/l_e)$ and $\gamma_c (= c/a)$ are the dimensionless length ratios of the unit cell; they are related to the porosity by

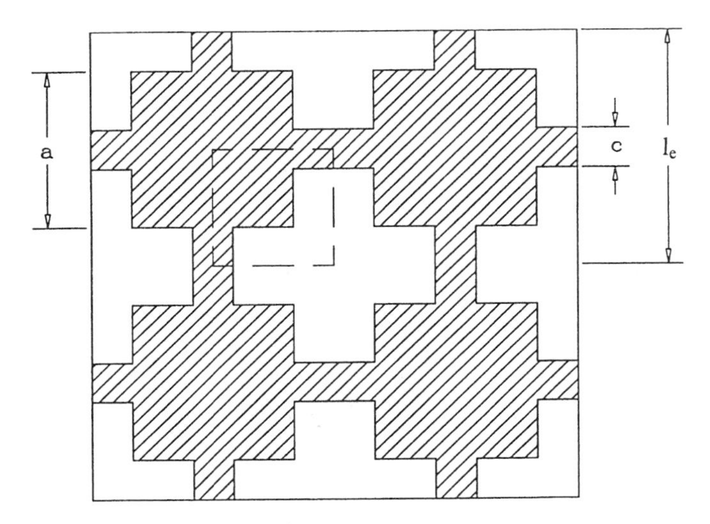

Figure 3. 2D in-line arrays of square cylinders and the unit cell used by Nozad et al. (1985).

$$1-\varepsilon=\gamma_a^2+2\gamma_c\gamma_a(1-\gamma_a) \tag{32}$$

For non-touching square cylinders where $\gamma_c=0$, $\gamma_a=(1-\varepsilon)^{1/2}$ and (31) reduces to

$$k_e/k_f=\left[1-(1-\varepsilon)^{1/2}\right]+\frac{(1-\varepsilon)^{1/2}\sigma}{\left[1-(1-\varepsilon)^{1/2}\right]\sigma+(1-\varepsilon)^{1/2}} \tag{33}$$

Since the unit cell shown in Figure 3 becomes identical to that of Kunii and Smith (1960) if $\gamma_c=0$, it is expected that (33) will behave very similarly to (30). This is indeed the case in terms of the functional form with respect to σ between (30) and (33); however, the coefficients in (33) are determinative, and are thus quite different from those in (30). The predictions of k_e/k_f based on (31)–(33) were in surprisingly good agreement with the numerical, as well as the experimental, results of Nozad et al. (1985).

C. Hsu et al.'s 3D Cube Model

With encouragement from the model for a 2D array of square cylinders, Hsu et al. (1995) further constructed a 3D model for in-line periodic arrays of cubes with the unit cell shown in Figure 4. They obtained the following expression for the determination of k_e/k_f

$$\begin{aligned}k_e/k_f&=1-\gamma_a^2-2\gamma_c\gamma_a+2\gamma_c\gamma_a^2+\sigma\gamma_c^2\gamma_a^2+\frac{\sigma\gamma_a^2(1-\gamma_c^2)}{\sigma+\gamma_a(1-\sigma)}\\&\quad+\frac{2\sigma\gamma_c\gamma_a(1-\gamma_a)}{\sigma+\gamma_c\gamma_a(1-\sigma)}\end{aligned} \tag{34}$$

where γ_a and γ_c are related to the porosity by

$$1-\varepsilon=(1-3\gamma_c^2)\gamma_a^3+3\gamma_c^2\gamma_a^2 \tag{35}$$

For non-touching cubes, $\gamma_a=(1-\varepsilon)^{1/3}$ and (34) reduces to

$$\frac{k_e}{k_f}=\left[1-(1-\varepsilon)^{2/3}\right]+\frac{(1-\varepsilon)^{2/3}\sigma}{\left[1-(1-\varepsilon)^{1/3}\right]\sigma+(1-\varepsilon)^{1/3}} \tag{36}$$

The predictions of k_e/k_f based on (34)–(36) of the 3D cube model with $\gamma_c=0$ and 0.13 are shown in Figure 5 for $\varepsilon=0.36$. For comparison, the predictions based on (31)–(33) of the 2D square cylinder model are also plotted in Figure 5. Evidently, the results from the 3D cube model are in better agreement with the experimental data of Nozad et al. (1985) than the results from the 2D square cylinder model.

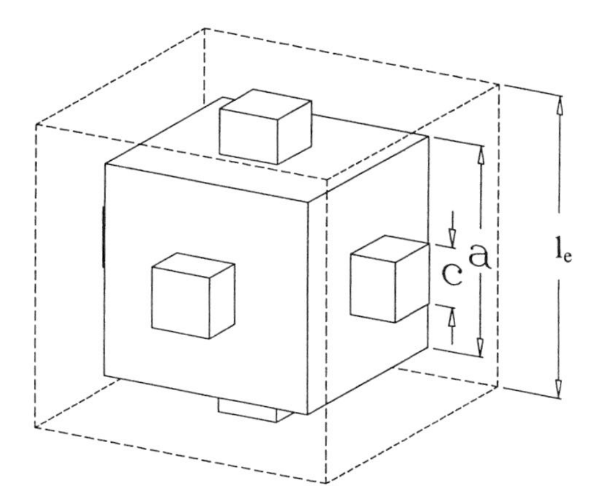

Figure 4. Unit cell of 3D in-line arrays of cubes used by Hsu et al. (1995).

VI. EVALUATION OF CLOSURE COEFFICIENTS G AND h_{fs}

We now return to the transient heat conduction problem of local thermal nonequilibrium. To render (22) and (23) applicable, analytical expressions for G and h_{fs} are required. They are provided below.

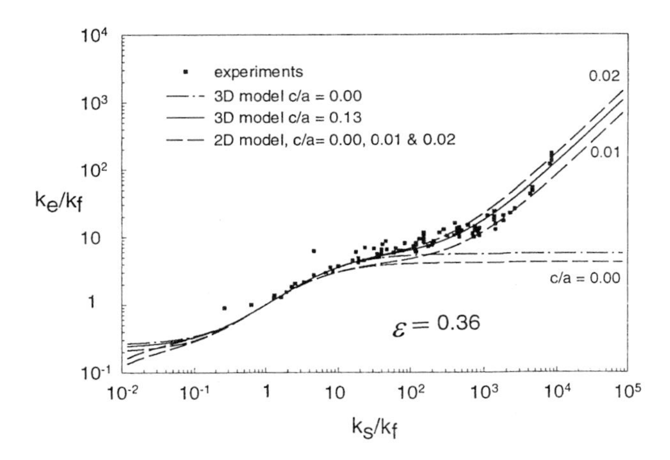

Figure 5. Predictions of effective thermal conductivity based on the 3D cube model and the 2D square cylinder model by Hsu et al. (1995), and their comparison with the experimental results of Nozad et al. (1985).

A. Tortuosity Parameter *G*

Since the value of G depends only on the local interfacial geometry and on the solid and fluid thermal properties, it is plausible to assume that the expression for G obtained under the local thermal equilibrium condition can be extended to the thermal nonequilibrium regime. This bears a resemblance to the closure modeling of turbulent flow, in which a simple flow is usually used to evaluate the closure coefficient in a closure relation. Therefore, G is obtained by solving (27) to yield

$$G = \left[\frac{k_e}{k_f} - \varepsilon - \sigma(1-\varepsilon)\right] \Big/ (1-\sigma)^2 \tag{37}$$

where k_e/k_f is evaluated by a lumped composite-layer model as described in the last section. To demonstrate the behavior of G, the 3D cube model of Hsu et al. (1995), described by Eqs. (34) and (35), is adopted here.

The values of G were calculated from (37) over a wide range of ε, γ_c, and σ. The results of G are shown in Figure 6 for different γ_c with ε fixed at 0.36. The distinctive feature is that G is always negative. This has to be so since $\varepsilon + \sigma(1-\varepsilon)$, which represents the result of the model of phase-segregated layers in parallel, is the upper limit of k_e/k_f. This implies that the tortuosity effect is to reduce the effective thermal conduction by increasing the thermal path, i.e., the phase-segregated parallel layer model has the shortest thermal path. As shown in Figure 4, the lower the thermal conductivity of solid is than that of fluid, the more the thermal path undulates. When $\sigma \gg 1$, the effect of touching is to enhance heat flow through the touching area. Most of the thermal lines pass through the touching area and the thermal lines are basically straight (i.e., $G \to 0$). These thermal lines are insensitive to the size of the touching area. Figure 7 shows the results of G when the touching parameter is fixed at $\gamma_c = 0.1$ while the porosity is varied. Although the value of G is always negative, the variation of G with porosity is not monotonic for different ranges of σ. In the range of $\sigma < 0.1$, maximal G occurs at ε near 0.7 and, when $\sigma > 10$, near $\varepsilon = 0.4$. Note that G becomes zero for cases of both $\varepsilon = 0$ and $\varepsilon = 1$, i.e., when the porous medium reduces to pure substance. Before moving to the interfacial heat transfer coefficient, we note that the results shown in Figures 6 and 7 apply specifically to media consisting of particulate solid particles. For spongy materials, with geometry quite symmetric between the solid and fluid phases, the values of k_e/k_f are quite different from those shown in Figure 5 (Hsu et al. 1994), and therefore the results of G will be quite different from those shown in Figures 6 and 7.

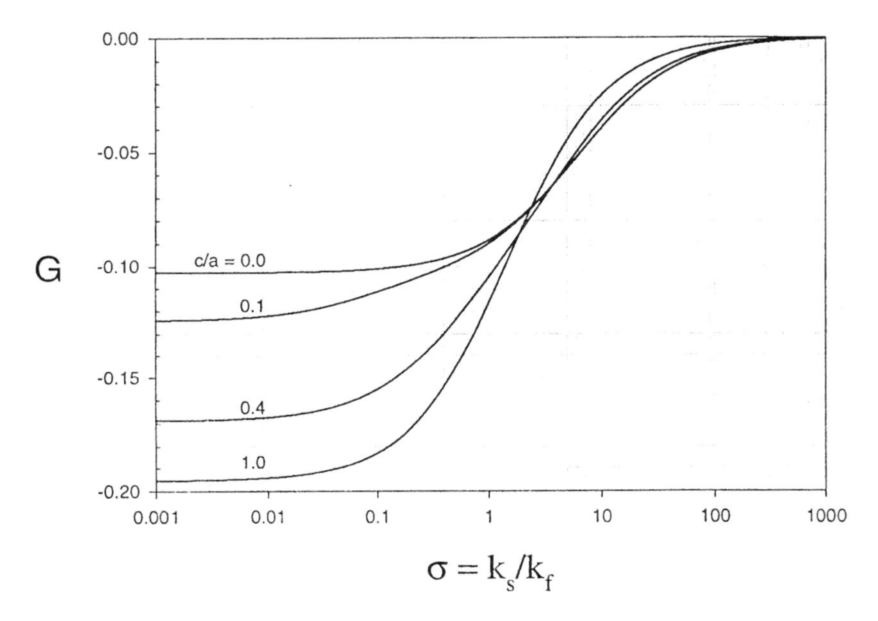

Figure 6. The effect of particle-touching on the interfacial thermal tortuosity coefficient when the porosity is fixed at 0.36.

B. Interfacial Heat Transfer Coefficient h_{fs}

The determination of the interfacial heat transfer coefficient h_{fs} has been one of the main research topics because of its significance in engineering applications, particularly in the design of reactors in chemical engineering. In mechanical engineering, one of the important applications is the use of a regenerator in thermal control. However, most of these early investigations were conducted under the condition of fluid flow. On the basis of experimental results for convective heat and mass transfer, Wakao et al. (1979) suggested the following correlation:

$$\hat{Nu} = \hat{h}_{fs} d_p / k_f = 2 + 1.1 Pr^{1/3} Re_p^{0.6} \tag{38}$$

where $\hat{Nu}$ is the Nusselt number for the convective interfacial heat transfer coefficient $\hat{h}_{fs}$, Pr the Prandtl number, and Re_p the Reynolds number based on the diameter of the solid particle. Amiri and Vafai (1994) adopted Eq. (38) for numerical computations of thermal nonequilibrium convective heat

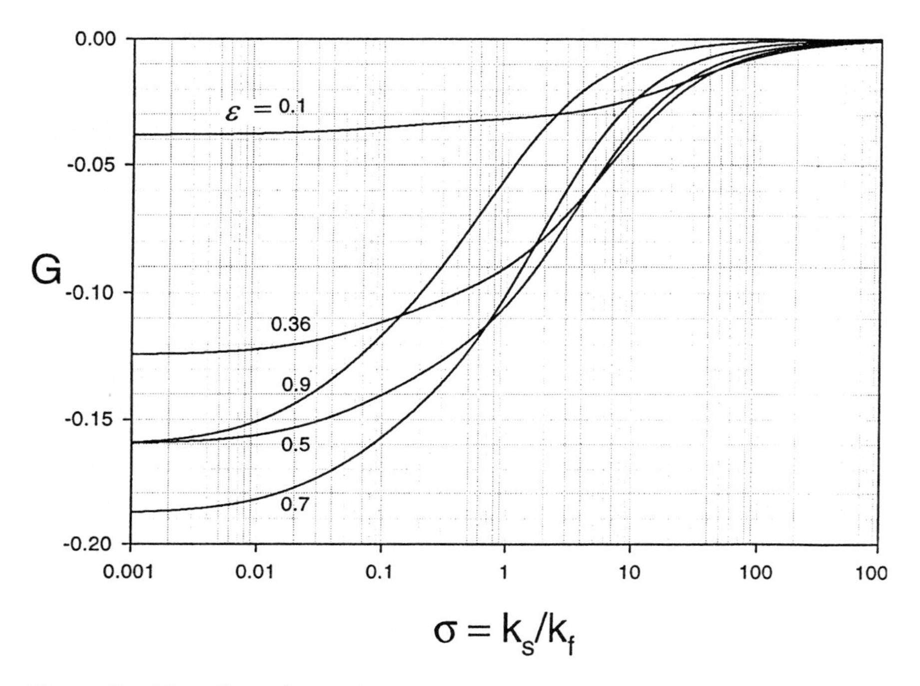

Figure 7. The effect of porosity on the interfacial thermal tortuosity coefficient when the particle-touching parameter is fixed at 0.1.

transfer in porous media. The correlation (38) suggests that $\hat{h}_{fs}d_p/k_f \rightarrow h_{fs}d_p/k_f = 2$ when $Re_p \rightarrow 0$. Therefore, $Nu = h_{fs}d_p/k_f = 2$ represents the simplest form of the interfacial heat transfer coefficient for stagnant heat conduction. However, this is not very meaningful physically since the Nusselt number Nu for the stagnant interfacial heat transfer coefficient has to depend on the thermal properties of both solid and fluid. In fact, the experimental results compiled later by Wakao and Kaguei (1982) showed that there exists considerable scattering in data of Nu when Re_p is small. As (38) was obtained by correlating the experimental data for large Re_p, the second term on the right-hand side of (38) is presumably dominant; this leads to a considerable error in the determination of the constant 2.0.

Quintard and Whitaker (1995) studied transient heat conduction in porous media and proposed a model for the interfacial heat transfer coefficient based on the unit cell of Chang (1983). Their expression of h_{fs} for spheres, in terms of our notation, reads

$$Nu = \frac{80}{3\pi} \frac{(\alpha^2 + \alpha + 1)\sigma}{(5 + \alpha - 3\alpha^2 - 2\alpha^3 - \alpha^4)\sigma + (1 + 2\alpha + 3\alpha^2 + 2\alpha^3 + \alpha^4)} \tag{39}$$

where $\alpha = (1 - \varepsilon)^{1/3}$. Evidently, (39) represents a model with two layers in series in a spherical geometry, with the layer thickness depending on porosity.

To investigate the dependency of h_{fs} on thermal diffusivity, Hsu (1999) proposed a quasi-steady model for dispersed spherical particles. The quasi-steady assumption is consistent with the quasi-steady closure modeling scheme given in Section III. By the same token, the validity of the quasi-steady model requires that the time scale of microscopic heat conduction is much smaller than the time scale of macroscopic heat conduction. Under this assumption, the stagnant interfacial heat transfer process was simplified into a model consisting of two heat conduction layers on the two sides of the interface, as shown in Figure 8. By solving the steady-state heat conduction equation in a spherical coordinate system satisfying the boundary conditions $T_f = \bar{T}_f$ at $r = r_0$ and $T_s = \bar{T}_s$ at $r = r_i$, Hsu obtained the interfacial heat flux from solid to fluid for each particle as

$$q = \frac{4\pi r_p k_f}{\dfrac{A}{1 + A} + \dfrac{B}{\sigma(1 - B)}} (\bar{T}_s - \bar{T}_f) \tag{40}$$

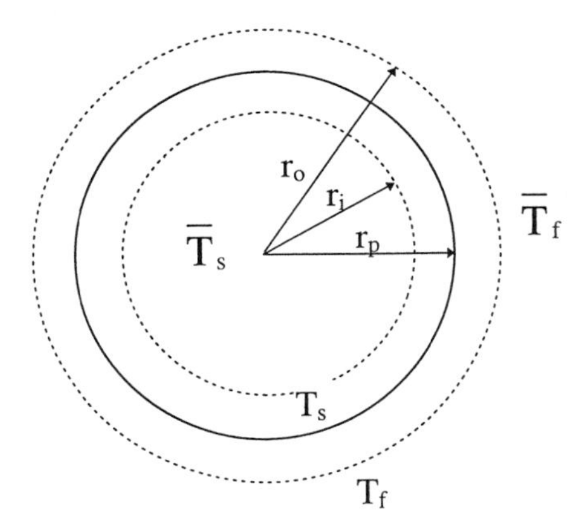

Figure 8. Schematic of a quasi-steady heat conduction for dispersed spherical particles immersed in fluid.

where $A = (r_0 - r_p)/r_p$ and $B = (r_p - r_i)/r_p$ are the nondimensional thicknesses of the conduction layers. Summing up (40) for n particles in REV gives the interfacial heat transfer per unit volume as $Q_{fs} = nq/V$, which, after invoking $a_{fs} = 6(1 - \varepsilon)/d_p$ and comparing with (23), results in

$$Nu = \frac{h_{fs}d_p}{k_f} = \frac{2\sigma}{\alpha_A\sigma + \alpha_B} \tag{41}$$

where $\alpha_A = A(1 + A)$ and $\alpha_B = B/(1 - B)$. The parameters A and B are related by

$$\beta A^2 \frac{3 + A}{1 + A} = B^2 \frac{3 - B}{1 - B} \tag{42}$$

as a result of thermal energy conservation within the two phases. Equations (41) and (42) indicate that the value of Nu depends on three parameters, A, σ, and β. While σ and β represent the material thermal properties, A depends also on how closely the medium is packed, i.e., on the porosity. The equivalent radius r_e of Chang's (1983) unit cell in porous media is given by $r_e/r_p = (1 - \varepsilon)^{-1/3}$. If the fluid outside the solid particle is assumed to reach a quasi-steady state when $r_0 = r_e$, then $1 + A = (1 - \varepsilon)^{-1/3}$, which gives $A = 0.16$ if ε takes a typical value of 0.36.

The results for Nu as evaluated by Hsu (1999), based on (41) and (42) as a function of σ for $A = 0.1$, 0.2, and 0.4, with varying β, are shown in Figure 9. Figure 9 indicates that Nu approaches zero and constant when σ approaches zero and infinity, respectively. When σ approaches zero, the solid particles behave as being nonconductive, i.e., little heat can conduct across the interface. On the other hand, when σ approaches infinity the solid particles become highly transmittable to heat flux and the heat flux across the interface then depends solely on the thermal conductivity of the fluids. In fact, the interfacial transfer coefficient h_{fs} becomes linearly proportional to the thermal conductivity of the fluids with the proportional constant depending solely on A. The dependence of the interfacial transfer coefficient on the thermal diffusivity ratio in the middle range of σ is interesting. The microscopic quasi-steady model depicted in Figure 8 suggests that the effective thermal conductivity of the conduction layers near the solid–fluid interface can be regarded as that of the two layers in series. Higher thermal diffusivity ratio β implies a thicker thermal layer inside the sphere, and therefore a lower value of Nu. It is expected that (41) behaves quite similarly to (39) as both were derived from the layer-in-series model. With a given value of α for (39), it is always possible to find the corresponding values of A and β so that (41) and (39) are identical. This is indeed the case, as shown in Figure 9, where the curve calculated from (39) with $\alpha = 0.86$ ($\varepsilon = 0.36$) is identical to that from (41) with $A = 0.2$ and $\beta = 5.4$. Apparently, Quintard

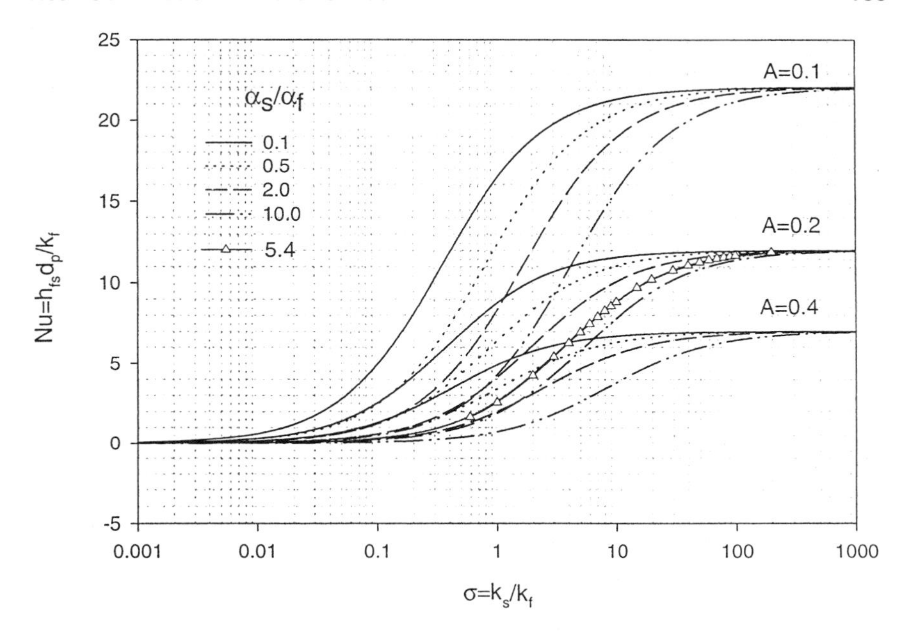

Figure 9. Dependence of the interfacial heat transfer coefficient on the local geometric (time scale) parameter and on the thermal property parameters of the solid–fluid phases.

and Whitaker's (1995) model, which does not depend on the thermal diffusivity ratio, represents a special case of Hsu's (1999) model.

VII. ONE-DIMENSIONAL TRANSIENT HEAT CONDUCTION

In this and the next sections, we shall discuss the characteristics of the model for transient heat conduction in porous media as described above and address some fundamental issues on the assumption of local thermal equilibrium. To this end, the calculations performed by Hsu and Wu (2000) for one-dimensional transient heat conduction through an infinitely extended composite slab of finite thickness are summarized below. The slab is assumed initially to be at uniform temperature T_i. At the left side of the slab the temperature is dropped suddenly to T_0 and at the right side it is insulated. The response of the two phases to the temperature perturbation is the main focus of the present investigation. The difference in the temperatures of the two phases during the transient process provides a measure of how the thermal equilibrium assumption holds. Since the validity of thermal equilibrium depends on the time scale chosen for observation, we shall normalize the problem by using the slab thickness l_m as the length scale, the temperature difference $T_i - T_0$ as the temperature scale, and the fluid diffusion time l_m^2/α_f as the time scale. After the substitution of (22) and (23) into (6) and (7), the normalization of the resultant equations in one-dimensional becomes

$$\varepsilon \frac{\partial \theta_f}{\partial t} = (\varepsilon + G)\frac{\partial^2 \theta_f}{\partial x^2} - G\sigma \frac{\partial^2 \theta_s}{\partial x^2} + Nu^*(\theta_s - \theta_f) \tag{43}$$

$$\lambda(1-\varepsilon)\frac{\partial \theta_s}{\partial t} = \sigma(1-\varepsilon+G\sigma)\frac{\partial^2 \theta_s}{\partial x^2} - G\sigma \frac{\partial^2 \theta_f}{\partial x^2} - Nu^*(\theta_s - \theta_f) \tag{44}$$

where $\theta_f = (\bar{T}_f - T_0)/(T_i - T_0)$ and $\theta_s = (\bar{T}_s - T_0)/(T_i - T_0)$ are the normalized non-dimensional temperatures, $\lambda = \sigma/\beta = (\rho_s Cp_s)/(\rho_f Cp_f)$ is the heat capacity ratio, and $Nu^* = Nu6(1-\varepsilon)(l_m/d_p)^2$ is the modified Nusselt number of the interfacial heat transfer. The initial condition for $t \leq 0$ is

$$\theta_f = \theta_s = 1, \qquad \text{in} \qquad 0 \leq x \leq 1 \tag{45}$$

and the boundary conditions for $t > 0$ are

$$\theta_f = \theta_s = 0 \qquad \text{at} \qquad x = 0 \tag{46a}$$

$$\frac{\partial \theta_f}{\partial x} = \frac{\partial \theta_s}{\partial x} = 0 \qquad \text{at} \qquad x = 1 \tag{46b}$$

Equations (43) and (44), subjected to the initial and boundary conditions (45) and (46), were solved numerically using a finite difference procedure. Central difference formulation was used for all spatial derivative terms. The Crank-Nicolson (1947) method, that had been widely used by others for solving transient parabolic diffusion equations was adopted.

The computations were carried out for different values of ε, λ, σ, and Nu^*. Here, λ takes an equivalent role to β. Note that G is not an independent parameter, since it can be determined from (34)–(37) once ε and σ are given. The time step of the computation was 10^{-4} with a spatial grid size of 0.01. Figure 10 shows the results of the transient response of the solid and fluid temperatures at different locations in the slab when $\varepsilon = 0.4$, $\lambda = 0.4$, $\sigma = 4.0$, and $Nu^* = 108$. For this case, where the solid has a conductivity four times that of the fluid, we see that the fluid temperature response lags behind the solid temperature, especially at $x = 0.25$. This temperature lag is initially equal to zero, increases with time, and decays to zero when t becomes of the order of one. Recall that t is scaled, with the time scale l_m^2/α_f. Therefore, both solid and fluid temperatures in the slab have become T_0 everywhere for $t > 1$. The temperature profiles at different times are shown in Figure 11. Evidently, the difference between the solid and fluid

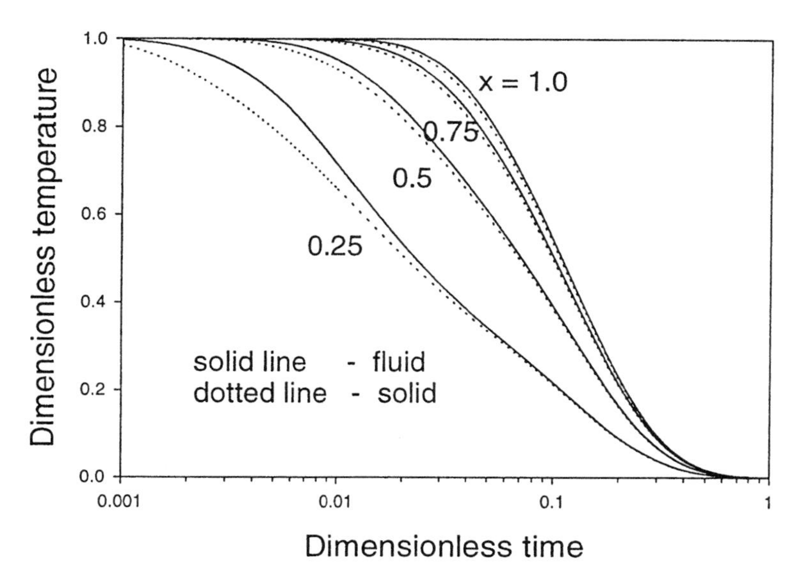

Figure 10. Time response of solid and fluid temperatures at different locations of an infinitely extended porous slab to a sudden drop of temperature at one wall, the other wall being insulated.

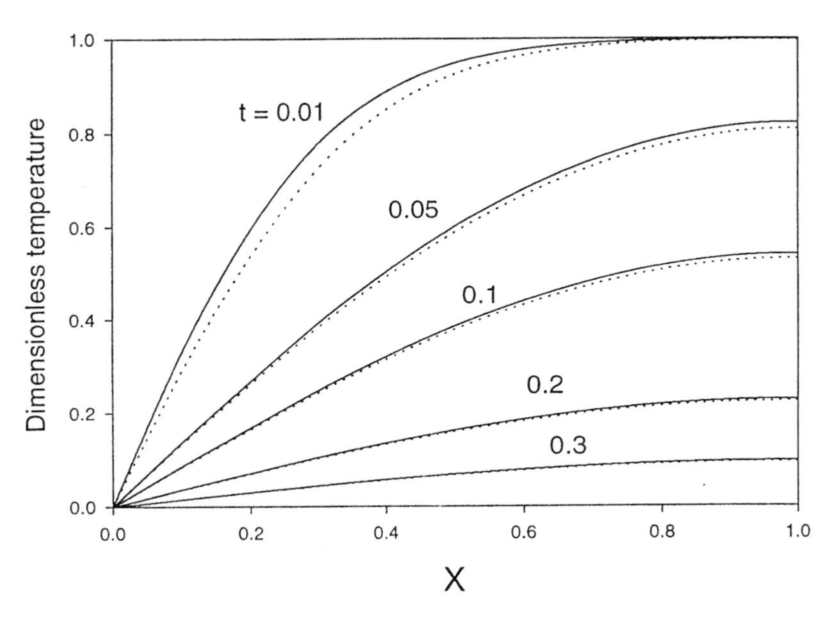

Figure 11. Profiles of solid and fluid temperatures in an infinitely extended porous slab at different times after a sudden drop of temperature at one wall, the other wall being insulated.

temperatures is large at small time. This difference diminishes when the time becomes large. Figures 10 and 11 clearly indicate that under the conditions of $\varepsilon = 0.4$, $\lambda = 0.4$, $\sigma = 4.0$, and $Nu^* = 108$, the validity of a local thermal equilibrium depends not only on the thermal properties, but also on the time scale and length scale used to measure the transient process. When the slab thickness is used as the length scale, the results shown in Figures 10 and 11 suggest that the transient conduction becomes locally thermal equilibrium when the time is greater than 0.1 l_m^2/α_f.

VIII. LOCAL THERMAL NONEQUILIBRIUM

In order to address the issue of local thermal nonequilibrium, we shall follow Quintard and Whitaker's (1993) concept of measuring the difference in thermal response between solid and fluid by calculating the temperature difference per unit volume, as follows:

$$\delta(t) = \int_0^1 (\theta_f - \theta_s)\mathrm{d}x = \frac{\sum_{n'=1}^{N_g}(\theta_{f,n'} - \theta_{s,n'})}{N_g} \tag{47}$$

where N_g is the grid number. Note that a positive value of $\delta(t)$ implies the fluid temperature lags behind the solid temperature during the cooling process, and vice versa. Instead, Quintard and Whitaker (1993) used $|\delta|^{1/2}$ as a measure which becomes harder to interpret when δ is negative. We notice that the temperature difference is constrained by the initial conditions and the boundary conditions, so that $\delta(0) = \delta(\infty) = 0$. To examine the effect of interfacial heat transfer on thermal response of the fluid and solid phases, numerical solutions for (43) and (44), subjected to initial condition (45) and boundary conditions (46a,b), were obtained for different values of Nu^* with other parameters fixed at $\sigma = 4.0$, $\lambda = 0.4$, $\varepsilon = 0.4$. The results are given in Figure 12 for $Nu^* = 60$, 108, 240, and 970. Figure 12 shows that δ approaches zero when t is very small or greater than one, and has different maximal values for different Nu^* at different times. The larger the values of Nu^*, the lower are the values of δ and the faster the porous medium reaches the condition of thermal equilibrium. Figure 12 shows that the transient time scale of a thermal nonequilibrium system decreases rapidly with increasing Nu^*. A large value of Nu^* will render the temperature difference relatively small so that the two-equation model can be lumped into a single-equation model for the solid–fluid mixture. Therefore, one can assume a local thermal equilibrium for $Nu^* > 1000$ if less than 1% in temperature difference is used as a threshold. Note that this may also depend on the values of λ, σ, and ε.

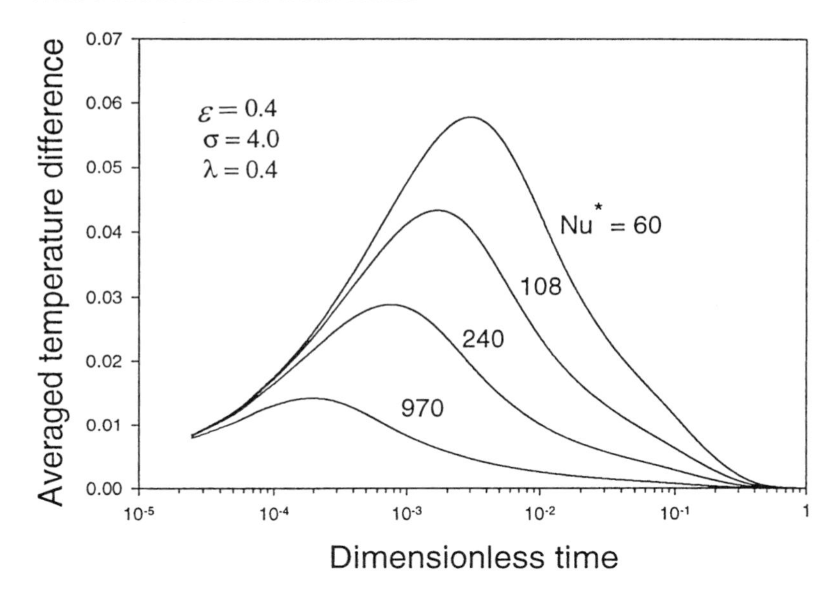

Figure 12. Effect of interfacial heat transfer on the total difference between macroscopic solid and fluid temperatures, indicating the establishment of a local thermal equilibrium.

The role of heat capacity (or thermal diffusivity) in controlling the local thermal equilibrium is played by the parameter λ. The effect of λ on δ is illustrated in Figure 13 for λ ranging from 0.01 to 100 while other parameters are fixed at $Nu^* = 108$, $\varepsilon = 0.4$, and $\sigma = 4.0$. At $\lambda = 0.01$, i.e., $\rho_f Cp_f \gg \rho_s Cp_s$, a maximal value of δ occurs when t is approximately 10^{-4}. Note that the positive value of δ implies a time lag of the fluid phase. The heat stored initially in the fluid phase is much greater than that in the solid phase; therefore, the fluid response is slower than that of the solid because of higher thermal inertia. For $\lambda = 1.0$, although there is no difference in the heat capacities between fluid and solid, there remains a departure from local thermal equilibrium because of the difference in thermal conductivities. The value of δ is almost equal to zero for all time when $\lambda = 1.84$, even though $\rho_f Cp_f < \rho_s Cp_s$. This represents the situation where the higher heat capacity in the solid is actually almost counterbalanced by the higher thermal conductivity of the solid. When λ is larger than 1.84, the high thermal mass of the solid phase results in a negative value of δ, i.e., the thermal response of the solid lags behind that of the fluid. When $\lambda = 4$, the ratio of the thermal diffusivity of the two phases is one and the diffusion time scales of the two phases are the same. Under this condition a departure from thermal equilibrium is caused by the thermal tortuosity. For higher $\lambda = 10$ and 100, the magnitude of negative δ increases monotonically with λ. The time when the maximum value of $|\delta|$ occurs increases with increasing λ. At a fixed value of σ, the solid has a time scale longer than the fluid if λ is large. Therefore, the solid requires a longer time than the fluid to respond to the sudden temperature change, and the time for $|\delta|$ to reach maximum is longer for larger λ. In fact, Figure 13 shows that the proper time scale becomes l_m^2/α_s for large λ.

The effect of σ on δ is illustrated in Figure 14 for σ ranging from 0.01 to 100 while other parameters are fixed at $\varepsilon = 0.4$, $Nu^* = 108$, and $\lambda = 0.4$. Figure 14 shows that the departure from local thermal equilibrium occurs when the ratio σ differs from 0.4. Again, the thermal equilibrium condition occurs when σ is not equal to one if λ is different from one, i.e., when the effects due to σ and λ are counterbalanced. The most interesting feature in Figure 14 is that the time when the maximum value of δ occurs remains almost the same when the value of σ varies. Apparently, σ does not produce a significant effect on the difference in response time of the two phases, as compared to the effect due to λ.

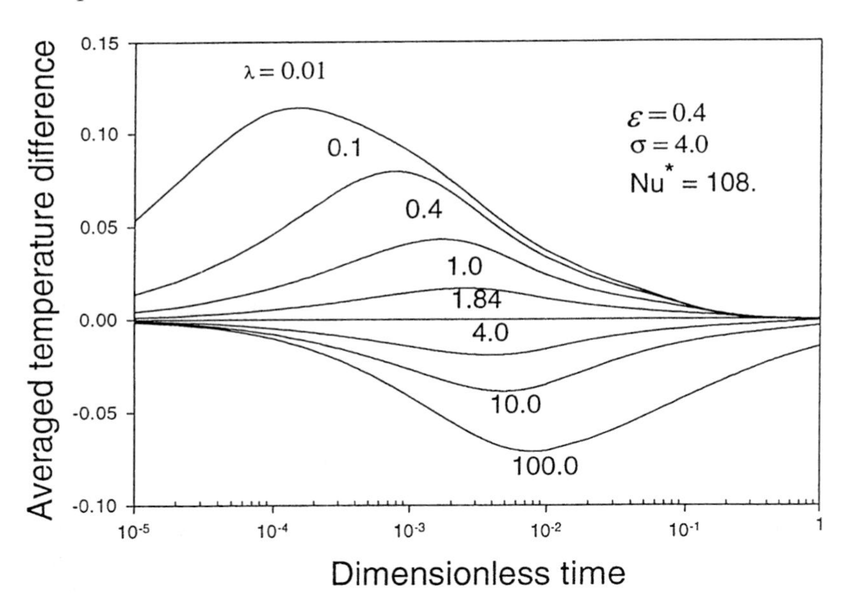

Figure 13. Effect of heat capacity ratio of solid to fluid on the total difference between macroscopic solid and fluid temperatures, indicating the establishment of a local thermal equilibrium.

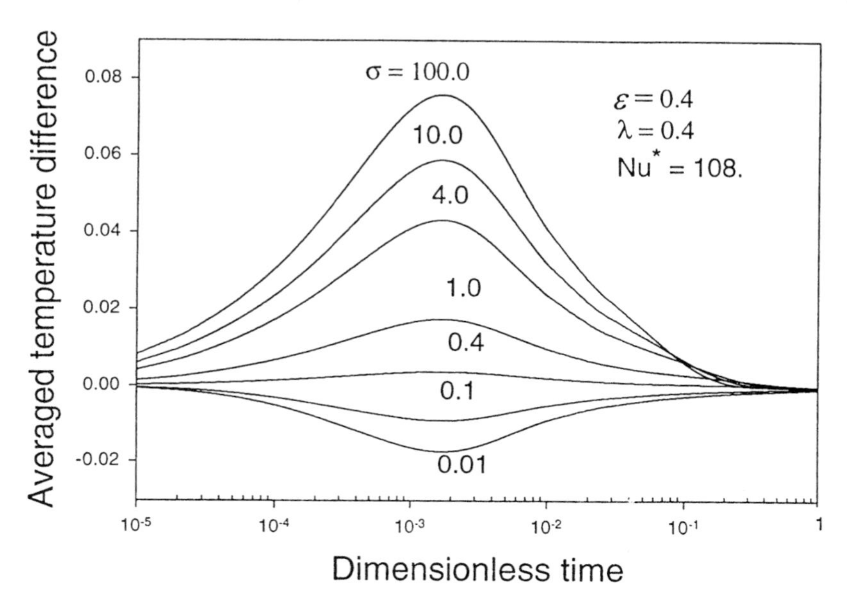

Figure 14. Effect of heat conductivity ratio of solid to fluid on the total difference between macroscopic solid and fluid temperatures, indicating the establishment of a local thermal equilibrium.

In Figure 15, the effect of porosity ε on δ is illustrated. For $\varepsilon = 0$ or 1.0, the porous medium reduces to a medium of pure solid or fluid, and δ has a zero value all the time. For $\varepsilon = 0.4$, 0.5, or 0.6, i.e., when the volumes of solid and fluid are about the same, the difference in thermal properties of the two phases gives rise to the largest thermal departure from local thermal equilibrium and δ has relatively larger values.

The above results indicate that the assumption of a local thermal equilibrium is valid when one of the following conditions is satisfied. These conditions are: (a) the porosity is close to 0 or 1; (b) the effects due to λ and σ are counterbalanced; (c) the value of Nu^* is very large; and (d) the time scale is large. These conditions, especially (b), imply that the commonly accepted condition of σ close to one is not sufficient to enforce the assumption of local thermal equilibrium.

IX. CONCLUDING REMARKS

Heat conduction in porous media under the local thermal nonequilibrium condition has been discussed in this chapter. The macroscopic transient heat

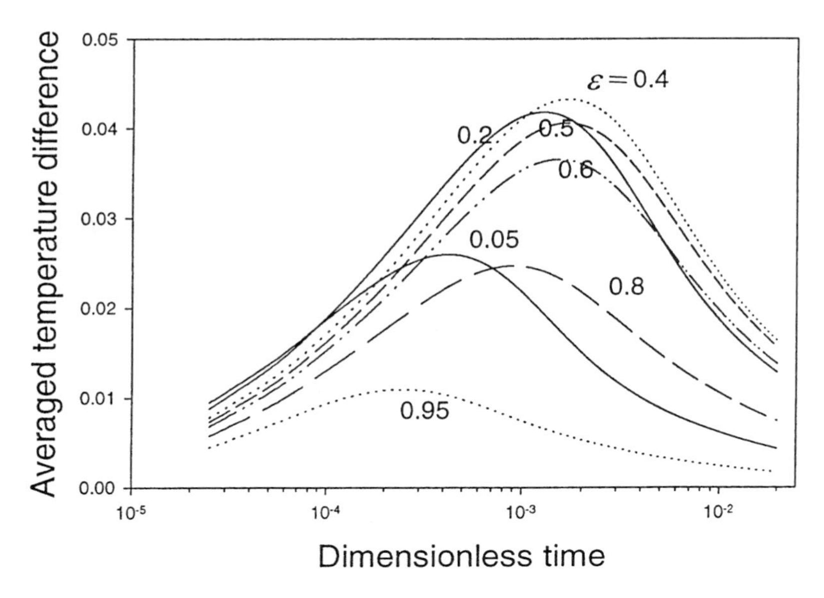

Figure 15. Effect of porosity on the total difference between macroscopic solid and fluid temperatures, indicating the establishment of a local thermal equilibrium.

conduction equations for fluid and solid phases were first obtained from microscopic equations by a phase average procedure. This leads to the closure problem due to thermal tortuosity and interfacial heat transfer between the two phases. The closure scheme of Hsu (1999) based on a microscopic quasi-steady assumption was adopted; this scheme appears to be consistent with, but mathematically more concise than, that of Quintard and Whitaker (1993). The macroscopic heat conduction equations were then combined into the single heat conduction equation for mixtures under the local thermal equilibrium assumption; this leads to the concept of effective stagnant thermal conductivity. The existing models for determining the effective stagnant thermal conductivity were reviewed. Then, the methods of evaluating the closure coefficients for thermal tortuosity and interfacial heat transfer into the two-equation model, especially the one proposed recently by Hsu (1999), were discussed. The characteristics of the two-equation model depends on five parameters—the porosity ε, the thermal conductivity ratio of solid to fluid σ, the heat capacity ratio of solid to fluid λ, the tortuosity parameter G, and the Nusselt number Nu^* for the interfacial heat transfer coefficient. It is noted that G is not an independent parameter since it is a function of ε and σ. The effect of these parameters on the local thermal nonequilibrium model, as investigated by Hsu and Wu (2000), are also presented.

This chapter has provided a complete and self-contained treatment of transient heat conduction in porous media under the assumption of local thermal nonequilibrium. However, it is noted that the results given in this chapter apply only to isotropic and homogeneous media. For media with anisotropic properties, the effective thermal conductivity will depend on the direction of thermal gradient. One such example was given by Hsu et al. (1996) for porous media made of wire screens. In this situation, the tensor forms for the effective stagnant thermal conductivity $\boldsymbol{K}_e$ should be used, and Eq. (37) in tensor form reads

$$\boldsymbol{G} = \frac{1}{(1-\sigma)^2}\left[\boldsymbol{K}_e/k_f - (\varepsilon + (1-\varepsilon)\sigma)\boldsymbol{I}\right] \tag{48}$$

For most layered media with organized solid structures, the tensors $\boldsymbol{K}_e$ and $\boldsymbol{G}$ are diagonal tensors. This has greatly simplified the determination of the thermal tortuosity coefficients.

Before closing, we want to emphasize that the treatment laid out in this chapter can be extended to media consisting of multi-component materials. The details of such extension require further investigations in the future.

ACKNOWLEDGMENT

This work was supported by the Hong Kong Government, through RGC Grant Nos. HKUST575/94E and HKUST815/96E.

NOMENCLATURE

Roman Letters

- a_{fs} specific area of fluid–solid interface per unit volume
- A conduction layer thickness outside a sphere, normalized by the radius of the sphere
- A_{fs} interfacial area between solid and fluid
- B conduction layer thickness inside a sphere, normalized by the radius of the sphere
- C_p specific heat
- G thermal tortuosity parameter
- h_{fs} interfacial heat transfer coefficient
- $\boldsymbol{I}$ unit matrix

M interfacial transfer parameter
n number of particles per unit volume
$\mathbf{n}_{fs}$ unit vector normal to the interface between fluid and solid
Q_{fs} interfacial heat transfer per unit volume
r radius
$\mathbf{s}$ area vector normal to the solid–fluid interface
V representative elementary volume (REV)

Greek Letters

β thermal diffusivity ratio of solid to fluid
δ total difference between macroscopic and fluid temperatures
γ_a length ratio of particle to unit cell
γ_c particle touching parameter
λ heat capacity ratio of solid to fluid
$\mathbf{\Lambda}_{fs}$ interfacial thermal tortuosity
σ thermal conductivity ratio of solid to fluid
τ thermal diffusion time scale

Subscripts

f fluid
p particle
s solid

REFERENCES

Amiri A, Vafai K. Analysis of dispersion effects and non-thermal equilibrium, non-Darcian, variable porosity incompressible flow through porous media. Int J Heat Mass Transfer 37:939–954, 1994.

Chang HC. Effective diffusion and conduction in two-phase media: a unified approach. AIChE Journal 29:846–853, 1983.

Cheng P, Hsu CT. Heat conduction. In: Ingham DB, Pop I, eds. Transport Phenomena in Porous Media. Oxford: Pergamon Press, Elsevier Science, 1998, p 57–76.

Crane RA, Vachon RI. A prediction of the bounds on the effective thermal conductivity of granular materials. Int J Heat Mass Transfer 20:711–723, 1977.

Crank J, Nicolson P. A practical method for numerical evaluation of solutions of partial differential equations of the heat-conduction type. Proc Cambridge Philos Soc. 43:50–67, 1947.

Deissler RG, Boegli JS. An investigation of effective thermal conductivities of powders in various gases. ASME Trans 80:1417–1425, 1958.

Hsu CT. A closure model for transient heat conduction in porous media. ASME J Heat Transfer, 121:733–739, 1999.

Hsu CT, Wu MW. A study on local thermal non-equilibrium based on one-dimensional transient heat conduction in porous media. 2000 (in manuscript).

Hsu CT, Cheng P, Wong KW. Modified Zehner–Schlunder models for stagnant thermal conductivity of porous media. Int J Heat Mass Transfer 37:2751–2759, 1994.

Hsu CT, Cheng P, Wong KW. A lumped parameter model for stagnant thermal conductivity of spatially periodic porous media. ASME J Heat Transfer 117:264–269, 1995.

Hsu CT, Wong KW, Cheng P. Effective thermal conductivity of wire screens. AIAA J Thermophys Heat Transfer 10:542–545, 1996.

Kaviany M. Principles of Heat Transfer in Porous Media. New York: Springer-Verlag, 1991.

Krupiczka R. Analysis of thermal conductivity in granular materials. Int Chem Eng 7:122–144, 1967.

Kunii D, Smith JM. Heat transfer characteristics of porous rocks. AIChE Journal 6:71–78, 1960.

Maxwell JC. A Treatise on Electricity and Magnetism. Oxford: Clarendon Press, 1873, p 365.

Nozad S, Carbonell RG, Whitaker S. Heat conduction in multiphase systems. I: Theory and experiments for two-phase systems. Chem Eng Sci 40:843–855, 1985.

Prasad V, Kladas N, Bandyopadhay A, Tian Q. Evaluation of correlations for stagnant thermal conductivity of liquid-saturated porous beds of spheres. Int J Heat Mass Transfer 32:1793–1796, 1989.

Quintard M, Whitaker S. One- and two-equation models for transient diffusion processes in two-phase systems. Adv Heat Transfer 23:369–367, 1993.

Quintard M, Whitaker S. Local thermal equilibrium for transient heat conduction: Theory and comparison with numerical experiments. Int J Heat Mass Transfer 38:2779–2796, 1995.

Rayleigh Lord. On the influence of obstacles arranged in rectangular order upon on the properties of a medium. Philos Mag 56:481–502, 1892.

Sahraoui M, Kaviany M. Slip and non-slip temperature boundary conditions at interface of porous, plain media: Conduction. Int J Heat Mass Transfer 36:1019–1033, 1993.

Swift, DL. The thermal conductivity of spherical metal powders including the effect of an oxide coating. Int J Heat Mass Transfer 9:1061-1073, 1966.

Tsotsas E, Martin H. Thermal conductivity of packed beds: A review. Chem Eng Prog 22:19–37, 1987.

Wakao N, Kaguei S. Heat and Mass Transfer in Packed Beds. New York: Gordon & Breach, 1982, p 294.

Wakao N, Kato K. Effective thermal conductivity of packed beds. J Chem Eng Jpn 2:24–32, 1969.

Wakao N, Vortmeyer D. Pressure dependency of effective thermal conductivity of packed beds. Chem Eng Sci 26:1753–1765, 1971.

Wakao N, Kaguei S, Funazkri T. Effect of fluid dispersion coefficients on particle-to-fluid heat transfer coefficients in packed beds. Chem Eng Sci 34:325–336, 1979.

Zehner P, Schlunder EU. Thermal conductivity of granular materials at moderate temperatures. Chem Ing-Tech 42:933–941, 1970 (in German).

5

Forced Convective Heat Transfer in Porous Media

Guy Lauriat and Riad Ghafir
Université de Marne-la-Vallée, Marne-la-Vallée, France

I. INTRODUCTION

Forced convective heat transfer in porous media has been the subject of many studies because of its important practical application in packed-bed chemical reactors, thermal insulation, compact heat exchangers, drying technology, electronic cooling, and numerous other applications. Up to the end of the 1970s, the overwhelming majority of existing studies were based on the Darcy flow model. For forced convection, this model leads to slug flow profiles so that the heat transfer results related to the assumption of slug flow can be directly transposed to the study of forced convection in porous media, provided that the assumption of local thermal equilibrium is valid. When the pore Reynolds number exceeds unity it has been established, beginning with Reynolds (1900) and Forchheimer (1901), that the inclusion of a quadratic term is required to account for flow recirculations, flow instabilities, and turbulence inside the pore volumes. However, the velocity profiles predicted by using the Darcy-extended Forchheimer momentum equation are still flat for boundary layer and duct flows for which the pressure does not depend on the normal coordinate to the wall. Similarly to clear fluids, the heat transfer results for external and confined flows are overestimated when using slug flow profiles, since viscous effects have been proved to be of significant importance to the convective transport of energy at the macroscopic scale. These effects should be considered in many practical applications in which porous media with relatively high porosities and permeabilities are used, especially for reducing pressure drop. For these reasons, the momentum diffusion caused by the friction of the boundary walls, modeled as a viscous term in the momentum equation, was included in a number of studies conducted in the past decades. In addition, the flow through boundary porous media become highly complex due to large-scale variations in porosity in the wall region, causing flow maldistribution and channeling in the boundary layer.

The above effects may be formulated using the volume-average technique of the continuity and momentum equations (Kaviany, 1991). For an incompressible Newtonian fluid flowing through an isotropic, homogeneous porous matrix, the continuity equation and Brinkman–Forchheimer–extended-Darcy equation of motion can be written as follows

$$\nabla \cdot \mathbf{v} = 0 \tag{1}$$

$$\frac{1}{\varepsilon}\frac{\partial \mathbf{v}}{\partial t} + \nabla \cdot \left(\frac{\mathbf{v} \otimes \mathbf{v}}{\varepsilon}\right) = -\frac{1}{\rho}\nabla p + \nu \nabla^2 \mathbf{v} - \left[\frac{\mathbf{v}}{K} + \frac{F}{K^{1/2}}|\mathbf{v}|\right]\varepsilon \mathbf{v} \tag{2}$$

where p and $\mathbf{v}$ are the volume-averaged pressure and velocity, respectively, ε is the porosity which may vary along the normal coordinate to the walls, K is the permeability, F the inertial coefficient which depends on the permeability as well as the microstructure of the porous matrix, ν is the fluid kinematic viscosity, and ρ the fluid density. It should be noted that the forces on the body due to gravity are neglected in Eq. (2) since we are considering forced convection only. The advective term in Eq. (2) results directly from the volume-averaging procedure applied to the momentum equation for the fluid phase. However, it will be shown in what follows that the development length is generally very short, so that the influence of the advective term is negligible for most of the flow fields that we will consider. At the end of the chapter, recent works on forced convective heat transfer in porous media saturated with non-Newtonian fluids are reviewed.

Most of the studies published in the heat transfer literature were conducted by assuming local thermal equilibrium (LTE). This assumption must be relaxed in many circumstances, and recent works on forced convective heat transfer in porous media (Amiri et al., 1995; Kuznetsov, 1997; Nield, 1998; Lee and Vafai, 1999) have considered the effects of local thermal nonequilbrium (LTNE). For an incompressible flow through a porous medium, by neglecting the viscous dissipation term and assuming that there is no local heat generation, the energy equations for the individual phases may be written as

Fluid phase energy equation

$$\varepsilon(\rho c_p)_f \frac{\partial T_f}{\partial t} + (\rho c_p)_f \mathbf{v}_f \cdot \nabla T_f = \nabla \cdot (k_{fe} \nabla T_f) + h_{fs} a_{sf} (T_s - T_f) \tag{3}$$

Solid phase energy equation

$$(1-\varepsilon)(\rho c_p)_s \frac{\partial T_s}{\partial t} = \nabla \cdot (k_{se} \nabla T_s) - h_{fs} a_{sf} (T_s - T_f) \tag{4}$$

where indexes s and f refer to the solid and fluid phases, respectively, T_f and T_s are the intrinsic phase averages of the fluid and solid temperatures, $\mathbf{v}_f$ the intrinsic phase-averaged velocity, h_{fs} is the fluid-to-solid heat transfer coefficient, a_{sf} the specific area of the porous bed, k_{fe} and k_{se} the fluid and solid effective thermal conductivities. The LTNE condition is considered in Section III.B.5. Under the LTE condition, Eqs. (3) and (4) can be matched into a phase-averaged energy equation as

$$(\rho c_p)_e \frac{\partial T}{\partial t} + (\rho c_p)_f \mathbf{v} \cdot (\nabla \mathbf{T}) = \nabla \cdot (\mathbf{k_e} \nabla \mathbf{T}) \tag{5}$$

In Eqs. (3)–(5), the effective conductivity includes thermal dispersion effects. In this chapter, the effect of thermal dispersion is viewed as a diffusive term added to the stagnant conductivity, and is discussed in Section III.B.4.

II. EXTERNAL FORCED CONVECTION

A. Forced Convection Boundary Layer along a Flat Plate

1. Boundary Layer Formulation

In the present analysis, the flow is assumed to be steady, incompressible, and two dimensional. The physical properties are constant and, at a first stage, the porosity is considered uniform. Assuming that the boundary layer approximation is also applicable, the governing equations can be written as

$$\frac{\partial u}{\partial x} + \frac{\partial v}{\partial y} = 0 \tag{6}$$

$$\frac{1}{\varepsilon^2}\left(u\frac{\partial u}{\partial x} + v\frac{\partial u}{\partial y}\right) = -\frac{1}{\varepsilon\rho}\frac{\mathrm{d}p}{\mathrm{d}x} + \frac{v}{\varepsilon}\frac{\partial^2 u}{\partial y^2} - \left[\frac{\mathrm{v}}{K} + \frac{F}{K^{1/2}}u\right]u \tag{7}$$

$$u\frac{\partial T}{\partial x} + v\frac{\partial T}{\partial y} = \alpha_e \frac{\partial^2 T}{\partial y^2} \tag{8}$$

where u, v are the volume-averaged velocity components and α_e is the effective thermal diffusivity, $\alpha_e = k_e/\rho c_p$. The x-axis has been chosen along the plate with the origin $x = 0$ at the leading edge, and the y-axis perpendicular to the flat plate. In writing Eq. (7), it has been assumed that $u^2 \gg v^2$ and, consequently, $|\mathbf{v}| \cong u$. The appropriate boundary conditions are

$$\left.\begin{array}{llll} u = u_\infty, & T = T_\infty, & x, y < 0 & \\ u = v = 0, & T = T_w, & x > 0, & y = 0 \\ u = u_\infty, & T = T_\infty, & x > 0, & y \to \infty \end{array}\right\} \quad (9)$$

This problem was extensively considered by Vafai and Tien (1981) for developing and for fully developed flows. Solutions of Eqs. (6)–(9) can be obtained through the application of the Kàrmàn–Pohlhausen integral method, similarity variables using, for example, the Falkner–Skan transformation, the method of matched asymptotic expansions, or numerical methods. The application of these methods was documented by Kaviany (1987), who was the first to apply the integral method. This method was found to yield meaningful and relatively accurate predictions for the skin friction coefficient and local Nusselt number. The application of transformed variables deduced from a scale analysis (Vafai and Tien, 1981; Bejan, 1984; Nakayama et al., 1990; Kumari et al., 1990; Nakayama and Pop, 1993) is considered next.

2. Momentum Boundary Layer over a Flat Plate

The pressure term in the x-momentum equation can be expressed as a function of free stream velocity and physical properties of the fluid-saturated porous medium by writing the momentum equation at a sufficiently large distance from the wall where $u = u_\infty$. This results in

$$\frac{1}{\varepsilon\rho}\frac{dp}{dx} = -\frac{\nu}{K}u_\infty - \frac{F}{K^{1/2}}u_\infty^2 \quad (10)$$

Equation (10) shows that the pressure gradient required for maintaining the free stream velocity at u_∞ is balanced by the solid matrix resistance. Therefore, the momentum equation can be written as

$$\frac{1}{\varepsilon^2}\left(u\frac{\partial u}{\partial x} + v\frac{\partial u}{\partial y}\right) = \frac{\nu}{K}(u_\infty - u) + \frac{F}{K^{1/2}}(u_\infty^2 - u^2) + \frac{\nu}{\varepsilon}\frac{\partial^2 u}{\partial y^2} \quad (11)$$

In the boundary layer developing region, the advective term is balanced by the Brinkman's term. Thus, one can conclude that the boundary layer thickness is of the order of

$$\delta \sim (\varepsilon\nu x/u_\infty)^{1/2} \quad (12)$$

In the fully developed region, the advective inertia can be neglected and a balance between the pressure gradient term and the Brinkman's term yields

$$\delta \sim \left(\frac{K/\varepsilon}{FRe_p + 1}\right)^{1/2} \quad (13)$$

where $Re_p = u_\infty \sqrt{K}/\nu$ is the pore Reynolds number. Therefore, the viscous boundary layer thickness increases as $x^{1/2}$ in the developing region and reaches a constant value in the developed region. If $FRe_p \ll 1$, the above momentum boundary-layer thickness is of the order of $(K/\varepsilon)^{1/2}$ (Vafai and Tien, 1981).

(a) Integral Method. Integrating the x-momentum equation with respect to y over the boundary layer thickness $\delta(x)$, eliminating the velocity component $v(x, y)$ by means of the continuity equation, and using a cubic approximation for the velocity profile (Kaviany, 1987) yields the following expression for the momentum boundary layer thickness

$$\left(\frac{\delta}{\sqrt{K/\varepsilon}}\right)^2 = \frac{140}{35 + 48FRe_p}\left[1 - e^{-\gamma X}\right] \qquad \text{for} \qquad 0 \le X \le (\varepsilon/Da)^{1/2} \tag{14}$$

where $Da = K/L^2$ is the Darcy number based on the plate length, L, $X = x/\sqrt{K/\varepsilon}$, and

$$\gamma = \left(\frac{70}{13}\frac{1}{Re_p} + \frac{96}{13}F\right)\varepsilon^{3/2} \tag{15}$$

The above solution shows that the thickness of the dimensionless boundary layer in the developed region depends on the group (FRe_p) only. It should be noted that the pore Reynolds number is related to the conventional fluid Reynolds number by $Re_p = Re_L\sqrt{Da}$. Equation (14) shows that the boundary layer thickness is almost constant for $X > 5/\gamma$. Hence, the dimensionless hydrodynamic development length is

$$X_e \cong \frac{65Re_p\varepsilon^{-3/2}}{70 + 96FRe_p} \tag{16}$$

In the case of negligible effects of the second-order flow resistance ($Re_p \ll 1$), the above expression reduces to

$$X_e \cong \frac{65}{70}Re_p\varepsilon^{-3/2} \tag{17}$$

or

$$X_e \cong 0.9\frac{Ku_\infty}{\nu\varepsilon^2} \tag{18}$$

These results are in agreement with the predictions of a scale analysis in Eq. (7) which shows that the advection term is significant over a length of the order of $(Ku_\infty/\nu\epsilon)$. It is clear that the thickness of the momentum boundary layer grows over a very short distance for most practical applications. However, this length may be quite large for a highly permeable medium and fluids with low kinematic viscosity. Beyond the developing length, the skin-friction coefficient becomes constant. This is a unique feature of flow through a porous medium. The skin friction coefficient is defined as

$$C_{fx} = \frac{\mu \frac{\partial u}{\partial y}\Big|_{y=0}}{\frac{1}{2}\rho u_\infty^2} = \frac{3\sqrt{K}}{\delta Re_p} \tag{19}$$

or

$$C_{fx} = \frac{3}{2\sqrt{35}} \frac{\sqrt{\varepsilon}}{Re_p} [35 + 48FRe_p]^{1/2} [1 - \mathrm{e}^{-\gamma X}]^{-1/2} \tag{20}$$

For negligible inertial effects, the developed friction coefficient is given as

$$C_f = \frac{3}{2} \frac{\varepsilon^{1/2}}{Re_p} \tag{21}$$

(b) Similarity Analysis for the Momentum Equation. The dimensional analysis of the momentum equation suggests the use of the following transformed variables (Nakayama et al., 1990):

$$\xi = b_1 x \qquad \text{and} \qquad \eta = \frac{y}{x}(Re_x/I)^{1/2} \tag{22}$$

where

$$b_1 = \varepsilon^2 \left(\frac{1}{FRe_p} + 1 \right) \frac{F}{\sqrt{K}}; \qquad Re_x = u_\infty x / \varepsilon \nu; \qquad I = \frac{1 - e^{-\xi}}{\xi} \tag{23}$$

Re_x is the local Reynolds number based on the pore velocity, u_∞/ε. The transformed streamwise coordinate ξ is deduced from Eq. (10) which shows that the pressure gradient at the boundary layer edge is linear. Therefore, ξ can be expressed as

$$\xi = (p'(0) - p'(x))/\rho(u_\infty/\varepsilon)^2 \tag{24}$$

where $p'(x)$ is the local averaged pressure related to the averaged intrinsic pressure p (i.e., the pressure read off a pressure gauge immersed into a porous medium) as $p'(x) = \varepsilon p(x)$. By defining a stream function $\psi(x, y)$ to satisfy the continuity equation such as

$$\psi(x, y) = \varepsilon \nu (Re_x I)^{1/2} f(\xi, \eta) \tag{25}$$

we can write

$$u = \frac{\partial \psi}{\partial y} = u_\infty f' \tag{26}$$

and

$$v = -\frac{\partial \psi}{\partial x} = \frac{\varepsilon \nu}{x}(Re_x/I)^{1/2}\left[\frac{1}{2}e^{-\xi}(\eta f' - f) - (1 - e^{-\xi})\frac{\partial f}{\partial \xi}\right] \tag{27}$$

where the prime denotes differentiation with respect to η. Substitution in Eq. (11) gives

$$f''' + \frac{1}{2}e^{-\xi}ff'' - (1 - e^{-\xi})\left[\frac{f' + FRe_p(f')^2}{1 + FRe_p} - 1\right] = (1 - e^{-\xi})\left(f'\frac{\partial f'}{\partial \xi} - f''\frac{\partial f}{\partial \xi}\right) \tag{28}$$

In terms of the transformed variables, the boundary conditions imposed on Eq. (28) are

$$f = f' = 0 \qquad \text{at} \qquad \eta = 0 \tag{29}$$

$$f' = 1 \qquad \text{for} \qquad \eta \to \infty \tag{30}$$

It can be seen that Eq. (28) possesses similarity in f only if special conditions are satisfied. From the definition of ξ, we can first write near the leading edge ($\xi \ll 1$)

$$f''' + \frac{1}{2}ff'' = 0 \tag{31}$$

This case reduces to the Blasius similarity solution for a semi-infinite flat plate embedded in a clear fluid, which leads to an increase in the boundary layer thickness as $x^{1/2}$. Second, for $\xi \gg 1$ Eq. (28) reduces to

$$f''' - \frac{f' + FRe_p(f')^2}{1 + FRe_p} + 1 = f'\frac{\partial f'}{\partial \xi} - f''\frac{\partial f}{\partial \xi} \tag{32}$$

If we assume that there exists a developed region in which f and f' are independent of ξ (constant boundary-layer thickness), the right-hand side term vanishes. Hence, this case also permits a similarity solution. Between these two asymptotic cases, a local similarity solution is achievable if it is assumed that the right-hand side term in Eq. (28) is small enough (Nakayama et al., 1990). Thus, the asymptotic similarity solutions for $\xi \ll 1$ and for $\xi \gg 1$ are connected with the local similarity solution of the following momentum equation

$$f''' + \frac{1}{2}e^{-\xi}ff'' - (1 - e^{-\xi})\left[\frac{f' + FRe_p(f')^2}{1 + FRe_p} - 1\right] = 0 \tag{33}$$

in which the streamwise coordinate ξ acts as a local parameter. The local skin friction coefficient can be evaluated from

$$C_{fx} = \varepsilon\nu\frac{\left.\dfrac{\partial u}{\partial y}\right|_{y=0}}{\frac{1}{2}u_\infty^2} = \frac{2f''(\xi, 0)}{(Re_x I)^{1/2}} \tag{34}$$

Nakayama et al. (1990) found $f''(\xi, 0)$ using a shooting method to solve numerically Eqs. (29), (30), and (33), and gave the following asymptotic values

$$\left.\begin{aligned} f''(0, 0) &= 0.332 \\ f''(\infty, 0)|_{Re_p=0} &= 1 \\ f''(\infty, 0)|_{Re_p\to\infty} &= 1.155 \end{aligned}\right\} \tag{35}$$

Boundary layer flows on a continuous moving solid surface at speed u_w through a porous medium find applications in many engineering domains. This problem was considered by Nakayama and Pop (1993) who employed the same transformed variables except that the wall velocity was taken as the velocity scale. Hence, the only change to bring into the above problem formulation is in Eq. (29). If the local Reynolds number is based on u_w instead of the free-stream velocity, the wall-boundary condition may be rewritten as

$$f = 0 \qquad \text{and} \qquad f' = 1 \qquad \text{at} \qquad \eta = 0 \tag{36}$$

Nakayama and Pop (1993) derived an exact solution for the case of $\xi \gg 1$ which permits a similarity solution as

$$f = \frac{3(1 + FRe_p)}{FRe_p}\frac{(A - 1)\left(1 - e^{-\eta/B}\right)}{1 - [(A - 1)/(A + 1)]e^{-\eta/B}} \tag{37}$$

where

$$A = \left(1 + \frac{2}{3}FRe_p\right)^{1/2} \qquad \text{and} \qquad B = (1 + FRe_p)^{1/2}$$

Thus, the dimensionless velocity is given by

$$f' = \frac{6}{FRe_p}\frac{[(A - 1)/(A + 1)]e^{-\eta/B}}{[1 - [(A - 1)/(A + 1)]e^{-\eta/B}]} \tag{38}$$

The local skin friction coefficient may be evaluated from

$$C_{fx} = -\varepsilon\nu \frac{\left.\frac{\partial u}{\partial y}\right|_{y=0}}{\frac{1}{2}u_w^2} = -\frac{2f''(\xi, 0)}{(Re_x I)^{1/2}} \tag{39}$$

$f''(\xi, 0)$ possesses the asymptotic values

$$f''(0, 0) = -0.444$$

$$f''(\infty, 0) = -\left[\frac{3 + 2FRe_p}{3(1 + FRe_p)}\right]^{1/2}$$

which leads to the asymptotic expressions for C_{fx}

$$C_{fx}/2 = 0.444/Re_x^{1/2} \qquad \text{for} \qquad \xi \ll 1 \tag{40}$$

and

$$C_{fx}/2 = \left(1 + \frac{2}{3}Re_x\right)^{1/2} \frac{\nu\varepsilon^{3/2}}{u_w K^{1/2}} \qquad \text{for} \qquad \xi \gg 1 \tag{41}$$

(c) Exact Solution for the Developed Region. Vafai and Thiyagaraja (1987) obtained an accurate solution using a singular perturbation analysis. The streamwise velocities were expanded in terms of powers of the porosity and results for up to the third order were given. Exact solutions were also obtained by Cheng (1987) and by Beckermann and Viskanta (1987) for a flow over a flat plate embedded in a porous medium. These closed-form solutions show that the velocity profile approaches asymptotically a simple exponential profile without inertial effects taken into consideration. This profile matches exactly the first-order term of the perturbation solution of Vafai and Thiyagaraja (1987). A closed-form analytical solution of Eq. (7) in the developed region is presented below.

The momentum equation can be put into dimensionless form by introducing the following dimensionless variables

$$U = u/u_\infty; \qquad Y = y/\sqrt{K/\varepsilon} \tag{42}$$

If the advection terms are neglected, substitution of (42) into (11) leads to

$$\frac{d^2U}{dY^2} + (1 - U) + FRe_p(1 - U^2) = 0 \tag{43}$$

Integration of Eq. (43) requires the following boundary conditions

$$\left.\begin{aligned} &U = 0 \qquad \text{at} \qquad Y = 0 \\ &\frac{dU}{dY} = 0 \qquad \text{and} \qquad U = 1 \qquad \text{for} \qquad Y \to \infty \end{aligned}\right\} \tag{44}$$

An analytical solution can be derived by multiplying (43) by $2\mathrm{d}U/\mathrm{d}Y$ and then integrating from Y to ∞ (Beckermann and Viskanta, 1987). Invoking the boundary conditions at $Y \to \infty$ yields

$$\frac{\mathrm{d}U}{\mathrm{d}Y} = \left(a(U-1)^2 + b(U-1)^3\right)^{1/2} \tag{45}$$

where

$$a = (1 + 2FRe_p); \qquad b = 2FRe_p/3 \tag{46}$$

After some algebraic manipulations, the fully developed profile is found to be

$$U = 1 - \frac{a}{b}\,\mathrm{sech}^2\left[a^{1/2}(Y + c_1)/2\right] \tag{47}$$

The constant of integration c_1 is obtained by applying the no-slip condition at the wall

$$c_1 = 2a^{-1/2}\mathrm{sech}^{-1}\left(\sqrt{a/b}\right) \tag{48}$$

The solution given by Eq. (47) matches the solution shown by Cheng (1987) for the same flow configuration and that given by Vafai and Kim (1989) for forced convection in a porous channel bounded by parallel plates, provided changes are made in the coordinate system. The solution given by Beckermann and Viskanta (1987) can be also transformed into the above expression.

The wall shear stress in dimensional form can be written as

$$\tau_w = \frac{\mu u_\infty}{\sqrt{K/\varepsilon}}[a - b]^{1/2} \tag{49}$$

The skin-friction coefficient for the fully developed velocity field is found to be

$$C_f = \frac{2\varepsilon^{1/2}}{Re_p}[a - b]^{1/2} \tag{50}$$

For the case where the inertial effects are neglected, we obtain

$$C_f = \frac{2}{Re_p}\varepsilon^{1/2} \tag{51}$$

Comparison between the above exact relation and the one obtained using the integral method shows that the correct coefficient for C_f is 2 instead of 3/2. Therefore, the integral solution underestimates the value of C_f by about 30% in the fully developed region. It should be noted that the integral treatment of the boundary layer leads to a more accurate solution in the developing rather than in the fully developed region (Kaviany, 1987). The boundary layer thickness, δ, can be deduced from the velocity profile by defining δ as the point where the velocity reaches 99% of its free stream value, i.e.

$$\frac{a}{b}\operatorname{sech}^2\left[a^{1/2}(Y+c_1)/2\right]=0.01 \tag{52}$$

This results in the following expression for the dimensional boundary layer thickness

$$\delta=\frac{2\sqrt{K/\varepsilon}}{\sqrt{1+2FRe_p}}\left\{\text{arc}\cosh\left[10\left(\frac{a}{b}\right)-\left(\left(100\frac{a}{b}-1\right)\left(\frac{a}{b}-1\right)\right)^{1/2}\right]\right\} \tag{53}$$

When $F=0$, a good approximation is

$$\delta\cong 5\sqrt{K/\varepsilon} \tag{54}$$

The results for the fully developed dimensionless velocity profile are plotted in Figure 1. Very similar profiles were shown by Cheng (1987) and by Beckermann and Viskanta (1987). As it can be seen in Figure 1, an

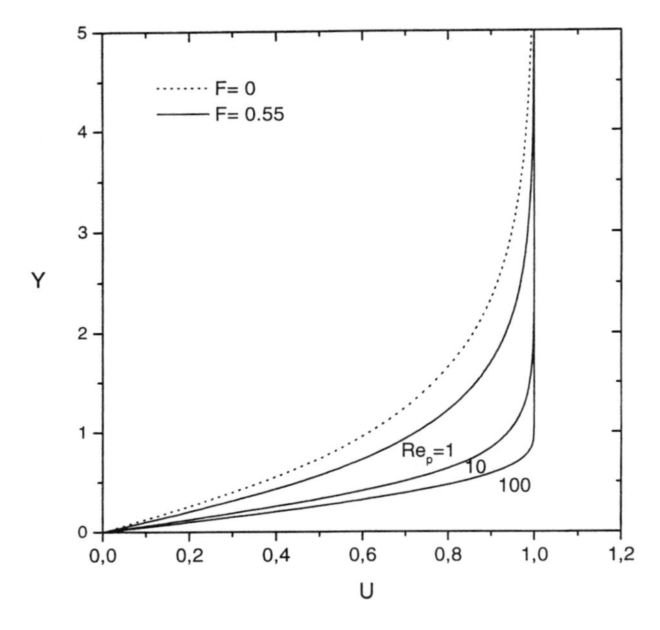

Figure 1. Effect of inertia on the fully developed dimensionless velocity profile over a flat plate.

increase of F and/or Re_p leads to a decrease in the boundary layer thickness. Hence, it can be concluded that the dimensional thickness of the momentum boundary layer depends not only on the permeability and porosity but also on the inertial effects. The same conclusions can be drawn regarding the developing length, which decreases when the inertial effects increase.

(d) Wall Effects Caused by Non-uniform Porosity. Forced convection from a flat isothermal plate embedded in variable porosity media was investigated experimentally by Vafai et al. (1985) and by Renken and Poulikakos (1989) for a porous matrix consisting of a packed bed of spheres. The measurements of Renken and Poulikakos (1989) were compared to closed-form, approximate, and numerical solutions (Vafai, 1984, 1986; Vafai et al., 1985; Beckermann and Viskanta, 1987; Cheng, 1987). The variation of porosity close to a solid boundary leads to flow maldistributions and channeling, which refer to the occurrence of a velocity overshoot in the vicinity of the wall. Such variations of porosity are caused by the nature of the contact between the wall and the solid matrix. Obviously, the observed effects depend on the structure of the porous matrix so that general conclusions are difficult to draw without specifying the microstructure of the porous matrix. Most of the investigations reported so far refer to unconsolidated porous media and, more precisely, to packed beds. Measurements of Roblee et al. (1958) and Benenati and Brosilow (1962) show a high-porosity region close to the boundary and variations of porosity as a damped oscillatory function of the distance from the boundary. In most of the theoretical and numerical studies, the oscillations of porosity were neglected and the variations of porosity were approximated by an exponential function of the form

$$\varepsilon = \varepsilon_\infty[1 + C\exp(-Ny/dp)] \tag{55}$$

where d_p is the particle diameter, C and N are empirical constants chosen so that the porosity approaches unity near the wall and decreases to ε_∞ at about four to five particle diameters away from the wall. It should be noted that Eq. (55) should be considered as a line-averaged porosity in the direction normal to the wall. For a packed-sphere bed, the Forchheimer constant and the permeability are related to the porosity and the particle diameter by the following equations

$$K = \frac{\varepsilon^3 d_p^2}{A_1(1-\varepsilon)^2} \tag{56}$$

$$F = \frac{B_1}{\sqrt{A_1}\varepsilon^{3/2}} \tag{57}$$

where the empirical constants A_1 and B_1 are determined from experimental results (from the results of Ergun (1952), $A_1 = 150$ and $B_1 = 1.75$). Using $\sqrt{K_\infty/\varepsilon}$ and u_∞ as scales for length and velocity and introducing the porosity variation into the momentum equation for the developed boundary layer regime yields

$$f_1(Y)(1-U) + (FRe_p)_\infty g_1(Y)(1-U^2) + \frac{d^2U}{dy^2} = 0 \tag{58}$$

where

$$f_1(Y) = \frac{\varepsilon_\infty^2(1-\varepsilon)^2}{\varepsilon^2(1-\varepsilon_\infty)^2} \qquad \text{and} \qquad g_1(Y) = \frac{\varepsilon_\infty^2(1-\varepsilon)}{\varepsilon^2(1-\varepsilon_\infty)}$$

The dependence of ε on Y can be deduced from Eq. (55). For uniform porosity, Eq. (58) reduces to (43). Due to the complicated nature of (58), a closed-form solution does not appear to be available. Vafai (1984, 1986) and Cheng (1987) have obtained analytical solutions for the velocity distribution, using the method of matched asymptotic expansions.

3. Thermal Boundary Layer

Unlike the momentum boundary layer, the thermal boundary layer does not become fully developed. Due to the complicated form of the velocity profile when boundary and inertia effects are accounted for, exact solutions for the energy equation have not yet been derived, except for the Darcy flow model.

(a) Similarity Solution Based on the Darcy Model. For the momentum equation based on the Darcy flow model, the heat transfer results are the same as for the asymptotic case of a flat plate embedded in a zero-Prandtl-number fluid, except that the Péclet number is based on the effective thermal diffusivity.

For an isothermal flat plate, this model yields the local and average Nusselt number as

$$Nu_x = \pi^{-1/2} Pe_x^{1/2} = 0.564 Pe_x^{1/2} \tag{59}$$

$$Nu = 1.13 Pe_L^{1/2} \tag{60}$$

If the wall is subjected to a uniform heat flux condition q_w, it can be shown that the temperature difference $(T_w(x) - T_\infty)$ varies as $x^{1/2}$ downstream. The local Nusselt number is given as

$$Nu_x = \frac{q_w}{k_e(T_w(x) - T_\infty)/x} = 0.886 Pe_x^{1/2} \tag{61}$$

The average Nusselt number, based on the average temperature difference between the leading edge and $x = L$, is

$$Nu_L = \frac{q_w}{k_e \overline{\Delta T}/L} = 1.329 Pe_L^{1/2} \tag{62}$$

where

$$\overline{\Delta T} = \frac{1}{L}\int_0^L (T_w(x) - T_\infty)\mathrm{d}x.$$

(b) Asymptotic Solutions. Assuming that the thermal boundary layer thickness is of the order of $\delta_t = \delta/Pr_e$, asymptotic analytical expressions for the temperature distribution can be obtained by accounting for the possible large differences in thickness of the thermal and the momentum boundary layers according to the value of the effective Prandtl number.

For low Prandtl number fluids, $\delta \ll \delta_t$ so that the temperature distribution in the limit $Pr \to 0$ is the Darcy limit given in Nield and Bejan (1992). For a more general case where the inertial effects are taken into account, and for a variable wall temperature in the form $\theta(x) = a_1 x^{p_1}$, an exact solution has been obtained by Vafai and Thiyagaraja (1987) for low Prandtl number fluids in terms of the gamma function and parabolic cylinder function $D_{-(2p_1+1)}$ as

$$\theta(x, y) = a_1 \Gamma(p_1 + 1)\left[2^{p_1+1/2}\pi^{-1/2}x^{p_1}\exp(-\kappa y^2/x)D_{-(2p_1+1)}\left(\sqrt{4\kappa y^2/x}\right)\right. \tag{63}$$

where $\kappa = u_\infty/8\alpha_e$. Hence, it can be deduced that

$$Nu_x = \frac{\Gamma(p_1 + 1)}{\Gamma(p_1 + 1/2)} Pe_L^{1/2} x^{1/2} \tag{64}$$

For very large Prandtl numbers, the thermal boundary layer is completely inside the momentum boundary layer. According to Levêque (1928), the velocity distribution very close to the wall can be assumed linear and is given by

$$u = \tau_w y/\mu \qquad (Pr \to \infty) \tag{65}$$

where τ_w is the wall shear stress given by Eq. (49) for the fully developed momentum boundary layer. Hence, the energy equation can be approximated by

$$y\frac{\partial T}{\partial x} = \frac{\alpha_e \mu}{\tau_w}\frac{\partial^2 T}{\partial y^2} \tag{66}$$

Introducing the similarity variable

$$\eta_1 = y\left(\frac{1}{9\xi_1 x}\right)^{1/3} \tag{67}$$

where

$$\xi_1 = \frac{\alpha_e \mu}{\tau_w} = \frac{K}{\left[\varepsilon(1 + 4FRe_p/3)\right]^{1/2} Re_p Pr_e]}$$

the energy equation and the boundary conditions for a developed momentum boundary layer and a developing thermal boundary layer can be written as

$$\left.\begin{aligned} &\frac{d^2\theta}{d\eta_1^2} + 3\eta_1^2 \frac{d\theta}{d\eta_1} = 0 \\ &\theta(0) = 0 \qquad \text{and} \qquad \theta(\infty) = 1 \end{aligned}\right\} \tag{68}$$

The solution of (68) has the form (Beckermann and Viskanta, 1987)

$$\theta = 1.12 \int_0^{\eta_1} \mathrm{e}^{-\zeta^3} \mathrm{d}\zeta \tag{69}$$

from which the following local Nusselt number can be obtained

$$Nu_x = x\frac{\partial\theta}{\partial y}\bigg|_{y=0} = 1.12\left[\frac{x^2}{9\xi_1}\right]^{1/3} \tag{70}$$

or

$$Nu_x = 0.5384\left[\varepsilon\left(1 + \frac{4}{3}FRe_p\right)\right]^{1/6}\left(\frac{Re_p Pre}{Da_x}\right)^{1/3} \tag{71}$$

where $Da_x = K/x^2$. The average Nusselt number is

$$Nu = \frac{3}{2} Nu_L \tag{72}$$

Equations (64) and (70) show that, for high effective Prandtl numbers, the Nusselt number is proportional to $Pr_e^{1/3}$ while Nu is proportional to $Pr_e^{1/2}$ for low effective Prandtl numbers.

(c) Similarity Solution for Non-Darcian Flows. Kaviany (1987) and Kumari et al. (1990) applied the Falkner–Skan transformation to the energy equation in order to achieve efficient numerical computation of the thermal boundary layer by using finite difference methods. Nakayama et al. (1990) and Nakayama and Pop (1993) employed a local similarity transformation

procedure for the stationary isothermal plate problem and for the case of forced convection over a continuous moving plate in a porous medium. The following new variables are introduced for the energy equation (8)

$$\theta(\xi, \eta_t) = \frac{T - T_w}{T_\infty - T_w} \qquad \eta_t = \frac{y}{x} Re_x^{1/2} = \sqrt{I}\eta \tag{73}$$

where η, ξ, and I are defined by Eqs. (22) and (23). It should be noted that the present definition of θ is not as in Nakayama et al. (1990). Therefore the following results differ slightly from those presented in their paper.

Thus, the energy equation is transformed into

$$\frac{1}{\varepsilon Pr_e}\theta'' + \frac{1}{2}\sqrt{I}f\theta' = \xi\left[\left(\sqrt{I}f\right)'\frac{\partial\theta}{\partial\xi} - \theta'\frac{\partial}{\partial\xi}\left(\sqrt{I}f\right)\right] \tag{74}$$

where the primes in (74) indicate differentiation with respect to η_t. The boundary conditions are

$$\theta = 0 \qquad \text{at} \qquad \eta_t = 0 \tag{75}$$

$$\theta = 1 \qquad \text{for} \qquad \eta_t \to \infty \tag{76}$$

where $Pr_e = \nu/\alpha_e$ is the effective Prandtl number. For $\xi \ll 1$, $I = 1$ and the right-hand side terms in (74) can be neglected. Thus, the similarity solution for a Newtonian fluid is recovered. For $\xi_1 \gg 1$, $I = 1/\xi$ and the terms $\partial\theta/\partial\xi$ and $\partial(\sqrt{I}f)/\partial\xi$ become small. From the consideration of these two limiting cases, Nakayama et al. (1990) suggested dropping the right-hand side of Eq. (74). The transformed energy equation is thus approximated as

$$\frac{1}{\varepsilon Pr_e}\theta'' + \frac{1}{2}\sqrt{I}f\theta' = 0 \tag{77}$$

Two integrations with respect to η_t and the use of the boundary conditions given by (75) and (76) lead to

$$\theta(\xi, \eta_t) = \frac{\int_0^{\eta_t} \exp\left[-\frac{1}{2}\varepsilon Pr_e\sqrt{I}\int_0^{\eta_t} f(\xi, \eta_t/\sqrt{I})\mathrm{d}\eta_t\right]\mathrm{d}\eta_t}{\int_0^{\infty} \exp\left[-\frac{1}{2}\varepsilon Pr_e\sqrt{I}\int_0^{\eta_t} f(\xi, \eta_t/\sqrt{I})\mathrm{d}\eta_t\right]\mathrm{d}\eta_t} \tag{78}$$

The local Nusselt number at the plate can be written as

$$\begin{aligned} Nu_x/Re_x^{1/2} &= x\frac{\partial\theta}{\partial y}\bigg|_{y=0} / Re_x^{1/2} = \theta'(\xi, 0) \\ &= \left[\int_0^{\infty} \exp\left(-\frac{1}{2}\varepsilon Pr_e I\int_0^{\eta_t} f(\xi, \eta_t)\mathrm{d}\eta_t\right)\mathrm{d}\eta_t\right]^{-1} \end{aligned} \tag{79}$$

By comparing the foregoing expression of $Nu_x/Re_x^{1/2}$ with the one found for forced convection over an isothermal flat plate in a Newtonian fluid, it can be concluded that the group $(\varepsilon Pr_e I)$ may be regarded as an equivalent Prandtl number. When $(\varepsilon Pr_e I)$ is large, the thermal boundary layer is much thinner than the velocity boundary layer and a linear velocity profile may be assumed across the boundary layer, such that (Beckermann and Viskanta, 1987)

$$f'(\xi, \eta) = f''(\xi, 0)\eta \tag{80}$$

Substitution of (80) into (77) yields

$$\theta(\xi, \eta_t) = \frac{3}{\Gamma(1/3)} \int_0^{\Omega} e^{-\Omega^3} d\Omega \bigg| \Omega = \left[\frac{1}{12}\frac{\varepsilon Pr_e}{\sqrt{I}} f''(\xi, 0)\right]^{1/3} \eta_t \tag{81}$$

where Γ is the gamma function such that $3/\Gamma(1/3) = 1.12$.

When $(\varepsilon Pr_e I)$ is small, the velocity boundary layer is much thinner than the thermal boundary layer and the slug flow approximation may be used. The numerical results for Eqs. (75)–(77) presented by Nakayama et al. (1990) for $\varepsilon Pr_e = 0.7$ show that the effects of Re_p are quite insensitive to θ (ξ, η_t) in the hydrodynamically developed region ($\xi \gg 1$) since the porous inertia effects on the dimensionless velocity profile are significant only for large ξ-values at which the thermal boundary layer thickness is much larger than the velocity boundary layer thickness, so that the Darcian flow model holds. This finding agrees with the results of Vafai and Tien (1981) who showed that the influences of boundary and inertial effects on the temperature profiles decrease with increasing downstream distance.

Forced convection over a continuous flat plate moving at speed u_w embedded in a motionless fluid-saturated porous medium was investigated by Nakayama and Pop (1993), using a local similarity solution procedure. They showed that the Nusselt number is substantially different from the case of a stationary plate embedded in a porous medium with a moving fluid. They presented asymptotic formulas for Nu_x according to the values of the dimensionless streamwise coordinate ξ and the group $(\varepsilon Pr_e I)$. For $(\varepsilon Pr_e I) \gg 1$

$$Nu(x)/Re_x^{1/2} = (\varepsilon Pr_e/\pi)^{1/2} \tag{82}$$

For $(\varepsilon Pr_e I) \ll 1$ and $\xi \ll 1$

$$Nu(x)/Re_x^{1/2} = 0.808 \varepsilon Pr_e \tag{83}$$

For $(\varepsilon Pr_e I) \ll 1$ and $\xi \gg 1$

$$Nu(x)/Re_x^{1/2} = \frac{3}{2FRe_p}(1 + FRe_p)^{1/2}\left[\left(1 + \frac{2}{3}FRe_p\right)^{1/2} - 1\right]\frac{\varepsilon Pr}{\xi^{1/2}} \qquad (84)$$

where $Re_x = u_w x/\varepsilon\nu$ is the Reynolds number based on the pore velocity u_w/ε. The foregoing equations suggest that the dependence of Nu_x on Pr_e is much higher for the moving plate than for the stationary plate.

Forced convection over a flat plate embedded in a porous medium saturated with non-Newtonian fluids (power-law model) was investigated by Hady and Ibrahim (1997). They used similarity-type variables to transform the momentum equation, including boundary friction and inertial effects, and the energy equation. The results indicate that the heat transfer from the plate decreases as the power-law index increases. Their results are, on the whole, in agreement with what has been found for non-Newtonian power-law fluids for the non-porous case. Vafai (1984) investigated the effect of variable porosity by using the method of matched asymptotic expansions and numerically solving the governing equations. Vafai et al. (1985) and Renken and Poulikakos (1989) conducted experimental studies to clarify the wall channeling effect on heat transfer in packed beds for constant wall heat flux and isothermal plate, respectively. Both of these studies shed light on the importance of variable porosity on heat transfer.

B. Transient External Forced Convection

Unsteady forced convection near heated or cooled bodies embedded in a porous medium has received little attention. Transient forced convection over an isothermal flat-plate was studied analytically by Nakayama and Ebinuma (1990) using the Darcy–Forchheimer model and time-dependent streamwise velocity for a flow that starts initially at rest. Bejan and Nield (1991) considered steady Darcian flows over a wall, both for suddenly imposed uniform temperature and for uniform heat flux. They performed a scale analysis for the energy equation which yielded time-dependent thickness of the thermal boundary layer and heat transfer rate for three regimes in the evolution of the temperature field for a wall with constant temperature as well as for a wall with constant heat flux. Details of this analysis can be found in the book by Nield and Bejan (1992).

In what follows we will refer to the work by Nakayama and Ebinuma (1990). Neglecting the advective and viscous terms in the momentum equation and axial conduction, the governing equations reduce to

$$\frac{\rho}{\varepsilon}\frac{\mathrm{d}u}{\mathrm{d}t} = -\frac{\mathrm{d}p}{\mathrm{d}x} - \frac{\mu}{K}u - \frac{\rho F}{\sqrt{K}}u^2 \qquad (85)$$

$$r_c \frac{\partial T}{\partial t} + u \frac{\partial T}{\partial x} = \alpha_e \frac{\partial^2 T}{\partial y^2} \tag{86}$$

where r_c is the solid–fluid heat capacity ratio. If we assume that a sudden uniform and constant pressure gradient is applied at $t = 0$ in order to induce a fluid flow, u is a function of t only. The appropriate initial and boundary conditions are

$$\begin{aligned} &u(0) = 0 \qquad T(0, x, y) = T_\infty \qquad T(t, 0, y) = T_0 \\ &T(t, x, 0) = T_w \qquad T(t, x, \infty) = T_\infty \end{aligned} \tag{87}$$

An exact solution to the momentum equation is obtained as

$$u = u_D \frac{2(1 - \mathrm{e}^{-(\gamma_1 t/t_D)})}{\gamma_1 + 1 + (\gamma_1 - 1)\mathrm{e}^{-(\gamma_1 t/t_D)}} \tag{88}$$

where

$$u_D = \frac{K}{\mu}\left(-\frac{\mathrm{d}p}{\mathrm{d}x}\right), \qquad t_D = \frac{\rho K}{\varepsilon \mu}, \qquad \gamma_1 = (1 + 4FRe_p)^{1/2} \tag{89}$$

For a Darcy flow ($F = 0$), Eq. (88) reduces to

$$u = u_D(1 - \mathrm{e}^{-(t/t_D)}) \tag{90}$$

Bejan and Nield (1991) determined the thickness of the thermal boundary layer from a scale analysis of the energy equation (86). At sufficiently short times, the transverse conduction is balanced by the thermal inertia. This yields the time-dependent boundary layer thickness scale

$$\delta \sim \left(\frac{\alpha_e t}{r_c}\right)^{1/2} \tag{91}$$

As time elapses, the longitudinal convection increases relative to the thermal inertia, and is balanced by the transverse conduction. In this second regime, the boundary layer scale becomes

$$\delta \sim \left(\frac{\alpha_e x}{u(t)}\right)^{1/2} \tag{92}$$

This occurs when $r_c \Delta T/t < u(t)\Delta T/x$. Thus, the boundary layer changes from a quasi-steady convective state in the upstream region to transient one-dimensional heat conduction farther downstream at $x = x_{s-t}$ where

$$x_{s-t} \sim \frac{u(t)t}{r_c} \tag{93}$$

The upstream region extends from the leading edge as time elapses and the downstream region is no longer present when $x_{s-t} \sim L$. Therefore, the solution becomes quasi-steady (or steady when u is time-independent) at the time $t_\infty \sim r_c L/u$.

In the case of the isothermal wall, analytical solutions were obtained by Bejan and Nield (1991) and by Nakayama and Ebinuma (1990) for these two asymptotic cases. For the upstream region extending from the leading edge, the quasi-steady-state solution is given by

$$\frac{T - T_w}{T_\infty - T_w} = \mathrm{erf}\left(\frac{y}{2(\alpha_e x/u(t))^{1/2}}\right) \tag{94}$$

For the downstream region, the transient solution is given by

$$\frac{T - T_w}{T_\infty - T_w} = \mathrm{erf}\left(\frac{y}{2(\alpha_e t/r_c)^{1/2}}\right) \tag{95}$$

Equation (95) is nothing else than the exact solution for transient heat conduction in a semi-infinite medium subjected to a sudden temperature rise at its bounding surface. As was noticed by Bejan and Nield (1991), Eqs. (94) and (95) are not exact at $x = x_{s-t}$ where thermal inertia, longitudinal convection, and transverse conduction are of same order of magnitude.

The following expressions for the local heat transfer coefficient are obtained from each of the asymptotic one-dimensional solutions

$$\frac{h(\alpha_e t_D/r_c)^{1/2}}{k_e} = \left\{ \begin{array}{ll} \dfrac{1}{\sqrt{\pi}}\left(\dfrac{t}{t_D}\right)^{-1/2} & \text{for } x > x_{s-t} \\ \dfrac{1}{\sqrt{\pi}}\left[\dfrac{2(1 - \mathrm{e}^{-(\gamma_1 t/t_D)})}{\gamma_1 + 1 + (\gamma_1 - 1)\mathrm{e}^{-(\gamma_1 t/t_D)}}\right]^{1/2}\left(\dfrac{r_c x}{u_D t_D}\right)^{-1/2} & \text{for } x < x_{s-t} \end{array} \right\} \tag{96}$$

The variation of the local heat transfer coefficient is shown in Figure 2 for a Darcy flow ($F = 0$ or $\gamma_1 = 1$) at various dimensionless times t/t_D. The horizontal lines are for the transient solution. It is clearly shown that the quasi-steady-state solution extends further downstream as time elapses, and the solution curve asymptotically approaches the steady-state solution curve.

The case of a wall with constant heat flux was considered by Bejan and Nield (1991). A few other shapes of bodies embedded in porous media was also investigated. Transient forced convection about a cylinder was first considered by Kimura (1989), who derived asymptotic analytical solutions

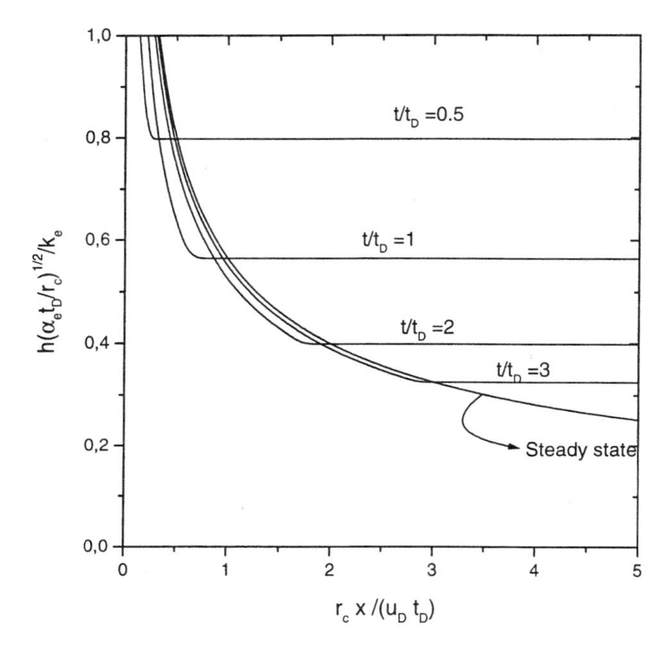

Figure 2. Variation of the local heat transfer coefficient for transient Darcy flow ($F = 0$).

for small and large times and presented numerical results valid for any time. Heat transfer from a circular cylinder with constant heat flux subjected to crossflow was studied analytically and experimentally by Kimura and Yongeya (1992). An order-of-magnitude analysis showed that the heat transfer is dominated by conduction at small times, but is eventually balanced by convection at larger times. Unsteady forced convection around a sphere was studied by Sano and Makizono (1995) for Darcy flows. Asymptotic solutions for large Péclet numbers were obtained for a sphere suddenly heated and, subsequently, maintained at constant temperature. They also observed that, at first stage, the heat transfer process is dominated mainly by conduction, and the convective effect becomes important as time elapses.

III. INTERNAL FORCED CONVECTION

Most of the earlier studies on convective heat transfer in ducts were conducted by assuming that the flow is fully developed. The hydrodynamic entrance length, which is usually short in porous media channel flows, is a function of geometry, permeability, porosity, and fluid viscosity. Numerical results displayed by Kaviany (1985) for the hydrodynamic characteristics of a low Reynolds number flow through a porous channel bounded by two parallel plates separated by a distance $2D$ have shown that the entrance length decreases rapidly as the dimensionless group $2D(\varepsilon/K)^{1/2}$ increases. Since the Darcy number is small in many practical applications, the assumption of fully developed velocity profile is much more justified in porous media flows than for Newtonian fluid flows.

Poulikakos and Renken (1987) considered channels packed with spherical beads. They presented detailed results of the thermal entry length for fully developed flow by using a general flow model for two channel geometries, parallel plates and circular pipe, with the walls maintained at constant temperature. In both configurations, they reported correlations for the dependence of the thermal entry length on the dimensionless pressure gradient and bead diameter. These correlations show that increases in flow rate, resulting from increase in pressure gradient and/or bead diameter, yield increases in thermal entry length. Cheng et al. (1988) analyzed thermally developing forced convection in an asymmetrically heated channel packed with spherical particles. Their study, which includes near-wall variations of porosity and transverse thermal dispersion effects, has shown that the thermal entrance length decreases as the dimensionless particle diameter decreases or as the Reynolds number increases. Marpu (1993) considered the effect of axial conduction in pipes filled with porous materials. The advective, Forchheimer, and Brinkman terms were included in the momentum equation. The results indicate that the inclusion of axial conduction plays a significant role at low Péclet numbers in the thermally developing region (i.e., for $x/(2DRePr_e) < 0.1$, where the Reynolds number is based on the pipe diameter and mean velocity). As for forced convection in clear fluids, it was found that axial conduction does not lead to any significant change in local Nusselt number for $Pe_e > 100$, even in the thermally developing region. It was also shown that the influence of axial conduction does not depend significantly on the type of flow description employed (Darcy's or modified-Darcy's models).

A. Darcy and Darcy–Forchheimer Flow Regimes

Hydrodynamically developed flows in a channel bounded by parallel plates, in a circular pipe, and in an annular duct, are considered in this section. The assumptions imposed are (a) steady laminar flows, (b) constant properties, (c) local thermal equilibrium, (d) thermal dispersion neglected, (e) inlet temperature known, and (f) axial conduction neglected.

In the Darcy–Forchheimer flow regime, the axial volume-averaged velocity is uniform and is described by the x-momentum equation

$$0 = \frac{1}{\varepsilon}\frac{\mathrm{d}p}{\mathrm{d}x} + \left[\frac{\mu}{K}u + \rho\frac{F}{\sqrt{K}}|u|\right]u \tag{97}$$

where $\mathrm{d}p/\mathrm{d}x$ is constant. When the porosity, permeability, and Forchheimer coefficient are assumed to be uniform a flat velocity profile results. The Darcy velocity in the x-direction is recovered for $F = 0$. The energy equation for $u =$ constant has the form

$$\rho c_p u\frac{\partial T}{\partial x} = k_e\frac{1}{y^m}\frac{\partial}{\partial y}\left(y^m\frac{\partial T}{\partial y}\right) \tag{98}$$

where $m = 0$ for the parallel-plate channel and $m = 1$ for the circular pipe and annular duct ($y = r$).

1. Thermally Developing Flow in Parallel Plate Channel and Circular Pipe

For the case of thermally developing flow in parallel plate channel and circular pipe, the thermal boundary conditions are

$$\left.\begin{aligned} &T(0, y) = T_i \\ &\frac{\partial T}{\partial y} = 0 \qquad \text{at} \qquad y = 0 \\ &T = T_w \qquad \text{or} \qquad -k_e\frac{\partial T}{\partial y} = q_w \qquad \text{at the duct walls} \end{aligned}\right\} \tag{99}$$

where T_w or q_w are constant. Therefore, the Graetz solutions for slug flows can be used to calculate the bulk temperature of the stream and the local Nusselt number at any distance x from the entrance of a porous filled channel. These solutions are reported extensively in the literature (Burmeister, 1993; Kakaç and Yener, 1995).

Defining the thermal entry length x_t as the distance between the entrance of the channel and the point at which $Nu_x = 1.05Nu_\infty$, the entry lengths for these two geometries are shown in Figure 3 by filled circles. It can be seen that the thermal entry length is about ten times shorter for constant wall temperature than for constant wall heat flux. The values of Nu in the fully developed region reported above are approximately 50 percent higher than the values predicted for laminar flows of Newtonian fluids in parallel plate channel or circular duct without a porous matrix. In addition, since the Nusselt number is based on the effective thermal conductivity, further increases in the heat transfer coefficient can be achieved for effective thermal conductivity higher than the fluid thermal conductivity. In conclusion, these

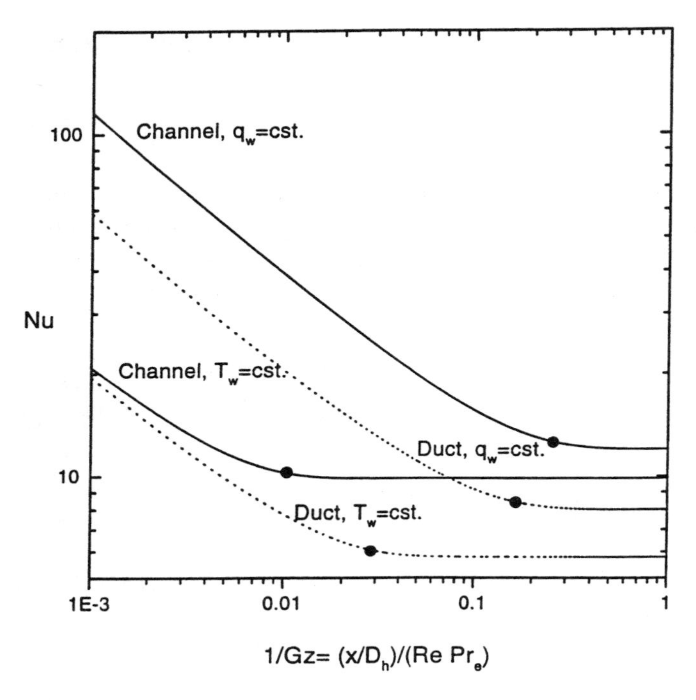

Figure 3. Variations of the local Nusselt number in the entrance region for parallel-plate channel and circular pipe (Darcy's flow).

results show that the insertion of a porous material within a duct may lead to a significant increase in the overall heat transfer from the wall to the fluid over the entrance length: this prediction is the basic concept used in the design of compact heat exchangers.

When the effects of flow inertia increase, the flow rate decreases for a fixed pressure drop while the profile is kept flat (slug flow). Since the above results are still valid, it can be concluded that both the thermal entry length and local Nusselt numbers decrease due to the decrease in the fluid Reynolds number. However, the fully developed results are not changed when the magnitude of Forchheimer inertia is increased. Consequently, the various curves corresponding to different values of the Forchheimer coefficient in the thermal entry region should converge toward the asymptotic value for fully developed heat transfer. Figure 4 presents the effect of *Re* on the Nusselt-number variations in the thermal entrance region of a circular duct with constant wall temperature.

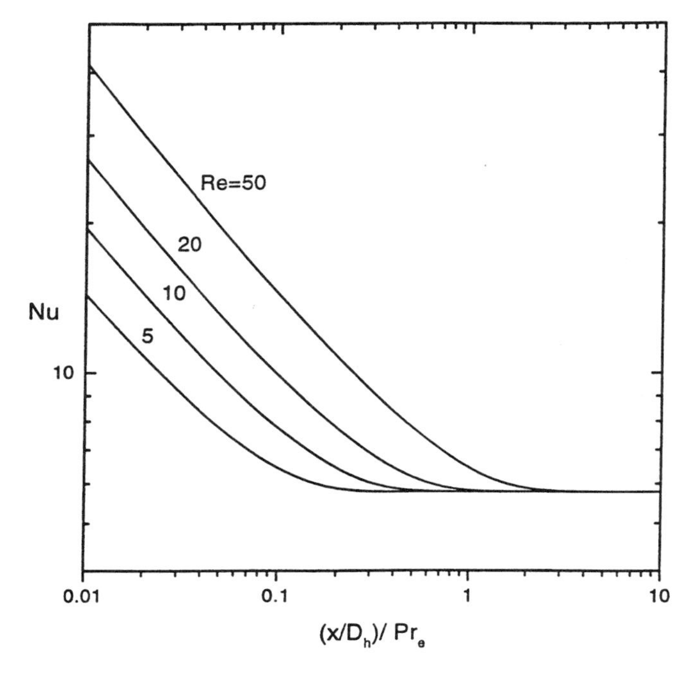

Figure 4. Effects of inertia on local Nusselt number in a circular pipe (Darcy–Forchheimer flow model).

2. Fully Developed Regime in a Porous Filled Annulus

We consider a porous filled annulus of inner and outer radii r_i and r_o, respectively. Under the conditions of constant heat rate to the fluid saturated porous medium, it is possible to derive exact solutions for the fully developed temperature profile and Nusselt number that are independent of Reynolds and Prandtl numbers. Since the two surfaces can be subjected to different heat fluxes, there are two Nusselt numbers of interest, one for each surface. We will employ the conventional notation for heat transfer in circular tube annulus (Kays and Perkins, 1985): subscript *ii* designates conditions on the inner surface when the inner surface alone is heated (the outer being insulated), and subscript *oo* designates conditions on the outer surface when the outer surface alone is heated (the inner being insulated). A single subscript (*i* or *o*) refers to the condition on the inner or outer surface, respectively, under any condition of simultaneous heating at both surfaces.

For the Darcy and Darcy–Forchheimer flow regimes, it can readily be shown that the fully developed temperature profile is of the form

$$T(x, r) = \gamma_2 x + \theta(r) \tag{100}$$

where

$$\theta(r) = \left(\frac{\gamma}{4\alpha_e} u_D\right)(r^2 - 2r_o^2 \ln r) + \frac{q_o r_o}{k_e} \ln r + \mathrm{a}_2 \tag{101}$$

and

$$\gamma_2 = \frac{2(r_i q_i + r_o q_o)}{\rho c_p u_D (r_o^2 - r_i^2)} \tag{102}$$

The mixing-cup temperature should be determined to evaluate a_2. The differences between the wall and mixing-cup temperatures are

$$T_i(x) - T_m(x) = \frac{2r_i}{k_e}(a_i q_i + b_i q_o) \tag{103}$$

$$T_o(x) - T_m(x) = \frac{2r_o}{k_e}(a_o q_i + b_o q_o) \tag{104}$$

where a_i, a_o, b_i, and b_o are functions of the radius ratio $R = r_i/r_o$ only, as given by

$$a_i = \frac{1}{8}\left(\frac{R^2 - 3}{1 - R^2}\right) - \frac{\ln R}{2(1 - R^2)^2} \tag{105}$$

$$b_i = \frac{R^4 - 4R^2 \ln R - 1}{8R(1 - R^2)^2} \tag{106}$$

$$a_o = \frac{R(1 - R^2) - 2R}{8(1 - R^2)} - \frac{R^3 \ln R}{2(1 - R^2)^2} \tag{107}$$

$$b_o = \frac{3(1 - R^2) - 2}{8(1 - R^2)} - \frac{R^4 \ln R}{2(1 - R^2)^2} \tag{108}$$

Use of Eqs. (103) and (104) gives the asymptotic Nusselt numbers based on the hydraulic diameter ($D_h = 2(r_o - r_i)$) as

$$Nu_i = \frac{h_i D_h}{k_e} = \frac{(1/R - 1)}{a_i + b_i(q_o/q_i)} \tag{109}$$

$$Nu_o = \frac{h_o D_h}{k_e} = \frac{(1 - R)}{a_o(q_i/q_o) + b_o} \tag{110}$$

For $q_o = 0$, it can be found that

$$Nu_{ii} = \frac{1}{a_i}\left(\frac{1}{R} - 1\right) \tag{111}$$

and, for $q_i = 0$

$$Nu_{oo} = \frac{1}{b_o}(1 - R) \tag{112}$$

so that

$$Nu_i = \frac{Nu_{ii}}{1 - (q_o/q_i)c_i} \tag{113}$$

$$Nu_o = \frac{Nu_{oo}}{1 - (q_i/q_o)c_o} \tag{114}$$

where $c_i = -b_i/a_i$ and $c_o = -a_o/b_o$ are called "influence coefficients." The temperature difference between the inner and outer surfaces is also derived as

$$T_i(x) - T_o(x) = \frac{D_h}{k_e}\left[q_i\left(\frac{1}{Nu_{ii}} + \frac{c_o}{Nu_{oo}}\right) - q_o\left(\frac{c_i}{Nu_{ii}} + \frac{1}{Nu_{oo}}\right)\right] \tag{115}$$

Values of the Nusselt numbers at the inner and outer walls are reported in Table 1 as a function of the radius ratio for constant heat rate on one surface with the other surface insulated. For comparison, the values given by Kays and Perkins (1985) for fully developed, laminar heat transfer to a Newtonian fluid flowing in a circular annulus are also reported in Table 1. The limiting cases $R = 0$ and $R = 1$ are for circular pipe and parallel plate channel, respectively.

Table 1. Values of Nusselt number in the case of an annulus filled with a porous medium

	Fully developed velocity profile		Slug flow	
R	Nu_{ii}	Nu_{oo}	Nu_{ii}	Nu_{oo}
0.00	∞	4.364	∞	8.00
0.05	17.81	4.792	16.82	7.638
0.10	11.91	4.834	11.29	7.341
0.20	8.499	4.883	8.201	6.898
0.40	6.583	4.979	6.617	6.383
0.60	5.912	5.099	6.176	6.136
0.80	5.58	5.24	6.032	6.028
1.00	5.385	5.385	6.00	6.00

3. Friction Factor

Obviously, such a heat transfer augmentation is obtained at the price of a dramatic increase of pressure drop, as can be readily demonstrated by using Darcy's law. The uniform velocity profile is given by

$$u = -\frac{K}{\mu}\frac{\mathrm{d}p}{\mathrm{d}x} = \text{constant} = u_D \tag{116}$$

For developed flow, the pressure drop over a length L is only due to the bulk frictional drag induced by the solid matrix (Darcy's pressure drop), which is

$$\Delta p_D = p(0) - p(L) = \mu u_D L/K \tag{117}$$

Defining a friction factor as

$$f_r = -\frac{(\mathrm{d}p/\mathrm{d}x)D_h}{\rho u_D^2/2} \tag{118}$$

where D_h is the hydraulic diameter of the duct, the result is

$$f_r = \frac{2}{ReDa} \tag{119}$$

where $Da = K/D_h^2$ and $Re = u_D D_h/v$ are the Darcy number and Reynolds number based on the hydraulic diameter, respectively. Since Da typically ranges from 10^{-8} to 10^{-3}, it is clear that the bulk frictional drag is very high compared to the boundary drag for flows of Newtonian fluids in the same duct but without porous material. Experiments (Ward, 1964) have shown that the Darcy equation is valid only at low velocities. Using $K^{1/2}$ as a length scale to define a pore Reynolds number as

$$Re_p = \frac{\rho u_D K^{1/2}}{\mu} \tag{120}$$

and a friction factor

$$f_r = \frac{(-\mathrm{d}p/\mathrm{d}x)K^{1/2}}{\rho u_D^2} \tag{121}$$

the Darcy friction factor for forced flow can be rewritten as

$$f_r = \frac{1}{Re_p} \tag{122}$$

If the pore Reynolds number exceeds $O(1)$ inertial effects, first as recirculating flows inside the volumes of the pores and later as turbulence, flatten the variation of f_r versus Re_p in a manner similar to the friction factor variations for turbulent flows in rough pipes. Ward (1964) noted that f_r should be rewritten as

$$f_r = \frac{1}{Re_p} + F \tag{123}$$

where F is an empirical constant which may be expressed as $F = K^{1/2}\beta_1$. The value of F depends both on the permeability and on the structure of the porous matrix (through β_1). From experimental results of isothermal water flows through particulate media, Ward (1964) determined that $F = 0.55$. Beavers and Sparrow (1969) performed experiments to determine the permeability of porous media consisting of a latticework of metallic fibers (foametal, wire screen assembly, and feltmetal). For materials belonging to the same structural family but showing large differences in permeability, they proposed $F = 0.074$. On the other hand, F is much larger for porous media made up of metallic fibers which have free fiber ends within the medium. For feltmetal, they measured an F-value of 0.132.

B. General Flow Model

1. Effects of Boundary Friction

This problem was documented by Kaviany (1985) who derived expressions for the pressure drop as a function of the Darcy number when the inertial effects are neglected. Using numerical integration, Kaviany also obtained the fully developed Nusselt number for laminar flow through a porous channel bounded by isothermal parallel plates. The solution of the momentum equation with the Forchheimer term omitted gives the following dimensionless velocity profile for constant $\mathrm{d}p/\mathrm{d}x$ in a channel of width $2D$

$$U(Y) = \frac{S}{\tanh S - S}\left(\frac{\cosh SY}{\cosh S} - 1\right) \tag{124}$$

where $Y = y/D$, $U = u/u_m$, and $S = 0.5(\varepsilon/Da)^{1/2}$ with $Da = (K/D^2)$.

Figure 5 shows the dimensionless velocity profile given by Eq. (124) for several values of the Darcy number. For small Da, the effects of the Brinkman term are restricted to a thin wall region, and a slug flow is obtained over most of the channel width. The parabolic velocity profile for Newtonian fluids is recovered at large values of Da. The overall frictional drag induced by both the solid matrix and the walls, as defined by Eq. (118), becomes

$$f_r = \frac{8}{Re\varepsilon}\left(\frac{S^3}{S - \tanh S}\right) \tag{125}$$

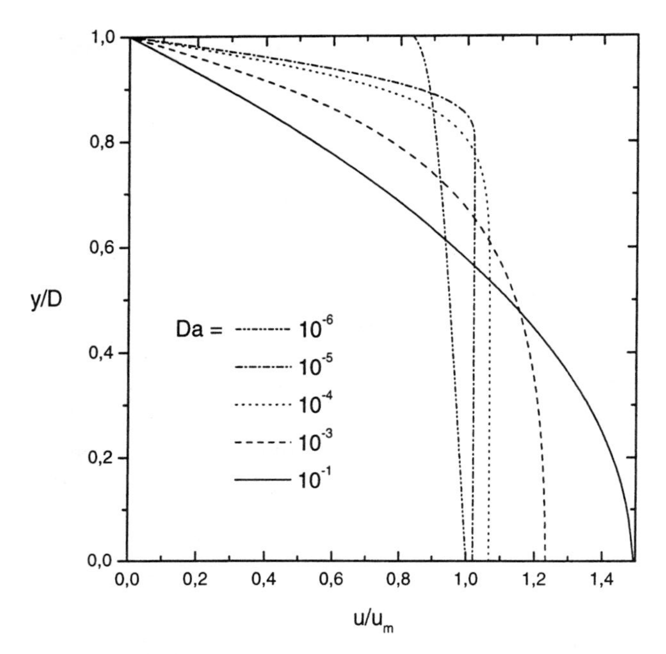

Figure 5. Fully developed velocity profile for several values of Da ($\epsilon = 1$) for a parallel-plate channel (Darcy–Brinkman model).

The total pressure drop over a distance x for the fully developed flow is found to be

$$\Delta p(x) = \Delta p_D(x)\left(\frac{S}{S - \tanh S}\right) \tag{126}$$

As expected, the total pressure loss reduces to Darcy's pressure drop for very low permeability (i.e., $S \to \infty$).

The fully developed temperature profile for constant wall heat flux was given by Lauriat and Vafai (1991) as

$$\theta(Y) = \frac{3(S - \tanh S)}{3\tanh S(S\tanh S + 5) + S(2S^2 - 15)} \left[2\left(\frac{\cosh SY}{coshS} - 1\right) - S^2(Y^2 - 1)\right] \tag{127}$$

where θ is defined as

$$\theta = \frac{T(x, y) - T_w(x)}{T_m(x) - T_w(x)} \tag{128}$$

The Nusselt number based on the hydraulic diameter is obtained from its definition as

$$Nu = \frac{24S(S - \tanh S)^2}{3 \tanh S(S \tanh S + 5) + S(2S^2 - 15)} \tag{129}$$

Equation (129) shows that the fully developed Nusselt number decreases from its asymptotic value for pure Darcian flow ($Nu = 12$) to the Newtonian fluid value $Nu = 8.24$ as (Da/ε) increases from 10^{-6} to 1. The numerical results presented by Kaviany (1985) for forced convection through a porous channel bounded by isothermal plates display similar trends. Therefore, these results indicate that the general effect of the Brinkman term is to decrease the heat transfer by reducing the flow velocity at the wall region.

2. Effects of Flow Inertia

A closed-form analytical solution for a fully developed flow in a porous channel bounded by parallel plates subject to a constant heat flux boundary condition was presented by Vafai and Kim (1989). The approach used for deriving the velocity profile is similar to the one used for solving the developed flow over a flat plate. In deriving this solution it has been inherently assumed that two distinct boundary layers exist at each boundary. As a consequence, this solution is not valid for channels filled with high porosity materials such as some metallic foams. The solution was derived for effective viscosity equal to fluid viscosity but the extension to classical models for effective viscosity is straightforward. Nield et al. (1996) presented a general analysis not based on boundary layer approximations, but of much more complicated form since it requires numerical integration of a transcendental equation.

The results for the velocity profile show that the momentum boundary layer thickness decreases as the inertial effect becomes more significant. Hence, it can be concluded that the thickness of the velocity boundary layer depends not only on the Darcy number but also on an inertial parameter defined as $\Lambda_1 = F\varepsilon^{1/2}Re$. These results are in agreement with those reported for the effect of the inertial parameter on the thickness of the boundary layer over a flat plate embedded in a porous medium. The dimensionless temperature profiles presented in Vafai and Kim (1989) and in Lauriat and Vafai (1991) show that the inertial parameter has less effect on the temperature profile than on the velocity profile. It was also found that an increase in the inertial parameter causes an increase in the Nusselt number. This is due to the augmented mixing of the fluid for larger inertial parameters, which leads to a more uniform velocity profile.

3. Effects of Variable Porosity

Variation in porosity near the walls was experimentally shown to have a significant influence on the velocity profiles for confined flows, and thus on heat transfer. Most of the studies on variable porosity media have concerned packed beds. Obviously, the effect of porosity variation along the normal to an impermeable wall is all the more important since the bed-to-particle ratio is relatively small, i.e., $D/d_p \leq 30$ for a packed pipe, as is often encountered in laboratory chemical reactors, for example. In a number of experimental and theoretical works, such as the pioneering studies of Schwartz and Smith (1958), Benenati and Brosilow (1962), and Schertz and Bischoff (1969), the flow maldistribution which arises in randomly packed beds was examined. When modeling flow and heat transfer, the properties of the fluid-saturated porous medium, such as permeability, inertia coefficient, effective viscosity, and thermal conductivity, need thus to be varied spatially within a very short distance from the walls. This strongly complicates analytical or numerical studies since sharp gradients in velocity and temperature occur in the near-wall region. Chandrasekhara and Vortmeyer (1979) were amongst the first to incorporate the effect of variation of porosity in a numerical work on isothermal forced flow in a circular pipe. They showed that the variation of porosity has a greater influence on channeling of velocity profiles near the walls than inertial effects. However, the computations were restricted to low wall porosity ($0.56 \leq \varepsilon_w \leq 0.588$). In Cohen and Metzner (1981) a theoretical model was proposed. This model divides a column randomly packed with spheres of uniform size into three regions: a wall region which extends a distance of one particle diameter away from the wall; a transition region in which appreciable porosity variations occur; and a bulk region in which the porosity may be considered as uniform. This work revealed essentially that the use of a single region model based on the average porosity leads to an over-prediction of the mass flux. Vortmeyer and Schuster (1983) examined analytically flow distributions within packed beds using a variational principle satisfying the no-slip condition and wall porosity equal to one, and approximated the porosity variation by an exponential expression. In this study, an explanation for the deviations between calculated velocity profiles inside a packed pipe and those which are often measured just above the packing is given. Fully developed forced convection through a packed-sphere bed between concentric cylinders maintained at different temperatures was considered by Cheng and Hsu (1986b). Radial variations of porosity, permeability, and stagnant thermal conductivity as well as transverse thermal dispersion were accounted for by using the Brinkman model. They derived an analytical solution based on the method of matched asymptotic expansions for the velocity profile, and showed that the magnitude of the velocity overshoots at the inner and outer cylinders increases as the dimensionless particle diameter is decreased. Hsu and Cheng (1988) obtained an analytical solution for the velocity profile of a fully developed flow in a packed parallel plate channel with the inclusion of the Forchheimer term. They demonstrated that the effect of variable porosity on the dimensionless pressure drop is more pronounced at low particle Reynolds numbers than at high Reynolds numbers because the Darcian resistance is proportional to $1/K$ while the Forchheimer resistance is proportional to $1/K^{1/2}$.

Cheng et al. (1988) and Chowdhury and Cheng (1989) examined thermally developing forced convection in a packed pipe and a packed channel. The flow, assumed to be hydrodynamically developed, was described by the Brinkman–Forchheimer extended-Darcy equation, with variable porosity approximated exponentially. Transverse thermal dispersion and variable stagnant thermal conductivity were taken into consideration in the energy equation, which was solved numerically. These papers present comparisons of theory and experiments for the temperature profiles in circular pipes and channels filled with glass spheres at different downstream locations.

Hunt and Tien (1988a) concentrated their study on packed chemical reactors. Non-Darcian formulation was used to predict the heat transfer rates with porosity varying form one at the wall to its bulk value after a few particle diameters. The stagnant thermal conductivity was allowed to have an exponential variation similar to that of the porosity so that its value close to the wall was the fluid conductivity. The results showed that the model used predicts well the heat transfer results of many experimental studies related to chemical reactors. Poulikakos and Renken (1987) and Renken and Poulikakos (1988) investigated numerically and experimentally the effect of flow channeling for forced convection at the thermal entry and fully developed regions in packed channels with the walls maintained at constant temperature. In these studies, it is shown that the thermal entry length increases almost proportionally to the imposed dimensionless axial pressure gradient or to the square of the bead diameter. A numerical and experimental investigation of forced convection in a packed-sphere square channel with the consideration of flow inertia, boundary friction, channeling, and thermal dispersion effects was presented by Chou et al. (1992). It was found that the values of the fully developed Nusselt number for constant wall heat flux are influenced mainly by the channeling effect with the Péclet number based on the sphere diameter is small, whereas the thermal dispersion effects become dominant when the Péclet number is large.

Kamiuto and Saitoh (1994, 1996) examined numerically fully developed forced convection in a cylindrical packed bed with constant wall temperature or with constant wall heat flux, and addressed the validity of their computations through comparisons with experimental data. They used the Ridgeway–Turback distribution to represent the local porosity variations. The asymptotic Nusselt number as $Re_d Pr \rightarrow 0$ was obtained as a function of the thermal conductivity ratio of the solid to that of the fluid and of the ratio of the radius of the pipe to the sphere diameter. For large $Re_d Pr$ they found that Nu depends mainly on particle Reynolds number and fluid Prandtl number. Amiri and Vafai (1994) employed a full general model for the momentum equation and a two-phase equation model for the energy equation to investigate numerically forced convection in a parallel plate channel packed with spherical particles of different sizes and materials with constant wall temperature. Their results permit one to determine the influence of a variety of effects and to present error maps for assessing the importance of various simplifying assumptions which are commonly used. Chou et al. (1994) re-examined experimental data from the literature and compared them to their measurements and numerical results for forced convection in packed pipes. They reconfirmed that the discrepancies among the various models used were mainly due to the modeling of channeling at low Péclet numbers. When using the flow model reported by Chandrasekhara and Vortmeyer (1979), their numerical predictions of Nu were found in agreement with data for both air and water flows at low Pe_d. Jiang et al. (1996a) presented a numerical study of convective heat transfer in a vertical porous annulus in which the effects of inertia,variable porosity, thermal dispersion, variable properties, fluid pressure, etc. were analyzed. The numerical results were compared with experimental data for water and R-318. They showed that the predicted values of the friction factor without consideration of variable porosity are 21 percent higher than those with variable porosity, which was found to agree well with experimental data. Jiang et al. (1996b, 1998) examined numerically and experimentally forced convection in a parallel plate channel. They considered variable porosity effects amongst a number of additional effects.

4. Effects of Thermal Dispersion

Thermal dispersion results from velocity gradients at the pore scale and the existence of the solid matrix which forces the flow to undergo a tortuous path due to the complex geometry of a porous medium. Most of the existing models assume that the effective conductivity k_e is the sum of a stagnant thermal conductivity, k_0, and a dispersion conductivity, k_d, which incorporates the additional transport due to transverse and longitudinal mixing within the pore (Cheng and Vortmeyer, 1988; Hunt and Tien, 1988b). The effect of thermal dispersion is thus viewed as a diffusive term, function of the phase-averaged velocity, porosity, geometry, etc. For confined flows, the thermal dispersion conductivity vanishes near the wall region where $k_0 \gg k_d$, while k_d is maximum at the regions where maxima of velocity occur. In the cases of hydrodynamically developed, axisymmetric channel flows, the transverse dispersion conductivity reaches its maximum at the channel axis, and the large variations in effective thermal conductivity with distance from the wall induce sharp temperature gradients in the wall region (as in turbulent flows). However, when solving the flow field in a porous medium by using the Darcy–Forchheimer model, all of the heat transfer results presented in Section III.A still hold provided that the expression of k_e in the definitions of Pr_e and Nu_x is written as $k_e = k_0 + k_d$, since k_d is constant under these circumstances. On the other hand, when wall effects are included into the problem formulation the expression of the thermal dispersion coefficient is a much more difficult task since it varies from the wall to the core region. Obviously, if wall channeling effects and viscous effects (Brinkman term) are shown to be restricted to a thin wall region, multilayer flow approaches should be introduced in order to simplify theoretical or numerical analyses. The mixing length concept proposed by Cheng and Vortmeyer (1988) for a packed bed provided some insight into the physics of the dispersion phenomenon, and was the first attempt to model the decrease in lateral mixing of fluid near solid boundary through the introduction of a wall function.

For a slug flow of velocity u_D in a constant porosity packed tube, Yagi and Wakao (1959), and Wakao and Kaguei (1982) gave the effective thermal conductivity as

$$k_e = k_0 + \gamma_d k_f Pr Re_d \tag{130}$$

where k_f and Pr are the thermal conductivity and Prandtl number of the fluid, respectively. $Re_d = u_D d_p / \nu$ and the constant γ_d was empirically determined to range between 0.09 and 0.1 according to the ratio of the tube diameter to the particle diameter. By rewriting Eq. (130) as

$$k_e = k_0(1 + \gamma_d Pe_e) \tag{131}$$

where $Pe_e = u_D d_p / \alpha_e$ is the effective Péclet number, it can be concluded that the transverse thermal dispersion is of the order of magnitude of the stagnant effective thermal conductivity if $Pe_e = 10$. When the wall effects are accounted for, the foregoing expression of k_e is applicable only in the core region. Hunt and Tien (1988b) modified Eq. (131) in order to introduce the non-uniform transverse velocity distribution as

$$k_e = k_0(1 + \gamma_d Pe_e U) \tag{132}$$

where $U = u/u_D$ and $u_D = -K_\infty/\mu(\mathrm{d}p/\mathrm{d}x)$. Consequently, $k_e \simeq k_0$ at the wall region as the velocity approaches zero. A similar relationship was also derived by Georgiadis and Catton (1988), who used a statistical model.

According to the mixing length concept introduced by Cheng and Vortmeyer (1988), the effective transverse thermal conductivity for forced convection in a porous medium bounded by a wall is given by

$$k_d/k_f = \gamma_d l_m Pr Re_{d,m} U \tag{133}$$

where $Re_{d,m}$ is the particle Reynolds number based on the mean velocity u_m, $U = u/u_m$, and l_m is a dimensionless mixing length (or dispersive length) for transverse thermal dispersion which depends on a dimensionless empirical constant obtained by matching analytical or numerical predictions with experimental data. Cheng (1987) showed that the mixing length is about a few particle diameters in length, that is to say, much larger than the thickness of the boundary friction layer ($\delta_v \sim K^{1/2}$) and of the order of the thickness of the non-uniform porosity layer for a packed-sphere bed.

The expression for the dispersive length for transverse thermal dispersion introduced by Hsu and Cheng (1988) for a fully developed flow of air in annular and cylindrical packed beds composed of glass spheres was $l_m = (1-\varepsilon)/\varepsilon^2$, where the porosity ε varies from the wall. The empirical constant $\gamma_d = 0.02$ was obtained by matching theory and experiments. For a two-layer mixing length, the expression of l_m in Eq. (133) is given by Cheng and Vortmeyer (1988) as

$$\begin{aligned} &l_m = y/\omega d_p \text{ in the wall region } (0 \le y \le \omega d_p) \qquad \text{and} \\ &l_m = 1 \text{ in the core region} \end{aligned} \tag{134}$$

whereas, for a Van Driest type of mixing length, the expression of l_m is given by

$$l_m = 1 - \exp(-y/\omega d_p) \tag{135}$$

where y is the normal distance from the wall. For forced convection through a packed-sphere bed between concentric cylinders of inner radius r_i, Cheng and Hsu (1986b) suggested the value of $\gamma_d = 0.2 \sim 0.25$, and $\omega = 2 \sim 2.5$ (d_p/r_i) in the definition of l_m.

In the paper by Cheng et al. (1988), Eq. (133) is rewritten as

$$k_d/k_f = \gamma_d l_m (Pr Re_{d,m} U)^{n_1} \tag{136}$$

where n_1 is an empirical constant obtained by matching the numerical results with experimental data. For a parallel plate channel of width $2H$, the expression for l_m is

$$l_m = \begin{Bmatrix} 1 - e^{-y/\omega}; & 0 \leq y \leq 1/\Gamma_1 \\ 1 - e^{-(\frac{2}{\Gamma_1} - y)/\omega}; & 1/\Gamma_1 \leq y \leq 2/\Gamma_1 \end{Bmatrix} \quad (137)$$

where $\Gamma_1 = d_p/H$. Using $C = 1$, $N = 2$, and $\varepsilon_\infty = 0.4$ in the exponential function for the porosity variation (Eq. 55), they found that $\omega = 1.5$, $\gamma_d = 0.17$, and $n_1 = 1$ give the best agreement with the experimental data.

Many other dispersion models have been reported in the literature since the pioneering work of Yagi and Wakao (1959). For a porous bed of particles, Kamiuto and Saitoh (1994, 1996) used the following expression

$$k_d/k_f = 0.3519(1 - \varepsilon)^{2.3819} Pr Re_d \quad (138)$$

where the variation of porosity is specified by the Ridgeway–Tarbuck distribution, and Re_d is the local particle Reynolds number based on the local axial velocity component.

In a recent experimental and numerical study of forced convection in parallel-plate channels filled with glass, stainless steel or bronze spherical particles, Jiang et al. (1998) used a modified version of the Hsu and Cheng (1990) model as

$$k_d = b_2(\rho c_p)_f d_p (u^2 + v^2)^{1/2}(1 - \varepsilon) \quad (139)$$

with

$$b_2 = 1.042\left[(\rho c_p)_f d_p u(1 - \varepsilon_m)\right]^{-0.8282} \quad (140)$$

where ε_m is the mean porosity.

Obviously, more sophisticated models of the dispersion conductivity must be used when the entrance length effect or/and axial conduction are considered. In the heat transfer literature, some investigations have considered the contribution of axial dispersion and noted that its importance is limited to low particle Reynolds number for porous-filled ducts with constant wall heat flux or constant wall temperature. This finding is due to the fact that convective heat transfer dominates axial diffusion at high flow rates. Wakao and Kaguei (1982) proposed an empirical correlation to model the thermal dispersion for transverse and longitudinal directions as

$$k_{d,x}/k_f = 0.5 Pr Re_d \quad (141)$$

$$k_{d,y}/k_f = 0.1 Pr Re_d \quad (142)$$

Chou et al. (1994) used a homogeneous version of the model of Cheng and Zhu (1987) for thermal dispersion conductivity in the case of tridimensional mixed convection in a cylindrical packed-sphere tube. That model is

$$k_d/k_f = \gamma_d l_m (PrRe_d)^{\lambda_1}\left[(U^2+V^2)/(PrRe_d)^2+W^2\right]^{0.5\lambda_1} \tag{143}$$

where $\gamma_d = 0.07$, $l_m = 1 - \exp[(n_2/d_p)^2]$, $\lambda_1 = 0.683$ for water flow through glass sphere beds, and $l_m = 1 - \exp[2(n_2/d_p)]$, $\lambda_1 = 1$ for air flows. In the expressions of l_m, n_2 is the normal distance from the wall. Using that model, the numerical heat transfer predictions of Chou et al. (1994) were found to be in good agreement with their experimental data.

A literature survey presented by Vafai and Amiri (1998) showed that the effect of longitudinal dispersion conductivity can be considered as insignificant for $(PrRe_d) > 10$ when wall heating is imposed perpendicular to the direction of flow. It is noteworthy that all of the models for dispersion effects depend on the fluid Prandtl number but that the Prandtl dependency was not found the same for all fluids: for air flow a linear dependence of both components of the dispersion conductivity with $(PrRe_d)$ is proposed in most of the existing models, whereas a nonlinear dependence has been suggested for water flows by Levec and Carbonell (1985).

5. Local Thermal Nonequilibrium

The local thermal equilibrium (LTE) condition assumes that the difference between the volume-averaged fluid and solid temperatures is negligible. This assumption has been widely employed in most of the studies on forced convection in porous filled ducts. It should be emphasized here that we are considering local thermal equilibrium at the macroscopic scale (scale of a representative elementary volume), and not local equilibrium at the pore scale. Amongst the conditions for LTE are relatively slow motion, dominant thermal diffusion, absence of temperature jump at the solid–fluid interfaces (caused by radiative exchanges between solid surfaces through a transparent fluid, for example), and absence of volumetric heating of one material but not the other, as happens for example when a gas flows through a metallic porous matrix which is inductively heated or if an exothermic reaction takes place inside the particles. The LTE assumption was examined by Carbonell and Whitaker (1984), who proposed criteria for the time and length scales required for the validity of LTE. In recent works (Vafai and Sözen, 1990a,b; Sözen and Vafai, 1990, 1991, 1993; Amiri and Vafai, 1994; Hwang et al., 1995; Branci et al., 1995; Amiri et al., 1995; Jiang et al., 1996a; You and Chang, 1997; Kuznetsov, 1997; Nield, 1998; Lee and Vafai, 1999), the effects of local thermal nonequilibrium (LTNE) have been considered for transient or steady forced convective flows in packed beds. Since the validity of the LTE condition is not directly linked to assumptions introduced for the flow modeling (LTNE may occur for conditions in which the Darcy model is valid as well as for conditions which require the use of a more general flow model), we present here only the volume-averaged energy equations for the solid and fluid phases, using the most commonly employed formulations.

Fluid phase energy equation:

$$\varepsilon(\rho c_p)_f \frac{\partial T_f}{\partial t} + (\rho c_p)_f \mathbf{v} \cdot \nabla T_f = \nabla \cdot (k_{fe} \nabla T_f) + h_{fs} a_{sf} (T_s - T_f) \tag{144}$$

Solid phase energy equation:

$$(1-\varepsilon)(\rho c_p)_s \frac{\partial T_s}{\partial t} = \nabla \cdot (k_{se} \nabla T_s) - h_{fs} a_{sf} (T_s - T_f) \tag{145}$$

where T_f and T_s are the intrinsic phase averages of the fluid and solid temperatures, respectively, h_{fs} is the fluid-to-solid heat transfer coefficient, a_{sf} the specific area of the porous bed, k_{fe} and k_{se} the fluid and solid effective thermal conductivities.

For a bed packed with spherical particles, the surface area per unit volume of representative elementary volume (REV) can be modeled as (Dullien, 1979)

$$a_{sf} = \frac{6(1-\varepsilon)}{d_p} \tag{146}$$

Such an expression can be readily demonstrated by considering a bed volume V_b of porosity ε packed with n_p spherical particles which have a volume V_p and a surface area A_p. Since the number of particles per unit volume is $n_p/V_b = (1-\varepsilon)/V_p$, the foregoing expression for a_{sf} is straightforward.

Hwang et al. (1995) used a surface area per unit volume as

$$a_{sf} = \frac{20.346(1-\varepsilon)\varepsilon^2}{d_p} \tag{147}$$

Obviously, the expression for a_{sf} is not universal since it depends on the porous matrix structure. From volume averaging techniques, it can be readily shown that the effective conductivities are, for stagnant conditions,

$$k_{fe} = \varepsilon k_f \tag{148}$$

$$k_{se} = (1-\varepsilon) k_s \tag{149}$$

where k_f and k_s are the fluid and solid thermal conductivities, respectively. For cases with fluid flow and when thermal dispersion effects are significant, the effective thermal conductivity can be estimated as

$$k_{fe} = \varepsilon k_f + k_d \tag{150}$$

where k_{fe} is generally direction dependent. It is often assumed that the dispersion conductivity (longitudinal as well as transverse component) is the same as for the one-phase or quasi-homogeneous model. This assumption is discussed in Sözen and Vafai (1993).

Determination of the fluid-to-particle heat transfer coefficient h_{fs} received extensive attention in the 1960s and early 1970s. Empirical correlations were established which show that the heat transfer coefficient depends on the shape and size of the particles of the bed, void fraction as well as the Prandtl number of the fluid, and the range of particle Reynolds number. A literature survey on h_{fs} has been made by Wakao et al. (1979) and Wakao and Kaguei (1982). Amongst the correlations widely used in recent works are those of Wakao et al. (1979), Kar and Dybbs (1982) for laminar flows, and Gamson et al. (1943) for turbulent flows. These correlations can be expressed in the following forms:

$$h_{fs} = \frac{k_f}{d_p}\left[2 + 1.1 Pr^{1/3}\left(\frac{\rho u d_p}{\mu}\right)^{0.6}\right] \qquad \text{(Wakao et al., 1979)} \tag{151}$$

$$h_{fs} = 0.004\left(\frac{d_v}{d_p}\right)^{0.35}\left(\frac{k_f}{d_p}\right) Pr^{0.33} Re_d^{1.35}$$

$$\text{for} \quad Re_d \leq 75 \qquad \text{(Kar and Dybbs, 1982)} \tag{152}$$

$$h_{fs} = 1.064\left(\frac{k_f}{d_p}\right) Pr^{0.33} Re_d^{0.59}$$

$$\text{for} \quad Re_d \geq 350 \qquad \text{(Gamson et al., 1943)} \tag{153}$$

where d_v is the average void diameter, which can be calculated as $d_v = 4\varepsilon / a_{sf}$.

For heat flux applied at the walls, it is not trivial to specify the boundary conditions for the two-energy equation model. This question was addressed by Amiri et al. (1995), but any definitive theory has been established on the best implementation of the boundary conditions. One possibility is to assume that each representative elementary volume at the wall surface receives a prescribed heat flux equal to the wall heat flux. As a consequence, the wall heat flux should be divided between the two phases according to their effective conductivities and wall temperature gradients. A second approach consists in assuming that the LTE condition prevails at the near-wall regions, and to consider that each of the phases receives an equal heat flux.

At first, we concentrate on the condition under which the effect of LTNE may be important for a steady, Darcian forced flow in the absence of longitudinal thermal dispersion. This problem has been addressed analytically by Nield (1998) and Lee and Vafai (1999) in the case of forced convection in a parallel-plate channel with constant boundary conditions (constant wall temperature or constant wall heat flux). The two energy equations for the volume-averaged fluid and solid temperatures are

$$0 = (1-\varepsilon)k_s \frac{\partial^2 T_s}{\partial y^2} + h'_{fs}(T_f - T_s) \tag{154}$$

$$(\rho c_p)_f u \frac{\partial T_f}{\partial x} = \varepsilon k_f \frac{\partial^2 T_f}{\partial y^2} + h'_{fs}(T_s - T_f) \tag{155}$$

where $h'_{fs} = h_{fs}a_{fs}$ is here the local convective heat transfer coefficient per unit volume between fluid and solid interfaces (Koh and Colony, 1974). Note that it is assumed that no volumetric heat sources are present, k_f and k_s are constant, and that axial conduction is negligible. Local thermal equilibrium at the boundaries is invoked so that $T_f = T_s = T_w$. For the purposes of this discussion, we will follow Nield (1998). The heat transfer coefficient is defined as

$$h = q_w/(T_w - T_{mf}) \tag{156}$$

where T_{mf} is the conventional mixing-cup fluid temperature and q_w is the sum of the heat fluxes transferred to the fluid and solid phases

$$q_w = \varepsilon k_f \left(\frac{\partial T_f}{\partial y}\right)_{y=D} + (1-\varepsilon)k_s\left(\frac{\partial T_s}{\partial y}\right)_{y=D} \tag{157}$$

The fully developed Nusselt number is defined as

$$Nu_D = 2Dh/k_e \tag{158}$$

where D is the channel half-spacing. The effective thermal conductivity of the fluid-saturated porous medium is approximated on the assumption that heat conduction in the solid and fluid phases occurs in parallel

$$k_e = \varepsilon k_f + (1-\varepsilon)k_s \tag{159}$$

(a) Constant Wall Temperature. Integration of the coupled system of differential equations for the phase temperatures, using the method of separation of variables, leads to the following Nusselt number based on the plate spacing

$$Nu_D = \frac{\pi^2}{2}\left\{1 - \frac{(1-\varepsilon)k_s}{k_e}\left[1 + \frac{4D^2 h'_{fs}}{\pi^2(1-\varepsilon)k_s}\right]^{-1}\right\} \tag{160}$$

For large values of h'_{fs}, Nu_D tends towards $\pi^2/2$, which corresponds to the LTE assumption. Therefore, the Nu_D relationship shows that the effect of LTNE is to reduce the value of the Nusselt number from its asymptotic value. The smallest Nu_D-value is attained when $D^2 h'_{fs}/k_s \ll 1$. In this case, Nu_D is given by

$$Nu_D = \left(\varepsilon \frac{k_f}{k_e}\right) \frac{\pi^2}{2} \tag{161}$$

(b) Constant Wall Heat Flux. Assuming that

$$T_f(x, y) = T_w(x) + \theta_f(y); \qquad T_s(x, y) = T_w(x) + \theta_s(y) \tag{162}$$

with $\theta_f = \theta_s = 0$ at $y = \pm D$ and

$$\frac{\partial T_s}{\partial x} = \frac{\partial T_f}{\partial x} = \frac{\mathrm{d}T_w}{\mathrm{d}x} = \frac{\mathrm{d}T_{mf}}{\mathrm{d}x} = \frac{q_w}{(\rho c_p)_f u D} \tag{163}$$

leads to the expression

$$Nu_D = 6\left[1 + \frac{3(1-\varepsilon)^2 k_s^2}{D^2 k_e h'_{fs}} \left(1 - \frac{\tanh(b_2 D)}{b_2 D}\right)\right]^{-1} \tag{164}$$

where b_2 is given by

$$b_2 = \left[h'_{fs}\left(\frac{k_e}{\varepsilon(1-\varepsilon)k_f k_s}\right)\right]^{1/2} \tag{165}$$

If $D^2 h_{fs}/k_s \gg 1$, the Nusselt number reduces to

$$Nu_D = 6\left[1 - \frac{3(1-\varepsilon)^2 k_s^2}{D^2 k_e h'_{fs}}\right] \leq 6 \tag{166}$$

When h'_{fs} tends to infinity, the Nu_D-value is 6, as was found by invoking the LTE assumption. On the other hand, Nu_D reduces to

$$Nu_D = 6\varepsilon\left(\frac{k_f}{k_e}\right) \tag{167}$$

as h'_{fs} tends to zero. In conclusion, the analytical solution of Nield (1998) demonstrates that the effect of LTNE is to decrease the heat transfer rate. It is shown that the important parameter is $N_1 = k_s/D^2 h'_{fs}$, and the LTNE effect is negligible if and only if $N_1 \ll 1$. For a porous bed of spherical particles, using the Kozeny relation for permeability and the correlation of Dixon and Cresswell (1979) for h_{fs} leads to

$$N_1 = \frac{30(1-\varepsilon)^2 Da(10k_s + Nu_{fs}k_f)}{\varepsilon^2 Nu_{fs}k_f} \tag{168}$$

where $Da = K/D^2$ and Nu_{fs} is the fluid-to-solid Nusselt number. Note that the Darcy–Forchheimer extended model must be used for $Re_p > 1$ but that the analytical solution of Nield is still valid. Also, since the transverse thermal dispersion conductivity is constant for slug flows, inclusion of dispersion effects does not change the foregoing conclusions.

Sözen and Vafai (1990, 1993) investigated numerically transient forced convective flows of a gas through a packed bed, using the Darcy–Forchheimer model. They concluded that the LTNE model is necessary to accurately compute the condensation rate in the packed bed. They showed that the LTE assumption is not valid for high particle Reynolds number and/or high Darcy flows. These conclusions were confirmed by Amiri and Vafai (1994), who investigated the influence of a variety of effects and presented comprehensive error maps for the applicability of simplified models. Amiri et al. (1995) considered forced convective flow in a parallel plate channel filled with a metallic porous material of uniform porosity for constant wall temperature and for constant wall heat flux. They employed a Darcy–Brinkman–Forchheimer flow model and accounted for LTNE. They compared their numerical results with the experimental data reported by Kuzay et al. (1991) for a square channel with constant wall heat flux. Their predictions of Nusselt number were found to be in good agreement with the experimental results. Hwang et al. (1995) investigated experimentally and numerically forced convection in an asymmetrically heated sintered porous channel filled with a highly conductive porous material. They found that the difference between the temperature of the solid and fluid phases decreases with an increase in particle Reynolds number. Therefore, they concluded that the assumption of local thermal equilibrium is more justified at large Re_d. Jiang et al. (1996a) studied numerically forced convection in a parallel-plate channel and concluded that the difference in the wall heat transfer coefficient predicted using the LTE assumption rather than LTNE is significant when the difference between the solid and fluid thermal conductivities is large, especially when the wall heat flux is high. Kuznetsov (1997) applied a perturbation technique for forced convection in a parallel-plate channel and concluded that the temperature difference between the fluid and solid phases is proportional to the ratio of the flow velocity to the mean velocity. Therefore, the LTE condition holds at the channel walls when the Brinkman term is included into the momentum equation. This review of recent works in which two energy equations were used allows us to conclude that additional works are needed in order to have a clear insight into the importance of LTNE conditions for forced convection in porous filled channels.

Vortmeyer and Schaeffer (1974) suggested replacing the energy equations for the two-phase model (also called the heterogeneous model) by an equivalent quasi-homogeneous model equation. By assuming that the second derivatives for the solid and fluid temperatures are equal (axial condition of equivalence), they obtained a one-phase equation in terms of temperature which is closer to the solid phase temperature than to the fluid phase temperature. Their theory, based on this equivalence assumption, allowed a very simple treatment of fixed bed heat regenerators, and also brought an easy physical interpretation for the axial thermal dispersion conductivity (Vortmeyer and Adam, 1984). Another approach, which allows one to solve the heterogeneous model with only one transport equation, was presented by Branci et al. (1995). By assuming that the fluid volumetric heat capacity is close enough to the solid one, i.e., $(\rho c_p)_f \cong (\rho c_p)_s$, Branci et al. showed that it is possible to replace the two-phase model for a packed bed with internal heat generation by a transport equation for an averaged temperature, defined as $T_{av} = \varepsilon T_f + (1 - \varepsilon)T_s$, and a heat balance which yields one of the intrinsic phase-averaged temperatures in terms of porosity, fluid-to-solid heat transfer coefficient, and particle diameter.

C. Forced Convection in Ducts Partially Filled with a Porous Medium

Forced convective heat transfer in a duct whose walls are layered by a porous medium has been analyzed in few research works only, although it is a common occurrence in many industrial, engineering, and environmental applications such as heat exchangers, heat pipes, electronic cooling, filtration, etc. As pointed out by Hadim (1994), the use of porous substrates may improve forced convective heat transfer in channels with a clear reduction in the pressure drop in comparison with the filling of the entire channel by a porous medium. Closed-form analytical solutions for parallel plates and circular pipe were first derived by Poulikakos and Kazmierczak (1987) for constant wall heat flux, and numerical results were obtained for constant wall temperature. Their study was based on the Brinkman-extended Darcy model which allows one to match shear stress and velocity at the fluid/porous interface. The use of the Darcy model in conjunction with the empirical relationship of Beavers and Joseph (1967) could be considered as an alternative solution to modeling such a problem. However, this relationship, which assumes a velocity gradient at the interface proportional to the difference between the slip and the Darcian velocity, involves an extra parameter depending on the structure of the interface, the value of which is controversial. Vafai and Thiyagaraja (1987) reconsidered this problem and obtained analytical solutions based on matched asymptotic expansions for velocity and temperature distributions for three classes of problems in porous media: the interface region between two different porous media, the interface region between a fluid region and a porous medium, and the interface region between an impermeable region and a porous medium. They showed that the correlation of Beavers and Joseph (1967) is valid for the linear regime only. Later, Vafai and Kim (1990) presented an exact solution for a fully developed flow over a flat plate, where the fluid is sandwiched between a porous medium from above and a solid flat plate from below. Forced convection in a parallel plate channel partially filled with a porous medium was numerically investigated by Jang and Chen (1992) by using the Darcy–Brinkman–Forchheimer model and by including thermal dispersion effects. They showed that the inertial phenomena have an important effect on Nusselt number calculations for pore Reynolds number larger than about 40. In agreement with the results presented by Poulikakos and Kazmierczak (1987), Jang and Chen (1992) showed that the Nusselt number curve exhibits a minimum for a critical thickness of the porous substrate which depends on the value of the permeability. Srinivasan et al. (1994) opted for the porous substrate approach for modeling the flow and heat transfer through spirally fluted tubes by introducing direction-dependent permeabilities to predict the swirl component in the tube. Comparison with experimental results have shown that this approach gives very good agreement with experiment. Chikh et al. (1995a,b, 1997) considered the annulus configuration and presented an analytical solution for the fully developed regime in the case of constant heat flux at the inner wall into the fluid-saturated porous medium. They presented numerical results for the constant temperature case and for the thermal entrance length by including the inertial term. Al-Nimr and Alkam (1994) presented numerical solutions for transient, developing, forced convection in annuli partially filled with a porous substrate, attached either to the inner cylinder or to the outer cylinder. In both cases, the boundary in contact with the porous substrate was exposed to a sudden change in its temperature while the other boundary was kept adiabatic. Using the Brinkman–Forchheimer extended Darcy model, they investigated the effects of various parameters regarding the geometry, the solid matrix, and the fluid on the hydrodynamic entrance length and Nusselt number. More recently, Ochoa-Tapia and Whitaker (1998) suggested modeling the heat transfer condition at the interface between a porous medium and a clear fluid as a jump condition for the heat flux, suggesting that accumulation, conduction, and convection of "excess surface thermal energy" might be introduced to account for the potential failure of local thermal equilibrium in the interface region. This theoretical work, based on the method of volume averaging and the generalized thermal energy transport equation, has not yet found application in analytical or numerical solutions.

Since the qualitative features of the flow and temperature fields are the same for all of these geometries and thermal boundary conditions, we document only the annular duct geometry. The results for a parallel plate channel can be found in the review paper by Lauriat and Vafai (1991).

1. Forced Convection in an Annular Duct Partially Filled with a Porous Medium.

Forced convection in an annular duct partially filled with a porous medium attached to the inner cylinder was studied analytically and numerically by Chikh et al. (1995a,b, 1997). In these studies, the outer cylinder was thermally insulated while both constant wall temperature and constant wall heat flux conditions at the inner cylinder were considered. The assumptions of constant fluid properties, homogeneous, isotropic porous medium, and local thermal equilibrium were introduced.

(a) Momentum Equation. The flow is assumed to be steady, laminar, incompressible, and fully developed. The equations governing the flow field are written separately for the porous and fluid regions as:

Porous region ($r_i \leq r \leq r_i + e$)

$$0 = -\frac{\mathrm{d}p}{\mathrm{d}x} + \mu \frac{1}{r}\frac{\mathrm{d}}{\mathrm{d}r}\left(r\frac{\mathrm{d}u_p}{\mathrm{d}r}\right) - \frac{\mu}{K}u_p - \frac{\rho\varepsilon F}{\sqrt{K}}u_p^2 \tag{169}$$

Fluid region ($r_i + e \leq r \leq r_o$)

$$0 = -\frac{\mathrm{d}p}{\mathrm{d}x} + \mu \frac{1}{r}\frac{\mathrm{d}}{\mathrm{d}r}\left(r\frac{\mathrm{d}u_f}{\mathrm{d}r}\right) \tag{170}$$

where e is the thickness of the porous substrate (Figure 6). The assumption that the effective viscosity of the fluid-saturated porous medium is equal to the fluid viscosity has been introduced in Eq. (169). No-slip conditions are applied at the walls, and the axial velocities and shear stresses are matched at the porous–fluid interface located at $r_i + e$

$$u_p = 0 \quad \text{at} \quad r = r_i \tag{171}$$

$$u_p = u_f; \quad \frac{\mathrm{d}u_p}{\mathrm{d}r} = \frac{\mathrm{d}u_f}{\mathrm{d}r} \quad \text{at} \quad r = r_i + e \tag{172}$$

$$u_f = 0 \quad \text{at} \quad r = r_o \tag{173}$$

(b) Energy Equation. Since the energy equations for fluid and porous layers look alike, it is convenient to develop a continuum approach by

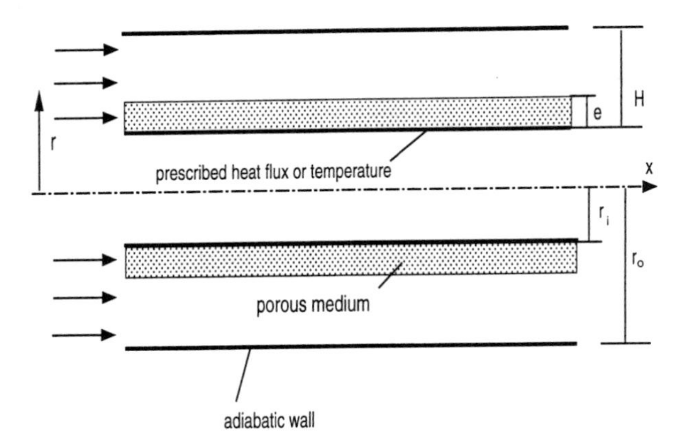

Figure 6. Schematic of the problem and coordinate system for an annulus partially filled with a porous medium.

considering the two layers in one domain which does not require any kind of matching at the interface. Hence the energy equation may be written in a general form, encompassing the fluid and porous layers, by introducing a binary parameter λ as

$$\rho c_p u \frac{\partial T}{\partial x} = \left\{ \lambda(k_e - k_f) + k_f \right\} \frac{1}{r} \frac{\partial}{\partial r} \left(r \frac{\partial T}{\partial r} \right) \tag{174}$$

This equation assumes negligible axial conduction. The values of λ are zero in the fluid layer and 1 in the porous layer, and k_f is the thermal conductivity of the fluid. The diffusion coefficient satisfies the compatibility conditions requiring that the temperatures and heat fluxes be continuous at the fluid–porous interface. The boundary conditions for the energy equation reduce then to

$$q_{wo} = 0 \qquad \text{at} \qquad r = r_0 \tag{175}$$

$$T = T_{wi} \qquad \text{at} \qquad r = r_i \tag{176}$$

for a constant inner wall temperature of T_{wi}, whereas

$$q_{wi} = -k_e \left. \frac{\partial T}{\partial r} \right|_{r=r_i} \qquad \text{at} \qquad r = r_i \tag{177}$$

for a constant wall heat flux of q_{wi}. The thermal entrance condition is

$$T = T_{in} \qquad \text{at} \qquad x = 0 \tag{178}$$

The governing equations may be rewritten in a dimensionless form by defining the following dimensionless quantities

$$R = \frac{r}{H} \qquad X = \frac{4x/D_h}{RePr} \qquad s = \frac{r_i + e}{H} \tag{179}$$

where $H = r_o - r_i$ is the annulus gap, $D_h = 2H$ the hydraulic diameter, Pr the fluid Prandtl number, and Re the Reynolds number based on D_h and reference velocity

$$u_r = -\frac{H^2}{\mu}\left(\frac{\mathrm{d}p}{\mathrm{d}x}\right) \tag{180}$$

$$\Lambda = \frac{k_e}{k_f}, \qquad Da = \frac{K}{H^2}, \qquad Re_p \varepsilon F = \frac{\varepsilon F \rho \sqrt{K}}{\mu} u_r \tag{181}$$

$$U_p = \frac{u_p}{u_r}, \qquad U_f = \frac{u_f}{u_r}, \qquad \theta = \frac{T - T_{in}}{\Delta T} \tag{182}$$

where $\Delta T = T_w - T_{in}$ for constant wall temperature and $\Delta T = q_{wi}H/k_e$ for constant wall heat flux into the fluid-saturated porous medium. Introducing the above dimensionless variables, Eqs. (169) and (170) become

(i) in the porous medium: $R_i \leq R \leq s$

$$0 = 1 + \frac{1}{R}\frac{\mathrm{d}}{\mathrm{d}R}\left(R\frac{\mathrm{d}U_p}{\mathrm{d}R}\right) - \frac{U_p}{Da} - \frac{Re_p \varepsilon F}{Da} U_p^2 \tag{183}$$

(ii) in the fluid region: $s \leq R \leq R_o$

$$0 = 1 + \frac{1}{R}\frac{\mathrm{d}}{\mathrm{d}R}\left(R\frac{\mathrm{d}U_f}{\mathrm{d}R}\right) \tag{184}$$

The associated boundary conditions are

$$\left.\begin{array}{llll} U_p = 0 & \text{at} & R = R_i & \\ U_f = 0 & \text{at} & R = R_o & \\ U_p = U_f; & \dfrac{\mathrm{d}U_p}{\mathrm{d}R} = \dfrac{\mathrm{d}U_f}{\mathrm{d}R} & \text{at} & R = s \end{array}\right\} \tag{185}$$

The dimensionless energy equation may be written as

$$U\frac{\partial \theta}{\partial X} = \{\lambda(\Lambda - 1) + 1\}\frac{1}{R}\frac{\partial}{\partial R}\left(R\frac{\partial \theta}{\partial R}\right) \tag{186}$$

The corresponding boundary conditions are, at the inner annulus wall $(R = R_i)$

$$\frac{\partial \theta}{\partial R} = -1 \qquad \text{for specified wall heat flux} \tag{187}$$

$$\theta = 1 \qquad \text{for specified wall temperature} \tag{188}$$

at the outer wall ($R = R_o$)

$$\frac{\partial \theta}{\partial R} = 0 \tag{189}$$

while the inlet condition reads $\theta = 0$ at $X = 0$.

(c) Exact Solution for Fully Developed Flow: Brinkman–Extended Darcy Model for Constant Wall Heat Flux. The momentum equation for the fluid region (184) can be easily integrated to yield the velocity profile in the fluid region, which is of the form

$$U_f(R) = -\frac{R^2}{4} + C_1 \ln R + C_2 \tag{190}$$

Neglecting the velocity quadratic term and introducing a new variable $Z = R/\sqrt{Da}$, Eq. (183) may be written as

$$0 = Da + \frac{\mathrm{d}^2 U_p}{\mathrm{d}Z^2} + \frac{1}{Z}\frac{\mathrm{d}U_p}{\mathrm{d}Z} - U_p \tag{191}$$

which is a modified Bessel equation of zeroth order, with a non-zero right-hand side. The solution of (191) is of the form

$$U_p(Z) = B_1 I_0(Z) + B_2 K_0(Z) + Da \tag{192}$$

where I_0 and K_0 are the modified Bessel functions of zeroth order of first and second kind, respectively. The constants B_1, B_2, C_1, and C_2, determined using the boundary conditions and properties of Bessel functions, are obtained as

$$B_1 = \frac{\frac{1}{4}(s^2 - R_o^2)K_{oi} - (s^2/2)\ln(s/R_o)K_{oi} + Da(K_{oi} - K_{os}) - (s/\sqrt{Da})\ln(s/R_o)K_{1s}Da}{I_{oi}K_{os} - I_{os}K_{oi} + (s/\sqrt{Da})\ln(s/R_o)(I_{1s}K_{oi} + K_{1s}I_{oi})} \tag{193}$$

$$B_2 = -\left(\frac{Da}{K_{oi}} + \frac{I_{oi}}{K_{oi}}B_1\right) \tag{194}$$

$$C_1 = \frac{s}{\sqrt{Da}}(B_1 I_{1s} - B_2 K_{1s}) + \frac{s^2}{2} \tag{195}$$

$$C_2 = -C_1 \ln(R_o) + \frac{R_o^2}{4} \tag{196}$$

the notations K_{oi}, I_{oi}, K_{os}, I_{os}, K_{1s}, and I_{1s} introduced in Eqs. (193)–(196) stand for the values of the modified Bessel functions K_o, I_o, K_1, and I_1 at the positions $R_i/\sqrt{Da}$ and $s/\sqrt{Da}$, respectively. From the velocity profiles for the fluid and porous regions, the mean velocity over the annulus section can be determined as

$$\bar{U} = \frac{2}{R_o^2 - R_i^2} \left\{ \begin{array}{l} B_1 s\sqrt{Da} I_{1s} - B_2 s\sqrt{Da} K_{1s} + Da\ s^2/2 - B_1 R_i \\ \sqrt{Da} I_{1i} + B_2 R_i \sqrt{Da} K_{1i} - Da\ R_i^2/2 - R_o^4/16 \\ -C_1 R_o^2/4 + C_1 (R_o^2/2) \ln R_o + C_2 R_o^2/2 + s^4/16 \\ -C_1 (s^2/2) \ln s + C_1 s^2/4 - C_2 s^2/2 \end{array} \right\} \tag{197}$$

Thermally fully developed flow implies that $\partial T/\partial x$ is constant and equal to the axial derivative of the bulk temperature, i.e., $\partial T/\partial x = \mathrm{d}T_B/\mathrm{d}x$, and that $T(x, r) = \lambda_2 x + \phi(r)$ for a constant wall heat flux prescribed on the inner cylinder. An energy balance yields

$$\frac{\mathrm{d}T_B}{\mathrm{d}x} = \frac{2 r_i q_{wi}}{\rho c_p \bar{u} (r_o^2 - r_i^2)} \tag{198}$$

As is well established, we need only the temperature difference ($T_{wi} - T$) to calculate the Nusselt number at the inner wall. Therefore, a dimensionless temperature difference defined as

$$\theta(x, r) = \frac{T_{wi}(x) - T(x, r)}{q_{wi} D_h / k_e} = \Phi(r) \tag{199}$$

where $\Phi(R) = \phi(r)/(q_{wi} D_h/k_e)$, is introduced in what follows. The energy equation (186) can thus be written as

$$\left(\frac{\Lambda}{\lambda(\Lambda - 1) + 1} \frac{R_i}{R_i^2 - R_o^2} \right) \frac{U}{\bar{U}} + \frac{1}{R} \frac{\mathrm{d}}{\mathrm{d}R} \left(R \frac{\mathrm{d}\Phi}{\mathrm{d}R} \right) = 0 \tag{200}$$

or

$$R^2 \frac{\mathrm{d}^2 \Phi}{\mathrm{d}R^2} + R \frac{\mathrm{d}\Phi}{\mathrm{d}R} = \Gamma_1 U R^2 \tag{201}$$

where Γ_1 is a constant defined as

$$\Gamma_1 = \left\{ \left(\frac{\Lambda}{\lambda(1 - \Lambda) - 1} \right) \left(\frac{R_i}{R_i^2 - R_o^2} \right) \right\} \frac{1}{\bar{U}} \tag{202}$$

The boundary conditions for (201) are

$$\Phi(R_i) = 0 \qquad \text{and} \qquad \left. \frac{\mathrm{d}\Phi}{\mathrm{d}R} \right|_{R_o} = 0 \tag{203}$$

Equation (201) is of the form of Cauchy's or Euler's equation. Its solution is given by

$$\Phi(R) = E + F \ln R + G\left(\ln R \int UR\mathrm{d}R - \int UR \ln R\mathrm{d}R\right) \tag{204}$$

where $G = -R_i/(R_i^2 - R_o^2)\bar{U}$ and U is given by Eq. (192) in the porous region, and $G = -R_i\Lambda/(R_i^2 - R_o^2)\bar{U}$ and U is given by Eq. (190) in the fluid region. The constants E and F for both regions are determined from the boundary conditions (203). Performing the integrations with the use of recurrence formulas for the Bessel functions yields the temperature profile as

in the porous region:

$$\begin{aligned}\Phi_p(R) = E_p + F_p \ln R - \frac{R_i}{(R_i^2 - R_o^2)\bar{U}}\Bigg\{&B_1 DaI_o\left(\frac{R}{\sqrt{Da}}\right) + \\ &B_2 DaK_o\left(\frac{R}{\sqrt{Da}}\right) - \frac{Da}{4}R^2\Bigg\}\end{aligned} \tag{205}$$

in the fluid region:

$$\Phi_f(R) = E_f + F_f \ln R - \frac{R_i\Lambda}{(R_i^2 - R_o^2)\bar{U}}\left\{\frac{R^2}{64} + C_1\frac{R^2}{4}\ln R - (C_1 - C_2)\frac{R^2}{4}\right\} \tag{206}$$

where

$$F_p = \frac{R_i\Lambda}{(R_i^2 - R_o^2)\bar{U}}\left\{\begin{array}{l}(R_o^4 - s^4)/16 + C_1(R_o^2 - s^2)/2\ln s + (C_2 - C_1/2) \\ (R_o^2 - s^2)/2 + B_1 s\sqrt{Da}I_1(s/\sqrt{Da}) - B_2 s\sqrt{Da}K_1(s/\sqrt{Da}) \\ -Da\ s^2/2\end{array}\right\} \tag{207}$$

$$E_p = -F_p \ln R_i + \frac{R_i}{(R_i^2 - R_o^2)\bar{U}}\left\{\begin{array}{l}B_1 DaI_o(R_i/\sqrt{Da}) + B_2 DaK_o(R_i/\sqrt{Da}) \\ -Da\ R_i^2/4\end{array}\right\} \tag{208}$$

$$F_f = \frac{R_o R_i \Lambda}{(R_i^2 - R_o^2)\bar{U}}\left\{\frac{R_o^3}{16} + C_1\frac{R_o}{2}\ln R_o + \left(C_2 - \frac{C_1}{2}\right)\frac{R_o}{2}\right\} \tag{209}$$

$$\begin{aligned}E_f = E_p + (F_p - F_f)\ln s - &\frac{R_i}{(R_i^2 - R_o^2)\bar{U}} \\ &\left\{\begin{array}{l}B_1 DaI_0(s/\sqrt{Da}) + B_2 DaK_0(s/\sqrt{Da}) - Da\ s^2/4 - \Lambda \\ (s^4/64 + C_1(s^2/4)\ln s - (C_1 - C_2)s^2/4)\end{array}\right\}\end{aligned} \tag{210}$$

The fully developed heat transfer coefficient at the inner wall, h_i, is defined as

$$q_{wi} = h_i(T_{wi} - T_m) \tag{211}$$

From this definition, the fully developed Nusselt number may be written as

$$Nu_i = \frac{h_i D_h}{k_e} = \frac{1}{\theta_{wi} - \theta_m} \tag{212}$$

Hence

$$Nu_i = \frac{(R_o^2 - R_i^2)\bar{U}}{2\int_{R_i}^{R_o} U\Phi(R)R\mathrm{d}R} \tag{213}$$

Performing the integral leads to the Nusselt number.

Velocity profiles and Nusselt number for $\Lambda = 1$, which may be considered as a typical conductivity ratio when using a porous material of high porosity (for insulation, for example), were discussed in Chikh et al. (1995a) for various values of the flow parameters. Figure 7 shows the variations in Nusselt number for $\Lambda = 1$ as a function of the dimensionless porous-layer thickness, and for different Darcy numbers. Here, the radius ratio is $r_o/r_i = 2$. However, the effects of the other parameters are qualitatively the same for any other value of the radius ratio considered. For a given Darcy number (or permeability), Figure 7 shows that the Nusselt number increases when the porous layer thickness increases up to a critical value beyond which Nu increases to end up at almost the same value as in the case of the completely porous channel. The physical explanation is that, when the porous layer thickness increases, the flow rate is reduced and hence the prescribed heat flux makes the wall temperature increase more than the mean temperature of the fluid. The Nusselt number, being inversely proportional to the temperature difference, decreases until a critical thickness is reached. Over this value, the inverse effect is produced, that is, the fluid mean temperature increases more than the wall temperature, and thus Nu is augmented. Similar results were shown by Poulikakos and Kazmierczak (1987) for a partially filled cylinder, and by Lauriat and Vafai (1991) for a parallel plate channel. It is worth noting that the limiting case of no porous medium ($Nu = 6.18$) agrees with the results given by Kays and Crawford (1993). The effect of permeability is also seen in this figure through the different values of Da. The more permeable the porous medium, the higher is the heat transfer and the lower is the critical thickness. One can deduce that

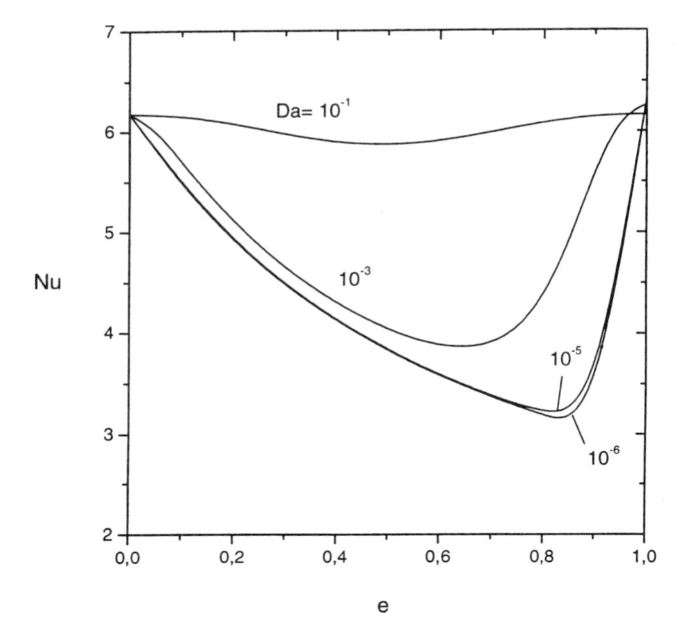

Figure 7. Values of Nusselt numbers for an annulus partially filled with a porous medium.

there is no need to fill the annular duct with the porous material to obtain the minimum heat transfer.

D. Forced Convection in Porous-filled Ducts Saturated with Non-Newtonian Fluids

Forced convection of non-Newtonian fluids through porous media has motivated increasing interest during recent years. This is because a number of working fluids in engineering applications exhibit non-Newtonian behavior. Examples may be found in fixed packed bed chemical reactors commonly used in industrial equipment, filtration, production of heavy crude oil by steam injection, biomechanics, ceramic processing, etc. Although external convection (mixed as well as forced convection) has been the topic of many studies during the past decade, as is evident form the review article of Shenoy (1994), very few research papers were published on forced convective heat transfer in porous filled ducts. In addition, the studies were restricted to purely viscous fluids, i.e., fluids for which an apparent viscosity can be defined.

1. Modified Darcy and Darcy–Forchheimer Equation for Non-Newtonian Fluids

(a) Power-law Fluids. The form of Darcy's law for inelastic time-independent non-Newtonian fluids described by the Ostwald-de-Waele power law model was proposed by Christopher and Middleman (1965), Kemblowski and Michniewicz (1979), and Dharmadhikari and Kale (1985) for fluid flow through packed particle beds. The modified Darcy's law for forced convection may be written in vectorial form as (Shenoy, 1992a)

$$\nabla p = \left(\frac{\mu^* |\langle \mathbf{v} \rangle|^{n-1}}{K^*} \right) \langle \mathbf{v} \rangle \tag{214}$$

where $\langle \mathbf{v} \rangle$ is the volume-averaged velocity, μ^* the consistency index of the fluid, n the power-law index, and K^* the modified permeability. For $n = 1$, the model reduces to Newton's law of viscosity with $\mu^* = \mu$. The deviation of n from unity is a measure of the degree of deviation from Newtonian behavior: when $n < 1$, the model describes pseudoplastic behavior (or shear thinning) whereas the behavior is dilatant (or shear thickening) for $n > 1$. The modified permeability is defined as follows

$$K^* = \frac{1}{2C_t} \left(\frac{n\varepsilon}{3n+1} \right)^n \left(\frac{50K}{3\varepsilon} \right)^{(n+1)/2} \tag{215}$$

where C_t is a modified tortuosity factor with the definition (Shenoy, 1994)

$$C_t = \begin{cases} (25/12) \\ \quad \text{Christopher and Middleman (1965)} \\ (2.5)^n 2^{(1-n)/2} \\ \quad \text{Kemblowski and Michniewicz (1979)} \\ \frac{2}{3} \left(\frac{8n}{9n+3} \right)^n \left(\frac{10n-3}{6n+1} \right) \left(\frac{75}{16} \right)^{(3(10n-3))/(10n+11)} \\ \quad \text{Dharmadhikari and Kale (1985)} \end{cases} \tag{216}$$

When using the expression of C_t given by Dharmadhikari and Kale (1985), the power-law index must be changed to $n' = n + 0.3(1 - n)$ in order to take into account the effects of the non-Newtonian behavior upon the path through the pore spaces. Since the model holds for beds of spherical particles of diameter d_p, the intrinsic permeability is given by the Ergun relation

$$K = \frac{\varepsilon^3 d_p^2}{150(1-\varepsilon)^2} \tag{217}$$

The validity of the modified Darcy's law for power-law fluids ends when a modified pore Reynolds number based on the interstitial velocity and pore scale exceeds a value of about unity. This Reynolds number is defined as

$$Re_K = \frac{\rho(K/\varepsilon)^{n-2}(u_D/\varepsilon)^{2-n}}{\mu^*} \tag{218}$$

where (u/ε) is the interstitial velocity. For power-law fluids, Shenoy (1992a) showed that the square velocity term introduced to model the inertia effects is the same for Newtonian as for non-Newtonian fluids. The reason is that the inertia term originates from advection terms and not from viscous terms when the Navier–Stokes equations are averaged over a representative elementary volume. Therefore, the Darcy–Forchheimer equation for power-law fluids can be expressed as

$$\nabla p = -\left(\frac{\mu^* |\langle \mathbf{v} \rangle|^{n-1}}{K^*} + \frac{\rho F |\langle \mathbf{v} \rangle|}{K^{1/2}}\right)\langle \mathbf{v} \rangle \tag{219}$$

For a unidirectional flow, the Darcy–Forchheimer slug velocity can be expressed as

$$u_{DF}[Re_K^* + 1]^{1/n} = \left\{\frac{K^*}{\mu^*}\left(-\frac{\mathrm{d}p}{\mathrm{d}x}\right)\right\}^{1/n} \tag{220}$$

where

$$Re_K^* = \frac{\rho F K^* u_{DF}^{2-n}}{\mu^* \sqrt{K}}$$

(b) Herschel–Bulkley Fluids. These complex fluids are of power-law type, but with a yield stress, τ_0. The apparent viscosity is a monotonic decreasing function when the shear stress increases (i.e., $n < 1$). For hydrodynamically developed confined flows along the x-axis, the modified Darcy's law for forced convection may now be written as

$$-\frac{\mathrm{d}p}{\mathrm{d}x} = \frac{\mu^*}{K^*}u^n + b^* \tag{221}$$

where

$$b^* = (0.24\varepsilon/K)^{1/2} C_t \tau_0 \tag{222}$$

and

$$C_t = (25/12)^{(n+1)/2} \tag{223}$$

When the yield stress τ_0 tends to zero, the power-law model is recovered, whereas a Bingham-plastic model is akin for $n = 1$. In the Darcy–Forchheimer flow regime, the slug velocity is given as

$$u_{DF}[Re_K^* + 1]^{1/n} = \left\{ \frac{K^*}{\mu^*} \left(-\frac{dp}{dx} - b^* \right) \right\}^{1/n} \tag{224}$$

where Re_K^* keeps the same definition as for a power-law fluid.

(c) Elastic Boger Fluids. The effect of elasticity can be accounted for by using the Boger constitutive equation. The first-order model was derived by Shenoy (1992b) to modify the Darcy's law for granular beds as follows

$$-\frac{dp}{dx} = \frac{\mu_o u}{K} - 0.37 \frac{C_E^{1/2} \mu_0 2\lambda_t u^2}{K^{3/2}} \tag{225}$$

where

$$C_E = C_0/\varepsilon^2 \tag{226}$$

$$K = C_0 d_p^2 \tag{227}$$

$$C_0 = \frac{\varepsilon^2}{150(1-\varepsilon)^2} \tag{228}$$

λ_t is a relaxation time whose physical meaning is that the shear stress decays as $\exp(-t/\lambda_t)$ after a sudden stop of the fluid motion. In the Darcy–Forchheimer flow regime, the velocity is (Shenoy, 1994)

$$u_{DF}\left[\,\overline{Re_K^*} + 1 - 0.37 We_{D,F}\right] = \frac{K}{\mu_0}\left(-\frac{dp}{dx}\right) \tag{229}$$

where $\overline{Re_K^*} = \rho F \sqrt{K} u_{DF}/\mu_0$ and $We_{D,F} = C_E^{1/2} 2\lambda_t u_{DF}/K^{1/2}$ is a Weissenberg number based on the pore scale.

2. Heat Transfer

Steady-state forced convection through a parallel plate channel subjected to a uniform heat flux boundary condition was investigated by Shenoy (1993) for porous media saturated by an elastic Boger fluid of constant viscosity and by using the Brinkman-extended Darcy model. The same geometry was also considered by Nakayama and Shenoy (1993) in the case of an inelastic power-law fluid, but the problem formulation was based on the Brinkman–Forchheimer–extended Darcy model. The approximate integral method employed in both studies leads to closed-form analytical expressions for the Nusselt number in the fully developed regime.

For a Boger fluid having a constant viscosity, the Brinkman–extended Darcy model leads to

$$Nu = \frac{12(\beta_2 + 3)(2\beta_2 + 3)}{2\beta_2^2 + 13\beta_2 + 17} \tag{230}$$

where β_2 is given by the following implicit equation

$$\beta_2 = \left\{(9 + 4\sigma^2[1 - 0.74We(2\beta_2/(2\beta_2 + 1))])^{0.5}\right\}/2 \tag{231}$$

with $\sigma = D/(K/\varepsilon)^{1/2}$, and $We = C_n^{1/2}\lambda u_c/K^{1/2}$ is the Weissenberg number based on the centerline velocity at the inlet and pore scale. A plot of Nu against σ (Shenoy, 1993) shows that Nu increases from $Nu = 8$ to $Nu = 12$ when σ increases from 0.1 to about 10^3, whatever We is. Obviously, an increase in elasticity produces a decrease in Nu.

For a power-law fluid, the Brinkman–Forchheimer–extended Darcy model yields, for the case of small Darcy number

$$Nu = \frac{15(2 - \delta_n)^2}{2\delta_n^3 - 5\delta_n^2 + 5} \tag{232}$$

where δ_n is the dimensionless thickness of the viscous sublayer, given as

$$\delta_n = \frac{Da^{*0.5}}{((n/1 + n) + \frac{2}{3}Re_K^* U_c^{2-n})^{1/(1+n)}} \tag{233}$$

with U_c the centerline dimensionless velocity which is scaled by the velocity u_D based on the Darcy law. U_c is given implicitly by

$$U_c^n + Re_K^* U_c^2 - 1 = 0 \tag{234}$$

where $Re_K^* = (\rho F K^* u_D^{2-n})/\mu^*$. Comparisons between the numerical results obtained for $Re_K^* = 0$, 1, and 10 with the results based on the integral method show reasonably good agreement for $Da \leq 10^{-2}$ (Nakayama and Shenoy, 1993). The main interest in the analytical results is that the effects of the power-law index and modified pore Reynolds number are well reflected, so that they can be used as a first approximation to readily calculate the heat transfer coefficient.

Chen and Hadim (1998) modeled heat transfer in a channel packed with an isotropic granular material using a modified Brinkman–Forchheimer–extended Darcy model for power-law fluids, which takes into account the effects of variable porosity and thermal dispersion. The momentum and energy equations were written as

$$\frac{\rho}{\varepsilon}(\mathbf{v}\cdot\nabla)\left(\frac{\mathbf{v}}{\varepsilon}\right) = -\nabla P - \frac{\mu^*}{K^*}|\mathbf{v}|^{n-1}\mathbf{v} - \frac{\rho C_F}{K^{1/2}}|\mathbf{v}|\mathbf{v} + \frac{\mu^*}{\varepsilon} \nabla\left[\frac{1}{\varepsilon^{n-1}}\left(\frac{\Delta:\Delta}{2}\right)^{(n-1)/2}\Delta\right]\mathbf{v} \tag{235}$$

$$(\rho c_p)_f(\mathbf{v}\cdot\nabla T) = \nabla\cdot(k_e\nabla T) \tag{236}$$

where

$$|\mathbf{v}| = \sqrt{u^2+v^2} \tag{237}$$

They used the correlation of Dharmadhikari and Kale (1985) for the modified permeability and tortuosity factor (Eqs. (215) and (216)). The variation of porosity was represented by an exponential function (55) where C and N were taken as $C = 1.4$ and $N = 5.0$. The transverse dispersion conductivity was accounted for by using a linear model (Hsu and Cheng, 1990).

Chen and Hadim (1998) have discussed the effects of particle diameter, power-law index, and modified particle Reynolds number on flow and heat transfer. As for a channel without a porous matrix, the results indicate that the Nusselt number increases when the power-law index is decreased below unity (shear thinning fluids), whereas the pressure drop decreases significantly if the inertia effects are not dominant ($Re_d^* \leq 50$ for the dimensionless particle diameters considered: $d_p/D_h = 0.1$ and 0.01). For large particle diameters, the effects of the power-law index on flow and heat transfer are significant, whereas for small values of the particle diameter the effects of the porous material are predominant and the power-law index has less effect at the same modified particle Reynolds number.

Forced convection with the assumption of hydrodynamically developed flows of power-law fluids in a saturated porous annulus was numerically examined by Alkam et al. (1998), who used a modified Brinkman–Forchheimer–extended Darcy model for solving transient thermal behaviors which result from sudden changes in thermal conditions at the boundaries. The problem formulation was basically the same as in Chen and Hadim (1998), but the porosity was assumed uniform and the thermal dispersion was neglected. Two types of thermal boundary conditions were considered: step temperature change at the inner wall (isothermal wall) with the outer wall kept adiabatic, and the opposite. The effects of modified Reynolds, Prandtl, and Darcy numbers based on the pore scale (i.e. $\sqrt{K^*}$), and power-law index were investigated for an annulus of radius ratio equal to 0.5. Nusselt number profiles in the thermal entry length are presented for various values of the flow and heat transfer parameters. It is shown that increasing Re_K^* or Pr^* leads to increases in local Nusselt number and thermal entry length, whereas increases in Da^* or in power-law index produce a decrease both in Nu_x and in thermal entry length.

NOMENCLATURE

Roman Letters

A_p	particle surface area
a_{sf}	specific area of the porous medium
C_f	friction coefficient
C_F	Forchheimer coefficient
C_t	modified tortuosity
c_p	specific heat
d_p	particle diameter
d_v	void average diameter
D	channel half-spacing
Da	Darcy number
D_h	hydraulic diameter
e	thickness of the porous substrate
F	empirical constant in the second-order flow resistance
f	dimensionless stream function
f_r	friction factor
g	acceleration of gravity
H	width of the annulus
h	heat transfer coefficient
$h_f s$	fluid-to-solid heat transfer coefficient
$h'_f s$	fluid-to-solid heat transfer coefficient per unit volume
I_0, I_1, K_0, K_1	Bessel functions
k_e	effective thermal conductivity of the fluid-saturated porous medium
k_0	stagnant thermal conductivity
k_d	dispersive thermal conductivity
K	permeability
K_∞	bulk permeability
K^*	modified permeability
l_m	mixing length
L	plate length
m	exponent defining the geometry
n	power law index
n'	modified power law index
n_2	normal distance from the wall
n_p	number of particles
Nu	Nusselt number
Nu_D	Nusselt number based on the plate spacing
Nu_{fs}	fluid-to-solid Nusselt number
Nu_∞	fully developed Nusselt number

p	pressure
p'	local averaged pressure
Δp_D	Darcy's pressure drop
Pe	Péclet number, RePr
Pr	Prandtl number
q	heat flux
r	radial coordinate
R	dimensionless radial coordinate
Re	Reynolds number
Re_d	particle Reynolds number
$Re_{d,m}$	particle Reynolds number based on the mean velocity
Re_K	Reynolds number based on permeability
Re_K^*	modified Reynolds number as $Re_K^* = \dfrac{\rho F K^* u_{DF}^{2-n}}{\mu^* \sqrt{K}}$
Re_p	pore Reynolds number
r_c	solid–fluid specific heat ratio
s	dimensionless radial position of the porous-fluid interface
t	time
t_∞	time at which the steady state is reached
T	temperature
T_{av}	average temperature
T_B	bulk temperature
T_m	mean temperature
T_{mf}	mixing-cup fluid temperature
u, v, w	Cartesian fluid velocity components
U, V, W	dimensionless Cartesian components of the fluid velocity
u_c	centerline velocity in the porous channel
u_D	Darcian velocity
u_{DF}	Darcy–Forchheimer velocity
$\bar{u}$	mean velocity
$\bar{U}$	dimensionless mean velocity
v	velocity vector
V_b	bed volume
V_p	particle volume
We_{DF}	Weinssenberg number based on the pore scale
x, y, z	Cartesian coordinates
X, Y, Z	dimensionless Cartesian coordinates
X_e	dimensionless hydrodynamic development length

Greek Letters

α	thermal diffusivity
β_1	empirical constant defining the structure of the porous matrix
δ	hydrodynamic boundary layer thickness

δ_n	dimensionless thickness of the viscous sublayer
δ_t	thermal boundary layer thickness
δ_v	thickness of boundary layer thickness
ε	porosity
ε_m	mean porosity
ε_∞	bulk porosity
$\eta, \eta_1, \eta_t, \xi, \xi_1, \zeta$	transformed variables
Γ	gamma function
λ_t	relaxation time
Λ	thermal conductivity ratio (k_e/k_f)
μ	dynamic viscosity
μ^*	consistency index
ν	kinematic viscosity
ψ	stream function
ρ	fluid density
σ	parameter equal to $D/(K/\varepsilon)^{1/2}$
θ	dimensionless temperature
τ	shear stress
τ_0	yield stress

Subscripts

e	effective
f	fluid
i	inner
in	entrance
L	number based on the plate length
o	outer
p	porous
s	solid
x	local variable
w	wall
∞	free stream property

REFERENCES

Al-Nimr MA, Alkam MK. Unsteady non-Darcian forced convection analysis in an annulus partially filled with a porous material. ASME J Heat Transfer 119:799–804, 1994.

Alkam MK, Al-Nimr MA, Mousa Z. Forced convection of non-Newtonian fluids in porous concentric annuli. Int J Numer Methods Heat Fluid Flow 8:703–716, 1998.

Amiri A, Vafai K. Analysis of dispersion effects and non-thermal equilibrium, non-Darcian, variable porosity incompressible flow through porous media. Int J Heat Mass Transfer 37:939–954, 1994.

Amiri A, Vafai K, Kuzay TM. Effects of boundary conditions on non-Darcian heat transfer through porous media and experimental comparisons. Numerical Heat Transfer, Part A 27:651–664, 1995.

Beavers GS, Joseph DD. Boundary conditions at naturally permeable wall. J Fluid Mech 13:197–207, 1967.

Beavers GS, Sparrow EM. Non-Darcy flow through fibrous porous media. J Appl Mech 36:711–714, 1969.

Beckermann C, Viskanta R. Forced convection boundary layer flow and heat transfer along a flat plate embedded in a porous medium. Int J Heat Mass Transfer 30:1547–1551, 1987.

Bejan A. Convection Heat Transfer. New York: Wiley, 1984.

Bejan A, Nield DA. Transient forced convection near a suddenly heated plate in a porous medium. Int Commun Heat Mass Transfer 18:83–91, 1991.

Benenati RF, Brosilow CB. Void fraction distribution in packed beds. AIChE J 8:359–361, 1962.

Branci F, Ghafir R, Chikh S, Si-Ahmed EK. Heat transfer of a non-Darcian flow in a porous medium with no local thermal equilibrium. In: Nguyen TH, Bilgen E, Mir A, Vasseur P, eds. Proceedings of Second International Thermal Energy Congress 2:445–450, ITEC, Ecole Polytechnique, Montréal 1995.

Burmeister LC. Convective Heat Transfer, 2nd ed. New York: Wiley, 1993, pp 15–124.

Carbonell RG, Whitaker S. Heat and mass transfer in porous media. In: Bear J, Corapcioglu MY, eds. Fundamentals of Transport Phenomena in Porous Media. Boston: Martinus Nijhof, 1984, pp 121–198.

Chandrasekhara BC, Vortmeyer D. Flow model for velocity distribution in fixed porous beds under isothermal conditions. Wärme- und Stoffübertragung 12:105–111, 1979.

Chen G, Hadim HA. Numerical study of non-Darcy forced convection in a packed bed saturated with a power law fluid. J Porous Media 1:147–157, 1998.

Cheng P. Wall effects on fluid flow and heat transfer in porous media. Proceedings of the 1987 ASME/JSME Thermal Engineering Joint Conference 2, 1987, pp 297–303.

Cheng P, Hsu CT. Fully-developed, forced convective flow through an annular packed-sphere bed with wall effects. Int J Heat Mass Transfer 29:1843–1853, 1986.

Cheng P, Vortmeyer D. Transverse thermal dispersion and wall channeling in a packed bed with forced convective flow. Chem Eng Sci 43:2523–2532, 1988.

Cheng P, Zhu CT. Effect of radial thermal dispersion on fully-developed forced convection in cylindrical packed tubes. Int J Heat Mass Transfer 30:2373–2383, 1987.

Cheng P, Hsu CT, Chowdhury A. Forced convection in the entrance region of a packed channel with asymmetric heating. ASME J Heat Transfer 110:946–954, 1988.

Chikh S, Boumedien A, Bouhadef K, Lauriat G. Analytical solution of non-Darcian forced convection in an annular duct partially filled with a porous medium. Int J Heat Mass Transfer 38:1543–1551, 1995a.

Chikh S, Boumedien A, Bouhadef K, Lauriat G. Non-Darcian forced convection analysis in an annulus partially filled with a porous medium. Numer Heat Transfer 28:707–722, 1995b.

Chikh S, Boumedien A, Bouhadef K, Lauriat G. Amélioration du transfert thermique par un dépôt poreux sur la paroi d'un échangeur tubulaire. Rev Gén Therm 36:41–50, 1997.

Chou FC, Lien WY, Lin SH. Analysis and experiment of non-Darcian convection in horizontal square packed-sphere channels—1. Forced convection. Int J Heat Mass Transfer 35:195–205, 1992.

Chou FC, Su JH, Lien SS. A re-evaluation of non-Darcian forced and mixed convection in cylindrical packed tubes. ASME J Heat Transfer 116:513–516, 1994.

Chowdhury A, Cheng P. Thermally developing flows in packed tubes and channels. Proceedings of International Conference on Mechanics of Two-phase Flows, Taipei, 1989, pp 421–427.

Christopher RH, Middleman S. Power-law flow through a packed tube. I and EC Fundamentals 4:422–426, 1965.

Cohen Y, Metzner AB. Wall effects in laminar flow of fluids through packed beds. AIChE J 27:705–715, 1981.

Dharmadhikari RV, Kale DD. Flow of non-Newtonian fluids through porous media. Chem Eng Sci 40:527–529, 1985.

Dixon AG, Cresswell DL. Theoretical predictions of effective heat transfer mechanisms in regular shaped packed beds. AIChE J 25:663–676, 1979.

Dullien FAL. Porous Media Fluid Transport and Pore Structure. New York: Academic Press, 1979.

Ergun S. Fluid flow through packed column. Chem Eng Prog 48:89–94, 1952.

Forchheimer F. Wasservevegung durch bioden. Z Ver Deutch Ing 45:1782–1788, 1901.

Gamson BW, Thodos G, Hougen OA. Heat, mass and momentum transfer in the flow gases through granular solids. Trans AIChE 39:1–35, 1943.

Georgiadis JG, Catton I. An effective equation governing convective transport in porous media. ASME J Heat Transfer 110:635–641, 1988.

Hadim A. Forced convection in a porous channel with localized heat sources. ASME J Heat Transfer 116:465–472, 1994.

Hady FM, Ibrahim FS. Forced convection heat transfer on a flat plate embedded in porous media for power-law fluids. Transp Porous Media 28:125–134, 1997.

Hsu ML, Cheng P. Closure schemes of the macroscopic energy equation for convective heat transfer in porous media. Int Commun Heat Mass Transfer 15:689–703, 1988.

Hsu ML, Cheng P. Thermal dispersion in porous media. Int J Heat Mass Transfer 33:1587–1597, 1990.

Hunt, ML, Tien CL. Non-Darcian convection in cylindrical packed beds. ASME J Heat Transfer 110, 378–384, 1988a.

Hunt ML, Tien CL. Effects of thermal dispersion forced convection in fibrous media. Int J Heat Mass Transfer 31:301–309, 1988b.

Hwang, GJ, Wu CC, Chao CH. Investigation of non-Darcian forced convection in an asymmetrically heated sintered porous channel. ASME J Heat Transfer 117:725–731, 1995.

Jang JY, Chen JL. Forced convection in a parallel plate channel partially filled with a high porosity medium. Int Commun Heat Mass Transfer 19:263–273, 1992.

Jiang PX, Ren ZP, Wang BX, Wang Z. Forced convective heat transfer in a plate channel filled with solid particles. J Therm Sci 5:43–53, 1996a.

Jiang PX, Wang BX, Luo DA, Ren ZP. Fluid flow and convective heat transfer in a vertical porous annulus. Numer Heat Transfer Part A 30:305–320, 1996b.

Jiang PX, Wang Z, Ren ZP. Fluid flow and convection heat transfer in a plate channel filled with solid particles. Proceedings of 11th International Heat Transfer Conference, Kyongju, 1998, 4:23–28.

Kakaç S, Yener Y. Convective Heat Transfer, 2nd ed. Boca Raton: CRC Press, 1995, pp 131–163.

Kamiuto K, Saitoh S. Fully developed forced-convection heat transfer in cylindrical packed beds with constant wall temperatures. JSME Int J Series B 37:554–559, 1994.

Kamiuto K, Saitoh S. Fully developed forced-convection heat transfer in cylindrical packed beds with constant wall heat fluxes. JSME Int J Series B 39:395–401, 1996.

Kar K, Dybbs A. Internal heat transfer coefficients of porous media. In: Beck JV, Yao LS, eds. Heat Transfer in Porous Media HTD-22, ASME, 1982.

Kaviany M. Laminar flow through a porous channel bounded by isothermal parallel plates. Int J Heat Mass Transfer 28:851–858, 1985.

Kaviany M. Boundary layer treatment of forced convection heat transfer from a semi-infinite plate embedded in porous media. ASME J Heat Transfer 109, 345–389, 1987.

Kaviany M. Principles of Heat Transfer in Porous Media. New York: Springer-Verlag, 1991.

Kays WM, Crawford ME. Convective Heat and Mass Transfer, 3rd ed. New York: McGraw-Hill, 1993.

Kays WM, Perkins HC. Forced convection, flow in ducts. In: Rohsenow WM, Hartnett JP, Ganić EN, eds. Handbook of Heat Transfer, Fundamentals. New York: McGraw-Hill, 1985, pp 7-46–7-67.

Kemblowski Z, Michniewicz M. A new look at the laminar flow of power-law fluids through granular beds. Rheol Acta 18:730–739, 1979.

Kimura S. Transient forced convection heat transfer from a circular cylinder in a saturated porous medium. Int J Heat Mass Transfer 32: 192–195, 1989.

Kimura S, Yoneya M. Forced convection heat transfer from a cylinder with constant heat flux in a saturated porous medium. Heat Transfer—J Res 21:250–258, 1992.

Koh JCY, Colony R. Analysis of cooling effectiveness for porous material in a coolant passage. ASME J Heat Transfer 96:81–91, 1974.

Kumari M, Nath G, Pop I. Non-Darcian effects on forced convection heat transfer over a flat plate in a highly porous medium. Acta Mech 84:201–207, 1990.

Kuzay TM, Collins JT, Khounsary AM, Morales G. Enhanced heat transfer with metal-wool-filled tubes. Proceedings of ASME/JSME Thermal Engineering Conference, 1991, pp 145–151.

Kuznetsov AV. Thermal nonequilibrium, non-Darcian forced convection in a channel filled with a fluid saturated porous medium—A perturbation solution. App Sci Res 57:119–131, 1997.

Lauriat G, Vafai K. Forced convection and heat transfer through a porous medium exposed to a flat plate or a channel. In: Kacaç S, Kilikiş B, Kulacki FA, Arinç F, eds. Convective Heat and Mass Transfer in Porous Media. Dordrecht: Kluwer Academic, 1991, pp 289–327.

Lee DY, Vafai K. Analytical characterization and conceptual assessment of solid and fluid temperature differentials in porous media. Int J Heat Mass Transfer 42:423–435, 1999.

Levec J, Carbonell RG. Longitudinal and lateral thermal dispersion in packed beds, AIChE J 31:591–602, 1985.

Lévêque DA. Les lois de la transmission de chaleur. Ann Mines 13:201–239, 1928.

Marpu DR. Non-Darcy flow and axial conduction effects on forced convection in porous material filled pipes. Wärme und Stoffübertragung 29:51–58, 1993.

Nakayama A, Ebinuma CD. Transient non-Darcian forced convective heat transfer from a flat plate embedded in a fluid-saturated porous medium. Int J Heat Fluid Flow 11:249–253, 1990.

Nakayama A, Pop I. Momentum and heat transfer over a continuous moving surface in a non-Darcian fluid. Wärme-und Stoffübertragung 28:177–184, 1993.

Nakayama A, Shenoy AV. Non-Darcy forced convective heat transfer in a channel embedded in a non-Newtonian inelastic fluid-saturated porous medium. Can J Chem Eng 71:168–173, 1993.

Nakayama A, Kokudai, Koyama H. Non-Darcian boundary layer flow and forced convective heat transfer over a flat plate in a fluid-saturated porous medium. ASME J Heat Transfer 112:157–162, 1990.

Nield DA. Effects of local thermal nonequilibrium in steady convective processes in a saturated porous medium: forced convection in a channel. J Porous Media 1:181–186, 1998.

Nield DA, Bejan A. Convection in Porous Media. New York: Springer Verlag, 1992.

Nield DA, Junqueira SLM, Lage JL. Forced convection in a fluid-saturated porous medium channel with isothermal or isoflux boundaries. J Fluid Mech 322:201–214, 1996.

Ochoa-Tapia JA, Whitaker S. Heat transfer at the boundary between a porous medium and a homogeneous fluid: the one-equation model. J Porous Media 1:31–46, 1998.

Poulikakos D, Kazmierczak M. Forced convection in a duct partially filled with a porous material. ASME J Heat Transfer 109:653–662, 1987.

Poulikakos D, Renken K. Forced convection in a channel filled with a porous medium, including the effects of flow inertia, variable porosity, and Brinkman friction. ASME J Heat Transfer 109:880–888, 1987.

Renken KJ, Poulikakos D. Experiments and analysis on forced convective heat transfer in a packed bed of spheres. Int J Heat Mass Transfer 31:1399–1408, 1988.

Renken KJ, Poulikakos D. Experiments on forced convection from a horizontal heated plate in a packed bed of glass spheres. J Heat Transfer 111:59–65, 1989.

Reynolds O. Papers on Mechanical and Physical Subjects. Cambridge, UK: Cambridge University Press, 1900.

Roblee LHS, Baird RM, Tierney JW. Radial porosity variations in packed beds. AIChE J 4:460–464, 1958.

Sano T, Makizono K. Unsteady forced convection around a sphere immersed in a porous medium at large Peclet number. Bull Univ Osaka Prefect 44:47–50, 1995.

Schertz WM, Bischoff KB. Thermal and material transport in non-isothermal packed beds. AIChE J 15:597–604, 1969.

Schwartz CE, Smith JM. Flow distribution in packed beds. Ind Eng Ch 45: 1209–1218, 1958.

Shenoy AV. Darcy–Forchheimer natural, forced and mixed convection heat transfer in non-Newtonian power-law fluid saturated porous media. Trans Porous Media 11:219–241, 1992a.

Shenoy AV. Darcy natural, forced and mixed convection heat transfer from an isothermal flat plate embedded in a porous medium saturated with an elastic fluid of constant viscosity. Int J Eng Sci 30:455–467, 1992b.

Shenoy AV. Forced convection heat transfer to an elastic fluid of constant viscosity flowing through a channel filled with a Brinkman–Darcy porous medium. Wärme-und Stoffübertragung 28:295–297, 1993.

Shenoy AV. Non-Newtonian fluid heat transfer in porous media. Adv Heat Transfer 24:101–190, 1994.

Sözen M, Vafai K. Analysis of non-thermal equilibrium condensing flow of a gas through a packed bed. Int J Heat Mass Transfer 33:1247–1261, 1990.

Sözen M, Vafai K. Analysis of oscillating compressible flow through a packed bed. Int J Heat Fluid Flow 12:130–136, 1991.

Sözen M, Vafai K. Longitudinal heat dispersion in porous beds with real gas flow. J Thermophys Heat Transfer 7:153–157, 1993.

Srinivasan V, Vafai K, Christensen RN. Analysis of heat transfer and fluid flow through a spirally fluted tube using a porous substrate approach. ASME J Heat Transfer 116:243–551, 1994.

Vafai K. Convective flow and heat transfer in variable porosity media. J Fluid Mech 147:233–259, 1984.

Vafai K. Analysis of the channeling effect in variable porosity media. ASME J Energy Resour Tech 108:131–139, 1986.

Vafai K, Amiri A. Non-Darcian effects in confined forced convective flows. In: Ingham DB, Pop I, eds. Transport Phenomena in Porous Media. Oxford: Pergamon, 1998.

Vafai K, Kim SJ. Forced convection in a channel filled with a porous medium: an exact solution. ASME J Heat Transfer 111:642–647, 1989.

Vafai K, Kim SJ. Fluid mechanics of interface region between a porous medium and a fluid layer—an exact solution. Int J Heat Fluid Flow 11:254–256, 1990.

Vafai K, Sözen M. Analysis of energy and momentum transport for fluid flow through a porous bed. ASME J Heat Transfer 112:690–699, 1990a.

Vafai K, Sözen M. An investigation of a latent heat storage porous bed and condensing flow through it. ASME J Heat Transfer 112:1014–1022, 1990b.

Vafai K, Thiyagaraja R. Analysis of flow and heat transfer at the interface region of a porous medium. Int J Heat Mass Transfer 30:1391–1405, 1987.

Vafai K, Tien CL. Boundary and inertia effects on flow and heat transfer in porous media. Int J Heat Mass Transfer 24:195–203, 1981.

Vafai K, Alkire RL, Tien CL. An experimental investigation of heat transfer in variable porosity media. ASME J Heat Transfer 105:642–647, 1985.

Vortmeyer D, Adam W. Steady-state measurements and analytical correlations of axial effective conductivities in packed beds at low gas flow rates. Int J Heat Mass Transfer 27:1465–1472, 1984.

Vortmeyer D, Schaeffer RJ. Equivalence of one- and two-phase flow models for heat transfer processes in packed beds: one dimensional theory. Chem Eng Sci 29:485–491, 1974.

Vortmeyer D, Schuster J. Evaluation of steady flow profiles in rectangular and circular packed beds by a variational method. Chem Eng Sci 38:1691–1699, 1983.

Wakao N, Kaguei S. Heat and mass transfer in packed beds. New York: Gordon & Breach, 1982.

Wakao N, Kaguei S, Funazkri T. Effect of fluid dispersion coefficients on particle-to-fluid heat transfer coefficients in packed beds. Chem Eng Sci 34:325–336, 1979.

Ward JC. Turbulent flow in porous media. J Hydraulic Div ASCE 90: HY5, 1964.

Yagi S, Wakao N. Heat and mass transfer from wall to fluid in packed beds. AIChE 5:79–158, 1959.

You HI, Chang CH. Numerical prediction of heat transfer coefficient for a pin fin channel flow. ASME J Heat Transfer 119:840–843, 1997.

6

Analytical Studies of Forced Convection in Partly Porous Configurations

A. V. Kuznetsov
North Carolina State University, Raleigh, North Carolina

I. INTRODUCTION

Forced convection in a composite region, part of which is occupied by a clear fluid and part by a fluid-saturated porous medium, has recently attracted considerable attention and become a subject for numerous investigations (Alkam and Al-Nimr, 1998; Al-Nimr and Alkam, 1997, 1998; Chikh et al., 1995a,b; Rudraiah, 1985). This interest is due to many important thermal engineering applications relevant to this problem. Solid matrix heat exchangers, the use of porous materials for heat transfer enhancement, fault zones in geothermal systems, and solidification of binary alloys are a few of these problems.

Momentum transport in the fluid/porous interface region was first investigated by Beavers and Joseph (1967). Proceeding from the results of their experimental investigation, Beavers and Joseph suggested the slip-flow boundary condition at the fluid/porous interface. According to this boundary condition, the velocity gradient on the fluid side of the interface is proportional to the slip velocity at the interface. The investigation by Beavers and Joseph was continued in works by Taylor (1971) and Richardson (1971). In all the papers mentioned above, fluid flow in the porous region was modeled by the Darcy equation. Considerable progress on this problem was made by Vafai and Kim (1990a), who modeled the flow in the porous region utilizing the so-called Brinkman–Forchheimer–

extended Darcy equation (Vafai and Kim, 1995b; Kuznetsov, 1998c). The Brinkman term in this equation represents viscous effects (Kaviany, 1991) and makes it possible to impose the no-slip boundary condition at the impermeable wall and also to match momentum equations at the fluid/porous interface, retaining continuity of the filtration (seepage) velocity. High flow velocities characterize many modern applications of porous media. In such cases it is also necessary to account for deviation from linearity in the momentum equation for porous media. This deviation is accounted for by the Forchheimer term representing the quadratic drag, which is essential for large particle Reynolds numbers (Vafai and Tien, 1981; Nakayama, 1995). From the physical point of view, quadratic drag appears in the momentum equation for porous media because, for large filtration velocities, the form drag due to the solid obstacles becomes comparable with the surface drag due to friction (Nield and Bejan, 1999).

Direct numerical simulation of the flow in the fluid/porous interface region was carried out in Sahraoui and Kaviany (1992). Further insight into this problem was gained in Ochoa-Tapia and Whitaker (1995a,b). Ochoa-Tapia and Whitaker, by means of complicated volume-averaging analysis of the momentum equations in the interface region, have shown that matching the Brinkman–Darcy and Stokes equations retains continuity of velocity but produces a jump in the shear stress. A heat transfer boundary condition at the fluid/porous interface for the case of thermal nonequilibrium between the fluid and solid phases was suggested by Ochoa-Tapia and Whitaker (1997). The boundary conditions suggested by Ochoa-Tapia and Whitaker were utilized in Kuznetsov (1996, 1997, 1998d), where solutions for different channels partly filled with a porous material were obtained.

The coupled fluid flow and heat transfer problem in a fully developed region of a parallel-plate channel filled with the Brinkman–Darcy porous medium was analytically investigated by Kaviany (1985) and Nakayama et al. (1988). Fully developed forced convection in channels partly filled with the Brinkman–Darcy porous medium was analytically studied in Poulikakos and Kazmierczak (1987). An important example of the perturbation analysis of heat transfer in the interface region between a clear fluid and a Brinkman–Forchheimer porous medium is given in Vafai and Thiyagaraja (1987). Among numerical works relevant to this problem we must mention the paper by Vafai and Kim (1990b), who studied forced convective flow and heat transfer in a composite fluid/porous system consisting of a fluid layer overlaying a porous substrate attached to the surface of the plate. Huang and Vafai (1993, 1994a, 1994b) studied forced convection in a composite fluid/porous system containing multiple emplaced porous blocks. Jang and Chen (1992) presented a numerical study of fully developed convection in a parallel-plate channel partly filled with a high porosity porous medium. Chen and Chen (1992) and Kim and Choi (1996) reported numerical investigations of convection in superposed fluid and porous layers. An insightful work by Hadim (1994) presents a numerical study of forced convection in a channel filled with a fluid-saturated porous medium and containing discrete heat sources on the bottom wall. Chang and Chang (1996) numerically analyzed the developing mixed convection in a vertical parallel-plate channel partly filled with a porous medium.

This chapter continues an analytical investigation of fluid flow and heat transfer in the interface region started in Vafai and Kim (1990a). Here the Brinkman–Forchheimer–extended Darcy equation is used to describe the momentum flow in the porous region. The Brinkman–Forchheimer extension of the Darcy law makes it possible to account for both viscous and nonlinear effects in the porous region. Utilizing the boundary layer technique, we will obtain solutions for the velocity and temperature distributions, as well as for the Nusselt number, for a number of composite geometrical configurations involving the fluid/porous interface.

II. FORCED CONVECTION IN PARALLEL-PLATE CHANNELS PARTLY FILLED WITH POROUS MEDIA

A. Problem Formulation and Governing Equations

In this approach we proceed from the boundary layer approach suggested in Vafai and Kim (1989, 1990a, 1995a). According to these papers, for most practical applications of porous media the value of the Darcy number is small; therefore the thickness of the momentum boundary layer in the porous region is usually smaller than the thickness of the porous region itself. In this chapter we adopt this approach and proceed from this idea. This assumption does not influence the governing equations for this problem. However, it will be utilized in obtaining the solution of the momentum equation in the porous region (Kuznetsov, 1999).

Figure 1 displays the schematic diagram of the problem. Four different geometrical configurations of parallel-plate channels partly filled with a fluid-saturated porous medium are considered. The parallel-plate channel displayed in Figure 1a has porous layers attached to the walls and a clear fluid region in the center. The walls of the channel are subject to a uniform heat flux. The parallel-plate channel displayed in Figure 1b has a porous core in the center and clear fluid layers at the walls. Again, the walls of the channel are subject to a uniform heat flux. The parallel-plate channel displayed in Figure 1c has a porous layer attached to one wall and a clear fluid layer at the other wall. The channel is subject to a uniform heat flux from the fluid side, while the porous side is adiabatic. The parallel-plate channel

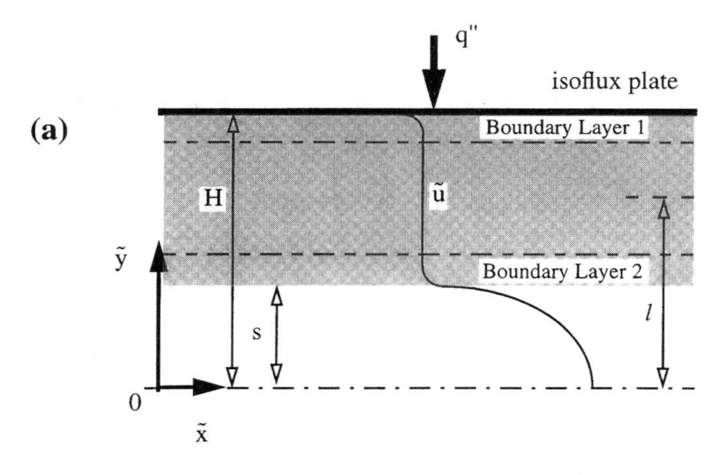

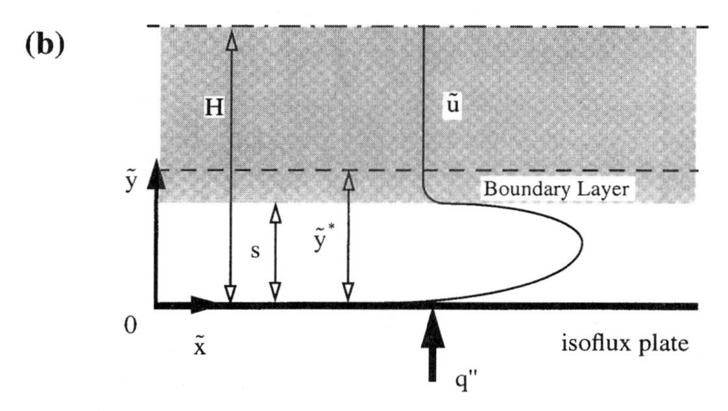

Figure 1. Different geometrical configurations of a parallel-plate channel partly filled with a porous medium: (a) parallel-plate channel with a porous layer at the walls; (b) parallel-plate channel with a porous core; (c) parallel-plate channel with a porous layer at one wall and a clear fluid layer at the other—heating from the fluid side; (d) parallel-plate channel with a porous layer at one wall and a clear fluid layer at the other—heating from the porous side.

displayed in Figure 1d is similar to the channel displayed in Figure 1c, the difference being that the channel in Figure 1d is subject to a uniform heat flux from the porous side while the fluid side is adiabatic.

For the problems displayed in Figures 1a, 1c, and 1d, there are two momentum boundary layers in the porous region. The first momentum boundary layer is near the upper plate, whereas the second boundary

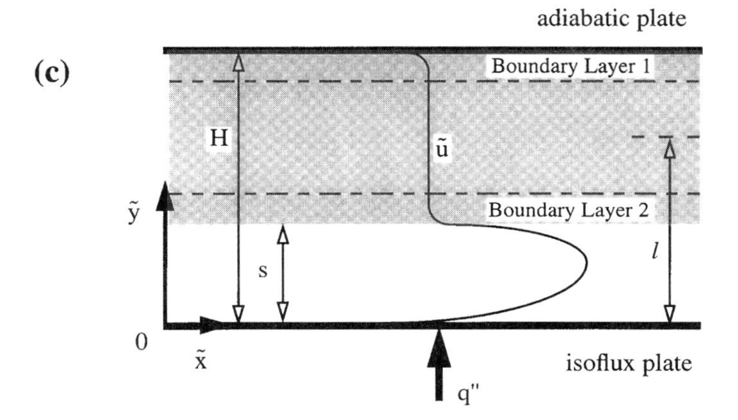

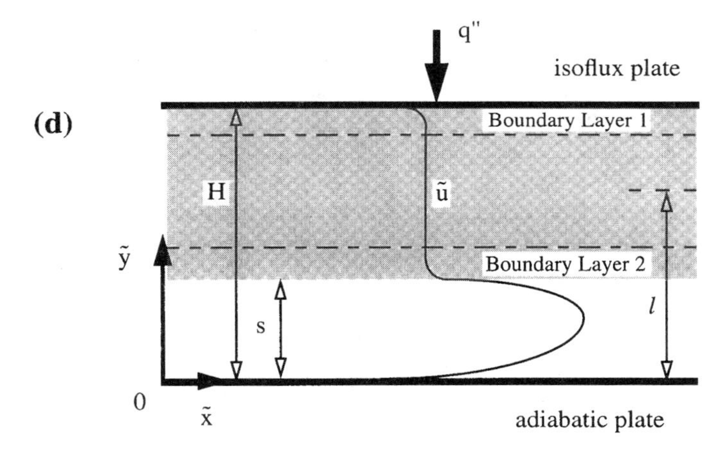

Figure 1 (*continued*)

layer is near the clear fluid/porous medium interface. For the problem displayed in Figure 1b there is only one momentum boundary layer, which is near the fluid/porous interface. The thickness of the boundary layers depends on the value of the Darcy number as well as on the value of the Forchheimer coefficient. The thickness of the boundary layers decreases with a decrease in the Darcy number and with an increase in the Forchheimer coefficient.

The set of governing equations for the problem shown in Figures 1a–d can be presented as

$$-\frac{d\tilde{p}}{d\tilde{x}} + \mu_f \frac{d^2\tilde{u}_f}{d\tilde{y}^2} = 0 \qquad 0 \le \tilde{y} \le s \tag{1}$$

$$-\frac{d\tilde{p}}{d\tilde{x}} + \mu_{\text{eff}} \frac{d^2\tilde{u}_f}{d\tilde{y}^2} - \frac{\mu_f}{K}\tilde{u}_f - \frac{\rho_f c_F}{K^{1/2}}\tilde{u}_f^2 = 0 \qquad s \le \tilde{y} \le H \tag{2}$$

$$\rho_f c_f \tilde{u}_f \frac{\partial \tilde{T}}{\partial \tilde{x}} = k_f \frac{d^2\tilde{T}}{d\tilde{y}^2} \qquad 0 \le \tilde{y} \le s \tag{3}$$

$$\rho_f c_f \tilde{u}_f \frac{\partial \tilde{T}}{\partial \tilde{x}} = k_{\text{eff}} \frac{d^2\tilde{T}}{d\tilde{y}^2} \qquad s \le \tilde{y} \le H \tag{4}$$

where c_f is the specific heat of the fluid, c_F is the Forchheimer coefficient, H is half the distance between the plates for the channels displayed in Figures 1a and 1b and the distance between the plates for the channels displayed in Figures 1c and 1d, k_f is the thermal conductivity of the fluid, k_{eff} is the effective thermal conductivity of the porous medium, K is the permeability of the porous medium, $\tilde{p}$ is the pressure, s is the thickness of the clear fluid region, $\tilde{T}$ is the temperature, $\tilde{u}_f$ is the filtration (seepage) velocity, $\tilde{x}$ is the streamwise coordinate, $\tilde{y}$ is the transverse coordinate, μ_f is the fluid viscosity, μ_{eff} is the effective viscosity in the Brinkman term for the porous region, and ρ_f is the density of the fluid. Equation (1) is the momentum equation for the clear fluid region whereas Eq. (2) is the momentum equation for the porous region, the Brinkman–Forchheimer–extended Darcy equation. Equations (3) and (4) are the energy equations for the clear fluid and porous regions, respectively. Following Kaviany (1985), Nakayama et al. (1988), Cheng et al. (1988), Vafai and Kim (1989), and Nield et al. (1996), longitudinal heat conduction is neglected in (3) and (4). This assumption is acceptable for large Péclet numbers.

In (2) we distinguish between the fluid viscosity, μ_f, and the effective viscosity in the Brinkman term, μ_{eff}. Most works which used the Brinkman model assumed that $\mu_{\text{eff}} = \mu_f$. However, recent direct numerical simulation (Martys et al., 1994) and recent experimental investigation (Givler and Altobelli, 1994) have demonstrated that there are situations when it is important to distinguish between these two coefficients. For example, in Givler and Altobelli (1994) a water flow through a tube filled with an open-cell rigid foam of high porosity was investigated. It was obtained that for this flow $\mu_{\text{eff}} = (7.5^{+3.4}_{-2.4})\mu_f$.

For the problems displayed in Figures 1a–d at the fluid/porous interface we utilize the form of boundary conditions suggested in Ochoa-Tapia and Whitaker (1995a,b)

$$\tilde{u}_f|_{\tilde{y}=s-0} = \tilde{u}_f|_{\tilde{y}=s+0} \qquad \mu_{\text{eff}}\frac{\mathrm{d}\tilde{u}_f}{\mathrm{d}\tilde{y}}\bigg|_{\tilde{y}=s+0} - \mu_f\frac{\mathrm{d}\tilde{u}_f}{\mathrm{d}\tilde{y}}\bigg|_{\tilde{y}=s-0}$$
$$= \beta\frac{\mu_f}{K^{1/2}}\tilde{u}_f|_{\tilde{y}=s} \qquad \text{at} \qquad \tilde{y} = s \tag{5}$$

$$\tilde{T}|_{\tilde{y}=s-0} = \tilde{T}|_{\tilde{y}=s+0} \qquad k_{\text{eff}}\frac{\partial\tilde{T}}{\partial\tilde{y}}\bigg|_{\tilde{y}=s+0} = k_f\frac{\partial\tilde{T}}{\partial\tilde{y}}\bigg|_{\tilde{y}=s-0} \qquad \text{at} \qquad \tilde{y} = s \tag{6}$$

where β is the adjustable coefficient in the stress jump boundary condition.

For the problem displayed in Figure 1a the boundary conditions in the center of the channel and at the peripheral wall are, respectively,

$$\frac{\partial\tilde{u}_f}{\partial\tilde{y}} = 0 \qquad \frac{\partial\tilde{T}}{\partial\tilde{y}} = 0 \qquad \text{at} \qquad \tilde{y} = 0 \tag{7}$$

$$\tilde{u}_f = 0 \qquad k_{\text{eff}}\frac{\partial\tilde{T}}{\partial\tilde{y}} = q'' \qquad \text{at} \qquad \tilde{y} = H \tag{8}$$

where q'' is the heat flux at the lower plate.

For the problem displayed in Figure 1b the boundary conditions at the peripheral wall and in the center of the channel are, respectively,

$$\tilde{u}_f = 0 \qquad k_f\frac{\partial\tilde{T}}{\partial\tilde{y}} = q'' \qquad \text{at} \qquad \tilde{y} = 0 \tag{9}$$

$$\frac{\partial\tilde{u}_f}{\partial\tilde{y}} = 0 \qquad \frac{\partial\tilde{T}}{\partial\tilde{y}} = 0 \qquad \text{at} \qquad \tilde{y} = H \tag{10}$$

For the problem displayed in Figure 1c the boundary conditions at the lower and upper plates are, respectively,

$$\tilde{u}_f = 0 \qquad k_f\frac{\partial\tilde{T}}{\partial\tilde{y}} = q'' \qquad \text{at} \qquad \tilde{y} = 0 \tag{11}$$

$$\tilde{u}_f = 0 \qquad \frac{\partial\tilde{T}}{\partial\tilde{y}} = 0 \qquad \text{at} \qquad \tilde{y} = H \tag{12}$$

For the problem displayed in Figure 1d the boundary conditions at the lower and upper plates are, respectively,

$$\tilde{u}_f = 0 \qquad \frac{\partial\tilde{T}}{\partial\tilde{y}} = 0 \qquad \text{at} \qquad \tilde{y} = 0 \tag{13}$$

$$\tilde{u}_f = 0 \qquad k_{\text{eff}}\frac{\partial \tilde{T}}{\partial \tilde{y}} = q'' \qquad \text{at} \qquad \tilde{y} = H \tag{14}$$

Outside the momentum boundary layers, the viscous term is negligible and the Brinkman–Forchheimer–extended Darcy equation, (2), reduces to

$$-\frac{d\tilde{p}}{d\tilde{x}} - \frac{\mu_f}{K}\tilde{u}_\infty - \frac{\rho_r c_F}{K^{1/2}}\tilde{u}_\infty^2 = 0 \tag{15}$$

where $\tilde{u}_\infty$ is the filtration velocity outside the momentum boundary layers.

B. Dimensionless Equations and Boundary Conditions

In the fully developed region of a channel with uniform wall heat flux, $\partial\tilde{T}/\partial\tilde{x}$ on the left side of Eqs. (3) and (4) is constant (Bejan, 1995). The value of $\partial\tilde{T}/\partial\tilde{x}$ can then be found from the following energy balance

$$\rho_f c_f H \tilde{U}\frac{\partial \tilde{T}}{\partial \tilde{x}} = q'' \tag{16}$$

where the mean flow velocity $\tilde{U}$ is defined by the following equation

$$\tilde{U} = \frac{1}{H}\int_0^H \tilde{u}_f d\tilde{y} \tag{17}$$

The Nusselt number is then defined as

$$Nu = \frac{Hq''}{k_f(\tilde{T}_w - \tilde{T}_m)} \tag{18}$$

where the mean temperature $\tilde{T}_m$ is defined by the following equation

$$\tilde{T}_m = \frac{1}{H\tilde{U}}\int_0^H \tilde{u}_f \tilde{T} d\tilde{y} \tag{19}$$

and $\tilde{T}_w$ is the wall temperature.

Introducing dimensionless variables, the momentum and energy equations (1)–(4), can be recast into the following dimensionless form

$$1 + \frac{d^2 u}{dy^2} = 0 \qquad 0 \le y \le S \tag{20}$$

$$1 + \gamma^2\frac{d^2 u}{dy^2} - \frac{1}{Da}u - Fu^2 = 0 \qquad S \le y \le 1 \tag{21}$$

$$\frac{d^2 T}{dy^2} = -Nu\frac{u}{U} \qquad 0 \le y \le S \tag{22}$$

$$R\frac{\mathrm{d}^2T}{\mathrm{d}y^2} = -Nu\frac{u}{U} \qquad S \le y \le 1 \tag{23}$$

The equation for the velocity outside the momentum boundary layer, (15), can be similarly recast into the dimensionless form as follows

$$1 - \frac{1}{Da}u_\infty - Fu_\infty^2 = 0 \tag{24}$$

The positive root of Eq. (24) is

$$u_\infty = \frac{-1 + [1 + 4Da^2F]^{1/2}}{2DaF} \tag{25}$$

In Eqs. (20)–(25) the following dimensionless variables are utilized:

$$Da = \frac{K}{H^2}, \qquad F = \frac{\rho_f c_F H^4}{K^{1/2}\mu_f^2}G, \qquad R = \frac{k_{\mathrm{eff}}}{k_f}, \qquad T = \frac{\tilde{T} - \tilde{T}_w}{\tilde{T}_m - \tilde{T}_w} \tag{26}$$

$$u = \frac{\tilde{u}_f\mu_f}{GH^2}, \qquad u_\infty = \frac{\tilde{u}_\infty\mu_f}{GH^2}, \qquad y = \frac{\tilde{y}}{H}, \qquad \gamma = \left(\frac{\mu_{\mathrm{eff}}}{\mu_f}\right)^{1/2} \tag{27}$$

where $G = -\mathrm{d}\tilde{p}/\mathrm{d}\tilde{x}$ is the applied pressure gradient.

The boundary conditions at the fluid/porous interface given by Eqs. (5) and (6), in the dimensionless variables take the following form

$$u|_{y=S+0} = u|_{y=S-0} = u_i \qquad \gamma^2\frac{\mathrm{d}u}{\mathrm{d}y}\bigg|_{y=S+0} - \frac{\mathrm{d}u}{\mathrm{d}y}\bigg|_{y=S-0} = \beta Da^{-1/2}u|_{y=S} \qquad \text{at} \qquad y = S \tag{28}$$

$$T|_{y=S+0} = T|_{y=S-0} = T_i \qquad R\frac{\mathrm{d}T}{\mathrm{d}y}\bigg|_{y=S+0} = \frac{\mathrm{d}T}{\mathrm{d}y}\bigg|_{y=S-0} \qquad \text{at} \qquad y = S \tag{29}$$

The dimensionless form of boundary conditions at the peripheral boundaries for the problem displayed in Figure 1a is

$$\frac{\mathrm{d}u}{\mathrm{d}y} = 0 \qquad \frac{\mathrm{d}T}{\mathrm{d}y} = 0 \qquad \text{at} \qquad y = 0 \tag{30}$$

$$u = 0 \qquad T = 0 \qquad \text{at} \qquad y = 1 \tag{31}$$

The dimensionless form of boundary conditions at the peripheral boundaries for the problem displayed in Figure 1b is

$$u = 0 \qquad T = 0 \qquad \text{at} \qquad y = 0 \tag{32}$$

$$\frac{du}{dy} = 0 \qquad \frac{dT}{dy} = 0 \qquad \text{at} \qquad y = 1 \tag{33}$$

The dimensionless form of boundary conditions at the peripheral boundaries for the problem displayed in Figure 1c is

$$u = 0 \qquad T = 0 \qquad \text{at} \qquad y = 0 \tag{34}$$

$$u = 0 \qquad \frac{dT}{dy} = 0 \qquad \text{at} \qquad y = 1 \tag{35}$$

The dimensionless form of boundary conditions at the peripheral boundaries for the problem displayed in Figure 1d is

$$u = 0 \qquad \frac{dT}{dy} = 0 \qquad \text{at} \qquad y = 0 \tag{36}$$

$$u = 0 \qquad T = 0 \qquad \text{at} \qquad y = 1 \tag{37}$$

Equations (22)–(23) determine the temperature distribution in the channel as a function of the Nusselt number. Equations (20)–(23) are solved subject to boundary conditions given by Eqs. (28)–(29) and either of Eqs. (30)–(31), (32)–(33), (34)–(35), or (36)–(37). Then the value of the Nusselt number can be found from the condition that the temperature distribution $T(y)$ must obey the definition for the mean temperature given by Eq. (19). This results in the following compatibility condition, which can be used for calculating the Nusselt number (Bejan, 1995).

$$\int_0^1 Tu \, dy = U \tag{38}$$

where

$$U = \frac{\tilde{U}\mu_f}{GH^2}$$

C. Composite Channel with a Fluid Core (Figure 1a)

1. Velocity Distribution

Velocity distribution in the channel displayed in Figure 1a can be found by integrating the dimensionless momentum equations, (20) and (21), along with boundary conditions given by Eq. (28) and by the first equations in (30) and (31).

(a) Velocity in the Clear Fluid Region ($0 \leq y \leq S$). For the clear fluid region ($0 \leq y \leq S$) the velocity distribution is obtained by integrating Eq.

(20) and utilizing boundary conditions given by the first equation in (28) and by the first equation in (30) as

$$u = u_i + \frac{S^2 - y^2}{2} \tag{39}$$

where u_i is the dimensionless velocity at the clear fluid/porous medium interface.

(b) Velocity in the Boundary Layer 2 Region ($S \leq y \leq L$). An exact solution for the velocity distribution in the porous region is not available. However, following the idea put forward in Vafai and Kim (1989, 1995a), it is possible to obtain a boundary layer solution. There are two momentum boundary layers in the porous region, one near the channel wall and the other near the clear fluid/porous medium interface. To obtain a boundary layer solution it is assumed that these boundary layers do not overlap in the center of the porous region. Between the momentum boundary layers, at $y = L$, the following conditions are utilized

$$u \to u_\infty \qquad \frac{du}{dy} \to 0 \tag{40}$$

The position L can be taken anywhere in the far-field region between two momentum boundary layers. For computations carried out in this chapter, it is assumed that L is the coordinate of the center of the porous layer, $L = S + (1 - S)/2$. Utilizing this approach, the following velocity distribution in the boundary layer 2 region is obtained

$$u = (u_\infty + u_1)\left[\frac{1 - z_2}{1 + z_2}\right]^2 - u_1 \tag{41}$$

where

$$z_2 = B\exp\{-D(y - S)\} \tag{42}$$

$$u_1 = 2u_\infty + \frac{3}{2DaF} \tag{43}$$

$$B = \frac{1 - [(u_i + u_1)/(u_\infty + u_1)]^{1/2}}{1 + [(u_i + u_1)/(u_\infty + u_1)]^{1/2}} \tag{44}$$

and

$$D = \frac{1}{\gamma}\left[\frac{2F(u_\infty + u_1)}{3}\right]^{1/2} \tag{45}$$

(c) Velocity in the Boundary Layer 1 Region ($L \le y \le 1$).

$$u = (u_\infty + u_1)\left[\frac{z_1 - 1}{z_1 + 1}\right]^2 - u_1 \tag{46}$$

where

$$z_1 = A \exp\{D(1 - y)\} \tag{47}$$

and

$$A = \frac{1 + [u_1/(u_\infty + u_1)]^{1/2}}{1 - [u_1/(u_\infty + u_1)]^{1/2}} \tag{48}$$

The dimensionless velocity at the clear fluid/porous medium interface, u_i, can be found from the jump in the shear stress condition given by the second equation in (28). This results in the following transcendental equation for u_i

$$-\gamma(u_i - u_\infty)\left[\frac{2}{3}F(u_i + u_1)\right]^{1/2} + S = \beta Da^{-1/2} u_i \tag{49}$$

(d) Mean Velocity. The dimensionless mean velocity U can then be found from the following equation

$$U = \int_0^S u\,\mathrm{d}y + \int_S^L u\,\mathrm{d}y + \int_L^1 u\,\mathrm{d}y \tag{50}$$

This results in the following equation for U

$$\begin{aligned} U = u_i S + \frac{S^3}{3} + u_\infty(1 - S) + \frac{2\sqrt{6}\gamma(u_\infty + u_1)^{1/2}}{F^{1/2}(1 + B)} \\ - \frac{2\sqrt{6}\gamma(u_\infty + u_1)}{[F(u_\infty + u_1)]^{1/2}(1 + B\exp\{-D(L - S)\})} - \frac{2\sqrt{6}\gamma(u_\infty + u_1)^{1/2}}{F^{1/2}(1 + A)} \\ + \frac{2\sqrt{6}\gamma(u_\infty + u_1)}{[F(u_\infty + u_1)]^{1/2}(1 + A\exp[\{D(1 - L)\})} \end{aligned} \tag{51}$$

2. Temperature Distribution

With the velocity distribution known, the temperature distribution in the channel can be obtained by integrating the energy equations, (22) and (23), subject to the boundary conditions given by Eq. (29) and by the second equations in (30) and (31). It is important to note that, as in Vafai and Kim (1989), the only assumption utilized in this process is the assumption that the momentum boundary layers do not overlap in the center of the porous region. No other approximations are needed for obtaining the temperature distribution in the channel and for determining the value of the Nusselt number.

(a) Dimensionless Temperature in the Clear Fluid Region $(0 \leq y \leq S)$. The dimensionless temperature in the clear fluid region can then be expressed as

$$T = T_i + \frac{Nu}{U}\left[\left(\frac{u_i}{2} + \frac{S^2}{4}\right)(S^2 - y^2) - \frac{1}{24}(S^4 - y^4)\right] \tag{52}$$

(b) Dimensionless Temperature in the Boundary Layer 2 Region $(S \leq y \leq L)$. The dimensionless temperature in the boundary layer 2 region is given by the following equation

$$T = T_i - \frac{1}{R}\frac{Nu}{U}\left\{Q(y - S) + \frac{u_\infty}{2}(y - S)^2 - \frac{6\gamma^2}{F} \times \ln\frac{1 + B\exp\{-D(y - S)\}}{1 + B}\right\} \tag{53}$$

where

$$Q = u_i S + \frac{S^3}{3} - \frac{2\sqrt{6}\gamma(u_\infty + u_1)^{1/2}B}{F^{1/2}(1 + B)} \tag{54}$$

(c) Dimensionless Temperature in the Boundary Layer 1 Region $(L \leq y \leq 1)$. Finally, the dimensionless temperature in the boundary layer 1 region is expressed as

$$T = Nu\frac{1}{RU}\left\{(1 - y)P + \frac{u_\infty}{2}(y + 1 - 2L)(1 - y) + \frac{6\gamma^2}{F} \times \ln\frac{1 + A\exp\{D(1 - y)\}}{1 + A}\right\} \tag{55}$$

where

$$P = Q + u_\infty(L - S) - \frac{2\sqrt{6}\gamma(u_\infty + u_1)^{1/2}}{F^{1/2}(1 + B\exp\{-D(L - S)\})} + \frac{2\sqrt{6}\gamma(u_\infty + u_1)^{1/2}}{F^{1/2}(1 + A\exp\{D(1 - L)\})} \tag{56}$$

The dimensionless temperature at the clear fluid/porous medium interface, T_i, can be found by matching the temperature distributions given by Eqs. (53) and (55) at $y = L$

$$T_i = Nu\Xi \tag{57}$$

where

$$\Xi = \frac{1}{RU}\left\{\frac{u_\infty}{2}(1-S)^2 + Q(1-S) - \frac{2\sqrt{6}\gamma(u_\infty + u_1)^{1/2}}{F^{1/2}(1 + B\exp\{-D(L-S)\})}\right.$$
$$\times (1-L) + \frac{2\sqrt{6}\gamma(u_\infty + u_1)^{1/2}}{F^{1/2}(1 + A\exp\{D(1-L)\})}(1-L) - \frac{6\gamma^2}{F}$$
$$\left.\times \ln\frac{1 + B\exp\{-D(L-S)\}}{1+B} + \frac{6\gamma^2}{F}\ln\frac{1 + A\exp\{D(1-L)\}}{1+A}\right\} \tag{58}$$

3. Nusselt Number

The Nusselt number can now be found from the compatibility condition, (38)

$$Nu = \frac{U}{\Psi_0 + \Psi_2 + \Psi_1} \tag{59}$$

where parameters Ψ_0, Ψ_2, and Ψ_1 are defined as the following integrals

$$\Psi_0 = \int_0^S (T/Nu)u\,\mathrm{d}y \tag{60}$$

$$\Psi_2 = \int_S^L (T/Nu)u\,\mathrm{d}y \tag{61}$$

and

$$\Psi_1 = \int_L^1 (T/Nu)u\,\mathrm{d}y \tag{62}$$

With the velocity and temperature distributions obtained, these integrals can also be calculated analytically. The value of Ψ_0 can be found as

$$\Psi_0 = \frac{S^3\Xi}{3} + \frac{17S^7}{315U} + S\Xi u_i + \frac{4S^5u_i}{15U} + \frac{S^3u_i^2}{3U} \tag{63}$$

The value of Ψ_2 can be found as

$$\Psi_2 = \Xi L u_\infty - \Xi S u_\infty + \frac{3D\gamma^2 L^2 u_\infty}{FRU} - \frac{L^2 Q u_\infty}{2RU} + \frac{S^3 u_\infty^2}{6RU}$$
$$- \frac{3\gamma^2(-8u_\infty + D^2S^2u_\infty + BD^2S^2u_\infty - 8u_1)}{DFRU(1+B)}$$
$$+ \frac{LQ(-4u_\infty + DSu_\infty - 4u_1)}{DRU} + \frac{4\Xi(u_\infty + u_1)}{D(1+B)}$$
$$- \frac{4\exp[DL]\Xi(u_\infty + u_1)}{D(\exp[DL] + B\exp[DS])} + \frac{4\exp[DL]Q(L-S)(u_\infty + u_1)}{DRU(\exp[DL] + B\exp[DS])}$$
$$- \frac{24\gamma^2(u_\infty + u_1)}{DFRU(1 + B\exp[-D(L-S)])}$$
$$+ [6DRU(1 + B\exp[-D(L-S)])]^{-1}$$
$$\times u_\infty\{12L^2u_\infty - DL^3u_\infty - BD\exp[-D(L-S)]L^3u_\infty - 24LSu_\infty$$
$$+3DL^2Su_\infty(1 + B\exp[-D(L-S)]) + 12S^2u_\infty - 3DLS^2u_\infty$$
$$\times (1 + B\exp[-D(L-S)]) + 12L^2u_1 - 24LSu_1 + 12S^2u_1\}$$
$$- \frac{Q}{2D^2RU}\{D^2S^2u_\infty - 8DS(u_\infty + u_1)\} - \frac{4Q(u_\infty + u_1)}{D^2RU}$$
$$\times \ln\frac{1 + B\exp[-D(L-S)]}{1+B} + [DFRU(1 + B\exp[-D(L-S)])]^{-1}$$
$$\times 6\gamma^2 \ln\left[\frac{1 + B\exp[-D(L-S)]}{1+B}\right]\{DLu_\infty(1 + B\exp[-D(L-S)])$$
$$-4(u_\infty + u_1)\} + \frac{4Su_\infty(u_\infty + u_1)}{DRU}\left\{L - S + \frac{1}{D}\right.$$
$$\left.\times \ln\frac{1 + B\exp[-D(L-S)]}{1+B}\right\} - \frac{8u_\infty(u_\infty + u_1)}{DRU}$$
$$\times \left\{\frac{L^2}{2} - \frac{S^2}{2} - \frac{S\ln[1+B]}{D} + \frac{L\ln[1 + B\exp\{-D(L-S)\}]}{D}\right.$$
$$\left. + \frac{\mathrm{Li}_2[-B]}{D^2} - \frac{\mathrm{Li}_2[-B\exp\{-D(L-S)\}]}{D^2}\right\} \tag{64}$$

where

$$\mathrm{Li}_2[\zeta] = -\int_0^{\zeta} \frac{\ln(1-\vartheta)}{\vartheta}\,\mathrm{d}\vartheta \tag{65}$$

is the dilogarithm function. To compute values of Li_2 for large negative values of the argument the following correlation is utilized

$$\mathrm{Li}_2[-\zeta] = \mathrm{Li}_2\left[\frac{1}{1+\zeta}\right] + \frac{1}{2}\ln^2\left[1+\frac{1}{\zeta}\right] - \frac{1}{2}\ln^2[\zeta] - \frac{\pi^2}{6}, \qquad \zeta > 0 \tag{66}$$

Finally, Ψ_1 can be found as

$$\begin{aligned}
\Psi_1 = \frac{1}{RU}\Bigg\{ & \frac{3D\gamma^2 u_\infty}{F}(1-L^2) - \frac{Pu_\infty}{2}(1-L^2) + \frac{(2-3L)u_\infty^2}{6} + \frac{L^3 u_\infty^2}{6} \\
& - \frac{P(4u_\infty - Du_\infty + 4u_1)}{D} + \frac{L^2 u_\infty(4u_\infty - DLu_\infty + 4u_1)}{2D} \\
& - \frac{24\gamma^2(u_\infty + u_1)}{DF(1+A)} + \frac{24\gamma^2(u_\infty + u_1)}{DF(1+A\exp[D(1-L)])} \\
& + \frac{4\exp[DL](1-L)P(u_\infty+u_1)}{D(A\exp[D]+\exp[DL])} + \frac{2(1-L)^2 u_\infty(u_\infty+u_1)}{D(1+A\exp[D(1-L)])} \\
& + \frac{2L^2 u_\infty(u_\infty+u_1)}{D} + \frac{LP(4u_\infty - Du_\infty + 4u_1)}{D} \\
& - \frac{Lu_\infty(Du_\infty + 8Lu_\infty - 2DLu_\infty + 8Lu_1)}{2D} \\
& + \frac{4P(u_\infty+u_1)}{D^2}\ln\frac{1+A\exp[D(1-L)]}{1+A} \\
& - \frac{6\gamma^2(-4u_\infty + DLu_\infty + ADLu_\infty\exp[D(1-L)] - 4u_1)}{DF(1+A\exp[D(1-L)])} \\
& \times \ln\frac{1+A\exp[D(1-L)]}{1+A} + \frac{4u_\infty(u_\infty+u_1)L}{D} \\
& \times \left\{1 - L - \frac{1}{D}\ln\frac{1+A\exp[D(1-L)]}{1+A}\right\} - \frac{8u_\infty(u_\infty+u_1)}{D} \\
& \times \left(\frac{1}{2} - \frac{L^2}{2} + \frac{\ln(1+A)}{D} - \frac{L\ln(1+A\exp[D(1-L)])}{D}\right) \\
& - \frac{1}{D^2}\mathrm{Li}_2[-A] + \frac{1}{D^2}\mathrm{Li}_2[-A\exp\{D[1-L]\}]\Bigg\}
\end{aligned} \tag{67}$$

Figure 2a displays velocity distributions between the center of the channel and the wall. This figure is computed for $F = 10^4$; $\gamma = 1$; $\beta = 0$; $R = 1$. As expected, velocity increases with an increase in the Darcy number. For $Da = 10^{-2}$ and $S = 0.5$ there is a small but visible jump in velocity at $y = L = 0.75$. This means that, for these parameter values, momentum boundary layers start to overlap in the center of the porous region and the boundary layer approximation fails. However, if we keep the same

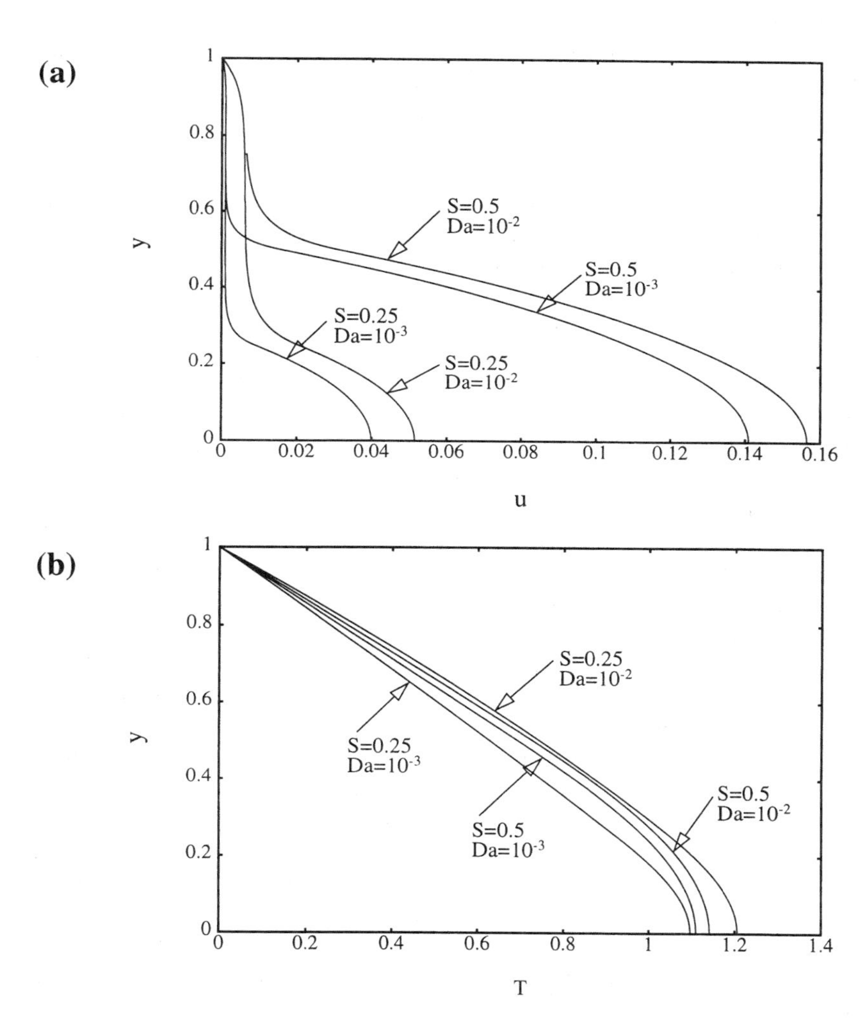

Figure 2. Velocity (a) and temperature (b) distributions in the channel.

value of the Darcy number but increase the thickness of the porous layer ($Da = 10^{-2}$ and $S = 0.25$), the boundary layer approximation and our solution are again valid. For the curves computed for $Da = 10^{-3}$ there are no jumps in velocity, which means that the boundary layer approximation is perfectly satisfied. Since the assumption that the momentum boundary layers do not overlap in the center of the porous region is the only assumption we utilized in obtaining our solution, the velocity profile computed according to Eqs. (39), (41), and (46) immediately shows whether our solution is applicable for particular parameter values. In general, we can say that, provided the thickness of the porous region near the wall is sufficient, our solution is valid for either relatively small Darcy numbers or relatively large values of the Forchheimer coefficient, or both. This probably covers the majority of practical situations. Our solution can fail for very small thickness of the porous layer, and also when the Darcy number takes on very large values (large in this case is of the order of unity) and the Forchheimer coefficient simultaneously takes on very small values.

Figure 2b displays dimensionless temperature computed for $F = 10^4$; $\gamma = 1$; $\beta = 0$; $R = 1$. These temperature distributions correspond to the velocity distributions depicted in Figure 2a.

Figure 3a displays a contour plot showing the dependence of the Nusselt number on the Darcy number and on the thickness of the clear fluid region, S. This figure is computed for $F = 10^4$; $\gamma = 1$; $\beta = 0$; $R = 1$. According to this figure, for a fixed value of the Darcy number the dependence $Nu = Nu(S)$ possesses a minimum. The value of S for which this minimum occurs and the magnitude of this minimum depend on the Darcy number. To give a physical interpretation of this phenomenon, let us start from the geometry where the porous medium completely occupies the channel ($S = 0$). A decrease in the width of the porous region results in appearance of the clear fluid region. Fluid velocity in the clear fluid region is much larger than in the porous region, and fluid flow shifts towards the center of the channel. Therefore, an increase in S first results in less intensive convective heat transfer at the walls of the channel and a decrease of the Nusselt number. However, further increase of the thickness of the clear fluid region results in more uniform velocity and temperature distribution along the thickness of the channel. This results in a larger temperature gradient at the wall of the channel and a larger value of the Nusselt number. The above gives a physical interpretation of the minimum possessed by the dependence of the Nusselt number on the thickness of the clear fluid region.

Figure 3b displays a contour plot showing the dependence of the Nusselt number on the Darcy number and on the Forchheimer coefficient. This figure is computed for $S = 0.25$; $\gamma = 1$; $\beta = 0$; $R = 1$. It can be seen that, for relatively large values of the Darcy number, the Nusselt number

(a)

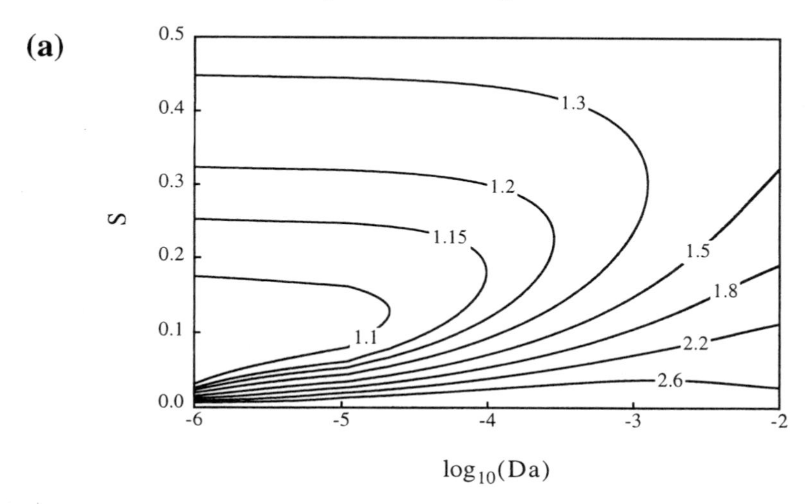

(b)

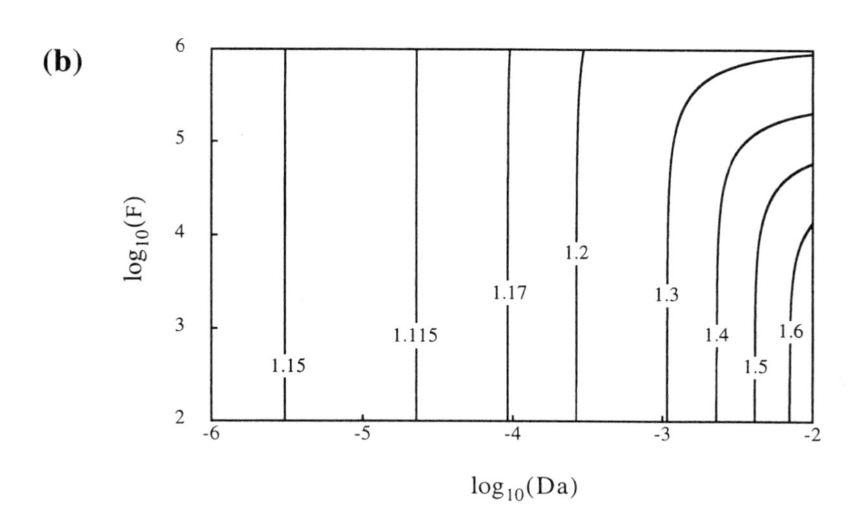

Figure 3. Contour plots showing the Nusselt number (a) as a function of the Darcy number and the thickness of the clear fluid region; (b) as a function of the Darcy number and the Forchheimer coefficient.

essentially decreases when the Forchheimer coefficient increases. This is because the Forchheimer term describes quadratic drag, and larger quadratic drag causes smaller filtration velocity in the porous region. This results in less intensive convective heat transfer at the channel walls, and in a smaller value of the Nusselt number. For small values of the Darcy number, however, filtration velocity becomes so small that the influence of the quadratic drag becomes insignificant and the Nusselt number does not visibly depend on the Forchheimer coefficient.

D. Composite Channel with a Porous Core (Figure 1b)

1. Velocity Distribution

Velocity distribution in the channel displayed n Figure 1b can be found by integrating the dimensionless momentum equations, (20) and (21), along with boundary conditions given by Eq. (28) and by the first equations in (32) and (33).

(a) Velocity in the Clear Fluid Region ($0 \leq y \leq S$). For the clear fluid region ($0 \leq y \leq S$) the velocity distribution is obtained by integrating Eq. (20) and utilizing boundary conditions given by the first equations in (28) and (32)

$$u = -\frac{y^2}{2} + \left(\frac{u_i}{S} + \frac{S}{2}\right)y \tag{68}$$

(b) Velocity in the Porous Region ($S \leq y \leq 1$). The velocity distribution in the porous region is obtained from the assumption that the momentum boundary layers do not overlap in the center of the porous region. If this is the case, in the center of the porous region (which coincides with the center of the channel), at $y = 1$, the conditions given by Eq. (40) must be satisfied. Integrating Eq. (21) and utilizing boundary conditions given by the first equation in (28) and by Eq. (40), the velocity distribution in the porous region ($S \leq y \leq 1$) is obtained as

$$u = (u_\infty + u_1)\left[\frac{1 - z}{1 + z}\right]^2 - u_1 \tag{69}$$

where

$$z = B\exp\{-D(y - S)\} \tag{70}$$

and B is defined by Eq. (44), D by Eq. (45), and u_1 by Eq. (43).

The dimensionless velocity at the fluid/porous interface, u_i, can be found from the second equation in (28) as

$$-\gamma(u_i - u_\infty)\left[\frac{2}{3}F(u_i + u_1)\right]^{1/2} + \frac{S}{2} - \frac{u_i}{S} = \beta Da^{-1/2}u_i \tag{71}$$

(c) Mean Velocity. The dimensionless mean velocity can now be obtained from the following equation

$$U = \int_0^S u\,\mathrm{d}y + \int_S^1 u\,\mathrm{d}y \tag{72}$$

Upon performing integration, the following equation for U is obtained

$$U = \frac{1}{12}S^3 + u_i\frac{S}{2} + u_\infty(1-S) + \frac{2\sqrt{6}\gamma(u_\infty + u_1)}{[F(u_\infty + u_1)]^{1/2}(1+B)} - \frac{2\sqrt{6}\gamma(u_\infty + u_1)}{[F(u_\infty + u_1)]^{1/2}(1 + B\exp\{-D(1-S)\})} \tag{73}$$

2. Temperature Distribution

With the velocity distribution known, the temperature distribution in the channel can be obtained by integrating the energy equations (22) and (23), along with boundary conditions given by Eq. (29) and by the second equations in (32) and (33).

(a) Dimensionless Temperature in the Clear Fluid Region $(0 \le y \le S)$. The dimensionless temperature distribution in the clear fluid region is

$$T = T_i\frac{y}{S} - \frac{Nu}{U}\left[-\frac{y^4}{24} + \left(\frac{u_i}{S} + \frac{S}{2}\right)\frac{y^3}{6}\right] + \frac{Nu}{U}\frac{y}{S}\left[-\frac{S^4}{24} + \left(\frac{u_i}{S} + \frac{S}{2}\right)\frac{S^3}{6}\right] \tag{74}$$

(b) Dimensionless Temperature in the Porous Region $(S \le y \le 1)$. The dimensionless temperature distribution in the boundary layer region $(S \le y \le 1)$ is obtained by integrating the dimensionless energy equation (23), and utilizing boundary conditions given by Eq. (29)

$$T = T_i + \frac{1}{R}\frac{T_i}{S}(y-S) - \frac{1}{R}\frac{Nu}{U}\left\{Q(y-S) + \frac{u_\infty}{2}(y-S)^2 - \frac{6\gamma^2}{F} \times \ln\frac{1 + B\exp\{-D(y-S)\}}{1+B}\right\} \tag{75}$$

where

$$Q = u_i\frac{S}{3} + \frac{S^3}{24} - \frac{2\sqrt{6}\gamma(u_\infty + u_1)^{1/2}B}{F^{1/2}(1+B)} \tag{76}$$

The dimensionless temperature at the fluid/porous interface, T_i, can then be obtained from the fact that the temperature distribution in the boundary layer region must satisfy the adiabatic boundary condition given by the second equation in (33). This results in the following equation for T_i

$$T_i = Nu\Xi \tag{77}$$

where

$$\begin{aligned}\Xi = \frac{S}{U}\Bigg\{ & u_i\frac{S}{3} + \frac{S^3}{24} + u_\infty(1-S) + \frac{2\sqrt{6}\gamma(u_\infty + u_1)^{1/2}}{F^{1/2}(1+B)} \\ & - \frac{2\sqrt{6}\gamma(u_\infty + u_1)^{1/2}}{F^{1/2}(1+B\exp\{-D(1-S)\})}\Bigg\}\end{aligned} \tag{78}$$

3. Nusselt Number

After the determination of the velocity and temperature distributions, the Nusselt number can be found by substituting these distributions into the compatibility condition given by Eq. (38). This results in the following equation for the Nusselt number

$$Nu = \frac{U}{\Psi_0 + \Psi_1} \tag{79}$$

where parameters Ψ_0 and Ψ_1 are defined as

$$\Psi_0 = \int_0^S (T/Nu)u\,\mathrm{d}y \tag{80}$$

$$\Psi_1 = \int_S^1 (T/Nu)u\,\mathrm{d}y \tag{81}$$

The value of Ψ_0 can then be found as

$$\Psi_0 = \frac{17S^7 + 168S^5u_i + 6720\Xi SUu_i + 56S^3(15\Xi U + 8u_i^2)}{20\,160U} \tag{82}$$

The value of Ψ_1 can be found as

$$
\begin{aligned}
\Psi_1 = {} & \Xi u_\infty - \Xi S u_\infty + \frac{3D\gamma^2 u_\infty}{FRU} - \frac{u_\infty}{2}\left(\frac{Q}{RU} - \frac{\Xi}{RS}\right) + \frac{S^3 u_\infty^2}{6RU} \\
& - \frac{3\gamma^2(-8u_\infty + D^2S^2u_\infty + BD^2S^2u_\infty - 8u_1)}{DFRU(1+B)} \\
& + \frac{(-4u_\infty + DSu_\infty - 4u_1)}{D}\left(\frac{Q}{RU} - \frac{\Xi}{RS}\right) + \frac{4\Xi(u_\infty + u_1)}{D(1+B)} \\
& - \frac{4\exp[D]\Xi(u_\infty + u_1)}{D(\exp[D] + B\exp[DS])} + \frac{4\exp[D](1-S)(u_\infty + u_1)}{D(\exp[D] + B\exp[DS])} \\
& \times \left(\frac{Q}{RU} - \frac{\Xi}{RS}\right) - \frac{24\gamma^2(u_\infty + u_1)}{DFRU(1 + B\exp[-D(1-S)])} \\
& + [6DRU(1 + B\exp[-D(1-S)])]^{-1} \\
& \times u_\infty\{12u_\infty - Du_\infty - BD\exp[-D(1-S)]u_\infty - 24Su_\infty \\
& + 3DSu_\infty(1 + B\exp[-D(1-S)]) + 12S^2u_\infty \\
& - 3DS^2u_\infty(1 + B\exp[-D(1-S)]) + 12u_1 - 24Su_1 + 12S^2u_1\} \\
& - \frac{1}{2D^2}\left(\frac{Q}{RU} - \frac{\Xi}{RS}\right)\{D^2S^2u_\infty - 8DS(u_\infty + u_1)\} - \frac{4(u_\infty + u_1)}{D^2} \\
& \times \left(\frac{Q}{RU} - \frac{\Xi}{RS}\right)\ln\frac{1 + B\exp[-D(1-S)]}{1+B} \\
& + [DFRU(1 + B\exp[-D(1-S)])]^{-1} 6\gamma^2 \\
& \times \ln\left[\frac{1 + B\exp[-D(1-S)]}{1+B}\right]\{Du_\infty(1 + B\exp[-D(1-S)]) \\
& - 4(u_\infty + u_1)\} + \frac{4Su_\infty(u_\infty + u_1)}{DRU} \\
& \times \left\{1 - S + \frac{1}{D}\ln\frac{1 + B\exp[-D(1-S)]}{1+B}\right\} - \frac{8u_\infty(u_\infty + u_1)}{DRU} \\
& \times \left\{\frac{1}{2} - \frac{S^2}{2} - \frac{S\ln[1+B]}{D} + \frac{\ln[1 + B\{-D(1-S)\}]}{D}\right. \\
& \left. + \frac{\mathrm{Li}_2[-B]}{D^2} - \frac{\mathrm{Li}_2[-B\exp\{-D(1-S)\}]}{D^2}\right\}
\end{aligned}
\tag{83}
$$

E. Composite Channel Heated from the Fluid Side (Figure 1c)

1. Velocity Distribution

Velocity distribution in the channel displayed in Figure 1c can be found by integrating the dimensionless momentum equations (20) and (21), along with boundary conditions given by Eq. (28) and by the first equations in (34) and (35).

(a) Velocity in the Clear Fluid Region ($0 \leq y \leq S$). For the clear fluid region ($0 \leq y \leq S$) the velocity distribution is obtained by integrating Eq. (20) and utilizing boundary conditions given by the first equations in (28) and (34) as

$$u = -\frac{y^2}{2} + \left(\frac{u_i}{S} + \frac{S}{2}\right) y \tag{84}$$

(b) Velocity in the Boundary Layer 2 Region ($S \leq y \leq L$). The procedure of finding the velocity distribution in this region is based on the boundary layer approximation discussed above, and is similar to the procedure of finding the velocity distribution in the boundary layer 2 region for the channel displayed in Figure 1a. The resulting velocity distribution is given by Eqs. (41)–(45). The difference from the case displayed in Figure 1a is that the dimensionless interfacial velocity u_i in Eqs. (41)–(45) now must be determined from Eq. (71).

(c) Velocity in the Boundary Layer 1 Region ($L \leq y \leq 1$). The procedure of finding the velocity distribution in this region is again similar to the procedure of finding the velocity distribution in the boundary layer 1 for the channel displayed in Figure 1a. The resulting velocity distribution is given by Eqs. (46)–(48).

The dimensionless velocity at the fluid/porous interface, u_i, in Eqs. (41)–(45) and (46)–(48) for the case displayed in Figure 1c must be found from Eq. (71).

(d) Mean Velocity. With the velocity distribution in the channel known, the dimensionless mean velocity can be obtained from Eq. (50). Upon performing integration, the following equation for U is obtained

$$U = \frac{1}{12}S^3 + u_i\frac{S}{2} + u_\infty(1-S) + \frac{2\sqrt{6}\gamma(u_\infty + u_1)^{1/2}}{F^{1/2}(1+B)}$$
$$- \frac{2\sqrt{6}\gamma(u_\infty + u_1)}{[F(u_\infty + u_1)]^{1/2}(1 + B\exp\{-D(L-S)\})} - \frac{2\sqrt{6}\gamma(u_\infty + u_1)^{1/2}}{F^{1/2}(1+A)}$$
$$+ \frac{2\sqrt{6}\gamma(u_\infty + u_1)}{[F(u_\infty + u_1)]^{1/2}(1 + A\exp\{D(1-L)\})} \quad (85)$$

2. Temperature Distribution

With the velocity distribution found, the temperature distribution in the channel can be obtained by integrating the energy equations (22) and (23), along with boundary conditions given by Eq. (29) and by the second equations in (34) and (35).

(a) Dimensionless Temperature in the Clear Fluid Region $(0 \le y \le S)$. The dimensionless temperature in the clear fluid region is

$$T = T_i\frac{y}{S} - \frac{Nu}{U}\left[-\frac{y^4}{24} + \left(\frac{u_i}{S} + \frac{S}{2}\right)\frac{y^3}{6}\right] + \frac{Nu}{U}\frac{y}{S}\left[-\frac{S^4}{24} + \left(\frac{u_i}{S} + \frac{S}{2}\right)\frac{S^3}{6}\right] \quad (86)$$

(b) Dimensionless Temperature in the Boundary Layer 2 Region *$(S \le y \le L)$*. The temperature distribution in the boundary layer 2 region is obtained by integrating Eq. (23) along with boundary conditions given by Eq. (29)

$$T = T_i + \frac{1}{R}\frac{T_i}{S}(y-S) - \frac{1}{R}\frac{Nu}{U}\left\{Q(y-S) + \frac{u_\infty}{2}(y-S)^2 - \frac{6\gamma^2}{F}\right.$$
$$\left. \times \ln\frac{1 + B\exp(-D(y-S))}{1+B}\right\} \quad (87)$$

where Q is defined by (76).

(c) Dimensionless Temperature in the Boundary Layer 1 Region $(L \le y \le 1)$. The dimensionless temperature distribution in the boundary layer 1 region $(L \le y \le 1)$ is obtained by integrating Eq. (23) and matching the temperature and the heat flux at $y = L$ with that which follows from Eq. (87). This results in the following temperature distribution in the boundary layer 1 region

$$T = T|_{y=1} - \frac{1}{R}\frac{T_i}{S}(1-y) + \frac{1}{R}\frac{Nu}{U}\left\{(1-y)P + \frac{u_\infty}{2}(y+1-2L)(1-y) + \frac{6\gamma^2}{F}\ln\frac{1+A\exp\{D(1-y)\}}{1+A}\right\} \tag{88}$$

where

$$T|_{y=1} = Nu\Omega \tag{89}$$

$$\Omega = \Xi + \frac{1}{R}\frac{\Xi}{S}(1-S) - \frac{1}{RU}\left\{(1-L)P + \frac{u_\infty}{2}(1-L)^2 + \frac{6\gamma^2}{F} \times \ln\frac{1+A\exp\{D(1-L)\}}{1+A} + Q(L-S) + \frac{u_\infty}{2}(L-S)^2 - \frac{6\gamma^2}{F} \times \ln\frac{1+B\exp\{-D(L-S)\}}{1+B}\right\} \tag{90}$$

and P is defined by Eq. (56).

The dimensionless temperature at the fluid/porous interface, T_i, can then be obtained from the fact that the temperature distribution in the boundary layer 1 region must satisfy the adiabatic boundary condition given by the second equation in (35). This results in the following equation for T_i

$$T_i = Nu\Xi \tag{91}$$

where

$$\Xi = \frac{S}{U}\left\{u_i\frac{S}{3} + \frac{S^3}{24} + u_\infty(1-S) + \frac{2\sqrt{6}\gamma(u_\infty+u_1)^{1/2}}{F^{1/2}(1+B)} - \frac{2\sqrt{6}\gamma(u_\infty+u_1)^{1/2}}{F^{1/2}(1+B\exp\{-D(L-S)\})} + \frac{2\sqrt{6}\gamma(u_\infty+u_1)^{1/2}}{F^{1/2}(1+A\exp\{D(1-L)\})} - \frac{2\sqrt{6}\gamma(u_\infty+u_1)^{1/2}}{F^{1/2}(1+A)}\right\} \tag{92}$$

3. Nusselt Number

The procedure of determining the Nusselt number for this case is similar to the procedure used for the channel displayed in Figure 1a. The value of the Nusselt number can be found from Eqs. (59)–(62), where Ψ_0, Ψ_2, and Ψ_1 are given by the following equations

$$\Psi_0 = \frac{17S^7 + 168S^5u_i + 6720\Xi SUu_i + 56S^3(15\Xi U + 8u_i^2)}{20\,160U} \tag{93}$$

$$
\begin{aligned}
\Psi_2 = {} & \Xi L u_\infty - \Xi S u_\infty + \frac{3D\gamma^2 L^2 u_\infty}{FRU} - \frac{L^2 u_\infty}{2}\left(\frac{Q}{RU} - \frac{\Xi}{RS}\right) + \frac{S^3 u_\infty^2}{6RU} \\
& - \frac{3\gamma^2(-8u_\infty + D^2 S^2 u_\infty + BD^2 S^2 u_\infty - 8u_1)}{DFRU(1+B)} \\
& + \frac{L(-4u_\infty + DSu_\infty - 4u_1)}{D}\left(\frac{Q}{RU} - \frac{\Xi}{RS}\right) + \frac{4\Xi(u_\infty + u_1)}{D(1+B)} \\
& - \frac{4\exp[DL]\Xi(u_\infty + u_1)}{D(\exp[DL] + B\exp[DS])} + \frac{4\exp[DL](L-S)(u_\infty + u_1)}{D(\exp[DL] + B\exp[DS])} \\
& \times \left(\frac{Q}{RU} - \frac{\Xi}{RS}\right) - \frac{24\gamma^2(u_\infty + u_1)}{DFRU(1 + B\exp[-D(L-S)])} \\
& + [6DRU(1 + B\exp[-D(L-S)])]^{-1} u_\infty \{12L^2 u_\infty - DL^3 u_\infty \\
& - BD\exp[-D(L-S)]L^3 u_\infty - 24LSu_\infty + 3DL^2 S u_\infty \\
& \times (1 + B\exp[-D(L-S)]) + 12S^2 u_\infty - 3DLS^2 u_\infty \\
& \times (1 + B\exp[-D(L-S)]) + 12L^2 u_1 - 24LSu_1 + 12S^2 u_1\} \\
& - \frac{1}{2D^2}\left(\frac{Q}{RU} - \frac{\Xi}{RS}\right)\{D^2 S^2 u_\infty - 8DS(u_\infty + u_1)\} \\
& - \frac{4(u_\infty + u_1)}{D^2}\left(\frac{Q}{RU} - \frac{\Xi}{RS}\right)\ln\frac{1 + B\exp[D(L-S)]}{1+B} \\
& + [DFRU(1 + B\exp[-D(L-S)])]^{-1} 6\gamma^2 \\
& \times \ln\left[\frac{1 + B\exp[-D(L-S)]}{1+B}\right]\{DLu_\infty(1 + B\exp[-D(L-S)]) \\
& - 4(u_\infty + u_1)\} \\
& + \frac{4Su_\infty(u_\infty + u_1)}{DRU}\left\{L - S + \frac{1}{D}\ln\frac{1 + B\exp[-D(L-S)]}{1+B}\right\} \\
& - \frac{8u_\infty(u_\infty + u_1)}{DRU}\left\{\frac{L^2}{2} - \frac{S^2}{2} - \frac{S\ln[1+B]}{D}\right. \\
& \left. + \frac{L\ln[1 + B\exp\{-D(L-S)\}]}{D} + \frac{\mathrm{Li}_2[-B]}{D^2} - \frac{\mathrm{Li}_2[-B\exp\{-D(L-S)\}]}{D^2}\right\}
\end{aligned}
\tag{94}
$$

and

$$
\begin{aligned}
\Psi_1 = \Omega\Bigg[& -Lu_\infty - \frac{4A\exp(D)(u_1+u_\infty)}{D[A\exp(D)+\exp(DL)]} + \frac{4A(u_1+u_\infty)+Du_\infty(1+A)}{D(1+A)}\Bigg] \\
& + \frac{1}{RU}\Bigg\{\frac{3D\gamma^2 u_\infty}{F}(1-L^2) - \left(P - \Xi\frac{U}{S}\right)\frac{u_\infty}{2}(1-L^2) \\
& + \frac{(2-3L)u_\infty^2}{6} + \frac{L^3u_\infty^2}{6} - \left(P - \Xi\frac{U}{S}\right)\frac{(4u_\infty - Du_\infty + 4u_1)}{D} \\
& + \frac{L^2u_\infty(4u_\infty - DLu_\infty + 4u_1)}{2D} - \frac{24\gamma^2(u_\infty+u_1)}{DF(1+A)} \\
& + \frac{24\gamma^2(u_\infty+u_1)}{DF(1+A\exp[D(1-L)])} + \left(P - \Xi\frac{U}{S}\right) \\
& \times \frac{4\exp[DL](1-L)(u_\infty+u_1)}{D(A\exp[D]+\exp[DL])} + \frac{2(1-L)^2u_\infty(u_\infty+u_1)}{D(1+A\exp[D(1-L)])} \\
& + \frac{2L^2u_\infty(u_\infty+u_1)}{D} + \left(P - \Xi\frac{U}{S}\right)\frac{L(4u_\infty - Du_\infty + 4u_1)}{D} \\
& - \frac{Lu_\infty(Du_\infty + 8Lu_\infty - 2DLu_\infty + 8Lu_1)}{2D} + \left(P - \Xi\frac{U}{S}\right) \\
& \times \frac{4(u_\infty+u_1)}{D^2}\ln\frac{1+A\exp[D(1-L)]}{1+A} \\
& - \frac{6\gamma^2(-4u_\infty + DLu_\infty + ADLu_\infty\exp[D(1-L)] - 4u_1)}{DF(1+A\exp[D(1-L)])} \\
& \times \ln\frac{1+A\exp[D(1-L)]}{1+A} + \frac{4u_\infty(u_\infty+u_1)L}{D}\Bigg\{1 - L - \frac{1}{D} \\
& \times \ln\frac{1+A\exp[D(1-L)]}{1+A}\Bigg\} - \frac{8u_\infty(u_\infty+u_1)}{D} \\
& \times \left(\frac{1}{2} - \frac{L^2}{2} + \frac{\ln(1+A)}{D} - \frac{L\ln(1+A\exp[D(1-L)])}{D}\right) \\
& - \frac{1}{D^2}\mathrm{Li}_2[-A] + \frac{1}{D^2}\mathrm{Li}_2\big[-A\exp\{D[1-L]\}\big]\Bigg\}
\end{aligned} \tag{95}
$$

F. Composite Channel Heated from the Porous Side (Figure 1d)

1. Velocity Distribution

Velocity distribution in the channel displayed in Figure 1d is exactly the same as in the channel displayed in Figure 1c. Consequently, velocity distribution is the same as that given in Section II.E.1. The difference from the channel displayed in Figure 1c is the heat transfer situation.

2. Temperature Distribution

The temperature distribution in the channel can be obtained by integrating the energy equations (22) and (23), along with boundary conditions given by Eq. (29) and by the second equations in (36) and (37).

(a) Dimensionless Temperature in the Clear Fluid Region ($0 \le y \le S$). The dimensionless temperature distribution in the clear fluid region is

$$T = T_i + \frac{Nu}{U}\left[\frac{S^4}{24} + u_i\frac{S^2}{6} + \frac{y^4}{24} - \left(\frac{u_i}{S} + \frac{S}{2}\right)\frac{y^3}{6}\right] \tag{96}$$

(b) Dimensionless Temperature in the Boundary Layer 2 Region ($S \le y \le L$). The dimensionless temperature distribution in the boundary layer 2 region ($S \le y \le L$) is given by Eq. (53), where Q is redefined as

$$Q = u_i\frac{S}{2} + \frac{S^3}{12} - \frac{2\sqrt{6}\gamma(u_\infty + u_1)^{1/2}B}{F^{1/2}(1+B)} \tag{97}$$

(c) Dimensionless Temperature in the Boundary Layer 1 Region ($L \le y \le 1$). The dimensionless temperature distribution in the boundary layer region ($L \le y \le 1$) is given by Eq. (55), where Q is redefined according to Eq. (97). The dimensionless temperature at the fluid/porous interface, T_i, can then be found from Eq. (58), where Q is according to Eq. (97) redefined again.

3. Nusselt Number

The procedure of determining the Nusselt number for this case is similar to the procedure used for the channel displayed in Figure 1a. The value of the Nusselt number can be found from Eqs. (59)–(62), where Ψ_0 is given by the following equation

$$\Psi_0 = \frac{13S^7 + 112S^5u_i + 2520\Xi SUu_i + 84S^3(5\Xi U + 3u_i^2)}{5040U} \tag{98}$$

Ψ_2 is given by Eq. (64) and Ψ_1 is given by Eq. (67). In Eqs. (64) and (67) Q must be redefined according to Eq. (97).

III. COUETTE FLOW IN PARTLY POROUS CONFIGURATIONS

A. Problem Formulation and Governing Equations

One of the important flow situations when both viscous (Brinkman) and nonlinear (Forchheimer) effects can be significant is Couette flow in a fluid-saturated porous medium. In an investigation by Nakayama (1992), analytical solutions for Couette flow of Newtonian and power-law non-Newtonian fluids through an inelastic porous medium are obtained. However, results of this reference are limited to fluid flow analysis only, and no investigation of heat transfer is made in this reference. Investigations of heat transfer in Couette flow through porous media are very limited. Daskalakis (1990) presented numerical solutions for an impulsively started Couette flow in a Brinkman–Darcy porous medium occupying the annular space between two concentric infinite cylindrical surfaces. At the initial moment, the outer cylinder is suddenly set in motion and rotates with a constant angular velocity while the inner cylinder stays motionless. Bhargava and Sacheti (1989) considered heat transfer in Couette flow of two immiscible fluids in a composite parallel-plate channel partly occupied by a clear fluid and partly by a Brinkman–Darcy porous medium. Recent papers by Kuznetsov (1998a,b) are the first attempts to analyze heat transfer in Couette flow through a Brinkman–Forchheimer–Darcy porous medium.

In this section, a problem of fluid flow and heat transfer in Couette flow through a parallel-plate channel partly filled with a fluid-saturated rigid porous medium is investigated. The fluid flow occurs due to a moving plane wall. Fluid flow in the porous region is described by the Brinkman–Forchheimer–extended Darcy equation. For the analysis of heat transfer, either isoflux fixed wall and insulated moving wall or insulated fixed wall and isoflux moving wall are considered. Analytical solutions for the flow velocity, temperature distribution, and for the Nusselt number are obtained.

Figure 4 depicts the schematic diagram of the problem. Two different situations of Couette flow in a partly porous channel are considered. For the case displayed in Figure 4a, a composite channel bounded by two infinite parallel plates is considered. The distance between the plates H. The lower part of the channel is occupied by a clear fluid while the upper part is occupied by a fluid-saturated porous medium with uniform permeability. The upper plate and the porous medium are fixed, while the lower plate moves with a constant velocity $\tilde{u}_w$. The fluid flow thus occurs due to a

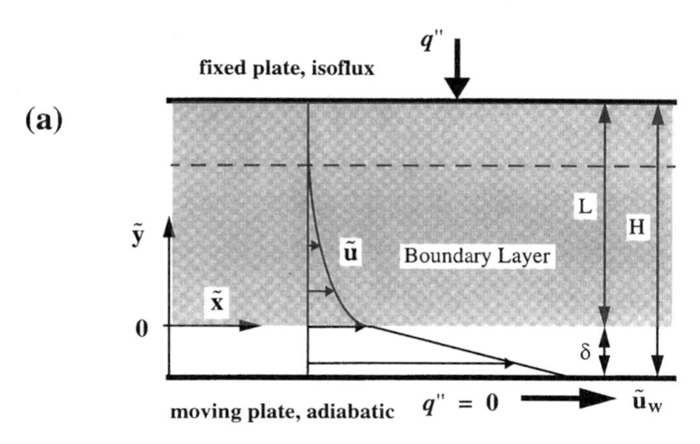

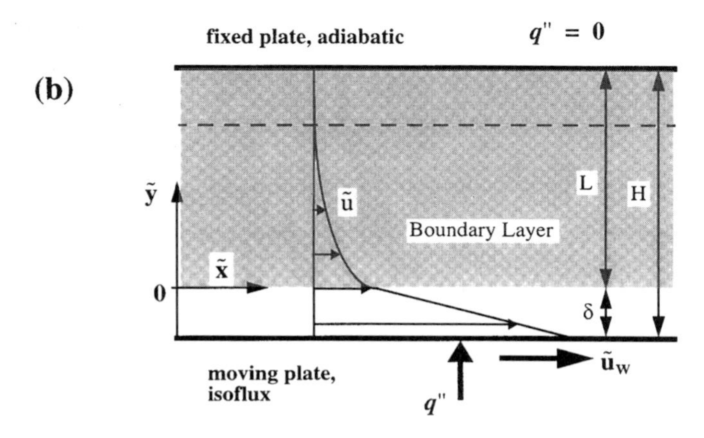

Figure 4. Couette flow in a parallel-plate channel party filled with a porous medium: (a) heating from the porous side; (b) heating from the fluid side.

moving plate which is separated from the porous medium by a gap of thickness δ filled with a clear fluid. A uniform heat flux is imposed at the fixed plate, while the moving plate is adiabatic. For the case displayed in Figure 4b the fluid flow situation is exactly the same, but in this case the uniform heat flux is imposed at the moving plate, while the fixed plate is adiabatic.

For the problems displayed in Figures 4a and 4b there is one momentum boundary layer in the porous region. This boundary layer is near the clear fluid/porous medium interface. As with the problems considered in Section II, the thickness of the boundary layer depends on the value of the Darcy number as well as on the value of the Forchheimer coefficient.

The set of governing equations for the problem shown in Figures 4a and b can be presented as

$$\mu_f \frac{\mathrm{d}^2 \tilde{u}_f}{\mathrm{d}\tilde{y}^2} = 0 \qquad -\delta \leq \tilde{y} \leq 0 \tag{99}$$

$$\mu_{\text{eff}} \frac{\mathrm{d}^2 \tilde{u}_f}{\mathrm{d}\tilde{y}^2} - \frac{\mu_f}{K}\tilde{u}_f - \frac{\rho_f c_F}{K^{1/2}}\tilde{u}_f^2 = 0 \qquad 0 \leq \tilde{y} \leq L \tag{100}$$

$$\rho_f c_f \tilde{u}_f \frac{\partial \tilde{T}}{\partial \tilde{x}} = k_f \frac{\partial^2 \tilde{T}}{\partial \tilde{y}^2} \qquad -\delta \leq \tilde{y} \leq 0 \tag{101}$$

$$\rho_f c_f \tilde{u}_f \frac{\partial \tilde{T}}{\partial \tilde{x}} = k_{\text{eff}} \frac{\partial^2 \tilde{T}}{\partial \tilde{y}^2} \qquad 0 \leq \tilde{y} \leq L \tag{102}$$

where L is the thickness of the porous layer and δ is the thickness of the fluid layer.

As in Section II, Eqs. (99) and (100) are the momentum equations for the clear fluid and porous regions, respectively. Equation (99) is the Stokes equation, while (100) is the Brinkman–Forchheimer–extended Darcy equation. Equations (101) and (102) are the energy equations for the clear fluid and porous regions, respectively.

As in Section II, for the problems displayed in Figures 4a and b at the fluid/porous interface we utilize the boundary conditions suggested in Ochoa-Tapia and Whitaker (1995a,b)

$$\tilde{u}_f|_{\tilde{y}=-0} = \tilde{u}_f|_{\tilde{y}=+0} \qquad \mu_{\text{eff}} \frac{\mathrm{d}\tilde{u}_f}{\mathrm{d}\tilde{y}}\bigg|_{\tilde{y}=+0} - \mu_f \frac{\mathrm{d}\tilde{u}_f}{\mathrm{d}\tilde{y}}\bigg|_{\tilde{y}=-0} = \beta \frac{\mu_f}{K^{1/2}} \tilde{u}_f|_{\tilde{y}=0} \qquad \text{at} \quad \tilde{y} = 0 \tag{103}$$

$$\tilde{T}|_{\tilde{y}=-0} = \tilde{T}|_{\tilde{y}=+0} \qquad k_{\text{eff}} \frac{\partial \tilde{T}}{\partial \tilde{y}}\bigg|_{\tilde{y}=+0} = k_f \frac{\partial \tilde{T}}{\partial \tilde{y}}\bigg|_{\tilde{y}=-0} \qquad \text{at} \quad \tilde{y} = 0 \tag{104}$$

For the problem displayed in Figure 4a the boundary conditions at the moving and fixed walls are, respectively,

$$\tilde{u}_f = \tilde{u}_w \qquad \frac{\partial \tilde{T}}{\partial \tilde{y}} = 0 \qquad \text{at} \qquad \tilde{y} = -\delta \tag{105}$$

$$\tilde{u}_f = 0 \qquad k_{\text{eff}} \frac{\partial \tilde{T}}{\partial \tilde{y}} = q'' \qquad \text{at} \qquad \tilde{y} = L \tag{106}$$

For the problem displayed in Figure 4b the boundary conditions at the moving and fixed walls are, respectively,

$$\tilde{u}_f = \tilde{u}_w \qquad k_{\text{f}} \frac{\partial \tilde{T}}{\partial \tilde{y}} = q'' \qquad \text{at} \qquad \tilde{y} = -\delta \tag{107}$$

$$\tilde{u}_f = 0 \qquad \frac{\partial \tilde{T}}{\partial \tilde{y}} = 0 \qquad \text{at} \qquad \tilde{y} = L \tag{108}$$

B. Dimensionless Equations and Boundary Conditions

Equations (99)–(102) can be recast into the following dimensionless form

$$\frac{\mathrm{d}^2 u}{\mathrm{d}y^2} = 0 \qquad -\frac{\delta}{H} \le y \le 0 \tag{109}$$

$$\gamma^2 \frac{\mathrm{d}^2 u}{\mathrm{d}y^2} - \frac{1}{Da} u - \frac{Re\, c_F}{Da^{1/2}} u^2 = 0 \qquad 0 \le y \le \frac{L}{H} \tag{110}$$

$$\frac{\mathrm{d}^2 T}{\mathrm{d}y^2} = -Nu \frac{\tilde{u}_w}{\tilde{U}} u \qquad -\frac{\delta}{H} \le y \le 0 \tag{111}$$

$$R \frac{\mathrm{d}^2 T}{\mathrm{d}y^2} = -Nu \frac{\tilde{u}_w}{\tilde{U}} u \qquad 0 \le y \le \frac{L}{H} \tag{112}$$

where

$$Da = \frac{K}{H^2}, \qquad Re = \frac{\rho_f \tilde{u}_w H}{\mu_f}, \qquad R = \frac{k_{\text{eff}}}{k_f}, \qquad T = \frac{\tilde{T} - \tilde{T}_w}{\tilde{T}_m - \tilde{T}_w} \tag{113}$$

$$u = \frac{\tilde{u}_f}{\tilde{u}_w}, \qquad y = \frac{\tilde{y}}{H}, \qquad \gamma = \left(\frac{\mu_{\text{eff}}}{\mu_f}\right)^{1/2} \tag{114}$$

The mean flow velocity $\tilde{U}$ is now determined by the following equation

$$\tilde{U} = \frac{1}{H} \int_{-\delta}^{L} \tilde{u}_f \mathrm{d}\tilde{y} \tag{115}$$

and the mean temperature $\tilde{T}_m$ is now determined by the following equation

$$\tilde{T}_m = \frac{1}{H\tilde{U}} \int_{-\delta}^{L} \tilde{u}_f \tilde{T}\, \mathrm{d}\tilde{y} \tag{116}$$

The value of the Nusselt number can be found by substituting u and T into the compatibility condition, which now takes the following form

$$\frac{\tilde{u}_w}{\tilde{U}} \int_{-\delta/H}^{L/H} Tu\, \mathrm{d}y = 1 \tag{117}$$

The boundary conditions at the fluid/porous interface given by Eqs. (103) and (104) in the dimensionless variables take the following form

$$u|_{y=+0} = u|_{y=-0} = u_i \qquad \gamma^2 \left.\frac{\mathrm{d}u}{\mathrm{d}y}\right|_{y=+0} - \left.\frac{\mathrm{d}u}{\mathrm{d}y}\right|_{y=-0} = \beta Da^{-1/2} u|_{y=0} \quad \text{at} \quad y = 0 \tag{118}$$

$$T|_{y=+0} = T|_{y=-0} = T_i \qquad R \left.\frac{\mathrm{d}T}{\mathrm{d}y}\right|_{y=+0} = \left.\frac{\mathrm{d}T}{\mathrm{d}y}\right|_{y=-0} \quad \text{at} \quad y = 0 \tag{119}$$

For the problem displayed in Figure 4a the dimensionless form of boundary conditions at the moving and fixed walls is

$$u = 1 \qquad \frac{\partial T}{\partial y} = 0 \qquad \text{at} \qquad y = -\frac{\delta}{H} \tag{120}$$

$$u = 0 \qquad T = 0 \qquad \text{at} \qquad y = \frac{L}{H} \tag{121}$$

For the problem displayed in Figure 4b the dimensionless form of boundary conditions at the moving and fixed walls is

$$u = 1 \qquad T = 0 \qquad \text{at} \qquad y = -\frac{\delta}{H} \tag{122}$$

$$u = 0 \qquad \frac{\partial T}{\partial y} = 0 \qquad \text{at} \qquad y = \frac{L}{H} \tag{123}$$

C. Composite Channel Heated from the Porous Side (Figure 4a)

The procedure of obtaining the solution for the channels displayed in Figure 4 is similar to the procedure utilized to obtain solutions for the channels displayed in Figure 1. Following the work by Vafai and Kim (1989), we proceed from the assumption that the momentum boundary layer, which is formed in the porous region near the fluid/porous interface, does not reach the fixed plate. For brevity, in this section we omit derivations and give only the results.

1. Velocity Distribution

(a) Velocity in the Clear Fluid Region $(-\delta/H \leq y \leq 0)$

$$u = u_i - (1 - u_i)\frac{H}{\delta}y \tag{124}$$

(b) Velocity in the Porous Region $(0 \leq y \leq L/H)$

$$u = \frac{3}{2}\frac{B}{A}\left\{\left[\frac{D\exp(B^{1/2}y/\gamma) - 1}{D\exp(B^{1/2}y/\gamma) + 1}\right]^2 - 1\right\} \tag{125}$$

where

$$A = \frac{Re\, c_F}{Da^{1/2}}, \qquad B = \frac{1}{Da}, \qquad D = \frac{1 + (1 + (2/3)(A/B)u_i)^{1/2}}{1 - (1 + (2/3)(A/B)u_i)^{1/2}} \tag{126}$$

The equation for the dimensionless velocity at the interface, u_i, follows from the second equation in (118) as

$$-\gamma u_i\left[\frac{2}{3}Au_i + B\right]^{1/2} + (1 - u_i)\frac{H}{\delta} = \beta B^{1/2}u_i \tag{127}$$

(c) Mean Velocity. The equation for the mean velocity $\tilde{U}$ can be found by integrating Eqs. (124) and (125) as

$$\frac{\tilde{U}}{\tilde{u}_w} = u_i\frac{\delta}{H} + \frac{1}{2}(1 - u_i)\frac{\delta}{H} + 6\gamma\frac{B^{1/2}}{A}\left[\frac{1}{1 + D\exp\left[\frac{B^{1/2}}{\gamma}\frac{L}{H}\right]} - \frac{1}{1 + D}\right] \tag{128}$$

2. Temperature Distribution

(a) Dimensionless Temperature in the Clear Fluid Region $(-\delta/H \leq y \leq 0)$

$$T = Nu\,\Lambda_1 \tag{129}$$

where

$$\Lambda_1 = -\frac{\tilde{u}_w}{\tilde{U}}\left[u_i\frac{y^2}{2} - (1-u_i)\frac{H}{\delta}\frac{y^3}{6}\right] - \frac{1}{2}\frac{\tilde{u}_w}{\tilde{U}}\frac{\delta}{H}[1+u_i]y + G_1 \tag{130}$$

(b) Dimensionless Temperature in the Porous Region $(0 \le y \le L/H)$

$$T = Nu\,\Lambda_2 \tag{131}$$

where

$$\Lambda_2 = -\frac{1}{R}\frac{\tilde{u}_w}{\tilde{U}}\frac{3}{2}\frac{B}{A}\frac{4\gamma}{B}\left[B^{1/2}y - \gamma\ln\left\{\frac{1+D\exp\left(\frac{B^{1/2}}{\gamma}y\right)}{1+D}\right\}\right]$$
$$-G_1\frac{H}{L}y + G_1 + \frac{1}{R}\frac{\tilde{u}_w}{\tilde{U}}\frac{3}{2}\frac{B}{A}\frac{4\gamma}{B}\frac{H}{L} \tag{132}$$
$$\times\left[B^{1/2}\frac{L}{H} - \gamma\ln\left\{\frac{1+D\exp\left(\frac{B^{1/2}}{\gamma}\frac{L}{H}\right)}{1+D}\right\}\right]y$$

The equation for the dimensionless temperature at the clear fluid/porous medium interface, T_i, can be found by matching heat fluxes at the clear fluid/porous medium interface

$$T_i = Nu\,G_1 \tag{133}$$

where

$$G_1 = \frac{1}{2R}\frac{\tilde{u}_w}{\tilde{U}}\frac{\delta}{H}[1+u_i]\frac{L}{H} - \frac{3}{2R}\frac{\tilde{u}_w}{\tilde{U}}\frac{B}{A}\frac{4\gamma}{B^{1/2}(1+D)} + \frac{3}{2R}\frac{\tilde{u}_w}{\tilde{U}}\frac{B}{A}\frac{4\gamma}{B}\frac{H}{L}$$
$$\times\left[B^{1/2}\frac{L}{H} - \gamma\ln\left\{\frac{1+D\exp\left(\frac{B^{1/2}}{\gamma}\frac{L}{H}\right)}{1+D}\right\}\right] \tag{134}$$

3. Nusselt Number

The Nusselt number can be found from the compatibility condition given by Eq. (117)

$$Nu = \frac{\tilde{U}}{\tilde{u}_w}[\Xi_1 + \Xi_2]^{-1} \tag{135}$$

where

$$\Xi_1 = \int_{-\delta/H}^{0} u\Lambda_1 \mathrm{d}y \tag{136}$$

and

$$\Xi_2 = \int_{0}^{L/H} u\Lambda_2 \mathrm{d}y \tag{137}$$

By performing integration, analytical expressions for Ξ_1 and Ξ_2 are obtained as

$$\Xi_1 = \frac{1}{60}\frac{\delta}{H}\left[30G_1(1+u_i) + \left(\frac{\delta}{H}\right)^2 \frac{\tilde{u}_w}{\tilde{U}}(8 + 9u_i + 3u_i^2)\right] \tag{138}$$

and

$$\begin{aligned}\Xi_2 = \frac{6\gamma}{A^2 R}\Bigg[&-\frac{B^{1/2}[AG_1 R + 6\gamma^2 \tilde{u}_w/\tilde{U}]}{1+D} - \gamma\frac{H}{L} \\ &\times \left(AG_1 R + 6\gamma^2 \frac{\tilde{u}_w}{\tilde{U}} \ln\frac{1 + D\exp[B^{1/2}L/(\gamma H)]}{1+D}\right) \\ &\times \ln\frac{1 + D\exp[B^{1/2}L/(\gamma H)]}{1+D} + \frac{6\gamma^2 B^{1/2}\tilde{u}_w/\tilde{U}}{1 + D\exp[B^{1/2}L/(\gamma H)]} + B^{1/2} \\ &\times \left\{AG_1 R + 6\gamma^2 \frac{\tilde{u}_w}{\tilde{U}} \ln\frac{1 + D\exp[B^{1/2}L/(\gamma H)]}{1+D}\right\}\Bigg]\end{aligned} \tag{139}$$

D. Composite Channel Heated from the Fluid Side (Figure 4b)

1. Velocity Distribution

Velocity distribution in the channel displayed in Figure 4b is exactly the same as in the channel displayed in Figure 4a. Consequently, velocity distribution is the same as given in Section III.C.1. The difference between this case and the channel displayed in Figure 4a is the heat transfer situation.

2. Temperature Distribution

(a) Dimensionless Temperature in the Clear Fluid Region $(-\delta/H \leq y \leq 0)$

$$T = Nu\frac{\tilde{u}_w}{\tilde{U}}(1-u_i)\frac{H}{\delta}\frac{y^3}{6} - Nu\frac{\tilde{u}_w}{\tilde{U}}u_i\frac{y^2}{2} - \left[\frac{1}{6}Nu\frac{\tilde{u}_w}{\tilde{U}}(1-u_i)\frac{\delta}{H} + \frac{1}{2} \times Nu\frac{\tilde{u}_w}{\tilde{U}}u_i\frac{\delta}{H} - T_i\frac{H}{\delta}\right]y + T_i \tag{140}$$

(b) Dimensionless Temperature in the Porous Region $(0 \leq y \leq L/H)$

$$T = \frac{6\gamma}{A}\frac{Nu}{R}\frac{\tilde{u}_w}{\tilde{U}}\left\{\gamma \ln\left[\frac{1 + D\exp\left[\frac{B^{1/2}}{\gamma}y\right]}{1+D}\right] - \frac{yB^{1/2}D\exp\left[\frac{B^{1/2}}{\gamma}\frac{L}{H}\right]}{1 + D\exp\left[\frac{B^{1/2}}{\gamma}\frac{L}{H}\right]}\right\} + T_i \tag{141}$$

where the dimensionless temperature at the clear fluid/porous medium interface, T_i, can be found as

$$T_i = Nu\frac{\tilde{u}_w}{\tilde{U}}G_2 \tag{142}$$

where

$$G_2 = 6\frac{B^{1/2}}{A}\gamma\frac{\delta}{H}\left[\frac{1}{1 + D\exp\left[\frac{B^{1/2}}{\gamma}\frac{L}{H}\right]} - \frac{1}{1+D}\right] + \frac{1}{6}(1-u_i)\left(\frac{\delta}{H}\right)^2 + \frac{1}{2}u_i\left(\frac{\delta}{H}\right)^2 \tag{143}$$

3. Nusselt Number

$$Nu = \left(\frac{\tilde{U}}{\tilde{u}_w}\right)^2\frac{1}{\Xi_3} \tag{144}$$

where

$$\Xi_3 = -\frac{1}{6}u_i\left(\frac{\delta}{H}\right)^3 + \frac{u_i}{2}\left(\frac{\delta}{H}\right)^2\left[\frac{1}{6}(1-u_i)\frac{\delta}{H} + \frac{1}{2}u_i\frac{\delta}{H} - G_2\frac{H}{\delta}\right] + u_i G_2\frac{\delta}{H}$$
$$-\frac{1}{30}(1-u_i)^2\left(\frac{\delta}{H}\right)^3 + \frac{(1-u_i)}{3}\left(\frac{\delta}{H}\right)^2$$
$$\times\left[\frac{1}{6}(1-u_i)\frac{\delta}{H} + \frac{1}{2}u_i\frac{\delta}{H} - G_2\frac{H}{\delta}\right] + \frac{1}{2}\frac{\delta}{H}G_2(1-u_i) + 36\frac{B}{A^2}\frac{\gamma^2}{R}\frac{L}{H}$$
$$\times\frac{D\exp[(B^{1/2}/\gamma)(L/H)]}{1+D\exp[(B^{1/2}/\gamma)(L/H)]}\left\{1 - \frac{1}{1+D\exp[(B^{1/2}/\gamma)(L/H)]}\right\}$$
$$-36\frac{B^{1/2}}{A^2}\frac{\gamma^3}{R}\left[\frac{1}{1+D} - \frac{1}{1+D\exp[(B^{1/2}/\gamma)(L/H)]}\right] - 6\gamma\frac{B^{1/2}}{A}G_2$$
$$\times\left[\frac{1}{1+D} - \frac{1}{1+D\exp[(B^{1/2}/\gamma)(L/H)]}\right] + 36\frac{B^{1/2}}{A^2}\frac{\gamma^3}{R}$$
$$\times\frac{1-D\exp[(B^{1/2}/\gamma)(L/H)]}{1+D\exp[(B^{1/2}/\gamma)(L/H)]}\ln\left[\frac{1+D\exp[(B^{1/2}/\gamma)(L/H)]}{1+D}\right] \tag{145}$$

IV. CONCLUSIONS

New boundary layer type solutions are obtained for a number of fundamentally important porous/fluid configurations. These new analytical solutions make it possible to investigate possibilities of enhancing heat transfer in composite channels. The new solutions are also valuable for gaining deeper insight toward understanding the transport process at the porous/fluid interface and for testing numerical codes.

ACKNOWLEDGMENT

The author acknowledges with gratitude the assistance of the North Carolina Supercomputing Center (NCSC) under an Advanced Computing Resources Grant and the support of the Alexander von Humboldt Foundation and the Christian Doppler Laboratory for Continuous Solidification Processes. The author is indebted to Professor W. Schneider for fruitful discussions and for motivating this research.

NOMENCLATURE

Roman Letters

A	parameter defined by Eq. (48) for the channels displayed in Figure 1. Also, parameter defined by Eq. (126) for the channels displayed in Figure 4
B	parameter defined by Eq. (44) for the channels displayed in Figure 1. Also, parameter defined by Eq. (126) for the channels displayed in Figure 4
c_f	specific heat of the fluid, $J\,\mathrm{kg}^{-1}\mathrm{K}^{-1}$
c_F	Forchheimer coefficient
D	parameter defined by Eq. (45) for the channels displayed in Figure 1. Also, parameter defined by Eq. (126) for the channels displayed in Figure 4
Da	Darcy number, K/H^2
F	scaled Forchheimer coefficient. $(\rho_f c_F/K^{1/2})(H^4/\mu_f^2)G$
G	applied pressure gradient, $-\mathrm{d}\tilde{p}/\mathrm{d}\tilde{x}$, $\mathrm{Pa\,m}^{-1}$
G_1	parameter defined by Eq. (134)
G_2	parameter defined by Eq. (143)
H	half of parallel plate separation distance for the channels displayed in Figures 1a and 1b, m. Also, parallel plate separation distance for the channels displayed in Figures 1c, 1d, 4a, and 4b, m
k_f	thermal conductivity of the fluid, $\mathrm{W\,m}^{-1}\,\mathrm{K}^{-1}$
k_{eff}	effective thermal conductivity of the porous medium, $\mathrm{W\,m}^{-1}\,\mathrm{K}^{-1}$
K	permeability of the porous medium, m^2
l	coordinate of the center of the porous region for the channels displayed in Figures 1a, 1c, and 1d, $(H+s)/2$, m
L	dimensionless coordinate of the center of the porous region for the channels displayed in Figures 1a, 1c, and 1d, l/H. Also, thickness of the porous layer for the channels displayed in Figure 4, m
$\mathrm{Li}_2[\zeta]$	the dilogarithm function, $-\int_0^{\zeta}\frac{\ln(1-\vartheta)}{\vartheta}\mathrm{d}\vartheta$
Nu	Nusselt number, $Hq''/[k_f(\tilde{T}_w-\tilde{T}_m)]$
$\tilde{p}$	intrinsic average pressure, Pa
P	parameter defined by Eq. (56)
q''	heat flux imposed at the isoflux plate, $\mathrm{W\,m}^{-2}$
Q	parameter defined by Eq. (54) for the channel displayed in Figure 1a. Also, parameter defined by Eq. (76) for the channels displayed in Figures 1b and 1c. Also, parameter defined by Eq. (97) for the channel displayed in Figure 1d
R	thermal conductivity ratio, k_{eff}/k_f
Re	Reynolds number, $\rho_f\tilde{u}_w H/\mu_f$
s	thickness of the clear fluid region for the channels displayed in Figure 1, m
S	dimensionless thickness of the clear fluid region for the channels displayed in Figure 1, s/H
T	dimensionless temperature, $(\tilde{T}-\tilde{T}_w)/(\tilde{T}_m-\tilde{T}_w)$
T_i	dimensionless temperature at the clear fluid/porous medium interface

$\tilde{T}$ intrinsic average temperature, K

$\tilde{T}_m$ mean temperature, K:
- for the channels displayed in Figure 1, $(1/H\tilde{U})\int_0^H \tilde{u}_f \tilde{T} d\tilde{y}$
- for the channels displayed in Figure 4, $(1/H\tilde{U})\int_{-\delta}^L \tilde{u}_f \tilde{T} d\tilde{y}$

$\tilde{T}_w$ temperature at the isoflux wall, K

u dimensionless velocity:
- for the channels displayed in Figure 1, $\tilde{u}_f \mu_f/(GH^2)$
- for the channels displayed in Figure 4, $\tilde{u}_f/\tilde{u}_w$

u_1 parameter defined by Eq. (43), $2u_\infty + 3/(2DaF)$

u_i dimensionless velocity at the clear fluid/porous medium interface

$\tilde{u}_f$ filtration (seepage) velocity, $m\,s^{-1}$

$\tilde{u}_w$ velocity of the lower plate, $m\,s^{-1}$

$\tilde{u}_\infty$ filtration velocity in the central part of the porous layer between the momentum boundary layers, $m\,s^{-1}$

u_∞ dimensionless filtration velocity in the central part of the porous layer between the momentum boundary layers, $\tilde{u}_\infty \mu_f/(GH^2)$

$\tilde{U}$ mean flow velocity, $m\,s^{-1}$
- for the channels displayed in Figure 1, $(1/H)\int_0^H \tilde{u}_f d\tilde{y}$
- for the channels displayed in Figure 4, $(1/H)\int_{-\delta}^L \tilde{u}_f d\tilde{y}$

U dimensionless mean flow velocity, $\tilde{U}\mu_f/(GH^2)$

$\tilde{x}$ streamwise coordinate, m

$\tilde{y}$ transverse coordinate, m

y dimensionless transverse coordinate $\tilde{y}/H$

z function of the coordinate y defined by Eq. (70), $B\exp\{-D(y-S)\}$

z_1 function of the coordinate y defined by Eq. (47), $A\exp\{D(1-y)\}$

z_2 function of the coordinate y defined by Eq. (42), $B\exp\{-D(y-S)\}$

Greek Letters

β adjustable coefficient in the stress jump boundary condition

δ thickness of the fluid layer for the channels displayed in Figure 4, m

γ constant, $(\mu_{eff}/\mu_f)^{1/2}$

Λ_1 parameter defined by Eq. (130)

Λ_2 parameter defined by Eq. (132)

μ_f fluid viscosity, $kg\,m^{-1}s^{-1}$

μ_{eff} effective viscosity in the Brinkman term for the porous region, $kg\,m^{-1}s^{-1}$

ρ_f density of the fluid, $kg\,m^{-3}$

Ω parameter defined by Eq. (90)

Ξ parameter defined by Eq. (58) for the channels displayed in Figures 1a and 1d. Also, parameter defined by Eq. (78) for the channel displayed in Figure 1b. Also, parameter defined by Eq. (92) for the channel displayed in Figure 1c

Ξ_1 parameter defined by Eq. (136)

Ξ_2 parameter defined by Eq. (137)

Ξ_3 parameter defined by Eq. (145)

Ψ_0 parameter defined by Eq. (60) for the channels displayed in Figures 1a, 1c, and 1d. Also, parameter defined by Eq. (80) for the channel displayed in Figure 1b

Ψ_1 parameter defined by Eq. (62) for the channels displayed in Figures 1a, 1c, and 1d. Also, parameter defined by Eq. (81) for the channel displayed in Figure 1b

Ψ_2 parameter defined by Eq. (61) for the channels displayed in Figures 1a, 1c, and 1d

REFERENCES

Alkam M, Al-Nimr M. Transient non-Darcian forced convection flow in a pipe partially filled with a porous material. Int J Heat Mass Transfer 41:347–356, 1998.

Al-Nimr M, Alkam M. Unsteady non-Darcian forced convection analysis in an annulus partially filled with a porous material. ASME J Heat Transfer 119:799–804, 1997.

Al-Nimr M, Alkam M. Unsteady non-Darcian fluid flow in parallel plates channel partially filled with porous material. Heat and Mass Transfer 33:315–318, 1998.

Beavers GS, Joseph DD. Boundary conditions at a naturally permeable wall. J Fluid Mech 30:197–207, 1967.

Bejan A. Convection Heat Transfer. 2nd ed. New York: Wiley, 1995.

Bhargava SK, Sacheti C. Heat transfer in generalized Couette flow of two immiscible Newtonian fluids through a porous channel: Use of Brinkman model. Indian J Tech 27, 211–214, 1989.

Chang W-J, Chang W-L. Mixed convection in a vertical parallel-plate channel partially filled with porous media of high permeability. Int J Heat Mass Transfer 39:1331–1342, 1996.

Chen F, Chen CF. Convection in superposed fluid and porous layers. J Fluid Mech 234:97–119, 1992.

Cheng P, Hsu CT, Chowdhury A. Forced convection in the entrance region of a packed channel with asymmetric heating. ASME J Heat Transfer 110:946–954, 1988.

Chikh S, Boumedian A, Bouhadef K, Lauriat G. Analytical solution of non-Darcian forced convection in an annular duct partially filled with a porous medium. Int J Heat Mass Transfer 38:1543–1551, 1995a.

Chikh S, Boumedian A, Bouhadef K, Lauriat G. Non-Darcian forced convection analysis in an annulus partially filled with a porous material. Numer Heat Transfer, Part A 28:707–722, 1995b.

Daskalakis J. Couette flow through a porous medium of high Prandtl number fluid with temperature-dependent viscosity. Int J Energy Res 14:21–26, 1990.

Givler RC, Altobelli SA. A determination of the effective viscosity for the Brinkman–Forchheimer flow model. J Fluid Mech 258:355–370, 1994.

Hadim A. Forced convection in a porous channel with localized heat sources. ASME J Heat Transfer 116:465–472, 1994.

Huang PC, Vafai K. Flow and heat transfer control over an external surface using a porous block arrangement. Int J Heat Mass Transfer 36:4019–4032, 1993.

Huang PC, Vafai K. Analysis of forced convection enhancement in a channel using porous blocks. J Thermophys Heat Transfer 8:563–573, 1994a.

Huang PC, Vafai K. Internal heat transfer augmentation in a channel using an alternative set of porous cavity-block obstacles. Numer Heat Transfer, Part A 25:519–539, 1994b.

Jang JY, Chen JL. Forced convection in a parallel plate channel partially filled with a high porosity medium. Int Commun Heat Mass Transfer 19:263–273, 1992.

Kaviany M. Laminar flow through a porous channel bounded by isothermal parallel plates. Int J Heat Mass Transfer 28:851–858, 1985.

Kaviany M. Principles of Heat Transfer in Porous Media. New York: Springer, 1991.

Kim SJ, Choi CY. Convective heat transfer in porous and overlying fluid layers heated from below. Int J Heat Mass Transfer 39:319–329, 1996.

Kuznetsov AV. Analytical investigation of the fluid flow in the interface region between a porous medium and a clear fluid in channels partially filled with a porous medium. Appl Sci Res 56:53–67, 1996.

Kuznetsov AV. Influence of the stress jump boundary condition at the porous-medium/clear-fluid interface on a flow at a porous wall. Int Commun Heat Mass Transfer 24:401–410, 1997.

Kuznetsov AV. Analytical investigation of heat transfer in Couette flow through a porous medium utilizing the Brinkman–Forchheimer–extended Darcy model. Acta Mech 129:13–24, 1998a.

Kuznetsov AV. Analytical investigation of Couette flow in a composite channel partially filled with a porous medium and partially with a clear fluid. Int J Heat Mass Transfer 41:2556–2560, 1998b.

Kuznetsov AV. Thermal nonequilibrium forced convection in porous media. In: Ingham DB, Pop I, eds. Transport Phenomena in Porous Media. Oxford: Elsevier, 1998c, pp 103–129.

Kuznetsov AV. Analytical study of fluid flow and heat transfer during forced convection in a composite channel partly filled with a Brinkman–Forchheimer porous medium. Flow, Turbulence and Combustion 60:173–192, 1998d.

Kuznetsov AV. Fluid mechanics and heat transfer in the interface region between a porous medium and a fluid layer—A boundary layer solution. J Porous Media 2:309–321, 1999.

Martys N, Bentz DP, Garboczi EJ. Computer simulation study of the effective viscosity in Brinkman's equation. Phys Fluids 6:1434–1439, 1994.

Nakayama A. Non-Darcy Couette flow in a porous medium filled with an inelastic non-Newtonian fluid. ASME J Fluids Eng 114:642–647, 1992.

Nakayama A. PC-aided Numerical Heat Transfer and Convective flow. Tokyo: CRC Press, 1995.

Nakayama A, Koyama H, Kuwahara F. An analysis on forced convection in a channel filled with a Brinkman–Darcy porous medium: Exact and approximate solutions. Wärme- und Stoffübertragung 23:291–295, 1988.

Nield DA, Bejan A. Convection in Porous Media. 2nd ed. New York: Springer, 1999.

Nield DA, Junqueira SLM, Lage JL. Forced convection in a fluid saturated porous medium channel with isothermal or isoflux boundaries. J Fluid Mech 322:201–214, 1996.

Ochoa-Tapia JA, Whitaker S. Momentum transfer at the boundary between a porous medium and a homogeneous fluid—I. Theoretical development. Int J Heat Mass Transfer 38:2635–2646, 1995a.

Ochoa-Tapia JA, Whitaker S. Momentum transfer at the boundary between a porous medium and a homogeneous fluid—II. Comparison with experiment. Int J Heat Mass Transfer 38:2647–2655, 1995b.

Ochoa-Tapia JA, Whitaker S. Heat transfer at the boundary between a porous medium and a homogeneous fluid. Int J Heat Mass Transfer 40:2691–2707, 1997.

Poulikakos D, Kazmierczak M. Forced convection in a duct partially filled with a porous material. ASME J Heat Transfer 109:653–662, 1987.

Richardson S. A model for the boundary conditions of a porous material. Part 2. J Fluid Mech 49 (Part 2):327–336, 1971.

Rudraiah N. Forced convection in a parallel plate channel partially filled with a porous material. ASME J Heat Transfer 107:322–331, 1985.

Sahraoui M, Kaviany M. Slip and no-slip velocity boundary conditions at interface of porous, plain media. Int J Heat Mass Transfer 35:927–943, 1992.

Taylor, GI. A model for the boundary conditions of a porous material. Part 1. J Fluid Mech 49 (part 2):319–326, 1971.

Vafai K, Kim SJ. Forced convection in a channel filled with a porous medium: an exact solution. ASME J Heat Transfer 111:1103–1106, 1989.

Vafai K, Kim SJ. Fluid mechanics of the interface region between a porous medium and a fluid layer—an exact solution. Int J Heat Fluid Flow 11:254–256, 1990a.

Vafai K, Kim SJ. Analysis of surface enhancement by a porous substrate. ASME J Heat Transfer 112:700–706, 1990b.

Vafai K, Kim SJ. Discussion of the paper by A. Hadim "Forced convection in a porous channel with localized heat sources". ASME J Heat Transfer 17:1097–1098, 1995a.

Vafai K, Kim SJ. On the limitations of the Brinkman–Forchheimer–extended Darcy equation. Int J Heat Fluid Flow 16:11–15, 1995b.

Vafai K, Thiyagaraja R. Analysis of flow and heat transfer at the interface region of a porous medium. Int J Heat Mass Transfer 30:1391–1405, 1987.

Vafai K, Tien CL. Boundary and inertia effects on flow and heat transfer in porous media. Int J Heat Mass Transfer 24:195–203, 1981.

7

Convective Boundary Layers in Porous Media: External Flows

I. Pop
University of Cluj, Cluj, Romania

D. B. Ingham
University of Leeds, Leeds, West Yorkshire, England

I. INTRODUCTION

During the last five decades buoyancy-driven flow through porous media has been the subject of many investigations. This is due to its many practical applications which can be modeled or approximated as transport phenomena in porous media, e.g., geothermal energy extraction, storage of nuclear waste material, groundwater flows, industrial and agricultural water distribution, oil recovery processes, thermal insulation engineering, pollutant dispersion in aquifers, cooling of electronic components, packed-bed reactors, food processing, casting and welding of manufacturing processes, the dispersion of chemical contaminants in various processes in the chemical industry and in the environment, to name just a few applications. This topic is of vital importance to these systems, thereby generating the need for a full understanding of transport processes through porous media. Comprehensive literature surveys concerning these processes can be found in the most recent books by Ingham and Pop (1998) and Nield and Bejan (1999), with reference to the review articles included therein.

Prandtl's (1904) boundary-layer theory proved to be of vital importance in Newtonian fluids as the Navier–Stokes equations can be converted into much simpler equations which are easier to handle. In the 1960s, with the increase of technological importance of transport phenomena through porous media, similar attempts were made by Wooding (1963) to solve equations which govern flow and heat transfer in porous media using the boundary-layer assumption. Several models were proposed in order to explain mathematical and physical aspects associated with boundary-layer flows and convective heat transfer in porous media. Among these, the Darcy law and a series of its modifications gained much acceptance; see, for example, Nakayama (1995, 1998). Boundary-layer assumptions were successfully applied to these models and much work has been done on them for different body geometries in the last three decades.

The present chapter is aimed at presenting a review of the updated papers which are relevant to convective boundary-layer flows over external surfaces such as flat surfaces and from the stagnation point of a two-dimensional cylindrical body. In particular, some of the details and the main results of these papers are reviewed in this chapter. However, it is beyond the scope of this chapter to discuss unsteady and conjugate convective flows in porous media since they were very recently reviewed in the excellent articles by Bradean et al. (1998), Kimura et al. (1997), Pop et al. (1998), and Pop and Nakayama (1999). It should be noted that much of the information which is available today on convective flows in porous media is restricted to very specialized theoretical studies, and a much more detailed knowledge of experimental data and theoretical analysis is further required in order to interpret the results correctly.

II. BASIC EQUATIONS

We start with the following assumptions: (i) at sufficient large Rayleigh numbers, and hence sufficient large velocities, one can expect that the fluid and porous medium are, in general, not in thermodynamical equilibrium, i.e., the temperatures T_s^* and T_f^* in the solid and fluid phases are not identically the same (see Nield and Bejan, 1999); (ii) the porous medium is isotropic; (iii) radiative effects, viscous dissipation, the work done by pressure changes, and the presence of heat sources are negligible; (iv) the Boussinesq approximation is valid, i.e., the fluid density ρ decreases linearly with the fluid temperature T_f^* according to $\rho = \rho_\infty[1 - \beta(T_f^* - T_\infty)]$. With these assumptions, the basic equations of convective flow in a porous medium are given by (Nield and Bejan 1999)

$$\nabla^* \cdot \mathbf{V}^* = 0 \tag{1a}$$

$$\frac{\mu}{K}\mathbf{V}^* = \rho_\infty[1 - \beta(T_f^* - T_\infty)]\mathbf{g}^* - \nabla^* p^* \tag{1b}$$

$$\varphi(\rho c)_f \frac{\partial T_f^*}{\partial t^*} + (\rho c)_f (\mathbf{V}^* \cdot \nabla^*) T_f^* = \varphi k_f \nabla^{*2} T_f^* + h^*(T_s^* - T_f^*) \tag{1c}$$

$$(1-\varphi)(\rho c_p)_s \frac{\partial T_s^*}{\partial t^*} = (1-\varphi) k_s \nabla^{*2} T_s^* - h^*(T_s^* - T_f^*) \tag{1d}$$

$$\varphi \frac{\partial C^*}{\partial t^*} + (\mathbf{V}^* \cdot \nabla^*) C^* = D_m \nabla^{*2} C^* \tag{1e}$$

where $\mathbf{V}^*$ is the mean fluid velocity vector, h^* is a heat transfer coefficient which is used to model the microscopic transfer of heat between the fluid and solid phases, and ∇^{*2} is the Laplacian in Cartesian coordinates (x^*, y^*).

III. FREE CONVECTION ALONG VERTICAL SURFACES

In this section we consider the steady convective boundary-layer flow induced by a vertical heated surface that is embedded in a fluid-saturated porous medium. In the case of steady ($\partial/\partial t^* \equiv 0$) convective flow, Eqs. (1) may be nondimensionalized using the variables

$$(x^*, y^*) = l(\hat{x}, \hat{y}), \qquad (u^*, v^*) = (\varphi \alpha_f / l)(\hat{u}, \hat{v})$$

$$p^* = (\varphi \alpha_f \mu / K)\hat{p}, \qquad T_f^* = T_\infty + T_f \Delta T$$

$$T_s^* = T_\infty + T_s \Delta T, \qquad C^* = C_\infty \hat{C}$$

where l is a suitably defined macroscopic length scale and $\alpha_f = \varphi k_f / (\rho c)_f$ is the thermal diffusivity. A further simplification is afforded by the introduction of a dimensionless stream function $\hat{\psi}$ according to $\hat{u} = \partial\hat{\psi}/\partial\hat{y}$ and $\hat{v} = -\partial\hat{\psi}/\partial\hat{x}$ along with the usual boundary-layer variables $\hat{x} = x$, $\hat{y} = Ra^{1/2} y$ and $\hat{\psi} = Ra^{1/2}\psi$, where $Ra = \rho_\infty g K \beta \Delta T l/(\varphi \mu \alpha_f)$ is the Rayleigh number based on the fluid properties. We can thus obtain from Eqs. (1) the basic equations which govern the steady free convection flow on a vertical surface embedded in a porous medium in the most general form (Rees and Pop, 2000)

$$\frac{\partial^2 \psi}{\partial y^2} = \frac{\partial T_f}{\partial y} S(x) \tag{2a}$$

$$\frac{\partial^2 T_f}{\partial y^2} = H(T_f - T_s) + \frac{\partial \psi}{\partial y}\frac{\partial T_f}{\partial x} - \frac{\partial \psi}{\partial x}\frac{\partial T_f}{\partial y} \tag{2b}$$

$$\frac{\partial^2 T_s}{\partial y^2} = H\gamma(T_s - T_f) \tag{2c}$$

$$\frac{1}{Le}\frac{\partial^2 C}{\partial y^2} = \frac{\partial \psi}{\partial y}\frac{\partial C}{\partial x} - \frac{\partial \psi}{\partial x}\frac{\partial C}{\partial y} \tag{2d}$$

where $S(x)$ is the sine of the angle between the outward normal from the two-dimensional cylindrical body surface and the downward vertical, and

$$H = \frac{h^* l^2}{Ra\varphi k_f}, \qquad \gamma = \frac{\varphi k_f}{(1-\varphi)k_s} \tag{3}$$

are constants with $H = \mathbf{O}(1)$ as $Ra \to \infty$. Such a scaling allows a detailed study of how the boundary layer undergoes the transition from strong thermal nonequilibrium near the leading edge of the plate to thermal equilibrium far from the leading edge.

We will now review some of the most representative boundary-layer papers for the vertical porous medium configuration.

A. Impermeable Surfaces

Here we consider a vertical surface embedded in a fluid-saturated porous medium which is in thermodynamic equilibrium ($T_f = T_s = T$) and for which there is no chemical reaction at the plate ($C = 0$). The governing equations (2), where $S(x) = 1$, reduce to

$$\frac{\partial^2 \psi}{\partial y^2} = \frac{\partial T}{\partial y} \tag{4a}$$

$$\frac{\partial^2 T}{\partial u^2} = \frac{\partial \varphi}{\partial y}\frac{\partial T}{\partial x} - \frac{\partial \psi}{\partial x}\frac{\partial T}{\partial y} \tag{4b}$$

Cheng and Minkowycz (1977) were the first to obtain similarity solutions to these equations for free convection from an impermeable surface when the wall temperature varies as x^m. Under this assumption, Eqs. (4a,b) have to be solved subject to the boundary conditions.

$$\begin{aligned} &\psi = 0, \qquad T = x^m \qquad \text{at} \qquad y = 0, \qquad x \geq 0 \\ &\frac{\partial \psi}{\partial y} \to 0, \qquad T \to 0 \qquad \text{as} \qquad y \to \infty, \qquad x \geq 0 \end{aligned} \tag{4c}$$

In order to solve Eqs. (4) we introduce the transformation

$$\psi = x^{(1+m)/2} f(\eta), \qquad \eta = y x^{(m-1)/2} \tag{5}$$

and T is found from Eq. (4a) and the boundary conditions (4c) to be given by

$$T = x^m f'(\eta)$$

Thus, Eqs. (4a,b) reduce to the following ordinary differential equation

$$f''' + \frac{m+1}{2} ff'' - mf'^2 = 0 \tag{6a}$$

which has to be solved subject to the boundary conditions

$$f(0) = 0, \qquad f'(0) = 1, \qquad f'(\infty) = 0 \tag{6b}$$

The numerical solution of Eqs. (6) has been performed by Cheng and Minkowycz (1977) for $-1/3 < m < 1$ and a further investigation of these equations was performed by Ingham and Brown (1986) for $-1/2 < m < \infty$. The latter authors have shown that a solution is possible only for $m > -1/2$, with the solution becoming singular as $m \to -1/2$. This may be verified by multiplying Eq. (6a) by f and integrating twice the resulting equation, using the boundary conditions (6b), to obtain

$$\int_0^\infty ff'^2 \mathrm{d}\eta = \frac{1}{2(2m+1)} \tag{7}$$

Since $f \geq 0$ for all $0 \leq \eta < \infty$ and $m < 0$ (Ingham and Brown, 1986), we see that the integral (7) must be positive, and hence $2m + 1 > 0$. Therefore, there are no solutions of Eqs. (6) for $m < -1/2$. It is easily seen that, for $m = -1/3$, Eqs. (6) have the exact analytical solution

$$f(\eta) = \sqrt{6}\tanh(\eta/\sqrt{6}), \qquad f''(0) = 0 \tag{8}$$

but for $m = 1$ the solution is

$$f(\eta) = 1 - \exp(-\eta), \qquad f''(0) = -1 \tag{9}$$

Ingham and Brown (1986) have also obtained asymptotic solutions of Eqs. (6) for m near $-1/2$, near 0, and for $m \gg 1$. The solution for m near $-1/2$ was obtained by taking $m = -1/2 + s$, where $s(> 0)$ is a small real quantity. It was found that

$$f''(0) = 0.078\,103\left(\frac{1}{2} + m\right)^{-3/4} + \cdots \tag{10}$$

for m near $-1/2$ and

$$f''(0) = -0.443\,75 - 0.856\,65m + 0.669\,43m^2 + \cdots \tag{11}$$

for m near 0, where $f''(0)$ is related to the heat flux at the plate, $(\partial T/\partial y)_w = x^{(3m-1)/2} f''(0)$. Further, Ingham and Brown (1986) have shown that dual solutions exist for $1 < m < \infty$. If m is very large, we introduce the variables $f = m^{-1/2}\tilde{f}(\tilde{\eta})$ and $\tilde{\eta} = m^{1/2}\eta$. Equation (6a) can then be reduced to

$$\tilde{f}''' + \frac{1}{2}\tilde{f}\tilde{f}'' - \tilde{f}'^2 = 0 \tag{12}$$

and this ordinary differential equation has to be solved subject to the boundary conditions (6b); primes now denote differentiation with respect to $\tilde{\eta}$. Solving this equation numerically, it was found that at least two possible solutions exist such that

$$f''(0) = -0.906\,38m^{1/2}, \qquad f(\infty) = 1.280\,77m^{-1/2} \tag{13a}$$

and

$$f''(0) = -0.913\,34m^{1/2} \qquad f(\infty) = 0.433\,65m^{-1/2} \tag{13b}$$

for $m \gg 1$. Although the values of $f''(0)$ in both solutions (13) are very close, the basic difference between them is that the second solution contains a region within the boundary layer where the fluid velocity $f'(\eta)$, or the temperature T, becomes negative. It was also reported by Ingham and Brown (1986) that, on this second branch of the numerical solution of Eq. (6) for $m \to 1^+$, $f(\infty) \to 0^+$ and $f(0) \to -1^-$.

The variation of $f''(0)$ as a function of m for the first solution of Eq. (12) was obtained by a direct numerical integration of Eqs. (6) for $-1/2 < m < 5$ and is shown in Figure 1 by the full line. Also shown is the exact solution (8), the solution (10) for $m \sim -1/2$, the two and three terms series solution (11) for $m \sim 0$, and the asymptotic solution (13a) for $m \gg 1$. This figure clearly shows that extreme care should be taken when dealing with free and mixed convection problems for vertical surfaces which are embedded in porous media when the wall temperature varies as a power of x. It appears that not all existing published papers have dealt with the existence of dual solutions correctly.

On the other hand, Banks and Zaturska (1986) have investigated the existence of eigensolutions for the present problem. When $m = -1/3$ and $m = 1$, for which closed form solutions (8) and (9) exist, a detailed analytical investigation of the eigensolutions was made. However, these authors worked this problem out in terms of the parameter

$$\beta = \frac{2m}{m+1} \tag{14}$$

In order to do this, let us introduce the following variables

$$\psi = \left(\frac{2x}{1+m}T_w(x)\right)^{1/2} F(x, \eta), \qquad \eta = y\left(\frac{1+m}{2x}T_w(x)\right)^{1/2} \tag{15}$$

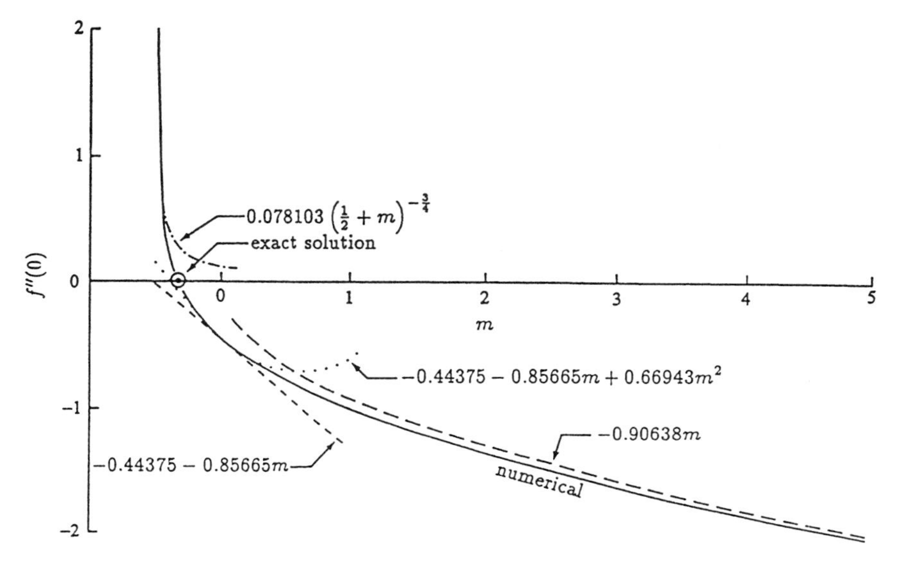

Figure 1. The variation of $f''(0)$ with m obtained from a numerical integration of Eqs. (6) and from some asymptotic solutions.

where $T_w(x) = x^m$. On substituting the transformation (15) into Eqs. (4a,b), we obtain the following equation for $F(x, \eta)$

$$\frac{\partial^3 F}{\partial \eta^3} + F\frac{\partial^2 F}{\partial \eta^2} - \beta\left(\frac{\partial F}{\partial \eta}\right)^2 = (2-\beta)x\left(\frac{\partial F}{\partial \eta}\frac{\partial^2 F}{\partial x \partial \eta} - \frac{\partial F}{\partial x}\frac{\partial^2 F}{\partial \eta^2}\right) \tag{16a}$$

along with the boundary conditions

$$\begin{aligned} &F = 0, \quad \frac{\partial F}{\partial \eta} = 1 \quad \text{at} \quad \eta = 0, \quad x \geq 0 \\ &\frac{\partial F}{\partial \eta} \to 0 \quad \text{as} \quad \eta \to \infty, \quad x \geq 0 \end{aligned} \tag{16b}$$

and an initial condition is also required of the form $F(x_0, \eta) = g(\eta)$, where $g(\eta)$ satisfies certain requirements. Here we assume that $\beta \neq 2$ since the case $\beta = 2$ corresponds to m being infinitely large. In a similar manner to the relation (7), an integral constraint for $F(x, \eta)$ also exists and is given explicitly as

$$(2+\beta)\int_0^\infty F\left(\frac{\partial F}{\partial \eta}\right)^2 \mathrm{d}\eta + (2-\beta)x\int_0^\infty \frac{\partial F}{\partial \eta}\left(2F\frac{\partial^2 F}{\partial x \partial \eta} + \frac{\partial F}{\partial x}\frac{\partial F}{\partial \eta}\right)\mathrm{d}\eta = \frac{1}{2} \tag{17}$$

Similarity solutions of equation (16a) may be obtained if we ignore the initial condition at x_0 and write $F(x, \eta) = f(\eta)$, where $f(\eta)$ satisfies the equation

$$f''' + ff'' - \beta f'^2 = 0 \tag{18a}$$

and the boundary conditions (16b) become

$$f(0) = 0, \qquad f'(0) = 1, \qquad f'(\infty) = 0 \tag{18b}$$

The problem posed by Banks and Zaturska (1986) is as follows: given that $g(\eta)$ is such that $F(x, \eta)$ differs only slightly from $f(\eta)$, as defined by Eq. (18), determine the leading-order term for $F(x, \eta) - f(\eta)$ as $x \to \infty$, depending on the value of the parameter β. Thus, $F(x, \eta)$ is taken to have the form

$$F(x, \eta) = f(\eta) + X(x)G(\eta) + \cdots \tag{19}$$

with

$$X(x) \sim x^{-e/(2-\beta)} \tag{20}$$

and $G(\eta)$ satisfies the equation

$$G_n''' + fG_n'' + (e_n - 2\beta)f'G_n' + (1 - e_n)f''G_n = 0 \tag{21a}$$

together with the boundary conditions

$$G_n(0) = G_n'(0) = 0, \qquad G_n'(\infty) = 0 \tag{21b}$$

It should be noted that a suffix n has been attached to the separation constant e and the function $G(\eta)$, since for each value of the parameter β in the range $-2 < \beta < 2 (m > -1.2)$ there exist solutions of Eqs. (18). It can be seen from expression (20) that, for these values of β and m of interest, the fluid flows are spatially stable as $x \to \infty$ if the minimum value in the set e_n is positive. With $f(\eta)$ known, Eqs. (21) constitute an eigenvalue problem for the eigenvalues e_n and eigenfunctions G_n. On the other hand, from (17) and (19) we obtain, on equating terms of $\mathbf{O}(1)$ and $\mathbf{O}(x)$, respectively,

$$2(2 + \beta) \int_0^\infty ff'^2 \mathrm{d}\eta = 1 \tag{22a}$$

$$(2 + \beta - e_n) \int_0^\infty f'(2fG_n' + f'G_n) \mathrm{d}\eta = 0 \tag{22b}$$

The condition (22a) shows that a similarity solution for $f(\eta)$ does not exist when $\beta = -2$ (i.e., $m = -1/2$). Further, the integrand in condition (22b) provides a constraint on the eigensolutions. One can infer from condition (22b) that, since $f(\eta)$ and $f'(\eta)$ are non-negative for each value of β in the range of interest, the integrand is of one sign if an eigenfunction exists which is of one sign in the interval $0 \leq \eta < \infty$. Thus, the first eigenvalue is given by

$$e_1 = 2 + \beta$$

which is real and positive for each $\beta > -2(m > -1/2)$ of interest. The determination of the first eigenfunction $G_1(\eta)$ is now reduced to a one-point numerical integration of Eqs. (21). Banks and Zaturska (1986) have also obtained analytical expressions for the first eigensolutions corresponding to the two analytical solutions of Eq. (18) for $\beta = 1$ $(m = 1)$ and $\beta = -1$ $(m = -1/3)$, respectively. It is also worth point out that for $\beta = 0$ $(m = 0)$ the first eigenvalue is $e_1 = 2$ with $G_1 = (\eta f' - f'')/f''(0)$ and that this corresponds to an origin shift in x in the boundary layer (Stewartson 1957). It should be noted that the value $e_1 = 2$ corresponds to the leading-edge-shift eigensolution of Daniels and Simpkins (1984) for free convective flow in a porous medium bounded by a uniformly heated vertical wall and a second thermally insulated wall which forms a corner of arbitrary angle. Further, it was stated by Banks and Zaturska (1986) that, for $\beta \to -2^+$ (or $m \to -1/2^+$), the present eigenvalue problem is similar to that of the wall jet which was studied by Riley (1962), where such a fluid flow may be described as neutrally stable. Typical results of the analytical and numerical investigations of Banks and Zaturska (1986) are presented in Figure 2.

Other aspects of this free convective boundary-layer flow have been documented by Na and Pop (1983), Merkin (1984), Zaturska and Banks (1985, 1987), Rees and Bassom (1991), and Wright et al. (1996).

B. Permeable Surfaces

Another important class of free convective boundary-layer flows in porous media is that of similarity solutions associated with permeable surfaces through which fluids can be injected into the porous medium or withdrawn from it. These problems were initially treated by Cheng (1977) and continued by Merkin (1978), Minkowycz and Cheng (1982), Govindarajulu and Malarvizhi (1987), and Chaudhary et al. (1995a,b). The situations we are interested in here give rise to the boundary conditions

$$\begin{aligned} &\psi = -\sigma x^{-(m+1)/2}, \qquad T = x^m \qquad \text{on} \qquad y = 0, \qquad x \geq 0 \\ &\frac{\partial \psi}{\partial y} \to 0, \qquad T \to 0 \qquad \text{as} \qquad y \to \infty, \qquad x \geq 0 \end{aligned} \tag{23}$$

where σ is called the mass transfer rate through the wall, with $\sigma > 0$ for injection and $\sigma < 0$ for suction (or withdrawal) of fluid, respectively. The introduction of the transformation

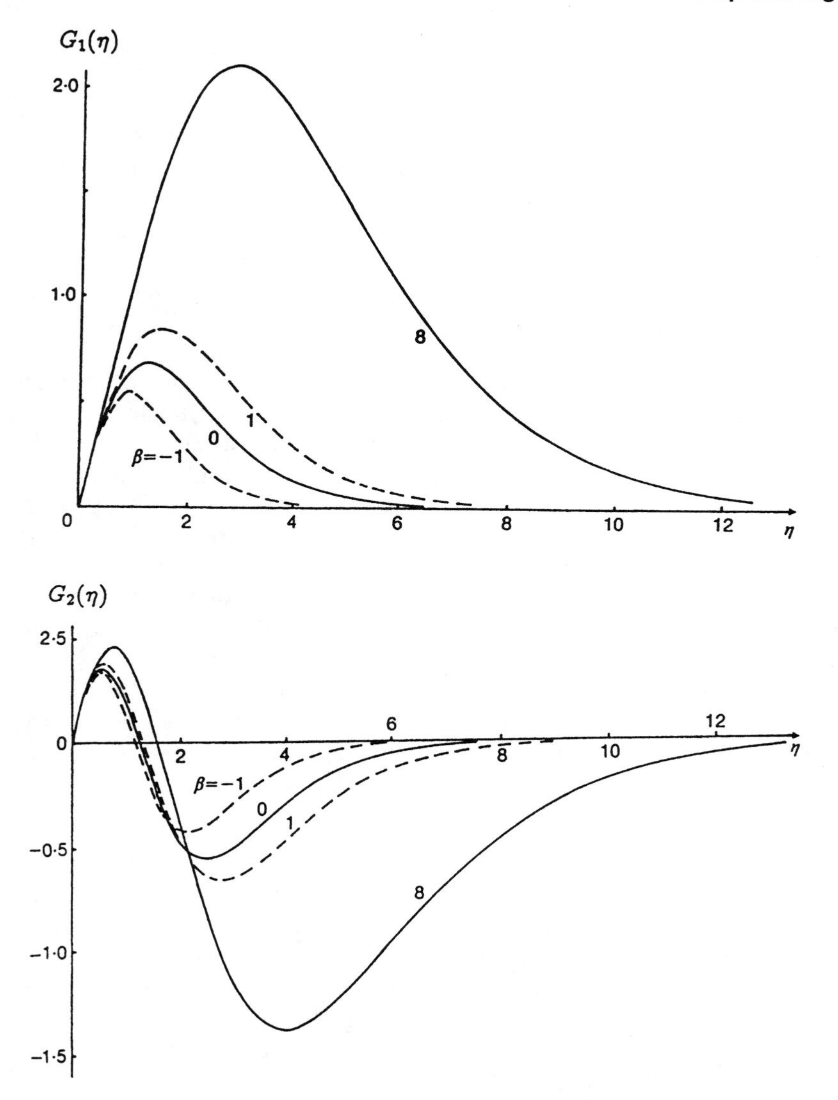

Figure 2. Profiles of the first (a) and second (b) eigensolutions obtained from the numerical integration of Eqs. (21).

$$\psi = \sqrt{2}x^{(m+1)/2}f(\eta), \qquad T = x^m\theta(\eta)$$
$$\eta = \frac{y}{\sqrt{2}}x^{(m-1)/2}$$

reduces Eqs. (4a,b) to the single equation

$$f''' + (m+1)ff'' - 2mf'^2 = 0 \tag{24a}$$

with the boundary conditions (23) requiring that

$$f(0) = -\sigma, \qquad f'(0) = 1, \qquad f'(\infty) = 0 \tag{24b}$$

It was shown by Chaudhary et al. (1995a) that Eqs. (24) have solutions only for $m > -1/2$, the same as for the case of $\sigma = 0$. However, an asymptotic solution has been obtained by these authors for $\sigma > 0$ when $m \sim -1/2$, and it was found that

$$f''(0) = 0.001\,19\sigma^3\left(\frac{1}{2} + m\right)^{-3} + \cdots \tag{25}$$

The variation of $f''(0)$ with m for $\sigma = 1$, as obtained from the numerical integration of Eq. (24) and from the asymptotic expansion (25), is shown in Figure 3. It is concluded from this figure that the two curves are

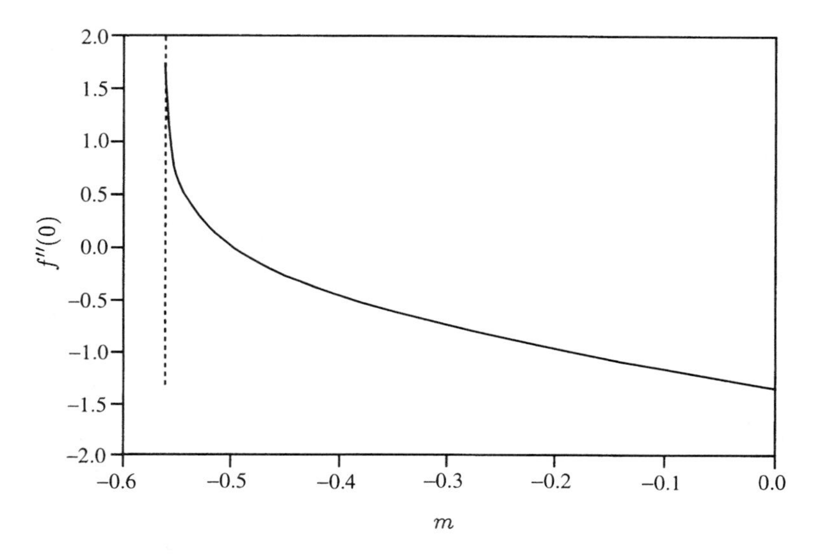

Figure 3. Variation of $f''(0)$ with m for $\sigma = 1$ obtained from a numerical integration of Eqs. (24) (shown by the full line) and from the asymptotic solution (25) (shown by the broken line).

in good agreement, with the difference becoming smaller as m decreases towards the singular value at $m = -1/2$. A numerical solution of Eq. (24) for $\sigma = 0$, 4, 8, and 12 with $m = 0$ and $m = 1$ has been also obtained by Chaudhary et al. (1995a), and a plot of the temperature profile $\theta(\eta)$ is given in Figure 4. We observe from Figure 4a that, for $m = 0$, a clear

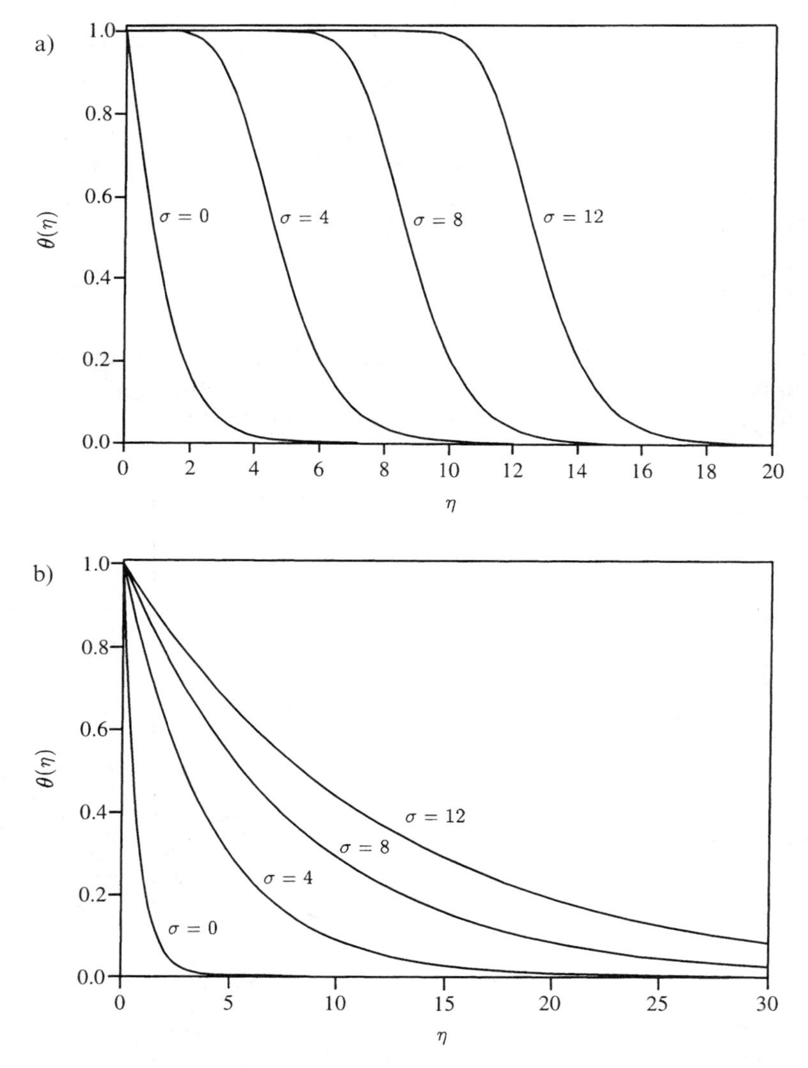

Figure 4. Temperature profiles for (a) $m = 0$ and (b) $m = 1$.

two-region structure emerges as σ increases. There is a thick inner region, where the temperature is constant (at its surface value), and a thinner shear layer at the outer edge where the ambient temperature is attained. These profiles are reminiscent of the temperature profiles seen at large distances from the leading edge of the plate in the constant surface temperature and fluid injection rate problem described by Merkin (1978). For $m = 1$, we observe from Figure 4b that, although the boundary layer becomes thicker as σ increases, there is no obvious two-region structure set up. These results suggest that the development of the solution for large σ could well depend on the value of m.

On the other hand, for $\sigma < 0$ (suction), Chaudhary et al. (1995a) have shown that Eqs. (24) have a solution even for values of $m < -1.2$, as can be seen in Figure 5 where the plot of $f''(0)$ obtained from a numerical solution of these equations for $\sigma = -1$ is given. This figure shows that a solution exists for $m > m_0(\sigma)$, where, for $\sigma = -1$, $m_0 \simeq -0.5619$, and that $f''(0)$ is approaching a constant value as $m \to m_0$ (shown by the broken line). Detailed solutions for strong suction, $\sigma \ll -1$, strong injection, $\sigma \gg 1$, and for $m \gg 1$ were also reported by Chaudhary et al. (1995a).

The same problem was also treated by Chaudhary et al. (1995b) for a vertical permeable surface in a porous medium but with a prescribed surface heat flux.

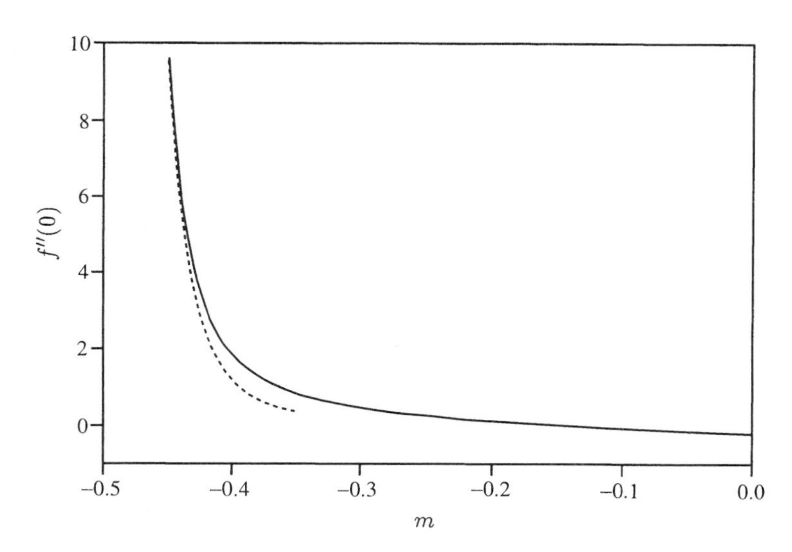

Figure 5. Variation of $f''(0)$ with m for $\sigma = -1$ obtained from a numerical integration of Eqs. (24) (shown by the full line). The value of $m_0 = -0.5619$ is shown by the broken line.

IV. FREE CONVECTION ON REACTING IMPERMEABLE SURFACES

A. Vertical Surfaces

Free convection boundary-layer flow along a vertical surface surrounded by a fluid-saturated porous medium which is driven by catalytic surface heating can be of importance in the design of equipment used in several types of engineering systems. Areas of research in this topic include tubular laboratory reactors, chemical vapor deposition systems, the oxidation of solid materials in large containers, the synthesis of ceramic materials by a self-propagating reaction, combustion in underground reservoirs for enhanced oil recovery, and the reduction of hazardous combustion products using catalytic porous beds, among others. Until recently this subject has received relatively little attention; see Ene and Pališevski (1987), Chao et al. (1996), and Minto et al. (1998). In the latter paper the authors have studied the free convection boundary layer on a vertical surface in a porous medium which is driven by an exothermic reaction, based on the assumption made by Chaudhary and Merkin (1994, 1995a,b, 1996) and Merkin and Chaudhary (1994), for the corresponding problem of viscous (non-porous) fluid, that the convective fluid contains a reactive species A which reacts to form some inert product when in contact with the surface. In particular, we assume that there is an exothermic catalytic reaction on this surface whereby reactant A is converted to an inert product B via the first-order, non-isothermal reaction scheme

$$\mathrm{A} \rightarrow \mathrm{B} + \text{heat}, \qquad \text{rate} = k_0 C^* \exp\left(-\frac{E}{RT_f^*}\right) \tag{26}$$

known as Arrhenius kinetics. The boundary-layer equations are (see, for example, Minto et al. 1998)

$$\theta = \frac{\partial \psi}{\partial y} \tag{27a}$$

$$\frac{\partial^2 \theta}{\partial y^2} = \frac{\partial \psi}{\partial y}\frac{\partial \theta}{\partial x} - \frac{\partial \psi}{\partial x}\frac{\partial \theta}{\partial y} \tag{27b}$$

$$\frac{1}{Le}\frac{\partial^2 \phi}{\partial y^2} = \frac{\partial \psi}{\partial y}\frac{\partial \phi}{\partial x} - \frac{\partial \psi}{\partial x}\frac{\partial \phi}{\partial y} \tag{27c}$$

which are to be solved subject to the boundary conditions

$$\left.\begin{aligned} &\psi = 0, \quad \left.\begin{aligned} \frac{\partial\theta}{\partial y} &= -\phi \exp\left(\frac{\theta}{1+\varepsilon\theta}\right) \\ \frac{\partial\phi}{\partial y} &= \lambda\phi \exp\left(\frac{\theta}{1+\varepsilon\theta}\right) \end{aligned}\right\} \quad \text{on} \quad y = 0, \quad x > 0 \\ &\frac{\partial\psi}{\partial y} \to 0, \quad \theta \to 0, \quad \phi \to 1 \quad \text{as} \quad y \to \infty, \quad x > 0 \\ &\psi = \theta = 0, \quad \phi = 1 \quad \text{on} \quad x = 0, \quad y > 0 \end{aligned}\right\} \tag{27d}$$

Here λ and ε are the reactant consumption and activation energy parameters defined as

$$\lambda = \frac{k_0 R T_\infty^2}{QDEC_\infty}, \qquad \varepsilon = \frac{RT_\infty}{E} \tag{28}$$

$\theta = (T - T_\infty)/T_R$, $C = \phi$, $T_R = RT_\infty^2/E$ is the scaling for the temperature and $Le = \varphi\alpha_f/D_m$ is the Lewis number. It is seen from Eqs. (27) that the case of $Le = 1$ leads to the relation

$$\phi(x, y) = 1 - \lambda\theta(x, y) \tag{29}$$

Equations (27) were solved numerically by Minto et al. (1998) for different values of the parameters λ and ε but all the results were obtained for $Le = 1$ only. Further, Minto et al. (1998) found asymptotic solutions of these equations for both x small and x large. Thus, the wall temperature, $\theta(x, 0) = \theta_w(x)$, and surface concentration, $\phi(x, 0) = \phi_w(x)$, can be expressed as follows

$$\left.\begin{aligned} \theta_w(x) &= 1.296\,18 x^{1/3}\{1 + 0.734\,29(1-\lambda)x^{1/3} + \cdots\} \\ \phi_w(x) &= 1 - 0.734\,29\lambda x^{1/3} + \cdots \end{aligned}\right\} \tag{30}$$

for small $x(\ll 1)$. For large values of $x(\gg 1)$ there are two cases to be considered, namely, when the reaction consumption parameter $\lambda = 0$ and $\varepsilon \neq 0$, and when $\lambda \neq 0$ and ε is arbitrary, giving rise to two essentially distinct types of asymptotic solutions. Thus, we have for $x \gg 1$

$$\begin{aligned} \theta_w(x) = x^{1/3}&\left(\exp\left(\frac{1}{\varepsilon}\right)\right)^{2/3} \\ &\left\{1.296\,18 - 0.853\,33\left(\left(\exp\left(\frac{1}{\varepsilon}\right)\right)^{-2/3}\varepsilon^{-2}x^{-1/3} + \cdots\right\}\right. \end{aligned} \tag{31}$$

which are valid only for $\lambda = 0$ and $\varepsilon \neq 0$. However, for $\lambda \neq 0$ and ε arbitrary we have, when $x \gg 1$,

$$\left.\begin{aligned}\theta_w(x) &= (\lambda)^{-1}\left\{1 - 0.443\,75(\lambda)^{-3/2}x\exp\left(-\frac{1}{\lambda+\varepsilon}\right)x^{-1/2} + \cdots\right\}\\ \phi_w(x) &= 0.443\,75(\lambda)^{-3/2}\exp\left(-\frac{1}{\lambda+\varepsilon}\right)x^{-1/2} + \cdots\end{aligned}\right\} \tag{32}$$

On the other hand, Minto et al. (1998) have solved numerically the two sets of equations corresponding to small and large values of x, using a method proposed by Mahmood and Merkin (1988), starting from $x = 0$. However, at $x = 0$ these equations are singular and, in order to remove this singularity, a new variable $\xi = x^{1/3}$ was used in all the numerical integrations. Figure 6 illustrates the variation of the dimensionless wall temperature $\theta_w(\xi)$ for $\lambda = 0$ (reactant consumption neglected) with $\varepsilon = 0.0$, 0.1, 0.2, and 3.0; here $\phi_w \equiv 1$, as can be seen from the relation (29). This figure clearly shows that, as $\xi \to \infty$, $\theta_w(\xi)$ does not approach constant values but instead increases towards infinite values. It is also observed that all the solutions exhibit a two-phase behavior. The initial reaction phase starts at the ambient temperature on the surface away from the leading edge. Initially there is a slow rate of increase in the wall temperature as we move along the surface and then the wall temperature suddenly starts to rise sharply. This sudden change in $\theta_w(\xi)$ can also be observed in Figure 7, where the dimensionless temperature (or velocity) is plotted as a function of η for $\xi = 0.60$, 0.65, 0.70, 0.71, and 0.72 when $\lambda = \varepsilon = 0$.

Further, we show in Figure 8 the variation of $\theta_w(\xi)$ and $\phi_w(\xi)$ for $\lambda \neq 0$ with $\varepsilon = 0.0$, 0.05, 0.1, and 0.2. Figures 8a and b show that $\theta_w(\xi) \to \lambda^{-1}$ and $\phi_w(\xi) \to 0$ as $\xi \to \infty$. It is also seen that the smaller value of ε, the higher is

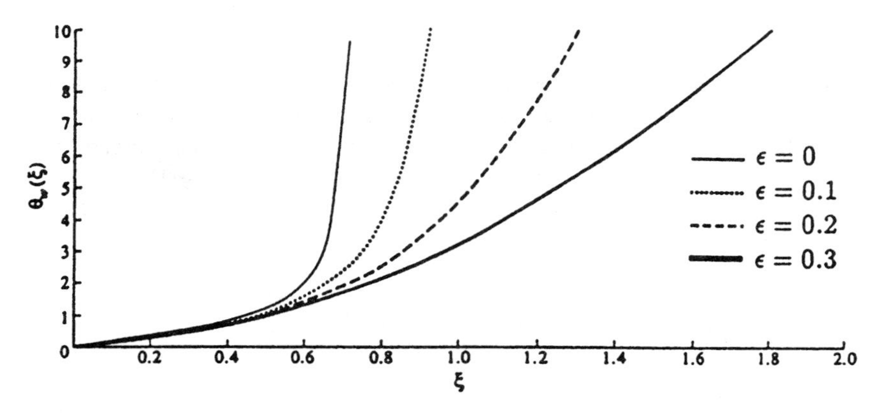

Figure 6. Variation of $\theta_w(\xi)$ with $\xi = x^{1/3}$ for $\lambda = 0$.

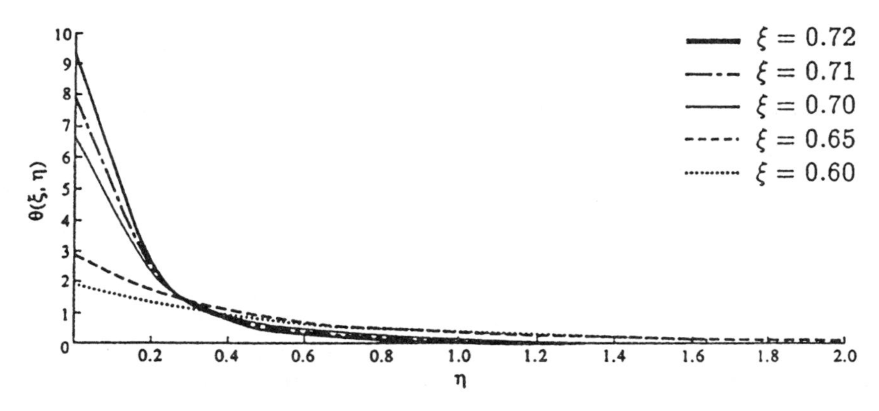

Figure 7. Temperature (or velocity) profiles for $\lambda = 0$ and $\epsilon = 0$.

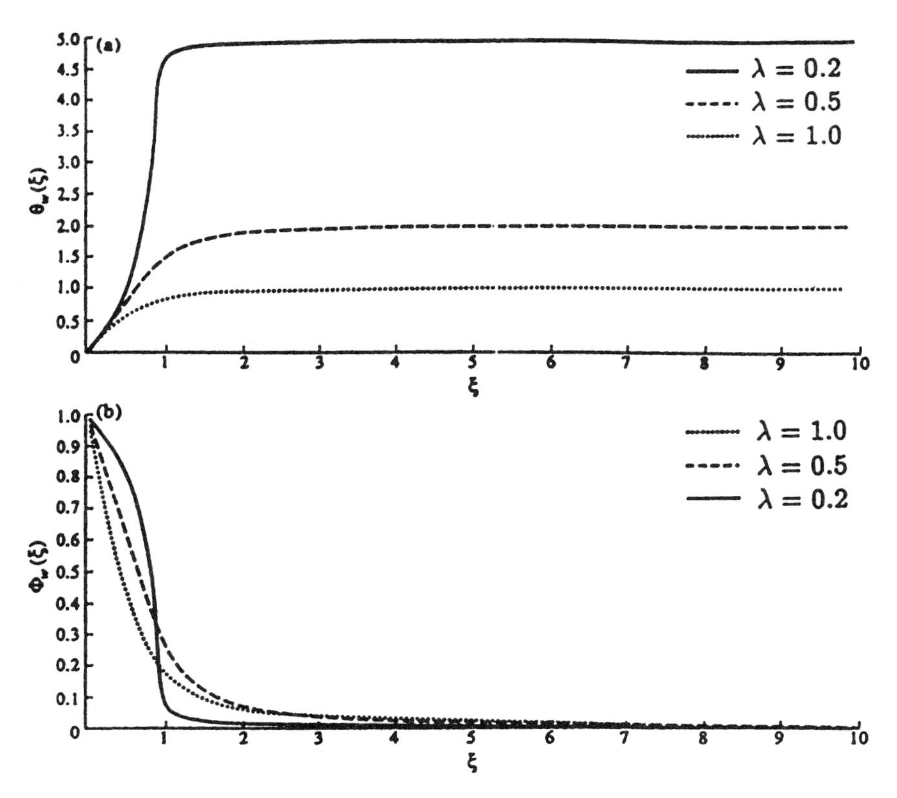

Figure 8. Variation of (a) $\theta_w(\xi)$ and (b) $\phi_w(\xi)$ with $\xi = x^{1/3}$ for $\epsilon = 0$.

the rate of increase or decrease of $\theta_w(\xi)$ and $\phi_w(\xi)$ after the initial phase of the reaction at low temperatures.

Finally, Figures 9 and 10 compare the numerical and asymptotic solutions in both the cases $\lambda = 0$ and $\lambda \neq 0$ when $\varepsilon \neq 0$. For this comparison a logarithmic scaling has been used to scale the x values, thereby producing a clearer picture of the behavior of the solution. Without going into further details, we should note that both the numerical and asymptotic solutions are in very good agreement.

Two very recent papers by Mahmood and Merkin (1998) and Merkin and Mahmood (1998) reported further results concerning the free convection boundary-layer flow on a vertical surface and near the stagnation point of a two-dimensional cylindrical body which is immersed in a porous medium, where the flow results from the heat released by an exothermic catalytic reaction on the bounding surface. We shall give below some results for the latter problem.

B. Stagnation Point

On noting that in this case $S(x) = x$ and taking $\psi = xf(y)$, Eqs. (27a–c) become

$$f' = \theta, \qquad \theta'' + f\theta' = 0 \tag{33a, b}$$

$$\phi'' + Lef\phi' = 0 \tag{33c}$$

which have to be solved subject to the boundary conditions

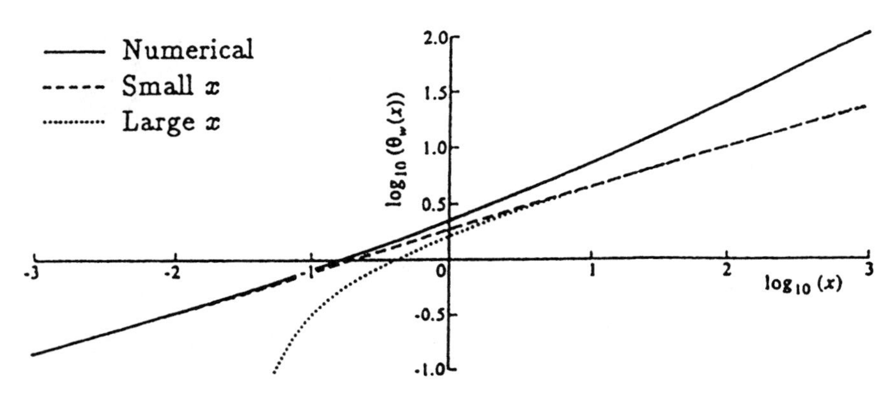

Figure 9. Variation of $\theta_w(x)$ with x for $\lambda = 0$ and $\epsilon = 1$.

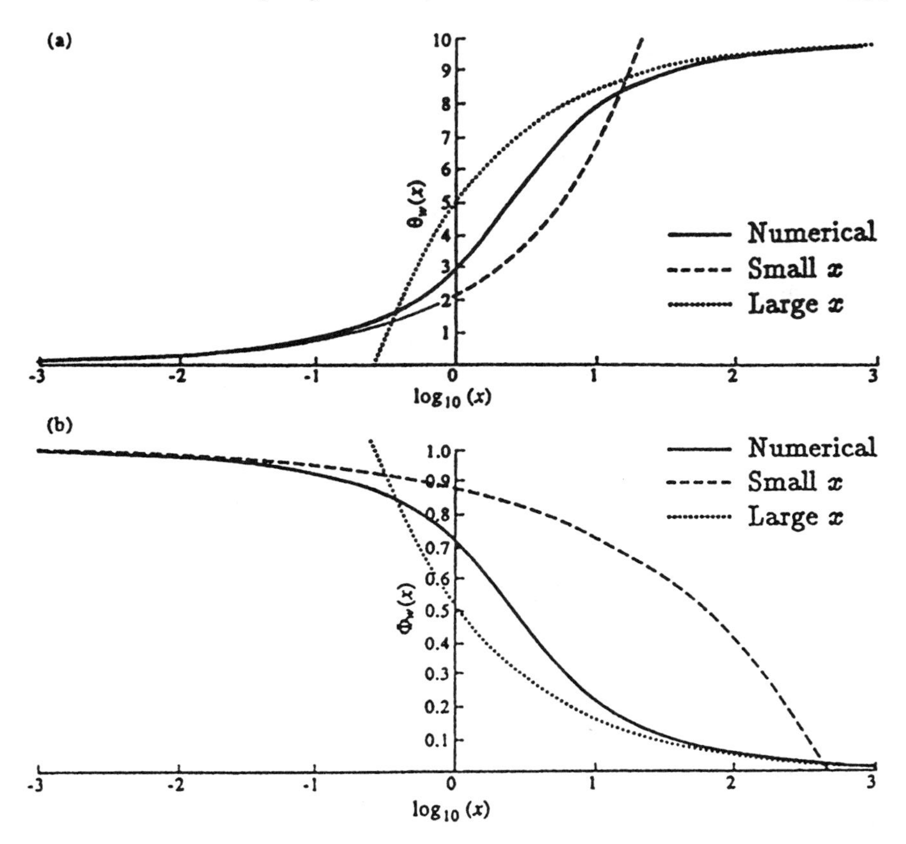

Figure 10. Variation of (a) $\theta_w(x)$ and (b) $\phi_w(x)$ with x for $\lambda = 0.1$ and $\epsilon = 0.2$.

$$\left.\begin{aligned} &f(0) = 0, \qquad \theta'(0) = -\lambda\phi_w \exp\left(\frac{\theta_w}{1+\theta_w}\right) \\ &\phi'(0) = \lambda\delta\phi_w \exp\left(\frac{\theta_2}{1+\varepsilon\theta_w}\right) \\ &f' \to 0, \qquad \theta \to 0, \qquad \phi \to 1 \qquad \text{as} \qquad y \to \infty \end{aligned}\right\} \tag{33d}$$

where δ is a dimensionless parameter defined in Merkin and Mahmood (1998), $\theta(0) = \theta_w$ and $\phi(0) = \phi_w$. If we now take

$$f = \theta^{1/2} M(Y), \qquad \phi = 1 - (1 - \phi_w) N(Y)$$
$$Y = \theta_w^{1/2} y$$

then Eqs. (33) reduce to

$$M''' + MM'' = 0, \qquad N'' + MN'' = 0 \tag{34a}$$

with the boundary conditions

$$\begin{aligned} &M(0) = 0, \qquad M'(0) = 1, \qquad N(0) = 1 \\ &M' \to 0, \qquad N \to 0 \qquad \text{as} \qquad Y \to \infty \end{aligned} \tag{34b}$$

On integrating Eqs. (34) it was found numerically by Merkin and Mahmood (1998) that

$$-M''(0) = C_0 = 0.627\,56$$

Further, on eliminating ϕ_w from the boundary conditions (33d) we find the following equation for θ_w

$$(1 - \alpha\theta_w)\lambda = C_0\theta_w^{3/2} \exp\left(-\frac{\theta_w}{1 + \varepsilon\theta}\right) \tag{35}$$

where $\alpha(Le) = \delta C_0 / C_1(Le)$ with $C_1(Le) = -\phi'(0)$, and expression (35) requires that $\theta_w < 1/\alpha$. On differentiating (35) with respect to θ_w, we find that the critical points (turning points where $d\lambda/d\theta_w = 0$) satisfy the equation

$$\alpha\varepsilon^2\theta_w^3 + (2\alpha\varepsilon - 3\varepsilon^2 - 2\alpha)\theta_w^2 + (\alpha + 2 - 6\varepsilon)\theta_w - 3 = 0 \tag{36}$$

On putting $\varepsilon = 0$ into this equation, we find that there are two critical points at

$$\theta_w^{(1,2)} = \frac{\alpha + 2 \pm \sqrt{\alpha^2 - 20\alpha + 4}}{4\alpha}, \qquad \alpha \neq 0$$

An important point to note from this expression is that there is a hysteresis phenomenon (coincident critical points) where $\alpha = 1 - 4\sqrt{6} = 0.2020$. To determine where there is a hysteresis phenomenon it is necessary to solve Eq. (36) simultaneously with the equation

$$3\alpha\varepsilon^2\theta_w^2 + 2(2\alpha\varepsilon - 3\varepsilon^2 - 2\alpha)\theta_w + \alpha + 2 - 6\varepsilon = 0 \tag{37}$$

Equations (36) and (37) were solved numerically by Merkin and Mahmood (1998) and the results are shown in Figure 11. For values of $\varepsilon > 0$ and α below the curve shown in Figure 11a, multiple solutions exist in the region between the upper $\theta_w^{(2)}$ and lower $\theta_w^{(1)}$ critical points. These are illustrated in Figure 11b for $\varepsilon = 0.02$ and, for this value of ε, $\theta_w^{(1)} = 1.5974$ and $\theta_w^{(2)} = 1565$.

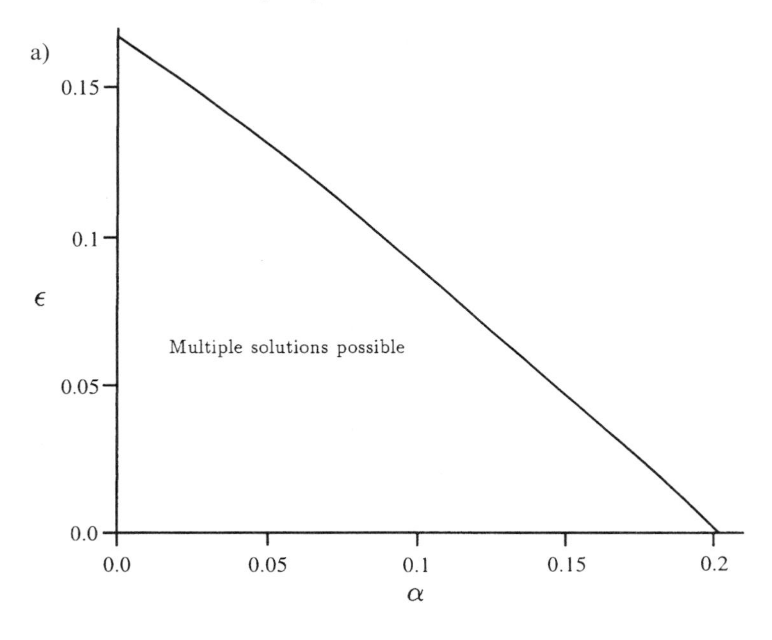

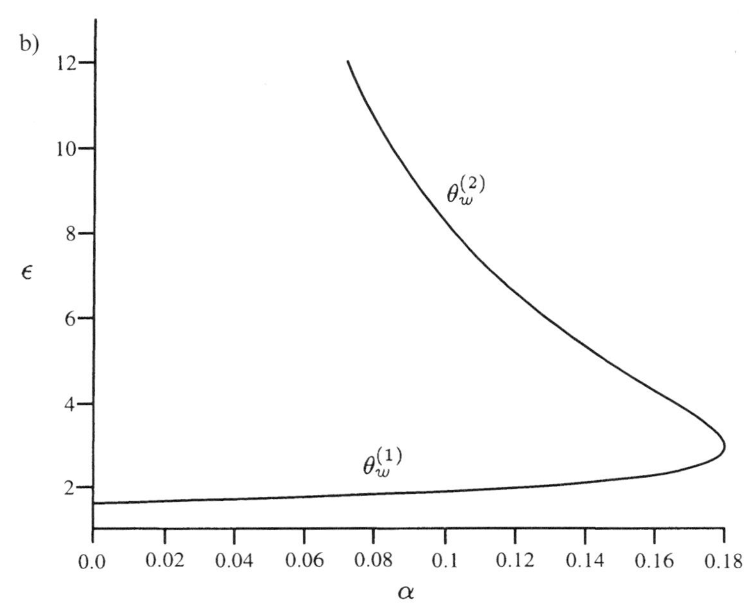

Figure 11. (a) Solution of Eqs. (36) in the (α, ϵ) plane; (b) graphs of $\theta_w^{(1)}$ and $\theta_w^{(2)}$ for $\epsilon = 0.02$.

V. FREE CONVECTION ON SURFACES USING A THERMAL NONEQUILIBRIUM MODEL

The subject of convective flow in a fluid-saturated porous medium when the solid and fluid phases are not in thermal equilibrium has its origin in two papers by Combarnous (1972) and Combarnous and Bories (1974) on the Darcy–Bénard problem. The review articles by Kuznetsov (1998) and Vafai and Amiri (1998) give very detailed information about the research on thermal nonequilibrium effects of the fluid flow through a porous packed bed. However, it appears that the problem of free convection on a body which is embedded in a fluid-saturated porous medium, using a two-temperature model, has only very recently been investigated. Rees (1998a,b) and Rees and Pop (1999, 2000) have studied the effect of adopting this model to the problem of the free convection boundary-layer from a vertical surface and near the stagnation point of a two-dimensional cylindrical body, both of which are embedded in a porous medium. We now report on some results of the two recent papers by Rees and Pop (1999, 2000).

A. Vertical Surfaces

The mathematical formulation of the problem of steady free convection boundary-layer flow along a vertical surface which is embedded in a porous medium consists of the following equations, as given by Eqs. (3) (Rees and Pop 2000)

$$\frac{\partial^2 \psi}{\partial y^2} = \frac{\partial T_f}{\partial y} \tag{38a}$$

$$\frac{\partial^2 T_f}{\partial y^2} = H(T_f - T_s) + \frac{\partial \psi}{\partial y}\frac{\partial T_f}{\partial x} - \frac{\partial \psi}{\partial x}\frac{\partial T_s}{\partial y} \tag{38b}$$

$$\frac{\partial^2 T_s}{\partial y^2} = H\gamma(T_s - T_f) \tag{38c}$$

subject to the boundary conditions

$$\left.\begin{array}{llllll} \psi = 0, & T_f = 1, & T_s = 1 & \text{at} & y = 0, & x \geq 0 \\ \dfrac{\partial \psi}{\partial y} \to 0, & T_f \to 0, & T_s \to 0 & \text{as} & y \to \infty, & x \geq 0 \end{array}\right\} \tag{38d}$$

To solve these equations, the usual boundary-layer transformation (5) with $m = 0$ can be used, namely

$$\psi = x^{1/2} f(x, \eta), \qquad T_f = \theta(x, \eta)$$
$$T_s = \phi(x, \eta), \qquad \eta = y/x^{1/2} \tag{39}$$

Applying the transformation (39) to Eqs. (38) and integrating Eq. (38a) once, we obtain

$$\frac{\partial f}{\partial \eta} = \theta, \qquad \frac{\partial^2 \phi}{\partial \eta^2} + H\gamma x(\phi - \theta) \tag{40a, b}$$

$$\frac{\partial^2 \theta}{\partial \eta^2} + \frac{1}{2} f \frac{\partial \theta}{\partial \eta} = Hx(\theta - \phi) + x\left(\frac{\partial f}{\partial \eta}\frac{\partial \theta}{\partial x} - \frac{\partial f}{\partial x}\frac{\partial \theta}{\partial \eta}\right) \tag{40c}$$

together with the boundary conditions

$$\left.\begin{array}{lllll} f = 0, & \theta = 1, & \phi = 1 & \text{at} & \eta = 0 \\ \theta \to 1, & \phi \to 0 & \text{as} & \eta \to \infty & \end{array}\right\} \tag{40d}$$

These equations form a system of parabolic equations whose solution is nonsimilar because of the x-dependent forcing induced by the terms proportional to H.

Normally, such a nonsimilar set of equations (40) is solved using a marching finite-difference scheme such as, for example, the Keller-box method (Cebeci and Bradshaw, 1984). Beginning at the leading edge, where the system reduces to an ordinary differential system, the solution at each streamwise station is obtained in turn at increasing distances from the leading edge. However, such solutions are typically supplemented by a series expansion for small values of x and by an asymptotic analysis for large values of x. The former often reveals no further information except perhaps a validation of the numerical scheme, but the latter often yields insights that may not be immediately obvious from the numerical solution. Rees and Pop (2000) have shown that the present problem is not of this general nature. However, when $x = 0$, Eq. (40b) for ϕ cannot be solved with both the boundary conditions (40d) satisfied. Therefore, this boundary layer has the rather unusual property of having a double-layer structure near the leading edge (small x), rather than far from it as is often the case encountered in other situations. This considerably complicates the numerical integration of Eqs. (40) since it is now essential to derive the small x solution very carefully before commencing the numerical integration of these equations. Rees and Pop (2000) have obtained the solution of Eqs. (40) in the form of series which are valid for small x by the use of the method of matched asymptotic expansions along with asymptotic matching between an inner and an outer layer. It was found that near the leading edge (small x) the temperature field has a two-layer structure, with the temperature profile of the solid phase appearing strongly in the outer layer. The fluid temperature profile appears only at $\mathbf{O}(x^{1/2})$ in the outer layer, being confined mainly within the inner layer. After a very careful analysis, Rees and Pop (2000) found that the rates of heat transfer for the two phases are given by

$$\frac{\partial \theta}{\partial \eta}(x,0) \sim b - \frac{\sqrt{H\gamma}}{7.060\,66}\left(\frac{a-\sqrt{a^2+8/\gamma}}{2}\right)x^{1/2} + \cdots$$
$$\frac{\partial \phi}{\partial \eta}(x,0) \sim -\sqrt{H\gamma}x^{1/2} + \cdots \tag{41a}$$

as $x \to 0$, where $a = 1.616\,13$ and $b = -0.443\,748$.

On the other hand, Rees and Pop (2000) have obtained an asymptotic solution for large x of Eqs. (40) in series of x^{-1}. They showed that θ and ϕ become almost identical as x increases and therefore it is quite possible that the difference between them is $\mathbf{O}(x^{-1})$ for large x, with no thin sublayer. The same is true for the solid and fluid rates of heat transfer, where

$$\frac{\partial \theta}{\partial \eta}(x,0) \sim \left(1+\frac{1}{\gamma}\right)^{-1/2}\left[b + 0.043\,689x^{-1}\ln x + \mathbf{O}(x^{-1})\right] \tag{41b}$$

as $x \to \infty$ and the term $\ln x$ was included because of the leading edge shift effect.

Having determined the correct asymptotic solutions (41) for both small and large values of x, Rees and Pop (2000) then solved Eqs. (40) numerically using the Keller-box method for $H = 1$ and various values of γ. The solutions obtained are summarized in Figures 12 and 13. Figure 12 displays the isotherms for the fluid (full lines) and solid (broken lines) phases in the coordinates (x, y) for $\gamma = 0.1$, 0.3, 1.0, 3.0, and 10.0. It is clearly seen from these figures that a state of local thermal equilibrium, indicated by the isotherms for the two phases being virtually coincident, is reached relatively close to the leading edge when γ is large. A large value of γ corresponds to the fluid having a high thermal conductivity relative to the solid, thereby allowing the fluid properties to dominate the development of the boundary-layer flow. Furthermore, it is important to note that the solid phase isotherms do not terminate at some point on the y-axis at $x = 0$. A full treatment of the flow, for which these solid phase isotherms would be closed or join onto an insulated surface, would necessarily entail a solution of the full elliptic equations at points within an $\mathbf{O}(Ra^{-1})$ distance of the origin. Figure 13 shows the variation of the rates of surface heat transfer for both the solid and fluid phases for some values of γ. The results are presented in terms of

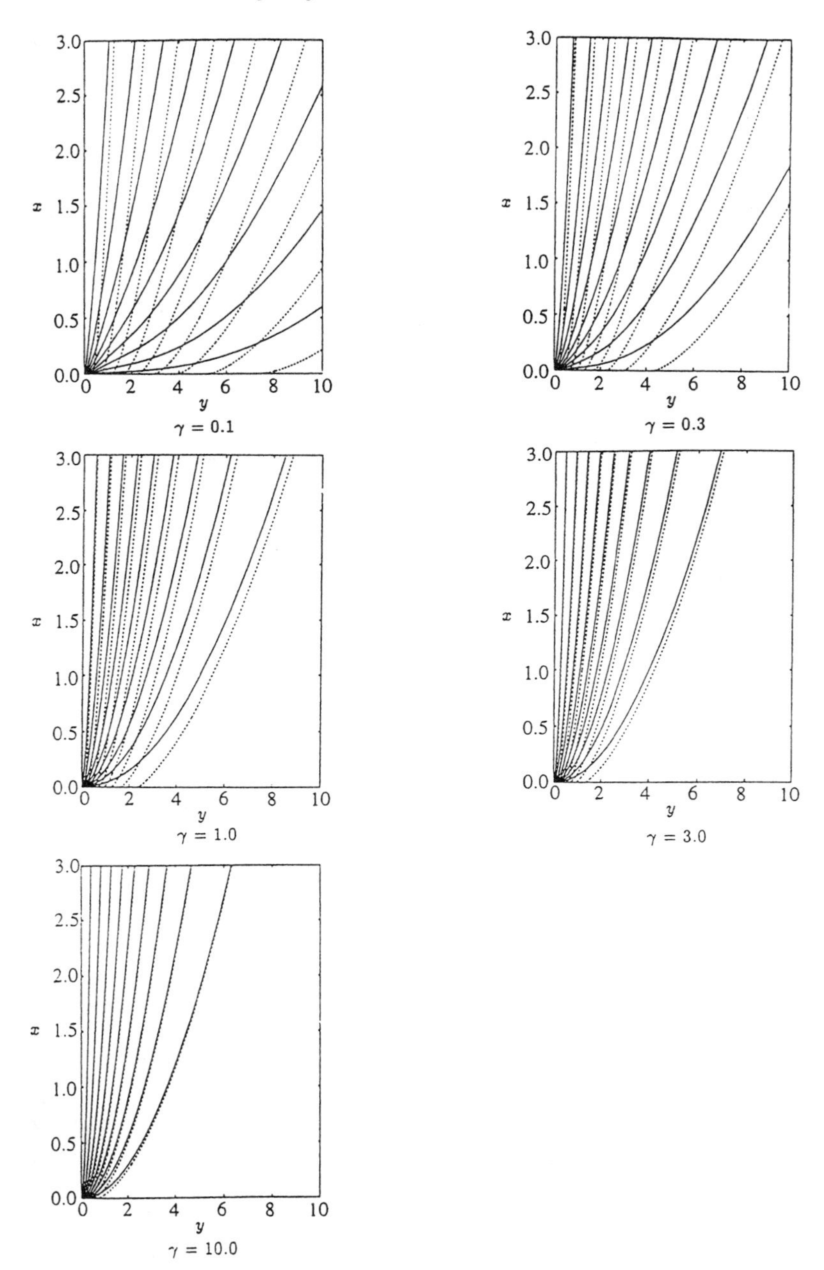

Figure 12. Isotherms for both the fluid (shown by full lines) and solid (shown by broken lines) phases.

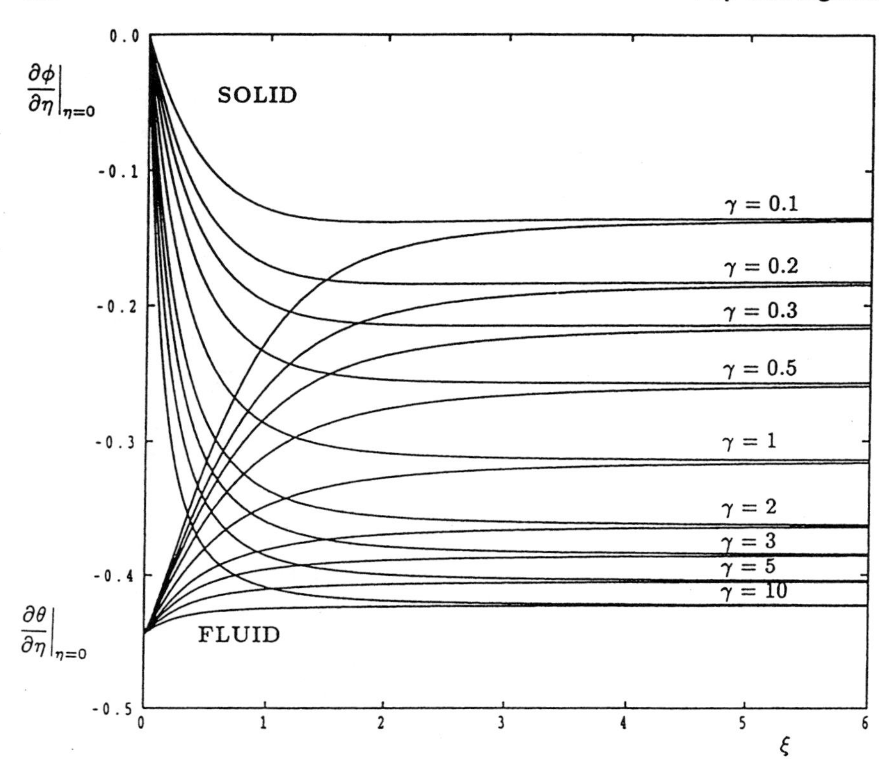

Figure 13. Variation of the fluid and solid surface rates of heat transfer with $\xi = x^{1/2}$.

$\xi = x^{1/2}$ in order to overcome the singularity in Eqs. (40) at $x = 0$. It can be seen from this figure that the solid rate of heat transfer increases in magnitude from zero until it becomes the same as that of the fluid phase. However, the fluid rate of heat transfer decreases in magnitude as ξ increases, varying from b at $\xi = 0$ to $b(1 + 1/\gamma)^{-1/2}$ as $\xi \to \infty$.

To this end, we draw the reader's attention to a very good paper by Rees (1999) in which he has analysed the effect of layering on the flow and the rate of heat transfer from a uniformly heated vertical surface in a porous medium. The sublayers comprising the porous medium are aligned such that the interfaces are parallel with the surface. The reported results indicate that it is highly likely that such a flow is stable and therefore will occur in practice.

B. Stagnation Point

The problem of steady free convection boundary-layer flow near the stagnation point of a two-dimensional body which is embedded in a porous medium, adopting a two-temperature model, has also been considered by Rees and Pop (1999). The basic model is the same as for the vertical surface described in the previous subsection. However, in this case $S(x) = x$ and we take $\psi = xf(y)$, $\theta = \theta(y)$, and $\phi = \phi(y)$. Then, Eqs. (2) transform into the following ordinary differential equations by using the fact that $f' = \theta$

$$f''' + ff'' = H(f' - \phi) \tag{42a}$$

$$\phi'' = H\gamma(\phi - f') \tag{42b}$$

which must be solved subject to the boundary conditions

$$\left.\begin{array}{llll} f(0) = 0, & f'(0) = 1, & \phi(0) = 1 & \\ f' \to 0 & \phi \to 0 & \text{as} & y \to \infty \end{array}\right\} \tag{42c}$$

The numerical results for the surface heat transfer rates of the fluid, $\theta'(0)$, and solid, $\phi'(0)$, phases obtained by solving Eqs. (42) are shown in Figure 14, where the parameters H and γ vary over the ranges $0.001 \leq H \leq 1$ and $0.1 \leq \gamma \leq 10$, respectively. It should be noted from this figure that when H is small there is a very substantial difference between the rates of heat transfer of the fluid and solid phases, indicating that nonequilibrium effects are strongest when H is small. The numerical solution also indicates that the thickness of the solid phase temperature field increases as H decreases, which is consistent with the decreasing rate of heat transfer in the limit. Furthermore, it can be seen from Figure 14 that the same qualitative effects are obtained as the parameter γ decreases. This confirms that the two-temperature model equations (42) can predict natural convective flow in porous media very accurately and efficiently.

In future studies it would be interesting to apply the model described in this section to other flow configurations, where nonequilibrium effects may prove to be of considerable importance.

VI. HORIZONTAL SURFACES

The buoyancy-induced flows in a fluid-saturated porous medium adjacent to horizontal surfaces also play an important practical role in heat transfer engineering. The buoyancy force in this situation is acting perpendicular to the direction of motion, which might create some mathematical difficulties. The first boundary-layer paper dealing with this problem appears to be

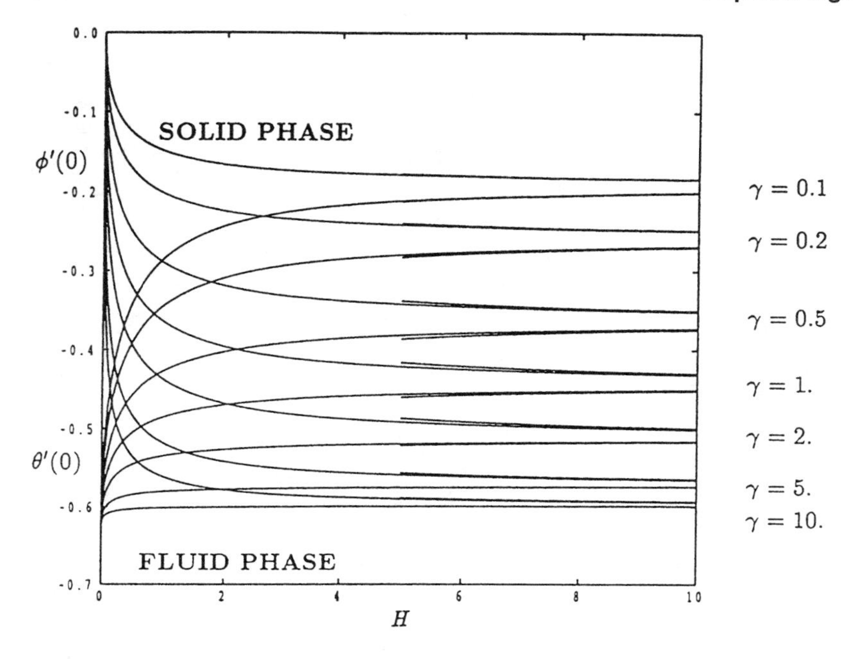

Figure 14. Variation of the fluid and solid surface rates of heat transfer with H.

that of McNabb (1965), who studied free convection in a saturated porous medium about a horizontal heated circular impermeable surface with wall temperature being a step function with respect to the radius.

In what follows, we review some of the recent work on free and mixed convection boundary-layer flow over horizontal surfaces in porous media.

A. Free Convection Case

Using appropriate scalings and assuming that the boundary-layer approximations hold, Eqs. (1) can be easily reduced to those which describe free convection flow on a horizontal surface in a porous medium (Chang and Cheng 1983). In nondimensional form these equations are as follows

$$\frac{\partial^2 \psi}{\partial y^2} = \mp \frac{\partial T}{\partial x} \tag{43a}$$

$$\frac{\partial^2 T}{\partial y^2} = \frac{\partial \psi}{\partial y}\frac{\partial T}{\partial x} - \frac{\partial \psi}{\partial x}\frac{\partial T}{\partial y} \tag{43b}$$

where the minus sign in Eq. (43a) corresponds to the boundary layer above a heated (or below a cooled) surface, while the plus sign is for the flow below a heated (or above a cooled) surface. Cheng and Chang (1976) were the first to have reduced Eq. (43) to a similarity form when the wall temperature is proportional to x^m. In this case, the boundary conditions for Eq. (43) are as follows

$$\begin{array}{llll} \psi = 0, & T = x^m & \text{as} \quad y = 0, & x \geq 0 \\ \dfrac{\partial \psi}{\partial y} \to 0, & T \to 0 & \text{as} \quad y \to \infty, & x \geq 0. \end{array} \tag{43c}$$

Equations (43) may be reduced to similarity form if we apply the transformation

$$\psi = x^{(m+1)/3} f(\eta), \qquad T = x^m h(\eta)$$
$$\eta = y x^{(m-2)/3}$$

where f and h are given by the following ordinary differential equations

$$f'' + mh + \frac{m-2}{3} \eta h' = 0 \tag{44a}$$

$$h'' + \frac{m+1}{3} f h' - m f' h = 0 \tag{44b}$$

which have to be solved subject to the boundary conditions

$$f(0) = 0, \qquad h(0) = 1, \qquad f'(\infty) = 0, \qquad h(\infty) = 0 \tag{44c}$$

The numerical solution of Eq. (44) has been performed by Cheng and Chang (1976) for $0.5 < m < 2$. These results have been confirmed by Merkin and Zhang (1990) and further calculations were performed by these authors for $-0.4 < m < \infty$. However, it was shown by Merkin and Zhang (1990) that there are no solutions of Eq. (44) for $m \leq -0.4$. This may be proved as follows. Suppose that there exists a solution of Eqs. (44) with $f'(0)$ bounded. Multiplying Eq. (44a) by h and then integrating from $\eta = 0$ to ∞ gives, after some manipulations,

$$f'(0) = \frac{5m+2}{6} \left(\int_0^\infty h^2 \mathrm{d}\eta + 2 \int_0^\infty f f' h \mathrm{d}\eta \right) \tag{45a}$$

On the other hand, by integrating Eq. (44a) directly, we obtain

$$f'(0) = -\frac{2(m+1)}{3} \int_0^\infty h \mathrm{d}\eta \tag{45b}$$

Since we require all the quantities in the integrands in (45) to be positive, we come to a contradiction when $m < -0.4$. Therefore, a solution of this problem is possible only for $m > -0.4$, with the solution becoming singular (unbounded) as $m \to -0.4$. Merkin and Zhang (1990) have made a detailed investigation of the nature of the solution near this singularity and found that

$$f'(0) = 0.135\,595(m + 0.4)^{-2/3} + \cdots \tag{46a}$$

$$h'(0) = 0.013\,961(m + 0.4)^{-4.3} + \cdots \tag{46b}$$

as $m \to -0.4$, where $f'(0)$ and $h'(0)$ are related to the wall velocity $(\partial\psi/\partial y)_w = x^{(2m-1)/3} f'(0)$ and the heat flux at the plate $(\partial T/\partial y)_w = x^{(4m-2)/3} h(0)$. Figures of $f'(0)$ and $h'(0)$, obtained by a direct numerical integration of Eqs. (44) for some values of m close to $m = -0.4$, are shown by the full lines in Figure 15. Also shown here by broken lines are the asymptotic expressions (46). These figures provide a clear confirmation of the asymptotic theory proposed by Merkin and Zhang (1990) for m close to -0.4 where the solution of Eq. (44) is singular.

It is worth mentioning that these authors have rigorously demonstrated that a boundary-layer solution of this problem does not exist for free convection flow below a heated (or above a cooled) isothermal horizontal plate in a porous medium. The case of a permeable horizontal surface in a porous medium has also been considered by Chaudhary et al. (1996), while Weidman and Amberg (1996) have studied the existence of similarity solutions for an inclined permeable surface in a porous medium.

Lesnic and Pop (1998a) have recently studied a problem of a different nature to that considered above; namely, they have assumed that the wall temperature is suddenly changed from that of the ambient fluid to a quadratic form in x. The governing equations can again be derived easily from Eqs. (1) and may be expressed in nondimensional form as (Lesnic and Pop 1998a)

$$\frac{\partial u}{\partial x} + \frac{\partial v}{\partial y} = 0 \tag{47a}$$

$$\frac{\partial u}{\partial y} = -\frac{\partial \theta}{\partial x} \tag{47b}$$

$$\frac{\partial \theta}{\partial t} + u\frac{\partial \theta}{\partial x} + v\frac{\partial \theta}{\partial y} = \frac{\partial^2 \theta}{\partial y^2} \tag{47c}$$

together with the boundary conditions

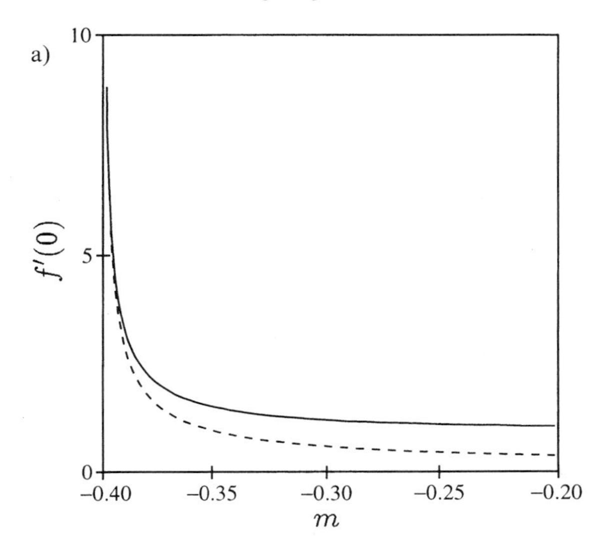

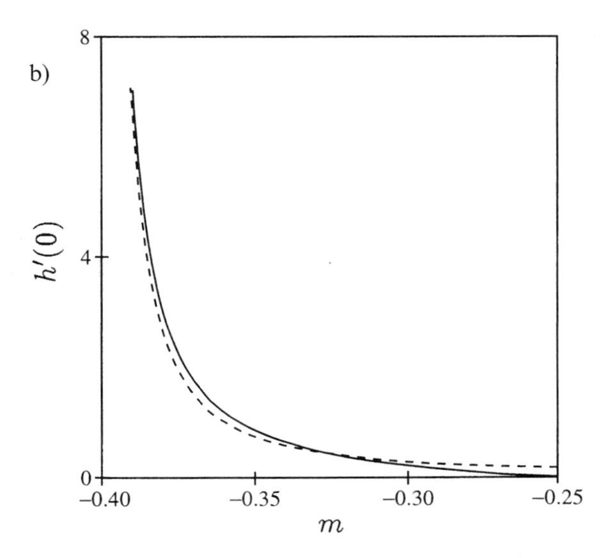

Figure 15. Variation of (a) $f'(0)$ and (b) $h'(0)$ with m obtained from a numerical integration of Eqs. (44) (shown by the full line) and form the asymptotic solution (46) (shown by the broken line).

$$\left.\begin{array}{llllll} t \leq 0: & u = v = \theta & \text{for} & x, y > 0 & & \\ t > 0: & v = 0, & \theta = \theta_w(x) & \text{at} & y = 0, & x > 0 \\ u \to 0, & \theta \to 0 & \text{as} & y \to \infty, & x > 0 & \end{array}\right\} \tag{47d}$$

It is assumed that $\theta_w(x)$ may take one of the following forms

$$\theta_w(x) = a_0 + a_1 x^2 \tag{48a}$$

$$\theta_w(x) = a_0 + 1 - \cos x \tag{48b}$$

where a_0 and a_1 are as yet unspecified constants. Guided by expression (48a), an exact solution of Eqs. (47) is sought in the form

$$u = x u_0(y, t), \qquad v = v_0(y, t)$$

$$\theta = \theta_0(y, t) + x^2 \theta_1(y, t)$$

where u_0, v_0, θ_0, and θ_1 satisfy the following equations, which are obtained from Eqs. (47a–c)

$$u_0 + \frac{\partial v_0}{\partial y} = 0, \qquad \frac{\partial u_0}{\partial y} = -2\theta_1 \tag{49a, b}$$

$$\frac{\partial \theta_0}{\partial t} + v_0 \frac{\partial \theta_0}{\partial y} = \frac{\partial^2 \theta_0}{\partial y^2} \tag{49c}$$

$$\frac{\partial \theta_1}{\partial t} + 2u_0\theta_1 + v_0 \frac{\partial \theta_1}{\partial y} = \frac{\partial^2 \theta_1}{\partial y^2} \tag{49d}$$

and have to be solved subject to the boundary conditions

$$\begin{array}{lllll} t \leq 0: & u_0 = v_0 = \theta_i = 0 & \text{for} & y > 0 & \\ t > 0: & v_0 = 0, & \theta_i = b_i, & \text{at} & y = 0 \\ u_0, \theta_i \to 0 & \text{as} & y \to \infty & & \end{array} \tag{49e}$$

where $i = 0, 1$. It was shown by Lesnic and Pop (1998a) that a steady-state solution of Eq. (49) can be determined for $a_0 = 1$ and $a_1 = 1/2$. Equations (49) were solved numerically for a range of values of t when $a_0 = 1$ and $a_1 = \pm 1/2$ and the steady-state solution was recovered when $a_1 = 1/2$. However, the situation is quite different when $a_1 < 0$. Namely, Eqs. (49) do not admit a steady-state solution and when $a_1 = -1/2$ and the time $t_s = 1.33$ a singular (unbounded) solution becomes apparent; this behavior is illustrated in Figure 16. The steady free convection flow case given by the periodic nature of $\theta_w(x)$, expressed in Eq. (48b), has been also investigated by Lesnic and

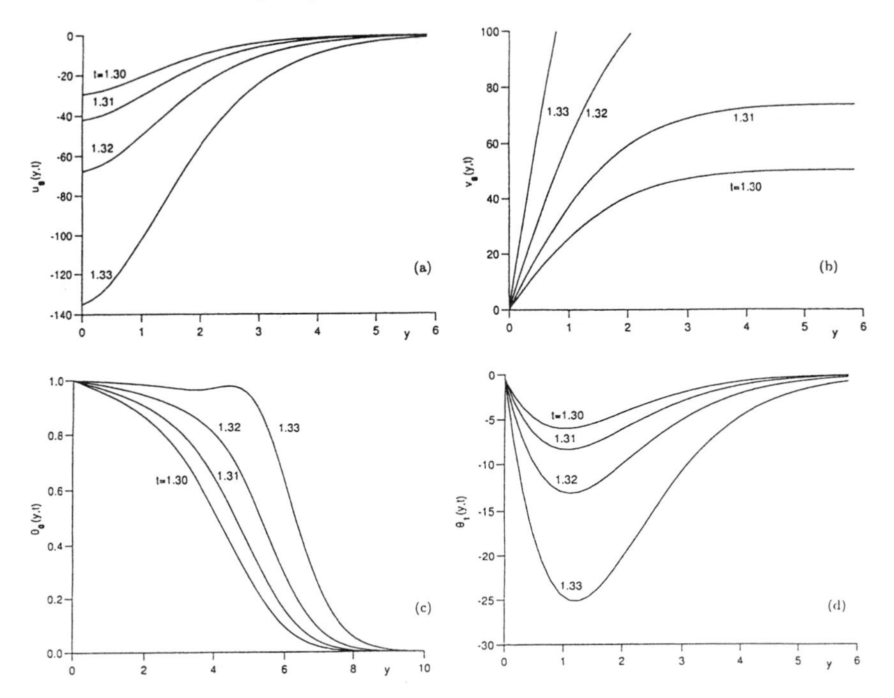

Figure 16. Velocity (a,b) and temperature (c,d) profiles obtained from a numerical integration of Eqs. (49) near the singularity times $t_s = 1.33$ for $a_0 = 1$ and $a_1 = -1/2$.

Pop (1998a). In this situation $\theta_w(x)$ is symmetrical about $x = 0$ and therefore it is sufficient to consider only the flow in the range $0 \leq x \leq \pi$. Expression (48b) shows that the flow starts at $x = 0$, close to which $\theta_w(x) \sim a_0 + \frac{1}{2}x^2$, as in the case when $a_1 = 1/2$ which was considered above, rather than at $x = \pi$, close to which

$$\theta_w(x) \sim a_0 + 2 - \frac{1}{2}(\pi - x)^2$$

In fact, this situation corresponds to the case $a_1 = -1/2$ (considered earlier) for which there is possibly no steady-state solution. Figure 17 shows the velocity and temperature profiles obtained from the full numerical solution of Eqs. (47) for values of $x \in \{0, \pi/4, \pi/2, 3\pi/4, \pi\}$. It can be seen from Figures 17 a and b that the velocity profiles $u(x, y)$ increase as x increases from 0 to $3\pi/4$ but decrease as x increases from $3\pi/4$ to π. These profiles are generally higher when $a_0 = 1$ than when $a_0 = 0$, since the initial profile $\theta_0(y)$ at $x = 0$ is $\theta_0(y) \equiv 0$ when $a_0 = 0$, and this is shown as a full line in Figure 3

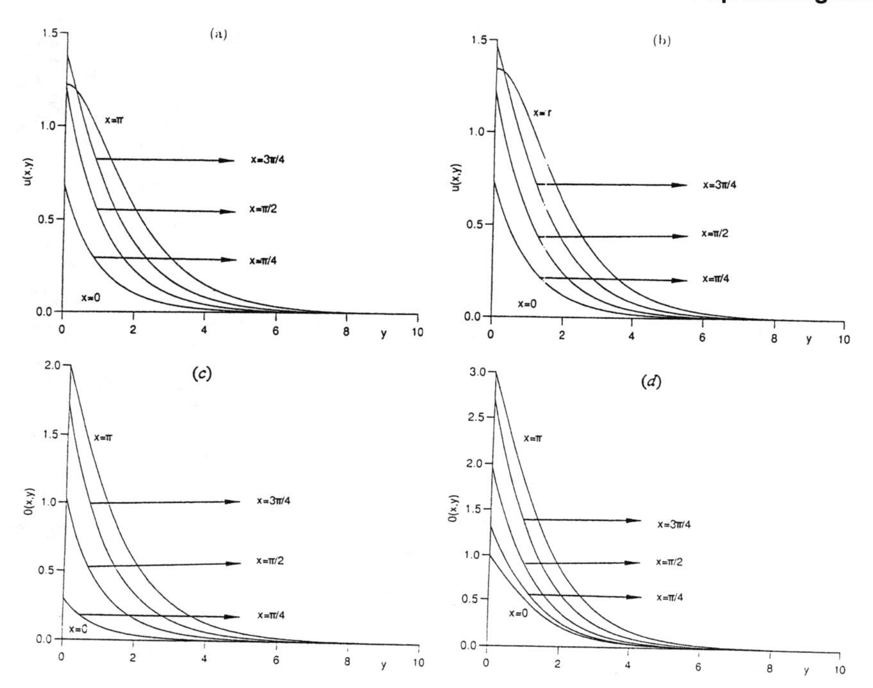

Figure 17. Velocity (a,b) and temperature (c,d) profiles obtained from a numerical integration of Eqs. (47) for (a,c) $a_0 = 0$ and (b,d) $a_0 = 1$.

in Lesnic and Pop (1998a). Further, Figures 17 c and d indicate that, as x increases from 0 to π, the temperature profiles $\theta(x, y)$ increase and are convex functions of y. Another important point to be noted from this figure is that $u(x, y)$ possesses an inflexion point when $x \in [3\pi/4, \pi]$ and this raises the question as to whether the solution obtained is a realistic physical situation over the range $x \in [3\pi/4, \pi]$.

We mention to this end the very recent paper by Angirasa and Peterson (1998) on the effect of the Rayleigh number on the free convection flow over a finite horizontal surface in a porous medium.

B. Mixed Convection Case

Mixed convection flows, or combined forced and free convection flows, arise in many transport processes in engineering devices and in nature. Such processes occur when the effect of the buoyancy forces in forced convection or the effect of forced flow in free convection becomes significant. The effect is especially pronounced in situations where the forced-flow velocity is low and the temperature difference is large. In mixed convection flows, the forced convection effects and the free convection effects are of comparable magnitude and there has been a large amount of work devoted to this topic. However, we shall limit our attention in this section to only two recent papers, by Merkin and Pop (1997) and Vynnycky and Pop (1997), in order to identify the most important mixed convection parameters for this horizontal flow configuration. Let us consider the mixed convection flow over a horizontal surface in a porous medium based on a similarity solution of the boundary-layer equations, assuming that the flow is driven by a heat source of constant heat flux Q (nondimensional) and the free stream is flowing in the positive x-direction with a velocity $U_0(x)$. The problem is governed by Eqs. (43a,b), which have to be solved subject to the boundary conditions

$$\left.\begin{array}{lllll} \psi = \dfrac{\partial T}{\partial y} = 0 & \text{at} & y = 0, & x \geq 0 & \\ \dfrac{\partial \psi}{\partial y} \to U_0(x), & T \to 0 & \text{as} & y \to \infty, & x \geq 0 \end{array}\right\} \tag{50a}$$

along with the heat flux integral

$$\int_0^\infty \psi \frac{\partial T}{\partial y} \mathrm{d}y = Q \tag{50b}$$

Equations (43a,b) subject to boundary conditions (50) allow a similarity solution provided $U_0(x)$ has the form

$$U_0(x) = Ux^{-1/2} \tag{51}$$

where U is a constant. It was shown by Merkin and Pop (1997) that the transformation

$$\psi = x^{1/4} f(\eta), \qquad T = x^{1/4}\theta(\eta), \qquad \eta = yx^{-3/4}$$

leads to the ordinary differential equations

$$4f'' - 3\eta\theta' - \theta = 0 \tag{52a}$$

$$4\theta' + f\theta = 0 \tag{52b}$$

which have to be solved subject to the boundary and integral constraints

$$\left.\begin{aligned} &f(0) = \theta'(0) = 0 \\ &f' \to q, \qquad \theta(\eta) \to \infty \qquad \text{as} \qquad \eta \to \infty \\ &\int_0^\infty f'\theta \mathrm{d}\eta = 1 \end{aligned}\right\} \tag{52c}$$

where

$$q = UQ^{-1/2} \tag{53}$$

We note that these equations are equivalent to the equations describing the horizontal wall plume which arises from a line thermal source embedded at the leading edge of an adiabatic horizontal surface in a porous medium (Shu and Pop 1997). Equations (52) were solved numerically by Merkin and Pop (1997) and the results are shown in Figure 18, where $f'(0)$ and $\theta(0)$ have been plotted as a function of q. The main point to note about these numerical results is that a solution exists only for $q \geq q_0$, where $q_0 = 1.406$, and that for $q > q_0$ there are two solutions for each value of q. However, on the upper solution branch $f'(0)$ and $\theta(0)$ behave as (Merkin and Pop 1997)

$$f'(0) \sim q - \frac{1}{2q} + \cdots, \qquad \theta(0) \sim -\frac{1}{\sqrt{2\pi q}} + \cdots \tag{54a, b}$$

as $q \to \infty$. The asymptotic expansions (54) are also shown in Figure 18 by the broken lines, and one can see that there is a good agreement with the numerically determined values. Profiles for the velocity $f'(\eta)$ and the temperature $\theta(\eta)$ fields on the upper and lower branches of the solution of Eqs.

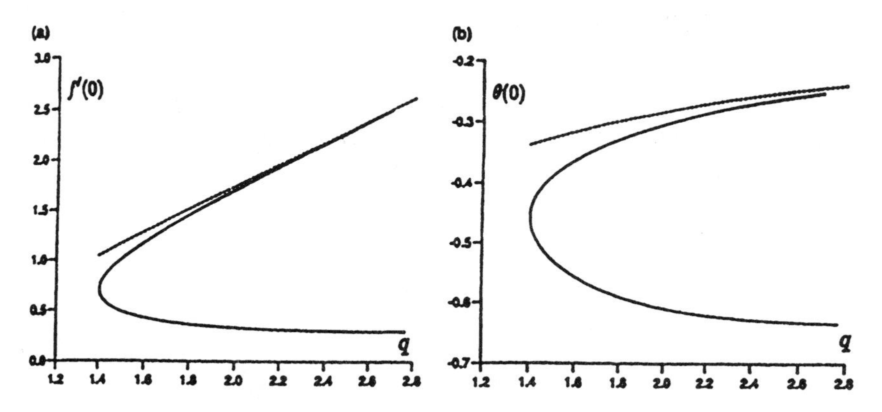

Figure 18. The variation of (a) $f'(0)$ and (b) $\theta(0)$ with q obtained from a numerical integration of Eqs. (52) (shown by the full lines) and from the asymptotic solution (54) (shown by the broken lines).

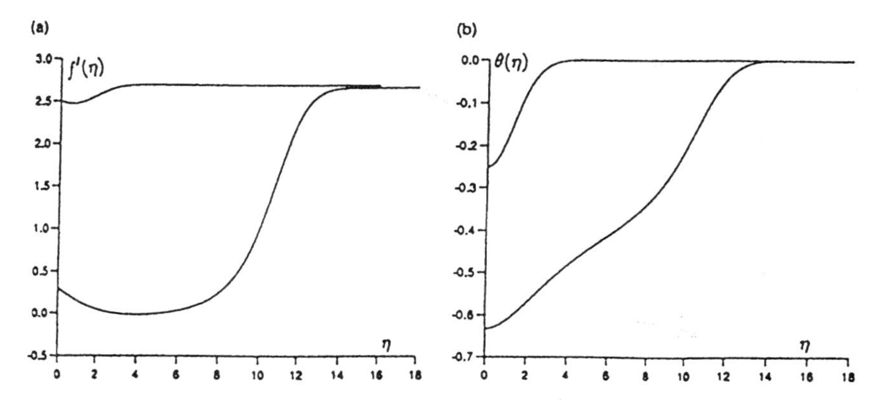

Figure 19. Velocity (a) and temperature (b) profiles on the upper and lower solution branches for $q = 2.7$.

(52) are shown in Figure 19 for the same value of $q = 2.7$. It is seen that there is a drop in the value of $f'(\eta)$ below $f'(0)$ before the asymptotic values are reached, and for the lower branch solutions this is much more pronounced and leads to a finite region of η over which $f'(\eta) < 0$.

Finally, we wish to mention that mixed convection flow over an isothermal finite horizontal surface buried in a fluid-saturated porous medium has been studied very recently by Vynnycky and Pop (1997). It has been found from this detailed analytical and numerical solution that, for high enough values of the Rayleigh number Ra, flow separation may occur for both heating and cooling surfaces. The predicted separation can be observed in Figure 20, where typical streamline and isotherm plots are illustrated for the Péclet number $Pe = 5$ and for some values of Ra/Pe increasing from negative to positive. It is seen that for $Ra/Pe > 0$ a separation bubble appears towards the trailing edge of the plate which reattaches to the x-axis further downstream beyond the trailing edge. However, separation in porous media mixed convection has been previously reported by Pop et al. (1995), albeit for a different flow configuration.

VII. CONCLUSIONS

In this chapter we have presented an exhaustive theoretical foundation which covers the recent developments in convective boundary-layer flow in porous media. The governing equations were presented, and we have shown how they can be transformed into a usable form by arguments

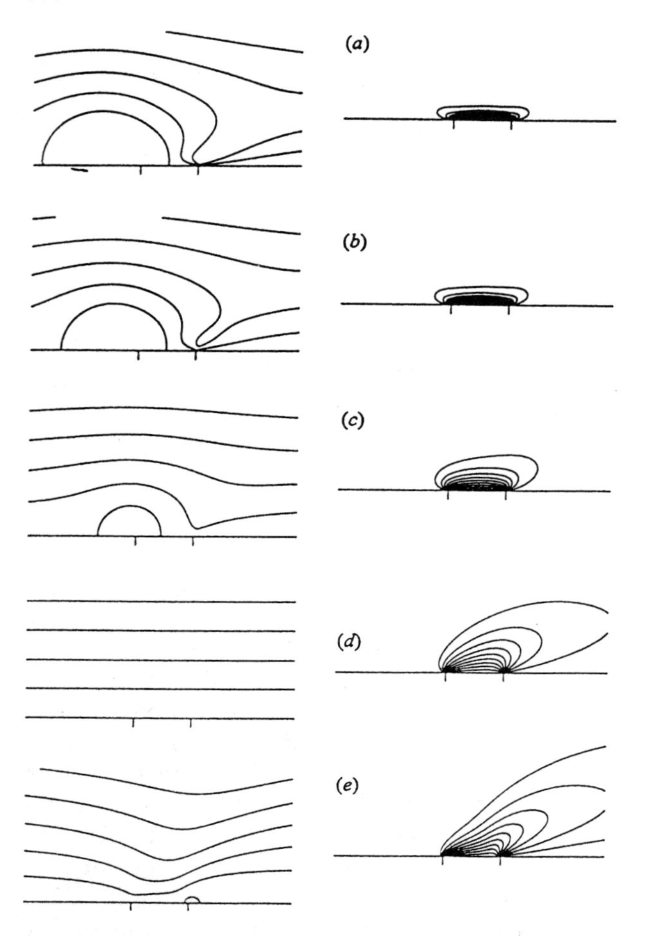

Figure 20. Streamlines, $0 \leq \psi \leq 5$ ($\Delta\psi = 0.5$) (left-hand plots), and isotherms, $0 \leq \theta \leq 1$ ($\Delta\theta = 0.1$) (right-hand plots) for $Pe = 5$: (a) $Ra/Pe = -200$; (b) $Ra/Pe = -100$; (c) $Ra/Pe = -50$; (d) $Ra/Pe = 0$; (e) $Ra/Pe = 10$.

derived from the physical aspects of the problem. Special attention has been given to free and mixed convection flows from vertical and horizontal surfaces, and the problems analysed demonstrate the validity, nature, and possible application to some heat transfer devices. However, due to space limitation we were not able to include the correlation expressions which were recently reported by Yu et al. (1991), Aldos et al. (1993a,b), and Lesnic et al. (1999). One of our objectives was also to describe two new theoretical models of convective flow in porous media. These models refer to flows which are driven by chemical reaction surface heating and to a flow which is affected by the use of a nonequilibrium model of microscopic transfer between the fluid and the solid phase of the porous medium.

We believe that the concepts presented in this chapter on convective boundary-layer flow in porous media should instigate new thoughts on some particular problems such as, for example, separation in mixed boundary layers, with particular emphasis on the nature of its asymptotic structure, and also the development of new turbulent models in porous media (Lage, 1998). With such models, one can not only gain a much deeper understanding of convective flows in porous media but also develop some new approaches to enhancement of heat transfer.

ACKNOWLEDGMENTS

The authors gratefully acknowledge the Royal Society for supporting some of the research described in this paper. They also wish to thank Dr Simon Harris for the assistance given during the preparation of this manuscript.

NOMENCLATURE

Roman Letters

a_0, a_1	constants defined in Eq. (48)
A	reactant species
B	product species
c	specific heat
c_p	specific heat at constant pressure
C^*	concentration
C	nondimensional concentration
D_m	solute diffusivity
e	eigenvalue defined in Eq. (20)
E	activation energy
$\mathbf{g}^*$	gravitational acceleration vector
h^*	solid/fluid heat transfer coefficient
H	scaled value of h^* defined in Eq. (3)
k	thermal conductivity
k_0	constant defined in Eq. (26)
K	porosity
l	length scale
Le	Lewis number, $= \varphi\alpha_f/D_m$

m	temperature exponent
p^*	pressure
Pe	Péclet number, $= U_\infty l/\alpha_f$
q	constant defined in Eq. (53)
Q	heat of reaction or rate of heat flux defined in Eq. (50b)
R	universal gas constant
Ra	Rayleigh number, $= \rho_\infty g K \beta_t \Delta T l/(\varphi \mu \alpha_f)$
$S(x)$	sine function
t^*	time
t	nondimensional time, $= (\alpha_f/l^2) Ra^{2/3} t^*$
t_s	time at which Eqs. (47) are singular
T^*	temperature
T	nondimensional temperature, $= (T^* - T_\infty)/\Delta T$
T_R	temperature scaling, $= RT_\infty^2/E$
ΔT	reference temperature, $= T_w - T_\infty$
u^*, v^*	velocity components along x^*, y^* axes, respectively
u, v	nondimensional boundary-layer velocity components along x, y axes, respectively
$U_0(x)$	nondimensional free stream velocity defined in Eq. (51)
U	constant in Eq. (51)
$\mathbf{V}^*$	velocity vector
x^*, y^*	Cartesian coordinates along and normal to the surface, respectively
x, y	nondimensional boundary-layer coordinates

Greek Letters

α	constant in Eq. (35)
α_f	thermal diffusivity, $= \varphi k_f/(\rho c)_f$
β	parameter defined in Eq. (14)
β_t	thermal expansion parameter
γ	parameter defined in Eq. (3)
δ	dimensionless parameter in Eq. (33d) representing the amount of heat released by the reaction
ε	activation energy parameter defined in Eq. (28)
η	independent variable
θ	nondimensional temperature function, $= (T - T_\infty)/\Delta T$ or $(T - T_\infty)/T_R$
λ	reactant consumption parameter defined in Eq. (28)
μ	fluid viscosity
ξ	reduced variable for x
ρ	density
σ	suction or injection parameter
φ	porosity
ϕ	nondimensional concentration or temperature of the solid phase, $= C^*/C_\infty$ or $(T_s^* - T_\infty)/\Delta T$
ψ	nondimensional stream function, $= \psi^*/(\varphi \alpha_f)$

Other Symbols, Subscripts and Superscripts

$*$	dimensional variables
$\wedge$	nondimensional variables
f	fluid
s	solid
w	wall condition
∞	ambient condition
$'$	differentiation with respect to the independent variables.

REFERENCES

Aldos TK, Chen TS, Armaly BF. Nonsimilarity solutions for mixed convection from horizontal surfaces in a porous medium—variable surface heat flux. Int J Heat Mass Transfer 36:463–470, 1993a.

Aldos TK, Chen TS, Armaly BF. Nonsimilarity solutions for mixed convection from horizontal surfaces in a porous medium—variable wall temperature. Int J Heat Mass Transfer 36:471–477, 1993b.

Angirasa D, Peterson GP. Upper and lower Rayleigh number bounds for two-dimensional natural convection over a finite horizontal surface situated in a fluid-saturated porous medium. Numer Heat Transfer, Part A 33:477–493, 1998.

Banks WHH, Zaturska MB. Eigensolutions in boundary-layer flow adjacent to a stretching wall. IMA J Appl Math 36:263–273, 1986.

Bradean R, Heggs PH, Ingham DB, Pop I. Convective heat flow from suddenly heated surfaces embedded in porous media. In: Ingham DB, Pop I, eds. Transport Phenomena in Porous Media. Oxford: Pergamon, 1998, pp 411–438.

Cebeci T, Bradshaw P. Physical and Computational Aspects of Convective Heat Transfer. New York: Springer, 1984.

Chang ID, Cheng P. Matched asymptotic expansions for free convection about an impermeable horizontal surface in a porous medium. Int J Heat Mass Transfer 26:163–173, 1983.

Chao BH, Wang H, Cheng P. Stagnation point flow of a chemically reactive fluid in a catalytic porous bed. Int J Heat Mass Transfer 39:3003–3019, 1996.

Chaudhary MA, Merkin JH. Free convection stagnation point boundary layers driven by catalytic surface reactions: I. The steady states. J Eng Math 28:145–171, 1994.

Chaudhary MA, Merkin JH. A simple isothermal model for homogeneous–heterogeneous reactions in boundary-layer flow. I. Equal diffusivities. Fluid Dyn Res 16:311–333, 1995a.

Chaudhary MA, Merkins JH. A simple isothermal model for homogeneous–heterogeneous reactions in boundary-layer flow. II. Different diffusivities for reactant and autocatalyst. Fluid Dyn Res 16:335–359, 1995b.

Chaudhary MA, Merkin JH. Free convection stagnation point boundary layers driven by catalytic surface reactions: II. Times to ignition. J Eng Math 30:403–415, 1996.

Chaudhary MA, Merkin JH, Pop I. Similarity solutions in free convection boundary-layer flows adjacent to vertical permeable surfaces in porous media. I. Prescribed surface temperature. Eur J Mech B/Fluids 14:217–237, 1995a.

Chaudhary MA, Merkin JH, Pop I. Similarity solutions in free convection boundary-layer flows adjacent to vertical permeable surfaces in porous media. II. Prescribed surface heat flux. Heat Mass Transfer 30:341–347, 1995b.

Chaudhary MA, Merkin JH, Pop I. Natural convection from a horizontal permeable surface in a porous medium—numerical and asymptotic solutions. Trans Porous media 22:327–344, 1996.

Cheng P. The influence of lateral mass flux on free convection boundary layers in a saturated porous medium. Int J Heat Mass Transfer 20:201–206, 1977.

Cheng P, Chang ID. Buoyancy induced flows in a saturated porous medium adjacent to impermeable horizontal surfaces. Int J Heat Mass Transfer 19:1267–1272, 1976.

Cheng P, Minkowycz WJ. Free convection about a vertical flat plate embedded in a porous medium with application to heat transfer from a dike. J Geophys Res 82:2040–2044, 1977.

Combarnous M. Description du transfert de chaleur par convection naturelle dans une couche poreuse horizontale à l'aide d'un coefficient de transfert solide–fluide. CR Acad Sci Paris A275:1375–1378, 1972.

Combarnous M, Bories S. Modelisation de la convection naturelle au sein d'une couche poreuse horizontal à l'aide d'un coefficient de transfert solide–fluide. Int J Heat Mass Transfer 17:505–515, 1974.

Daniels PG, Simpkins PG. The flow induced by a heated vertical wall in a porous medium. Quart J Mech Appl Math 37:339–354, 1984.

Ene HI, Poliševski D. Thermal Flow in Porous Media. Dordrecht: Reidel, 1987.

Govindarajulu T, Malarvizhi G. A note on the solution of free convection boundary-layer flow in a saturated porous medium. Int J Heat Mass Transfer 30:1769–1771, 1987.

Ingham DB, Brown SN. Flow past a suddenly heated vertical plate in a porous medium. Proc R Soc London A403:51–80, 1986.

Ingham DB, Pop I, eds. Transport Phenomena in Porous Media. Oxford: Pergamon, 1998.

Kimura S, Kiwata T, Okajima A, Pop I. Conjugate natural convection in porous media. Adv Water Resour 20:111–126, 1997.

Kuznetsov AV. Thermal non-equilibrium forced convection in porous media. In: Ingham DB, Pop I, eds. Transport Phenomena in Porous Media. Oxford: Pergamon, 1998, pp 103–129.

Lage JL. The fundamental theory of flow through permeable media from Darcy to turbulence. In: Ingham DB, Pop I, eds. Transport Phenomena in Porous Media. Oxford: Pergamon, 1998, pp 1–30.

Lesnic D, Pop I. Free convection in a porous medium adjacent to horizontal surfaces. J Appl Math Mech (ZAMM) 78:197–205, 1998a.

Lesnic D, Pop I. On the mixed convection over a horizontal surface embedded in a porous medium. In: Tupholme GE, Wood AS, eds. Mathematics of Heat Transfer. Bradford, 1998b, pp 219–224.

Lesnic D, Ingham DB, Pop I. Free convection boundary-layer flow along a vertical surface in a porous medium with Newtonian heating. Int J Heat Mass Transfer 42:2621–2627, 1999.

McNabb A. On convection in a saturated porous medium. In: Proceeding of 2nd Australasian Conference on Hydraulics and Fluid Mechanics. New Zealand: University of Auckland Press, 1965.

Mahmood T, Merkin JH. Mixed convection on a vertical circular cylinder. J Appl Math Phys (ZAMP) 39:186–203, 1988.

Mahmood T, Merkin JH. The convective boundary-layer flow on a reacting surface in a porous medium. Transp Porous Media 32:285–298, 1998.

Merkin JH. Free convection boundary layers in a saturated porous medium with lateral mass flux. Int J Heat Mass Transfer 21:1499–1504, 1978.

Merkin JH. A note on the solution of a differential equation arising in boundary-layer theory. J Eng Math 18:31–36, 1984.

Merkin JH, Chaudhary MA. Free-convection boundary layers on vertical surfaces driven by an exothermic surface reaction. Q J Mech Appl Math 47:405–428, 1994.

Merkin JH, Mahmood T. Connective flow on reactive surfaces in porous media. Transp Porous Media 33:279–293, 1998.

Merkin JH, Pop I. Mixed convection on a horizontal surface embedded in a porous medium: the structure of a singularity. Transp Porous Media 18:625–631, 1997.

Merkin JH, Zhang G. On the similarity solutions for free convection in a saturated porous medium adjacent to impermeable horizontal surfaces. Wärme- und Stoffübertr 25:179–184, 1990.

Minkowycz WJ, Cheng P. Local non-similar solutions for free convection flow with uniform lateral mass flux in a porous medium. Lett Heat Mass Transfer 9:159–168, 1982.

Minto BJ, Ingham DB, Pop I. Free convection driven by an exothermic reaction on a vertical surface embedded in porous media. Int J Heat Mass Transfer 41:11–23, 1998.

Na TY, Pop I. Free convection flow past a vertical flat plate embedded in a saturated porous medium. Int J Eng Sci 21:517–526, 1983.

Nakayama A. PC-aided Numerical Heat Transfer and Convective Flow. Tokyo: CRC Press, 1995.

Nakayama A. A unified treatment of Darcy–Forchheimer boundary-layer flows. In: Ingham DB, Pop I, eds. Transport Phenomena in Porous Media. Oxford: Pergamon, 1998, pp 179–204.

Nield DA, Bejan A. Convection in Porous Media. 2nd ed. New York: Springer, 1999.

Pop I, Nakayama A. Conjugate free and mixed convection heat transfer from vertical fins embedded in porous media. In: Sunden B, Heggs PJ, eds. Recent Advances in Analysis of Heat Transfer for Fin Type Surfaces. Southampton: Computational Mechanics Publications, 1999, pp 67–96.

Pop I, Lesnic D, Ingham DB. Conjugate mixed convection on a vertical surface in a porous medium. Int J Heat Mass Transfer 38:1517–1525, 1995.

Pop I, Ingham DB, Merkin JH. Transient convection heat transfer in a porous medium: external flows. In: Ingham DB, Pop I, eds. Transport Phenomena in Porous Media. Oxford: Pergamon, 1998, pp 205–231.

Prandtl L. Über Flüssigkeitsbewegung bei sehr kleiner Reibung. In: Proceedings of 3rd International Mathematics Congress, Heidelberg, 1904; also NACA TM 452, 1928.

Rees DAS. Vertical free convective boundary-layer in a porous medium using a thermal non-equilibrium model: elliptical effects. In: Proceedings of Bangladesh Mathematics Conference, 1998a, pp 13–22.

Rees DAS. Onset of Darcy–Bénard convection using a thermal non-equilibrium model. Int Commun Heat Mass Transfer, 1998b.

Rees DAS. Free convective boundary-layer flow from a heated surface in a layered porous medium. J Porous Media, 2:39–58, 1999.

Rees DAS, Bassom AP. Some exact solutions for free convective flows over heated semi-infinite surfaces in porous media. Int J Heat Mass Transfer 34:1564–1567, 1991.

Rees DAS, Pop I. Free convection boundary-layer flow in a porous medium using a thermal non-equilibrium model. Int Commun Heat Mass Transfer, 26:945–954, 1999.

Rees DAS, Pop I. Vertical free convective boundary-layer flow in a porous medium using a thermal non-equilibrium model. J Porous Media, 3:31–43, 2000.

Riley N. Asymptotic expansions in radial jets. J Math Phys 41:132–146, 1962.

Shu JJ, Pop I. Inclined wall plumes in porous media. Fluid Dyn Res 21:303–317, 1997.

Stewartson K. On asymptotic expansions in the theory of boundary layers. J Math and Phys 36:173–191, 1957.

Vafai K, Amiri A. Non-Darcian effects in combined forced convective flows. In: Ingham DB, Pop I, eds. Transport Phenomena in Porous Media. Oxford: Pergamon, 1998, pp 313–329.

Vynnycky M, Pop I. Mixed convection due to a finite horizontal flat plate embedded in a porous medium. J Fluid Mech 351:359–378, 1997.

Weidman PD, Amberg MF. Similarity solutions for steady laminar convection along heated plates with variable oblique suction: Newtonian and Darcian fluid flow. QJ Mech Appl Math 49:373–403, 1996.

Wooding RA. Convection in a saturated porous medium at large Rayleigh or Péclet numbers. J Fluid Mech 15:527–544, 1963.

Wright SD, Ingham DB, Pop I. On natural convection from a vertical plate with a prescribed surface heat flux in a porous medium. Transp Porous Media 22:183–195, 1996.

Yu WS, Lin HT, Lu CS. Universal formulations and comprehensive correlations for non-Darcy natural convection and mixed convection in porous media. Int J Heat Mass Transfer 34:2859–2868, 1991.

Zaturska MB, Banks WHH. A note concerning free-convective boundary-layer flows. J Eng Math 19:247–249, 1985.

Zaturska MB, Banks WHH. On the spatial stability of free-convection flows in a saturated porous medium. J Eng Math 21:41–46, 1987.

8

Porous Media Enhanced Forced Convection Fundamentals and Applications

José L. Lage and Arunn Narasimhan
Southern Methodist University, Dallas, Texas

I. INTRODUCTION

What is *enhanced heat transfer*? According to Webster's Dictionary (1995), the word *enhance* means "to raise to a higher degree, to intensify, to increase the value, attractiveness, or quality of, to improve." Hence, enhanced heat transfer is a heat transfer that has been improved. An important subsequent question is: *How can heat transfer be improved*? The answer to this question is simple: *Heat transfer can be improved by reducing the thermal resistance of the transfer process.* This is what *enhancing heat transfer* is all about!

It is well known that the heat transfer relation between heat flow (or *current*) and temperature difference (or *potential difference*) depends on the heat transfer mode, i.e., it depends on heat being transferred by diffusion, convection, and/or radiation. In convection heat transfer the general relation between heat q and the driving temperature difference is

$$q = hA(T_w - T_{ref}) \tag{1}$$

where the parameters are defined in the nomenclature. In Eq. (1), the potential driving the heat transfer is $(T_w - T_{ref})$ and the thermal resistance is $1/(hA)$.

Keep in mind that Eq. (1) is the definition of the thermal resistance $(1/hA)$, or, more specifically, the definition of the convection heat transfer coefficient h. Another important observation is that q and $(T_w - T_{ref})$ are not totally independent quantities, as Eq. (1) might seem to suggest. Recall that the relationship between q and T_w is defined by Fourier's law at the interface, namely

$$q = -k_f A \frac{\partial T}{\partial n}\bigg|_w \tag{2}$$

where the derivative is the local fluid temperature variation along a direction n perpendicular to the solid–fluid interface.

Equations (1) and (2) can be combined into a single equation

$$\frac{h}{k_f} = \frac{-\partial T/\partial n|_w}{(T_w - T_{\text{ref}})} \tag{3}$$

Equation (3) can be used as another definition of h, in this case independent of the surface area A. In fact, multiplying both sides of (3) by a representative length D results in a dimensionless equation

$$Nu = \frac{hD}{k_f} = \frac{-D(\partial T/\partial n)|_w}{(T_w - T_{\text{ref}})} \tag{4}$$

defining the Nusselt number Nu, i.e., a dimensionless representative of h.

Consider now the case in which q is fixed. Decreasing $(T_w - T_{\text{ref}})$ reduces the thermal resistance $1/(hA)$ by the same factor, according to Eq. (1). Sometimes, reducing $(T_w - T_{\text{ref}})$ might be the main objective of the design engineer because a lower temperature difference translates into a lower interface temperature (recall that T_{ref} is invariant), alleviating the temperature requirement for selecting the convection fluid or the solid material for the surface. If T_w is the fixed design parameter instead, then $(T_w - T_{\text{ref}})$ becomes invariant. In this case, increasing the heat flow q would decrease the thermal resistance $1/(hA)$.

An enhanced convection heat transfer configuration can be evaluated by comparison with a reference configuration using the factor-of-merit E_f, defined as

$$E_f = \frac{hA}{(hA)_{\text{ref}}} \tag{5}$$

Two ways to reduce the thermal resistance $(1/hA)$ of a certain reference configuration are: (1) to increase the heat transfer coefficient h; and (2) to increase the area A crossed by the heat flow.

The convection heat transfer coefficient h, as indicated in Eq. (3), can be modified by changing the surface temperature T_w, the fluid thermal conductivity k_f, or the fluid temperature variation at the solid–fluid interface $-\partial T/\partial n|_w$. Very often, a change in the heat transfer area A indirectly yields a change in the convection coefficient h by modifying the surface temperature or the fluid temperature variation. Sometimes the change is beneficial, increasing h; sometimes the change is detrimental.

This chapter focuses on alternative ways to enhance the convection heat transfer process by increasing A, and also on the indirect effects that this increase might have on h.

Frequently, the heat transfer area A is increased by including within the fluid path a geometrically complex solid structure. It so happens that many of these structures are permeable. One of the objectives of this chapter, therefore, is to highlight that the fundamentals of convection through porous media can be used to analyze several enhanced configurations, even when the geometry of the enhanced structure does not necessarily resemble the geometry of what is generally perceived as a porous medium. An excellent example of how beneficial a porous medium analysis might be to understand the behavior of complex permeable structure was illustrated by Bejan and Morega (1993), who considered the cross-flow through arrays of staggered cylinders, a structure which frequently would not be considered as a porous medium.

Ways by which a thermal design engineer can modify the thermal resistance of a heat transfer apparatus, by modifying A or h, constitute nowadays an entire body of research and technology included in the general theme of *enhanced heat transfer*. Even though much has been accomplished in the past several years, much more is still to come as supporting activities, such as materials processing and manufacturing, provide new advanced concepts and techniques for carrying out design ideas set forth by the thermal engineers.

Notwithstanding the appropriateness of acknowledging and studying the *past*, the applications section of this chapter considers only contemporary (post-1993) discoveries in the field of enhanced convection heat transfer and porous media. This is deliberately done because the recent book by Webb (1994), a landmark in the area of enhanced heat transfer, already provides an excellent broad review of the scientific and technical advances of enhanced heat transfer realized prior to 1993. His book is strongly recommended.

II. FUNDAMENTALS

Observe in Eq. (1) that the reference temperature T_{ref} is loosely defined, i.e., in principle T_{ref} can be any invariant temperature, one that does not change. It is, therefore, fundamental that the same T_{ref} be used when comparing the thermal performances of two distinct systems.

Heat transfer enhancement by surface modification and/or insertion of permeable obstructions to the flow field is quantified by using Eq. (5). The case of unimpeded fluid flow along a smooth surface is usually considered as a reference configuration. An enhanced configuration could be built by brazing a wire-mesh to the smooth surface, for instance. In this case, the thermal resistance of the surface is affected directly by changing the heat transfer area A because the heat can now flow into the fluid region by following the solid wire-mesh path. Moreover, the heat transfer coefficient h is also affected indirectly by modifying the flow configuration. This new flow configuration will result in a new temperature or temperature variation at the fluid–solid interface, ultimately affecting h. Observe that this latter effect is independent of contact between the wire-mesh and the surface. Even if the wire-mesh is placed away from the surface (without touching it), but close enough to affect the velocity field near the heated surface, the thermal resistance will be modified. However, in this case, the heat transfer area A is exactly the same as the heat transfer area of the reference configuration.

Predicting the variation of h by varying A requires, invariably, the prediction of the new fluid velocity field and the new temperature distribution by solving the momentum and energy balance equations. This is so because it is practically very difficult to modify the heat transfer area without affecting the nearby velocity and temperature variations. These variations affect the surface temperature and/or the temperature variation at the surface, which in turn affect h.

A. Balance Equations

Considering the heat transfer definition, Eq. (3), it is clear that the fluid temperature and the temperature variation at the solid–fluid interface are fundamental parameters to determine h. These local parameters are affected by the energy transport within the moving fluid, which is affected by the fluid velocity. For an unobstructed flow system the energy transport equation, assuming incompressible fluid with constant and uniform properties and zero viscous dissipation, is

$$\rho c_p \left[\frac{\partial T}{\partial t} + (\boldsymbol{v} \cdot \nabla) T \right] = k_f \nabla^2 T \tag{6}$$

where ∇ is the gradient operator, and ∇^2 is the Laplacean operator.

It is, therefore, imperative that the fluid velocity be known before the fluid temperature variation can be determined. To this end, the balance of momentum for an incompressible Newtonian fluid with constant and uniform properties is frequently invoked, namely

$$\rho\left[\frac{\partial \boldsymbol{v}}{\partial t} + (\boldsymbol{v} \cdot \nabla)\boldsymbol{v}\right] = -\nabla p + \mu \nabla^2 \boldsymbol{v} \tag{7}$$

With the additional mass balance equation for incompressible fluid, i.e.

$$\nabla \cdot \boldsymbol{v} = 0 \tag{8}$$

the system of Eqs. (6)–(8) can be solved with suitable initial and boundary conditions for several flow configurations (see Bejan 1995, for instance). The surface temperature and/or temperature variation, determined from the temperature solution, can be used in (3) to find h.

Now, consider the case in which the flow field is modified by the insertion of a porous medium within the fluid path. Equations (6)–(8) can still be used to determine the temperature distribution necessary to calculate h. However, in most cases the new heat flow surface has such a complicated shape that even the mapping of the solid–fluid interface would require a tremendous effort. Therefore, general analytic solutions for predicting the temperature or temperature variation at the fluid–solid interface are unattainable. Even a direct numerical simulation using Eqs. (6)–(8) is precluded by the complicated internal geometry of most porous media. The option then is to create a model to circumvent this difficulty.

It is instructive at this point to recall a similar difficulty for determining the fluid temperature variation in the case of an unobstructed flow configuration. Even though Eqs. (6)–(8) are frequently taken for granted, a more natural approach to find the fluid velocity and temperature fields would be to follow each fluid molecule in order to obtain their collective effect on the transport of momentum and energy. The number of molecules, however, is frequently too large for this approach to be feasible. The alternative is to create a *continuum approach*, which considers a fluid volume in space containing a large number of molecules. It is only the overall (net) effect of all molecules within this volume, represented by continuum quantities such as T, p, or $\boldsymbol{v}$, which is considered when deriving (6)–(8).

The only requirement of the continuum approach to be accurate is that the mean free-path distance traveled by the molecules should be much smaller than the dimensions of the control volume where the transport processes take place. Even though this restriction seems very clear, an important step for deriving Eqs. (6)–(8) is to take the limit when the volume of fluid goes to zero (see Bejan 1995:6). This apparent contradiction is resolved by recalling that the limit as the volume *tends* to zero does not necessarily mean that the volume *is* zero. Or, even though the volume is considered very small, it is still large enough as compared to the mean free-path length of the molecules included in it.

A similar continuum approach can be used when deriving the transport of momentum and energy within a porous medium. Consider now another continuum, called *porous-continuum*, one in which the mean pore and solid obstacle dimensions of the porous medium are much smaller than the dimensions of the control volume where the transport processes take place. With this in mind, balance equations can be derived for the transport of momentum and energy in a porous medium in the very same way as the balance equations were derived for the unobstructed flow.

The presence of solid obstacles within the fluid path modifies the convection process. New constitutive equations become necessary to close the derivation of the transport equations. New transport quantities representing the net (volume-averaged) continuum quantities T, p, and $\boldsymbol{v}$ emerge from the derivation of the balance equations, such as $\langle T \rangle$ (representing the solid and fluid local thermal equilibrium temperature), $\langle p \rangle$, and $\langle \boldsymbol{v} \rangle$. The resulting energy, momentum, and mass balance equations, for a rigid, isotropic, and homogeneous porous medium, are respectively (Lage 1998; Nield and Bejan 1999)

$$\left[\varepsilon(\rho c_p)_f + (1-\varepsilon)(\rho c_p)_s\right]\frac{\partial \langle T \rangle}{\partial t} + (\rho c_p)_f(\langle \boldsymbol{v} \rangle \cdot \nabla)\langle T \rangle = k_e \nabla^2 \langle T \rangle \tag{9}$$

$$\frac{\rho_f}{\varepsilon}\left[\frac{\partial \langle \boldsymbol{v} \rangle}{\partial t} + \frac{1}{\varepsilon}(\langle \boldsymbol{v} \rangle \cdot \nabla)\langle \boldsymbol{v} \rangle\right] = -\nabla \langle p \rangle + \mu_e \nabla^2 \langle \boldsymbol{v} \rangle - \frac{\mu}{K}\langle \boldsymbol{v} \rangle - \rho_f C |\langle \boldsymbol{v} \rangle| \langle \boldsymbol{v} \rangle \tag{10}$$

$$\nabla \cdot \langle \boldsymbol{v} \rangle = 0 \tag{11}$$

Equation (10) is sometimes referred to as the *general* momentum equation for flow through a porous medium. Except for the brackets $\langle \ \rangle$, which denote quantities averaged over a representative porous volume (a formal discussion of the volume-averaging approach is presented by Kaviany [1991], and a more recent and extensive discussion by Whitaker [1999], the volume porosity ε (considered equal to the isotropic surface porosity of the matrix), and the subscripts e for effective, f for fluid, and s for solid, Eqs. (9) and (10) are essentially the same as (6) and (8). The only difference between (6) and (9), for a steady-state process, is the utilization of an *effective* thermal conductivity k_e in place of the fluid thermal conductivity k.

Equation (10) differs from Eq. (7) by the inclusion of the two last terms, where K represents the permeability of the medium and C represents the form coefficient of the medium. These two extra terms emerge from the momentum balance effects of the solid permeable structure placed within the fluid path, respectively the viscous and form drag effects.

The fluid velocity $\langle \boldsymbol{v} \rangle$, averaged over a representative elementary porous volume, is usually called the *Darcy* or *seepage velocity*. The Darcy velocity is related to the velocity averaged over the pore-volume $\langle \boldsymbol{v} \rangle_p$, the volume occupied by the fluid only, by the equation $\langle \boldsymbol{v} \rangle = \varepsilon \langle \boldsymbol{v} \rangle_p$.

It is important to recognize that the effect of inserting a porous medium within a fluid path is to obstruct (hinder) the flow, increasing the pressure-drop for the same flow rate. This is, of course, detrimental to the thermohydraulic behavior of the system. However, there are situations in which a porous medium enhances the energy transport by increasing the thermal conductivity of the medium (when $k_e > k$). Moreover, considering internal convection, when the fluid path is filled with a homogeneous and isotropic porous medium the resulting fluid velocity profile gets flattened, i.e., the fluid tends to move faster near the solid–porous medium interface leading to a more effective convection near the solid bounding surface. Hence, an enhanced heat transfer configuration to compensate the increase in pumping power can be achieved.

Equations (9)–(11), with suitable initial and boundary conditions, can be solved to find the temperature variation at the boundary of the porous medium in contact with the solid surface of the channel (porous–solid interface). The porous medium (fluid and solid permeable matrix) is now seen as a continuum medium, much like the fluid medium.

B. Convection Heat Transfer Coefficient

The effect of increasing the surface area A of the system by inserting a porous medium is now reflected in the local effective heat transfer coefficient of the porous medium, defined in terms of a local nondimensional Nusselt number, as

$$Nu_e = \frac{h_e D}{k_e} = \frac{-D(\partial \langle T \rangle / \partial n)|_w}{(\langle T_w \rangle - T_{\text{ref}})} \tag{12}$$

The use of angle brackets when writing Eq. (12) is an important reminder of the special meaning of the temperature in Eq. (9). Even though T_w is defined along the solid and porous medium interface, the porous-continuum approach used for deriving Eqs. (9)–(11) must also be extended to this interface, for consistency. This means that the interface temperature $\langle T_w \rangle$ is pointwise only in a porous-continuum sense.

Observe that T_w has two components along the porous–solid interface. One is the average temperature along the region bathed by the fluid (the temperature of the fluid within the pores of the porous medium next to the interface). The other component is the average temperature of the solid matrix of the porous medium in contact with the solid. Equation (9), derived by following a porous-continuum approach, assumes these two local temperature components to be locally the same (thermal equilibrium).

A common distinct definition of the Nusselt number is

$$Nu_m = \frac{h_e D}{k_f} = \frac{k_e}{k_f}\left(\frac{-D(\partial\langle T\rangle/\partial n)|_w}{(\langle T_w\rangle - \langle T\rangle_m)}\right) \tag{13}$$

Observe some striking differences between Eqs. (12) and (13). First, the use of the fluid thermal conductivity k_f in the denominator of (13) instead of the effective thermal conductivity k_e of the porous medium. This results in the inclusion of the group k_e/k_f in the right-hand side of (13). Second, the use of a local bulk-temperature $\langle T\rangle_m$, defined as the speed-weighted temperature cross-sectional average

$$\langle T\rangle_m = \frac{\int_{A_c}\langle u\rangle\langle T\rangle \mathrm{d}A_c}{\int_{A_c}\langle u\rangle \mathrm{d}A_c} \tag{14}$$

where A_c is the flow cross-sectional area of the channel, in place of the reference temperature $\langle T\rangle_{\text{ref}}$ shown in (12). Even though there are some theoretical advantages in using $\langle T\rangle_m$ as a reference temperature, it is important to recognize that $\langle T\rangle_m$ is not easy to measure accurately in a porous medium. Moreover, in most systems, $\langle T\rangle_m$ varies along the main flow direction.

Consider now the classical problem of convection through a regular confined channel of perimeter w and length L, with isoflux surfaces, filled with an isotropic and homogeneous porous medium. The longitudinal flow direction is parallel to the x-direction. It is possible to find the relation between Nu_e and Nu_m in this case by starting with Eq. (13), rewritten as

$$\langle T_w\rangle - \langle T\rangle_m = \frac{Dq_w''}{Nu_m k_f} \tag{15}$$

where the uniform surface heat flux is $q_w'' = -k_e\partial\langle T\rangle/\partial n|_w$. Equation (15) can be integrated along the surface A of the channel, as

$$\frac{1}{A}\int_A (\langle T_w\rangle - \langle T\rangle_m)\mathrm{d}A = \frac{1}{A}\int_A \left(\frac{Dq_w''}{Nu_m k_f}\right)\mathrm{d}A \tag{16}$$

The integrand on the right-hand side of Eq. (16) is constant when the flow is thermally fully developed (because Nu_m is constant in this case). Observe, also in reference to (16), that the integral of $\langle T_w\rangle$ along A results in the surface-area-averaged temperature $\langle T_w\rangle_{\text{avg}}$; therefore

$$\langle T_w\rangle_{\text{avg}} - \frac{1}{A}\int_A \langle T\rangle_m \mathrm{d}A = \frac{Dq_w''}{Nu_m k_f} \tag{17}$$

To find the surface-averaged fluid bulk temperature, the fluid longitudinal heat diffusion is neglected and the steady-state version of the first law of thermodynamics for a control volume is invoked,

$$\langle T\rangle_m = \frac{(wx)q''_w}{\dot{m}c_{p_f}} + \langle T\rangle_{\text{in}} \tag{18}$$

where w is the channel perimeter, x is the distance from the inlet to the point where $\langle T\rangle_m$ is measured, $\langle T\rangle_{\text{in}}$ is the fluid inlet temperature (cross-sectional-area-averaged temperature in the case of nonuniform $\langle T\rangle_{\text{in}}$), and $\dot{m}$ is the fluid mass flow rate. Observe that $\langle T\rangle_m$ in (18) is the transverse-averaged fluid bulk temperature (varying along the x-direction only). Equation (18) can be integrated from $x = 0$ to $x = L$, multiplied by (w/A), and simplified using the relations $(w\mathrm{d}x) = \mathrm{d}A$ and $(wL) = A$, to yield

$$\frac{1}{A}\int_A \langle T\rangle_m \mathrm{d}A = \frac{wLq''_w}{2\dot{m}c_p} + \langle T\rangle_{\text{in}} \tag{19}$$

Now, Eqs. (19) and (17) can be combined into a single equation

$$\frac{Dq''_w}{k_e(\langle T_w\rangle_{\text{avg}} - \langle T\rangle_{\text{in}})} = Nu_{e\text{-avg}} = \frac{1}{\left(\dfrac{k_e}{k_f Nu_m} + \dfrac{wL}{2D}\dfrac{k_e}{\dot{m}c_p}\right)} \tag{20}$$

Equation (20) presents the relation between the Nusselt number Nu_m defined on the basis of the wall temperature (surface-averaged in the present case) and the fluid bulk temperature, Eq. (13), and the surface-averaged Nusselt number $Nu_{e\text{-avg}}$ defined on the basis of the surface-averaged wall temperature and a reference temperature (in the present case equal to the inlet fluid temperature). Keep in mind that Eq. (20) was derived assuming: (a) thermally fully developed flow; and (b) negligible longitudinal diffusion.

The last term of Eq. (20) can be rewritten in dimensionless form using the Reynolds number $Re = \dot{m}D/(\mu A_c)$ and the Prandtl number $Pr = \mu c_p/k_f$. Therefore, (20) becomes

$$Nu_{e\text{-avg}} = \frac{(k_f/k_e)}{\left[\dfrac{1}{Nu_m} + \left(\dfrac{wL}{2A_c}\right)\dfrac{1}{RePr}\right]} \tag{21}$$

The group within parentheses in the denominator of Eq. (21) is dependent only on the geometry of the channel. The main advantage of Eq. (21) is to provide a direct verification of theoretical results predicting Nu_m by comparing the value of $Nu_{e\text{-avg}}$ from (21) with the experimental value (it is easier to measure $Nu_{e\text{-avg}}$ than to measure Nu_m). For instance, considering the case of plug flow (Darcy flow) through a porous medium sandwiched between two parallel isoflux plates, of length L and set apart by a distance $H = D$, for which theory predicts Nu_m equal to $6k_e/k_f$ (see Nield and Bejan 1999:62), the predicted $Nu_{e\text{-avg}}$ is simply

$$Nu_{e\text{-avg}} = \frac{1}{\left[\frac{1}{6} + \left(\frac{L}{H}\right)\frac{1}{RePr_e}\right]} \tag{22}$$

where an effective Prandtl number is introduced as $Pr_e = \mu c_p/k_e$. Clearly, from Eq. (22), when the flow rate is very high $Nu_{e\text{-avg}}$ tends to 6 as expected because the bulk temperature approaches the inlet temperature of the fluid (see Eqs. (12) and (13) to verify that Nu_e approaches $k_f Nu_m/k_e$).

If the channel surface is held at a constant and uniform temperature $\langle T_w \rangle$, then the local Nusselt number defined in terms of $\langle T \rangle_{\text{ref}}$ is a function of x,

$$Nu_e(x) = (k_f/k_e)\frac{[\langle T_w \rangle - \langle T \rangle_m(x)]}{(\langle T_w \rangle - \langle T \rangle_{\text{in}})} Nu_m \tag{23}$$

according to Eqs. (12) and (13), if the reference temperature chosen when defining Nu_e is $\langle T \rangle_{\text{in}}$.

For cases in which the local bulk temperature $\langle T \rangle_m$ is too difficult to measure, an optional equation is made available by invoking the first law of thermodynamics. The result, in nondimensional form, is

$$Nu_{e\text{-avg}} = (k_f/k_e)\left[1 - \frac{wL}{2A_c}\frac{1}{RePr}\frac{Dq''_{\text{avg}}}{(\langle T_w \rangle - \langle T \rangle_{\text{in}})}\right] Nu_m \tag{24}$$

where q''_{avg} represents the surface-averaged heat flow crossing the channel surface. Again, for a plug flow through a porous medium sandwiched between two parallel isothermal plates, of length L and separated by a distance $H = D$, the theory predicts Nu_m uniform and equal to 4.93 k_e/k_f (Nield and Bejan 1999:62). Therefore, $Nu_{e\text{-avg}}$ becomes

$$Nu_{e\text{-avg}} = 4.93\left[1 - \frac{wL}{2A_c}\frac{1}{RePr}\frac{Dq''_{\text{avg}}}{(\langle T_w \rangle - \langle T \rangle_{\text{in}})}\right] \tag{25}$$

C. Effective Thermal Conductivity

It is clear from the previous section that one of the important effects of porous medium enhancing technique is related to the effective thermal conductivity of the porous medium.

The effective thermal conductivity of a porous medium depends on the thermal conductivities of the fluid and solid matrix, on the structure of the solid matrix (i.e., the paths available for diffusion through the fluid and through the solid), and on the amount of fluid and solid present in the porous medium. Unfortunately, exact analytical representations for the effective thermal conductivity can be derived only in two specific cases. One case is that of a porous medium in which fluid and solid are distributed parallel to each other and in parallel to the heat flow direction. In this case

$$k_e = \varepsilon k_f + (1 - \varepsilon)k_s \tag{26}$$

The other case involves a porous medium in which the fluid and solid are in series (perpendicular to the heat flow direction), in which case

$$\frac{1}{k_e} = \frac{\varepsilon}{k_f} + \frac{(1 - \varepsilon)}{k_s} \tag{27}$$

Equation (26) provides an upper bound value for the effective thermal conductivity of a porous medium, and Eq. (27) provides a lower bound value. Cheng and Hsu (1998) proposed a correlation similar to (26), that is

$$k_e = \varepsilon k_f + (1 - \varepsilon)k_s + (k_f - k_s)G \tag{28}$$

where the new parameter G represents the structural (heat flow path, or tortuosity of the matrix) effect on the effective thermal conductivity of the porous medium. Even though the parameter G is extremely difficult to calculate in most cases, Eq. (28) is a valuable reminder of the importance of considering the internal geometry of a porous medium when modeling k_e.

Nield (1991) suggested a simplified correlation for the effective thermal conductivity of a porous medium in which the fluid and solid thermal conductivities are not too dissimilar,

$$k_e = k_f^{\varepsilon} k_s^{(1-\varepsilon)} \tag{29}$$

Equations (26)–(29) suggest that the effective diffusivity of a porous medium tends to the fluid thermal conductivity when the porosity is increased and to the thermal conductivity of the solid when the porosity is decreased.

When enhanced heat transfer is the objective behind the use of a porous insert, then the thermal conductivity of the solid should be higher than that of the fluid for the effective thermal conductivity of the porous medium to be higher than the fluid thermal conductivity. It is interesting to note that, even though the precise characterization of the effective diffusivity of a porous medium is a precondition for finding the temperature distribution and the enhanced heat transfer coefficient, no accurate predictive general correlation for k_e is available.

Chapter 3 of Kaviany (1991) presents a very good review of analytical representations and experimental data available for determining k_e. In it, the possibility of having a porous medium consisting of contacting or noncontacting particles is also analyzed. The case of contacting particles is complicated by the contact thermal resistance effect. Cheng and Hsu (1998) performed another more recent review of the subject. They indicated that the majority of experiments to determine k_e were done with fluids having thermal conductivity smaller than that of the solid matrix, $1 < k_s/k_f < 10^3$. These experimental results are then useful for the case of porous enhanced heat transfer.

Some additional important implications of the porous-continuum approach in regard to the solid–porous interface in conduction heat transfer were considered by Pratt (1989, 1990) and more recently by Lage (1999).

D. Dispersion

An additional effect of including a solid obstruction along the flow path in convection heat transfer is the mixing brought about when the fluid is forced to deviate from a straight path. It is well known from fluid mechanics that a fluid flowing around a single solid obstacle is prone to mixing by the vortices created at the trailing edge of the obstacle. The strength of these vortices depends on the fluid properties, flow characteristics (speed, velocity variation), and ultimately on the shape of the solid obstacle. Another characteristic configuration inducing mixing is the confluence of two flows, originally following distinct paths.

When considering a permeable medium, vortices and confluence of fluid streams are very common. They are caused by the geometrical irregularities of the fluid path available in the solid matrix. Observe that this analysis is not limited to what is generally understood as a porous medium (small porosity, packed particles), but also applies to flows through tube bundles, microchannels, fins, and many other configurations found in enhanced convection heat transfer.

The extra mixing caused by the tortuous fluid path imposed by a solid obstacle is not specifically accounted for by the model of Eq. (9). This extra mixing, called *dispersion*, has thermal consequences which are similar to the extra mixing caused by turbulence. It is, therefore, common to include the dispersion effect into the effective thermal conductivity k_e of the porous medium, mimicking what is usually done when modeling turbulence.

Dispersion is expected to be direction dependent, i.e., for instance, the mixing effect caused by vortices and/or confluence of fluid streams along the main flow direction might differ from the mixing effect caused in the transverse flow direction. Experimental results reported in the literature (Nield and Bejan 1999) indicate a predominance of transverse dispersion.

It is also reasonable to believe that dispersion depends on the local fluid velocity, on the characteristic dimension of the flow path, and also on the local characteristic dimension of the solid obstruction. The latter dependency is borrowed from the studies performed with flow around bluff bodies.

It is now clear that not only the approach for taking dispersion into account is similar to the approach used in turbulence. The modeling of dispersion, i.e., the effect of dispersion on the diffusivity, is also extremely complex, like the turbulence effect. Simple models are available, and can be summarized into a single equation by introducing an extension to Eq. (16) written in terms of the transverse effective conductivity k_e^T, namely

$$k_e^T = \varepsilon k_f + (1 - \varepsilon)k_s + (k_f - k_s)G + k_f PeD \tag{30}$$

where Pe is a local Péclet number, $Pe = vd/\alpha_f$, D is a positive parameter to be determined experimentally (function of the internal structure, porosity, contact resistance, etc.), d is a characteristic length scale, and α_f is the thermal diffusivity of the fluid, equal to $k_f/(\rho c_p)$. Because the majority of experiments are performed with a packed bed of spheres, the length scaled d is usually set equal to the sphere diameter (or particle diameter) d_p.

At low flow speeds, when the Péclet number is small, the dispersion effect can be neglected. Moreover, when the dispersion effect is present but not accounted for, the resulting heat transfer enhancement obtained by solving the balance equations will be conservative as the effect of dispersion is always to increase k_e.

III. APPLICATIONS

A review of publications dealing with enhanced heat transfer techniques that are related to porous medium enhanced heat transfer is now presented. References are loosely classified into any one of the following categories: porous inserts and cavities, microsintering, porous coating and porous fins, microchannels, permeable ribs and perforated baffles, and cylinder arrays.

The plethora of recent (post-1993) enhanced heat transfer research publications makes a broad review very difficult. So, broadness is attempted only in the variety of problems reviewed and not in the accumulation of references that pertain to any particular topic. Comprehensive and recent reviews on the general topic of heat transfer enhancement, like those by Andrews and Fletcher (1996), Somerscales and Bergles (1997), Bergles (1997), Sathe and Sammakia (1998), and Kalinin and Dreitser (1998), are cited for their wealth of information and recommended as additional references.

A. Porous Inserts and Cavities

The simplest configuration pertaining to convection heat transfer enhancement using porous media is that of a channel filled with a porous insert. The majority of work in this area was done numerically, with limited experimental model verification.

Hadim (1994) considered the problem of single-phase forced convection in a porous channel with localized heat sources. Hadim and Bethancourt (1995) extended the analysis, considering the forced convection in a partially porous channel with discrete heat sources.

Tong et al. (1993) considered the enhancement achieved by including a porous medium within an isothermal parallel plates channel (Figure 1). They studied numerically several configurations by varying the width s of the porous insert placed symmetrically at the center of the channel. The flow was assumed steady, laminar, and hydrodynamically fully developed (unidirectional). Accordingly, the simplified flow, Eq. (10)

$$0 = -\frac{1}{\rho}\frac{\mathrm{d}\langle p\rangle}{\mathrm{d}x} + \nu\frac{\mathrm{d}^2\langle u\rangle}{\mathrm{d}y^2} - \frac{\nu}{K}\langle u\rangle - \frac{\varepsilon F}{K^{1/2}}\langle u\rangle^2 \tag{31}$$

was used, in which x is a Cartesian coordinate aligned with the longitudinal flow direction, u is the longitudinal fluid speed, and y is the corresponding perpendicular Cartesian coordinate (transverse to the flow direction). Tong et al. (1993) assumed μ_e equal to μ. They also replaced the form coefficient C in Eq. (10) by the group $\varepsilon F/K^{1/2}$, as shown in Eq. (31), where F is usually termed the *inertia* or *Forchheimer* coefficient. Although very common in the literature, this substitution has no physical basis (see Lage 1998 for a detailed historical account and a physical justification for preserving the momentum equation written in terms of C).

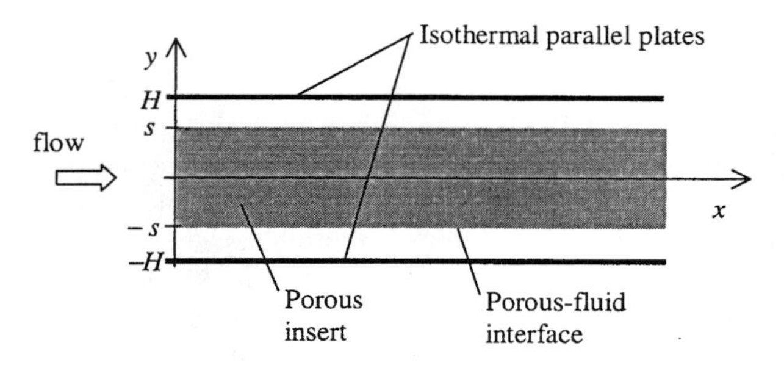

Figure 1. Isothermal parallel plates channel with enhancing porous insert configuration studied by Tong et al. (1993).

The last two terms of Eq. (31) are zero only in the gap-region near the channel surfaces occupied by fluid when the porous insert does not fill completely the channel, i.e., $s < H$. This configuration is referred to as the *partial-fill* configuration, in contrast to the *full-fill* configuration when $s = H$. No distinction is made between the porous-continuum quantities $\langle P \rangle$ and $\langle u \rangle$ of Eq. (31) and the fluid-continuum quantities P and u when solving the equations within a porous medium region or within a fluid region, respectively. Boundary conditions $\langle u \rangle = u$ and $\mathrm{d}\langle u \rangle/\mathrm{d}y = \mathrm{d}u/\mathrm{d}y$ are imposed at the porous–fluid interface (Figure 1).

The energy equation used by Tong et al. (1993) was

$$(\rho c_p)_f \langle u \rangle \frac{\partial \langle T \rangle}{\partial x} = k_e \frac{\partial^2 \langle T \rangle}{\partial y^2} \tag{32}$$

for the porous region, and similarly for the fluid region with k_e replaced by k_f. Again, no distinction is made between the porous-continuum quantities of Eq. (32) and their fluid-continuum counterparts. The porous–fluid interface boundary conditions are: $\langle T \rangle = T$, and $k \partial T/\partial y = k_e \partial \langle T \rangle / \partial y$.

Before considering the results obtained by Tong et al. (1993), it is important to highlight the Nusselt number definition used in their work, namely

$$Nu_m = \frac{h_e H}{k_f} = \kappa \left(\frac{-H(\partial \langle T \rangle / \partial y)|_{y=H}}{(\langle T_w \rangle - \langle T \rangle_m)} \right) \tag{33}$$

The parameter κ is equal to unity for the partial-fill configuration (i.e., when $s < H$, Figure 1). Otherwise, when the porous insert fills the entire channel, i.e., $s = H$, then $\kappa = k_e/k_f$. The Nusselt number definition of Eq. (33) is a slightly modified version of the Nusselt number already presented in Eq. (13). When s is zero (no porous insert), then the surface-averaged Nusselt number becomes equal to 1.885 (Bejan 1995:298)—keep in mind that the characteristic dimension D used by Tong et al. (1993) when defining the Nusselt number was one-half the plate spacing.

Tong et al. (1993) found conditions under which Nu_m could be maximized by filling the channel partially with the porous insert. This enhancement is the result of the extra hydrodynamic drag imposed by the porous medium, which forces the fluid to flow faster through the unimpeded channel region, i.e., the gap-region next to the isothermal channel surfaces. However, when the porous insert fills the entire channel ($s = H$), it is assumed to become a direct participant of the heat transfer process, with zero thermal-contact resistance between the porous medium and the surfaces of the channel. Therefore, a porous insert with high effective thermal conductivity k_e could yield a better heat transfer coefficient when in contact with the channel surface than when partially filling the channel.

When $F = 0$, $Da = K/H^2 = 1 \times 10^{-3}$ and κ is smaller than 2.5, Tong et al. found that Nu_m can be maximized using the *partial-fill* configuration (the value of s which optimizes Nu_m in this case varies almost linearly from $0.7H$ to $0.5H$ as the value of κ decreases from 2.5 to 0.1). If κ is larger than 2.5, the *full-fill* configuration yields the highest Nu_m. The penalty for enhancing the heat transfer is a higher pressure-drop.

One can argue that the option of placing a porous insert in the middle of a channel, without contacting the surfaces of the channel, as a heat transfer enhancing device is an inefficient thermal design. The increased fluid speed near the channel surface, responsible for enhancing the heat transfer in this case, can be obtained more effectively (with a smaller pressure-drop) by simply increasing the flow rate of the clear (no insert) channel. In practice, however, this might not be true because a porous insert can induce flow-mixing inside the channel at a laminar flow rate, therefore yielding a better heat transfer coefficient without the high pressure-drop penalty characteristic of an equivalent turbulent flow case in a clear channel.

Moreover, when radiation is involved the porous core can perform as a re-radiating surface, as in a convection-to-radiation converter configuration. This was demonstrated by the numerical study of Zhang et al. (1997). Their model considered fully developed (unidirectional) laminar flow in the region between the porous insert and the channel (pipe) surface, and in the region occupied by the insert with $\mu_e = \mu$. The energy equation uses k_e, calculated from Eq. (26), and an extra term modeling the radiation effect inside the porous medium. Because of the radiation effect, the porous insert tends to cool more than the fluid within the unobstructed region. High permeability inserts would then be desirable to allow more fluid to flow through the insert, leading to better heat transfer enhancement. However, higher permeability is generally linked to higher porosity, in which case the re-radiating surface area of the porous insert would be smaller, hindering the radiation heat transfer to the channel surface. An optimization problem seems to exist in this case.

A very interesting numerical investigation of forced convection in an isothermal parallel-plates channel was presented by Huang and Vafai (1994). The configuration chosen is depicted in Figure 2. The objective of alternating inserted porous blocks and porous-filled cavities is to enhance the heat transfer from the plates to the fluid. Because of the nonuniformity of the channel geometry with alternating porous blocks and cavities, the flow is inherently nondeveloped. Accordingly, the flow model consisted of using a general momentum equation (similar to (10)) written in a vorticity-

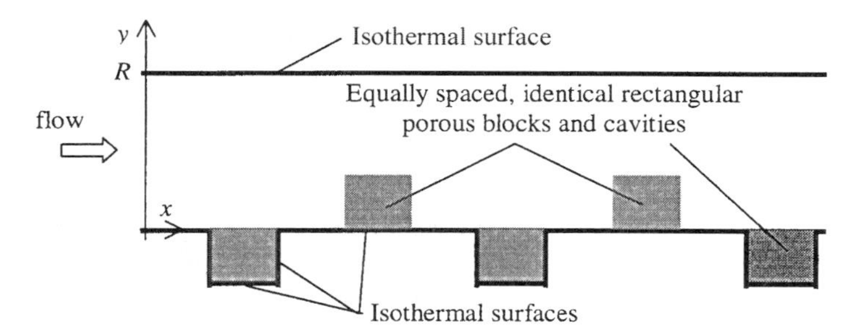

Figure 2. Isothermal parallel plates channel with enhancing configuration of porous blocks and porous-filled cavities studied by Huang and Vafai (1994).

stream function formulation. The Nusselt number was defined similarly to Eq. (13), with the plate spacing in place of the parameter D.

Results were obtained for $\mu_e = \mu$, $k_e = k_f$, and a wide range of flow parameters. The enhancing effect of alternating porous blocks and porous cavities is to promote flow recirculation and mixing, enhancing the heat transfer from the channel surfaces. The enhancing effect is obviously dependent on the flow parameters. Huang and Vafai (1994) found that the interactions between the vortices residing inside the porous-filled cavities and the vortices after the porous blocks have a significant effect on the flow and thermal characteristics of the channel.

Heat transfer enhancement in cooling channels by the use of rolled copper mesh-type porous inserts, brazed to the surfaces of the channel, was investigated by Sozen (1996). This configuration is similar to that presented in Figure 1, with $s = H$, the channel being circular, and the surface of the channel being isoflux instead of isothermal. A finite-differences code, developed and benchmarked against experimental data obtained previously at Argonne National Laboratory, was used for numerical simulations (Sozen and Kuzay 1996).

Comparisons between the average heat transfer performance of plain tubes and tubes with porous inserts were presented. The results of Sozen (1996) showed that the ratio of the average Nusselt number for tubes with porous inserts to the average Nusselt number for plain tubes decreases with increasing Reynolds number. The average Nusselt number of enhanced tubes increased by a factor within the range 20–30 for a Reynolds number in the range 200–2000. On the other hand, the hydraulic pressure drop of the porous-enhanced tubes was 2–3 orders of magnitude higher than that of the plain tubes, for similar Reynolds numbers.

A recent study in this area was presented recently by Ould-Amer et al. (1998). Their configuration is depicted in Figure 3, and is seen as another interesting variation of the previous two configurations.

The use of porous inserts for heat transfer enhancement in recirculating flows, specifically flow over a backward-facing step, was investigated by Abu-Hijleh (1997) and Martin et al. (1998a). This configuration is similar to the ones studies by Huang and Vafai (1994) and Vafai and Huang (1994). The difference is that the flow is made inherently recirculating by the presence of a sudden channel expansion, such as the configuration studied by Ould-Amer et al. (1998), following a backward-facing step.

Abu-Hijleh (1997) considered a configuration in which part of the channel surface, downstream of the step, was porous (Figure 4). The objective was to mitigate the detrimental effect of the recirculating flow by drafting the recirculation within the porous surface. The drag imposed by the porous medium tends to suppress the recirculation.

Martin et al. (1998a) considered inserting a porous block within the flow path immediately past the backward facing step, as shown in Figure 5. Numerical simulations were performed considering laminar flow through high porosity (from 0.9 to 0.99 porosity) inserts, composed of small-diameter (150 μm) silicon carbide fibers aligned transverse to the stream-wise flow direction. The insert length and porosity were varied to determine the most favorable combination for maximum temperature reduction with minimum head-loss.

One important detail distinguishes the work of Martin et al. (1998a) from the majority of studies considering convection in porous medium. The energy transport model invoked by Martin et al. (1998a) did not consider the solid porous matrix in thermal equilibrium with the saturating fluid. In this case, a two-equation energy model must be used, involving a heat transfer coefficient between the solid matrix and the fluid. Interestingly enough, the solid matrix-to-fluid heat transfer coefficient was established

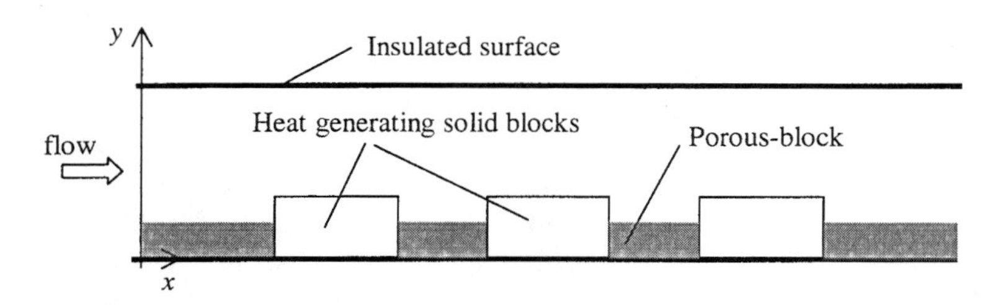

Figure 3. Enhancing configuration of porous blocks between heat generating solid blocks studied by Ould-Amer et al. (1998).

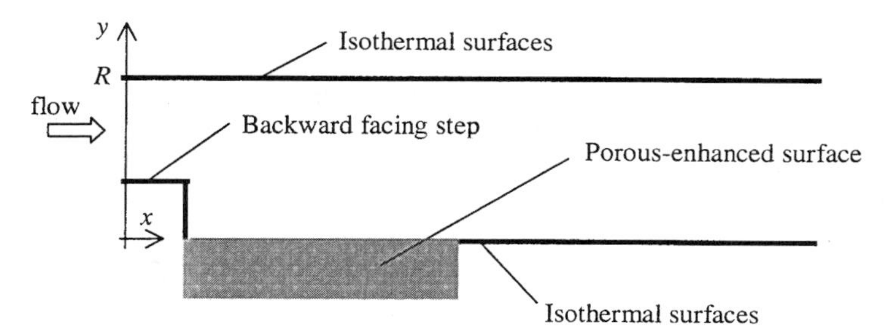

Figure 4. Configuration of porous-enhanced channel downstream of a backward-facing step studied by Abu-Hijleh (1997).

from results of convection through a bundle of parallel cylinders; i.e., results from a configuration generally not perceived as a porous configuration were used to simulate convection heat transfer through a porous medium. Martin et al. (1998a) indicated that the assumption of thermal equilibrium seems to be valid only when long inserts (more than one-tenth of the plate spacing) with porosity less than 0.98 are used.

In general, Martin et al. (1998a) found that the porous insert reduces or eliminates the lower wall recirculation zone characteristic of flow past a backward-facing step. A maximum wall temperature reduction of more than 30%, compared to the maximum wall temperature found in a channel without porous insert, is obtained when the porous insert length is more than one-fifteenth of the plate spacing and the insert porosity is less than 0.98. In this case the pressure-drop is always more than six times the pressure drop of the channel configuration with no porous insert. Only incremental maximum wall-temperature reductions are possible when the porous inserts are

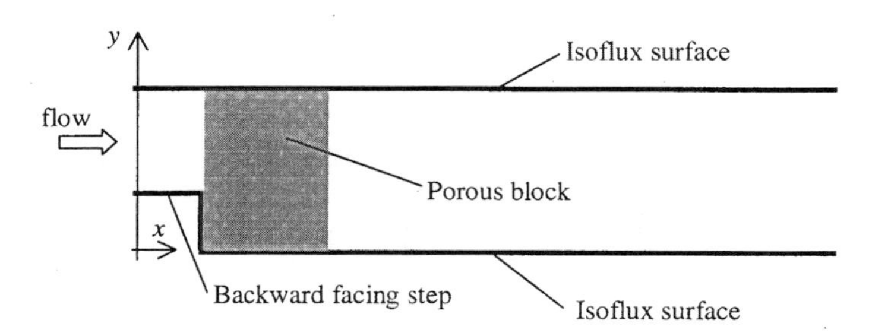

Figure 5. Configuration of porous block placed downstream of a backward-facing step studied by Martin et al. (1998).

longer than one-fifteenth of the plate spacing. However, the head-loss can be as high as 100 times that of the no-insert configuration.

Even though excellent heat transfer characteristics were observed within the inserts themselves, due to the high-conductivity fiber material, there are some cases in which the recirculation zone is lengthened. This happens, for instance, when the inserts are short (about 15/100 of the channel plate spacing or less) and extremely porous (0.99 porosity). That is, the inclusion of a porous block might be detrimental to the thermal performance of the channel when not properly designed.

The single-phase, forced convection cooling of a thin rectangular channel heated uniformly at the top and bottom surfaces and filled with a porous medium, as shown in Figure 6, was studied theoretically and numerically by Lage et al. (1996). This microporous enhanced channel was designed for cooling high-frequency microwave phased-array radar antennas used in military aircraft. The porous insert has two vital functions for this particular application: (1) to enhance the convection heat transfer mechanism; and (2) to induce a uniform flow along the x-direction, shown in Figure 6, leading to a uniform transverse temperature along the y-direction.

The thermohydraulic performance of the enhanced channel was obtained through numerical simulations considering a porous layer insert

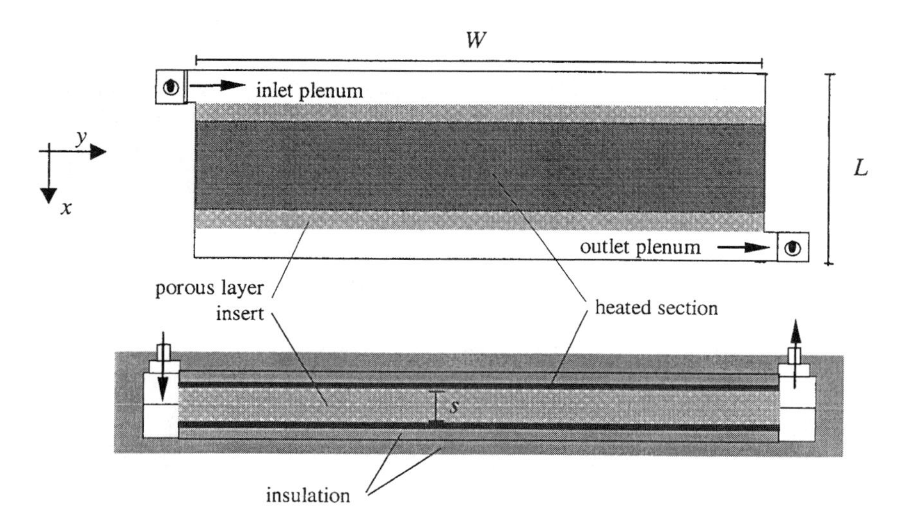

Figure 6. Microporous enhanced cold plate configuration studied by Porneala et al. (1999).

made of aluminum, and air, water, and a synthetic oil as coolants. Energy, momentum, and mass transport equations used for the simulations were similar to the steady-state versions of Eqs. (9)–(11). The very small dimension s of the flow channel, as compared to W and L, warranted the application of a simplified transport model in which the original three-dimensional equations were reduced to a set of two-dimensional equations. Consequently, a volumetric heat source term was included in the energy equation simulating the heat transfer from the surfaces of the channel. Results were compared against the performance of the plain-channel configuration, i.e., a channel without the porous enhancing insert.

A general theoretical expression for predicting the total pressure-drop within the enhanced channel, fundamental for preliminary design considerations, was developed by Lage et al. (1996). Flow mapping based on streamline distribution indicated that a high degree of flow uniformity is achieved along the width of the enclosure when a porous insert is in place. The penalty for the more uniform flow is an increase of the pressure-drop from inlet to outlet of the channel when compared against the results obtained with a plain-channel configuration. The more uniform flow clearly reduces the transverse temperature variation. Lage et al. (1996) found also that the aluminum porous insert always enhances the cooling of the channel by reducing the maximum temperature and reducing the maximum transversal temperature difference.

Even though Lage et al. (1996) showed that a porous enhanced channel configuration could yield much better cooling performance, the practical question of identifying a suitable porous medium insert was not considered. A subsequent specific experimental study was performed by Antohe et al. (1997), who reported results on the hydraulic characteristics of a suitable candidate for the porous insert. The material examined, aluminum alloy 6101-O porous matrix, was chosen for its unique properties, namely low density (for small-to-negligible impact on the total weight of the system), intrinsic mechanical strength (for structural rigidity), and good material compatibility (for brazing the porous layer to the channel surfaces). In addition, the thermal conductivity of this solid porous matrix is much higher than that of most coolants.

The original uncompressed porous matrix can be manufactured in densities (ratio of solid material volume to total volume) varying from 2% to 15%. The void (pore) size can be independently selected from 400 to 1600 pores per meter. The matrix density can be artificially increased by mechanical compression. This is very advantageous as it provides a means to broaden the range of hydraulic parameters of the microporous insert. Moreover, the mechanical compression enlarges the contact (interface) area between the porous insert and the solid surfaces of the channel, facilitating the brazing process, improving the heat transfer across the interface, and enhancing the structural rigidity of the channel.

The rate of fluid flow through the channel is determined by the hydraulic characteristics of the flow passage. The pressure-drop across the porous insert, for a given flow rate and fluid dynamic viscosity, depends on the structure of the insert. Antohe et al. (1997) found that the compressed microporous inserts behave like a porous medium. Indeed, the hydraulic behavior of the compressed inserts can be characterized by the permeability and form coefficient of the medium. These two properties of the metallic microporous insert had to be obtained experimentally as predictive models for this type of porous layer do not exist.

On the basis of the experimental hydraulic parameters obtained by Antohe et al. (1997), Antohe et al. (1996) obtained further numerical simulations using the numerical model introduced by Lage et al. (1996). The results by Antohe et al. (1996) indicated that the hydraulic efficiency of the system, in terms of pressure-drop, could vary by a factor of five, depending on the characteristics of the porous insert used.

Porneala et al. (1999) presented experimental results of thermohydraulic tests performed with the microporous-enhanced design shown in Figure 6. With guidance from previous studies, several microporous-enhanced channels were manufactured for testing. The thermohydraulic performances of these devices were compared against two other configurations using lanced-offset finstock inserts (Manglik and Bergles 1995; Hu and Herold 1995a,b) one of them having a split-flow arrangement. Single-phase convective tests were performed using polyalphaolefin (PAO), a synthetic oil commonly used for cooling military avionics. Their experimental results indicated the feasibility of manufacturing microporous-enhanced channels with very thin porous inserts, yielding excellent thermal performance. An additional original analysis for predicting the pressure-drop penalty when using a coolant with temperature-dependent viscosity was presented recently by Nield et al. (1999)

Pulsating forced flow as a means to enhance the heat transfer from a heated surface has been given more attention in recent years. Flow pulsation enhances the effective axial diffusion in the presence of an axial gradient of concentration or temperature. The enhanced thermal diffusion can be thousands of times larger than transport by molecular diffusion alone.

Kim et al. (1994) and Guo et al. (1997) dealt with different aspects of the problem of pulsating flow through a porous medium-enhanced channel, considering flow channels that are filled completely or partially with a porous medium. The numerical study presented by Guo et al. (1997), for instance, focused on the heat transfer characteristics of pulsating flow in a circular pipe partially filled with a porous medium attached to the channel surface. This configuration is shown schematically in Figure 7. The flow model assumed unidirectional (fully developed) transient momentum balance, with zero convective inertia and $\mu_e = \mu$. No special consideration was given to the porous–fluid interface, nor to the differences between porous-continuum quantities and fluid-continuum quantities. The impacts of the permeability, the thickness of the porous layer, the ratio of effective thermal conductivity of porous material to fluid, as well as the pulsating flow frequency and amplitude, were investigated.

Main emphasis was placed on the minimization of the pressure-drop and the enhancement of the heat transfer by partially filling the channel. The results indicated that the effective thermal diffusivity (thermal molecular diffusion augmented by the mixing effect of the pulsating flow) of a channel partially filled with a porous material is generally higher than the effective diffusivity achieved when the channel is filled entirely with porous medium, and also higher than the diffusivity of the channel without enhancing porous medium. That is, a maximum effective diffusivity is shown to exist for a porous medium thickness between zero and the radius of the channel. This maximum, and the corresponding optimum thickness of porous medium, depends on several thermohydraulic and geometric parameters.

When the permeability of the porous medium increases, the dependence of the effective diffusivity on the porous medium thickness weakens considerably. Generally, as the porous thickness increases beyond the optimal value, the effective diffusivity tends to be smaller than the thermal diffusivity of the channel without a porous insert.

For highly conductive porous media (i.e., $k_e \geq 10k_f$), the bulk-temperature-based Nusselt number (Eq. (13)) increases with increasing thickness of the porous medium, whereas for $k_e = k_f$, a minimum exists in the Nu_m versus porous-thickness relation.

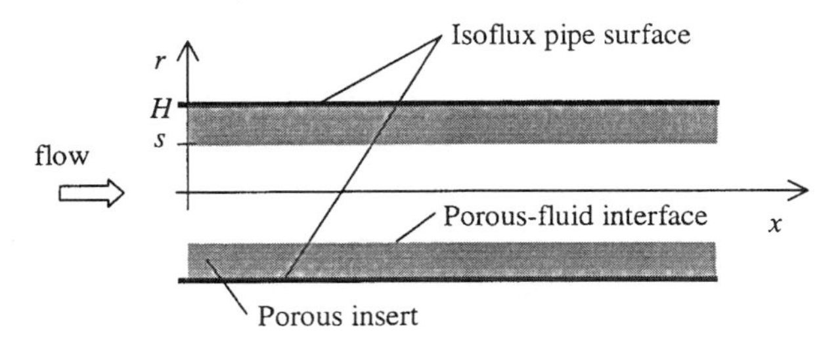

Figure 7. Porous-enhanced isoflux circular channel configuration studied by Guo et al. (1997).

In an extension of the work by Im and Ahluwalia (1994), which discussed the problem of radiation within a pipe filled with small diameter ($\sim 100\ \mu$m) silicon carbide fibers, Martin et al. (1998b) studied numerically the coupling of convective and radiative heat transfer modes on laminar flow. Employing fiber arrays within the flow path alters the velocity profile in the same way as a porous medium does; i.e., it flattens the velocity profile within an entry length, which becomes very short as the porosity of the fiber array is reduced.

The results confirm that convective and radiative internal heat transfer augmentation with fiber arrays has great potential as an enhancing technique. For specified parameters, the heat exchange effectiveness is doubled with the fiber arrays as compared to the no-fibers flow, resulting in a 30% reduction in the wall outlet temperature. Although local thermal equilibrium is evident in the case of conduction and convection alone, a significant departure from the thermal equilibrium assumption is seen when the radiation effect is included and becomes comparable to convection and conduction

B. Microsintering, Porous Fins, and Porous Coating

Microsintering is a useful technique for building porous-enhanced channels by assembling small-diameter metal particles next to a surface. After partial (surface) melting and re-solidification, the particles form a porous flow passage, which is beneficial for the heat transfer process (Lindemuth et al. 1994).

Two-phase heat dissipation utilizing porous-enhanced channels made of high conductivity material was reported recently by Peterson and Chang (1998). Porous channels of various sizes were fabricated using sintered copper particles inside rectangular copper channels with base dimensions of 25 mm × 25 mm, either 3 mm or 10 mm in height. The experiments were conducted using subcooled water as the working fluid, and test conditions ranged from an inlet temperature of 85 to 95°C, inlet pressures of 1.062 to 1.219 bar, flow rates of 22.5 to 150 ml/min, and heat fluxes of 10 to 25 W/cm^2.

Using water with inlet subcooling from 6.6 to 10.8°C, the heat transfer coefficient was 1.25, 1.94, 1.79, and 3.33 W/cm^2 °C when porous channels were used with mean particle diameters of 0.97, 0.54, 0.39, and 0.33 mm, respectively. Hence porous channels made of smaller diameter spheres tended to yield a higher heat transfer coefficient, probably because of the larger heat transfer area. Also, the pressure-drop increased with the heat flux, and this increase was more pronounced at high flow rates.

Some limitations of the sintering technique are: (1) material limitations; (2) practicality (cost-effectiveness) on small dimensions; (3) limited range of hydraulic parameters (mostly limited by the spherical shape of the sintered particles).

Two relatively new enhancing techniques involving the applications of porous media consider the use of fins made of porous media and the coating of heat transfer surfaces with a very thin porous coating paint. Studies describing these two techniques involve phase-change heat transfer, as described next.

The influence of surface condition on boiling heat transfer has been proven to be considerable. Earlier studies have observed decreased superheats due to increased surface roughness during boiling tests with *n*-pentane. Webb (1994), in his review, addressed three types of surface enhancement techniques developed for use in pool boiling. They are: (1) attached nucleation promoters; (2) metal or nonmetal coatings; (3) nucleation sites formed by mechanical working or chemical etching of the base surface. Of these, porous metallic coating is considered to be the most viable enhancement technique; it is formed by sintering, brazing, flame spraying, foaming, or electrolytic deposition. Enhancement in boiling by porous metallic coating is influenced by various geometric parameters such as particle size, particle shape, coating thickness, and porosity.

Microporous enhancement coatings, developed and studied by O'Connor and You (1995), O'Connor et al. (1995), and Chang and You (1997a), have been proven to augment pool boiling heat transfer. These coatings effectively produce multilayered microporous structures over the heated surface. In a more recent study by Chang and You (1997b), experiments were conducted on pool boiling heat transfer from tubes immersed in FC-87 and R-123. Under increasing heat flux conditions, six configurations were tested, such as plain, microporous enhanced, integral fins, micro-porous enhanced low fin, Turbo-B, and high flux.

As a variation of the theme, an experimental study of pool boiling heat transfer from a flat microporous surface immersed in a highly wetting fluid was also performed by Chang and You (1997b). Here, microscale enhancement techniques with five different coatings are used to augment boiling heat transfer performance characteristics such as incipient superheat, nucleate boiling heat transfer, and CHF. The experimental results confirm that the surface microgeometry is a dominant factor for enhancing the nucleate boiling heat transfer performance over the surface area increase (macrogeometry).

A parametric study of nucleate boiling on structured surfaces was reported in a two-part experimental work by Chien and Webb (1998a,b) in which the effect of geometric dimensions on the boiling performance of tunneled enhanced boiling surfaces is identified.

Structured surfaces are made by reforming the base surface to make fins of a standard or special configuration. The structured surfaces all consist of interconnected tunnels and pores (or narrow gaps at the surface). Two key geometric characteristics of the surfaces are the subsurface tunnels and surface pores or fin gaps. These key features are further defined by several dimensional parameters. A description of the test apparatus and a survey of prior work to investigate the effect of pore diameter and pore pitch were given by Chien and Webb (1998b). The results of the detailed parametric study to define the effects of pore diameter and pore pitch lead to the following conclusions:

1. The dry-out heat flux increases with the increase of the total open area. For a given pore pitch, a higher dry-out heat flux will be obtained using larger, rather than smaller, pore sizes.
2. At a certain reduced heat flux, part of the tunnel will become flooded and the performance will be reduced. Smaller pore size will inhibit flooding at reduced heat fluxes.
3. It is possible to select the preferred pore diameter and pore pitch for operation over a specific heat flux range.
4. The boiling coefficient is strongly controlled by the pore size at a given heat flux. The optimum pore diameter is $d_p \sim 0.23\,\mathrm{mm}$ for 1378 fins/m tube at $q'' < 30\,\mathrm{kW/m^2}$.

Renken and Mueller (1993) experimentally demonstrated the potential of condensation enhancement utilizing a porous metallic coating. The experimental conditions could produce observable dropwise, transition, and filmwise condensation on the surface. Comparison of experimental heat transfer results with analytical and numerical predictions was presented. Porosity and permeability values are necessary for the analytical model and to calculate the effective thermal conductivities of the porous coatings. The analytical prediction indicates a very strong dependence of the heat transfer enhancement on coating thickness. The results clearly demonstrate that the employment of a porous metallic coating on a condensing surface is a viable alternative for heat transfer enhancement in condensation problems.

C. Microchannels

Large heat transfer coefficients from microchannel enhanced convective devices are possible because of the relatively large heat transfer surface area. Walpole and Missagia (1993) provided a very interesting review of this subject. It is interesting to observe that if the microchannels were built with irregularities that were randomly oriented, the final solid structure would resemble that of an isotropic porous medium.

Obviously, when the microchannel density (number of channels per volume) is high, a porous medium approach to the numerical simulation of a microchannel heat sink is possible and advantageous. An important particular characteristic of microchannel heat sinks, when seen as porous media, is the anisotropy of the medium. Therefore, when modeling fluid flow through microchannel heat sinks, the transport equations (9)–(11) must be written considering varying thermohydraulic properties such as permeability K, form coefficient C, effective thermal conductivity k_e and porosity ε.

Notwithstanding this apparent difficulty, microchannel heat sinks are generally of simple construction with parallel channels of regular shape (often rectangular) machined (or etched) through the surface of a flat plate. The geometrical regularity of most microchannel heat sinks warrants a simplified porous medium model. Fluid flow will take place only along the axis-direction of the microchannels. Consequently, Eqs. (9)–(11) can in fact be used as long as they are written in scalar form using distinct properties in each direction (when the channel longitudinal axis is parallel to the axis of the microchannels, then the flow through the heat sink can be described by a single equation; otherwise at least two equations may be necessary). Keep in mind that *surface porosity* becomes important, and distinct from volumetric porosity, in this case.

Recent experimental work by Peng et al. (1994a,b) discussed the heat transfer and pressure-drop characteristics of rectangular microchannels convectively cooled with water. Peng et al. (1995) indicated that flat-plate rectangular microchannels are capable of dissipating more than $10\,\mathrm{kW/m^2}$ with single-phase convective cooling.

Single-phase miniature heat sinks operating under high surface heat fluxes tend to present a large streamwise increase in coolant temperature and a corresponding streamwise increase in the heat sink surface temperature. This increase is detrimental to temperature-sensitive devices such as electronic chips. To lessen this detrimental effect, it is often necessary to have the heat sink operating with a high flow rate.

Two-phase flow heat sinks are much less prone to presenting a large temperature variation by relying upon the latent heat of the coolant. Experimental two-phase flow results using mini-channel and microchannel heat sinks were presented by Bowers and Mudawar (1994a). Optimum heat sink geometry, in terms of channel spacing and overall thickness, was ascertained on the basis of a heat diffusion analysis.

The disadvantages of two-phase flow heat sinks are the critical heat flux (CHF) and the density variation experienced by the fluid during phase change. The latter phenomenon seems to play a critical role in increasing the total pressure-drop of the system (Bowers and Mudawar 1994b). Moreover, when the CHF is approached by a flow boiling system, there is a sudden dry-out at the heat transfer surface, accompanied by a drastic reduction in heat transfer coefficient and a corresponding rise in surface temperature.

The small hydraulic diameters of mini- and microchannels suggest an increased frequency and effectiveness of droplet impact with the channel wall in regions of high values of equilibrium quality of the liquid-gas phase at the end of the heated length. This may result in increase of the heat transfer coefficient in those regions and enhanced CHF compared to droplet flow regions in channels with large hydraulic diameter.

The forced convection heat transfer and flow characteristics of water flowing through microchannel plates with extremely small rectangular channels having hydraulic diameters of 0.133–0.367 mm and different geometric configurations were investigated experimentally by Peng and Peterson (1996). The geometric configuration of the microchannel plate and the individual microchannels has a critical effect on single-phase convection heat transfer, and this effect is different for laminar and turbulent flow conditions.

A new dimensionless parameter is introduced to help define the effect of the aspect ratio,

$$Z = \frac{\min(H, W)}{\max(H, W)} \tag{34}$$

where H and W are the height and width of the microchannel plate, respectively. Peng and Peterson (1996) indicated the existence of an optimum Z-value which maximizes the turbulent convective heat transfer. This value is approximately equal to 0.5, regardless of the ratio H/W or W/H. Empirical heat transfer correlations were proposed by Peng and Peterson (1996) for laminar flow

$$Nu = 0.1165\left(\frac{D_h}{W_c}\right)^{0.81}\left(\frac{H}{W}\right)^{-0.79} Re^{0.62} Pr^{1/3} \tag{35}$$

and for turbulent flow

$$Nu = 0.072\left(\frac{D_h}{W_c}\right)^{1.15}\left[1 - 2.421(Z - 0.5)^2\right] Re^{0.8} Pr^{0.33} \tag{36}$$

where Nu is the Nusselt number based on a log-mean temperature difference and the hydraulic diameter of the channel D_h, W_c is the center-to-center microchannel distance, H is the microchannel height, W is the width of the microchannel, Re is the Reynolds number based on the mean fluid speed and hydraulic diameter, and Pr is the fluid Prandtl number.

D. Permeable Fences and Perforated Baffles

Recent experimental articles by Hwang and Liou (1994, 1995a,b) considered the use of solid ribs, placed perpendicular to the main flow direction, to enhance the heat transfer of a convectively cooled channel. When not in contact with each other, the ribs form a fence which is permeable to the fluid flow. Fence surface porosity varying from zero to 50% has been tested. Operating as a turbulating device, the permeable fence affects the flow field by generating mixing, enhancing the heat transfer from the surfaces of the channel.

Another interesting recent study by Hwang et al. (1998) considered the effect of fence thickness on the pressure-drop and heat transfer coefficient of an air-cooled heated channel. The effects of varying the fence thickness-to-height ratio (from 0.16 to 1.0) were experimentally investigated, as well as the effects of varying the fence surface porosity (from 10% to 44%). One of the main contributions of this work was to identify the link between fence thickness and fence permeability. That is, when the fence thickness increases, the fence permeability decreases and the fence thermohydraulic performance tends to that of an impermeable fence. The experimental results indicate that fences with thickness-to-height ratio smaller than 0.16 and surface porosity smaller than 10% yield thermohydraulic results identical to the results obtained with impermeable fences.

An important observation from the experimental results of Hwang et al. (1998) is that the surface-averaged heat transfer coefficient of a fence-enhanced channel, with fence surface porosity greater than 10%, is consistently larger than the heat transfer coefficient of a channel with an impermeable fence. Also, the pressure-drop is consistently lower, decreasing with an increase in the surface porosity. Perhaps unexpectedly, the length-to-height ratio has a very small influence on the average heat transfer coefficient when the surface porosity is greater then 10%. The same is true regarding the pressure-drop.

The use of perforated baffles as flow-straightening devices is very common in many engineering applications. However, the use of perforated baffles as heat transfer enhancing devices is a relatively new idea. When placed inclined to a heated surface (Figure 8), perforated baffles can force the flow to impinge directly onto the heat transfer surface. The presence of baffles within a channel can also disturb the flow, generating additional mixing. Both characteristics can be explored to enhance the heat transfer from a heated surface.

Perforated baffles can be seen as porous media of very short length (most perforated baffles are very thin), and as such they can be analyzed using the fundamental concepts of flow through porous media. Similar to the case of microchannel heat sinks, when modeling fluid flow through perforated baffles the transport equations (9)–(11) must be written considering varying thermohydraulic properties such as permeability K, form coefficient C, effective thermal conductivity k_e, and porosity ε.

Perforated baffles are generally of simple construction with parallel holes of regular shape (often circular) drilled through the surface of a flat plate. Hence, fluid flow will take place only along the axial direction of the holes. Consequently, Eqs. (9)–(11) can be used as long as they are written in scalar form using distinct properties in each direction (when the channel longitudinal axis is parallel to the axis of the holes, then the flow through the baffle is described by a single equation; otherwise at least two equations may be necessary).

Surface porosity becomes important in this case also, and distinct from volumetric porosity. Baffle surface porosity ε_s can be defined as the ratio of the total cross-sectional area of the holes to the entire surface area of the baffle.

Experimental thermohydraulic results of a baffle-enhanced channel placed perpendicular to the main flow direction were reported in Habib et al. (1994). Dutta et al. (1998) conducted experiments using a perforated baffle inclined to 5 degrees. Experiments were conducted using four different baffle orientations (as described in Figure 8). The heat transfer enhancement

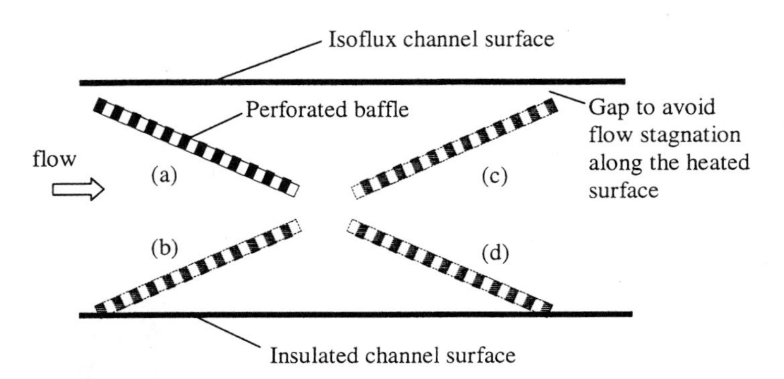

Figure 8. Inclined baffle-enhanced channel configurations (a), (b), (c), and (d) studied by Dutta et al. (1998). Configuration (a) was studied by Dutta and Dutta (1998), using different baffles and baffle inclination.

is quantified by comparing the heat transfer coefficient using solid or perforated baffle to the heat transfer coefficient of a fully developed pipe flow with the same Reynolds number. Results show that the location along the heated channel and the orientation of the baffle play a crucial role in enhancing the surface heat transfer. Configuration (d) in Figure 8 results in better thermal performance when using solid baffle than for perforated baffle, a result almost entirely independent of the baffle position along the channel (comparable results between solid and perforated baffles are found when the baffle is placed near the beginning of the heated section of the channel). Noteworthy is the fact that the local heat transfer enhancement achieved by the perforated baffle used by Dutta et al. (1998) could be as high as 650%. The maximum local enhancement of a solid baffle was 400%.

Dutta and Dutta (1998) performed experiments with a baffle-enhanced rectangular channel heated from the top. Several baffles and baffle inclinations (from 3.8° to 9.6°), positioned as configuration (a) in Figure 8, were used. All baffles were positioned at the same distance from the leading edge of the heated surface, i.e., at $0.013\,L$, where L is the heated surface length. Baffle surface porosity varied from 1.49 to 19.86%.

Results of local heat transfer coefficient seem to indicate that baffles with small surface porosity yield a higher heat transfer coefficient downstream of the baffle, with the solid baffle being the most efficient of all. However, the average Nusselt number results indicate that, for comparable pressure-drop, the perforated baffle yields better thermal performance than the solid baffle, as a direct consequence of the much better heat transfer along the heated surface length covered by the baffle. This better heat transfer coefficient is a consequence of the redirection of the flow—by the holes of the perforated baffle—towards the heated surface. Unfortunately, the experimental results of Dutta and Dutta (1998) and Dutta et al. (1998) are difficult to parameterize because too many parameters are allowed to vary from one experimental configuration to another.

The effect of baffle surface porosity on the thermohydraulic performance of inclined perforated baffle suggests the possibility of an optimum perforation density. This is because of the two competing effects of decreasing the surface porosity of the baffle, thereby enhancing the heat transfer by promoting a stronger flow towards the heated surface, resulting in an increased pressure-drop due to the more restricted flow passage.

E. Cylinder Arrays

Frictional losses and convection heat transfer in sparse, periodic cylinder arrays under laminar air cross-flow were evaluated in the numerical study by Martin et al. (1998c). Interesting in their approach is the desire to correlate microscopic results with macroscopic quantities, e.g., to infer from local information of fluid velocity and pressure-drop inside the pores the correct value of the permeability of the medium. This approach is based on a clear distinction between fluid-continuum and porous-continuum.

The cylinders were arranged in square or triangular patterns, with a fluid volume fraction (porosity) ε ranging from 0.80 to 0.99. The particle Reynolds number, based on the volume-averaged fluid speed and on the diameter of the cylinders, was varied from 3 to 160.

A comparison of the pressure-drop versus flow rate results with the Ergun and Forchheimer relations was made to examine their validity. Results suggest that the Ergun equation (which is known to be valid for a packet bed of spheres only) and the Forchheimer equation do not correlate well with the numerical results. Regarding the comparison with the Forchheimer equation, it is important to realize the difficulty in determining the inertia coefficient in the present case. This difficulty might have influenced the comparison. A modified pressure-drop versus fluid-speed relation, following a power-law function of the permeability-based Reynolds number and with strong dependency on the porosity, was proposed.

The effective heat transfer coefficient between cylinder surfaces and fluid, defined using volume-averaged temperatures, was also calculated and presented in terms of a Nusselt number. A correlation between this Nusselt number and the particle Reynolds number was also provided. No conclusive evidence was found in support of using the permeability to obtain a representative length scale for correlating the heat transfer results.

IV. CONCLUSIONS

The design of porous enhanced heat exchangers for convection heat transfer is more art than science. Even though relatively complex transport equations are available, the transport properties necessary to solve these equations (such as effective thermal conductivity and effective viscosity) are mostly unknown.

Numerical simulations at the microscale are useful for estimating the transport properties, but this approach is limited to simple geometries at the microscale.

Fundamental theory or experimentally derived correlations for determining these transport properties exist only for porous media of specific (simple) internal geometry—mainly packed bed of spheres—operating under specific configurations. There is an overwhelming tendency to apply the few existing predictive models to porous media having distinct characteristics, a consequence of the lack of general predictive models.

Even when using particular models, complications known to affect the heat transfer performance of porous enhanced systems are usually not accounted for. The anisotropy of the conductivity tensor (caused by the thermal dispersion effect), for instance, is often neglected.

Frequently neglected also is the dimensional difference between quantities defined by the classical continuum approach and quantities defined by the porous-continuum approach. This oversight leads to inconsistencies when modeling a porous bounded interface, for instance. The frequently invoked thermal equilibrium assumption has limited validity in several applications. The alternative two-equations energy transport model is not commonly used because of the difficulty in measuring the local heat transfer coefficient between the solid matrix and the saturating fluid.

The application of porous medium concepts to the analysis of irregular, anisotropic, and nonuniform permeable devices used in enhancing convection heat transfer is very positive and should be pursued more vigorously. For one thing it would help to highlight the limitations of existing porous medium formulations and theories. It would help also to determine when a particular enhancing geometry falls outside the porous continuum range. At the same time, it broadens the applicability horizon of porous media concepts.

ACKNOWLEDGMENTS

The first author is indebted to the Institute of Process Engineering and to the Institute of Energy Engineering of the Swiss Federal Institute of Technology for the hospitality and support provided during the fall of 1998 when most of this work was accomplished. Financial support provided by SMU is also gratefully acknowledged.

NOMENCLATURE

Roman Letters

A area
A_c cross-sectional area
C form-drag coefficient
D representative length
E_f enhancement parameter
F Forchheimer parameter
G representative structural parameter
h_e effective convection heat transfer coefficient

H	height, separation distance
k_e^T	effective transverse thermal conductivity
L	length
$\dot{m}$	mass flow rate
n	coordinate measured along the perpendicular direction of a solid–fluid interface
Nu_e	effective Nusselt number
$Nu_{e\text{-avg}}$	surface-averaged Nu_e
Nu_m	effective Nusselt number based on mean temperature
$\langle p \rangle$	volume averaged pressure
P	nondimensional pressure
Pr_e	effective Prandtl number based on k_e
q_w''	surface heat flux
T_b	bulk temperature
T_w	surface temperature
T_{ref}	reference (fixed) temperature
$\langle T \rangle$	volume-averaged temperature
$\langle T_w \rangle$	volume-averaged surface temperature
$\langle T_w \rangle_{\text{avg}}$	surface-averaged $\langle T_w \rangle$
$\langle T \rangle_m$	mean volume-averaged temperature
$\langle T \rangle_{\text{in}}$	inlet $\langle T \rangle$
$\langle u \rangle$	volume-averaged (Darcy) speed
$\mathbf{v}$	vector velocity
$\langle \mathbf{v} \rangle$	volume-averaged (Darcy) velocity
$\langle \mathbf{v} \rangle_p$	pore-volume (intrinsic) averaged velocity
W	width
Z	dimensionless aspect ratio

Greek Letters

Θ	heat exchange effectiveness
κ	ratio of thermal conductivities
μ_e	effective dynamic viscosity

REFERENCES

Abu-Hijleh B. Convection heat transfer from a laminar flow over a 2-D backward facing step with asymmetric and orthotropic porous floor segments. Num Heat Transfer A 31:325–335, 1997.

Andrews MJ, Fletcher LS. Comparison of several heat transfer enhancement technologies for gas heat exchangers. ASME J Heat Transfers 118:897–902, 1996.

Antohe BV, Lage JL, Price DC, Weber RM. Numerical characterization of micro heat exchangers using experimentally tested porous aluminum layers. Int J Heat Fluid Flow 17:594–603, 1996.

Antohe BV, Lage JL, Price DC, Weber RM. Experimental determination of permeability and inertia coefficients of mechanically compressed alumiunum porous matrices. ASME J Fluids Eng 119:404–412, 1997.

Bejan A. Convection Heat Transfer. 2nd ed. New York: John Wiley, 1995.

Bejan A, Morega AM. Optimal arrays of pin fins and plate fins in laminar forced convection. ASME J Heat Transfer 115:75–81, 1993.

Bergles AE. Heat transfer enhancement—the encouragement and accommodation of high heat fluxes. ASME J Heat Transfer 119:8–19, 1997.

Bowers MB, Mudawar I. Two-phase electronic cooling using mini-channel and microchannel heat sinks: Part 1—design criteria and heat diffusion constraints. ASME J Electron Packag 116:290–297, 1994a.

Bowers MB, Mudawar I. Two-phase electronic cooling using mini-channel and microchannel heat sinks: Part 2—flow rate and pressure drop constrains. ASME J Electron Packag 116:298–305, 1994b.

Chang JY, You SM. Enhanced boiling heat transfer from micro-porous surfaces: effects of coating composition and method. ASME J Heat Transfer 119:319–325, 1997a.

Chang JY, You SM. Enhanced boiling heat transfer from micro-porous cylindrical surfaces in saturated FC-87 and R-123. Int J Heat Mass Transfer 40:4449–4460, 1997b.

Cheng P, Hsu C-T. Heat conduction. In: Ingham DB, Pop I, eds. Transport Phenomena in Porous Media. Kidlington: Pergamon, 1998, pp 57–76.

Chien L-H, Webb RL. A parametric study of nucleate boiling on structured surfaces, part 1: effect of tunnel dimensions. ASME J Heat Transfer 120:1042–1048, 1998a.

Chien L-H, Webb RL. A parametric study of nucleate boiling on structured surfaces, part 2: effect of pore diameter and pore pitch. ASME J Heat Transfer 120:1049–1054, 1998b.

Dutta P, Dutta S. Effect of baffle size, perforation, and orientation on internal heat transfer enhancement. Int J Heat Mass Transfer 41:3005–3013, 1998.

Dutta S, Dutta P, Jones RE, Khan JA. Heat transfer coefficient enhancement with perforated baffles. ASME J Heat Transfer 120:795–797, 1998.

Guo Z, Kim SY, Sung HJ. Pulsating flow and heat transfer in a pipe partially filled with a porous medium. Int J Heat Mass Transfer 40:4209–4218, 1997.

Habib A, Mobarak AM, Sallak MA. Abdel Hadi EA, Affify RI. Experimental investigation of heat transfer and flow over baffles of different heights. ASME J Heat Transfer 116:363–368, 1994.

Hadim, A. Forced convection in a porous channel with localized heat sources. ASME J Heat Transfer 116:465–472, 1994.

Hadim A, Bethancourt A. Numerical study of forced convection in a partially porous channel with discrete heat sources. ASME J Electron Packag 117:46–51, 1995.

Hu S, Herold KE. Prandtl number effect on offset fin heat exchanger performance: predictive model for heat transfer and pressure drop. Int J Heat Mass Transfer 38:1043–1051, 1995a.

Hu S, Herold KE. Prandtl number effect on offset fin heat exchanger performance: experimental results. Int J Heat Mass Transfer 38:1053–1061, 1995b.

Huang PC, Vafai K. Internal heat transfer augmentation in a channel using an alternate set of porous cavity–block obstacles. Numer Heat Transfer A 25:519–539, 1994.

Hwang JJ, Liou T-M. Augmented heat transfer in a rectangular channel with permeable ribs mounted on the wall. ASME J Heat Transfer 116:912–920, 1994.

Hwang JJ, Liou T-M. Effect of permeable ribs on heat transfer and friction in a rectangular channel. ASME J Turbomach 117:265–271, 1995a.

Hwang JJ, Liou T-M. Heat transfer in a rectangular channel with perforated turbulence promoters using holographic interferometry. Int J Heat Mass Transfer 38:3197–3207, 1995b.

Hwang JJ, Lia TY, Liou T-M. Effect of fence thickness on pressure drop and heat transfer in a perforated-fence channel. Int J Heat Mass Transfer 41:811–816, 1998.

Im KH, Ahluwalia RK. Radiative enhancement of tube side heat transfer. Int J Heat Mass Transfer 37:2635–2646, 1994.

Kalinin EK, Dreitser GA. Heat transfer enhancement in heat exchangers. Adv Heat Transfer 31:159–322, 1998.

Kaviany M. Principles of Heat Transfer in Porous Media. New York: Springer-Verlag, 1991.

Kim SY, Kang BH, Hyun JM. Heat transfer from pulsating flow in a channel filled with porous media. Int J Heat Mass Transfer 37:2025–2033, 1994.

Lage JL. The fundamental theory of flow through permeable media from Darcy to turbulence. In: Ingham DB, Pop I, eds. Transport Phenomena in Porous Media. Kidlington: Pergamon, 1998, pp 1–30.

Lage JL. The implications of the thermal equilibrium assumption for surrounding-driven steady conduction within a saturated porous medium layer. Int J Heat Mass Transfer 42:477-485, 1999.

Lage JL, Weinert AK, Price DC, Weber RM. Numerical study of low permeability microporous heat sink for cooling phased-array radar systems. Int J Heat Mass Transfer 39:3622–3647, 1996.

Lindemuth JE, Johnson DM, Rosenfeld JH. Evaluation of porous metal heat exchangers for high heat flux applications. ASME, Heat Transfer Div 301:93–98, 1994.

Manglik RM, Bergles AE. Heat transfer and pressure drop correlations for rectangular offset strip fin compact heat exchangers. Exp Therm Fluid Sci 10:171–180, 1995.

Martin AR, Saltiel C, Shyy W. Heat transfer enhancement with porous inserts in recirculating flows. ASME J Heat Transfer 120:458–467, 1998a.

Martin AR, Saltiel C, Shyy W. Convective and radiative internal heat transfer augmentation with fiber arrays. Int J Heat Mass Transfer 41:3431–3440, 1998b.

Martin AR, Saltiel C, Shyy W. Frictional losses and convective heat transfer in sparse periodic cylinder arrays in cross flow. Int J Heat Mass Transfer 41:2383–2397, 1998c.

Nield DA. Estimation of the stagnant thermal conductivity of saturated porous media. Int J Heat Mass Transfer 34:1575–1576, 1991.

Nield DA, Bejan A. Convection in Porous Media. 2nd ed. New York: Springer Verlag, 1999.

Nield DA, Porneala DC, Lage JL. A theoretical study, with experimental verification, of the temperature-dependent viscosity effect on the forced convection through a porous medium channel. ASME J Heat Transfer 121:365–369, 1999.

O'Connor JP, You SM. A painting technique to enhance pool boiling heat transfer in saturated FC-72. ASME J Heat Transfer 117:387–393, 1995.

O'Connor JP, You SM, Price DC. Thermal management of high power microelectronics via immersion cooling. IEEE Trans CPMT 18:656–663, 1995.

Ould-Amer Y, Chikh S, Bouhadef K, Lauriat G. Forced convection cooling enhancement by use of porous materials. Int J Heat Fluid Flow 19:251–258, 1998.

Peng XF, Peterson GP. Convective heat transfer and flow friction for water flow in microchannel structures. Int J Heat Mass Transfer 39:2599–2608, 1996.

Peng XF, Peterson GP, Wang BX. Frictional flow characteristics of water flowing through rectangular microchannels. Exp Heat Transfer 7:249–264, 1994a.

Peng XF, Peterson GP, Wang BX. Heat transfer characteristics of water flowing through microchannels. Exp Heat Transfer 7:265–283, 1994b.

Peng XF, Wang BX, Peterson GP, Ma HB. Experimental investigation of heat transfer in flat plates with rectangular microchannels. Int J Heat Mass Transfer 38:127–137, 1995.

Peterson GP, Chang CS. Two-phase heat dissipation utilizing porous-channels of high-conductivity material. ASME J Heat Transfer 120:243–252, 1998.

Porneala DC, Lage JL, Price DC. Experiments of forced convection through microporous enhanced cold plates for cooling airborne microwave antennas. Proceedings of 1999 ASME National Heat Transfer Conference, Albuquerque, NM, 1999.

Pratt M. On the boundary conditions at the macroscopic level. Transp. Porous Media 4:259–280, 1989.

Pratt M. Modeling of heat transfer by conduction in a transition region between a porous medium and an external fluid. Transp Porous Media 5:71–95, 1990.

Renken KJ, Mueller CD. Measurements of enhanced film condensation utilizing a porous metallic coating. AIAA J Thermophysics Heat Transfer 7:148–152, 1993.

Sathe S, Sammakia B. A review of recent developments in some practical aspects of air-cooled electronic packages. ASME J Heat Transfer 120:830–839, 1998.

Somerscales EFC, Bergles AE. Enhancement of heat transfer and fouling mitigation. Adv Heat Transfer 30:197–249, 197.

Sozen M. Use of porous inserts in heat transfer enhancement in cooling channels. Proceedings of 1996 ASME International Mechanical Engineering Congress and Exhibition, Atlanta, GA, 1996.

Sozen M, Kuzay TM. Enhanced heat transfer in round tubes with porous inserts. Int J Heat Fluid Flow 17:124–129, 1996.

Tong TW, Sharatchandra MC, Gdoura Z. Using porous inserts to enhance heat transfer in laminar fully-developed flows. Int Commun Heat Mass Transfer 20:761–770, 1993.

Vafai K, Huang PC. Analysis of heat transfer regulation and modification employing intermittently emplaced porous cavities. ASME J Heat Transfer 116:604–613, 1994.

Walpole JN, Missagia LJ. Microchannel heat sinks for two-dimensional diode laser array. In: Evans GA, Hammer JM, eds. Surface Emitting Semiconductor Lasers and Arrays. New York: Academic Press, 1993.

Webb RL. Principles of Enhanced Heat Transfer. New York: John Wiley, 1994.

Webster's Unabridged Dictionary, 1995.

Whitaker S. The Method of Volume Averaging. Dordrecht: Kluwer Academic, 1999.

Zhang JM, Sutton WH, Lai FC. Enhancement of heat transfer using porous convection-to-radiation converter for laminar flow in a circular duct. Int J Heat Mass Transfer 40:39–48, 1997.

9
Flow and Thermal Convection in Rotating Porous Media

P. Vadasz
University of Durban-Westville, Durban, South Africa

I. INTRODUCTION

The study of flow in rotating porous media is motivated by its practical applications in geophysics and engineering. Among the applications of rotating flow in porous media to engineering disciplines, one can find the food processing industry, chemical process industry, centrifugal filtration processes, and rotating machinery. Detailed discussion on particular applications has been presented by Nield and Bejan (1999), Bejan (1995), and Vadasz (1997).

Very limited research is available on *isothermal* flow in *rotating* porous media, whereas some results are available for natural convection in rotating porous media; e.g., Palm and Tyvand (1984), Rudraiah et al. (1986), Patil and Vaidyanathan (1983), and Jou and Liaw (1987). Nield (1991b), while presenting a comprehensive review of the stability of convective flows in porous media, found also that the effect of rotation on convection in a porous medium attracted limited interest. The lack of experimental results was particularly noticed.

The main reason behind the lack of interest for this type of flow is probably the fact that isothermal flow in homogeneous porous media following Darcy's law is irrotational (Bear, 1972), hence the effect of rotation on this flow is not significant. However, for a heterogeneous medium with spatially dependent permeability or for free convection in a nonisothermal homogeneous porous medium the flow is no longer irrotational, and hence the effects of rotation become significant. In some applications these effects can be small, e.g., when the porous media Ekman number is high. Nevertheless, the effect of rotation is of interest as it may generate secondary flows in planes perpendicular to the main flow direction. Even when these secondary flows are weak, it is essential to understand their source as they might be detectable in experiments. The latter claim can be substantiated by looking at the corresponding problems in rotating flows in pure fluids (non-porous domains). There, the Coriolis effect and secondary motion in planes perpendicular to the main flow direction are controlled by the Ekman number. Experiments (Hart 1971; Johnston et al. 1972; Lezius and Johnston, 1976) showed that this secondary motion is detectable, even for very low or very high Ekman numbers, although the details of this motion may vary considerably according to the pertaining conditions. One can therefore expect to obtain secondary motion when a solid porous matrix is present in a similar geometric configuration, although the details of the motion cannot be predicted *a priori* on the basis of physical intuition only. This creates a strong motivation to investigate the effect of rotation in isothermal heterogeneous porous media. For high angular velocity of rotation or high permeability, conditions which pertain to some engineering applications, the Ekman number can become of order of magnitude unity or lower and then the effect of rotation becomes even more significant. The same motivation applies for investigating the effect of free convection in porous media.

The sequence adopted in this chapter consists of a presentation of the dimensionless equations governing the flow and transport phenomena in a rotating frame of reference followed by a classification of problems, i.e., isothermal rotating flows in heterogeneous porous media and free convection in rotating homogeneous porous systems. The available results to the different types of problems are then presented and, finally, recommendations are proposed regarding further research needed on topics which fall short of having available reliable results.

II. GOVERNING EQUATIONS

The dimensionless equations governing the flow and heat transfer in a rotating heterogeneous (but isotropic) porous medium, following Darcy's law but extended to include the Coriolis and centrifugal terms are presented in the form

1. *Continuity equation*

$$\nabla \cdot V = 0 \tag{1}$$

2. *Darcy's law* (extended to include rotation effects)

$$V = -K\left[\nabla p_r + Ra_g T\nabla(\hat{e}_g \cdot X) - Ra_\omega T\hat{e}_\omega \times (\hat{e}_\omega \times X) + \frac{1}{Ek}\hat{e}_\omega \times V\right] \tag{2}$$

3. *Energy equation*

$$\frac{\partial T}{\partial t} + V \cdot \nabla T = \nabla^2 T + \dot{Q} \tag{3}$$

where it was assumed that local thermal equilibrium between the solid and fluid phases applies, and provision was made in Eq. (3) for the possibility of internal heat generation $\dot{Q}$. For cases without heat generation, $\dot{Q} = 0$. In Eqs. (1)–(3) V is the dimensionless filtration velocity (Darcy's flux), p_r is the dimensionless reduced pressure generalized to include the constant components of the centrifugal as well as the gravity terms, T is the dimensionless temperature, $K(X)$ is the dimensionless permeability function, $\hat{e}_\omega$ is a unit vector in the direction of the imposed angular velocity, and $X = x\hat{e}_x + y\hat{e}_y + z\hat{e}_z$ is the position vector. The values $u_c = v_o/l_c$ and $\Delta p_c = \mu_o v_o/K_o$ were used to scale the filtration velocity and pressure, respectively, and K_o was used to scale the permeability function. For heat convection problems a rescaling of the equations by using $u_c = \alpha_{eo}/l_c$ and $\Delta p_c = \mu_o \alpha_{eo}/K_o$ for scaling the filtration velocity and pressure, respectively, was applied. The characteristic length l_c, used to scale the space variables x_*, y_*, and z_*, needs to be defined in the context of the particular problem under consideration, and the characteristic time $\Delta t_c = l_c/u_c$ was used to scale the time variable. A linear approximation was assumed for the relationship between the density and temperature in the form

$$\rho = 1 - \beta_T T \tag{4}$$

Furthermore, by using the Oberbeck–Boussinesq approximation, the density was assumed constant everywhere except in a body force term of Eq. (2). In Eq. (4) $\beta_T = \beta_* \Delta T_c$, where β_* is the thermal expansion coefficient and ΔT_c is a characteristic temperature difference (typically $\Delta T_c = (T_H - T_C)$) used to scale the temperature in the equations. The dimensionless groups in (2) are the gravity-related Rayleigh number in porous media, the porous media Rayleigh number related to the centrifugal body force, and the porous media Ekman number, defined as

$$Ra_g = \frac{\beta_* \Delta T_c g_* l_c K_o}{\nu_o \alpha_{eo}} \qquad Ra_\omega = \frac{\beta_* \Delta T_c \omega_*^2 l_c^2 K_o}{\nu_o \alpha_{eo}} \qquad Ek = \frac{\varepsilon \nu_o}{2\omega_* K_o} \tag{5}$$

where ε is the porosity, ω_* is the angular velocity of rotation, g_* is the acceleration of gravity, K_o is a reference value of permeability, ν_o is the kinematic viscosity, α_{eo} is the effective thermal diffusivity extended to account for the ratio between the heat capacity of the fluid and the effective heat capacity of the porous domain, l_c is a characteristic length scale, β_* is the thermal expansion coefficient, and ΔT_c is a characteristic temperature difference. These dimensionless groups control the significance of the different phenomena. Therefore, the value of Ekman number (Ek) controls the significance of the Coriolis effect, and the ratio between the gravity-related Rayleigh number (Ra_g) and the Rayleigh number related to the centrifugal body force (Ra_ω) controls the significance of gravity with respect to centrifugal forces as far as free convection is concerned. This ratio is $Ra_g/Ra_\omega = g_*/\omega_*^2 l_c$.

Equations (1)–(3) include three distinct mechanisms of coupling. Two of these couplings are linear. The first is the coupling between the pressure terms and the filtration velocity components, i.e., a coupling between Eq. (1) and the three components of Eq. (2). The second linear coupling is due to the Coriolis acceleration acting on the filtration velocity components in the plane perpendicular to the direction of the angular velocity. The third coupling is nonlinear as it involves the gravity or centrifugal buoyancy terms in Eq. (2) and the energy in Eq. (3). As $\boldsymbol{V}$ in Eq. (2) is dependent on temperature (T) because of the buoyancy terms, it follows that the convection term $\boldsymbol{V} \cdot \nabla T$ in the energy equation causes the nonlinearity. Therefore the coupling between the temperature T and the filtration velocity $\boldsymbol{V}$ is the only source of nonlinearity in the Darcy's formulation of flow and heat transfer in rotating porous media.

Decoupling the equations is difficult without losing generality, although the linear couplings can be resolved. Resolving the Coriolis-related coupling is particularly useful. In doing so, a Cartesian coordinate system is used and, without loss of generality, one can choose $\hat{\boldsymbol{e}}_\omega = \hat{\boldsymbol{e}}_z$. A further choice is made, which reduces the generality of the problem, namely that the gravity acceleration is collinear with the z-axis and directed downwards, i.e., $\hat{\boldsymbol{e}}_g = -\hat{\boldsymbol{e}}_z$. Some generality is lost as problems where the directions of rotation and gravity are not collinear will not be represented by the resulting equations Nevertheless, it is of interest to demonstrate this particular choice of system as it represents a significant number of practical cases. Subject to these choices ($\hat{\boldsymbol{e}}_\omega = \hat{\boldsymbol{e}}_z$ and $\hat{\boldsymbol{e}}_g = -\hat{\boldsymbol{e}}_z$), Eq. (2) can be expressed in the following form (with the r dropped from p_r for convenience)

$$\left[1 + KEk^{-1}\hat{\boldsymbol{e}}_z \times\right]\boldsymbol{V} = -K\left[\nabla p + Ra_\omega T(x\hat{\boldsymbol{e}}_x + y\hat{\boldsymbol{e}}_y) - Ra_g T\hat{\boldsymbol{e}}_z\right] \tag{6}$$

An important observation can be made by presenting Eq. (6) explicitly in terms of the three scalar components

$$u - KEk^{-1}v = -K\left[\frac{\partial p}{\partial x} + Ra_\omega Tx\right] \tag{7}$$

$$KEk^{-1}u + v = -K\left[\frac{\partial p}{\partial y} + Ra_\omega Ty\right] \tag{8}$$

$$w = -K\left[\frac{\partial p}{\partial z} + Ra_g T\right] \tag{9}$$

where u, v, and w are the corresponding x, y, and z components of the filtration velocity vector $\boldsymbol{V}$. It is now observed that the first two equations (7) and (8) for the horizontal components of $\boldsymbol{V}$ are "Coriolis coupled," while the third is not. Since this type of coupling is linear, it is possible to decouple them to obtain

$$\boldsymbol{V_H} = -\mathbb{E}_\mathrm{H} \cdot [\nabla_H p + Ra_\omega T \boldsymbol{X}_H] \tag{10}$$

where $\boldsymbol{V_H} = u\hat{\boldsymbol{e}}_x + v\hat{\boldsymbol{e}}_y$ is the horizontal filtration velocity, $\boldsymbol{X}_H = x\hat{\boldsymbol{e}}_x + y\hat{\boldsymbol{e}}_y$ is the horizontal position vector, $\nabla_H \equiv \partial/\partial x\hat{\boldsymbol{e}}_x + \partial/\partial y\hat{\boldsymbol{e}}_y$ is the horizontal gradient operator, and $\mathbb{E}_\mathrm{H}$ is the following operator

$$\mathbb{E}_\mathrm{H} = \frac{K}{[1 + Ek^{-2}K^2]}\begin{bmatrix} 1 & Ek^{-1}K \\ -Ek^{-1}K & 1 \end{bmatrix} \tag{11}$$

The definition of the operator $\mathbb{E}_\mathrm{H}$ can be extended to include the z component of $\boldsymbol{V}$, thus leading to

$$\boldsymbol{V} = -\mathbb{E} \cdot [\nabla p + \boldsymbol{B}T] \tag{12}$$

where $\boldsymbol{B} = Ra_\omega x\hat{\boldsymbol{e}}_x + Ra_\omega y\hat{\boldsymbol{e}}_y - Ra_g\hat{\boldsymbol{e}}_z$ is the buoyancy coefficient vector and the operator $\mathbb{E}$ is defined in the form

$$\mathbb{E} = \frac{K}{[1 + Ek^{-2}K^2]}\begin{bmatrix} 1 & Ek^{-1}K & 0 \\ -Ek^{-1} & 1 & 0 \\ 0 & 0 & (1 + Ek^{-2}K^2) \end{bmatrix} \tag{13}$$

It is interesting to observe from Eqs. (12) and (13) that the Coriolis effect due to rotation is equivalent to a particular form of anisotropic porous medium. The anisotropy is represented by the tensor operator $\mathbb{E}$. It should be pointed out that the analogy between the Coriolis effect and anisotropy was identified by Palm and Tyvand (1984) for the problem of gravity-driven thermal convection in a rotating porous layer at marginal stability. It has been shown here that this analogy is more general and applies also to centrifugally driven convection and to isothermal flows in rotating heterogeneous porous media ($Ra_\omega = Ra_g = 0 \rightarrow \boldsymbol{B} = 0$ in Eq. (12)).

The non-Darcy models presented in this chapter deal predominantly with an extension of Eq. (2) which includes a time derivative term. For homogeneous porous media $K = 1$, and this model is presented in the form

$$\frac{Da}{\varepsilon Pr}\frac{\partial V}{\partial t} + V = -\Big[\nabla p_r + Ra_g T \nabla(\hat{e}_g \cdot X) - Ra_\omega T \hat{e}_\omega \times (\hat{e}_\omega \times X) - \frac{1}{Ek}\hat{e}_\omega \times V\Big] \tag{14}$$

From these equations it is clear that for very small values of the Prandtl number (i.e., $Pr = \mathbf{O}(Da)$), which are typical for liquid metals, the time derivative term in Eq. (14) should be retained. Even when $Da \ll Pr$ there might be instances when the time derivative in (14) cannot be neglected especially when oscillatory or chaotic flows are of interest. In such circumstances, neglecting the time derivative term is equivalent to the neglect of the higher derivative from the equation, and might prevent all boundary (or initial) conditions from being satisfied. Other non-Darcy models and their appropriate implementation are discussed by Nield (1983, 1991a).

III. CLASSIFICATION OF ROTATING FLOWS IN POROUS MEDIA

Rotating flows in porous media can be dealt with by classifying them first into three major categories:

1. Isothermal flows in heterogeneous porous media subject to rotation
2. Convective flows in non-isothermal homogeneous porous media subject to rotation
3. Convective flows in non-isothermal heterogeneous porous media subject to rotation

This chapter is concerned with the first two categories. The third category is not covered, simply because there are no reported research results on free convection in rotating heterogeneous porous media. Regarding category (1) above, it is evident from Eq. (2), subject to isothermal conditions, i.e., $Ra_g = Ra_\omega = 0$, that heterogeneity of the medium is an important condition for the effects of rotation to be significant. This can be observed, for example, from the basic Darcy's law which under homogeneous conditions (i.e., $K = \text{const.} = 1$) yields irrotational types of flows, i.e., the vorticity

$\nabla \times V = 0$. The heterogeneity of the medium, $K \equiv K(X)$, introduces a nonvanishing vorticity, i.e., $\nabla \times V \neq 0$. Nonisothermal flows, as a result of free convection, allow also a nonvanishing vorticity.

Convective flows in rotating porous media are further classified into three categories, each one of which can be separated into three cases. In order to justify this classification we proceed first by defining the phenomenon of free convection. Free convection is the phenomenon of fluid flow driven by density variations in a fluid subject to body forces. Therefore, there are two necessary conditions to be met in order to obtain convective flow: (i) *density variations exist within the fluid;* and (ii) *the fluid must be subjected to body forces.* Density is in general a function of pressure, temperature, and solute concentration (in the case of a binary mixture), i.e., $\rho \equiv \rho(p, T, S)$, and its variation with respect to pressure is much smaller than that with respect to temperature or concentration (i.e., $\beta_p \ll \beta_T$, leading to Eq. (4) for example). Hence the convection can be driven either by temperature variations or by variations in solute concentrations or by both. In the two latter cases, Eq. (4) should be extended to include S in the form $\rho = 1 - \beta_T + \beta_S S$, the solute transport equation should be included in the model, and Eq. (2) should be extended as well to account for the effect of solute on density. Gravity, centrifugal forces, electromagnetic forces (in the case of liquid metals subject to an electric field) are only examples of body forces which represent the second necessary condition for free convection to occur. Some of these body forces are constant, such as gravity for example, while others can vary linearly with the perpendicular distance from the axis of rotation, like the centrifugal force. The lack of body forces, for example under microgravity conditions in outer space, prevents occurrence of convection. The two conditions mentioned above are indeed necessary for free convection to occur; however, they are not sufficient. The relative orientation of the density gradient with respect to the body force is an important factor for providing the sufficient conditions for convection to occur. This is shown graphically in Figure 1 for the particular case of thermal convection, where $\boldsymbol{B}$ represents the body force and $\nabla\rho = -\beta_T \nabla T$ is the direction of the density gradient.

Given this basic introduction to the causes of the setup of convection, the following classification of convective flows is introduced:

1. Convection due to thermal buoyancy
2. Convection due to thermo-solutal buoyancy

A third category related to convection due to solutal buoyancy alone could have been introduced but it would not bring any significant contribution to (1) above as the results obtained for thermal buoyancy alone can be easily converted to the third case by analogy.

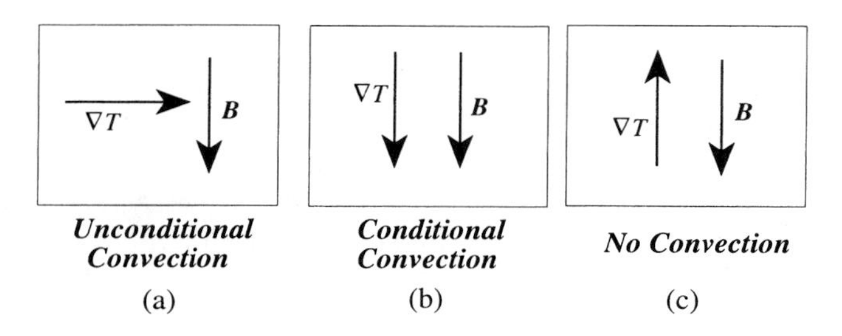

Figure 1. Effect of the relative orientation of the temperature gradient with respect to the body force on the setup of convection.

Three separate cases for each one of the above categories can be considered, depending on the driving body force, i.e., *convection driven by gravity, convection driven by the centrifugal force,* and *convection driven by both gravity and centrifugal force.*

IV. ISOTHERMAL FLOW IN HETEROGENEOUS POROUS MEDIA SUBJECT TO ROTATION

A. General Background

Considering Darcy's regime under isothermal conditions ($Ra_g = Ra_\omega = 0$) yields the following equations, from (6) and (12)

$$[1 + KEk^{-1}\hat{e}_z \times]\boldsymbol{V} = -K\nabla p \qquad \text{or} \qquad \boldsymbol{V} = -\mathbb{E}\cdot\nabla p \tag{15}$$

where $\boldsymbol{B} = 0$ in Eq. (12) and $\mathbb{E}$ is the tensor operator defined by Eq. (13). Further decoupling between the pressure and the filtration velocity is made possible by applying the divergence operator ($\nabla\cdot$) on (15) and making use of (1), leading to

$$\nabla^2 p + Ek^{-2}K^2\frac{\partial^2 p}{\partial z^2} + \gamma_1\frac{\partial K}{\partial z}\frac{\partial p}{\partial z} + \gamma_2\nabla_H K\cdot\nabla_H p + Ek^{-1}\gamma_3 J_H(K,p) = 0 \tag{16}$$

where γ_1, γ_2, and γ_3 are the coefficients

$$\gamma_3 = \frac{2}{[1 + Ek^{-2}K^2]}; \qquad \gamma_1 = \frac{2}{\gamma_3 K}; \qquad \gamma_2 = \frac{\gamma_e[1 - Ek^{-2}K^2]}{2} \tag{17}$$

and $J_H(K, p)$ is the horizontal Jacobian, defined as

$$J_H(K, p) = \frac{\partial K}{\partial x}\frac{\partial p}{\partial y} - \frac{\partial K}{\partial y}\frac{\partial p}{\partial x} \tag{18}$$

Let us consider now a few cases of interest to demonstrate the application of the theory to some particular examples. When the heterogeneity can be represented by a vertical stratification $K \equiv K(z)$ and therefore is independent of the horizontal coordinates x and y, the horizontal Jacobian and the horizontal permeability gradient vanish, i.e., $J_H = 0$ and $\nabla_H K = 0$, which yields

$$\nabla^2 p + Ek^{-2}K^2\frac{\partial^2 p}{\partial z^2} + \gamma_1\frac{\partial K}{\partial z}\frac{\partial p}{\partial z} = 0 \tag{19}$$

For a homogeneous medium $K = 1$, and the pressure equation takes the form $\nabla^2 p + Ek^{-2}\partial^2 p/\partial z^2 = 0$. Compressing the vertical coordinate by using the transformation $\zeta = z/(1 + Ek^{-2})^{1/2}$ yields, for this case, a Laplace equation in the form $\partial^2 p/\partial x^2 + \partial^2 p/\partial y^2 + \partial^2 p/\partial \zeta^2 = 0$. One can notice that, except for the compression of the vertical coordinate, the rotation does not have any significant effect in homogeneous media.

B. Taylor–Proudman Columns and Geostrophic Flow in Rotating Porous Media

To present the topic of Taylor–Proudman columns we consider Eq. (15) multiplied by $[Ek/K]$ for $\hat{e}_\omega = \hat{e}_z$ and rescale the pressure in the form $p_R = Ek\, p$ to yield

$$\left[\frac{Ek}{K} + \hat{e}_z \times\right] V = -\nabla p_R \tag{20}$$

Given typical values of viscosity, porosity, and permeability, one can evaluate the range of variation of Ekman number in some engineering applications. There, the angular velocity may vary from 10 rpm to 10 000 rpm, leading to Ekman numbers in the range from $Ek = 1$ to $Ek = 10^{-3}$. The latter value is very small, pertaining to the conditions considered in this section. Therefore, in the limit of $Ek \to 0$, say $Ek = 0$, Eq. (20) takes the simplified form

$$\hat{e}_z \times V = -\nabla p_R \tag{21}$$

and the effect of permeability variations disappears. Taking the "curl" of (21) leads to

$$\nabla \times (\hat{e}_z \times V) = 0 \tag{22}$$

Evaluating the curl operator on the cross product of the left-hand side of (22) gives

$$(\hat{e}_z \cdot \nabla)\boldsymbol{V} = 0 \tag{23}$$

Equation (23) is identical to the Taylor–Proudman condition for pure fluids (non-porous domains); it thus represents the proof of the Taylor–Proudman theorem in porous media and can be presented in the following simplified form

$$\frac{\partial \boldsymbol{V}}{\partial z} = 0 \tag{24}$$

The conclusion expressed by Eq. (24) is that $\boldsymbol{V} = \boldsymbol{V}(x, y)$, i.e., it cannot be a function of z, where z is the direction of the angular velocity vector. This means that all filtration velocity components can vary only in the plane perpendicular to the angular velocity vector.

The consequence of this result can be demonstrated by considering a particular example (see Greenspan (1980) for the corresponding example in pure fluids). Figure 2 shows a closed cylindrical container filled with a fluid-saturated porous medium. The topography of the bottom surface of the container is slightly changed by fixing securely a small solid object. The container rotates with a fixed angular velocity ω_*. Any forced horizontal flow in the container is expected to adjust to its bottom topography. However, since Eq. (24) applies to each component of $\boldsymbol{V}$ it applies in parti-

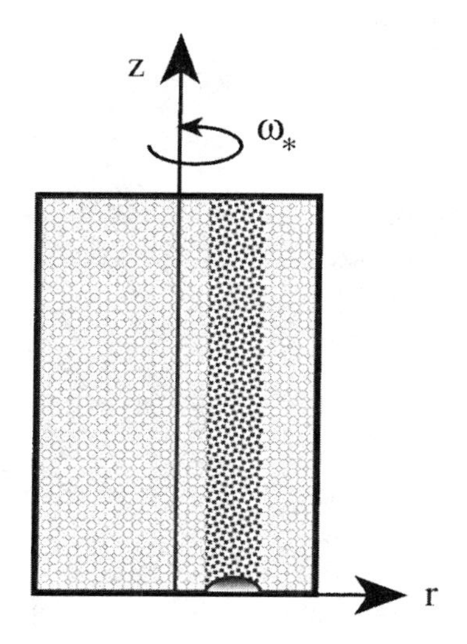

Figure 2. A closed cylindrical container filled with a fluid-saturated porous medium. A solid object is fixed at the bottom and a qualitative description of a Taylor–Proudman column in porous media is given.

cular to w, i.e. $\partial w/\partial z = 0$. But the impermeability conditions at the top and bottom solid boundaries require $\boldsymbol{V} \cdot \hat{e}_n = 0$ at $z = h(r, \theta)$ and at $z = 1$, where $h(r, \theta)$ represent the bottom topography in polar coordinates (r, θ). The combination of these boundary conditions with the requirement that $\partial w/\partial z = 0$ yields $w = 0$ anywhere in the field. Hence, a flow over the object, as described qualitatively in Figure 3a, becomes impossible as it introduces a vertical component of velocity. Therefore, the resulting flow may adjust around the object, as presented qualitatively in Figure 3b. However, since this flow pattern is also independent of z, it extends over the whole height of the container, resulting in a fluid column above the object which rotates as a solid body. This will demonstrate a Taylor–Proudman column in porous media, as presented qualitatively in Figure 2. It should be pointed out that the column in porous media should be expected in the macroscopic sense, i.e., the diameter of the column (and naturally the diameter of the obstacle) must have macroscopic dimensions which are much greater than the microscopic characteristic size of the solid phase, or any other equivalent pore scale size.

Experimental confirmation of these theoretical results is necessary. However, because of the inherent difficulty of visualizing the flow, i.e., the Taylor–Proudman column through a grandular porous matrix, an indirect method was adopted. The rationale behind the proposed method was to place a porous layer in between two pure fluid layers inside a cylinder (see Figure 4a). A disturbance in the form of a small fixed obstacle was created on the bottom wall in the fluid layer located below the porous layer, causing the appearance of a Taylor–Proudman column in this layer. As the fundamental property of a Taylor–Proudman column is that it does not allow for

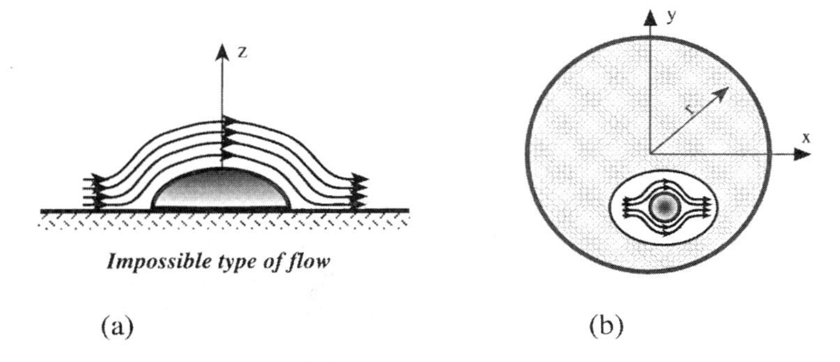

Figure 3. (a) An impossible type of flow over the object. (b) The flow adjusts around the object (as seen from above) and extends at all heights, creating a column above the object which behaves like a solid body.

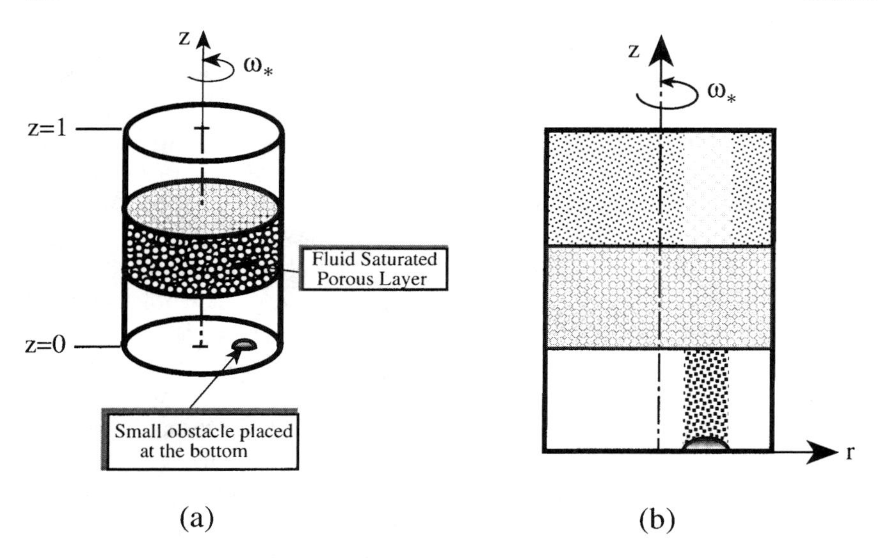

Figure 4. (a) A closed cylindrical container divided into a porous layer and two fluid layers above and below the porous layer. A solid object is fixed to the bottom surface of the container. (b) Qualitative description of the Taylor–Proudman column as observed in the preliminary runs of the experiment.

variations of velocity in the vertical direction, the proposed experimental set-up should allow us to detect Taylor–Proudman columns in the top undisturbed fluid layer located above the porous layer, if the column exists in the porous layer as a result of the disturbance in the bottom fluid layer. This means that a small object located at the bottom of the lower fluid layer should create a Taylor–Proudman column which extends upwards through the porous layer into the upper fluid layer. It is because of this expectation, that the top fluid layer will "feel" through the buffer porous layer the small object located at the bottom of the lower fluid layer, that some colleagues of the author named the effect resulting from the experiment "*The Princess and the pea.*" Of course, if experiments will not allow us to detect Taylor–Proudman columns in the top undisturbed fluid layer another possibility exists which is consistent with Eq. (24). This could, for example, imply that the filtration velocity is zero in a rotating porous layer, meaning that the whole fluid in a rotating porous layer rotates as a solid body instantaneously (almost).

The experimental apparatus consists of a record player turntable adjusted to allow variable angular velocity in the designed range and to provide a better dynamic balance when the cylindrical container is placed securely on the rotating plate. Preliminary results confirm the appearance of Taylor–Proudman columns in the top fluid layer, as shown qualitatively in Figure 4b; however, the degree of confidence in these results should be improved by improving the visualization technique and cross-checking the appearance of the columns to exclude sources other than the obstacle at the bottom fluid layer.

A further significant consequence of Eq. (24) is represented by a geostrophic type of flow. Taking the z-component of (24) yields $\partial w/\partial z = 0$, and the continuity equation (1) becomes

$$\frac{\partial u}{\partial x} + \frac{\partial v}{\partial y} = 0 \tag{25}$$

Therefore, the flow at high rotation rates has a tendency towards two-dimensionality and a stream function, ψ, can be introduced for the flow in the x–y plane

$$u = -\frac{\partial \psi}{\partial y}; \qquad v = \frac{\partial \psi}{\partial x} \tag{26}$$

which satisfies identically the continuity equation (1). Substituting u and v with their stream function representation given by (26) in (21) yields

$$\frac{\partial \psi}{\partial x} = \frac{\partial p_R}{\partial x} \tag{27}$$

$$\frac{\partial \psi}{\partial y} = \frac{\partial p_R}{\partial y} \tag{28}$$

As both the pressure and the stream function can be related to an arbitrary reference value, the conclusion from Eqs. (27) and (28) is that the stream function and the pressure are the same in the limit of high rotation rates ($Ek \to 0$). This type of geostrophic flow means that *isobars* represent *streamlines* at the leading order, for $Ek \to 0$.

C. Flow through Heterogeneous Porous Media in a Rotating Square Channel

In the previous section the case of high rotation rates (i.e., $Ek \ll 1$) was considered. Let us consider now the case when small rotation rates are of interest, i.e., $Ek \gg 1$. This is particularly useful in many practical applications. The problem of axial flow through a long rotating square channel filled with fluid-saturated porous material is considered. The axial flow is imposed through an axial pressure gradient while the channel rotates about an axis perpendicular to the horizontal walls (Figure 5). With a homoge-

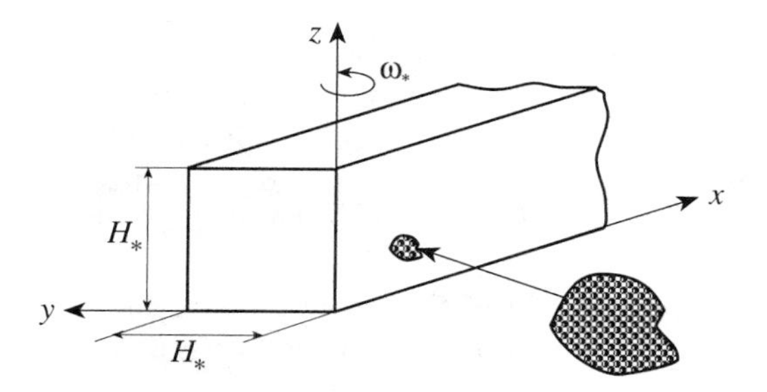

Figure 5. A heterogeneous fluid-saturated porous medium in a rotating square channel.

neous porous medium the permeability is constant throughout the flow domain, resulting in a uniform distribution of the filtration velocity, and the effect of rotation does not affect the flow. However, for heterogeneous porous media the permeability is spatially dependent, thus leading to secondary circulation in a plane perpendicular to the imposed fluid motion. The particular case where the permeability varies only along the vertical coordinate is considered. This assumption is compatible with the assumption of developed flow in pure fluids (nonporous domains).

As $Ek \gg 1$, an expansion of the dependent variables is used in the form

$$\boldsymbol{V} = \boldsymbol{V}_0 + Ek^{-1}\boldsymbol{V}_1 + \mathbf{O}(Ek^{-2}) \tag{29}$$

$$p = p_0 + Ek^{-1}p_1 + \mathbf{O}(Ek^{-2}) \tag{30}$$

The substitution of these expansions in Eqs. (1) and (15) yields a hierarchy of partial differential equations at different orders. At the leading order a basic flow solution is found in the form

$$u_0 = K(z), \qquad v_0 = w_0 = 0 \tag{31}$$

To the first order, a two-dimensional continuity equation for the y and z components of $\boldsymbol{V}_1$ in the y–z plane allows for the introduction of a stream function defined as $v_1 = \partial\psi_1/\partial z$, $w_1 = -\partial\psi_1/\partial y$. Hence, the equation governing the flow in the y–z plane takes the form

$$\frac{\partial^2\psi_1}{\partial y^2} + \frac{\partial^2\psi_1}{\partial z^2} - \frac{\mathrm{d}(\ln K)}{\mathrm{d}z}\frac{\partial\psi_1}{\partial z} = -K\frac{\mathrm{d}K}{\mathrm{d}z} \tag{32}$$

which has to be solved subject to the impermeability boundary conditions at the walls, $\psi_1 = 0$ at $y = 0, 1$ and at $z = 0, 1$.

An analytical solution to Eq. (32) subject to the rigid walls boundary conditions was presented by Vadasz (1993a) for a particular monotonic form of the permeability function, i.e., $K = e^{\gamma z}$. The solution is a single vortex rotating counter-clockwise in the y–z plane, as presented in Figure 6a. The forcing term $(-K\mathrm{d}K/\mathrm{d}z)$ in (32) is responsible for the single vortex and the resulting flow direction. A monotonic function $K(z)$ keeps the sign of the forcing term unchanged for all values of z. Therefore, the single vortex turns counter-clockwise when K increases in the z-direction and clockwise when K decreases, as presented qualitatively in Figure 6b. When $K(z)$ is not monotonic e.g., when it decreases in the first half of the domain ($z \in [0, 0.5]$) and increases in the second half ($z \in [0.5, 1]$), the sign of the forcing term in Eq. (32) changes accordingly, leading to two vortices as presented qualitatively in Figure 6c. This secondary circulation, in a plane perpendicular to the main flow direction, is a result of the Coriolis effect of the main forced flow.

Extending the problem to a case which applies to all values of Ekman number requires the introduction of a scaling for $\boldsymbol{V}$ and K in the form $\boldsymbol{V}^* = \boldsymbol{V}/Ek$ and $K^* = K/Ek$. Then, subject to the same assumptions of developed flow, implying also $K^* \equiv K^*(z)$, the following equation is obtained from Eq. (15)

$$\frac{\partial^2 \psi^*}{\partial y^2} + \left[1 + (K^*)^2\right]\frac{\partial^2 \psi^*}{\partial z^2} + \left[(K^*)^2 - 1\right]\frac{\mathrm{d}(\ln K^*)}{\mathrm{d}z}\frac{\partial \psi^*}{\partial z} = -K^*\frac{\mathrm{d}K^*}{\mathrm{d}z} \tag{33}$$

$$u^* = k^*\left[1 + \frac{\partial \psi^*}{\partial z}\right] \tag{34}$$

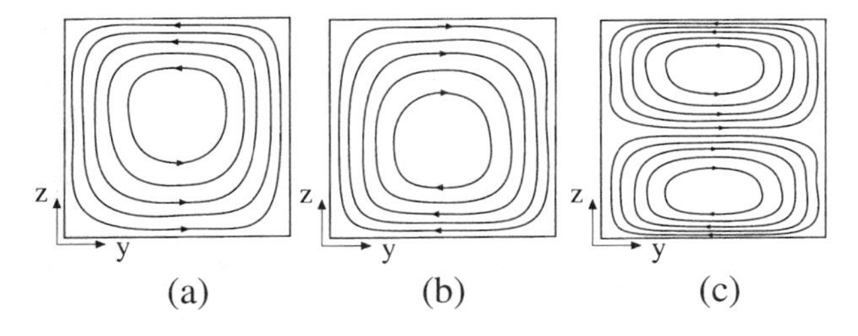

Figure 6. The flow field in the y–z plane of a rotating heterogeneous porous channel: (a) results of analytical solution for a permeability increasing upwards; (b) qualitative results for a permeability decreasing upwards; (c) qualitative results for a permeability decreasing in the bottom half and increasing in the top half of the domain.

where the stream function was redefined in terms of v^* and w^* in the form $v^* = \partial\psi^*/\partial z$, $w^* = -\partial\psi^*/\partial y$. Monotonic, non-monotonic antisymmetric, as well as non-monotonic but symmetric variations of permeability were investigated by Vadasz and Havstad (1999) by using a finite-differences numerical method (Havstad and Vadasz 1999) to solve the three-dimensional fluid flow problem formulated by Eqs. (33) and (34). The accuracy of the method was established by Havstad and Vadasz (1999) and was accomplished by using a coordinate transform which allows mesh refinement to be tailored to the boundary layer position and thickness for a wide range of Ekman number values. The results, corresponding to the three different classes of permeability functions for $Ek = 1$, provide a confirmation of the analytical solutions regarding the nature of the secondary flows. In addition, the significant back-impact of this secondary flow on the basic axial flow via an axial flow deficiency was identified.

V. ONSET OF FREE CONVECTION DUE TO THERMAL BUOYANCY CAUSED BY CENTRIFUGAL BODY FORCES

A. General Background

Considering Darcy's regime for a homogeneous porous medium ($K = 1$) subject to a centrifugal body force and neglecting gravity ($Ra_g/Ra_\omega \ll 1$), Eqs. (1), (2), and (3) with $K = 1$ and $Ra_g = 0$ represent the mathematical model for this case. The objective in the first instance is to establish the convective flow under small rotation rates, thus $Ek \gg 1$, and as a first approximation the Coriolis effect can be neglected, i.e., $Ek \to \infty$. Following these conditions, the governing equations when heat generation is absent (i.e., $\dot{Q} = 0$) become

$$\nabla \cdot \boldsymbol{V} = 0 \tag{35}$$

$$\boldsymbol{V} = -[\nabla p - Ra_\omega T \hat{\boldsymbol{e}}_\omega \times (\hat{\boldsymbol{e}}_\omega \times \boldsymbol{X})] \tag{36}$$

$$\frac{\partial T}{\partial t} + \boldsymbol{V} \cdot \nabla T = \nabla^2 T \tag{37}$$

The three cases corresponding to the relative orientation of the temperature gradient with respect to the centrifugal body force as presented in Figure 1 are considered. Case 1(a) in Figure 1 corresponds to a temperature gradient which is perpendicular to the direction of the centrifugal body force and leads to unconditional convection. The solution representing this convection pattern is the objective of the investigation. Cases 1(b) and 1(c) in Figure 1 corresponding to temperature gradients collinear with the centrifugal body force represents a stability problem. The objective is then to establish the stability condition as well as the convection pattern when this stability condition is not satisfied.

B. Temperature Gradients Perpendicular to the Centrifugal Body Force

An example of a case when the imposed temperature is perpendicular to the centrifugal body force is a rectangular porous domain rotating about the vertical axis, heated from above and cooled from below. For this case the centrifugal buoyancy term in Eq. (36) becomes $Ra_{\omega} Tx\hat{\boldsymbol{e}}_x$, leading to

$$V = -\nabla p - Ra_{\omega} Tx\hat{\boldsymbol{e}}_x \tag{38}$$

An analytical two-dimensional solution to this problem (see Figure 7) for a small aspect ratio of the domain was presented by Vadasz (1992). The solution to the nonlinear set of partial differential equations was obtained through an asymptotic expansion of the dependent variables in terms of a small parameter representing the aspect ratio of the domain.

The convection in the core region far from the sidewalls was the objective of the investigation. To first-order accuracy, the heat transfer coefficient represented by the Nusselt number was evaluated in the form

$$Nu = -\left[1 + \frac{Ra_{\omega}}{24} + \mathbf{O}(H^2)\right] \tag{39}$$

where Nu is the Nusselt number and the length scale used in the definition of Ra_{ω}, Eq. (5), is $l_c = H_*$.

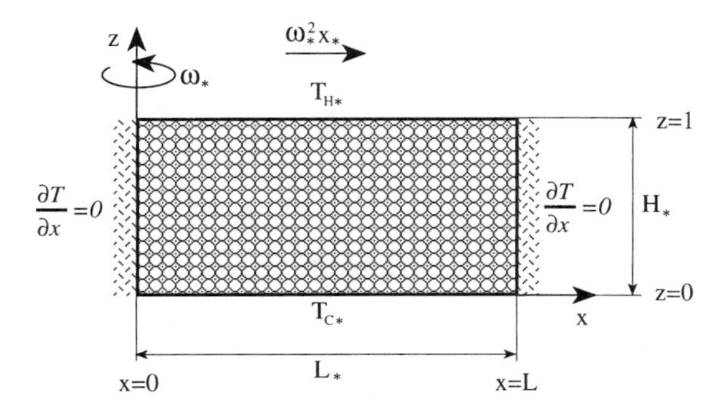

Figure 7. A rotating rectangular porous domain heated from above and cooled from below.

A different approach was used by Vadasz (1994a) to solve a similar problem without the restriction of a small aspect ratio. A direct extraction and substitution of the dependent variables was found to be useful for decoupling the nonlinear partial differential equations, resulting in a set of independent nonlinear ordinary differential equations which was solved analytically. To obtain an analytical solution to the nonlinear convection problem we assume that the vertical component of the filtration velocity, w, and the temperature T are independent of x, i.e., $\partial w/\partial x = \partial T/\partial x = 0$ $\forall x \in (0, L)$, being functions of z only. It is this assumption that will subsequently restrict the validity domain of the results to moderate values of Ra_ω (practically, $Ra_\omega < 5$). Subject to the assumptions of two-dimensional flow, $v = 0$ and $\partial(\cdot)/\partial y = 0$, and that w and T independent of x, the governing equations take the form

$$\frac{\partial u}{\partial x} + \frac{\mathrm{d}w}{\mathrm{d}z} = 0 \tag{40}$$

$$u = -\frac{\partial p}{\partial x} - Ra_\omega x T \tag{41}$$

$$w = -\frac{\partial p}{\partial z} \tag{42}$$

$$\frac{\mathrm{d}^2 T}{\mathrm{d}z^2} - w\frac{\mathrm{d}T}{\mathrm{d}z} = 0 \tag{43}$$

The method of solution consists of extracting T from Eq. (41) and expressing it explicitly in terms of u, $\partial p/\partial x$, and x. This expression of T is then introduced into Eq. (43) and the derivative $\partial/\partial x$ is applied to the result. Then, the continuity equation (40) is substituted in the form $\partial u/\partial x = -\mathrm{d}w/\mathrm{d}z$ and Eq. (42) yields a nonlinear ordinary differential equation for w in the form

$$\frac{\mathrm{d}^3 w}{\mathrm{d}z^3} = -w\frac{\mathrm{d}^2 w}{\mathrm{d}z^2} = 0 \tag{44}$$

An interesting observation regarding Eq. (44) is the fact that it is identical to the Blasius equation for boundary layer flows of pure fluids (nonporous domains) over a flat plate. To observe this, one simply has to substitute $w(z) = -f(z)/2$ to obtain $2f''' + ff'' = 0$, which is the Blasius equation. Unfortunately, no further analogy to the boundary layer flow in pure fluids exists, mainly because of the quite different boundary conditions and because the derivatives $[\mathrm{d}(\cdot)/\mathrm{d}z]$ and the flow (w) are in the same direction. The solutions for the temperature T and the horizontal component of the filtration velocity u are related to the solution of the ordinary differential equation

$$\varphi'\varphi''' - \varphi''^2 + \varphi\varphi'^2 = 0 \tag{45}$$

where $(\cdot)'$ stands for $\mathrm{d}(\cdot)/\mathrm{d}z$ and

$$u = x\varphi(z) \tag{46}$$

$$T(z) = -\frac{1}{Ra_\omega}[P + \varphi(z)] \tag{47}$$

where P is a constant defined by

$$P = -Ra_\omega \int_0^1 T(z)\mathrm{d}z \tag{48}$$

The relationship (48) is a result of imposing a condition of no net flow through any vertical cross-section in the domain, stating that $\int_0^1 u\mathrm{d}z = 0$.

The following boundary conditions are required to the solution of (44) for w: $w = 0$ at $z = 0$ and $z = 1$, representing the impermeability condition at the solid boundaries, and $T = 0$ at $z = 0$ and $T = 1$ at $z = 1$. Since $\partial u/\partial x = \varphi$ according to Eq. (46), then, following the continuity equation (40), $\varphi = -\mathrm{d}w/\mathrm{d}z$ and the temperature boundary conditions can be converted into conditions in terms of w by using Eq. (47), leading to the following complete set of boundary conditions for w:

$$z = 0: \qquad w = 0 \qquad \text{and} \qquad \frac{\mathrm{d}w}{\mathrm{d}z} = P \tag{49}$$

$$z = 1: \qquad w = 0 \qquad \text{and} \qquad \frac{\mathrm{d}w}{\mathrm{d}z} = P + Ra_\omega \tag{50}$$

Equations (49) and (50) represent four boundary conditions, whereas only three are necessary to solve the third-order equation (44). The reason for the fourth condition comes from the introduction of the constant P, whose value remains to be determined. Hence, the additional two boundary conditions are expressed in terms of the unknown constant P and the solution subject to these four conditions will determine the value of P as well. A method similar to Blasius's method of solution was applied to solve (44). Therefore, $w(z)$ was expressed as a finite power series and the objective of the solution was to determine the power series coefficients. Once the solution for $w(z)$ and P was obtained, u and T were evaluated by using $\varphi(z) = -\mathrm{d}w/\mathrm{d}z$ and Eqs. (46) and (47).

Then, for presentation purposes, a stream function ψ was introduced to plot the results ($u = \partial\psi/\partial z$, $w = -\partial\psi/\partial x$). An example of the flow field represented by the streamlines is presented in Figure 8 for $Ra_\omega = 4$ and for an aspect ratio of 3 (excluding a narrow region next to the sidewall at $x = L$). Outside this narrow region next to $x = L$ the streamlines remain open on the right-hand side. They are expected to close in the end region. Nevertheless, the streamlines close on the left-hand side throughout the domain. The reason for this is the centrifugal acceleration, which causes u to vary linearly with x, thus creating (due to the continuity equation) a nonvanishing vertical component of the filtration velocity w at all values of x. The local Nusselt number Nu, representing the local vertical heat flux, was evaluated as well by using the definition $Nu = | - \partial T/\partial z|_{z=0}$ and the solution for T. A comparison between the heat flux results obtained from this solution and the results obtained by Vadasz (1992) using an asymptotic method is presented in Figure 9. The two results compare well as long as Ra_ω is very small. However, for increasing values of Ra_ω the deviation from the linear relationship pertaining to the first-order asymptotic solution ($Nu = 1 + Ra_\omega/24$, according to Vadasz [1992] is evident.

C. Temperature Gradients Collinear with the Centrifugal Body Force

The problem of stability of free convection in a rotating porous layer when the temperature gradient is collinear with the centrifugal body force was treated by Vadasz for a narrow layer adjacent to the axis of rotation (Vadasz, 1994b) and distant from the axis of rotation (Vadasz, 1996a), respectively. The problem formulation corresponding to the latter case is presented in Figure 10. In order to include explicitly the dimensionless offset distance from the axis of rotation x_0, and to keep the coordinate system linked to the porous layer, Eq. (38) was presented in the form

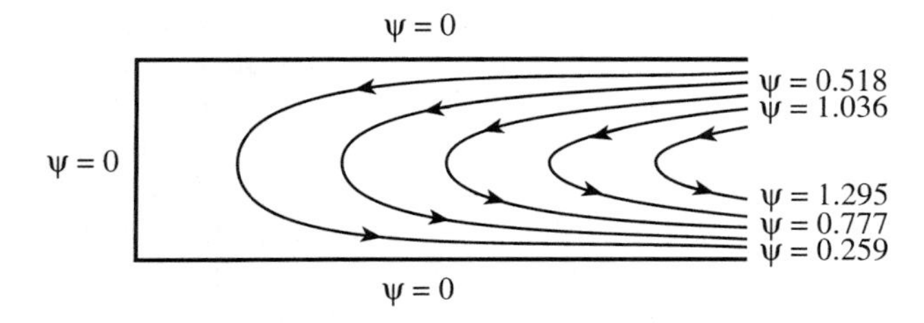

Figure 8. Graphical description of the resulting flow field; five streamlines equally spaced between their minimum value $\psi_{\min} = 0$ at the rigid boundaries and their maximum value $\psi_{\max} = 1.554$. The values in the figure correspond to every other streamline.

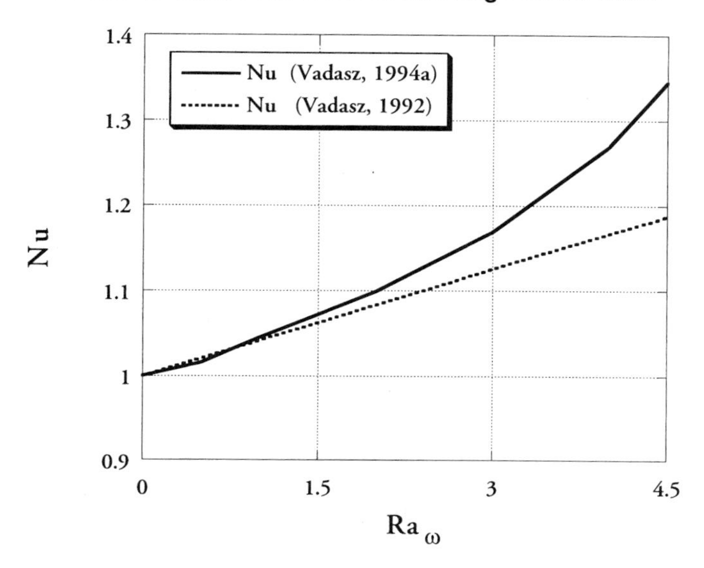

Figure 9. Graphical representation of the local Nusselt number (Nu) versus the centrifugal Rayleigh number (Ra_ω).

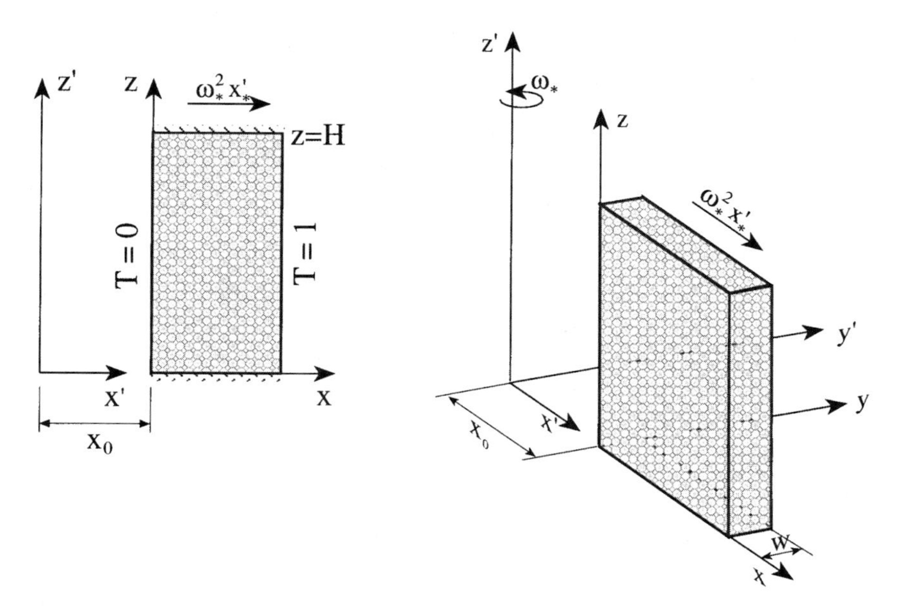

Figure 10. A rotating fluid-saturated porous layer distant from the axis of rotation and subject to different temperatures at the sidewalls.

$$V = -\nabla p - [Ra_{\omega o} + Ra_{\omega} x] T \hat{e}_x \tag{51}$$

Two centrifugal Rayleigh numbers appear in Eq. (51); the first one, $Ra_{\omega} = \beta_{T*} \Delta T_c \omega_*^2 x_{0*} L_* K_0 / \alpha_{eo} v_o$, represents the contribution of the offset distance from the rotation axis to the centrifugal buoyancy, while the second, $Ra_{\omega} = \beta_{T*} \Delta T_c \omega_*^2 L_*^2 K_o / \alpha_{eo} v_o$, represents the contribution of the horizontal location within the porous layer to the centrifugal buoyancy. The ratio between the two centrifugal Rayleigh numbers is

$$\eta = \frac{Ra_{\omega}}{Ra_{\omega o}} = \frac{1}{x_0} \tag{52}$$

and can be introduced as a parameter in the equations transforming Eq. (51) in the form

$$V = -\nabla p - Ra_{\omega o}[1 + \eta x] T \hat{e}_x \tag{53}$$

From Eq. (53) it is observed that when the porous layer is far away from the axis of rotation then $\eta \ll 1$ ($x_0 \gg 1$) and the contribution of the term ηx is not significant, whereas for a layer close enough to the rotation axis $\eta \gg 1$ ($x_0 \ll 1$) and the contribution of the first term becomes insignificant. In the first case the only controlling parameter is $Ra_{\omega o}$ while in the latter case the only controlling parameter is $Ra_{\omega} = \eta Ra_{\omega o}$. The flow boundary conditions are $V \cdot \hat{e}_n = 0$ on the boundaries, where $\hat{e}_n$ is a unit vector normal to the boundary. These conditions stipulate that all boundaries are rigid and therefore nonpermeable to fluid flow. The thermal boundary conditions are: $T = 0$ at $x = 0$, $T = 1$ at $x = 1$, and $\nabla T \cdot \hat{e}_n = 0$ on all other walls, representing the insulation condition on these walls.

The governing equations accept a basic motionless conduction solution in the form

$$[V_b, T_b, p_b] = \left[0, x, \left(-Ra_{\omega o}(x^2/2 + \eta x^3/3) + \text{const.}\right)\right] \tag{54}$$

The objective of the investigation was to establish the condition when the motionless solution (54) is not stable and consequently a resulting convection pattern appears. Therefore a linear stability analysis was employed, representing the solution as a sum of the *basic solution* (54) and *small perturbations* in the form

$$[V, T, p] = [V_b + V', T_b + T', p_b + p'] \tag{55}$$

where $(\cdot)'$ stands for perturbed values. Solving the resulting linearized system for the perturbations by assuming a normal modes expansion in the y and z direction and $\theta(x)$ in the x direction, i.e., $T' = A_{\kappa} \theta(x) \exp[\sigma t + i(\kappa_y y + \kappa_z z)]$, and using the Galerkin method to solve for $\theta(x)$, one obtains at marginal stability, i.e., for $\sigma = 0$, a homogeneous set of linear algebraic equations. This homogeneous linear system accepts a nonzero solution only for particular values of $Ra_{\omega o}$ such that its determinant vanishes. The solution of this system was evaluated up to order 7 for different values of η, representing the offset distance from the axis of rotation. However, useful information was obtained by considering the approximation of order 2. At this order the system reduces to two equations which lead to the characteristic values of $Ra_{\omega o}$ in the form

$$R_{0,c} = \frac{\beta[(1+\alpha)^2 + (4+\alpha)^2]}{2\alpha(\beta^2 - \gamma^2)} \pm \frac{\sqrt{\beta^2[(1+\alpha)^2 + (4+\alpha)^2]^2 - 4(\beta^2 - \gamma^2)(1+\alpha)^2(4+\alpha)^2}}{2\alpha(\beta^2 - \gamma^2)} \tag{56}$$

where the following notation was used

$$R_o = \frac{Ra_{\omega o}}{\pi^2}; \qquad R = \frac{Ra_{\omega}}{\pi^2}; \qquad \alpha = \frac{\kappa^2}{\pi^2}; \qquad \beta = 1 + \frac{\eta}{2}; \qquad \gamma^2 = \frac{256\eta^2}{81\pi^4} \tag{57}$$

and κ is the wavenumber such that $\kappa^2 = \kappa_y^2 + \kappa_z^2$ while the subscript c in Eq. (56) represents characteristic values. A singularity in the solution for $R_{o,c}$, corresponding to the existence of a single root for $R_{o,c}$, appears when $\beta^2 = \gamma^2$. This singularity persists at higher orders as well. Resolving for the value of η when this singularity occurs shows that it corresponds to negative η values, implying that the location of the rotation axis falls within the boundaries of the porous domain (or to the left side of the cold wall—a case of little interest due to its inherent unconditional stability). This particular case will be discussed later in this section.

The critical values of the centrifugal Rayleigh number as obtained from the solution up to order 7 are presented graphically in Figure 11a in terms of both $R_{o,cr}$ and R_{cr} as a function of the offset parameter η. The results presented in Figure 11 are particularly useful in order to indicate the stability criterion for all positive values of η. It can be observed from the figure that as the value of η becomes small, i.e., for a porous layer far away from the axis of rotation, the critical centrifugal Rayleigh number approaches a limit value of $4\pi^2$. This corresponds to the critical Rayleigh number in a porous layer subject to gravity and heated from below. For high values of η it is appropriate to use the other centrifugal Rayleigh number R, instead of R_o, by introducing the relationship $R = \eta R_o$ (see Eqs. (52) and (57)) in order to establish and present the stability criterion. It is observed from Figure 11a that as the value of η becomes large, i.e. for a

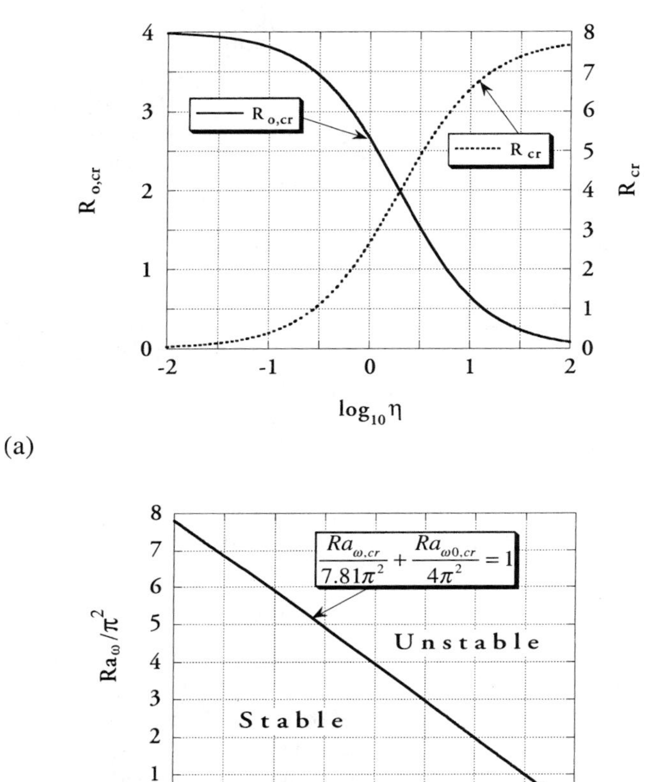

Figure 11. (a) Variation of the critical values of the centrifugal Rayleigh numbers as a function of η; (b) stability map on the Ra_{ω}–$Ra_{\omega o}$ plane, showing the division of the plane into stable and unstable zones.

porous layer close to the axis of rotation, the critical centrifugal Rayleigh number approaches a limit value of $7.81\pi^2$. This corresponds to the critical Rayleigh number for the problem of a rotating layer adjacent to the axis of rotation as presented by Vadasz (1994b). The stability map on the Ra_{ω}–$Ra_{\omega o}$ plane is presented in Figure 11b, showing that the plane is divided between the stable an unstable zones by the straight line $(Ra_{\omega,cr}/7.81\pi^2)+(Ra_{\omega o,cr}/4\pi^2) = 1$.

The results for the convective flow field were presented graphically by Vadasz (1996a), where it was concluded that the effect of the variation of the centrifugal acceleration within the porous layer is definitely felt when the box is close to the axis of rotation, corresponding to an eccentric shift of the convection cells towards the sidewall at $x = 1$. However, when the layer is located far away from the axis of rotation (e.g., $x = 50$) the convection cells are concentric and symmetric with respect to $x = 1/2$, as expected for a porous layer subject to gravity and heated from below (here "below" means the location where $x = 1$).

Although the linear stability analysis is sufficient for obtaining the stability condition of the motionless solution and the corresponding eigenfunctions describing qualitatively the convective flow, it cannot provide information regarding the values of the convection amplitudes, nor regarding the average rate of heat transfer. To obtain this additional information, Vadasz and Olek (1998) analyzed and provided a solution to the nonlinear equations by using Adomian's decomposition method to solve a system of ordinary differential equations for the evolution of the amplitudes. This system of equations was obtained by using the first three relevant Galerkin modes for the stream function and the temperature, while including the time derivative term in Darcy's equation in the form $(1/\chi)\partial \mathbf{V}/\partial t$, where $\chi = \varepsilon Pr/Da$, and Da, Pr are the Darcy and Prandtl numbers, respectively, defined as $Da = K_o/L_*^2$ and $Pr = \boldsymbol{\nu}_o/\alpha_{eo}$ (see Eq. (14) with $Ra_g = 0$ and $Ek \to \infty$). Then the following equations were obtained for the time evolution of the amplitudes X, Y, Z

$$\dot{X} = a(Y - X) \tag{58}$$

$$\dot{Y} = RX - Y - (R - 1)XZ \tag{59}$$

$$\dot{Z} = 4b(XY - Z) \tag{60}$$

where a is a parameter proportional to χ, b depends on the aspect ratio, and R is a rescaled Rayleigh number. The results obtained are presented in Figure 12 in the form of projection of trajectories data points onto the Y–X and Z–X planes. Different transitions as the value of R varies are presented and relate to the convective fixed point which is a stable simple node in Figure 12a, a stable spiral in Figures 12b and c, and loses stability via a Hopf bifurcation in Figure 12d, where the trajectory describes a limit cycle, moving towards a chaotic solution presented in Figures 12e and f. A further transition from chaos to a periodic solution was obtained at a value of R slightly above 100, which persists over a wide range of R values. This periodic solution is presented in Figures 12g and h for $R = 250$.

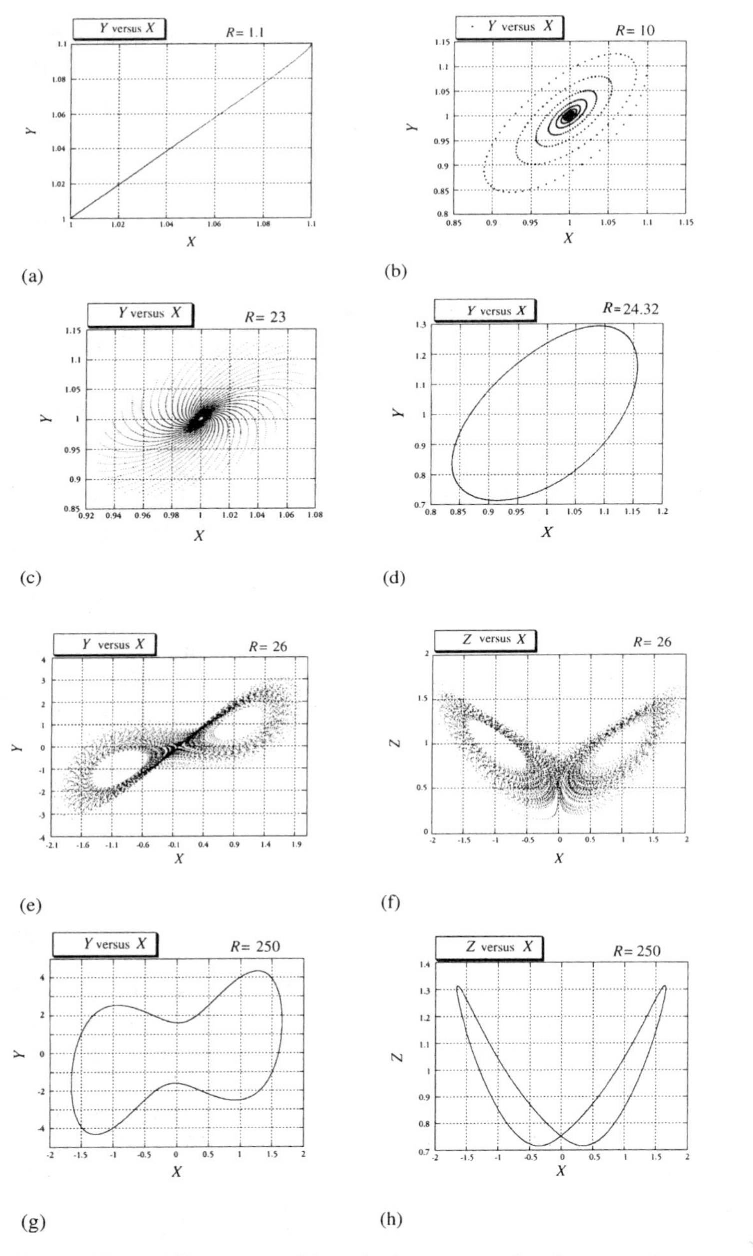

Figure 12. Different transitions in free convection in a rotating porous layer.

Previously in this section (see Eq. (56)) a singularity in the solution was identified and associated with negative values of the offset distance from the axis of rotation. It is this resulting singularity and its consequences which were investigated by Vadasz (1996b) and are the objective of the following presentation. As this occurs at negative values of the offset distance from the axis of rotation, it implies that the location of the rotation axis falls within the boundaries of the porous domain, as presented in Figure 13. This particular axis location causes positive values of the centrifugal acceleration on the right side of the rotation axis and negative values on its left side. The rotation axis location implies that the value of x_0 is not positive. It is therefore convenient to introduce this fact in the problem formulation, specifying explicitly that $x_0 = -|x_0|$. As a result, Eq. (53) can be expressed in the form

$$V = -\nabla p - Ra_{\omega}[x - |x_0|]T\hat{e}_x \tag{61}$$

The solution for this case is similar to the previous case, leading to the same characteristic equation for R_c at order 2 as obtained previously in (56) for $R_{o,c}$, with the only difference appearing in the different definition of β and γ as follows

$$\beta = \left(\frac{1}{2} - |x_0|\right); \qquad \gamma^2 = \frac{256}{81\pi^4} \tag{62}$$

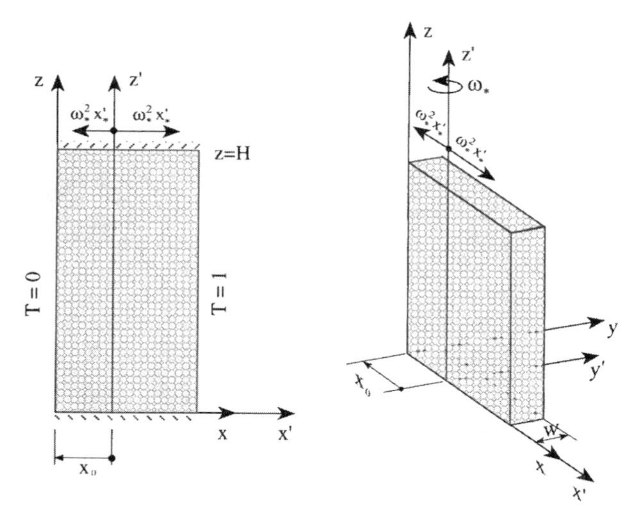

Figure 13. A rotating porous layer having the rotation axis within its boundaries and subject to different temperatures at the side walls.

The singularity is obtained when $\beta^2 = \gamma^2$, corresponding to $\beta = \gamma$ or $\beta = -\gamma$. Since β is uniquely related to the offset distance $|x_0|$ and $\gamma = 16/9\ \pi^2$ is a constant, one can relate the singularity to specific values of $|x_0|$. At order 2 this corresponds to $|x_0| = 0.3199$ and $|x_0| = 0.680$. It was shown by Vadasz (1996b) that the second value $|x_0| = 0.680$ is the only one which has physical consequences. This value corresponds to a transition beyond which, i.e., for $|x_0| \geq 0.68$, no positive roots of R_c exist. It therefore implies an unconditional stability of the basic motionless solution for all values of R if $|x_0| \geq 0.68$. The transitional value of $|x_0|$ was investigated at higher orders, showing $|x_0| \geq 0.765$ at order 3, and the value increases with increasing order. The indications are that as the order increases the transition value of $|x_0|$ tends towards the limit value of 1.

The results for the critical values of the centrifugal Rayleigh number, expressed in terms of R_{cr} versus $|x_0|$, are presented in Figure 14. From the figure it is clear that increasing the value of $|x_0|$ has a stabilizing effect. The results for the convective flow field as obtained by Vadasz (1996b) are presented in Figures 15–17 for different values of $|x_0|$. Keeping in mind that to the right of the rotation axis the centrifugal acceleration has a destabilizing effect while to its left a stabilizing effect is expected, the results presented in Figures 15b and c reaffirm this expectation, showing an eccentric shift of the convection cells towards the right side of the rotation axis. When the rotation axis is moved further towards the hot wall, say at $|x_0| = 0.6$ as presented in Figure 16a, weak convection cells appear even to the left of the rotation axis. This weak convection becomes stronger as $|x_0|$ increases, as observed in Figure 16b for $|x_0| = 0.7$, and formation of boundary layers

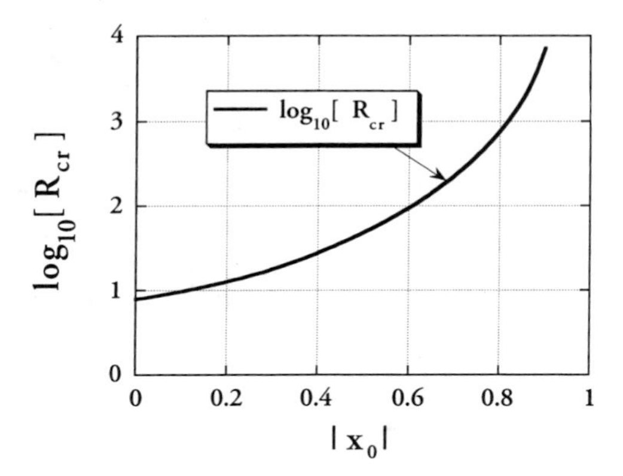

Figure 14. Variation of the critical values of the centrifugal Rayleigh number as a function of $|x_0|$.

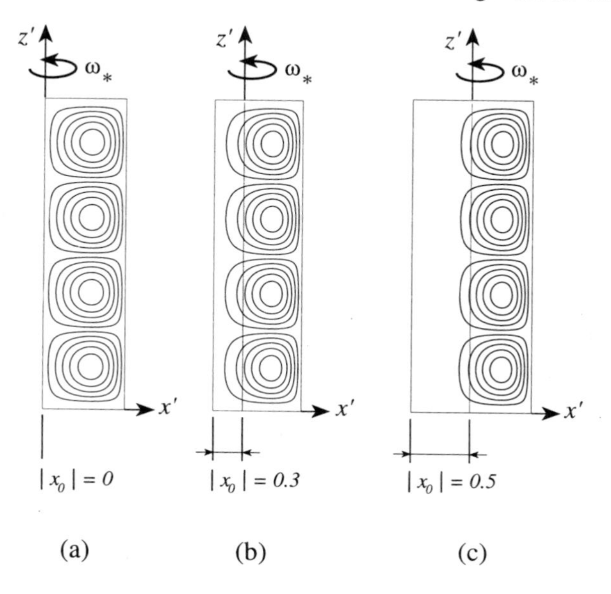

Figure 15. The convective flow field at marginal stability for three different values of $|x_0|$; 10 streamlines equally divided between $\psi_{\min}$ and $\psi_{\max}$.

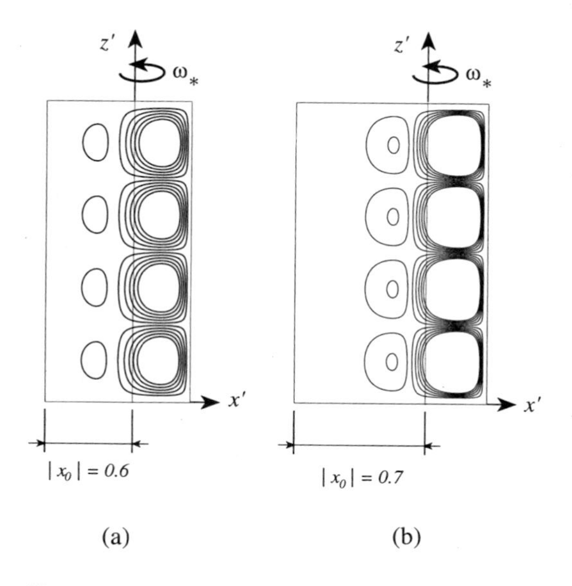

Figure 16. The convective flow field at marginal stability for two different values of $|x_0|$; 10 streamlines equally divided between $\psi_{\min}$ and $\psi_{\max}$.

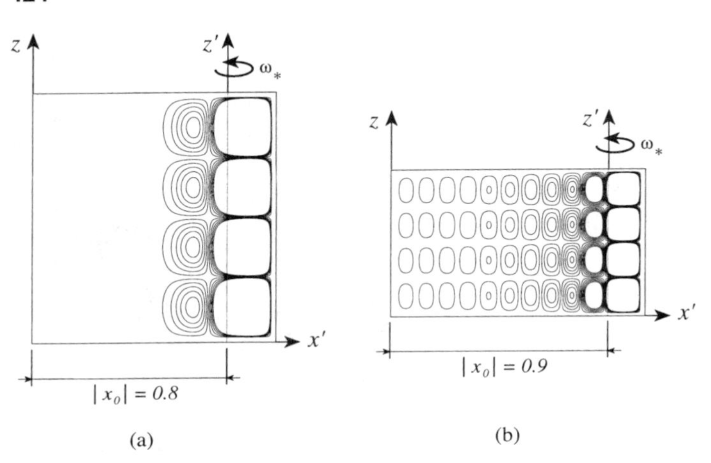

Figure 17. The convective flow field at marginal stability for two different values of $|x_0|$; 10 streamlines equally divided between $\psi_{\min}$ and $\psi_{\max}$.

associated with the primary convection cells is observed to the right of the rotation axis. These boundary layers become more significant for $|x_0| = 0.8$, as represented by sharp streamline gradients in Figure 17a. When $|x_0| = 0.9$, Figure 17b shows that the boundary layers of the primary convection are well established and the whole domain is filled with weaker secondary, tertiary, and further convection cells. The results for the isotherms corresponding to values of $|x_0| = 0$, 0.5, 0.6, and 0.7 are presented in Figure 18, where the effect on the temperature of moving the axis of rotation within the porous layer is evident.

VI. CORIOLIS EFFECT ON FREE CONVECTION DUE TO THERMAL BUOYANCY CAUSED BY CENTRIFUGAL BODY FORCES

In the previous section centrifugally driven free convection was discussed under conditions of small rotation rates, i.e., $Ek \gg 1$. Then, as a first approximation the Coriolis effect was neglected. It is the objective of the present section to show the effect of the Coriolis acceleration on free convection even when this effect is small, i.e., $Ek \gg 1$. A long rotating porous box is considered as presented in Figure 19. The possibility of internal heat generation is included but the case without heat generation, i.e., when the box is heated from above and cooled from below, is dealt with separately. This case was solved by Vadasz (1993b) by using an asymptotic expansion in

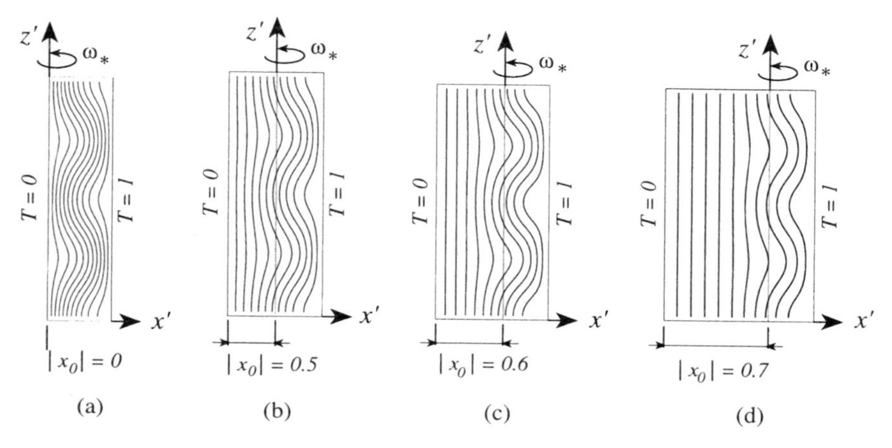

Figure 18. The convective temperature field at marginal stability for four different values of $|x_0|$; 10 isotherms equally divided between $T_{\min} = 0$ and $T_{\max} = 1$.

terms of two small parameters representing the reciprocal Ekman number in porous media and the aspect ratio of the domain. Equations (7), (8), and (9) were used with $Ra_g = 0$, $K = 1$, and neglecting the component of the centrifugal acceleration in the y direction (a small aspect ratio). Then an expansion of the form

$$[\boldsymbol{V}, T, p] = \sum_{m=0}^{\infty} \sum_{n=0}^{\infty} H^m Ek^{-n} [\boldsymbol{V}_{mn}, T_{mn}, p_{mn}] \tag{63}$$

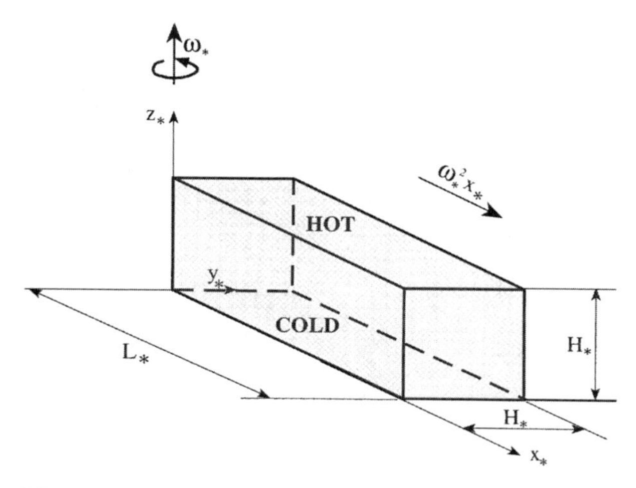

Figure 19. Three-dimensional convection in a long rotating porous box.

is introduced in the equations, where H is the aspect ratio, i.e., $H = H_*/L_*$.

To leading order the zero powers of the aspect ratio H^m and Ek^{-n} are used, which yields

$$v_{00} = w_{00} = 0; \qquad T_{00} = z; \qquad u_{00} = -\frac{Ra_\omega x}{2}[2z - 1] \tag{64}$$

This solution holds for the core region of the box and is presented in Figures 20c and f. To orders 1 in Ek^{-n} and 0 in H^m the Coriolis effect is first detected in a plane perpendicular to the leading-order free convection flow, leading to the following analytical solution for the stream function ψ_{01} in the y–z plane

$$\psi_{01} = -\frac{16 Ra_\omega x}{\pi^4} \sum_{i=1}^{\infty} \sum_{j=1}^{\infty} \frac{\sin[(2i-1)\pi y] \sin[(2j-1)\pi z]}{(2i-1)(2j-1)[(2i-1)^2 + (2j-1)^2]} \tag{65}$$

and the corresponding temperature solution is

$$T_{01} = -\frac{16 Ra_\omega x}{\pi^5} \sum_{i=1}^{\infty} \sum_{j=1}^{\infty} \frac{\cos[(2i-1)\pi y] \sin[(2j-1)\pi z]}{(2j-1)[(2i-1)^2 + (2j-1)^2]^2} \tag{66}$$

From the solutions it was concluded that the Coriolis effect on free convection is controlled by the combined dimensionless group

$$\sigma = \frac{Ra_\omega}{Ek} = \frac{2\beta_{T^*} \Delta T_c \omega_*^3 L_* H_* K_o^2}{\alpha_{eo} v_o^2 \varepsilon} \tag{67}$$

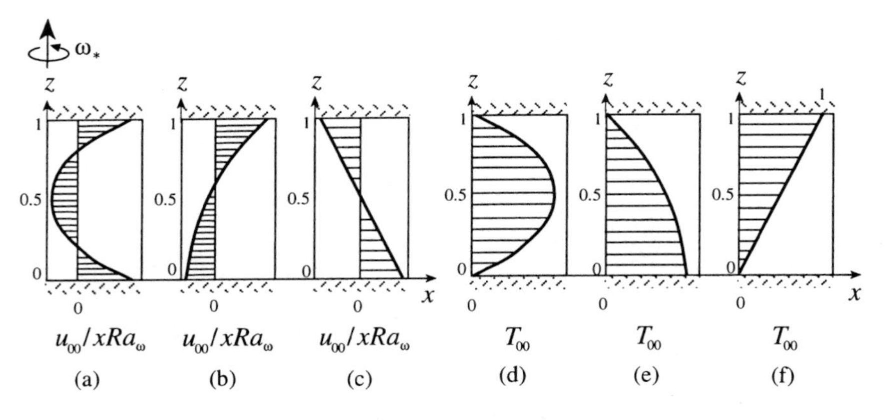

Figure 20. The leading-order convection for the core region: (a) filtration velocity, with heat generation, perfectly conducting top and bottom walls; (b) filtration velocity, with heat generation, perfectly conducting top wall, insulated bottom wall; (c) filtration velocity, no heat generation, heating from above; (d) temperature, as in (a); (e) temperature, as in (b); (f) temperature, as in (c).

The flow and temperature fields in the plane y–z, perpendicular to the leading-order free convection plane as evaluated through the analytical solution, are presented in Figure 21c in the form of streamlines and isotherms. The single vortex in this plane is a consequence of the monotonic variation of the free convection flow field in the core region, i.e., by the sign of $\partial u_{00}/\partial z$. By extending this argument to evaluate the free convection flow and temperature field in a similar box to that in Figure 19 but including internal heat generation and subject to different top and bottom boundary conditions, the following cases were considered by Vadasz (1995):

1. A uniform rate of internal heat generation and perfectly conducting top and bottom walls, i.e., $Q = 1$, $z = 0$: $T = 0$, and $z = 1$: $T = 0$
2. A uniform rate of internal heat generation, perfectly conducting top wall and adiabatic bottom wall, i.e., $Q = 1$, $z = 0$: $\partial T/\partial z = 0$, and $z = 1$: $T = 0$

As a result, a basic flow u_{00}, at the leading order, was evaluated in the form

(a) For boundary conditions set 1

$$u_{00} = \frac{Ra_{\omega}x}{2}\left[z^2 - z + \frac{1}{6}\right] \tag{68}$$

(b) For boundary conditions set 2

$$u_{00} = \frac{Ra_{\omega}x}{2}\left[z^2 - \frac{1}{3}\right] \tag{69}$$

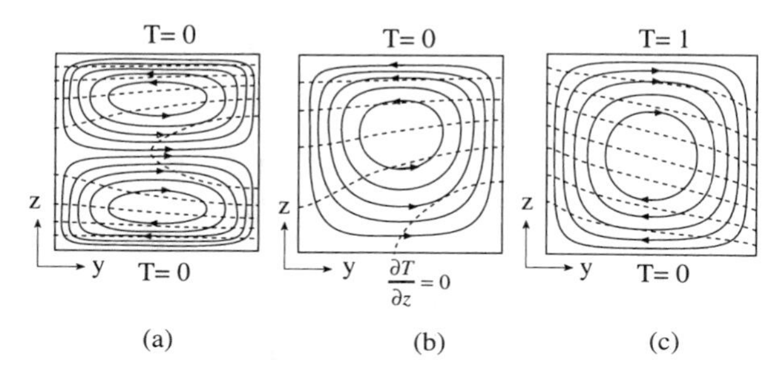

Figure 21. The flow and temperature fields in the y–z plane: (a) with heat generation, perfectly conducting top and bottom walls; (b) with heat generation, perfectly conducting top wall, insulated bottom wall; (c) no heat generation, heating from above.

The graphical description of the leading-order core solutions in terms of $u_{00}/Ra_{\omega}x$ is presented in Figures 20a and b and the solutions for T_{00} in Figures 20d and e. Figures 20a and d correspond to the boundary conditions set 1, while Figures 20b and e correspond to the boundary conditions set 2. Because of the change of sign of the gradient $\partial u_{00}/\partial x$ in case 1 (see Eq. (68) and Figure 20a), a double vortex secondary flow as presented in Figure 21a was obtained in the y–z plane. Since the gradient $\partial u_{00}/\partial z$ for case 2 does not change sign, a single vortex secondary flow is the result obtained in the y–z plane for this case, as shown in Figure 21b. A comparison of the flow direction of this vortex with the direction of the vortex resulting from the analytical solution for the case without heat generation while heating from above (see Figure 21c) shows again the effect of the basic flow vertical gradient $\partial u_{00}/\partial z$ on the secondary flow, i.e., a positive gradient is associated with a counter-clockwise flow and a negative gradient favors a clockwise flow. The resulting effect of these secondary flows on the temperature is presented by the dashed lines in Figure 21, representing the isotherms.

VII. CORIOLIS EFFECT ON FREE CONVECTION DUE TO THERMAL BUOYANCY CAUSED BY GRAVITY FORCES

The problem of a rotating porous layer subject to gravity and heated from below was originally investigated by Friedrich (1983) and by Patil and Vaidyanathan (1983). Both studies considered a non-Darcy model, which is probably subject to the limitations as shown by Nield (1991a). Friedrich (1983) focused on the effect of Prandtl number on the convective flow resulting from a linear stability analysis as well as a nonlinear numerical solution, while Patil and Vaidyanathan (1983) dealt with the influence of variable viscosity on the stability condition. The latter concluded that variable viscosity has a destabilizing effect while rotation has a stabilizing effect. Although the non-Darcy model considered included the time derivative in the momentum equation, the possibility of convection setting in as an oscillatory instability was not explicitly investigated by Patil and Vaidyanathan (1983). It should be pointed out that for a pure fluid (non-porous domain) convection sets in as oscillatory instability for a certain range of Prandtl number values (Chandrasekhar, 1961). This possibility was explored by Friedrich (1983), who presented stability curves for both monotonic and oscillatory instability. Jou and Liaw (1987) investigated a similar problem of gravity-driven thermal convection in a rotating porous layer subject to transient heating from below. By using a non-Darcy model they established the stability conditions for the marginal state without considering the possibility of oscillatory convection.

An important analogy was discovered by Palm and Tyvand (1984) who showed, by using a Darcy model, that the onset of gravity-driven convection in a rotating porous layer is equivalent to the case of an anisotropic porous medium. The critical Rayleigh number was found to be

$$Ra_{g,cr} = \pi^2\left[(1 + Ta)^{1/2} + 1\right]^2 \tag{70}$$

where Ta is the Taylor number, defined here as

$$Ta = \left(\frac{2\omega_* K_o}{\varepsilon v_o}\right)^2 \tag{71}$$

and the corresponding critical wave number is $\pi(1 + Ta)^{1/4}$. The porosity is missing in the Palm and Tyvand (1984) definition of Ta. Nield (1999) has pointed out that these authors and others have omitted the porosity from the Coriolis term. This result, Eq. (70) (amended to include the correct definition of Ta), was confirmed by Vadasz (1998) for a Darcy model extended to include the time derivative term (see Eq. (14) with $Ra_\omega = 0$), while performing linear stability as well as a weak nonlinear analysis of the problem to provide differences as well as similarities with the corresponding problem in pure fluids (nonporous domains). As such, Vadasz (1998) found that, in contrast to the problem in pure fluids, overstable convection in porous media at marginal stability is not limited to a particular domain of Prandtl number values (in pure fluids the necessary condition is $Pr < 1$). Moreover, it was also established by Vadasz (1998) that in the porous media problem the critical wave number in the plane containing the streamlines for stationary convection is not identical to the critical wave number associated with convection without rotation, and is therefore not independent of rotation, a result which is quite distinct from the corresponding pure-fluids problem. Nevertheless, it was evident that in porous media, just as in the case of pure fluids subject to rotation and heated from below, the viscosity at high rotation rates has a destabilizing effect on the onset of stationary convection, i.e., the higher the viscosity, the less stable is the fluid. An example of stability curves for overstable convection is presented in Figure 22 for $Ta = 5$, where κ is the wave number. The upper bound of these stability curves is represented by a stability curve corresponding to stationary convection at the same particular value of the Taylor number, whereas the lower bound was found to be independent of the value of Taylor number and corresponds to the stability curve for overstable convection associated with $\chi = 0$.

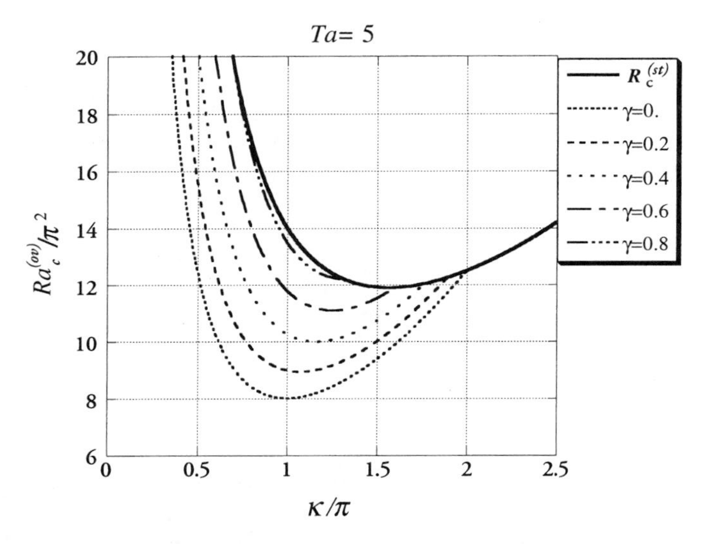

Figure 22. Stability curves for overstable gravity-driven convection in a rotating porous layer heated from below ($\gamma = \chi/\pi^2$, $R = Ra_g/\pi^2$).

Two conditions have to be fulfilled for overstable convection to set in at marginal stability, i.e., (i) the value of Rayleigh number has to be higher than the critical Rayleigh number associated with overstable convection, and (ii) the critical Rayleigh number associated with overstable convection has to be greater than the critical Rayleigh number associated with stationary convection. The stability map obtained by Vadasz (1998) is presented in Figure 23, which shows that the Ta–γ ($= \chi/\pi^2$) plane is divided by a continuous curve (almost a straight line) into two zones, one for which convection sets in as stationary, and the other where overstable convection is preferred. The dotted curve represents the case when the necessary condition (i) above is fulfilled but condition (ii) is not. Weak nonlinear stationary as well as oscillatory solutions were derived, identifying the domain of parameter values consistent with supercritical pitchfork (in the statitionary case) and Hopf (in the oscillatory case) bifurcations. The identification of the critical point corresponding to the transition from supercritical to subcritical bifurcations was presented on the γ–Ta parameter plane. The possibility of a codimension-2 bifurcation, which is expected at the intersection between the stationary and overstable solutions, although identified as being of significant interest for further study, was not investigated by Vadasz (1998).

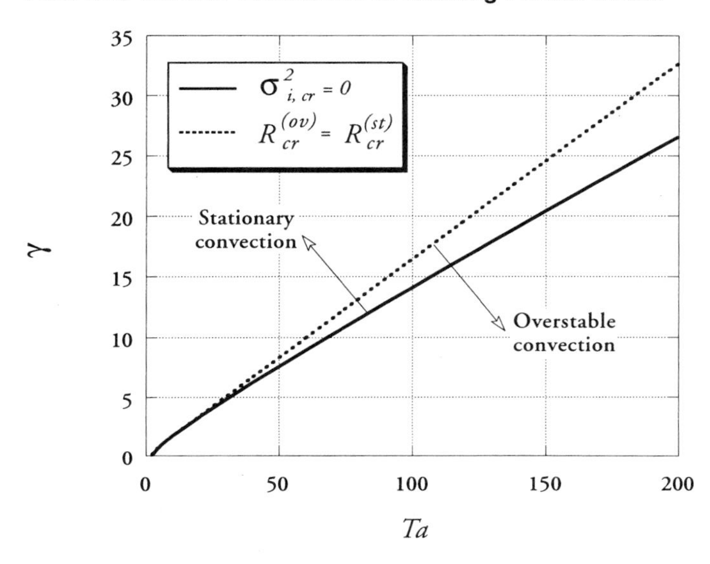

Figure 23. Stability map for gravity-driven convection in a rotating porous layer heated from below ($\gamma = \chi/\pi^2$, $R = Ra_g/\pi^2$).

VIII. ONSET OF FREE CONVECTION DUE TO THERMOHALINE BUOYANCY CAUSED BY GRAVITY FORCES

Limited research results are available on free convection in a rotating porous medium due to thermohaline buoyancy caused by gravity forces. Chakrabarti and Gupta (1981) investigated a non-Darcy model which includes the Brinkman term as well as a nonlinear convective term in the momentum equation (in the form $(\boldsymbol{V} \cdot \nabla)\boldsymbol{V}$). Therefore the model's validity is subject to the limitations pointed out by Nield (1991a). Both linear and nonlinear analyses were performed and overstability was particularly investigated. Overstability is affected in this case by both the presence of a salinity gradient and by the Coriolis effect. Apart from the thermal and solutal Rayleigh numbers and the Taylor number, two additional parameters affect the stability. These are the Prandtl number $Pr = v_o/\alpha_{ei}$, and the Darcy number $Da = K_o/H_*^2$, where H_* is the layer's height. The authors found that, in the range of values of the parameters which were considered, the linear stability results favor setting-in of convection through a mechanisms of overstability. The results for nonlinear steady convection show that the system becomes unstable to finite-amplitude steady disturbances before it becomes unstable to disturbances of infinitesimal amplitude. Thus the porous layer may exhibit subcritical instability in the presence of rotation.

These results are surprising, at least in the sense of their absolute generality, and the authors mention that further confirmation is needed in order to increase the degree of confidence in these results.

A similar problem was investigated by Rudraiah et al. (1986) while focusing on the effect of rotation on linear and nonlinear double-diffusive convection in a sparsely packed porous medium A non-Darcy model identical to the one used by Chakrabarti and Gupta (1981) was adopted; however, the authors spelled out explicitly that the model validity is limited to high porosity and high permeability, which makes it closer to the behavior of a pure fluid system (nonporous domain). It is probably for this reason that the authors preferred to use the nonporous medium definitions for Rayleigh and Taylor numbers which differ by a factor of Da and Da^2, respectively, from the corresponding definitions for porous media. It is because of these definitions that the authors concluded that for small values of Da number the effect of rotation is negligible for values of $Ta < 10^6$. This means that rotation has a significant effect for large rotation rates, i.e., $Ta > 10^6$. If the porous media Taylor number had been used instead, i.e., the proper porous media scales, then one could have significant effects of rotation at porous media Taylor numbers as small as $Ta > 10$. Hence, the results presented by Rudraiah et al. (1986) are useful provided $Da = \mathbf{O}(1)$ which is applicable for high permeability (or sparsely packed) porous layers. Marginal stability and overstability were investigated and the results show different possibilities of existence of neutral curves by both mechanisms, i.e., monotonic as well as oscillatory instability. In this regard the results appear more comprehensive in the study by Rudraiah et al. (1986) than in Chakrabarti and Gupta (1981). The finite amplitude analysis was performed by using a severely truncated representation of a Fourier series for the dependent variables. As a result, a seventh-order Lorenz model of double diffusive convection in a porous medium in the presence of rotation was obtained. From the study of steady, finite amplitude analysis the authors found that subcritical instabilities are possible, depending on the parameter values. The effect of the parameters on the heat and mass transport was investigated as well, and results presenting this effect are discussed in Rudraiah et al. (1986).

IX. FREE CONVECTION DUE TO THERMAL BUOYANCY CAUSED BY COMBINED CENTRIFUGAL AND GRAVITY FORCES

Section V dealt with the onset of free convection due to thermal buoyancy caused by centrifugal body forces. Consequently, the effect of gravity was neglected, i.e., $Ra_g = 0$. For this assumption to be valid the following condition has to be satisfied: $Ra_g/Ra_\omega = g_*/\omega_*^2 L_* \ll 1$. It is of interest to investigate the case when $Ra_g \sim Ra_\omega$ and both centrifugal and gravity forces are of the same order of magnitude. Vadasz and Govender (1998, 1999) present an investigation of cases corresponding to the ones presented in Section V.C, including the effect of gravity in a direction perpendicular to the centrifugal force. Figure 10 is still applicable to the present problem, with the slight modification of drawing the gravity acceleration g_* in the negative z direction. The notation remains the same and Eq. (53) becomes.

$$V = -\nabla p - Ra_{\omega o}[1 + \eta x]T\hat{e}_x + Ra_g T\hat{e}_z \tag{72}$$

where $\eta = 1/x_0 = Ra_\omega/Ra_{\omega o}$ represents the reciprocal of the offset distance from the axis of rotation. The approach being the same as in Section V.C, the solution is expressed as a sum of a basic solution and small perturbations, as presented in Eq. (55). However, because of the presence of the gravity component in Eq. (72), a motionless conduction solution is no longer possible. Therefore, the basic solution far from the top and bottom walls is obtained in the form

$$\begin{aligned} &u_b = v_b = 0; \qquad w_b = Ra_g\left(x - \frac{1}{2}\right); \qquad T_b = x; \\ &p_b = \frac{1}{2} Ra_g z - Ra_{\omega o} x^2 \left[\frac{1}{2} + \frac{1}{3}\eta x\right] + \text{const.} \end{aligned} \tag{73}$$

Substituting this basic solution into the governing equations, and linearizing the result by neglecting terms that include products of perturbations which are small, yields a set of partial differential equations for the perturbations. Assuming a normal modes expansion in the y and z directions in the form

$$T' = A_\kappa \theta(x) \exp\left[\sigma t + i(\kappa_y y + \kappa_z z)\right] \tag{74}$$

where κ_y and κ_z are the wave numbers in the y and$z$ directions respectively, i.e., $\kappa^2 = \kappa_y^2 + \kappa_z^2$, and using the Galerkin method, the following set of linear algebraic equations is obtained at marginal stability (i.e., for $\sigma = 0$)

$$\begin{aligned} \sum_{m=1}^{M} \Bigg\{ &[2(m^2\pi^2 + \kappa^2)^2 - \kappa^2 Ra_{\omega o}(2 + \eta)]\delta_{ml} + \Bigg[\frac{16ml\kappa^2 Ra_{\omega o}}{\pi^2(l^2 - m^2)^2} \\ &- i\frac{8ml\kappa_z Ra_g}{\pi^2(l^2 - m^2)^2}\left[\pi^2(l^2 + m^2) + 2\kappa^2\right]\Bigg]\delta_{m+l,2p-1}\Bigg\} a_m = 0 \end{aligned} \tag{75}$$

for $l = 1, 2, 3, \ldots, M$ and $i = \sqrt{-1}$. In Eq. (75) δ_{ml} is the Kronecker delta function and the index p can take arbitrary integer values, since it stands only for setting the second index in the Kronecker delta function to be an odd integer. A particular case of interest is the configuration when the layer is placed far away from the axis of rotation, i.e., when the length of the layer L_* is much smaller than the offset distance from the rotation axis x_{0*}. Therefore, for $x_0 = (x_{0*}/L_*) \to \infty$ or $\eta \to 0$, the contribution of the term ηx in Eq. (72) is not significant. Substitution of this limit into Eq. (75) and solving the system at the second order, i.e. $M = 2$, yields a quadratic equation for the characteristics values of $Ra_{\omega o}$. This equation has no real solutions for values of $\alpha = \kappa_z^2/\pi^2$ beyond a transitional value $\alpha_{tr} = (27\pi^3/16Ra_g)^2$. This value was evaluated at higher orders as well, showing that for $M = 10$ the transitional value varies very little with Ra_g, beyond a certain Ra_g value around 50π.

The critical values of $Ra_{\omega o}$ were evaluated for different values of R_g $(= Ra_g/\pi)$ and the corresponding two-dimensional convection solutions in terms of streamlines are presented graphically for the odd modes in Figure 24a, showing the perturbation solutions in the x–y plane as skewed convection cells when compared with the case without gravity. The corresponding convection solutions for the even modes are presented in Figure 24b, where it is evident that the centrifugal effect is felt predominantly in the central region of the layer, whereas the downwards and upwards basic gravity-driven convection persists along the left and right boundaries, respectively, although not in straight lines. Beyond the transition value of α, the basic gravity-driven convective flow (Eq. (73)) is unconditionally stable. These results were shown to have an analogy with the problem of gravity-driven convection in a nonrotating, inclined porous layer (Govender and Vadasz, 1995). Qualitative experimental confirmation of these results was presented by Vadasz and Heerah (1998), who used a thermosensitive liquid-crystal tracer in a rotating Hele–Shaw cell.

When the layer is placed at an arbitrary finite distance from the axis of rotation no real solutions exist for the characteristic values of $Ra_{\omega o}$ corresponding to any values of γ other than $\gamma = \kappa_z Ra_g = 0$. In the presence of gravity $Ra_g \neq 0$, and $\gamma = 0$ can be satisfied only if $\kappa_z = 0$. Therefore the presence of gravity in this case has no other role but to exclude the vertical modes of convection. *The critical centrifugal Rayleigh numbers and the corresponding critical wave numbers are the same as in the corresponding case without gravity, as presented in Section V.C.* However, the eigenfunctions representing the convection pattern are different as they exclude the vertical modes, replacing them with a corresponding horizontal mode in the y direction. Therefore, a cellular convection in the x–y plane is superimposed to the basic convection in the x–z plane.

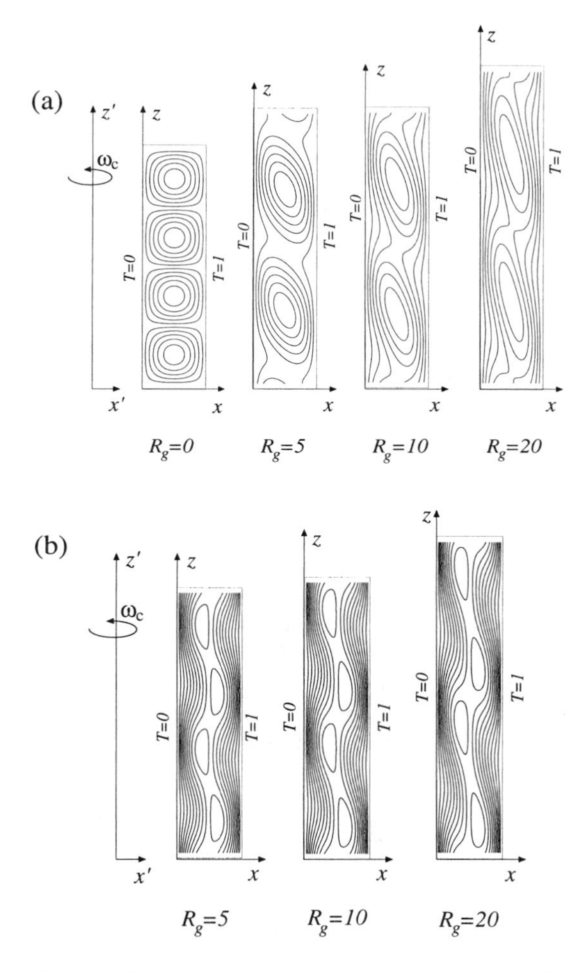

Figure 24. The convective flow field (streamlines) at marginal stability for different values of R_g ($= Ra_g/\pi$): (a) the odd modes; (b) the even modes.

X. CONCLUDING REMARKS

The effect of rotation was shown to have a significant impact on the flow in porous media. In isothermal systems this effect is limited to the effect of the Coriolis acceleration on the flow. It was shown that, similarly to nonporous domains, Taylor–Proudman columns and geostrophic flows may exist in porous media as well. The effect of rotation on the flow in a heterogeneous porous channel was presented, showing the existence of a secondary flow in a plane perpendicular to the axial imposed flow. In nonisothermal systems the effect of rotation is expected in free convection. Then the rotation may affect the flow through two distinct mechanisms, namely thermal buoyancy caused by centrifugal forces and the Coriolis force (or a combination of both). Since free convection may be driven also by gravity force and the orientation of the buoyancy force with respect to the imposed thermal gradient has a distinctive impact on the resulting convection, a significant number of combinations of different cases arise in the investigation of the rotation effects in nonisothermal systems. Results pertaining to some of these cases were presented. However, the lack of extensive experimental confirmation is particularly noticed. Although some experimental results were reported, it is recommended that the effort of getting experimental confirmation of theoretical results be expanded. This recommendation is valid for the investigation of isothermal systems as well.

NOMENCLATURE

Roman Letters

$\hat{\mathbf{e}}_x, \hat{\mathbf{e}}_y, \hat{\mathbf{e}}_z$	unit vectors in the x, y, and z direction, respectively
$\hat{\mathbf{e}}_n$	unit vector normal to the boundary, positive outwards
$\hat{\mathbf{e}}_g$	unit vector in the direction of the acceleration of gravity
$\hat{\mathbf{e}}_\omega$	unit vector in the direction of the angular velocity of rotation
Ek	porous media Ekman number, defined by Eq. (5)
g_*	acceleration of gravity
H	front aspect ratio of the porous layer, $= H_*/L_*$
H_*	height of the layer
K	dimensionless permeability function
K_o	reference value of permeability, dimensional
L_*	length of the porous layer
p	pressure, dimensionless
p_r	dimensionless reduced pressure, generalized to include the constant components of the centrifugal and gravity terms
$\dot{Q}$	internal heat generation, dimensionless
Ra_g, Ra_ω	gravity and centrifugal Rayleigh numbers, defined by Eq. (5)
S	solute concentration, dimensionless
T	temperature, dimensionless
T_C, T_H	coldest and hottest wall temperatures, respectively, dimensional
Ta	porous media Taylor number (reciprocal of Ekman number), defined by Eq. (71)
u, v, w	horizontal x and y, and vertical components of the filtration velocity, respectively
V	filtration velocity, dimensionless
W_*	width of the layer

W	top aspect ratio of the porous layer, $= W_*/L_*$
$\boldsymbol{X}$	position vector, $= x\hat{\boldsymbol{e}}_x + y\hat{\boldsymbol{e}}_y + z\hat{\boldsymbol{e}}_z$
x_0	dimensionless offset distance from rotation axis
x	horizontal length coordinate
y	horizontal width coordinate
z	vertical coordinate

Greek Letters

α_{eo}	effective thermal diffusivity
β_*	thermal expansion coefficient
β_T	dimensionless thermal expansion coefficient, $= \beta_* \Delta T_c$
δ_{ij}	Kronecker delta function
ε	porosity
η	reciprocal of the offset distance from the rotation axis, $= 1/x_0 = Ra_\omega / Ra_{\omega o}$
μ_o	fluid dynamic viscosity
ν_o	fluid kinematic viscosity
ρ	density, dimensionless
ψ	stream function
ω_*	angular velocity of rotating porous domain

Subscripts

$*$	dimensional values
c	characteristic values
cr	critical values

REFERENCES

Bear J. Dynamics of Fluids in Porous Media. New York: Elsevier, 1972, pp 131–132.

Bejan A. Convection Heat Transfer. 2nd ed. New York: Wiley, 1995.

Chakrabarti A, Gupta AS. Nonlinear thermohaline convection in a rotating porous medium. Mech Res Commun 8:9–22, 1981.

Chandrasekhar S. Hydrodynamic and Hydromagnetic Stability. Oxford: Oxford University Press, 1961.

Friedrich R. Einfluβ der Prandtl-Zahl auf die Zellularkonvektion in einem rotierenden, mit Fluid gesättigten porösen Medium. Z Angew Math Mech 63:246–249, 1983.

Govender S, Vadasz P. Centrifugal and gravity driven convection in rotating porous media—an analogy with the inclined porous layer. ASME HTD 309:93–98, 1995.

Greenspan HP. The Theory of Rotating Fluids. Cambridge: Cambridge University Press, 1980.

Hart JE. Instability and secondary motion in a rotating channel flow. J. Fluid Mech 45:341–351, 1971.

Havstad MA, Vadasz P. Numerical solution of the three dimensional fluid flow in rotating heterogeneous porous channel. Int J Numer Meth Fluids, 31:411–429, 1999.

Johnston JP, Haleen RM, Lezius DK. Effects of spanwise rotation on the structure of two-dimensional fully developed turbulent channel flow. J Fluid Mech 56:533–557, 1972.

Jou JJ, Liaw JS. Thermal convection in a porous medium subject to transient heating and rotation. Int J Heat Mass Transfer 30:208–211, 1987.

Lezius DK, Johnston JP. Roll-cell instabilities in a rotating laminar and turbulent channel flow. J Fluid Mech 77:153–175, 1976.

Nield DA. The boundary correction for the Rayleigh–Darcy problem: limitations of the Brinkman equation. J Fluid Mech 128:37–46, 1983.

Nield DA. The limitations of the Brinkman–Forchheimer equation in modeling flow in a saturated porous medium and at an interface. Int J Heat Fluid Flow 12:269–272, 1991a.

Nield DA. The stability of convective flows in porous media. In Kakaç S, Kilkis B, Kulacki FA, Arniç F eds. Convective Heat and Mass Transfer in Porous Media. Dordrecht: Kluwer, 1991b, pp 79–122.

Nield DA. Modeling the effects of a magnetic field or rotation on flow in a porous medium: momentum equation and anisotropic permeability analogy. Int J Heat Mass Transfer, 42:3715–3718, 1999.

Nield DA, Bejan A. Convection in Porous Media. 2nd ed. New York: Springer Verlag, 1999.

Palm E, Tyvand A. Thermal convection in a rotating porous layer. J Appl Math Phys (ZAMP) 35:122–123, 1984.

Patil PR, Vaidyanathan G. On setting up of convection currents in a rotating porous medium under the influence of variable viscosity. Int J Eng Sci 21:123–130, 1983.

Rudraiah N, Shivakumara IS, Friedrich R. The effect of rotation on linear and non-linear double-diffusive convection in a sparsely packed porous medium. Int J Heat Mass Transfer 29:1301–1317, 1986.

Vadasz P. Natural convection in porous media induced by the centrifugal body force—the solution for small aspect ratio. J Energy Resour Technol 114:250–254, 1992.

Vadasz P. Fluid flow through heterogeneous porous media in a rotating square channel. Transp Porous Media 12:43–54, 1993a.

Vadasz P. Three-dimensional free convection in a long rotating porous box: analytical solution. J Heat Transfer 115:639–644, 1993b.

Vadasz P. Centrifugally generated free convection in a rotating porous box. Int J Heat Mass Transfer 37:2399–2404, 1994a.

Vadasz P. Stability of free convection in a narrow porous layer subject to rotation. Int Commun Heat Mass Transfer 21:881–890, 1994b.

Vadasz P. Coriolis effect on free convection in a long rotating porous box subject to uniform heat generation. Int J Heat Mass Transfer 38:2011–2018, 1995.

Vadasz P. Stability of free convection in a rotating porous layer distant from the axis of rotation. Transp Porous Media 23:153–173, 1996a.

Vadasz P. Convection and stability in a rotating porous layer with alternating direction of the centrifugal body force. Int J Heat Mass Transfer 39:1639–1647, 1996b.

Vadasz P. Flow in rotating porous media. In du Plessis P, ed; Rahman M, series ed. Fluid Transport in Porous Media, from the series Advances in Fluid Mechanics 13. Southampton: Computational Mechanics Publications, 1997, pp 161–214.

Vadasz P. Coriolis effect on gravity-driven convection in a rotating porous layer heated from below. J Fluid Mech 376:351–375, 1998.

Vadasz P, Govender S. Two-dimensional convection induced by gravity and centrifugal forces in a rotating porous layer far away from the axis of rotation. Int J Rotat Mach 4:73–90, 1998.

Vadasz P, Govender S. Stability and stationary convection induced by gravity and centrifugal forces in a rotating porous layer distant from the axis of rotation, 1999 (submitted for publication).

Vadasz P, Havstad MA. The effect of permeability variations on the flow in a heterogeneous porous channel subject to rotation. ASME J Fluids Eng 121:568–573, 1999.

Vadasz P, Heerah A. Experimental confirmation and analytical results of centrifugally-driven free convection in rotating porous media. J Porous Media 1:261–272, 1998.

Vadasz P, Olek S. Transitions and chaos for free convection in a rotating porous layer. Int J Heat Mass Transfer 41:1417–1435, 1998.

10

Numerical Modeling of Convective Heat Transfer in Porous Media Using Microscopic Structures

A. Nakayama and F. Kuwahara
Shizuoka University, Hamamatsu, Japan

I. INTRODUCTION

The complexity associated with the geometric structure of a porous medium does not allow us to treat the detailed velocity and temperature fields inside each individual porous structure. Thus, it has been a common practice to introduce volume-averaged quantities and concentrate on the overall aspects of mass, momentum, and energy conservation principles. Accordingly, a number of heuristic and semi-heuristic models have been introduced to describe Darcy and non-Darcy flows and dispersion in heat and mass transfer through a porous medium. The model constants in these models are usually determined on the basis of exhaustive experimental data.

Detailed flow and temperature fields inside a microscopic structure may be investigated using a microscopic structure (such as a lattice structure) rather than treating complex porous media in reality. The microscopic numerical results obtained at pore scale can be processed to extract the macroscopic hydrodynamic and thermal characteristics in terms of the volume-averaged quantities. A great deal of effort has been directed towards this endeavor. In this chapter, we shall review a numerical modeling strategy of Newtonian and non-Newtonian fluid flows and heat transfer in porous media on the basis of microscopic structures. The idea of microscopic simulations using a structural model to determine macroscopic flow and heat transfer characteristics will be elucidated, and a series of numerical investigations will be examined. Turbulence in porous media is also treated from the microscopic view, which eventually leads us to establish a set of macroscopic turbulence model equations.

II. PERIODIC STRUCTURES

In the periodic models, the matrix is envisioned as a collection of objects, as shown in Figure 1. By virtue of the structural periodicity, the set of the governing equations for steady flow and heat transfer can be solved only for a single structural cell. Figures 1a and b show two distinct three-dimensional models, namely a collection of spheres and one of cubes. The minimum porosity for the collection of spheres is limited to about 0.3, while that for the collection of cubes can be reduced to virtually zero. The two-dimensional structural models, namely the collections of squares and circular cylinders, are also presented with their possible porosity ranges in Figures 1c–f.

In reality, each fluid particle inside a natural (or synthetic) porous medium experiences complex three-dimensional motion. In this sense,

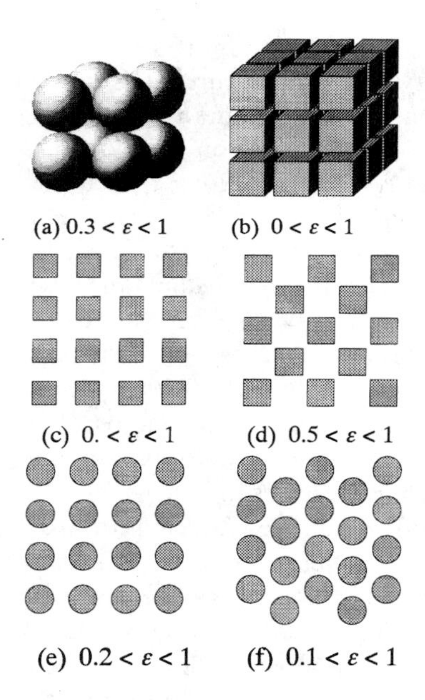

Figure 1. Periodic models.

three-dimensional models are more relevant than two-dimensional ones for simulating flows through porous media. However, such three-dimensional computations are extremely expensive and time consuming, even when using a periodic structural model. Fortunately, a series of our numerical investigations using both two- and three-dimensional models reveals that the two-dimensional models lead to expressions for the permeability almost identical to those obtained using the three-dimensional models. Furthermore, the thermal dispersion predicted using a two-dimensional model is found to be very close to what has been experimentally observed. Although there is a certain limitation to it, a two-dimensional model can be exploited to elucidate complex flow and heat transfer characteristics associated with a porous medium. Such two-dimensional models are shown in Figure 2.

Figure 2a shows a collection of square rods, placed randomly, whereas Figure 2b shows a collection of "regular" structural elements, oriented and placed randomly. The flow through the collection of randomly placed square rods (Figure 2a) may be approximated by the flow through the collection of randomly oriented structural elements (Figure 2b) without much loss of generality. However, neither of the models is practical for direct numerical computations of flows through a porous medium, since the number of grid nodes and the amount of computer time required for such calculations would be tremendous, even with the model consisting of randomly oriented structural elements (as shown in Figure 2b). In the next section, we shall consider a macroscopically uniform flow with a certain macroscopic flow angle, flowing through an infinite number of square rods placed in a regular fashion. Then, only a single structural unit needs to be taken for direct numerical computations, since periodic boundary conditions can be imposed along the unit boundaries. (There exists the

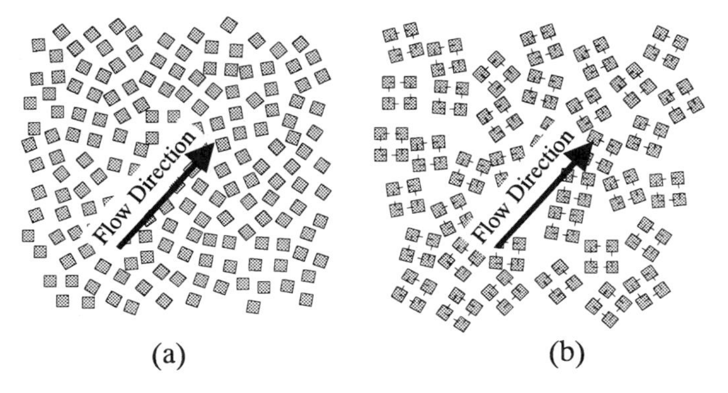

Figure 2. Two-dimensional models.

possibility that more than one unit may be involved in recurring patterns when the flow is fast and unsteady. However, the time-averaged macroscopic quantities may be estimated by performing a steady flow calculation using a single unit.) We may carry out such direct computations for various flow angles and take ensemble averages of the results, since such averaged results are believed to follow closely those of the macroscopic flows through the porous medium consisting of randomly oriented structural elements (Figure 2b), and are even capable of approximating those of the flows through the more realistic arrangement (Figure 2a).

III. MATHEMATICAL MODEL

Let us consider a macroscopically uniform flow with an angle θ through an infinite number of square rods placed in a regular fashion, as shown in Figure 3. The macroscopic velocity field follows

$$\langle \boldsymbol{u} \rangle = |\langle \boldsymbol{u} \rangle|(\cos\theta \boldsymbol{i} + \sin\theta \boldsymbol{j}) \tag{1}$$

where $\langle\ \rangle$ denotes volume average, namely

$$\langle \phi \rangle = \frac{1}{V}\int_V \phi \mathrm{d}V \tag{2}$$

The control volume V is much smaller than a macroscopic characteristic length and can be taken as H^2 for this periodic structure. Due to the periodicity of the model, only one structural unit, as indicated by dashed lines in Figure 3, may be taken as a calculation domain.

The governing equations for the detailed flow field, namely the equations of continuity and momentum, are given as follows:

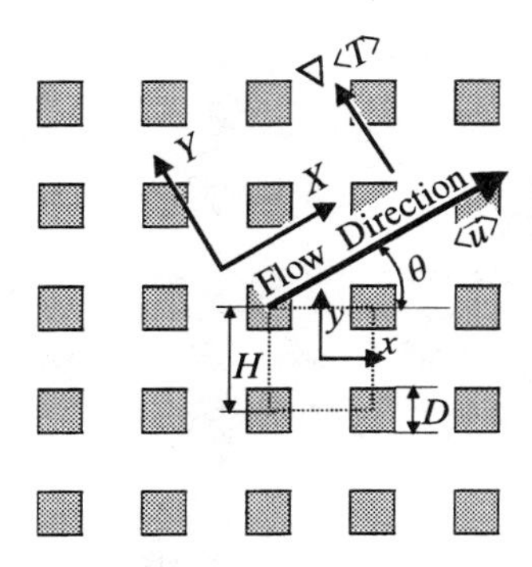

Figure 3. Physical model and its coordinate system.

$$\nabla \cdot \boldsymbol{u} = 0 \tag{3}$$

$$\nabla \cdot \boldsymbol{uu} = -\frac{1}{\rho}\nabla p + \nu\nabla^2 \boldsymbol{u} \tag{4}$$

The boundary and compatibility conditions are given by

on the solid walls:

$$\boldsymbol{u} = \boldsymbol{0} \tag{5}$$

on the periodic boundaries

$$\boldsymbol{u}|_{x=-\frac{H}{2}} = \boldsymbol{u}|_{x=\frac{H}{2}} \tag{6a}$$

$$\boldsymbol{u}|_{y=-\frac{H}{2}} = \boldsymbol{u}|_{y=\frac{H}{2}} \tag{6b}$$

$$\int_{-\frac{H}{2}}^{\frac{H}{2}} u\,\mathrm{d}y\bigg|_{x=-\frac{H}{2}} = \int_{-\frac{H}{2}}^{\frac{H}{2}} u\,\mathrm{d}y\bigg|_{x=\frac{H}{2}} = H|\langle \boldsymbol{u}\rangle|\cos\theta \tag{7a}$$

$$\int_{-\frac{H}{2}}^{\frac{H}{2}} v\,\mathrm{d}x\bigg|_{y=-\frac{H}{2}} = \int_{-\frac{H}{2}}^{\frac{H}{2}} v\,\mathrm{d}x\bigg|_{y=\frac{H}{2}} = H|\langle \boldsymbol{u}\rangle|\sin\theta \tag{7b}$$

We shall define the Reynolds number based on the Darcian velocity $|\langle \boldsymbol{u}\rangle|$ and length of structural unit H as

$$Re = |\langle \boldsymbol{u}\rangle|H/\nu \tag{8}$$

whereas the porosity for the case of this two-dimensional array of square rods is given by

$$\varepsilon = 1 - \left(\frac{D}{H}\right)^2 \tag{9}$$

The microscopic velocity field within the structural model can be determined by solving the governing equations (3) and (4) with the appropriate boundary and compatibility conditions as given above. The corresponding microscopic temperature field will be discussed later in connection with the thermal dispersion mechanism.

IV. METHOD OF COMPUTATION

The foregoing governing equations are readily discretized by integrating them over a grid volume. The SIMPLE algorithm for the pressure–velocity coupling, as proposed by Patankar and Spalding (1972), can be adopted to correct the pressure and velocity fields. The calculation starts by solving the two momentum equations, and subsequently the estimated velocity field is corrected by solving the pressure correction equation reformulated from the discretized continuity and momentum equations, such that the velocity field fulfills the continuity principle. Then the energy conservation equation (if the temperature field is sought) is solved to find the corresponding temperature field. This iteration sequence must be repeated until convergence is achieved. Convergence can be measured in terms of the maximum change in each variable during an iteration. The maximum change allowed for the convergence check may be set to an arbitrarily small value (such as 10^{-5}), as the variables are normalized by appropriate references. A fully implicit scheme may be adopted, with the hybrid differencing scheme for the advection terms. Calculations are carried out using the grid system (45×45). Further details on this numerical procedure can be found in Patankar (1980) and Nakayama (1995).

V. DETERMINATION OF PERMEABILITY

A. Related Work

A series of numerical and analytical attempts has been made to determine the permeability purely from a theoretical basis. Eidsath et al. (1983), Coulaud et al. (1988), Sahraoui and Kaviany (1991), and Fowler and Bejan (1994) carried out two-dimensional numerical simulations for flows across banks of circular cylinders, whereas the authors (Kuwahara et al. 1994) investigated a collection of square rods to cover a wide range of porosity, virtually from zero to unity.

Three-dimensional analyses were also conducted by Larson and Higdon (1989) for Stokes flows through a lattice of spheres, and by the authors (Nakayama et al. 1995) for fully elliptic flows through a lattice of cubes to study not only the Darcy contribution but also the porous inertial contribution to the macroscopic pressure drop. In what follows, we shall review some of recent work carried out by our research group.

B. Microscopic Flow Fields

Two distinct velocity vector plots obtained at $\theta = 0^\circ$ and 45° for three different Reynolds numbers, namely $Re = 10^{-1}$, 10, and 10^3, are compared in Figure 4. When the Reynolds number is comparatively small, say $Re \leq 10$, the velocity profiles for both $\theta = 0^\circ$ and 45° exhibit parabolic profiles, as in a fully developed channel flow, such that the viscous force contribution to the pressure drop predominates over the inertial contribu-

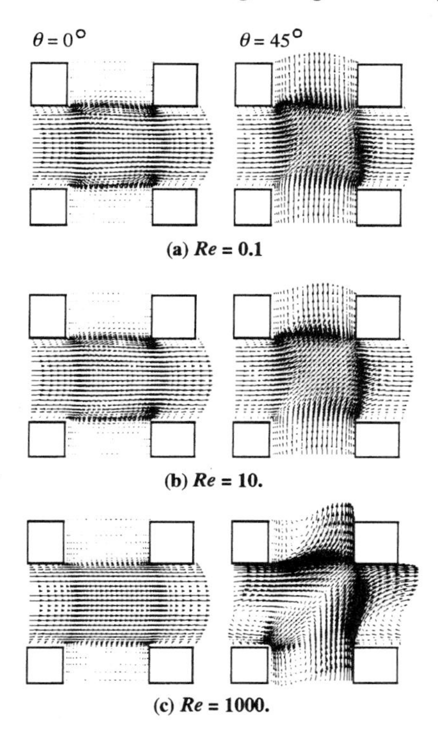

Figure 4. Velocity vector plots.

tion. As the Reynolds number increases, a distinct difference appears between the velocity field of $\theta = 0^\circ$ and that of $\theta = 45^\circ$. Flow separation takes place for $\theta = 45^\circ$ such that the inertial contribution to the pressure drop becomes significant, whereas the flow field for $\theta = 0^\circ$ remains as the channel flow type such that the inertial contribution is negligibly small. Accordingly, we may expect that the resulting macroscopic pressure drop for a fixed mass flow rate would be fairly insensitive to the macroscopic flow direction for low Reynolds number flows, while it becomes sensitive to the flow direction as the Reynolds number increases.

C. Macroscopic Pressure Gradient and Permeability

The macroscopic pressure gradient (i.e., the gradient of the intrinsic average pressure measured along the macroscopic flow direction) may readily be evaluated using the microscopic numerical results as

$$-\frac{\mathrm{d}\langle p\rangle^f}{\mathrm{d}X} = \frac{\cos\theta}{H(H-D)}\int_{-(H-D)/2}^{(H-D)/2}\left(p|_{x=-(\frac{H-D}{2})} - p|_{x=\frac{H-D}{2}}\right)\mathrm{d}y + \frac{\sin\theta}{H(H-D)}\int_{-(H-D)/2}^{(H-D)/2}\left(p|_{y=-(\frac{H-D}{2})} - p|_{y=\frac{H-D}{2}}\right)\mathrm{d}x \tag{10}$$

The pressure gradient results are assembled in Figure 5 in terms of the dimensionless pressure gradient against the Reynolds number. All data show that the dimensionless pressure gradient stays constant for $Re \leq 10$, irrespective of the flow angle, as we expected from the velocity vector plots in Figure 4. The pressure gradient increases drastically as Re goes beyond 10, in which the porous inertial contribution becomes appreciable as compared with the viscous (Darcian) contribution.

The Forchheimer-extended Darcy's law may be written as

$$-\frac{\mathrm{d}\langle p\rangle^f}{\mathrm{d}X} = \frac{\mu}{K}|\langle \boldsymbol{u}\rangle| + \rho b|\langle \boldsymbol{u}\rangle|^2 \tag{11a}$$

or, in dimensionless form, as

$$-\frac{\mathrm{d}\langle p\rangle^f}{\mathrm{d}X}\frac{H}{\rho|\langle \boldsymbol{u}\rangle|^2}Re = \frac{H^2}{K} + bH\ Re \tag{11b}$$

Thus the permeability K may readily be determined by reading the intercept of the ordinate variable in Figure 5.

The permeability, thus obtained for various porosities, is presented in Figure 6 as a function of the porosity. Constancy of the dimensionless

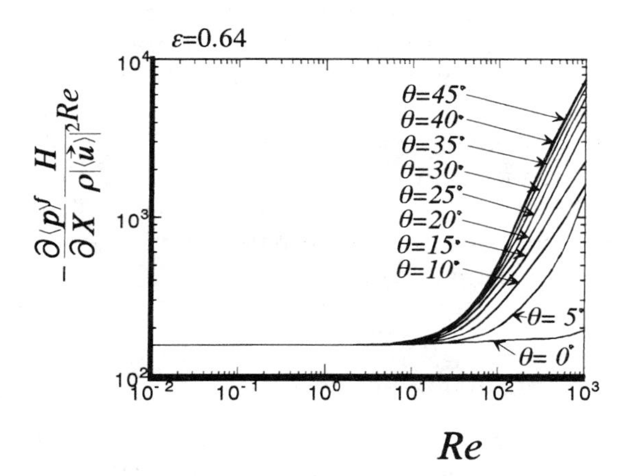

Figure 5. Effect of Reynolds number on pressure gradient.

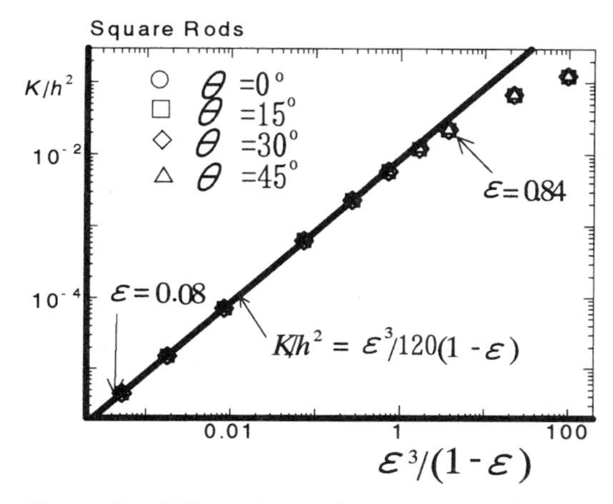

Figure 6. Effect of porosity on permeability.

pressure gradient, irrespective of θ, substantiates the validity of the periodic numerical models to determine K from first principles. From the figure, the following functional relation can be extracted

$$K = \varepsilon^3 D^2/120(1-\varepsilon)^2: \qquad \text{square cylinders} \tag{12}$$

whose functional form is identical to that of Ergun's empirical expression, namely

$$K = \varepsilon^3 D^2/150(1-\varepsilon)^2: \qquad \text{Ergun (1952)} \tag{13}$$

where D is the average diameter of spherical particles. Similar low Reynolds number calculations have been performed for a collection of circular cylinders in square arrangement. The resulting expression for the permeability, which is very close to the foregoing expressions, is given by

$$K = \varepsilon^3 D^2/144(1-\varepsilon)^2: \qquad \text{circular cylinders} \tag{14}$$

Exhaustive three-dimensional numerical calculations have also been conducted for low Reynolds number flows, using collections of cubes (Nakayama et al. 1995) and spheres (Kuwahara et al. 1994) in simple cubic arrangements. The resulting expressions, listed below, turn out to be very close to those obtained from the two-dimensional models

$$K = \varepsilon^3 D^2/147(1-\varepsilon)^2: \qquad \text{spheres} \tag{15}$$

$$K = \varepsilon^3 D^2/152(1-\varepsilon)^2: \qquad \text{cubes} \tag{16}$$

For the cases of circular cylinders and spheres, D stands for the diameter. Thus, all two- and three-dimensional models lead to almost identical expressions for the permeability, which are in good accord with the empirical expression proposed by Ergun (1952). It is rather surprising that a two-dimensional model as simple as the square cylinder model is capable of predicting the permeability, which is in almost perfect agreement with the well-known empirical relationship.

The second term on the right-hand side of (11b) represents the porous inertial contribution which increases with Re. The coefficient b for the inertial contribution may be determined by fitting the numerical results obtained at high Reynolds numbers into (11b). Such an attempt, and its extension to a three-dimensional model, are discussed in the next section.

VI. DETERMINATION OF POROUS INERTIA

A. Two-dimensional Model

As shown in Figure 5, the macroscopic pressure gradient increases drastically as Re goes beyond unity, since the porous inertial contribution takes over that of viscous force. The Forchheimer constant b, describing the porous inertia effects on the pressure drop, can be determined by fitting the numerical results obtained at high Reynolds numbers into Eq. (11b). Extensive calculations were performed, using the two-dimensional numerical model of square cylinders, to investigate the porous inertial contribution. Figure 7 shows a typical variation of the Forchheimer constant b predicted using the structural model of square cylinders. The abscissa variable is chosen anticipating Ergun's functional form (Ergun 1952) for the Forchheimer term, namely

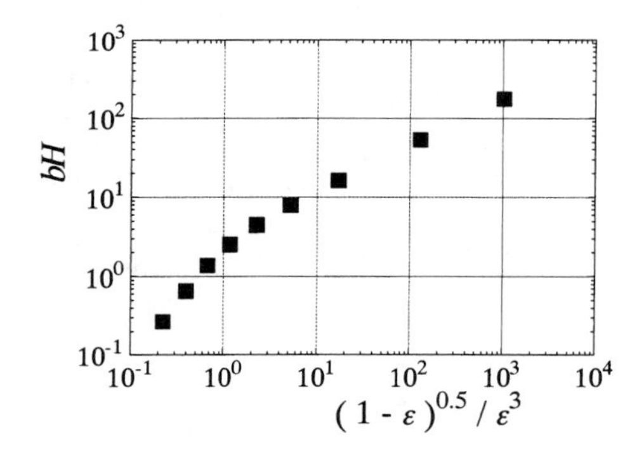

Figure 7. Forchheimer constant based on two-dimensional model.

$$b = 1.75\frac{(1-\varepsilon)}{\varepsilon^3}\frac{1}{D} \tag{17}$$

However, no proportional relationship between b and $(1-\varepsilon)/\varepsilon^3$ can be seen from these two-dimensional results. This suggests that we must resort to a three-dimensional model to simulate the porous inertial contribution, since flow separation and wake are strictly of a three-dimensional nature.

B. Three-dimensional Model

Thus, we consider a macroscopically uniform flow through a three-dimensional array of obstacles, namely an infinite number of cubes placed in a regular fashion in an infinite space, as shown in Figure 8a. Only one structural unit of $H \times H \times H$, as shown in Figure 8b can be taken as a calculation domain in consideration of the geometric periodicity. The direction of

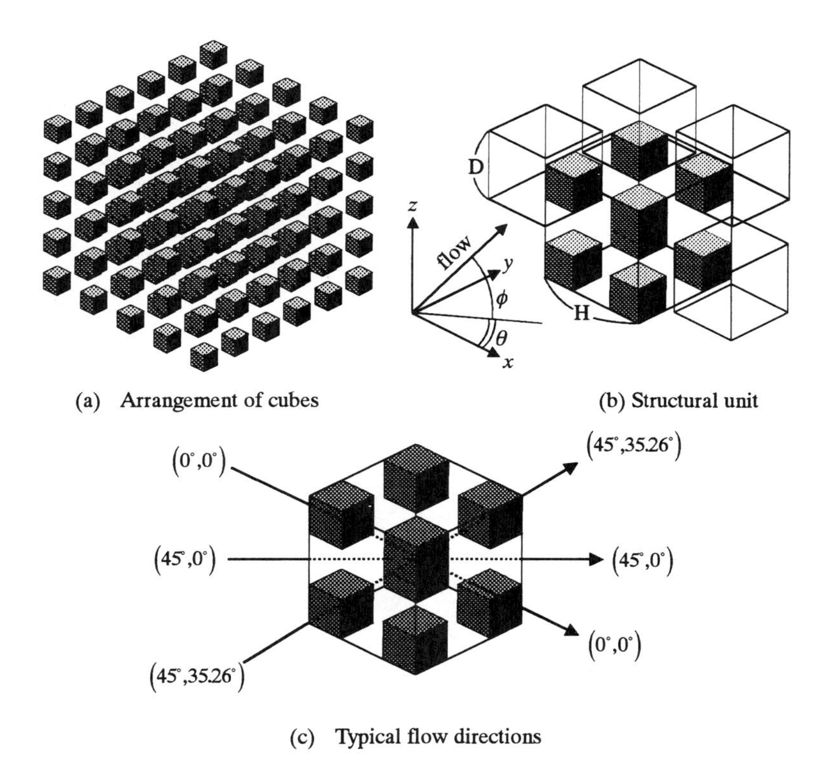

Figure 8. Numerical model.

the microscopically uniform flow is expressed in terms of (θ, ϕ), as illustrated in Figure 8c, such that

$$\langle \boldsymbol{u} \rangle = |\langle \boldsymbol{u} \rangle|(\cos\theta\cos\phi\boldsymbol{i} + \sin\theta\cos\phi\boldsymbol{j} + \sin\phi\boldsymbol{k}) \tag{18}$$

The porosity may be varied by changing the ratio of the size of cube D to the unit cell size H, as

$$\varepsilon = 1 - \left(\frac{D}{H}\right)^3 \tag{19}$$

A three-dimensional computer program can be constructed by a straightforward extension of the two-dimensional program discussed earlier. Computations have been carried out distributing $45 \times 45 \times 45$ nodes within one structural unit to acquire sufficient accuracy.

The values of Forchheimer constant b based on the three-dimensional results are presented for three different flow directions in Figure 9, which clearly shows that the three-dimensional results possess the same functional relationship $b \propto (1 - \varepsilon)/\varepsilon^3$ as in Ergun's formula. The multiplicative constant, however, is rather sensitive to the macroscopic flow direction, as shown in the figure. In reality, a fluid particle travels through a porous medium, changing its direction freely. Thus, any numerical model seems to have its limitations for simulating such random flow motions. For the first approximation, we may average the multiplicative constant for the Forchheimer constant over the flow angles as

$$\frac{4}{\pi\tan^{-1}(1/\sqrt{2})}\int_0^{\tan^{-1}(1/\sqrt{2})}\int_0^{\pi/4} bD\left(\frac{\varepsilon^3}{1-\varepsilon}\right)\mathrm{d}\theta\mathrm{d}\phi \cong 0.5 \tag{20}$$

such that

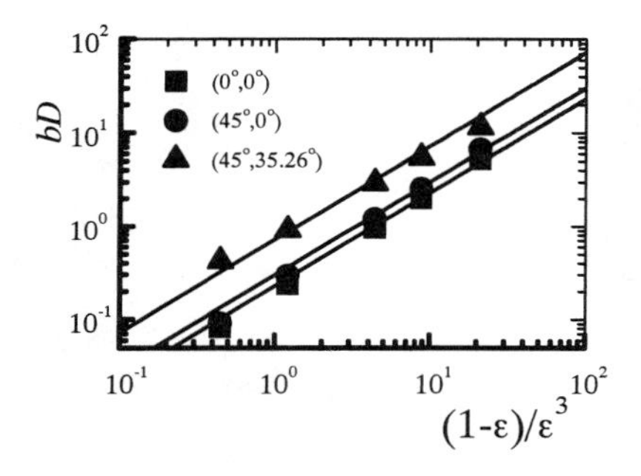

Figure 9. Forchheimer constant based on three-dimensional model.

$$b \approx 0.5 \frac{(1-\varepsilon)}{\varepsilon^3} \frac{1}{D}: \qquad \text{present model} \tag{21}$$

The predicted multiplicative constant 0.5 is about one-third the Ergun multiplicative constant, 1.75. This shows the limitation of the model based on a regular arrangement. In order to make a realistic prediction of the porous inertial effects, irregularity of porous structures, along with possible flow unsteadiness, must be taken into full consideration.

VII. DETERMINATION OF THERMAL DISPERSION

A. Related Work

Howells (1974), Acrivos et al. (1980), Larson and Higdon (1989), Kim and Russel (1985), and many others carried out analytical investigations on flows through porous media. They all considered Stokes equations under the assumption of low Reynolds (or Péclet) number. The Stokes model, however, cannot be adopted to study the dispersion and porous inertial contributions, which dominate over the diffusive contributions at high velocities. Dispersion in porous media was studied by Koch and Brady (1985) and Koch et al. (1989), who obtained closed-form expressions for the dispersion tensor. However, in their analysis, extremely dilute suspension of particles (i.e., high porosity) having the same thermal conductivity as the fluid was assumed, using Stokes flow approximation along with the point force approximation. Thus, no Reynolds number or boundary layer effects were implemented in their analysis. Full Navier–Stokes and energy equations were solved by Eidsath et al. (1983) and Edwards et al. (1991) for flows through a periodic structure of circular cylinders with in-line and staggered arrangements. In their models, the thermal conductivity of particles is assumed to be zero, such that no coupling of the energy equations for the fluid and solid phases was present. Kuwahara et al. (1994) studied the conjugate heat transfer problems of Pop et al. (1985, 1986), and proposed an idea to determine the transverse dispersion coefficient by fitting the numerical results against the similarity solution for forced convection from a line heat source in a porous medium, and conducted exhaustive numerical calculations for a large computational domain made of a lattice of square rods. Arquis et al. (1991, 1993) extended the numerical model proposed by Coulaud et al. (1988) to the coupling of momentum and heat transfer in order to study both axial and transverse dispersion coefficients. Arquis and his group imposed a macroscopic temperature gradient either normal or parallel to a macroscopically uniform flow such that the microscopic temperature field within only one structural unit was needed, as in the velocity field, to determine the corresponding dispersion coefficient. However, in these numerical studies, computations were carried out only for a limited number of sets of the parameters such as the porosity, macroscopic flow direction and Péclet number. No general functional relationship for the dispersion coefficient as a function of these parameters were drawn from these studies, due to a rather narrow porosity range.

Kuwahara et al. (1996) followed the numerical approach proposed by Arquis et al. (1991, 1993) to determine the transverse dispersion coefficient purely from a theoretical basis. A macroscopically uniform flow was assumed to pass through a lattice of square rods placed regularly in an infinite space where a macroscopically linear temperature gradient was imposed perpendicularly to the flow direction. The macroscopic results were integrated over a unit structure to evaluate the transverse thermal dispersion coefficient. Two distinct expressions for the transverse dispersion, as a function of porosity and Péclet number, have been established for the low and high Péclet number ranges.

Yagi et al. (1960) were the first to measure the effective longitudinal (axial) thermal conductivities of a packed bed. A numerical experiment to investigate longitudinal thermal dispersion was conducted by Kuwahara and Nakayama (1999). This time, a macroscopically linear temperature gradient was imposed in parallel to the macroscopic flow direction. This numerical experiment, which agrees well with available experiments, confirms that the longitudinal dispersion is substantially higher than the transverse dispersion.

B. Numerical Model

Again, we shall consider a macroscopically uniform flow with an angle θ, as given by Eq. (1), through an infinite number of square rods placed in a regular fashion, as illustrated in Figure 3. Two distinct macroscopic temperature fields are considered to obtain the transverse and longitudinal dispersions. Thus, we impose a macroscopically linear temperature gradient perpendicularly to the macroscopic flow direction, namely

$$\nabla\langle T\rangle = \frac{\Delta T}{H}(-\sin\theta \boldsymbol{i} + \cos\theta \boldsymbol{j}) \tag{22}$$

to determine the transverse thermal dispersion, and a macroscopically linear temperature gradient in parallel to the macroscopic flow direction, namely

$$\nabla\langle T\rangle = \frac{\Delta T}{H}(\cos\theta \boldsymbol{i} + \sin\theta \boldsymbol{j}) \tag{23}$$

to determine the longitudinal thermal dispersion.

Again, only one structural unit, as indicated by dashed lines in Figure 3, may be taken as a calculation domain. In addition to the continuity and Navier–Stokes equations, as given by Eqs. (3) and (4), we consider the energy conservation equation for the whole domain (i.e. fluid and solid phases)

$$\rho_f C_{pf} \nabla \cdot (\boldsymbol{u}T) = k_f \nabla^2 T: \qquad \text{fluid phase} \tag{24}$$

$$k_s \nabla^2 T = 0: \qquad \text{solid phase} \tag{25}$$

where the subscripts f and s denote fluid and solid phases, respectively. The thermal boundary and periodic constraints are given as follows:

on the solid walls:

$$T|_s = T|_f \tag{26a}$$

$$k_s \frac{\partial T}{\partial n}\bigg|_s = k_f \frac{\partial T}{\partial n}\bigg|_f \tag{26b}$$

on the periodic boundaries:

$$T|_{x=-\frac{H}{2}} = T|_{x=\frac{H}{2}} + \Delta T \sin\theta \tag{27a}$$

$$T|_{y=-\frac{H}{2}} = T|_{y=\frac{H}{2}} - \Delta T \cos\theta \tag{27b}$$

as we determine the transverse thermal dispersion, and

$$T|_{x=-\frac{H}{2}} = T|_{x=\frac{H}{2}} - \Delta T \cos\theta \tag{28a}$$

$$T|_{y=-\frac{H}{2}} = T|_{y=\frac{H}{2}} - \Delta T \sin\theta \tag{28b}$$

as we determine the longitudinal thermal dispersion.

C. Microscopic Expressions for Tortuosity and Thermal Dispersion

Following Cheng (1978), Nield and Bejan (1992), and Nakayama (1995), we integrate the microscopic energy equation (24) for the incompressible fluid over an elementary control volume, and obtain

$$\rho_f C_{p_f} \langle \boldsymbol{u} \rangle \cdot \nabla \langle T \rangle^f = \nabla \cdot \left[k_f \nabla \varepsilon \langle T \rangle^f + \frac{1}{V} \int_{A_{\text{int}}} k_f T \mathrm{d}\mathbf{A} - \rho_f C_{p_f} \langle T' \boldsymbol{u}' \rangle \right] + \frac{1}{V} \int_{A_{\text{int}}} k_f \nabla T \cdot \mathrm{d}\mathbf{A} \tag{29}$$

Similarly, the microscopic energy equation for the solid (25) may be integrated to give

$$\nabla \cdot \left[k_s \nabla (1-\varepsilon) \langle T \rangle^s - \frac{1}{V} \int_{A_{\text{int}}} k_s T \mathrm{d}\mathbf{A} \right] - \frac{1}{V} \int_{A_{\text{int}}} k_f \nabla T \cdot \mathrm{d}\mathbf{A} = 0 \tag{30}$$

where A_{int} is the total interface between the fluid and solid, and $\mathrm{d}\mathbf{A}$ is its vector element pointing outward from the fluid side to solid side. $\langle T \rangle^f$ and $\langle T \rangle^s$ are the intrinsic averages of the fluid temperature and the solid phase temperature, respectively. The continuity of heat flux at the interface is implemented in the above equation. We shall assume that thermal equilibrium exists between the fluid and solid matrix, namely

$$\langle T \rangle^f = \langle T \rangle^s = \langle T \rangle \tag{31}$$

Adding (29) and (30), we have

$$\rho_f C_{p_f} \langle \boldsymbol{u} \rangle \cdot \nabla \langle T \rangle = \nabla \cdot \left\{ \left[\varepsilon k_f + (1-\varepsilon) k_s \right] \nabla \langle T \rangle + \frac{1}{V} \int_{A_{\text{int}}} (k_f - k_s) T \mathrm{d}\mathbf{A} - \rho_f C_{p_f} \langle T' \boldsymbol{u}' \rangle \right\} \tag{32a}$$

which can be arranged as

$$\rho_f C_{p_f} \langle \boldsymbol{u} \rangle \cdot \nabla \langle T \rangle = \nabla \cdot \left\{ \left(k_e \overline{\overline{I}} + \overline{\overline{k}}_{\text{tor}} + \overline{\overline{k}}_{\text{dis}} \right) \cdot \nabla \langle T \rangle \right\} \tag{32b}$$

where

$$k_e \equiv \varepsilon k_f + (1-\varepsilon) k_s \tag{33}$$

$$\frac{1}{V} \int_{A_{\text{int}}} (k_f - k_s) T \mathrm{d}\mathbf{A} \equiv \overline{\overline{k}}_{\text{tor}} \cdot \nabla \langle T \rangle \tag{34}$$

$$-\rho_f C_{p_f} \langle T' \boldsymbol{u}' \rangle \equiv \overline{\overline{k}}_{\text{dis}} \cdot \nabla \langle T \rangle \tag{35}$$

The first two terms $-(k_e \overline{\overline{I}} + \overline{\overline{k}}_{\text{tor}}) \cdot \nabla \langle T \rangle$ on the right-hand side of (32b) account for molecular diffusion, whereas the third term $-\overline{\overline{k}}_{\text{dis}} \cdot \nabla \langle T \rangle$ accounts for contributions from thermal (mechanical) dispersion. k_e, defined in (33), is the stagnant thermal conductivity. The apparent conductivity tensors $\overline{\overline{k}}_{\text{tor}}$ and $\overline{\overline{k}}_{\text{dis}}$ are introduced to model the tortuosity molecular diffusion term and the thermal dispersion term, respectively, by a gradient-type diffusion hypothesis.

We shall determine both tortuosity and thermal dispersion conductivities purely from a theoretical basis by substituting the microscopic numerical results in to Eqs. (34) and (35). Let us set one coordinate along the macroscopic flow direction. When the macroscopically linear temperature gradient is imposed along the Y direction normal to the X direction of the macroscopic flow, only diagonal components of the tortuosity and dispersion conductivity tensors are non-zero components. Thus, from Eqs. (34) and (35), the YY components of $\bar{\bar{k}}_{\text{tor}}$ and $\bar{\bar{k}}_{\text{dis}}$ can readily be determined from

$$(k_{\text{tor}})_{YY} = \frac{\left(\dfrac{k_s - k_f}{V} \int_{A_{\text{int}}} T \mathrm{d}\mathbf{A}\right) \cdot (-\sin\theta \boldsymbol{i} + \cos\theta \boldsymbol{j})}{(\Delta T/H)} \tag{36}$$

$$(k_{\text{dis}})_{YY} = \frac{-(\rho_f C_{p_f}/H^2)}{(\Delta T/H)} \int_{-H/2}^{H/2} \int_{-H/2}^{H/2} (T - \langle T \rangle)(\boldsymbol{u} - \langle \boldsymbol{u} \rangle^f) \, \mathrm{d}x \mathrm{d}y \cdot (-\sin\theta \boldsymbol{i} + \cos\theta \boldsymbol{j}) \tag{37}$$

respectively.

Similarly, when the macroscopically linear temperature gradient is imposed along the X direction of the macroscopic flow, the XX components of $\bar{\bar{k}}_{\text{tor}}$ and $\bar{\bar{k}}_{\text{dis}}$ can readily be determined from

$$(k_{\text{tor}})_{XX} = \frac{\left(\dfrac{k_s - k_f}{V} \int_{A_{\text{int}}} T \mathrm{d}\mathbf{A}\right) \cdot (\cos\theta \boldsymbol{i} + \sin\theta \boldsymbol{j})}{(\Delta T/H)} \tag{38}$$

$$(k_{\text{dis}})_{XX} = \frac{-(\rho_f C_{p_f}/H^2)}{(\Delta T/H)} \int_{-H/2}^{H/2} \int_{-H/2}^{H/2} (T - \langle T \rangle)(\boldsymbol{u} - \langle \boldsymbol{u} \rangle^f) \, \mathrm{d}x \mathrm{d}y \cdot (\cos\theta \boldsymbol{i} + \sin\theta \boldsymbol{j}) \tag{39}$$

D. Microscopic Temperature Field and Thermal Conductivity due to Tortuosity

Typical temperature fields obtained at $\theta = 0^\circ$ and 45° for two different Reynolds numbers, namely $Re = 10$ and 10^3, are compared in Figures 10 and 11. (Note that the corresponding velocity fields are shown in Figure 4.) Figure 10 shows the cases in which the macroscopic temperature gradient is imposed perpendicularly to the macroscopic flow direction, while Figure 11 shows the cases in which the gradient is imposed in parallel to the macro-

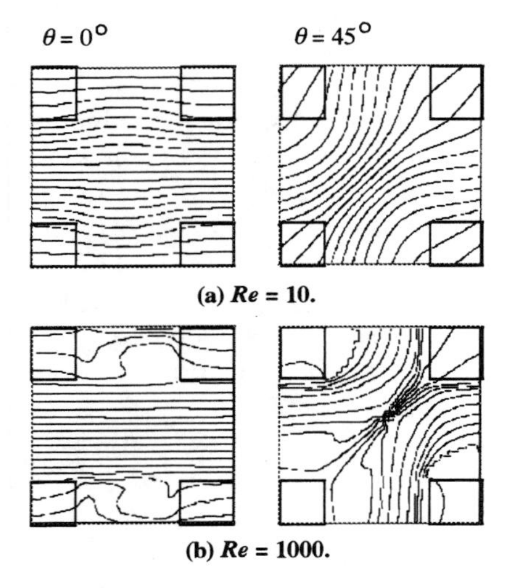

Figure 10. Isotherms (perpendicular).

scopic flow direction. The isotherms obtained at low Reynolds numbers exhibit a typical pattern for the case of pure thermal conduction. For the case of high Reynolds number flows, the temperature pattern becomes very complex as a result of thermal dispersion, as seen from Figure 11b, where comparatively uniform temperature regions exist within recirculation

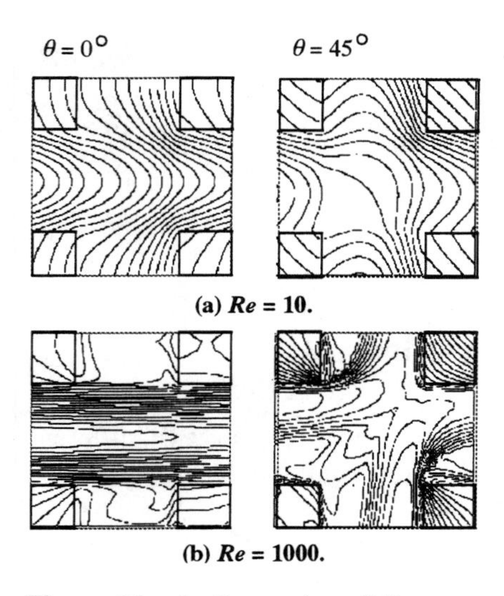

Figure 11. Isotherms (parallel).

regions. In Figure 12, the tortuosity conductivity $(k_{tor})_{YY}$, determined by feeding the microscopic numerical results in Eq. (36), is plotted against the Péclet number Pe for the case of $\varepsilon = 0.64$. It is interesting to note that all data obtained for different macroscopic flow angles fall onto a horizontal asymptote as Pe decreases. An approximate analytical formula for this asymptotic value may be found by approximating the temperature distribution in a structural unit at low Péclet number (see Figure 10a) by a piecewise linear temperature distribution within a one-dimensional composite slab, as

$$\frac{(k_{tor})_{YY}}{k_f} = \left(\frac{k_s}{k_f} - 1\right)\frac{(1-\varepsilon)}{(1-\varepsilon)^{1/2} + (k_s/k_f)(1-(1-\varepsilon)^{1/2})} \tag{40}$$

The tortuosity conductivity results obtained for a low Péclet number range with changing porosity and thermal conductivity ratio are plotted together with the curves generated by the foregoing approximate formula, in Figure 13 with the abscissa $(1-\varepsilon)$. An excellent agreement can be seen between the numerical results and the approximate formula for the entire range of porosity.

Figure 12 clearly shows that the tortuosity conductivity decreases with Pe increases. However, the contribution from thermal dispersion predominates over the tortuosity contribution with increasing Pe, such that the tortuosity conductivity may no longer be important and may well be neglected in such a high Péclet number range.

E. Thermal Conductivity due to Thermal Dispersion

From the microscopic temperature fields shown in Figure 11, it can be seen that the isotherms for the case in which the macroscopic temperature gradient aligns with the macroscopic flow are distorted considerably along the macroscopic flow direction even when the Reynolds number is as low as 10

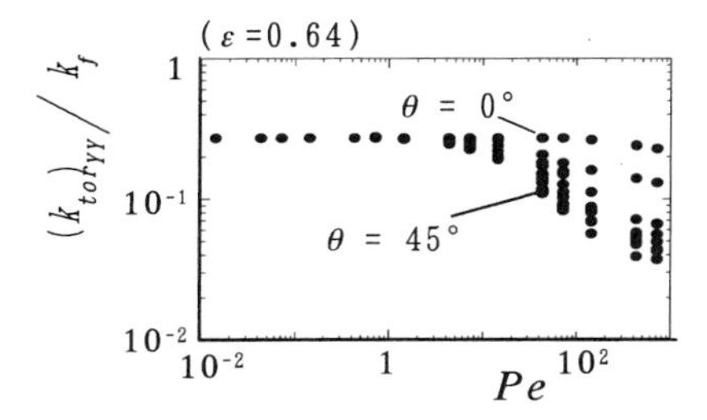

Figure 12. Effect of Péclet number on apparent thermal conductivity due to tortuosity.

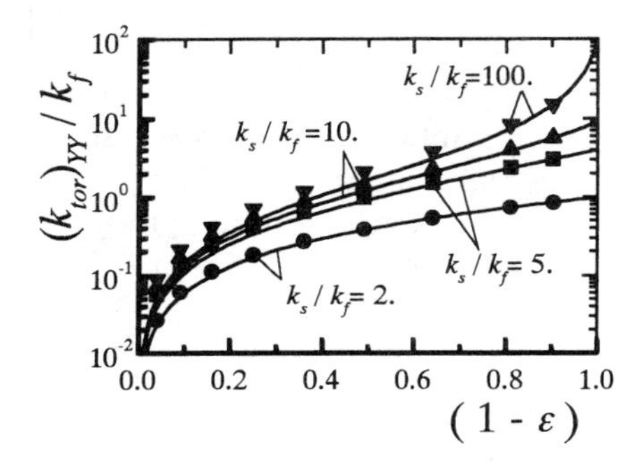

Figure 13. Apparent thermal conductivity due to tortuosity.

(see the upper part of Figure 11), whereas the distortions in the isotherms for the case of the macroscopic temperature gradient perpendicular to the macroscopic flow are not significant even for cases of high Reynolds number (see the lower part of Figure 10) in which flow reversals take place.

This marked difference in the distortions, observed for the isotherms in the two distinct cases, is closely associated with the fact that the macroscopic net flow does not exist in the direction perpendicular to the macroscopic flow direction, and that only weak mixing actions due to microscopic wake and flow reversals are solely responsible for the transverse mechanical dispersion. Naturally, we expect that the longitudinal effective thermal conductivity (dispersion coefficient) is much more significant than the transverse one. This unique feature associated with convective heat transfer in porous media often puzzles those who are accustomed to applying the boundary layer theory to pure fluid flow problems. One should bear in mind that the boundary layer approximations may fail even for the case of macroscopically unidirectional flow through a porous medium.

The microscopic temperature results obtained for various flow angles are processed using Eq. (37), and the resulting thermal conductivity $(k_{\mathrm{dis}})_{YY}$ due to the transverse thermal dispersion is plotted for the case of $\varepsilon = 0.64$ in Figure 14. The figure suggests that the lower and higher Péclet number data follow two distinct limiting lines. The lower Péclet number data vary in proportion to Pe^{17} where as the high Péclet number data vary in proportion to Pe. The thermal conductivity $(k_{\mathrm{dis}})_{XX}$ due to the longitudinal thermal dispersion is plotted for the case $\varepsilon = 0.64$ in Figure 15, where the low Péclet number data vary in proportion to Pe^2, as suggested by Taylor (1953) in the classical study of solute in a fluid flowing in a tube, whereas the high Péclet number data vary in proportion to Pe, as in the case of the transverse thermal dispersion. As expected from the microscopic tempera-

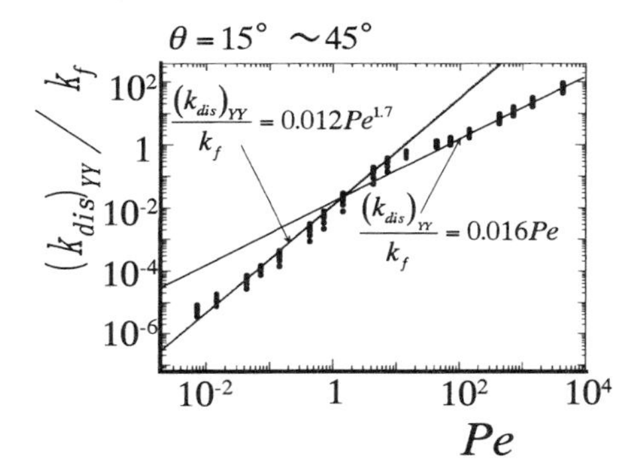

Figure 14. Apparent thermal conductivity due to dispersion (perpendicular).

ture fields, the level of the transverse thermal dispersion is found to be substantially lower than that of the longitudinal thermal dispersion.

Exhaustive computations were conducted to extract functional relationships for the thermal dispersion conductivities at low and high Péclet number ranges, in terms of the Péclet number and porosity. The thermal dispersion conductivities obtained for different ε were plotted with the abscissa variable $(1-\varepsilon)$ to investigate the porosity dependency. The resulting expressions for the low and high Péclet number ranges are given as follows.

For the transverse dispersion:

$$\frac{(k_{\mathrm{dis}})}{k_f} = 0.022\frac{Pe_D^{1.7}}{(1-\varepsilon)^{1/4}} \qquad \text{for} \qquad (Pe_D < 10) \tag{41a}$$

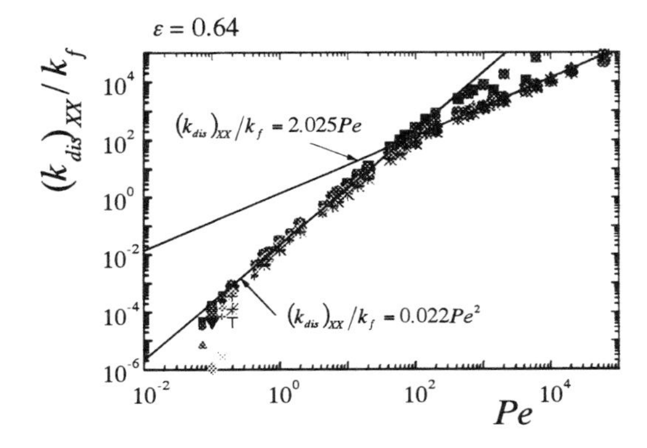

Figure 15. Apparent thermal conductivity due to dispersion (parallel).

$$\frac{(k_{\text{dis}})}{k_f} = 0.052(1-\varepsilon)^{1/2} Pe_D \qquad \text{for} \qquad (Pe_D > 10) \tag{41b}$$

for the longitudinal dispersion:

$$\frac{(k_{\text{dis}})_{XX}}{k_f} = 0.022 \frac{Pe_D^2}{(1-\varepsilon)} \qquad \text{for} \qquad (Pe_D < 10) \tag{42a}$$

$$\frac{(k_{\text{dis}})_{XX}}{k_f} = 2.7 \frac{Pe_D}{\varepsilon^{1/2}} \qquad \text{for} \qquad (Pe_D > 10) \tag{42b}$$

where the relationship $Pe_D = Pe(1-\varepsilon)^{1/2}$ is used.

In Figure 16, the foregoing correlations (41b) and (42b) for the high Péclet number range are compared against the experimental data obtained by Fried and Combarnous (1971) for the longitudinal and transverse thermal dispersion coefficients in a packed bed. The figure shows that the present correlation (42b), obtained from the numerical experiment, agrees very well with the experimental data for the axial thermal dispersion coefficient, whereas the correlation (41b) for the transverse thermal dispersion underpredicts the corresponding experimental data. Both the predicted correlations and experimental data clearly show that the level of the longitudinal dispersion is much higher than that of the transverse dispersion. Overall, the agreement between the present correlation and the experimental data can be seen to be good, in light of the simplicity of the present two-dimensional model.

Throughout this chapter, we assume that local thermal equilibrium prevails within fluid-saturated porous media. However, when there is a strong heat generation in the fluid phase or in the solid phase, the intrinsic

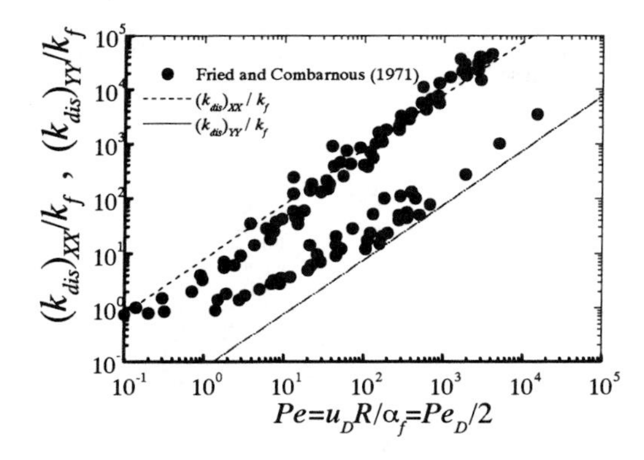

Figure 16. Comparison of prediction and experiment.

average of one phase will be significantly different from that of the other phase. Similarly, when the hot fluid enters a porous medium, the temperature of the entering fluid can be much higher than that of the medium, such that heat transfer takes place from the fluid phase to the solid phase. Under these situations, the assumption of local thermal equilibrium is no longer valid. Thus, we must allow for heat transfer between the phases. The theoretical approaches along this line may be found in the excellent book on convection in porous media written by Kaviany (1991).

VIII. MATHEMATICAL MODEL FOR NON-NEWTONIAN FLUID FLOWS IN POROUS MEDIA

A. Related Work

A number of industrially important fluids, including fossil fuels, which may saturate underground beds, exhibit non-Newtonian fluid behavior. Naturally, the understanding of non-Newtonian fluid flows through porous media represents interesting challenges in geophysical systems, chemical reactor design, certain separation processes, polymer engineering, or in petroleum production. Many of the inelastic non-Newtonian fluid flows encountered in such engineering processes are known to follow the so-called "power-law model" in which the pressure drop is in proportion to a power function of the mass flow rate.

Darcy's law for such power-law fluid flows in porous media was first proposed by Christopher and Middleman (1965) and Dharmadhikari and Kale (1985). The model was used to attack the problem of free convection of non-Newtonian fluid from a vertical plate in a porous medium by Chen and Chen (1988) and the corresponding problem of mixed convection by Wang et al. (1990). Later, dealing with free convection from an arbitrary body shape in a non-Newtonian fluid-saturated porous medium, Nakayama and Koyama (1991) introduced the modified permeability to unify the two distinct expressions proposed by Christopher and Middleman (1965) and Dharmadhikari and Kale (1985).

Since Darcy's law accounts only for a balance of the viscous force and pressure drop, it fails to describe the flow inertia effects on the pressure drop, which become significant when the Reynolds number is high. Shenoy (1993) introduced a non-Newtonian version of the Forchheimer extension of Darcy's law to investigate various aspects associated with convective flows in power-law fluid-saturated porous media. Nakayama and Shenoy (1993) pointed out that the inertial drag force is independent of a particular stress–strain relationship, such that the same initial drag expression as the Newtonian one should be used for all cases of non-Newtonian fluids. This point, however, has not been substantiated either experimentally or numerically. A comprehensive review on non-Newtonian fluid flow and heat transfer in porous media may be found in Shenoy (1994).

In this section, non-Newtonian power-law fluid flows passing through porous media are investigated numerically, using a three-dimensional periodic array, as shown in Figure 8. A full set of three-dimensional momentum equations along with the power-law strain rate–shear stress relationship are numerically treated at pore scale. The resulting microscopic results are processed to extract the macroscopic pressure gradient–Darcian velocity relationship and compared with the Christopher and Middleman (1965) formula and its Forchheimer modification.

B. Governing Equations and Constitutive Equation

The governing equations, namely the continuity and momentum equations for the steady incompressible non-Newtonian fluid flows, are given in terms of Cartesian tensors as

$$\frac{\partial u_j}{\partial x_j} = 0 \tag{43}$$

$$\rho \frac{\partial}{\partial x_j}(u_j u_i) = -\frac{\partial p}{\partial x_j} + \frac{\partial}{\partial x_j}\left(\mu_{\mathrm{ap}}\left(\frac{\partial u_i}{\partial x_j} + \frac{\partial u_j}{\partial x_i}\right)\right) \tag{44}$$

where

$$\left.\begin{aligned} \mu_{\mathrm{ap}} &= \mu^* \Phi^{(n-1)/2} && \text{for} \quad \Phi^{1/2} \geq \frac{|\langle \boldsymbol{u} \rangle|}{H} \times 10^{-3} \\ \mu_{\mathrm{ap}} &= \mu^* \left(\frac{|\langle \boldsymbol{u} \rangle|}{H} \times 10^{-3}\right)^{n-1} && \text{for} \quad \Phi^{1/2} \leq \frac{|\langle \boldsymbol{u} \rangle|}{H} \times 10^{-3} \end{aligned}\right\} \tag{45}$$

and

$$\Phi = \frac{\partial u_k}{\partial x_l}\left(\frac{\partial u_k}{\partial x_l} + \frac{\partial u_l}{\partial x_k}\right) \tag{46}$$

We employ the constitutive equation (45) such that the Ostwald–de Waele model, the so-called "power-law model," holds when the square root of the dissipation function $\Phi^{1/2}$ is greater than a threshold value, namely $10^{-3}|\langle \boldsymbol{u} \rangle|/H$. When $\Phi^{1/2}$ is less than the threshold value, on the other hand, the fluid is assumed to be Newtonian with the apparent viscosity μ_{ap} set to a constant value, so as to account for the fact that all fluids behave like Newtonian fluids as the strain rate becomes sufficiently small (see Figure 17).

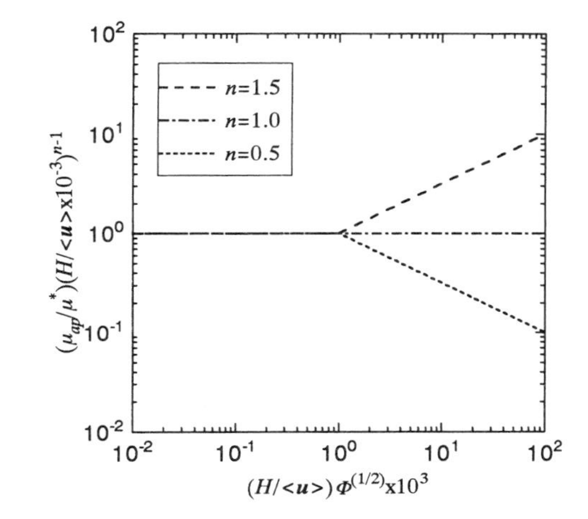

Figure 17. Apparent viscosity.

Preliminary calculations on laminar flows in ducts revealed that the constitutive equation (45) leads to profiles which are almost exact for the cases of fully developed flows. Many of the inelastic non-Newtonian fluids encountered in chemical engineering processes are known to follow this empirical power-law model. For the power-law fluids, we may define the Reynolds number based on the Darcian velocity and the size of the structural unit as

$$Re_H = \frac{\rho |\langle \boldsymbol{u} \rangle|^{2-n} H^n}{\mu^*} \tag{47}$$

C. Determination of Modified Permeability

The gradient of the intrinsic average pressure $\langle p \rangle^f$ may readily be evaluated, using the microscopic results, as

$$\begin{aligned} -\frac{\mathrm{d}\langle p \rangle^f}{\mathrm{d}X} = & \frac{\cos\theta\cos\phi}{H(H^2-D^2)} \iint_{Af} (p|_{x=0} - p|_{x=H}) \mathrm{d}y\,\mathrm{d}z \\ & + \frac{\sin\theta\cos\phi}{H(H^2-D^2)} \iint_{Af} (p|_{y=0} - p|_{y=H}) \mathrm{d}z\,\mathrm{d}x \\ & + \frac{\sin\phi}{H(H^2-D^2)} \iint_{Af} (p|_{z=0} - p|_{z=H}) \mathrm{d}x\,\mathrm{d}y \end{aligned} \tag{48}$$

where X is the coordinate taken in the macroscopic flow direction, whereas $A^f(= H^2 - D^2)$ is the cross-sectional area of the flow passage. The Forchheimer modification of Darcy's law for power-law fluids may be written as

$$-\frac{\mathrm{d}\langle p\rangle^f}{\mathrm{d}s} = \frac{\mu^*}{K^*}|\langle \boldsymbol{u}\rangle|^n + \rho b|\langle \boldsymbol{u}\rangle|^2 \tag{49a}$$

or, in dimensionless form, as

$$-\frac{\mathrm{d}\langle p\rangle^f}{\mathrm{d}s}\frac{H}{\rho|\langle \boldsymbol{u}\rangle|^2}Re_H = \frac{H^{n+1}}{K^*} + bHRe_H \tag{49b}$$

Thus, the modified permeability $K^*[m^{n+1}]$ and the Forchheimer constant $b[1/m]$ may readily be determined by fitting the numerical results for the macroscopic pressure gradient against Eq. (49b).

Typical pressure gradient results are plotted against the Reynolds number in Figure 18 for the case of $\varepsilon = 0.488$, $(\theta, \phi) = (45°, 35.26°)$, and $n = 0.5$, 1.0, and 1.5 for pseudoplastic, Newtonian, and dilatant fluids, respectively. It is clearly seen that the dimensionless pressure gradient stays constant for the low Reynolds number range, where the viscous force dominates, and then increases steeply for the high Reynolds number range, where the porous inertial contribution becomes predominant. All three curves merge together, which substantiates the assumption made by Nakayama and Shenoy (1993), namely that the inertial drag force is independent of a particular stress–strain relationship, so that the same inertial

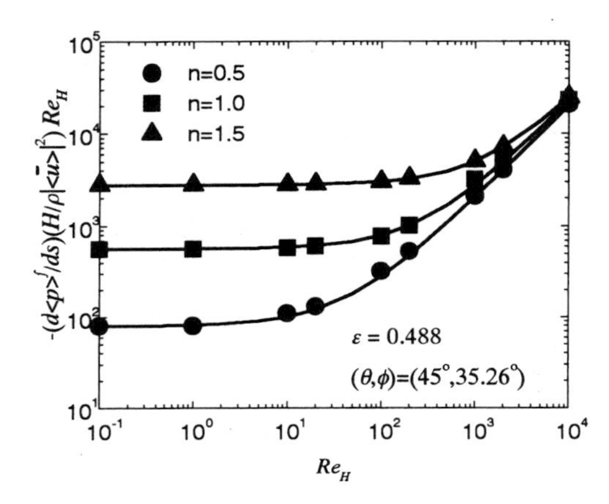

Figure 18. Effect of Reynolds number on pressure gradient.

drag expression as the Newtonian one can be used for all cases of non-Newtonian fluids under these conditions.

The modified permeability K^* may be determined by reading the intercept on the ordinate axis in the figure. The permeability, thus obtained for three distinct flow angles $(\theta, \phi) = (0^\circ, 0^\circ), (45^\circ, 0^\circ)$, $(45^\circ, 35.26^\circ)$, with $\varepsilon = 0.488$, are plotted in Figure 19 for $n = 0.5$, 1.0, and 1.5. The figure shows that the modified permeability K^* is virtually independent of the macroscopic flow direction. This suggests that the average mean shear rate within a unit cell, for the case of low Reynolds number, is fairly insensitive to the macroscopic flow direction. (Note that the velocity field within a unit cell changes markedly with the flow direction.) Thus, we expect the dimensionless permeability to be a function of the porosity ε and power-law index n.

Figure 20 shows the effects of porosity ε and power-law index n on the dimensionless permeability. The lines generated from Christopher and Middleman's (1965) formula based on a simple hydraulic radius concept are also indicated in the same figure, where the abscissa variable is chosen following their formula

$$K^* = \frac{6}{25}\left(\frac{n\varepsilon}{1+3n}\right)^n \left(\frac{\varepsilon D}{3(1-\varepsilon)}\right)^{1+n} = \frac{6}{25}\left(\frac{n}{1+3n}\right)^n \left(\frac{H}{3}\right)^{1+n} \frac{\varepsilon^{1+2n}}{(1-\varepsilon)^{2(1+n)/3}} \tag{50}$$

It is interesting to note that the numerical results essentially follow the Christopher and Middleman formula, which suggests that the permeability for a collection of cubes of size D is nearly the same as that of a packed bed of spheres of diameter D, under the same porosity. For the case of a

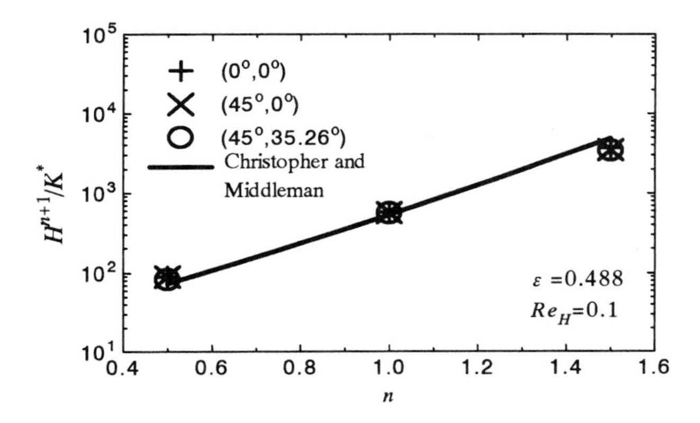

Figure 19. Effects of flow angle and power-law index n on permeability.

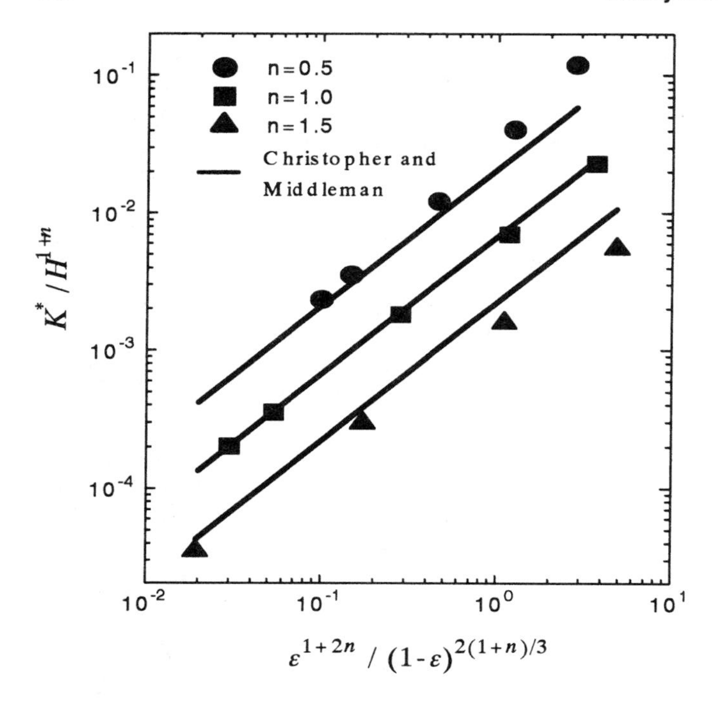

Figure 20. Effects of porosity ϵ and power-law index n on permeability.

Newtonian fluid, Kuwahara et al. (1994) conducted three-dimensional numerical calculations using a collection of spheres in a simple cubic arrangement, and obtained results which closely follow those of the cubes. Numerical calculations using spheres for the case of a non-Newtonian fluid have not been reported yet. The amount of CPU time required for such calculations would be tremendous.

IX. MATHEMATICAL MODEL FOR TURBULENT FLOW IN POROUS MEDIA

A. Related Work

Turbulence may become appreciable in porous media when the pore Reynolds number becomes sufficiently high. There exist some experimental reports, such as Mickeley et al. (1965), Kirkham (1967), Macdonald et al. (1979), and Dybbs and Edwards (1984), which confirm the existence of turbulence within a porous medium at a high Reynolds number. Fand et al. (1986) refer to the high Reynolds number regime, where turbulence effects become significant, as the post-Forchheimer flow regime, showing the curve representing transition from the laminar to turbulent regimes in the graph of the macroscopic pressure gradient versus the Reynolds number. An excellent review on high Reynolds number flows in porous media can be found in Kaviany (1991).

There is only a limited number of sets of experimental data which cover the turbulent flow regime in porous media. It has been pointed out by Antohe and Lage (1997) that the pressure drop data of Kececioglu and Jiang (1994) indicate a marked transition from the Forchheimer regime to the turbulence regime, whereas no such pronounced effects of turbulence are reflected in the data of Fand et al. (1987).

Kuwahara et al. (1998) proposed a spatially periodic array of square rods as a numerical model of microscopic porous structure, so as to simulate the microscopic details of turbulent flow through a porous medium of regular arrangement in a two-dimensional space of infinite extent. Extensive numerical calculations based on a low Reynolds number k–ε model were carried out for a wide range of porosity and Reynolds number, so as to elucidate the hydrodynamic behavior of turbulent flow (post-Forchheimer flow) in porous media. The microscopic numerical results thus obtained at pore scale were processed to extract the macroscopic hydrodynamic characteristics in terms of the volume-averaged quantities. It has been clarified numerically how a transition from Forchheimer to turbulent flow reflects on the relationship between the macroscopic pressure gradient and the Reynolds number.

B. Numerical Model and Boundary Conditions

As long as the pore Reynolds number is sufficiently high (such that the turbulence length scale is much smaller than the pore scale), any reliable turbulence models designed for pure fluid flows (without a porous matrix) may be used to resolve microscopic details of turbulent flow fields within a microscopic porous structure.

Again, a macroscopically uniform flow with a macroscopic flow angle θ through an infinite number of square rods is assumed, as illustrated in Figure 3. The microscopic governing equations are given as follows:

$$\frac{\partial \bar{u}_j}{\partial x_j} = 0 \tag{51}$$

$$\frac{\partial \bar{u}_j \bar{u}_i}{\partial x_j} = -\frac{1}{\rho}\frac{\partial \bar{p}}{\partial x_i} + \frac{\partial}{\partial x_j}\left(\nu\left(\frac{\partial \bar{u}_i}{\partial x_j} + \frac{\partial \bar{u}_j}{\partial x_i}\right) - \overline{u_i' u_j'}\right) \tag{52}$$

$$\frac{\partial \bar{u}_j k}{\partial x_j} = \frac{\partial}{\partial x_j}\left(\left(\nu + \frac{\nu_t}{\sigma_k}\right)\frac{\partial k}{\partial x_j}\right) - \overline{u_i' u_j'}\frac{\partial \bar{u}_i}{\partial x_j} - \varepsilon \tag{53}$$

$$\frac{\partial \bar{u}_j \varepsilon}{\partial x_j} = \frac{\partial}{\partial x_j}\left(\left(\nu + \frac{\nu_t}{\sigma_\varepsilon}\right)\frac{\partial \varepsilon}{\partial x_j}\right) + \left(-c_1\overline{u_i' u_j'}\frac{\partial \bar{u}_i}{\partial x_j} - c_2 f_e \varepsilon\right)\frac{\varepsilon}{k} \tag{54}$$

where

$$-\overline{u_i' u_j'} = \nu_t\left(\frac{\partial \bar{u}_i}{\partial x_j} + \frac{\partial \bar{u}_j}{\partial x_i}\right) - \frac{2}{3}k\delta_{ij} \tag{55}$$

and

$$\nu_t = c_D f_\mu \frac{k^2}{\varepsilon} \tag{56}$$

In this section, the symbol ε is reserved for the dissipation rate of the kinetic energy k, whereas ϕ is assigned for the porosity. The barred quantities represent ensemble-averaged components, whereas the primes indicate fluctuating components. We use a low Reynolds number form of the $k-\varepsilon$ model since the pore Reynolds number in the case of turbulent flow in a porous medium can be as low as a few thousands, to which the standard (high Reynolds number) version cannot be applied. The damping functions proposed by Abe et al. (1992) are chosen to account for the low Reynolds number effects

$$f_\mu = \left(1 - \exp\left(-\frac{(\nu\varepsilon)^{1/4} n}{14\nu}\right)\right)^2 \left(1 + \frac{5}{(k^2/\nu\varepsilon)^{3/4}}\exp\left(-\left(\frac{(k^2/\nu\varepsilon)}{200}\right)^2\right)\right) \tag{57}$$

$$f_\varepsilon = \left(1 - \exp\left(-\frac{(\nu\varepsilon)^{1/4} n}{3.1\nu}\right)\right)^2 \left(1 - 0.3\exp\left(-\left(\frac{(k^2/\nu\varepsilon)}{6.5}\right)^2\right)\right) \tag{58}$$

where n is the coordinate normal to the wall. Abe et al. (1992) use the Kolmogorov velocity $(\nu\varepsilon)^{1/4}$ as a velocity scale in place of the conventional friction velocity u_τ, and claim that their model performs well even for the cases of flow separation and reattachment, where conventional low Reynolds number models may fail to produce reasonable results since the friction velocity becomes vanishingly small there. Such turbulent flow separation and reattachment may take place within the porous media under consideration. The model constants are given as follows:

$$c_D = 0.09, \qquad c_1 = 1.55, \qquad c_2 = 1.9, \qquad \sigma_k = 1.4, \qquad \sigma_\varepsilon = 1.3 \tag{59}$$

The boundary, compatibility, and periodic constraints are given as follows:

on the solid walls:

$$\bar{\boldsymbol{u}} = \boldsymbol{0}, \qquad k = 0, \qquad \varepsilon = \nu \frac{\partial^2 k}{\partial n^2} \tag{60}$$

on the periodic boundaries:

$$\bar{\boldsymbol{u}}|_{x=2H} = \bar{\boldsymbol{u}}|_{x=0}, \qquad \bar{\boldsymbol{u}}|_{y=H} = \bar{\boldsymbol{u}}|_{y=0} \tag{61}$$

$$\int_0^H \bar{u}\mathrm{d}y|_{x=2H} = \int_0^H \bar{u}\mathrm{d}y|_{x=0} = H|\langle\bar{\boldsymbol{u}}\rangle|\cos\theta \tag{62}$$

$$\int_0^{2H} \bar{v}\mathrm{d}x|_{y=H} = \int_0^{2H} \bar{v}\mathrm{d}x|_{y=0} = 2H|\langle\bar{\boldsymbol{u}}\rangle|\sin\theta \tag{63}$$

$$\begin{aligned} &k|_{x=0} = k|_{x=2H}, \qquad k|_{y=0} = k|_{y=H}, \qquad \varepsilon|_{x=0} = \varepsilon|_{x=2H}, \\ &\varepsilon|_{y=0} = \varepsilon|_{y=H} \end{aligned} \tag{64}$$

C. Microscopic Velocity and Pressure Fields

The velocity vectors, pressure, and turbulence kinetic energy distributions in a microscopic porous structure obtained at $Re_H = 5000$ for the cases of porosity $\phi = 0.64$ and 0.84 are presented in Figures 21 and 22, respectively. The flow accelerates around the windward corners and separates over the leeward corners to form recirculation bubbles, as in the case of the backward facing step. The pressure increases at the front stagnation face of the square rod and decreases drastically around the corner, as can be seen from the pressure contours. The turbulence kinetic energy is high around the corner where a strong flow acceleration takes place and, subsequently, a strong shear layer is formed downstream of the corner. The two sets of results for high and low porosities are very much alike, except that separation bubbles are somewhat suppressed for the case of the lower porosity. A series of calculations was carried out using the low Reynolds number model for a wide range of Reynolds number, 5000 to 10^7. The resulting velocity and pressure patterns obtained at the same porosity are found to be very similar for all cases. Thus, the effects of the Reynolds number on the microscopic velocity and pressure fields diminish as the Reynolds number becomes sufficiently large.

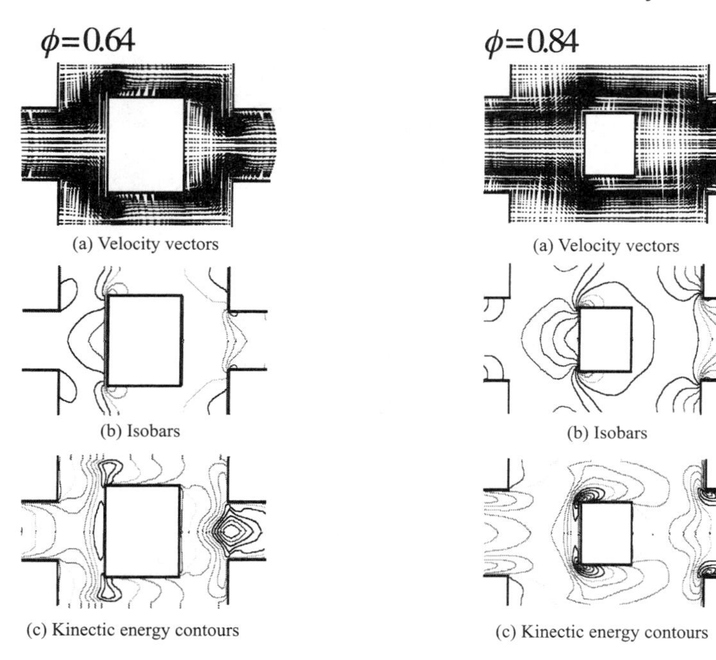

Figure 21. Microscopic results at $\phi = 0.64$.

Figure 22. Microscopic results at $\phi = 0.84$.

D. Determination of Macroscopic Pressure Gradient

The pressure gradient results obtained with $\phi = 0.84$ for various Reynolds numbers are assembled in Figure 23 in terms of the dimensionless pressure gradient $(-\mathrm{d}\langle\bar{p}\rangle^f/\mathrm{d}X)(H^2/\mu|\bar{\boldsymbol{u}}|)$. In the same figure, the pressure results of laminar flow calculations ($Re_H \leq 10^4$) and those of turbulent flow calculations based on the standard k–ε model ($10^5 \leq Re_H \leq 10^7$) are also presented for comparison. It is clearly seen that the dimensionless pressure gradient stays virtually constant for the low Reynolds number laminar range, say $Re \leq 10$, where the viscous force dominates, and then increases steeply as the porous inertial contribution becomes predominant. For these laminar flow regimes, the well-known Forchheimer-modified Darcy law holds. The permeability K may be readily determined by reading the value of the intercept on the ordinate axis in the figure.

It is interesting to note that all the pressure gradient results from the turbulent flow calculations follow a straight line extrapolated along Forchheimer-modified Darcy's law in the laminar regime. In the range of $5000 \leq Re \leq 10^5$, the results based on the low Reynolds number k–ε model

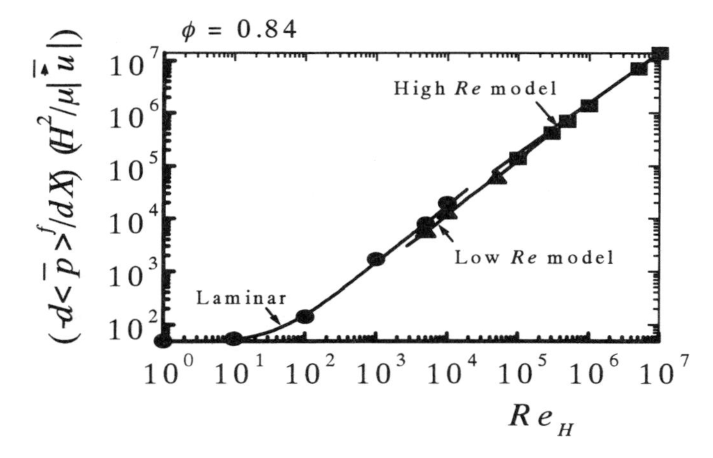

Figure 23. Effect of Reynolds number on pressure gradient.

lie slightly lower than those based on the standard k–ε model. The difference between the two models, however, diminished in the high Reynolds number range, where the velocity and pressure patterns within a porous structure are essentially frozen, as is observed in the microscopic results. Thus, no remarkable changes due to a transition from the laminar to a turbulent regime, as in the case of pure fluid flows (in the absence of porous matrix), are predicted in the pressure gradient versus Reynolds number relationship, which conforms with the experimental data obtained by Fand et al. (1987). Hence, the Forchheimer-extended Darcy model for the laminar regime may well be used to estimate the pressure drop in the turbulent regime. The numerical results are compared against the two distinct experimental correlations reported by Fand et al. (1987) and Kececioglu and Jiang (1994) for the turbulent flow regime. These two experimental correlations differ significantly from each other as shown in Figure 24. It is interesting to note that the curve based on the present prediction extends between those of the experimental correlations. Further experimental investigation is needed to clarify this divergence in the experimental data.

X. MACROSCOPIC TURBULENCE MODEL

A. Related Work

Rudraiah (1983) introduced the Reynolds decomposition for the macroscopic governing equations to treat turbulent flows in porous media. A comprehensive review of the Reynolds decomposition and turbulence modeling using modified Darcy's equations has been provided by Rudraiah (1988). In his work, however, only zero-equation models based on a gradient diffusion model for closure were investigated, in order to treat comparatively simple free convective turbulent flows in porous media.

Perhaps Lee and Howell (1987) were the first to introduce a set of transport equations for the turbulence kinetic energy and its rate of dissipation for analyzing turbulent flows in porous media. However, no account was taken of possible production and dissipation due to the presence of porous matrix, since they considered only highly porous media.

Recently, two distinct two-equation turbulence models have been proposed for turbulent flows in porous media. Antohe and Lage (1997) chose to carry out the Reynolds averaging over the volume-averaged macroscopic equations to obtain a two-equation turbulence model equation, whereas Masuoka and Takatsu (1996) derived a set of macroscopic turbulence model equations by spatially averaging the turbulence transport equations of the two-equation turbulence model equations. Antohe and Lage (1997) examined their model equations for the decay of the turbulence kinetic energy and its dissipation rate, assuming a unidirectional fully developed flow through an isotropic porous medium. Their model demonstrates that the only possible steady-state solution for the case is "zero" macroscopic turbulence kinetic energy. This solution should be re-examined, since the macroscopic turbulence kinetic energy in a forced flow through a porous medium must stay at a certain level as long as the presence of porous matrix keeps on generating it. (The situation is analogous to that of turbulent fully developed flow in a conduit.) Masuoka and Takatsu (1996) modeled the dissipation rate inherent in porous matrix such that it varies in proportion to the turbulence kinetic energy. None of these models has been verified experimentally.

Since the small eddies must be modeled first, we must start with the Reynolds averaged set of the governing equations and integrate them over a representative control volume to obtain the set of macroscopic turbulence model equations. Therefore, the procedure based on the Reynolds averaging of the spatially averaged continuity and momentum equations, such as that proposed by Antohe and Lage (1997), is questionable, since eddies larger than the scale of the porous structure are not likely to survive long enough to be detected.

Nakayama and Kuwahara (1999) purposed a comprehensive set of macroscopic two-equation turbulence model equations which is sufficiently general and capable of simulating most turbulent flows in porous media. The macroscopic transport equations for the turbulence kinetic energy and its dissipation rate are derived by spatially averaging the Reynolds-averaged transport equations along with the k–ε turbulence model. For the closure problem, various unknown terms describing the production and dissipation rates inherent in porous matrix are modeled collectively. In order to establish the unknown model functions, we conduct an exhaustive numerical experiment for turbulent flows through a periodic array, directly solving the microscopic governing equations, namely the Reynolds-averaged set of continuity, Navier–Stokes, turbulence kinetic energy, and its dissipation rate equations The microscopic results obtained from the numerical experiment are integrated spatially over a unit porous structure to determine the unknown model functions.

The macroscopic turbulence model, thus established, is tested for the case of macroscopically unidirectional turbulent flow. The streamwise variations of the turbulence kinetic energy and its dissipation rate predicted by the present macroscopic model are compared against those obtained from a large-scale direct computation over an entire field of saturated porous medium, to substantiate the validity of the macroscopic turbulence model. In this section, we shall examine the macroscopic turbulence model proposed by Nakayama and Kuwahara (1999).

B. Microscopic Governing Equations

In addition to the microscopic equations listed in Eqs. (51)–(54), we consider the energy equations for both fluid and solid structures as

$$\rho_f c_{pf} \frac{\partial \bar{T}}{\partial t} + \rho_f c_{pf} \frac{\partial \bar{u}_j \bar{T}}{\partial x_j} = \frac{\partial}{\partial x_j}\left(k_f \frac{\partial \bar{T}}{\partial x_j} - \rho_f c_{pf} \overline{T' u_j'} \right) \tag{65}$$

where the turbulent heat flux tensors are given by

$$-\rho_f c_{pf} \overline{T' u_j'} = \rho_f c_{pf} \frac{\nu_t}{\sigma_T} \frac{\partial \bar{T}}{\partial x_j} \tag{66}$$

and

$$\rho_s c_s \frac{\partial \bar{T}}{\partial t} = \frac{\partial}{\partial x_j}\left(k_s \frac{\partial \bar{T}}{\partial x_j} \right) \tag{67}$$

The barred quantities represent ensemble-averaged components, whereas the primes indicate fluctuating components. The subscripts f and s refer to fluid phase and solid phase, respectively. (Note that k_f here is not the turbulent kinetic energy but the thermal conductivity of the fluid phase.)

Kuwahara et al. (1998) numerically investigated the microscopic turbulence fields within a fluid-saturated periodical array, using both low- and high-Reynolds number versions of the k–ε model, and found that the difference in the volume-averaged quantities predicted by both of the models is insignificant. Thus, we choose the standard (high Reynolds number) version of the k–ε model along with conventional wall functions to save the number of grid nodes.

C. Macroscopic Continuity, Momentum, and Energy Equations

We integrate the Reynolds-averaged equation (51) over a control volume V, which is much larger than a microscopic (pore structure) characteristic size but much smaller than a macroscopic characteristic size. Thus, the macroscopic equation of continuity is given by

$$\frac{\partial}{\partial x_j}\langle \bar{u}_j \rangle^f = 0 \tag{68}$$

In the numerical study of turbulent flow through a periodic array, Kuwahara et al. (1998) concluded that the Forchheimer-extended Darcy's law holds even in the turbulent flow regime in porous media. The experimental data, provided by Fand et al. (1987), also support the validity of this law. Thus, we follow Vafai and Tien (1981) and introduce Forchheimer's modification of the last two terms on the right-hand side, representing the intrinsic volume average of the total surface force (acting on the fluid inside the pore) and the inertial dispersion, respectively. Hence, we obtain the macroscopic momentum equation as follows

$$\begin{aligned} &\frac{\partial \langle \bar{u}_i \rangle^f}{\partial t} + \frac{\partial}{\partial x_j}\langle \bar{u}_j \rangle^f \langle \bar{u}_i \rangle^f \\ &= -\frac{1}{\rho}\frac{\partial}{\partial x_i}\left(\langle \bar{p} \rangle^f + \frac{2}{3}\rho_f \langle k \rangle^f\right) + \frac{\partial}{\partial x_j}\left[2(\nu + \nu_t)\langle s_{ij} \rangle^f\right] \\ &\quad - \phi\left(\frac{\nu}{K} + \frac{\phi C}{\sqrt{K}}\left(\langle \bar{u}_j \rangle^f \langle \bar{u}_j \rangle^f\right)^{1/2}\right)\langle \bar{u}_i \rangle^f \end{aligned} \tag{69}$$

where

$$s_{ij} = \frac{1}{2}\left(\frac{\partial \bar{u}_i}{\partial x_j} + \frac{\partial \bar{u}_j}{\partial x_i}\right)$$

is the mean rate of strain. The permeability K and Forchheimer constant C are empirical constants which depend on the porosity $\phi = V_f/V$.

Similarly, spatial integration of the two microscopic energy equations (65) and (67) yields a set of macroscopic energy equations. We shall assume that thermal equilibrium exists between the fluid and solid matrix, namely, $\langle \bar{T} \rangle^f = \langle \bar{T} \rangle^s$. Adding the two macroscopic energy equations, we have

$$\left[\phi\rho_f c_{p_f} + (1-\phi)\rho_s c_s\right]\frac{\partial\langle\bar{T}\rangle^f}{\partial t} + \phi\rho_f c_{p_f}\frac{\partial}{\partial x_j}\langle\bar{u}_j\rangle^f\langle\bar{T}\rangle^f$$
$$= \frac{\partial}{\partial x_i}\left[\left(k_e + \phi\frac{\rho_f c_{p_f}\nu_t}{\sigma}\right)\delta_{ij} + (k_{\text{tor}})_{ij} + (k_{\text{dis}})_{ij}\right]\frac{\partial\langle\bar{T}\rangle^f}{\partial x_j} \tag{70}$$

where

$$k_e \equiv \phi k_f + (1-\phi)k_s \tag{71a}$$

$$\frac{1}{V}\int_{A_{\text{int}}}(k_f - k_s)\bar{T}n_j \mathrm{d}A \equiv (k_{\text{tor}})_{ij}\frac{\partial\langle\bar{T}\rangle^f}{\partial x_j} \tag{71b}$$

and

$$-\phi(\rho c_p)_f\langle\bar{u}_j''\bar{T}''\rangle^f \equiv (k_{\text{dis}})_{ij}\frac{\partial\langle\bar{T}\rangle^f}{\partial x_j} \tag{71c}$$

k_e is the stagnant thermal conductivity. The double prime denotes the deviation from the intrinsic average such that $a'' \equiv a - \langle a\rangle^f$. The apparent conductivity tensors $(k_{\text{tor}})_{ij}$ and $(k_{\text{dis}})_{ij}$ are introduced to model the tortuosity molecular diffusion term and the thermal dispersion term, respectively. The empirical and theoretical expressions for $(k_{\text{tor}})_{ij}$ and $(k_{\text{dis}})_{ij}$ may be found elsewhere (Kuwahara et al. 1996; Kaviany 1991). When the Reynolds number is high, both k_e and $(k_{\text{tor}})_{ij}$ may be ignored in comparison with the thermal dispersion and turbulent diffusion.

D. Macroscopic Transport Equations for Turbulence Kinetic Energy and its Dissipation Rate

To describe turbulent diffusion, we may recast the eddy diffusivity formula (56) using the intrinsically averaged values of the turbulence quantities as

$$\nu_t = c_D\frac{\left(\langle k\rangle^f\right)^2}{\langle\varepsilon\rangle^f} \tag{72}$$

The macroscopic transport equations for $\langle k\rangle^f$ and $\langle\varepsilon\rangle^f$ may be obtained by integrating the microscopic transport equations (53) and (54). The presence of the porous matrix yields two additional terms in the macroscopic turbulence kinetic energy equation, namely the production term $2\nu_t\langle s_{ij}''s_{ij}''\rangle^f$ and the dissipation term $(\nu/V_f)\int_{A_{\text{int}}}(\partial k/\partial x_j)n_j\mathrm{d}A$. (Note, it is $2\nu_t\langle s_{ij}''s_{ij}''\rangle^f$ that is solely responsible for kinetic energy production in the case of macroscopically uniform flow with zero macroscopic mean shear, and also that $(\partial k/\partial x_j)n_j$ in the dissipation term is always negative due to the no-slip

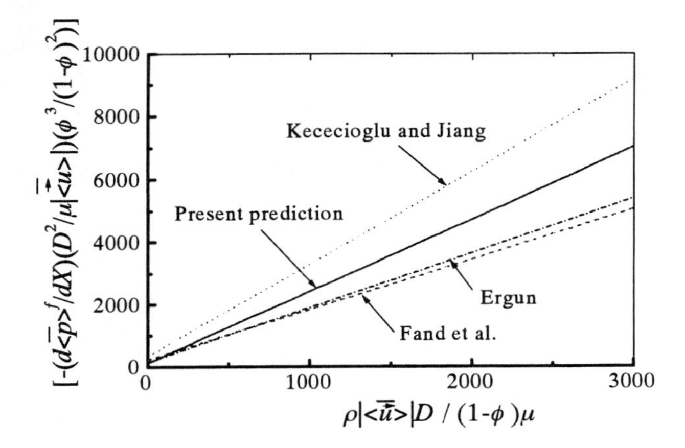

Figure 24. Correlations for turbulent flows in porous media.

requirements. See also the discussion by Masuoka and Takatsu (1996). The sum of these two terms corresponds to the net production rate due to the presence of the porous matrix, which balances the dissipation rate for the case of fully developed macroscopically unidirectional flow with zero macroscopic mean shear through a porous medium. Thus, we model these two terms collectively as

$$\varepsilon_\infty \equiv 2\nu_t \langle s''_{ij} s''_{ij} \rangle^f + \frac{\nu}{V_f} \int_{A_{\rm int}} \frac{\partial k}{\partial x_j} n_j {\rm d}A \tag{73}$$

A similar argument can be made for the corresponding macroscopic equation for the dissipation rate. Thus, we set

$$c_2 \frac{\varepsilon_\infty^2}{k_\infty} \equiv 2c_1 \nu_t \langle s''_{ij} s''_{ij} \rangle^f \frac{\langle \varepsilon \rangle^f}{\langle k \rangle^f} + \frac{\nu}{V_f} \int_{A_{\rm int}} \frac{\partial \varepsilon}{\partial x_j} n_j {\rm d}A^f \tag{74}$$

Hence, we obtain the following set of macroscopic transport equations for $\langle k \rangle^f$ and $\langle \varepsilon \rangle^f$

$$\begin{aligned} &\frac{\partial \langle k \rangle^f}{\partial t} + \frac{\partial}{\partial x_j} \langle \bar{\boldsymbol{u}}_j \rangle^f \langle k \rangle^f \\ &\quad = \frac{\partial}{\partial x_i} \left[\left(\nu + \frac{\nu_t}{\sigma_k} \right) \delta_{ij} + \frac{(k_{\rm dis})_{ij}}{Le_k \phi \rho_f c_{pf}} \right] \frac{\partial \langle k \rangle^f}{\partial x_j} \\ &\quad + 2\nu_t \langle s_{ij} \rangle^f \langle s_{ij} \rangle^f - \langle \varepsilon \rangle^f + \varepsilon_\infty \end{aligned} \tag{75}$$

and

$$\frac{\partial\langle\varepsilon\rangle^f}{\partial t}+\frac{\partial}{\partial x_j}\langle\bar{u}_j\rangle^f\langle\varepsilon\rangle^f$$
$$=\frac{\partial}{\partial x_i}\left[\left(\nu+\frac{\nu_t}{\sigma_\varepsilon}\right)\delta_{ij}+\frac{(k_{\text{dis}})_{ij}}{Le_\varepsilon\phi\rho_f c_{pf}}\right]\frac{\partial\langle\varepsilon\rangle^f}{\partial x_j}$$
$$+\left(2c_1\nu_t\langle s_{ij}\rangle^f\langle s_{ij}\rangle^f-c_2\langle\varepsilon\rangle^f\right)\frac{\langle\varepsilon\rangle^f}{\langle k\rangle^f}+c_2\frac{\varepsilon_\infty^2}{k_\infty} \tag{76}$$

where Le_k and Le_ε are the Lewis numbers for the mechanical dispersions. The model functions, namely ε_∞ and k_∞, should be determined from either turbulence measurements or numerical experiments.

E. Preliminary Consideration for Turbulence Model Functions

Experimental determination of the unknown model functions, namely ε_∞ and k_∞, requires detailed turbulence measurements on the pore scale, which may not easily be done even with artificially consolidated porous media. Alternatively, we may choose to conduct a numerical experiment to determine these functions, using the set of microscopic governing equations.

For macroscopically uniform flow with zero mean shear, the macroscopic model equations (75) and (76) reduce to

$$\langle\bar{u}_1\rangle^f\frac{\mathrm{d}\langle k\rangle^f}{\mathrm{d}x}=-\langle\varepsilon\rangle^f+\varepsilon_\infty \tag{77}$$

and

$$\langle\bar{u}_1\rangle^f\frac{\mathrm{d}\langle\varepsilon\rangle^f}{\mathrm{d}x}=-c_2\frac{\left(\langle\varepsilon\rangle^f\right)^2}{\langle k\rangle^f}+c_2\frac{\varepsilon_\infty^2}{k_\infty} \tag{78}$$

respectively. When the flow is periodically fully developed, the foregoing equations yield

$$\langle k\rangle^f=k_\infty \qquad \text{and} \qquad \langle\varepsilon\rangle^f=\varepsilon_\infty \tag{79}$$

Thus, we can determine the unknown model functions ε_∞ and k_∞ from the intrinsic volume-average values of the turbulence kinetic energy and its dissipation rate attained in a periodically fully developed flow through a porous medium.

Kuwahara et al. (1994) and Nakayama et al. (1995) conducted a series of numerical experiments for the laminar flow regime, using various two- and three-dimensional numerical models for porous media, such as arrays of square rods, circular rods, spheres, and cubes. They found that all these models lead to almost identical expressions for the permeability which are in good accord with Ergun's empirical formula (Ergun 1952). The numerical study of thermal dispersion by Kuwahara et al. (1996) also supports the possibility of utilizing the results based on a two-dimensional numerical model to estimate the thermal dispersion in packed spheres. Thus, we assume a two-dimensional periodic array of square rods placed in an infinite space, as illustrated in Figure 3.

F. Determination of Turbulence Model Functions

Having established the microscopic turbulence fields, the intrinsic volume-average values can readily be evaluated by integrating the microscopic turbulence quantities over the fluid phase domain within the structural unit. As can be expected from the microscopic results, the effect of the Reynolds number on the intrinsic average quantities, $\langle k \rangle^f = k_\infty$ and $\langle \varepsilon \rangle^f = \varepsilon_\infty$, is negligibly small, whereas that of the porosity is substantial, as can be seen from Figures 25a and b, plotted for k_∞ and ε_∞, respectively. These figures suggest the following correlations

$$k_\infty = 3.7 \frac{1-\phi}{\sqrt{\phi}} |\langle \bar{\boldsymbol{u}} \rangle|^2 = 3.7(1-\phi)\phi^{3/2} \langle \bar{u}_j \rangle^f \langle \bar{u}_j \rangle^f \tag{80}$$

and

$$\varepsilon_\infty = 39 \frac{(1-\phi)^2}{\phi} \frac{|\langle \bar{\boldsymbol{u}} \rangle|^3}{H} = 39\phi^2 (1-\phi)^{5/2} \frac{1}{D} \left(\langle \bar{u}_j \rangle^f \langle \bar{u}_j \rangle^f \right)^{3/2} \tag{81}$$

Nakayama et al. (1995) and Kuwahara et al. (1996) showed through numerical experiments that the results based on a two-dimensional numerical model can be used to estimate the pressure drop and thermal dispersion in packed spheres. Thus, the present formulas (80) and (81) for the model constants k_∞ and ε_∞, determined numerically using the microscopic results based on the square rod array, may be used to investigate the turbulent flow through packed spheres, as the square rod size D is taken as the average diameter of the packed spheres.

G. Assessment of Macroscopic Turbulence Model

In order to assess the present macroscopic turbulence model, we shall investigate a steady unidirectional highly turbulent flow entering a semi-finite periodic array of square rods, from both microscopic and macroscopic points of view. From the microscopic view, we carry out a large-scale direct computation using the set of microscopic equations for a row of periodic structural units, as shown in Figure 26, and then integrate the microscopic

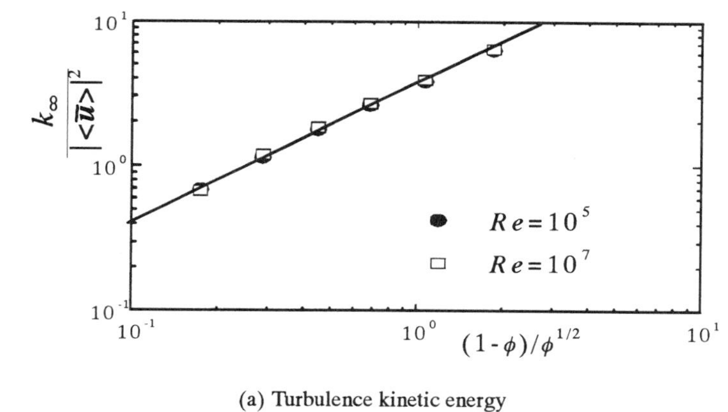

(a) Turbulence kinetic energy

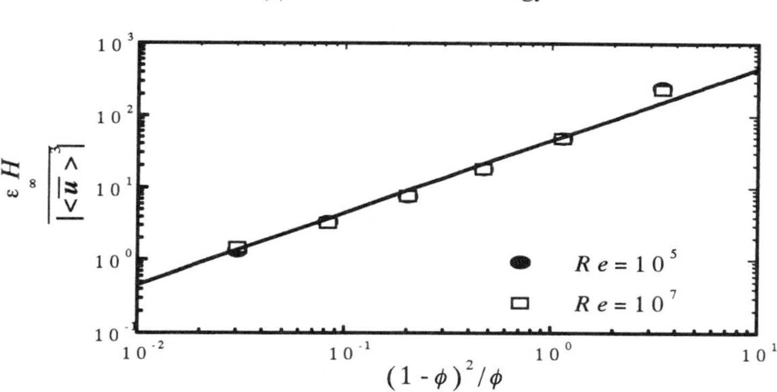

(b) Dissipation rate of turbulence kinetic energy

Figure 25. Effect of porosity on turbulence.

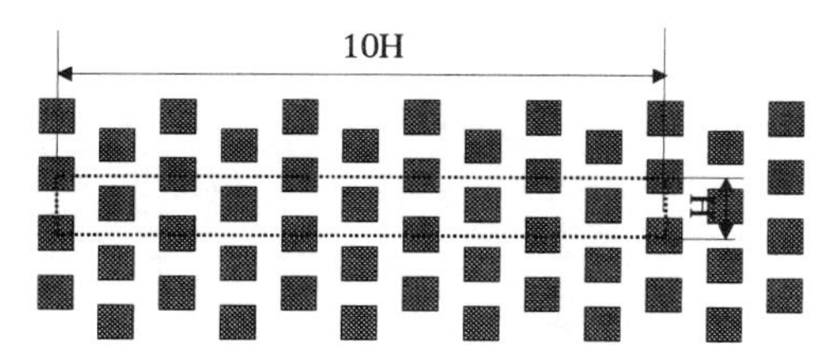

Figure 26. Physical model for a large-scale direct computation.

results of turbulence quantities over every unit to obtain the macroscopic decay of turbulence. (The periodic boundary conditions are used only for the upper and lower boundaries.) From the macroscopic view, we solve the macroscopic turbulence model equations (75) and (76), with the model constants as given by (80) and (81), to predict the streamwise decay of turbulence, and compare the results against the foregoing results based on large-scale direct computation.

Figures 27a and b show the microscopic fields of turbulence kinetic energy and its dissipation rate obtained from the large-scale direct computation for the case of $Re_H = 10^5$, $\phi = 0.75$, the inlet turbulence kinetic energy $\langle k \rangle^f = 10k_\infty$, and its dissipation rate $\langle \varepsilon \rangle^f = 30\varepsilon_\infty$. It can be seen that the turbulence decays drastically over a short distance from the entrance, and that periodically fully developed patterns are established as the fluid passes downstream. The streamwise decay of the macroscopic turbulence kinetic energy and that of the dissipation rate predicted by the present macroscopic turbulence model are presented in Figures 28a and b, along with the intrinsically averaged values based on large-scale direct computation. Good agreement between the two sets of results reveals the validity of the present macroscopic turbulence model.

Finally, let us estimate the effect of turbulence on scalar transport as compared with that of mechanical dispersion. From Eq. (70), the apparent thermal conductivity tensor is given by

$$(k_{\text{ap}})_{ij} = \left(k_e + \phi \frac{\rho_f c_{p_f} \nu_t}{\sigma} \right) \delta_{ij} + (k_{\text{tor}})_{ij} + (k_{\text{dis}})_{ij} \tag{82}$$

For the case of high Reynolds number, the molecular diffusion terms are negligible. Hence, we may approximate the apparent thermal conductivity tensor as

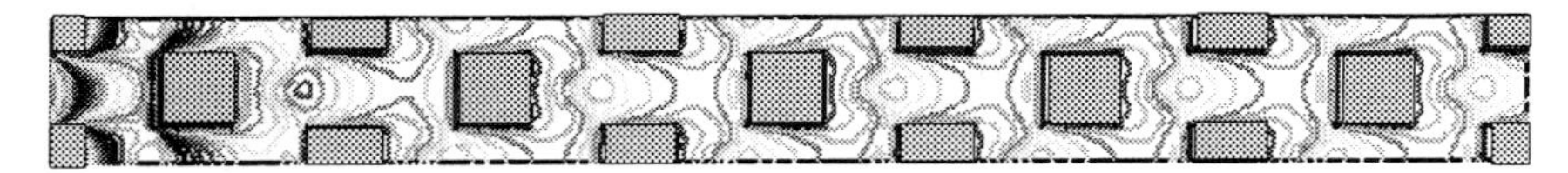

(a) Turbulence kinetic energy

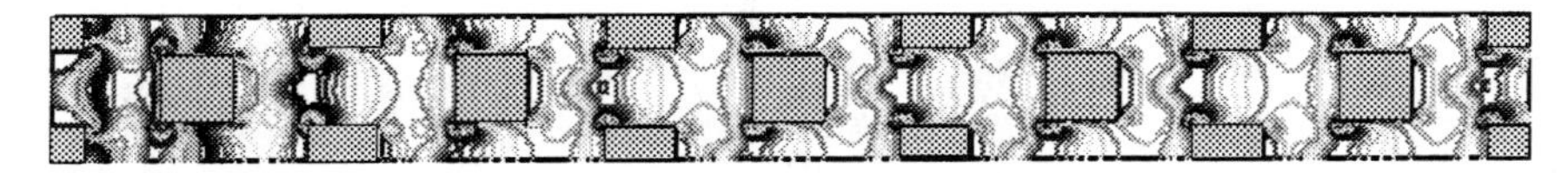

(b) Dissipation rate of turbulence kinetic energy

Figure 27. Macroscopic results from a large-scale direct computation.

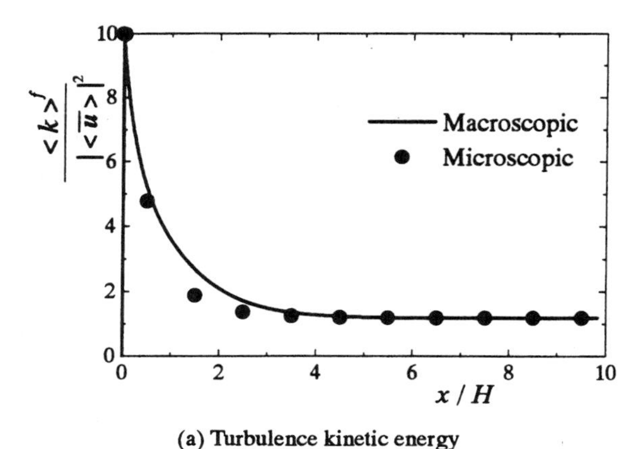

(a) Turbulence kinetic energy

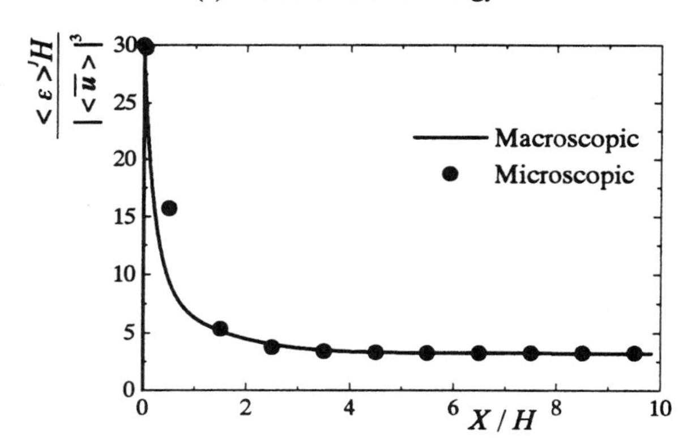

(b) Dissipation rate of turbulence kinetic energy

Figure 28. Decay of turbulence.

$$(k_{\text{ap}})_{ij} \approx \phi \frac{\rho_f c_{pf} \nu_t}{\sigma} \delta_{ij} + (k_{\text{dis}})_{ij} \tag{83}$$

Its transverse component (normal to the macroscopic flow direction) may be estimated as

$$(k_{\text{ap}})_{22} \approx \left(0.03 \frac{\phi}{\sigma_T \sqrt{1-\phi}} + 0.05\sqrt{1-\phi} \right) \phi \rho_f c_{pf} \langle \bar{u}_1 \rangle^f D \tag{84}$$

Equations (80), (81), and (72) are used to evaluate the fully developed value of ν_t whereas the correlation based on the numerical experiment by Kuwahara et al. (1996) is used to estimate $(k_{\text{dis}})_{22}$. The foregoing equation suggests that the contribution from turbulent mixing is comparable to that from mechanical dispersion. It is also interesting to note that, for given mass flow rate $\phi\langle\bar{\boldsymbol{u}}\rangle^f$ and particle size D, the increase in the porosity ϕ enhances turbulent mixing and, at the same time, damps mechanical dispersion within porous media. Thus, even in a sufficiently highly porous medium where the effect of mechanical dispersion on scalar transport is insignificant, that of turbulence mixing may not be negligible.

XI. CONCLUDING REMARKS

In this chapter, the recent developments in numerical modelling of convective flows in porous media have been reviewed. Regular arrays of obstacles are used to describe microscopic porous structures. A series of new numerical experiments using regular arrays of obstacles reveal that both two- and three-dimensional models are quite effective for determining the permeability and thermal dispersion coefficients, purely from the theoretical basis without any empiricisms. For determining the porous inertia, however, it is essential to use a three-dimensional model, since only three-dimensional models are capable of capturing the complex three-dimensional fluid motion associated with porous inertia. Turbulent transport in porous media has been also treated from a microscopic point of view, which eventually has led to a complete set of macroscopic two-equation turbulence model equations for analyzing high Reynolds number convective flows.

A further numerical experiment using an irregular array of obstacles is strongly recommended to provide a more realistic prediction of the effects of porous inertia, irregularity of porous structure, and possible flow unsteadiness on the macroscopic flow and heat transfer characteristics. Finally, it must be stressed that a DNS (direct numerical simulation) study is essential for further exploration of one of the most challenging topics in porous media, namely, turbulence.

NOMENCLATURE

Roman Letters

A	surface area
A_{int}	total interface between the fluid and solid
C	Forchheimer constant
c_1, c_2, c_D	turbulence model constants
c_p	specific heat at constant pressure
D	size of obstacle
H	size of unit cell

k	turbulence kinetic energy; thermal conductivity
K	permeability
Le	Lewis number
p	pressure
Re	Reynolds number
T	temperature
u_i, $\boldsymbol{u}$	velocity vector
u, v	velocity components
x, y	Cartesian coordinates

Greek Letters

ε	porosity, dissipation rate of turbulence kinetic energy
ν	kinematic viscosity
ν_t	effective viscosity
σ_k, σ_ε	effective Prandtl number
ρ	density

Special Symbols

$\bar{a}$	ensemble mean
a'	turbulent fluctuation
a''	deviation from intrinsic average
$\langle a \rangle$	volume average
$\langle a \rangle^{f,s}$	intrinsic average
$\lvert\alpha\rvert$	absolute value

Subscripts and Superscripts

ap	apparent
dis	dispersion
f	fluid
int	interface
s	solid
tor	tortuosity

REFERENCES

Abe K, Nagano Y, Kondoh T. An improved k–ε model for prediction of turbulent flows with separation and reattachment. Trans JSME, Ser B58:3003–3010, 1992.

Acrivos A, Hinch EJ, Jeffrey DJ. Heat transfer to a slowly moving fluid from a dilute fixed bed of heated spheres. J Fluid Mech 101:403–421, 1980.

Antohe BV, Lage JL. A general two-equation macroscopic turbulence model for incompressible flow in porous media. Int J Heat Mass Transfer 13:3013–3024, 1997.

Arquis E, Caltagirone JP. Le Breton P. Détermination des propriétés de dispersion d'un milieu périodique à partir de l'analyse locale des transferts. CR Acad Sci Paris II 313:1087–1092, 1991.

Arquis E, Caltagirone JP, Delmas A. Derivation in thermal conductivity of porous media due to high interstitial flow velocity. Proceedings of the 22nd International Thermal Conductivity Conference, Arizona, 1993.

Chen HT, Chen CK. Free convection of non-Newtonian fluids along a vertical plate embedded in a porous medium. Trans ASME J Heat Transfer 110:257–260, 1988.

Cheng P. Heat transfer in geothermal systems. Adv Heat Transfer 14:1–105, 1978.

Christopher RH, Middleman S. Power-law flow through a packed tube. Ind Eng Chem Fundam 4:422–429, 1965.

Coulaud O, Morel P, Caltagirone JP. Numerical modeling of nonlinear effects in laminar flow through a porous medium. J Fluid Mech 190:393–407, 1988.

Dharmadhikari RV, Kale DD. Flow of non-Newtonian fluids through porous media. Chem Eng Sci 40:527–529, 1985.

Dybbs A, Edwards RV. A new look at porous media fluid mechanics—Darcy to turbulent. In: Bear J, Corapcioglu A, ed. Fundamentals of Transport Phenomena in Porous Media. Netherlands: Martinus Nijhoff, 1984, pp 199–254.

Edwards DA, Shapiro M, Brenner H, Shapira M. Dispersion of inert solutes in spatially periodic two-dimensional model porous media. Transp Porous Media 6:337–358, 1991.

Eidsath A, Carbonell RG, Whitaker S, Herman LR. Dispersion in pulsed systems III: Comparison between theory and experiment for packed beds. Chem Eng Sci 38:1803–1816, 1983.

Ergun S. Fluid flow through packed columns. Chem Eng Prog 48:89–94, 1952.

Fand RM, Steinberger TE, Cheng P. Natural convection heat transfer from a horizontal cylinder embedded in a porous medium. Int J Heat Mass Transfer 29:119–133, 1986.

Fand RM, Kim BYK, Lam ACC, Phan RT. Resistance to flow of fluids through simple and complex porous media whose matrices are composed of randomly packed spheres. Trans ASME J Fluids Eng 109:268–274, 1987.

Fowler AJ, Bejan A. Forced convection in banks of inclined cylinders at low Reynolds number. Int J Heat Fluid Flow 15:90–99, 1994.

Fried JJ, Combarnous MA. Dispersion in porous media. Adv Hydro Sci 7:169–282, 1971.

Howells ID. Drag due to the motion of a Newtonian fluid through a sparse random array of small fixed rigid objects. J Fluid Mech 64:449–475, 1974.

Kaviany M. Principles of Heat Transfer in Porous Media. New York: Springer-Verlag, 1991, pp 42–49.

Kececioglu I, Jiang Y. Flow through porous media of packed spheres saturated with water. Trans ASME J Fluids Eng 116:164–170, 1994.

Kim S, Russel WB. Modeling of porous media by renormalization of the Stokes equations. J Fluid Mech 154:269–286, 1985.

Kirkham CE. Turbulent flow in porous media—An analytical and experimental study. Melbourne, Australia: Department of Civil Engng, University of Melbourne, 1967.

Koch DL, Brady JF. Dispersion in fixed beds. J Fluid Mech 154:399–427, 1985.

Koch DL, Cox RG, Brenner H, Brady JF. The effect of order on dispersion in porous media. J Fluid Mech 200:173–188, 1989.

Kuwahara F, Nakayama A. Numerical determination of thermal dispersion coefficients using a periodic porous structure. Trans ASME J Heat Transfer 121:160–163, 1999.

Kuwahara F, Nakayama A, Koyama H. Numerical modelling of heat and fluid flow in a porous medium. Proceedings of 10th International Heat Conference, vol 5, 1994, pp 309–314.

Kuwahara F, Nakayama A, Koyama H. A numerical study of thermal dispersion in porous media. Trans ASME J Heat Transfer 118:756–761, 1996.

Kuwahara F, Kameyama Y, Yamashita S, Nakayama A. Numerical modeling of turbulent flow in porous media using a spatially periodic array. J Porous Media 1:47–55, 1998.

Larson RE, Higdon JJL. A periodic grain consolidation model of porous media. Phys Fluids A1:38–46, 1989.

Lee K, Howell JR. Forced convective and radiative transfer within a highly porous layer exposed to a turbulent external flow field. Proceedings of 2nd ASME/JSME Thermal Engineering Joint Conference, vol 2, 1987, pp 377–386.

Macdonald IF, El-Sayed MS, Mow K, Dullien FAL. Flow through a porous media–Ergun equation revisited. Ind Eng Chem Fund 18:199–208, 1979.

Masuoka T, Takatsu Y. Turbulence model for flow through porous media. Int J Heat Mass Transfer 39:3803–2809, 1996.

Mickeley HS, Smith KA, Korchak EI. Fluid flow in packed beds. Chem Eng Sci 23:237–246, 1965.

Nakayama A. PC-aided Numerical Heat Transfer and Convective Flow. Boca Raton: CRC Press, 1995, pp 177–250.

Nakayama A, Koyama H. Buoyancy-induced flow of non-Newtonian fluids over a non-isothermal body of arbitrary shape in a fluid-saturated porous medium. Appl Sci Res 48:55–70, 1991.

Nakayama A, Kuwahara F. A macroscopic turbulence model for flow in a porous medium. Trans ASME J Fluids Engineering 121:427–433, 1999.

Nakayama A, Kuwahara F. Convective flow and heat transfer in porous media. Recent Res Dev Chem Eng 3:121–177, 1999.

Nakayama A, Shenoy AV. Non-Darcy forced convective heat transfer in a channel embedded in a non-Newtonian inelastic fluid-saturated porous medium. Can J Chem Eng 71:168–173, 1993.

Nakayama A, Kuwahara F, Kawamura Y, Koyama H. Three-dimensional numerical simulation of flow through a microscopic porous structure.

Proceedings of ASME/JSME Thermal Engineering Conference, vol 3, 1995, pp 313–318.
Nield DA, Bejan A. Convection in Porous Media. New York: Springer-Verlag, 1992.
Patankar SV. Numerical Heat Transfer and Fluid Flow. Washington, DC: Hemisphere, 1980.
Patankar SV, Spalding DB. A calculation procedure for heat, mass and momentum transfer in three-dimensional parabolic flows. Int J Heat Mass Transfer 15:1787–1806, 1972.
Pop I, Sunada JK, Cheng P, Minkowycz WJ. Conjugate heat transfer from long vertical plate fins embedded in a porous medium at high Rayleigh numbers. Int J Heat Mass Transfer 28:1629–1636, 1985.
Pop I, Ingham DB, Heggs PJ, Gardner D. Conjugate heat transfer from a downward projecting fin immersed in a porous medium. In: Heat Transfer 1986. Washington, DC: Hemisphere, vol 5, 1986, pp 2635–2640.
Rudraiah N. Turbulent convection in a porous medium using spectral method. Proceedings, II Asian Congress on Fluid Mechanics. Beijing, China: Science Press, 1983, pp 1021–1027.
Rudraiah N. Turbulent convection in porous media with non-Darcy effects. ASME HTD 96:747–754, 1988.
Sahroui M, Kaviany M. Slip and no-slip boundary condition at interface of porous, plain media. ASME/JSME Therm Eng Proc 4:273–286, 1991.
Shenoy AV. Darcy–Forchheimer natural, forced and mixed convection heat transfer in non-Newtonian power-law fluid-saturated porous media. Transp Porous Media 11:219–241, 1993.
Shenoy AV. Non-Newtonian fluid heat transfer in porous media. Adv Heat Transfer 24:101–190, 1994.
Taylor GI. Dispersion of solute matter in solvent flowing slowly through a tube. Proc R Soc London A219:186–203, 1953.
Vafai K, Tien CL. Boundary and inertia effects on flow and heat transfer in porous media. Int J Heat Mass Transfer 24:195–203, 1981.
Wang C, Tu C, Zhang X. Mixed convection of non-Newtonian fluids from a vertical plate embedded in a porous medium. Acta Mech Sin 6:214–220, 1990.
Yagi, S, Kunii D, Wakao N. Studies on axial effective thermal conductivities in packed beds. AIChE J 6:543–546, 1960.

11

Natural Convective Heat Transfer in Porous-Media-Filled Enclosures

P. H. Oosthuizen
Queen's University, Kingston, Ontario, Canada

I. INTRODUCTION

Heat transfer by natural convection across porous-media-filled enclosures (or cavities)—here termed porous enclosures—will be considered in this chapter. In such flows, the porous medium, in which there is a buoyancy-driven flow, is entirely enclosed by solid walls along which differential heating is applied, the resulting temperature differences leading to the generation of the buoyancy forces that cause the flow. Some examples of the various types of enclosure being considered are shown in Figure 1. General discussions of the area are given by Bejan (1995), Carbonell and Whitaker (1984), Cheng (1979), Kaviany (1991), and Oosthuizen and Naylor (1998).

Flow in porous enclosures has received considerable attention, due in large part to the fact that flows of this type occur in a number of practically important situations. Flows of this type arise, for example, in some building and similar applications in which heat is transferred across an insulation-filled enclosure. Other examples occur in geothermal and oil extraction applications.

II. VERTICAL AND INCLINED RECTANGULAR ENCLOSURES

The most widely considered situation involving a porous enclosure is that of a rectangular enclosure completely filled with a saturated porous medium and with one wall heated to a uniform temperature, T_H, the opposite wall

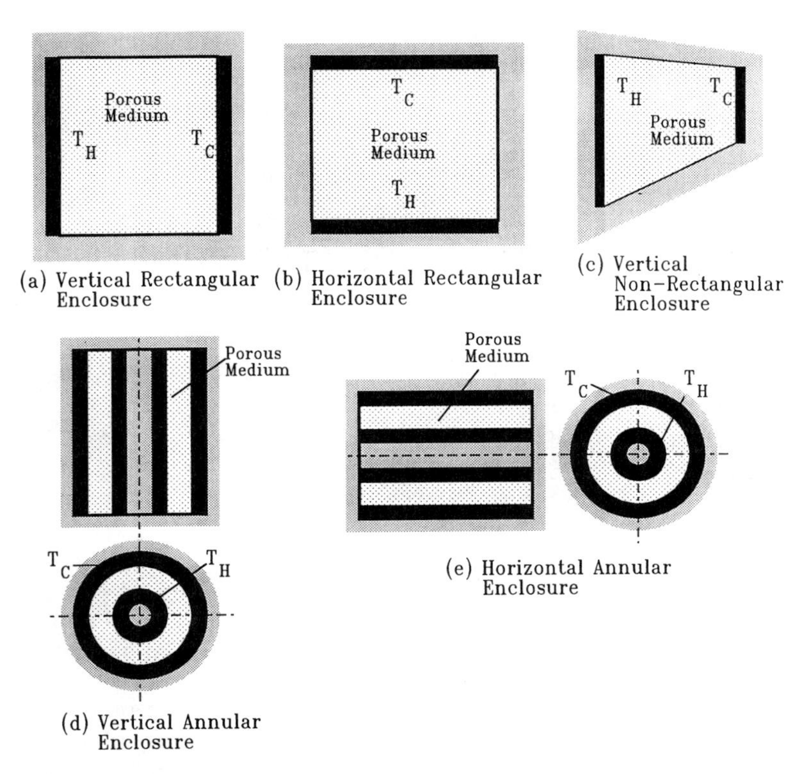

Figure 1. Examples of types of enclosure being considered in the present chapter.

cooled to a uniform temperature, T_C, and with the remaining two walls adiabatic. In general, the enclosure is inclined to the vertical. This flow situation is therefore as shown in Figure 2.

Most studies of this type of flow have assumed that the enclosure is vertical, i.e., that the heated and cooled walls are vertical and that the adiabatic walls are horizontal. Most studies have also used the Darcy flow model and the Boussinesq approximation; e.g., see Ansari and Daniels (1994), Beck (1972), Beckermann et al. (1986a,b), Bejan (1979), Chan et al. (1970), Llagostera and Figueiredo (1998), Manole and Lage (1992, 1993), Misra and Sarkar (1995), Rao and Glakpe (1992), Sai et al. (1997), Seki et al. (1978), Simpkins and Blythe (1980), Walker and Homsy (1978), and Weber (1975). In this case, using the coordinate system shown in Figure 3, the governing equations are

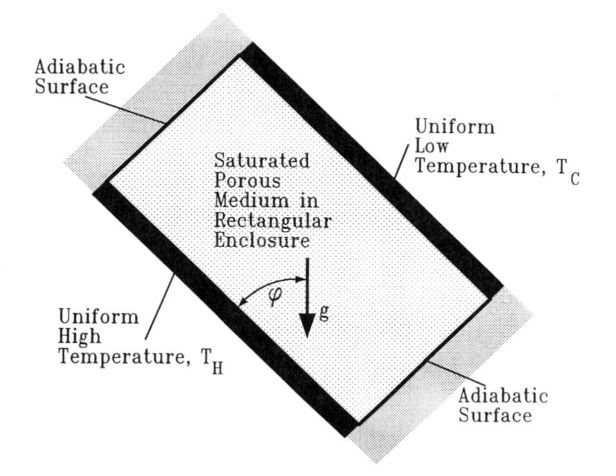

Figure 2. Inclined rectangular enclosure.

$$\frac{\partial u}{\partial x} + \frac{\partial v}{\partial y} = 0 \tag{1}$$

$$\frac{\partial p}{\partial x} = -\frac{\mu_f u}{K}, \qquad \frac{\partial p}{\partial y} = -\frac{\mu_f v}{K} + \beta g \rho_f (T - T_C) \tag{2}$$

$$u\frac{\partial T}{\partial x} + v\frac{\partial T}{\partial y} = \left(\frac{k_a}{\rho_f c_f}\right)\left(\frac{\partial^2 T}{\partial x^2} + \frac{\partial^2 T}{\partial z^2}\right) \tag{3}$$

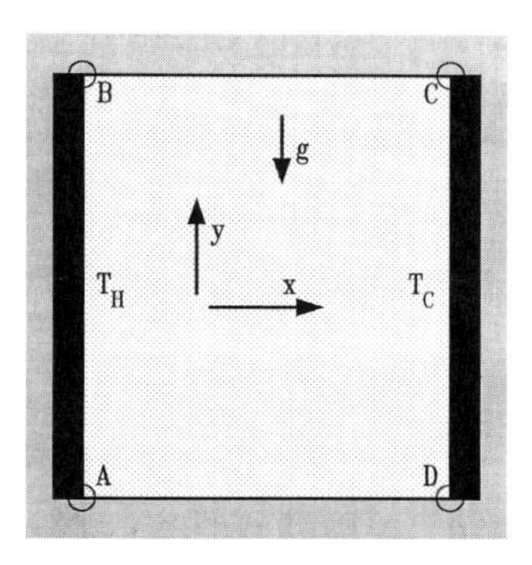

Figure 3. Coordinate system and wall segments used.

The temperature of the cold wall T_C has, it will be noted, been used as the reference temperature in expressing the change in density. The subscript f refers to the properties of the fluid with which the porous material is saturated, and k_a is the effective thermal conductivity of the saturated porous medium.

The above equations are conventionally written in dimensionless form. For this purpose, if the enclosure width W is used as the reference length scale and if the following dimensionless variables are then introduced

$$X = x/W \qquad Y = y/W \tag{4}$$

$$U = uW/\alpha_a \qquad V = uW/\alpha_a \tag{5}$$

$$P = pK/\mu_f\alpha_f \qquad \theta = (T - T_C)/(T_H - T_C) \tag{6}$$

where α_a is the apparent thermal diffusivity of the porous material, i.e., $\alpha_a = k_a/\rho_f c_f$, the governing equations become

$$\frac{\partial U}{\partial X} + \frac{\partial V}{\partial Y} = 0 \tag{7}$$

$$\frac{\partial P}{\partial X} = -U, \qquad \frac{\partial P}{\partial Y} = -V + Ra^*\theta \tag{8}$$

$$U\frac{\partial \theta}{\partial X} + V\frac{\partial \theta}{\partial Y} = \frac{\partial^2 \theta}{\partial X^2} + \frac{\partial^2 \theta}{\partial Z^2} \tag{9}$$

Here

$$Ra^* = \frac{\beta g K \rho_f (T_H - T_C) W}{\alpha_a \mu_f} \tag{10}$$

is the Darcy-modified Rayleigh number.

The boundary conditions on the solution are, using the wall segments defined in Figure 3.

1. On wall AB, i.e., at $X = 0$

$$u = 0, \qquad T = T_H, \qquad \text{i.e., } U = 0, \qquad \theta = 1$$

2. On wall BC, i.e., at $Y = A$

$$v = 0, \qquad \frac{\partial T}{\partial y} = 0, \qquad \text{i.e., } V = 0, \qquad \frac{\partial \theta}{\partial Y} = 0$$

3. On wall CD, i.e., at $X = 1$

$$u = 0, \qquad T = T_C, \qquad \text{i.e., } U = 0, \qquad \theta = 0$$

4. On wall DA, i.e., at $Y = 0$

$$v = 0, \qquad \frac{\partial T}{\partial y} = 0, \qquad \text{i.e., } V = 0, \qquad \frac{\partial \theta}{\partial Y} = 0$$

Here $A = H/W$ is the aspect ratio of the enclosure.

Once the above set of equations has been solved to give the temperature distribution, the local heat transfer rate q_w on the heated and cooled walls can be determined by using Fourier's law, i.e., by using

$$q_w = -k_a \frac{\partial T}{\partial y}$$

Using these results, the mean Nusselt number Nu can be determined, where

$$Nu = \frac{\overline{q_w} W}{k_a(T_H - T_C)} \tag{11}$$

Here $\overline{q_w}$ is the mean heat transfer rate.

It should be noted that, if the convective motion has a negligible effect on the heat transfer rate, i.e., if there is effectively pure conduction, $Nu = 1$.

A consideration of the above governing equations and the boundary conditions shows that the parameters on the solution are Ra^* and A, i.e.

$$Nu = \text{function}(Ra^*, A) \tag{12}$$

Although approximate analytical solutions to the above equations can be obtained (see later), full solutions have to be obtained numerically. Such solutions have been obtained using finite-difference, finite-element, and finite-volume methods. Some typical results showing the variation of Nu with Ra^* and A are given in Figures 4 and 5. The results in these figures are largely based on those given by Prasad and Kulacki (1984).

It will be seen from Figures 4 and 5 that at low values of Ra^*, Nu tends to its pure conduction value of 1. It is to be expected that at large values of the Darcy-modified Rayleigh number the flow in the enclosure will consist essentially of a boundary layer flow along the walls with a central core that is essentially isothermal at a temperature of $(T_H + T_C)/2$, this flow situation being shown schematically in Figure 6. Now, for a boundary layer flow over a plane vertical surface the mean Nusselt number based on the height of the surface is proportional to the square root of the Darcy-modified Rayleigh number based on the height of the surface. It is to be expected, therefore, that at large values of $Ra^*(H/W)$, i.e., of Ra^*_H, where Ra^*_H is the Darcy-modified Rayleigh number based on the height of the enclosure, i.e.,

$$Ra^*_H = \frac{\beta g K \rho_f (T_H - T_C) H}{\alpha_a \mu_f} \tag{13}$$

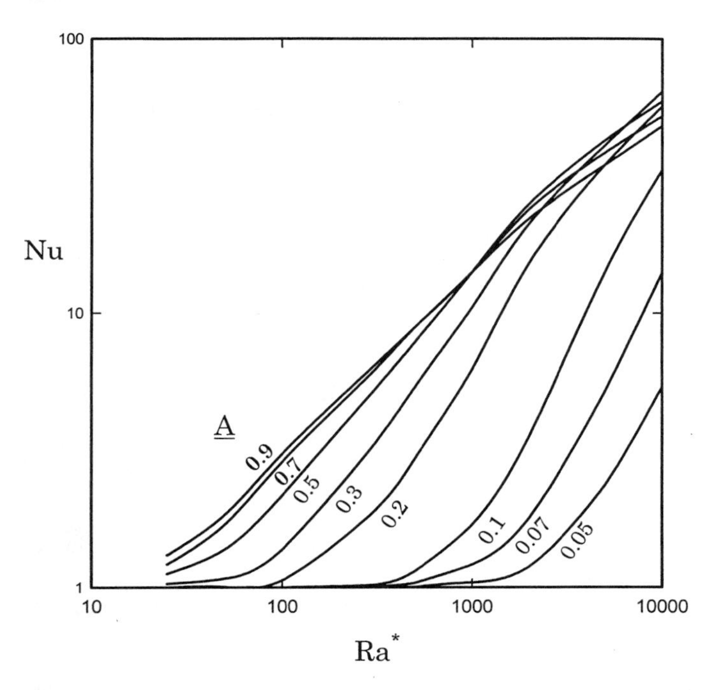

Figure 4. Variation of mean Nusselt number with Darcy-modified Rayleigh number for $A < 1$.

Nu will be proportional to $(W/H)Ra_H^{*\,0.5}$, i.e., proportional to $(Ra^*/A)^{0.5}$. It will be seen from Figures 4 and 5 that this is indeed the case, these results indicating that, when this flow condition exists, the results can be approximately described by

$$Nu = 0.51\left(\frac{Ra^*}{A}\right)^{0.5} \tag{14}$$

Now, the boundary layer thickness on the walls will be of the order of $H/Ra_H^{*\,0.5}$. Hence this type of flow is expected to exist when $H/Ra_H^{*\,0.5}$ is much less than W, i.e., when $Ra_H^{*\,0.5}/A$ is much greater than the order of 1, i.e., when $(Ra^*/A)^{0.5}$ is much greater than the order of 1. If the aspect ratio is very small, however, the heat transfer rate will again approach its conduction limit, as shown by the results given in Figure 5.

Of course, if the aspect ratio is high enough the boundary layers on the hot and cold walls will merge and a type of fully developed flow will exist between the two walls. Since the boundary layer thickness on the walls will

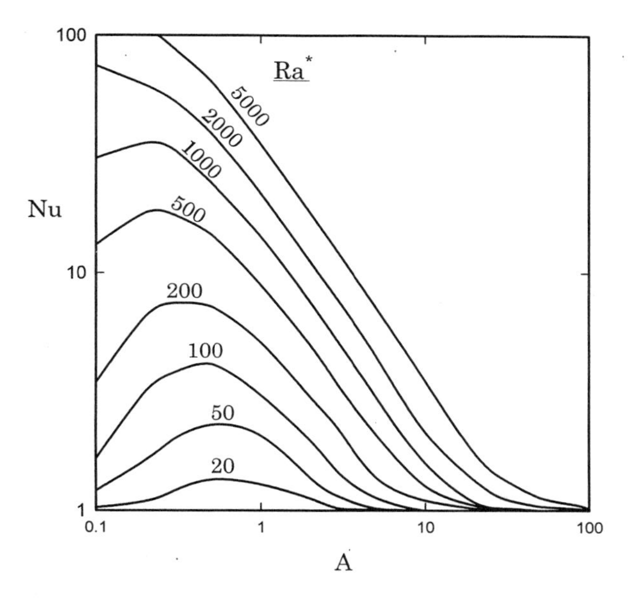

Figure 5. Variation of mean Nusselt number with A for various values of the Darcy-modified Rayleigh number.

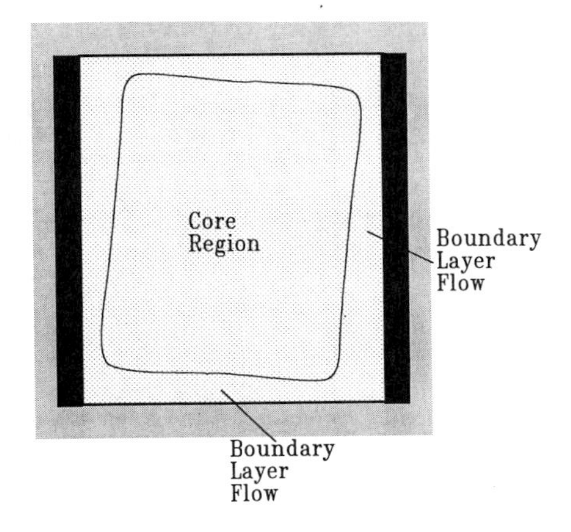

Figure 6. Assumed flow pattern at high values of the Darcy-modified Rayleigh number.

be of the order of $H/Ra_H^{*\,0.5}$, this type of flow is expected to exist when $H/Ra_H^{*\,0.5}$ is of the order of W or larger, i.e., when $Ra_H^{*\,0.5}/A$ is less than the order of 1, i.e., when $(Ra^*/A)^{0.5}$ is less than the order of 1. When this type of flow exists, there will be no change in the temperature distribution with distance up the wall and the Nusselt number will have its pure conduction value. This is clearly shown by the results given in Figure 5.

The above discussion was for vertical rectangular enclosures. When the enclosure is inclined, the hot and cold walls making an angle ϕ to the vertical, ϕ being as defined in Figure 2,

$$Nu = \text{function}(Ra^*, A, \phi) \tag{15}$$

Studies of this situation are described by Hsiao (1998), Inaba et al. (1987), Mbaye et al. (1993), Mihir et al. (1987, 1988), Rees and Bassom (1998), Vasseur et al. (1990), Wang and Zhang (1990), and Zhang (1992).

Some typical numerically and experimentally determined results given by Inaba et al. (1987), showing the effect of ϕ on the Nusselt number, are shown in Figure 7. It will be seen from this figure that the Nusselt number has a maximum value when ϕ is greater than 0°, the effective height over

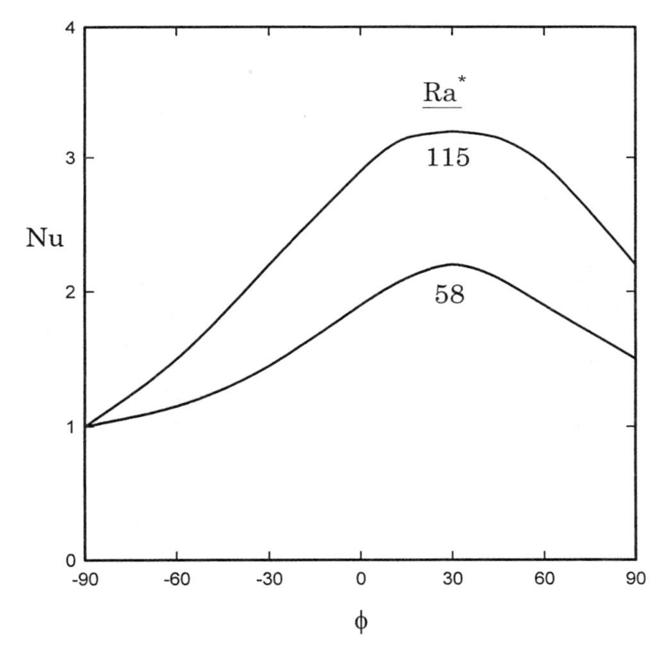

Figure 7. Typical effect of inclination angle on the Nusselt number.

which the buoyancy forces act then being greater than H, with the result that the intensity of the fluid motion is increased, resulting in a higher Nusselt number. The experimental results indicate that the maximum Nu occurs when ϕ is approximately 30°, whereas the numerical results indicate that it occurs at a somewhat greater value of ϕ. Also, when $\phi = -90°$, the fluid in the enclosure is stably stratified and there is no convective motion, and consequently $Nu = 1$. The effect of inclination angle when ϕ is near 90° will be discussed in the next section.

On the basis of their experimental results, Inaba et al. (1987) suggest the following equation for the Nusselt number for $75° > \phi > -30°$

$$Nu = 0.22 Pr_a^{0.13} A^{-0.34} [Ra^* \cos(30° - \phi)]^{0.53} \tag{16}$$

this equation applying when $30 < Ra^* \cos(30° - \phi) < 1030$, $1.1 < Pr_a < 45$, and $2.5 < A < 7.9$. Here, Pr_a is the effective Prandtl number of the porous medium, i.e., $\mu_f c_f / k_a$. This parameter does not, of course, occur in Darcy model-based analyses.

The results discussed above were all relevant to the case where the heated wall is maintained at a uniform temperature. In a number of practical situations there is, however, essentially a uniform heat flux q_w rather than a uniform temperature at the hot wall. In this situation, the following Darcy-modified Rayleigh number is used

$$Ra_q^* = \frac{\beta g K \rho_f q_w W^2}{\alpha_a \mu_f k_a} \tag{17}$$

Solutions for this boundary condition are also available for the vertical enclosure case; e.g., see Bejan (1983). At high values of the Darcy-modified Rayleigh number, where there are distinct boundary layers on the vertical walls, these solutions indicate that the heat transfer rate is approximately given by

$$Nu = 0.5 Ra_q^{*0.5} \tag{18}$$

III. HORIZONTALLY HEATED RECTANGULAR ENCLOSURES

Attention is given in this section to the flow in a rectangular enclosure that is heated from below, the situation considered being shown in Figure 8. This situation corresponds to the inclined enclosure case discussed in the previous section with $\phi = 90°$. Studies of this situation are given, for example, by Elder (1967), Georgiadis and Catton (1986), Lai and Kulacki (1991a), and Naylor and Oosthuizen (1995).

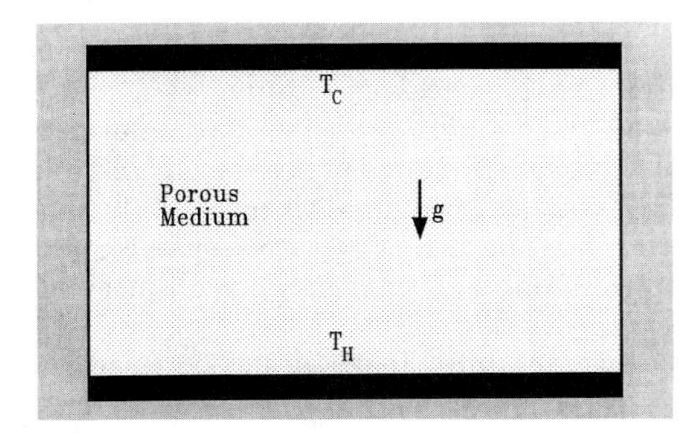

Figure 8. Rectangular enclosure heated from below.

In a given situation, if the temperature difference across the enclosure, $(T_H - T_C)$, is small there will be no convective motion in the enclosure and $Nu = 1$. However, with heating from below, this situation is unstable and, therefore, once $(T_H - T_C)$ and hence Ra^* exceeds a certain critical value convection develops and Nu rises above 1. This is the same situation as that associated with Bénard convection when the enclosure is filled with a pure fluid. Studies of the stability of enclosures with very large values of A, i.e., effectively in infinite horizontal porous layers, indicate that disturbances in the flow grow with time, i.e., convection develops, when Ra^* exceeds $4\pi^2$, i.e., exceeds 39.5. Once the Darcy-modified Rayleigh number exceeds this value, a cellular flow develops that is similar to the Rayleigh–Bénard flow that occurs with a pure fluid. There have been a number of studies of the heat transfer rate that exists once the convection develops. As shown in Figure 9, these results can be approximately represented by

$$Ra^* < 40\text{: } Nu = 1; \qquad Ra^* > 40\text{: } Nu = \frac{Ra^*}{40} \tag{19}$$

The results discussed above were all relevant to an effectively infinite horizontal porous layer. The actual aspect ratio of the enclosure will affect both the value of Ra^* at which convective flow develops and the variation of Nu when there is convective flow.

When the enclosure is inclined slightly to the horizontal, i.e., ϕ (see Figure 2) is less than but near 90°, the value of Ra^* at which the flow develops is less than 39.5; e.g., see Kaneko et al. (1974). Once ϕ becomes less than about 80°, the flow in the enclosure becomes unicellular, i.e., becomes similar to that which occurs in a vertical enclosure.

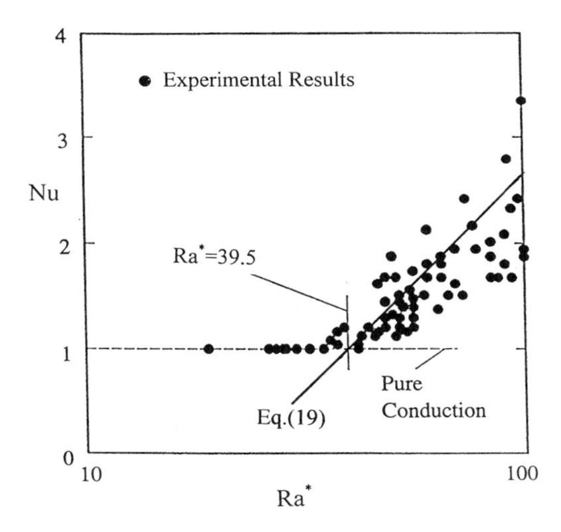

Figure 9. Variation of mean Nusselt number with Darcy-modified Rayleigh number for flow in an infinite horizontal porous layer.

IV. NONRECTANGULAR ENCLOSURES

The discussion given in the previous two sections was concerned with flow in rectangular porous enclosures. However, in many practical situations the flow occurs in enclosures with other shapes. A number of studies of flow in nonrectangular enclosures have therefore been undertaken; e.g., see Bejan and Poulikakos (1982), Burns and Tien (1979), and Poulikakos and Bejan (1983a). Some of the enclosure shapes considered in these studies are shown in Figures 1c, d, and e.

One situation that is of significant practical importance is that shown in Figure 1e, i.e., two-dimensional flow in a saturated porous medium contained in the annulus between two concentric horizontal cylinders, the inner and outer cylinders being at temperatures T_H and T_C respectively and having radii r_i and r_o respectively. Attention will here be restricted to the case where $T_H > T_C$. This general type of flow has been considered, for example, by Arnold et al. (1991), Barbosa Mota and Saatdjian (1994), Bejan and Tien (1979), Kaviany (1986), Muralidhar and Kulacki (1986), and Rao et al. (1985, 1987, 1988).

For this type of flow it is usual to define the Darcy-modified Rayleigh number as

$$Ra^* = \frac{\beta g K \rho_f (T_H - T_C) r_i}{\alpha_a \mu_f} \tag{20}$$

and to define

$$R = \frac{r_o}{r_i} \tag{21}$$

At low values of Ra^* a flow with a single cell on each side of the vertical center-line will exist, as shown schematically in the left-side sketch in Figure 10. However, in the upper portion of the enclosure there is essentially a near-horizontal heated surface below a near-horizontal cold surface, a situation that was discussed in the preceding section. When the Darcy-modified Rayleigh number exceeds a certain value, a secondary cellular motion therefore develops in this upper portion of the enclosure, i.e., a multicellular flow develops. This is illustrated schematically in the right-side sketch in Figure 10.

The value of Ra^* at which the multi-cellular pattern first develops depends on the value of R. For example, Arnold et al. (1991) found that a multicellular flow developed when Ra^* exceeded about 120 when $R = 2$ but did not develop until Ra^* exceeded about 260 when $R = 1.2$. If a Darcy-modified Rayleigh number that is based on the size of the gap between the two surfaces, i.e., on $(r_o - r_i)$, is used, i.e., if the following is used,

$$Ra_g^* = \frac{\beta g K \rho_f (T_H - T_C)(r_o - r_i)}{\alpha_a \mu_f} \tag{22}$$

the two values are 120 and approximately 50 respectively. This illustrates that when R is near 1, i.e., when the gap size is small compared to the radius of the inner surface and curvature effects in the flow are therefore small, the multicellular flow starts at a value of the Darcy-modified Rayleigh number

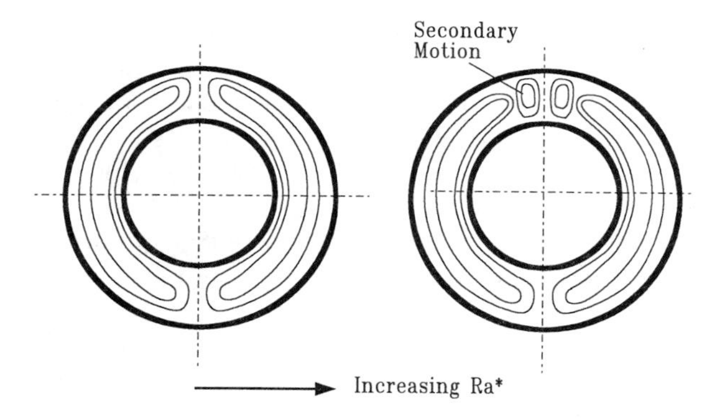

Figure 10. Streamline patterns in a horizontal annular enclosure filled with a porous medium.

based on the gap size that is close to 40. This is the value at which Rayleigh–Bénard type flow develops in a horizontal plane porous layer.

Typical variations of mean Nusselt number based on the conditions at the inner surface, Nu_i, with Ra_g^* are shown in Figure 11. Here, because the Nusselt number is the ratio of the actual heat transfer rate to the heat transfer rate that would exist with pure conduction, Nu_i is defined by

$$Nu_i = \frac{\overline{q_{wi}} 2\pi r_i}{2\pi k_a (T_H - T_C)/\ln(r_o/r_i)} = \frac{\overline{q_{wi}} r_i \ln R}{k_a (T_H - T_C)} \tag{23}$$

where $\overline{q_{wi}}$ is the mean heat transfer rate over the inner surface. Caltagirone (1976) proposed the following equation for $1.19 \leq R \leq 4$

$$Nu_i = 0.44 Ra^{*0.5} \frac{\ln R_o}{1 + 0.916/R^{0.5}} \tag{24}$$

Here, as before, $R = r_o/r_i$.

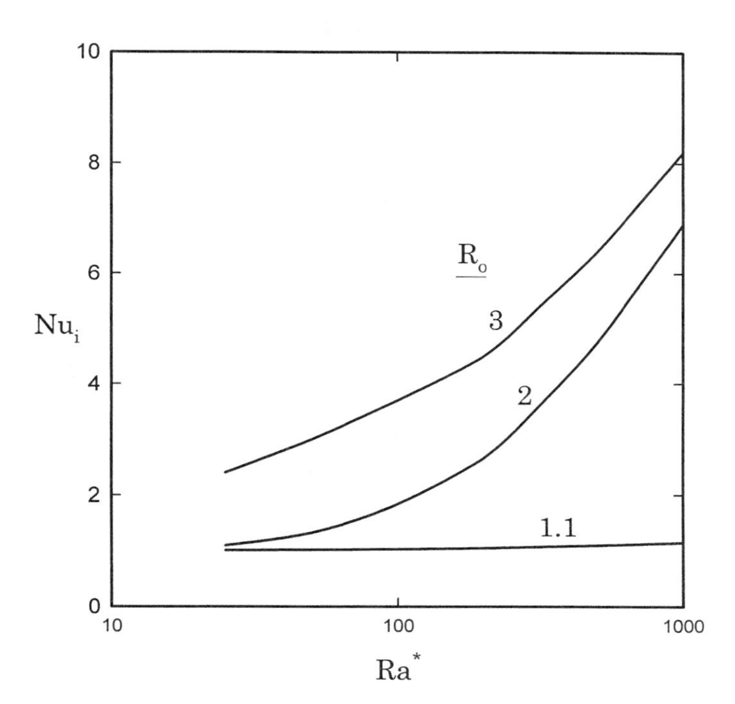

Figure 11. Typical Nusselt number–Darcy-modified Rayleigh number variations for flow in a horizontal annular enclosure filled with a porous medium.

The above discussion concerned flow in the enclosure between two concentric horizontal cylinders. In some practical situations the two cylinders are not concentric, i.e., the inner cylinder is eccentrically positioned relative to the outer cylinder. The case where the center-lines of the two cylinders lie on the same vertical line but are displaced by a distance of εr_i along this line, the center-line of the inner cylinder being above that of the outer cylinder, has been considered by Bau (1984), Himasekhar and Bau (1986), and Barbosa Mota et al. (1994), for example. Typical variations of mean Nusselt number with ε for various Darcy-modified Rayleigh numbers based on the inside radius are shown in Figure 12. The sharp changes in the slope of the curve, indicated by the dotted line in Figure 12, are associated with a change in the secondary flow pattern near the top of the annulus.

Studies of heat transfer across a vertical porous annulus have also been undertaken; e.g., see Bau and Torrance (1981), Campos et al. (1990), David et al. (1989), and Havstad and Burns (1982). Havstad and Burns

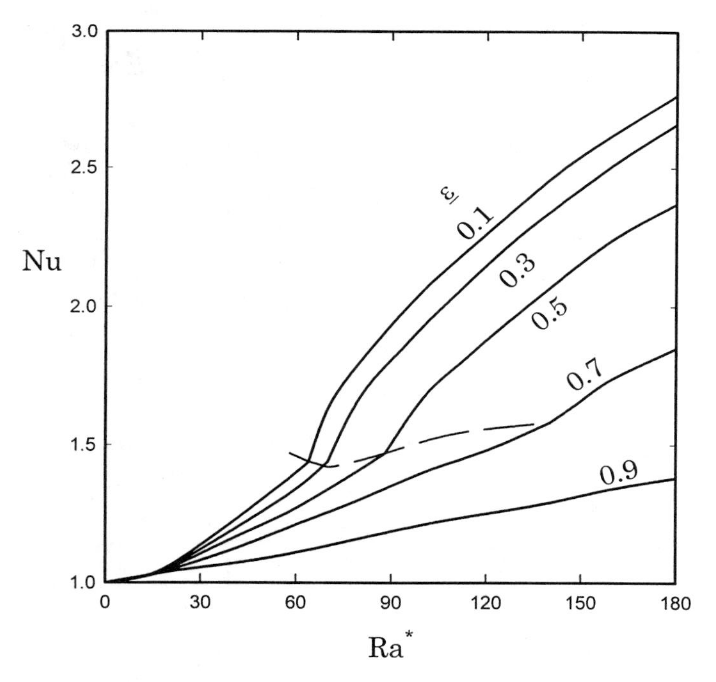

Figure 12. Typical variations of Nusselt number with dimensionless eccentricity for various Darcy-modified Rayleigh numbers for flow in a horizontal eccentric annular enclosure filled with a porous medium.

(1982) found that their numerical results could be correlated using the following equation

$$Nu = 1 + 0.2196\left[\frac{1}{R_o}\left(1 - \frac{1}{R_o}\right)\right]^{1.334}(Ra^* R_o)^{0.9296}\left(\frac{H}{r_o}\right)^{1.168} \exp\left(-\frac{3.702}{R_o}\right) \tag{25}$$

Here H is the height of the enclosure. This equation is applicable for $1 \leq H/r_o \leq 20$, $0 < Ra^* R_o < 150$.

Some studies of the heat transfer rate across the porous enclosure formed between two concentric spherical surfaces of radii r_i and r_o maintained at temperatures T_H and T_C respectively are also available. Defining

$$Nu_i = \frac{q_{wi}}{q_{\text{conduction}}} = \frac{q_{wi} 4\pi r_i^2}{4\pi k_a (T_H - T_C)/(1/r_i - 1/r_o)} = \frac{q_{wi} r_i (1 - 1/R_o)}{k_a (T_H - T_C)} \tag{26}$$

the following equation applies at high Darcy-modified Rayleigh numbers (approximately when $Nu > 1.5$)

$$Nu = 0.76 Ra^{*0.5} \frac{1 - 1/R_o}{1 + 1.42/R_o^{1.5}} \tag{27}$$

where Ra^* is again the Darcy-modified Rayleigh number based on the inner surface radius and R_o is again r_o/r_i.

V. ENCLOSURES PARTLY FILLED WITH POROUS MEDIA

In a number of situations that occur in practice, natural convection occurs in an enclosure that is only partly filled with a porous medium. Studies of this type of problem are discussed, for example, by Naylor and Oosthuizen (1994, 1996a,b), Oosthuizen (1987, 1995, 1996), Oosthuizen and Paul (1987, 1993, 196a,b), Poulikakos and Bejan (1983b), and Song and Viskanta (1991, 1994). Some of the situations that have been considered for a rectangular enclosure are shown in Figure 13. In some cases there is an impermeable barrier between the porous and fluid layers which prevents flow between the layers. In such cases it is often adequate to assume that the barrier offers no thermal resistance. In other cases there is effectively no barrier between the porous and fluid layers. In such cases there is a continuity of normal velocity component, pressure, temperature, and heat flux across the interface. A typical streamline pattern for the case where there is no barrier between

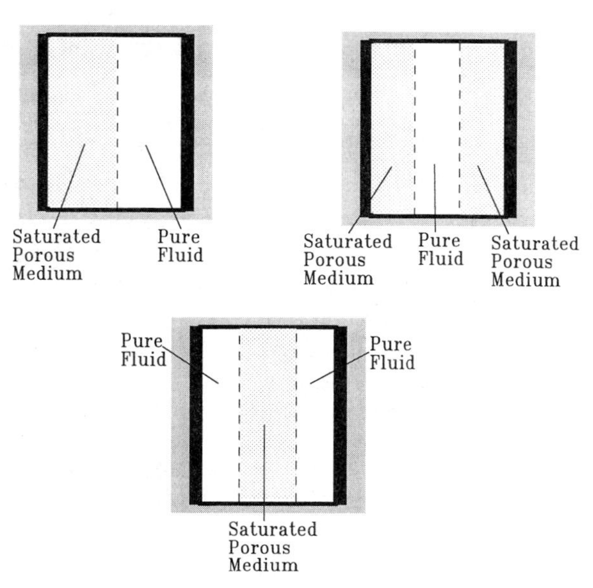

Figure 13. Rectangular enclosures partly filled with a porous medium.

the layers and where there is a centrally positioned fluid layer between two porous layers adjacent to the walls is shown in Figure 14.

One important result that follows from studies of the heat transfer rate across enclosures partly filled with a porous medium is that the heat transfer rate can be kept near to the pure conduction value even when the enclosure contains a relatively small amount of porous material. The amount of porous material required will be largely dependent in a given situation on the conductivity ratio $k_r = k_a/k_f$. This effect is illustrated by the typical results presented in Figure 15. This result has significant consequences in some thermal insulation situations because it indicates that less insulation material may be needed than is conventionally assumed.

The above discussion was concerned with situations involving vertical porous layers. Some studies of flows in which the porous layer is horizontal are also available; e.g., see Naylor and Oosthuizen (1994, 1995).

VI. ENCLOSURES FILLED WITH ANISOTROPIC POROUS MEDIA

Most studies of natural convection in porous enclosures assume that the porous medium is isotropic, i.e., that it has the same properties in all direc-

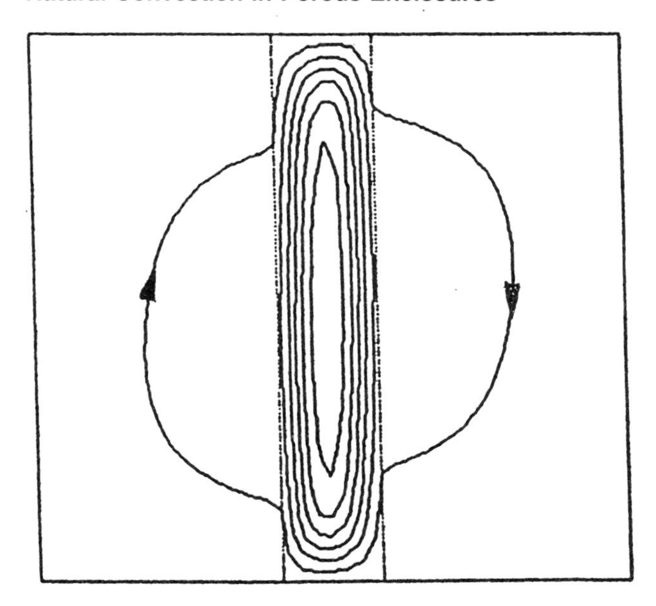

Figure 14. Typical streamline pattern for flow in a rectangular enclosure partly filled with two porous layers and with a fluid layer in the center of the enclosure. The width of the fluid layer is $0.15W$.

tions. However, in many real situations, both the permeability K and the apparent thermal conductivity k_a have values that depend on the direction considered. Studies of the effect of anisotropy on the flow in a rectangular enclosure containing a porous medium have been undertaken by Kimura et al. (1993), Ni and Beckerman (1991), Degan et al. (1995, 1998), Chang and Lin (1994), Degan and Vasseur (1996, 1997), and Zhang et al. (1993). In most of the studies, it has been assumed that the principal axes of the apparent thermal conductivity coincide with the horizontal and vertical coordinate directions. In general, however, the principal directions will be oriented at an arbitrary angle ϕ to the horizontal axes, as shown in Figure 16. In many cases, the nature of the anisotropy for the flow and the heat transfer is very different and in many cases it is possible to assume that K is the same in all directions even when the variation of k_a with direction has to be accounted for; in other words, only the anisotropy in the apparent thermal conductivity has to be accounted for, i.e., it is only necessary to consider the thermal anisotropy. The situation shown in Figure 16 was investigated by Degan et al. (1998) using this assumption. They obtained both numerical and approximate boundary type solutions. The boundary layer analysis of Degan et al. (1998) indicates that the heat transfer rate is given by

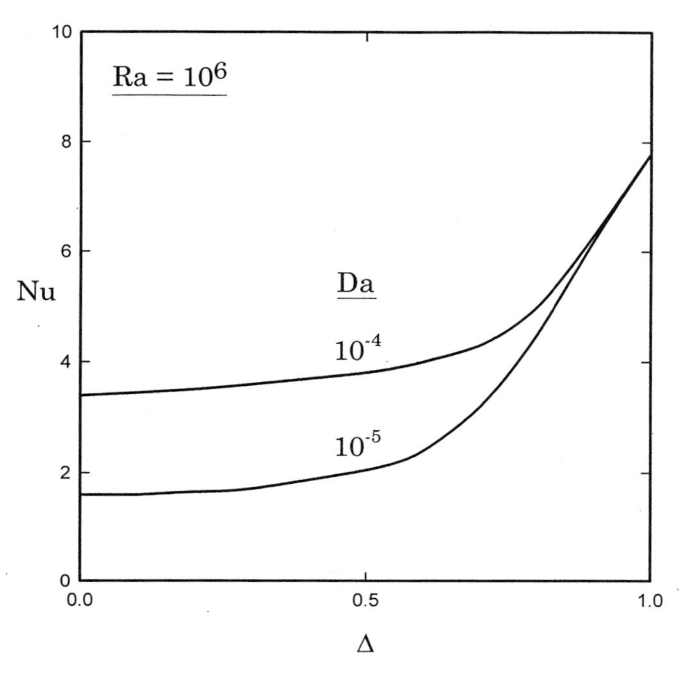

Figure 15. Effect of dimensionless thickness of fluid layer on heat transfer rate across a square enclosure with a porous layer adjacent to the heated and cooled vertical walls (see Figure 13, top right).

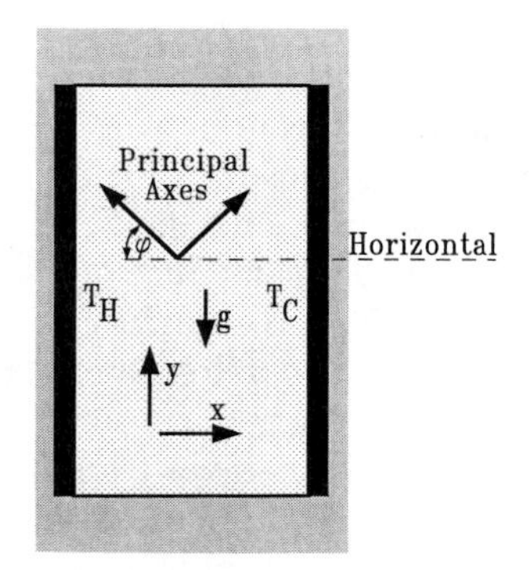

Figure 16. Arrangement of principal axes of thermal conductivity.

$$Nu = 0.51\left(\frac{Ra_2^*}{aA}\right)^{1/2} \tag{28}$$

where

$$Ra_2^* = \frac{\beta g K \rho_f (T_H - T_C) W}{\alpha_{2a}\mu_f} \tag{29}$$

$$\alpha_{2a} = \frac{k_{a2}}{\rho_f c_f}, \qquad a = \cos^2\phi + k_a^* \sin^2\phi \tag{30}$$

$$k_a^* = \frac{k_{a1}}{k_{a2}}, \qquad Nu = \frac{\overline{q_w} W}{k_{a2}(T_H - T_C)} \tag{31}$$

Here $\overline{q_w}$ is the mean heat transfer rate and k_{a1} and k_{a2} are the values of the apparent thermal conductivity in the two principal directions. A typical result for the effect of anisotropy on the heat transfer rate, obtained by Degan et al. (1998), is shown in Figure 17. Both the numerical analysis and the boundary layer analysis give essentially the same results under the conditions considered in this figure, in which

$$Ra^* = \frac{\beta g K \rho_f (T_H - T_C) W}{\alpha_{ma}\mu_f}, \qquad \alpha_{ma} = \frac{\sqrt{k_{a1}k_{a2}}}{\rho_f c_f} \tag{32}$$

A study of the effect of anisotropy of both permeability and apparent thermal conductivity on the flow in a horizontal annular enclosure filled with a porous medium was undertaken by Aboubi et al. (1998).

Layered porous media are a special case of anisotropic porous media. Flow in a layered porous medium has received quite wide attention, typical of the studies in this area being those of Gjerde and Tyvand (1984), Lai and Kulacki (1988), Masuoka et al. (1978), McKibbin and O'Sullivan (1980 and 1981), McKibbin and Tyvand (1982, 1983, 1984), Pan and Lai (1996), Poulikakos and Bejan (1983b), Rana et al. (1979), and Rees and Riley (1990).

VII. OTHER ENCLOSURE FLOWS

In some situations heat transfer occurs from a body buried in a porous medium that is contained within an enclosure, a typical situation being shown in Figure 18. Situations of this type can occur with a heated pipe

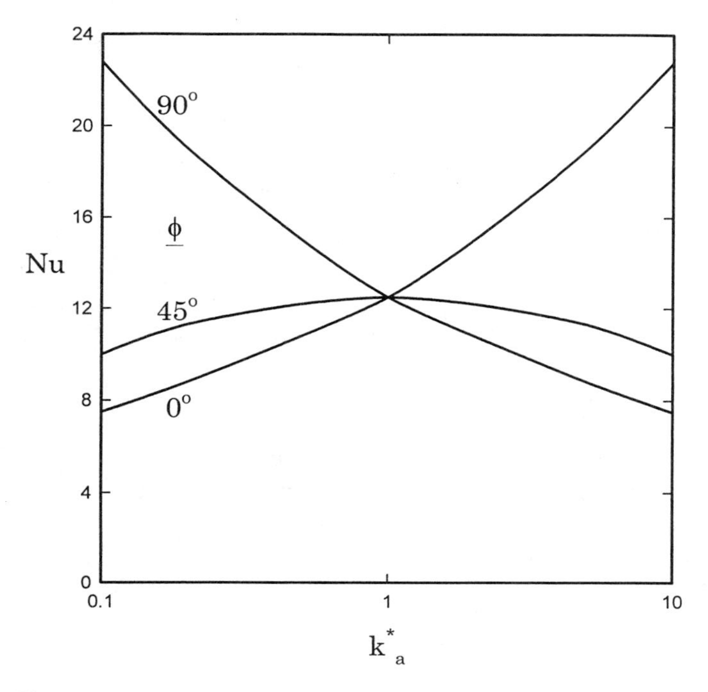

Figure 17. Effect of thermal anisotropy on Nu for $Ra^* = 2500$ and $A = 4$.

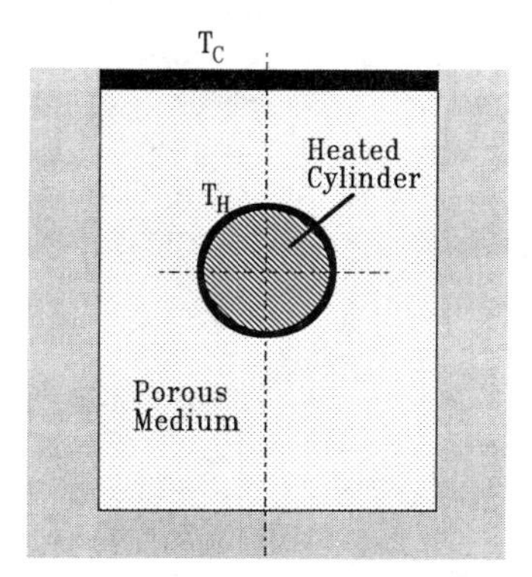

Figure 18. Heated cylinder in a rectangular enclosure filled with a porous medium and with a cooled upper surface.

buried in a trench that is filled with water-saturated soil. Studies of this general type of flow are described by Hsiao (1995), Hsiao and Chen (1994), Hsiao et al. (1992), Nag et al. (1994) and Oosthuizen and Naylor (1995, 1996a,b), for example.

Several studies of time-dependent flows in rectangular porous enclosures are available; e.g., see Antohe and Lage (1997), Holst and Aziz (1972), and Rajen and Kulacki (1992). Attention has been given both to the transient flow that follows the sudden heating (or cooling) of one wall and to the flow that results when the wall temperature varies periodically with time. In the latter case, consideration has been given to the conditions under which resonance occurs, i.e., to the conditions under which there is an increase in the heat transfer rate as a result of the coincidence of the period of the wall temperature variation with the natural frequency of the enclosure flow.

All of the studies discussed above assumed that there was a linear relationship between the density and the temperature. For some liquids, however, there is a density maximum, as illustrated in Figure 19. The most commonly encountered liquid of this type is water, which has a density maximum at a temperature of 3.98°C. For water in the region of the density maximum, the density variation is approximately described by

$$\rho_{\max} - \rho = a\rho_{\max}(T - T_{\max})^2 \tag{33}$$

This equation gives adequate accuracy for temperatures between 0°C and about 10°C. Studies of flows in which the density maximum is important have been undertaken, for example, by Blake et al. (1984), Chang and Yang (1995), Oosthuizen and Nguyen (1992), Poulikakos (1984), and Yen (1974).

When there is a density maximum the flow can be down or up the heated surface, depending on the temperature of the surface relative to the maximum density temperature. Bejan (1995) used a scale analysis to derive

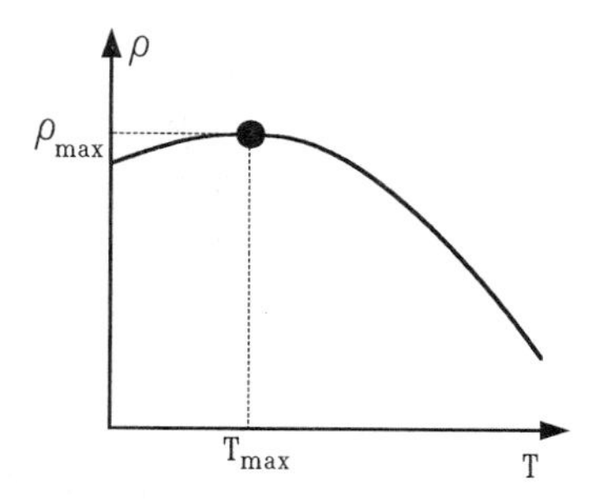

Figure 19. Variation of density with temperature for a liquid having a density maximum.

the following form of equation for high Rayleigh number flow in a vertical rectangular enclosure containing a porous medium that is saturated with a liquid whose density can be described by the parabolic form of the equation that was discussed above, i.e., by Eq. (33)

$$Nu = 0.26C \frac{1/A}{Ra_H^{*-0.5} + (C-1)Ra_C^{*-0.5}} \tag{34}$$

where C is a constant and

$$Ra_H^* = \frac{\beta g K \rho_f (T_H - T_{\max})}{\alpha_a \mu_f} \tag{35}$$

and

$$Ra_C^* = \frac{\beta g K \rho_f (T_{\max} - T_C)}{\alpha_a \mu_f} \tag{36}$$

Most previous studies of flow in porous-media-filled enclosures that take account of the density maximum have been concerned with cases in which there is freezing in the porous medium; e.g., see Oosthuizen and Paul (1991), Sasaguchi (1995), and Sasaki et al. (1990a,b). Situations involving a phase change will not, however, be discussed in this chapter.

The discussion given thus far in this chapter has been concerned entirely with situations in which the buoyancy forces that cause the flow result from density changes in the enclosure that in turn result from temperature changes. However, in some cases there are concentration differences in the fluid that saturates the porous medium and these concentration differences can also result in density changes in the enclosure. Natural convective flows that result from density differences caused by both temperature and concentration differences are termed double-diffusive flows. There have been a number of studies of such double-diffusive flows in rectangular enclosures filled with a porous medium; e.g., see Alavyoon et al. (1994), Bejan and Khair (1985), Bejan and Tien (1978), Charrier-Mojtabi et al. (1997), Gobin et al. (1998), Goyeau et al. (1996), Karimi-Fard et al. (1997, 1998), Lai and Kulacki (1991b), Mamou et al. (1994, 1995a,b, 1998a,b), Nguyen et al. (1994), Nield (1968), Nield and Bejan (1992), Nithiarasu et al. (1996, 1997a,b), Rudraiah et al. (1986), and Trevisan and Bejan (1985, 1986, 1990). Most of these studies have been concerned with the bifurcations that can occur in the flow as a result of changes in the relative importance of the buoyancy forces resulting from the temperature differences and from the concentration differences.

All of the above discussion has been concerned with flows that are the result of density changes in the gravitational force field. However, convection can also result from density changes in a porous medium exposed to a centrifugal force field. In many such cases the forces resulting from both the gravitational and the centrifugal force fields are important. Enclosed porous media flows in the presence of a centrifugal force field have been considered by Jou and Liaw (1987a,b), Palm and Tyvand (1984), Rudraiah et al. (1986), Vadasz (1993, 1994), and Zhao et al. (1998), for example. Convection resulting from electromagnetic forces can also sometimes occur.

VIII. NON-DARCY AND OTHER EFFECTS

The discussions given in the previous sections were all based on the assumption that the Darcy model applied. There are, however, a number of situations involving flow in a porous enclosure in which the basic Darcy model does not provide an adequate description of the flow. For this reason a number of studies have been undertaken that try to determine when the Forchheimer and Brinkman extensions of the basic Darcy model, when thermal dispersion, and when tunneling near the wall, for example, are important; e.g., see Aboubi et al. (1998), Beckerman et al. (1986a), Bejan and Poulikakos (1984), Hadim and Chen (1995), Hsiao et al. (1992), Kaviany (1986), Lage (1992), Lauriat and Prasad (1986, 1989), Marpu (1995), Muralidhar and Kulacki (1986), Nithiarasu et al. (1996, 1997b), Poulikakos and Bejan (1985), Prasad and Tuntomo (1987), Sai et al. (1997), Tong and Subramanian (1985), Vafai and Tien (1981), Vasseur and Robillard (1987), and Vasseur et al. (1989, 1990).

In general, the Nusselt number for flow in a porous enclosure will depend on the fluid Rayleigh number, $Ra_f = \beta g K \rho_f (T_H - T_C) W / \alpha_f \mu_f$, the fluid Prandtl number, $Pr_f = v_f / \alpha_f$, the Darcy number, $Da = K/W^2$, the Forchheimer number, $Fs = b/W$, the conductivity ratio, $\lambda = k_f/k_a$, the specific heat ratio, $S = (\rho c)_a/(\rho c)_f$, and the viscosity ratio, $\Lambda = \mu_a/\mu_f$.

Results obtained numerically with various models for conditions for which experimental results are available indicate that, under some conditions, the use of the extended Darcy–Brinkman–Forchheimer model does give results that agree better with experiment than do the results given by the Darcy model, and that the inclusion of thermal diffusion effects does not seem to bring about a significant improvement in the agreement between numerical and experimental results.

REFERENCES

Aboubi K, Robillard L, Vasseur P. Natural convection in horizontal annulus filled with an anisotropic porous medium. Int J Numer Methods Heat Fluid Flow 8:689–702, 1998.

Alavyoon F, Masuda Y, Kimura S. On natural convection in vertical porous enclosures due to opposing fluxes of heat and mass prescribed at the vertical walls. Int J Heat Mass Transfer 37:195–206, 1994.

Ansari A, Daniels PG. Thermally driven tall cavity flows in porous media: the convective regime. Proc R Soc Lond A 444:375–388, 1994.

Antohe BV, Lage JL. Prandtl number effect on the optimum heating frequency of an enclosure filled with fluid or with a saturated porous medium. Int J Heat Mass Transfer 40:1313–1323, 1997.

Arnold F, Blehaut J, Clavier JY, Gorges D, Lohou F, Saatdjian E. Natural convection in a porous medium between concentric, horizontal cylinders. Proceedings of 7th International Conference on Numerical Methods in Thermal Problems, Stanford, 1991, Vol 2 (Pt 2), pp 1084–1091.

Barbosa Mota JP, Saatdjian E. Natural convection in porous, horizontal cylindrical annulus. J Heat Transfer 116:621–626, 1994.

Barbosa Mota JP, Le Prevost JF, Puons E, Saatdjian E. Natural convection in porous, horizontal eccentric annuli. Proceedings of 10th International Heat Transfer Conference, Brighton, 1994, Vol 5, pp 435–440.

Bau HH. Thermal convection in a horizontal, eccentric annulus containing a saturated porous medium—an extended perturbation expansion. Int J Heat Mass Transfer 27:2277–2287, 1984.

Bau HH, Torrance KE. Onset of convection in a permeable medium between vertical coaxial cylinders. Phys Fluids 24:382–385, 1981.

Beck JL. Convection in a box of porous material saturated with fluid. Phys Fluids 15:1377–1383, 1972.

Beckermann C, Viskanta R, Ramadhyani S. A numerical study of non-Darcian natural convection in a vertical enclosure filled with a porous medium. Numer Heat Transfer 10:557–570, 1986a.

Beckermann C, Ramadhyani S, Viskanta R. Natural convection flow and heat transfer between a fluid layer and a porous layer inside a rectangular enclosure. Proceedings of AIAA/ASME 4th Thermophysics and Heat Transfer Conference, Boston, 1986b, pp 1–12.

Bejan A. On the boundary layer regime in a vertical enclosure filled with a porous medium. Lett Heat Mass Transfer 6:93–102, 1979.

Bejan A. The boundary layer regime in a porous layer with uniform heat flux from the side. Int J Heat Mass Transfer 26:1339–1346, 1983.

Bejan A. Convection Heat Transfer. 2nd ed. New York: Wiley, 1995.

Bejan A, Khair KR. Heat and mass transfer by natural convection in a porous medium. Int J Heat Mass Transfer 28:909–918, 1985.

Bejan A, Poulikakos D. Natural convection in an attic-shaped space filled with porous material. J Heat Transfer 104:241–247, 1982.

Bejan A, Poulikakos D. The nonDarcy regime for vertical boundary layer natural convection in a porous medium. Int J Heat Mass Transfer 27:717–722, 1984.

Bejan A, Tien CL. Laminar natural convection heat transfer in a horizontal cavity with different end temperatures. J Heat Transfer 100:641–647, 1978.

Bejan A, Tien CL. Natural convection in horizontal space bounded by two concentric cylinders with different end temperatures. Int J Heat Mass Transfer 22:919–927, 1979.

Blake KR, Bejan A, Poulikakos D. Natural convection near 4°C in a water saturated porous layer heated from below. Int J Heat Mass Transfer 27:2355–2364, 1984.

Burns PJ, Tien CL. Natural convection in porous media bounded by concentric spheres and horizontal cylinders. Int J Heat Mass Transfer 22:929–939, 1979.

Caltagirone JP. Thermoconvective instabilities in a porous medium bounded by two concentric horizontal cylinders. J Fluid Mech 76:337–362, 1976.

Campos H, Morales JC, Lacoa U, Campo A. Thermal aspects of a vertical annular enclosure divided into a fluid region and a porous region. Int Commun Heat Mass Transfer 17:343–354, 1990.

Carbonell RG, Whitaker S. Heat and mass transfer in porous media. In: Bear J, Corapciolglu MY, eds. Fundamentals of Transport Phenomena in Porous Media. Dordrecht: Martinus Nijhoff, 1984.

Chan BKC, Ivey CM, Barry JM. Natural convection in enclosed porous media with rectangular boundaries. J Heat Transfer 92:21–27, 1970.

Chang WJ, Lin HC. Natural convection in a finite wall rectangular cavity filled with an anisotropic porous medium. Int J Heat Mass Transfer 37:303–312, 1994.

Chang WJ, Yang DF. Transient natural convection of water near its density extremum in a rectangular cavity filled with porous medium. Numer Heat Transfer A: Appl 28:619–633, 1995.

Charrier-Mojtabi MC, Karimi-Fard M, Azaiez M, Mojtabi A. Onset of a double-diffusive convective regime in a rectangular porous cavity. J Porous Media 1:104–118, 1997.

Cheng P. Heat transfer in geothermal systems. Adv Heat Transfer 14:1–105, 1979.

David E, Lauriat G, Prasad V. Non-Darcy natural convection in packed-sphere beds between concentric vertical cylinders. AIChE Symp Ser 85:90–95, 1989.

Degan G, Vasseur P. Natural convection in a vertical slot filled with an anisotropic porous medium with oblique principal axes. Numer Heat Transfer A: Appl 30:392–412, 1996.

Degan G, Vasseur P. Boundary-layer regime in a vertical porous layer with anisotropic permeability and boundary effects. Int J Heat Fluid Flow 18:334–343, 1997.

Degan G, Vasseur P, Bilgen E. Convective heat transfer in a vertical anisotropic porous layer. Int J Heat Mass Transfer 38:1975–1987, 1995.

Degan G, Beji H, Vasseur P. Natural convection in a rectangular cavity filled with an anisotropic porous medium. Proceedings of 11th International Heat Transfer Conference, Kyongju, 1998, Vol 4, pp 441–446.

Elder JW. Steady free convection in a porous medium heated from below. J Fluid Mech 27:29–48, 1967.

Georgiadis J, Catton I. Prandtl number effect on Bénard convection in porous media. J Heat Transfer 108:284–290, 1986.

Gjerde KM, Tyvand PA. Thermal convection in a porous medium with continuous periodic stratification. Int J Heat Mass Transfer 27:2289–2295, 1984.

Gobin D, Goyeau B, Songbe JP. Double diffusive natural convection in a composite fluid-porous layer. J Heat Transfer 120:234–242, 1998.

Goyeau B, Songbe JP, Gobin D. Numerical study of double-diffusive natural convection in a porous cavity using the Darcy–Brinkman formulation. Int J Heat Mass Transfer 39:1363–1378, 1996.

Hadim HA, Chen G. Numerical study of non-Darcy natural convection of a power-law fluid in a porous cavity. Proceedings of the ASME International Mechanical Engineering Congress and Exposition, San Francisco, 1995, Vol 317, pp 301–307.

Havstad MA, Burns PJ. Convective heat transfer in vertical cylindrical annuli filled with a porous medium. Int J Heat Mass Transfer 25:1755–1766, 1982.

Himasekhar K, Bau HH. Large Rayleigh number convection in a horizontal eccentric annulus containing a saturated porous medium. Int J Heat Mass Transfer 29:703–712, 1986.

Holst PH, Aziz K. Transient three-dimensional natural convection in confined porous media. Int J Heat Mass Transfer 15:73–90, 1972.

Hsiao SW. A numerical study of transient natural convection about a corrugated plate embedded in an enclosed porous medium. Int J Numer Methods Heat Fluid Flow 5:629–645, 1995.

Hsiao SW. Natural convection in an inclined porous cavity with variable porosity and thermal dispersion effects. Int J Numer Methods Heat Fluid Flow 8:97–117, 1998.

Hsiao SW, Chen CK. Natural convection heat transfer from a corrugated plate embedded in an enclosed porous medium. Numer Heat Transfer A: Appl 25:331–345, 1994.

Hsiao SW, Cheng P, Chen CK. Non-uniform porosity and thermal dispersion effects on natural convection about a heated horizontal cylinder in an enclosed porous medium. Int J Heat Mass Transfer 35:3407–3418, 1992.

Inaba H, Fukuda T, Sugawara M, Chen CF. Natural convection heat transfer in an inclined air layer packed with spherical particles. Proceedings of ASME/JSME Thermal Engineering Joint Conference, Honolulu, 1987, pp 85–89.

Jou JJ, Liaw JS. Transient thermal convection in a rotating porous medium confined between two rigid boundaries. Int Commun Heat Mass Transfer 14:147–153, 1987a.

Jou JJ, Liaw JS. Thermal convection in a porous medium subject to transient heating and rotation. Int J Heat Mass Transfer 30:208–211, 1987b.

Kaneko T, Mohtadi MF, Aziz K. An experimental study of natural convection in inclined porous media. Int J Heat Mass Transfer 17:485–496, 1974.

Karimi-Fard M, Charrier-Mojtabi MC, Vafai K. Non-Darcian effects on double-diffusive convection within a porous medium. Numer Heat Transfer A: Appl 31:837–852, 1997.

Karimi-Fard M, Charrier Mojtabi MC, Mojtabi A. Analytical and numerical simulation of double-diffusive convection in a tilted cavity filled with porous medium. Proceedings of 11th International Heat Transfer Conference, Kyongju, 1998, Vol 4, pp 453–458.

Kaviany M. Non-Darcian effects on natural convection in porous media confined between horizontal cylinders. Int J Heat Mass Transfer 29:1513–1519, 1986.

Kaviany M. Principles of Heat Transfer in Porous Media. 1st ed. New York: Springer-Verlag, 1991.

Kimura S, Masuda Y, Kazuo Hayashi T. Natural convection in an anisotropic porous medium heated from the side (Effects of anisotropic properties of porous matrix). Heat Transfer Jpn Res 22:139–153, 1993.

Lage JL. Comparison between the Forchheimer and the convective inertia terms for Bérnard convection with a fluid saturated porous medium. Proceedings of 28th National Heat Transfer Conference and Exhibition, San Diego, 1992, ASME HTD-193, pp 49–55.

Lai FC, Kulacki FA. Natural convection across a vertical layered porous cavity. Int J Heat Mass Transfer 31:1247–1260, 1988.

Lai FC, Kulacki FA. Experimental study of free and mixed convection in horizontal porous layers locally heated from below. Int J Heat Mass Transfer 34:525–541, 1991a.

Lai FC, Kulacki FA. Coupled heat and mass transfer by natural convection from vertical surfaces in porous media. Int J Heat Mass Transfer 34:1189–1194, 1991b.

Lauriat G, Prasad V. Natural convection in a vertical cavity: a numerical study for Brinkman-extended Darcy formulation. Proceedings of AIAA/ASME 4th Thermophysics and Heat Transfer Conference, Boston, 1986, pp 13–22.

Lauriat G, Prasad V. Non-Darcian effects on natural convection in a vertical porous enclosure. Int J Heat Mass Transfer 32:2135–2148, 1989.

Llagostera J, Figueiredo JR. Natural convection in porous cavity: application of UNIFAES discretization scheme. Proceedings of 11th International Heat Transfer Conference, Kyongju, 1998, pp. 465–470.

McKibbin R, O'Sullivan MJ. Onset of convection in a layered porous medium heated from below. J. Fluid Mech 96:375–393, 1980.

McKibbin R, O'Sullivan MJ. Heat transfer in a layered porous medium heated from below. J Fluid Mech 111:141–173, 1981.

McKibbin R, Tyvand PA. Anisotropic modeling of thermal convection in multilayered porous media. J Fluid Mech 118:315–339, 1982.

McKibbin R, Tyvand PA. Thermal convection in a porous medium composed of alternating thick and thin layers. Int J Heat Mass Transfer 26:761–780, 1983.

McKibbin R, Tyvand PA. Thermal convection in a porous medium with horizontal cracks. Int J Heat Mass Transfer 27:1007–1023, 1984.

Mamou M, Vasseur P, Bilgen E, Gobin D. Double-diffusive convection in a shallow porous layer. Proceedings of 10th International Heat Transfer Conference, Brighton, 1994, Vol 5, pp 339–344.

Mamou M, Vasseur P, Bilgen E, Gobin D. Double-diffusive convection in an inclined slot filled with porous medium. Eur J Mech B: Fluids 14:629–652, 1995a.

Mamou M, Vasseur P, Bilgen E. Multiple solutions for double-diffusive convection in a vertical porous enclosure. Int J Heat Mass Transfer 38:1787–1798, 1995b.

Mamou M, Vasseur P, Bilgen E. Double-diffusive convection instability in a vertical porous enclosure. J Fluid Mech 368:263–289, 1998a.

Mamou M, Vasseur P, Bilgen E. A Galerkin finite-element study of the onset of double-diffusive convection in an inclined porous enclosure. Int J Heat Mass Transfer 41:1513–1529, 1998b.

Manole DM, Lage JL. Numerical benchmark results for natural convection in a porous medium cavity. Proceedings of ASME Winter Annual Meeting, Anaheim, 1992, ASME HTD-216, pp 55–60.

Manole DM, Lage JL. Nonuniform grid accuracy test applied to the natural-convection flow within a porous medium cavity. Numer Heat Transfer B: Fundam 23:351–368, 1993.

Marpu DR. Forchheimer and Brinkman extended Darcy flow model on natural convection in a vertical cylindrical porous annulus. Acta Mech 109:41–487, 1995.

Masouka T, Katsuhara T, Nakazono Y, Isozaki S. Onset of convection and flow patterns in a porous layer of two different media. Heat Transfer: Jpn Res 7: 39–52, 1978.

Mbaye M, Bilgen E, Vasseur P. Natural convection heat transfer in an inclined porous layer boarded by a finite thickness wall. Int J Heat Fluid Flow 14:284–291, 1993.

Mihir S, Vasseur P, Robillard L. Multiple steady states for unicellular natural convection in an inclined porous layer. Int J Heat Mass Transfer 30:2097–2113, 1987.

Mihir S, Vasseur P, Robillard L. Parallel flow convection in a tilted two-dimensional porous layer heated from all sides. Phys Fluids 31:3480–3487, 1988.

Misra D, Sarkar A. A comparative study of porous media models in a differentially heated square cavity using a finite element method. Int J Numer Methods Heat Fluid Flow 5:735–752, 1995.

Muralidhar K, Kulacki FA. Non-Darcy natural convection in a saturated horizontal porous annulus. Proceedings of AIAA/ASME Fourth Thermophysics and Heat Transfer Conference, Boston, 1986, pp 23–31.

Nag A, Sarkar A, Sastri VMK. On the effect of porous thick horizontal partial partition attached to one of the active walls of a differentially heated square cavity. Int J Numer Methods Heat Fluid Flow 4:399–411, 1994.

Naylor D, Oosthuizen PH. Free convection in an enclosure partly filled with a porous medium and partially heated from below. Proceedings of 10th International Heat Transfer Conference, Brighton, 1994, Vol 5, pp 351–356.

Naylor D, Oosthuizen PH. Free convection in a horizontal enclosure partly filled with a porous medium. J Thermophys Heat Transfer 9:797–800, 1995.

Naylor D, Oosthuizen PH. Natural convective heat transfer in an enclosure partly filled with a non-porous insulation. Int J Numer Methods Heat Fluid Flow 6:37–48, 1996a.

Naylor D, Oosthuizen PH. A numerical study of the optimum thermal insulation of air spaces. Proceedings of CSME Forum, Hamilton, 1996b, pp 183–190.

Nguyen HD, Paik S, Douglass RW. Study of double-diffusive convection in layered anisotropic porous media. Numer Heat Transfer B: Fundam 26:489–505, 1994.

Ni J, Beckerman C. Natural convection in a vertical enclosure filled with an anisotropic porous media. J Heat Transfer 113:1033–1037, 1991.

Nield DA. Onset of thermohaline convection in a porous medium. Water Resour Res 4:551–560, 1968.

Nield, DA, Bejan A. Convection in Porous Media. 1st ed. New York: Springer-Verlag, 1992.

Nithiarasu P, Seetharamu KN, Sundararajan T. Double diffusive natural convection in an enclosure filled with fluid-saturated porous medium: a generalized non-Darcy approach. Numer Heat Transfer A: Appl 30:413–426, 1996.

Nithiarasu P, Sundararajan T, Seetharamu KN. Double-diffusive natural convection in a fluid saturated porous cavity with a freely convecting wall. Int Commun Heat Mass Transfer 24:1121–1130, 1997a.

Nithiarasu P, Seetharamu KN, Sundararajan T. Non-Darcy double-diffusive natural convection in axisymmetric fluid saturated porous cavities. Heat Mass Transfer (Waerme und Stoffuebertragung) 32:427–433, 1997b.

Oosthuizen PH. Natural convection in an inclined partitioned square cavity half-filled with a porous medium. ASME National Heat Transfer Conference, Pittsburgh, 1987, ASME 87-HT-15.

Oosthuizen PH. Natural convection in an inclined square enclosure partly filled with a porous medium and with a partially heated wall. Proceedings of ASME Energy Sources Technology Conference and Exhibition, Houston, 1995, ASME HTD-302, pp 29–42.

Oosthuizen PH. Natural convection in a square enclosure partly filled with two layers of porous material. Proceedings of 4th International Conference on Advanced Computational Methods in Heat Transfer, Udine, 1996, pp 63–72.

Oosthuizen PH, Naylor D. A numerical study of natural convective heat transfer from a cylinder in an enclosure partly filled with a porous medium. Proceedings of 30th National Heat Transfer Conference, Portland, 1995, ASME HTD-309, Vol. 7, pp 57–64.

Oosthuizen PH, Naylor D. Natural convective heat transfer from a cylinder in an enclosure partly filled with a porous medium. Int J Numer Methods Heat Fluid Flow 6:51–63, 1996a.

Oosthuizen PH, Naylor D. A numerical study of natural convective heat transfer from a cylinder buried in a porous medium that partly fills an enclosure. Proceedings of CSME Forum, Hamilton, 1996b, pp 160–166.

Oosthuizen PH, Naylor D. An Introduction to Convective Heat Transfer Analysis. 1st ed. New York: McGraw-Hill, 1998.

Oosthuizen PH, Nguyen TH. Maximum density effects on heat transfer from a heated cylinder buried in a porous medium filled enclosure with a cool top. Proceedings of 3rd UK National Heat Transfer Conference incorporating 1st European Conference on Thermal Sciences, Birmingham, 1992, Vol 2, pp 1123–1129.

Oosthuizen PH, Paul JT. Natural convective flow in a square cavity partly filled with a porous medium. Proceedings ASME/JSME Thermal Engineering Joint Conference, Honolulu, 1987, Vol 2, pp 407–412.

Oosthuizen PH, Paul JT. Maximum density effects on heat transfer from a heated cylinder buried in a frozen porous medium in an enclosure. Proceedings of ASME Winter Annual Meeting, Atlanta, 1991, ASME HTD-177, pp 13–23.

Oosthuizen PH, Paul JT. Natural convection in a rectangular enclosure with a partially heated wall and partly filled with a porous medium. Proceedings of 8th International Conference on Numerical Methods in Thermal Problems, Swansea, 1993, Vol VIII, Part I, pp 467–478.

Oosthuizen PH, Paul JT. Natural convection in a square enclosure partly filled with two layers of porous material. Proceedings of 4th International Conference on Advanced Computational Methods in Heat Transfer, Udine, pp 63–71, 1996a.

Oosthuizen PH, Paul JT. Natural convection in a square enclosure partly filled with a centrally positioned porous layer and with a partially heated wall. Proceedings of 2nd Thermal-Sciences and 14th UIT National Heat Transfer Conference, Rome, 1996b, Vol 2, pp 851–856.

Palm, E. Tyvand A. Thermal convection in a rotating porous layer. J Appl Math Phys (ZAMP) 35:122–123, 1984.

Pan CP, Lai FC. Reexamination of natural convection in a horizontal layered porous annulus. J Heat Transfer 118:990–992, 1996.

Poulikakos D. Maximum density effects on natural convection in a porous layer differentially heated in the horizontal direction. Int J Heat Mass Transfer 27:2067–2075, 1984.

Poulikakos D, Bejan A. Numerical study of transient high Rayleigh number convection in an attic-shaped porous layer. J Heat Transfer 105:476–484, 1983a.

Poulikakos D, Bejan A. Natural convection in vertically and horizontally layered porous media headed from the side. Int J Heat Mass Transfer 26:1805–1814, 1983b.

Poulikakos D, Bejan A. The departure from Darcy flow in natural convection in a vertical porous layer. Phys Fluid 28:3477–3483, 1985.

Prasad V, Kulacki FA. Convective heat transfer in a rectangular porous cavity—effect of aspect ratio on flow structure and heat transfer. J Heat Transfer 106:158–165, 1984.

Prasad V, Tuntomo A. Inertial effects on natural convection in a vertical porous cavity. Numer Heat Transfer 11:295–320, 1987.

Rajen G, Kulacki FA. Vibration-induced thermal convection in fluid-saturated porous media. Proceedings of 28th National Heat Transfer Conference and Exhibition, San Diego, 1992, ASME HTD-193, pp 87–90.

Rana R, Horne RN, Cheng P. Natural convection in a multi-layered geothermal reservoir. J Heat Transfer 101:411–416, 1979.

Rao YF, Fukuda K, Hasegawa S. Flow patterns of natural convection in horizontal cylindrical annuli. Int J Heat Mass Transfer 28:705–714, 1985.

Rao YF, Fukuda K, Hasegawa S. Steady and transient analyses of natural convection in a horizontal porous annulus with the Galerkin method. J Heat Transfer 109:919–927, 1987.

Rao YF, Fukuda K, Hasegawa S. A numerical study of three-dimensional natural convection in a horizontal porous annulus with the Galerkin method. Int J Heat Mass Transfer 31:695–707, 1988.

Rao YF, Glakpe EK. Natural convection in a vertical slot filled with porous medium. Int J Heat Fluid Flow 13:97–99, 1992.

Rees DAS, Bassom AP. The onset of convection in an inclined porous layer heated from below. Proceedings of 11th International Heat Transfer Conference, Kyongju, 1998, pp 497–502.

Rees DAS, Riley DS. The three-dimensional stability of finite-amplitude convection in a layered porous medium heated from below. J Fluid Mech 211:437–461, 1990.

Rudraiah N, Shivakumara IS, Friedrich R. The effect of rotation on linear and nonlinear double diffusive convection in sparsely packed porous medium. Int J Heat Mass Transfer 29:1301–1317, 1986.

Sai BVK Satya, Seetharamu KN, Aswathanarayana PA. Finite element analysis of heat transfer by natural convection in porous media in vertical enclosures: Investigations in Darcy and non-Darcy regimes. Int J Numer Methods Heat Fluid Flow 7:367–400, 1997.

Sasaguchi K. Effect of density inversion of water on the melting process of frozen porous media. Proceedings of 1995 ASME/JSME Thermal Engineering Joint conference, Maui, 1995, Vol 3, pp 371–378.

Sasaki A, Aiba S, Fukusako S. Freezing heat transfer within water-saturated porous media. JSME Int J Ser 2 33:296–304, 1990a.

Sasaki A, Aiba S, Fukusako S. Numerical study on freezing heat transfer in water-saturated porous media. Numer Heat Transfer A: Appl 18:17–32, 1990b.

Seki N, Fukusako S, Inaba H. Heat transfer in a confined rectangular cavity packed with porous media. Int J Heat Mass Transfer 21:985–989, 1978.

Simpkins PG, Blythe PA. Convection in a porous layer. Int J Heat Mass Transfer 23:881–887, 1980.

Song M, Viskanta R. Natural convection flow and heat transfer between a fluid layer and an anisotropic porous layer within a rectangular enclosure. Proceedings of ASME Winter Annual Meeting, Atlanta, 1991, ASME HTD-177, pp 1–12.

Song M, Viskanta R. Natural convection flow and heat transfer within a rectangular enclosure containing a vertical porous layer. Int J Heat Mass Transfer 37:2425–2438, 1994.

Tong TW, Subramanian E. A boundary-layer analysis for natural convection in vertical porous enclosures—use of the Brinkman-extended Darcy model. Int J Heat Mass Transfer 28:563–571, 1985.

Trevisan OV, Bejan A. Natural convection combined heat and mass transfer buoyancy effects in a porous medium. Int J Heat Mass Transfer 28:1597–1611, 1985.

Trevisan OV, Bejan A. Mass and heat transfer by natural convection in a vertical slot filled with a porous medium. Int J Heat Mass Transfer 29:403–415, 1986.

Trevisan O, Bejan A. Combined heat and mass transfer by natural convection in porous medium. Adv Heat Transfer 20:315–352, 1990.

Vadasz P. Three-dimensional free convection in a long rotating porous box: analytical solution. J Heat Transfer 115:639-644, 1993.

Vadasz P. Fundamentals of flow and heat transfer in rotating porous media. Proceedings of 10th International Heat Transfer Conference, Brighton, 1994, Vol 5, pp 405–410.

Vafai K, Tien CL. Boundary and inertia effects on flow and heat transfer in porous media. Int J Heat Mass Transfer 24:195–203, 1981.

Vasseur P, Robillard L. The Brinkman model for boundary layer regime in a rectangular cavity with uniform heat flux from the side. Int J Heat Mass Transfer 30:717–727, 1987.

Vasseur P, Wang CH, Mihir S. The Brinkman model for natural convection in a shallow porous cavity with uniform heat flux. Numer Heat Transfer A: 15:221–242, 1989.

Vasseur P, Wang CH, Sen M. Natural convection in an inclined rectangular porous slot. The Brinkman-extended Darcy model. J Heat Transfer 112:507–511, 1990.

Walker KL, Homsy GM. Convection in a porous cavity. J Fluid Mech 87:449–474, 1978.

Wang BX, Zhang X. Natural convection in liquid-saturated porous media between concentric inclined cylinders. Int J Heat Mass Transfer 33:827–833, 1990.

Weber JE. The boundary layer regime for convection in a vertical porous layer. Int J Heat Mass Transfer 18:569–573, 1975.

Yen YC. Effects of density inversion on free convective heat transfer in porous layer heated from below. Int J Heat Mass Transfer 17:1349–1356, 1974.

Zhang X. Natural convection in an inclined water-saturated porous cavity: the duality of solutions. Proceedings of ASME Winter Annual Meeting, Anaheim, 1992, ASME HTD-216, pp 47–54.

Zhang X, Nguyen TH, Kahawita R. Connective flow and heat transfer in an anisotropic porous layer with principal axes non-coincident with the gravity vector. Proceedings of ASME Winter Annual Meeting, New Orleans, 1993, ASME HTD-264, pp 79–86.

Zhao M, Robillard L, Vasseur P. Mixed convection in a low radiation horizontal cylinder. Int Commun Heat Mass Transfer 25:1031–1041, 1998.

12

The Stability of Darcy–Bénard Convection

D. A. S. Rees
University of Bath, Bath, England

I. INTRODUCTION

Many years have passed since Horton and Rogers (1945) and Lapwood (1948) published their pioneering studies into the onset of convection in porous layers heated from below, the well-known Horton–Rogers–Lapwood or Darcy–Bénard problem. This configuration, a horizontal layer of either finite or infinite extent, has been studied in very great detail and especially so since the weakly nonlinear study of Palm et al. (1972) and the detailed numerical stability analysis of Straus (1974). The main reason for such activity lies in the fact that there are many extensions to the governing Darcy–Boussinesq equations. Commonly cited examples are boundary, inertia, local thermal non-equilibrium, anisotropic, and thermal dispersion effects. The presence of one or more of these or other modifications serves to change the nature of the resulting flow and accounts for the huge research literature. I will refer to such modified problems by the generic term 'Darcy–Bénard', even though further words such as 'Forchheimer' or 'Brinkman' could quite legitimately claim a place in that term.

The Darcy–Bénard problem is striking in its simplicity: a uniform-thickness horizontal porous layer with uniform temperature and impermeable upper and lower surfaces. The macroscopic equations governing the filtration of the fluid in the case where Darcy's law applies, the fluid is Boussinesq, the matrix is rigid and isotropic, and the solid and fluid phases are in local thermal equilibrium, are so straightforward that the linearized stability analysis proceeds analytically within the space of a few lines. Even the associated weakly-nonlinear analysis may be undertaken analytically within a few pages. However, this is where the simplicity ends. More complicated scenarios either involve the use of numerical methods or sentence the researcher to heavy-duty algebraic work which, though tractable, will nevertheless take much longer than a directly numerical approach!

The present author is very aware of how much space was set aside in the excellent book by Nield and Bejan (1992) to cover the topic of Darcy–Bénard convection, even though the treatment given even there could not have been comprehensive, especially in terms of presenting detailed results. Indeed, this topic is now sufficiently well advanced and mature to merit a book of its own, much as Koschmeider's (1993) book describes the analogous subject of Bénard convection and takes a more pedagogical stance. With this in mind, it has been necessary to restrict the scope of this chapter so that some material other than references and the author's name and affiliation should fit into the allotted space! Therefore the present review will concentrate exclusively on flows in horizontal layers, rather than on inclined or vertical layers, and it will avoid multi-diffusive flows. The important topic of convection in rotating systems has been dealt with admirably in two very recent reviews by Vadasz (1997, 1998), and therefore little will be mentioned here. The author hopes that offense will not be taken by those whose papers have not been cited.

It seems appropriate to split the review into two main sections, the first of which describes briefly the classical Darcy–Bénard problem where no extensions to Darcy's law or "variations on a theme" are included. This is organized systematically and describes in turn linearized theory, weakly nonlinear theory and strongly nonlinear computations. The second section deals with those aspects which arise from properties of the porous medium (e.g., boundary and inertia effects, anisotropy, and thermal dispersion), arise from fluid properties alone (e.g. temperature-dependent viscosity and chemical reactions), and external influences such as unsteady heating, different types of thermal boundary conditions, and mixed convective effects. These are arranged according to the different type of extension considered. The majority of papers cover various aspects of linearized theory, the onset problem. Decreasing numbers deal in turn with weakly nonlinear theory and strongly nonlinear flows.

II. THE CLASSICAL DARCY–BÉNARD PROBLEM

A. Linearized Theory

In apparently independent investigations, Horton and Rogers (1945) and Lapwood (1948) considered the onset of convection in a horizontal saturated porous layer heated from below. The two bounding surfaces are held at uniform temperatures (i.e., infinitely or perfectly conducting), the flow is assumed to be governed by Darcy's law, and the Boussinesq approximation is taken to apply. If the dimensional horizontal coordinates are $\hat{x}$ and $\hat{z}$ and the dimensional upwards vertical coordinate is $\hat{y}$, then the governing equations of motion are

$$\frac{\partial \hat{u}}{\partial \hat{x}} + \frac{\partial \hat{v}}{\partial \hat{y}} + \frac{\partial \hat{w}}{\partial \hat{z}} = 0 \tag{1}$$

$$(\hat{u}, \hat{v}, \hat{w}) = -\frac{K}{\mu}\nabla\hat{p} + \frac{\rho\tilde{g}\beta K(T - T_c)}{\mu}(0, 1, 0) \tag{2}$$

$$\sigma\frac{\partial T}{\partial \hat{t}} + \hat{u}\frac{\partial T}{\partial \hat{x}} + \hat{v}\frac{\partial T}{\partial \hat{y}} + \hat{w}\frac{\partial T}{\partial \hat{z}} = \kappa\left(\frac{\partial^2 T}{\partial \hat{x}^2} + \frac{\partial^2 T}{\partial \hat{y}^2} + \frac{\partial^2 T}{\partial \hat{z}^2}\right) \tag{3}$$

where all the terms have their usual meanings: K is permeability, ρ is a reference density, β the coefficient of cubical expansion, μ the fluid viscosity, and

$$\sigma = \frac{(1-\phi)(\rho c)_s + \phi(\rho c)_f}{(\rho c)_f} \tag{4}$$

is the heat capacity ratio of the saturated medium to that of the fluid. Hence c is the heat capacity and ϕ is the porosity of the medium. These equations may be nondimensionalized, using the following substitutions

$$(x, y, z) = (\hat{x}, \hat{y}, \hat{z})/d, \qquad (u, v, w) = \frac{d}{\kappa}(\hat{u}, \hat{v}, \hat{w}) \tag{5}$$

$$p = \frac{K}{\kappa\mu}\hat{p}, \qquad t = \frac{\kappa}{d^2\sigma}\hat{t}, \qquad \theta = \frac{T - T_c}{T_h - T_c} \tag{6}$$

where T_c and T_h are the upper (cold) and lower (hot) boundary temperatures, respectively, and d is the uniform depth of the layer. Equations (1)–(3) become

$$\frac{\partial u}{\partial x} + \frac{\partial v}{\partial y} + \frac{\partial w}{\partial z} = 0 \tag{7}$$

$$(u, v, w) = -\nabla p + Ra(0, \theta, 0) \tag{8}$$

$$\sigma\frac{\partial \theta}{\partial t} + u\frac{\partial \theta}{\partial x} + v\frac{\partial \theta}{\partial y} + w\frac{\partial \theta}{\partial z} = \frac{\partial^2 \theta}{\partial x^2} + \frac{\partial^2 \theta}{\partial y^2} + \frac{\partial^2 \theta}{\partial z^2} \tag{9}$$

where

$$Ra = \frac{\rho \tilde{g} \beta d K (T_h - T_c)}{\mu \kappa} \tag{10}$$

is the Darcy–Rayleigh number; this value plays the same role in porous convection as does the usual Rayleigh number for convecting clear fluids, as it expresses a balance between buoyancy and viscous forces. These equations are to be solved subject to the boundary conditions

$$v = 0, \qquad \theta = 1 - y \qquad \text{at} \qquad y = 0, 1 \tag{11}$$

and it is straightforward to verify that

$$u = v = w = 0, \qquad p_y = Ra(1 - y), \qquad \theta = 1 - y \tag{12}$$

is a solution of equations (7)–(9); this is referred to as the conduction solution. The topic of this chapter is the stability and evolution of disturbances to this and similar basic profiles.

The determination of conditions governing the onset of convection is *usually* undertaken using a linearized theory. Here the solution is perturbed about the basic solution and then linearized. A cellular convective planform is typically assumed for the disturbance profile which involves one or two wavenumbers, and a critical Rayleigh number is derived by solving an ordinary or partial differential eigenvalue problem. In the present case, the onset of two-dimensional disturbances proceeds initially by defining the streamfunction ψ according to $u = \psi_y$, $v = -\psi_x$, $w = 0$. If we also subtract out the basic profile (for which $\psi = 0$) and linearize, there results the system

$$\psi_{xx} + \psi_{yy} = Ra\theta_x, \qquad \theta_t = \theta_{xx} + \theta_{yy} + \psi_x \tag{13}$$

subject to $\psi = 0$ and $\theta = 0$ at $y = 0, 1$. If we assume that convection consists of rolls in the x-direction with wavenumber k, then the substitution

$$\psi = e^{\lambda t} f(y) \sin kx, \qquad \theta = e^{\lambda t} g(y) \cos kx \tag{14}$$

yields

$$f'' - k^2 f + kg = 0, \qquad \lambda g = g'' - k^2 g + kf = 0 \tag{15}$$

subject to $f = g = 0$ at $y = 0, 1$. Here λ is the exponential growth rate. When $\lambda = 0$, disturbances are neutrally stable and neither grow nor decay, and the solution of (15) yields the criterion for the onset of convection. For the present classical problem, the solutions are, simply, that

$$f(y) = A(k^2 + \pi^2) \sin \pi y, \qquad g(y) = Ak \sin \pi y, \qquad Ra = \frac{(k^2 + \pi^2)^2}{k^2} \tag{16}$$

where A is an arbitrary amplitude. A graph of Ra against k is depicted in Figure 1. It is easily shown that the minimum value is $Ra = 4\pi^2$ and this occurs when $k = \pi$, which corresponds to rolls with square cross-section. In Eq. (14) it has been assumed that exchange of stabilities holds (see Drazin and Reid 1981 for details of this concept), which is true for this flow and means that $\lambda = 0$ corresponds to neutral stability. Otherwise, it would be necessary to represent the x-dependencies in (14) in complex exponential form and impose $Re(\lambda) = 0$ for neutral stability; in such cases nonzero values of the imaginary part of λ would correspond to a translational wave-speed of the disturbance. The neutral curve given in Figure 1 should be interpreted as meaning that infinitesimal disturbances will decay whenever both the Rayleigh number and the wavenumber correspond to points below the curve.

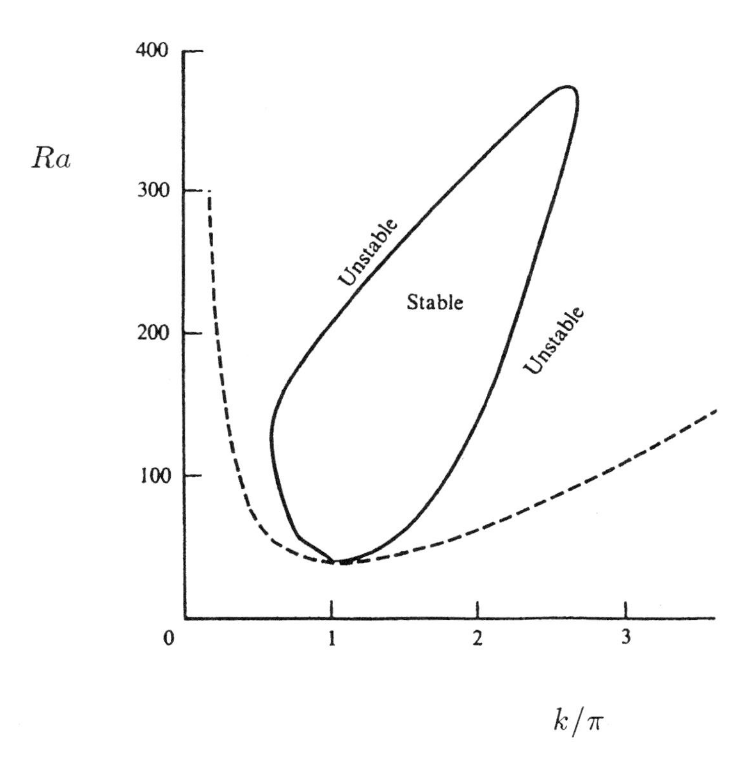

Figure 1. Regions of stable and unstable rolls. The dashed line is the neutral stability curve for the onset of convection and the solid line, from Straus (1974), delineates the region of stable steady rolls. Reproduced by permission of Cambridge University Press.

The above analysis covers only that case with perfectly conducting, impermeable bounding surfaces. Other cases involve different combinations of impermeable (IMP) and constant pressure (FRE), and conducting (CON) and constant heat flux (CHF) for both surfaces (but note that Nield and Bejan (1992) refer to the CHF case as "insulating" and label it INS). Values of the critical Rayleigh number and the associated wavenumber are given in Table 1. It is interesting to note that when a constant heat flux is specified on both boundaries the preferred wavenumber is zero; the only other case where this happens is when both perfectly conducting surfaces are at constant pressure. It is to be presumed that a smooth transition takes place when any thermal boundary condition is varied smoothly from one extreme case to the another. Finally, we note that not all the theoretical cases presented in Table 1 form practical configurations for experimental work.

B. Weakly Nonlinear Analyses

Linearized analyses only yield information on the conditions for which small-amplitude disturbances have a zero growth rate. They do not give information on whether the flow above the critical Rayleigh number is stable, on whether there exists flow at subcritical Rayleigh numbers, or to decide between competing instabilities. These are the roles, in turn, of weakly nonlinear and strongly nonlinear studies. Weakly nonlinear analyses are asymptotic analyses and are therefore valid asymptotically close to the point of onset given by linearized theory. However, much of the qualitative nature of their results carries over to the strongly nonlinear regime.

Table 1. Values of the critical Rayleigh number and wavenumber for various combinations of boundary conditions. After Nield (1968)

Upper surface		Lower surface		Ra_c	k_c
IMP	CON	IMP	CON	$4\pi^2$	π
IMP	CON	IMP	CHF	27.10	2.33
IMP	CHF	IMP	CHF	12	0
IMP	CON	FRE	CON	27.10	2.33
IMP	CHF	FRE	CON	17.65	1.75
IMP	CON	FRE	CHF	π^2	$\pi/2$
IMP	CHF	FRE	CHF	3	0
FRE	CON	FRE	CON	12	0
FRE	CON	FRE	CHF	3	0
FRE	CHF	FRE	CHF	0	0

The seminal paper by Palm et al. (1972) concentrates on the supercritical flow and heat transfer of unit-aspect-ratio convection using a power series expansion in $(1 - Ra_c/Ra)^{1/2}$. Steady convection at Rayleigh numbers up to about 5 times the critical value is described well using this power series, even though it is, strictly speaking, an asymptotic analysis.

Although a rigorous weakly nonlinear analysis for Bénard convection by Newell and Whitehead (1969) was published thirty years ago, the first journal paper to appear from which the analogous Darcy–Bénard stability results may be inferred is Rees and Riley (1989a) (although Joseph (1976) reports the results directly). In these types of analysis, Ra is assumed to be within $O(\varepsilon^2)$ of $Ra_c = 4\pi^2$ and k within $O(\varepsilon)$ of $k_c = \pi$. The temperature field, for example, and the Rayleigh number are expanded in power series in ε as follows

$$\theta = \theta_0 + \varepsilon\theta_1 + \varepsilon^2\theta_2 + \varepsilon^3\theta_3 + \ldots, \qquad Ra = R_0 + \varepsilon^2 R_2 + \ldots \tag{17}$$

where $R_0 = 4\pi^2$. Here θ_0 is the conduction solution and θ_1 comprises one or more disturbances satisfying the linearized equations of motion. Normally the $O(\varepsilon^2)$ equations may be solved fairly easily, either analytically or numerically. However, the nonlinear interactions between the modes introduced at $O(\varepsilon)$ give rise to generally insoluble equations at $O(\varepsilon^3)$ since the inhomogeneous terms contain components which are proportional to solutions of the corresponding homogeneous equations. A solvability condition may then be derived by choosing an appropriate value of R_2. For example, if we set $\theta_1 = \pi^{-2}A(\tau)\cos\pi x \sin\pi y$ in Eq. (16), where $\tau = \frac{1}{2}\varepsilon^2 t$ is a slow timescale, then the solvability condition is

$$A_\tau = R_2 A - A^3 \tag{18}$$

It is straightforward to show that $A = 0$ is the only steady solution when $R_2 < 0$, which corresponds to $Ra < Ra_c$, and that this solution is unstable when $R_2 > 0$. For $R_2 > 0$ there also exist the solutions $A = \pm R_2^{1/2}$. More complicated stability analyses involve allowing for small wavenumber changes, small changes in orientation of the basic roll, and the inclusion of other rolls. Such stability analyses are known, respectively, as the Eckhaus, zigzag, and cross-roll instabilities. The first two may be accommodated by defining slow x and z variables: $X = \varepsilon x$ and $Z = \varepsilon^{1/2} z$. A three-dimensional analysis yields the amplitude equation

$$A_\tau = R_2 A + \left(2\frac{\partial}{\partial X} - \frac{i}{\pi}\frac{\partial^2}{\partial Z^2}\right)^2 A - A^2\bar{A} \tag{19}$$

Details of the ensuing analysis may be gleaned from Rees and Riley (1989a) and more directly from Rees (1996). In short, if the basic solution takes the form of

$$A = (R_2 - 4K^2)e^{iKX} \tag{20}$$

this corresponds to a roll with a wavenumber $\pi + \varepsilon K$. Solutions with $K < 0$ are unstable to zigzag disturbances, whereas solutions with $4K^2 < R_2 < 12K^2$ are subject to the Eckhaus instability. In the former case two rolls aligned at equal but opposite small angles to the original roll grow at the expense of the basic roll, forming a sinuous roll boundary. One of these two rolls eventually dominates, leaving a stable steady-state configuration. In the latter case a pair of perfectly aligned rolls with wavenumbers either side of the original roll grow, again with one dominating whose wavenumber lies within the stable domain.

Finally, the cross-roll instability may be studied by considering a roll at a finite angle χ to the basic roll; if we let

$$\theta_1 = \pi^{-2} A \cos \pi x \sin \pi y + \pi^{-2} B \cos \pi (x \cos \chi - z \sin \chi) \sin \pi y \tag{21}$$

then A and B may be shown to satisfy the equations

$$A_\tau = R_2 A + 4A_{XX} - A[A\bar{A} + \Omega(\chi) B\bar{B}] \tag{22a}$$

$$B_\tau = R_2 B + 4A_{X_B X_B} - B[B\bar{B} + \Omega(\chi) A\bar{A}] \tag{22b}$$

Here we have suppressed Z-variations in A, and $X_B = \varepsilon(\cos \chi - z \sin \chi)$ is the slow variable perpendicular to the direction of the B-roll. A general analysis of (22) shows that single rolls are the favored mode when $\Omega(\chi) > 1$ for all values of χ, whereas square or rectangular cells are preferred if the minimum value of $\Omega(\chi)$ is less than 1. For the present problem we have

$$\Omega(\chi) = \frac{70 + 28 \cos^2 \chi - 2 \cos^4 \chi}{49 - 2 \cos^2 \chi + \cos^4 \chi} \tag{23}$$

which varies monotonically from 2 at $\chi = 0$ to its minimum, 10/7 at $\chi = \pi/2$ (Rees and Riley 1989b). Therefore rolls are preferred, and it may be shown that the roll given by (20) is stable when $R_2 > 40K^2/3$. This criterion is more restrictive than the Eckhaus criterion, and therefore there exists a band of stable wavenumbers lying between $K = 0$ and $K = +\sqrt{3R_2/40}$, the corresponding domain of existence of the basic roll being $|K| \leq \sqrt{R_2/4}$. We note that Riahi (1983) considered the same problem but with finitely conducting boundaries. There is a region in parameter-space wherein square cells are preferred.

As mentioned earlier, a weakly nonlinear analysis is an asymptotic analysis and it is therefore valid only in the $\varepsilon \rightarrow 0$ limit. The present analysis corresponds to the bottom of the Straus (1974) stability envelope shown in Figure 1 and which is discussed later. We note that Joseph (1976) presents a more accurate version of the zigzag instability. Similar weakly nonlinear analyses have been undertaken for more complicated versions of the classical Darcy–Bénard problem and a wide variety of phenomena ensue, but we defer discussion of these cases to the next section. Energy stability analyses also exist for the onset of convection, but for the classical Darcy–Bénard problem they yield identical results to those of linearized stability theory.

C. Strongly Nonlinear Convection

One advantage of weakly nonlinear theory is that layers of infinite extent may be considered easily. When the strongly nonlinear regime is encountered, progress is gained through numerical computation. Although spectral methods may, by their nature, be said to apply to an infinite domain, finite difference or finite element techniques must be applied to a finite computational domain. When Darcy flow is considered and sidewalls are taken to be insulated and impermeable, then sidewalls coincide with planes of symmetry in the infinite domain and much may then be said about the infinite case, even though whole classes of possible disturbances are absent due to the presence of the side walls.

At Rayleigh numbers slightly higher than that given by linearized theory, the realized flow is steady and two-dimensional. Straus (1974) undertook a detailed stability analysis of this strongly nonlinear convection and determined the whole regime in *Ra–k* space within which convection is stable; this region is displayed in Figure 1. It is important to be able to interpret correctly the information presented in this figure. First, it is essential to note that this represents only the stable regime; it does not give any indication as to the relative probabilities of different stable flows given a random initial condition in a time-dependent simulation. Second, there is no information as to what the final stable planform will be, for this depends on the initial disturbance; on the right-hand part of the stability envelope we know that the most dangerous disturbance is one aligned at right-angles to the basic roll, but the graph does not give the wavenumber of that disturbance. Finally, when *Ra* is larger than the maximum value on that neutral curve, the resulting flow may not be deduced without further knowledge—one could guess that it is three-dimensional, but not whether it is steady.

A long series of papers by Straus, Schubert, Steen and others have sought to extend our knowledge of the flow at increasing Rayleigh numbers; see Holst and Aziz (1972), Horne and O'Sullivan (1974), Horne (1979),

Straus and Schubert (1979, 1981), Schubert and Straus (1979, 1982), Kimura et al. (1986, 1987), Aidun (1987), Aidun and Steen (1987), Steen and Aidun (1988), and Caltagirone et al. (1987). In a square box with insulating side walls, the primary transition from a motionless state to steady convection takes place when $Ra = 4\pi^2$, and the second transition to a time-periodic flow occurs at roughly $Ra = 391$. This value was determined very accurately by Riley and Winters (1991), who used a finite element steady flow solver and examined the iteration matrix to find when the linearized disturbance equations admit a Hopf bifurcation. They recorded a value of $Ra = 390.72$ 01 which should now be regarded as being definitive. In the same paper Riley and Winters also computed how this second critical value of Ra varies with the box aspect ratio.

At increasing Rayleigh numbers there follows a sequence of transitions between time-periodic flow and aperiodic flow; the precise demarcation of the points of transition are, in the present author's experience, highly dependent on the level of resolution of the numerical scheme. The papers by Kimura et al. (1986, 1987) are devoted to determining this sequence and find that the flow undergoes a short sequence of transitions between periodicity and quasi-periodicity before becoming nonperiodic. A very thorough analysis has been presented by Graham and Steen (1992), who have managed to uncover much of the bifurcation structure of flows at such values of Ra. In particular, they show that quasi-periodic motion sets in at $Ra = 505$ where the periodic flow is destabilized by a disturbance at an incommensurate frequency. Other unsteady flows are computed, points of bifurcation examined with a diagram of possible bifurcation relating each solution to one another and inferring the existence of unstable quasi-periodic flows. Furthermore, this bifurcation structure is shown to be able to explain the observations of Kimura et al. (1986) that there are two stable quasi-periodic solutions in the range $500 < Ra < 560$. Other box aspect ratios are considered by Graham and Steen; in particular, the aspect ratio 2.495 yields a double Hopf bifurcation where two periodic solutions bifurcate off the steady solution branch at the same Rayleigh number.

Other details of modal competition between possible steady-state solutions may be found in Riley and Winters (1989), where attention is focused on how the bifurcation structure varies with the box aspect ratio. In Figure 2 we display a sample bifurcation diagram for a box of unit aspect ratio. At increasing values of Ra after the primary 1-cell destabilization at $Ra = 4\pi^2$, a 2-cell state and then a 3-cell flow bifurcate off the conduction solution, but are initially unstable. The 2-cell solution eventually gains stability and this point is marked by the bifurcation of a mixed-mode branch of the 2-cell branch. A similar scenario occurs for the 3-cell solution, except that two bifurcations to mixed mode solutions must occur before it is stabilized. We

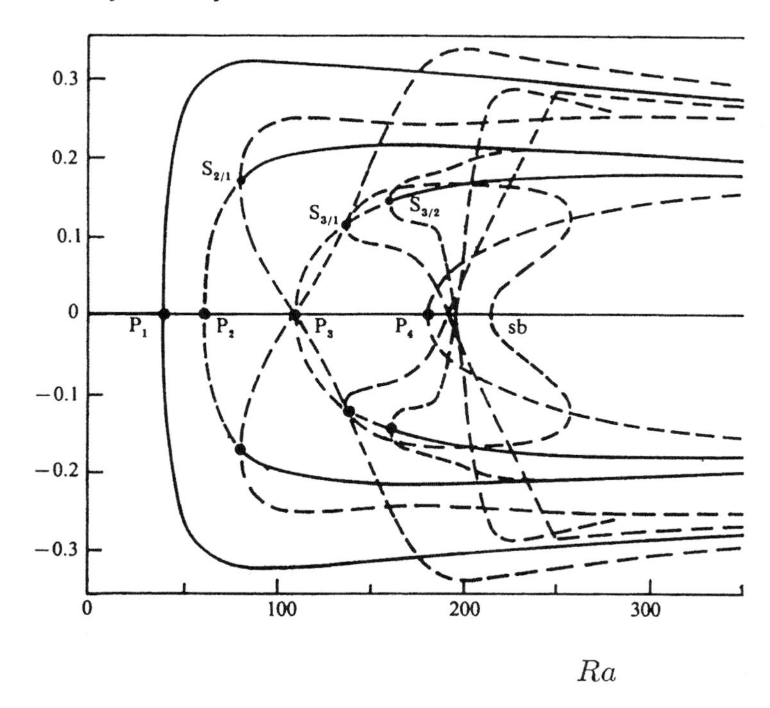

Figure 2. Computed bifurcation structure for two-dimensional flow in a square box. The stable and unstable branches are denoted by full and broken curves respectively. The vertical axis is the temperature at the midpoint of the left-hand sidewall. From Riley and Winters (1989). Reproduced by permission of Cambridge University Press.

note that a similar study of the bifurcation structure in three dimensions, even for a cubic box, has not yet been undertaken, for although three-dimensional flows are unstable in an infinite layer at sufficiently low Rayleigh numbers, the spatial constriction in a cube or a cuboid is sufficient to stabilize such flows; see Steen (1983) and Kimura et al. (1990). A very detailed account of the onset of oscillatory convection is presented in Kimura (1998).

Other early papers are concerned with the effect of a net mass flux through the layer. Sutton (1970) considered vertical flow through permeable boundaries; in this case stabilization (an increased critical Rayleigh number) or destabilization depends on the strength of the throughflow and the aspect ratio of the finite cavity. Prats (1967), on the other hand, studied the effect of a horizontal flow; here the criterion for the onset of convection is unchanged since the Darcy equations retain their precise form when expressed in the frame of reference which is moving with the mean flow. Thus, in an infinite layer, all the above–quoted results of Palm et al. (1972) and of Straus and co-workers are unaffected.

III. VARIATIONS ON THE THEME

A. Inertia Effects

In this subsection we discuss how the presence of fluid inertia (form drag or Forchheimer drag) influences the onset and development of convection. The governing dimensionless equations are given by

$$u_x + v_y + w_z = 0, \qquad \underline{u}(1 + G|\underline{u}|) = -\nabla p + Ra(0, \theta, 0),$$
$$\theta_t + \underline{u} \cdot \nabla\theta = \nabla^2\theta \tag{24}$$

where $G = \tilde{K}\rho\kappa/\mu d$ is an inertia parameter and $\tilde{K}$ is a material parameter related to the microscopic lengthscale of the medium (L) and the porosity (ϕ) via Ergun's (1952) relation

$$\tilde{K} = \frac{1.75L}{150(1-\phi)} \tag{25}$$

for example. Given that the additional term over and above the normal Darcy-flow problem is nonlinear and given that there is no basic flow, it is very straightforward to show that the presence of inertia has no effect on the onset of convection in a horizontal porous layer. This point may also be gleaned from the paper by He and Georgiadis (1990), who also considered the extra effects of internal heating and hydrodynamic dispersion. When there is a basic flow in the porous medium, the presence of inertia serves to modify the stability characteristics from those corresponding to Darcy flow. One example is that of the inclined Darcy–Bénard problem, where unpublished numerical simulations by the present author have shown that convection is delayed by the presence of inertia when compared with that of the corresponding Darcy-flow case given in Rees and Bassom (1998). A more pertinent example is that of the combined effects of a horizontal pressure gradient with inertia, studied by Rees (1998). In this analytical study the author showed that two-dimensional cells (whose axes are perpendicular to the direction of the pressure gradient) have a critical Rayleigh number which rises with both G and the strength Q of the horizontal flow:

$$Ra_c = \pi^2\left[(1 + G|Q|)^{1/2} + (1 + 2G|Q|)^{1/2}\right]^2 \tag{26a}$$

The critical wavenumber is

$$k_c = \pi \left[\frac{1 + 2G|Q|}{1 + G|Q|} \right]^{1/4} \tag{26b}$$

We note that it is the combined effect of inertia and the horizontal pressure gradient which causes this rise; when inertia is absent, Prats (1967) showed that the dynamics even of strongly nonlinear convection are unchanged when viewed in the frame of reference moving with the mean flow. However, when the rolls are aligned with the mean flow direction the critical Rayleigh is $Ra_c = 4\pi^2(1 + G|Q|)$ with $k_c = \pi$, leading to a clear preference for such rolls in an unbounded layer.

A weakly nonlinear stability analysis was undertaken by He and Georgiadis (1990), who found that the standard pitchfork bifurcation is modified to one with straight lines intercepting the zero amplitude axis; this aspect is a consequence of the Forchheimer term. They also found that the positive amplitude flow arises at a different value of Ra from that of negative amplitude flow. This was attributed to the combined effects of inertia, internal heating, and dispersion. More recently, Georgiadis (personal communication, 1996) has indicated that these combined effects actually cause a splitting of the bifurcation "curve" into two, corresponding to different modes of convection, in much the same way as Rees and Riley (1986) found two different modes in the case of Darcy flow in a layer with symmetrical boundary imperfections. The one arising at the lower value of Ra is generally stable in an infinite layer, while the other is unstable Such results are easily proved using the appropriate amplitude equations.

In the work of He and Georgiadis (1990) the imperfections to the classical problem were sufficiently strong to cause the weakly nonlinear analysis to progress only to second order. A suitable third-order analysis was undertaken by Rees (1996), who assumed that inertia effects were weak. The straight line behavior of the solution curves near onset are reproduced, but at higher Rayleigh numbers the usual square root behavior is re-established. This counter-intuitive result, which says that inertia effects are strongest when the flow is weakest and vice versa, is a direct consequence of the fact that the destabilizing effect of buoyancy is balanced at high Rayleigh numbers by the most nonlinear term in the amplitude equation

$$A_\tau = R_2 A + \alpha A|A| - A^2 \bar{A} \tag{27}$$

where α is a scaled inertia parameter. The author also developed a full weakly nonlinear stability analysis and found that inertia causes some wavenumbers less than the critical value to regain stability, but the cross-roll instability is more effective and reduces the stable wavenumber range.

Strongly convecting flow was considered by Strange and Rees (1996), who undertook a finite difference computation of flow in a square cavity.

Using a 32×32 grid and a streamfunction/temperature formulation they found that unsteady convection is delayed by the presence of inertia, and that, for values of G less than 0.05, the critical value is given by

$$Ra = Ra_{c2} + 2.95 \times 10^4 G \tag{28}$$

where Ra_{c2} is the value corresponding to incipient unsteady flow in the absence of inertia. The same qualitative result was found by Kladias and Prasad (1989, 1990), who also used the Brinkman terms in their fundamental model.

B. Brinkman Effects and Advective Inertia

The appropriate momentum equation is now taken to be

$$\begin{aligned} \phi^{-1}(\underline{u} \cdot \nabla)\underline{u} = \nabla p + (Pr/Ra)^{1/2}\nabla^2 \underline{u} - (F\phi Da^{-1})|\underline{u}|\underline{u} \\ - (\phi/Da)(Pr/Ra)^{1/2}\underline{u} + \phi(0, \theta, 0) \end{aligned} \tag{29}$$

(Lage et al. 1992) where the first term is the advective inertia term and the second term on the right-hand side is the Brinkman term. The paper by Lage et al. (1992) discusses at length the role of the Prandtl number on the onset of convection, concluding that the criterion for onset is unaffected by the value of Pr. In this regard these authors show that earlier work by Kladias and Prasad (1989) is incorrect in stating that the criterion is dependent on Pr, and therefore verify much earlier work by Walker and Homsy (1977). In fact, the nondimensionalization used to derive the above equation obscures this result: if we replace all occurrences of $\underline{u}$ by $\underline{u}Da(RaPr)^{-1/2}$, then Eq. (29) becomes

$$\begin{aligned} (Da^2\phi^{-2}Pr^{-1})(\underline{u} \cdot \nabla)\underline{u} = (Ra/\phi)\nabla p + (Da/\phi)\nabla^2 \underline{u} \\ - (FDa/Pr)|\underline{u}|\underline{u} - \underline{u} + (0, \theta, 0) \end{aligned} \tag{30}$$

The Prandtl number is now seen to appear only in nonlinear terms and therefore will not affect linearized analyses based on a no-flow solution. However, we see that the criterion for onset will depend only on (Da/ϕ), the coefficient of the Brinkman term, rather than on Da as suggested by Lage et al. (1992). Walker and Homsy (1977) computed the neutral curve shown in Figure 3 which depicts the transition between Darcy flow when Da is very small and clear fluid flow at sufficiently large values (10^{-3} and greater). The reader is referred to the very detailed introductions in Lage et al. (1992) and Georgiadis and Catton (1988) for a description of earlier work on this topic.

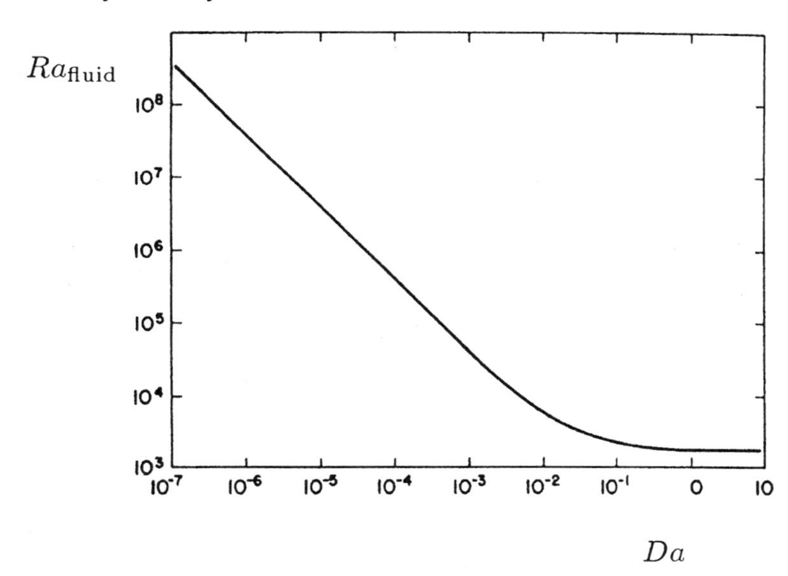

Figure 3. Critical fluid Rayleigh number as a function of the Darcy number. From Walker and Homsy (1977). Reproduced by permission of ASME.

A very recent paper by Néel (1998) considers how a horizontal pressure gradient affects convection in the presence of inertia and Brinkman effects. The results presented therein correspond to a cubic inertia term replacing the usual quadratic term, and therefore they cannot be compared directly with the above papers. In her work, the author does not include advective inertia, but performs a small parameter expansion in terms of an inertia parameter and an equivalent to the Darcy number to show that these effects cause an increase in the critical Rayleigh number. The effects of a horizontal mean flow are also considered.

C. The Effects of Throughflow

Throughflow has been studied in two different types of case: (i) horizontal flow and (ii) vertical flow. We have already discussed briefly the papers by Prats (1967) and Néel (1998) on the effects of horizontal flow, but a further paper by Dufour and Néel (1998) considers Darcy flow in finite horizontal channels. Here various end-wall boundary conditions are imposed and the resulting flow patterns investigated. Weakly nonlinear stability analyses are performed, as are numerical simulations. Given a reasonable entry condition (uniform flow and an imposed temperature profile), an entry-region effect is obtained whereby increasing flow rates yield increasing distances before strong traveling-wave convection is obtained.

A somewhat neglected paper by Sutton (1970) was the first to consider the effect on Ra_c of a vertical throughflow of fluid. On taking the nondimensional upward mean velocity as a small parameter, a perturbation analysis is undertaken which shows that the correction to the critical Rayleigh number is positive, i.e., that the effect of very small throughflow is stabilizing. Some numerical analysis of the perturbation equations was also presented for a small range of values of the throughflow parameter. Sutton showed that, although destabilization occurs for small vertical velocities, stabilization is re-established at higher flow rates. This is consistent with the fact that the overall temperature drop occurs in a decreasingly thick region as the flow rate increases. The small destabilization may be argued as being a consequence of a nonlinear temperature profile. Nield (1987) considered the analogous case of convection between insulating boundaries and showed that throughflow is always stabilizing. In the case of mixed-type boundaries, stabilization or destabilization depends on whether the throughflow is towards or away from the more restrictive boundary.

Riahi (1989) performed a weakly nonlinear analysis using the method of Schlüter et al. (1965). At a sufficiently high Péclet number (equivalent to a scaled throughflow velocity) the resulting finite-amplitude flow takes the form of hexagons (three rolls aligned at 60° to one another) initially with subcritical motions occurring. At higher convection amplitudes, square cells, originally unstable, are stabilized. The existence of multiple stable states implies that the realized flow depends on the initial disturbance.

D. Hydrodynamic Dispersion

Dispersion in the porous medium context is the local mixing which takes place due to the complex paths the fluid takes through the porous medium; the reader is referred to Section 2.2.3 of Nield and Bejan (1992) for a detailed discussion of the modeling. In the papers by Neichloss and Degan (1975) and Kvernvold and Tyvand (1980), dispersion was modeled using an energy equation of the form

$$\theta_t + \underline{u} \cdot \nabla\theta = \nabla \cdot (\mathbf{D} \cdot \nabla\theta) \tag{31}$$

where $\mathbf{D} = (1 + \varepsilon_2 \underline{u} \cdot \underline{u})\mathbf{I} + (\varepsilon_1 - \varepsilon_2)\underline{uu}$ is the dispersion tensor with the values of $\varepsilon_1 = D/15$ and $\varepsilon_2 = D/40$ chosen. Here D is a dispersion factor which takes small values. On the other hand, Georgiadis and Catton (1988) use $\mathbf{D} = 1 + Di|\underline{u}|$ where Di is also a small parameter. It is clear that dispersion is a nonlinear effect and, in the presence of a no-flow basic state, has no influence on the onset problem.

Neichloss and Degan (1975) performed an analysis using the same series expansion technique as Palm et al. (1972) to obtain steady supercritical flows. They show that increasing dispersion decreases the overall Nusselt number, and the influence is significant for coarse materials. Although these authors undertake a two-dimensional stability analysis, they conclude that the flow is stable within the limitations of the theory. A more comprehensive analysis was presented by Kvernvold and Tyvand (1980), who extended the results of Straus (1974) to flows with dispersion. At values of D as small as 1/150 the stability envelope, similar to that shown in Figure 1, extends to much higher values of Ra and indicates that stable steady convection can persist for a considerably greater range of Rayleigh numbers.

E. Internal Heating

When the porous medium is subject to a uniform distribution of heat sources the nondimensional energy transport equation takes the form

$$\theta_t + \underline{u} \cdot \nabla\theta = \nabla^2\theta + 1 \tag{32}$$

the final term representing the steady uniform generation of heat. Often two Rayleigh numbers are used to categorize the flow: an external Rayleigh number, which is based on any physically imposed temperature difference in the system (i.e., heating from below), and an internal Rayleigh number, Ra_I, which depends on the rate of heat generation. Problems involving internal heat generation are frequently referred to as "penetrative convection," although such a term is used quite legitimately in a wider context such as those flows presented in Section 7.4 of Nield and Bejan (1992).

When the horizontal bounding surfaces are held at identical constant temperatures the critical value of Ra_I is approximately 470. As the temperature of the lower surface is raised, this value decreases until it reaches zero when $Ra = 4\pi^2$. The detailed Ra/Ra_I map is given in Figure 4, where the stable and unstable regions are delineated. A nonlinear stability analysis of the case $Ra = 0$ was carried out by Tveitereid (1977), who showed that down-hexagons (those with fluid flowing downwards in the centre of the hexagonal pattern) are stable up to approximately 8 times the critical value (i.e., 8×470), whereas up-hexagons are always stable. Initially rolls are unstable, but regain their stability at about 3 times the critical value. No information is given about the relative stability of square or rectangular planform flows. Other aspects involving different boundary conditions are discussed briefly in Nield and Bejan (1992).

A substantially different problem ensues when considering a rectangular porous cavity. If all four bounding walls are maintained at the same

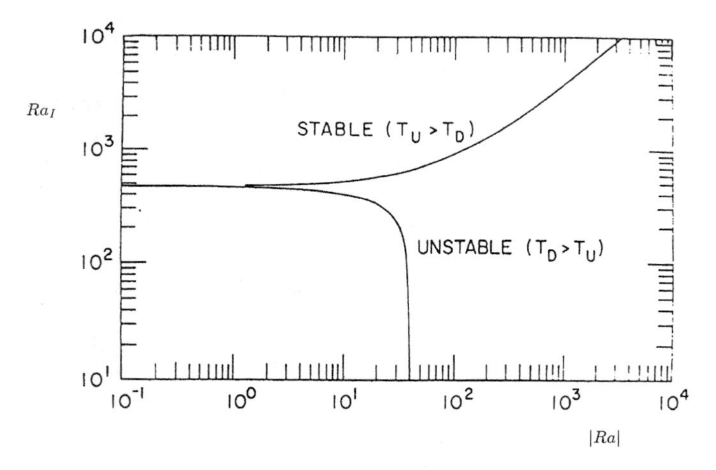

Figure 4. Critical internal Rayleigh number against the external Rayleigh number for stabilizing and destabilizing temperature differences. From Gasser and Kazimi (1976). Reproduced by permission of ASME.

uniform temperature, then there is flow at all nonzero values of Ra_I. Indeed, at high values of Ra_I (with $Ra = 0$) convection at the vertical sidewalls becomes of boundary layer type; see Blythe et al. (1985). However, in such a situation, the upper part of the cavity is unstably stratified and the flow described by Blythe et al. (1985) is unlikely to be realized in practice. A very recent numerical study by Banu et al. (1998) has sought to investigate the onset of unsteady convection in rectangular cavities; the authors have found that incipient unsteady flow occurs at values of Ra_I which are highly dependent on the aspect ratio of the cavity. Since convective instabilities at the time-dependent motion are confined to the top of the cavity, it is possible to show that the critical value is proportional to the inverse third power of the aspect ratio for tall thin cavities. On the other hand, the detailed dynamics becomes very complicated at large aspects ratios where the cavity is shallow. In this case the flow may become chaotic and it loses left/right symmetry; snapshots of a chaotic sequence are shown in Figure 5, where the downward pointing plumes may be seen to be generated whenever there is sufficient room near the top of the cavity, and subsequently travel towards the nearer side wall. A detailed numerical study of possible flow regimes in Ra_I-aspect ratio space is needed.

Nield (1995) considered the case of internal heating which decreases with depth and varies periodically with time, such as would occur in solar energy collectors or solar ponds. The author assumed that the time-depen-

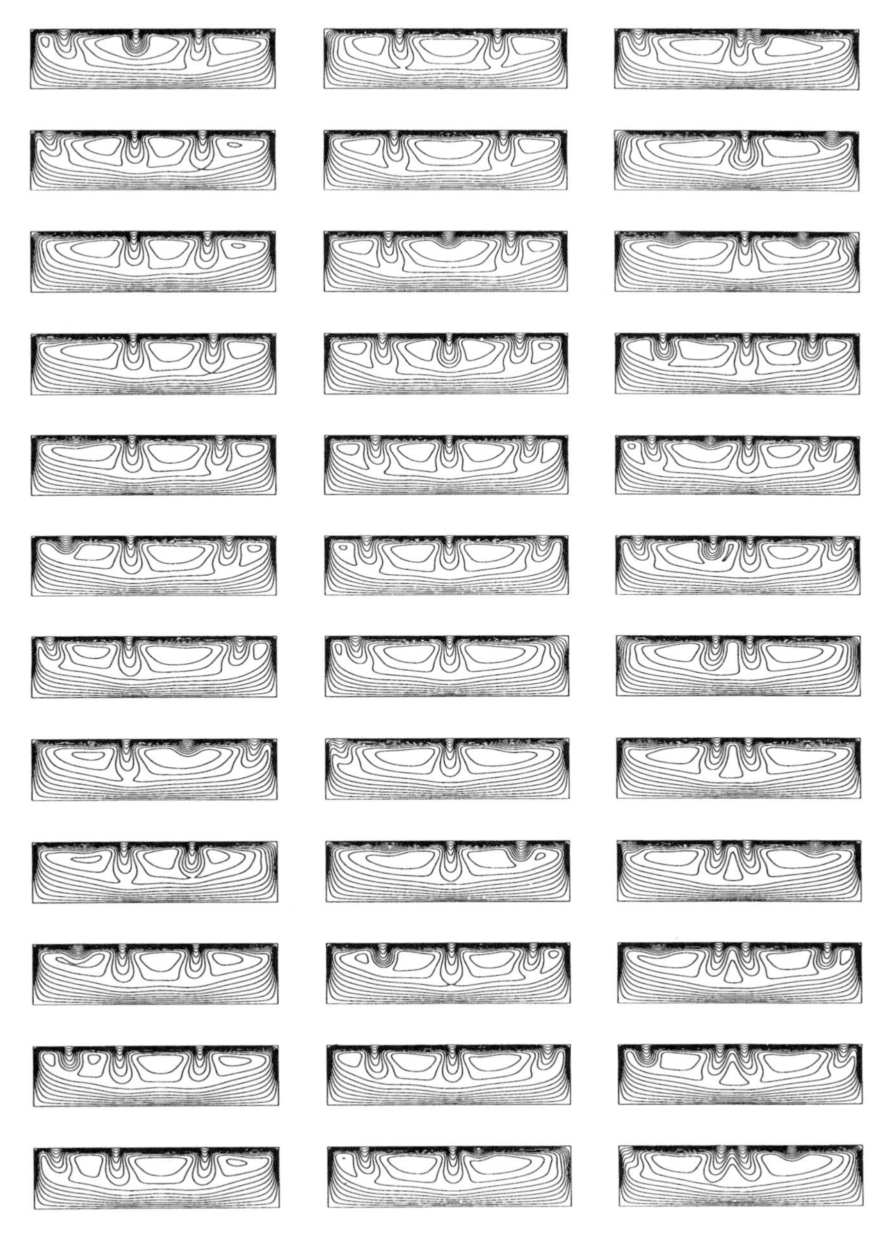

Figure 5. Instantaneous isotherms displaying a chaotic solution for a 4×1 cavity with uniform internal heating at $Ra = 3000$. The individual frames represent snapshots at equal time increments. Time progresses downwards from top left.

dence is quasi-static, in line with the physics of such applications. A linearized analysis using a one-term Galerkin approximation is presented since the flow inhabits a very large parameter space.

Ames and Cobb (1994) consider the onset of convection in a layer with a temperature drop between the infinite horizontal surfaces and where the saturating fluid is water close to its density minimum at approximately 4°C. The results are necessarily detailed, but the authors also use a nonlinear energy theory. In many circumstances the linear and nonlinear onset criteria are different, suggesting that subcritical motions may occur. The extra effect of an anisotropic permeability is examined in Straughan and Walker (1996a); they find that convection arises with a complex growth rate, a Hopf bifurcation to unsteady flow.

The paper by Rionera and Straughan (1990) considers the effect of a y-dependent gravity (to use the present notation) on convection in a layer with a linear density–temperature relationship and nonzero Ra. Again, linearized and nonlinear energy analyses were used, and since the critical value of the Rayleigh number using the energy method lies below that obtained using a standard linearized analysis, it is again deduced that subcritical motions arise.

The addition of Brinkman effects was considered by Vasseur and Robillard (1993), who also assumed that the layer (without internal heat generation) is heated using a uniform heat flux at the lower horizontal surface. The other three bounding surfaces are insulated. As there is no heat sink, the mean temperature rises. On subtracting out this rise, the resulting mathematical problem resembles one involving heat generation. As the thermal boundary conditions are different from those assumed by the above-quoted authors, no direct comparison can be made with their conclusions. Analytical solutions are presented for various values of the critical Rayleigh number.

F. Local Thermal Nonequilibrium

The great majority of papers dealing with convective flows in porous media assume that the solid matrix and the saturating fluid are in local thermal equilibrium (LTE), meaning that the difference between the fluid and solid temperatures is negligible throughout the medium. Vafai and Amiri (1998), in their review article on forced convective flows in porous media, state that sufficiently slow dynamic processes allow the use of the LTE assumption. Although we believe that this is true for the Darcy–Bénard problem, recent boundary-layer analyses by Rees and Pop (2000) and Rees (1999) show that LTE is not always recovered in steady-state flows.

If we use θ and Φ to denote the fluid phase and solid phase temperature fields, respectively, then the two energy transport equations may be written in the dimensionless form

$$\theta_t + \underline{u} \cdot \nabla\theta = \nabla^2\theta + H(\Phi - \theta), \qquad \alpha\Phi_t = \nabla^2\Phi + \gamma H(\theta - \Phi) \tag{33}$$

The three parameters, α, γ, and H, are defined according to

$$\alpha = \frac{(\rho c)_s k_f}{(\rho c)_f k_s}, \qquad \gamma = \frac{\varepsilon k_f}{(1-\varepsilon)k_s}, \qquad H = \frac{hd^2}{\varepsilon k_f} \tag{34}$$

respectively, and may be described as a diffusivity ratio, a porosity-scaled conductivity ratio, and a dimensionless inter-phase heat transfer coefficient. Discussions on typical values for h, the dimensional inter-phase heat transfer coefficient, may be found in Kuznetsov (1998) and Vafai and Amiri (1998). For such problems the Darcy–Rayleigh number is based upon the thermal properties of the fluid phase, rather than on the average properties, and is also inversely proportional to the porosity. LTE corresponds to conditions where $|\theta - \Phi|$ is negligible compared with typical values of θ and Φ.

Combarnous (1972) presented a numerical study of steady large-amplitude convection using the above model. At a fixed value of $Ra = 200$ with a unit aspect ratio convection cell, he showed that the Nusselt number is independent of γ when H is sufficiently large (which recovers LTE), and that LTE is established at decreasing values of H as γ increases. Much more recently, Banu and Rees (2000) have sought to determine conditions for the onset of convection. These authors obtain the same qualitative results as Combarnous in the sense that LTE corresponds either to large values of H or to sufficiently large values of γ. In Figure 6 we show the detailed neutral curves for various values of γ. Comparison with the classical Darcy–Bénard case for which LTE applies is afforded by plotting $\gamma Ra/(1+\gamma)$, which corresponds to the usual Darcy–Bénard Rayleigh number. It is assumed that the layer is infinite in horizontal extent and therefore $\gamma Ra/(1+\gamma)$ has been minimized with respect to the wavenumber. We note that large values of γ or sufficiently large values of H correspond to LTE. When H is very small, the neutral curves correspond to $\gamma Ra/(1+\gamma) = 4\pi^2\gamma/(1+\gamma)$.

G. Anisotropy, Layering, and Heterogeneity

Storesletten (1998) has very recently presented a thorough and comprehensive review of convection in anisotropic porous media, taking into account anisotropy in both the permeability and thermal diffusivity. Most of that review concentrates on the Darcy–Bénard problem. In the very short time

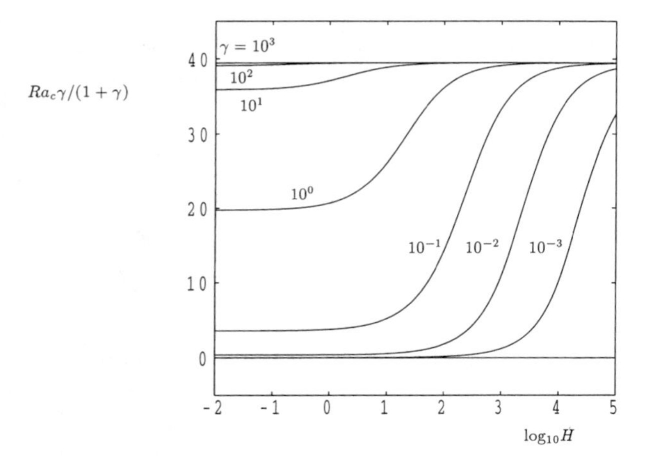

Figure 6. Values of $Ra_c\gamma/(1+\gamma)$ against $\log_{10} H$ for various values of γ. From Banu and Rees (2000).

since the submission of the manuscript of Storesletten (1998) only two further papers have appeared. Mamou et al. (1998) consider how anisotropy affects the onset of convection in a layer where the upper and lower surfaces are cooled and heated by uniform heat fluxes rather than having constant temperatures imposed. In the isotropic case the critical Rayleigh number for an infinite layer is 12 (Nield 1968). Mamou et al. restrict attention to media where one of the principal axes of the permeability tensor is horizontal. It is shown that the criterion for the onset of convection depends very strongly on the ratio of the principal permeabilities, as does the detailed pattern of convection. The second paper, by Parthiban and Patil (1997), considers the combined effects of anisotropy, internal heating, and an inclined temperature gradient.

Frequent mention is made of the connection between anisotropy and layering in the Darcy–Bénard context. McKibbin (1985) gives a comprehensive account of work in the area of layering, and further comments may be found in McKibbin (1991). Rees and Riley (1990) provide an extension to the earlier weakly nonlinear (two-dimensional) heat transfer analysis of McKibbin and O'Sullivan (1981). In this paper Rees and Riley consider whether or not two-dimensional rolls are stable in layers composed of isotropic sublayers of different permeabilities and thicknesses, and conclude that in some circumstances square cells are to be preferred. The necessary modification to the overall rate of heat transfer is given. Of particular interest is the fact that layering allows the possibility of multimodal neutral curves; these allow for sudden changes in the preferred wavelength and pattern of convection when a parameter (such as a sublayer permeability) changes. Comprehensive details may be found in McKibbin and O'Sullivan (1980). In a three-layer medium Rees and Riley (1990) found a parameter set with the critical Rayleigh number corresponding to three preferred wavenumbers simultaneously, suggesting that increasingly complicated modal interactions may be uncovered as the number of sublayers increases.

The effects of continuous variations in the permeability and thermal diffusivity (as opposed to the discontinuous changes corresponding to layering) were considered in quite general terms in Braester and Vadasz (1993). In general, weak heterogeneities (i.e., small-amplitude variations) modify the manner in which convection ensues. Convection is maintained at all nonzero Rayleigh numbers and a smooth transition to a convecting regime is obtained as the Rayleigh number increases.

H. Inclined Temperature Gradient

Now we consider the case where there is an additional horizontal temperature gradient superimposed on the standard vertical gradient. In nondimensional terms, the horizontal surfaces have the following linear temperature variations

$$y = 0: \qquad \theta = 1 - \tilde{\beta}x, \qquad y = 1: \qquad \theta = -\tilde{\beta}x \tag{35}$$

where $\tilde{\beta}$ is the ratio of the horizontal to the vertical temperature gradients. Unlike the classical Darcy–Bénard problem in which there is a basic flow whose stability is analyzed, there is always a horizontal flow when $\tilde{\beta} \neq 0$. This is referred to as a Hadley circulation, but there is no net horizontal flow.

Weber (1974) considered small values of $\tilde{\beta}$ and determined criteria for the onset of convection in the form of longitudinal rolls (i.e., with axes in the x-direction): $Ra_c = 4\pi^2(1 + \tilde{\beta}^2)$ at a wavenumber of π. Later, Nield (1991) relaxed the restriction of small values of $\tilde{\beta}$ and used a low-order Galerkin expansion to confirm the qualitative result that a horizontal temperature increases the critical Rayleigh number. He also presented details of stationary and traveling transverse modes. In another paper Nield (1994) discovered that Ra_c for longitudinal rolls does not increase indefinitely as $\tilde{\beta}$ increases, but reaches a maximum before decreasing to zero. The implication is that the vertical temperature difference, which is a characteristic of the Darcy–Bénard problem, is not essential to induce instability in this type of problem. More detailed linear analyses are presented in Lage and Nield (1998), where the authors also provide strongly nonlinear computations (for transverse modes using a variant of the well-known SIMPLE algorithm) and extend the work to doubly diffusive convection. In the strongly nonlinear regime a complicated bifurcation structure involving both stationary and travelling rolls is found. Further references may be found in Lage and Nield (1998). Of some considerable interest is the nonlinear energy stability analysis of Kaloni and Qiao (1997); their results suggest that strongly nonlinear convection may in many circumstances arise at Rayleigh numbers well below that predicted by the standard linearized theory as given by Nield (1991, 1994).

Nield (1990) also considered the additional effect of introducing a horizontal pressure gradient to induce an overall fluid flux in the positive-x direction. As the surface temperature gradient is negative, this additional effect destabilizes the basic flow relative to when there is no mean horizontal flow. Again, it is found that a local vertical temperature difference is not always essential for instability to arise.

The additional effects of anisotropy and an internal heat source were considered by Parthiban and Patil (1997). The authors conclude that longitudinal modes form the preferred pattern of convection at onset, even with the strong tendency towards hexagonal convection brought about by the presence of internal heating, as shown by Tveitereid (1977).

I. Boundary Imperfections

Here we consider how the classical Darcy–Bénard problem is modified by the presence of perturbed boundary conditions such as wavy boundaries and nonuniform bounding temperatures (both steady and unsteady). We note that these two types of boundary imperfection have identical qualitative effects within the weakly nonlinear regime when the imperfection is steady and applies to the upper and lower surfaces, although the quantitative results are different. Here the nonuniformity is described by sinusoidal variations in the x-direction.

Perturbation analyses of temperature variations were first carried out by O'Sullivan and McKibbin (1986), Rees and Riley (1989a,b) and Rees (1990), using weakly nonlinear theory. The general conclusion from these studies is that the resulting flow depends very strongly on the wavelength and symmetry of the boundary imperfection. When the imperfection has the critical wavelength (i.e., with a wavenumber of π), and is not symmetric, then the transition to convection is smooth and the concept of a critical Rayleigh number is inapplicable (O'Sullivan and McKibbin 1986; Rees and Riley 1989a). When the wavenumber is greater than 3π longitudinal modes are preferred, causing convection to be three-dimensional (Rees and Riley 1989b). At smaller wavenumbers it is possible for rectangular planforms to arise abruptly, and these are composed of a pair of rolls aligned at equal but opposite angles from the direction of the x-axis. At slightly higher Rayleigh numbers these cells destabilize and one of the pair dominates the other. This is a mutual resonance effect mediated by the boundary imperfection. At wavenumbers close to the critical value rolls with spatially deformed axes or spatially varying wavenumbers may arise (Rees and Riley 1989a), and when the wavenumber is very small the pattern of convection at onset may then be quasi-periodic (Rees 1990).

Further and more detailed weakly nonlinear analyses have been undertaken by Riahi (1993, 1995, 1996, 1998) and a centre manifold analysis by Néel (1992). The first three of these papers and that of Néel form an extension of the work of O'Sullivan and McKibbin (1986) and Rees and Riley (1989a,b), who consider the boundary imperfection to have only one sinusoidal component, to the multicomponent case. Given the complexity of the possible patterns when given one component (Rees and Riley 1989b), the number of different possibilities multiplies enormously in the more general cases. Generally, convection is three-dimensional and nonperiodic, although many special cases exist for which this is not true. Examples are cited where the flow may undertake a transition between different patterns as the imperfection amplitude is altered. Riahi (1998) considers the same type of fundamental problem, but assumes that the boundaries are poorly conducting, an extension of earlier work contained in Riahi (1983). Other aspects are studied, using reduction methods, by Néel (1990a,b).

Limited results are available for strongly nonlinear convection in the presence of boundary imperfections. As part of their study, Rees and Riley (1986) undertook a two-dimensional numerical simulation of convection in a symmetric layer with wavy boundaries. The onset of convection is abrupt and is delayed by the presence of the nonuniformity. However, the onset of time-periodic flow takes place at much smaller Rayleigh numbers than those corresponding to the uniform layer. The mechanism generating unsteady flow is no longer a thermal boundary layer instability, but appears to be a cyclical interchange between two distinct modes which support each other via the imperfection, and its onset is not a Hopf bifurcation. At relatively high amplitudes of the wavy surface, the basic flow may bifurcate directly to unsteady flow. Rathish Kumar et al. (1998) have also studied the effects of large-amplitude undulations of the lower surface of a porous cavity. They investigate the effects of varying the wave phase, the wave amplitude, and the number of waves, using a finite element technique.

In the above papers the boundary imperfections were held stationary in space and in most cases the preferred mode of convection at onset is steady. Indeed, rolls show no tendency to exhibit a horizontal mean motion in the absence of moving boundary conditions or a horizontal pressure gradient, although the presence of an inclined temperature gradient may sometimes cause such motion, as mentioned earlier. Just as stationary small-amplitude imperfections cause weak fluid motion whose presence affects the onset of convective instability, so a moving imperfection will cause the basic fluid motion to be unsteady, which may influence the location of the convective cells.

The first paper in this subtopic deals with moving thermal boundary conditions of the form

$$y = 0: \qquad \theta = 0, \qquad y = 1: \qquad \theta = a\cos[k(x + Ut)] \tag{36}$$

In their paper, Ganapathy and Purushothaman (1992) assume that the wavenumber of the imperfection, k, is large and also include Brinkman and advective inertia effects. This is not a stability problem as the mean temperature gradient across the layer is zero. They perform a long-wave approximate analysis to determine the subsequent motion. Mamou et al. (1996) use the boundary conditions

$$y = 0: \qquad \theta = 1 + a\sin[k(x - Ut)], \qquad y = 1: \qquad \theta = 0 \tag{37}$$

in a numerical simulation of strong two-dimensional convection. These authors assume that $k = \pi$, the critical value for the classical Darcy–Bénard problem, that the flow is spatially periodic, and that both a and U take finite (rather than infinitesimal) amplitudes. Additionally, they compute the flow in a frame of reference which is moving with the imperfection, and reduce the problem to one in which the imperfection is stationary but which has an overall mean horizontal flow. The main result of their computational work is to demonstrate that there are two main flow regimes: (i) where the convective motion follows the thermal wave, and (ii) where the convective pattern drifts at a slower speed than the thermal wave. In particular they deduce that there is a critical value of Ra at which one behavior gives way to the other. Much of this qualitative behavior has been explained by Banu and Rees (1999), who perform a weakly nonlinear stability analysis based on the boundary conditions

$$\begin{aligned} y = 0: &\qquad \theta = 1 + a\cos[\pi(x - Ut)], \\ y = 1: &\qquad \theta = a\cos[\pi(x - Ut)] \end{aligned} \tag{38}$$

The amplitude of convection is taken to be $O(\varepsilon)$ and the transitional case, mimicking the computations of Mamou et al. (1996), requires $a = O(\varepsilon^3)$ and $U = O(\varepsilon^2)$. For a given wave speed, convection follows the boundary wave until a certain value of R_2 (see Section II) is exceeded. The resulting nonlinear motion is governed by a straightforward ordinary differential system and has to be computed numerically. As R_2 increases the flow may be either periodic (taking all possible multiples of the forcing period) or quasi-periodic, but the overall mean horizontal motion of the convective pattern decreases. An asymptotic analysis, using the technique of multiple scales analysis for large values of R_2, shows that the horizontal drift is proportional to R_2^{-1} in this limit; a typical flow for such a regime is given in Figure 7.

An entirely different type of imperfection has been investigated by Impey et al. (1990) and Vadasz and Braester (1992). These authors consider the effect of imperfectly insulated sidewalls on convection in a horizontal

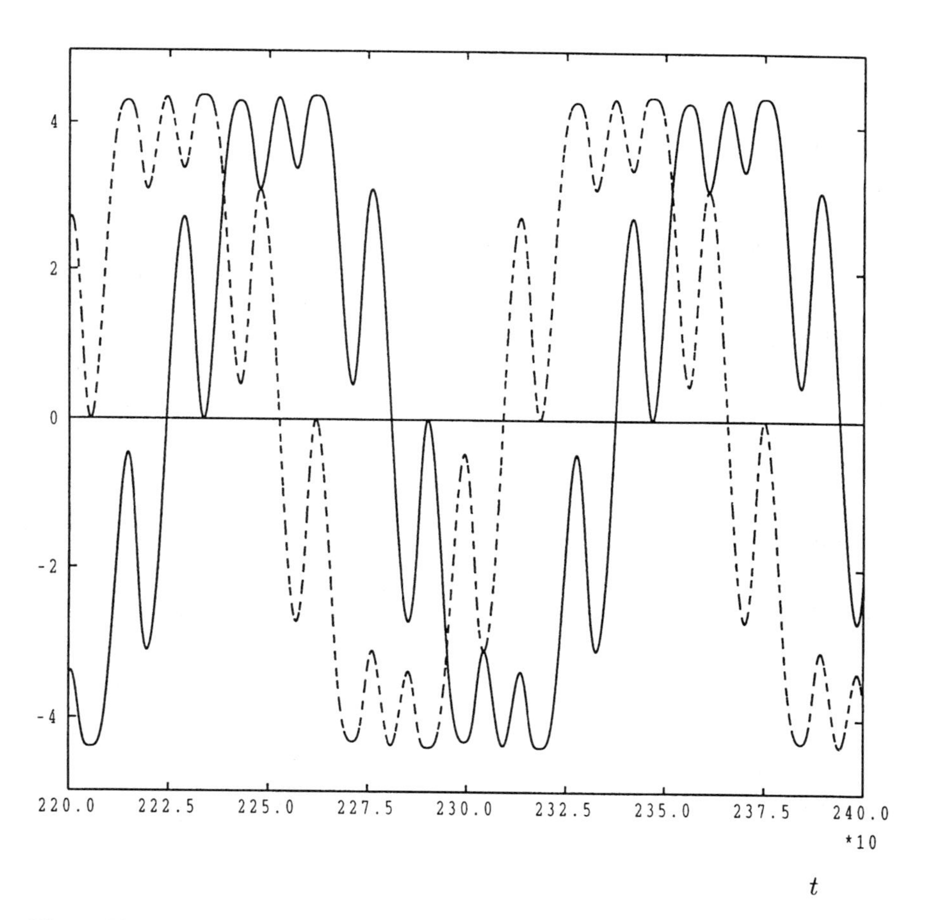

Figure 7. Evolution of the flow for $R_2 = 19.057\,07$ with $\alpha = 0.5$. The real part of the complex amplitude is given by the solid curve, and the imaginary part corresponds to the dashed curve. This flow has 9 times the forcing period. From Banu and Rees (1999).

cavity with upper and lower surfaces at a uniform temperature. Vadasz and Braester (1992) use weakly nonlinear theory to show that weak imperfections modify the usual pitchfork bifurcation to a so-called imperfect bifurcation. Thus weak convection persists at relatively low Rayleigh numbers and grows sharply in strength near the critical Rayleigh number corresponding to the classical Darcy–Bénard problem. A second stable flow also exists within the weakly nonlinear regime if the Rayleigh number is sufficiently large. Impey et al. (1990) provide a comprehensive account of the effects of different shapes and symmetries of the sidewall imperfections, and extend their results well outside the weakly nonlinear regime by using numerical continuation techniques drawn from bifurcation theory. Great emphasis is laid upon how different solution branches gain or lose stability as the Rayleigh number increases.

Vadasz et al. (1993), while also summarizing the results of Vadasz and Braester (1992), also consider strong convection in a cavity with perfectly conducting sidewalls. Conditions are determined under which a conducting no-flow state may be obtained. For general sidewall temperature profiles, flow exists at all nonzero Rayleigh numbers. In shallow cavities convection is confined to the regions near the sidewalls at subcritical Rayleigh numbers, but the patterns migrate towards the middle and fill the cavity as the Rayleigh number increases. Cases with heating from above as well as heating from below are considered.

J. Large Wavenumber Effects

In two separate recent studies the flow within a tall cavity heated from below has been considered, but different sidewall boundary conditions are imposed and this leads to very great qualitative differences between the two cases. Lewis et al. (1997) allow the sidewalls to be insulated, and perform a large-wavenumber asymptotic analysis of nonlinear convection. Three separate nonlinear regimes appear as the Rayleigh number increases, but convection remains unicellular. On the other hand, Rees and Lage (1997) use perfectly conducting boundary conditions by setting a linearly decreasing temperature profile up the sidewalls. For all cell aspect ratios the onset problem is degenerate in the sense that any combination of an odd and an even mode is destabilized simultaneously at the critical Rayleigh number. This degeneracy persists even into the nonlinear regime. A large-wavenumber analysis of the tall cavity problem shows that strong boundary layers form on the top and the bottom of the cavity and a third may exist in the interior.

K. Recent Experimental Advances

A primary difficulty in experimental work involving porous media is the determination of the detailed macroscopic flowfield. Conventional visualization techniques suffer due to the random nature of most media and the difference in refractive indices of the fluid and the matrix of those media which are transparent. Invasive techniques yield only a small amount of information, and others provide information only at boundaries. Howle et al. (1993) overcame these difficulties using suitably regular media (stacked square cross-section bars or punched disks) and a shadowgraph technique. Despite the relatively coarse nature of the media, a remarkably sharp transition to convection was observed for the barred medium. A more comprehensive account is given in Howle et al. (1997).

The magnetic resonance imaging technique was used by Shattuck et al. (1995, 1997) to provide accurate imaging and heat transport measurements for both ordered and disordered media consisting of spheres. It was found that pattern selection in disordered media is influenced strongly by the presence of packing defects. On the other hand, well-defined convective patterns appear when the spheres are uniformly packed. The authors provide a very comprehensive account of the effects of different container geometries and conclude that theoretical modeling of macroscopic laws may well need more detailed information on the pore structure.

L. Miscellaneous Topics

Finally, we present brief descriptions of various studies which cannot be categorized easily into the above subsections.

Parthiban and Patil (1996) consider the onset of convection of a rarefied gas for which the Knudsen number is close to unity. Compressibility effects are included, as are thermal dissipation effects (in the usual fluid-dynamical sense of heat being generated by sufficiently high fluid velocities). The authors study both the Darcy and Darcy–Brinkman models. It is found that rarefaction serves to increase both the critical Rayleigh number and the wavenumber. A simpler model was used by Stauffer et al. (1997) in an independent study of strong convection using numerical methods. We also note that the time derivative of the density in the continuity equations used in these papers is multiplied by 1 in Parthiban and Patil (1996) and by the porosity in Stauffer et al. (1997), and that the Rayleigh numbers are defined differently. Therefore it is difficult to compare these studies.

Malashetty et al. (1994) performed a linear stability analysis of a layer saturated with a chemically reacting fluid. The basic state consists of a nonlinear temperature profile whose degree of nonlinearity depends on the size of the Frank–Kamenetskii parameter. A low-order Galerkin expansion of the linearized stability equations was used to yield criteria for the onset of convection. It is found that this type of chemical reaction enhances convection and destabilizes the layer in comparison with the classical Darcy–Bénard problem.

The presence of a vertical baffle occupying part of the vertical extent of a porous cavity was considered by Chen and Wang (1993). The primary effect of the presence of the baffle is to inhibit the onset of convection by restricting the space within which convection can occur, although a suitably placed baffle will not influence onset criteria.

Sometimes it is advantageous to delay the onset of convection in order to reduce the rate of heat transfer across a porous layer. With this in mind, Tang and Bau (1993) have presented and analyzed a particular feedback control stabilization strategy. The controller modifies the boundary temperature using (i) a measured deviation of the temperature from the conductive state, and (ii) its time derivative. It is found that the critical Rayleigh number may be increased by a factor of 4. The conduction state is maintained through small modulations in the boundary temperatures.

The possibility of combining surface tension effects in a highly porous medium with standard instability mechanisms in order to explain the frequent occurrence of hexagonal planforms was postulated by Hennenberg et al. (1997). The Darcy flow model with a linear equation of state is insufficient to cause stable hexagonal patterns, and the addition of Brinkman effects does not change this qualitative result as the symmetry of the layer has not been changed. Hennenberg et al. refer to their paper as tentative, and detailed comments are given in Nield (1998) about the modeling of the free surface and about other mechanisms which could explain the presence of hexagonal cells.

Skeldon et al. (1997) present a method for investigating the bifurcation structure associated with competition between hexagonal patterns and rolls. Although the paper uses directional solidification as the application, the primary aim is to devise suitable finite element grids which will capture correctly all the symmetries involved in hexagonal/roll interactions. The computation of points of bifurcation is of great importance in understanding the roles played by the many competing solutions of a convection problem. In this regard much information may be gained from analyzing the eigenvalues of the Newton–Raphson iteration matrix when computing steady-state solutions. At relatively low Rayleigh numbers this is quite inexpensive, but efficiency and accuracy are of increasing importance at increasing Rayleigh numbers. Both may be achieved by means of two different methods described in detail in Straughan and Walker (1996b), who also give applications in porous medium convection. The first is a modified version of the compound matrix method, and the second is the Chebyshev tau method.

Porous rocks may exhibit large-scale imperfections such as layering or faulting. Although much work has been presented in the literature on the onset problem for layered media, little exists for faulted media, such as is presented in Joly et al. (1996). In that paper the authors devise a method for analyzing stability which involves the finite element technique. The method is able to test whether complicated two-dimensional flows are unstable to three-dimensional disturbances. It is illustrated by an application to a layered medium, the middle sublayer of which has vertical faulting modeled by a relatively high permeable porous medium.

NOMENCLATURE

Roman Letters

a	amplitude of thermal wave
A, B	amplitudes of weakly nonlinear convection
c	heat capacity
d	macroscopic length scale
D, Di	dispersion factors
Da	Darcy number
$\mathbf{D}$	dispersion tensor
f	reduced streamfunction
F	inertial coefficient
g	temperature
G	inertia parameter
$\tilde{g}$	gravity
h	inter-phase coefficient of heat transfer
H	nondimensional inter-phase coefficient of heat transfer
k	wavenumber of convection
K	permeability, small perturbation to wavenumber
$\tilde{K}$	dimensional inertia parameter
L	microscopic length scale
p	pressure
Pr	Prandtl number
Q	fluid flux speed
Ra	Darcy–Rayleigh number
t	time
T	temperature
u, v, w	fluid flux velocities in the x, y, and z directions
U	nondimensional velocity of thermal wave
x, y, z	Cartesian coordinates
X, X_B, Z	slow spatial variables

Greek Letters

α	scaled inertia parameter, diffusivity ratio
β	coefficient of cubical expansion
$\tilde{\beta}$	ratio of temperature gradients
γ	porosity-scaled conductivity ratio
ε	small expansion parameter
θ	scaled temperature
κ	thermal diffusivity
λ	exponential growth rate
μ	viscosity
ρ	fluid density
σ	heat capacity ratio
τ	slow time scale
ϕ	porosity
Φ	temperature of the solid phase
χ	angle between rolls
ψ	streamfunction
Ω	inter-roll coupling coefficient

Superscripts and Subscripts

$\hat{}$	dimensional
c	cold, critical
f	fluid
h	hot
s	solid
0,1,2,3	terms in weakly nonlinear theory

REFERENCES

Aidun CK. Stability of convection rolls in porous media. ASME J Heat Transfer 94:331–336, 1987.

Aidun CK, Steen, PH. Transition of oscillatory convective heat transfer in a fluid-saturated porous medium. AIAA J Thermophys Heat Transfer 1:268–273, 1987.

Ames KA, Cobb SS. Penetrative convection in a porous medium with internal heat sources. Int J Eng Sci 32:92–105, 1994.

Banu N, Rees DAS. The effect of a travelling thermal wave on weakly nonlinear convection in a porous layer heated from below. Proceedings of Workshop on Fluid Mechanics, Sylhet, Bangladesh (September 1998), 1999.

Banu N, Rees DAS. Onset of Darcy–Bénard convection using a thermal nonequilibrium mode. Submitted for publication, 2000.

Banu N, Rees DAS, Pop I. Steady and unsteady free convection in porous cavities with internal heat generation. Proceedings of 11th International Heat Transfer Conference, Kyongju-ju, Korea, 1998, Vol 4, pp 375–380.

Blythe PA, Daniels PG, Simpkins PG. Convection in a fluid-saturated porous medium due to internal heat generation. Int Commun Heat Mass Transfer 12:493–504, 1985.

Braester C, Vadasz P. The effect of a weak heterogeneity of a porous medium on natural convection. J Fluid Mech 254:345–362, 1993.

Caltagirone JP, Fabrie P, Combarnous M. De la convection naturelle oscillante en milieu poreux au chaos temporel? C R Acad Sci Paris II 305:549–553, 1987.

Chen F, Wang CY. Convective instability in saturated porous enclosures with a vertical insulating baffle. Int J Heat Mass Transfer 36:1897–1904, 1993.

Combarnous M. Description du transfert de chaleur par convection naturelle dans une couche poreuse horizontale à l'aide d'un coefficient de transfert solide–fluide. C R Acad Sci Paris II A275:1375–1378, 1972.

Drazin PG, Reid WH. Hydrodynamic Stability. Cambridge: Cambridge University Press, 1981.

Dufour F, Néel M-C. Numerical study of instability in a horizontal porous channel with bottom heating and forced horizontal flow. Phys Fluids 10:2198–2207, 1998.

Ergun S. Fluid flow through packed columns. Chem Eng Prog 48:89–94, 1952.

Ganapathy R, Purushothaman R. Free convection in a saturated porous medium due to a travelling thermal wave. Z Anger Math Mech 72:142–145, 1992.

Gasser RD, Kazimi MS. Onset of convection in a porous medium with internal heat generation. ASME J Heat Transfer 98:49–54, 1976.

Georgiadis JG, Catton I. Dispersion in cellular thermal convection in porous media. Int J Heat Mass Transfer 31:1081–1091, 1988.

Graham MD, Steen PH. Strongly interacting traveling waves and quasiperiodic dynamics in porous medium convection. Physica D 54:331–350, 1992.

He XS, Georgiadis JG. Natural convection in porous media: effect of weak dispersion on bifurcation. J Fluid Mech 216:285–298, 1990.

Hennenberg M, Ziad Saghir M, Rednikov A, Legros JC. Porous media and the Bénard–Marangoni problem. Transp Porous Media 27:327–355, 1997.

Holst PH, Aziz K. Transient three-dimensional natural convection in confined porous media. Int J Heat Mass Transfer 15:73–90, 1972.

Horne RN. Three-dimensional natural convection in a confined porous medium heated from below. J Fluid Mech 92:751–766, 1979.

Horne RN, O'Sullivan MJ. Oscillatory convection in a porous medium heated from below. J Fluid Mech 66:339–352, 1974.

Horton CW, Rogers FT. Convection currents in a porous medium. J Appl Phys 16:367–370, 1945.

Howle L, Behringer RP, Georgiadis J. Visualization of convective fluid flow in a porous medium. Nature 362:230–232, 1993.

Howle L, Behringer RP, Georgiadis JG. Convection and flow in porous media, Part 2, Visualization by shadowgraph. J Fluid Mech 332:247–262, 1997.

Impey MD, Riley DS, Winters KH. The effect of sidewall imperfections on pattern formation in Lapwood convection. Nonlinearity 3:197–230, 1990.

Joly N, Bernard D, Ménégazzi P. ST2D3D: A finite-element program to compute stability criteria for natural convection in complex porous structures. Numer Heat Transfer B29:91–112, 1996.

Joseph DD. Stability of fluid motions II. Berlin: Springer, 1976.

Kaloni PN, Qiao Z. Non-linear stability of convection in a porous medium with inclined temperature gradient. Int J Heat Mass Transfer 40:1611–1615, 1997.

Kimura S. Onset of oscillatory convection in a porous medium. In: Ingham DB, Pop I, eds. Transport Phenomena in Porous Media. Oxford: Pergamon, 1998, pp 77–102.

Kimura S, Schubert G, Straus JM. Route to chaos in porous-medium thermal convection. J Fluid Mech 166:305–324, 1986.

Kimura S, Schubert G, Straus JM. Instabilities of steady, periodic and quasi-periodic modes of convection in porous media. ASME J Heat Transfer 109:350–355, 1987.

Kimura S, Schubert G, Straus JM. Time dependent convection in fluid-saturated porous cube heated from below. J Fluid Mech 207:153–189, 1990.

Kladias N, Prasad V. Natural convection in horizontal porous layers: effects of Darcy and Prandtl numbers. ASME J Heat Transfer 112:675–684, 1989.

Kladias N, Prasad V. Flow transitions in buoyancy-induced non-Darcy convection in a porous medium heated from below. ASME J Heat Transfer 112:675–684, 1990.

Koschmeider EL. Bénard cells and Taylor vortices. Cambridge: Cambridge Monographs on Mechanics and Applied Mathematics, 1993.

Kuznetsov AV. Thermal nonequilibrium forced convection in porous media. In: Ingham DB, Pop I, eds. Transport Phenomena in Porous Media. Oxford: Pergamon, 1998, pp 103–130.

Kvernvold O, Tyvand PA. Dispersion effects on thermal convection in porous media. J Fluid Mech 99:673–686, 1980.

Lage JL, Nield DA. Convection induced by inclined gradients in a shallow porous medium layer. J Porous Media. 1:57–69, 1998.

Lage JL, Bejan A, Georgiadis JG. The Prandtl number effect near the onset of Bénard convection in a porous medium. Int J Heat Fluid Flow 13:408–411, 1992.

Lapwood ER. Convection of a fluid in a porous medium. Proc Camb Phil Soc 44:508–521, 1948.

Lewis S, Rees DAS, Bassom AP. High wavenumber convection in tall porous containers heated from below. Q J Mech Appl Math 50:545–563, 1997.

McKibbin R. Thermal convection in layered and anisotropic porous media: a review. In: Wooding RA, White I, eds. Convective Flow in Porous Media. Wellington, New Zealand: DSIR, 1985, pp 113–127.

McKibbin R. Convection and heat transfer in layered and anisotropic porous media. Proceedings of ICHMT International Seminar on Heat and Mass Transfer in Porous Media, Dubrovnic, Yugoslavia, 1991, pp 327–336.

McKibbin R, O'Sullivan MJ. Onset of convection in a layered porous medium heated from below. J Fluid Mech 96:375–393, 1980.

McKibbin R, O'Sullivan MJ. Heat transfer in a layered porous medium heated from below. J Fluid Mech 111:141–173, 1981.

Malashetty MS, Cheng P, Chao BH. Convective instability in a horizontal porous layer saturated with a chemically reacting fluid. Int J Heat Mass Transfer 37:2901–2908, 1994.

Mamou M, Robillard L, Bilgen E, Vasseur P. Effects of a moving thermal wave on Bénard convection in a horizontal saturated porous layer. Int J Heat Mass Transfer 39:347–354, 1996.

Mamou M, Mahidjiba A, Vasseur P, Robillard L. Onset of convection in an anisotropic porous medium heated from below by a uniform heat flux. Int Commun Heat Mass Transfer 25:799–808, 1998.

Néel M-C. Convection in a horizontal porous layer of infinite extent. Eur J Mech B 9:155–176, 1990a.

Néel M-C. Convection naturelle dans une couche poreuse horizontale d'extension infinie: chauffage inhomogène. C R Acad Sci Paris II 309:1863–1868, 1990b.

Néel M-C. Inhomogeneous boundary conditions and the choice of the convective patterns in a porous layer. Int J Eng Sci 30:507–521, 1992.

Néel M-C. Convection forcée en milieux poreux: écarts à la loi de Darcy. C R Acad Sci Paris II 326:615-620, 1998.

Neichloss H, Dagan G. Convective currents in a porous layer heated from below: the influence of hydrodynamic dispersion. Phys Fluids 18:757–761, 1975.

Newell AC, Whitehead JA. Finite bandwidth, fine amplitude convection. J Fluid Mech 38:279–303, 1969.

Nield DA. Onset of thermohaline convection in a porous medium. Water Resour Res 11:553–560, 1968.

Nield DA. Convective instability in porous media with throughflow. AIChemEng J 33:1222–1224, 1987.

Nield DA. Convection in a porous medium with inclined temperature gradient and horizontal mass flow. Heat Transfer 1990, Hemisphere, 1990, Vol 5, pp 153–158.

Nield DA. Convection in a porous medium with inclined temperature gradient. Int J Heat Mass Transfer 34:87–92, 1991.

Nield DA. Convection in a porous medium with inclined temperature gradient: additional results. Int J Heat Mass Transfer 37:3021–3025, 1994.

Nield DA. Onset of convection in a porous medium with nonuniform time-dependent volumetric heating. Int J Heat Fluid Flow 16:217–222, 1995.

Nield DA. Modelling the effect of surface tension on the onset of natural convection in a saturated porous medium. Transp Porous Media 31:365–368, 1998.

Nield DA, Bejan A. Convection in Porous Media. New York: Springer-Verlag, 1992.

O'Sullivan MJ, McKibbin R. Heat transfer in an unevenly heated porous layer. Transp Porous Media 1:293–312, 1986.

Palm E, Weber JE, Kvernvold O. On steady convection in a porous medium. J Fluid Mech 54:153–161, 1972.

Parthiban C, Patil PR. Convection in a porous medium with velocity slip and temperature jump boundary conditions. Heat Mass Transfer 32:27–31, 1996.

Parthiban C, Patil PR. Thermal instability in an anisotropic porous medium with internal heat source and inclined temperature gradient. Int Commun Heat Mass Transfer 24:1049–1058, 1997.

Prats M. The effect of horizontal fluid motion on thermally induced convection currents in porous mediums. J Geophys Res 71:4835–4838, 1967.

Rathish Kumar BV, Murthy PVSN, Singh P. Free convection heat transfer from an isothermal wavy surface in a porous enclosure. Int J Numer Meth Fluids 28:633–661, 1998.

Rees DAS. The effect of long-wave thermal modulations on the onset of convection in an infinite porous layer heated from below. Q J Mech Appl Math 43:189–214, 1990.

Rees DAS. The effect of inertia on the stability of convection in a porous layer heated from below. J Theor Appl Fluid Mech 1:154–171, 1996.

Rees DAS. The effect of inertia on the onset of mixed convection in a porous layer heated from below. Int Commun Heat Mass Transfer 24:277–283, 1998.

Rees DAS. Vertical thermal boundary layer flow in a porous medium using a thermal nonequilibrium model: elliptic effects. Proceedings of Workshop on Fluid Mechanics, Sylhet, Bangladesh (September 1998), 1999.

Rees DAS, Bassom AP. The onset of convection in inclined porous layers. Proceedings of International Conference on Heat Transfer, Kyong-ju, Korea, 1998, Vol 4, pp 497–502.

Rees DAS, Lage JL. The effect of thermal stratification on natural convection in a vertical porous insulation layer. Int J Heat Mass Transfer 40:111–121, 1997.

Rees DAS, Pop I. Vertical thermal boundary layer flow in a porous medium using a thermal nonequilibrium model. J Porous Media, 3: 31–44, 2000.

Rees DAS, Riley DS. Convection in a porous layer with spatially periodic boundary conditions: resonant wavelength excitation. J Fluid Mech 166:503–530, 1986.

Rees DAS, Riley DS. The effects of boundary imperfections on convection in a saturated porous layer: near-resonant wavelength excitation. J Fluid Mech 199:133–154, 1989a.

Rees DAS, Riley DS. The effects of boundary imperfections on convection in a saturated porous layer: non-resonant wavelength excitation. Proc R Soc Lond A421:303–339, 1989b.

Rees DAS, Riley DS. The three-dimensional stability of finite-amplitude convection in a layered porous medium heated from below. J Fluid Mech 211:437–461, 1990.

Riahi DN. Nonlinear convection in a porous layer with finite conducting boundaries. J Fluid Mech 129:153–171, 1983.

Riahi DN. Nonlinear convection in a porous layer with permeable boundaries. Int J Non-linear Mech 24:459–463, 1989.

Riahi DN. Preferred pattern of convection in a porous layer with a spatially non-uniform boundary temperature. J Fluid Mech 246:529–543, 1993.

Riahi DN. Finite amplitude thermal convection with spatially modulated boundary temperatures. Proc R Soc Lond 449:459–478, 1995.

Riahi DN. Modal package convection in a porous layer with boundary imperfections. J Fluid Mech 318:107–128, 1996.

Riahi DN. Finite bandwidth, long wavelength convection with boundary imperfections: near-resonant wavelength excitation. Int J Math Sci 21:171–182, 1998.

Riley DS, Winters KH. Modal exchange mechanisms in Lapwood convection. J Fluid Mech 215:309–329, 1989.

Riley DS, Winters KH. Time-periodic convection in porous media: the evolution of Hopf bifurcations with aspect ratio. J Fluid Mech 223:457–474, 1991.

Rionera S, Straughan B. Convection in a porous medium with internal heat source and variable gravity effects. Int J Eng Sci 28:497–503, 1990.

Schlüter A, Lortz D, Busse FH. On the stability of finite-amplitude convection. J Fluid Mech 23:129–144, 1965.

Schubert G, Straus JM. Three-dimensional and multicellular steady and unsteady convection in fluid-saturated porous media at high Rayleigh numbers. J Fluid Mech 94:25–38, 1979.

Schubert G, Straus JM. Transitions in time-dependent thermal convection in fluid-saturated porous media. J Fluid Mech 121:301–303, 1982.

Shattuck MD, Behringer RP, Johnson GA, Georgiadis JG. Onset and stability of convection in porous media: visualization by magnetic resonance imaging. Phys Rev Lett 75:1934–1937, 1995.

Shattuck MD, Behringer RP, Johnson GA, Georgiadis JG. Convection and flow in porous media, Part 1, Visualization by magnetic resonance imaging. J Fluid Mech 332:215–246, 1997.

Skeldon AC, Cliffe KA, Riley DS. Grid design for the computation of a hexagon–roll interaction using a finite element method. J Comput Phys 133:18–26, 1997.

Stauffer PH, Auer LH, Rosenberg ND. Compressible gas in porous media: a finite amplitude analysis of natural convection. Int J Heat Mass Transfer 40:1585–1589, 1997.

Steen PH. Pattern selection for finite-amplitude convection states in boxes of porous media. J Fluid Mech 136:219–241, 1983.

Steen PH, Aidun CK. Time-periodic convection in porous media: transition mechanism. J Fluid Mech 196:263–290, 1988.

Storesletten L. Effects of anisotropy on convective flow through porous media. In; Ingham DB, Pop I, eds. Transport Phenomena in Porous Media. Oxford: Pergamon, 1998, pp 261–283.

Strange R, Rees DAS. The effect of fluid inertia on the stability of free convection in a saturated porous medium heated from below. Proceedings of International Conference on Porous Media and their Applications in Science, Engineering and Technology, Kona, Hawaii, 1996, pp 71–84.

Straughan B, Walker DW. Anisotropic porous penetrative convection. Proc R Soc Lond 452:97–115, 1996a.

Straughan B, Walker DW. Two very accurate and efficient methods for computing eigenvalues and eigenfunctions in porous convection problems. J Comp Phys 127:128–141, 1996b.

Straus JM. Large amplitude convection in porous media. J Fluid Mech 64:51–63, 1974.

Straus JM, Schubert G. Three-dimensional convection in a cubic box of fluid-saturated porous material. J Fluid Mech 91:155–165, 1979.

Straus JM, Schubert G. Modes of finite-amplitude three-dimensional convection in rectangular boxes of fluid-saturated porous material. J Fluid Mech 103:23–32, 1981.

Sutton FM. Onset of convection in a porous channel with throughflow. Phys Fluids 13:1931–1934, 1970.

Tang J, Bau HH. Feedback control stabilization of the no-motion state of a fluid confined in a horizontal porous layer heated from below. J Fluid Mech 257:485–505, 1993.

Tveitereid M. Thermal convection in a horizontal porous layer with internal heat sources. Int J Heat Mass Transfer 20:1045–1050, 1977.

Vadasz P. Flow in rotating porous media. Fluid transport in porous media. In: Duplessis P, ed. Advances in Fluid Mechanics. Southampton: Computational Mechanics Publications, 1997.

Vadasz P. Free convection in rotating porous media. In: Ingham DB, Pop I, eds. Transport Phenomena in Porous Media. Oxford: Pergamon, 1998, pp 285–312.

Vadasz P, Braester C. The effect of imperfectly insulated sidewalls on natural-convection in porous-media. Acta Mech 91:215–233, 1992.

Vadasz P, Braester C, Bear J. The effect of perfectly conducting side walls on natural convection in porous media. Int J Heat Mass Transfer 36:1159–1170, 1993.

Vafai K, Amiri A. Non-Darcian effects in confined forced convective flows. In: Ingham DB, Pop I, eds. Transport Phenomena in Porous Media. Oxford: Pergamon, 1998, pp 313–329.

Vasseur P, Robillard L. The Brinkman model for natural convection in a porous layer: effects of nonuniform thermal gradient. Int J Heat Mass Transfer 36:4199–4206, 1993.

Walker K, Homsy GM. A note on convective instabilities in Boussinesq fluids and porous media. ASME J Heat Transfer 99:338–339, 1977.

Weber JE. Convection in a porous medium with horizontal and vertical temperature gradients. Int J Heat Mass Transfer 17:241–248, 1974.

13
Double-Diffusive Convection in Porous Media

Abdelkader Mojtabi and Marie-Catherine Charrier-Mojtabi
Université Paul Sabatier, Toulouse, France

I. INTRODUCTION

A. Definitions

Natural convective flow in porous media, due to thermal buoyancy alone, has been widely studied (Combarnous and Bories 1975) and well documented in the literature (Cheng 1978; Bejan 1984; Nield and Bejan 1992), whereas only a few works have been devoted to double-diffusive convection in porous media. This type of convection concerns the processes of combined (simultaneous) heat and mass transfer which are driven by buoyancy forces. Such phenomena are usually referred to as thermohaline, thermosolutal, double-diffusive, or combined heat and mass transfer natural convection; in this case the mass fraction gradient and the temperature gradient are independent (no coupling between the two). Double-diffusive convection frequently occurs in seawater flow and mantle flow in the earth's crust, as well as in many engineering applications.

Soret-driven thermosolutal convection results from the tendency of solute to diffuse under the influence of a temperature gradient. The concentration gradient is created by the temperature field and is not the result of a boundary condition; see De Groot and Mazur (1962), Patil and Rudraiah (1980). For saturated porous media, the phenomenon of cross-diffusion is further complicated because of the interaction between fluid and porous matrix, and accurate values of cross-diffusion coefficients are not available. This makes it impossible to proceed to a practical quantitative study of cross-diffusion effects in porous media. The Dufour coefficient is an order of magnitude smaller than the Soret coefficient in liquids, and the corresponding contribution to the heat flux can be ignored. Knobloch (1980) and Taslim and Narusawa (1986) demonstrated in a fluid medium and a porous medium respectively that there exists a close relationship between cross-diffusion problems (taking into account the Dufour effect and Soret effect) and double-diffusion problems.

Recent interest in double-diffusive convection through porous media has been motivated by its importance in many natural and industrial problems. Some examples of thermosolutal convection can be found in astrophysics, metallurgy, electrochemistry, and geophysics. Double-diffusive flows are also of interest with respect to contaminant transport in groundwater and exploitation of geothermal reservoirs.

Two regimes of double-diffusive convection are commonly distinguished. When the faster diffusing component is destabilizing, as it is when stably stratified saltwater is heated from below in a horizontal cell, the system is in the diffusive regime. When the slower diffusing component is destabilizing, as is the case when cold fresh water is overlain by hot salty water, the system is in the fingering regime. In such binary fluids, the diffusivity of heat is usually much higher than the diffusivity of salt; thus, a displaced particle of fluid loses any excess heat more rapidly than it loses any excess solute. The resulting buoyancy force may tend to increase the displacement of the particle from its original position, causing instability. The same effect may cause overstability, involving oscillatory motions of large amplitude since heat and solute diffuse at widely different rates.

The current state of knowledge concerning double-diffusive convection in a saturated porous medium is summarized in the overviews of Nield and Bejan (1998). Of the many works in the literature related to double-diffusive convection in a cavity, the majority can be classified into two categories: cavities with imposed uniform heat and mass fluxes and cavities with imposed uniform temperature and concentration. It is important to note that most of these works are theoretical.

B. Experimental Studies

We consider, here, the most significant experimental studies in thermosolutal convection in porous media. The first was carried out by Griffith (1981), who used both a Hele–Shaw cell and a sand-tank model with salt and sugar or heat and salt as the diffusing components and porous medium of glass spheres to study the "diffusive" configuration (a thin diffusive interface). He measured salt–sugar and heat–salt fluxes through two-layer convection systems and compared the results with predictions from a model. This was applied to the Wairakei geothermal system, and the observed values were consistent with those found in laboratory experiments.

The second work was carried out by Imhoff and Green (1988), who studied double-diffusive groundwater fingers, using a sand-tank model and a salt–sugar system. They observed that double-diffusive groundwater fingers can transport solutes at rates as much as two orders of magnitude larger than those associated with molecular diffusion in motionless groundwater. This could play a major role in the vertical transport of near-surface pollutants in groundwater.

The third experimental work, by Murray and Chen (1989), is closer to our study, Charrier-Mojtabi et al. (1998), and concerns the onset of double-diffusive convection in a finite box filled with porous medium. The experiments were performed in a horizontal layer consisting of 3 mm diameter glass beads contained in a box 24 cm × 12 cm × 4 cm rigid. The rigid top and bottom walls of the box provide a linear basic-state temperature profile but only allow a nonlinear time-dependent basic-state profile for salinity. They observed that, when a porous medium is saturated with a fluid having a stabilizing salinity gradient, the onset of convection was marked by a dramatic increase in heat flux at the critical ΔT, and the convection pattern was three-dimensional, whereas two-dimensional rolls are observed for single-component convection in the same apparatus. They also observed a hysteresis loop reducing the temperature difference from supercritical to subcritical values.

C. Linear Stability Analysis

Concerning the theoretical studies, various modes of double-diffusive convection can be developed, depending not only on how both thermal and solutal gradients are imposed relative to each other but also on the numerous nondimensional parameters involved.

Most of the published work regarding double-diffusive convection in porous media concerns linear stability analysis. The linear stability characteristics of flow in horizontal layers with imposed vertical temperature and concentration gradients has been the subject of many studies. The onset of thermosolutal convection was predicted by Nield (1968) on the basis of linear stability analysis. This flow configuration was later studied by many investigators. Taunton et al. (1972) extended Nield's analysis and considered salt-fingering convection in a porous layer. Trevisan and Bejan (1985) studied mass transfer in the case where buoyancy is entirely due to temperature gradients. Rudraiah et al. (1986) applied linear and nonlinear stability analysis and showed that subcritical instabilities are possible in the case of two-component fluids. Brand et al. (1983) obtained amplitude equations for the convective instability of a binary fluid mixture in a porous medium. They found an experimentally feasible example of a codimension-two bifurcation (intersection of stationary and oscillatory bifurcation lines).

With regard to porous layers heated from the side, the focus has been on the double-diffusive instability of double boundary-layer structures that form near a vertical wall immersed in a temperature and concentration stratified porous medium. The stability of this problem was studied by Gershuni et al. (1976) and independently by Khan and Zebib (1981). The occurrence of both monotonic and oscillatory instability was predicted. Raptis et al. (1981) constructed similarity solutions for the boundary-layer near a vertical wall immersed in a porous medium with constant temperature and concentration. Nield et al. (1981) analyzed the convection induced by inclined thermal and solutal gradients in a shallow horizontal layer of a porous medium. The onset of double-diffusive convection in situations where the buoyancy forces induced by the thermal and solutal effects oppose each other and are of equal intensity was recently predicted by Mamou et al. (1998) and Karimi-Fard et al. (1999).

D. Numerical and Analytical Studies

1. Vertical Porous Layer Subjected to Constant Heat and Mass Fluxes

Although the most basic geometry for the study of simultaneous heat and mass transfer from the side is the vertical wall, most of the available studies dealing with double-diffusion convection are in confined porous media and concern rectangular cavities subjected to constant heat and mass fluxes at their vertical walls.

For a vertical wall immersed in an infinite porous medium, Bejan and Khair (1985) studied the vertical natural convective flows due to the combined buoyancy effects of thermal and species diffusion. They presented an order-of-magnitude analysis of boundary layer equations which yields functional relations for the Nusselt and Sherwood numbers in limiting cases. This fundamental problem was re-examined by Lai and Kulacki (1991). Their solutions cover a wide range of governing parameters.

The similarity approach employed by Bejan and Khair (1985) was generalized by Jang and Chang (1988a,b) to consider the effect of wall inclination on a two-layer structure. More recently, Nakayama and Hossain (1995) obtained an integral solution for aiding-flow adjacent to vertical surfaces. Rastogi and Poulikakos (1995) considered non-Newtonian fluid-saturated porous media and presented similarity solutions for aiding-flows with constant wall temperature and concentration as well as constant wall flux conditions.

Rectangular cavities with imposed uniform heat and mass fluxes have been the subject of numerous studies. Trevisan and Bejan (1986) developed an analytical Oseen-linearized solution for boundary-layer regimes for $Le = 1$, and proposed a similarity solution for heat transfer driving flows for $Le > 1$. They also performed an extensive series of numerical experiments that validate the analytical results and provide heat and mass transfer data in the domain not covered by analytical study. The same configuration was considered by Alavyoon (1993) for co-operative ($N > 0$) buoyancy forces and by Alavyoon et al. (1994) for opposing ($N < 0$) buoyancy forces. They presented an analytical solution valid for stratified flow in slender enclosures ($A \gg 1$) and a scale analysis that agrees with the heat-driven and solute-driven limits, using numerical and analytical methods and scale analysis. Comparisons between fully numerical and analytical solutions are presented for a wide range of parameters. They also show the existence of oscillatory convection with opposing buoyancy forces. Transient heat and mass transfer in a square porous enclosure has been studied numerically by Lin (1993). He showed that an increase of the buoyancy ratio N improves heat and mass transfer and causes the flow to approach steady-state conditions in a short time. An extension of these studies to the case of the inclined porous layer subject to transverse gradients of heat and solute was carried out by Mamou et al. (1995a). Their results are presented for $10^{-3} \leq Le \leq 10^3$, $0.1 \leq Ra_T \leq 10^4$, $-10^4 \leq N \leq 10^4$, $2 \leq A \leq 15$, and $-180^\circ \leq \phi \leq 180^\circ$, where ϕ corresponds to the inclination of the enclosure. They obtained an analytical solution by assuming parallel flow in the core region of the tilted cavity. The existence of multiple steady-state solutions for opposing buoyancy forces has been demonstrated numerically. Mamou et al. (1995b) have also shown numerically that, in square cavities where the thermal and solutal buoyancy forces counteract each other ($N = -1$), a purely diffusive (motionless) solution is possible even for Lewis numbers different from unity.

2. Vertical Cavities with Imposed Temperature and Concentration

The configuration of a vertical cavity with imposed temperature and concentration along the vertical side-walls was considered by Trevisan and Bejan (1985), Charrier-Mojtabi et al. (1998), and by Angiraza et al. (1997). Trevisan and Bejan (1985) considered a square cavity submitted to horizontal temperature and concentration gradients. Their numerical simulations are compared to scaling analysis. They found that the onset of the convective regime depends on the cell aspect ratio A, the Lewis number, the thermal and solutal Rayleigh numbers Ra_T and Ra_S or the buoyancy ratio N. Their numerical simulations were carried out for the range $0.01 \leq Le \leq 100$, $50 \leq Ra_T \leq 10^4$, and $-5 \leq N \leq +3$ for $A = 1$. Angiraza et al. (1997), without making approximations of boundary layer character, numerically solved the Darcy-type equation. They found that, for high Rayleigh number aiding flows, the numerical solutions match the similarity solutions very closely. However, they differ substantially for opposing flows and for low Rayleigh numbers. Flow and transport follow complex patterns depending on the interaction between the diffusion coefficients and the buoyancy ratio $N = Ra_S/Ra_T$. The Nusselt and Sherwood numbers reflect this complex interaction.

E. Other Configurations

The double diffusive case of natural convection in a vertical annular porous layer under the condition of constant heat and mass fluxes at the vertical boundaries was analyzed by Marcoux et al. (1999). The system of governing equations was solved numerically to obtain a detailed description of the velocity, temperature, and concentration within the cavity in order to emphasize the influence of the dimensionless parameters Ra_T, Le, N, and curvature on steady and unsteady convective flows. For the case of high aspect ratios ($A \gg 5$), an analytical solution is proposed on the basis of a parallel flow model. The good agreement of this solution with numerical results shows that the analytical model can be faithfully used to obtain a concise description of the problem for these cases, as seen in Figures 1 and 2.

Double-diffusive convection over a sphere was analyzed by Lai and Kulacki (1990), while Yucel (1990) similarly treated the flow over a vertical cylinder. Flow over a horizontal cylinder, with the concentration gradient being produced by transpiration, was studied by Hassan and Mujumdar (1985). All the above studies (Sections I.C, I.D, I.E) describe the momentum conservation in the porous medium, using the Darcy model.

F. Other Formulations and Physical Problems

1. Brinkman and Brinkman–Forchheimer Model

Poulikakos (1986) studied the criterion of the onset of double-diffusive convection using the Darcy–Brinkman model to describe momentum conservation in the porous medium: the results clearly show the influence of Darcy number. Chen and Chen (1993) also used the Brinkman and Forchheimer terms to consider nonlinear two-dimensional horizontally per-

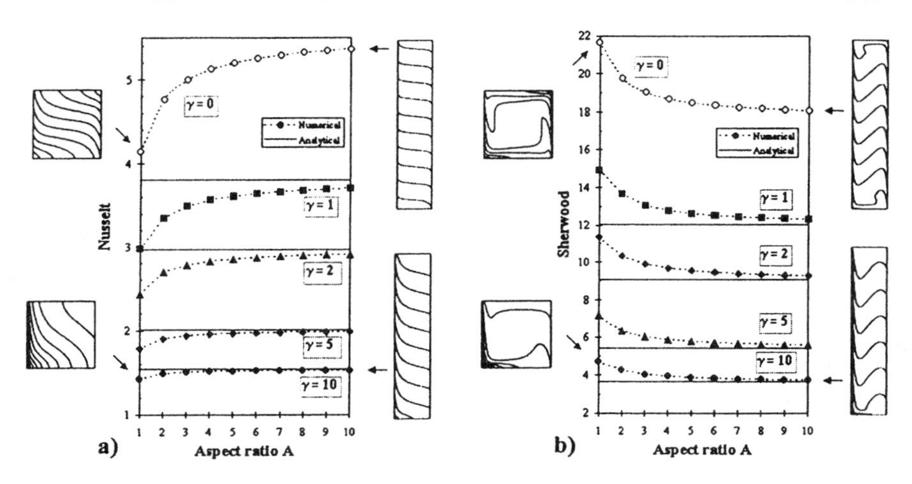

Figure 1. Influence of the aspect ratio A on the Nu_i (a) and Sh_i (b) numbers for different values of γ. Isotherms (a) and isohalines (b) at steady-state for $Ra_T = 100$, $Le = 10$, $A = 1$ and 10, $\gamma = 0$ and 10.

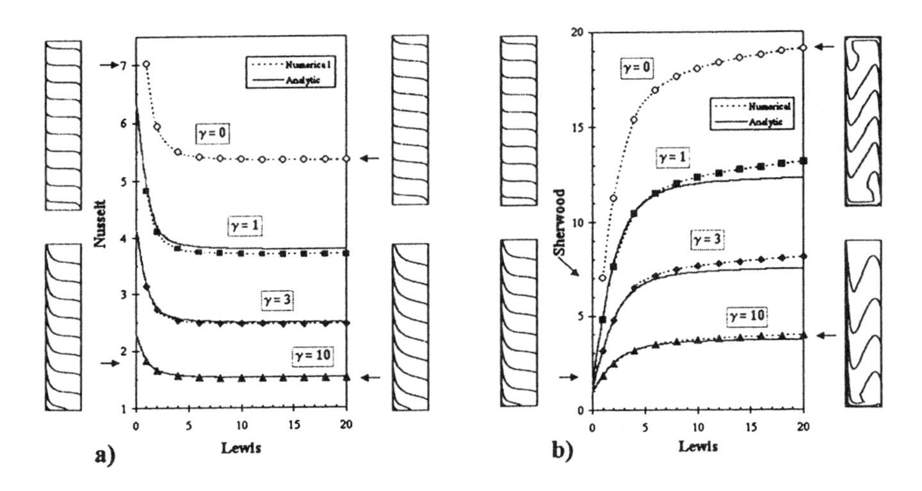

Figure 2. Influence of the Lewis number on the Nu_i (a) and Sh_i (b) numbers for different values of γ. Isotherms (a) and isohalines (b) at steady-state for $Ra_T = 100$, $N = 1$, $A = 10$, $Le = 1$ and 20, $\gamma = 0$ and 10.

iodic, double-diffusive fingering convection. The stability boundaries which separate regions from different regimes of convection are identified. The Darcy–Brinkman formulation was adopted recently by Goyeau et al. (1996) for a vertical cavity with imposed temperature and concentration along the vertical side-walls. This study deals with natural convection driven by co-operating thermal and solutal buoyancy forces. The numerical simulations presented span a wide range of the main parameters (*Ra* and Darcy number, *Da*) in the domain of positive buoyancy numbers, N and $Le > 1$. This contribution completes certain observations on the Darcy regime already mentioned in the previous studies. It is shown that the numerical results for mass transfer are in excellent agreement with scaling analysis over a very wide range of parameters.

Multiphase transport is another aspect of double-diffusive convection. Vafai and Tien (1989), Tien and Vafai (1990) studied phase change effects and multiphase transport in porous materials. They used the Darcy law for flow motion without the Boussinesq approximation. The problem was modeled by a system of transient intercoupled equations governing the two-dimensional multiphase transport process in porous media. It should be noted that (aside from non-Darcian effects) the problem of double-diffusive convection within a porous medium will then be a special case of multiphase transport in porous media as analyzed in Vafai and Tien (1989) and Tien and Vafai (1990). The more recent work by Karimi-Fard et al. (1997) studied double-diffusive convection in a square cavity filled with a porous medium. Several different flow models for porous media, such as Darcy flow, Forchheimer's extension, Brinkman's extension, and generalized flow are considered. The influence of boundary and inertial effects on heat and mass transfer is analyzed to determine the validity of Darcy's law in this configuration. It is shown that the inertial and boundary conditions have a profound effect on the double-diffusive convection.

A comparison between different models is presented in Figure 3. The plots clearly show that the difference between the models increases with an increase in *Da*. Figure 4 shows the influence of *Le* on heat transfer for $Pr = 1$, 10, and 20. Boundary and inertial effects are also shown in Figure 4. It can be seen that the use of the Darcy results induces an overestimation for *Nu* in comparison with models based on Forchheimer extension and Brinkman extension. The essential non-Darcian effect is the boundary effect. The plots clearly show that the generalized model and the Brinkman extension of the Darcy model give almost the same *Nu*. An interesting effect is observed for double-diffusive convection. As seen in Figure 4, heat transfer is maximized for a critical value of the Lewis number. This behavior exists for all models, but is more significant for the Darcy

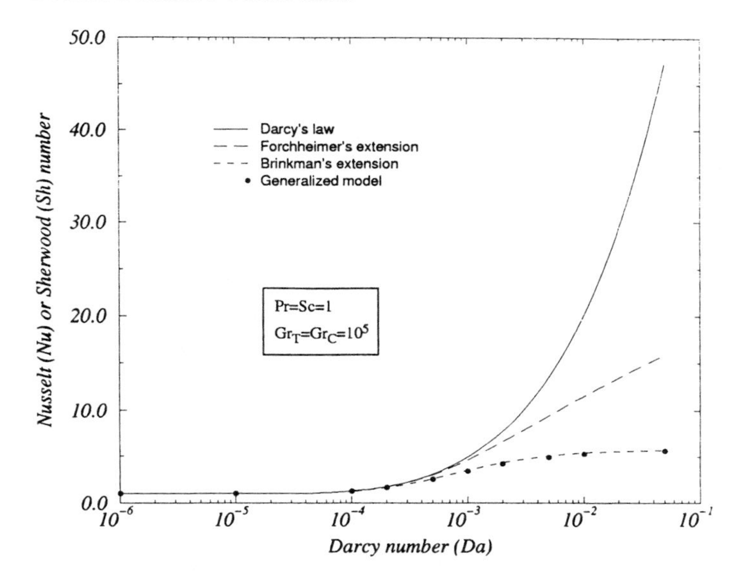

Figure 3. Variations of Nusselt or Sherwood number as a function of Darcy number for different flow models.

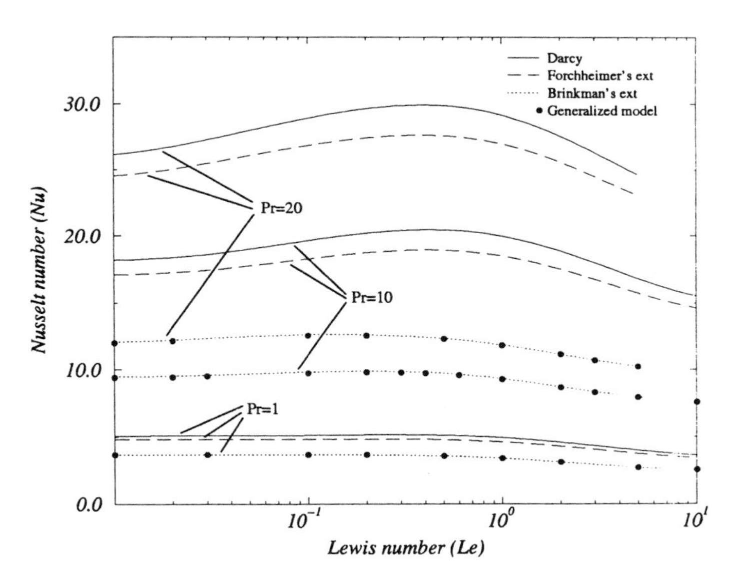

Figure 4. Variations of Nusselt number as a function of Lewis and Prandtl numbers for $Gr_T = Gr_C = 10^{-5}$, $Da = 10^{-3}$, and $\Lambda = 2.34$.

model and Forchheimer's extension of the Darcy model than for Brinkman's extension and the generalized models.

2. Double-Diffusive Convection in an Anisotropic or Multidomain Porous Medium

Tyvand (1980) was the first to study double-diffusive convection in an anisotropic porous medium. He considered a horizontal layer which retains horizontal isotropy with respect to permeability, thermal diffusivity, and solute diffusivity. It was shown that for porous media, with a thermally insulating solid matrix, the stability diagram has the same shape as in the case of isotropy. The onset of double-diffusive convection in a rotating porous layer of infinite horizontal extent was investigated numerically by Patil et al. (1989) for anisotropic permeability and horizontal isotropy. Double-diffusive convection in layered anisotropic porous media was studied numerically by Nguyen et al. (1994). A rectangular enclosure, consisting of two anisotropic porous layers with dissimilar hydraulic and transport properties, was considered. The problem was solved numerically. Four different sets of boundary constraints were imposed on the system, including aiding diffusion, opposing diffusion, and the two modes of cross diffusion. The results show that each set of boundary conditions produces distinct flow, temperature, and concentration fields. The overall heat transfer rates may or may not be sensitive to the Rayleigh numbers, depending on the orientation of the boundary conditions of the temperature and concentration fields. Recently, double-diffusive convection in dual permeability, dual porosity media was studied by Saghir and Islam (1999). The Brinkman model is used as the momentum balance equation and solved simultaneously with mass and energy balance equations in the two-dimensional domain. Special emphasis is given to the study of double-diffusive phenomena in a layered porous bed with contrasting permeabilities. The study is completed for a wide range of permeability contrasts.

II. MATHEMATICAL FORMULATION

A. Governing Equations Describing the Conservation Laws

1. Momentum Equation

The basic dynamic equations for the description of the flow in porous media have been the subject of controversial discussion for several decades. Most of the analytical and numerical work presented in the literature is based on the Darcy–Oberbeck–Boussinesq formulation. Darcy's law is valid only when the pore Reynolds number *Re* is of the order of 1. Lage (1992) studied the effect of the convective inertia term for Bénard convection in porous media. He concluded that the convective term, included in the general momentum equation, has no significant effect on the calculation of overall heat transfer. Chan et al. (1970) utilized Brinkman's extension to study natural convection in porous media with rectangular impermeable boundaries. However, they essentially concluded that non-Darcian effects have very little influence on heat transfer results. For many practical applications, however, Darcy's law is not valid, and boundary and inertial effects need to be accounted for. A fundamental study of boundary and inertial effects can be found in the work of Vafai and Tien (1981) and Hsu and Cheng (1985). A systematic study of the non-Darcian effects in natural convection is presented in the work of Ettefagh et al. (1991). These authors report a formal derivation of a general equation for fluid flow through an isotropic, rigid, and homogeneous porous medium.

The general final equation for an incompressible fluid is

$$\rho\left[\frac{\partial \boldsymbol{V}'}{\varepsilon \partial t}+\frac{1}{\varepsilon^2}\boldsymbol{V}'\cdot\nabla\boldsymbol{V}'\right]=-\nabla P'+\rho\boldsymbol{g}+\mu_e\nabla^2\boldsymbol{V}'-\frac{\mu}{K}\boldsymbol{V}'-\frac{b\rho}{K^{1/2}}\|\boldsymbol{V}'\|\boldsymbol{V}' \tag{1}$$

where ρ, μ, μ_e, K, b, and ε are fluid density, dynamic viscosity, effective viscosity, permeability, form coefficient, and porosity respectively. We suppose that the medium is homogeneous and spatially invariant and the viscosity is taken as a constant.

Double-diffusive convection is often studied using the Darcy formulation and Boussinesq approximation, provided that the fluid moves slowly so that the inertial effects are negligible, and one can usually drop the time derivative term completely on the basis of the analysis given by Nield and Bejan (1992), as

$$\boldsymbol{V}'=\frac{K}{\mu}(-\nabla P'-\rho g\boldsymbol{k}) \tag{2}$$

where $\boldsymbol{V}'=(u',v',w')$ and P' are the seepage (Darcy) velocity and pressure respectively. $\boldsymbol{k}=-\sin(\varphi)\boldsymbol{x}+\cos(\varphi)\boldsymbol{y}$ defines the tilt of the cavity.

2. Continuity Equation

Conservation of fluid mass, assuming an incompressible fluid and no sources or sinks, can be expressed as

$$\nabla\cdot\boldsymbol{V}'=0 \tag{3}$$

3. Energy Equation

The macroscopic description of heat transfer in porous media by a single energy equation implies the assumption of local thermal equilibrium between the moving fluid phase and the solid phase ($T_s = T_f = T$). This hypothesis has been investigated by several authors (Sözen and Vafai 1990; Kaviany 1995; Gobbé and Quintard 1994; Quintard and Whitaker 1996a,b). For situations in which local thermal equilibrium is not valid, models have been proposed based on the concept of two macroscopic continua, one for the fluid phase and the other for the solid phase; see Quintard et al. (1997).

The temperature differences imposed across the boundaries are small, and consequently the Boussinesq approximation is valid. The single-energy equation is

$$\frac{(\rho c)_m}{(\rho c)_f}\frac{\partial T'}{\partial t'} + V' \cdot \nabla T' = \alpha_e \nabla^2 T' \tag{4}$$

where c is the specific heat, α_e is the effective thermal conductivity of saturated porous medium divided by the specific heat capacity of the fluid. Subscript f refers to fluid properties, whereas subscript m refers to the fluid–solid mixture and s to the solid matrix, where

$$(\rho c)_m = \varepsilon(\rho c)_f + (1-\varepsilon)(\rho c)_s \tag{5}$$

$$\alpha_e = \varepsilon\frac{k_f}{(\rho c)_f} + (1-\varepsilon)\frac{k_s}{(\rho c)_f} = \varepsilon\alpha_f + (1-\varepsilon)\frac{k_s}{(\rho c)_f} \tag{6}$$

which correspond to effective thermal conductivity obtained as the weighted arithmetic mean of the conductivities k_s and k_f. In general, the effective thermal conductivity depends in a complex fashion on the geometry of the medium. Many other expressions of k^* do exist, such as geometric mean $k^* = k_s^{\varepsilon} k_f^{1-\varepsilon}$, and many others are listed in the book by Kaviany (1995).

4. Mass Transfer Equation

For a porous solid matrix saturated by a fluid mixture, we have

$$\varepsilon\frac{\partial C'}{\partial t'} + V' \cdot \nabla C' = D_m \nabla^2 C' \tag{7}$$

Parameter D_m represents the diffusivity of a constituent through the fluid-saturated porous matrix.

5. Combined Heat and Mass Transfer

Generally, the transport of heat and mass are not directly coupled and Eqs. (4) and (6) hold without change. In thermosolutal convection, coupling takes place because the density ρ of the binary fluid depends on both temperature T' and mass fraction C'. For small density variations due to temperature and mass fraction changes at constant pressure, the density variations can be expressed as

$$\rho(T, C) = \rho_r(1 - \beta_T(T - T_r) - \beta_c(C - C_r)) \tag{8}$$

where T_r and C_r are taken as the reference state, and the coefficients of volumetric expansion with temperature $\beta_T = -\left(\frac{1}{\rho}\right)\left(\frac{\partial\rho}{\partial T}\right)_C$ or with concentration $\beta_C = -\left(\frac{1}{\rho}\right)\left(\frac{\partial\rho}{\partial C}\right)_T$ are assumed constant. It is noted that the expansion coefficient β_T is usually positive and the expansion coefficient β_C is negative if C corresponds to the mass fraction of the denser component.

In some circumstances there is direct coupling. This occurs when cross-diffusion (Soret and Dufour effects) is not negligible. The Soret effect refers to the mass flux produced by temperature gradients, and the Dufour effect refers to the heat flux produced by a concentration gradient. With no heat or mass sources we have, instead of Eqs. (4) and (6),

$$\frac{(\rho c)_m}{(\rho c)_f}\frac{\partial T'}{\partial t'} + V' \cdot \nabla T' = \alpha_e\left(\nabla^2 T' + \frac{\alpha_m}{\alpha_e}\nabla^2 C'\right) \tag{9}$$

$$\varepsilon\frac{\partial C'}{\partial t'} + V' \cdot \nabla C' = D_m\left(\nabla^2 C' + \frac{D_T}{D_m}\nabla^2 T'\right) \tag{10}$$

where $\alpha_m/\alpha_e = D_d$ and $D_T/D_m = S_T$ are, respectively, the Dufour and Soret dimensional coefficient of the porous medium.

B. Nondimensional Equations (Case of Darcy Model)

The fluid flow within the porous medium is assumed to be incompressible and governed by Darcy's law. The contribution to the heat flux by the Dufour effect is assumed negligible in liquids. The Oberbeck–Boussinesq approximation is applicable in the range of temperatures and concentrations expected. We introduce nondimensional variables with the help of the following scales: L for distance, $L^2(\rho c)_m/k_e$ for time, α_e/L for velocity, ΔT for temperature, ΔC for concentration, $k_e\mu/K(\rho c_p)_f$ for pressure. Thus we obtain the system of governing equations for nondimensional variables

$$\left.\begin{aligned} &\nabla \cdot \boldsymbol{V} = 0 \\ &\boldsymbol{V} = -\nabla P + (Ra_T T + Ra_S C)\boldsymbol{k} \\ &\frac{\partial T}{\partial t} + \boldsymbol{V} \cdot \nabla T = \nabla^2 T \\ &\varepsilon \frac{\partial C}{\partial t} + \boldsymbol{V} \cdot \nabla C = \frac{1}{Le}(\nabla^2 C + S_T^* \nabla^2 T) \end{aligned}\right\} \tag{11}$$

$\boldsymbol{k} = -\sin(\varphi)\boldsymbol{x} + \cos(\varphi)\boldsymbol{y}$ defines the tilt of the cavity.

The problem formulated involves the following nondimensional parameters: the thermal Rayleigh number Ra_T, the solutal Rayleigh number Ra_S, the Lewis number Le, the parameter of Soret effect S_T^*, the normalized porosity ε; these five dimensionless parameters governing the convective dynamics are defined by

$$Ra_T = \frac{Kg\beta_T(\rho c)_f \Delta TL}{k^* \nu}, \qquad Ra_S = \frac{Kg\beta_C(\rho c)_f \Delta CL}{k^* \nu},$$

$$Le = \frac{a}{D}, \qquad S_T^* = \frac{D_T}{D}\Delta T_r, \qquad \varepsilon = \varepsilon^* \frac{(\rho c)_f}{(\rho c)_m}$$

If we introduce the buoyancy ratio

$$N = \frac{Ra_S}{Ra_T} = \frac{\beta_C \Delta C_r}{\beta_T \Delta T_r}$$

N is positive for cooperative buoyancy forces and negative for opposing buoyancy forces. The Darcy equation becomes

$$\boldsymbol{V} = -\nabla P + Ra_T(T + NC)\boldsymbol{k} \tag{12}$$

III. ONSET OF DOUBLE-DIFFUSIVE CONVECTION IN A TILTED CAVITY

A. Linear Stability Analysis

The purpose of this subsection is to analyze the linear stability of a purely diffusive solution in a tilted rectangular or infinite box with porous medium saturated by binary fluid. We complete the previous results obtained for horizontal layers by Nield (1968) and by Charrier-Mojtabi et al. (1998). The influence of the tilt of the cavity on the bifurcation points is analyzed.

We show the existence of oscillatory instability even for the case where $Le = 1$, and for various tilts of the cavity.

With reference to Figure 5, we consider a Cartesian frame with an angle of tilt φ with respect to the vertical axis. We assume that the rectangular porous cavity (height H, width L, aspect ratio $A = H/L$) is bounded by two walls at different but uniform temperatures and concentrations, respectively T_1 and T_2 (C_1 and C_2); the two other walls are impermeable and adiabatic. We assume that the medium is homogeneous and isotropic, that Darcy's law is valid, and that the Oberbeck–Boussinesq approximation is applicable. The Soret and Dufour effects are assumed to be negligible (see Section IV).

The dimensionless thermal, species, and velocity boundary conditions are given by the equations

$$\left.\begin{array}{llll} \dfrac{\partial C}{\partial y} = \dfrac{\partial T}{\partial y} = V = 0 & \text{for} & y = 0, A & \forall x \\ T = C = U = 0 & \text{for} & x = 0 & \forall y \\ T = C = 1;\ U = 0 & \text{for} & x = 1 & \forall y \end{array}\right\} \tag{13}$$

The motionless double-diffusive solution ($\boldsymbol{V}_0 = 0$, $T_0 = x$, $C_0 = x$) is a particular solution of the set of Eqs. (11) and (13) for a horizontal cell. To study the stability of this solution we introduce infinitesimal 3D perturbations ($\boldsymbol{v}$, θ, c) defined by: $\boldsymbol{v} = \boldsymbol{V}^* - \boldsymbol{V}_0$; $\theta = T^* - T_0$; $c = C^* - C_0$, where $\boldsymbol{V}^*$, T^*, C^* indicate the disturbed flow and $\boldsymbol{V}_0$, T_0, C_0 indicate the basic flow. We assume that the perturbation quantities ($\boldsymbol{v}, \theta, c$) are small and we ignore the smaller second-order quantities. After linearization, we obtain the following system of equations for small disturbances

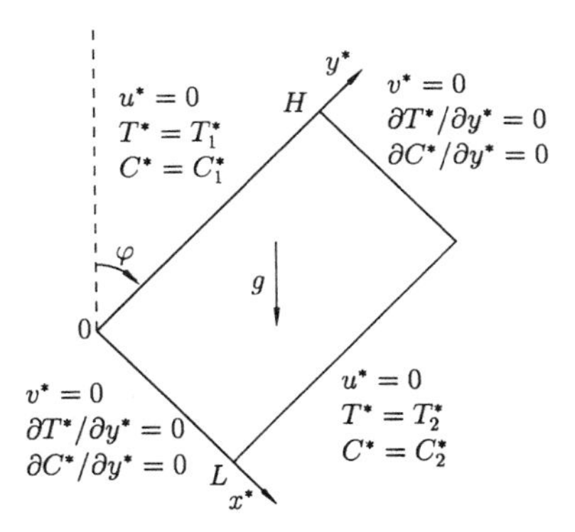

Figure 5. Definition sketch.

$$\left.\begin{aligned} \mathbf{v} &= -\nabla p + Ra_T(\theta + Nc)\mathbf{k} \\ \frac{\partial \theta}{\partial t} &= \nabla^2 \theta - u \\ \varepsilon \frac{\partial c}{\partial t} &= \frac{\nabla^2 c}{Le} - u \end{aligned}\right\} \tag{14}$$

Operating on the first equation of (14) twice with curl, using the continuity equation and taking only the x component of the resulting equation, we obtain

$$\nabla^2 u = -Ra_T\left(\frac{\partial^2(\theta + Nc)}{\partial x \partial y}\cos(\varphi) + \left(\frac{\partial^2(\theta + Nc)}{\partial y^2} + \frac{\partial^2(\theta + Nc)}{\partial z^2}\right)\sin(\varphi)\right) \tag{15}$$

with the following boundary conditions

$$\left.\begin{aligned} &\frac{\partial u}{\partial y} = \frac{\partial c}{\partial y} = \frac{\partial \theta}{\partial y} = 0; \qquad \text{for} \qquad y = 0, A \qquad \forall x, \forall z, \forall t \\ &u = c = \theta = 0; \qquad \text{for} \qquad x = 0, 1 \qquad \forall y, \forall z, \forall t \end{aligned}\right\} \tag{16}$$

1. Linear Stability Analysis for an Infinite Horizontal Cell

We first consider the two limit cases of horizontal cells ($\varphi = \pm\pi/2$). In this situation the cross-derivative term in Eq. (15) is simplified. The problem can be solved by direct calculation and no numerical approximation is needed. Equations (14) and (15) become

$$\left.\begin{aligned} \nabla^2 u &= -JRa_T\left(\frac{\partial^2(\theta + Nc)}{\partial y^2} + \frac{\partial^2(\theta + Nc)}{\partial z^2}\right) \\ \frac{\partial \theta}{\partial t} &= \nabla^2 \theta - u \\ \varepsilon \frac{\partial c}{\partial t} &= \frac{\nabla^2 c}{Le} - u \end{aligned}\right\} \tag{17}$$

where J is defined by

$$\begin{cases} \varphi = +\frac{\pi}{2} \rightarrow J = 1 \\ \varphi = -\frac{\pi}{2} \rightarrow J = -1 \end{cases}$$

The boundary conditions associated with this problem are

$$\left.\begin{aligned}&\frac{\partial u}{\partial y}=\frac{\partial c}{\partial y}=\frac{\partial \theta}{\partial y}=0; \qquad \text{for} \qquad y=0, A \qquad \forall x, \forall z, \forall t\\ &u=c=\theta=0; \qquad \text{for} \qquad x=0,1 \qquad \forall y, \forall z, \forall t\end{aligned}\right\} \tag{18}$$

When we consider a cell of infinite extension in directions y and z, the perturbation functions are written as

$$(u(x,y,z,t), \theta(x,y,z,t), c(x,y,z,t)) = (u(x), \theta(x), c(x))\mathrm{e}^{\sigma t + I(ky+\ell z)} \tag{19}$$

where $u(x)$, $\theta(x)$, $c(x)$ are the amplitude, k and ℓ are the wavenumbers in directions y and z, respectively, I is the imaginary unit and σ is defined by $\sigma = \sigma_r + I\omega$. The marginal state corresponds to $\sigma_r = 0$.

We substitute expansions (19) into (17) and then obtain the following linear differential equations for amplitude

$$\left.\begin{aligned}(\mathrm{D}^2-\alpha^2)u &= JRa_T\alpha^2(\theta+Nc)\\ (\mathrm{D}^2-\alpha^2-\sigma)\theta - u &= 0\\ (\mathrm{D}^2-\alpha^2-\varepsilon\sigma Le)c - Leu &= 0\end{aligned}\right\} \tag{20}$$

where D is the operator $\mathrm{D} = \mathrm{d}/\mathrm{d}x$ and $\alpha^2 = k^2 + \ell^2$.

In these equations α is an overall horizontal wavenumber. The system of equations (20) must be solved subject to the boundary conditions

$$u(x) = c(x) = \theta(x) = 0 \qquad \text{for} \qquad x=0 \qquad \text{and} \qquad x=1 \tag{21}$$

Solutions of the form

$$(u, c, \theta) = (u_0, c_0, \theta_0)\sin(i\pi x) \tag{22}$$

are possible if

$$B(B+\sigma)(B+\sigma\varepsilon Le) - JRa_T\alpha^2[NLe(B+\sigma) + B + \varepsilon\sigma Le] \tag{23}$$

where $B = (i\pi)^2 + \alpha^2$.

At marginal stability, $\sigma = I\omega$, where ω is real. The real and imaginary parts of Eq. (23) become

$$\left.\begin{aligned}(B^2-\varepsilon Le\omega^2) - JRa_T\alpha^2(NLe+1) &= 0\\ \omega\big[(1+\varepsilon Le)B^2 - JRa_T\alpha^2 Le(N+\varepsilon)\big] &= 0\end{aligned}\right\} \tag{24}$$

Two solutions are possible

$$\begin{cases} \omega = 0 \\ Ra_T = \dfrac{JB^2}{\alpha^2(NLe+1)} \end{cases} \quad \text{and} \quad \begin{Bmatrix} Ra_T = \dfrac{JB^2(1+\varepsilon Le)}{\alpha^2 Le(N+\varepsilon)} \\ \omega^2 = -\dfrac{B^2(1+\varepsilon NLe^2)}{\varepsilon Le^2(N+\varepsilon)} \end{Bmatrix} \tag{25}$$

since B^2/α^2 has the minimum value $4\pi^2$, attained when $i = 1$ and $\alpha = \pi$.

(a) Case $\varphi = +\pi/2$ ($J = 1$). The saturated porous medium is heated from below, where the highest concentration is imposed. The two critical solutions are

$$\begin{cases} \omega = 0 \\ Ra_{Tc} = \dfrac{4\pi^2}{NLe+1} \end{cases} \quad \text{and} \quad \begin{Bmatrix} Ra_{Tc} = \dfrac{4\pi^2(1+\varepsilon Le)}{Le(N+\varepsilon)} \\ \omega_c^2 = -\dfrac{4(1+\varepsilon NLe^2)\pi^4}{\varepsilon Le^2(N+\varepsilon)} \end{Bmatrix} \tag{26}$$

For co-operative buoyancy forces ($N > 0$), $\omega_c^2 < 0$, the motionless solution loses is stability via stationary bifurcation with $Ra_{Tc} = 4\pi^2/(NLe+1)$. For opposing buoyancy forces ($N < 0$), stationary bifurcation is possible if $N > -1/Le$ and Hopf bifurcation is possible if $N \in]-\varepsilon, -1/\varepsilon Le^2[$ (the pulsation ω_c must be positive); this latter relation is acceptable for $Le > 10$, i.e., for liquids. We can verify that if the Hopf bifurcation occurs it will appear before the stationary bifurcation.

For $N < -\varepsilon$, the motionless double-diffusive solution is infinitely linearly stable for all values of ε and Le.

(b) Case $\varphi = -\pi/2$ ($J = -1$). The saturated porous medium is now heated from the top, where the highest concentration is imposed. The two critical solutions are

$$\begin{cases} \omega = 0 \\ Ra_{Tc} = -\dfrac{4\pi^2}{NLe+1} \end{cases} \quad \text{and} \quad \begin{cases} Ra_{Tc} = -\dfrac{4\pi^2(1+\varepsilon Le)}{Le(N+\varepsilon)} \\ \omega_c^2 = -\dfrac{4(1+\varepsilon NLe^2)\pi^4}{\varepsilon Le^2(N+\varepsilon)} \end{cases}$$

For cooperative buoyancy forces ($N > 0$), $\omega_c^2 < 0$, the motionless double-diffusive solution is infinitely linearly stable for all values of ε and Le. For opposing buoyancy forces ($N < 0$), the stationary bifurcation is possible if $N < -1/Le$ and Hopf bifurcation is possible only if $N \in]-\varepsilon, -1/\varepsilon Le^2[$ (the pulsation ω_c must be positive); this latter relation is acceptable for $Le > 10$, i.e., for liquids. In this case, if the Hopf bifurcation occurs it will appear after the stationary bifurcation.

2. Linear Stability Analysis for a General Case

In the general case, for any tilt, the motionless double-diffusive steady-state linear distribution ($V_0 = 0,\ T_0 = x,\ C_0 = x$) is not a solution of Eqs. (11) with $S_T^* = 0$. When the thermal and solutal buoyancy forces are of the same order but have opposite signs ($Ra_T = -Ra_S \Leftrightarrow N = -1$), the steady linear distribution ($V_0 = 0,\ T_0 = x,\ C_0 = x$) is a particular solution of (11) for any aspect ratio and for any tilt. To study the stability of this solution, we use a numerical approach based on the Galerkin method; analytical resolution of the stability problem is not possible. Three situations are considered, $Le = 1$, $Le > 1$, and $Le < 1$.

(a) Case $Le = 1$. A complete analysis of this situation shows that the motionless solution can lose its stability via a Hopf bifurcation. Figures 6 and 7 show the influence of normalized porosity on the critical Rayleigh number and the pulsation corresponding to the Hopf bifurcation for a square cavity and $Le = 1$. We can see that the critical Rayleigh number increases with the normalized porosity. This means that ε has a stabilizing effect. In this case, the mass and thermal diffusion coefficients are identical and they do not cause instability. The cause of instability is the difference between the unsteady temperature and concentration profiles. The difference increases when ε decreases, which is consistent with the results presented in Figure 6. Moreover, for $\varepsilon = 1$, the temperature and concentration

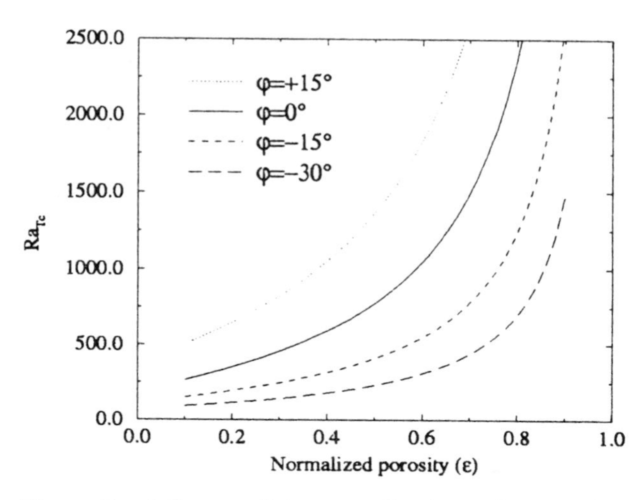

Figure 6. Influence of the normalized porosity ϵ on the critical Rayleigh number Ra_{Tc} of the Hopf bifurcation for $A = 1$ and $Le = 1$.

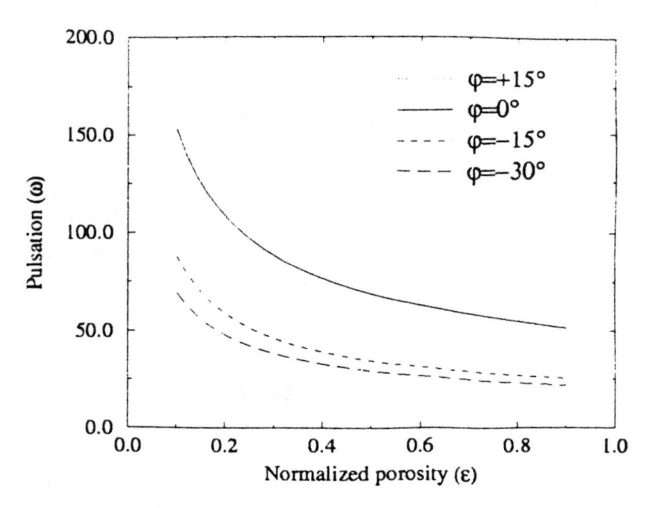

Figure 7. Influence of the normalized porosity ϵ on the pulsation ω_c for $A = 1$ and $Le = 1$.

profiles are identical and there are no sources of instability. The motionless double-diffusive solution is then infinitely linearly stable.

(b) $Le > 1$. In this case, the thermal diffusivity is higher than the mass diffusivity, which means that the concentration perturbations have the most destabilizing effect. Thus the stability of the motionless solution depends directly on the destabilizing effects of the concentration. Karimi-Fard et al. (1999) have shown that the lowest critical parameter is obtained for $\varphi = -90^\circ$ (the upper wall is maintained at the highest concentration), which corresponds to the case where the concentration field is the most destabilizing. This destabilizing effect decreases with φ, which induces the increase of the critical parameter. These authors demonstrated that he first primary bifurcation creates either branches of steady solutions or time-dependent solutions via Hopf bifurcation. They identified two types of steady bifurcation: transcritical or pitchfork bifurcations, depending on the aspect ratio of the box, as seen in Figure 8. The nature of bifurcation depends on ε, Le and A. The porosity of the porous medium was found to have a strong influence on the nature of the first bifurcation and there exists a threshold for convective motion even when $Le = 1$. These results agree with those obtained by Mamou et al. (1998) for a vertical cavity subjected to constant fluxes of heat and solute on the vertical walls when the two horizontal walls are impermeable and adiabatic. Trevisan and Bejan (1985),

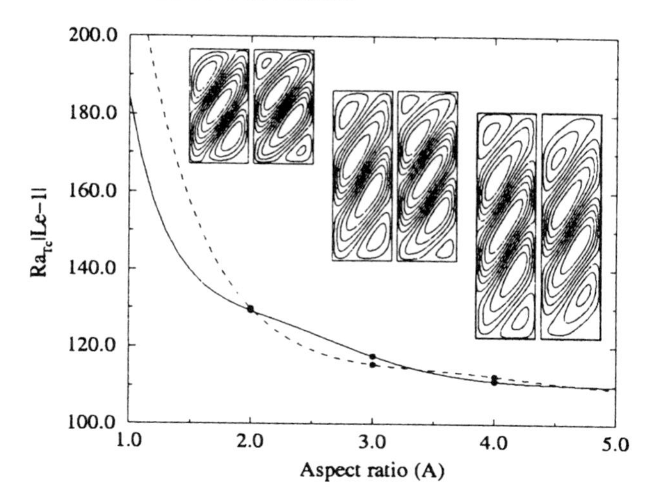

Figure 8. Evolution of transcritical (solid line) and pitchfork (dashed line) bifurcations with respect to the aspect ratio for $\varphi = 0$. The streamfunctions associated with the first bifurcation are drawn on the left side ($A = 2$, 3, and 4).

however, found that convection was strongly attenuated in the vicinity of $N = -1$ and that the flow disappeared completely if $Le = 1$ and $N = -1$.

The numerical resolution of the perturbation equations shows the existence of two zones in the (Le, ε) parameter space separated by the curve $\varepsilon Le^2 = 1$. When $\varepsilon Le^2 > 1$, the first primary bifurcation creates steady-state branches of the solution and, for $\varepsilon Le^2 < 1$, the first bifurcation is a Hopf bifurcation. It is important to observe that these results do not depend on either the aspect ratio or the tilt of the cavity. As can be observed in Figure 9, the same curve (solid line) was obtained for all tested angles of tilt.

(c) $Le < 1$. For Lewis numbers lower than one, the stability of the solution will depend on the destabilizing effect of the temperature. In this case the situation is more complicated. There are still two zones in the (Le, ε) parameter space, but they are separated by a curve depending on both the angle of tilt and the aspect ratio. Figure 9 shows the results obtained for a square cavity and for three angles of tilt ($\varphi = -15^\circ$, $\varphi = 0^\circ$ and $\varphi = 15^\circ$). Each discontinuous line represents a codimension-two bifurcation curve and delimits, with the curve defined by $\varepsilon Le^2 - 1 = 0$, the zone where the first bifurcation, which is a Hopf one, occurs. A section of Figure 9 for $\varepsilon = 0.5$ is presented in Figure 10. This figure shows the evolution of critical Rayleigh numbers associated with transcritical and Hopf bifurcation as a function of

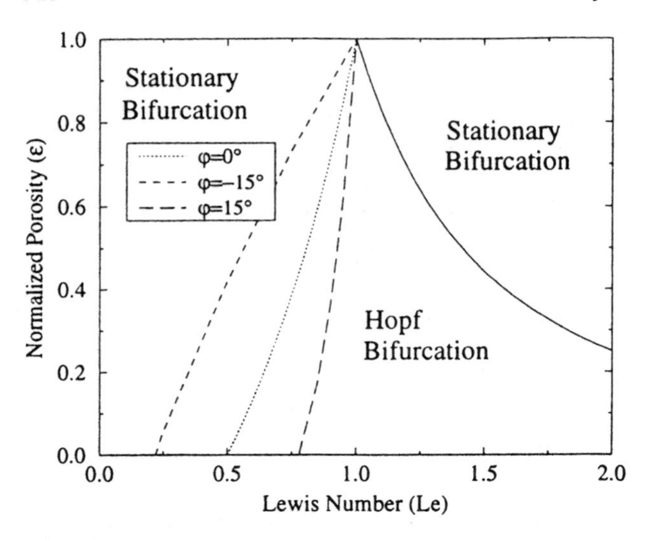

Figure 9. Domains of the existence of stationary and Hopf bifurcation in (Le, ϵ) parameter space for $A = 1$.

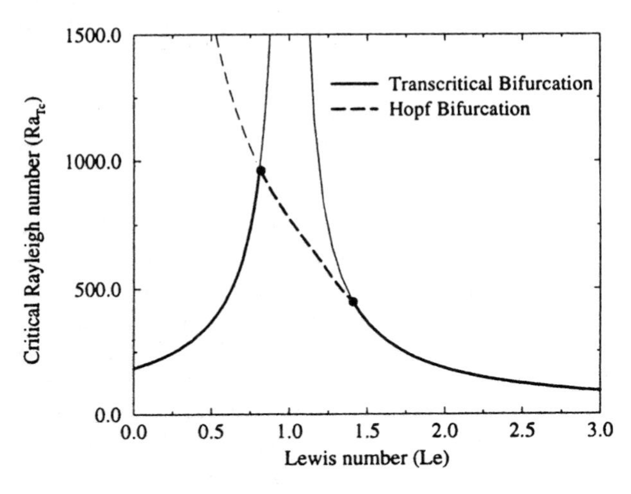

Figure 10. Critical Rayleigh number versus Lewis number, for $A = 1$, $\varphi = 0$, and $\epsilon = 0.5$.

Lewis number for $A = 1$ and $\varphi = 0^\circ$. The curve of Hopf bifurcation crosses the transcritical curve at a codimension-two bifurcation point.

3. Comparisons between Fluid and Porous Medium

These recent papers have been published in Physics of Fluids (Gobin and Bennacer 1994; Ghorayeb and Mojtabi 1998; Xin et al. 1998) on the same problem in a fluid medium with the same boundary conditions and for $N = -1$. In a fluid medium, the first primary bifurcation is never a Hopf one.

The existence of a Hopf bifurcation in a porous medium may be explained through normalized porosity. This parameter induces different evolutions in time between the temperature and the concentration. The difference is enhanced when the normalized porosity decreases. Indeed, diffusion and advection of concentration can only be carried out in space occupied by fluid and thus both diffusion and advection are magnified by ε^{-1}, compared to diffusion and advection of heat. On the other hand, results concerning the bifurcations which lead to steady states are very similar to the ones obtained in a fluid medium: the bifurcations are transcritical or pitchfork, depending on the aspect ratio A and the tilt of the cavity. The perturbation equations also have centro-symmetry.

B. Weakly Nonlinear Analysis

The purpose of this subsection is to get the normal form of the amplitude equation and to determine the characteristics of supercritical solutions (stream function, Nusselt number, and Sherwood number) near the bifurcation point for a square vertical cavity. The weakly nonlinear analysis that we are going to carry out is based on the multiscale technique. The nonlinear stability problem, formulated n terms of (ψ, θ, c), for $N = -1$, gives

$$\left.\begin{aligned} 0 &= \nabla^2\psi - Ra\left(\frac{\partial c}{\partial x} - \frac{\partial \theta}{\partial x}\right) \\ \frac{\partial \theta}{\partial t} &= \nabla^2\theta - \frac{\partial \psi}{\partial y} - \frac{\partial \psi}{\partial y}\frac{\partial \theta}{\partial x} + \frac{\partial \psi}{\partial x}\frac{\partial \theta}{\partial y} \\ \varepsilon\frac{\partial c}{\partial t} &= \frac{\nabla^2 c}{Le} - \frac{\partial \psi}{\partial y} - \frac{\partial \psi}{\partial y}\frac{\partial c}{\partial x} + \frac{\partial \psi}{\partial x}\frac{\partial c}{\partial y} \end{aligned}\right\} \tag{27}$$

Let us rewrite Eq. (27) in the form

$$\frac{\partial \tilde{\boldsymbol{u}}}{\partial t} = L(\boldsymbol{u}) - N(\boldsymbol{u}, \boldsymbol{u}) \tag{28}$$

where $\boldsymbol{u} = (\psi, \theta, c)$, $\tilde{\boldsymbol{u}} = (0, \theta, \varepsilon c)$,

$$L = \begin{bmatrix} \nabla^2 & Ra\partial/\partial x & -Ra\partial/\partial x \\ -\partial/\partial y & \nabla^2 & 0 \\ -\partial/\partial y & 0 & \nabla^2/Le \end{bmatrix} \tag{29}$$

and

$$N(\boldsymbol{u}, \boldsymbol{u}) = \left(0, \frac{\partial\psi}{\partial y}\frac{\partial\theta}{\partial x} - \frac{\partial\psi}{\partial x}\frac{\partial\theta}{\partial y}, \frac{\partial\psi}{\partial y}\frac{\partial c}{\partial x} - \frac{\partial\psi}{\partial x}\frac{\partial c}{\partial y}\right)$$

L and N represent the linear and nonlinear parts of the evolution operator, respectively.

In order to study the onset of convection near the critical Rayleigh number, we expand the linear operator and the solution into power series of the positive parameter η defined by

$$\eta = \frac{Ra - Ra_c}{Ra_c} \Rightarrow Ra = Ra_c(1 + \eta) \qquad \text{with} \qquad \eta \ll 1 \tag{30}$$

Thus

$$\left.\begin{aligned} L &= L_0 + \eta L_1 \\ \boldsymbol{u} &= \eta \boldsymbol{u}_1 + \eta^2 \boldsymbol{u}_2 \end{aligned}\right\} \tag{31}$$

where

$$L_0 = \begin{bmatrix} \nabla^2 & Ra_c\frac{\partial}{\partial x} & -Ra_c\frac{\partial}{\partial x} \\ -\partial/\partial y & \nabla^2 & 0 \\ -\partial/\partial y & 0 & \nabla^2/Le \end{bmatrix} \quad \text{and} \quad \begin{bmatrix} 0 & Ra_c\frac{\partial}{\partial x} & -Ra_c\frac{\partial}{\partial x} \\ 0 & 0 & 0 \\ 0 & 0 & 0 \end{bmatrix}$$

It may be noted that L_0 is the operator which governs the linear stability.

By introducing Eqs. (30) and (31) into (28), with the classical transformation of time $\tau = \eta t$, we obtain, after equating like powers of η, the sequential system of equations

$$\left.\begin{aligned} 0 &= L_0(\boldsymbol{u}_1) \qquad \text{at order } \eta \\ \frac{\partial\tilde{\boldsymbol{u}}_1}{\partial\tau} &= L_0(\boldsymbol{u}_2) + L_1(\boldsymbol{u}_1) - N(\boldsymbol{u}_1, \boldsymbol{u}_1) \qquad \text{at order } \eta^2 \\ \frac{\partial\tilde{\boldsymbol{u}}_2}{\partial\tau} &= L_0(\boldsymbol{u}_3) + L_1(\boldsymbol{u}_2) - N(\boldsymbol{u}_1, \boldsymbol{u}_2) - N(\boldsymbol{u}_2, \boldsymbol{u}_1) \qquad \text{at order } \eta^3 \end{aligned}\right\} \tag{32}$$

etc. The first-order equation leads us to solve the linear system

$$\left.\begin{aligned}
0 &= \nabla^2\psi_1 - Ra_c\left(\frac{\partial c_1}{\partial x} - \frac{\partial \theta_1}{\partial x}\right)\\
0 &= \nabla^2\theta_1 - \frac{\partial \psi_1}{\partial y}\\
0 &= \frac{\nabla^2 c_1}{Le} - \frac{\partial \psi_1}{\partial y}
\end{aligned}\right\} \tag{33}$$

Taking into account the boundary conditions (18), Eq. (33) yields $c_1 = Le\theta_1$ such that we have

$$\boldsymbol{u}_1 = A(\tau)(\psi_1, \theta_1, c_1 = Le\theta_1) = A(\tau)\boldsymbol{\phi}$$

where $\boldsymbol{\phi}$ is the eigenmode of the linear stability problem and A its amplitude. The solution of system (33) to the first order of approximation does not allow us to determine the amplitude (A). Only the minimum value of Ra_C is found. The eigenmode $\boldsymbol{\phi}$ may be written for a square cavity

$$\left.\begin{aligned}
\psi_1 &= \sum_{i=1}\sum_{j=1} a^1_{i,j}\sin(i\pi x)\sin(j\pi y),\\
\theta_1 &= \sum_{i=1}\sum_{j=0} b^1_{i,j}\sin(i\pi x)\cos(j\pi y), \qquad \text{and} \qquad c_1 = Le\theta_1
\end{aligned}\right\} \tag{34}$$

Substituting (34) into (33) we obtain, by direct identification

$$b^1 n, 0 = 0 \qquad \forall n$$

The amplitude $A(\tau)$ of the first-order solution is known by using the solvability of the Fredholm alternative or compatibility condition. Before solving the problem for each $\boldsymbol{u}_i$, it is necessary to determine the eigenmode of the adjoint operator L_0^* of L_0 defined by

$$L_0^* = \begin{bmatrix} \nabla^2 & \partial/\partial y & \partial/\partial y \\ -Ra_c\partial/\partial x & \nabla^2 & 0 \\ Ra_c\partial/\partial x & 0 & \nabla^2/Le \end{bmatrix}$$

The second equation of system (32) leads to

$$\frac{\mathrm{d}(A(\tau))}{\mathrm{d}\tau}\tilde{\boldsymbol{\phi}} = L_0(\boldsymbol{u}_2) + A(\tau)L_1(\boldsymbol{\phi}) - A^2(\tau)N(\boldsymbol{\phi}, \boldsymbol{\phi}) \tag{35}$$

The existence of a solution for Eq. (35) requires the compatibility equation to be satisfied such that

$$\frac{\mathrm{d}(A(\tau))}{\mathrm{d}\tau}\left\langle\boldsymbol{\phi}^*, \tilde{\boldsymbol{\phi}}\right\rangle = A(\tau)\langle\boldsymbol{\phi}^*, L_1(\boldsymbol{\phi})\rangle - A^2(\tau)\langle\boldsymbol{\phi}^*, N(\boldsymbol{\phi}, \boldsymbol{\phi})\rangle \tag{36}$$

where $\boldsymbol{\phi}^*$ is the eigenvector of L_0^* adjoint of L_0 and the inner product is defined as

$$\langle \psi, \theta \rangle = \int_0^1 \int_0^1 \psi\theta \mathrm{d}x\mathrm{d}y$$

To determine the coefficients of the amplitude equation (36) we must first solve the adjoint linear problem

$$\begin{aligned} 0 &= \nabla^2 \psi_1^* + \left(\frac{\partial c_1^*}{\partial y} + \frac{\partial \theta_1^*}{\partial y} \right) \\ 0 &= \nabla^2 \theta_1^* - Ra_c \frac{\partial \psi_1^*}{\partial x} \\ 0 &= \frac{\nabla^2 c_1^*}{Le} + Ra_c \frac{\partial \psi_1^*}{\partial x} \end{aligned} \tag{37}$$

Taking into account the boundary condition relative to the adjoint problem we obtain

$$c_1^* = -Le\theta_1^*$$

The eigenmode $\boldsymbol{\phi}^*$ for the adjoint problem may be written as

$$\left. \begin{aligned} \psi_1^* &= \sum_{i=1} \sum_{j=1} a_{i,j}^{1*} \sin(i\pi x) \sin(j\pi y), \\ \theta_1^* &= \sum_{i=1} \sum_{j=0} b_{i,j}^{1*} \sin(i\pi x) \cos(j\pi y) \qquad \text{and} \qquad c_1^* = -Le\theta_1^* \end{aligned} \right\} \tag{38}$$

After introducing the expression of the two eigenmodes $\boldsymbol{\phi}$ and $\boldsymbol{\phi}^*$ into (36) one obtains

$$\left\langle \boldsymbol{\phi}^*, N(\boldsymbol{\phi}, \boldsymbol{\phi}) \right\rangle \neq 0 \qquad \text{and} \qquad \left\langle \boldsymbol{\phi}^*, L_1(\boldsymbol{\phi}) \right\rangle \neq 0$$

when $\mathrm{d}A(\tau)/\mathrm{d}\tau = 0$, the two steady solutions for the square cavity are

$$A = 0 \qquad \text{and} \qquad A = \frac{\langle \boldsymbol{\phi}^*, L_1(\boldsymbol{\phi}) \rangle}{\langle \boldsymbol{\phi}^*, N(\boldsymbol{\phi}, \boldsymbol{\phi}) \rangle}$$

The stationary bifurcation is then transcritical. If we consider that $\theta_1^* = \bar{\theta}_1/(Le - 1)$, then, after some algebraic manipulations, we obtain the amplitude A

$$A = \frac{\langle \boldsymbol{\phi}, L_1(\boldsymbol{\phi}) \rangle}{\langle \boldsymbol{\phi}^*, N(\boldsymbol{\phi}, \boldsymbol{\phi}) \rangle} = \frac{Ra_c(Le - 1)}{Le + 1} \frac{\left\langle \psi_1^*, \frac{\partial \theta_1}{\partial x} \right\rangle}{\left\langle \bar{\theta}_1, \left(\frac{\partial \theta_1}{\partial x} \frac{\partial \psi_1}{\partial y} - \frac{\partial \theta_1}{\partial y} \frac{\partial \psi_1}{\partial x} \right) \right\rangle}$$

We verify that near the bifurcation point the stream function and temperature are proportional to the following: $(Ra - Ra_c)(Le - 1)/(Le + 1)$, which is in good agreement with the numerical results (Figure 11a).

The importance of thermal and mass exchange are given by the overall Nusselt and Sherwood numbers, respectively, at the vertical walls. The dimensionless Nu and Sh numbers are defined in a square cavity by

$$Nu = \int_0^1 -\frac{\partial T}{\partial x}\bigg|_{x=0 \text{ or } 1} dy \qquad \text{and} \qquad Sh = \int_0^1 -\frac{\partial C}{\partial x}\bigg|_{x=0 \text{ or } 1} dy \tag{39}$$

Substituting T and C by their expressions into Eq. (39), we obtain

$$\begin{aligned} Nu &= 1 + \eta \int_0^1 -\frac{\partial \theta_1}{\partial x}\bigg|_{x=0,1} dy + \eta^2 \int_0^1 -\frac{\partial \theta_2}{\partial x}\bigg|_{x=0,1} dy + \cdots \\ Sh &= 1 + \eta \int_0^1 -\frac{\partial c_1}{\partial x}\bigg|_{x=0,1} dy + \eta^2 \int_0^1 -\frac{\partial c_2}{\partial x}\bigg|_{x=0,1} dy + \cdots \end{aligned} \tag{40}$$

If we introduce θ_1 and c_1 given by (34) into (40), we verify that

$$\int_0^1 \frac{\partial \theta_1}{\partial x}\bigg|_{x=0,1} = \frac{1}{Le}\int_0^1 \frac{\partial c_1}{\partial x}\bigg|_{x,0,1} = \sum_{n=1} b^1_{n,0} = 0 \qquad \text{since:} \qquad b^1_{n,0} = 0 \qquad \forall n$$

The final expressions of Nu and Sh are then:

$$\left.\begin{aligned} Nu &= 1 + \eta^2 \int_0^1 -\frac{\partial \theta_2}{\partial x}\bigg|_{x=0,1} dy + \cdots \\ Sh &= 1 + \eta^2 \int_0^1 -\frac{\partial c_2}{\partial x}\bigg|_{x=0,1} dy + \cdots \end{aligned}\right\} \tag{41}$$

These results show that $(Nu - 1)$ and $(Sh - 1)$ are proportional to η^2. The numerical simulation performed in this study confirms this analytical result (Figures 11b and c).

C. Numerical Results

1. Numerical Procedures

Two numerical models based on formulation with primitive variables, one with a spectral collocation method and the second one with a finite volume method, have been performed (Charrier-Mojtabi et al. 1997).

The validity of the two codes was first established by comparing our results to those obtained by Goyeau et al. (1996) and Trevisan and Bejan (1985). For fluxes of heat and mass prescribed at vertical walls, we also

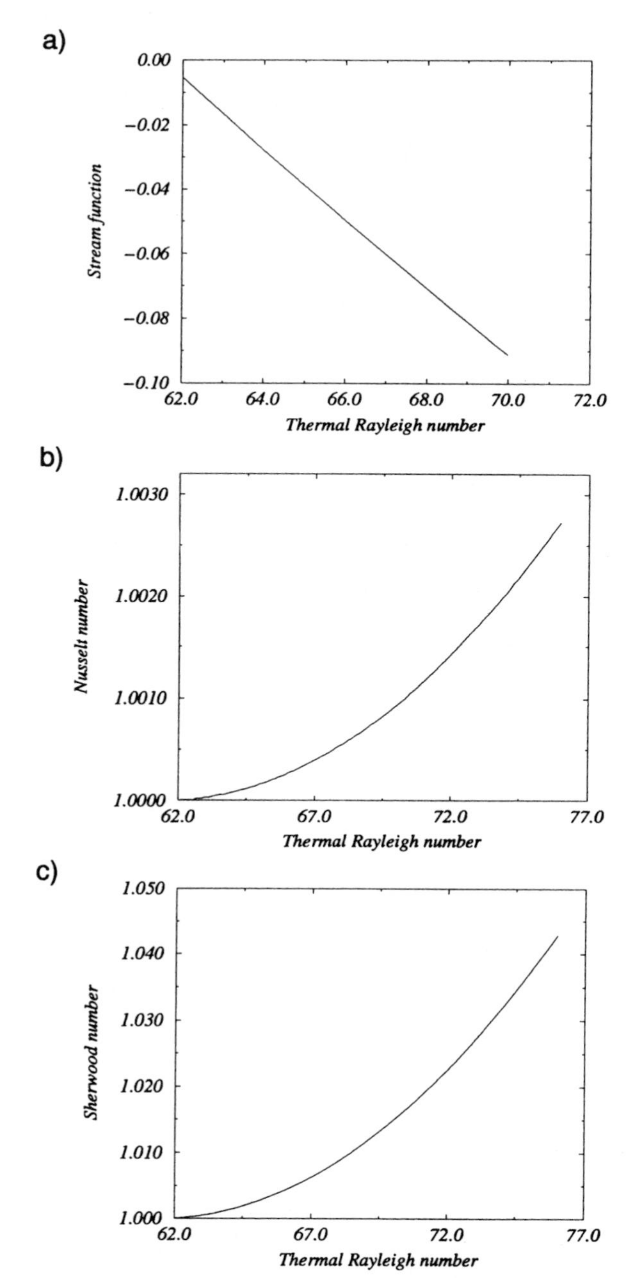

Figure 11. $Le = 4$, $N = -1$, $A = 1$: (a) stream function at the center of the cavity versus Ra, near Ra_c; (b) average Nusselt number versus Ra, near Ra_c; (c) average Sherwood number versus Ra, near Ra_c.

compared our results to those obtained by Alavyoon et al. (1994). We found, like these authors, that oscillatory flows occur for sufficiently large values of the Rayleigh number

2. Numerical Determination of the Critical Rayleigh Number Ra_c for Different Values of the Lewis Number

For the present case (constant temperatures and concentrations imposed at the vertical walls), the study of the transition between the purely diffusive regime and the thermosolutal convective regime, obtained for $N = -1$, was carried out for $Le = 0.1, 0.2, 0.3, 2, 3, 4, 7, 11$, in a square cavity ($A = 1$) and for $\varphi = 0°$.

The transition between the equilibrium solution and the convective regime systematically occurs for a critical thermal Rayleigh number satisfying the relation

$$Ra_c|Le - 1| = 184.06$$

This is in very good agreement with the stability analysis performed in Section III.A, as indicated in Table 1 and Figure 12. The thermosolutal supercritical convective regime obtained just after the transition is symmetrical with respect to the center of the cavity, as shown in Figure 13. For $Le = 4$, $A = 1$, $N = -1$, the stream function at the center of the cavity, the global Nusselt number, and the Sherwood number are plotted as functions of the Rayleigh number (Figure 11) near the bifurcation point. We observe that the stream function depends linearly on the Rayleigh number, whereas the global Nusselt number and Sherwood number vary quadratically with the Rayleigh number. These variations are in good agreement with the results obtained by nonlinear stability analysis (Eq. (41)). The bifurcation diagrams for the value of the stream function in the center of the cavity, the

Table 1. $Ra_c|Le - 1|$ as a function of the aspect ratio A

$A = 0.5$	$N \times M$	6×6	7×7	8×8	20×10	40×20
	$Ra_c|Le - 1|$	517.36	517.12	517.01	516.87	516.85
$A = 1$	$N \times M$	6×6	7×7	8×8	20×20	30×30
	$Ra_c|Le - 1|$	184.33	184.15	184.13	184.06	184.06
$A = 2$	$N \times M$	6×6	7×7	8×8	8×16	20×40
	$Ra_c|Le - 1|$	129.34	129.38	129.25	129.22	129.21
$A = 5$	$N \times M$	6×6	7×7	8×8	5×25	14×70
	$Ra_c|Le - 1|$	109.71	109.55	109.31	109.21	109.16
$A = 10$	$N \times M$	7×7	8×8	3×30	6×60	10×100
	$Ra_c|Le - 1|$	117.75	111.01	106.77	106.37	106.35

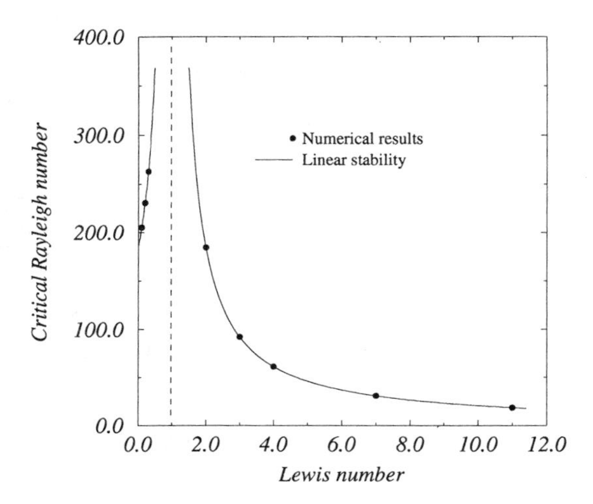

Figure 12. $Ra_c = f(Le)$ for $A = 1$, $\varphi = 0$: analytical and numerical results.

global Nusselt, and Sherwood number are presented in Figures 13a–c, respectively, for $Le = 4$, $N = -1$ and $A = 1$. One can observe the presence of two other branches of the solution (branches II and III), different to the one corresponding to the transition described in the previous section (branch I); see Figure 14.

D. Scale Analysis

Scale analysis was applied to double-diffusive convection in order to determine the heat and mass transfer at the wall.

1. Boundary Layer Flow

Bejan and Khair (1985) studied the phenomenon of naturally convective heat and mass transfer near a vertical surface embedded in a fluid-saturated porous medium. The vertical surface is maintained at a constant temperature T_0 and constant concentration C_0, different to the porous medium temperature T_∞ and concentration C_∞ observed sufficiently far from the wall. The scale of the flow, temperature, and concentration fields near the vertical wall are determined on the basis of order-of-magnitude analysis.

This study shows that the vertical boundary-layer flux is driven by heat transfer when ($|\beta_T \Delta T| \gg |\beta_C \Delta C| \Leftrightarrow |N| \ll 1$) or by mass transfer when ($|\beta_C \Delta C| \gg |\beta_T \Delta T| \Leftrightarrow |N| \gg 1$), or by a combination of heat and mass

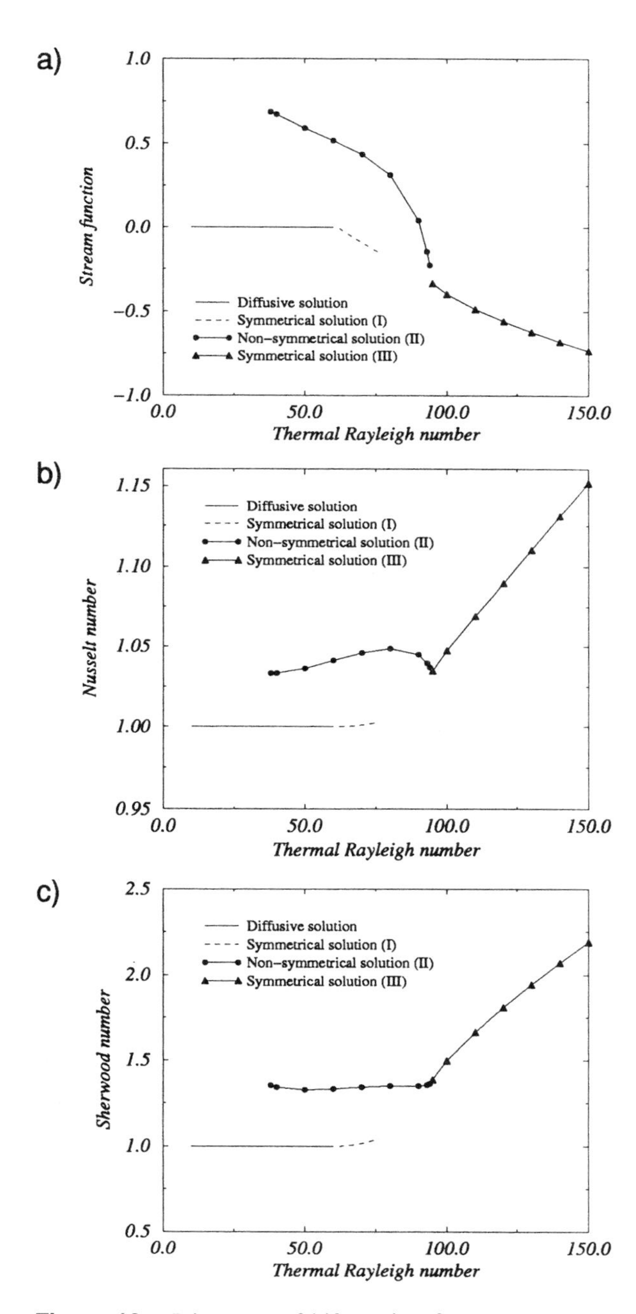

Figure 13. Diagrams of bifurcation for $Le = 4$, $N = -1$, $A = 1$: (a) stream function $= f(Ra)$, at the center of the cavity; (b) Nusselt number $= f(Ra)$; (c) Sherwood number $= f(Ra)$.

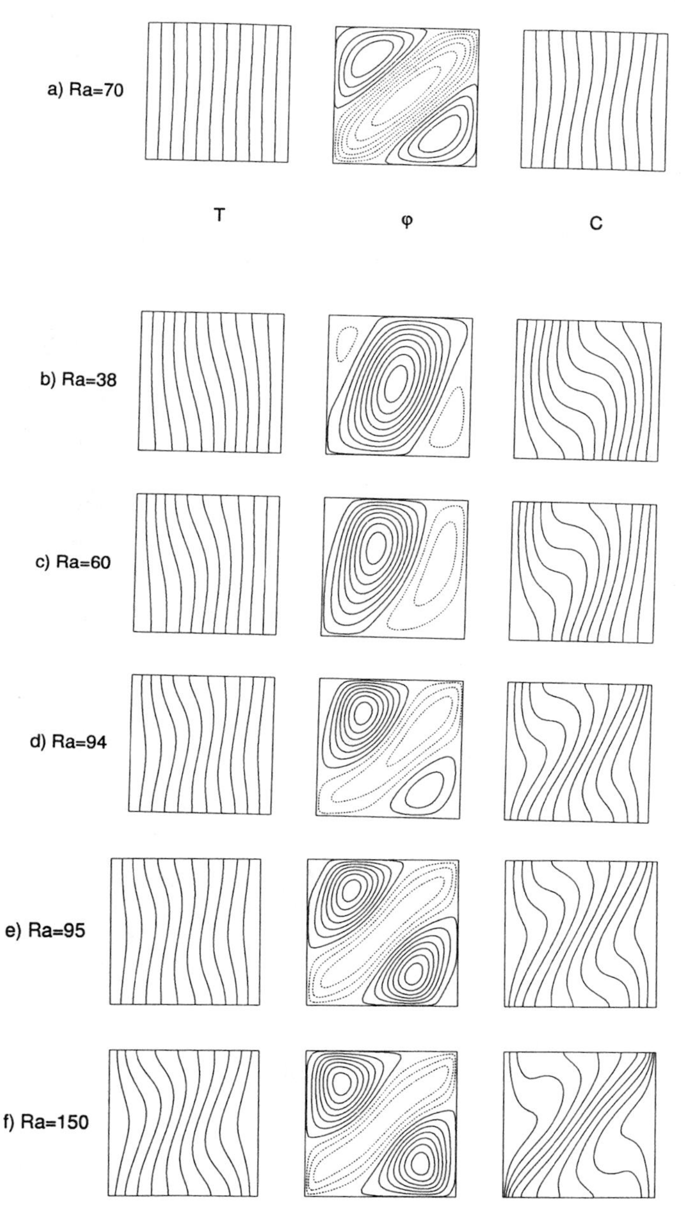

Figure 14. Isotherms, streamlines and isoconcentrations for $Le = 4$, $N = -1$, $A = 1$: branch I: (a) $Ra = 70$; branch II: (b) $Ra = 38$, (c) $Ra = 60$, (d) $Ra = 94$; branch III: (e) $Ra = 95$, (f) $Ra = 150$ (dashed lines correspond to clockwise rotations).

transfer effects. These authors have distinguished four limiting regimes, depending on N and Le numbers:

For heat transfer driven flow ($|N| \ll 1$) they found for $Le \gg 1$: $Nu \approx Ra^{1/2}$ and $Sh \approx (RaLe)^{1/2}$; and for $Le \ll 1$: $Nu \approx Ra^{1/2}$ and $Sh \approx Ra^{1/2}Le$.

For mass transfer driven flow ($|N| \gg 1$) they found for $Le \gg 1$: $Nu \approx (Ra|N|/Le)^{1/2}$ and $Sh \approx (RaLe|N|)^{1/2}$ and for $Le \ll 1$: $Nu \approx (Ra|N|)^{1/2}$ and $Sh \approx (RaLe|N|)^{1/2}$.

2. Effect of the Buoyancy Ratio *N* on the Heat and Mass Transfer Regimes in a Vertical Porous Enclosure

Previous studies have dealt with vertical boxes with either imposed temperatures and concentrations along the vertical side-walls (Trevisan and Bejan 1985; Charrier-Mojtabi et al. 1997; Karimi-Fard et al. 1999), or prescribed heat and mass fluxes across the vertical side-walls (Trevisan and Bejan 1986; Alavyoon et al. 1994; Mamou et al. 1995b, 1998). For both of these boundary conditions, when the ratio of the solutal to thermal buoyancy forces, N, is equal to -1, a purely diffusive state (equilibrium solution) can be obtained at low thermal Rayleigh numbers and any Lewis number (Karimi-Fard et al. 1999).

In general, flow and transport follow complex patterns depending on the aspect ratio of the cell, the interaction between the diffusion coefficients (Le), and the buoyancy ratio (N). These groups account for the many distinct heat and mass transfer regimes that can exist. Trevisan and Bejan (1985) identified these regimes on the basis of scale analysis and numerical experiments.

For heat driven flows ($|N| \ll 1$) there are five distinct regimes and, in each subdomain of the two-dimensional domain (Le, Ra_T/A^2) they give the overall heat and mass transfer rates as follows:

$$\text{subdomain 1:} \qquad Sh \approx \frac{1}{A}(Ra_T Le)^{1/2}, \qquad Nu \approx \frac{1}{A} Ra_T^{1/2}$$

In the case of $N = 0$ and $A = 1$, the numerical simulations conducted by Goyeau et al. (1996) show that the Nusselt number does not depend on the Lewis number for a given Ra_T, since the flow is totally driven by the thermal buoyancy force. On the other hand, the Sherwood number increases with Le and Ra_T. The power law deduced from the computed values of the Sherwood number gives $Sh = 0.40\ (Ra_T Le)^{0.51}$, which is in close agreement with the preceding scaling law.

subdomain 2: $Sh \approx \frac{1}{A} Le Ra_T^{1/2}, \qquad Nu \approx \frac{1}{A} Ra_T^{1/2}$

subdomain 3: $Sh \approx 1, \qquad Nu \approx \frac{1}{A} Ra_T^{1/2}$

subdomain 4: $Sh \approx 1, \qquad Nu \approx 1$

subdomain 5: $Sh \approx \frac{1}{A} (Ra_T Le)^{1/2}, \qquad Nu \approx 1.$

For mass-driven flows ($|N| \gg 1$) five distinct regimes are also possible, and in each subdomain of the two-dimensional domain (Le, $Ra_T |N| / A^2$) the authors give the overall heat and mass transfer rates as

subdomain 1: $Sh \approx \frac{1}{A} (Ra_T |N| Le)^{1/2},$

$$Nu \approx \frac{1}{A Le^{1/2}} (Ra_T |N|)^{1/2}$$

A regression of numerical results obtained by Goyeau et al. (1996) for higher values of N and $A = 1$ leads to the following correlation: $Sh = 0.75 (Ra_T Le N)^{0.46}$, where the exponent is in fairly good agreement with the value 0.5 assessed by the scale analysis.

subdomain 2: $Sh \approx \frac{1}{A} (Ra_T |N| Le)^{1/2}, \qquad Nu \approx \frac{1}{A} (Ra_T |N|)^{1/2}$

subdomain 3: $Sh \approx 1, \qquad Nu \approx \frac{1}{A} (Ra_T |N|)^{1/2}$

subdomain 4: $Sh \approx 1, \qquad Nu \approx 1$

subdomain 5: $Sh \approx \frac{1}{A} (Ra_T |N| Le)^{1/2}, \qquad Nu \approx 1$

It is shown by Goyeau et al. (1996) that numerical results for mass transfer are in good agreement with the scaling analysis over a wide range of parameters. As a conclusion to the analysis presented by these authors, it is clear that more investigations are required in order to derive the appropriate scaling laws in the domains where the flow is fully dominated neither by the thermal nor by the solutal component of the buoyancy force.

IV. SORET EFFECT AND THERMOGRAVITATIONAL DIFFUSION IN MULTICOMPONENT SYSTEMS

A. Soret Effect

A review of these studies may be found in the papers by Platten and Legros (1984) and Turner (1985). Binary fluids in a horizontal porous cell, initially homogeneous in composition, heated from below, will, in the steady state, display a concentration gradient due to the so-called thermal diffusion or Soret effect. Therefore, depending on the sign of the Soret coefficient, the onset of convection can be delayed or anticipated. The Soret coefficient is strongly dependent on composition of the binary fluids. In the last decade, a reviewed interest was given to this problem due to the rich dynamic behavior involved in the stabilizing concentration gradient. The first instability sets in as oscillations of increasing amplitude, while the first bifurcation is stationary in horizontal cells saturated by a pure fluid in the Rayleigh–Bénard configuration. Finite amplitude convection is characterized by traveling waves, and sometimes by localized traveling waves, etc. Next, with increasing Rayleigh number, there is a bifurcation towards steady overturning convection.

The critical Rayleigh number, deduced from linear stability theory, for the marginal state of stationary instability, in the absence of an imposed solutal gradient, is given by

$$Ra_c = \frac{4\pi^2}{1 + S_T^*(1 + Le)} \tag{42}$$

We find for free, permeable, and conductive boundaries in a fluid medium a similar relation

$$Ra_c = \frac{27\pi^4/4}{1 + S_T^*(1 + Le)}$$

where Ra_c is the critical Rayleigh number corresponding to exchange of stability. Marginal oscillatory instability occurs for

$$Ra_c = \frac{4\pi^2(\sigma + \varepsilon^* Le)}{Le(\varepsilon^* + \sigma S_T^*)} \tag{43}$$

The general situation, with both cross-diffusion and double diffusion (thermal and solutal gradients imposed), was studied by Patil and Rudraiah (1980). Brand and Steinberg (1983) pointed out that with the Soret effect it is possible to have oscillatory convection induced by heating from above. Generally, the mass and heat fluxes are given respectively by

$$j_C = -\rho D \nabla C' - \rho C'(1 - C') D_T \nabla T' \tag{44}$$

$$j_T = -\lambda \nabla T' - \rho C' T' D_T \frac{\partial \mu}{\partial C} \nabla C' \tag{45}$$

where μ is the chemical potential of the solute.

The ratio $S_T = D_T/D =$ (thermal diffusion coefficient)/(isothermal diffusion coefficient) is commonly referred to as the Soret coefficient (in K^{-1}). Its magnitude and sign may vary to a large extent from one chemical to another, and even, for a given chemical, S_T is a complicated function of state variables. When we impose a concentration gradient ($C = 0$ at $x = 0$ and $C = 1$ at $x = 1$) in the dimensionless form, it is usual to ignore the Soret effect (i.e., the concentration gradient induced by the temperature gradient). This is due to the low values of the Soret coefficient; for classical binary mixtures S_T is between 10^{-4} and $10^{-2}\,K^{-1}$. We have what is called the thermosolutal problem. In this case the concentration gradient exists even in the absence of a thermal gradient.

Knobloch (1980) and Taslim and Narusawa (1986) demonstrated in a fluid medium and porous medium respectively that a close relationship exists between cross-diffusion problems (taking into account the Dufour effect and the Soret effect) and double-diffusion problems. In fact, they demonstrated that these two problems are mathematically identical.

B. Thermogravitational Diffusion

Thermogravitational diffusion denotes a physical process occurring when a thermal gradient is applied on a fluid mixture. It might contribute to large numbers of natural physical processes.

A fluid mixture saturating a vertical porous cavity under a gravity field, and exposed to a uniform horizontal thermal gradient, is subject not only to convective transfer but also to thermodiffusion, corresponding to a concentration gradient associated with the Soret effect. The coupling of these two phenomena is called thermogravitational diffusion and leads to species separation. The convective steady state obtained in this case is characterized by a large concentration contrast between the top and the bottom of the cell. This contrast is measured by the separation factor, which is defined as the ratio of the mass fraction of the denser component at the bottom of the cell to its mass fraction at the top ($q = C_{\text{bottom}}/C_{\text{top}}$).

This phenomenon, well known for more than a hundred years, has been lately under investigation owing to its involvement in several natural physical processes in geophysics and mineralogy, where a fluid saturates a porous medium (Jamet et al. 1992; Benano-Melly et al. 1999). Industrial projects using this thermogravitational diffusion phenomenon coupled to convection in order to separate or to concentrate species have been developed.

The different analytical studies by Fury et al. (1939) and by Estebe and Schott (1970) into this phenomenon have shown the existence of a maximum separation ratio obtained for the corresponding optimum permeability. Marcoux and Charrier-Mojtabi (1998) consider a thermogravitational cell bounded by temperature-imposed vertical walls and adiabatic horizontal walls and filled with homogeneous isotropic porous medium saturated by a two-component incompressible fluid. The dimensionless form of the equations considered in that work lead to five parameters: the thermal Rayleigh number, the buoyancy ratio N, the normalized porosity, the Lewis number, and the dimensionless Soret number. These authors have numerically studied the influence of each of these parameters in species separation. The numerical results show the expected existence of a maximum separation corresponding to an optimal Rayleigh number. But up to now there is no agreement between numerical and experimental results, already observed by Jamet et al. (1992) and Marcoux and Charrier-Mojtabi (1998), and, as seen in Figure 15, this remains an open question still to be resolved. The numerical curves in Figure 16 show the influence, on the optimal Ra number, of vertical separation of the Lewis number and the Soret number. They confirm rather simple previous analytical results established by Estebe and Schott (1970), who evaluated the maximum separation ratio and the

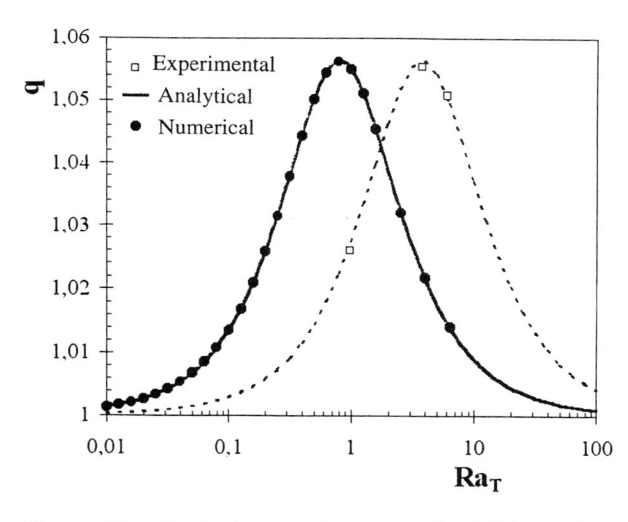

Figure 15. Vertical separation versus Rayleigh number.

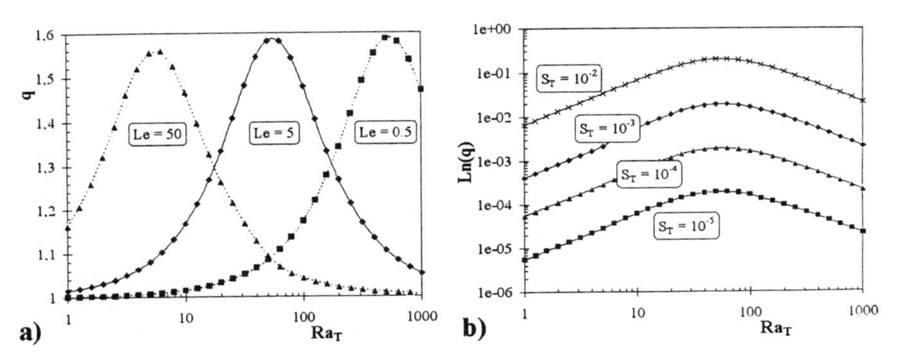

Figure 16. Vertical separation versus Rayleigh number, for various values of Lewis number (a) and Soret number (b).

corresponding optimum permeability as functions of the different physical parameters.

V. CONCLUSIONS AND OUTLOOK

Several modern engineering processes can benefit from a better understanding of double-diffusive convection in saturated porous medium where the flow and transport follow complex patterns that depend on the interaction between diffusion coefficients and buoyancy ratio.

In geophysics, recent efforts are focused more on heat and mass transfer flows in regions below geothermal reservoirs in order to provide better understanding of the processes which transfer heat and chemicals from deep magmatic sources to the base of reservoirs and to surface discharge features (McKibbin 1998).

Another important area of practical interest is one in materials science, namely in the casting and solidification of metal alloy where double-diffusive convection in the mushy zone, characterized by high variation of porosity, can have an important effect on the quality of the final product (Sinha and Sundararajan 1992; Gobin et al., 1998).

The double-diffusive convection phenomena described in this chapter depend essentially on the gravity field, but they can also be observed in the case of pure weightlessness in a cavity filled with a saturated porous medium subjected to vibration (Khallouf et al. 1995). It is the coupling between these two external force fields (gravity and inertia) and the diffusion that organizes the flow into a form which permits its control (Gershuni and Lyubimov, 1998).

In microgravity conditions, the surface tension effect can induce stable convective motions when the conductive situation becomes unstable. Both linear and nonlinear stability analyses of Marongoni double-diffusive convection in binary mixtures, saturated porous media, subjected to the Soret effect, are needed for a better understanding and better control of fluid motions in microgravity.

Comprehensive predictions, made possible by means of the thermogravitational diffusion model, require experimental values of the Soret coefficient. For most binary mixtures the Soret coefficient is unknown. To date, quantitative experimental data suitable for model validation are quite scarce and, thus, coordinated efforts between modeling and experimentation are needed to provide an ultimate understanding of double-diffusive convection in porous media.

NOMENCLATURE

Roman Letters

A	aspect ratio of the cell, H/L
a	thermal diffusivity
C	mass fraction
C_1 (C_2)	mass fraction at cold (hot) vertical wall, $\Delta C = C_2 - C_1$
C_r	reference mass fraction, $= C_1$
c	disturbance concentration
H	height of the cavity
L	width of the cavity
Le	Lewis number; $Le = a/D$
D	mass diffusivity of the constituent through the fluid mixture
g	intensity of gravity ($\boldsymbol{g} = -g\boldsymbol{e}_y$)
k	wavenumber
k_c	critical wavenumber
k^*	effective thermal conductivity of the porous medium
I	$\sqrt{-1}$
M, N	orders of approximation
N	Buoyancy ratio ($N = Ra_S/Ra_T$)
Nu	average Nusselt number
q	vertical separation
Ra_c	critical thermal Rayleigh number
Ra_S	solutal Rayleigh number based on L; $Ra_s = Kg\beta_C(\rho c)_f \Delta CL/k^*\nu$
Ra_T	thermal Rayleigh number based on L; $Ra_T = Kg\beta_T(\rho c)_f \Delta TL/k^*\nu$
r_i	inner cylinder radius

r_o	outer cylinder radius
S_T	dimensionless Soret number
Sh	average Sherwood number
T	dimensionless temperature
T_1 (T_2)	temperature at cold (hot) vertical wall, $\Delta T = T_2 - T_1$
T_r	reference temperature, $= T_1$
U	dimensionless horizontal component of the velocity
V	dimensionless vertical component of the velocity

Greek Letters

α	wavenumber
α_e	effective thermal diffusivity of the porous medium
β_T	coefficient of volumetric expansion with respect to the temperature
β_c	coefficient of volumetric expansion with respect to the concentration
γ	curvature parameter, $= (r_o - r_i)/r_i$
ε	normalized porosity, $= \varepsilon^*(\rho c)_f/(\rho c)^*$
ε^*	porosity of the porous matrix
θ	disturbance temperature
ν	kinematic viscosity of fluid
$(\rho c)_f$	heat capacity of fluid
$(\rho c)^*$	heat capacity of saturated porous medium
σ	heat capacity ratio, $\sigma = (\rho c)_f/(\rho c)^*$
ψ	disturbance stream function

REFERENCES

Alavyoon F. On natural convection in vertical porous enclosures due to prescribed fluxes of heat and mass at the vertical boundaries. Int J Heat Mass Transfer 36:2479–2498, 1993.

Alavyoon F, Masuda Y, Kimura S. On natural convection in vertical porous enclosures due to opposing fluxes of heat and mass prescribed at the vertical walls. Int J Heat Mass Transfer 37:195–206, 1994.

Angirasa D, Peterson GP, Pop I. Combined heat and mass transfer by natural convection with opposing buoyancy effects in a fluid saturated porous medium. Int J Heat Mass Transfer 40:2755–2773, 1997.

Bejan A. Convection Heat Transfer. New York: Wiley, 1984 (2nd ed, 1995).

Bejan A, Khair KR. Heat and mass transfer by natural convection in a porous medium. Int J Heat Mass Transfer 28:909–918, 1985.

Benano-Melly L, Caltagirone JP, Faissat B, Montel F. Etude sur un modèle de transport dans les mélanges en milieux poreux. J Entropie, 17: 41–45, 1999.

Brand H, Steinberg V. Convective instabilities in binary mixtures in a porous medium. Physica A 119:327–338, 1983.

Brand HR, Hohenberg PC, Steinberg V. Amplitude equation near a polycritical point for the convective instability of binary fluid mixture in a porous medium. Phys Rev A27:591–593, 1983.

Chan BKC, Ivey CM, Arry JM. Natural convection in enclosed porous media with rectangular boundaries. J Heat Transfer 92:21–27, 1970.

Charrier-Mojtabi MC, Karimi-Fard M, Azaiez M, Mojtabi A. Onset of a double diffusive convective regime in a rectangular porous cavity. J Porous Media 1:104–118, 1997.

Chen F, Chen CF. Double-diffusive fingering convection in porous medium. Int J Heat Mass Transfer 36:793–807, 1993.

Cheng P. Heat transfer in geothermal systems. Adv Heat Transfer 14:1–105, 1978.

Combarnous M, Bories S. Hydrothermal convection in saturated porous media. In: Advances in Hydroscience. Academic Press, 1975, pp 231–307.

De Groot SR, Mazur P. Nonequilibrium thermodynamics. Amsterdam: North-Holland Pub. Co., 1962.

Estebe J, Schott J. Concentration saline et cristallisation dans un milieu poreux par effet thermogravitationnel. CR Acad Sci Paris 271:805–807, 1970.

Ettefagh J, Vafai K, Kim SJ. Non-Darcian effects in open-ended cavities filled with a porous medium. J Heat Transfer 113:747–756, 1991.

Fury WH, Jones RC, Onsager L. On the theory of isotope separation by thermal diffusion. Phys Rev 55:1083–1095, 1939.

Gershuni DZ, Zhukhovisskii EM, Lyubimov DV. Thermal concentration instability of a mixture in a porous medium. Sov Phys Dokl 21:375–377, 1976.

Gershuni GZ, Luybimov DV. Thermal vibrational convection. New York: John Wiley and Sons, 1998.

Ghorayeb, K, Mojtabi A. Double-diffusive convection in vertical rectangular cavity. Phys Fluids 9:2339–2348, 1997.

Gobbé C, Quintard M. Macroscopic description of unsteady heat transfer in heterogeneous media. High Temp High Press 26:1–14, 1994.

Gobin D, Bennacer R. Double diffusion in a vertical fluid layer: Onset of the convective regime. Phys Fluids 6:59–67, 1994.

Gobin D, Goyeau B, Songbe JP. Double diffusive convection in a composite fluid-porous layer. J Heat Transfer 120:234–242, 1998.

Goyeau B, Songbe JP, Gobin D. Numerical study of double-diffusive natural convection in a porous cavity using the Darcy–Brinkman formulation. Int J Heat Mass Transfer 39:1363–1378, 1996.

Griffith RW. Layered double-diffusive convection, in porous media. J Fluid Mech 102:221–248, 1981.

Hassan M, Mujumdar AS. Transpiration-induced buoyancy effect around a horizontal cylinder embedded in porous medium. Int J Energy Res 9:151–163, 1985.

Hsu CT, Cheng P. The Brinkman model for natural convection about a semi-infinite vertical flat plate in porous medium. Int J Heat Mass Transfer 28:683–697, 1985.

Imhoff BT, Green T. Experimental investigation in a porous medium. J Fluid Mech 188:363–382, 1988.

Jamet P, Fargue D, Costesèque P, de Marsily G, Cernes A. The thermogravitational effect in porous media: a modelling approach. Transp Porous Media 9:223–240, 1992.

Jang JY, Chang WJ. Buoyancy-induced inclined boundary layer flows in a porous medium resulting from combined heat and mass buoyancy effects. Int Commun Heat Mass Transfer 15:17–30, 1988a.

Jang JY, Chang WJ. The flow and vortex instability of horizontal natural convection in a porous medium resulting from combined heat and mass buoyancy effects. Int J Heat Mass Transfer 31:769–777, 1988b.

Karimi-Fard M, Charrier-Mojtabi MC, Vafai K. Non-Darcian effects on double-diffusive convection within a porous medium. Numer Heat Transfer A 31:837–852, 1997.

Karimi-Fard M, Charrier-Mojtabi MC, Mojtabi A. Onset of stationary and oscillatory convection in a tilted porous cavity saturated with a binary fluid. Phys Fluids 11:1346–1358, 1999.

Kaviany M. Principles of Convective Heat Transfer in Porous Media. 2nd ed. New York: Springer, 1995.

Khallouf H, Gershuni GZ, Mojtabi A. Numerical study of two-dimensional thermovibrational convection in rectangular cavities. Numer Heat Transfer 27:297–305, 1995.

Khan AA, Zebib A. Double-diffusive instability in a vertical layer of a porous medium. J Heat Transfer 103:179–181, 1981.

Knobloch E. Convection in binary fluids. Phys Fluids 23:1918–1920, 1980.

Lage JL. Effect of the convective inertia term on Bénard convection in a porous medium. Numer Heat Transfer A 22:469–485, 1992.

Lage JL. The fundamental theory of flow through permeable media from Darcy to turbulence. Transport Phenomena in Porous Media. Oxford: Pergamon/Elsevier Science, 1998, pp 1–30.

Lai FC, Kulacki FA. Coupled heat, and mass transfer from a sphere buried in an infinite porous medium. Int J Heat Mass Transfer 33:209–215, 1990.

Lai FC, Kulacki FA. Coupled heat, and mass transfer by natural convection from vertical surfaces in porous medium. Int Heat Mass Transfer 34:1189–1194, 1991.

Lin KW. Unsteady natural convection heat and mass transfer in saturated porous enclosure, wäm-und Stoffübertragung 28:49–56, 1993.

Lin TF, Huang CC, Chang TS. Transient binary mixture natural convection in square enclosures. Int J Heat Mass Transfer 33:287–299, 1990.

McKibbin R. Mathematical models for heat and mass in geothermal systems. Transport Phenomena in Porous Media. Oxford: Pergamon/Elsevier Science, 1998, pp 131–154.

Mamou M, Vasseur P, Bilgen E, Gobin D. Double-diffusive convection in an inclined slot filled with porous medium. Eur J Mech B/Fluids 14:629–652, 1995.

Mamou M, Vasseur P, Bilgen E. Multiple solutions for double-diffusive convection in a vertical porous enclosure. Int J Heat Mass Transfer 38:1787–1798, 1995.

Mamou M, Vasseur P, Bilgen E. Double-diffusive convection instability in a vertical porous enclosure. J Fluid Mech 368:263–289, 1998.

Marcoux M, Platten JK, Chavepeyer G, Charrier-Mojtabi MC. Diffusion thermogravitationnelle entre deux cylindres coaxiaux; effets de la courbure. Entropie 198:89–96, 1996.

Marcoux M, Charrier-Mojtabi MC. Etude paramétrique de la thermogravitation en milieu poreux. Comptes Rendus de l'Academie des Sciences 326:539–546, 1998.

Marcoux M, Charrier-Mojtabi MC, Azaiez M. Double-diffusive convection in an annular vertical porous layer. Int J Heat Mass Transfer, 2313–2325, 1999.

Murray BJ, Chen CF. Double-diffusive convection in a porous medium. J Fluid Mech 201:147–166, 1989.

Nakayama A, Hossain MA. An integral treatment for combined heat and mass transfer by natural convection in a porous medium. Int J Heat Mass Transfer 38:761–765, 1995.

Nguyen HD, Paik S, Douglass RW. Study of double diffusion convection in layered anisotropic porous media. Numer Heat Transfer B 26:489–505, 1994a.

Nguyen HD, Paik S, Venkatakrishnan KS. Thermohaline instability in rotating anisotropic porous media. Numer Heat Transfer B 26:489–505, 1994a.

Nield DA. Onset of thermohaline convection in a porous medium. Water Resour Res 4:553–560, 1968.

Nield DA, Bejan A. Convection in Porous Media. New York: Springer-Verlag, 1992 (2nd ed 1998).

Nield DA, Manole DM, Lage JL. Convection induced by inclined thermal and solutal gradients in a shallow horizontal layer of porous medium. J Fluid Mech 257:559–574, 1981.

Patankar SV. Numerical Heat Transfer and Fluid Flow. London: McGraw-Hill, 1980.

Patil P, Rudraiah N. Linear convective stability and thermal diffusion of a horizontal quiescent layer of two-component fluid in a porous medium. Int J Eng Sci 18:1055–1059, 1980.

Patil P, Paravathy CP, Venkatakrishnan KS. Thermohaline instability in a rotating anisotropic porous medium. Appl Sci Res 46:73–88, 1989.

Platten JK, Legros JC. Convection in Liquids. New York: Springer-Verlag, 1984.

Poulikakos D. Double-diffusive convection in a horizontally sparsely packed porous layer. Int Commun Heat Mass Transfer 13:587–598, 1986.

Quintard M, Whitaker S. Transport in chemically and mechanically heterogeneous porous media I: theoretical development of region averaged equations for slightly compressible single-phase flow. Adv Water Resour 19:29–47, 1996a.

Quintard M, Whitaker S. Transport in chemically and mechanically heterogeneous porous media II: theoretical development of region averaged equations for slightly compressible single-phase flow. Adv. Water Resour 19:47–60, 1996b.

Quintard M, Kaviany M and Whitaker S. Two-medium treatment of heat transfer in porous media: numerical results for effectives properties. Advances in water resources 20(2-3):77–94, 1997.

Raptis A, Tzivanidis G, Kafousias N. Free convection and mass transfer flow through a porous medium bounded by infinite vertical limiting surface with constant suction. Lett Heat Mass Transfer 8:417–424, 1981.

Rastogi SK, Poulikakos D. Double-diffusion from a vertical surface in a porous region saturated with a non-Newtonian fluid. Int J Heat Mass Transfer 38:935–946, 1995.

Rudraiah N, Srimani PK, Friedrich R. Amplitude convection in a two-component fluid porous layer. Int Commun Heat Mass Transfer 3:587–598, 1986.

Saghir MZ, Islam MR. Double-diffusive convection in dual-permeability, dual-porosity porous media. Int J Heat Mass Transfer 42:437–454, 1999.

Sinha SK, Sundararajan T. A variable property analysis of alloy solidification using the anisotropic porous approach. Int J Heat Mass Transfer 35:2865–2877, 1992.

Sözen M, Vafai K. Analysis of the non-thermal equilibrium condensing flow of gas through a packed bed. Int J Heat Mass Transfer 33:1247–1261, 1990.

Taslim ME, Narusawa U. Binary fluid convection and double-diffusive convection in a porous medium. J Heat Transfer 108:221–224, 1986.

Taunton JW, Lightfoot EN, Green T. Thermohaline instability and salt fingers in a porous medium. Phys Fluids 15:748–753, 1972.

Tien HC, Vafai K. Pressure stratification effects on multiphase transport across a vertical slot porous insulation. J Heat Transfer 112:1023–1031, 1990.

Trevisan O, Bejan A. Natural convection with combined heat and mass transfer buoyancy effects in a porous medium. Int J Heat Mass transfer 28:1597–1611, 1985.

Trevisan O, Bejan A. Mass and heat transfer by natural convection in a vertical slot filled with porous medium. Int J Heat Mass Transfer 29:403–415, 1986.

Trevisan O, Bejan A. Combined heat and mass transfer by natural convection in porous medium. Adv Heat Transfer 20:315–352, 1990.

Tyvand PA. Thermohaline instability in anisotropic porous media. Water Resour Res 16:325–330, 1980.

Turner JS. Multicomponent Convection. Ann Rev Fluid Mech 17:11–44, 1985.

Vafai K, Tien HC. A numerical investigation of phase change effects in porous materials. Int J Heat Mass Transfer 32:195–203, 1981.

Vafai K, Tien HC. Boundary and inertia effects on flow and heat transfer in porous media. Int J Heat Mass Transfer 24:195–1277, 1989.

Xin S, Le Quéré P, Tuckerman L. Bifurcation analysis of double-diffusive convection with opposing horizontal thermal and solutal gradients. Phys Fluids 10:850–857, 1998.

Yucel A. Natural convection heat and mass transfer along a vertical cylinder in porous layer heated from below. Int J Heat Mass Transfer 33:2265–2274, 1990.

14

Mixed Convection in Saturated Porous Media

F. C. Lai
University of Oklahoma, Norman, Oklahoma

I. INTRODUCTION

Early studies on convection in porous media were largely devoted to buoyancy-induced flows and forced convection. The interaction mechanism between these two modes of convection was given very little attention. Initial research on mixed convection was motivated by the desire to provide a basic understanding of the upward convective drift of subsurface ground water due to buoyancy caused by high temperatures in the geothermal region of Wairakei, New Zealand (Wooding 1960, 1963). Research on mixed convection in porous media was also performed to study the formation of an island geothermal reservoir after a magma chamber was intruded from below (Cheng and Teckchandani 1977; Cheng and Lau 1977; Cheng 1978).

In addition, the design of a saltless solar pond requires a basic understanding of continuous withdrawal of warm water from the bottom of the pond for distribution to the point of use while cool make-up water is continuously added at the top. This creates a gentle flow downward in the direction of increasing temperature, i.e., an opposing forced flow in a volumetrically heated packed bed (Hadim and Burmeister 1988). In another solar space heating application, a horizontal packed bed is charged by passing hot air from solar collectors through the bed horizontally (Salt and Mahoney 1987).

Motivation to study mixed convection in porous media has also come from the need to characterize the convective transport processes around a deep geological repository for the disposal of high-level nuclear waste, e.g., spent fuel rods from nuclear reactors. Presently proposed repositories would cover an area of up to 5 km^2 and be 600 m below ground level. The parameters which are expected to affect the temperature field around a repository include the natural stratigraphy of the site, groundwater flow caused by the hydrostatic head of the water table, layout of tunnels and rooms, and the variation in waste heat generation with time.

Research on mixed convection in fluid-saturated porous media which is more relevant to geothermal reservoir applications has been reviewed in detail by Cheng (1978), and hence will be given less attention in this chapter. On the other hand, research conducted over the last twenty years will be reviewed in more detail in order to cover recent progress. Although the details of the non-Darcy formulation and its implications are discussed in other chapters, non-Darcy effects are included here whenever the need for a complete discussion of the subject is called for.

II. EXTERNAL FLOWS

A. Inclined and Vertical Surfaces

Cheng (1977a) has considered the problem of mixed convection in porous media along inclined surfaces. His analysis has been based on the Darcy formulation and boundary-layer approximations. Both aiding and opposing flows have been considered. Assuming a power law variation of the wall temperature ($T_w = T_\infty \pm Ax^\lambda$), similarity solutions are obtained for two cases: (1) a uniform flow along a vertical isothermal plate ($\lambda = 0$); and (2) an accelerating flow over a 45° inclined plate of constant heat flux ($\lambda = 1/3$). The governing parameter for the problem is Ra/Pe. The heat transfer rate is found to approach asymptotically the forced and free convection limits as the value of the governing parameter approaches zero and infinity, respectively (Figure 1).

Merkin (1980) followed Cheng's (1977a) study with an analysis of mixed convection boundary-layer flow in porous medium adjacent to a vertical, uniform heat flux surface. As pointed out by Cheng, a similarity solution does not exist. Two different coordinate perturbations were applied, for small and large downstream distances. These expansions were then matched for intermediate downstream distances, using numerical integration of the governing equations. Joshi and Gebhart (1985) extended Merkin's analysis by using the method of matched asymptotic expansions. It was shown that the first correction to the boundary-layer theory (neglected by Merkin) occurs at the same level as the second correction

(a)

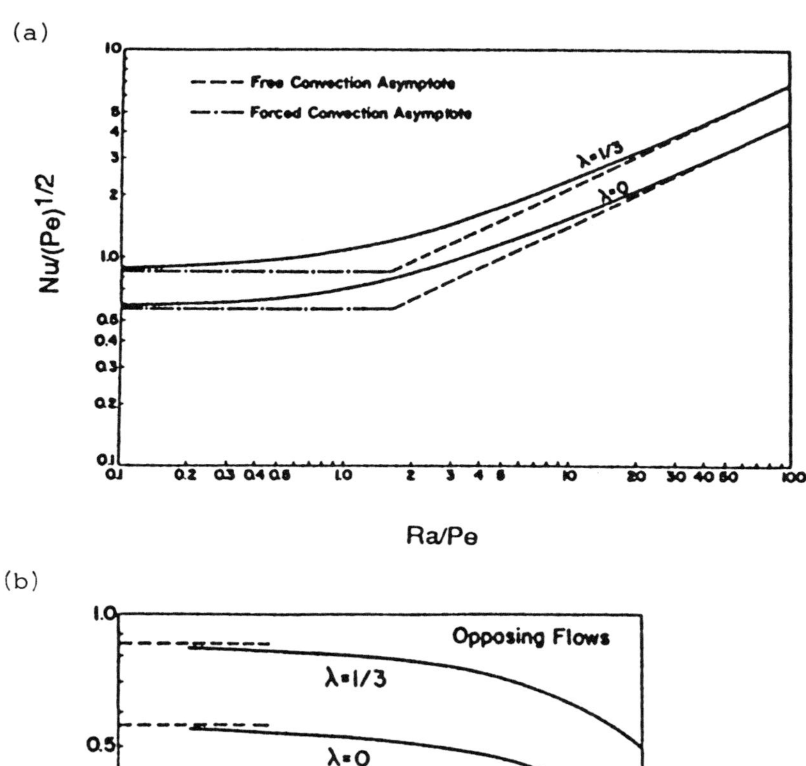

(b)

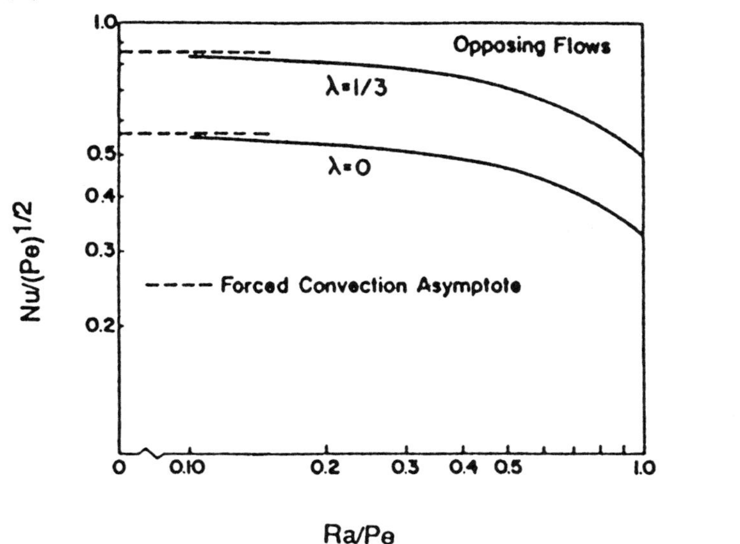

Figure 1. Nusselt numbers for mixed convection about vertical ($\lambda = 0$) and inclined ($\lambda = 1/3$) surfaces in saturated porous media: (a) aiding flows; (b) opposing flows (Cheng 1977a).

due to mixed convection. Therefore, both of these effects must be included in a physically consistent analysis.

For geothermal applications, the effect of flow injection and withdrawal at the wall is of considerable importance for heat transfer. Lai and Kulacki (1990a) have presented similarity solutions for the special case where the wall temperature, the free stream velocity, and the injection or suction velocity at the wall are prescribed power functions of distance. As in the case of free convection (Cheng 1977c), heat transfer coefficients are found to increase with suction velocity and decrease with injection velocity.

For other heating conditions at the wall (e.g., variable wall temperature $T_w = T_\infty + ax^n$, and variable wall heat flux $q_w = bx^m$), Hsieh et al. (1993a,b) have obtained nonsimilarity solutions for mixed convection along a vertical plate in a porous medium. They proposed two sets of correlations for the average Nusselt numbers; one applies to the forced convection dominated regime and free convection dominated regime (Eqs. (1) and (3)), while the other is good for the entire regime (Eqs. (2) and (4)).
Variable wall temperature (for $0 \le n \le 1$)

$$\frac{\overline{Nu}Pe_L^{-1/2}}{2G_1(n)} = \left[1 + \left[\frac{2G_2(n)(Ra_L/Pe_L)^{1/2}}{2(n+1)G_1(n)}\right]^2\right]^{1/2} \tag{1}$$

$$\frac{\overline{Nu}(Pe_L^{1/2} + Ra_L^{1/2})^{-1}}{G_1(n)} = \left[(2\chi_L)^2 + \left[(1-\chi_L)\frac{2G_2(n)}{(n+1)G_1(n)}\right]^2\right]^{1/2} \tag{2}$$

Variable wall heat flux (for $-0.5 \le m \le 1$)

$$\frac{\overline{Nu}Pe_L^{-1/2}}{2G_3(m)} = \left[1 + \left[\frac{2G_4(m)(Ra_L^*/Pe_L^{3/2})^{1/3}}{2(m+2)G_3(m)}\right]^3\right]^{1/3} \tag{3}$$

$$\frac{\overline{Nu}(Pe_L^{1/2} + Ra_L^{*\,1/3})^{-1}}{G_3(m)} = \left[(2\chi_L^*)^3 + \left[(1-\chi_L^*)\frac{3G_4(m)}{(m+2)G_3(m)}\right]^3\right]^{1/3} \tag{4}$$

where

$$G_1(n) = 0.5650 + 0.7631n - 0.2813n^2 + 0.0821n^3 \tag{5}$$

$$G_2(n) = 0.4457 + 0.8099n - 0.3831n^2 + 0.1286n^3 \tag{6}$$

$$G_3(m) = 0.8864 + 0.5488m - 0.1559m^2 + 0.0516m^3 \tag{7}$$

$$G_4(m) = 0.7718 + 0.3043m - 0.1189m^2 + 0.0444m^3 \tag{8}$$

$$\chi_L = \left[1 + \left(\frac{Ra_L}{Pe_L}\right)^{1/2}\right]^{-1} \tag{9}$$

$$\chi_L^* = \left[1 + \left(\frac{Ra_L^*}{Pe_L^{3/2}}\right)^{1/3}\right]^{-1} \tag{10}$$

For the studies discussed above, boundary-layer approximations have been employed. It is recognized, however, that this is valid only at high Rayleigh numbers. Under such conditions, the validity of Darcy's law then becomes questionable. It is expected that non-Darcy effects, especially inertia, will become important at high Rayleigh numbers. In addition, thermal dispersion effects may also become important when inertial effects are prevalent.

The effects of flow inertia and thermal dispersion on mixed convection heat transfer over a vertical surface in a porous medium have been studied by Lai and Kulacki (1991a). In their study, the inertial effects are expressed in terms of the Forchheimer number Fo which is defined as $(Kb/d)(\alpha/\nu)$, where b is the Forchheimer coefficient. The thermal dispersion effect is introduced by assuming the effective thermal diffusivity α_e to have two components: α, the molecular diffusivity, and α', the diffusivity due to thermal dispersion. The dispersive diffusivity is assumed to be proportional to the streamwise velocity component with the proportional constant C defined as the thermal dispersion coefficient. Similarity solutions have been obtained for the problem under consideration. While inertial effects tend to reduce the heat transfer rate (Figure 2), it is found that dispersion effects can increase the heat transfer rate significantly (Figure 3).

The inertial and viscous effects on mixed convection about a vertical surface have been studied by Ranganathan and Viskanta (1984). Their analysis also included the effects of porosity variation and of low blowing velocity at the surface. It was shown that the effects of boundary friction and inertia are quite significant and cannot be ignored. It was also shown that porosity variation has a negligible effect on heat transfer results, but blowing at the surface produces large changes in the velocity and temperature distribution.

Chen et al. (1996) extended the work of Ranganathan and Viskanta (1984) to include the effects of thermal dispersion. Their results show that the inertia and viscous effects tend to reduce heat transfer whereas the channeling effect, caused by the porosity variation near the wall, enhances

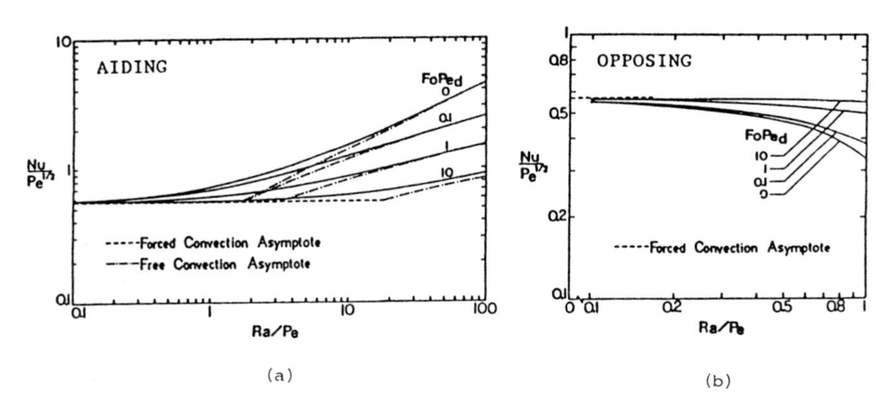

Figure 2. Nusselt numbers on a vertical plate with inertia effects: (a) aiding flows; (b) opposing flows (Lai and Kulacki 1991a).

heat transfer. Whether the heat transfer is increased or decreased depends on the competition between these effects. They also reported a significant increase in heat transfer due to thermal dispersion effects.

For coupled heat and mass transfer by mixed convection, Lai (1991a) has obtained similarity solutions for a vertical plate maintained at a constant temperature and constant concentration. The governing parameters for the problem were identified to be the buoyancy ratio $N(= \beta_C(C_w - C_\infty)/\beta_T(T_w - T_\infty))$, Lewis number Le, and mixed convection

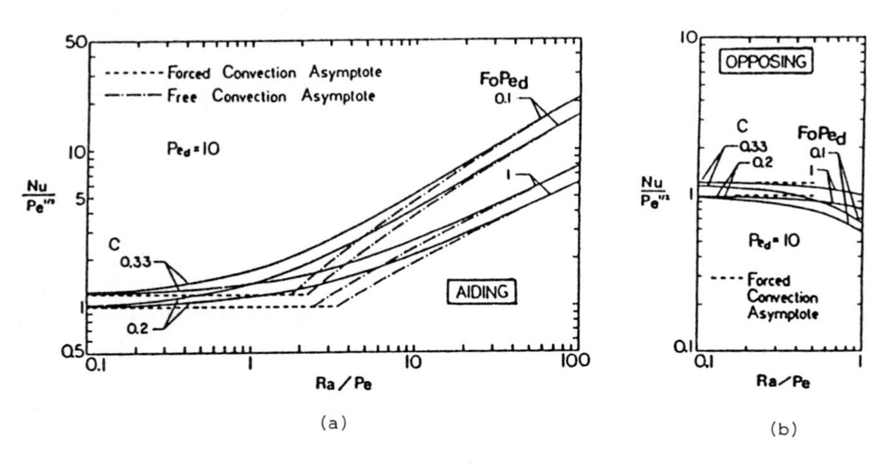

Figure 3. Nusselt numbers on a vertical plate with inertia and dispersion effects: (a) aiding flows; (b) opposing flows (Lai and Kulacki 1991a).

parameter Ra/Pe. It has been shown that, for a given Le, the buoyancy dominated regime extends toward a smaller value of Ra/Pe as N increases (Figure 4). However, for a given N, the buoyancy dominated regime retreats toward a larger value of Ra/Pe as Le increases. The study has been extended by Yih (1998) to include the case when the plate is maintained at constant heat and mass fluxes.

B. Horizontal Surfaces

Cheng (1977b) has also presented similarity solutions for mixed convection over horizontal surfaces in porous media. Similarity solutions have been obtained for the special case where the free stream velocity ($u = Bx^m$) and wall temperature ($T_w = T_\infty \pm Ax^\lambda$) vary according to the power function of distance. Under this restricted condition, solutions are reported for two cases: parallel flow and stagnation-point flow. The governing parameter in this case is $Ra/Pe^{3/2}$. Heat transfer coefficients are also found to approach asymptotically the forced and free convection limits as the value of the governing parameter approaches zero and infinity, respectively (Figure 5).

Minkowycz et al. (1984) extended Cheng's work to the case of an arbitrary power law variation of wall temperature (i.e., $T_w = T_\infty + ax^n$) for both parallel and stagnation-point flows where similarity solutions are not possible. The local non-similarity method was employed, and the analysis was carried out to the third level of truncation. It was found that all three levels of truncation gave identical results for $n = 0.5$.

For other heating conditions at a horizontal surface in a porous medium (e.g., variable wall temperature $T_w = T_\infty + ax^n$, and variable wall heat flux $q_w = bx^m$), Aldoss et al. (1993a,b, 1994) also reported nonsimilarity solutions. They proposed two sets of correlations for the average Nusselt numbers; one applies to the forced convection dominated regime and free convection dominated regime (Eqs. (11) and (13)) while the other is good for the entire regime (Eqs. (12) and (14)).

Variable wall temperature (for $0.5 \le n \le 2$)

$$\frac{\overline{Nu}Pe_L^{-1/2}}{2f_1(n)} = \left[1 + \left[\frac{3f_2(n)(Ra_L/Pe_L^{3/2})^{1/3}}{2(n+1)f_1(n)}\right]^3\right]^{1/3} \tag{11}$$

$$\frac{\overline{Nu}(Pe_L^{1/2} + Ra_L^{1/3})^{-1}}{g_1(n)} = \left[(2\chi_L)^3 + \left[\frac{3}{n+1}(1-\chi_L)\frac{g_2(n)}{g_1(n)}\right]^3\right]^{1/3} \tag{12}$$

Variable wall heat flux (for $0 \le m \le 2$)

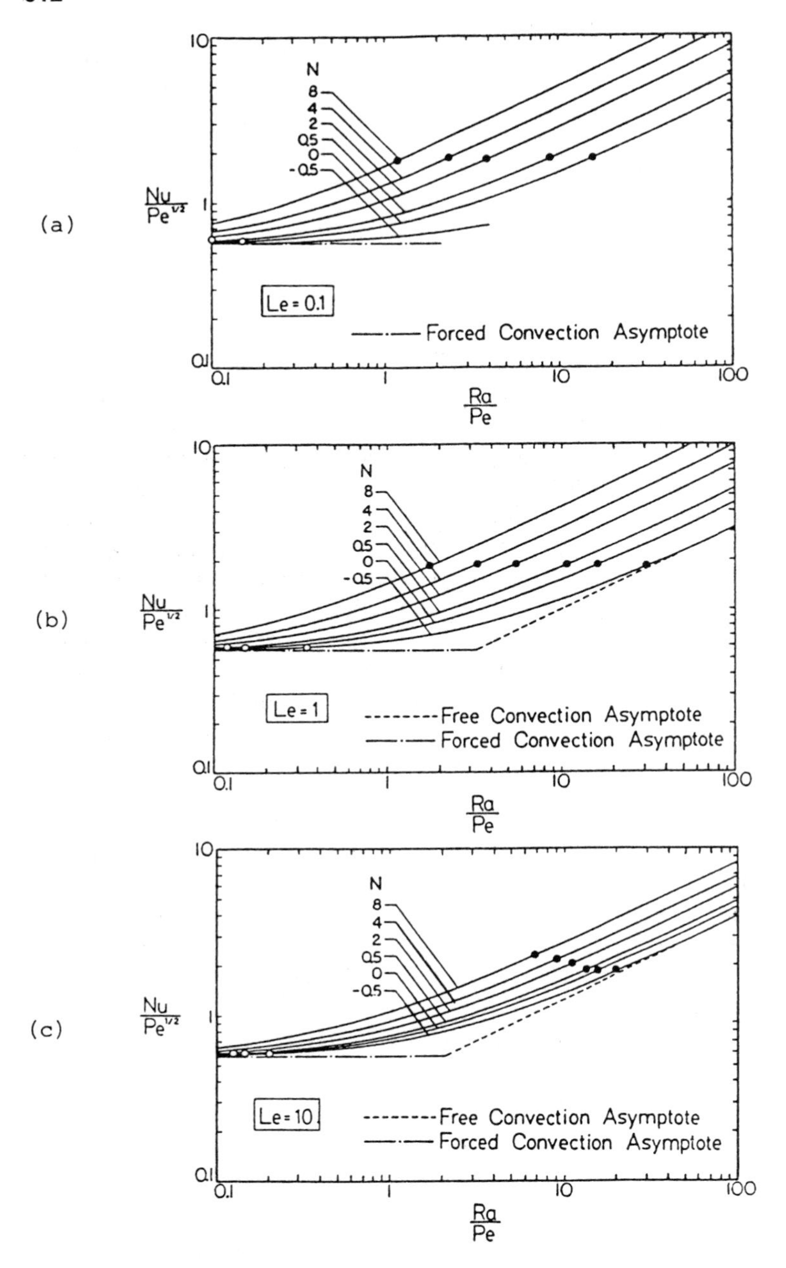

Figure 4. Nusselt numbers for coupled heat and mass transfer by mixed convection along a vertical surface in a saturated porous medium (Lai 1991a).

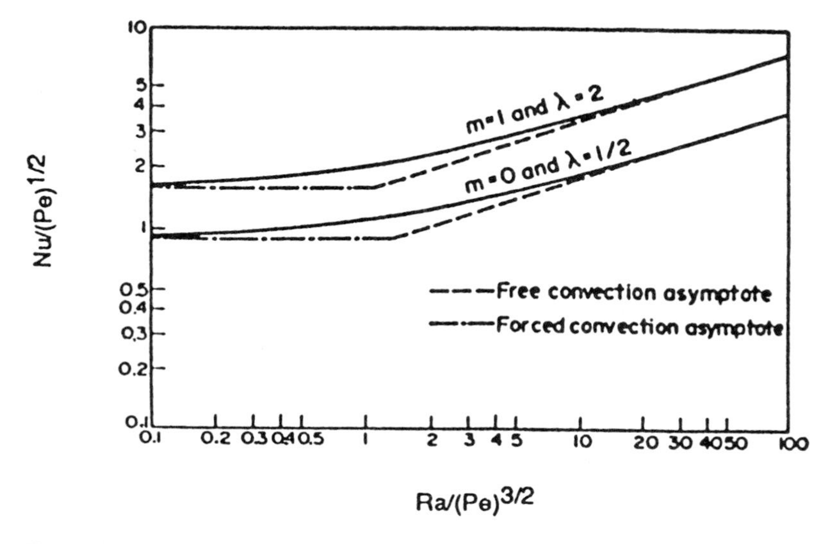

Figure 5. Nusselt numbers for mixed convection over a horizontal plate in a saturated porous medium: parallel flow with constant wall heat flux ($m = 0$, $\lambda = 1/2$); stagnation point flow with $T_w \propto x^2$ ($m = 1$, $\lambda = 2$) (Cheng 1977b).

$$\frac{\overline{Nu}Pe_L^{-1/2}}{2f_3(m)} = \left[1 + \left[\frac{2f_4(m)(Ra_L^*/Pe_L^2)^{1/4}}{(m+2)f_3(m)}\right]^3\right]^{1/3} \tag{13}$$

$$\frac{\overline{Nu}(Pe_L^{1/2} + Ra_L^{*\,1/4})^{-1}}{g_3(m)} = \left[(2\chi_L^*)^3 + \left[\frac{4}{m+2}(1-\chi_L^*)\frac{g_4(m)}{g_3(m)}\right]^3\right]^{1/3} \tag{14}$$

where

$$f_1(n) = 0.582 + 0.685n - 0.167n^2 + 0.0285n^3 \tag{15}$$

$$f_2(n) = 0.474 + 0.762n - 0.168n^2 + 0.0319n^3 \tag{16}$$

$$g_1(n) = 0.597 + 0.640n - 0.0124n^2 + 0.0155n^3 \tag{17}$$

$$g_2(n) = 0.493 + 0.703n - 0.111n^2 + 0.0144n^3 \tag{18}$$

$$f_3(m) = 0.8872 + 0.5298m - 0.1034m^2 + 0.0163m^3 \tag{19}$$

$$f_4(m) = 0.8597 + 0.3596m - 0.0641m^2 + 0.0103m^3 \tag{20}$$

$$g_3(m) = 0.889 + 0.520m - 0.091m^2 + 0.0118m^3 \tag{21}$$

$$g_4(m) = 0.861 + 0.353m - 0.053m^2 + 0.0072m^3 \tag{22}$$

$$\chi_L = \left[1 + \left(\frac{Ra_L}{Pe_L^{3/2}}\right)^{1/3}\right]^{-1} \tag{23}$$

$$\chi_L^* = \left[1 + \left(\frac{Ra_L^*}{Pe_L^2}\right)^{1/4}\right]^{-1} \tag{24}$$

Chandrasekhara (1985) has also extended Cheng's work to include the effect of variable permeability. Similarity solutions have been obtained for the same two cases presented by Cheng (1977b). It is shown that the variation of permeability, and in turn the variation of conductivity of the porous medium, has increased the heat transfer rate appreciably (Figure 6).

To refine the analysis further, Chandrasekhara and Namboodiri (1985) have included the effect of both variable permeability and a no-slip boundary, i.e., the Brinkman effect, on mixed convection about inclined surfaces. Similarity solutions for both aiding and opposing flows were obtained for the case of uniform wall temperature and a linear variation of wall temperature with distance from the leading edge. The governing parameters are Gr/Re^2 and σ^2/Re, where $\sigma = x/\sqrt{K_0}$ is a "local porosity" parameter and K_0 is the permeability of the porous medium at the edge of the boundary layer. Their results have clearly shown flow channeling near

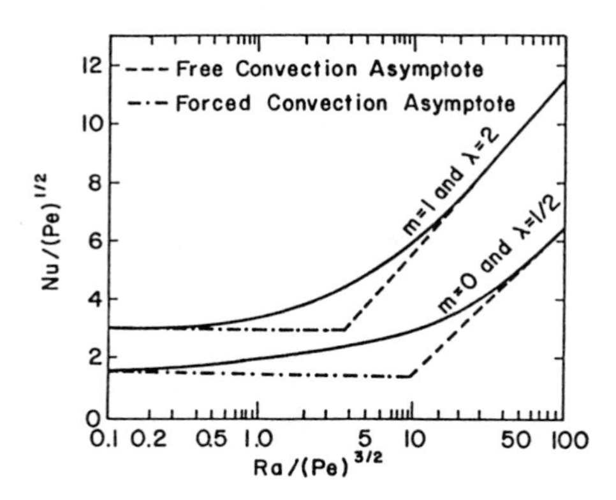

Figure 6. Nusselt numbers for a flat plate in a medium with variable permeability near the wall: parallel flow with constant wall heat flux ($m = 0$, $\lambda = 1/2$); stagnation point flow with $T_w \propto x^2$ ($m = 1$, $\lambda = 2$) (Chandrasekhara 1985).

the wall (Figure 7). It is due to flowing channeling that heat transfer coefficients increase for all values of σ^2/Re.

For the cases of flow injection and withdrawal, Lai and Kulacki (1990b) also reported similarity solutions for the special case where the wall velocity is a prescribed power function of distance. It is found that the heat transfer coefficient is enhanced by the withdrawal of fluid from the surface, whereas it is decreased by injection. Brouwers (1994) applied the classical film model to the same problem. The results he obtained are in good agreement with those of Lai and Kulacki (1990b) using the boundary-layer model.

For non-Darcy flows, Lai and Kulacki (1987) considered inertial effects. On the basis of the Ergun model, they have presented similarity solutions for the case of a uniform flow over a horizontal surface where the surface temperature varies with $x^{1/2}$. The limiting cases of free and forced convection have also been included. In their study, the inertial effects are characterized by the Ergun number Er which is defined as $K'\alpha/d\nu$, where K' is the inertial coefficient in the Ergun equation. They show that heat transfer coefficients are decreased as a result of flow inertia (Figure 8). Since thermal dispersion effects may become very important when inertia is prevalent, Lai and Kulacki (1989) extended their earlier study to include such effects. Similarity solutions again exist for the problem under consideration and it is found that the heat transfer coefficient is significantly increased as a result of thermal dispersion (Figure 9).

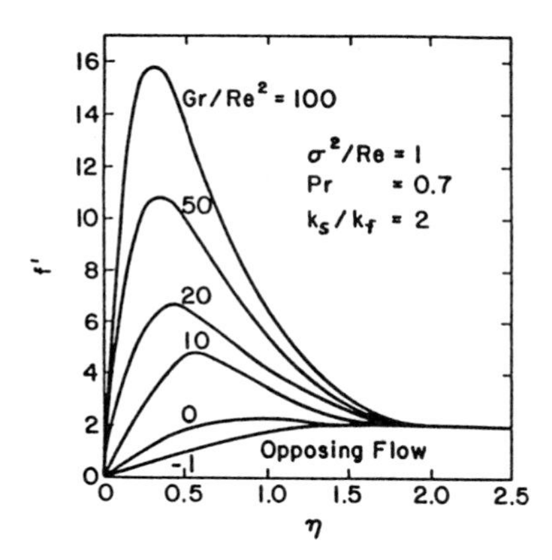

Figure 7. Dimensionless velocity f' for aiding ($Gr/Re^2 > 0$) and opposing ($Gr/Re^2 < 0$) flows for an inclined plate, including variable permeability and the no-slip wall condition (Chandrasekhara and Namboodiri 1985).

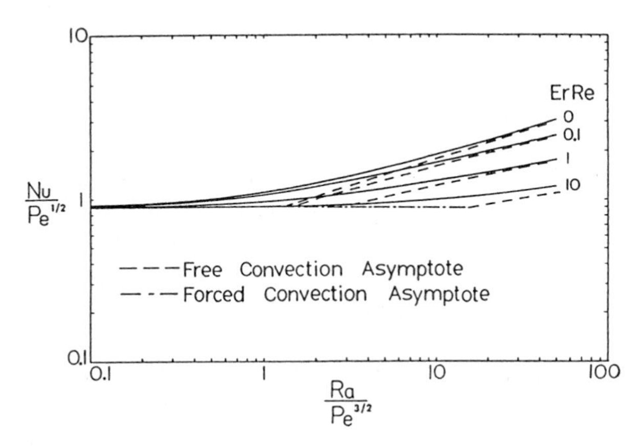

Figure 8. Nusselt numbers for non-Darcy flow over a horizontal surface with inertia effects (Lai and Kulacki 1987).

For the case of a heated plate near an impermeable surface, where the boundary-layer approximation is no longer applicable, Oosthuizen (1988) has performed a numerical study. Qualitatively, his results are in good agreement with those presented by Prasad et al. (1988a) and will be elaborated upon in the next section.

Hsu and Cheng (1980a,b) have applied a linear stability analysis to determine the condition of the onset of vortex instability for flow over

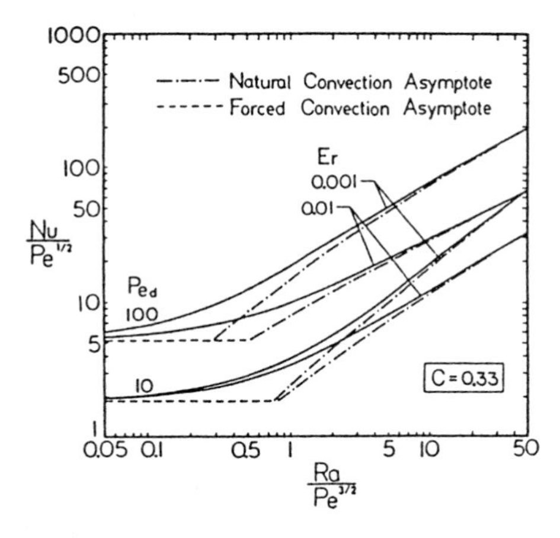

Figure 9. Nusselt number of non-Darcy flow over a horizontal surface with inertia and thermal dispersion effects (Lai and Kulacki 1989).

inclined and horizontal impermeable surfaces. At a given value of $Ra/Pe^{3/2}$, it is found that aiding flows are more stable than opposing flows for inclined surfaces. However, stagnation-point flows are more stable than parallel aiding flows for horizontal surfaces. In both cases, it is noted that the effect of increasing external flow is to suppress the growth of vortex disturbances.

The work of Hsu and Cheng (1980b) has been extended by several investigators to include various non-Darcy effects. For example, the boundary and inertia effects have been considered by Jang and Lie (1993), while the thermal dispersion and inertia effect as well as the variable porosity effect have been studied by Jang and Chen (1993a,b). The results from these studies show that the thermal dispersion effect enhances heat transfer and stabilizes the flow whereas the inertia effect exhibits the opposite trend. The variable porosity effect tends to enhance heat transfer but destabilize the flow. As with thermal dispersion, the effect of variable porosity is more pronounced at a higher value of mixed convection parameter $\xi(= Pe_x^{3/2}/Ra_x)$. In addition, Jang et al. (1995) have reported the effect of surface mass flux on vortex instability in mixed convection flow over a horizontal surface in a porous medium. Their results show that blowing reduces the heat transfer rate and destabilizes the flow, while suction displays the opposite trend.

When solute transport is involved, Li and Lai (1997) have obtained a similarity solution for coupled heat and mass transfer by mixed convection from a horizontal surface subject to constant heat and mass fluxes. They also reported that the similarity solution for a horizontal surface maintained at a constant temperature and constant concentration is not physically plausible. The governing parameters for the problem have been identified to be the buoyancy ratio $N(= \beta_C(C_w - C_\infty)/\beta_T(T_w - T_\infty))$, Lewis number Le, and mixed convection parameter $Ra/Pe^{3/2}$. For a given Le, it is found that the buoyancy dominated regime extends toward a smaller value of $Ra/Pe^{3/2}$ as N increases (Figure 10). However, for a given N ($N > 1$), the buoyancy dominated regime retreats toward a larger value of Ra/Pe as Le increases, whereas the opposite trend is observed for $|N| < 1$.

C. Cylinders and Spheres

By employing a generalized similarity transformation similar to that of Merkin (1979) for the free convection problem, Cheng (1982) was able to obtain the similarity solution for mixed convection over a horizontal cylinder and a sphere. It was found that the transformed equations are identical to those for mixed convection about a vertical isothermal surface (Cheng

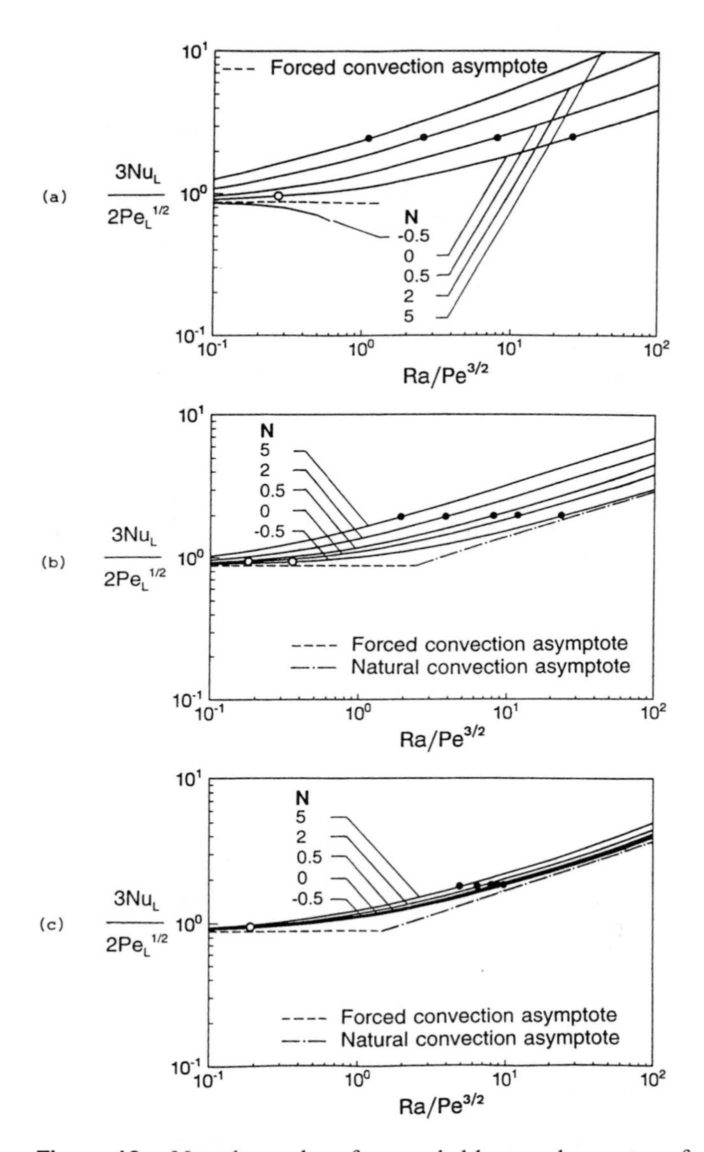

Figure 10. Nusselt numbers for coupled heat and mass transfer by mixed convection over a horizontal surface in a saturated porous medium: (a) $Le = 0.1$; (b) $Le = 1$; (c) $Le = 10$ (Li and Lai 1997).

1977a). Following the same approach, but with a different transformation, Huang et al. (1986) have obtained the solution for the constant heat flux case.

Minkowycz et al. (1985) have extended Cheng's analysis to consider aiding mixed convection about a nonisothermal cylinder and a sphere. Based on a curvilinear orthogonal coordinate system, together with the boundary-layer simplifications, approximate solutions are obtained by the local nonsimilarity method.

Badr and Pop (1988) have also examined the problem of mixed convection over a horizontal isothermal cylinder via a series expansion in combination with a finite-difference scheme. Both aiding and opposing flows have been considered. They reported that the solutions based on the boundary-layer approximation (Minkowycz et al. 1985) are in good agreement with their results, even at small Grashof numbers. For opposing flows, they found that there exists a recirculating zone near the upper half of the cylinder (Figure 11).

Zhou and Lai (1998) have re-examined the problem of mixed convection over a horizontal cylinder in a saturated porous medium, using a finite-difference method with body-fitted coordinates. For flows at a small Reynolds number and a small mixed convection parameter Gr/Re, their results agree very well with those reported by Badr and Pop (1988). However, for flows at a large Reynolds number or a large mixed convection parameter, significant discrepancies have been found between these two studies (Figures 11 and 12). More importantly, some steady solutions reported by Badr and Pop were found to be oscillatory (e.g., at $Re = 100$ and $Gr = 400$). Although no conclusion has been reached at this time, it is suspected that the condition which permits a series solution in the study by Badr and Pop (1988) may not hold at a large mixed convection parameter (Figure 13).

For crossflow over a horizontal cylinder near an impermeable surface, Oosthuizen (1987) has performed a numerical study. His results show that the presence of the surface has a negligible effect on heat transfer when the dimensionless buried depth of the cylinder is greater than three times its diameter. In addition, the presence of the surface tends to increase local heat transfer coefficients on the upper upstream quarter of the cylinder and decrease it on the upper downstream side of the cylinder.

For mixed convection along a vertical cylinder, Hooper et al. (1994) have obtained nonsimilarity solutions for the case of uniform wall temperature. They proposed the following correlation for the average Nusselt number

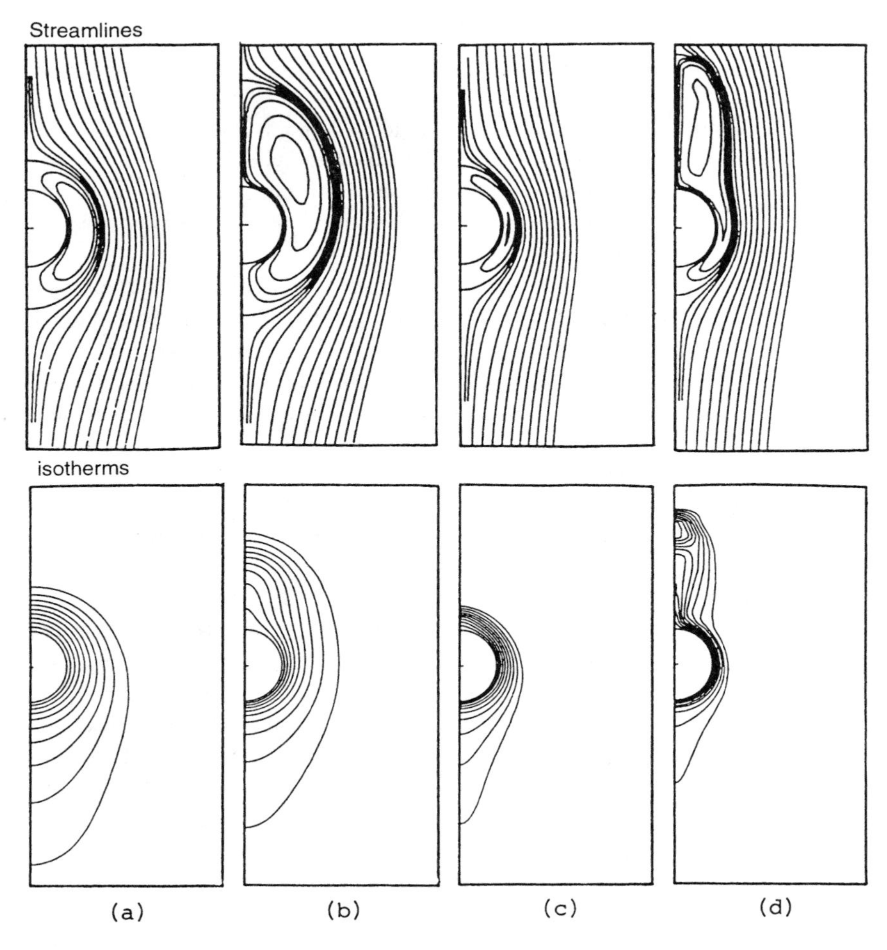

Figure 11. Flow and temperature fields for opposing mixed convective flow past a horizontal cylinder in a saturated porous medium: (a) $Re = 20$, $Gr = 60$; (b) $Re = 20$, $Gr = 80$; (c) $Re = 100$, $Gr = 300$; (d) $Re = 100$, $Gr = 400$ (Badr and Pop 1988).

$$\frac{\overline{Nu}(Pe_L^{1/2} + Ra_L^{1/2})^{-1}}{2F_1(\xi_L)} = \left[\chi^{m_2(\xi_L)} + \left[(1-\chi)\frac{F_2(\xi_L)}{F_1(\xi_L)} \right]^{m_2(\xi_L)} \right]^{1/m_2(\xi_L)} \tag{25}$$

where

$$F_1(\xi_L) = 0.565 + 0.2212\xi_L - 0.0126\xi_L^2 + 0.000\,73\xi_L^3 \tag{26}$$

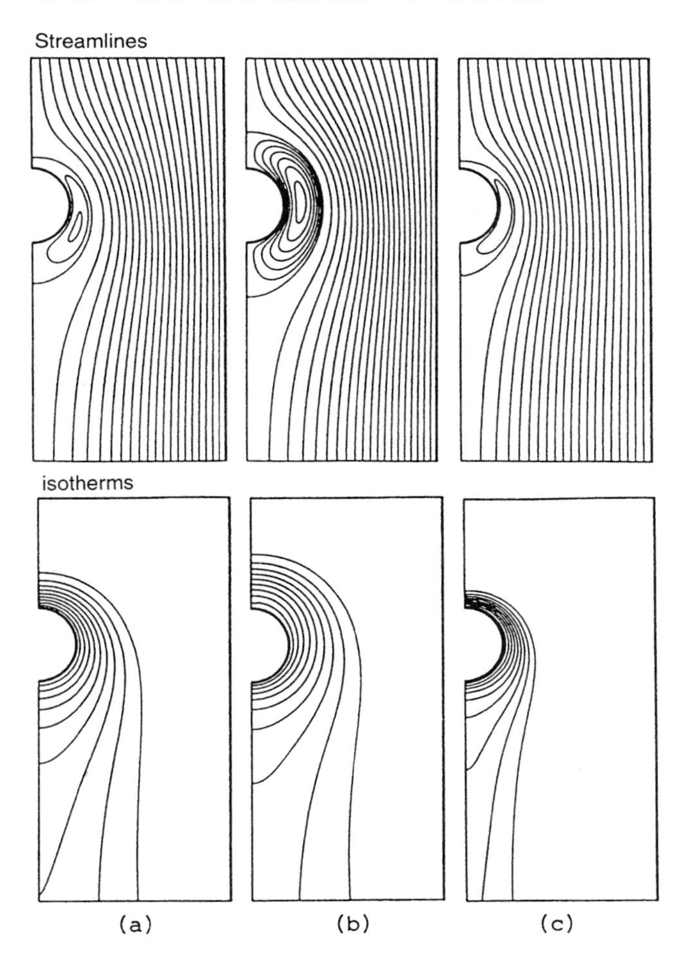

Figure 12. Flow and temperature fields for opposing mixed convective flow past a horizontal cylinder in a saturated porous medium: (a) $Re = 20$, $Gr = 60$; (b) $Re = 20$, $Gr = 80$; (c) $Re = 100$, $Gr = 300$ (Zhou and Lai 1998).

$$F_2(\xi_L) = 0.4457 + 0.1754\xi_L - 0.0827\xi_L^2 + 0.000\,48\xi_L^3 \tag{27}$$

$$m_2(\xi_L) = 2 - 0.1199\xi_L \tag{28}$$

$$\xi_L = \frac{(L/r_0)}{Pe_L^{1/2} + Ra_L^{1/2}} \tag{29}$$

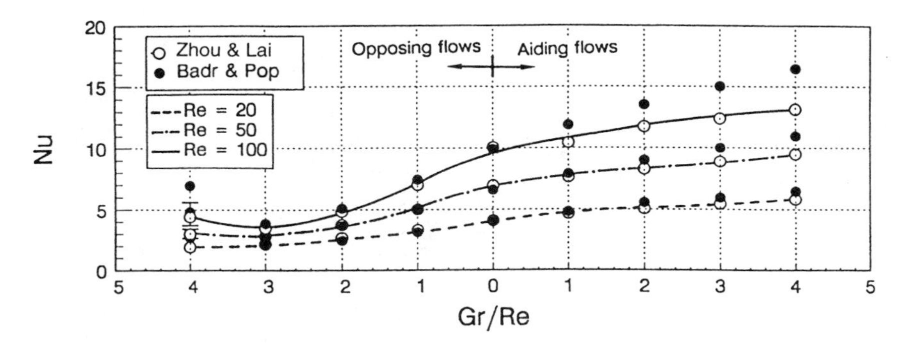

Figure 13. Comparison of heat transfer results for mixed convection over a horizontal cylinder in a saturated porous medium (Zhou and Lai 1998).

$$\chi = \left[1 + \left(\frac{Ra_L}{Pe_L}\right)^{1/2}\right]^{-1} \tag{30}$$

For non-Darcy flows, Kumari and Nath (1989) have studied inertia effects on mixed convection heat transfer along a vertical cylinder. Both forced flow and buoyancy dominated regimes have been considered. Their results show that the heat transfer coefficient increases with buoyancy for aiding flows, whereas it decreases for opposing flows. Other non-Darcy effects, which include boundary, inertia, variable porosity, and thermal dispersion effects, have been considered by Chen et al. (1992). As in other studies of non-Darcy convection, their results also show that the inertia effects reduce the heat transfer coefficients considerably. The channeling effect, on the other hand, enhances heat transfer. Whether heat transfer is enhanced or reduced depends on the relative magnitude of these effects. For other heating conditions (e.g., variable wall temperature and variable wall heat flux), nonsimilarity solutions have been reported by Aldoss et al. (1996) for the entire mixed convection regime.

Nguyen and Paik (1994) have obtained numerical solutions for mixed convection about a sphere buried in a saturated porous medium, using the Chebyshev–Legendre spectral method. Two types of heating conditions were considered; variable surface temperature and variable surface heat flux. As in the case of cylinders, recirculating secondary cells appear in opposing flow conditions.

D. Other Geometries

So far, analytical studies of mixed convection have been limited to simple geometries such as flat plates, cylinders, and spheres. Two-dimensional or axisymmetric bodies of arbitrary shape have received rather little attention, despite the fact that they are encountered frequently in applications.

By using a general transformation, Nakayama and Koyama (1987) were able to obtain similarity solutions for a vertical wedge, vertical cone, horizontal cylinder, and sphere. For the first two cases, similarity solutions exist only when the surface temperature varies with the same power function as that of the free stream velocity, while for the last two cases the surface temperature is constant. For a flat plate, they have shown that the resulting equation is reduced to that obtained by Cheng (1977a).

Under the assumption of a slender body, Lai et al. (1990b) have reported similarity solutions for a vertical cylinder and a paraboloid of revolution. The first case corresponds to accelerating flow past a vertical cylinder with a linear temperature variation along the axis, whereas the second corresponds to uniform flow over a paraboloid of revolution at a constant temperature. For the limiting case of free convection, the results are in good agreement with those presented by Minkowycz and Cheng (1976), although different transformations have been used. For cylinders, it is reported that the heat transfer coefficient decreases with an increase in the dimensionless radius. For paraboloids of revolution, however, it is found that the heat transfer coefficient increases with the dimensionless radius at higher values of Ra/Pe.

For many engineering applications, a heat source buried underground can be approximated by a point or a line source of heat in a porous medium. For mixed convection over a horizontal line source in an infinite porous medium (i.e., a free plume), Cheng and Zheng (1986) have obtained local similarity solutions. Their study has been extended to include the inertial and thermal dispersion effects by Lai (1991b), using a local nonsimilarity method. It has been shown that the inertial effect tends to reduce the mixed convective flow and thicken the thermal boundary layer whereas thermal dispersion further enhances this influence (Figure 14).

Numerical solutions for mixed convection from a line source of heat located at the leading edge of an adiabatic vertical wall in a saturated porous medium (i.e., a wall plume) have been obtained by Kumari et al. (1988). Their study has been extended by Jang and Shiang (1997) to include non-Darcy effects. It is shown that non-Darcy effects reduce the peak velocity and increase the maximum temperature.

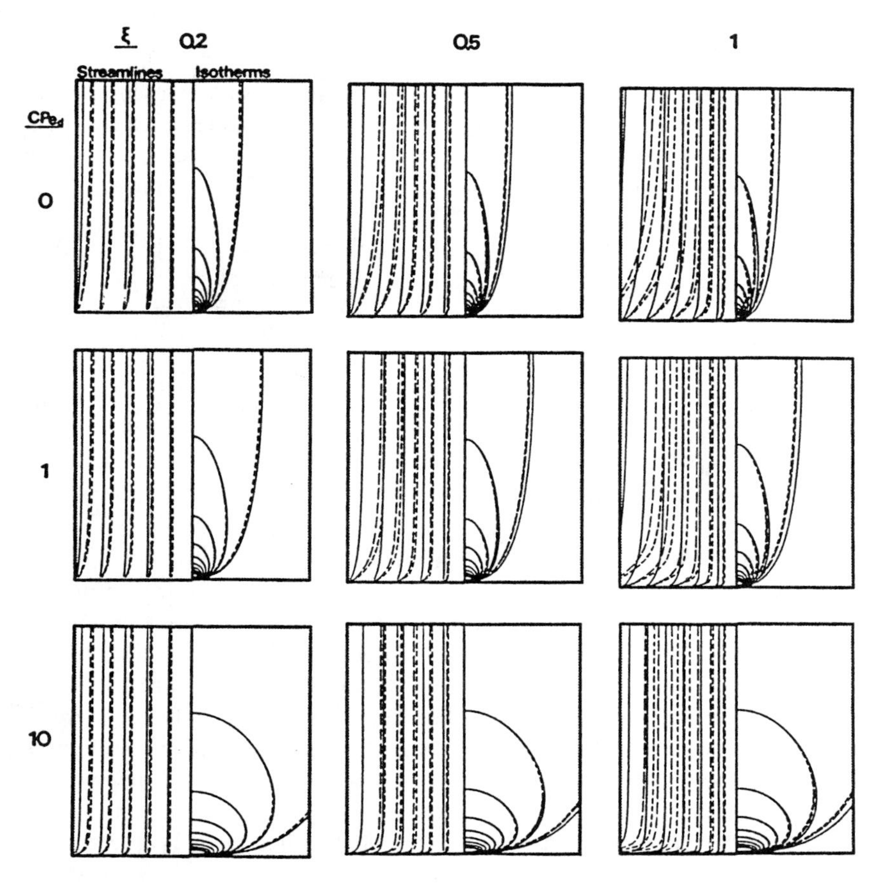

Figure 14. Flow and temperature fields for non-Darcy mixed convection past a horizontal line source of heat in a saturated porous medium (······ $ErPe_d = 0.01$; ------- $ErPe_d = 0.1$; ——— $ErPe_d = 1$) (Lai 1991b).

E. Effects of Variable Properties

In all of the studies discussed above, the thermophysical properties of the fluid are assumed to be constant. However, it is known that these properties may change with temperature, especially fluid viscosity. To accurately predict the heat transfer rate, it is necessary to take into account this variation of viscosity. In spite of its importance in many applications, this effect has been reported only recently by Lai and Kulacki (1990c). Similarity solutions have been obtained for an isothermally heated plate with fluid viscosity varying as an inverse function of temperature. A new parameter, θ_e, is introduced to take into account temperature-dependent viscosity. Its value is determined by the first temperature coefficient of viscosity and the operating temperature difference. For liquids, θ_e is negative, and it is positive for gases. The results show that, for liquids, heat transfer coefficients are greater than those for the constant-viscosity case, whereas they are less for gases.

III. INTERNAL FLOWS

A. Horizontal Layers

1. Onset of Thermal Instability

Extensive research has been conducted on mixed convection in horizontal porous layers. Both horizontal and vertical through flows in the presence of a buoyancy-induced velocity field have been considered for the theoretical and experimental studies. The conditions for the onset of thermal instability are determined and the heat transfer rates for a wide range of governing parameters are reported. Mixed convection effects have also been considered for flow through differentially heated horizontal concentric cylinders.

Wooding (1960, 1963) made the first attempt to examine the effect of a heated liquid rising through a semi-infinite porous medium. His analyses were closely related to the geothermal systems in Wairakei, New Zealand. It was shown that the thermal boundary layer is table provided that the modified Rayleigh number of the system does not exceed a critical value, and that the wavenumber of the critical neutral disturbance is finite. In the absence of permeability difference and viscosity variation, the theory developed for the analysis of thermal boundary layers at large Rayleigh or Péclet number is equivalent to that for a laminar incompressible flow in a two-dimensional half-plane or round jet.

The effect of vertical through flow on the onset of cellular convection in a two-dimensional horizontal porous layer bounded laterally by insulated walls was studied by Sutton (1970). For weak forced flow, the solutions to the eigenvalue problem were obtained analytically by a series expansion method. For large values of aspect ratio and no through flow, Sutton's solution approaches $Ra_c = 4\pi^2$ for free convection. At large aspect ratios, the critical Rayleigh number monotonically increases with the strength of the vertical through flow, which is in agreement with the results of Homsy and Sherwood (1976). Homsy and Sherwood's analysis is an extension of the earlier work of Wooding (1960). Interestingly, they also found that the direction of through flow (upward or downward) has no effect on the critical Rayleigh number, provided that the Saffman–Taylor mode is negligible.

On the other hand, the effect of a horizontal through flow on the onset of thermal instability in a horizontal porous layer heated from below was first investigated by Prats (1966). It was shown that the eigenvalue problem is identical to that of Lapwood (1948) if a linear transformation of coordinates is made before the stability analysis is performed. The total horizontal velocity in his problem is the sum of the base (uniform) velocity and a perturbed term due to thermally induced currents. Thus the shape of the streamlines depends on the relative values of forced flow and thermally induced velocity.

Combarnous and Bia (1971) performed experiments in horizontal porous layers with a low-velocity horizontal through flow, heated from below. The criterion for the onset of convection was unaffected by the existence of a horizontal mean flow, which is in agreement with the linear theory (Prats 1966). Their experimental data indicate that although the convective flows depend on both *Ra* and *Pe*, the influence of mean flow is not apparent on the measured temperatures in the conduction regime, $Ra_c < 4\pi^2$.

The effect of thermal dispersion on stability of the flow field was studied by Rubin (1974, 1975) by employing a method similar to that of Prats (1966). However, there exist two major differences: (1) because Prats ignored the flow field accelerations, his analysis did not require the assumption that the solid matrix should be a poor conductor; and (2) in Prat's analysis, the frame of coordinates moves with the velocity of the heat wave, whereas, in Rubin's work, it moves with the steady-state barycentric flow velocity. It was found that thermal dispersion enhances the stability of the flow field and may inhibit the appearance of convective currents, which would appear if the dispersion effects were not included. Rubin also found that both the longitudinal and the lateral dispersivity affect stability and the dimensions of the convection cells.

Recently, Hadim and Burmeister (1988) determined the criteria for the onset of convection in a volumetrically heated horizontal porous layer with a downward flow. They considered an isothermal free boundary at the top and an isothermal rigid wall at the bottom ($T_b > T_t$). The internal heating was considered to be due to the solar radiation, q, at the top, whose absorption is generally determined in terms of extinction coefficient. In their analysis, they used a perturbation technique to linearize the equations of motion, energy, and continuity which were then solved numerically. The stability criteria obtained by them are functions of the internal Rayleigh number, $Ra_I = (g\beta qKH^2)/(k_m \nu \alpha_m)$, external Rayleigh number, $Ra_E = (g\beta KH\Delta T)/(\nu\alpha_m)$, Péclet number $Pe = w_0 H/\alpha_m$ (where w_0 is the downward flow velocity), and extinction parameter N (Figure 15).

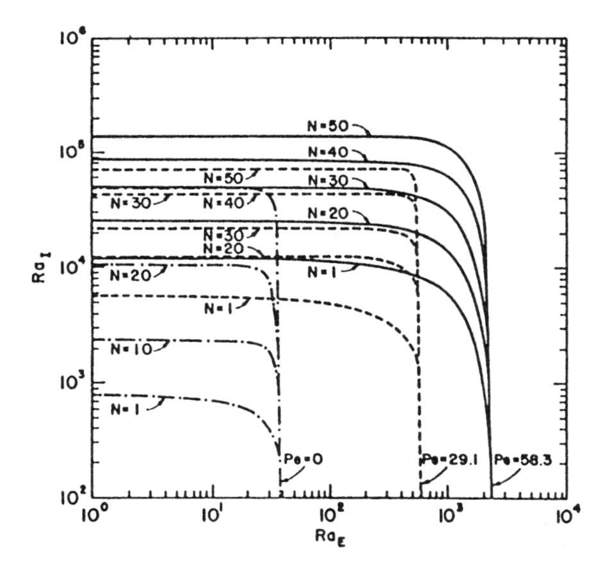

Figure 15. Stability curves in a horizontal porous layer with volumetric heating and downflow at the top (Hadim and Burmeister 1988).

2. Uniform Heating

Islam and Nandkumar (1986) and Nandkumar et al. (1987) investigated the problem of buoyancy-induced secondary flows in a rectangular duct filled with saturated porous media through which an axial flow is maintained by an imposed pressure gradient. They considered the Darcy–Brinkman flow model for steady, fully developed flow in the duct and assumed negligible axial conduction and a constant rate of heat transfer per unit length, q'. Also, the definition of dimensionless temperature depends on the case being considered:

Case I. Top half insulated and bottom half heated by uniform flux

$$\theta = \frac{T - T_b}{q'/k_m} \tag{31}$$

Case II. Uniform wall temperature around the periphery

$$\theta = \frac{T - T_w}{q'/k_m} \tag{32}$$

A finite-difference numerical scheme was used to solve the above equations. Since axial conduction is neglected, their solutions are valid only for large Péclet numbers. The results show a secondary flow pattern of two counter-rotating cells in a plane perpendicular to the axial flow when the modified Grashof number is low. The flow field is thus helicoid in nature. With an increase in Grashof number, a gradual restructuring of the two-vortex pattern takes place in a region close to the lower part of the duct. Finally, at an upper critical Grashof number, the twin counter-rotating vortices become unstable and give rise to a stable four-vortex pattern (Figure 16). This flow pattern is then sustained for Grashof numbers up to 10^4, the upper limit of the computational results. The cellular flow pattern and bifurcation reported in this study are similar to those observed for Bénard convection in porous media (Combarnous and Bories 1975; Caltagirone 1975; Schubert and Straus 1979; Kladias and Prasad 1988, 1989a,b), except that the critical Grashof number has been modified by somewhat different thermal boundary conditions and the forced flow.

It is also interesting to note that, with decreasing Grashof number, the four-cell pattern sustains itself until a lower critical point is reached below which the flow changes abruptly in to a two-vortex structure displaying a hysteresis phenomenon. As is evident from Figure 16, both two- and four-vortex solutions are possible in the range of $1000 < Gr < 3250$. The multi-cellular nature of helicoid flow and the hysteresis behavior were also observed for isothermal heating (Case II). The helicoid flow reported by Islam and Nandkumar (1986) is in agreement with the observation by Combarnous and Bia (1971). However, the alternating flow pattern of moving rolls for $Pe < 0.75$, as observed by these investigators, could not have been predicted by Islam and Nandkumar (1986) because their formulation was valid for large Péclet numbers only.

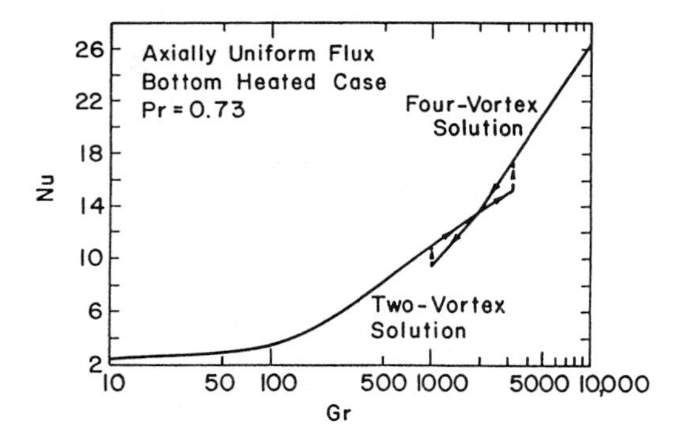

Figure 16. Nusselt numbers for a square channel with an insulated upper half and constant flux lower half. Grashof number is based on a temperature modulus defined in terms of wall flux (Islam and Nandkumar 1986).

As can be expected, the bifurcation from two- to four-cell pattern and range of Grashof number for multiple solutions are strong functions of aspect ratio ($Da \to 0$ in the work of Islam and Nandkumar). The effect of aspect ratio A on two- and four-vortex flow pattern and hysteresis was studied for $0.6 < A < 3$. Figure 17 shows that the upper and lower critical Grashof numbers increase significantly with a decrease in aspect ratio, and also that the range of Grashof number for dual solutions reduces with A. Similar multiple solutions have been reported for a fluid-filled duct (Nandkumar et al., 1984), and the porous media results suggest that the nonlinearity of the convective terms is not the only cause of multiple solutions in the pure fluid case.

Nandkumar et al. (1987) extended the above work to trace the entire path of symmetric solutions. The limit points on the solution path were calculated precisely by using the extended system formulation (Weinitschke 1985) which showed that the pattern of solutions may be much more complicated than that reported above by Islam and Nandkumar (1986). For example, seven limit points were obtained for a square channel, out of which only two (L1 and L7 in Figure 18) could have been predicted in the earlier study. The five additional limit points were detected using the arc length procedure and the extended system. Indeed, the number of limit points depends strongly on the aspect ratio, e.g., 2 for $A = 0.6$, 7 for $A = 1$, and 9 for $A = 1.4$. these calculations also showed that, for any given Grashof number, there always exists an odd number of solutions. With the increasing aspect ratio, multiple solutions occur at low values of Gr. This is demonstrated by the increasing number

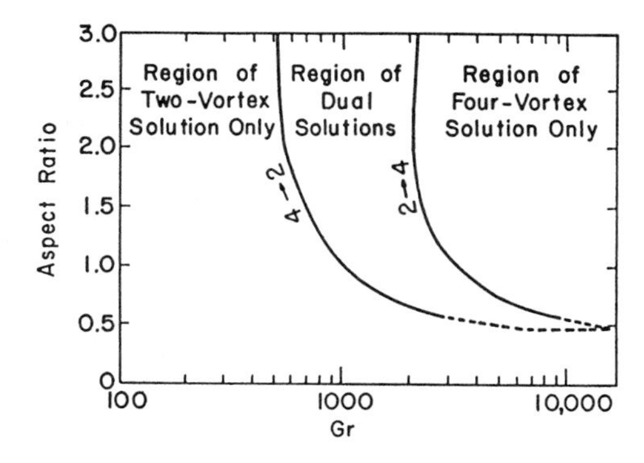

Figure 17. Region of dual (2-cell and 4-cell) solutions in the aspect ratio–Grashof number space for a channel with the upper half insulated and the lower half at constant flux (Islam and Nandkumar 1986).

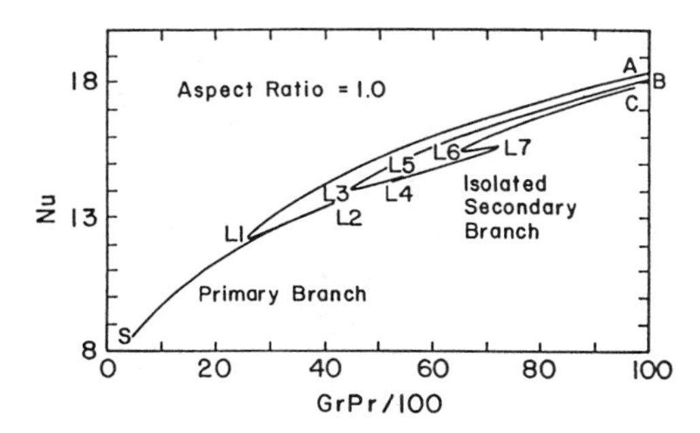

Figure 18. Limit points of solutions for a square channel with the upper half insulated and the lower half at constant flux. Solutions at L1 and L7 were predicted by Islam and Nandkumar (1986) (Nandkumar et al. 1987).

of folds in the *Nu* number curve at low *Gr* when the aspect ratio is relatively large. As is evident from the experimental work of Combarnous and Bia (1971), the flow at large Grashof numbers may not remain symmetric due to oscillatory behavior; hence, further computations are needed to verify these results. Also, under these conditions, the boundary, viscous, and inertia effects may influence the solutions significantly, as shown by the recent studies on natural convection in horizontal porous layers heated from below (Kladias and Prasad 1988, 1989a,b).

Chou et al. (1992) conducted a numerical and experimental study on fully developed non-Darcy mixed convection in horizontal porous channel. Their analysis included flow inertia, no-slip boundary, channeling, and thermal dispersion effects. Their experiments used two sets of channels; one was made of stainless steel with a dimension of $4.75 \times 4.75 \times 85\,\text{cm}$ while the other was made of aluminum with a dimension of $9.5 \times 9.5 \times 60\,\text{cm}$. Electric heaters were wrapped around both channels to provide a constant-flux heating condition at the walls. The numerically predicted flow and temperature fields agree qualitatively well with those of Islam and Nandkumar(1986). Their numerical prediction of Nusselt number is in good agreement with the experimental data for the stainless steel channel, but overpredicts for the aluminum channel. They attributed the discrepancy for the latter case to the short test section (i.e., the entrance effect) and the overestimation of the thermal dispersion effect. In the follow-up study (Chou et al., 1994), when they used a nonlinear correlation between the thermal dispersive conductivity and the Péclet number (for water), as suggested by Levec and Carbonell (1985), they were able to improve the agreement between the numerical prediction and experimental data.

3. Discrete Heating

A series of numerical studies have been conducted by the author and his colleagues to examine the effects of horizontal flow on buoyancy-induced velocity and temperature fields in a horizontal porous layer discretely heated at the bottom and isothermally cooled at the top. Both isothermal and uniform flux heating were considered for varying sizes of one or multiple heat sources. The formulation was based on Darcy flow and the computations were carried out in the range of $1 \leq Ra \leq 500$ and $0.01 \leq Pe \leq 50$. Rayleigh and Péclet numbers were based on the layer height H. The dimensionless length of heat source, $A = D/H$, was considered a geometric parameter in these studies. For computational purposes, the domain was considered long enough such that the flow could be assumed parallel and the axial conduction could be neglected at the exit.

As a base case, the heat source of unit length, $A = 1$, was considered and the bottom surface was treated as insulated except for the isothermally heated segment (Prasad et al. 1988a). The results for this configuration indicate that when the forced flow is weak, i.e., at a small Péclet number, a thermal plume rises above the heat source and a pair of counter-rotating cells is generated in a region directly above the source (Figure 19). Qualitatively, these two recirculating cells are very similar to what have been observed in the case of natural convection (Prasad and Kulacki 1986, 1987), but, because of the through flow, the upstream cell is lifted upward and the downstream cell is moved downward. Thus, symmetry at the midpoint of the heated segment, as observed in the case of natural convection, is destroyed. However, the temperature fields remain symmetric as long as the Péclet number is low. With an increase in Pe, the relative strength of the forced flow increases, the symmetric nature of the isotherms vanishes, the strength of the two recirculating cells become weaker, and the buoyancy-induced convective rolls and thermal plume move downstream. Also, the downstream recirculating cell is always weaker than the upstream one in the mixed convection regime.

The velocity profile for $Ra = 500$ and $Pe = 0.1$ indicates the existence of two pairs of convective rolls. The appearance of these two additional slender cells is very special since such a four-cell pattern had not been observed in the corresponding case of natural convection. The results of Elder (1967), Horne and O'Sullivan (1974a, 1978), and Prasad and Kulacki (1986, 1987) indicate that the flow is bicellular as long as $D/H \leq 2$. It was suspected that this may be due to false diffusion introduced by upwind differencing in the numerical scheme. To verify this, the supposedly more accurate QUICK scheme was used which predicted a bicellular flow field (Lai et al. 1987a). Further calculations showed that the difference

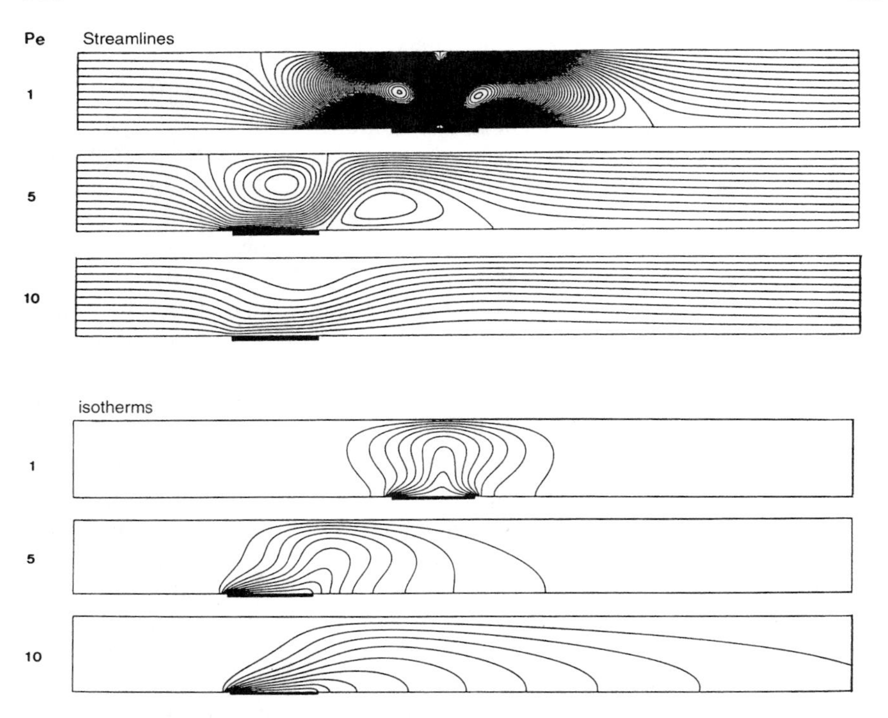

Figure 19. Streamline and temperature contours for a porous layer with a finite, isothermal heat source on the lower surface. $Ra = 100$ (Prasad et al. 1988a).

in heat transfer results between the predictions made by upwind and QUICK schemes is small as long as Ra is not very high and Pe not very low, such as $Ra = 500$ and $Pe = 0.1$.

Plots of overall Nusselt number exhibit that, for $Pe > 1$, Nusselt number increases monotonically with the Péclet number as long as $Ra \leq 10$ (Figure 20). However, this behavior is altered when the modified Rayleigh number is high. For example, Nu for $Ra = 100$ decreases with an increase in Pe beyond unity, reaching a minimum at $Pe > 5$. This is primarily because the enhancement in heat transfer by an increase in forced flow is not able to compensate for the reduction due to diminishing buoyant effects (or recirculation). When forced convection effects take over, Nusselt number increases monotonically with Pe. Forced convection becomes the dominant mode of heat transfer when $Ra \leq 100$, $Pe \geq 50$. A comparison between Nusselt numbers in forced convection in a channel and those reported by Cheng (1977b) for horizontal plates show that the flat plate solutions can be

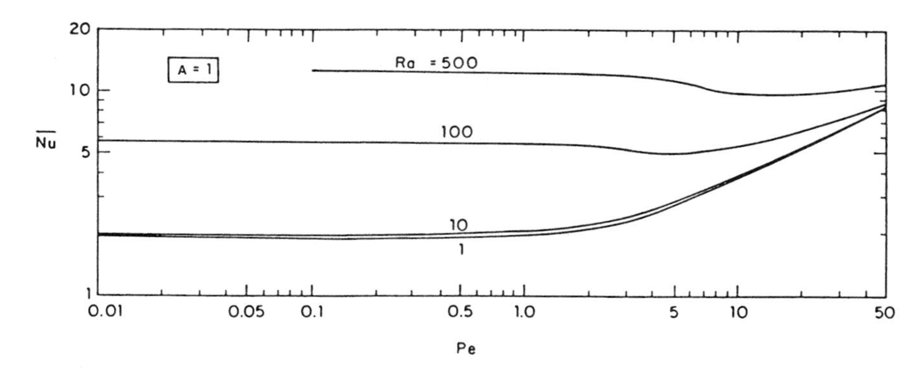

Figure 20. Average Nusselt numbers for an isothermal heat source on the lower wall of a channel (Prasad et al. 1988a).

used with reasonable accuracy as long as the free convection effects are negligible.

It was also found that flow structure, temperature field, and heat transfer coefficients change significantly with the size of the heat source (Lai et al. 1987b). If the Rayleigh number is low, only two recirculating cells are produced, one near the leading edge and the other at the trailing edge of the heat source. However, at high Rayleigh numbers the number of cells increases with the size of the heat source. For example, two pairs of convective cells are observed when $Ra = 100$, $Pe = 0.1$, and $A = 3$. These flow structures at low Péclet numbers are very similar to the natural convection flow field reported by Prasad and Kulacki (1986, 1987). As seen earlier, an increase in forced flow again destroys the multicellular flow pattern. The Nusselt number curves (Figure 21) show similar competitive effects of buoyant and forced flows, as displayed in Figure 20 for $A = 1$.

The interaction mechanisms between the forced flow and the buoyancy-induced velocity fields in the presence of multiple heat sources has also been examined (Lai et al. 1990a). In the free convection regime, a direct consequence of the existence of multiple heat sources is the appearance of multiple pairs of convective cells and reversed thermal plumes. For every additional heat source one pair of cells is generated. The flow and temperature fields thus obtained are very similar to those reported by Prasad and Kulacki (1986) for a long heat source. However, the basic difference between the two cases is that the convective cells are produced at much lower modified Rayleigh numbers than in the case of a long heated segment. The symmetric nature of the middle cells and thermal plumes further suggests that the heat sources (except those at each end) behave independently.

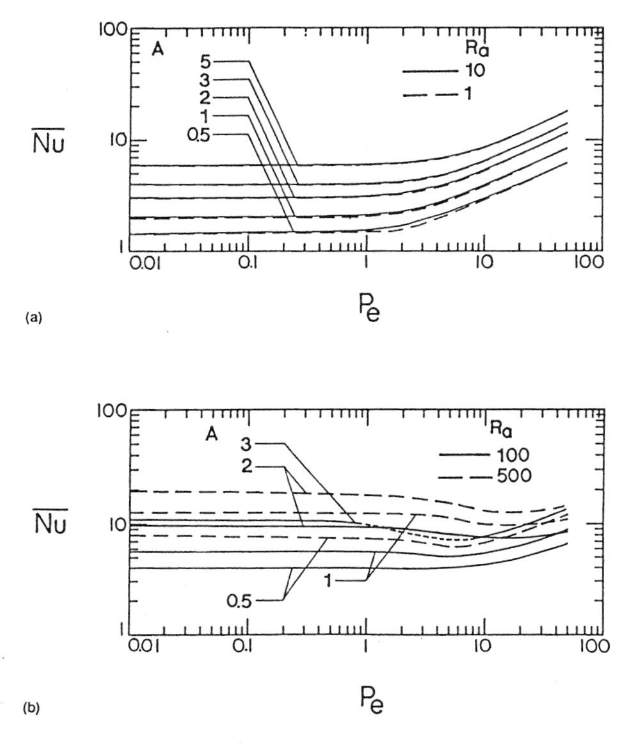

Figure 21. Average Nusselt numbers for an isothermal heat source on the lower wall of a channel: effect of heater size (Lai et al. 1987b).

With the introduction of forced flow, the buoyancy-driven recirculation loses its strength and finally disappears, as observed in the case of a single heat source. The most interesting finding is that at certain values of Rayleigh and Péclet numbers, e.g., $Ra = 100$ and $Pe = 5$, the flow becomes unstable, as has been observed in the case of a long heat source. The temperature field, flow field, and heat transfer rate also vary with time. As shown in Figure 22, the oscillation in the flow field is due to the destruction and regeneration of the recirculating cells. Initially, two pairs of cells form in the flow field, but with time, these cells become weaker. Finally, the first inner cell is decomposed into two cells and the second cell is swept away by the forced flow. The newly generated first inner cell then replaces the last cell; in the meantime, another two inner cells grow. Shortly after these two cells are developed, the whole process repeats itself. Similar behavior is observed in the flow field when the number of heat sources is more than two.

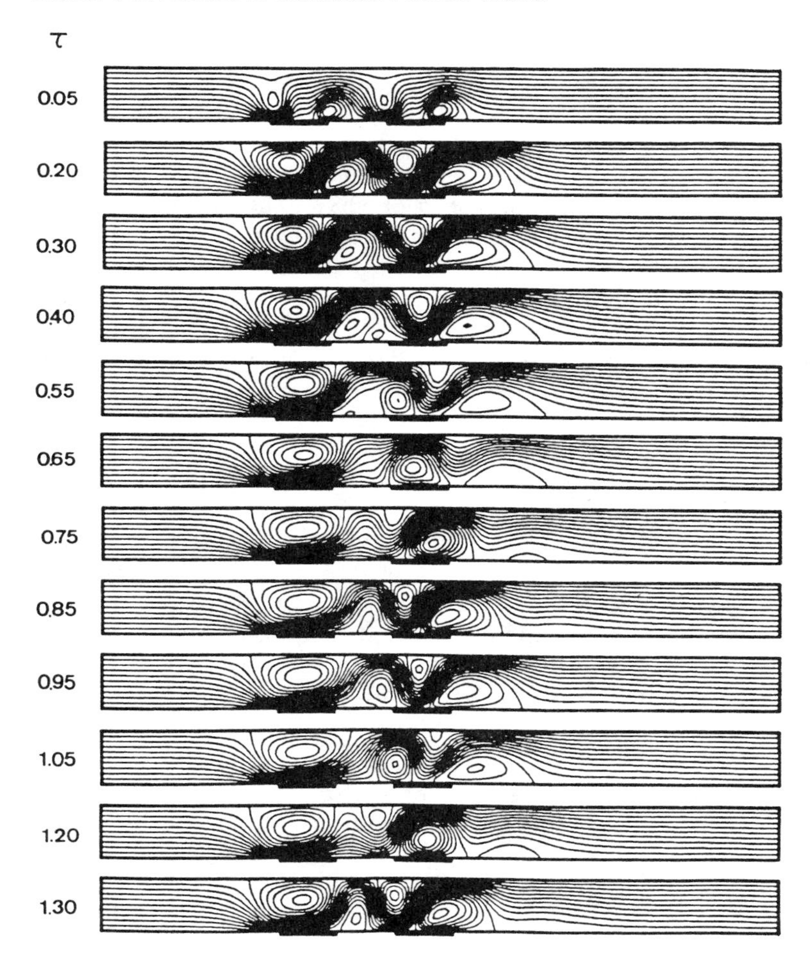

Figure 22. Unsteady flow in a channel with two isothermal heaters at $Ra = 100$ and $Pe = 5$. Heater length/layer height = 1 (Lai et al. 1990a).

In general, the overall Nusselt number for multiple heat sources exhibits a similar dependence on Ra and Pe as is observed for a single heat source. However, the reduction in Nu from the free convection asymptote with increasing Péclet number is more significant when the source number increases. Also, oscillatory behavior seems to dominate in the range of Péclet number where the heat transfer rate has been significantly reduced by the forced flow. In general, the multicellular and oscillatory flows reported in these studies need further investigation to understand completely the bifurcation phenomena, to examine the effects of inertia and boundary viscous forces, and to reveal the multiplicity of solutions, if any.

Experiments have also been conducted to validate the numerical results on mixed convection in a discretely heated horizontal porous layer (Lai and Kulacki, 1991b). Glass beads of 3 mm diameter were used with water to produce a saturated porous layer of 50.8 mm height. A rectangular test section 1067 mm long and 305 mm wide was used to simulate the two-dimensional flow. The strip heaters on the lower surface produced a constant heat flux boundary condition and permitted a wide range of dimensionless heater lengths. Temperature was measured at several locations in the porous bed to characterize the cellular flow structure. Experiments were conducted for a large range of Péclet number, $0 < Pe < 200$, to cover the free, mixed, and forced convection regimes.

For the three sizes of heat source selected, it was possible to obtain a heat transfer correlation in terms of $\overline{Nu}_D$ and Ra_D for free convection,

$$\overline{Nu}_D = 0.269 Ra_D^{0.451}, \qquad Ra_D \geq 40 \tag{33}$$

where the Nusselt and Rayleigh numbers are based on the heater length. A modified version of this correlation in terms of effective thermal conductivity, $k_e = \omega k_f + (1-\omega)k_m$, where ω is the weighting convection parameter (Prasad et al. 1985), was obtained as

$$\overline{Nu}_{D,e} = 0.516 Ra_{D,e}^{0.357}, \qquad Ra_{D,e} \geq 40 \tag{34}$$

which is very close to the correlation obtained from the numerical solutions

$$\overline{Nu}_D = 0.520 Ra_D^{0.354}, \qquad Ra_D \geq 40 \tag{35}$$

Excellent agreement between the predicted and measured values was also achieved for mixed convection. The experimental data are generally lower than the numerical predictions when $D/H = 1$, whereas they are higher for $D/H = 5$. The largest difference is about 20% but mostly they are well within 10%. It is observed that the maximum difference between the experimental and numerical results occurs when $Ra/Pe^{3/2} \gg 1$ and $Ra/Pe^{3/2} \ll 1$, i.e., in the asymptotic free and forced convection regimes, respectively. A large scattering of experimental data in these two regimes may be due to the use of an inappropriate thermal conductivity, k_m, when the fluid is convected most. A much better agreement was achieved when the experimental data were presented in terms of the effective thermal conductivity, k_e.

For these studies, the heat transfer data could be best correlated in terms of $\overline{Nu}_D/Pe_D^{1/2}$ and $Ra_D/Pe_D^{3/2}$. Thus

$$\frac{\overline{Nu_D}}{Pe_D^{1/2}} = \left[1.274 + 0.079\left(\frac{Ra_D}{Pe_D^{3/2}}\right)\right]^{0.506} \tag{36}$$

This correlation was modified to define the governing parameters in terms of k_e, or

$$\frac{\overline{Nu_D}}{Pe_D^{1/2}} = \left[1.895 + 0.200\left(\frac{Ra_D}{Pe_D^{3/2}}\right)\right]^{0.375} \tag{37}$$

which is very close to the correlation obtained from the numerical solutions

$$\frac{\overline{Nu_D}}{Pe_D^{1/2}} = \left[1.917 + 0.210\left(\frac{Ra_D}{Pe_D^{3/2}}\right)\right]^{0.372} \tag{38}$$

Figure 23 shows the agreement between the experimental data and Eqs. (36) and (37). Also, Eqs. (36)–(38) permit one to estimate the Péclet numbers for which $\overline{Nu_D}$ is a minimum as a function of Rayleigh number. In general, $Pe_{D,\min} \propto Ra_D^{2/3}$ for $Ra_D > 40$.

The possibility of flow transition to the oscillatory mixed convection flow regime has been studied by recording the fluctuations in temperatures (Lai 1988) and numerical simulation (Lai and Kulacki 1991c). It was, however, not feasible to determine from the numerical and experimental work performed, the conditions for the onset of oscillatory flow and the range of Ra and Pe for such flows to exist. On the other hand, Hsu and Cheng (1980a,b) have studied the onset of vortex instability of mixed convective flow over a horizontal surface. A "critical" Péclet number at which the secondary flow is initiated is reported to be a function of $Ra/Pe^{3/2}$. Although not directly related, it is expected that the second Pe_{critical} for oscillations will have a similar functional form. To elaborate further, the plot presented by Hsu and Cheng (1980a) can be transformed and reproduced as shown in Figure 24. For a given Ra_D, an increase in Péclet number will bring the flow field from the asymptotic free convection to the asymptotic forced convection regime, which is represented by a solid straight line. The dashed line, on the other hand, separates the stable and unstable convection regimes. Clearly, the flow is always stable for $Ra_D = 10$ whereas, for higher Rayleigh numbers, a transition from stable to unstable flows is certainly possible. Hence, the analytical solutions of Hsu and Cheng explain successfully the oscillatory phenomena observed in the case of a bounded horizontal porous layer.

The problem of mixed convection resulting from horizontal forced flow which is interacting with natural convection driven by a horizontal temperature gradient has been studied by Bejan and Imberger (1979).

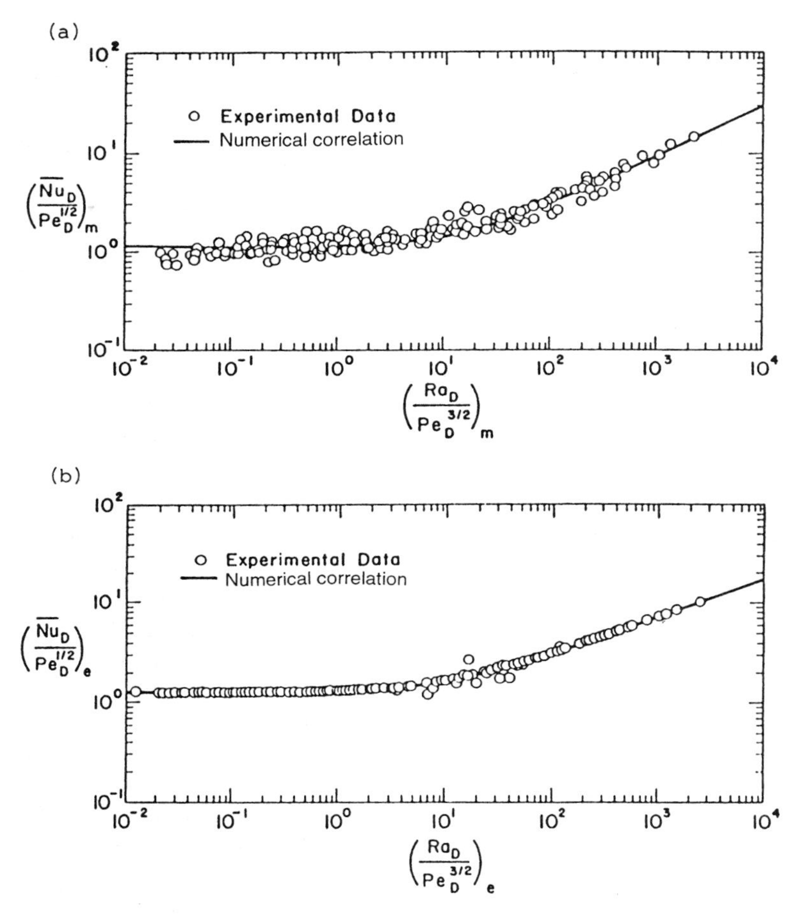

Figure 23. Comparison of experimental Nusselt numbers with numerical correlation using: (a) mean thermal conductivity; (b) effective thermal conductivity (Lai and Kulacki 1991b).

They used an asymptotic expansion method to study this mode of convection in a porous channel of large aspect ratios. Other than aspect ratio (W/H), this problem is governed by Rayleigh number and a pressure parameter P which is equivalent to the Péclet number. It was shown that a net discharge leads to a nonparallel flow structure characterized by a variable longitudinal temperature gradient. Haajizadeh and Tien (1984) also studied this problem in three parameter ranges: (a) $Ra \to 0$ for a fixed aspect ratio and $P = O(1)$; (b) an infinitely large aspect ratio for a fixed Ra and

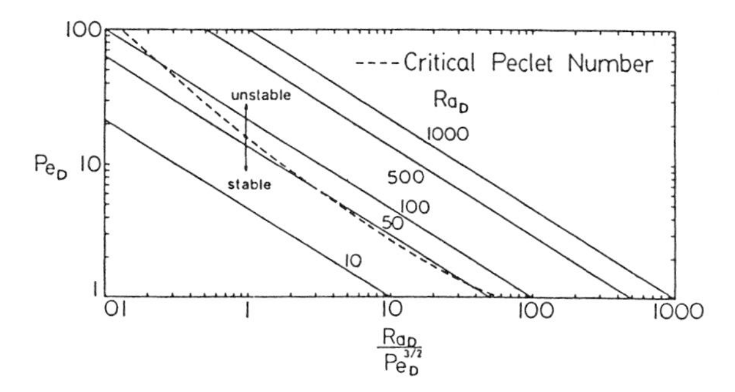

Figure 24. Critical Péclet numbers for the onset of secondary (oscillatory) flow over a horizontal surface of length D (Lai 1988).

$P = O(H/W)$; and (c) $Ra_H \leq 80$, $2 \leq W/H \leq 5$ and $0 \leq P/Ra_H \leq 0.5$. Both asymptotic and finite-difference numerical solutions were obtained, which showed that the asymptotic solutions are valid for $Ra_H^2/(W/H)^3 \leq 50$ and $P \leq 1.5$. For the range of parameters considered by Haajizadeh and Tien, the heat transfer rates due to natural convection and forced flow can be simply added together despite the nonlinear interaction between these two modes of convection. They also found that even a small rate of through flow has a significant effect on temperature distribution and heat transfer across the channel. Also, for $P/Ra_H \geq 0.2$, the contribution of buoyancy to the overall heat transport process is almost negligible. These studies further suggest that the interaction mechanism between the horizontal through flow and the flow induced by the horizontal temperature gradient is very different from that between the horizontal forced flow and free convection due to a vertical temperature gradient.

B. Vertical Layers

Research on mixed convection in porous layers confined between two vertical walls has been rather limited. Lai et al. (1988) have considered the vertical channel with a discrete wall heat source, and Hadim (1994) has carried out a numerical study of the development of laminar mixed convection in a vertical channel with uniform wall temperature and uniform wall heat flux.

1. Uniform Heating

Hadim (1994) carried out a numerical study of buoyancy-aided mixed convection in an isothermally heated vertical channel filled with a fluid-saturated porous medium. He employed the Darcy–Brinkman–Forchheimer model and presented the results in terms of modified Grashof number ($Gr = g\beta KH\Delta T/\nu^2$), Darcy number ($Da = K/H^2$), Reynolds number ($Re = VH/\nu$), inertia coefficient ($b = 0.55$), and $Pr = 0.72$. The objective of this study was to examine the evolution of mixed convection in the entrance region.

He has also reported that the slope of the curve of bulk temperature versus axial length increases as the Darcy number is decreased. As a result, the thermal entry length for fixed values of *Gr* and *Re* decreases with *Da*. The increased velocities near the walls enhance the heat transfer rate within a short length from the entrance. The effect of decreasing Darcy number is, however, important only at low values of *Da*, i.e., in the Darcy regime. At large Darcy number, the flow in this region is dominated by forced convection and the Nusselt number is almost independent of *Da*. The mixed convection flows, which start a short distance away from the inlet, exhibit a significant enhancement in heat transfer at low Darcy numbers (Muralidhar, 1989). When the flow becomes fully developed, the Nusselt number does not vary with *Da* in the Darcy regime. However, in the non-Darcy flow regime, the Nusselt number decrease with *Da* and asymptotically approaches the fluid value, a phenomenon discussed in detail by Prasad et al. (1988b).

The non-Darcy numerical results of Hadim (1994) further show that the effects of Grashof and Reynolds numbers for a given *Da* remain qualitatively similar to what have been reported for the pure fluid case, that is, very close to the entrance; and in the fully developed region, forced convection dominates the overall heat transfer. In the Darcy flow regime, the heat transfer rate is governed by the sole parameter Gr/Re. The Nusselt number increases with Gr/Re in the inlet region but is independent of this parameter in the fully developed region (Figure 25). However, when the Reynolds number is very low, axial conduction is not negligible, and Gr/Re does not remain the sole parameter to govern the mixed convection.

Mixed convection in a vertical porous channel subject to asymmetric wall heating has been studied by Hadim and Chen (1994). Two heating conditions have been considered: uniform wall temperature and uniform wall heat flux. Their results show that, for fixed Grashof and Reynolds numbers, the average Nusselt number increases as the Darcy number decreases. Also, the heat transfer enhancement in the mixed convection regime is more pronounced for the heating condition by uniform wall temperature.

Chang and Chang (1996) have obtained numerical solutions for mixed convection in a vertical channel partially filled with porous medium. On the basis of the Brinkman–Forchheimer–extended Darcy formulation, they have taken into account the inertial and viscous effects. Their results

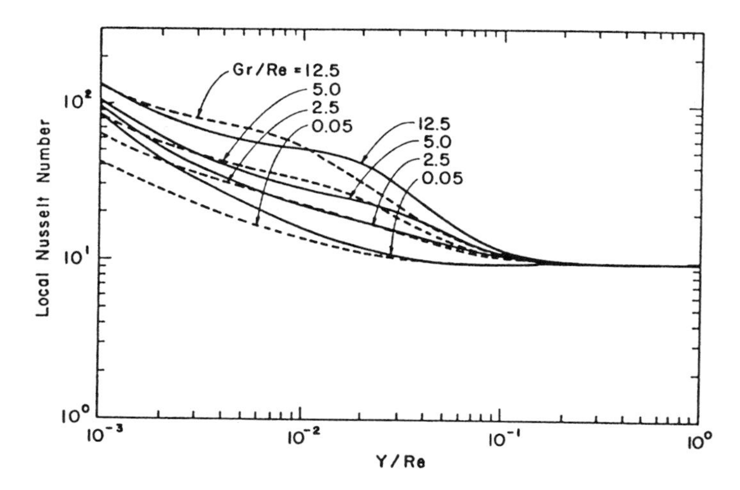

Figure 25. Local Nusselt numbers for non-Darcy flows in a channel: $Da = 10^{-6}$; $Re = 20$ (——) and $Re = 100$ (---) (Hadim 1994).

show that an increase in the Darcy number or Grashof number leads to an increase in the local Nusselt number, friction coefficient, and the hydrodynamic entrance length, but a reduction in the thermal entrance length. However, an increase in the porous layer thickness produces the opposite results.

2. Discrete Heating

Lai et al. (1988) have carried out a numerical study to examine the effects of forced flow on free convection induced by a finite heat source on an otherwise adiabatic vertical wall of a two-dimensional wall channel. The channel was considered to be isothermally cooled. The results for Darcy flow were obtained for both aiding and opposing flows and for $Ra = 0$, 10, 50, and 100, and $0.01 \leq Pe \leq 100$, when the length of the heated segment was equal to the layer width, $D = H$.

In the absence of a forced flow, it is expected that a recirculatory flow will be induced by the isolated heat source. This is also evident from the work of Prasad et al. (1986) on free convection from a finite wall heat source in a vertical cavity. The buoyancy-induced flow pattern is significantly modified if an external pressure gradient causes upflow. When the forced flow is weak, buoyant effects dominate the velocity field generally. However, the acceleration caused by buoyancy forces deflects the main flow toward the heat source, causing a recirculation zone at the cold wall side (Figure 26).

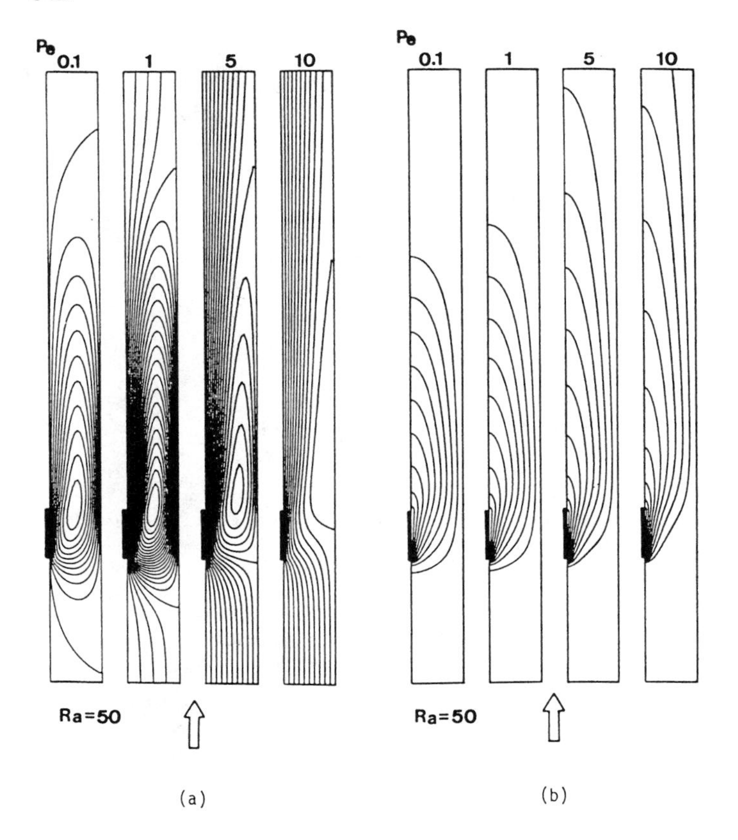

Figure 26. Aiding mixed convection in a vertical channel with an isothermal heat source: (a) streamline contours: $Pe = 0.1$, $\Delta\psi = 5$; $Pe = 1$, $\Delta\psi = 0.25$; $Pe = 5$, $\Delta\psi = 0.1$; and $Pe = 10$, $\Delta\psi = 0.1$; (b) isotherm contours: $\Delta\theta = 0.1$ for all Pe (Lai et al. 1988).

An increase in the externally induced velocity, or Péclet number, moves the convective cell upward. This delays the separation of the main flow from the cold wall. When the Péclet number is large, the enhanced thermal convection in the upward direction weakens the opposing buoyant effects on the cold wall. As a result, the strength of the secondary flow decreases substantially.

Figure 27 demonstrates the effect of Péclet number on the opposing mixed convective flow. When the main flow is weak, buoyancy causes an upflow in a region close to the heated segment. A recirculating flow is then produced in the hot wall region and the downward main flow is deflected

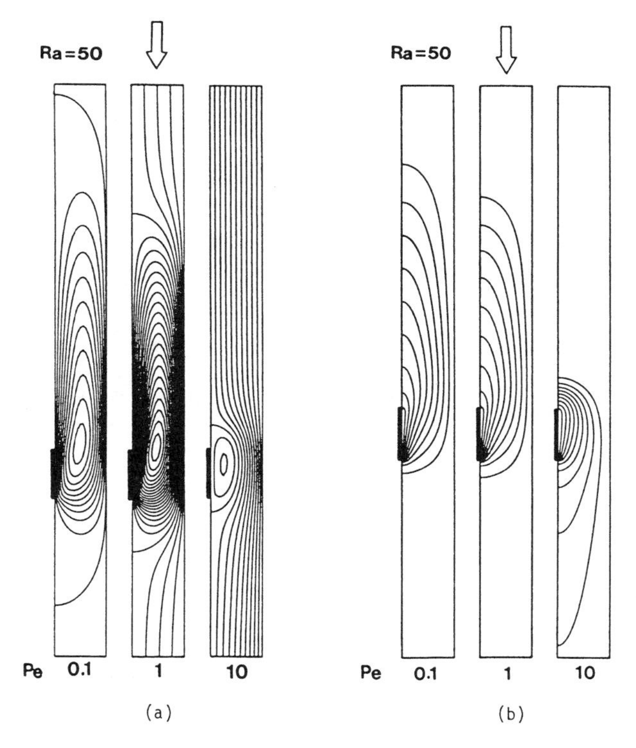

Figure 27. Opposing mixed convection in a vertical channel with an isothermal heat source: (a) streamline contours: $Pe = 0.1$, $\Delta\psi = 5$; $Pe = 1$, $\Delta\psi = 0.25$; $Pe = 10$, $\Delta\psi = 0.1$; (b) isotherm contours: $\Delta\theta = 0.1$ for all Pe (Lai et al. 1988).

toward the cold wall. This structure is contrary to what has been observed in the case of aiding flow. When the Péclet number is small, secondary flows for both aiding and opposing mixed convection are close to each other in strength and extent (Figures 26 and 27). However, an increase in Pe for opposing mixed flow substantially reduces the strength and size of the convective cell. It also moves the thermal plume downward, close to the heat source, rather than moving it upward as in the case of aiding flow. As a result, the region of intense thermal activity shrinks substantially with the increase in Péclet number and the heat transfer rate is reduced from its free convection value for $Pe > 0$.

Figure 28 shows that, as long as the forced flow aids buoyancy, $\overline{Nu}$ increases with Ra and/or Pe. An increase in Péclet number beyond a certain value results in an enhancement in heat transfer rate, and the slope of the

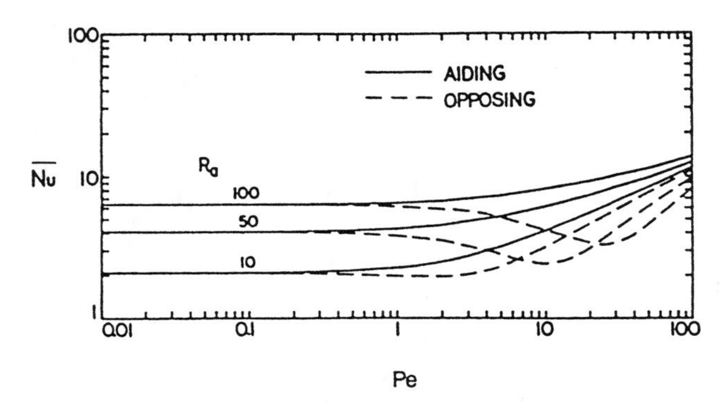

Figure 28. Average Nusselt numbers for aiding and opposing mixed convection in a vertical channel with a finite heat source (Lai et al. 1988).

Nusselt number curve starts increasing with *Pe*. With a further increase in Péclet number, the slope reaches a value close to 0.5, i.e., the slope reported for forced convection. The smaller the modified Rayleigh number, the earlier (in terms of *Pe*) the Nusselt number curve deviates from the free convection asymptote and the sooner it joins the forced convection curve.

Figure 28 further indicates that the deviation from the free convection asymptote is a weak function of the direction of main flow. However, the heat transfer rate for the opposing flow decreases with an increase in Péclet number as soon as the forced flow starts to contribute to heat transfer significantly. The negative slope of the Nusselt number curve, which is initially small, also increases with external flow rate or *Pe*. The overall heat transfer rate finally reaches a minimum before it starts increasing with *Pe*. The higher the modified Rayleigh number, the larger is the Péclet number required for the inflection point. Another interesting feature of Figure 28 is that, in the case of opposing flow, the Nusselt number for a lower *Ra* may exceed that for larger values. For example, $\overline{Nu}$ for $Ra = 10$ is greater than that for $Ra = 50$ and 100 if $Pe > 20$. This trend continues unless the Péclet number is very high such that buoyancy effects become negligible. Similar behavior has been demonstrated by the boundary layer solutions for a vertical plate (Cheng 1977a).

It is of interest to compare these results with those reported for natural convection in a vertical layer with a finite heat source on one wall (Prasad et al. 1986), and for mixed flow on a fully heated vertical plate (Cheng 1977a). Nusselt numbers reported by Prasad et al. for aspect $D/H < 5$ are significantly higher (by about 25% at $Ra = 100$) than the present asymptotic values for free convection. However, results discussed above support the conclusion of Prasad et al. that the Nusselt number for free convection in the vertical channel decreases from a peak to an asymptotic value as the heater length/channel width approaches infinity, i.e., a fully heated wall. Evidently, the asymptotic Nusselt number for the geometry considered by them is the one obtained here for free convection. A comparison between the present results and the boundary layer solutions for an isothermally heated vertical plate embedded in a saturated porous medium (Cheng 1977a) shows that the heat transfer rate from a flat plate is higher than that for a vertical channel if the forced flow is upward. The behavior is just the reverse in the case of opposing flow. The difference is largest for a free convection regime, and decreases with an increase in Péclet number. The values of the Nusselt numbers are very close to each other in the forced convection regime. It may thus be concluded that, in the boundary-layer forced convection regime, the flat plate solution of Cheng (1977a) can make reasonable predictions for the vertical channel problem.

C. Vertical Annuli

1. Discrete Heating

Reda (1988) considered a finite aspect ratio annulus (height/gap ~ 4) discretely heated on the inner wall and isothermally cooled at the outer surface. The porous layer between the concentric cylinders were superimposed with a fluid layer to obtain a permeable top, and downflow was permitted in order to study mixed convection heat transfer. Since the radius ratio was large, $r_o/r_i \sim 23$, the effects of outer wall on the temperature and flow fields were minimal and the geometry was closer to a cylinder embedded in a porous medium (Prasad and Kulacki 1984; Prasad 1986). This is also evident from the measured temperature profiles reported by Reda. With the total disappearance of buoyancy-induced upflow as the criterion, the transition from mixed to forced convection for opposing flows was predicted to occur for $Ra/Pe \sim 0.5$, independent of the heat source length or power input. This is in full agreement with the numerical data reported by Choi et al. (1989), as will be elaborated upon in the following discussion.

Choi et al. (1989) have extended the work of Lai et al. (1988) on mixed convection in vertical porous channels to examine the effects of both curvature and uniform flux heating. Figure 29 presents the effects of curvature on overall Nusselt number and shows that Nu for both aiding and opposing flows increases with the radius ratio parameter $\gamma = (r_R - r_i)/r_i)$. The higher the Péclet number, the larger is the rate of enhancement at all radius ratios. However, solutions for low Péclet number show that Nu decreases with an

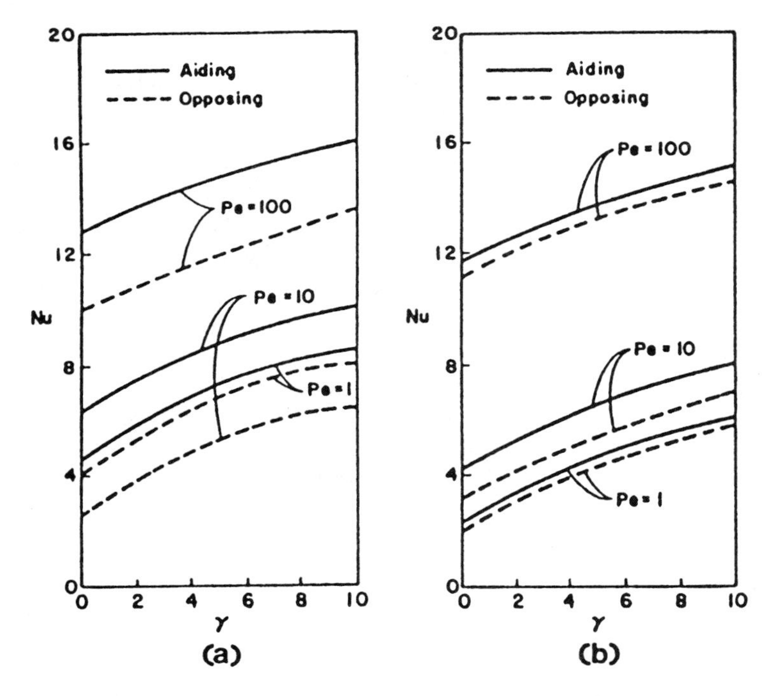

Figure 29. Average Nusselt numbers for mixed convection in a vertical annulus, $Ra = 50$: (a) isothermal heat source; (b) constant flux heat source (Choi et al. 1989).

increase in radius ratio, indicating that the solution asymptotically approaches that for a cylinder, which was first observed by Prasad and Kulacki (1984) in the case of free convection in a vertical annulus. When the heat transfer process is dominated by forced convection, the outer cylinder has no influence on Nusselt number (Figure 29). Also, Nusselt numbers for uniform flux heating are generally higher than that for the isothermal case.

For design purposes, the correlations of Nusselt number (based on the gap width) have been sought in the form

$$\frac{\overline{Nu}}{Pe^a} = \left[C_1 + C_2 \left(\frac{Ra}{Pe^b} \right) \right]^{C_3} \tag{39}$$

Through nonlinear regression, the values of a and b are obtained as 1/2 and 1 for isothermal heating, and 1/2 and 2 for the uniform flux heating, respectively. In the limit of natural convection ($Pe \rightarrow 0$), the heat transfer correlations for a vertical channel ($r_o/r_i \rightarrow 1$) are

Isothermal walls:

$$\overline{Nu} = 0.472 Ra^{0.5412} \tag{40}$$

where Ra is based on the gap width and the wall temperature differences.

Uniform flux on one wall:

$$\overline{Nu} = 1.021 Ra^{*\,0.2504} \tag{41}$$

where Ra^* is based on the gap width and the temperature modules, qH/k_m.

When curvature effects are included and the mixed convection regime is considered, Nusselt number correlations are

Isothermal wall, aiding flow:

$$\frac{\overline{Nu}}{Pe^{1/2}} = (3.373 + \gamma^{0.566})\left[0.068 + 0.032\left(\frac{Ra}{Pe}\right)\right]^{0.489)} \tag{42}$$

Isothermal wall, opposing flow:

$$\frac{\overline{Nu}}{Pe^{1/2}} = (2.269 + \gamma^{0.511})\left[0.047 + 0.047\left(\frac{Ra}{Pe}\right)\right]^{0.509} \tag{43}$$

Constant heat flux, aiding flow:

$$\frac{\overline{Nu}}{Pe^{1/2}} = (7.652 + \gamma^{0.892})\left[0.0004 + 0.0005\left(\frac{Ra^*}{Pe^2}\right)\right]^{0.243} \tag{44}$$

Constant heat flux, opposing flow:

$$\frac{\overline{Nu}}{Pe^{1/2}} = (4.541 + \gamma^{0.787})\left[0.017 + 0.0021\left(\frac{Ra^*}{Pe^2}\right)\right]^{0.253} \tag{45}$$

Choi and Kulacki (1992) have conducted experiments to verify the numerical results presented earlier (Choi et al. 1989). However, their study was limited to the case of constant flux heat source and $\gamma = 1$. Under these conditions, their experimental data were best correlated by (Figure 30)

Aiding flows:

$$\frac{\overline{Nu}}{Pe^{1/2}} = \left[2.924 + 0.668\left(\frac{Ra^*}{Pe^{3/2}}\right)\right]^{0.333} \tag{46}$$

Opposing flows:

$$\frac{\overline{Nu}}{Pe^{1/2}} = \left[2.082 + 1.886\left(\frac{Ra^*}{Pe^{3/2}}\right)\right]^{0.300} + \left[0.312 - 0.814\left(\frac{Ra^*}{Pe^{3/2}}\right)\right.$$

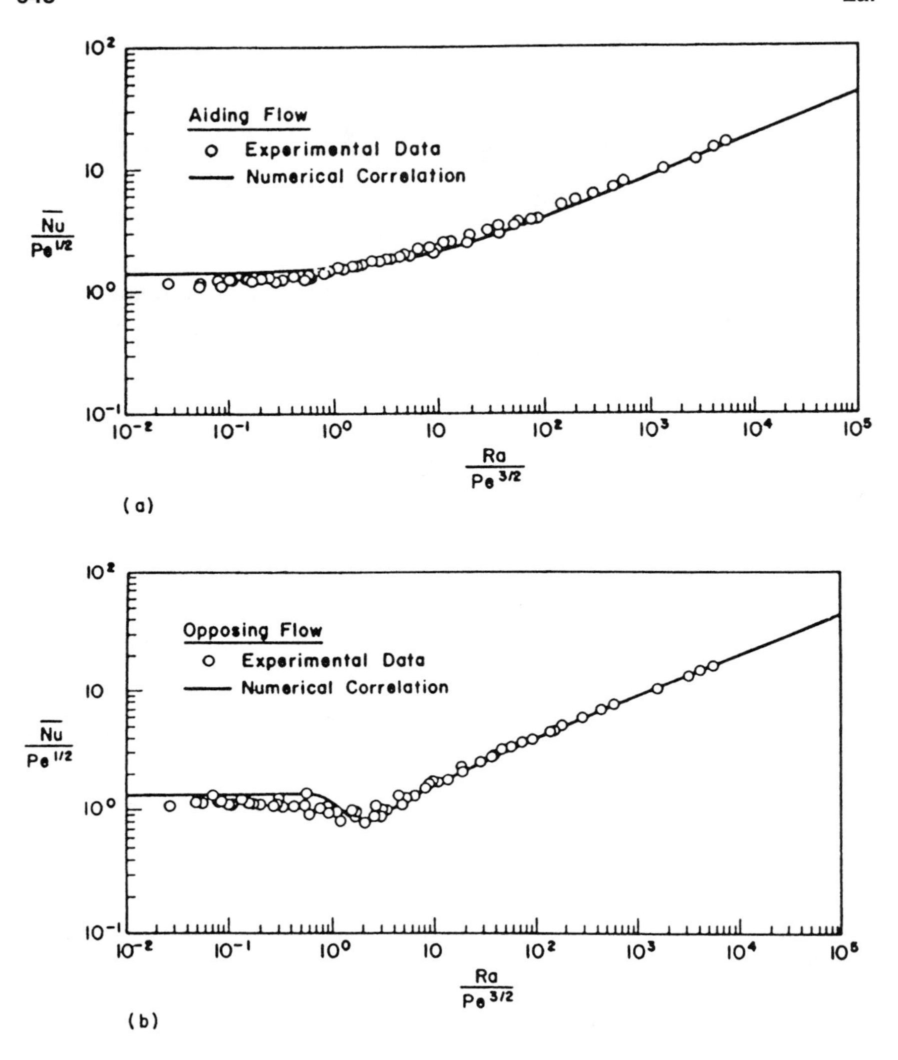

Figure 30. Comparison of experimental Nusselt numbers with numerical correlation: (a) aiding flows; (b) opposing flows (Choi and Kulacki 1992).

$$+ 1.895 \times 10^{-5}\left(\frac{Ra^*}{Pe^{3/2}}\right)^2]\exp\left[-\left|\frac{(Ra^*/Pe^{3/2})}{1.3}\right|\right] \tag{47}$$

For opposing flows, the second term is included for a better fit of data in the mixed convection rate, $0.1 < Ra/Pe^{3/2} < 10$.

Choi and Kulacki (1993) have extended their earlier study (Choi et al. 1989) to include non-Darcy effects (i.e., inertial and viscous effects). Their results show that the reduction of the Nusselt number is pronounced as the Rayleigh number increases in the natural convection dominated regime and as the Péclet number increases in the forced convection dominated regime. However, it is interesting to find that the Nusselt number increases substantially in the mixed convection regime for opposing flows, which is contrary to other studies in non-Darcy flows.

2. Differential Heating

Muralidhar (1989) has extended the work of Prasad and Kulacki (1984) on natural convection in a vertical annulus to examine the effects of superimposed aiding flow. Only Darcy flow was considered in the range of $0 < Ra < 500$ and $0 < Pe < 10$ for height/gap $= 10$ and $r_o/r_i = 2$. Thermal boundary conditions considered were isothermal heating and cooling on the inner and outer walls, respectively. His results indicate that forced convection dominates in the entry length defined by $0 < z/(r_o - r_i) < 1$. At $Z = z/(r_o - r_i) = 0.275$, the change from free convection is minimal, but it increases with the axial distance, and at $Z = 10.675$, the variation in Nu due to free convection is recovered. His numerical data (Figure 31) show that the Nusselt number increases with Ra and/or Pe. The incremental change in Nu as Ra increases is about the same for all Péclet numbers ($Pe \leq 10$). Hence, the percentage change in the average heat transfer rate drops at higher Péclet numbers. A surprising effect exhibited in Muralidhar's results is the sharp change in Nu as Pe changes from 0 to 1. For $Ra = 100$, the free convection motion generated can be expected to be more vigorous than the flow at $Pe = 1$. However, as Pe is increased beyond zero, the recirculation pattern is replaced by thin thermal boundary layers, which give rise to large heat transfer rates. In general, for $Pe >$

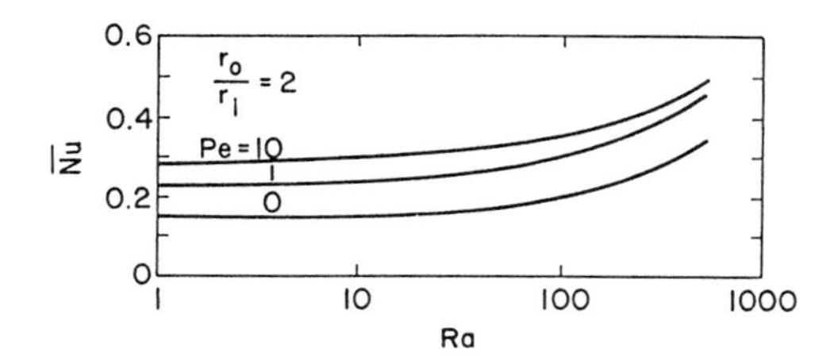

Figure 31. Numerically predicted average Nusselt numbers for aiding mixed convection in a vertical annulus (Muralidhar 1989).

100 and $0 < Ra < 100$, the heat transfer is reported to be forced convection, whereas for $Ra > 500$ and $0 < Pe < 10$, buoyancy forces determine the Nusselt number.

Parang and Keyhani (1987) have obtained analytical solutions for an asymmetrically heated annulus with uniform heat flux q_i and q_o on the inner and outer walls, respectively. The solutions for the Darcy and Darcy–Brinkman models were presented in terms of modified Bessel functions. Their solutions show that wall effects are more pronounced at the outer wall; that is, for a given radius ratio $\kappa(= r_o/r_i)$ and Gr/Re, the exclusion of the Brinkman term will result in a larger error in the heat transfer coefficient on the outer cylinder. This difference increases with an increase in radius ratio as the heat flux ratio $Q(= q_i/q_o)$ approaches zero. As is expected, boundary effects are strengthened with an increase in Gr/Re. The solutions also indicate that, for certain combinations of κ and Q, the heat transfer coefficient on one of the walls of the annulus may diminish as a result of mixed convection effects. For example, for $Q = 100$, $\kappa = 10$, and $Gr/Re = 10$, buoyancy forces enhance the heat transfer coefficient by 48.6% on the inner surface, while on the outer wall it is decreased by 30%. However, the opposite effect is observed if the heat flux ratio is reduced to, say, $Q = 2$.

Clarksean et al. (1988) have recently conducted an experimental and numerical study of mixed convection between concentric vertical cylinders for slightly different boundary conditions. Motivated by the feasibility of tapping into a liquid magma region located near the earth's surface to extract thermal energy, they considered an adiabatic inner cylinder and isothermally heated outer wall. Their geometry was closer to an open duct since the radius ratio of their experimental apparatus was close to 12. As a result, the radial temperature distributions have zero slope in a region close to the inner wall. The numerical and experimental data were presented for $0.05 < Ra/Pe < 0.5$ and showed the Nusselt number to be proportional to $(Ra/Pe)^{-1/2}$. It is evident from the work of other investigators, discussed above, that the heat transfer in this case is dominated by forced convection.

Experiments in a horizontal porous annulus were conducted by Vanover and Kulacki (1987), who simulated an inner cylinder heated by a constant flux and the outer cylinder isothermally cooled, for $r_o/r_i = 2$. Experimental data were obtained for fully developed flow in the range of $0 < Pe < 520$ and $Ra < 830$ using 1 and 3 mm diameter glass beads with water. In their study, the Péclet and Rayleigh numbers were based on the gap width $(r_o - r_i)$ and the temperature modulus $q(r_o - r_i)/k_m$. They found that, when the Rayleigh number is large, Nusselt numbers in mixed convection may be lower than the free convection values. This "crossover" in the heat transfer coefficient is attributed to a restructuring of the flow as forced convection begins to play a dominant role. Muralidhar (1989), however, did not observe this phenomenon since he considered Péclet numbers only up to 10 and focused his attention on the developing flows. Heat transfer correlations obtained by Vanover and Kulacki are

Mixed convection:

$$\overline{Nu_D} = 0.619 Pe_D^{0.177} Ra^{0.092}, \qquad 6 < Pe < 82 \tag{48}$$

Forced convection:

$$\overline{Nu_D} = 0.117 Pe^{0.657}, \qquad Pe > 180 \tag{49}$$

where the overall Nusselt number is normalized with its conduction value, $Nu_{\text{cond}} = 1.44$, for an annulus with $r_o/r_i = 2$.

D. Other Configurations

Several other studies on mixed convection in porous media reported in the literature do not fall into the above categories. Another class of problem, which has been studied extensively, is the effect of fluid discharge/recharge on buoyancy-induced motion in a geothermal reservoir. Cheng and coworkers (Cheng and Lau 1977; Cheng and Teckchandani 1977; Cheng 1978), Horne and O'Sullivan (1974b), Schrock and Laird (1976), Troncoso and Kassoy (1983), and many other investigators have worked on this problem to simulate the Wairakei geothermal field under production, the natural hot springs discharge at Long Valley, California, the coastal geothermal reservoir with application to liquid waste disposal in the Florida acquifier, and several other recharge/discharge phenomena. The readers are referred to Cheng (1978), who has reviewed these studies in great detail.

IV. CONCLUDING REMARKS

In this chapter, the recent literature on mixed convective heat transfer in porous media has been reviewed. It appears that research on this topic, although intensive in recent years, is still in its early stage. For theoretical development, most problems studied earlier are of simple geometry, and the solutions obtained are based on some simplified assumptions; e.g., most commonly, the Darcy flow model and boundary-layer approximations. These studies, nevertheless, lay the foundation for future development. With the accumulation of fundamental understanding, recent research has focused on correcting or improving the Darcy flow model. Non-Darcy effects have been identified in a variety of flows, and, qualitatively, their influences on the heat transfer results are well established now.

Despite the increased volume of research in this field, experimental results are still very few. For published measurements, there exist discrepancies between numerical predictions and experimental data. Although this kind of discrepancy is often found in the literature of heat transfer in porous media, it once again reflects the urgent need for more fundamental experimental studies. So far, no satisfactory theory is available for resolving this problem. However, it is felt that the so-called "divergence" observed in earlier measurements of Nusselt number in a porous layer could be attributed to the incorrect value of thermal conductivity. Simply put, the role of conduction within the solid matrix and that of convection within the saturating fluid produce a value of conductivity different from stagnant values as the driving potential for convection increases. It is important to realize that most numerical research on convective transport in porous media can neatly avoid the issue of property values because of the use of the volume averaging technique and dimensionless variables. Admittedly, no theoretical argument has yet been offered on which to base the use of the effective thermal conductivity. Nevertheless, the proposed effective thermal conductivity model has the comforting results of (1) reducing the considerable scatter which is usually found in measurements of Nusselt numbers over a wide range of Rayleigh (for natural convection) and Péclet (for mixed convection) numbers, and (2) improving agreement between experimental and calculated Nusselt numbers.

At this time, it is postulated that there is a connection between the concept of effective thermal conductivity and that of thermal dispersion. Unfortunately, the experimental data on thermal dispersion are very scarce in the literature. Most analytical studies have used hydrodynamic dispersion for approximation. It is important to note that there exists a basic difference between these two dispersion mechanisms: heat does transfer through the solid phase whereas fluid cannot. Therefore, the analytical results thus obtained are valid only qualitatively. The functional form of relation between the effective thermal conductivity and thermal dispersion will then require more research of a decidedly fundamental nature. It is hoped that this review will provide useful information to thermal engineers and will stimulate further research interest in a very fundamental and yet very challenging problem.

NOMENCLATURE

Roman Letters

A	aspect ratio or heater size parameter
b	Forchheimer coefficient
C	thermal dispersion coefficient
D	heater length for horizontal layer case and gap width for annular case, [m]
d	particle diameter or mean pore diameter, [m]
Da	Darcy number, K/H^2 or K/D^2
Er	Ergun number, $K'\alpha/d\nu$
Fo	Forchheimer number, Kb/x
f	dimensionless stream function
Gr	Grashof number
g	acceleration due to gravity, [m/s^2]
H	height of porous layer, [m]
h	heat transfer coefficient, [W/m^2 K]
K	permeability, [m^2]
K'	inertial coefficient of the Ergun equation
k	thermal conductivity, [W/m K]
Le	Lewis number
m	constant defined in the power law variation of free stream velocity or wall heat flux
N	buoyancy ratio or extinction parameter
Nu	local Nusselt number
$\overline{Nu}$	average Nusselt number
n	constant defined in the power law variation of wall temperature
Pe	Péclet number
Pr	Prandtl number
q	heat flux, [W/m^2]
Ra	Rayleigh number based on constant temperature difference
Ra^*	Rayleigh number based on constant flux
Re	Reynolds number
r	radius, [m]
T	temperature, [K]

Greek Letters

α_e	effective thermal diffusivity, [m^2/s]
α	molecular thermal diffusivity, [m^2/s]
α'	diffusivity due to thermal dispersion, [m^2/s]
β	coefficient of volume expansion

γ radius ratio parameter, $r_o/r_i - 1$
η similarity variable
θ dimensionless temperature
κ radius ratio, r_o/r_i
λ constant defined in the power law variation of wall temperature
μ dynamic viscosity, $[\mathrm{N\,s/m^2}]$
ν kinematic viscosity, $[\mathrm{m^2/s}]$
ρ density, $[\mathrm{kg/m^3}]$
σ local porosity parameter

Subscripts

C concentration-driven convection
c cold wall
D quantities defined on the basis of length of heat source or gap width
d quantities defined on the basis of particle diameter or mean pore diameter
e quantities defined on the basis of effective thermal conductivity
H quantities defined on the basis of height of porous layer
h hot wall
m mean value, or quantities defined on the basis of stagnant thermal conductivity
T temperature-driven convection
w condition at the wall
∞ condition at infinity

REFERENCES

Aldoss TK, Chen TS, Armaly BF. Nonsimilarity solutions for mixed convection from horizontal surfaces in a porous medium: variable surface heat flux. Int J Heat Mass Transfer 36:463–470, 1993a.

Aldoss TK, Chen TS, Armaly BF. Nonsimilarity solutions for mixed convection from horizontal surfaces in a porous medium: variable wall temperature. Int J Heat Mass Transfer 36:471–478, 1993b.

Aldoss TK, Chen TS, Armaly BF. Mixed convection over nonisothermal horizontal surfaces in a porous medium: the entire regime. Numer Heat Transfer A 25:685–702, 1994.

Aldoss TK, Jarrah MA, Al-Sha'er BJ. Mixed convection from a vertical cylinder embedded in a porous medium: non-Darcy model. Int J Heat Mass Transfer 39:1141–1148, 1996.

Badr HM, Pop I. Combined convection from an isothermal horizontal rod buried in a porous medium. Int J Heat Mass Transfer 31:2527–2541, 1988.

Bejan A, Imberger J. Heat transfer by forced and free convection in a horizontal channel with differentially heated ends. J Heat Transfer 101:417–421, 1979.

Brouwers HJH. Heat transfer between a fluid-saturated porous medium and a permeable wall with fluid injection or withdrawal. Int J Heat Mass Transfer 37:989–996, 1994.

Caltagirone JP. Thermoconvective instabilities in a horizontal porous layer. J Fluid Mech 72:269–287, 1975.

Chandrasekhara BC. Mixed convection in the presence of horizontal impermeable surfaces in saturated porous media with variable permeability. Wärme- und Stoffübertragung 19:195–201, 1985.

Chandrasekhara BC, Namboodiri PMS. Influence of variable permeability on combined free and forced convection about inclined surfaces in porous media. Int J Heat Mass Transfer 28:199–206, 1985.

Chang WJ, Chang WL. Mixed convection in a vertical parallel-plate channel partially filled with porous media of high permeability. Int J Heat Mass Transfer 39:1331–1342, 1996.

Chen CH, Chen TS, Chen CK. Non-Darcy mixed convection along nonisothermal vertical surfaces in porous media. Int J Heat Mass Transfer 39:1157–1164, 1996.

Chen CK, Chen CH, Minkowycz WJ, Gill US. Non-Darcian effects on mixed convection about a vertical cylinder embedded in a saturated porous medium. Int J Heat Mass Transfer 35:3041–3046, 1992.

Cheng P. Combined free and forced boundary layer flows about inclined surfaces in a porous medium. Int J Heat Mass Transfer 20:807–814, 1977a.

Cheng P. Similarity solutions for mixed convection from horizontal impermeable surfaces in saturated porous media. Int J Heat Mass Transfer 20:893–898, 1977b.

Cheng P. The influence of lateral mass flux on free convection boundary layers in a saturated porous medium. Int J Heat Mass Transfer 20:201–206, 1977c.

Cheng P. Heat transfer in geothermal systems. In: Advances in Heat Transfer. New York: Academic Press, 1978, Vol 14, pp 1–105.

Cheng P. Mixed convection about a horizontal cylinder and a sphere in a fluid-saturated porous medium. Int J Heat Mass Transfer 25:1245–1247, 1982.

Cheng P, Lau KH. The effect of steady withdrawal of fluid in geothermal reservoirs. Proceedings of the Second United Nations Symposium on the Development and Use of Geothermal Resources, 1977, Vol 3, pp 1591–1598.

Cheng P, Teckchandani L. Numerical solutions for transient heating and fluid withdrawal in a liquid-dominated geothermal reservoir. American Geophysical Union Monograph No. 20, *The Earth's Crust—Its Natural and Physical Properties*, 1977, pp 705–721.

Cheng P, Zheng TM. Mixed convection in the thermal plume above a horizontal line source of heat in a porous medium of infinite extent. Proceedings of the 8th International Heat Transfer Conference, 1986, Vol 5, pp 2671–2675.

Choi CY, Kulacki FA. Mixed convection through vertical porous annuli locally heated from the inner cylinder. J Heat Transfer 114:143–151, 1992.
Choi CY, Kulacki FA. Non-Darcian effects on mixed convection in a vertical packed-sphere annulus. J Heat transfer 115:506–509, 1993.
Choi CY, Lai FC, Kulacki FA. Mixed convection in vertical porous annuli. AIChE Symposium Series 269, 1989, Vol 85, pp 356–361.
Chou FC, Cheng CJ, Lien WY. Analysis and experiment of non-Darcian convection in horizontal square packed-sphere channels – II mixed convection. Int J Heat Mass Trasnfer 35:1197–1208, 1992.
Chou FC, Su JH, Lien SS. A re-evaluation of non-Darcian forced and mixed convection in cylindrical packed tubes. J Heat Transfer 116:513–516, 1994.
Clarksean R, Kwendakwema N, Boehm R. A study of mixed convection in a porous medium between vertical concentric cylinders. Proceedings of 1988 ASME/AIChE National Heat Transfer Conference, 1988, Vol 2, pp 339–344.
Combarnous MA, Bia P. Combined free and forced convection in a porous media. Soc Petrol Eng J 11:399–405, 1971.
Combarnous MA, Bories S. Hydrothermal convection in a saturated porous media. Adv Hydrosci 10:231–307, 1975.
Elder JW. Steady free convection in a porous medium heated from below. J Fluid Mech 27:29–48, 1967.
Haajizadeh M, Tien CL. Combined natural and forced convection in a horizontal porous channel. Int J Heat Mass Transfer 27:799–813, 1984.
Hadim HA. Numerical study of non-Darcy mixed convection in a vertical porous channel. J Thermophys Heat Transfer 8:371–372, 1994.
Hadim A, Burmeister LC. Onset of convection in a porous medium with internal heat generation and downward flow. J Thermophys Heat Transfer 2:343–349, 1988.
Hadim HA, Chen G. Non-Darcy mixed convection in a vertical porous channel with asymmetric wall heating. J Thermophys Heat Transfer 8:805–807, 1994.
Homsy GM, Sherwood AE. Convective instabilities in porous media with through flow. AIChE J 22:168–174, 1976.
Hooper WB, Chen TS, Armaly BF. Mixed convection along an isothermal vertical cylinder in porous media. J Thermophys Heat Transfer. 8:92–99, 1994.
Horne RN, O'Sullivan MJ. Oscillatory convection in a porous medium heated from below. J Fluid Mech 66:339–352, 1974a.
Horne RN, O'Sullivan MJ. Oscillatory convection in a porous medium: the effect of throughflow. Proceedings of 5th Australian Conference on Hydrology and Fluid Mechanics. New Zealand: University of Canterbury Press, 1974, Vol 2, pp 231–236.
Horne RN, O'Sullivan MJ. Convection in a porous medium heated from below: the effect of temperature dependent viscosity and thermal expansion coefficient. J Heat Transfer 100:448–452, 1978.
Hsieh JC, Chen TS, Armaly BF. Nonsimilarity solutions for mixed convection from vertical surfaces in porous media: variable surface temperature of heat flux. Int J Heat Mass Transfer 36:1485–1494, 1993a.

Hsieh JC, Chen TS, Armaly BF. Mixed convection along a nonisothermal vertical flat plate embedded in a porous medium: the entire regime. Int J Heat Mass Transfer 36:1819–1826, 1993b.

Hsu CT, Cheng P. The onset of longitudinal vortices in mixed convective flow over an inclined surface in a porous medium. J Heat Transfer 102:544–549, 1980a.

Hsu CT, Cheng P. Vortex instability of mixed convective flow in a semi-infinite porous medium bounded by a horizontal surface. Int J Heat Mass Transfer 23:789–798, 1980b.

Huang MJ, Yih KA, Chou YL, Chen CK. Mixed convection flow over a horizontal cylinder or a sphere embedded in a porous medium. J Heat Transfer 108:469–471, 1986.

Islam RM, Nandkumar K. Multiple solutions for buoyancy-induced flow in saturated porous media for large Péclet numbers. J Heat Transfer 108:866–871, 1986.

Jang JY, Chen JL. Thermal dispersion and inertia effects on vortex instability of a horizontal mixed convection flow in a saturated porous medium. Int J Heat Mass Transfer 36:383–390, 1993a.

Jang JY, Chen JL. Variable porosity effect on vortex instability of a horizontal mixed convection flow in a saturated porous medium. Int J Heat Mass Transfer 36:1573–1582, 1993b.

Jang JY, Lie KN. Boundary and inertia effects on vortex instability of a horizontal mixed convection flow in a porous medium. Numer Heat Transfer A 23:361, 1993.

Jang JY, Shiang CT. The mixed convection plume along a vertical adiabatic surface embedded in a non-Darcian porous medium. Int J Heat Mass Transfer 40: 1693–1699, 1997.

Jang JY, Lie KN, Chen JL. Influence of surface mass flux on vortex instability of a horizontal mixed convection flow in a saturated porous medium. Int J Heat Mass Transfer 38:3305–3312, 1995.

Joshi Y, Gebhart B. Mixed convection in porous media adjacent to a vertical uniform heat flux surface. Int J Heat Mass Transfer 28:1783–1786, 1985.

Kladias N, Prasad V. Non-Darcy oscillating convection in horizontal porous layers heated from below. Proceedings of 1st National Fluid Dynamics Congress, 1988, Vol 3, pp 1757–1764.

Kladias N, Prasad V. Natural convection in horizontal porous layers: effects of Darcy and Prandtl numbers. J Heat Transfer 111:926–935, 1989a.

Kladias N, Prasad V. Convective instabilities in horizontal porous layers heated from below: effects of grain size and its properties. New York: American Society of Mechanical Engineers, 1989b, ASME HTD—Vol 107, pp 369–379.

Kumari M, Pop I, Nath G. Darcian mixed convection plumes along vertical adiabatic surfaces in a saturated porous medium. Therm Fluid Dyn 22:173–178, 1988.

Kumari M, Nath G. Non-Darcy mixed convection boundary layer flow on a vertical cylinder in a saturated porous medium. Int J Heat Mass Transfer 32:183–187, 1989.

Lai FC. Free and mixed convection in porous media. PhD dissertation, University of Delaware, 1988.

Lai FC. Coupled heat and mass transfer by mixed convection from a vertical plate in a saturated porous medium. Int Commun Heat Mass Transfer 18:83–106, 1991a.

Lai FC. Non-Darcy mixed convection from a line source of heat in saturated porous medium. Int Commun Heat Mass Transfer 18:875–887, 1991b.

Lai FC, Kulacki FA. Non-Darcy convection from horizontal impermeable surfaces in saturated porous media. Int J Heat Mass Transfer 30:2189–2192, 1987.

Lai FC, Kulacki FA. Thermal dispersion effects on non-Darcy convection over horizontal surfaces in saturated porous media. Int J Heat Mass Transfer 32:971–976, 1989.

Lai FC, Kulacki FA. The influence of lateral mass flux on mixed convection over inclined surfaces in saturated porous media. J Heat Transfer 112:515–518, 1990a.

Lai FC, Kulacki FA. The influence of surface mass flux on mixed convection over a horizontal plate in saturated porous medium. Int J Heat Mass Transfer 33:576–579, 1990b.

Lai FC, Kulacki FA. The effects of variable viscosity on convective heat transfer along a vertical surface in a saturated porous medium. Int J Heat Mass Transfer 33:1028–1031, 1990c.

Lai FC, Kulacki FA. Non-Darcy mixed convection along a vertical wall in a saturated porous medium. J Heat Transfer 113:252–255, 1991a.

Lai FC, Kulacki FA. Experimental study of free and mixed convection in porous media. Int J Heat Mass Transfer 34:525–541, 1991b.

Lai FC, Kulacki FA. Oscillatory mixed convection in horizontal porous layers locally heated from below. Int J Heat Mass Transfer 34:887–890, 1991c.

Lai FC, Kulacki FA, Prasad V. Numerical study of mixed convection in porous media. In: Numerical Methods in Thermal Problems. Swansea: Pineridge Press. 1987a, Vol V, Part 1, pp 784–796.

Lai FC, Prasad V, Kulacki FA. Effects of the size of heat source on mixed convection in horizontal porous layers heated from below. Proceedings of 2nd ASME/JSME Thermal Engineering Joint Conference. New York: American Society of Mechanical Engineers, 1987b, Vol 2, pp 413–419.

Lai FC, Prasad V, Kulacki FA. Aiding and opposing mixed convection in a vertical porous layer with a finite wall heat source. Int J Heat Mass Transfer 31:1049–1061, 1988.

Lai FC, Choi CY, Kulacki FA. Free and mixed convection in horizontal porous layers with multiple heat sources. J Thermophys Heat Transfer 4:221–227, 1990a.

Lai FC, Pop I, Kulacki FA. Free and mixed convection from slender bodies of revolution in porous media. Int J Heat Mass Transfer 33:1767–1769, 1990b.

Lapwood ER. Convection of a fluid in a porous medium. Proc Cam Phil Soc 44:508–521, 1948.

Levec J, Carbonell RG. Longitudinal and lateral thermal dispersion in packed beds, part II: comparison between theory and experiment. AIChE J 31:591–602, 1985.

Li CT, Lai FC. Coupled heat and mass transfer by mixed convection from a horizontal surface in saturated porous medium. Proceedings of the 1997 National Heat Transfer Conference, 1997, Vol 11, pp 169–176.

Merkin JH. Free convection boundary layers on axisymmetric and two-dimensional bodies of arbitrary shape in a saturated porous medium. Int J Heat Mass Transfer 22:1461–1462, 1979.

Merkin JH. Mixed convection boundary layer flow on a vertical surface in a saturated porous medium. J Eng Math 14:301–313, 1980.

Minkowycz WJ, Cheng P. Free convection about a vertical cylinder embedded in a porous medium. Int J Heat Mass Transfer 19:805–813, 1976.

Minkowycz WJ, Cheng P, Hirschberg RN. Nonsimilar boundary layer analysis of mixed convection about a horizontal heated surface in a fluid-saturated porous medium. Int Commun Heat Mass Transfer 11:127–141, 1984.

Minkowycz WJ, Cheng P, Chang CH. Mixed convection about non-isothermal cylinders and spheres in a porous medium. Numer Heat Transfer 8:349–359, 1985.

Muralidhar K. Mixed convection flow in a saturated porous annulus. Int J Heat Mass Transfer 32:881–888, 1989.

Nakayama A, Koyama H. A general similarity transformation for combined free and forced convection flows within a fluid-saturated porous medium. J Heat Transfer 109:1041–1045, 1987.

Nandkumar K, Masliyah JH, Law HS. Multiple steady state and hysteresis in mixed convection flow in horizontal square tubes. In: Numerical Methods for Nonlinear problems. Swansea: Pineridge Press, 1984, Vol 2, pp. 935–947.

Nandkumar K, Weinitschke HJ, Sankar SR. The calculation of singularities in steady mixed convection flow through porous media. In: Mixed Convection Heat Transfer—1987. New York: American Society of Mechanical Engineers, 1987, HTD—Vol 84, pp 67–73.

Nguyen HD, Paik S. Unsteady mixed convection from a sphere in water-saturated porous media with variable surface temperature–heat flux. Int J Heat Mass Transfer 37:1783–1794, 1994.

Oosthuhizen PH. Mixed convection heat transfer from a cylinder in a porous medium near an impermeable surface. In: Mixed Convection Heat Transfer—1987. New York: American Society of Mechanical Engineers, 1987, HTD—Vol 84, pp 75–82.

Oosthuhizen PH. Mixed convection heat transfer from a heated horizontal plate in a porous medium near an impermeable surface. J Heat Transfer 110:390–394, 1988.

Parang M, Keyhani M. Boundary effects in laminar mixed convection flow through an annulus porous medium. J Heat Transfer 109:1039–1041, 1987.

Prasad V. Numerical study of natural convection in a vertical porous annulus with constant heat flux on the inner wall. Int J Heat Mass Transfer 29:841–853, 1986.

Prasad V, Kulacki FA. Natural convection in a vertical porous annulus. Int J Heat Mass Transfer 27:207–219, 1984.

Prasad V, Kulacki FA. Effects of the size of heat source on natural convection in horizontal porous layers heated from below. Proceedings of 8th International Heat Transfer Conference. New York: Hemisphere Press, 1986, pp 2677–2682.

Prasad V, Kulacki FA. Natural convection in horizontal porous layers with localized heating from below. J Heat Transfer 109:795–798, 1987.

Prasad V, Kulacki FA, Keyhani M. Natural convection in porous media. J Fluid Mech 150:89–119, 1985.

Prasad V, Kulacki FA, Stone K. Free convection in a porous cavity with a finite wall heat source. In: Natural Convection in Enclosures—1986. New York: American Society of Mechanical Engineers, 1986 HTD—Vol 63, pp 91–98.

Prasad V, Lai FC, Kulacki FA. Mixed convection in horizontal porous layers heated from below. J Heat Transfer 110:395–402, 1988a.

Prasad V, Lauriat G, Kladias N. Reexamination of Darcy–Brinkman solutions for free convection in porous media. Proceedings of the 1988 National Heat Transfer Conference. New York: American Society of Mechanical Engineers, Vol 1, pp 569–580.

Prats M. The effects of horizontal fluid flow on thermally induced convection currents in porous mediums. J Geophys Res 71:4835–4837, 1966.

Ranganathan P, Viskanta R. Mixed convection boundary-layer flow along a vertical surface in a porous medium. Numer Heat Transfer 7:305–317, 1984.

Reda DC. Mixed convection in a liquid-saturated porous medium. J Heat Transfer 110:147–154, 1988.

Rubin H. Heat dispersion effect on thermal convection in a porous medium layer. J Hydrol 21:173–185, 1974.

Rubin H. A note on the heat dispersion effect on thermal convection in a porous medium layer. J Hydrol 25:167–168, 1975.

Salt H, Mahoney KJ. Mixed convection in horizontal-flow packed beds. In: Mixed Convection Heat Transfer—1987. New York: American Society of Mechanical Engineers, 1987, HTD—Vol 84, pp 83–89.

Schrock VE, Laird ADK. Physical modelling of combined forced and natural convection in wet geothermal formations. J Heat Transfer 98:213–220, 1976.

Schubert G, Straus JM. Three-dimensional and multi-cellular steady and unsteady convection in fluid-saturated porous media at high Rayleigh numbers. J Fluid Mech 94:25–38, 1979.

Sutton FM. Onset of convection in a porous channel with net through flow. Phys Fluids 13:1931–1934, 1970.

Troncoso J, Kassoy DR. An axisymmetric model for the charging of a liquid-dominated geothermal reservoir. Int J Heat Mass Transfer 26:1389–1401, 1983.

Vanover DE, Kulacki, FA. Experimental study of mixed convection in a horizontal porous annulus. In: Mixed Convection Heat Transfer—1987. New York: American Society of Mechanical Engineers, 1987, HTD—Vol 84, pp 83–89.

Weinitschke HJ. On the calculation of limit and bifurcation points in stability problem of elastic shells. Int J Solids Struct 21:79–95, 1985.

Wooding RA. Rayleigh instability of a thermal boundary layer in flow through a porous medium. J Fluid Mech 9:183–192, 1960.

Wooding RA. Convection in a saturated porous medium at large Rayleigh number or Péclet number. J Fluid Mech 15:527–544, 1963.

Yih KA. Coupled heat and mass transfer in mixed convection over a vertical flat plate embedded in saturated porous media: PST–PSC or PHF–PMF. Heat Mass Transfer 34:55–62, 1998.

Zhou MJ, Lai FC. Aiding and opposing mixed convection from a cylinder in a saturated porous medium. 7th AIAA/ASME Joint Thermophysics and Heat Transfer Conference, 1998, AIAA Paper 98–2674.

15
Radiative Transfer in Porous Media

John R. Howell
University of Texas at Austin, Austin, Texas

I. INTRODUCTION

Radiative transport in porous media has important engineering applications in: combustion; heat exchangers for high temperature applications, including solar collectors; regenerators and recuperators; insulation systems; packed and circulating bed combustors and reactors; manufacturing and materials processing; and proposed energy storage and conversion methods.

In many of these applications, the porous medium acts as a means to absorb or emit radiant energy that is transferred to or from a fluid. Generally, the fluid itself (particularly if it is a gas) can be assumed to be transparent to radiation, because the dimensions for radiative transfer among the solid structure elements of the porous medium are usually much less than the radiative mean free path for scattering or absorption in the fluid. In other applications, no fluid is present and the heat transfer at high temperatures is by a combination of radiative and conductive transfer. There are also cases where the energy transfer among the particles and with the surrounding is by radiation alone, which is the case for some designs of space radiator where fluid particles are transmitted from a nozzle to a collection device, and are allowed to cool by radiative loss before being recycled.

Reviews by Vortmeyer (1978), Tien (1988), Kaviany and Singh (1993), Dombrovsky (1996), and Baillis-Doermann and Sacadura (1998) provide excellent recent overviews of radiative transfer in porous and dispersed media, along with extensive bibliographies covering pertinent literature. This chapter concentrates on recent developments and applications where radiative transfer in porous materials is an important effect.

Vortmeyer (1978) reviews the models of radiation/conduction interaction in packed beds that were available at that time. Comparisons of model predictions among themselves and with available experimental data are presented. A careful reading indicates that a designer or practitioner desiring to predict the behavior of a packed bed would have a difficult time in choosing a method that could be used with confidence. Many of the models were arbitrarily adjusted to agree with experiment, and they require input data on bed characteristics (usually porosity, particle diameter, particle spacing) and the radiative properties of the bed materials at the bed operating temperature. Many of these data are poorly known and must be obtained by experiment on a particular system. This remains the case.

Tien (1988) concentrates on radiative transfer in packed beds of spherical particles, providing a detailed review of the literature up to the time of publication. He discusses the effect of dependent and anisotropic scattering on the prediction of radiative transfer, and notes the lack of experimental data available for verification of the various proposed models. Comparisons of the models with the experiments of Chen and Churchill (1963) show poor agreement, for the most part.

Contemporary research and literature recognize the complexity of the interaction of radiation with closely packed elements such as is the case in many porous materials. This complexity is the reason why many of the earlier simplified models of this interaction have failed to provide agreement with experiment over extended ranges of parameters. In particular, many early models neglected the effect of anisotropic and/or dependent scattering. Newer models have thus moved toward numerical and statistical approaches that incorporate the detailed structure and properties of a particular packing and material at the expense of simplified correlations.

Kaviany and Singh (1993) concentrate on the use of Monte Carlo techniques to include the effects of dependent scattering, and quantify the difference in radiative transfer that results from values calculated under the independent scattering assumption. This work is probably the most complete exposition of radiative transfer in porous media. They examine the various correlations for radiative conductivity outlined in Vortmeyer (1978), as well as predictions based on two-flux and discrete ordinates calculations, and show that the predictions are in poor agreement with more exact Monte Carlo calculations except for bed materials with very high emissivity.

Dombrovsky (1996) provides material on the effect of anisotropic scattering in porous beds for a wide variety of particles. Most attention is given to highly disperse systems, where independent scattering governs.

The review by Baillis-Doermann and Sacadura (1998) succinctly summarizes recent publications on radiation in porous materials, and reaches similar conclusions to other contemporary reviews: the effect of dependent scattering is very important in porous media; criteria for when to treat a medium as continuous or discontinuous have not been established; experimental data for verification of analytical approaches is lacking; polarization effects have largely been ignored; and improved inversion techniques for property determination are needed.

References to other work that provides the details of more complex models are also given here; unfortunately, the results of much of this work cannot readily be reduced to formulae or correlations. Where contemporary understanding has led to accurate and useable correlations, these are included.

A. Applications

Some systems have extremely large local temperature gradients, which may violate an assumption inherent in many early models of transfer through packed beds, which is that the dimensionless temperature change over a bed particle diameter or model cell is small. For example, combustion in liquid- and gas-fueled submerged combustion burners occurs within the pore structure of a reticulated ceramic matrix rather than at or above the burner surface. The combustion energy release is localized, and the reaction zone temperature within the porous medium can increase by over 1000 K over a few pore diameters. Radiation from the incandescent matrix is an important mechanism for transferring energy upstream in these burners. The radiant energy is absorbed by the cooler upstream ceramic matrix, and is transferred to the incoming premixed fuel and air by convection. This mechanism provides an inherent preheater for these burners. Radiant heating from the downstream burner face is also important. Reviews of these burners have been given by Howell et al. (1995, 1996).

Bohn and Mehos (1990) examined the use of ceramic foams as an absorbing medium for high-performance solar collectors in concentrating collector systems. A fluid passes through the porous foam to remove the absorbed solar energy. The radiative absorption characteristics of the medium were measured by Hale and Bohn (1993) for use in analysis of collector efficiency.

Echigo (1982, 1991) and Echigo et al. (1986) examined industrial applications of porous media as a means of converting the enthalpy of a flowing gas into radiation for redirection into other uses. This is in a sense the inverse of the solar collector application.

Kudo et al. (1990), Taniguchi et al. (1993), and Mohamad et al. (1994), among others, have investigated the influence of radiative transfer on the performance of packed and fluidized beds. Tien (1988) reviews earlier literature.

A number of manufacturing and materials processing techniques involve interactions between radiation and porous media. Examples include: desktop manufacturing systems that use a high-intensity laser to sinter or melt a powder in a layer-by-layer buildup of a three-dimensional object (Kandis et al. 1998; Das et al. 1998; Wu and Lee, 1999); radiation-enhanced *in situ* processing of filament wound structures (Chern et al. 1995a,b), where energy from infrared lamps penetrates the matrix of thermoset resin and fibers to begin the polymerization of the resin during processing; thermal processing of laid-up preimpregnated thermoplastic composites; and the radiative drying of paper and other materials where radiation penetrates into the porous material and cannot be treated as a purely surface boundary condition (Fernandez and Howell, 1997).

Hoffmann et al. (1996) proposed a porous matrix for storing energy from the combustion of low-energy content gases by using a cyclic flow through a porous bed with high thermal capacity. Radiative transfer becomes an important loss mechanism as the bed achieves high temperatures.

Yano et al. (1991) proposed a fire shelter for people or devices exposed for short periods to harsh fire environments. The shelter uses stored water evaporating from a porous surface to provide thermal protection to the inside of the shelter from the external radiation field.

As these and other new technologies take advantage of the special characteristics of the interaction of radiation with porous media, the nature of this interaction must be well understood so that complete advantage can be gained.

B. Fundamentals

There are two common approaches to treating radiative transfer in porous media. If the structure of the medium is well defined, then it is possible to model the radiative transfer among the discrete structural elements. Such an approach can provide detailed radiative transfer information, and the characteristics of both the physical structure and surface radiative properties of the porous material can be included. This approach requires more information about the radiative properties of the solid elements than is often available, and usually is applied only when the structural elements have a simple geometric form. The data required for this approach include the directional and spectral emissivity and reflectivity of the structural elements and their detailed geometry. If the elements of the structure are not opaque, then the problem is complicated by the need to consider refraction at the solid–fluid interface of the elements as well as the external and internal reflections that occur at these interfaces. An even more difficult problem arises if the elements of the porous medium are so closely spaced that reflected radiation from one element is affected by that from adjacent elements. Such effects put the radiative transfer problem into the realm of dependent scattering, which is a most difficult area for accurate analysis or simple correlations. Singh and Kaviany (1991) flatly conclude that "independent theory is shown to fail for systems with low porosity and is not suitable for packed beds..." and note that deviations are significant from independent scattering theory for porosities as high as $\varepsilon = 0.935$.

The criterion for determining whether the porous material can be treated as a continuum depends on two factors. The first is the dimensionless bed size (L/d) in terms of the minimum bed dimension L and the particle or pore diameter d_p or d_{pore}, and the second is the particle size relative to the important wavelengths λ of the radiation. The latter criterion is usually expressed in terms of the size parameter $\xi = \pi d_p/\lambda$.

For most industrial processes and practical systems, the porous material is contained within a system that is many pore or particle diameters in extent ($L/d \geq\sim 10$), and the pores or particles are large relative to the wavelengths carrying the important radiative energy ($\xi \geq\sim 5$). In this case, the porous medium may be treated as continuous, and the effective radiative properties are measured by averaging over the pore structure. In this way, traditional radiative transfer analysis methods for participating media can be applied without resort to detailed element-by-element modeling. This is not to say that element-by-element modeling is not required to predict the radiative properties themselves for use in the continuum analysis.

Singh and Kaviany (1994) point out that when temperature gradients are large and/or the thermal conductivity of the elements is small, the assumption of isothermal elements may fail, and the analysis of radiative transfer among elements must account for this fact.

C. Treatment of the Porous Medium as a Continuum

If the porous medium can be treated as a continuum for purposes of radiative transfer, then the standard radiative transfer equation (RTE) is appropriate to describe the propagation of radiative intensity through the medium. Detailed derivation of the RTE and the variable solution techniques cannot be completely covered in the space available here. However, a very brief overview is now given, along with reference to detailed treatments.

The RTE can be written as

$$\frac{\partial i_\lambda(\Omega)}{\partial \tau_\lambda} = I_\lambda(\Omega) - i_\lambda(\Omega) \tag{1}$$

The RTE shows that the attenuation of radiation intensity along a given direction depends on the difference between the *source function* I_λ and the local *intensity* i_λ. Each of these quantities is spectrally (wavelength) dependent. The variables are in turn defined in terms of properties and more familiar variables by

$$d\tau_\lambda = (a_\lambda + \sigma_\lambda)ds = \kappa_\lambda ds; \qquad \beta_\lambda = \frac{\sigma_\lambda}{\kappa_\lambda};$$
$$I_\lambda(\Omega) = (1 - \beta_\lambda)i_{\lambda b} + \beta_\lambda \int_{\Omega_i = 4\pi} i_\lambda(\Omega_i)\Phi(\Omega_i, \Omega)d\Omega_i \tag{2}$$

where $d\tau_\lambda$ is the differential optical thickness, which indicates the ability of the medium to attenuate the intensity across a differential length ds as caused by the medium attenuation coefficient κ_λ, which is the sum of the absorption coefficient a_λ and the scattering coefficient $\sigma_{R\lambda}$. The quantity $\beta_\lambda = \sigma_{R\lambda}/(a_\lambda + \sigma_{R\lambda})$ is the *single scattering albedo* of the medium, and is a measure of the fraction of attenuation that is due to scattering. The source function I_λ is composed of two terms: the first is the contribution to intensity by radiation emission from the medium. The second is the contribution from intensity traveling in other directions and then scattered into directions within the solid angle Ω.

The RTE can be formally integrated to give

$$i_\lambda(\tau_\lambda, \Omega) = i_\lambda(0, \Omega)\exp(-\tau_\lambda) + \int_0^{\tau_\lambda} I_\lambda(\tau_\lambda^*, \Omega)\exp[-(\tau_\lambda - \tau_\lambda^*)]d\tau_\lambda^* \tag{3}$$

To solve either Eq. (1) or (3), the local temperature of the medium must be known so that the local blackbody intensity $i_{\lambda b}$ can be specified in the RTE, allowing solution for the local intensity i_λ. Thus, the energy equation must be solved for the medium temperature distribution along with the RTE.

D. The Energy Equation

The form of the energy equation depends on the complexity of the particular problem to be solved. If energy transfer between the porous medium and a flowing fluid is to be calculated, then it may be necessary to write energy equations on both the porous medium and the fluid, with a convective transfer term providing the coupling between the two equations. This approach allows calculation of differing bed and fluid temperatures. If the fluid is transparent to radiation, the radiative flux divergence will then appear only in the equation for the porous material. In this case, the energy equations (in one dimension) become

Fluid phase energy equation:

$$\rho u \frac{\mathrm{d}T_f}{\mathrm{d}x} - \frac{1}{c_p}\frac{\mathrm{d}}{\mathrm{d}x}\left(k_f \frac{\mathrm{d}T_f}{\mathrm{d}x}\right) + \frac{1}{c_p} S + \frac{1}{c_p} h_v (T_f - T_s) = 0 \tag{4}$$

Solid phase energy equation:

$$\frac{\mathrm{d}}{\mathrm{d}x}\left(k_s \frac{\mathrm{d}T_s}{\mathrm{d}x}\right) - \frac{\mathrm{d}}{\mathrm{d}x}(q_R) + h_v (T_f - T_s) = 0 \tag{5}$$

Here, h_v is the volumetric heat transfer coefficient and S is the volumetric source strength that results from combustion or other internal energy source. The properties are the effective properties (i.e., they are dependent upon the structure of the medium for the solid and the flow configuration for the fluid). If multiple species are present, as in a combustion problem, additional terms for species diffusion must be included in the fluid-phase equation.

It is clear from the coupling between Eqs. (4) and (5) through the convective transfer terms that the presence of radiation will affect both the solid and fluid temperature distributions even though the radiation term appears only in the energy equation for the solid. In many papers before 1975 (Vortmeyer 1978), linearization of the radiation terms was a common approach, especially when the medium was optically thick. Better computational ability has made this unnecessary.

If the porous medium has no flowing fluid associated with it, then the porous medium temperature alone is sufficient and a single energy equation can be written that describes the net radiative and conductive flux divergence in the medium. For steady state and with no energy sources or sinks in the medium, Eqs. (4) and (5) reduce to

$$\nabla \cdot q_R + \nabla \cdot q_C = 0 \tag{6}$$

Alternatively, if the flow rate of the fluid is sufficiently large, then the volumetric heat transfer coefficient becomes large and the local fluid and solid temperatures become equal. Equations (4) and (5) can then be combined to give

$$\rho u \frac{\mathrm{d}T}{\mathrm{d}x} - \frac{1}{c_p}\frac{\mathrm{d}}{\mathrm{d}x}\left(k_{\mathrm{eff}} \frac{\mathrm{d}T}{\mathrm{d}x}\right) + \frac{1}{c_p}\frac{d}{d_x}(q_R) + \frac{1}{c_p} S = 0 \tag{7}$$

In this equation, k_{eff} is the effective thermal conductivity of the fluid-saturated porous medium (see the chapter on conduction in porous media in this Handbook), and the other properties are corrected for the porosity and are on a per unit of total volume basis.

The radiative flux divergence appearing in the energy equations (5)–(7), $\nabla \cdot q_R$, can be determined from the local radiative intensity by the relation

$$\nabla \cdot q_R = \int_{\lambda=0}^{\infty} \left[\int_{\Omega=4\pi} i_\lambda(\Omega)\mu \mathrm{d}\Omega \right] \mathrm{d}\lambda \tag{8}$$

The RTE and the porous medium energy equation in each case are fully coupled and must be solved simultaneously to find the solid temperature distribution, which may in turn be coupled to the fluid temperature distribution through Eqs. (4) and (5). More details of the methods of solution of the RTE for this class of problems are available in the standard radiation texts (Siegel and Howell 1992; Modest 1993; Howell and Mengüç 1998).

A major difficulty in application of RTE solution methods to porous media is in determining the appropriate property values to use in the solutions. Values of the properties determine to some extent the choice of appropriate solution technique for the RTE. Howell and Mengüç (1998) provide a table of solution method characteristics to aid in the choice of solution technique for a particular problem.

It is clear from the form of the RTE (Eq. (1) or (3)) that the necessary radiative properties under the continuum assumption are the spectral absorption and scattering coefficients of the medium (a_λ and $\sigma_{R,\lambda}$) and the single-scattering phase function $\Phi(\Omega_i, \Omega)$. The phase function may also be spectrally dependent; however, data on its spectral behavior is seldom available, and simple models (isotropic scattering or simple angular dependencies) are usually assumed to apply and to be independent of wavelength.

II. RADIATIVE PROPERTY MEASUREMENT AND PREDICTION

With the exception of porous media used as absorbers in solar collectors, where a very-high-temperature external radiation source (the Sun) is present, almost all radiative transfer applications in porous media require knowledge of the radiative properties at wavelengths in the infrared region of the spectrum. Internal medium temperatures are limited by the material properties of the porous medium, and, for applications where bed temperatures are of the order of 1500 K or below, the important radiative transfer will be at wavelengths of 2 μm and greater. For solar collectors, where the effective incident energy distribution is proportional to that of a blackbody source at the effective solar temperature of 5780 K, the important wavelengths are in the visible and near-IR portions of the spectrum. In such a case, the spectral dependence of the radiative properties must be considered and averaged properties using Kirchhoff's law are inappropriate.

A. Dependent Scattering

When particles, fibers, or other bodies are in close proximity, the scattering and absorption of radiation is affected in comparison with predictions based on the properties of individual particles. This arises from two effects. The first is that the internal radiation field in non-opaque particles is affected by the scattering from surrounding particles (the near-field effect). Secondly, scattered radiation from one particle can constructively or destructively interfere with that from another particle (the far-field effect). As the volume fraction occupied by particles increases (porosity decreases), these effects become more important. A considerable amount of attention has been paid to these effects in recent years. The computations are difficult, and usually require simplifying assumptions that make the absolute accuracy of the predictions somewhat suspect.

For spherical particles, Kaviany and Singh (1993) give a comprehensive discussion of the effect of dependent scattering, and recommend that independent scattering can be assumed when the criterion

$$C + 0.1d_p > \frac{\lambda}{2} \tag{9}$$

is met, where C is the interparticle clearance distance. This criterion can also be written as

$$\xi_C + 0.1\xi > \frac{\pi}{2} \tag{10}$$

where the clearance parameter $\xi_C = \pi C/\lambda$. If this criterion is not met, the more detailed dependent scattering approach should be used. This result is derived for porosities typical of rhombohedral packing, $\varepsilon \approx 0.26$. The authors state that dependent scattering in packed beds remains an important effect for bed porosities as high as 0.935, and the effect is most pronounced for opaque particles. They note that both the porosity requirement and the relation between C and λ (Eq. (9) or (10)) must be satisfied before the assumption of independent scattering can be used with confidence. Earlier, Brewster and Tien (1982) provided the criterion for independent scattering as

$$\xi > \pi \frac{C}{\lambda} \frac{(1-\varepsilon)^{1/2}}{0.905 - (1-\varepsilon)^{1/2}} \tag{11}$$

where the independent scattering region (i.e., deviation of more than 5% from independent scattering results) is demarcated when $C/\lambda = 0.5$ is inserted into the equation (Tien 1988). This relation is based on far-field effects.

For example, for a packed bed of spherical particles with an average bed temperature near 500 K, Wien's displacement law (Howell and Mengüç 1998) gives the wavelength at the peak of the blackbody spectrum as $\lambda_{max} =$

$C_3/T = 2898$ (μm K)/500 K = 5.8 μm. This wavelength exactly splits the energy in the blackbody spectrum such that 25% of the radiant power is at shorter wavelengths and the remainder is at longer wavelengths. It provides a reasonable criterion for an approximate determination of whether dependent scattering will be important. If the bed is closely packed so that the clearance distance can be taken as near zero, then the particle diameter below which dependent scattering becomes important according to Eq. (9) is $d_p = 5\lambda_{max} = 29\,\mu m$. Equation (11), for the same temperature, λ_{max}, and with $\varepsilon = 0.26$, gives the minimum particle diameter for assuming dependent scattering as $d_p = 30.6\,\mu m$. Lower temperature beds will have correspondingly greater values of λ_{max} and the particle diameters included in the dependent scattering region will be proportionally larger. Note that dependent scattering will occur for all particle diameters in the part of the spectrum where the wavelength becomes large enough to violate the criterion of Eqs. (9, 10). Quite often, the fraction of energy at these long wavelengths is small enough to be neglected.

Kavany and Singh (1993) also present a relation for a scaling factor that gives the relation between transmission through a packed bed of opaque particles with independent scattering and the result assuming dependent scattering, and show that this result is only dependent on the bed porosity (i.e., independent of particle radiative emissivity). For nonopaque particles, they find the scaling approach to be unfeasible.

Closely spaced parallel cylinders are of interest in applications such as *in-situ* curing by an infrared source during filament winding and the determination of nanomaterial properties. Lee (1994) and Chern et al. (1995c) used practically identical assumptions and approaches to determine the radiative properties of arrays of parallel cylinders subject to normal irradiation and dependent scattering effects. They show the ranges of cylinder volume fraction and size parameter where dependent scattering becomes important for various values of the fiber effective refractive index [e.g., see Figures 1–3]. Lee and Grzesik (1995) later extended the analysis to coated and uncoated fibers subject to obliquely incident radiation. Chern et al. (1995b) show that inclusion of dependent scattering effects for parallel cylinders is important in treating radiative transfer within filament-wound epoxy/glass-fiber composites because of the large fiber volume fraction and small size parameter of the fibers typical of these systems. For fiber layers, the interaction among wavelength, refractive index, fiber diameter, and fiber spacing is complex enough that no simple criterion is available for deciding whether dependent scattering will be important, and maps similar to Figures 1–3 must be referred to. However, they do note that, in general, for ξ (based on cylinder diameter) > 2, the dependent scattering is relatively independent

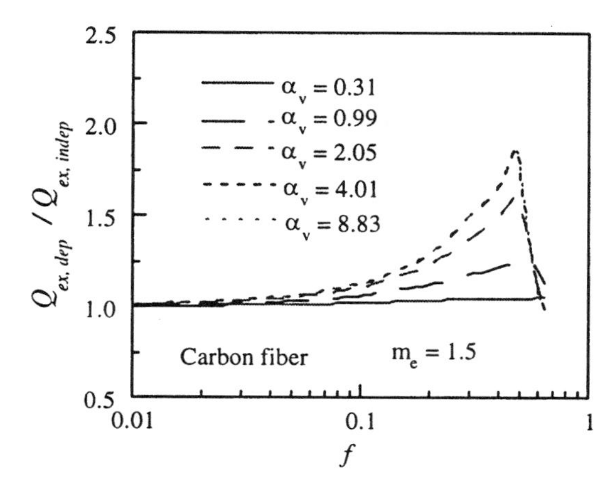

Figure 1. Variation of the extinction efficiency for normal incidence with fiber size and concentration. Carbon fibers dispersed in epoxy resin (Chern et al. 1995c).

of the refractive index of the fibers and the size parameter, and dependent scattering is present for $\varepsilon < 0.925$.

B. Measured Properties of the Equivalent Homogeneous Material

Because most treatments of radiative transfer in porous media rely on solution of the radiative transfer equation, it is necessary to measure or predict the effective continuum radiative properties of the particular porous med-

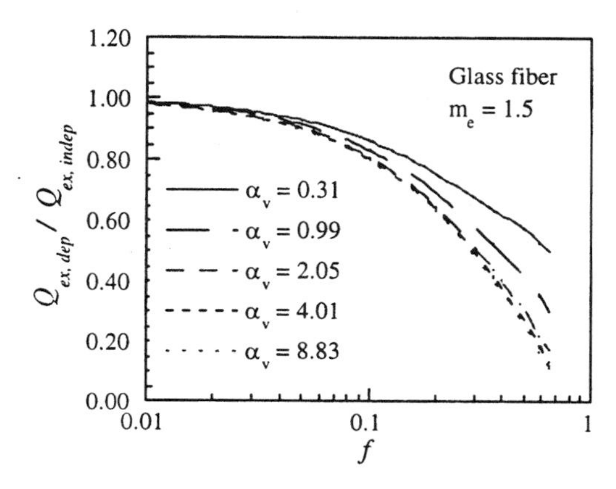

Figure 2. Variation of the extinction efficiency for normal incidence with fiber size and concentration. S-glass fibers dispersed in epoxy resin (Chern et al. 1995c).

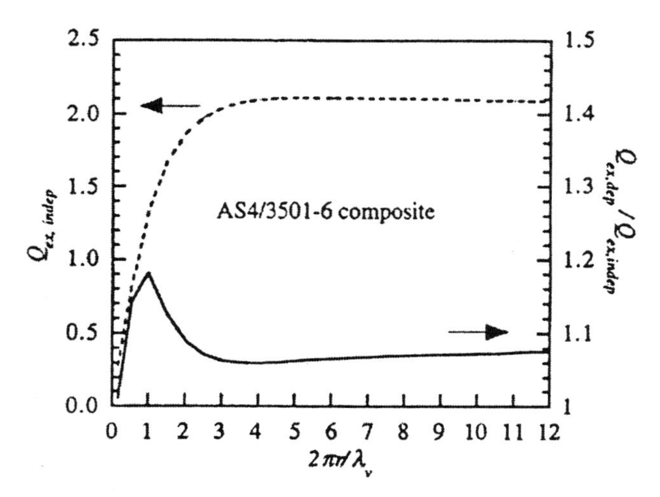

Figure 3. The extinction efficiency with and without consideration of dependent scattering effects for normal incidence, with fiber size and concentration. S-glass fibers dispersed in epoxy resin (Chern et al. 1995c).

ium. Various approaches to finding these properties have been attempted. They generally can be separated into two classes: direct or indirect measurement of the required properties, or prediction of the properties from models of the geometrical structure and surface properties of the porous material. The properties, from either measurements or predictions, are expected to be dependent on the characteristics of the particular porous material; i.e., the particle scale, material, and geometry. In addition, the properties may be functions of temperature and wavelength and, depending on the structure of the material, anisotropic. Depending on the criteria for particle size and clearance relative to important wavelengths (Eqs. (9–11)), dependent scattering and absorption effects may also become important. The question of adequately describing particle geometry is the most daunting.

First, some available measured properties are viewed to illustrate the methods and difficulties of obtaining accurate and useful radiative property values for porous materials.

Most measurements of radiative properties are made by inferring the detailed properties from measurements of transmission or reflection of radiation from the porous material. Inversion of these measurements is then used to find the best set of scattering coefficient, absorption coefficient, and phase function that will predict the measured effects.

Two problems are inherent in these measurements. First, inverse techniques are susceptible to large uncertainties in the inferred values that depend on the experimental uncertainty in the measured values, and there are questions of uniqueness as well. Second, the model of radiative transfer that is used in the inversion may have particular assumptions embedded within it. For example, a simple phase function behavior (isotropic, linearly anisotropic, etc.) may be assumed to simplify the radiation model used in the inversion. If this is done, then the resulting values of inferred absorptions and scattering coefficient will depend on that assumed type of phase function. It follows that the radiative model used in describing radiation in porous material must use the same assumptions, or the absorbing and scattering coefficients will not be compatible. This is sometimes overlooked when reported properties are taken from the literature and then used in an incompatible model of radiative transfer.

1. Beds of Spheres

Chen and Churchill (1963) measured the transmission of radiation through an isothermal bed of randomly packed equal-diameter spheres, as well as beds of cylinders and irregular grains. They included materials of glass, aluminum oxide, steel, and silicon carbide. They used this data to determine the bed effective scattering and absorption cross-sections based on a two-flux model. Their data have served as a benchmark test for comparison of predictions by many researchers.

Kamiuto et al. (1990) measured the angularly dependent reflectance of planar beds of glass and alumina spheres contained between glass plates due to normally incident radiation. The radiation source was an expanded 0.6238 μm He–Ne beam. An inversion technique was used to find the extinction coefficient, albedo, and Henyey–Greenstein asymmetry factor.

Kudo et al. (1991) measured the transmittance through a randomly packed bed of polished ball bearings of uniform diameter. Diameters of 7.94, 4.76, and 3.18 mm were contained within a 6.35 cm square duct by a fine-mesh screen. A He–Ne laser (0.6328 μm) beam was expanded and made normally incident and of uniform intensity across the bed face. The spheres had a measured surface absorptivity of 0.28 at this wavelength. The authors observed significant transmittance through the bed in regions near the duct surfaces, whereas the transmittance near the bed centerline decreased rapidly with increasing bed thickness.

2. Fibers and Fiber Layers

Glass-fiber insulations generally have very low densities, so that, even though the fiber diameter is in the range where dependent scattering might be important, the fiber spacing is so large that dependent effects can be neglected.

Tong et al. (1983) measured the transmittance of fibrous insulation layers using radiation from a 1300 K blackbody source. A monochromator was used to obtain spectral data. A two-flux and a linear anisotropic scattering model were used to invert the transmittance data and find the effective radiative conductivity of the fiber layers for use in multi-mode heat transfer calculations.

Nicolau et al. (1994) investigated methods for experimentally determining the optical thickness, albedo, and four parameters describing a scattering phase function. They present data for fiberglass insulation and silica fiber–cellulose insulation. The data are spectrally dependent, using relatively wide (1 μm) spectral bands.

3. Open-celled Materials

The use of reticulated ceramic foams in submerged-combustion burners has shown significant advantages in allowing low-pollutant combustion, burning of low-energy-value fuels, and compact combustion systems. Prediction of burner behavior requires accurate thermophysical and transport properties for these materials, so considerable effort has been expended in determining them. Radiative transfer is very important, and the reticulated structure allows long radiation paths. These materials are therefore of particular interest when examining radiation in porous materials.

Howell et al. (1996) review the radiative properties of commercially available reticulated porous ceramic foams. These materials have an open pore structure in a random dodecahedral matrix created by the interconnected struts or webs that make up the foam. For these materials it is difficult to specify a particle shape or size that is representative of the material. Instead, most researchers and manufacturers specify an effective pore diameter, d_{pore}. Thus, the scale of the structure is described by d_{pore} and the porosity. The porosity is typically about 85%, and pore sizes range from approximately 25 to 2 pores per centimeter (ppcm) [65 to 8 pores per inch (ppi)]. This is well outside the range where dependent scattering should be of importance. The foam composition is usually specified by the manufacturer in terms of a base material (silicon carbide, silicon nitride, mullite, cordierite, alumina, and zirconia) plus a stabilizing binder (e.g., magnesia and yttrium).

Most investigators report properties by simply specifying the material and the manufacturer's specification of "pores per centimeter" or ppcm. However, Hsu and Howell (1992), Younis and Viskanta (1993), and Hendricks (1993) show that the actual pore diameter varies even among materials with the same specified ppcm supplied by the same manufacturer. Thus, when using experimental property data, significant error may be introduced into the model results unless reliable property data are available on exactly the material being modeled.

When the porous material is treated as a homogeneous absorbing and scattering medium, Eqs. (1) and (2) show that it is necessary to know the effective absorption and scattering coefficients, as well as the scattering phase function which describes the angular distribution of scattered energy relative to the direction of incidence. Hsu and Howell (1992) used a two-flux radiation model to infer the effective radiative extinction coefficient κ from experimental heat transfer measurements. They obtained the result as a function of pore size for partially stabilized zirconia reticulated ceramic foam in the form of a data correlation

$$\kappa = 1340 - 1540 d_{pore} + 527 d_{pore}^2 \tag{12}$$

where κ is the mean (spectrally averaged for long wavelength radiation) extinction coefficient in m^{-1} and d_{pore} is the actual pore diameter in mm. This correlation applies for pore sizes in the range $0.3\,mm < d_{pore} < 1.5\,mm$. The data were collected for temperatures in the range $290 < T < 890\,K$, although no significant temperature dependence was observed. Additionally, they present a relation, based on geometric optics, that predicted the trend of the data very well

$$\kappa = (3/d_{pore})(1 - \varepsilon) \tag{13}$$

Equation (13) applies for $d_{pore} > 0.6\,mm$. Here, ε is the porosity of the sample, which varied over a narrow range from 0.87 at large pore diameters to 0.84 at the smallest diameters. The method used to obtain these correlations from the experimental data includes the assumption of isotropic scattering, and it was not possible to determine independent values of albedo or scattering or absorption coefficients.

Hale and Bohn (1993) measured scattered radiation from a sample of reticulated alumina that resulted from an incident laser beam at 488 nm. They then matched Monte Carlo predictions of the scattered radiation, calculated from various values of extinction coefficient and scattering albedo, and chose the values that best matched the experimental data. They carried out this approach for reticulated alumina samples of 4, 8, 12, and 26 ppcm. It was found that a scattering albedo of 0.999 and an assumed isotropic scattering phase function reproduced the measured data for all pore sizes. These data, along with the large reported albedo value, indicates that alumina is very highly scattering, and that radiative absorption is extremely small for this material.

Hendricks and Howell (1994, 1996a) measured the normal spectral transmittance and normal-hemispherical reflectance of three sample thicknesses each of reticulated partially stabilized zirconia and silicon carbide at pore sizes of 4, 8, and 26 ppcm. The measurements covered a spectral range of 400–5000 nm. They used an inverse discrete ordinates method to find the spectrally dependent absorption and scattering coefficients as well as the constants appropriate for use in either the Henyey–Greenstein approximate phase function or a composite isotropic/forward scattering phase function. Some of the data are shown in Figures 4 and 5. The data shown in these two figures are compatible; i.e., the absorptions and scattering coefficient data were obtained along with data for the Henyey–Greenstein phase function; the data for these coefficients will differ if another form is chosen for the phase function. Compatible sets are also given in Hendricks and Howell (1996a) for the choice of a modified diffuse-forward scattering phase function.

Both materials showed best agreement with experimental data when this anisotropic scattering phase function was used. The spectral data for absorption and scattering coefficient for silicon carbide were fairly independent of wavelength, which is the same as the spectral behavior of pure silicon

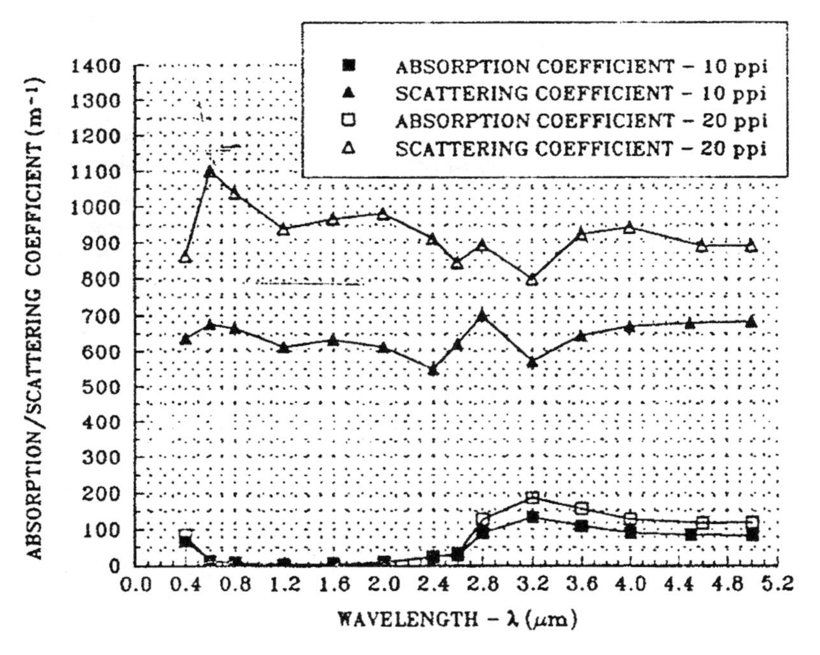

Figure 4. Spectral absorption and scattering coefficients for partially stabilized zirconium oxide reticulated ceramic foam, based on an assumed Henyey–Greenstein phase function (Hendricks and Howell 1996a).

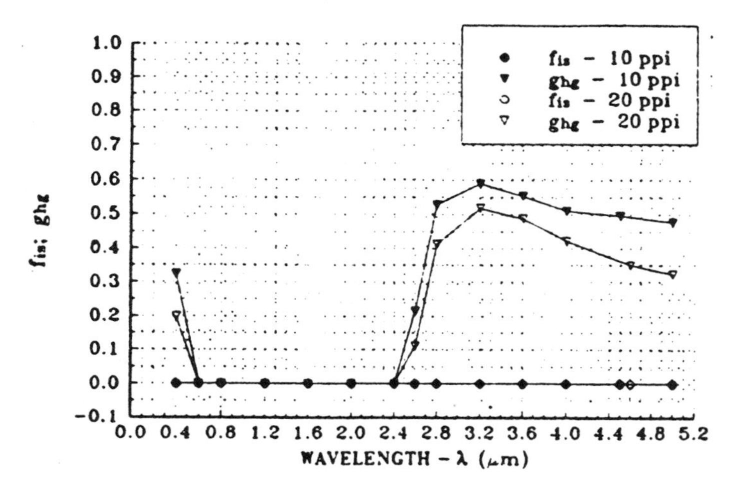

Figure 5. Spectrally dependent parameters for Henyey-Greenstein scattering phase function for partially stabilized zirconium oxide reticulated ceramic foam (Hendricks and Howell 1996a).

carbide in solid form. For partially stabilized zirconia the properties showed a significant change across the range 2500–3000 nm, but were independent of wavelength on each side of this value. The spectral characteristics of pure and stabilized zirconia do not show this behavior. The phase function in particular shows a radical change in behavior for the reticulated ceramics across this spectral range for PSZ. Scattering albedos were in the range of 0.81–0.999 and varied with wavelength for zirconia, with some variation with pore diameter, and were in the range 0.55–0.81 for silicon carbide.

An important point in interpreting data obtained by inverse analysis of experimental data is that the predicted properties are interrelated. For example, if a given form is assumed for the phase function (e.g., isotropic or Henyey–Greenstein) in the inverse analysis, then the values of absorption and scattering coefficient depend on the form chosen. It is important, therefore, to use the same form of phase function as used in the data reduction when carrying out radiative analysis using measured absorption and scattering properties.

Integration over wavelength of the spectral results of Hendricks and Howell provides mean extinction coefficients data that can be compared with those of Hsu (1991) and Hsu and Howell (1992). The Hale and Bohn (1993) data are for the single laser-source wavelength of 488 nm. Hendricks and Howell (1996a) found that a modified geometrical optics relation fits the data for the integrated extinction coefficient of both zirconia and silicon carbide. They recommend the relation

$$\kappa = (4.4/d_{pore})(1 - \varepsilon) \tag{14}$$

where κ is in m^{-1} and d_{pore} is the actual pore diameter in mm. The correlations of Hsu and Howell (1992) and Hendricks and Howell (1996a), along with the 488 nm data of Hale and Bohn (1993), are shown in Figure 6.

The data presented in Figure 6 obviously have similar characteristics, regardless of the material. The Hale and Bohn data are plotted on the basis of pore diameter calculated as the inverse of the ppcm values reported; however, measured pore sizes are generally smaller than nominal sizes computed in this way (see Table 1). It may be possible to collapse all of the data for extinction coefficient onto a single curve; however, actual pore size data were not reported by Hale and Bohn. Even if the data can be collapsed in this way, it will be necessary to have additional data for the scattering albedo of the material so that the individual scattering and absorption coefficients can be recovered from the extinction coefficient data.

High-temperature experiments (1200–1400 K) were performed by Mital et al. (1996) on reticulated ceramic samples of mullite, silicon carbide, cordierite, and yttria–zirconia–alumina. They reduced the data using a gray two-flux approximation assuming isotropic scattering. Temperature dependence was found to be small. They present the relations for absorption and scattering coefficients for these materials as

$$a = (2 - \varepsilon_r)\frac{3}{2d_{pore}}(1 - \varepsilon) \tag{15}$$

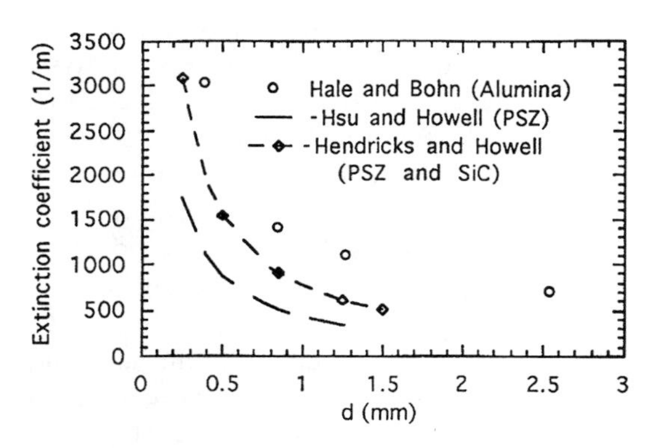

Figure 6. Extinction coefficient versus pore diameter for various reticulated ceramics (Howell et al. 1996).

Table 1. Comparison of nominal and actual pore characteristics in reticulated ceramic media (from Howell et al. 1996)

	Mean pore diameter (mm)					Mean web diameter (mm)	
ppcm	Alumina*	Cordierite†	Partially stabilized ZrO_2†	Partially stabilized ZrO_2‡	Silicon carbide‡	Partially stabilized ZrO_2‡	Silicon carbide‡
4	1.52	–	1.450	1.346	1.725	0.506	0.686
8	0.94	1.25	0.967	0.737	1.015	0.356	0.508
12	0.76	–	0.708	0.610	0.800	0.254	0.279
18	0.42	–	0.467	0.467	–	0.15	–
26	0.29	–	0.335	0.335	–	0.075	–

* Younis and Viskanta (1993)
† Hsu and Howell (1992)
‡ Hendricks (1993)

$$\sigma_r = \varepsilon_r \frac{3}{2d_{pore}}(1 - \varepsilon) \tag{16}$$

When summed to obtain the extinction coefficient, the result agrees with Eq. (13).

Moura et al. (1998) examined various experimental techniques that can be used to generate data that can then be inverted to find the fundamental radiative properties of porous materials. Four generic methods for generating data were examined, and the accuracy of each in estimating properties was delineated. Generally, better inversions were obtained from experimental measurement techniques that give directional rather than hemispherical data, as would be expected. Also, it was pointed out that a major difficulty in experimental data inversion is the need to assume a simple form for the scattering phase function, which can limit the accuracy of radiative transfer calculations based on the inverted data. This paper provides useful information for planning the appropriate measurement technique to find accurate radiative properties of a given material.

C. Radiative Property Prediction

If a simple geometry can be specified as representative of a porous material, it is possible to predict the equivalent radiative properties of a homogeneous medium using radiative transfer analysis based on surface–surface interactions.

1. Cell Models

Most early models of radiative properties in porous materials have been based on simplified models of the structure. For example, random packings of spheres were assumed to have similar characteristics to regular packings. This clearly allows much simpler analysis because of geometric regularity. Usually, a unit cell of the regular packing was analyzed, and the characteristics of this cell were then assumed to apply to the entire porous medium. Radiation and conduction are assumed to occur in some parallel–series mode that is dependent on the assumed cell structure. These models suffer from the same problem that exists in predicting the thermal conductivity of a two- or more component mixture. That is, the bounds between a pure parallel conductivity and a pure series conductivity can be very far apart, and so the conductivity of a particular structure becomes very dependent on the exact structure and properties of each component. Any assumed simplification in the structure thus may produce averaged properties that are far from those of the real structure. Thus, cell models may predict agreement for a limited range of parameters but will fail outside of that range. Further, the radiative behavior of the cell components is usually assumed to be governed by diffuse reflections, and this is often far from the case. Finally, most early cell models did not recognize the importance of dependent scattering effects. Vortmeyer (1978) reviews many early cell models, and compares their predictions with experiment and with each other. The models agree best with each other when the bed particles are radiatively black, as might be expected because radiative transfer for each model becomes less dependent on the reflective behavior of the assumed cell structure.

2. Beds of Spheres

The structure of the porous material affects its radiative properties. A porous medium composed of regularly or randomly packed equal-diameter spheres has a well-determined size parameter, $\xi = \pi d_p/\lambda$, based on the diameter d_p of the spheres. The absorption and scattering characteristics for such a medium are discussed in the classic texts (Bohren and Huffman 1983). If the spheres are close-packed and have small diameters relative to the important wavelengths, then the potential for dependent scattering should be checked by using Eqs. (10, 11).

Ray tracing and Monte Carlo techniques have successfully predicted radiation transfer through regular and random arrays of particles and fibers. Howell (1998) reviewed the literature dealing with this technique, and included a section on packed beds and fiber arrays.

Yang (1981) and Yang et al. (1983) derived the statistical attributes of a randomly packed bed of identical cold (non-emitting) spheres (distribution of number of contact points between spheres and the conditional distributions of contact angles for spheres with a given number of contact points). Distributions were generated from a zero-momentum random packing model modified from that of Jodrey and Tory (1979). Figure 7 shows the distribution of the number of contact points for spheres away from the system boundary and near the boundary as computed from this program, which has been used by many other researchers.

The upper layer of spheres in a randomly packed container is not level in either real or simulated packed beds. Thus, defining the bed thickness is somewhat arbitrary, especially for beds that are only a few sphere diameters in depth. Wavy container boundaries were assumed so that the radial dependence of packing characteristics present in rigid containers could be ignored. Yang (1981) and Yang et al. (1983) used the bed packing distributions to construct a ray-tracing algorithm for specularly reflecting spheres, and com-

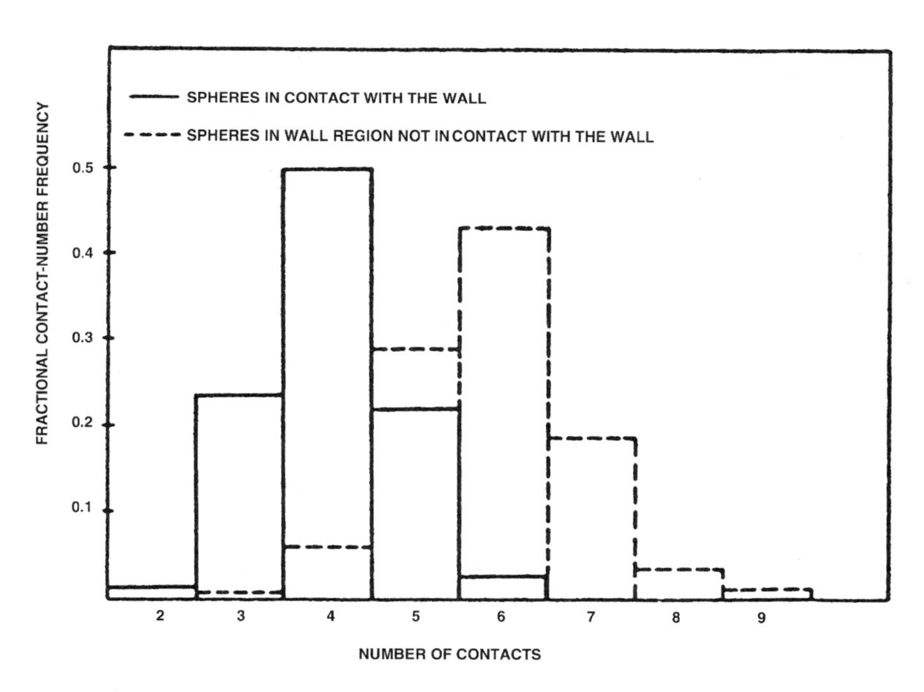

Figure 7. Frequency distribution of number of contacts with adjacent spheres in a randomly packed bed: near boundary, solid lines; bed interior, dashed lines (Yang 1981).

puted the fraction of incident diffuse radiation that was transmitted through the bed. Comparisons with the experimental results of Chen and Churchill (1963) showed similar trends of transmittance versus bed thickness, but not exact agreement.

Abbasi and Evans (1982) adapted a Monte Carlo-based Knudsen gas diffusion code to compute the transfer of photons through a plane layer of poorly conducting, diffusely reflecting porous material. They predicted the radiative transmittance T (fraction of incident radiation transmitted) through a bed of thickness L with pore diameter $d_{pore} \pm \bar{\sigma}$ and porosity $0.1 < \varepsilon < 0.5$, where $\bar{\sigma}$ is the standard deviation of the distribution around d_{pore}. The result is

$$T = \frac{d_{pore}}{L}\left(0.0496 + 0.5333\varepsilon - 0.096\,53\frac{\bar{\sigma}}{d_{pore}}\right) \tag{17}$$

When particle sizes are large such that geometric optics is applicable, Saatdjian (1987) provides a model for the extinction behavior of uniformly dispersed spheres, and shows that an extinction coefficient can be predicted that conforms to exponential attenuation in the limit of small spheres. A porosity range of $0.70 < \varepsilon < 1.0$ is covered by the theory. This work is applicable to cases where the particles are separated, such as in particle-laden flows or in radiating-droplet space radiators, where only radiation transfer governs interparticle energy transfer.

Tien (1988) discusses all of the above references and others in detail, and shows the comparisons of the various models with the experimental data for transmission through a bed of spheres generated by Chen and Churchill (1963).

Kudo et al. (1991) used a model similar to that of Yang for determining transmittance, and added further experimental results for comparison. They simulated random packing by perturbing regular packing within a bounded region and then removing those spheres that overlapped the boundary after perturbation. This produced a low-density region near the boundary, as observed in real packed beds. This was found to introduce a considerable increase in bed radiative transmissivity. Both specularly and diffusely reflecting spheres were modeled.

Singh and Kaviany (1991, 1992, 1994) and Kaviany and Singh (1993) extended the Monte Carlo analysis of packed beds of spheres to consider semitransparent and emitting spheres, and in particular to show that dependent scattering effects are quite important even for quite large bed porosities. Near-field dependent scattering effects are inherently included in these Monte Carlo analyses of packed beds, but are difficult to include in many other methods. Kaviany and Singh (1993) pointed out that the results of the earlier analyses by Yang and Kudo did not account for the nondiffuse nature of the incident radiation in the Chen and Churchill experiments, and thus tended to underpredict the experimental bed transmittance.

Xia and Strieder (1994a) derive the effective emissivity $\varepsilon_{R,\text{eff}}$ of an isothermal semi-infinite slab of randomly packed spheres in terms of the sphere solid surface emissivity $\varepsilon_{R,s}$. They predict the values of $\varepsilon_{R,\text{eff}}$ to lie within the bounds

$$\varepsilon_{R,s} + 0.3804\varepsilon_{R,s}(1-\varepsilon_{R,s})\varepsilon \leq \varepsilon_{R,\text{eff}} \leq \varepsilon_{R,s} + 0.6902(1-\varepsilon_{R,s})\varepsilon \tag{18}$$

This relation is valid for $\varepsilon_{R,s} > 0.6$. Based on these results, an estimate for $\varepsilon_{R,\text{eff}}$ is

$$\varepsilon_{R,\text{eff}} = \left\{\frac{(1+\varepsilon_{R,s})}{2} + \frac{(1-\varepsilon_{R,s})}{2}\left[\varepsilon(0.6902 + 0.3803\varepsilon_{R,s}) - 1\right]\right\} \pm \varepsilon(1-\varepsilon_{R,s})(0.3451 - 0.1902\varepsilon_{R,s}) \tag{19}$$

In a companion paper, Xia and Strieder (1994b) provide a more complex but more accurate relation for the upper bound on bed emissivity. Equations (18) and (19) contain the assumption of independent scattering, so the bed characteristics must conform to the constraints of Eqs. (9–11).

Argento and Bouvard (1996) revisited the work of Yang et al. (1983) and were able to use Monte Carlo results evaluated in the bed interior, thus eliminating the problem of indeterminate bed height in random packing.

Göbel et al. (1998) have further extended Monte Carlo analysis to account for the effect of inhomogeneities inside nonopaque spherical particles on the radiative scattering from individual spheres. Results are compared with the limiting cases of transparent refracting spheres and purely reflecting solid spheres. Independent scattering must be assumed in the approach used, and the spheres are also assumed to be large compared with the wavelength of the incident radiation.

3. Fiber Layers

Kudo et al. (1995) and Li et al. (1996) analyzed transfer through a bed of randomly oriented fibers, and a bed of fibers that are randomly oriented but lie in planes parallel to the bed surface. Mie scattering was assumed to occur from individual fibers. Comparison with available experimental results for transmission through fiber beds was quite good.

The anisotropic extinction coefficient $e^*(T) = \kappa^*(T)/\rho$ of rigid fiber insulation with various anisotropic fiber orientations was calculated for independent scattering by Marschall and Milos (1997). The authors used basic electromagnetic theory to predict values of $e^*(T)$ and compared the difference in radiative transfer that results from the use of the assumption of isotropic properties in the radiative conductivity for the diffusion equation. Differences of over 20% were found in a number of cases. The results for the anisotropic case are shown to be scalable within 5% of the exact calculation by superposing the results for a matrix of anisotropic fibers and the results for fibers normal to the incident radiation. No data for $e^*(T)$ are given.

When fibers are coated with a thin surface of metal, their radiative properties are greatly changed. Dombrovsky (1998) has provided predictions of the scattering efficiency and extinction coefficient of fibers with various coatings in the infrared and microwave regions of the spectrum. Cases for fibers randomly aligned within a plane and randomly aligned in space are given for monodisperse systems and for systems with fiber diameter distributions. All results are under the assumption of independent scattering.

Lee and Cunnington (1998) review models for transfer through fibrous media, including discussion of the effect of dependent scattering.

4. Open-celled Materials

Fu et al. (1997) used a unit cell model to predict the extinction coefficient and single scattering albedo of reticulated ceramics, and compared their results with available data. Comparison was generally satisfactory, being based on assumed ranges of surface properties for many of the comparison materials, since the actual surface properties were not known.

Kamiuto and Matsushita (1998) predicted the extinction coefficient of various open-celled structures simulated by cubic unit-cells, based on the work of Kamiuto (1997). They compared the predictions with experimental measurements obtained by an inversion technique using emission data from a heated plane layer of porous material. They assumed a Henyey–Greenstein phase function, and obtained good agreement between experiment and prediction for the extinction coefficient and asymmetry factor for Ni–Cr and cordierite porous plates. They also compared predictions with the data of Hsu and Howell (1992) and Hendricks and Howell (1996a), including the spectrally dependent data of the latter. Comparison was good, although differences were observed between the magnitudes of the predicted and measured spectral asymmetry factors in the Henyey–Greenstein phase function.

III. SOLVING THE RTE FOR POROUS MEDIA

A. Radiative Transfer in Optically Thick Porous Media

If the radiation travels a short mean free path before attenuation in comparison with the dimensions of the porous medium, then the RTE reduces considerably. Radiation in that case can be treated as a diffusion process (Deissler, 1964), and the local radiative energy flux is expressed as

$$q_R(S) = -\frac{16n^2\sigma T^3}{3\kappa}\frac{\partial T}{\partial S} = -k_R\frac{\partial T}{\partial S} \tag{20}$$

where k_R is the *radiative conductivity* of the medium and n is the refractive index. The requirement for application of this form is that the bed optical thickness $\tau = \kappa L \gg 1$, where L is the smallest dimension of the bed.

B. Radiative Transfer in Porous Media with Significant Mean Free Path and Homogeneous Effective Properties

The most difficult case to treat is when the radiative mean free path is of the order of the overall bed dimensions. In this case, complete solution of the RTE may be required. The standard methods include the *P–n* method, in which the intensity in the differential form of the RTE is expanded in a series of spherical harmonics; the discrete ordinates method, in which the sphere of solid angles about an element is divided into discrete solid angular increments, and the integrated RTE is solved along each ordinate direction; the Monte Carlo technique, which simulates the behavior of the radiation by sampling a large set of radiative energy packets throughout their lifetime; and others. The two-flux method is often used when one-dimensional radiative transfer is assumed; however, Singh and Kaviany (1991) show that this method does not give satisfactory results even for isotropic scatterers if the particles in the bed are absorbing. Each of these methods is discussed at length in Siegel and Howell (1992) and Modest (1993).

C. Detailed Treatment of Radiative Transfer Accounting for Structure

If the structure of the medium is known or can be determined, then it is possible to use conventional surface–surface interchange analysis to determine the radiative transfer. This approach is used by Antoniak et al. (1996) and Palmer et al. (1996). They used a Monte Carlo cell-by-cell analysis to simulate radiative transfer among fixed geometric shapes (rectangles, triangles, circles in two-dimensional arrays). Simulations were performed for both diffuse and specular reflections from the array elements. These are an extension of earlier cell-based models, in that the actual radiative characteristics of the cells are computed by detailed radiative analysis using Monte Carlo rather than by the use of approximate radiant interchange relations. Antoniak et al. (1996) provide correlations of their detailed results for diffusely incident and parallel normally incident radiation onto the face of a triangular array of parallel tubes or rectangles. The correlations are approximate, and are found to be insensitive to the pitch of the array. They do account for the surface absorptivity of the array elements. A portion of their results is given in Table 2. For collimated radiation not normal to the bed face, additional information appears in Antoniak et al. (1996).

A less conventional radiative analysis was introduced by Hendricks and Howell (1996b). They noted that radiative transfer in porous materials may propagate through a considerable path length in an interconnected open-celled medium before encountering a solid surface. These path lengths may extend to several cell diameters, depending upon the particular geometry of the reticulated web structure of the solid that defines the cells and the web size. They proposed that the radiative intensity for such a case be divided into two parts: the portion that is transmitted through the cells without interaction, and the portion that has interacted with the solid structure and then continued on after reflection They rewrote the RTE to account for the intensity in two portions, resulting in an effective extinction coefficient κ_{eff} given by

$$\kappa_{\mathrm{eff}} = (a + \sigma_R)(1 - A) + AB \qquad (21)$$

where A and B are found from experimental measurements of the direct transmission $T(s)$ through a porous material, which they found to be of the form

$$T(s) = A \exp(-Bs) \qquad (22)$$

Figure 8 shows data on transmittance through some particular porous media samples, from which values of A and B can be deduced by curve fit. They showed that this approach had some advantages in describing radiative transfer in porous media. They measured the directly transmitted fraction to be used in the analysis by simply observing the illumination passing through samples that were sliced to thinner and thinner sections. A filtering program was used to remove the transmitted radiation that occurred because of reflection/scattering within the medium, so that only the radiation transmitted without encountering an obstruction was considered. The measurement of the other necessary properties was carried out as described earlier.

Jones et al. (1996) measured the directional distribution of radiative intensity leaving a randomly packed bed of heated stainless steel spheres with a porosity of 0.371. They used the combined-mode radiative/conductive formulation based on discrete ordinates formulation for radiation. Bed radiative properties were derived on the basis of dependent scattering theory and used in the continuum model. Agreement was good normal to the bed surface, but deviated at near-grazing angles to the surface. The authors indicate that the RTE based on average bed radiative properties may not be valid for use near the bed surface, where a discrete model of radiative transfer among elements may be necessary.

Table 2. Correlation of transmittance and reflectance for a triangular array of n rows of parallel tubes or rectangles (Antoniak et al. 1996)

Element geometry	Transmittance T_n for $n > 1$ rows*	D_0†	Array reflectance
Diffuse incident radiation			
Cylinder	$T_n = (1 - D_0)^n[1 + (1 - \alpha_R)D_0]$ for $n \le 5$	$D_{0.5} = 0.622$	$R_n = (1 - \alpha_R)D_0[1 + \sum_{i=2}^{n}(1 - D_0)^{2(i-1)}]$
	$T_n = T_5 \exp(-n/10)$ for $n > 5$	$D_{0.2} = 0.265$	where $D_0 = D$ (cylinder diameter)
Rectangle	$T_n = (1 - D_0)^k[1 + (1 - \alpha_R)D_0]$ for $n \le 5$	$D_{0.5} = 0.845$	$R_n = 0.6(1 - \alpha_R)D_0[1 + \sum_{i=2}^{n}(1 - D_0)^{2(i-1)}]$
	where $k = 1.28n^{0.51}$	$D_{0.2} = 0.644$	where $D_0 =$ from this table
	$T_n = T_5 \exp(-n/10)$ for $n > 5$		
Collimated normally incident radiation‡			
Cylinder	$T_n = (1 - D_0)^{1.4n}[1 + (1 - \alpha_R)D_0]$	$1 - D_{0.5} = 0.441$	$R_{0.5} = 0.58\rho_R$
		$1 - D_{0.2} = 0.735$	$R_{0.2} = 0.52\rho_R$
Rectangle	$T_n = (1 - D_0)^k[1 + (1 - \alpha_R)D_0]$ where $k = 1.28n^{0.51}$	$1 - D_{0.2} = 1.2036$	$R_{0.2} = 0.26\rho_R$

* For 1 row, $T_1 = 1 - D_0$.

† D_0 is the fraction of incident energy striking the first row of an array. For example, $D_{0.5}$ indicates the fraction or incident energy intercepted by the first row when the elements in the first row cover 50% of the face area of the bed.

‡ For collimated incident radiation, the correlations apply for three or more rows in the array. For two rows, $T_2 = 1 - 2D$.

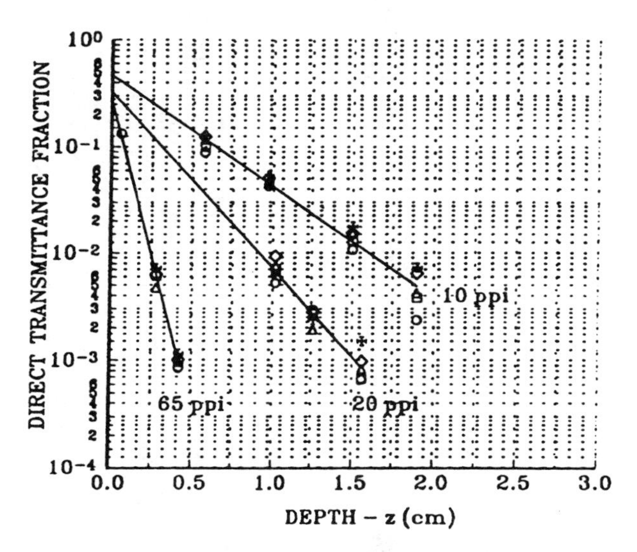

Figure 8. Experimental transmittance versus depth through reticulated partially stabilized ZrO_2 (Hendricks and Howell 1996b).

An entirely different approach has been proposed by Galaktionov (1998). He defines a universal linear radiative transfer operator G, and derives its properties through thermodynamic and mathematical requirements. By examining the behavior of this operator when applied to particular cases of radiative transfer, he shows that the RTE is recovered and that the spherical harmonics approximation also can be derived. The great merit of this approach is that it can be applied to the case of a porous medium without resort to detailed prescription of the medium radiative properties; however, considerable work remains in defining the appropriate boundary conditions to be applied to formulation in terms of G.

IV. COUPLING OF RADIATION WITH OTHER HEAT TRANSFER MODES THROUGH THE ENERGY EQUATION

For a system without convective motion, the total heat flux can be expressed as

$$q(S) = q_R(S) + q_C(S) = -k_R \frac{\partial T}{\partial S} - k_{\text{eff}} \frac{\partial T}{\partial s} = -(k_R + k_{\text{eff}}) \frac{\partial T}{\partial S} = -k_{\text{tot}} \frac{\partial T}{\partial S} \tag{23}$$

The total thermal conductivity k_{tot} accounts for both radiative and conductive transfer, but is highly nonlinear because of its dependence on T^3 (Eq. (20)). The value of k_{eff} for conventional conductivity is the composite value for conduction through the solid porous material and the material in the interparticle space if it is present. The advantage of this approach is that the energy equation terms involving conduction and radiation become considerably simpler than if the complete RTE had to be solved to determine the radiative flux divergence. Equation (7) reduces, in this case, to

$$\nabla \cdot q_R + \nabla \cdot q_C = \nabla \cdot (q_R + q_C) = -\nabla \cdot (k_{tot} \nabla T) = 0 \tag{24}$$

which can be solved with conventional numerical tools for handling conduction problems with temperature-dependent thermal conductivity. This approach is used in the great majority of studies of multi-mode heat transfer in porous materials.

The radiative transfer through silica fibers randomly oriented within a plane layer was compared on a spectral basis using this approach by Jeandel et al. (1993) in the spectral range 4–40 μm. They predicted the effective conductivity using a two-flux model, and included the effect of a distribution of cylinder radii. Spectral absorption and scattering coefficients were predicted, and these were used to predict radiative transfer through a layer of fibers. The results compared well with measurements.

Lee and Cunnington (1998) extended this approach, for the case of a fibrous insulation, to include a forward-scattering fraction, the effect of a distribution of cylinder diameters, and some assumptions about the temperature-dependent thermal conductivity of the bed. Independent scattering was assumed to exist. They found it necessary to use the rigorously derived scattering phase function to predict the forward scattering effect. They used the modified diffusion solution to predict the heat transfer through the fibrous material used in space-shuttle thermal protection tiles, and found good agreement between experiment and prediction.

Im and Ahluwalia (1994) and Martin et al. (1998) proposed using small diameter fibers inside a duct to augment the heat transfer between a duct surface and a flowing fluid. Radiation from the duct surface to the fibers provided the mechanism for augmentation. The fiber energy balance coupled the net radiative exchange to conduction within the fibers and convective transfer between the fluid flowing in the duct and the fibers. Reductions in dimensionless duct wall temperature of 30% were found in comparison with the case with no fiber inserts. The tradeoff between heat transfer augmentation and pressure drop was important, and it was found that pressure drop increased monotonically with decreasing porosity.

The difficulty with the use of a radiative conductivity is in predicting accurate values for its use. Clearly, the radiative conductivity should depend on particle size and spacing, on the surface emissivity of the particles in the bed, and probably on the particle thermal conductivity which affects the temperature distribution within the particles. Each system may require individual determination of its effective radiative properties for use in Eq. (24). For some simple systems, correlations are available. Kaviany (1991), Kaviany and Singh (1993) and Singh and Kaviany (1994) note that the radiative conductivity for use in optically thick systems with porosity between 0.4 and 0.5, composed of opaque solid spherical particles with surface emissivity ε_R and thermal conductivity k_s, can be approximated by

$$k_r = 4d\sigma T^3\{0.5756\varepsilon_R \tan^{-1}[1.5353(k_s^*)^{0.8011}] + 0.1834\} \tag{25}$$

where

$$k_s^* = \frac{k_s}{4d\sigma T^3} \tag{26}$$

Kaviany and Singh (1994) state that this formulation is reasonably accurate for both diffuse and specular particles and is weakly dependent on bed porosity.

V. UNRESOLVED PROBLEMS AND FUTURE RESEARCH

Considerable progress has been made in the last decade in defining methods for treating radiative transfer in porous media. Properties necessary for radiative analysis can be estimated from those available presently, and predictive techniques have improved greatly. However, it is still difficult to predict properties accurately for new classes of porous materials, especially for cases where dependent scattering may occur and the geometry of the porous solid is complex. The high-temperature processing of porous material such as in the selective laser sintering technique will require better methods and data as radiation becomes the dominant heat transfer mode, and presently available data and methods need further improvement.

Many researchers have recognized the inaccuracy of simple approaches, and have gone to quite detailed modeling to allow inclusion of the effects of anisotropic scattering, dependent scattering, etc. Methods such as Monte Carlo are successful and can be quite accurate; however, they provide results that are particular to the system and geometry under study. By the nature of such results, they do not provide simple handbook relations, and, for this reason, few are presented in this chapter. As the understanding increases of radiative transfer in nonhomogeneous systems such as porous media, this may change; however, the very nature of the complex interaction of radiation with complex geometries makes the realization of such hopes doubtful in the near future.

NOMENCLATURE

Roman Letters

a	linear radiation absorption coefficient
A, B	coefficients in curve fit, Eq. (22)
C	interparticle clearance distance
d	diameter
e^*	anisotropic mass extinction coefficient, κ/ρ
i	intensity of radiation
I	source function, Eq. (2)
L	packed bed minimum dimension
n	refractive index
q	energy flux
s	distance along path
S	volumetric source strength

Greek Letters

β	single-scattering albedo, σ_R/κ
ε_R	radiative emissivity
κ	extinction coefficient, $= a + \sigma_R$
λ	wavelength
μ	cosine of angle relative to surface normal or direction of travel
ξ	size parameter, $\pi d_p/\lambda$
ξ_C	clearance parameter, $\pi C/\lambda$
σ	Stefan–Boltzmann constant
σ_R	linear radiation scattering coefficient
$\bar{\sigma}$	standard deviation
τ	optical thickness, κs
Φ	scattering phase function
Ω	solid angle

Subscripts

b	blackbody
f	fluid
i	incident or incoming
max	value at peak of blackbody spectrum

R	radiative
s	solid
tot	value for combined conduction and radiation transfer
v	volumetric
λ	wavelength

REFERENCES

Abbasi MH, Evans JW. Monte Carlo simulation of radiant transport through an adiabatic packed bed or porous solid. AIChE J 28:853–854, 1982.

Antoniak ZI, Palmer BJ, Drost MK, Welty JR. Parametric study of radiative heat transfer in arrays of fixed discrete surfaces. J Heat Transfer 118:228–230, 1996.

Argento C, Bouvard D. A ray tracing method for evaluating the radiative heat transfer in porous media. Int J Heat Mass Transfer 39:3175–3180, 1996.

Baillis-Doermann D, Sacadura JF. Thermal radiation properties of dispersed media: Theoretical prediction and experimental characterization. In: Menguç MP, ed. Radiative Transfer II; Proceedings of 2nd International Symposium on Radiation Transfer. New York: Begell House, 1998, pp 1–38.

Bohn MS, Mehos MS. Radiative transport models for solar thermal receiver/reactors. In: Beard JT, Ebadian MA, eds. Proceedings of 12th ASME International Solar Energy Conference. New York: ASME, 1990, pp 175–182.

Bohren C, Huffman D. Absorption and Scattering of Light by Small Particles. New York: Wiley–Interscience, 1983.

Brewster MQ, Tien CL. Radiative transfer in packed fluidized-beds: Dependent vs independent scattering. J Heat Transfer 104:573–579, 1982.

Chen JC, Churchill SW. Radiant heat transfer in packed beds. AIChE J 9:35–41, 1963.

Chern BC, Moon TJ, Howell JR. Thermal analysis of in-situ curing for thermoset, hoop-wound structures using infrared heating: Part I—numerical predictions using independent scattering. J Heat Transfer 117:674–680, 1995a.

Chern BC, Moon TJ, Howell JR. Thermal analysis of in-situ curing for thermoset, hoop-wound structures using infrared heating: Part II—dependent scattering effects. J Heat Transfer 117:681–686, 1995b.

Chern BC, Moon TJ, Howell JR. Dependent radiative transfer regime for unidirectional fiber composites exposed to normal incident radiation. Proceedings of 4th ASME/JSME Joint Symposium, Maui, 1995c.

Das S, Beaman JJ, Wohlert M, Bourell DL. Direct laser freeform fabrication of high performance metal components. Rapid Prototyp J 4: 1998.

Deissler RG. Diffusion approximation for thermal radiation in gases with jump boundary condition. J Heat Transfer 86:240–246, 1964.

Dombrovsky LA. Radiation Heat Transfer in Disperse Systems. New York: Begell House, 1996.

Dombrovsky LA. Calculation of infrared and microwave radiative properties of metal coated microfibers. In: Mengüç MP, ed. Radiative Transfer II; Proceedings of 2nd International Symposium on Radiation Transfer. New York: Begell House, 1998, pp 355–366.

Echigo R. Effective energy conversion method between gas enthalpy and thermal radiation and application to industrial furnaces. Proceedings of 7th International Heat Transfer Conference, Munich. Washington DC: Hemisphere, 1982, Vol VI, pp 361–366.

Echigo R. Radiation enhanced/controlled phenomena of heat and mass transfer in porous media. Proceedings of ASME/JSME Thermal Engineering Joint Conference, 1991,Vol 4, pp xxi–xxxii.

Echigo R, Yoshizawa Y, Hanamura K, Tomimura T. Analytical and experimental studies on radiative propagation in porous media with internal heat generation. Proceedings of 8th International Heat Transfer Conference, San Francisco, CA. Washington DC: Hemisphere, 1986, Vol II, pp 827–832.

Fernandez M, Howell JR. Radiative drying model of porous materials. Drying Tech 15: 1997.

Fu X, Viskanta R, Gore JP. A model for the volumetric radiation characteristics of cellular ceramics. Int Commun Heat Mass Transfer 24:1069–1082, 1997.

Galaktionov AV. Radiative heat transfer in dispersed media: New phenomenological approach. In: Mengüç MP, ed. Radiative Transfer II; Proceedings of 2nd International Symposium on Radiation Transfer. New York: Begell House, 1998, pp 39–51.

Göbel G, Lippek A, Wreidt T, Bauckhage K. Monte Carlo simulation of light scattering by inhomogeneous spheres. In: Mengüç MP, ed. Radiative Transfer II: Proceedings of 2nd International Symposium on Radiation Transfer. New York: Begell House, 1998, pp 367–376.

Hale MJ, Bohn MS. Measurement of the radiative transport properties of reticulated alumina foams. Proceedings of ASME/ASES Joint Solar Energy Conference, Washington, DC, 1993, pp 507–515.

Hendricks TJ. Thermal radiative properties and modelling of reticulated porous ceramics. PhD dissertation, The University of Texas at Austin, 1993.

Hendricks TJ, Howell JR. Inverse radiative analysis to determine spectral radiative properties using the discrete ordinates method. Proceedings of 10th International Heat Transfer Conference, Brighton. Levittown: Taylor and Francis, 1994, Vol 2, pp. 75–80.

Hendricks TJ, Howell JR. Absorption/scattering coefficients and scattering phase functions in reticulated porous ceramics. J Heat Transfer 118:79–87, 1996a.

Hendricks TJ, Howell JR. New radiative analysis approach for reticulated porous ceramics using discrete ordinates. J Heat Transfer 118:911–917, 1996b.

Hoffmann JG, Echigo R, Tada S, Yoshida H. Analytical study on flame stabilization in reciprocating combustion in porous media with high thermal conductivity. Twenty-sixth Symposium (International) on Combustion, 1996, pp 2709–2716.

Howell JR. The Monte Carlo method in radiative transfer. J Heat Transfer 120:547–560, 1998.

Howell JR, Mengüç MP. Radiation. In: Hartnett JP, Irvine T, eds. Handbook of Heat Transfer Fundamentals. New York: McGraw-Hill, 1998, Chap 7.

Howell JR, Hall MJ, Ellzey JL. Combustion of liquid hydrocarbon fuels within porous inert media. Proceedings of Combined Central/Western States Meeting. The Combustion Institute, 1995.

Howell JR, Hall MJ, Ellzey JL. Combustion of hydrocarbon fuels within porous inert media. Prog Energy Combust Sci 22:121–145, 1996.

Hsu PF. Analytical and experimental study of combustion in porous inert media. PhD dissertation, The University of Texas at Austin, 1991.

Hsu PF, Howell JR. Measurements of thermal conductivity and optical properties of porous partial stabilized zirconia. Exp Heat Transfer 5:293–313, 1992.

Im KH, Ahluwalia RK. Radiative enhancement of tube-side heat transfer. Int J Heat Mass Transfer 37:2635–2646, 1994.

Jeandel J, Boulet P, Morlot G. Radiative transfer through a medium of silica fibres oriented in parallel planes. Int J Heat Mass Transfer 36:531–536, 1993.

Jodrey WS, Tory EM. Simulation of random packing of spheres. Simulation 1–12, 1979.

Jones PD, McLeod DG, Dorai-Raj DE. Correlation of measured and computed radiation intensity exiting a packed bed. J Heat Transfer 118:94–102, 1996.

Kamiuto K. Study of Dul'nev's model for the thermal and radiative properties of open-cellular porous materials. JSME Int J, Series B 40:577–582, 1997.

Kamiuto K, Matsushita T. High-temperature radiative properties of open-cellular porous materials. In: Lee JS, ed. Heat Transfer 1998: Proceedings of 11th International Heat Transfer Conference. Levittown: Taylor & Francis, 1998, Vol 7, pp 385–390.

Kamiuto K, Sato M, Iwamoto M. Radiative properties of packed-sphere systems. In Hetsroni G, ed. Heat Transfer 1990: Proceedings of 9th International Heat Transfer Conference, Jerusalem, 1990, Vol 6, pp 427–432.

Kandis M, Buckley W, Bergman TL. A model of polymer powder sintering induced by laser irradiation. In: Lee JS, ed. Heat Transfer 1998: Proceedings of 11th International Heat Transfer Conference. Levittown: Taylor & Francis, 1998, Vol 5, pp 205–210.

Kaviany M, Principles of Heat Transfer in Porous Media. New York: Springer-Verlag, 1991.

Kaviany M, Singh BP. Radiative heat transfer in porous media. In: Hartnett JP, Irvine T, eds. Advances in Heat Transfer. San Diego: Academic Press, 1993, Vol 23, pp 133–186.

Kudo K, Taniguchi H, Kaneda H, Yang W, Zhang YZ, Guo KH, Matsumura M. Flow and heat transfer simulation in circulating fluidized beds. In: Basu P, Horio M, Hasatani M, eds. Circulating Fluidized Bed Technology III: Proceedings of 3rd International Conference on Circulating Fluidized Beds. Oxford: Pergamon Press, 1990, pp 269–274.

Kudo K, Taniguchi H, Kim YM, Yang WJ. Transmittance of radiative energy through three-dimensional packed spheres. Proceedings of 1991 ASME/JSME Thermal Engineering Joint Conference, 1991, pp 35–42.

Kudo K, Li B, Kuroda A. Analysis of radiative energy transfer through fibrous layer considering fibrous orientation. ASME HTD—Vol 315:37–43, 1995.

Lee SC. Dependent vs independent scattering in fibrous composites containing parallel fibers. J Thermophys Heat Transfer 8:641–646, 1994.

Lee SC, Cunnington GR. Heat transfer in fibrous insulations: Comparison of theory and experiment. J Thermophys Heat Transfer 12:297–303, 1998.

Lee SC, Cunnington GR. Theoretical Models for Radiative Transfer in Fibrous Media. In: Tien CL, ed. Annual Review of Heat Transfer. New York: Begell House, 1998, Vol IX, pp. 159–218.

Lee SC, Grzesik JA. Scattering characteristics of fibrous media containing closely spaced parallel fibers. J Thermophys Heat Transfer 9:403–409, 1995.

Li B, Kudo K, Kuroda A. Study on radiative heat transfer through fibrous layer. Proceedings of 3rd KSME–JSME Thermal Engineering Conference, KyongJu, Korea, 1996, Vol III, pp 279–284.

Marschall J, Milos FS. The calculation of anisotropic extinction coefficients for radiation diffusion in solid fibrous ceramic insulations. Int J Heat Mass Transfer 40:627–634, 1997.

Martin AR, Saltiel C, Chai J, Shyy W. Convective and radiative internal heat transfer augmentation with fiber arrays. Int J Heat Mass Transfer 41:3431–3440, 1998.

Mital R, Gore JP, Viskanta R. Measurements of radiative properties of cellular ceramics at high temperatures. J Thermophys Heat Transfer 10:33–38, 1996.

Modest MF. Radiative Heat Transfer. New York: McGraw-Hill, 1993.

Mohamad AA, Ramadhyani S, Viskanta R. Modelling of combustion and heat transfer in a packed bed with embedded coolant tubes. Int J Heat Mass Transfer 37:1181–1191, 1994.

Moura LM, Baillis D, Sacadura JF. Identification of thermal radiation properties of dispersed media: Comparison of different strategies. In: Lee JS, ed. Heat Transfer 1998: Proceedings of 11th International Heat Transfer Conference, Levittown: Taylor & Francis, 1998, Vol 7, pp 489–414.

Nicolau VP, Raynaud M, Sacadura JF. Spectral radiative properties identification of fiber insulating materials. Int J Heat Mass Transfer 37, Suppl 1:311–324, 1994.

Palmer BJ, Drost MK, Welty JR. Monte Carlo simulation of radiative heat transfer in arrays of fixed discrete surfaces using cell-to-cell photon transport. Int J Heat Mass Transfer 39:2811–2819, 1996.

Saatdjian E. A cell model that estimates radiative heat transfer in a particle-laden flow. J Heat Transfer 109:256–258, 1987.

Siegel R, Howell JR. Thermal Radiation Heat Transfer. 3rd ed. Washington: Taylor & Francis, 1992.

Singh BP, Kaviany M. Independent theory versus direct simulation of radiative heat transfer in packed beds. Int J Heat Mass Transfer 34:2869–2881, 1991.

Singh BP, Kaviany M. Modeling radiative heat transfer in packed beds. Int J Heat Mass Transfer 35:1397–1405, 1992.

Singh BP, Kaviany M. Effect of solid conductivity on radiative heat transfer in packed beds. Int J Heat Mass Transfer 37:2579–2583, 1994.

Taniguchi H, Kudo K, Kuroda A, Kaneda H, Fukuchi T, Song KK. Effects of several parameters on heat transfer in circulating fluidized bed boiler. In: Lee JS, Chung SH, Kim KH, eds. Proceedings of 6th International Symposium on Transport Phenomena in Thermal Engineering, Seoul: KSME, 1993, Vol IV, p 54.

Tien CL. Thermal radiation in packed and fluidized beds. J Heat Transfer 110:1230–1242, 1988.

Tong TW, Yang QS, Tien CL. Radiative heat transfer in fibrous insulations—Part II: Experimental study. J Heat Transfer 105:76–81, 1983.

Vortmeyer D. Radiation in packed solids. In: Rogers JT, ed. Heat Transfer 1978: Proceedings of 6th International Heat Transfer Conference. Washington, DC: Hemisphere, 1978, Vol 6, pp 525–539.

Wu AKC, Lee SH-K. Computation of Radiative Properties in 1-D Sphere Packings for Sintering Applications, ASME paper NHTC99–60, 1999.

Xia Y, Strieder W. Complementary upper and lower truncated sum, multiple scattering bounds on the effective emissivity. Int J Heat Mass Transfer 37:443–450, 1994a.

Xia Y, Strieder W. Variational calculation of the effective emissivity for a random bed. Int J Heat Mass Transfer 37:451–460, 1994b.

Yang YS. Heat transfer through a randomly packed bed of spheres. PhD dissertation, The University of Texas at Austin, 1981.

Yang YS, Klein DE, Howell JR. Radiative heat transfer through a randomly packed bed of spheres by the Monte Carlo method. J Heat Transfer 105:325–332, 1983.

Yano T, Ochi M, Enya S. Protection against fire and high temperature by using porous media and water. Proceedings of ASME/JSME Thermal Engineering Joint Conference, 1991, pp 213–218.

Younis LB, Viskanta RV. Experimental determination of the volumetric heat transfer coefficient between stream of air and ceramic foam. Int J Heat Mass Transfer 36:1425–1434, 1993.

16
Weak Turbulence and Transitions to Chaos in Porous Media

P. Vadasz
University of Durban-Westville, Durban, South Africa

I. INTRODUCTION

The fundamental understanding of the transition from laminar to turbulent convection in porous media is far from being conclusive. While major efforts are under way, there is still a significant challenge in front of the scientist and engineer to uncover the complex behavior linked to this transition. In pure fluid (nonporous domain) shear flows the time-dependent and three-dimensional form of turbulence is well established experimentally and numerically. It is caused by the nonlinear terms in the isothermal Navier–Stokes equations. In isothermal flow in porous media no experiments identifying the three-dimensional nature of a transition from the Darcy regime, via an inertia-dominated regime, towards turbulence are available. In particular, this detailed description of turbulence is missing in the problem of porous media convection where an additional nonlinear interaction appears as a result of the coupling between the equations governing the fluid flow and the energy equation. The latter can typically cause a transition to a nonsteady and nonperiodic regime at much lower values of the parameter controlling the flow when compared to the corresponding isothermal system. It is for this reason that the nonsteady and nonperiodic convective regime associated with a reduced set of equations is referred to as weak turbulence (Choudhury 1997).

The wide variety of engineering applications of transport phenomena in porous media provide the solid practical modification for this investigation. The distinction between modern applications of convection in porous media, which include small values of Prandtl number and consequently moderate values of the Darcy–Prandtl number, and traditional applications, which are typically associated with moderate values of Prandtl number (high values of the Darcy–Prandtl number), is of particular interest in the present review. The resulting effect of high values of the Darcy–Prandtl number is to yield a very small coefficient to the time derivative term in Darcy's equation, hence allowing the neglect of this term. On the other hand, small Prandtl number convection can be associated with liquid metals ($Pr = O(10^{-3})$). The process of solidification of binary alloys can be selected in particular as an appropriate example of small Prandtl number convection in porous media as the mushy layer which forms on the interface between the solid phase and the liquid metal to be solidified can be regarded as a porous medium for all practical purposes. Even for moderate values of Prandtl number (and consequently high values of Darcy–Prandtl number) it is essential to retain the time derivative term in Darcy's equation when investigation of wave effects is being considered. Gheorghita (1966) showed that retaining the time derivative term for investigating wave effects is important even in isothermal systems. Comprehensive reviews of the fundamentals of heat convection in porous media are presented by Nield and Bejan (1999) and Bejan (1995).

The objective of the present chapter is to review the results of transitions to chaos in a fluid-saturated porous layer subject to gravity and heated from below. The nonsteady and nonperiodic convective regime associated with a reduced set of equations is referred to as weak turbulence (Choudhury 1997). This is related to "turbulence," manifesting itself in the time domain while the space domain does not exhibit irregularities in the solution. In practice, the situation is more complex; however, weak turbulence could very well describe the phenomenon occurring over a transitional range of Rayleigh numbers, whereas temporal-spatial turbulence (strong turbulence) is expected to manifest itself beyond this transitional range. Moreover, the weak-turbulent regime can be anticipated to persist even beyond the transitional range within the core of the convective domain, while boundary layers may incorporate the temporal-spatial effect. The results reported show that the transition from steady convection to chaos is sudden for small Prandtl number convection, and follows a period-doubling sequence for moderate Prandtl number convection (Vadasz and Olek 1999b,c). Both transitions occur initially via a Hopf bifurcation, producing a "solitary limit cycle" which may be associated with a homoclinic explosion. Work on obtaining analytical results suggesting an explanation for the appearance of this solitary limit cycle and explaining a well-known phenomenon of hysteresis is presented by Vadasz (1999).

II. GOVERNING EQUATIONS

A. Original System

A fluid-saturated porous layer subject to gravity and heated from below, as presented in Figure 1, is considered. A Cartesian coordinate system is used such that the vertical axis z is collinear with gravity, i.e., $\mathbf{e}_g = -\mathbf{e}_z$. The time derivative term is not neglected in Darcy's equation, but, other than that, Darcy's law is assumed to govern the fluid flow, while the Boussinesq (1903) approximation is applied for the effects of density variations. Under these conditions the following dimensionless set of governing equations applies

$$\nabla \cdot V = 0 \tag{1}$$

$$\left[\frac{1}{Pr_D}\frac{\partial}{\partial \hat{t}} + 1\right] V = -\nabla p + RaT\mathbf{e}_z \tag{2}$$

$$\frac{\partial T}{\partial \hat{t}} + V \cdot \nabla T = \nabla^2 T \tag{3}$$

Equations (1)–(3) are presented in a dimensionless form. The values $\alpha_{eo}/H_* M_f$, $\mu_* \alpha_{eo}/K_o M_f$, and $\Delta T_c = (T_H - T_C)$ are used to scale the filtration velocity components (u_*, v_*, w_*), pressure (p_*), and temperature variations $(T_* - T_C)$, respectively, where α_{eo} is the effective thermal diffusivity, μ_* is the fluid viscosity, K_o is the permeability of the porous matrix, and M_f is the ratio between the heat capacity of the fluid and the effective heat capacity of the porous domain. The height of the layer H_* was used for scaling the variables x_*, y_*, z_*, and H_*^2/α_{eo} was used for scaling the time t_*.

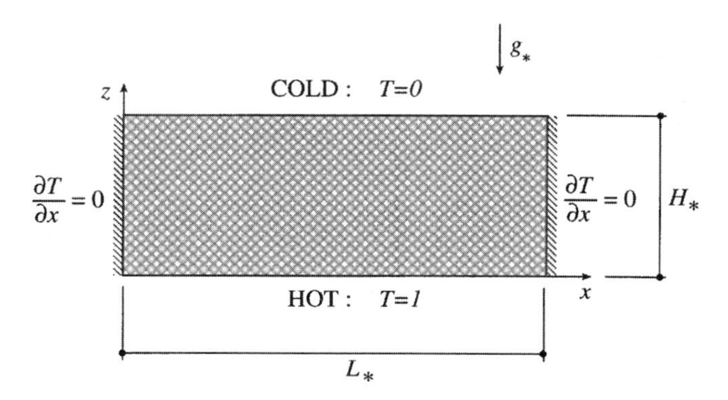

Figure 1. A fluid-saturated porous layer heated from below. (Courtesy: Kluwer Academic Publishers.)

Accordingly, $x = x_*/H_*$, $y = y_*/H_*$, $z = z_*/H_*$, and $\hat{t} = t_*\alpha_{eo}/H_*^2$. In Eq. (2) Ra is the gravity-related Rayleigh number defined in the form $Ra = \beta_* \Delta T_c g_* H_* K_o M_f / \alpha_{eo} \nu_*$.

The time derivative term was included in Darcy's equation (2), where Pr_D is a dimensionless group which includes the Prandtl and Darcy numbers as well as the porosity of the porous domain and is defined by $Pr_D = \phi Pr/Da$. Hence, while Pr can take values as small as 10^{-3} for liquid metals and up to 10^3 for oils, the corresponding values for Pr_D will be magnified by a factor of ϕ/Da, which is typically a large number. This factor can take values from 10 to 10^{20}. Therefore the values of Pr_D can be expected in the range from 10^{-2} to 10^{23}. In traditional applications of transport phenomena in porous media typical values of Pr_D are quite large, a fact that provides the justification for neglecting the time derivative term in Darcy's equation. However, when wave phenomena are of interest, the time derivative term should be retained in order to prevent the reduction of the order of the system in the time domain (Gheorghita 1966). Vadasz and Olek (1999c) demonstrated the effect of including this term in the equations on the convection results for moderate Prandtl number values and consequently high values of Darcy–Prandtl number. When small Prandtl numbers are considered (corresponding to Darcy–Prandtl numbers around $Pr_D \sim 100$), that are applicable in modern applications of convection in porous media such as the solidification of binary alloys, there is even less justification for neglecting the time derivative term in Darcy's equation.

As all the boundaries are rigid the solution must follow the impermeability conditions there, i.e., $\mathbf{V} \cdot \mathbf{e}_n = 0$ on the boundaries, where $\mathbf{e}_n$ is a unit vector normal to the boundary. The temperature boundary conditions are: $T = 1$ at $z = 0$, $T = 0$ at $z = 1$, and $\Delta T \cdot \mathbf{e}_n = 0$ on all other walls, representing the insulation condition on these walls.

For convective rolls having axes parallel to the shorter dimension (i.e., y) $v = 0$, and the governing equations can be presented in terms of a stream function defined by $u = \partial\psi/\partial z$ and $w = -\partial\psi/\partial x$, which, upon applying the curl ($\nabla\times$) operator on Eq. (2), yields the following system of partial differential equations from Eqs. (1), (2), and (3)

$$\left[\frac{1}{Pr_D}\frac{\partial}{\partial \hat{t}} + 1\right]\left[\frac{\partial^2 \psi}{\partial x^2} + \frac{\partial^2 \psi}{\partial z^2}\right] = -Ra\frac{\partial T}{\partial x} \tag{4}$$

$$\frac{\partial T}{\partial \hat{t}} + \frac{\partial \psi}{\partial z}\frac{\partial T}{\partial x} - \frac{\partial \psi}{\partial x}\frac{\partial T}{\partial z} = \frac{\partial^2 T}{\partial x^2} + \frac{\partial^2 T}{\partial z^2} \tag{5}$$

where the boundary conditions for the stream function are $\psi = 0$ on all solid boundaries.

The set of partial differential equations (4) and (5) forms a nonlinear coupled system which, together with the corresponding boundary conditions, accepts a basic motionless conduction solution.

B. Reduced System for Weak Turbulence

To obtain the solution to the nonlinear coupled system of partial differential equations (4) and(5), we represent the stream function and temperature in the form

$$\psi = A_{11} \sin\left(\frac{\pi x}{L}\right) \sin(\pi z) \tag{6}$$

$$T = 1 - z + B_{11} \cos\left(\frac{\pi x}{L}\right) \sin(\pi z) + B_{02} \sin(2\pi z) \tag{7}$$

This representation is equivalent to a Galerkin expansion of the solution in both x and z directions, truncated when $i + j = 2$, where i is the Galerkin summation index in the x direction and j is the Galerkin summation index in the z direction. Substituting (6) and (7) into (4) and (5), multiplying the equations by the orthogonal eigenfunctions corresponding to (6) and (7), and integrating them over the domain, i.e. $\int_0^L \mathrm{d}x \int_0^1 \mathrm{d}z(\cdot)$, yields a set of three ordinary differential equations for the time evolution of the amplitudes, in the form

$$\frac{\mathrm{d}A_{11}}{\mathrm{d}t} = -\frac{Pr_D \gamma}{\pi^2}\left[A_{11} + \frac{Ra}{\eta\pi} B_{11}\right] \tag{8}$$

$$\frac{\mathrm{d}B_{11}}{\mathrm{d}t} = -B_{11} - \frac{1}{\pi\eta} A_{11} - \frac{1}{\eta} A_{11} B_{02} \tag{9}$$

$$\frac{\mathrm{d}B_{02}}{\mathrm{d}t} = \frac{1}{2\eta} A_{11} B_{11} - 4\gamma B_{02} \tag{10}$$

where the time was rescaled and additional notation was introduced ($t = \eta\pi^2 \hat{t}/L$, $\eta = (L^2 + 1)/L$, and $\gamma = L/\eta$). While the severe truncation may seem to limit significantly the validity of the model in connection with the original problem, Malkus (1972) showed that the set of three equations associated with the three modes in (6) and (7) decouple from the rest (with exact closure), at least in the sense of weighted residuals. Therefore, their solution is relevant even when solving at higher truncation levels as this set of three equations needs to be solved separately before one attempts to solve the rest of the set corresponding to higher modes. In general, we will see that the dynamics related to this reduced system is so rich that it is important to understand it prior to attempting to solve at higher orders.

The nonsteady and nonperiodic convection regime associated with this reduced set of equations is referred to as weak turbulence (Choudhury 1997).

The fixed (i.e., stationary) points of the system (8), (9), and (10) are obtained by setting all the time derivatives equal to zero and solving the resulting algebraic equations. They yield the following possible solutions: $A_{11} = B_{11} = B_{20} = 0$, representing the motionless conduction solution, and

$$A_{11} = \mp 2\sqrt{2}(L^2 + 1)^{1/2}\left(\frac{Ra}{\pi^2\eta^2} - 1\right)^{1/2} \tag{11}$$

$$B_{11} = \pm \frac{2\sqrt{2}\pi L\eta^2}{(L^2 + 1)^{1/2} Ra}\left(\frac{Ra}{\pi^2\eta^2} - 1\right)^{1/2} \tag{12}$$

$$B_{02} = -\frac{(Ra - \pi^2\eta^2)}{\pi Ra} \tag{13}$$

representing the steady convective solutions. It is convenient to introduce the following further notation $R = Ra/\pi^2\eta^2$, $\alpha = Pr_D\gamma/\pi^2$, and rescale the amplitude with respect to their convective fixed points in the following form, $X = -A_{11}/2\eta\sqrt{2\gamma(R-1)}$, $Y = \pi R B_{11}/2\sqrt{2\gamma(R-1)}$, and $Z = -\pi R B_{02}/(R-1)$, to provide the following set of scaled equations, which are equivalent to Eqs. (8), (9), and (10)

$$\dot{X} = \alpha(Y - X) \tag{14}$$

$$\dot{Y} = RX - Y - (R-1)XZ \tag{15}$$

$$\dot{Z} = 4\gamma(XY - Z) \tag{16}$$

where the rescaled amplitudes X, Y, and Z are defined in the text above and in the Nomenclature, and the dots $(\cdot)$ denote time derivatives $\mathrm{d}(\,)/\mathrm{d}t$.

Equations (14), (15), and (16) are equivalent to Lorenz equations (Lorenz 1963; Sparrow 1982), although with different coefficients. The value of γ has to be consistent with the wavenumber at the convection threshold, which is required for the convection cells to fit into the domain and fulfil the boundary conditions. This requirement, which in pure fluids (nonporous domains) results in $\gamma = 2/3$, yields for porous media convection a value of $\gamma = 0.5$. With this value of γ, the definitions of R and α are explicitly expressed in the form $R = Ra/4\pi^2$ and $\alpha = Pr_D/2\pi^2$.

C. Linear Analysis

The fixed points of the rescaled system are $X_1 = Y_1 = Z_1 = 0$, corresponding to the motionless solution, and $X_{2,3} = \pm 1$, $Y_{2,3} = \pm 1$, $Z_{2,3} = 1$, corresponding to the convection solution.

The next step is to perform a stability analysis of the stationary solutions in order to determine the nature of the dynamics about the fixed points. The system (14), (15), and (16) has the general form $\dot{\boldsymbol{X}} = \boldsymbol{f}(\boldsymbol{X})$, and the equilibrium (stationary or fixed) points $\boldsymbol{X}_s$ are defined by $\boldsymbol{f}(\boldsymbol{Xs}) = \boldsymbol{0}$. The stability matrix is established by evaluating the Jacobian $(\partial f_i/\partial X_j)_{Xs}$ at the fixed point of interest $\boldsymbol{X}_s$. The eigenvalues of the stability matrix, evaluated by solving the zeros of the characteristic polynomial associated with the stability matrix, provide the stability conditions. A fixed point is stable if all eigenvalues corresponding to its stability matrix are negative (or, in the case of complex eigenvalues, they have negative real parts) and it is not stable if at least one eigenvalue becomes positive (or, in the case of complex eigenvalues, it has a positive real part).

The stability of the fixed point associated with the motionless solution ($X_1 = Y_1 = Z_1 = 0$) is controlled by the zeros of the following characteristic polynomial equation for the eigenvalues σ_i $(i = 1, 2, 3)$

$$(4\gamma + \sigma)[\alpha R - (\alpha + \sigma)(1 + \sigma)] = 0 \tag{17}$$

The first eigenvalue $\sigma_1 = -4\gamma$ is always negative as $\gamma > 0$. The other two eigenvalues are always real and are given by $\sigma_{2,3} = [-(\alpha + 1) \pm \sqrt{(\alpha + 1)^2 + 4\alpha(R - 1)}]/2$. σ_3 is always negative and σ_2 provides the stability condition for the motionless solution in the form $\sigma_2 < 0 \Leftrightarrow R < 1$. Therefore the critical value of R, where the motionless solution loses stability and the convection solution (expressed by the other two fixed points) takes over, is obtained as $R_{c1} = R_{cr} = 1$, which corresponds to $Ra_{cr} = 4\pi^2$.

The stability of the fixed points associated with the convection solution ($X_{2,3}$, $Y_{2,3}$, $Z_{2,3}$) is controlled by the following cubic equation for the eigenvalues σ_i $(i = 1, 2, 3)$

$$\sigma^3 + (4\gamma + \alpha + 1)\sigma^2 + 4\gamma(\alpha + R)\sigma + 8\gamma\alpha(R - 1) = 0 \tag{18}$$

Equation (18) yields three eigenvalues. The smallest eigenvalue σ_1 is always real and negative over the whole range of parameter values. The other two are real and negative at slightly supercritical values of R; therefore the convection fixed points are stable, i.e., simple nodes. As the value of R increases these two roots move on the real axis towards the origin, the smaller of the two chasing the other one and reducing the distance between them. For $\alpha = 5$ (which is consistent with $Pr_D \cong 98.7$) and $\gamma = 0.5$ these roots become equal when $R \simeq 1.28$, while for $\alpha = 50$ and $\gamma = 0.5$ they become equal when $R \simeq 1.26$. It is exactly at this point that these two roots become complex conjugate. However, they still have negative real parts; therefore the convection fixed points are stable, i.e., spiral nodes. As the value of R increases further, both the imaginary and real parts of these two complex conjugate eigenvalues increase and, on the complex plane, they cross the imaginary axis; i.e., their real part becomes non-negative at a value of R given by

$$R_{c2} = \frac{\alpha(\alpha + 4\gamma + 3)}{(\alpha - 4\gamma - 1)} \tag{19}$$

At this point the convection fixed points lose their stability and other, periodic or chaotic, solutions take over. However, just prior to the first transition happening, at the point when the complex eigenvalues cross the imaginary axis, a Hopf bifurcation occurs; i.e., at R_{c2} these eigenvalues are purely imaginary, leading to a solitary limit cycle (we call this a "solitary limit cycle" because it is obtained only at this value of R and is distinct from any periodic solution beyond the transition point). The evolution of the complex eigenvalues is presented in Figure 2a for $\alpha = 5$, and in Figure 2b for $\alpha = 50$, providing a graphical description of the sequence of events leading to the loss of stability of the convection fixed points. For $\alpha = 5$ and $\gamma = 0.5$ the loss of stability of the convection fixed points is evaluated, using (19), to be $R_{c2} = 25$, whereas for $\alpha = 50$ and $\gamma = 0.5$ it yields $R_{c2} = 58.51$.

An interesting observation can be made by investigating the behavior of the value of R_{c2} as α becomes very big. When $\alpha \gg 1$ the time derivative term in Darcy's equation and consequently in Eq. (14) is very small, and this fact provides the justification for neglecting this term. Then (14) yields the solution $Y = X$ which is valid for all time values. By taking the limit of R_{c2} in Eq. (19) one obtains that $R_{c2} \to \alpha$ as $\alpha \to \infty$. Therefore, although a much higher value for the first transition is required when the values of α are very big, this transition still exists, and may eventually lead to chaos, while neglecting the time derivative term in Eq. (14) wipes out this possibility as the dimensionality of the system reduces to two.

III. NONLINEAR METHODS OF SOLUTION

A. Weak Nonlinear Method (Local Analytical Solution)

This method was applied by Vadasz (1999) to provide analytical insight into the transition observed computationally by Vadasz and Olek (1999b,c). The stationary (fixed) points of the system (14), (15), and (16) are the convective solutions $X_S = Y_S = \pm 1$, $Z_S = 1$, and the motionless solution $X_S = Y_S =$

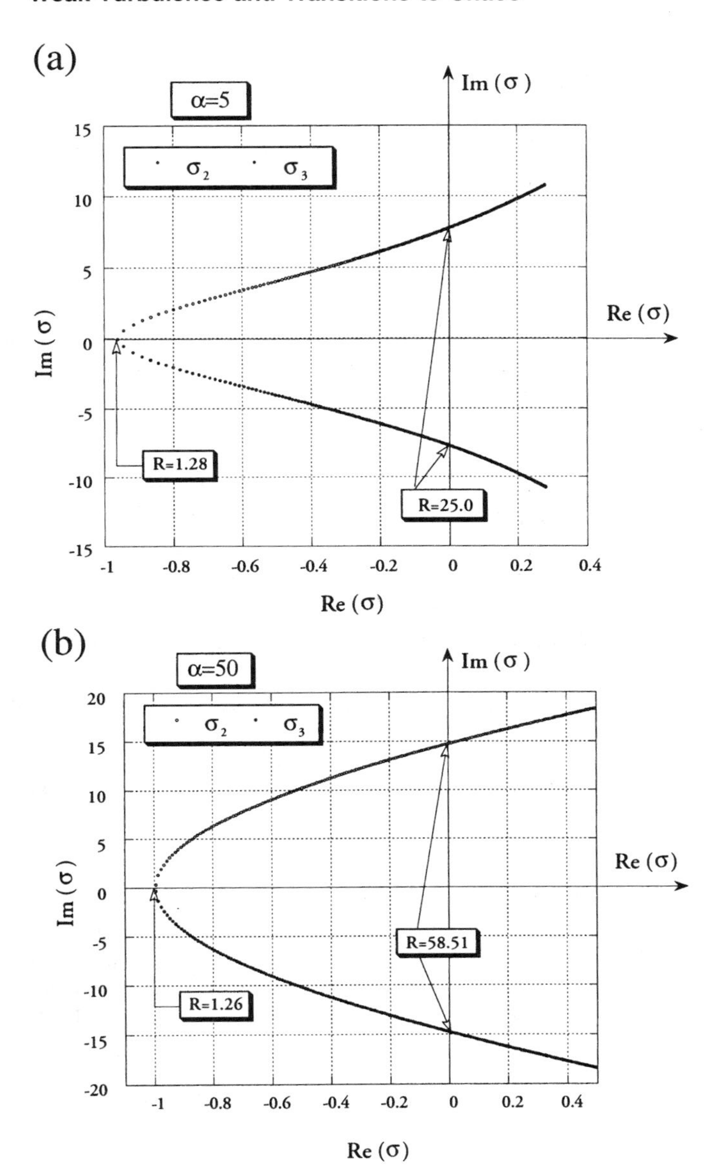

Figure 2. The evolution of the complex eigenvalues with increasing Rayleigh number, for $\gamma = 0.5$ and: (a) $\alpha = 5$ ($Pr_D = 98.7$); (b) $\alpha = 50$ ($Pr_D = 987$). (Courtesy: Kluwer Academic Publishers.)

$Z_S = 0$ (occasionally referred to as the origin). The expansion around the motionless stationary solution yields the familiar results of a pitchfork bifurcation from a motionless state to convection at $R = 1$. We expand now the dependent variables around the convection stationary points in the form

$$X = X_S + \varepsilon X_1 + \varepsilon^2 X_2 + \varepsilon^3 X_3 + \ldots \tag{20}$$

$$Y = Y_S + \varepsilon Y_1 + \varepsilon^2 Y_2 + \varepsilon^3 Y_3 + \ldots \tag{21}$$

$$Z = Z_S + \varepsilon Z_1 + \varepsilon^2 Z_2 + \varepsilon^3 Z_3 + \ldots \tag{22}$$

We also expand R in a finite series of the form $R = R_o(1 + \varepsilon^2)$ which now defines the small expansion parameter as $\varepsilon^2 = (R - R_o)/R_o$, where $R_o = R_{c2}$ is the value of R where the stationary convective solutions lose their stability in the linear sense (see Vadasz and Olek 1999b,c). Therefore the present weak nonlinear analysis is expected to be restricted to initial conditions sufficiently close to *any one* of the convective fixed points. Introducing a long time scale $\tau = \varepsilon^2 t$ and replacing the time derivatives in Eqs. (14), (15), and (16) with $\mathrm{d}/\mathrm{d}t \to \mathrm{d}/\mathrm{d}t + \varepsilon^2 \mathrm{d}/\mathrm{d}\tau$ yields a hierarchy of ordinary differential equations at the different orders. The leading order provides the stationary solutions, while at order ε we get the familiar homogeneous linearized system

$$\dot{X}_1 - \alpha(Y_1 - X_1) = 0 \tag{23}$$

$$\dot{Y}_1 - [X_1 - Y_1 \mp (R_o - 1)Z_1] = 0 \tag{24}$$

$$\dot{Z}_1 - 4\gamma(\pm X_1 \pm Y_1 - Z_1) = 0 \tag{25}$$

where the $\pm$ upper- or under-sign corresponds to the selected stationary convective point, either $X_S = Y_S = 1$, $Z_S = 1$ corresponding to the upper-sign, or $X_S = Y_S = -1$, $Z_S = 1$ corresponding to the under-sign. The following stages will focus on the solution around $X_S = Y_S = 1$, $Z_S = 1$; hence the upper-sign holds. Similar equations hold for the other stationary point. The solutions to the linear set (23)–(25) have the form

$$X_1 = a_1 \mathrm{e}^{i\sigma_o t} + a_1^* \mathrm{e}^{-i\sigma_o t}; \qquad Y_1 = b_1 \mathrm{e}^{i\sigma_o t} + b_1^* \mathrm{e}^{-i\sigma_o t};$$
$$Z_1 = c_1 \mathrm{e}^{i\sigma_o t} + c_1^* \mathrm{e}^{-i\sigma_o t} \tag{26}$$

where the coefficients $a_1(\tau)$, $a_1^*(\tau)$, $b_1(\tau)$, $b_1^*(\tau)$, $c_1(\tau)$, and $c_1^*(\tau)$ are allowed to vary over the long time scale τ and $\pm i\sigma_o$ are the imaginary parts of the complex eigenvalues corresponding to the linear system at marginal stability (i.e., the real part o the eigenvalues is 0). They are related to α and γ by the equation $\sigma_o^2 = 8a\gamma(\alpha + 1)/(\alpha - 4\gamma - 1)$, which can be established by working out the relationships between the $O(\varepsilon)$ coefficients in the solution (26). These relationships are obtained by substituting the solutions (26) into the linear equations (23)–(25), and yield

$$b_1 = \frac{(\alpha + i\sigma_o)}{\alpha} a_1; \qquad b_1^* = \frac{(\alpha - i\sigma_o)}{\alpha} a_1^* \tag{27}$$

$$c_1 = \frac{\sigma_o[\sigma_o - i(\alpha + 1)]}{\alpha(R_o - 1)} a_1; \qquad c_1^* = \frac{\sigma_o[\sigma_o + i(\alpha + 1)]}{\alpha(R_o - 1)} a_1^* \tag{28}$$

The relationship for R_o is also obtained in the form $R_o = \alpha(\alpha + 4\gamma + 3)/(\alpha - 4\gamma - 1)$, recovering Eq. (19). At higher orders of hierarchy of nonhomogeneous ordinary differential equations sharing the same homogeneous operator as Eqs. (23)–(25) are obtained. At order $O(\varepsilon^3)$ the forcing functions belonging to the nonhomogeneous part of the equations resonate the homogeneous operator unless a solvability condition is fulfilled (for details see Vadasz 1999).

The solvability condition takes the form of a constraint on the amplitude of the solution at order $O(\varepsilon)$ and yields an amplitude equation of the form

$$\frac{\mathrm{d}a}{\mathrm{d}t} = h_{21}[\varepsilon^2 - h_{32}aa^*]a \tag{29}$$

and a similar equation for a^*, where in Eq. (29) $a = \varepsilon a_1$, $a^* = \varepsilon a_1^*$, and h_{21}, h_{32} are complex coefficients which depend on α, γ, σ_o, and R_o. Their lengthy expressions are skipped here. The coefficient of the nonlinear term in Eq. (29) plays a role of particular importance as it controls the direction of the Hopf bifurcation which results from the post-transient solution to (29). Its impact on the solution and the corresponding analysis is presented later in this chapter (see Section IV.A).

B. Adomian's Decomposition Method (Computational Solution)

Adomian's decomposition method (Adomian 1988, 1994), is applied to solve the system of Eqs. (14), (15), and (16). The method provides, in principle, an analytical solution in the form of an infinite power series for each dependent variable, and its excellent accuracy in solving nonlinear equations was demonstrated by Olek (1994, 1997). The solution follows Olek (1997) and considers the following more general dynamical system of equations

$$\frac{\mathrm{d}X_i}{\mathrm{d}t} = \sum_{j=1}^{m} b_{ij}X_j + \sum_{l=1}^{m}\sum_{j=1}^{m} a_{ijl}X_jX_l, \qquad \forall i = 1, 2, \ldots, m \tag{30}$$

given the initial conditions $X_i(0)$, $i = 1, 2, \ldots, m$. It can be easily observed that the system of equations (14), (15), and (16) is just a particular case of Eq. (30).

Olek (1994, 1997) used the decomposition method to solve a variety of nonlinear problems, some of which have closed-form analytical solutions, and a comparison was provided between the results obtained via the decomposition method and either exact analytical or numerical results. The conclusion from the comparison was that the decomposition method provided results which were accurate up to 14 significant digits. Even when only three terms were kept in the decomposition series solution of the Lotka–Volterra equations, the results agreed by at least five significant digits with a corresponding numerical solution. The problem can actually be solved to the desired accuracy by including more terms in the computation of the series.

For the system of equations (30) the nonlinear terms are of the rather simple X^2 form, so that very simple symmetry rules for decomposition polynomials can be used. If we denote $L \equiv \mathrm{d}/\mathrm{d}t$, the formal solution of (30) may be presented in the form

$$X_i(t) = X_i(0) + \mathrm{L}^{-1}\left[\sum_{j=1}^{m} b_{ij}X_j + \sum_{l=1}^{m}\sum_{j=1}^{m} a_{ijl}X_jX_l\right] \qquad \forall i = 1, 2, \ldots, m \tag{31}$$

where $\mathrm{L}^{-1} \equiv \int_0^t [\cdot]\mathrm{d}t$. According to the decomposition method, an expansion of the following form is assumed

$$X_i(t) = \sum_{n=0}^{\infty} \tilde{X}_{in} \qquad \forall i = 1, 2, \ldots, m \tag{32}$$

Substituting (32) into (31) yields, after rearranging the products,

$$X_i(t) = X_i(0) + \mathrm{L}^{-1}\left[\sum_{j=1}^{m} b_{ij}\sum_{n=0}^{\infty}\tilde{X}_{jn} + \sum_{l=1}^{m}\sum_{j=1}^{m} a_{ijl}\sum_{n=0}^{\infty}\sum_{k=0}^{n}\tilde{X}_{jk}\tilde{X}_{l(n-k)}\right] \qquad \forall i = 1, 2, \ldots, m \tag{33}$$

The solution is ensured by requiring

$$\tilde{X}_{i0} = X_i(0) \qquad \forall i = 1, 2, \ldots, m \tag{34}$$

$$\tilde{X}_{i1} = \mathrm{L}^{-1}\left[\sum_{j=1}^{m} b_{ij}\tilde{X}_{j0} + \sum_{l=1}^{m}\sum_{j=1}^{m} a_{ijl}\sum_{k=0}^{0}\tilde{X}_{jk}\tilde{X}_{l(0-k)}\right] \qquad \forall i = 1, 2, \ldots, m \tag{35}$$

$$\tilde{X}_{i2} = \mathrm{L}^{-1}\left[\sum_{j=1}^{m} b_{ij}\tilde{X}_{j1} + \sum_{l=1}^{m}\sum_{j=1}^{m} a_{ijl}\sum_{k=0}^{1}\tilde{X}_{jk}\tilde{X}_{l(1-k)}\right] \qquad \forall i = 1, 2, \ldots, m \tag{36}$$

$$\vdots$$

$$\tilde{X}_{in} = \mathrm{L}^{-1}\left[\sum_{j=1}^{m} b_{ij}\tilde{X}_{j(n-1)} + \sum_{l=1}^{m}\sum_{j=1}^{m} a_{ijl}\sum_{k=0}^{n-1}\tilde{X}_{jk}\tilde{X}_{l(n-k-1)}\right] \qquad \forall i = 1, 2, \ldots, m \tag{37}$$

After carrying out the integrations, the following solution is obtained

$$X_i(t) = \sum_{n=0}^{\infty} c_{i,n}\frac{t^n}{n!} \qquad \forall i = 1, 2, \ldots, m \tag{38}$$

where $c_{i0} = X_i(0)\ \forall i = 1, 2, \ldots, m$ and the general term for $n \geq 1$ is defined through the following recurrence relationship

$$c_{i,n} = \sum_{j=1}^{m} b_{ij}c_{j,(n-1)} + \sum_{l=1}^{m}\sum_{j=1}^{m}\sum_{k=0}^{n-1} a_{ijl}\frac{(n-1)!c_{j,k}c_{l,(n-k-1)}}{k!(n-k-1)!} \qquad \forall i = 1, 2, \ldots, m \tag{39}$$

The convergence of the series (38) is difficult to assess *a priori*. Irrespective of this difficulty, the practical need to compute numerical values for the solution at different values of time requires the truncation of the series, and therefore its convergence needs to be established in each particular case. To achieve this goal, the decomposition method can be used as an algorithm for the approximation of the dynamical response in a sequence of time intervals $[0, t_1), [t_1, t_2), \ldots, [t_{n-1}, t_n)$ such that the solution at t_p is taken as the initial condition in the interval $[t_p, t_{p+1})$ which follows. This approach has the following advantages: (i) in each time interval one can apply a theorem, proved by Répaci (1990), which states that the solution obtained by the decomposition method converges to a unique solution as the number of terms in the series becomes infinite; and (ii) the approximation in each interval is continuous in time and can be obtained with the desired accuracy corresponding to the desired number of terms.

The latter procedure is adopted in the computation of the solution to Eqs. (14), (15), and (16). One can easily observe that this set of equations is just a particular case of Eqs. (30) with $m = 3$. The set of equations (14), (15), and (16) provides the following nonzero coefficients for substitution in Eq.

(30): $b_{11} = -\alpha$; $b_{12} = \alpha$; $b_{21} = R$; $b_{22} = -1$; $b_{33} = -4\gamma$; $a_{213} = -(R-1)$; $a_{312} = 4\gamma$. Except for these coefficients, all others are identically zero. Therefore the coefficients $c_{i,n}$ in Eq. (38) take the particular form

$$c_{1,n} = -\alpha(c_{1,(n-1)} - c_{2,(n-1)}) \tag{40}$$

$$c_{2,n} = Rc_{1,(n-1)} - c_{2,(n-1)} - (R-1)\sum_{k=0}^{n-1} \frac{(n-1)!c_{1,k}c_{3,(n-k-1)}}{k!(n-k-1)!} \tag{41}$$

$$c_{3,n} = -4\gamma c_{3,(n-1)} + 4\gamma \sum_{k=0}^{n-1} \frac{(n-1)!c_{1,k}c_{2,(n-k-1)}}{k!(n-k-1)!} \tag{42}$$

C. Pseudo-spectral Method (Numerical Solution)

Kimura et al. (1986) used a psuedo-spectral method to identify the route to chaos in two-dimensional, single-cell, porous-medium thermal convection. They used the Darcy's equation without including the time derivative term and derived a general set of spectral equations for N modes. Using a previously derived model for three-dimensional convection in porous media, Straus and Schubert (1979) suggest the introduction of a function ζ, defined by $u = \partial^2\zeta/\partial x\partial z$, $v = \partial^2\zeta/\partial y\partial z$, which satisfies identically the condition of vanishing vertical vorticity. The latter condition is obtained by taking the *curl* ($\nabla\times$) of Eq. (2) without the time derivative term (i.e., $Pr_D \to \infty$). Then, by introducing the further definition $w = -(\partial^2\zeta/\partial x^2 + \partial^2\zeta/\partial y^2)$, the continuity equation (1) is also satisfied identically. Substituting these definitions into the vorticity equations in the horizontal plane and into the energy equation yields, after some mathematical manipulations, one nonlinear, fourth-order, partial differential equation for ζ. When the problem is reduced to two dimensions there is a relationship between ζ and the stream function, in the form $\psi = \partial\zeta/\partial x$. The equation for ζ in the two-dimensional case considered by Kimura et al. (1986) is presented in the form

$$\frac{\partial(\nabla^2\zeta)}{\partial t} + \frac{\partial^2\zeta}{\partial x\partial z}\nabla^2\left(\frac{\partial\zeta}{\partial x}\right) - \frac{\partial^2\zeta}{\partial x^2}\nabla^2\left(\frac{\partial\zeta}{\partial z}\right) = \nabla^2\zeta + Ra\frac{\partial^2\zeta}{\partial x^2} \tag{43}$$

that needs to be solved subject to the boundary conditions $\partial^2\zeta/\partial x^2 = \partial^2\zeta/\partial z^2 = 0$ at $z = 0,1$ and $\partial^2\zeta/\partial x\partial z = \nabla^2(\partial\zeta/\partial x) = 0$ at $x = 0,1$. Expanding the solution in terms of eigenfunctions obtained from the solution of the corresponding linear problem in the form

$$\zeta(t, x, z) = \sum_{n=1}^{\infty} \sum_{j=0}^{\infty} A_{nj}(t) \sin(n\pi z) \cos(j\pi x) \tag{44}$$

and substituting it into Eq. (43) yields an infinite set of coupled, nonlinear, first-order, ordinary differential equations for the spectral coefficients A_{nj}. These equations are truncated with $n + j \leq N$, where N is a positive integer referred to as the truncation number, and were solved numerically. The main difficulty (as reported by Kimura et al. (1986) in applying the Galerkin method to (43) is the evaluation of convolution products arising from the nonlinear terms; overall computation time is proportional to N^4. The authors report that the pseudo-spectral method (or the collocation method) bypasses the evaluation of the convolution products and is thereby more efficient in computing the nonlinear terms. The results were presented in terms of the power spectrum versus frequencies for different values of N. The highest value of N used was $N = 42$, and different transitions to periodic, multi-periodic, quasi-periodic, and chaos were identified.

The neglect of the time derivative term in Darcy's equation, as well as the significantly higher number of terms used by Kimura et al. (1986) in the spectral representation of the solution, prevents a comparison of these results with the others presented in this chapter that correspond to the weak nonlinear convection regime.

IV. ANALYSIS AND RESULTS

A. Investigation of the Weak Nonlinear Results

As indicated in Section IV.A, the coefficient of the nonlinear term in Eq. (29) plays a role of particular importance as it controls the direction of the Hopf bifurcation which results from the post-transient solution to (29). To observe this point further we follow Vadasz (1999) and represent Eq. (29) for the complex amplitude a as a set of two equations for the absolute value of the amplitude, $r = |a|$, and its phase, θ, in the form: $a = r\exp(i\theta)$, $a^* = r\exp(-i\theta)$, with $aa^* = r^2$. Substituting this representation in (29) yields

$$\beta \frac{\mathrm{d}r}{\mathrm{d}t} = [\varepsilon^2 - \varphi r^2] r \tag{45}$$

$$\frac{\mathrm{d}\theta}{\mathrm{d}t} = m_{21}\varepsilon^2 - m_{31} r^2 \tag{46}$$

where the following notation was introduced to separate the real and imaginary parts of the coefficients in Eq. (29): $h_{21} = h_{21}^{\circ} + im_{21}$, $h_{31} = h_{31}^{\circ} + im_{31}$, and $\beta = 1/h_{21}^{\circ}$, $\varphi = h_{31}^{\circ}/h_{21}^{\circ}$.

Now, it can be observed that when the coefficient of the nonlinear term φ is positive the Hopf bifurcation is forward (i.e., supercritical) whereas a negative value of φ yields an inverse bifurcation (i.e., subcritical). The coefficient φ was evaluated for $\gamma = 0.5$ as a function of α and was presented graphically by Vadasz (1999), where it was observed that its value was negative over the whole domain of validity of the Hopf bifurcation. The relaxation time β was also evaluated by Vadasz (1999) for $\gamma = 0.5$ as a function of α and was shown to be positive for $\alpha \geq 3$. Note that values of $\alpha < 3$ are not consistent with the Hopf bifurcation and with the solutions considered here, as can be observed from Eq. (19) for R_o. For such values of α the solution of this system decays to the stationary points. We can therefore confirm previously suggested results (Sparrow 1982; Wang et al. 1992; Yuen and Bau 1996), related to the Lorenz equations for other applications, that the Hopf bifurcation in this system is indeed subcritical. A further analysis of the periodic solution at slightly subcritical values of R shows that this solution is unstable for $\varepsilon^2 < 0$ (i.e., $R < R_o$). We are therefore faced with a periodic solution which exists only for $\varepsilon^2 < 0$ (i.e., $R < R_o$) but it is not stable in this domain. However, it is at this point that the further investigation of the amplitude equation provides a marked insight into the details of this Hopf bifurcation at the point where the steady convective solutions lose their stability but the resulting periodic solution unfolding from the amplitude equation is unstable for $R < R_o$ and does not exist when $R > R_o$. The post-transient amplitude solution is obtained from Eq. (45) in the form $r^2 = \varepsilon^2/\varphi$, which clearly yields a real value for the amplitude only when $\varepsilon^2 \leq 0$ (i.e., $R \leq R_o$) given the already established fact that $\varphi < 0$. The post-transient frequency correction, $\dot{\theta}$, can be obtained for $R \leq R_o$ by substituting the post-transient solution for r^2 into (46) in the form $\dot{\theta} = (m_{21} - m_{31}/\varphi)\varepsilon^2$.

To gain more insight into the nature of the limit cycle and the subcriticality of the Hopf bifurcation resulting from the amplitude equation, we undertake further investigation of this equation, the major point being the question of where exactly the supercritical solution disappeared, what is the reason for its disappearance, and what more can we learn about the unstable subcritical limit cycle. This is done by working out the transient solution of Eq. (45), which can be easily obtained by a simple integration. Prior to that, it is convenient to introduce the following notation, which simplifies the analysis; $\chi = \varphi/\beta$ and $\xi = \varepsilon^2/\varphi$. Clearly, $\chi < 0$ over all the cases to be considered in this analysis, while $\xi > 0$ for $\varepsilon^2 < 0$ (subcritical conditions), $\xi < 0$ for $\varepsilon > 0$ (supercritical conditions), and $\xi = 0$ for $\varepsilon^2 = 0$ (critical conditions). Using this notation, the amplitude equation (45) takes the form

$$\frac{\mathrm{d}r}{\mathrm{d}t} = \chi[\xi - r^2]r \tag{47}$$

The solution to Eq. (47) is obtained by direct integration in the following form: $r^2 = \xi \exp(2\xi\chi t)/[D + \exp(2\xi\chi t)]$ for $\xi \neq 0$, ($\varepsilon^2 \neq 0$); and $r^2 = 1/[2(\chi t - D)]$ for $\xi = 0$, ($\varepsilon^2 = 0$); where D is a constant of integration to be determined from the initial conditions. Introducing the initial conditions $r = r_o$ at $t = 0$, the transient solutions take the form

$$r^2 = \frac{\xi}{\left[1\left(1 - \frac{\xi}{r_o^2}\right)\exp(-2\xi\chi t)\right]} \qquad \text{for} \qquad \xi \neq 0\ (\varepsilon^2 \neq 0) \tag{48}$$

$$r^2 = \frac{r_o^2}{[1 + 2r_o^2\chi t]} \qquad \text{for} \qquad \xi = 0\ (\varepsilon^2 = 0) \tag{49}$$

Clearly, both solutions (48) and (49) are valid at $t = 0$ and then yield $r^2 = r_o^2$, which can be easily recovered by substituting $t = 0$ in (48) and (49). Therefore the question which arises is what happens at a later time $t > 0$ which causes these solutions to disappear when $\varepsilon^2 > 0$ (i.e., when $\xi < 0$). To answer this question, we separate the discussion into three cases as follows: (1) the supercritical case, $\varepsilon^2 > 0$ ($\xi < 0$); (2) the subcritical case, $\varepsilon^2 < 0$ ($\xi > 0$); and (3) the critical case, $\varepsilon^2 = 0$ ($\xi = 0$). In all three cases it should be remembered that the value of χ is always negative, whereas ξ is negative, zero, or positive, depending on whether the conditions are supercritical, critical, or subcritical, respectively.

1. The Supercritical Case, $\varepsilon^2 > 0$ ($\xi < 0$)

For this case the argument of the exponent in the denominator of Eq. (48) and the value of ξ/r_o^2 are always negative. The denominator has, at $t = 0$, a negative value and this value increases until such time that the denominator vanishes, causing the solution of r^2 to become infinite. To evaluate this critical time t_{cr} when the solution diverges we check the condition when the denominator becomes equal to 0, providing the following result

$$t_{cr} = \frac{1}{2\chi\xi}\ln\left[1 - \frac{\xi}{r_o^2}\right] \tag{50}$$

Obviously, since $\xi < 0$ in this case the argument of the natural logarithm is always greater than 1; therefore, this critical time exists unconditionally in the supercritical regime, causing the amplitude solution to diverge. The significance of this will be discussed below.

2. The Subcritical Case, $\varepsilon^2 < 0$ ($\xi > 0$)

For this case the argument of the exponent in the denominator of Eq. (48) and the value of ξ/r_o^2 are always positive. The denominator has, at $t = 0$, a positive value and this value decreases or increases depending on the value of the ratio ξ/r_o^2. To establish the conditions for the denominator of the solution (48) to become zero and cause the solution to diverge, we evaluate the critical time in a similar way as for the supercritical case, providing an identical result for t_{cr} as presented in Eq. (50). However, this time, under subcritical conditions $\xi > 0$, the coefficient of the natural logarithm is negative and its argument can be greater or less than 1 depending on the value of ξ/r_o^2.

It is easier to notice the behavior of the solution in the subcritical case by presenting the equation for the critical time corresponding to the subcritical case in the following form, where we express explicitly the fact that $\chi < 0$

$$t_{cr} = -\frac{1}{2|\chi|\xi}\ln\left[1 - \frac{\xi}{r_o^2}\right] \tag{51}$$

From this equation it is clear that the condition for existence of a critical time is that the argument of the natural logarithm in Eq. (51) be positive and less than one, i.e., $0 < (1 - \xi/r_o^2) < 1$. Otherwise, either the $\ln(\cdot)$ function does not exist or the critical time becomes negative, having no physical significance. The right-hand side of this inequality yields: $\xi/r_o^2 > 0$, which is unconditionally satisfied under subcritical conditions ($\xi > 0$). The left-hand side provides the following condition for existence of a critical time: $\xi/r_o^2 < 1$. The interesting fact coming out from this result is that a critical time when the limit cycle solution diverges exists in the subcritical case as well, subject to the condition $\xi/r_o^2 < 1$.

3. The Critical Case, $\varepsilon^2 = 0$ ($\xi = 0$)

For the critical case we use Eq. (49) and evaluate the condition for its denominator to vanish in order to establish the existence of a critical time in this case also. One may expect it to exist on continuity arguments as it exists conditionally in the subcritical case and unconditionally in the supercritical case. From Eq. (49) it can be easily evaluated as

$$t_{cr} = -\frac{1}{2\chi r_o^2} \qquad \text{for} \qquad \xi = 0 \tag{52}$$

The same result as presented in Eq. (52) can be obtained by applying the limit as $\xi \to 0$ and using the L'Hôpital rule on Eq. (50) which is valid for both the subcritical and supercritical conditions. This indicates that the critical time t_{cr} as a function of ξ varies smoothly as it passes through the critical point $\xi = 0$ ($\varepsilon^2 = 0$).

The significance of the existence of a critical time when the limit cycle solution diverges is explained simply in terms of the breakdown of the asymptotic expansion which implicitly assumes (a) that the solution is local around *any one (but only one)* of the fixed points, and (b) that the expansion is valid around the critical value of R, i.e., around R_o. The second assumption does not seem to be violated, at least not for the slightly sub-/supercritical case; however, the first assumption is strongly violated by a solution which tends to infinity, starting at subcritical conditions when $\xi/r_o^2 < 1$. This condition implies $r_o^2 > \xi$; i.e., at a given subcritical value of R, as long as the initial conditions for r_o^2 are smaller than ξ ($r_o^2 < \xi$), the solution decays, spiraling towards the corresponding fixed point around which we applied the expansion. When the initial conditions satisfy $r_o^2 = \xi$ a solitary limit cycle solution around this fixed point exists (the terminology "solitary limit cycle" is used to indicate that this limit cycle can be obtained only at $r_o^2 = \xi$). As the initial conditions move away from this fixed point and $r_o^2 > \xi$, the other fixed point may affect the solution as well; however, the asymptotic expansion used does not allow it, and this is why the solution diverges, indicating the breakdown of the expansion used. Therefore the divergence of the solution for $r_o^2 > \xi$ should be interpreted in this light only, i.e., the breakdown of the expansion used. While the divergence of the transient solution as $t \to t_{cr}$ indicates the breakdown of the assumed expansion, it is sensible to suggest a physical interpretation of this result as the tendency of the solution to be repelled away from the neighborhood of the present fixed point ($X_S = Y_S = Z_S = 1$) towards the other fixed point ($X_S = Y_S = -1$; $Z_S = 1$), or alternatively its tendency to orbit around both fixed points. We can imagine a process of gradually increasing the value of R towards R_o (i.e., decreasing the value of ξ towards $\xi = 0$). As we do so and get closer to R_o, a wider range of initial conditions falls into the category which satisfies the solution's divergence condition. To the question of what solution would therefore exist when this condition is fulfilled, one can anticipate (with hindsight of the computational results which are presented in the next section) that the solution may move towards the other fixed point, indicating a homoclinic explosion, or wander around both fixed points, suggesting a chaotic solution. Transforming the condition for this transition to occur, from $r_o^2 > \xi$ to the original physical parameters of the system, by substituting the definition of ξ, one can observe that there is a value of $R \leq R_o$, say R_t, beyond which the transition occurs; this can be expressed in the form

$$R_t = R_o(1 - |\varphi| r_o^2) \tag{53}$$

where the minus sign and the absolute value of φ appear in order to show explicitly that $\varphi < 0$ ($\varphi = -2.4$ in the present case). If $R < R_t$ the solution decays, spiraling towards the corresponding fixed point; at $R = R_t$ we expect the solitary limit cycle solution, and beyond this transitional value of R, i.e., $R > R_t$, the solution moves away from this fixed point either (a) towards the other fixed point, or (b) wanders around both fixed points before it stabilizes towards one of them, or (c) yields a chaotic behavior by being attracted to the nonwandering set (Lorenz attractor). The present expansion cannot provide an answer to select between these three possibilities. However, it is important to stress that for any initial condition r_o^2, which we choose, we can find a value of $R \leq R_o$ which satisfies Eq. (53). At that value of R we expect to obtain a limit cycle solution and beyond it a possible chaotic solution.

To present the analytical solutions graphically, the following rescaled variables relevant to Eqs. (48)–(52) are introduced: $\tilde{r} = r/r_o$, $\tilde{t} = r_o^2|\chi|t$, and $\tilde{\xi} = \xi/r_o^2$. Substituting these rescaled variables into Eqs. (48)–(52) transforms the solutions to the form in which they are plotted in Figures 3 and 4. Figure 3 shows an example of the amplitude solution (48) for three values of $\tilde{\xi}$,

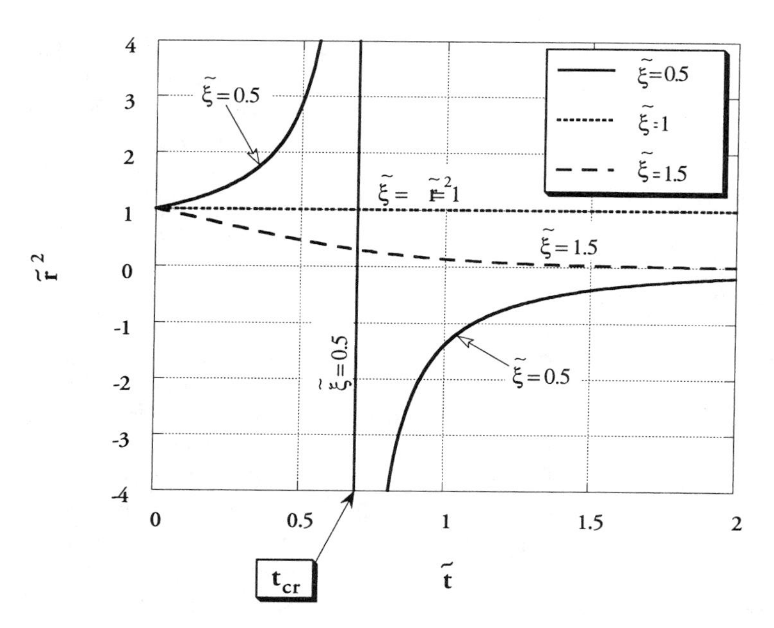

Figure 3. The amplitude solution for three values of $\tilde{\xi}$ corresponding to: (i) sub-transitional conditions, $\tilde{\xi} = 1.5$; (ii) transitional conditions, $\tilde{\xi} = \tilde{r}^2 = 1$; and (iii) super-transitional conditions, $\tilde{\xi} = 0.5$. (Courtesy: Kluwer Academic Publishers.)

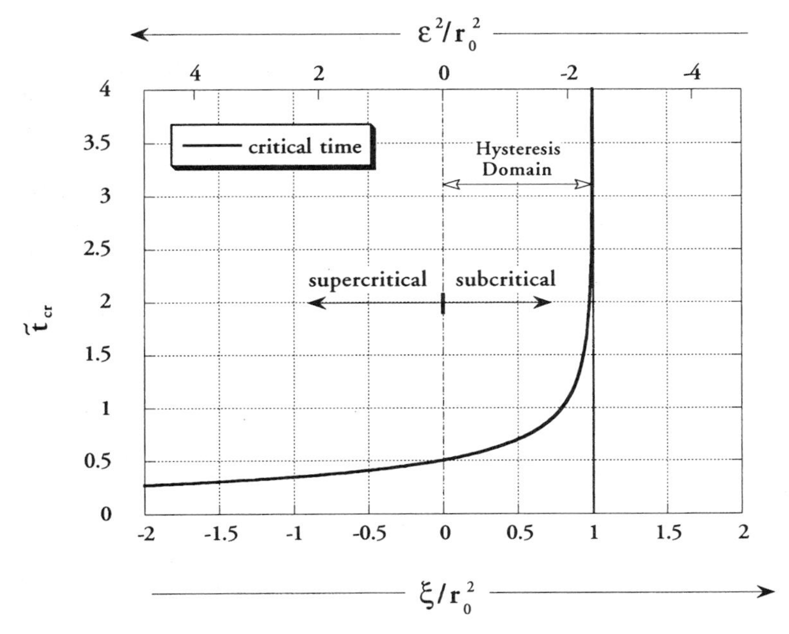

Figure 4. The variation of the critical time $\tilde{t}_{cr}$ when the amplitude solution diverges, as a function of $\tilde{\xi}$. Super- or sub-critical domains as well as the hysteresis domain are identified. (Courtesy: Kluwer Academic Publishers.)

corresponding to sub-transitional conditions (when $\tilde{\xi} > 1$), transitional conditions (when $\tilde{\xi} = \tilde{r}^2 = 1$), and super-transitional conditions (when $\tilde{\xi} < 1$ and the solution diverges at $\tilde{t} = \tilde{t}_{cr}$). The variation of the critical time $\tilde{t}_{cr}$ with $\tilde{\xi}$ is presented in Figure 4, where one observes that the critical time tends to infinity as $\tilde{\xi} \to 1$. This fact suggests that it should not be difficult to obtain numerically the solitary limit cycle solution around $\tilde{\xi} \approx 1$ as the time needed for this solution to be destabilized becomes very large in this neighborhood, at least as long as the present analysis is valid, i.e., for values of r_o^2 not too far away from one of the stationary points ($r_o^2 \ll 1$).

B. Hysteresis

Experimental and numerical results of transitions to chaos in the Lorenz system (Wang et al. 1992; Yuen and Bau 1996; Sparrow 1982) suggest the existence of a hysteresis mechanism which is described as follows: when increasing the value of R gradually by approaching R_o from below, the transition to chaos occurs at $R = R_o$; while repeating the same procedure, but approaching R_o from above, the transition from chaos to the stationary solution occurs at a value of $R < R_o$. Vadasz (1999) provides an explanation of the hysteresis phenomenon in connection with the transitional value of R which is presented in Eq. (53). When approaching R_o from below, say $R < R_t$, the initial conditions lead the solution to one of the convective fixed points, i.e., $r = 0$ (the fixed points represent the steady solutions of convective rolls, moving clockwise or counterclockwise). As we gradually increase the value of R by staring the next experiment (or numerical procedure) with initial conditions taken from the post-transient previous solution obtained at the slightly lower value of R, the new initial conditions are very close to the fixed point, i.e., $r_o^2 \approx 0$ (they are not exactly at the fixed point because the post-transient values of the previous solution are reached asymptotically, and at any finite time there is a slight departure between the solution and the steady state), and, according to Eq. (53), the corresponding transitional value R_t is very close to R_o. However, when one approaches R_o from above, the initial conditions taken from the previous solution at a value of $R > R_o$ are quite large and far away from the fixed point. Therefore, in such a case one may expect to obtain a chaotic solution for subcritical values of R until the value of R_t is reached from above, which this time could be quite far away from R_o. In graphical terms, this process can be observed on Figure 4 where, by moving towards R_o form the left with not negligible initial conditions, the transition is expected when the critical time disappears, i.e., at values of $R < R_o$, whereas by moving towards R_o from the right and gradually decreasing the value of ξ and consequently of r_o^2 (because of the gradual process), one could keep the ratio ξ / r_o^2 greater than one, as both $\xi \to 0$ and $r_o^2 \to 0$ simultaneously. To make this explanation more transparent, the variation of the explicit critical time as a function of $(R - R_o)$, using Eq. (50) and the definitions of ξ and ε^2, is presented in Figure 5 for different values of r_o^2. The disappearance of the critical time at $R = R_t$ is an indication that the amplitude does not diverge, and therefore a steady convective solution can be obtained. From the figure it is evident that for small values of r_o^2 (e.g., $r_o^2 = 0.001$), corresponding to the forward transition from steady convection to chaos, the asymptote of t_{cr} (i.e., the point when the critical time disappears) occurs very close to $R = R_o$, whereas for values of r_o^2 which are not so small, corresponding to the reverse transition from chaos to steady convection (e.g., $r_o^2 = 0.5$), the asymptote of t_{cr} occurs at values of $R < R_o$ which are quite far away from R_o. This explains the reason for observing hysteresis in the transition from steady convection to chaos and backwards, using initial conditions corresponding to a previous solution at a slightly different value of R.

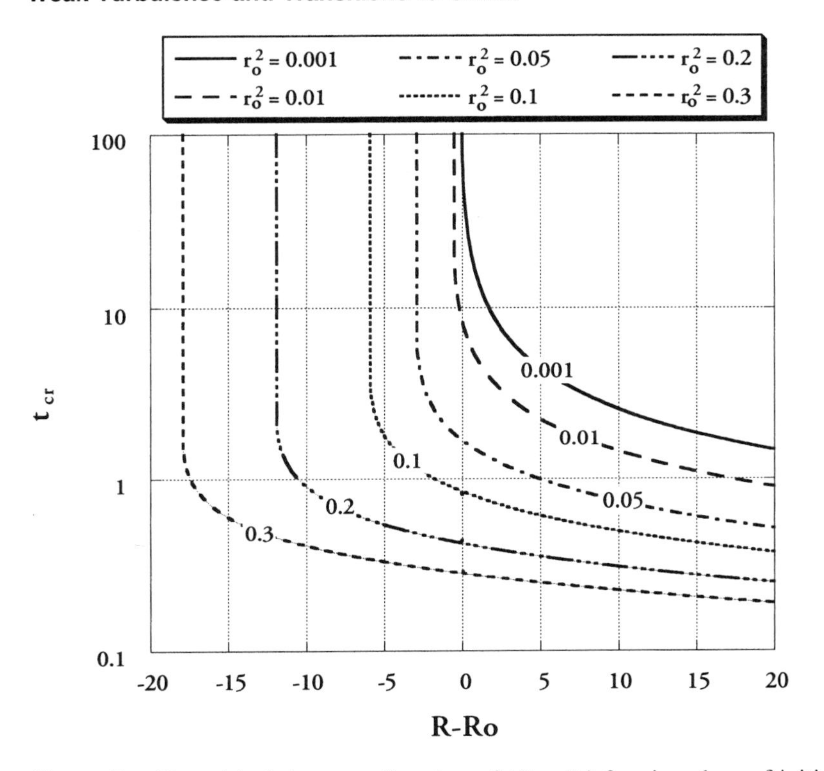

Figure 5. The critical time as a function of $(R - R_o)$ for six values of initial conditions in terms of r_o^2. The transition from steady convection to chaos (or backwards) is linked to the existence (disappearance) of this critical time, explaining the mechanism for hysteresis. (Courtesy: Kluwer Academic Publishers.)

C. Computational Results and Poincaré Maps

Adomian's decomposition method of solution was applied to obtain the sets of results presented in this section. They follow Vadasz and Olek (1999b,c), using the method developed by Vadasz and Olek (1998) for the solutions of the corresponding problem of convection in a rotating porous layer. All solutions in Subsections IV.C.1 and IV.C.2 were obtained by using the same initial conditions, which were selected to be in the neighborhood of the positive convective fixed point. As such, the common initial conditions are at $t = 0$: $X = Y = Z = 0.9$. Different initial conditions were used in Subsection IV.C.3. The value of γ used in all computations was consistent with the critical wavenumber at the marginal stability in porous media convection, i.e., $\gamma = 0.5$. All computations were carried out up to a value of maximum time $\tau_{\max} = 210$. The time step used for the computational procedure described in Section IV.B was constant for all cases and set to be $\Delta\tau = 10^{-3}$, while 15 terms in the series were used throughout. In what follows, the reference to projections on $Y - X$, $Z - X$, and $Z - Y$ planes corresponds to the planes $Z = 0$, $Y = 0$, and $X = 0$, respectively. In all the results presented *the data points are not connected*, unless indicated otherwise.

1. Small Prandtl Numbers

For small Prandtl number convection we considered a value of $\alpha = 5$ which is consistent with a Darcy–Prandtl number of $Pr_D \cong 98.7$. The solution data points were post-processed for graphical representation of the results in the form of Poincaré maps and state space projections of trajectories onto the Y–X, Z–X, and Z–Y planes.

The sequence of events leading to the loss of stability of the steady convection regime is presented in Figure 6 in terms of projections of trajectories (including the transient solution) onto the Y–X plane. In Figure 6a we can observe that for a Rayleigh number slightly above the loss of stability of the motionless solution ($R = 1.1$) the trajectory moves to the steady convection point $X = Y = Z = 1$ on a straight line. At $R = 5$, beyond the point where the eigenvalues become a complex conjugate pair (i.e., $R > 1.28$), the trajectories approach the fixed point on a spiral (Figure 6b). The spiraling approach towards the fixed point persists, increasing the rate of motion in the angular direction with increasing value of R, as can be observed in Figures 6c, d, and e. In Figures 6e, f we observe a white hole around the convection fixed point, indicating that the maximum time allocated for the computation was not sufficient for the trajectory to reach the fixed point. It is evident that it moves towards the fixed point by observing the behavior in the time domain (not shown here). This result is the first evidence of approaching the transition point. At a value of $R = 24.647\,752$ we obtain a solitary limit cycle, signifying the loss of stability of the steady convective fixed points. The post-transient results corresponding to this solitary limit cycle are presented in Figure 7 as projections of the trajectory's data points on the Y–X, Z–X, and Z–Y planes (Figures 7a, b, and c, respectively) and the corresponding post-transient Poincaré section (Figure 7d). We notice that this transition occurs at a subcritical value of R ($R_{c2} = 25.0$). From the Poincaré section (Figure 7d) we can see that the periodic regime associated with this limit cycle has one period only. As soon as the value of R is slightly higher than this transitional value of $R_t = 24.647\,752$, i.e., for $R > R_t$, but lower than R_{c2}, we observe that a homoclinic explosion has occurred and a chaotic regime takes over. This is shown in Figure 8 for

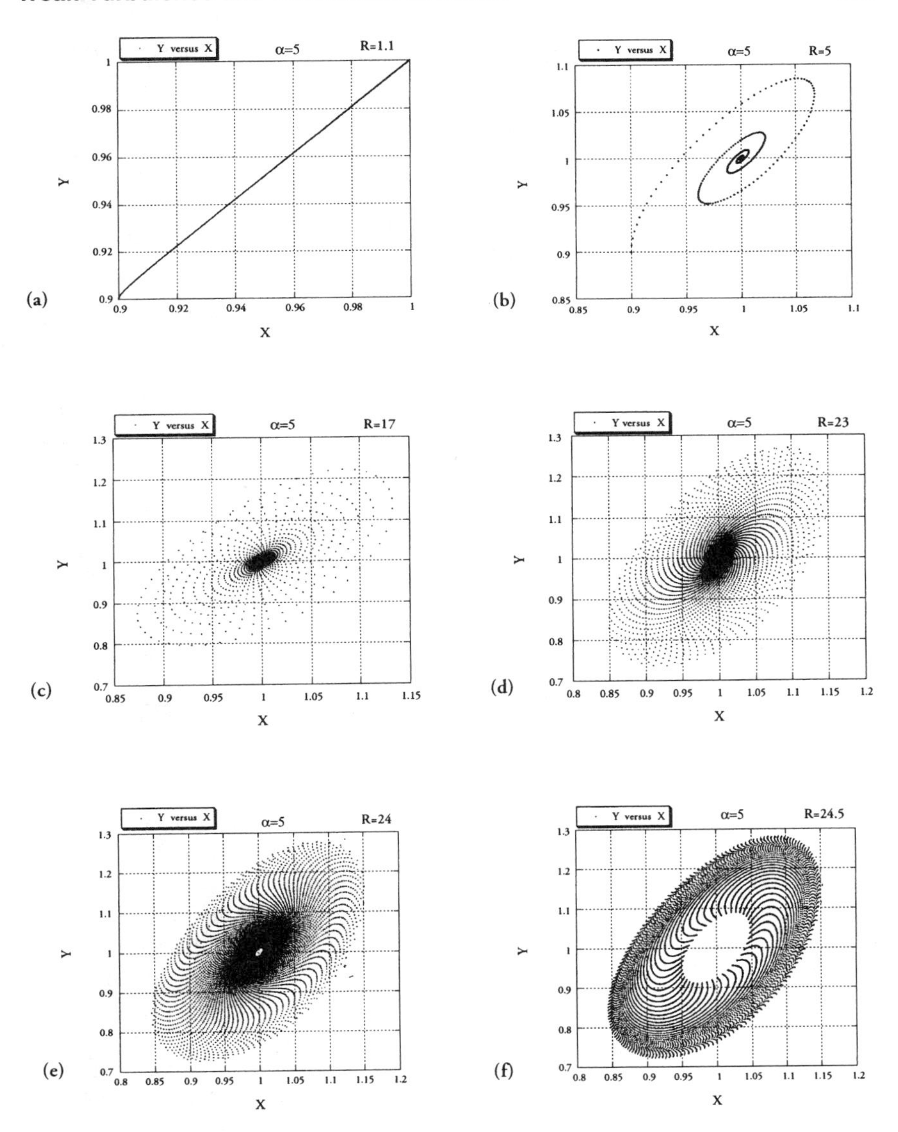

Figure 6. The evolution of trajectories over time for $\alpha = 5$ ($Pr_D = 98.7$), and for increasing values of Rayleigh number (in terms of R) corresponding to $R \geq 1.1$ and $R \leq 57.53$. The graphs represent the projection of the solution data points (not connected) onto the Z–X plane. (Courtesy: Kluwer Academic Publishers.)

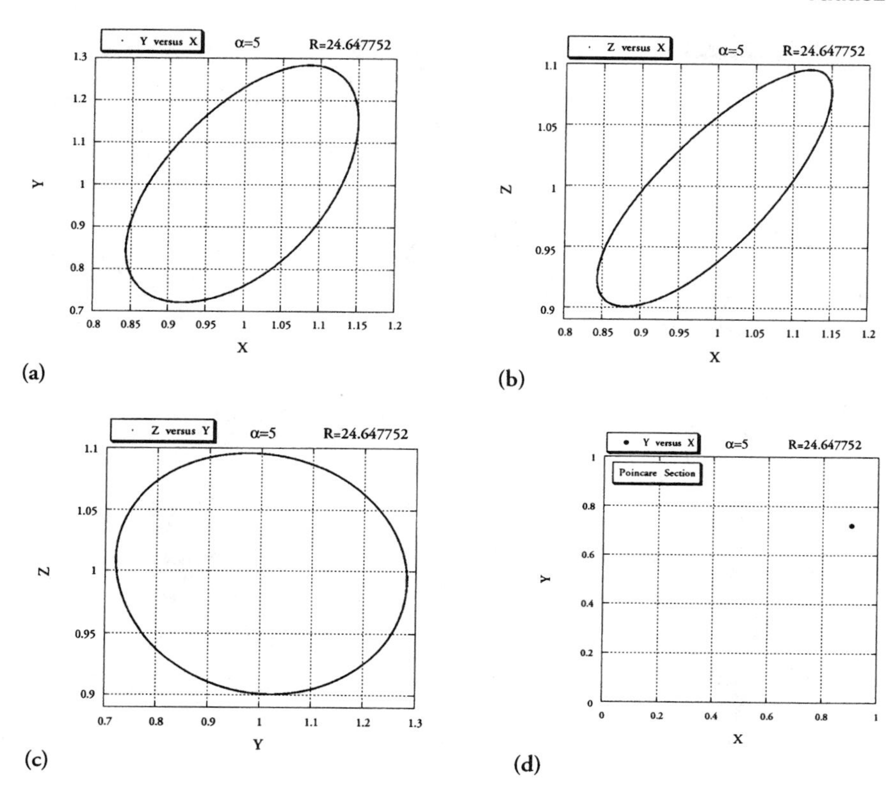

Figure 7. The evolution of trajectories over time for $\alpha = 5$ ($Pr_D = 98.7$), and for a value of Rayleigh number (in terms of R) associated with the solitary limit cycle at the transition to a nonperiodic regime and corresponding to $R = 24.647\,752$. The graphs (a)–(c) represent the projection of the solution data points (not connected) onto the (a) Y–X plane, (b) Z–X plane, and (c) Z–Y plane. (d) Poincaré section at $Z = 1$. (Courtesy: Kluwer Academic Publishers.)

$R = 24.9 < R_{c2}$, where the strange attractor can be identified on the Poincaré section in Figure 8d. The results for the same values of R in the time domain are presented in Figure 8 for X as a function of t. The decay of the solution corresponding to $R = 23$ towards the steady-state value of $X = 1$ is clearly identified in Figure 9a, and the inset presented in Figure 9b highlights its oscillatory behavior. On the other hand, for $R = 24.9$, Figure 9c shows a typical chaotic result and the inset presented in Figure 9d focuses on the post-transient time domain $170 < t < 210$. It is worth re-emphasizing the fact that the computational results show a transition to

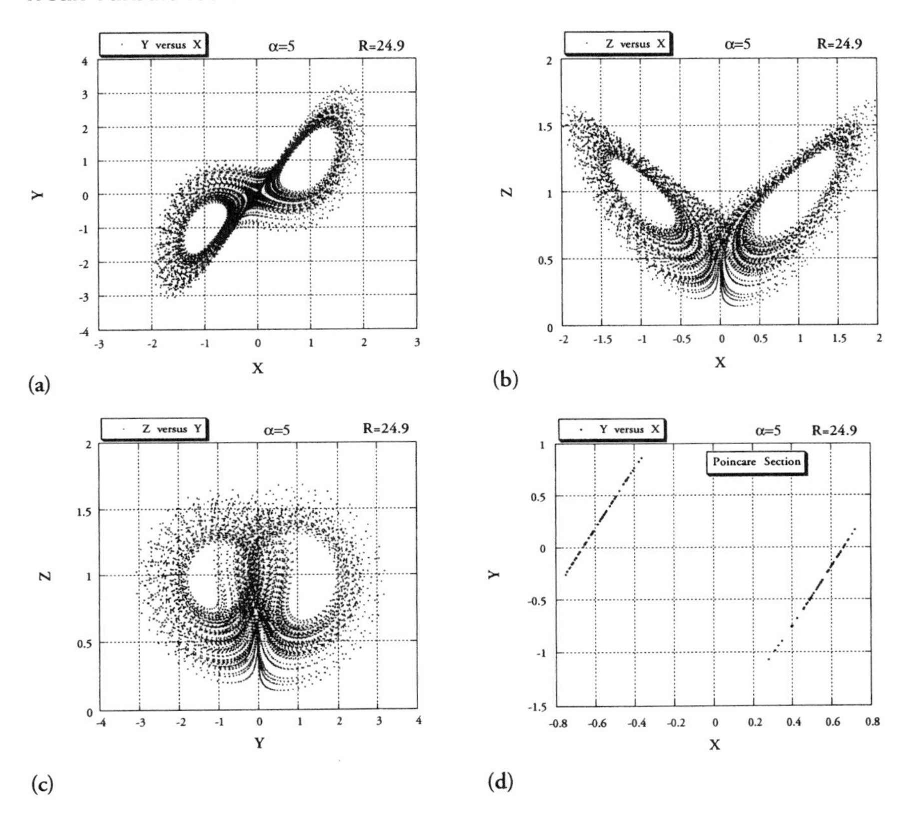

Figure 8. The evolution of trajectories over time for $\alpha = 5$ ($Pr_D = 98.7$), and for a value of Rayleigh number (in terms of R) just beyond the transition to a nonperiodic regime corresponding to $R = 24.9$. The graphs (a)–(c) represent the projection of the solution data points (not connected) onto the (a) Y–X plane, (b) Z–X plane, and (c) Z–Y plane. (d) Poincaré section at $Z = 1$. (Courtesy: Kluwer Academic Publishers.)

chaos at a subcritical value of R (the critical value is $R_o = 25$). A comparison between Figures 9a and 9c at a common transient time domain $0 < t < 50$ shows that the envelope of the function $X(t)$ converges for $R = 23$ (Figure 9a) and diverges for $R = 24.9$ (Figure 9c). This suggests that somewhere in between $R = 23$ and $R = 24.9$ the envelope of the function $X(t)$ will neither converge nor diverge, producing a typical limit cycle. Looking for this limit cycle provides the result presented in Figure 9e, where it is evident that the envelope of the function $X(t)$ does not converge nor diverge, and the inset presented in Figure 9f (where the data points are connected)

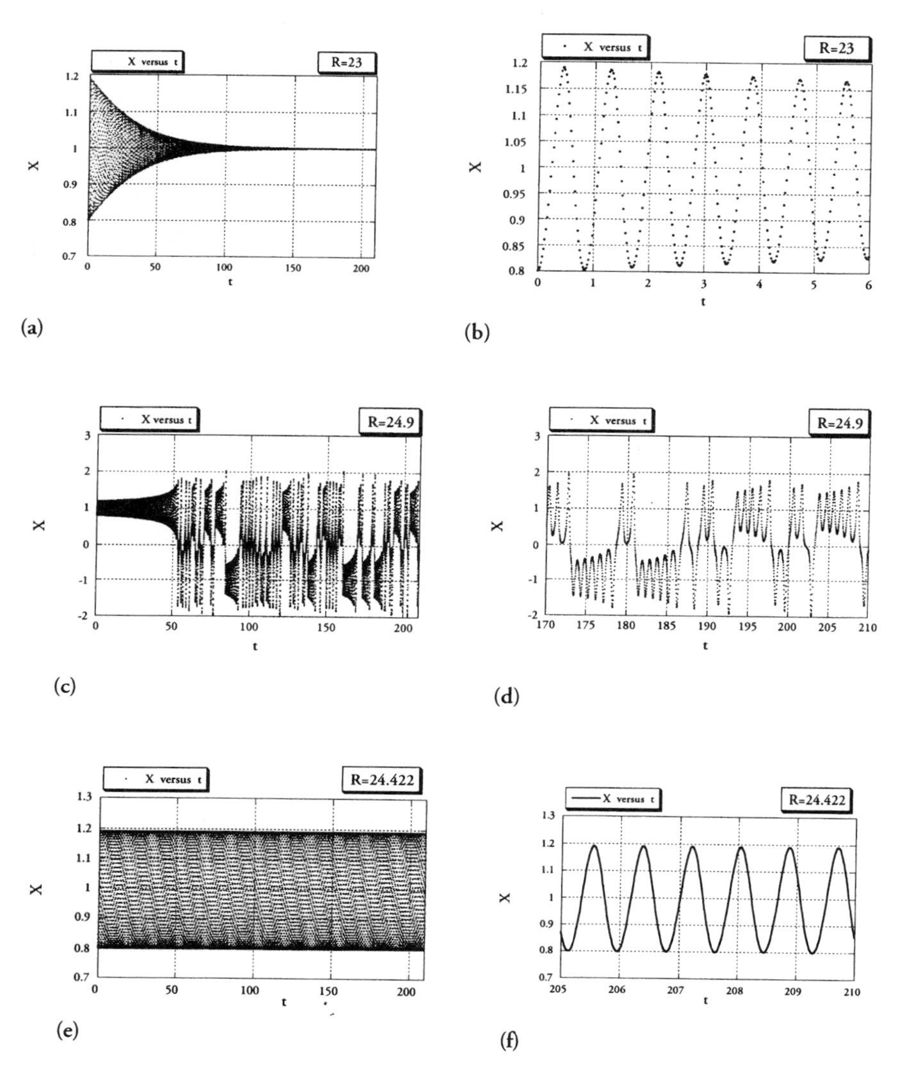

Figure 9. The computational results for the evolution of $X(t)$ in the time domain for $\alpha = 5$ ($Pr_D = 98.7$), and for three values of Rayleigh number (in terms of R): (a) X as a function of time for $R = 23$; the solution stabilizes to the fixed point; (b) the inset of Figure 9a detailing the oscillatory decay of the solution; (c) X as a function of time for $R = 24.9$; the solution exhibits chaotic behavior; (d) the inset of Figure 9c detailing the chaotic solution; (e) X as a function of time for $R = 24.422$; the solution is periodic; (f) the inset of Figure 9e detailing the periodic solution (data points are connected). (Courtesy: Kluwer Academic Publishers.)

demonstrates the periodic behavior of the solution. The chaotic regime persists up to $R \sim 74.5$, where two hooks appear in the Poincaré map (not shown here). Typically, appearance of hooks in return maps is associated with homoclinic explosions giving birth to a stable periodic orbit which should be detected as a slightly higher value of R. Sparrow (1982) describes this process in detail for the solution of the Lorenz equations in pure fluids (nonporous domains). He explains that the contraction of nearby trajectories on the Poincaré map which is associated with the bend in the hook may lead to one (or more) of the unstable periodic orbits becoming stable. While Sparrow (1982) had difficulty in detecting the hook in numerical experiments and needed to introduce a "tent-type" of return map, our results clearly indicate that for convection in porous media the hooks are easily detectable on the Poincaré section at $Z = 1$ (to be shown later in Figure 11). Figure 10 shows that, as indeed expected, a period-8 periodic solution is obtained at $R = 75$. At $R = 77$ the solution is chaotic again, as presented in Figure 11, and the two hooks on the Poincaré map indicate another periodic window, which is to be expected. The results of the projection of trajectories data points onto the Y–X, Z–X, and Z–Y planes and their corresponding Poincaré section for $R = 86$ are presented in Figure 12 and confirm the expected periodic window, which is identified with a period-4 periodic solution. Increasing the Rayleigh number further shows a return to a chaotic solution, as presented in Figure 13 for $R = 100$ where the hooks still appear in the corresponding Poincaré section (Figure 13d). The transition to a high Rayleigh number periodic regime occurs at $R \approx 105$. The results corresponding to this regime are presented in Figure 14 for $R = 150$, where a period-2 periodic solution is identified on the Poincaré section in Figure 14d. The solution at higher values of R continues to belong to a period-2 periodic type. We can therefore conclude the observation around these regimes of periodic windows within the broad band of chaotic solutions by pointing out a sequence of period halving as one increases the Rayleigh number. Originally, a period-8 periodic solution was obtained at $R = 75$, which became of period-4 at $R = 88$ and eventually period-2 at $R = 109$. It can be suggested, therefore, that the transition from chaos to a high Rayleigh number periodic convection occurs via a cascade of period-halving bifurcations. This result is consistent with a similar route from chaos to a periodic regime in the corresponding problem in pure fluids (nonporous domains), as presented by Sparrow (1982).

2. Moderate Prandtl Numbers

For moderate Prandtl number convection we considered a value of $\alpha = 50$, which is consistent with a Darcy–Prandtl number of $Pr_D \cong 987$. Prior to

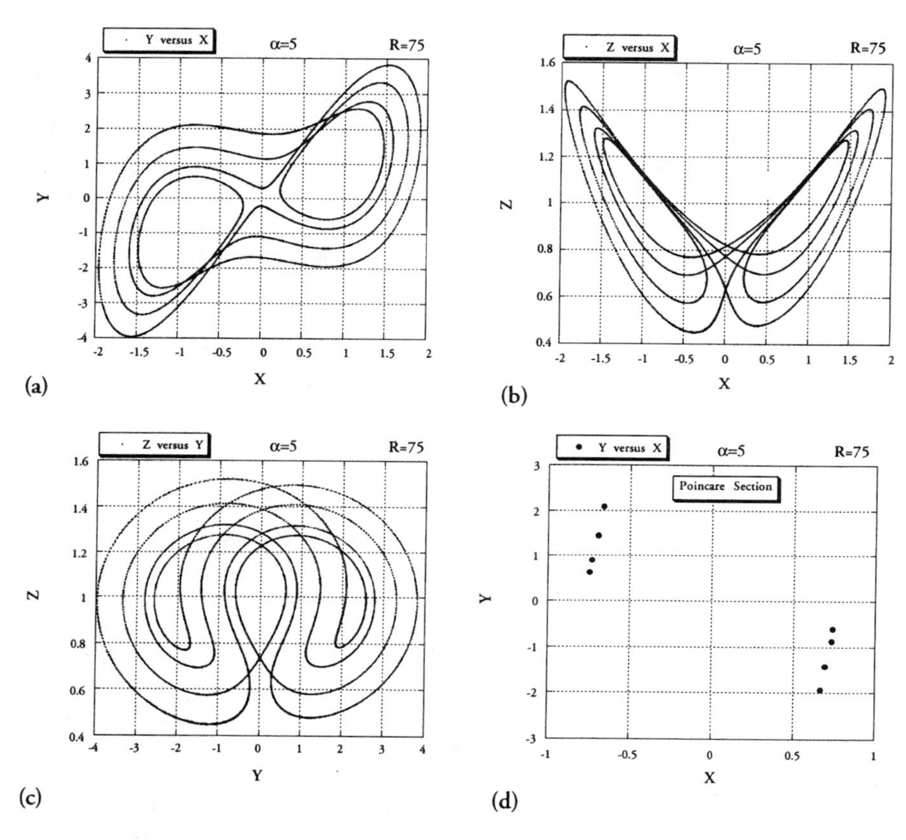

Figure 10. The evolution of trajectories over time for $\alpha = 5$ ($Pr_D = 98.7$), and for a value of Rayleigh number (in terms of R) corresponding to a period-8 periodic window at $R = 75$. The graphs (a)–(c) represent the projection of the solution data points (not connected) onto the (a) Y–X plane, (b) Z–X plane, and (c) Z–Y plane. (d) Poincaré at $Z = 1$. (Courtesy: Kluwer Academic Publishers.)

presenting the results, we would like to consider two particular limit cases which will be referred to during the results presentation. First, it is evident that for large values of α, leading to the motivation for ignoring the time derivative term in Eq. (14) (which is equivalent to neglecting this term in the original Darcy's equation (2)), the solution lies on the plane $Y = X$ for all values of t. We will observe the behavior of the computational results relative to this plane. Second, although we used the same initial conditions (in the neighborhood of the positive convective fixed point) for all computations, we would like to stress that computing the results with initial condi-

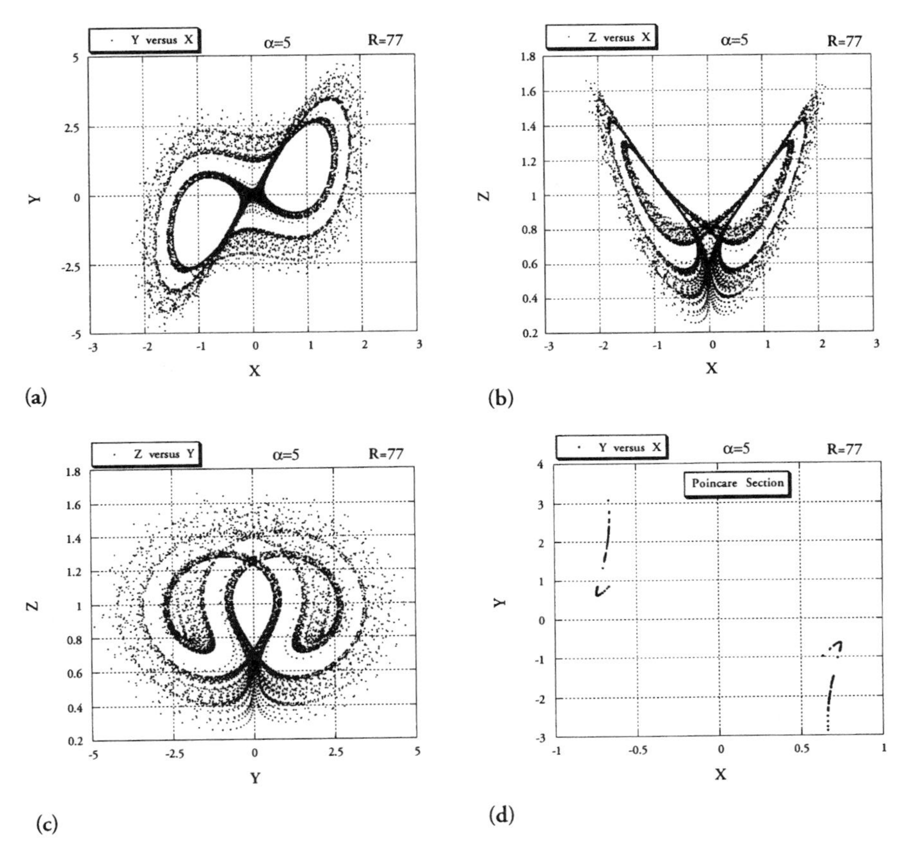

Figure 11. The evolution of trajectories over time for $\alpha = 5$ ($Pr_D = 98.7$), and for a value of Rayleigh number (in terms of R) corresponding to $R = 77$. The graphs (a)–(c) represent the projection of the solution data points (not connected) onto the (a) Y–X plane, (b) Z–X plane, and (c) Z–Y plane. (d) Poincaré section at $Z = 1$. (Courtesy: Kluwer Academic Publishers.)

tions located on the Z-axis (e.g., $X(0) = Y(0) = 0$ and $Z(0) = 0.5$) leads the solution towards the origin ($X = Y = Z = 0$), producing the motionless solution. The reason for this is the fact that the Z-axis lies on the stable manifold of the origin. We will see that this fact is relevant to some of the results.

The sequence of events leading to the loss of stability of the steady convection regime is presented in Figures 15 and 16 in terms of projections of trajectories (including the transient solution) onto the Z–X and Y–X planes, respectively. In Figure 15a we can observe that for a Rayleigh num-

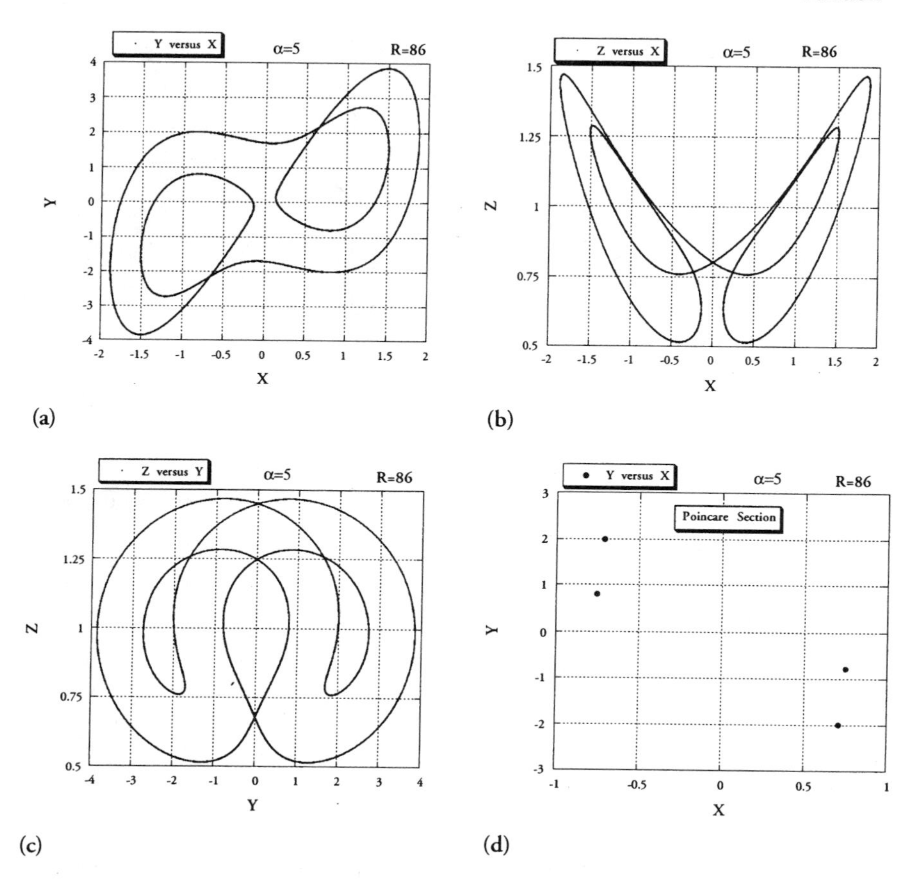

Figure 12. The evolution of trajectories over time for $\alpha = 5$ ($Pr_D = 98.7$), and for a value of Rayleigh number (in terms of R) corresponding to a period-4 periodic window at $R = 86$. The graphs (a)–(c) represent the projection of the solution data points (not connected) onto the (a) Y–X plane, (b) Z–X plane, and (c) Z–Y plane. (d) Poincaré section at $Z = 1$. (Courtesy: Kluwer Academic Publishers.)

ber slightly above the loss of stability of the motionless solution ($R = 1.1$) the trajectory moves to the steady convection point $X = Y = Z = 1$ on a straight line (except for a slight initial overshooting). At $R = 30$, beyond the point where the eigenvalues became a complex conjugate pair (i.e., $R > 1.26$), the trajectories approach the fixed point on a spiral (Figure 15b). The spiraling approach towards the fixed point persists, increasing the rate of motion in the angular direction with increasing value of R, as can be observed in Figure 15c, d, and e. In Figure 15f we observe that the

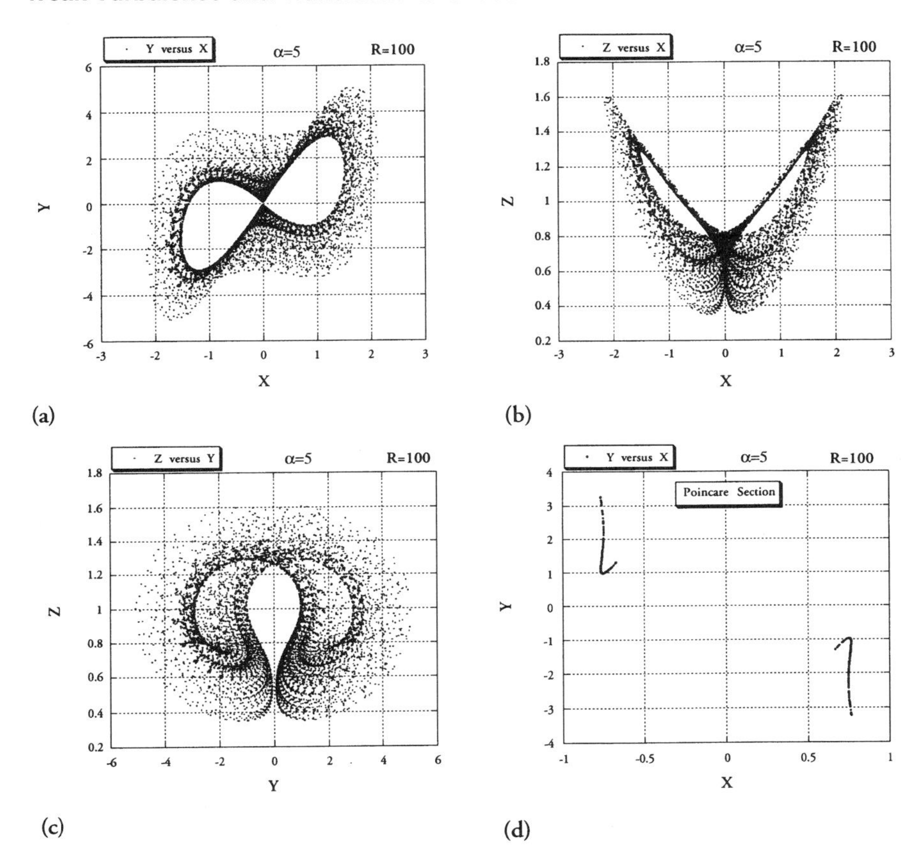

Figure 13. The evolution of trajectories over time for $\alpha = 5$ ($Pr_D = 98.7$), and for a value of Rayleigh number (in terms of R) corresponding to $R = 100$. The graphs (a)–(c) represent the projection of the solution data points (not connected) onto the (a) Y–X plane, (b) Z–X plane, and (c) Z–Y plane. (d) Poincaré section at $Z = 1$. (Courtesy: Kluwer Academic Publishers.)

maximum time allocated for the computation was not sufficient for the trajectory to reach the fixed point. It is evident that it moves towards the fixed point by observing the behavior in the time domain (not shown here). This result is the first evidence of approaching the transition point. We can observe the Prandtl number effect on the solution in Figure 16. The relatively high value of α corresponding to a Darcy–Prandtl number of $Pr_D \sim 1000$ suggested that the solution should lie on the plane $Y = X$. We can observe the projection of the solution data points on the Y–X plane (i.e., the

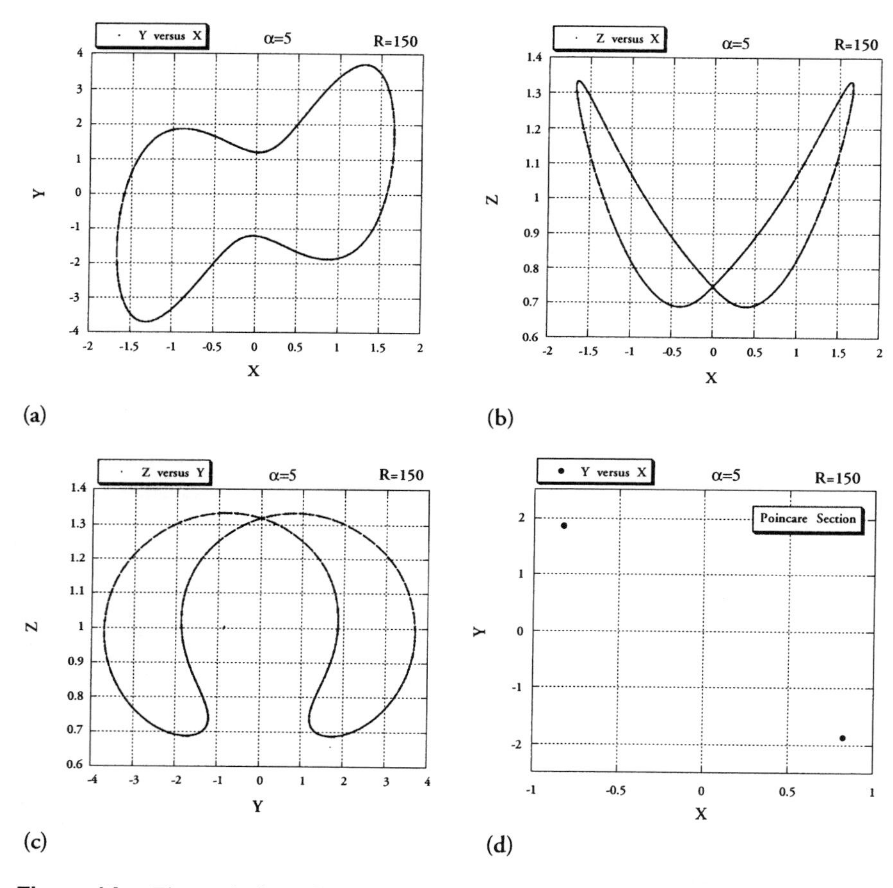

Figure 14. The evolution of trajectories over time for $\alpha = 5$ ($Pr_D = 98.7$), and for a value of Rayleigh number (in terms of R) corresponding to a period-2 periodic regime at $R = 150$. The graphs (a)–(c) represent the projection of the solution data points (not connected) onto the (a) Y–X plane, (b) Z–X plane, and (c) Z–Y plane. (d) Poincaré section at $Z = 1$. (Courtesy: Kluwer Academic Publishers.)

plane corresponding to $Z = 0$) in Figure 16. Initially, for $R = 1.1$ (Figure 16a), the solution indeed lies on the plane $Y = X$ as it moves towards the steady fixed point. However, as the value of R increases the solution departs from this plane because of the oscillatory fashion (spiraling due to the complex pair of eigenvalues) in which the trajectory approaches the fixed point. The $Y = X$ plane is represented by a straight line on this plane. As the value of R approaches the point of transition (i.e., around $R \approx 57$), the

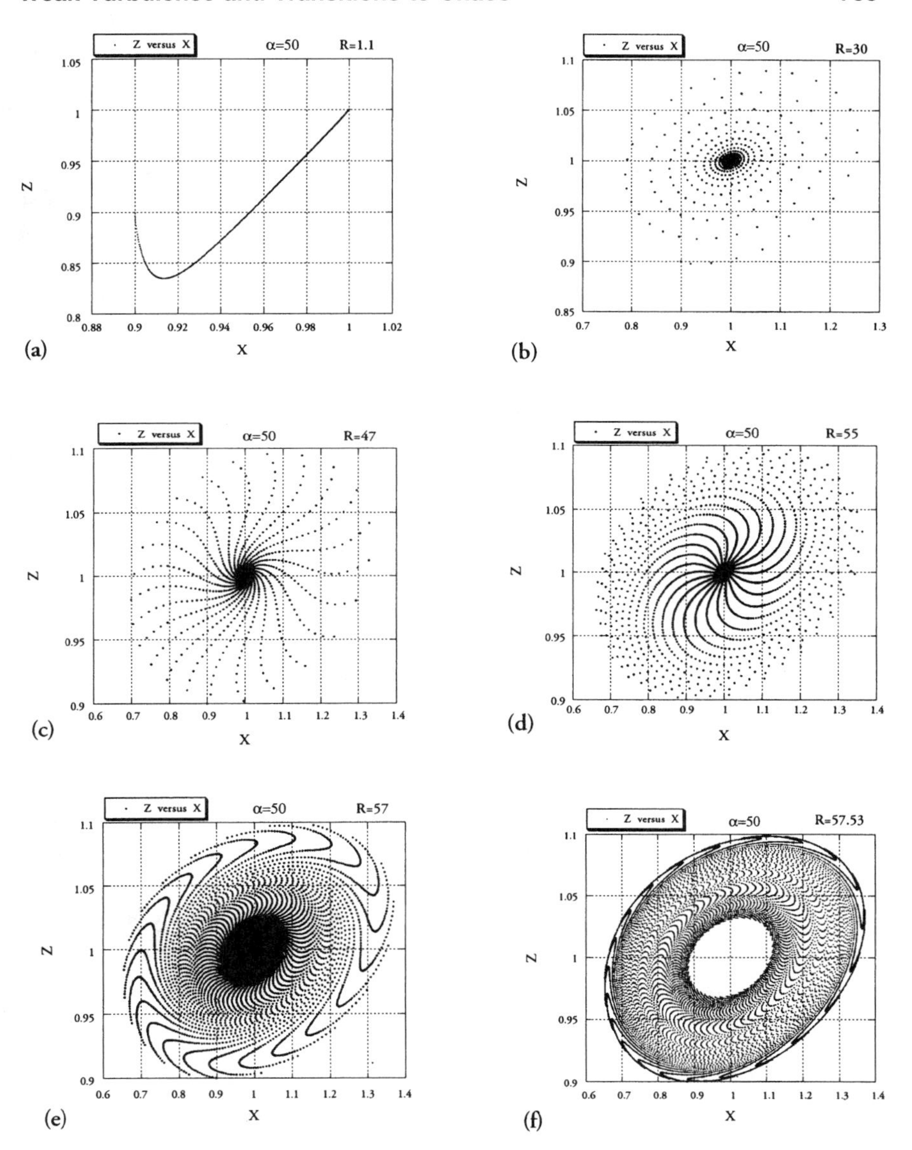

Figure 15. The evolution of trajectories over time for $\alpha = 50$ ($Pr_D = 987$), and for increasing values of Rayleigh number (in terms of R) corresponding to $R \geq 1.1$ and $R \leq 57.73$. The graphs represent the projection of the solution data points (not connected) onto the Z–X plane. (Courtesy: Kluwer Academic Publishers.)

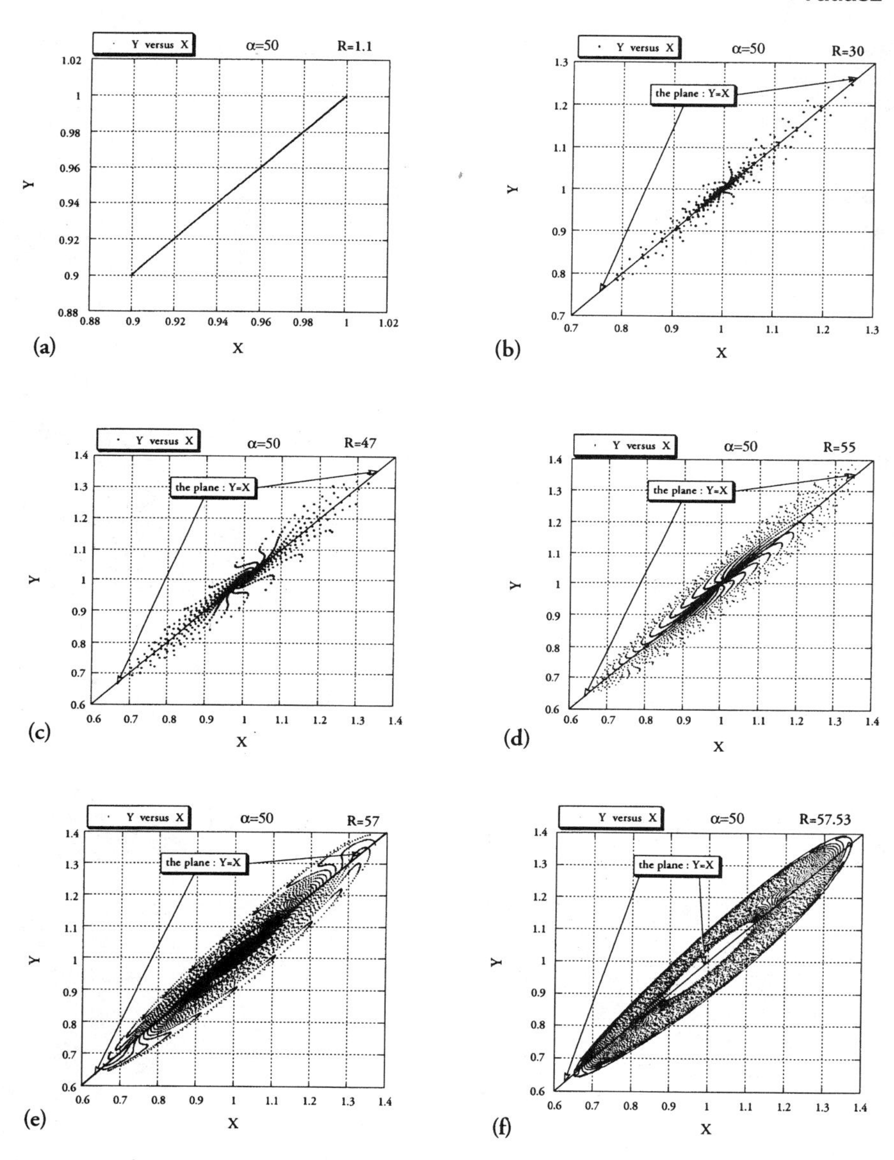

Figure 16. The evolution of trajectories over time for $\alpha = 50$ ($Pr_D = 987$), and for increasing values of Rayleigh number (in terms of R) corresponding to $R \geq 1.1$ and $R \leq 57.53$. The graphs represent the projection of the solution data points (not connected) onto the Y–X plane. The deviation of the solution from the plane $Y = X$ as the value of R increases is evident. (Courtesy: Kluwer Academic Publishers.)

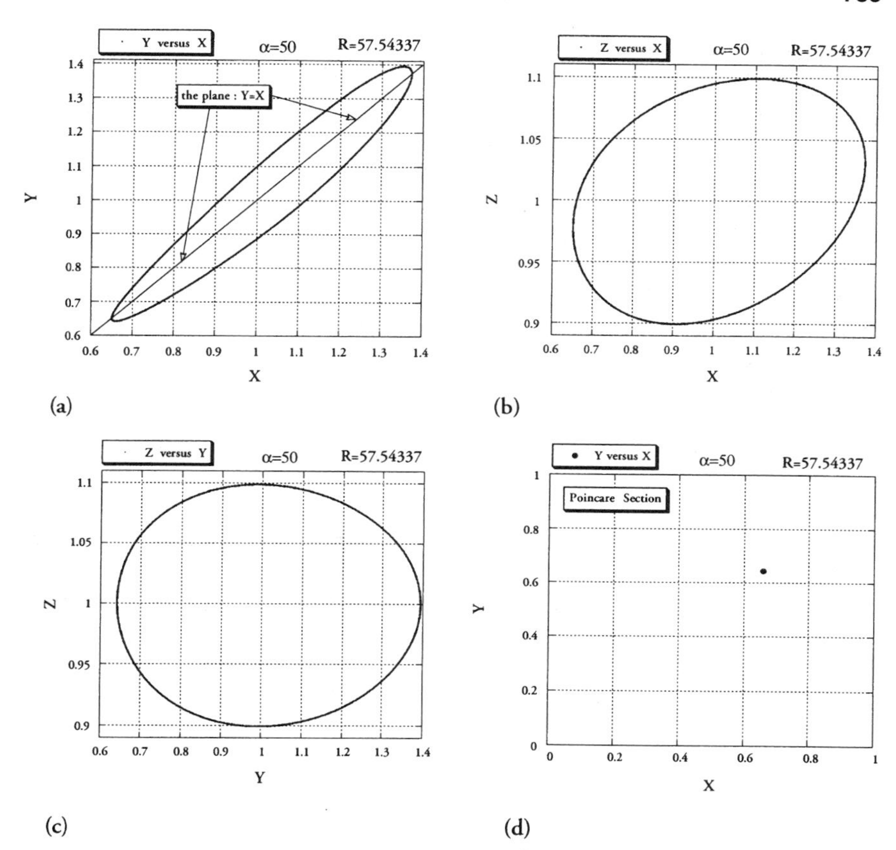

Figure 17. The evolution of trajectories over time for $\alpha = 50$ ($Pr_D = 987$), and for a value of Rayleigh number (in terms of R) associated with the solitary limit cycle at the first transition from steady convection, and corresponding to $R = 57.543\,37$. The graphs (a)–(c) represent the projection of the solution data points (not connected) onto the (a) Y–X plane, (b) Z–X plane, and (c) Z–Y plane. (d) Poincaré section at $Z = 1$. (Courtesy: Kluwer Academic Publishers.)

departure from the $Y = X$ plane is quite significant, as can be observed in Figure 16e, f. At a value of $R = 57.543\,37$ we obtain a solitary limit cycle signifying the loss of stability of the steady convective fixed points. The post-transient results corresponding to this solitary limit cycle are presented in Figure 17 as projections of the trajectory's data points on the Y–X, Z–X, and Z–Y planes (Figures 17a, b, and c, respectively) and the corresponding post-transient Poincaré section (Figure 17d). First, we notice that this transition occurs at a subcritical value of R ($R_{c2} = 58.51$). One can observe also in Figure 17a that the post-transient solution is already quite distant from the plane $Y = X$. From the Poincaré section (Figure 17d) we can see that the periodic regime associated with this limit cycle has one period only. As soon as the value of R is slightly higher than this transitional value $R_t = 57.543\,37$, i.e., for $R > R_t$, we observe that a homoclinic explosion has occurred, introducing a second period to the solution. The sequence of events following this first transition is presented in Figure 18 as projections of the trajectory's data points on the Z–X plane and its corresponding Poincaré section for three different values of R. At $R = 100$ the Poincaré section (Figure 18b) shows the period-2 periodic regime which took over beyond $R = R_t$. The results for the other two values of R, presented in Figure 18c, d and Figure 18e, f, show clearly a sequence of period doubling, i.e., a period-4 periodic regime at $R = 275$ and a period-8 periodic regime at $R = 285$. Two of the frequencies in the period-8 regime (Figure 18e, f) are very close to each other. Following this period-doubling cascade, at a value of R above ~ 290 we obtained the first transition to chaos which is presented in Figure 19a, b. We can observe that the return map on the Poincaré section in Figure 19b shows a hook at the lower left corner of the section. As previously mentioned, the appearance of hooks in return maps is associated with homoclinic explosions giving birth to a stable periodic orbit, which should be detected at a slightly higher value of R. Figure 19c, d shows that, as indeed expected, a new period-6 periodic solution is obtained at $R = 300$. This corresponds to the first periodic window within the chaotic regime. Above $R \sim 300$ the solution becomes again chaotic, as presented in Figure 20 for $R = 315$; however, it is evident that the strange attractor has changed its shape. Another important observation can be made by noticing the Poincaré section on Figure 20d. A second hook appeared in the return map, providing a hint regarding another periodic window which is expected. This second hook appeared first are $R \sim 310$ (not shown here). Indeed so, around $R \sim 338$ another periodic window appears, as observed in Figure 21, corresponding to $R = 338$, where it is evident that this is another period-6 periodic solution.

Further increasing the Rayleigh number shows a repetition of the same scenario of appearance of periodic windows, although symmetry-breaking phenomena occur as well. A periodic window associated with the value of R at $R = 374.7$ identifies a period-3 periodic solution. At $R = 399$ we obtained another period-6 periodic window, as presented in Figure 22. An additional period-3 solution is identified at $R = 437$ (not shown graphically). An interesting result appears in Figure 23, which shows a chaotic solution at $R = 452.8$. We can observe on Figure 23a that some point on the Z-axis, which belong to the stable manifold of the origin at the first transition,

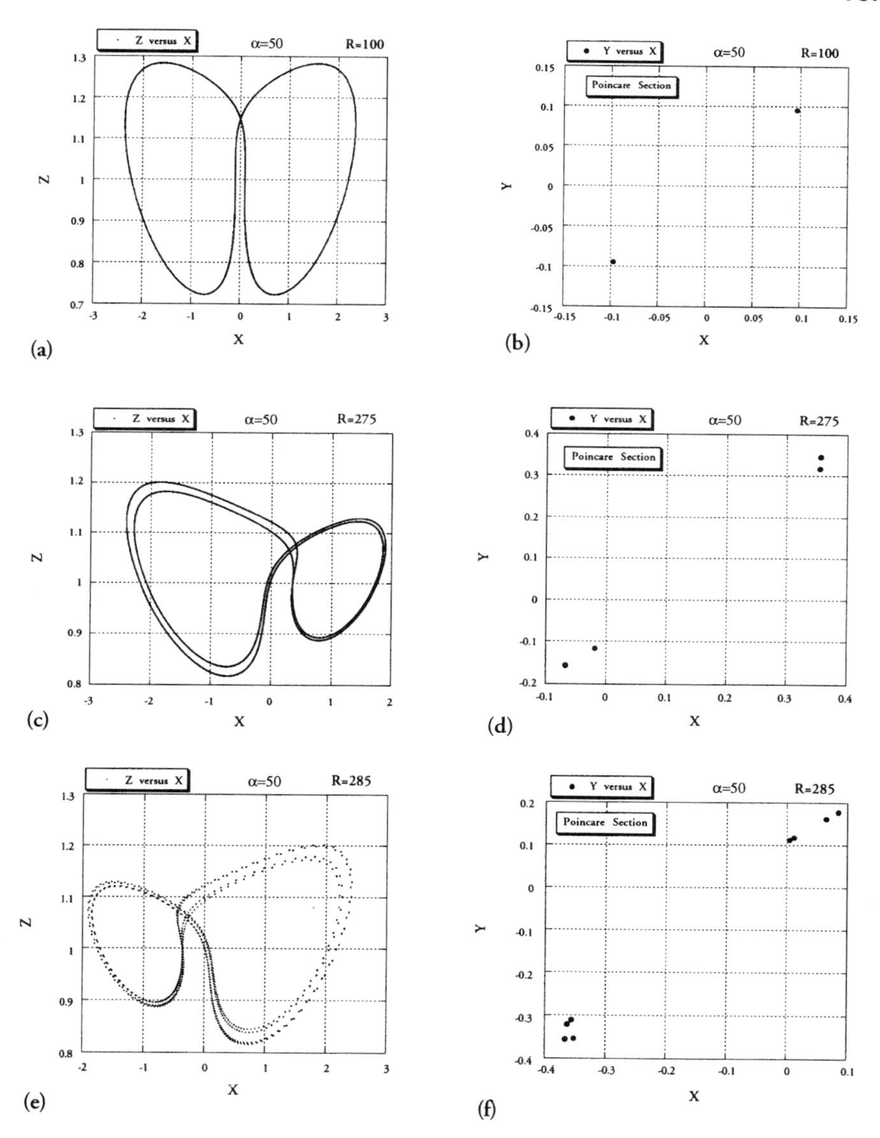

Figure 18. The evolution of trajectories over time for $\alpha = 50$ ($Pr_D = 987$), and for a value of Rayleigh number (in terms of R) corresponding to three different values of R, identifying a period-doubling route to chaos: (a) projection of the solution data points (not connected) onto the Z–X plane, for $R = 100$; (b) Poincaré section at $Z = 1$ for $R = 100$; (c) projection of the solution data points (not connected) onto the Z–X plane, for $R = 275$; (d) Poincaré section at $Z = 1$ for $R = 275$; (e) projection of the solution data points (not connected) onto the Z–X plane for $R = 285$; (f) Poincaré section at $Z = 1$ for $R = 285$. (Courtesy: Kluwer Academic Publishers.)

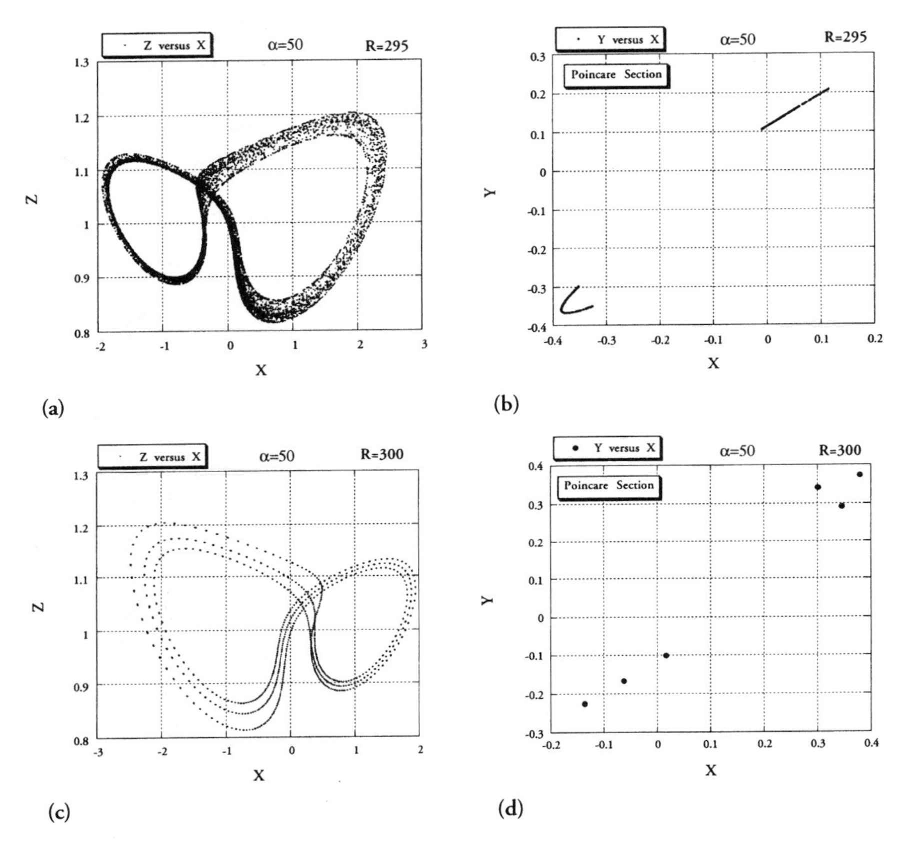

Figure 19. The evolution of trajectories over time for $\alpha = 50$ ($Pr_D = 987$), and for two values of Rayleigh number (in terms of R) associated with the neighborhood of the first periodic window within the chaotic regime: (a) projection of the solution data points (not connected) onto the Z–X plane, for $R = 295$; (b) Poincaré section at $Z = 1$ for $R = 295$; (c) projection of the solution data points (not connected) onto the Z–X plane, for $R = 300$; (d) Poincaré section at $Z = 1$ for $R = 300$, identifying a period-6 periodic window. (Courtesy: Kluwer Academic Publishers.)

became part of the strange attractor. Tracking down this result, we have been able to establish that this happened first at a value of $R \sim 376$. Other than that, we can observe also two additional hooks in the Poincaré section (Figure 23d). As expected, this leads to another periodic window around $R \sim 452.9$, identifying a period-5 periodic solution. Beyond a value of $R \sim 535$ a period-4 periodic regime is established, as shown in Figure 24, corresponding to $R = 548$.

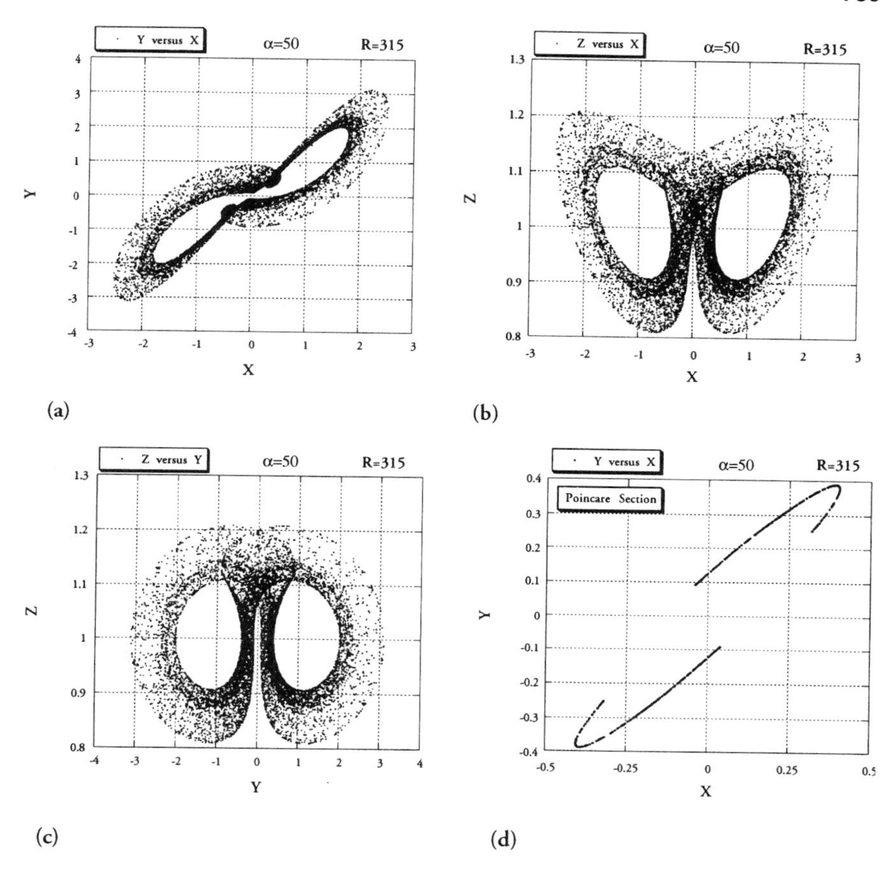

Figure 20. The evolution of trajectories over time for $\alpha = 50$ ($Pr_D = 987$), and for a value of Rayleigh number (in terms of R) corresponding to $R = 315$. The graphs (a)–(c) represent the projection of the solution data points (not connected) onto the (a) Y–X plane, (b) Z–X plane, and (c) Z–Y plane. (d) Poincaré section at $Z = 1$. (Courtesy: Kluwer Academic Publishers.)

3. Comparison between Computational and Analytical Results

The objective in the presentation of the following results is to demonstrate the appearance of the solitary limit cycle at a particular value of R prior to the first transition and to compare the computational values (Adomian's decomposition results) of $R = R_t$, where this transition occurs, with its corresponding analytical values which were presented in Eq. (53), Section IV.A, for different initial conditions (consistent with the weak nonlinear solution). While, in these computations, the value of R as well as the initial conditions

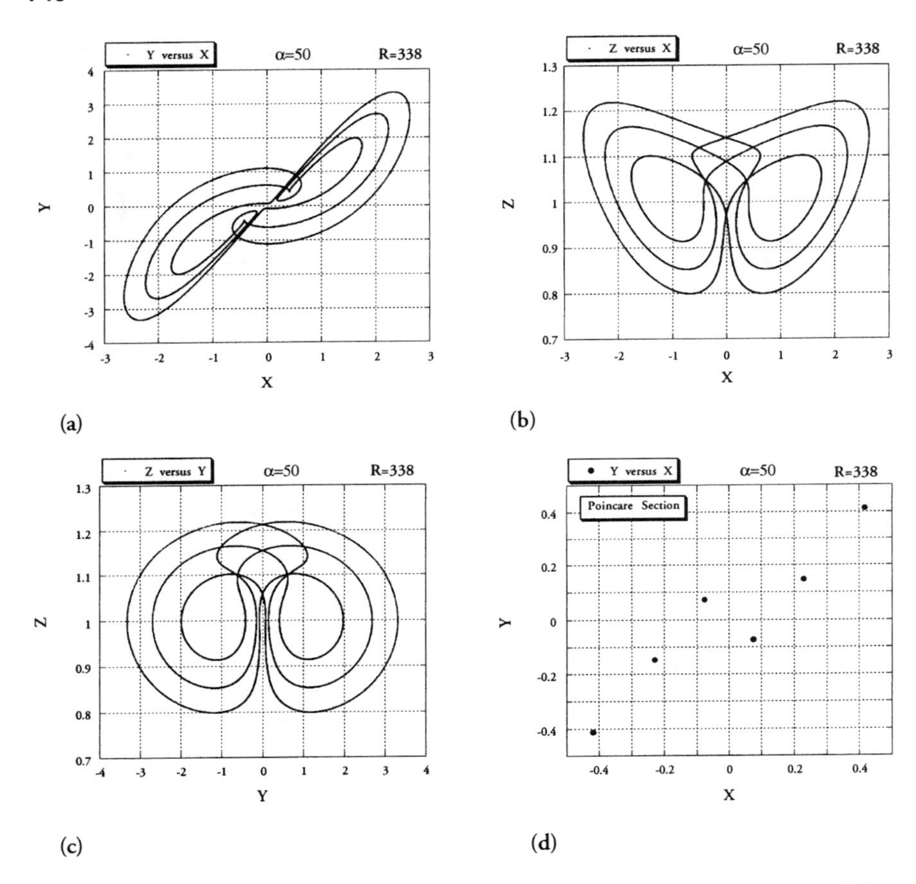

Figure 21. The evolution of trajectories over time for $\alpha = 50$ ($Pr_D = 987$), and for a value of Rayleigh number (in terms of R) corresponding to a period-6 periodic window at $R = 338$. The graphs (a)–(c) represent the projection of the solution data points (not connected) onto the (a) Y–X plane, (b) Z–X plane, and (c) Z–Y plane. (d) Poincaré section at $Z = 1$. (Courtesy: Kluwer Academic Publishers.)

vary from one computation to another, the values of γ and α are kept constant at $\gamma = 0.5$ and $\alpha = 5$. These values of $\gamma = 0.5$ and $\alpha = 5$ yield the critical values: $R_o = 25$, $\sigma_o^2 = 60$, and the following coefficients for the amplitude equation (45): $\varphi = -2.4$ and $\beta = 0.403\,226$.

In order to compare the computational results to the analytical ones obtained in Section IV.A via the weak nonlinear theory, we have to make sure that the initial conditions for the computations are consistent with the initial conditions corresponding to the weak nonlinear solution. It should be

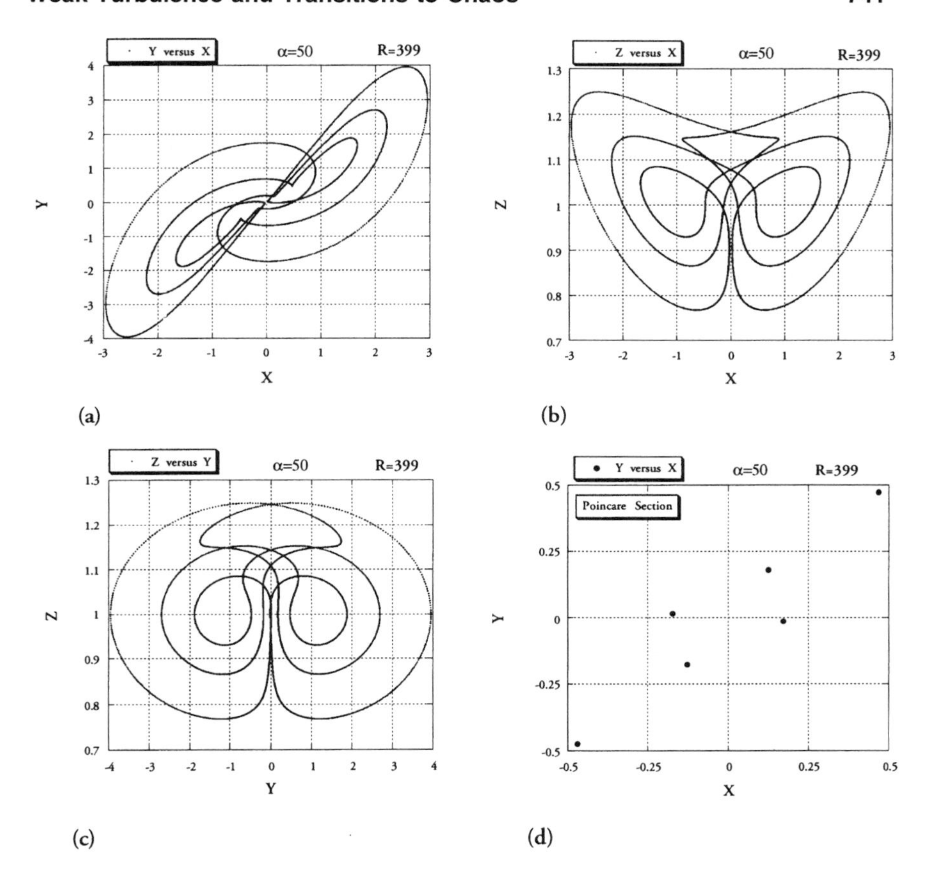

Figure 22. The evolution of trajectories over time for $\alpha = 50$ ($Pr_D = 987$), and for a value of Rayleigh number (in terms of R) corresponding to a period-6 periodic window at $R = 399$. The graphs (a)–(c) represent the projection of the solution data points (not connected) onto the (a) Y–X plane, (b) Z–X plane, and (c) Z–Y plane. (d) Poincaré section at $Z = 1$. (Courtesy: Kluwer Academic Publishers.)

pointed out that the set of possible initial conditions in the weak nonlinear solutions (48) and (49) is constrained because we did not include the decaying solutions of the form $a_{12}(\tau)\exp(\sigma_3 t)$ (with $\sigma_3 < 0$ and real) in Eq. (26). Therefore, this constraint, which is equivalent to setting $(a_{12})_{\tau=0} = 0$ and $(\theta)_{\tau=0} = 0$, is kept valid for all computational results as well. The weak nonlinear solution provides the following conditions, which are necessary and sufficient to ensure the consistency of the initial conditions between the weak nonlinear and the computational solutions

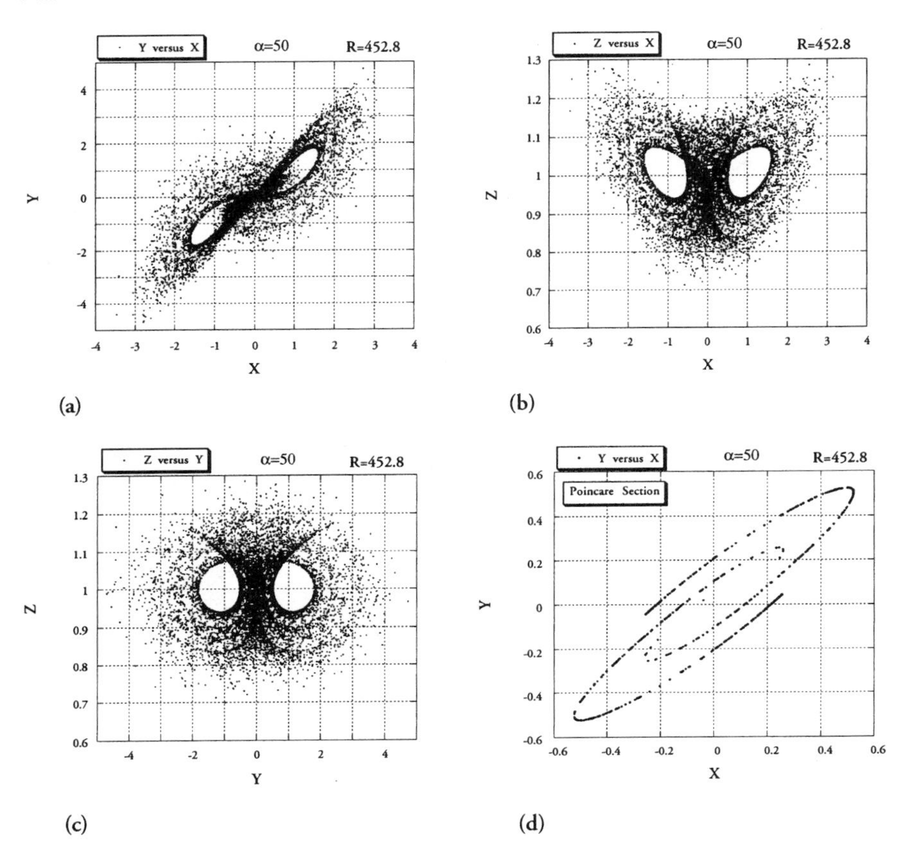

Figure 23. The evolution of trajectories over time for $\alpha = 50$ ($Pr_D = 987$), and for a value of Rayleigh number (in terms of R) corresponding to $R = 452.8$. The graphs (a)–(c) represent the projection of the solution data points (not connected) onto the (a) Y–X plane, (b) Z–X plane, and (c) Z–Y plane. (d) Poincaré section at $Z = 1$. (Courtesy: Kluwer Academic Publishers.)

$$
\begin{aligned}
X^{(o)} &= Y^{(o)} = 1 + 2r_o; \\
Z^{(o)} &= 1 + \frac{\sigma_o^2}{\alpha(R_o - 1)}[X^{(o)} - 1] = 1 + \frac{2\sigma_o^2}{\alpha(R_o - 1)} r_o
\end{aligned}
\tag{54}
$$

where $X^{(o)}$, $Y^{(o)}$, and $Z^{(o)}$ are the initial conditions for X, Y, and Z, respectively, and r_o is the initial condition for r, as used in the weak nonlinear solution in Section IV.A. furthermore, using Eq. (54), it is noted that $r_o = (X^{(o)} - 1)/2 = (Y^{(o)} - 1)/2$. Clearly, this yields negative

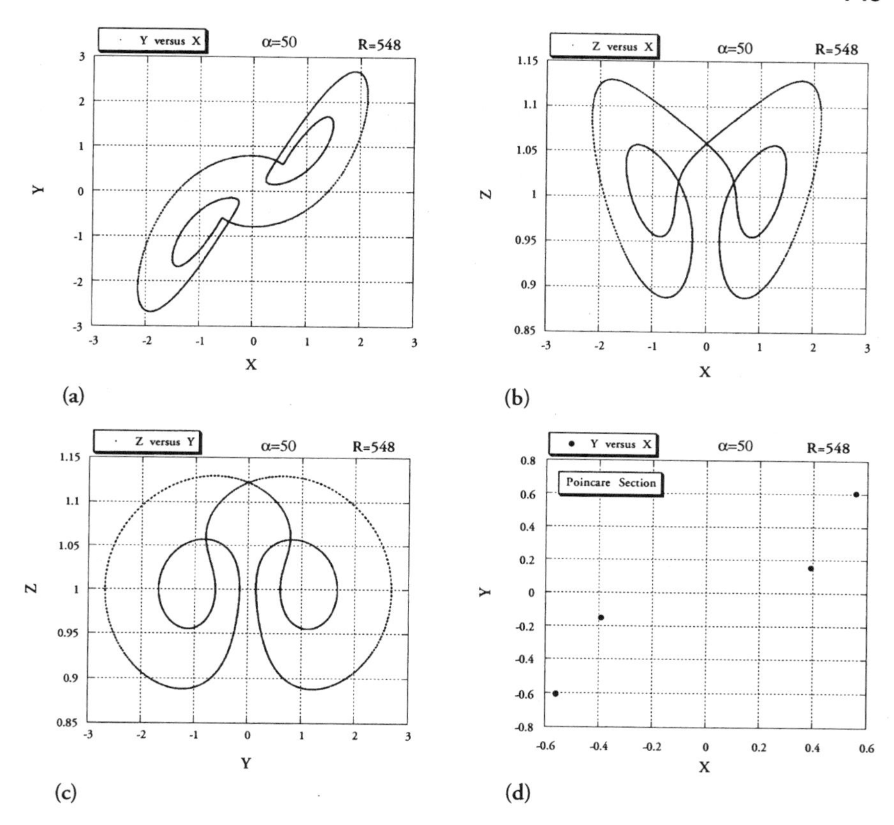

Figure 24. The evolution of trajectories over time in the state space for $\alpha = 50$ ($Pr_D = 987$), and for a value of Rayleigh number (in terms of R) corresponding to a period-4 periodic regime at $R = 548$. The graphs (a)–(c) represent the projection of the solution data points (not connected) onto the (a) Y–X plane, (b) Z–X plane, and (c) Z–Y plane. (d) Poincaré section at $Z = 1$. (Courtesy: Kluwer Academic Publishers.)

values of r_o if $X^{(o)} < 1$ or $Y^{(o)} < 1$. We therefore extend the definition of r and allow it to take negative values. This is equivalent to a phase shift in the limit cycle solution of the form $\tilde{\theta} = \theta + \pi$. For r_o this corresponds to a phase shift $\tilde{\theta}_o = \theta_o + \pi = \pi$, because $\theta_o = (\theta)_{\tau=0} = 0$, implicitly in the present case. With $\alpha = 5$, Eq. (54) yields $Z^{(o)} = (X^{(o)} + 1)/2$. It should be stressed that the solitary limit cycle was obtained computationally irrespective of whether the initial conditions were consistent with the weak nonlinear solution or not; the consistency is imposed only for quantitative comparison purposes.

A sequence of numerous computations was performed in order to evaluate these transitional R values. The results were presented by Vadasz (1999) and are reproduced in Figure 25, where the continuous curve represents the analytical solution expressed by Eq. (53) while the dots represent the computational results. The very good agreement between the analytical and computational solutions in the neighborhood of the convective fixed point (i.e., $|r_o| \ll 1$) is evident from Figure 25. As the initial conditions move away from the convective fixed point and the value of $|r_o|$ increases, the computational results depart from the analytical solution, which reconfirms the validity of the weak nonlinear solution in the neighborhood of a convective fixed point and its breakdown far away from this point. The departure of the computational results from the analytical ones is clearly not symmetrical with respect to $r_o = 0$. While the weak nonlinear solution is

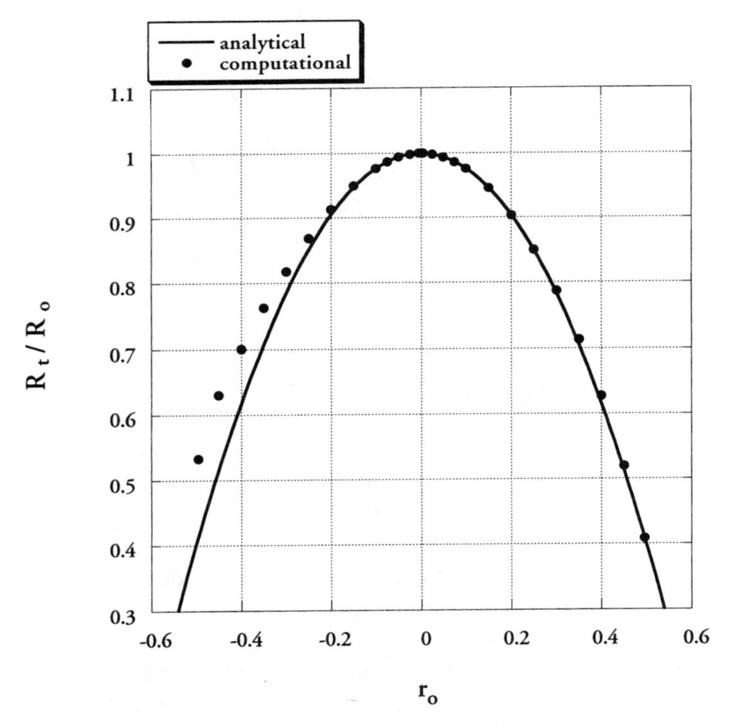

Figure 25. Transitional sub-critical values of Rayleigh number in terms of R_t/R_o as a function of the initial conditions r_o. A comparison between the weak nonlinear solution (___ analytical) and the Adomian's decomposition solution (• computational), for $\alpha = 5$ ($Pr_D = 98.7$). (Courtesy: Kluwer Academic Publishers.)

symmetrical with respect to $r_o = 0$, there is no reason to expect this symmetry from the computational solution as ones moves away from the fixed point. The maximum value of $|r_o|$ for which we could obtain results and still be consistent with the weak nonlinear solution around one fixed point was $r_o = \pm 0.5$. Actually, at $r_o = -0.5$ the corresponding initial conditions are $X^{(o)} = Y^{(o)} = 0$ and $Z^{(o)} = 0.5$, which lie on the Z-axis that is included on the stable manifold of the origin. Therefore, the computational results obtained for this set of initial conditions lead naturally to the origin, producing the motionless solution. In order to evaluate the solitary limit cycle as we approach $r_o = -0.5$, we evaluated the computational solution at $r_o = 0.495$.

Another interesting result from the computations is the fact that it was relatively very easy to detect the solitary limit cycle when the initial conditions were close to the convective fixed point, i.e., around $r_o = 0$. There, the critical time is very large, as established via the weak nonlinear analysis, and if the value of R is sufficiently close to R_t the limit cycle appears and persists for a very long time. Naturally, in this neighborhood the accuracy in estimating the value of R_t is somewhat compromised, for the same reason. Nevertheless, if the maximum time for presenting the solution (i.e., $t_{\max}$) is sufficiently large, this accuracy problem around $r_o = 0$ can be resolved. In our case, with $t_{\max} = 210$, the results provided accurate values of R_t. However, as we move away from the neighborhood of $r_o = 0$ by using initial conditions further away from the convective fixed point, it becomes more and more difficult to detect the solitary limit cycle, and more computations are needed by modifying the value of R closer and closer to R_t. While detection of the solitary limit cycle becomes more difficult as the initial conditions move away from the convective fixed point, the accuracy of the value of R_t once the limit cycle was detected becomes extremely high. For example, at $r_o = -0.25$ the limit cycle appeared at $R = R_t = 21.715\,916\,620\,635$. The need to use 13 significant digits in order to obtain the limit cycle solution over the whole time domain just emphasizes the difficulty of detecting the limit cycle as one move away from the convective fixed point and the associated accuracy in evaluating R_t. This is also an indication that this limit cycle becomes less and less table as one departs from the neighborhood of the convective fixed point. As one moves even further away, more significant digits are required in order to establish the value of R_t and detect the limit cycle over the whole time domain, and naturally, when this process reaches the limit of over 14 digits, which corresponds to the double-precision computation, the limit cycle cannot be detected for the whole time domain selected.

A further comparison between the computational and analytical results was presented by Vadasz (1999). The comparison between the shapes of the analytical and computational limit cycles for $r_o = -0.1$ and $r_o = -0.125$ shows that they are quite close. The corresponding limit cycle solutions corresponding to values of $|r_o|$ smaller than 0.1 show that analytical and computational results overlap. On the other hand, the comparison between the analytical and computational limit cycle results corresponding to $r_o = -0.495$, i.e., quite far away from the convective fixed point, shows a marked departure between the two solutions, not only in the quantitative sense, as presented by the deviation of the computational value of R_t from its predicted analytical value (see Figure 25), but also qualitatively since their shapes are substantially different. While the analytical periodic orbit maintains its elliptical shape similar, regardless of r_o, the computational results show that as the solitary limit cycle approaches conditions consistent with the homoclinic orbit its shape is altered considerably (the shape of the homoclinic orbit is far different from that of an ellipse; see Section IV.E and Figure 28, later).

D. Bifurcation Diagrams

In order to establish the relevant regimes of interest on which to focus our attention in detail, Vadasz and Olek (1999b) computed the complete solutions for a wide range of R values between $1 < R \leq 150$ with a resolution of $\Delta R = 0.25$. The computational results identified peaks and valleys (i.e., maxima and minima) in the post-transient solution for each value of R; these were plotted as a function of R, producing the bifurcation diagram presented in Figure 26 in terms of peaks and valleys in the post-transient values of Z versus R, for $\alpha = 5$ ($Pr = 98.7$). Avoidance of the use of a Poincaré map for the bifurcation diagram was motivated by reported difficulties (Sparrow 1982) with selecting only one appropriate Poincaré section to capture the results over a wide range of R values for the corresponding problem in pure fluids (nonporous domains). On the bifurcation diagram in Figure 26 we observe a sudden bifurcation from steady convection to chaos in the neighborhood of the critical value $R_{c2} = 25$. From the weak nonlinear analysis (Vadasz 1999) it was established that this is a subcritical Hopf bifurcation. We have shown in the detailed computational as well as in the analytical presentation that the transition occurs at a subcritical value of R. We cannot associate yet the bifurcations seen on the bifurcation diagram at higher values of R with the number of periods in the corresponding periodic regimes as this bifurcation diagram is not plotted on a Poincaré section. Its major objective is to guide us regarding windows of parameter values which are worth further detailed investigations. As such, we can identify possible wide periodic windows within the broad chaotic regime around $R \approx 75$, 85, and transition to a high Rayleigh number periodic

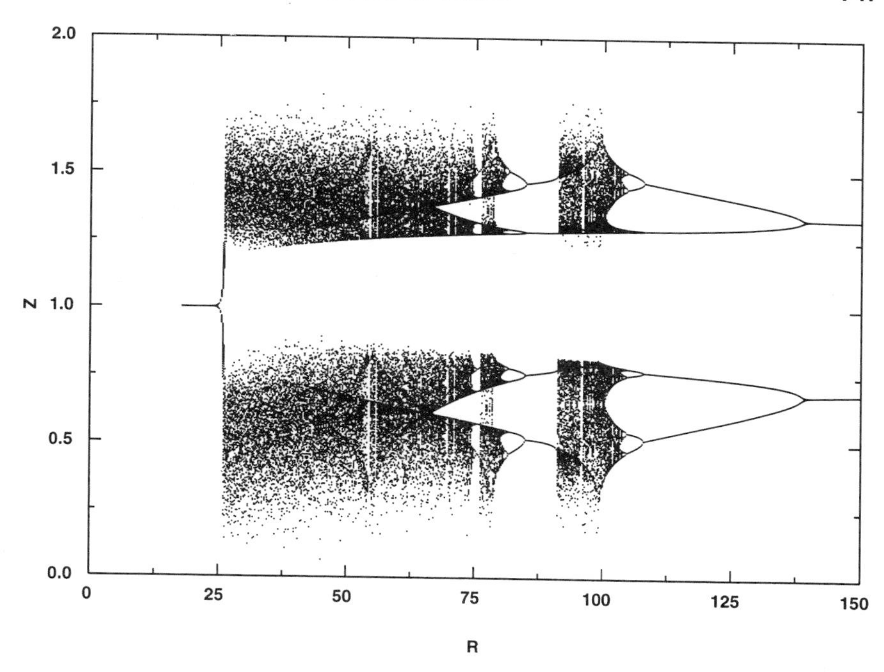

Figure 26. Bifurcation diagram of Z versus R, representing peaks and valleys of the post-transient solution of $Z(t)$ for $\alpha = 5$ ($Pr_D = 98.7$). Periodic windows can be clearly identified. (Courtesy: Kluwer Academic Publishers.)

regime about $R \approx 105$. There are narrower windows as well which we did not search for at this stage. These results were used in the further computational investigation presented in Subsection IV.C.1.

For the case of moderate Prandtl numbers, i.e., $\alpha = 50$ ($Pr = 987$), Vadasz and Olek (1999c) computed the complete solutions for a wider range of R values between $1 < R \leq 570$ with a resolution of $\Delta R = 1$ in most of the parameter domain and smaller resolutions in particular domains of interest. Similarly, as for small Prandtl numbers, the computational results identified peaks and valleys (i.e., maxima and minima) in the post-transient solution for each value of R, which were plotted as a function of R, producing the bifurcation diagram presented in Figure 27 in terms of peaks and valleys in the post-transient values of Z versus R. On the bifurcation diagram in Figure 27 we observe a sudden bifurcation from steady convection to a periodic regime in the neighborhood of the critical value $R_{c2} = 58.51$. This transition occurs also at a subcritical value of R. We can identify the first transition to chaos at around $R \approx 290$, and further

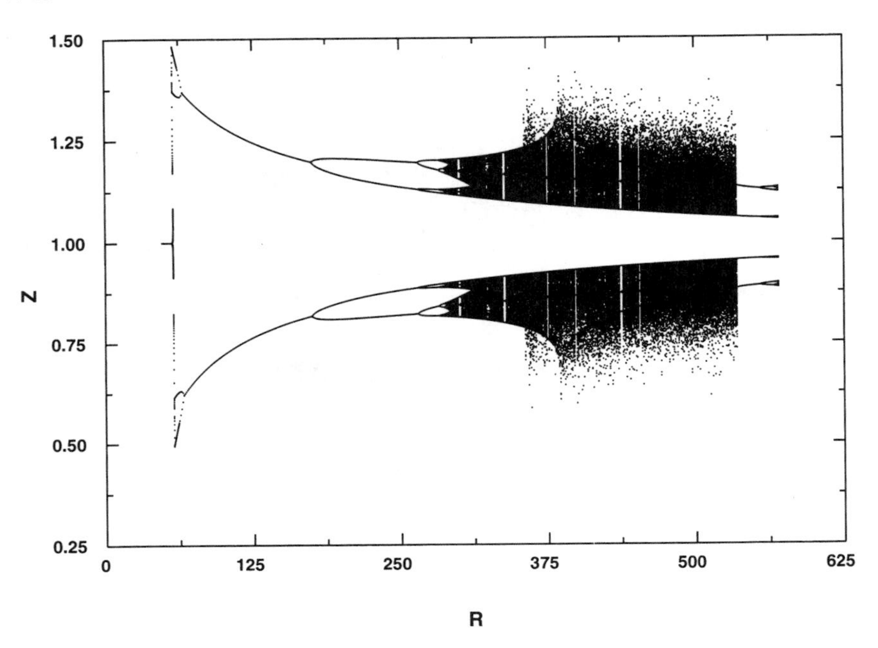

Figure 27. Bifurcation diagram of Z versus R, representing peaks and valleys of the post-transient solution of $Z(t)$ for $\alpha = 50$ ($Pr_D = 987$). Periodic windows can be clearly identified. (Courtesy: Kluwer Academic Publishers.)

possible periodic windows around $R \approx 300$, 338, 375, 400, 437, 453, and transition to a high Rayleigh number periodic regime above $R \approx 535$. When comparing these results with the small Prandtl number results we can conclude that increasing the Prandtl number not only delays the loss of stability of the steady convection from $R_t = 24.647\,752$ corresponding to $\alpha = 5$, to $R_t = 57.543\,37$ corresponding to $\alpha = 50$, but also that the transition in the latter case is not moving directly towards chaos but rather towards a multi-periodic regime. The first transition to chaos in the latter case is delayed even further by the moderate Prandtl number and occurs following a period-doubling sequence at a value of R above ~ 290.

E. Homoclinic Orbit and Homoclinic Explosion

The transitions from steady convection to chaos in a fluid-saturated porous layer heated from below are typically associated with a homoclinic explosion when the trajectory which originally moves around one fixed point departs towards the other fixed point. The trajectory's behavior depends on the Rayleigh number and on the initial conditions. The existence of a homoclinic orbit in this convective flow was already established (Sparrow 1982). However, the difficulty to recover computationally this orbit is also evident because of its unstable nature (Sparrow 1982). A simple procedure allowing the computational recovery of the homoclinic orbit was presented by Vadasz and Olek (1999a), which combines the Adomian's decomposition computational solution with the weak nonlinear analytical results to provide effective guidelines.

The results of $X(t)$, $Y(t)$, and $Z(t)$ are presented in terms of projections of the trajectory's data points onto the planes $Y = 0$, $Z = 0$, and $X = 0$, for a value of $\alpha = 5$. As long as the initial conditions are not too far away from one of the fixed points $X = Y = Z = 1$ or $X = Y = -1$, $Z = 1$, it is not difficult to recover the orbit associated with the Hopf bifurcation relevant to the transition from steady convection to chaos, as presented in Section IV.C. However, as the initial conditions depart significantly from the fixed points, the orbit becomes more and more unstable and its recovery becomes more difficult. The region around the origin poses particular difficulties because (i) it is far away from both fixed points, and (ii) the Z-axis, being part of the stable manifold of the origin, prevents the choice of initial conditions on the axis, as the solution then will naturally converge towards the origin ($X = Y = Z = 0$) and prevent the recovery of the homoclinic orbit. The method adopted by Vadasz and Olek (1999a) is the use of quantitative approximations based on Vadasz (1999) and successive computations for initial conditions around the origin but not necessarily too close to it. This procedure allows us to establish that the homoclinic orbit in the neighborhood of the origin lies on the plane $Y \approx 2X$, for non-negative values of Z. Accordingly, we choose the initial conditions as close as possible to the origin on this plane. The results presented here correspond to the initial conditions $X_o = 0.01$, $Y_o = 0.02$, and $Z_o = 0$. Then, for $R = 12.200\,139\,732\,122$ we obtain the one-sided (single branch) homoclinic orbit, observed when the trajectory makes one single loop around the fixed point. The accuracy (number of significant digits) required for the value of R in order to recover the periodic orbit far away from both fixed points was analyzed and discussed by Vadasz (1999). There is no great novelty in recovering the one-sided homoclinic orbit; this has been done previously and reported by Sparrow (1982). The interesting part is the ability to use this procedure to recover the complete two-sided (both branches) homoclinic orbit. This is accomplished by a very slight variation of R. For $R = 12.200\,139\,732\,123$ Vadasz and Olek (1999a) obtained the complete homoclinic orbit, as presented in Figure 28, following the trajectory to make a complete loop around both fixed points. The direction of the trajectory on Figure 28a, representing the projection of the trajectory's data points onto

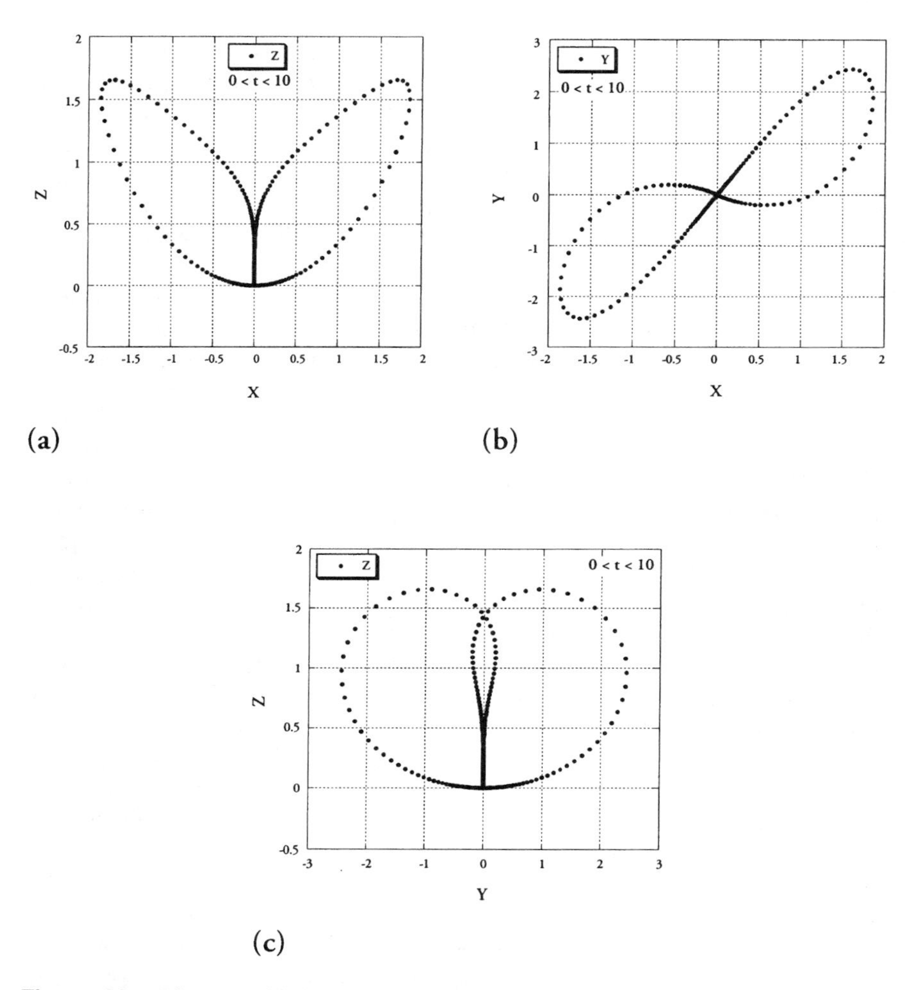

Figure 28. The two-sided homoclinic orbit (trajectory data points are not connected): (a) projection of trajectory data points on the plane $Y = 0$; (b) projection of trajectory data points on the plane $Z = 0$; (c) projection of trajectory data points on the plane $X = 0$. (Courtesy: Elsevier Science Ltd.)

the plane $Y = 0$, is of particular interest. It is evident from the results in the time domain (not shown here) that the trajectory, starting from the initial conditions at $Z_o = 0$ and $X_o = 0.01$ (and $Y_o = 0.02$, not observed on this projection), moves counter-clockwise on the bottom side of the right branch of the homoclinic orbit and returns straight to the origin from above. At the origin the trajectory makes a sudden right turn, switching to the bottom side of the left branch of the orbit, where it moves clockwise, making a complete loop by returning to the origin from above (actually very close to it but not exactly to the origin). Figures 28b and 28c represent the projection of the same trajectory on the planes $Z = 0$ and $X = 0$, respectively. The closer we choose the value of X_o to the origin, the closer the trajectory will recover the accurate homoclinic orbit (recall that the value of Y_o is selected to lie on the plane $Y \approx 2X$). Naturally, we cannot choose the origin itself as the initial conditions, because it lies on the Z-axis which, being part of the stable manifold of the origin, will cause the solution to remain at the origin, recovering the trivial solution.

V. CONCLUSIONS

The rich dynamics associated with the transition from steady convection to chaos in porous media convection was demonstrated by using analytical as well as computational methods. A brief description of the different methods was presented and selected results were illustrated.

NOMENCLATURE

Roman Letters

Da	Darcy number, $= K_o/H_*^2$
$\mathbf{e}_x$	unit vector in the x direction
$\mathbf{e}_y$	unit vector in the y direction
$\mathbf{e}_z$	unit vector in the z direction
$\mathbf{e}_n$	unit vector normal to the boundary, positive outwards
$\mathbf{e}_g$	unit vector in the direction of gravity acceleration
g_*	gravity acceleration
H	front aspect ratio of the porous layer, $= H_*/L_*$
H_*	height of the layer
K	dimensionless permeability function
K_o	a reference value of permeability, dimensional
L_*	length of the porous layer
L	reciprocal of the front aspect ratio, $= 1/H = L_*/H_*$
M_f	heat capacity ratio, $= \rho_f c_{p,f}/[\phi\rho_f c_{p,f} + (1-\phi)\rho_s c_s]$
p	reduced pressure, dimensionless
Pr_D	Darcy–Prandtl number, $= \phi Pr/Da$
Pr	Prandtl number, $= \nu_*/\alpha_{e^*}$
Ra	porous media gravity-related Rayleigh number, $= \beta_* \Delta T_c g_* H_* K_o M_f/\alpha_{e*}\nu_*$

Ra_o	critical value of Rayleigh number for the transition from steady convection to chaos
R	scaled Rayleigh number, $= Ra/4\pi^2$
R_o	critical value of R for the transition from steady convection to chaos
r	absolute value of the complex amplitude
r_o	initial condition of r
t_*	time, dimensional
T	temperature, dimensionless, $= (T_* - T_C)/(T_H - T_C)$
T_C	coldest wall temperature, dimensional
T_H	hottest wall temperature, dimensional
V	filtration velocity, dimensionless
u	horizontal x component of the filtration velocity
v	horizontal y component of the filtration velocity
w	vertical component of the filtration velocity
x	horizontal length coordinate
y	horizontal width coordinate
z	vertical coordinate
X	rescaled amplitude A_{11}, $= -A_{11}/2\eta\sqrt{2\gamma(R-1)}$
Y	rescaled amplitude B_{11}, $= \pi R B_{11}/2\sqrt{2\gamma(R-1)}$
Z	rescaled amplitude B_{20}, $= \pi R B_{02}/(R-1)$

Greek Letters

α	parameter related to the time derivative term in Darcy's equation
α_{eo}	effective thermal diffusivity, dimensional, $= [\phi k_f + (1-\phi)k_s]/[\phi \rho_f c_{p,f} + (1-\phi)\rho_s c_s]$
β_*	thermal expansion coefficient
β	relaxation time parameter in Eq. (45)
ε	asymptotic expansion parameter, defined by $\varepsilon^2 = (R - R_0)/R_0$
ζ	function related to the stream function by $\psi = \partial\zeta/\partial x$
η	parameter, $= (L^2 + 1)/L$
θ	higher order frequency correction
μ_*	dynamic viscosity of fluid, dimensional
ν_*	kinematic viscosity of fluid, dimensional
ξ	parameter, $= \varepsilon^2/\varphi$
τ	long time scale
ϕ	porosity
φ	coefficient of the nonlinear term in the amplitude equation (45)
χ	parameter, $= \varphi/\beta$
ψ	stream function

Subscripts

$*$	dimensional value
c	characteristic value
cr	critical value
C	related to the coldest wall
f	related to the fluid phase
H	related to the hottest wall
o	reference value
s	related to the solid phase
t	transitional value

Superscripts

$*$	complex conjugate
(o)	initial conditions

REFERENCES

Adomian G. A review of the decomposition method in applied mathematics. J Math Anal Appl 135:501–544, 1988.

Adomian G. Solving Frontier Problems in Physics: the Decomposition Method. Dordrecht: Kluwer Academic Publishers, 1994.

Bejan A. Convection Heat Transfer. 2nd ed. New York: Wiley, 1995.

Boussinesq J. Theorie Analitique de la Chaleur. 2. Paris: Gauthier-Villars, 1903, p 172.

Choudhury RS. Stability conditions for the persistence, disruption and decay of two-dimensional dissipative three-mode patterns in moderately extended nonlinear systems and comparisons with simulations. In: Debnath L, Choudhury SR, eds. Nonlinear Instability Analysis, Advances in Fluid Mechanics. Southampton: Computational Mechanics Publications, 1997, pp 43–91.

Gheorghita StI. Mathematical Methods in Underground Hydro-Gaso-Dynamics. Bucharest: Romanian's Academy Editions, 1966 (in Romanian).

Kimura S, Schubert G, Straus JM. Route to chaos in porous-medium thermal convection. J Fluid Mech 166:305–324, 1986.

Lorenz EN. Deterministic on-periodic flows. J Atmos Sci 20:130–141, 1963.

Malkus WVR. Non-periodic convection at high and low Prandtl number. Mem Soc R Sci Liège IV 6:125–128, 1972.

Nield DA, Bejan A. Convection in porous media. 2nd ed. New York: Springer-Verlag, 1999.

Olek S. An accurate solution to the multispecies Lotka–Volterra equations. SIAM Rev 36:480–488, 1994.

Olek S. Solution to a class of nonlinear evolution equations by Adomian's decomposition method. 1997 (manuscript in preparation).

Répaci A. Non-linear dynamical systems: on the accuracy of Adomian's decomposition method. Appl Math Lett 3:35–39, 1990.

Sparrow C. The Lorenz Equations: Bifurcations, Chaos, and Strange Attractors. New York: Springer-Verlag, 1982.

Straus JM, Schubert G. Three-dimensional convection in a cubic box of fluid-saturated porous material. J Fluid Mech 91:155–165, 1979.

Vadasz P. Local and global transitions to chaos and hysteresis in a porous layer heated from below. Transp Porous Media 37:213–245,1999.

Vadasz P, Olek S. Transitions and chaos for free convection in a rotating porous layer. Int J Heat Mass Transfer 14:1417–1435, 1998.

Vadasz P, Olek S. Computational recovery of the homoclinic orbit in porous media convection. Int J Nonlinear Mech 43:89–93, 1999a.

Vadasz P, Olek S. Weak turbulence and chaos for low Prandtl number gravity driven convection in porous media. Transp Porous Media 37:69–91, 1999b.

Vadasz P, Olek S. Route to chaos for moderate Prandtl number convection in a porous layer heated from below. Transp Porous Media, 1999c (in press).

Wang Y, Singer J, Bau HH. Controlling chaos in a thermal convection loop. J Fluid Mech 237:479–498, 1992.

Yuen P, Bau HH. Rendering a subcritical Hopf bifurcation supercritical. J Fluid Mech 317:91–109, 1996.

17

Transport Phenomena in Porous Media: Modeling the Drying Process

O. A. Plumb
Washington State University, Pullman, Washington

I. INTRODUCTION

Drying has been practiced since it was first discovered that foods could be preserved and that clay, through drying, could be converted to a useful structural material. Today, many products are dried for a variety of reasons including preservation, reduction of weight or volume for shipping and packaging, improved dimensional stability, or as a step during processing. The drying process has been the focus of considerable fundamental and applied research because the processing variables have a direct effect on energy consumption, product quality, and productivity. Many of the products that are dried are porous materials or particulate materials or powders dried in such a manner that they behave as a porous medium. As an example of the economic importance of the problem, Franzen et al. (1987) estimated that in excess of 2×10^8 kJ are consumed annually in drying food products alone. Strumillo et al. (1995) estimate that 12% of all industrial energy expenditures are for drying.

The early fundamental work on drying of porous materials is based on the theories for simultaneous heat and mass transfer in porous media developed by Luikov (1966) and Phillip and de Vries (1957) in the 1950s. The more recent modeling efforts are typically based on the work of Whitaker (1977), who applied volume averaging to the well-known transport equations for continua in an attempt to arrive at a general theory for drying of porous materials. The more recent reviews include those by Hartley (1987), which focuses on soil, Bories (1991), Waananen et al. (1993), Turner and

Perre (1997), and Perre and Turner (1997a). In addition to these there are a number of reviews related to specific materials, classes of materials, and types of dryers (see Mujumdar 1995, for example).

For a review of the state of the art from the more applied viewpoint, including the design of drying equipment, one can refer to the monographs by Keey (1978) and Cook and DuMont (1991) and the handbook edited by Mujumdar (1995). It is important to note that the link between theory and practice for drying and the design of drying equipment has not been closed to the same extent that theory and practice have been linked in the design of heat exchangers. This is because the transport phenomena involved in drying porous materials are quite complex and fundamental questions remain to be resolved. Thus, the design of dryers is still more art than science (Cook and DuMont 1991).

Porous materials that are traditionally dried during processing can be classified in a number of ways. First, different processes might be used, depending on whether the product is monolithic, such as bricks or wood, or granular as in the case of grains or particulate ceramics produced in an aqueous phase process. Other classifications based on geometry could include sheets—paper and textiles—and fibrous materials, including wood fibers. Many chemicals produced in the powder form are not in themselves porous; however, they may be dried as a packed bed or a thin tape or sheet, thus creating a porous medium. Second, materials might be classified according to whether they are hygroscopic, containing both bound and free water (biological products, including wood and many food products), or nonhygroscopic. For nonhygroscopic materials free water is of primary interest despite the fact that there may exist small amounts of absorbed water that is not necessarily removed during the drying process.

The drying process can be classified according to the technology used to remove the liquid. Convective drying in air is the most common drying process. Others include radiative drying, electromagnetic drying (typically microwaves) or infrared radiation, contact drying, supercritical drying (above the critical point for the liquid to be removed), freeze drying, or a combination of these and possibly others. For high temperature drying, particularly for materials for which oxidation is a concern, superheated steam can be used in place of air as the drying medium (Perre et al. 1993, for example). From the more practical standpoint, a drying process can be classified according to the type of equipment to be utilized. Typical possibilities include drum dryers, flash dryers, kilns, press dryers, and fluidized bed dryers.

This chapter will focus on modeling the transport phenomena that occur during the drying of porous materials. First, the basic transport equations, which are developed in detail elsewhere, will be reviewed. This will be followed by a discussion of the boundary conditions necessary for application to a wide range of drying processes and a discussion of the transport properties necessary to model actual drying processes. Simplifying assumptions that are of value in modeling specific processes will then be discussed. A brief section on freeze drying and electromagnetic drying is included as these are commonly practiced. Numerical procedures useful in solving the model equations will not be covered. These are discussed in considerable detail elsewhere (for example, Turner and Perre 1997).

II. REVIEW OF TRANSPORT EQUATIONS AND BOUNDARY CONDITIONS

A. Transport Equations

In drying porous materials we are interested in the simultaneous transport of heat, liquid (free and bound), vapor, and air (unless the drying environment is controlled), which results from the application of heat at the surface by conduction, convection, and/or radiation, or internally by electromagnetic waves. In most processes we are interested in liquid water and water vapor; however, it could be any volatile liquid. Although the drying process can be modeled using a microscopic approach (Prat 1993), we will take a macroscopic approach based on the presumption that the appropriate macroscopic balance equations can be obtained through a volume averaging process. The macroscopic balance equations are assumed to be valid over an averaging volume or representative elementary volume (REV) which includes a representative sample of the geometry including void volume and flow channels and the distribution of phases—solid, liquid, gas. Within the averaging volume the phases will be assumed to be in thermal and mass equilibrium. Hence, the computed quantities, including temperature, velocities of the various phases, and mass fractions of the various phases, are average values for the REV. In addition, we will assume that the solid phase is not deformable. This assumption leads to a significant simplification but may not be appropriate if shrinkage occurs during the drying process. Drying models for wood that include shrinkage have been developed by Lartigue et al. (1989), Salin (1992), Perre and Passard (1995), and Ferguson (1997). For food products, models that include shrinkage have been developed by Suarez and Viollaz (1988).

With the above assumptions and other restrictions outlined by Whitaker (1977), the basic balance equations can be written as follows liquid phase:

$$\frac{\partial}{\partial t}(\rho_l\phi_l) + \nabla \cdot (j_l) = -\dot{m}_l \tag{1}$$

bound liquid:

$$\frac{\partial}{\partial t}(\rho_b\phi_b) + \nabla \cdot (j_b) = -\dot{m}_b \tag{2}$$

gas phase:

$$\frac{\partial}{\partial t}(\rho_g\phi_g) + \nabla \cdot (j_g) = \dot{m}_l + \dot{m}_b \tag{3}$$

vapor phase:

$$\frac{\partial}{\partial t}(\rho_v\phi_v) + \nabla \cdot (j_v) = \dot{m}_l + \dot{m}_b \tag{4}$$

air phase:

$$\frac{\partial}{\partial t}(\rho_a\phi_a) + \nabla \cdot (j_a) = 0 \tag{5}$$

energy:

$$\begin{aligned} &\left[\rho_s\phi_s c_{ps} + (\rho_l\phi_l + \rho_b\phi_b)c_{pl} + \rho_g\phi_g c_{pg}\right]\frac{\partial T}{\partial t} \\ &+ \left[(\rho_l v_l + \rho_b v_b)c_{pl} + \rho_g c_{pg} v_g\right] \cdot \nabla T \\ &= \nabla \cdot (k_{\text{eff}}\nabla T) - h_{fg}(\dot{m}_l + \dot{m}_b) - h_s\dot{m}_b + \dot{s} \end{aligned} \tag{6}$$

In Eqs. (1)–(6) the ρs refer to densities and ϕs to volume fractions. Note that the volume fractions, porosity, and liquid phase saturation are related through the following constraints

$$\phi_a + \phi_v = \phi_g \tag{7a}$$

$$\phi_g + \phi_l + \phi_s = 1 \tag{7b}$$

$$\phi_g + \phi_l = \varepsilon = 1 - \phi_s \tag{7c}$$

$$S = \phi_l/\varepsilon \tag{7d}$$

The vs in the above equations represent mass averaged velocities and the js represent fluxes. The effective thermal conductivity k_{eff} arises from the formal volume averaging procedure (Whitaker 1977) and will be assumed to be a predictable property which depends strongly upon liquid phase saturation. For convenience, the product $\rho_i\phi_i$, which is the mass fraction of component i, will be written as ρ_i^*.

Fourier's law for the transfer of heat has already been invoked in Eq. (6) and it is assumed that the gas phase can be treated as incompressible. For Eqs. (1)–(5), closure requires Darcy's law (multiphase) for the mass-averaged velocities and Fick's law for the diffusive fluxes:

Darcy's law:

$$j_l = \rho_l^* v_l = -\frac{\rho_l^* K k_{rl}}{\mu_l}(\nabla p_l - \rho_l g) \tag{8a}$$

$$j_g = \rho_g^* v_g = -\frac{\rho_g^* K k_{rg}}{\mu_g}\nabla p \tag{8b}$$

Fick's law:

$$j_v = \rho_v^* v_v = -\frac{\rho_v^* K k_{rg}}{\mu_g}\nabla p - \rho_g D_{va}\nabla\left(\frac{\rho_v^*}{\rho_g}\right) \tag{9a}$$

$$j_b = -\rho_s D_{bl}\nabla\left(\frac{\rho_b^*}{\rho_s}\right) \tag{9b}$$

Here it has been assumed that the free liquid flux is totally convective, the bound water flux is diffusive, and both diffusion and convection contribute to the vapor phase flux. Thermally driven diffusion is neglected for both vapor and bound water. In addition, it has been assumed that gravitational effects are not important in the gas phase (no natural convection). To complete the closure we will assume that the gas phase is incompressible and behaves as a mixture of ideal gases, and utilize the following relationships which result from the assumption of local equilibrium:

ideal gas:

$$p = \rho_g R_g T, \qquad p_v = \rho_v R_v T, \qquad p_a = \rho_a R_a T \qquad \text{and} \qquad p_v + p_a = p \tag{10}$$

capillary pressure:

$$p_c = p - p_l \tag{11}$$

vapor–liquid equilibrium (Clausius–Clapeyron):

$$p_v = p^0 \exp\left\{\frac{-p_c}{\rho_l R T} + \frac{h_{fg}}{R_v}\left(\frac{1}{T} - \frac{1}{T^0}\right)\right\} \tag{12}$$

bound liquid–vapor equilibrium (sorption curve):

$$p_v = p^0 f(\rho_b^*, T) \tag{13}$$

Note that the effect of curvature (Kelvin effect) on the vapor–liquid equilibrium is included. This effect becomes important in very small pores. Neglecting the Kelvin effect is probably consistent with the assumption that the gas phase behaves as a continuum. Thus, at pressures near one atmosphere, we are restricted to pore sizes greater than the order of $10\mu m$ (Rohsenow and Choi 1961). A detailed discussion of the thermodynamics associated with the assumption of local equilibrium can be found in Baggio et al. (1997).

The system of Eqs. (1)–(13) is typically solved after reduction to three nonlinear partial differential equations with three unknowns. One choice of dependent variables is moisture content or total moisture content, defined as

$$M = \frac{\rho_l \phi_l + \rho_b \phi_b + \rho_v \phi_v}{\rho_s \phi_s} \tag{14}$$

temperature, and gas phase pressure (Bories 1991; Hartley 1987). For many drying processes the vapor phase makes a negligible contribution to the total moisture content as defined in Eq. (14) because the density of water vapor is three orders of magnitude smaller than that of liquid water. We begin with this definition for the sake of generality, recognizing that for high pressure and supercritical drying and for liquids other than water the contribution of the vapor phase to the total moisture content may be significant.

Turner and Perre (1997) suggest the use of different dependent variables depending on the nature of the problem. For example, for a nonhygroscopic porous material they suggest liquid saturation, temperature, and air density (Turner and Perre 1995), or moisture content, temperature, and gas phase pressure. Couture et al. (1995) suggest moisture content, air density, and the total enthalpy (in place of temperature). Pruess (1987) utilizes pressure, temperature, and gas phase saturation. In what follows, the moisture content as defined in Eq. (14), temperature, and gas phase pressure are selected as the primary dependent variables. It should be noted, however, that if the drying takes place above the boiling point for the liquid, the enthalpy formulation (in place of temperature) eliminates the need to track drying fronts. This offers great advantage if the problem is multidimensional.

To obtain a final drying model that carries a minimal set of assumptions to make it applicable to a wide range of drying problems, we first sum Eqs. (1), (2), and (4) and eliminate the mass-averaged velocities and diffusive fluxes by substituting Eqs. (8a), (8b), (9a), and (9b) into both the conservation equation for total moisture and the energy equation. The results are

$$\rho_s\phi_s\frac{\partial M}{\partial t}+\nabla\cdot\left\{-\frac{\rho_l^*Kk_{rl}}{\mu_l}(\nabla p_l-\rho_l g)-\rho_s D_{bl}\nabla\left(\frac{\rho_b^*}{\rho_s}\right)-\frac{\rho_v^*Kk_{rg}}{\mu_g}\right.$$
$$\left.\nabla p-\rho_g D_{va}\nabla\left(\frac{\rho_v^*}{\rho_g}\right)\right\}=0 \tag{15}$$

and

$$\overline{\rho c_p}\frac{\partial T}{\partial t}+\left[-\frac{\rho_l^*Kk_{rl}}{\mu_l}(\nabla p_l-\rho_l g)c_{pl}-\frac{\rho_g^*Kk_{rg}}{\mu_g}(\nabla p)c_{pg}\right]\cdot\nabla T$$
$$=\nabla\cdot(k_{\text{eff}}\nabla T)-h_{fg}(\dot m_l+\dot m_b)-h_s\dot m_b+\dot s \tag{16}$$

In Eq. (16) the convective heat fluxes resulting from the diffusion of vapor and bound water have been assumed to be negligible and the mass-averaged heat capacity for the wet solid is defined as

$$\overline{\rho c_p}=\rho_s^*c_{ps}+(\rho_l^*+\rho_b^*)c_p+\rho_g^*c_{pg} \tag{17}$$

We next write the liquid phase pressure in terms of the total pressure, moisture content, and temperature, using Eq. (11) coupled with the assumption that the capillary pressure can be expressed in terms of temperature and pressure through the Leverett function (Kaviany 1995)

$$p_c=\frac{\sigma F(S)}{\sqrt{K/\varepsilon}} \tag{18a}$$

$$\mathrm{d}p_c=\frac{\sigma}{\sqrt{K/\varepsilon}}\frac{\partial F}{\partial S}\mathrm{d}S+\frac{F}{\sqrt{K/\varepsilon}}\frac{\partial\sigma}{\partial T}\mathrm{d}T=D_{\sigma S}\mathrm{d}S+D_{\sigma T}\mathrm{d}T \tag{18b}$$

where $\sqrt{K/\varepsilon}$ is used to characterize the pore level length scale.

Writing the saturation in terms of moisture content and combining the convective terms results in

moisture:

$$\rho_s^*\frac{M}{\partial t}+\nabla\cdot\left\{-D_{p_M}\nabla p-D_{pl}\left[-D_{\sigma S}\left(\frac{\rho_s^*}{\rho_l\varepsilon}\nabla M-\frac{1}{\rho_l\varepsilon}\nabla\rho_b^*\right.\right.\right.$$
$$\left.\left.\left.-\frac{1}{\rho_l\varepsilon}\nabla\rho_v^*\right)-D_{\sigma T}\nabla T-\rho_e g\right]-D_{bl}\nabla\rho_b^*-\rho_g D_{va}\nabla\left(\frac{\rho_v^*}{\rho_g}\right)\right\}=0 \tag{19}$$

energy:

$$\overline{\rho c_p}\frac{\partial T}{\partial t} + \left\{ D_{pl}\left[\nabla p - D_{\sigma S}\left(\frac{\rho_s^*}{\rho_l \varepsilon}\nabla M - \frac{1}{\rho_l \varepsilon}\nabla \rho_b^* - \frac{1}{\rho_l \varepsilon}\nabla \rho_v^* \right) \right.\right.$$
$$\left.\left. - D_{\sigma T}\nabla T - \rho_l g \right] c_{pl} - D_{pg}(\nabla p)c_{pg} \right\} \cdot \nabla T \tag{20}$$
$$= \nabla \cdot (k_{\text{eff}}\nabla T) - h_{fg}(\dot{m}_l + \dot{m}_b) - h_s \dot{m}_b + \dot{s}$$

where the transport coefficients are defined as

$$D_{p_M} = \rho_l^* \frac{Kk_{rl}}{\mu_l} + \frac{\rho_v^* Kk_{rg}}{\mu_g} = D_{pl} + D_{p_g} \tag{21}$$

If we make the assumption that

$$\rho_b^* = \begin{cases} \text{constant} & \text{if} \quad S > 0 \\ \rho_s^* M - \rho_v^* & \quad S = 0 \end{cases}$$

then

$$\nabla \rho_b^* = \begin{cases} 0 & \text{for} \quad S > 0 \\ \rho_s^* \nabla M - \nabla \rho_v^* & \text{for} \quad S = 0 \end{cases}$$

Following Hartley (1987) the gradient of the vapor phase mass fraction, ρ_v^*, is next expressed in terms of the selected primary variables M, T, and p. We first note that, using Eq. (10),

$$\nabla\left(\frac{\rho_v}{\rho_g}\right) = \frac{\rho_a \nabla \rho_v - \rho_v \nabla \rho_a}{\rho_g^2} \tag{22}$$

and again, from Eq. (10),

$$\nabla \rho_a = \frac{1}{R_a T}\nabla p - \frac{p}{R_a T^2}\nabla T - \frac{R_v}{R_a}\nabla \rho_v \tag{23}$$

combining Eqs. (22) and (23) results in

$$\nabla\left(\frac{\rho_v}{\rho_g}\right) = \frac{1}{R_a T \rho_g^2}\left[p \nabla \rho_v - \rho_v \nabla p + \frac{p\rho_v}{T}\nabla T \right] \tag{24}$$

Since $\rho_v = f(M, T, p)$,

$$\mathrm{d}\rho_v = \left(\frac{\partial \rho_v}{\partial M}\right)_{T,P} \mathrm{d}M + \left(\frac{\partial \rho_v}{\partial T}\right)_{M.P} \mathrm{d}T + \left(\frac{\partial \rho_v}{\partial p}\right)_{M.T} \mathrm{d}p \tag{25}$$

Substituting (25) into (24) and combining terms leads to the final result

$$\nabla\left(\frac{\rho_v}{\rho_g}\right) = D_{v_M}\nabla M + D_{v_T}\nabla T + D_{v_p}\nabla p \tag{26}$$

where the diffusion coefficients are defined as

$$D_{v_M} = \frac{p}{R_a T \rho_g^2} \frac{\partial \rho_v}{\partial M} \tag{27a}$$

$$D_{v_T} = \frac{p}{R_a T \rho_g^2} \left(\frac{\partial \rho_v}{\partial T} + \frac{\rho_v}{T} \right) \tag{27b}$$

$$D_{v_P} = \frac{p}{R_a T \rho_g^2} \left(\frac{\partial \rho_v}{\partial p} - \frac{\rho_v}{p} \right) \tag{27c}$$

Substituting this result into Eqs. (19) and (20) yields the total moisture and energy equations in terms of the three primary variables M, T, and p.

To obtain the transport equation for pressure we begin with the continuity equation for air (Eq. (5)) and note that the combined convective and diffusive flux of air can be written as

$$j_a = \rho_a v_a = \frac{-\rho_a K k_{rg}}{\mu_g} \nabla p - \rho_g D_{av} \nabla \left(\frac{\rho_a}{\rho_g} \right) \tag{28}$$

For the case of binary diffusion the flux of air can be written in terms of the vapor flux

$$\nabla \rho_a = \frac{1}{R_a T} \nabla p - \frac{1}{R_a T^2} \nabla T - \frac{R_v}{R_a} \nabla \rho_v \tag{29}$$

which has been established in Eqs. (24) and (25). Making use of (7a), (7b), and (7c), the left-hand side of (5) can be written

$$\frac{\partial}{\partial t}(\rho_a \phi_a) = \frac{\partial}{\partial t} \left[\frac{p - p_v}{R_a T} \left(\varepsilon - \frac{\rho_l \phi_l}{\rho_l} - \frac{\rho_v \phi_v}{\rho_v} \right) \right] \tag{30}$$

Using Eq. (14), this becomes

$$\frac{\partial}{\partial t} \left\{ \frac{p - p_v}{R_a T} \left[\varepsilon - \frac{\rho_s \phi_s M}{\rho_l} + \frac{\rho_b \phi_b}{\rho_l} + \rho_v \phi_v \left(\frac{1}{\rho_l} + \frac{1}{\rho_v} \right) \right] \right\} \tag{31}$$

For typical drying problems the time evolution of the mass fraction of the vapor phase is very small compared to that for the liquid phase or bound water. Therefore, the last term in the brackets in (31) may be dropped. Making use of the chain rule to expand $p_v(M, T)$ results in

$$
\begin{aligned}
&\left(\frac{\varepsilon}{R_a T} - \frac{\rho_s \phi_s M}{\rho_l R_a T} + \frac{\rho_b \phi_b}{\rho_l R_a T}\right)\frac{\partial p}{\partial t} + \frac{p - p_v}{\rho_l R_a T}\frac{\partial}{\partial t}(\rho_b \phi_b) \\
&+ \left(-\frac{\varepsilon p_a}{R_a T^2} - \frac{\varepsilon}{R_a T}\left(\frac{\partial p_v}{\partial T}\right)_M + \frac{p_a}{R_a T^2}\frac{\rho_s \phi_s M}{\rho_l} - \frac{p_a}{R_a T^2}\frac{\rho_b \phi_b}{\rho_l}\right. \\
&\left. + \frac{\rho_s \phi_s M}{\rho_l R_a T}\left(\frac{\partial p_v}{\partial T}\right)_M - \frac{\rho_b \phi_b}{\rho_l R_a T}\left(\frac{\partial p_v}{\partial T}\right)_M\right)\frac{\partial T}{\partial t} + \left(-\frac{\varepsilon}{R_a T}\left(\frac{\partial p_v}{\partial M}\right)_T\right. \\
&\left. + \frac{\rho_s \phi_s}{\rho_l}\left(\frac{p - p_v}{R_a T}\right) + \frac{\rho_s \phi_s M}{\rho_l R_a T}\left(\frac{\partial p_v}{\partial M}\right)_T - \frac{\rho_b \phi_b}{\rho_l R_a T}\left(\frac{\partial p_v}{\partial M}\right)_T\right)\frac{\partial M}{\partial t} \\
&+ \nabla \cdot \left[\frac{\rho_a^* K k_{rg}}{\mu_g}\nabla p - \rho_g D_{av}\left(\frac{1}{\rho_g R_a T}\left(1 - \frac{\rho_a}{\rho_g}\right)\nabla p\right.\right. \\
&\left.\left. + \frac{p}{\rho_g R_a T^2}\left(\frac{R_a \rho_a}{R_g \rho_g} - 1\right)\nabla T - \frac{R_v}{\rho_g R_a}\nabla \rho_v\right)\right] = 0
\end{aligned}
\tag{32}
$$

In Eq. (32) we first note that it is generally assumed that the mass fraction of bound water is constant if free water exists. This, along with the assumption that the mass fraction of bound water is much greater than that of the vapor, leads to

$$
\rho_b \phi_b = \begin{cases} \text{constant} & \text{if} \quad \phi_l > 0 \\ \rho_s \phi_s M & \text{if} \quad \phi_l = 0 \end{cases} \tag{33}
$$

Finally, we note that the Kelvin effect, $(\partial p_v / \partial M)$, determined from Eq. (12) is negligible except for very small pores, and the terms involving $(\partial p / \partial T)$ can be determined directly from Eq. (12)

$$
\left(\frac{\partial p_v}{\partial T}\right)_M = \frac{p_v h_{fg}}{R_v T^2} \quad \text{for} \quad v_v \approx v_{fg} \tag{34}
$$

As a result, Eqs. (19), (20), and (32) represent a general model for the drying problem.

B. Boundary Conditions

The boundary conditions of practical interest for most drying problems are convective or radiative or, in some cases, a combination of the two. The case of direct contact appears occasionally in situations where the porous medium is in direct contact with a heated surface. Roll dryers for paper and textiles and press dryers for veneer and wood-based composites are examples. For the general case of a combined convective/radiative boundary condition we can write

$$[\rho_l v_l - \rho_l D_{bl} \nabla(\rho_b \phi_b) + \rho_v v_g \phi_v - \rho_g D_{va} \nabla\left(\frac{\rho_v \phi_v}{\rho_g}\right)]_{\text{surface}} = h_m(p_{v_{\text{surface}}} - p_{v,\infty}) \tag{35}$$

$$[-k_{\text{eff}} \nabla T + (\rho_l v_l + \rho_b v_b) h_{fg} - \rho_b v_b h_s]_{\text{surface}} = h(T_{\text{surface}} - T_\infty) + \sigma^* \varepsilon^* \left(T^4_{\text{surface}} - T^4_\infty\right) \tag{36}$$

$$p = p_\infty \tag{37}$$

where h_m and h are the convective mass and heat transfer coefficients respectively, σ^* is the Stefan–Boltzmann constant, and ε^* is the surface emissivity. The convective heat transfer coefficient is typically determined from the existing empirical correlations for the appropriate geometry and flow conditions characterized by the Reynolds number or from the exact solutions that arise from the laminar boundary layer theory. The general form is

$$Nu = C\, Re^m Pr^n \tag{38}$$

where C is dependent on the geometry, m is for simple geometries, dependent on whether the flow is laminar or turbulent, and Pr is the Prandtl number, which is the ratio of the diffusivity of momentum (kinematic viscosity) to the diffusivity of heat (thermal diffusivity). The power, n, can be assumed to be 1/3 for most applications. The convective mass transfer coefficient can then be determined from the Chilton–Colburn analogy

$$\frac{h_m}{h} = \frac{D_{av} Le^n}{k} \tag{39}$$

where Le is the Lewis number, which is the ratio of the diffusivity of heat to the mass diffusivity.

Some notes of caution are associated with the boundary conditions as expressed in Eqs. (35), (37), (38), and (39). First, in the case of high intensity drying, where the mass flux in the gas phase at the surface of the porous medium is significant, the resulting blowing effect alters the convective heat and mass transfer coefficients (Kays and Crawford 1980). Furthermore, the pressure adjusts from the surface pressure to p_∞ through a loss coefficient analogous to that which would be used to predict the effluent from a small hole in a pipe. In addition, Plumb et al. (1985), Chen and Pei (1989), Lee et al. (1992), and Gong and Plumb (1994b) have questioned the use of analogy to determine the mass transfer coefficient from knowledge of the heat transfer coefficient. In all four of these studies it was found that the mass transfer coefficient had to be adjusted to obtain good agreement between model predictions and the experimental results.

Gong and Plumb (1994b) conclude that the convective mass transfer coefficient (for softwood) agrees with that predicted from analogy at high moisture content when a significant portion of the surface is covered with free water. It then decreases to a minimum near the fiber saturation point where the free liquid disappears. At this point the convective mass transfer coefficient is on the order of 30% of that predicted from analogy. It then increases, again approaching that predicted from analogy, as the moisture content approaches zero. This increase at low moisture content is presumably a result of the evaporation of bound water from a nearly continuous solid surface. Earlier experiments with both wood (Plumb et al. 1985) and packed beds of glass beads (Lee et al. 1992) led to similar conclusions.

In the case of direct contact drying (Plumb et al. 1986) the boundary condition for the energy equation is written

$$\left[-k_{\text{eff}}\nabla T + (\rho_l v_l + \rho_b v_b)h_{fg} + \rho_b v_b h_s\right]_{\text{surface}} = h_c\left(T_{\text{surface}} - T_p\right) \tag{40}$$

where h_c is the contact conductance, which is dependent on the materials involved as well as the pressure.

C. Transport Properties

The transport properties which appear in the set of transport equations developed in Section II.A (Eqs. (19), (20), and (32)) include the permeability, K; the liquid phase permeability, k_{rl}; the gas phase permeability, k_{rg}; the liquid and gas phase viscosities, μ_l and μ_g; the capillary pressure function, $F(S)$; the effective vapor phase diffusion coefficient, D_{va}; the diffusion coefficient for bound liquid, D_{bl}; and the effective thermal conductivity, k_{eff}. Except for the viscosities, these properties are all strongly dependent on moisture content (liquid phase saturation) and the porosity and pore structure of the solid phase (geometry). In addition, even in the absence of thermally driven transport the diffusion coefficients and the effective thermal conductivity are temperature dependent. The diffusivities are also weakly dependent on pressure.

A detailed discussion of the above transport properties for a wide range of materials that are traditionally dried is beyond the scope of this review. What will be presented are some of the empirical relations that are traditionally used to estimate transport properties including their dependence on geometry, temperature, moisture content, and pressure. For a more in-depth discussion of transport properties the reader is referred to Kaviany (1995), Dullien (1992), and Adler (1992). Tabulations of thermal conductivities and diffusion coefficients for a wide range of materials can be found in Mujumdar (1995).

1. Permeability and Relative Permeability

The most commonly used relation for the permeability is the well-known Kozeny equation

$$K = \frac{d_p^2 \varepsilon^3}{\beta(1-\varepsilon)^2} \tag{41}$$

In Eq. (41) d_p is the mean particle diameter (for packed beds) or a mean pore size (for monolithic materials), and the suggested value of the constant β, which was originally recommended to be 150, should be 180 on the basis of more recent measurements by MacDonald et al. (1979) and Crawford and Plumb (1986). Experiments leading to Eq. (41) were made using packed beds of relatively uniform sand or glass beads; thus, the Kozeny equation should be taken to be only a rough estimate for monolithic materials (food products or wood) or packed beds containing nonspherical particles or composed of particles having a wide distribution of sizes. Furthermore, this relation is restricted to flows dominated by viscous effects and, hence, is limited to Reynolds numbers based on the mean particle or pore diameter of less than unity.

A number of empirical relations for the relative permeabilities have been developed (see Kaviany 1995, for example). These have been developed primarily by soil scientists and groundwater hydrologists, so they are most appropriate for granular materials having an aspect ratio close to unity and a relatively uniform particle size distribution. The relations developed by Fatt and Klikoff (1959), for example, are commonly used

$$k_{rl} = S^{*3} \tag{42}$$

$$k_{rg} = (1 - S^*)^3 \tag{43}$$

where S^* is the normalized saturation, defined as

$$S^* = \frac{S - S_{ir}}{1 - S_{ir}} \tag{44}$$

In Eq. (44) S_{ir} is the irreducible saturation (saturation below which the liquid is no longer a continuous phase) for the porous material in question. The irreducible saturation is typically in the range of 10–15%. Other commonly used correlations include those of van Genuchten (1980) and Verma et al. (1985).

2. Capillary Pressure

We have assumed in the derivation in Section II.A (Eq. (17)) that the capillary function takes the general form proposed by Leverett (1941) and the characteristic length r^* is taken to be $\sqrt{K/\varepsilon}$. The function $F(S)$ has been represented as a polynomial fit to Leverett's original data (Udell 1985)

$$F(S) = 1.417(1 - S^*) - 2.120(1 - S^*)^2 + 1.263(1 - S^*)^3 \tag{45}$$

or as a power law (van Genuchten 1980)

$$F(S) = -p_0[S^{*-1/\lambda} - 1]^{1-\lambda} \tag{46}$$

It is important to note that the capillary function exhibits hysteresis (Dullien 1998). Hysteresis can come into play in some drying problems (Lee et al. 1992; Molenda et al. 1992). In these cases it is essential to have a capillary pressure relationship for both drying and wetting (imbibition) (see Luckner et al. 1989).

3. Diffusivities of the Vapor Phase and Bound Water

The diffusivity of water vapor in a porous medium in the presence of immobile liquid remains a subject of investigation (Gu et al. 1998). We begin with the binary diffusion coefficient for water vapor in air, which can be expressed (Vargaftik 1975) as

$$D_c = D_c^0 \frac{p_0}{p}\left(\frac{T}{273}\right)^n \tag{47}$$

where $p_0 = 1$ bar, $D_c^0 = 2.13 \times 10^{-5}$ m^2/s, and $n = 1.8$. In a dry porous medium this diffusion coefficient must be corrected for the effect of the presence of the solid phase on the diffusion process. This is done through a tortuosity coefficient, τ. Eidsath et al. (1983) have shown how the tortuosity evolves theoretically through volume averaging. Through comparison with experimental data a value of 0.7 is recommended. Measurements for soil have resulted in tortuosities of around 0.66 (Cass et al. 1984). The measurements of Gu et al. (1998) in packed beds of uniform glass beads resulted in $\tau = 0.78$. Hence, for a dry porous material the effective diffusivity is written

$$D_{va} = \tau\varepsilon D_c \tag{48}$$

If immobile (below the irreducible saturation) liquid is present, then it is logical that the effective diffusivity would be further reduced since the tortuosity and the gas phase volume fraction will be smaller, leading to

$$D_{va} = \tau\varepsilon S_g D_c \tag{49}$$

Indeed, the model developed by Millington and Quirk (1961) suggests a decrease in D_{va} as the liquid saturation increases (S_g decreases). To the contrary, Phillip and de Vries (1957) proposed that the diffusion of water vapor should be enhanced in the presence of immobile liquid if a temperature gradient is imposed. The proposed enhancement is a result of condensation and evaporation at the liquid–vapor interfaces. The evaporation and condensation result in a short circuit for the vapor transport, thus essentially increasing the tortuosity coefficient. This enhancement has been verified experimentally by several investigators including Jackson (1964), Cass et al. (1984), and Gu et al. (1998). Webb (1998) suggest that the enhancement can take place even in the absence of temperature gradients. Although the data on enhanced diffusion are not extremely consistent, they do agree that there is very little enhancement at low liquid saturation ($S_l < 0.05$) and that the enhancement increases significantly (values as high as four or greater) as the liquid saturation increases to 0.2. With enhancement included, the effective vapor phase diffusion coefficient can be written

$$D_{va} = \eta \tau \varepsilon S_g D_c \tag{50}$$

where η is the enhancement factor, which depends on the liquid phase saturation.

4. Effective Thermal Conductivity

Numerous models have been proposed for the prediction of the effective thermal conductivity of complex mixtures, including porous materials partially saturated with liquid. For dry granular porous media (packed beds) empirical models have been developed that can be used successfully to predict the effective thermal conductivity (Kaviany 1995, pp 144–145). For packed beds partially saturated with liquid the general correlations that are available are less conclusive (Plumb 1991). The simple model given by Eq. (51), due to Somerton et al. (1974), is widely used

$$k_{\text{eff}} = k_0 + S^{*1/2}(k_l - k_0) \tag{51}$$

where k_0 is the effective thermal conductivity of the dry porous medium. The limited experimental results presented in Plumb (1991) indicate that this relationship underpredicts the effective thermal conductivity. For wood, a hygroscopic porous material, both Turner and Perre (1997) and Plumb and Gong (1997) utilize an effective thermal conductivity that is a linear function of moisture content.

III. MODELING DRYING PROCESSES

For many applications it is neither desirable nor necessary to solve the complete set of equations developed in Section II.A along with the boundary conditions of Section II.B. The model equations can be greatly simplified through the judicious use of assumptions appropriate to the application of interest. Judicious assumptions are developed through the use of scale analysis (Bejan 1984), which is used to determine the relative magnitudes of the contribution due to the various physical phenomena that affect the drying process. In the following discussion scale analysis will be used to illustrate the justification for simplifying assumptions commonly utilized in analyzing or modeling practical drying problems. Before presenting these assumptions, a brief discussion of the results of experimental observations is appropriate.

Experimental observations of the drying of porous materials are often presented in the format illustrated in Figure 1, which shows drying rate as a function of moisture content. Although there are many drying scenarios that would differ significantly, Figure 1 illustrates qualitatively what one might observe if a porous material is dried convectively in air at temperature less than 100°C. After an initial transient period the free water at the surface will reach the saturation temperature. The drying rate is constant while the surface remains at high moisture content. This is because the mass transfer process is externally controlled. That is, the resistance to mass transfer by convection from the surface is much greater than the internal resistance to liquid water flow by capillary action to the surface. Hence, the gradient in moisture content (or saturation) in the material being dried is small. At some point in the process the internal resistance begins to dominate since

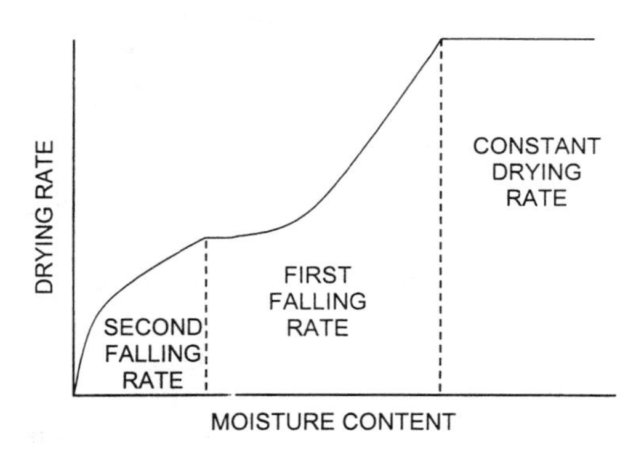

Figure 1. Drying rate as a function of moisture content for a porous material.

the relative permeability for the liquid phase decreases significantly as the saturation (moisture content) decreases. This leads to the first falling rate period. A second falling rate period is often observed. This is the result of the switch from liquid phase transport to vapor phase transport becoming the dominant mechanism for the movement of moisture to the drying surface. As will be illustrated in the next section, the effective diffusivity for vapor is typically less than that for liquid (a factor of 26 for the specific numbers used for illustration). Thus, the likelihood of the lumped capacitance or constant drying rate model being valid for a drying problem at low moisture content (no free liquid) is not high.

In order to illustrate scale analysis quantitatively it is necessary to have knowledge of the transport properties discussed in Section II.C. Some values that may be typical have been selected and are listed in Table 1. The most important property is the permeability, which is selected to be $10^{-11}\,\mathrm{m}^2$. For a material having a porosity of 0.35, such as a packed bed of sand or other uniformly sized particulate material, this implies, from Eq. (41), a particle diameter of 133 μm. The fluid to be removed is assumed to be water and the dimension of the material in the drying direction is assumed to be 2 cm. We will assume that the material is a slab exposed to the drying medium on both sides; thus the characteristic dimension used in the calculations is the half thickness of 1 cm. The convective heat transfer coefficient is assumed to be $20\,\mathrm{W/m^2.K}$, which is on the high side for many applications. The Chilton–Colburn analogy (Incropera and DeWitt 1990) can then be used to estimate the mass transfer coefficient

$$h_m = \frac{h D_c L^{1/3}}{k} \tag{52}$$

Table 1. Parameter values used to illustrate scale analysis

Particle/pore diameter	$d_p = 133\,\mu\mathrm{m}$
Permeability	$K = 10^{-11}\,\mathrm{m}^2$
Porosity	$\epsilon = 0.35$
Characteristic length scale	$L = 0.01\,\mathrm{m}$
Convective heat transfer coefficient	$h = 20\,\mathrm{W/m^2.K}$
Diffusion coefficient for water vapor in air	$D_c = 2 \times 10^{-5}\,\mathrm{m^2/s}$
Thermal conductivity of air	$k = 0.025\,\mathrm{W/m.K}$
Density	$\rho_2 = 1000\,\mathrm{kg/m^3}$
Mixture specific heat	$c_p = 1000\,\mathrm{J/kg.K}$
Tortuosity	$\tau = 0.78$

Assuming that $Le^{1/3} \approx 1$ for water vapor in air, and using the values for the thermal conductivity of air and the diffusivity of water vapor in air given in Table 1, results in a convective mass transfer coefficient of 0.016 m/s.

A. The Thermal Transient

As indicated above, the drying process begins with a thermal transient where the material to be dried is heated from its initial temperature to the saturation temperature (which is approximately equal to the wet bulb temperature) of water. It is important to estimate the length of this transient and compare it to the total drying time to determine whether or not the drying that takes place during this transient period contributes significantly to the total drying process. The time scale associated with the propagation of this thermal front from the surface to the center of the slab of material under consideration is given by

$$t_T = L^2/\alpha = (.01\,\mathrm{m})^2/1.\mathrm{E}\text{-}06\,\mathrm{m}^2/\mathrm{s} = 100\,\mathrm{s} \tag{53}$$

The total drying time can be estimated from the energy requirement to evaporate all of the water. If the slab is initially saturated then

$$t_{\mathrm{dry}} = \frac{mh_{fg}}{h(T_S - T_\infty)} \tag{54}$$

where m is the initial mass of liquid per unit area and T_s is the surface temperature. If the average temperature difference during the drying process is 10°C (with $h = 20\,\mathrm{W/m^2.K}$), then the total drying time is 25 hours. Hence, consistent with observation, the initial transient usually represents an insignificant portion of the total drying time.

B. Constant Drying Rate

The assumption necessary to model a drying process as a constant rate process is analogous to the lumped capacitance assumption used to model transient heat conduction. As indicated above, the lumped capacitance assumption will be valid when the resistance to internal flow of moisture to the surface is much less than the resistance to mass transfer by convection from the surface. For situations where the primary internal mechanism is liquid transport via capillary action the ratio of these two resistances evolves through the nondimensionalization of the boundary condition given by Eq. (36), considering liquid transport only and utilizing Eqs. (8a) and (18a)

$$-\frac{\partial S}{\partial x^*} = \frac{h_m L}{D_l}(p_v^* - 1) \tag{55}$$

where

$$D_l = \frac{Kk_{rl}\sigma}{\mu_l\sqrt{K/\varepsilon}}\frac{\partial F(S)}{\partial S} \tag{56}$$

can be viewed as an effective diffusion coefficient for liquid transport due to capillary action. The quantity $h_m L/D_l(S)$ is the mass transfer Biot number (Bi_m). The general rule of thumb is that for $Bi_m < 0.1$ (Incropera and DeWitt 1990) saturation gradients within the material being dried can be neglected, leading to the lumped capacitance model for drying during the constant rate period. Because the effective liquid phase diffusion coefficient $D_l(S)$ decreases with saturation, the Biot number increases with time, inevitably leading to a point where the lumped capacitance model is no longer valid and internal gradients in moisture content must be considered. Using the properties given in Table 1, the Biot number for liquid phase transport resulting from capillary action can be estimated using Eq. (56). If we take the liquid phase relative permeability to be unity, which is appropriate at the beginning of the drying process if the material is nearly saturated, and assume $\mathrm{d}F/\mathrm{d}S$ is of the order unity (for Udell's polynomial at $S^* = 0.9$, $\mathrm{d}F/\mathrm{d}S \simeq -1$), then, using properties from Table 1, the effective diffusivity for the liquid phase can be estimated to be

$$D_l = 1.3E\text{-}04\,\mathrm{m}^2/\mathrm{s} \tag{57}$$

The Biot number for mass transfer is then

$$Bi_m = \frac{h_m L}{D_l} = 1.23 \tag{58}$$

which implies that, for this case, the lumped capacitance model is not appropriate and a constant rate drying period will not exist. From Eqs. (56) and (58) it can be readily seen that a thin material (small L) or a material having a higher permeability would have a value for Bi_{ml} less than 0.1.

For the case of diffusive drying, where transport in the vapor phase is dominant (low initial liquid saturation), it is more difficult to imagine a situation that would involve a constant drying rate. From Table 1 we estimate the effective diffusivity for the vapor phase by multiplying the binary diffusion coefficient for water vapor in air by the tortuosity and the porosity

$$D_v = \varepsilon\tau D_c = (0.35)(0.78)(2\times 10^{-5}\,\mathrm{m}^2/\mathrm{s}) = 5\times 10^{-6}\,\mathrm{m}^2/\mathrm{s} \tag{59}$$

leading to

$$Bi_v = \frac{h_m L}{D_v} = 32 \tag{60}$$

which rules out the lumped capacitance solution except for those cases involving extremely thin materials (paper or textiles) or situations involving a very small convective mass transfer coefficient.

C. Low Intensity Drying

Several investigators (Spolek and Plumb 1980; Bories 1991; Turner and Perre 1997) have illustrated that if a material having sufficiently high permeability is dried at a temperature below the boiling point the transport of liquid and vapor driven by gradients in the total pressure can be neglected. Utilizing this assumption reduces the number of primary variables from three to two, moisture content and temperature.

$$\rho_s\phi_s\frac{\partial M}{\partial t}+\nabla\cdot\left\{\frac{-\rho_l^*Kk_{rl}}{\mu_l}\nabla p_l-D_{bl}\nabla\rho_b^*-\rho_gD_{va}\nabla\left(\frac{\rho_v}{\rho_g}\right)\right\}=0 \tag{61}$$

$$\overline{\rho c_p}\frac{\partial T}{\partial t}+\left(\frac{-\rho_l^*Kk_{rl}}{\mu_l}\nabla p_l\right)c_{pl}\cdot\nabla T=\nabla\cdot(k_{\text{eff}}\nabla T)-h_{fg}(\dot{m}_l+\dot{m}_b)-h_s\dot{m}_b \tag{62}$$

Note that in the energy equation convective transport due only to the capillary-driven flow is retained. One note of caution against dropping the pressure-driven (convective) transport must be included. Both the experimental results of Lee et al. (1992) for packed beds of glass beads and the numerical predictions of Turner and Perre (1997) for wood show that if the drying process is initiated at high moisture content an underpressure (pressure less than that in the drying medium) will be developed by the "sucking action" resulting from capillarity. The underpressure works against the capillary pressure, thus impeding the flow of liquid to the surface. The importance of this effect as a function of the transport properties and drying parameters has not been well established.

The effects of gravity are discussed in considerable detail by Puiggali and Quintard (1992). They illustrate that gravitational effects are negligible when the gravity number (capillary number times the bond number) is small and the capillary number large. These two parameters are expressed as

gravity number:

$$N_g=\frac{K\rho_l^2g}{\mu_l\dot{m}_s} \tag{63}$$

capillary number:

$$N_c = \frac{\sigma}{\rho_l g d_p L} \tag{64}$$

where $\dot{m}_s$ is the surface mass flux. If the particle size, and thus the permeability, is sufficiently small, then gravity is negligible. However, a specific range of particle size or permeability cannot be identified because of the dependence on the drying rate and the depth of the medium being dried.

In many cases, heat diffuses into the material much more rapidly than liquid (or total moisture) diffuses out. Thus, after an initial transient which is relatively short compared to the total drying time, the material remains nearly isothermal during the entire drying period (Gong and Plumb 1994b). For wood, during the constant rate period and much of the first falling period, the temperature remains at the saturation temperature as long as free liquid remains at the surface. Once the free liquid disappears from the surface (the second falling rate period) the temperature of the wood undergoes a second relatively short transient as the temperature approaches the temperature of the drying medium. Turner and Perre (1997) illustrate similar behavior for low temperature drying. From this discussion it is possible to conclude that if the initial and intermediate transient is short compared to the total drying time, then the drying process can be adequately approximated using an isothermal model. The scale analysis associated with this assumption is illustrated in Section III.A.

One further simplifying assumption in modeling low intensity drying concerns the relative magnitude of the vapor flux compared to the liquid flux at high moisture content. Bories (1991) shows that the effective diffusivity for the liquid phase is much higher than that for the vapor phase at high moisture content. Plumb et al. (1985) have invoked this assumption in a model for wood drying that achieved reasonable agreement with experimental results. When modeling the drying process at low moisture content, where transport in the vapor phase or transport of bound water is dominant, this assumption offers little advantage because the diffusive transport of vapor or bound water must be included in the model.

The relative importance of the liquid phase transport as compared to the diffusive transport in the vapor phase can be determined by comparing the time scales associated with the two processes. The effective diffusivities for the two processes are given in Eqs. (56) and (59). The time scales for the processes can be computed by dividing the square of the length scale by the diffusivity. Since the length scale is the same for both processes the ratio of the time scales becomes

$$\frac{D_l}{D_v} = \frac{1.3 \times 10^{-4}}{5 \times 10^{-6}} = 26 \tag{65}$$

indicating that the time scale for the vapor transport is much larger than that for the liquid transport. This justifies neglecting the transport in the vapor phase. We should note that if the permeability were to be reduced by two orders of magnitude to 10^{-13} m^2 this assumption would not be justified. For convective transport in the vapor phase (driven by pressure gradients) the time scale can be developed by noting that

$$q_s = \dot{m}h_{fg} = \rho U_c h_{fg} \tag{66}$$

Therefore the time scale can be derived

$$t_{vc} = L/U_c = L\rho h_{fg}/q_s \tag{67}$$

This allows the following ratios of time scales to be written

$$\frac{t_{vD}}{t_{vc}} = \frac{Lq_s}{D_v \rho h_{fg}} = 0.77 \tag{68}$$

$$\frac{t_{lD}}{t_{vc}} = \frac{Lq_s}{D_l \rho h_{fg}} = 3 \times 10^{-6} \tag{69}$$

These ratios of time scale illustrate that the liquid phase transport dominates until the surface dries out, resulting in the effective diffusivity of the liquid phase approaching zero as the relative permeability approaches zero.

D. High Intensity Drying

For the sake of discussion, drying above the boiling point or under vacuum will be termed high intensity drying. For this case the dominant physical phenomena can be much different from those for the case of low intensity drying. In the most extreme case, when convection driven by the gas phase pressure gradient dominates, diffusion in the gas phase and capillary transport of liquid can be neglected. For this situation a drying front develops and propagates into the dried material. Thus, the general equations (19), (20), (32) become

$$\rho_s^* \frac{\partial M}{\partial t} + \nabla \cdot \{-(D_{Pl} + D_{Pg})\nabla p - D_{bl}\nabla \rho_b^*\} = 0 \tag{70}$$

$$\overline{\rho c_p}\frac{\partial T}{\partial t} + (D_{Pl}c_{pl} + D_{Pg}c_{pg})\nabla p \cdot \nabla T = \nabla \cdot (k_{\text{eff}}\nabla T) - h_{fg}(\dot{m}_l + \dot{m}_b) - h_s \dot{m}_b \tag{71}$$

$$\frac{\rho_g \rho_s}{\rho_l}\frac{\partial M}{\partial t} + \nabla \cdot \left[-\frac{\rho_g \phi_g}{\mu_g} K k_{rg} \nabla p\right] = \dot{m}_l + \dot{m}_b \tag{72}$$

It should be noted that it is necessary to correct the convective boundary conditions (Eqs. (35) and (36)) for the nonzero velocity normal to the drying surface for high intensity drying.

Examples of studies of high intensity drying involving a model similar to that discussed above can be found in Bories (1991) and Plumb et al. (1986). In both of these models liquid transport via capillary action is retained in the model, so they might be classified as medium intensity instead of high intensity drying as described in Eqs. (70)–(72).

IV. OTHER COMMON DRYING PROCESSES

A. Electromagnetic Drying

Recent discussions of the use of electromagnetic waves in drying can be found in Turner and Jolly (1991), Schmidt et al. (1992), Jones (1992), Chen and Schmidt (1997), and Perre and Turner (1997b). Electromagnetic waves at four different wavelengths have been found to be useful to some extent for drying applications. Ultraviolet (UV) energy is used for curing inks and coatings. Infrared (IR), which, like UV, is absorbed at the surface, is used extensively for the drying of paper and coatings or thin films. Radio frequency (RF) and microwave energy are of greater interest for drying porous materials because dielectric heating provides a volumetric heat source internal to the medium to be dried. They are of particular interest because water has a high loss factor for both microwave and RF, resulting in significant heat generation concentrated in the areas where liquid water is present. Microwave or RF drying is usually coupled with conventional convective drying; thus the model equations and boundary conditions are the same as those discussed in Sections II.A and II.B, except for the addition of a volumetric heat source in the energy equation (20).

The papers on dielectric drying referenced above use an approximation to the solution to Maxwell's equations to estimate the strength of the local electric field. A few papers, for example Li et al. (1994) and Perre and Turner (1997b), have solved Maxwell's equations jointly with the transport equations in addressing a drying problem. The source term is typically approximated as

$$S = \tfrac{1}{2}\omega\varepsilon_0\varepsilon'' E^2 \tag{73}$$

where E is the electric field strength at the material surface, ω the frequency, ε_0 the permittivity of free space and ε'' an effective loss factor. The loss factor is a function not only of the particular material being dried, but also of frequency, temperature, and moisture content. Chen and Schmidt (1997) show the variation of both the dielectric constant and the loss factor for several materials as a function of temperature and moisture content.

As indicated above, the use of dielectric heating offers the advantage of producing a volumetric heat source which is strongest where the moisture content is highest. This can result in internal temperatures which are higher than the temperature of the drying medium. In fact, internal boiling can be easily produced, resulting in pressure gradients which can drive liquid and vapor toward the surface. The model and experimental results that are available indicate that convective drying, enhanced dielectrically, can result in more rapid increase of internal temperature early in the drying process and the sustenance of a constant rate, or at least a higher drying rate, for a significant portion of the drying process.

B. Freeze Drying

Materials that are temperature sensitive, including certain biological materials, pharmaceuticals, and food products, are often freeze dried. Reviews of the freeze drying process can be found in Mellor (1978) and Liapis and Bruttini (1995, 1997). The freeze drying process is generally considered in three stages. First the material is frozen rapidly, usually to a temperature well below the freezing point of water or solvent to be removed. The free water is then removed by sublimation in the second stage. In the final stage, the bound or adsorbed water is removed. The process is generally carried out at low pressure in a chamber from which the vapor is continuously removed. Aside from the advantage of low temperature processing, freeze drying also eliminates or at least greatly reduces moisture transport in the liquid phase. This offers some advantage for a certain materials. For example, ceramics processed in the aqueous phase typically have a very small particle size resulting in a very high capillary pressure (Scherer 1992). The capillary effects can produce cracks through the action of surface tension forces on individual particles or clusters of particles. It is important to note that the freezing process can also produce structural damage due to the significant decrease in the water density near the freezing point. For solvents other than water that exhibit a small change in density, freeze drying can be a viable alternative.

In modeling the freeze drying process we need to consider only the continuity equations for the water vapor, the air, and the bound water along with the energy equation. For this case Eq. (32) remains and Eqs. (15) and (16) can be written as follows

$$\rho_s \phi_s \frac{\partial M}{\partial t} + \nabla \cdot \left[-D_{bl} \nabla \rho_b^* - D_{pg} \nabla p - \rho_g D_{va} \nabla \left(\frac{\rho_v}{\rho_g} \right) \right] = 0 \tag{74}$$

$$\overline{\rho c_p}\frac{\partial T}{\partial t} + \left[\frac{\rho_g}{\rho_v} D_{pg} c_{pg} \nabla p\right] \cdot \nabla T = \nabla \cdot (k_{\text{eff}} \nabla T) - h_{ls}(\dot{m}_l + \dot{m}_b) - h_s \dot{m}_b \tag{75}$$

V. CONCLUDING REMARKS

The advances in computing power that have taken place over the past 20 years have allowed the development and solution of complex drying models that were previously not possible to solve. Still much remains to be done before we can actually take the knowledge of transport phenomena that occur during drying of porous materials and apply it directly to the design of drying equipment.

One of the complicating factors is the paucity of data that is available on the necessary transport properties. As indicated in Section II.C, these include permeability, relative permeabilities, capillary pressure, effective thermal conductivity, and effective diffusivities for both bound water and water vapor. Because of the wide variety of materials of interest, it is unlikely that this problem will be overcome in the near future. The ultimate solution would be the development of empiricism or theory that allows the computation of properties from knowledge of structure—one of the fundamental issues of materials science.

A second complicating factor is that our ability to compute far exceeds the availability of experimental results to validate the existing models. Experimental progress is hampered by the difficulty in making local measurements of moisture content *in situ* during drying. The number of papers in the literature reporting numerical results far exceeds the number reporting experimental results. Much remains to be done in carefully validating models, particularly some of the simplifying assumptions that are developed through scale analysis in Section IV.D. It is these simplified models that offer the potential to contribute in a significant way to the design of actual drying equipment.

Two areas that deserve additional study, that have been touched upon only recently, are those of dealing with real porous materials that are not homogeneous and coupling the transport problem with the mechanical problem so that the stresses and ultimately the damage to products that occurs during drying can be predicted. This coupled problem, involving shrinkage, also gives rise to the question of the effect of shrinkage on the transport properties.

Finally, the ultimate goal of modeling is, at least in part, to develop tools that are useful in the design of actual drying equipment. Currently there remains a considerable gap. Real drying involves materials having complex geometry, e.g., wood chips, food particles, and granular materials, having a wide size distribution, etc. These types of geometries do not lend themselves well to simple one- or two-dimensional models. In addition, the boundary conditions can be complex. For example, in a rotary drum dryer the material receives heat through contact with the walls of the dryer, through contact with other particles, and from convection. In this type of situation the specification of the surface heat flux is not simple and straightforward.

NOMENCLATURE

Roman Letters

c_p	specific heat (J/kg.K)
D_{va}	effective diffusion coefficient for vapor in air (m^2/s)
D_{bl}	effective diffusion coefficient for bound liquid (m^2/s)
$F(S)$	capillary function
g	gravitational acceleration (m/s^2)
h	convective heat transfer coefficient ($W/m^2.K$)
h_{es}	latent heat of evaporation (J/kg)
h_{ls}	latent heat of sublimation (J/kg)
h_m	convective mass transfer coefficient (kg/s)
h_s	heat of adsorption (J/kg)
j	mass flux ($kg/m^2.s$)
k_{eff}	effective thermal conductivity (W/m.K)
k_r	relative permeability
K	permeability (m^2)
Le	Lewis number
$\dot{m}$	mass generation ($kg/m^3.s$)
M	moisture content, dry basis
Nu	Nusselt number
p	pressure (N/m^2)
Pr	Prandtl number
R	gas constant (kJ/kg.K)
Re	Reynolds number
S	saturation
$\dot{s}$	energy source term (W/m^3)
t	time (s)
T	temperature (K)
v	velocity (m/s)

Greek Letters

ε porosity
μ dynamic viscosity ($N.s/m^2$)
ρ density (kg/m^3)
σ surface tension (N/m)
ϕ volume fraction

Subscripts

a air
b bound water
c capillary
g gas phase
l liquid phase
s solid phase, or saturation
t temperature
v vapor
σ surface tension

REFERENCES

Adler PM. Porous Media: Geometry and Transport. Massachusetts: Butterworth–Heinemann, 1992.

Baggio P, Bonacina C, Schrefler BA. Some considerations of modeling heat and mass transfer in porous media. Transp Porous Media 28:233–251, 1997.

Bejan A. Convective Heat Transfer. New York: John Wiley & Sons, 1984.

Bories SA. Fundamentals of drying capillary-porous bodies. In: Kakac S, Kilkis B, Kulacki FA, Arinc F, eds. Convective Heat and Mass Transfer in Porous Media, NATO ASI Series E, Vol 196. Dordrecht: Kluwer, 1991, pp 391–434.

Cass A, Campbell GS, Jones TL. Enhancement of thermal water vapor diffusion in soil. Soil Sci Soc Am J 48:25–32, 1984.

Chen P, Pei DCT. A mathematical model of drying processes. Int J Heat Mass Transfer 32:297–310, 1989.

Chen P, Schmidt PS. Mathematical modeling of dielectrically-enhanced drying. In: Turner I, Mujumdar AS, eds. Mathematical Modeling and Numerical Techniques in Drying Technology. New York: Marcell Dekker, 1997, pp 439–479.

Cook EM, DuMont HD. Process Drying Practice. New York: McGraw-Hill, 1991.

Couture F, Fabrie P, Puiggali J-R. An alternative choice for the drying variables leading to a mathematically and physically well described problem. Drying Technol 13:519–550, 1995.

Crawford CW, Plumb OA. The influence of surface roughness on resistance to flow through packed beds. J Fluids Eng 108:343–347, 1986.

Dullien FAL. Porous Media: Fluid Transport and Pore Structure. 2nd ed. New York: Academic Press, 1992.

Dullien FAL. Capillary effects and multiphase flow in porous media. J Porous Media 1:1–29, 1998.

Eidsath A, Carbonell RG, Whitaker S, Hermann LR. Dispersion in pulsed systems—III. Comparison between theory and experiments for packed beds. Chem Eng Sci 38:1803–1816, 1983.

Fatt I, Klikoff, WA. Effect of fractional wettability on multiphase flow through porous media. AIME Trans 216:246, 1959.

Ferguson WJ. A numerical prediction of the effect of airflow and wet bulb temperature on the stress development during convective wood drying. In: Turner I, Mujumdar AS, eds. Mathematical Modeling and Numerical Techniques in Drying Technology. New York: Marcel Dekker, 1997, pp 239–277.

Franzen K, Liang H, Litchfield B, Murakami E, Nichols C, Waananen K, Miles G, Okos M. Design and Control of Energy Efficient Drying Processes with Specific Reference to Foods, Vol I—Literature Review. DOE/ID/12608-2(Vol 1), 1987, p iii.

Gong L, Plumb OA. The effect of heterogeneity on wood drying, Part I: Model development and predictions. Drying Technol 12:1983–2001, 1994a.

Gong L, Plumb OA. The effect of heterogeneity on wood drying, Part II: Experimental results. Drying Technol 12:2003–2026, 1994b.

Gu L, Ho CK, Plumb OA, Webb SW. Diffusion with condensation and evaporation in porous media. Proceedings of 7th AIAA/ASME Joint Thermophysics and Heat Transfer Conference, Albuquerque, NM, 1998.

Hartley JG. Coupled heat and moisture transfer in soils: a review. In: Mujumdar AS, ed. Advances in Drying, Vol 4. Washington, DC: Hemisphere, 1987, pp 199–245.

Incropera FP, DeWitt DP. Fundamentals of Heat and Mass Transfer. 3rd ed. New York: John Wiley & Sons, 1990.

Jackson, RD. Water vapor diffusion in relatively dry soil: I. Theoretical considerations and sorption experiments. Soil Sci Soc Proc 172–176, 1964.

Jones PL. Electromagnetic wave energy in drying processes. In: Mujumdar AS, ed. Drying '92. Amsterdam: Elsevier Science, 1992, pp 114–136.

Kaviany M. Principles of Heat Transfer in Porous Media. 2nd ed. New York: Springer, 1995.

Kays WM, Crawford ME. Convective Heat and Mass Transfer. 2nd ed. New York: McGraw-Hill, 1980.

Keey RB. Introduction to Industrial Drying Operations. New York: Pergamon, 1978.

Lartigue C, Puiggali JR, Quintard M. A simplified model of moisture transport and shrinkage in wood. In: Mujumdar AS, Roques M, eds. Drying '89. Washington, DC: Hemisphere, 1989, pp 169–175.

Lee WC, Plumb OA, Gong L. An experimental study of heat and mass transfer during drying of packed beds. Trans ASME J Heat Transfer 114:727–734, 1992.

Leverett MC. Capillary behavior in porous solids. AIME Trans 142:152–169, 1941.

Li W, Ebadian MA, White TL, Grubb RG, Foster D. Temperature and pore pressure distribution in a concrete slat during the microwave decontamination process. In: Hewitt GF, ed. Heat Transfer 1994, New York: Hemisphere, Vol 5. 1994, pp 321–326.

Liapis AI, Bruttini R. Freeze drying. In: Mujumdar AS, ed. Handbook of Industrial Drying. 2nd ed. New York: Marcel Dekker, 1995, pp 309–343.

Liapis AI, Bruttini R. Mathematical models for the primary and secondary drying states of the freeze-drying of pharmaceuticals on trays and in vials. In: Turner I, Mujumdar AS, eds. Mathematical Modeling and Numerical Techniques in Drying Technology. New York: Marcel Dekker, 1997, pp 481–535.

Luckner L, van Genuchten M Th, Nielsen DR. A consistent set of parametric models for the two-phase flow of immiscible fluids in the subsurface. Water Resour Res 25:2187–2193, 1989.

Luikov AV. Heat and Mass Transfer in Capillary Porous Bodies. Oxford: Pergamon, 1966.

MacDonald IF, El-Sayed MS, Mow K, Dullien FAL. Flow through porous media—the Ergun equation revisited. I&EC Fund 18:199–208, 1979.

Mellor JD. Fundamentals of Freeze-Drying. London: Academic Press, 1978.

Milllington RJ, Quirk JP. Permeability of porous solids. Trans Faraday Soc 57:1200–1207, 1961.

Molenda CHA, Crausse P, Lemarchand D. The influence of capillary hysteresis effects on the humidity and heat coupled transfer in a non-saturated porous medium. Int J Heat Mass Transfer 35:1385–1396, 1992.

Mujumdar AL. handbook of Industrial Drying. 2nd ed. New York: Marcel Dekker, 1995.

Perre P, Passard J. A control volume procedure compared with the finite element method for calculating stress and strain during wood drying. Drying Technol 13:635–660, 1995.

Perre P, Turner I. the use of macroscopic equations to simulate heat and mass transfer in porous media. In: Turner I, Mujumdar AS, eds. Mathematical Modeling and Numerical Techniques in Drying Technology. New York: Marcel Dekker, 1997a, pp 83–156.

Perre P, Turner IW. Microwave drying of softwood in an oversized waveguide: Theory and experiment. AIChE J 43:2579–2595, 1997b.

Perre P, Moser M, Martin M. Advances in transport phenomena during convective drying with superheated steam and moist air. Int J Heat Mass Transfer 36:2725–2746, 1993.

Phillip JR, de Vries DA. Moisture movement in porous materials under temperature gradients. Trans Am Geophys Union 38:222–232, 1957.

Plumb OA. Heat transfer during unsaturated flow in porous media. In: Kakac S, Kilkis B, Kulacki FA, Arinc F, eds. Convective Heat and Mass Transfer in

Porous Media. NATO ASI Series E, Vol 196. Dordrecht: Kluwer, 1991, pp 435–464.

Plumb OA, Gong L. Modeling the effect of heterogeneity of wood drying. In: Turner I, Mujumdar AS, eds. Mathematical Modeling and Numerical Techniques in Drying Technology. New York: Marcel Dekker, 1997, pp 221–258.

Plumb OA, Spolek GA, Olmstead BA. Heat and mass transfer in wood during drying. Int J Heat Mass Transfer 28:1669–1678, 1985.

Plumb OA, Couey LM, Shearer D. Contact drying of wood veneer. Drying Technol 4:387–413, 1986.

Prat M. Percolation model of drying under isothermal conditions in porous media. Int J Multiphase Flow 19:691–704, 1993.

Pruess K. TOUGH User's Guide. LBL-20700. California: Lawrence Berkeley Laboratory, 1987.

Puiggali JR, Quintard M. Properties and simplifying assumptions for classical drying models. In: Mujumdar AS, ed. Advances in Drying, Vol 5. Washington, DC: Hemisphere, 1992, pp 109–143.

Rohsenow WM, Choi H. Heat, Mass and Momentum Transfer, Chap 11. New Jersey: Prentice-Hall, 1961.

Salin JG. Numerical prediction of checking during timber drying and a new mechano-sorptive creep model. Holz als Roh- und Werkstoff 50:195–200, 1992.

Scherer GW. Drying of ceramics made by sol–gel processing. In: Mujumdar, AS, ed. Drying '92. Amsterdam: Elsevier Science, 1992, pp 92–113.

Schmidt PS, Bergman TL, Pearce JA, Chen P. Heat and mass transfer considerations in dielectrically-enhanced drying. In: Mujumdar AS, ed. Drying '92. Amsterdam: Elsevier Science, 1992, pp 137–160.

Somerton WH, Keese JA, Chu SC. Thermal behaviour of unconsolidated oil sands. SPE J 14:513–521, 1974.

Spolek GA, Plumb OA. A numerical model of heat and mass transport in wood during drying. In Mujumdar AS, ed. Drying 80, Vol 2. Washington, DC: Hemisphere, 1980, pp 84–92.

Strumillo C, Jones PL, Romuald Z. Energy aspects of drying. In: Mujumdar AS, ed. Handbook of Industrial Drying, 2nd ed. New York: Marcel Dekker, 1995.

Suarez C, Viollaz PE. Effect of shrinkage on drying behavior of potato slabs. Proceedings of Sixth International Drying Symposium, IDS '88, Versailles, France, Vol I. OP 385–389. 1988.

Turner, IW, Jolly PG. Combined microwave and convective drying of a porous material. Drying Technol 9:1209–1269, 1991.

Turner IW, Perre P. A comparison of the drying simulation codes TRANSPORE and WOOD2D which are used for the modelling of two-dimensional wood drying processes. Drying Technol 13:695–735, 1995.

Turner I, Perre P. A synopsis of strategies and efficient resolution techniques used for modelling and numerically simulating the drying process. In: Turner I, Mujumdar AS, eds. Mathematical Modeling and Numerical Techniques in Drying Technology. New York: Marcel Dekker, 1997, pp 1–82.

Udell KS. Heat transfer in porous media considering phase change and capillarity—the heat pipe effect. Int J Heat Mass Transfer 28:485–495, 1985.

van Genuchten MTh. A closed-form equation for predicting the hydraulic conductivity of unsaturated soils. Soil Sci Soc Am J 44:892–898, 1980.

Vargaftik NB. Tables on the Thermophysical Properties of Liquids and Gases. 2nd ed. Washington, DC: Hemisphere, 1975.

Verma AK, Preuess K, Tsang CF, Witherspoon PA. A study of two-phase concurrent flow of steam and water in an unconsolidated porous medium. Proceedings of 23rd ASME/AIChE National heat Transfer Conference, Denver, CO, 1985.

Waananen KM, Litchfield JB, Okos MR. Classification of drying models for porous solids. Drying Technol 11:1–40, 1993.

Webb SW. Pore-scale modeling of transient and steady-state vapor diffusion in partially saturated porous media. Proceedings of AIAA/ASME Joint Thermophysics and Heat Transfer Conference, Albuquerque, NM, 1998.

Whitaker S. Simultaneous heat, mass and momentum transfer in porous media: a theory of drying. Adv Heat Transfer 13:119–200, 1977.

18
Remediation of Soils Contaminated with Hydrocarbons

V. K. Dhir
University of California, Los Angeles, California

I. INTRODUCTION

Contamination of subsurface soil and groundwater can occur as a result of release of contaminants from leaky underground storage tanks, ruptured pipe lines, chemical waste disposal, and oil spills. If left untreated, the contaminated soil and groundwater can pose a significant environmental risk. Aside from contamination of groundwater and the subsurface soil, the release of petroleum products in the subsurface environment can lead to explosion hazards due to accumulation of hydrocarbon vapors in building basements, and the degradation of utility lines which come in contact with leaked hydrocarbons.

After a liquid contaminant is released into the subsurface environment, it percolates downward as well as spreading laterally because of capillary and gravity forces. When the amount of contaminant released is small, the contaminant may be held in the interstices of the soil particles as a discontinuous phase and may not reach the water table. However, if the leak is large, the contaminant may continue to move downward until it reaches a low permeability soil layer or groundwater. Figure 1a shows a typical scenario for an underground leakage source. The variation of contaminant saturation in the soil is depicted qualitatively in Figure 1b. Depending upon the saturation of contaminant in the soil, several zones such as unsaturated zone, capillary zone, and free hydrocarbon layer can be identified. In the unsaturated zone, the contaminant is held in the interstices by capillary and adhesion forces, and exists as a discontinuous phase. The contaminant

(a)

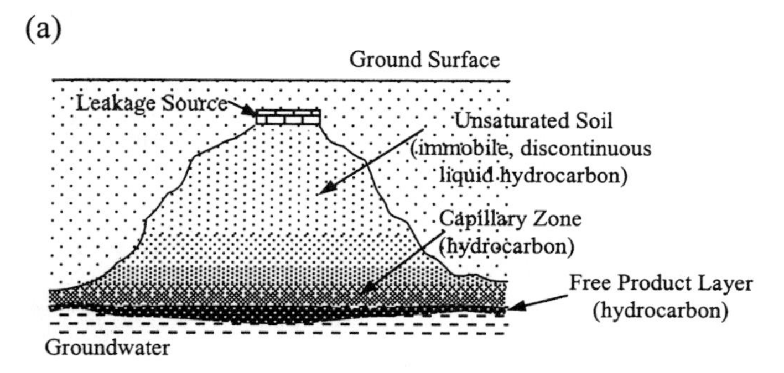

(b)

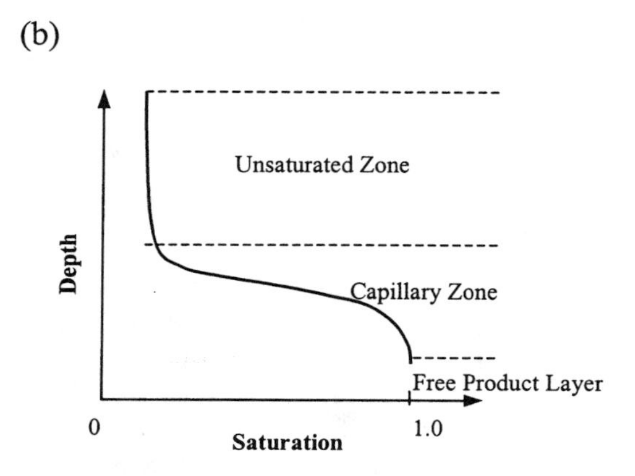

Figure 1. (a) Schematic of a typical contaminated site; (b) variation of contaminant saturation with depth.

is immobile and generally the saturation is less than 0.2. The capillary zone lies above the water table and, in this zone, the saturation of the contaminant very rapidly increases and attains a value close to unity. Now the contaminant becomes a continuous phase. Excess contaminant forms as a free product layer just above the water table.

There are several regulatory acts that govern the management of hazardous waste. In 1976, Congress passed the Resource Conservation and Recovery Act (RCRA). This act (see, e.g., LaGrega et al. 1994) held generators of hazardous waste responsible from "cradle to grave" for the wastes they produce. The act is directed at the waste currently being created rather than at correcting the undesirable practices of the past. The act also allows "citizen suits" against any entity that is alleged to be in violation of permits, standards, regulations, requirements, etc. This act was amended in

1984 as Hazardous and Solid Waste Amendments of 1986 (HSWA). This was a much more detailed act than RCRA. It specified standards for issuance of permits, penalties for violation of the law, and controls on underground storage tanks. It also included a minimal technical requirement for land disposal facilities and prohibited the disposal of uncontained liquid hazardous waste in landfills or surface impoundments.

In order to clean up the existing waste sites, Congress passed, in 1980, the Comprehensive Environmental Response, Compensation, and Liability Act (CERCLA). This act established a $1.6 billion fund to implement a waste site clean-up program over a five-year period. The generators of hazardous waste were required to report to the Environmental Protection Agency (EPA) any facility at which hazardous wastes were present or had been treated, stored, or disposed of (LaGrega et al. 1994). The act was amended in 1986 with the Superfund Amendments and Reauthorization Act (SARA). The act (SARA) created a fund of $8.5 billion to clean abandoned waste disposal sites and $500 million to clean soils contaminated as a result of leaking underground petroleum tanks. The act also had a provision for the "community's right to know" and for industries to inform the public of hazardous substances being used and plans for emergencies.

In response to these regulations, several ex-situ and in-situ strategies for remediation of soil have been proposed and/or implemented in the past. The ex-situ strategies have included excavation and transportation of the contaminated soil to a clean-up location. This option can be quite expensive, as large amounts of soil need to be hauled away. Also, during the transportation of soil, the public can be unnecessarily exposed to hazardous materials. Additionally, the possibility of contamination of additional soil at the clean-up site exists. In some cases, the contaminated soil has been disposed of at commercial landfill sites. But this solution cannot be considered remediation as it merely involves transfer of contamination from one location to another. Site isolation, using physical barriers, has also been considered. The success of this strategy depends on the ability of the barriers to provide protection over long periods of time. The possibility of failure of these barriers with time cannot be ruled out, and, as such, this strategy merely postpones the task of clean-up of contamination to the future.

In-situ remediation strategies have several advantages over the ex-situ remediation options, including non-removal of the contaminated soil. As a result, the risk to public health and safety due to excavation, transportation, and handling of soil is eliminated. No additional land areas for implementation of treatment are needed and the volume of the contaminated soil is not increased, as would be the case if the soil were transported to an ex-situ clean-up site. Various in-situ remediation techniques that have been pursued in the past for soil contaminated with hydrocarbons are:

1. Air stripping and vapor extraction (soil venting)
2. Bioremediation
3. In-situ detoxification
4. In-situ multiphase fluidization
5. Joule resistance heating
6. Radio-frequency soil heating
7. Soil washing
8. Steam injection
9. Vitrification

A. Air Stripping and Vapor Extraction (Soil Venting)

In this scheme, ambient or preheated air is forced into the contaminated soil through injection wells. Evaporation of contaminants occurs as air moves through the soil. The air, after passing through the contaminated soil, is withdrawn through extraction wells. The contaminants in the exiting air are either combusted, condensed, or are removed by other means before the air is discharged into the atmosphere. The method is low cost, easy to implement, and has been field tested for clean-up of soils in the unsaturated zone. The method is applicable for removal of high- to medium-volatility contaminants (moderate vapor pressures at temperature 100–150°C). Proper location of supply and withdrawal wells requires a good knowledge of the subsurface geology and of the physics of the hydrodynamic, thermal, and mass transport processes that occur in the soil. The method is not suited for soils with low permeabilities. The heterogeneities in soil can lead to channeling and uneven removal of contaminants from the soil.

The vapor extraction process is very similar to air stripping except that in this case vacuum is applied to the subsurface soil and air is sucked through the soil containing the contaminant in the unsaturated state. The induced flow of air causes the liquid contaminant to evaporate. The exit air, laden with contaminants, is treated before being discharged to the atmosphere. The desirable features and limitations of this process are similar to those for air stripping. However, if the air pressure in the subsurface soils is very low and the pore size is relatively small, slip conditions (little or no shear) in the pore space may occur. The slip in the pore space will increase the flow rate of gases for the same pressure drop.

B. Bioremediation

This process involves degradation of hydrocarbon contaminants by microbial activity (Smith and Hinchee 1992). Microbes (e.g., bacteria, fungi) are added to the contaminated soil or are activated in situ. Oxygen and other nutrients are also added to the soil to support the microbial activity. The biological activity of the microbes strongly depends on the pH value and the concentration of nutrients, toxic compounds, and the soil temperature. This activity has been quantified by Brock (1970) at temperatures varying from −12 to 100°C. The optimum range of temperature depends on the type of micro-organisms. It has been found that degradation rates of contaminants almost double for every 10°C temperature rise up to a certain maximum before declining again. Over a limited temperature range, the biodegradation rate can be described by the van't Hoff–Arrhenius relationship. Generally the degradation period can be divided into two phases. In the first phase, the degradation is very rapid, whereas in the second phase, degradation slows down because of diffusion-limited transport of pollutants to the interface. Heavy metals cannot be removed by biological techniques and must be extracted by chemical techniques.

A single type of micro-organism cannot degrade all types of pollutants. As such, it is important that a proper type of microbe and environmental condition for growth of the microbes be present. To bioremediate several types of pollutants, a two-step remediation process is required (e.g., Ferdinandy and van Vlerken, 1998). As an example, to remediate chlorinated compounds (chlorobenzenes), initially no oxygen is allowed so that dechlorination can take place. Thereafter, oxygen is added for aerobic mineralization of the intermediate products into water and carbon dioxide. In remediating TNT (tri-nitro-toluene), initially aerobes are used to consume oxygen from soil–water slurries. Once the anerobic conditions are established, anerobic microbes are added to metabolize TNT. Roberts et al. (1998) have studied the effect of the presence of metals such as copper, iron, lead, and zinc on bioremediation of soils containing TNT. It was found that the presence of lead generally inhibited the metabolism. Copper also tended to be an inhibitor to microbes that were required to create anerobic conditions in the aqueous phase. Iron did not have any effect on the metabolization. Zinc had some effect, especially in delaying the complete removal of the second intermediate product formed during TNT degradation. Use of foams has been proposed (e.g. Chowdiah et al. 1998) for delivering nutrients and oxygen to the subsurface soil. The surfactants used to generate foam can improve soil wettability and contaminant desorption. Also, foams provide large fluid–gas interfacial areas in comparison to those without foam.

Bioremediation is a simple and low-cost technology that has been used in the field by itself or in conjunction with other technologies such as air stripping. The technology has several limitations in that it is difficult to control and monitor the microbial activity. Nonuniformities in the distribution of oxygen and nutrients can lead to uneven biodegradation. The remediation process can take long periods of time (many months to years) and its application is limited to regions that are not very deep (less than 7 m).

C. In-situ Detoxification

This process is an extension of the drilling technology in which treatment agents such as steam, air, and other oxidants (in solid or liquid form) are added to the soil, which is then mixed thoroughly. After treatment, the off-gases are vented to the ambient air or are recycled. Thorough mixing of the soil and the treatment agents results in even remediation of the soil. By monitoring the composition of the off-gases, the treatment conditions can be adjusted to obtain a desired degree of remediation. Only very narrow columns of soil can be remediated at a given time, and the depth of treatment is limited to about 6 m. The feasibility of the technology has been shown on a field site containing hydrocarbons (Ghassemi 1988).

D. In-situ Multiphase Fluidization

In this method (e.g., Niven and Khalili 1998) fluidization of the contaminant granular soil is achieved by a jet of co-currently flowing gas and liquid inserted into the soil (upflow washing). The fluidization of soil causes the release of nonaqueous phase liquid (NAPL) ganglia trapped in the interstitials of the soil particles. Fluidization also leads to selective removal of fines. This in turn causes the removal of contaminants sorbed or linked to the fines such as heavy metal (lead) contaminants. The released or upwashed contaminants are collected into waste water storage tanks for onsite or offsite treatment. Figure 2 (Niven and Khalili 1998) shows a schematic for field implementation of the method for cleanup of saturated or unsaturated granular soils (sands). In column and tank test sections 96 to 99% reductions in the contaminant concentration has been obtained in soils contaminated with diesel. After remediation, final diesel levels of less than 1000 mg/kg in uniform fine sand and less than 200 mg/kg have been obtained in clayey sands under laboratory conditions. Fluidization of the soil can be obtained with a jet of liquid alone. However, it has been found that jets of liquid alone are less efficient than air–liquid jets in removing contaminants from uniform particle size sands.

The method has the advantage that both hydrocarbon and metal contaminants can be removed simultaneously. It can be used along with other remediation technologies. However, prior to treatment of the contaminated soil, it may be necessary to isolate the site by installation of a slurry well.

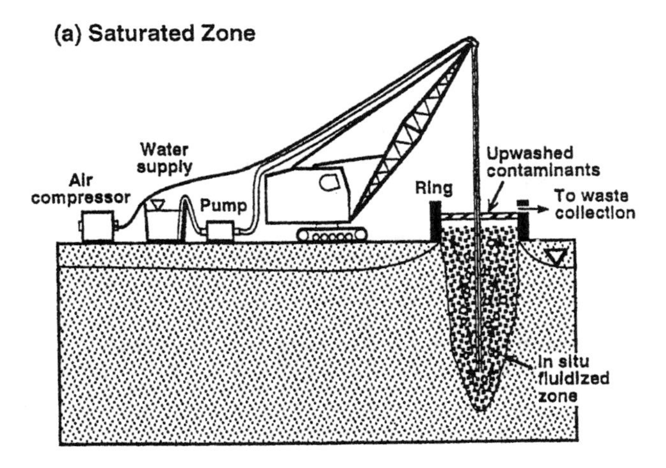

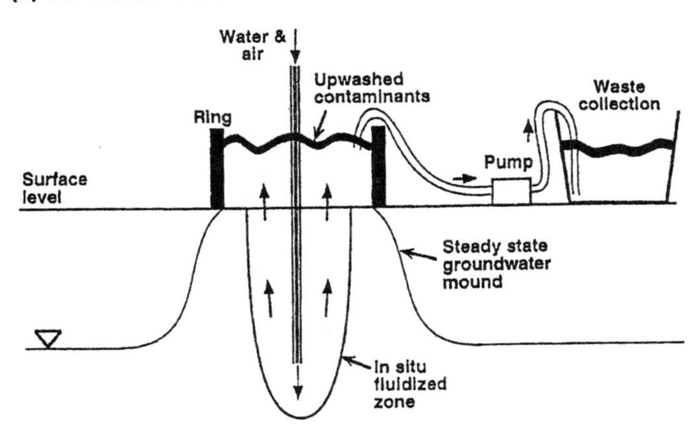

Figure 2. A schematic diagram of the field implementation of in-situ multiphase fluidization: (a) saturated zone; (b) unsaturated zone (as proposed by Niven and Khalili 1998).

E. Joule Resistance Heating

In this method, the soil mass is directly heated by using it as a conductor of electrical current. In implementing this technology, an array of electrodes is formed by inserting metal pipes into the contaminated soil. Vapor of volatile organics and water are collected through the electrodes, which act as extraction wells, and are treated in the same manner as in air stripping or vacuum venting. Bench and field tests using this technique have been performed by

Heath (1990). It has been shown that the method is applicable to a wide range of soils including sand, silty loam, and bentonite clay. On bench-scale tests, up to 99% of contaminants present in the soil have been removed. Moisture in the soil contributes significantly to the electrical conductivity of the soil. As a result, when moisture is driven off, the conductivity of the soil matrix drops significantly and to maintain the same heat generation rate the voltage must be increased.

F. Radio-frequency Soil Heating

In implementation of this technology, an exciter array or electrodes are used to create an electromagnetic energy field in the radio frequency band in the contaminated soil. The electrodes are placed into holes drilled in the soil and the energy field heats the soil, leading to release of vapor of volatile and semi-volatile organics (Ghassemi 1988). The released vapor and gases can be removed and collected for further treatment by using hollow electrodes and applying vacuum to the electrodes. The electrodes are energized with power from a modified radio transmitter. The energy in the soil is generated because of ohmic and dielectric heating. Ohmic heating results from movement of electrons through the soil, whereas dielectric heating is caused by distortion of molecular structure due to an imposed electric field. Power input is proportional to the dielectric properties of the soil, the applied radio frequency, and the strength of the applied field. The dielectric properties strongly depend on the moisture content of the soil. As such, the choice of operating frequency is based on the dielectric properties of the soil and the extent of the contaminated soil. The frequency range of remediation applications is found to be between 6.8 MHz and 2.5 GHz.

Radio-frequency heating is most suitable for remediation of soils containing semi-volatile organics as it provides rapid heating of soils in the temperature range of 100–150°C. In addition to the system for radio-frequency heating, the implementation of the technology requires a vapor barrier to limit escape of vapor to untreated soil or ambient and an off-gas treatment system. Although the technology has been tested in the field, the cost of implementation of the technology can be quite high. The high temperatures can adversely affect the humic matter in the soil and microbial activity.

G. Soil Washing

In this process (Ghassemi 1988) the contaminants are removed from the soil by flushing or rinsing the soil with surfactants or other reagents. Gravity head is used to supply the reagents to the contaminant zone, and excess solution and contaminants are collected in ditches or drains placed around the region to be remediated. Lee (1998) reported results of one-dimensional column and two-dimensional experiments in which water-saturated and unsaturated sands contaminated with tetrachloro-ethylene were flushed with a 1% solution of sorbitan monooleate (a surfactant). It was found that in both the saturated and unsaturated cases more than 99.5% of the original mass of the contaminant was removed after flushing 12 pore volumes. A model for surfactant-enhanced remediation of non-aqueous phase liquids (NAPL) has been developed by White and Oostrom (1998). The model is based on nonlinear mass conservation for two immiscible liquids (water, NAPL) and aqueous-phase organic and aqueous-phase surfactant. Surfactant concentrations in the aqueous phase affect interfacial tension and, indirectly, NAPL mobility. The application of the technique is limited to shallow soil regions or to surface spills. The method is not applicable to a mixture of contaminants having different solubility characteristics and it is difficult to control the clean-up process. The contaminant with the collected excess reagent is in a very dilute form and, as such, it is cumbersome and expensive to handle and treat large volumes. The reagents can be sorbed in the soil and, because of the non-homogeneities of the soil, reagents may not reach evenly in all regions of the soil.

H. Steam Injection

In this process low pressure steam is injected into the soil containing organic contaminants through several wells located near the outer boundary of the zone of contamination. Vacuum is applied through an extraction well located in the middle of a contamination zone. The heat released as a result of condensation of steam causes soil and contaminant temperature to rise and, in turn, volatilizes the contaminants. The steam exiting the extraction well and containing the condensate liquid droplets and organic vapor is further condensed and organic contaminants are separated for disposal. The off-gases are also treated before they are discharged to the atmosphere. The technology has been implemented in the field (e.g., Udell and Stewart 1989). Removal efficiencies as high as 99.9% (Nunno et al. 1989) have been reported. However, the removal efficiencies depended on several factors such as type of soil (e.g., sand or clay), the polarity of the organic compounds, and the vapor pressure–temperature relationship of the organics.

The technology is applicable to removal of volatile and semi-volatile compounds. Because of the thermal energy supplied by the steam, the treatment times can be significantly reduced as compared to air stripping carried out at ambient temperature. Inhomogeneities in the soil can limit the usefulness of this technique as steam channeling will occur around less permeable soil. In such a case, the contaminant must diffuse out of the less permeable soil. The time constant associated with the diffusion process can be very long. The implementation of this scheme depends on the availability of steam, which can add to the cost of the clean-up process. Also high temperatures can inhibit or destroy microbial activity.

I. Vitrification

This technology involves heating the contaminated soil to such a temperature that melting of the soil occurs. The heating is accomplished by passing current across electrodes embedded in the soil. On cooling, the soil is converted into glass and gases exiting the soil are contained and passed through a treatment system before being discharged to the atmosphere. The technology is more suited for radioactive waste. The implementation of the technology is expensive and only a limited region of the soil can be remediated at a given time.

Next, two of the most promising technologies—soil venting (vapor extraction or air stripping) and remediation by injecting steam—are described in detail.

II. SOIL VENTING

Since the first study of soil venting, conducted by the American Petroleum Institute (1980), several bench-scale and field studies documenting the effectiveness of soil vapor extraction as a remediation technology for soils contaminated with organics have been reported in the literature. Figure 3 shows a schematic diagram of a typical soil venting system utilizing pre-heated air. Several parameters, such as flow rate, well location, permeability of the soil, the composition and volatility of the contaminant, and the temperature of the soil and incoming air, affect the venting process. Crow et al. (1985) studied the effect of air flow rate on vapor removal rate at a gasoline spill site. The results indicated that vapor recovery rate increased with increase in air flow rate. Also, the reduction in soil vapor concentrations was higher in the region near the centerline between the air-inlet wells and the vapor recovery well, where higher air flow rate existed. A similar case study to remediate gasoline-contaminated soil and spill sites was conducted by Fall and Pickens (1988). The system was operated primarily during the day time and was shut off at night to allow the subsurface vapor pressures to re-equilibrate. This pulsed venting was found to increase the efficiency of the remediation process and maximize hydrocarbon removal from the soil. The observed vapor recovery rates in the pulse mode were at least two orders of

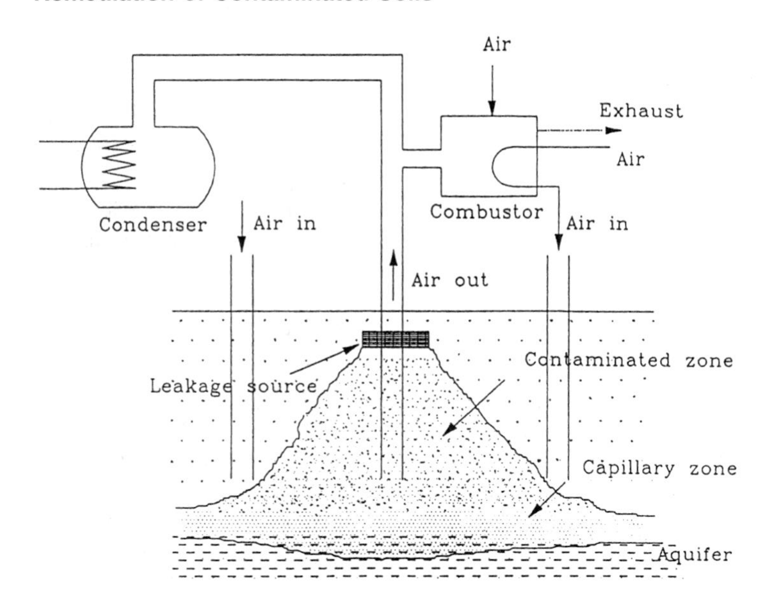

Figure 3. Schematic of a soil venting system.

magnitude greater than those reported by Crow et al. (1985). It was noted that decrease in soil vapor concentrations at any location depended on the flow rate of air rather than on the induced subsurface vacuum at that location. The effect of air flow rate on removal of volatiles from a simulated gasoline site has also been investigated by Thornton and Wootan (1982). They found that after very high initial gasoline concentrations in the effluent air the removal rate leveled off, and the amount of gasoline removed increased with air flow rate.

Texas Research Institute (1984) has studied the effect of venting well geometry on the remediation process. It was found that using a short slotted section at the bottom of the inlet vent pipes for release of incoming air provided more efficient vapor removal than that obtained when the pipes were slotted all along from the soil surface to the bottom of the well.

Through laboratory-scale experiments in which residual saturation of gasoline was established in sand columns, Marley and Hoag (1984) and Marley (1985) have studied the mechanisms that control the soil venting process. Gasoline vapor removal rate was found to decrease with time, and it was rationalized that the reduction was caused by a decrease in total vapor pressure of the gasoline as more volatile fractions of gasoline were removed. Also, vapor concentration in the effluent air was found to be near the saturation concentration corresponding to the soil temperature, even at high flow rates. This observation implied that liquid was evaporating at a rate faster than the maximum rate at which it could be swept away by inflowing air. Since saturation vapor pressure limits the evaporation rate, the evaporation rate can be increased by elevating the temperature of the incoming air. Katsumata (1992) conducted experiments similar to those of Marley and Hoag (1984), using preheated air, and found that remediation times could be reduced significantly by heating the incoming air.

As shown in Figure 4, the contaminant within the contamination zone can be partitioned into four main phases: (1) vapor phase (air phase); (2) dissolved phase in pure water (water phase); (3) sorbed phase on soil particles (solid phase); and (4) as a nonaqueous phase liquid (NAPL) or immiscible phase. Several models to describe the venting process have been proposed in the literature. Baehr et al. (1989) developed a one-dimensional model to describe the venting process. In the model, liquid phases were assumed to be immobile and the air flow distribution was obtained for a radially symmetric region with a single vacuum cell. Species concentration in all phases was obtained from local equilibrium considerations. The model included volatilization from NAPL, dissolution from the air phase into the aqueous phase, and adsorption on soil from the aqueous phase. For high gas flow rates, the model predictions were found to be insensitive to the degree of partitioning into water and adsorbed phases. The model results were compared with one-dimensional laboratory experiments in which gasoline was used as the contaminant. Gasoline was divided into five constituents whose vapor pressures were obtained by calibrating the model with the data from an experiment conducted at low air flow rate. Good agreement between the predicted and experimentally observed hydrocarbon flow rates at the exit

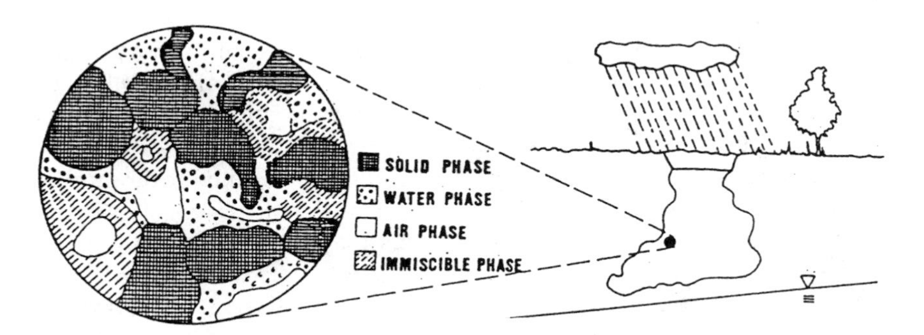

Figure 4. Partitioning of contaminant into various phases.

was found. However, the comparison did not provide a rigorous verification of the assumption of local equilibrium for various constituents.

Johnson et al. (1990) also developed a model similar to that of Baehr et al. (1989), using an equilibrium assumption for interphase mass transfer. They obtained analytical expressions for the time required to achieve steady vapor flow and the distance over which air becomes saturated upon entering the contaminated soil. For typical extraction parameters and soil properties in the field, the air flow was found to reach steady state within a time of one day to one week. For regions through which air flow does not occur, a diffusion-limited model was used. The effect of temperature on gas flow was accounted for by allowing changes in gas density and viscosity. However, the temperature affects the vapor pressure as well as species concentration at the liquid–vapor interface and in the bulk.

A model for soil vapor extraction has also been developed by Rathfelder et al. (1991). In this model, nonequilibrium with respect to phase partitioning was considered and first-order rate expressions with overall transfer factors were used both for the NAPL phases and for the aqueous phase. No rationalization was given for the choice of magnitude of transfer factors. Soil moisture was assumed to be the preferentially wetting fluid and solids only adsorbed the aqueous phase. The model was exercised to study the influence of contaminant vapor pressures.

Sepehr and Samani (1993) developed a three-dimensional model for gas flow to study the effect of partial penetration and partial screening of vapor extraction wells. The model accounted for non-homogeneity and anistropy of the soil, and provided pressure distribution around vapor extraction wells. The model mainly focused on gas flow in the subsurface environment and did not include mass and heat transfer associated with evaporation. Several field situations were simulated numerically, and the predicted pressures were found to be in good agreement with data.

Sleep and Sykes (1989) developed a model for transport of organics that allowed for nonequilibrium in mass transport. An overall mass transfer rate that included a product of mass transfer coefficient and interfacial area was used. The dissolution rate of organics in the aqueous phase, evaporation rate of organics into gas phase, and evaporation rate of organics into gas phase from the aqueous phase were modeled using the overall mass transfer rate. Fluid motion, as a result of buoyancy in the gas phase, was modeled but no modeling of the temperature distribution in the soil was included. Adsorption of hydrophobic organic chemicals on low organic-carbon aquifer materials has been experimentally investigated by Brusseau et al. (1991). In the experiments, chemical concentrations in water varied from 30 to 60 μg/l and it was found that sorption was rate limited. The results of experiments with higher pore velocity of water could only be explained with a nonequilibrium mass transfer model. However, the equilibrium model was found to give satisfactory results for low velocities. Armstrong et al. (1994) have also considered equilibrium advective-dispersive transport of organic contaminants contained in air, dissolved in aqueous phase, and adsorbed on the soil particles. Again, the product of mass transfer coefficient and interfacial area was used. The authors suggested that an accurate value of mass transfer rates can be obtained only by calibrating the model results with data from laboratory or field tests. It was also noted that, for a given set of physical parameters, a critical flow rate existed. Beyond this flow rate, no significant decrease in remediation time was found when the flow rate was increased. Furthermore, for a given average flow rate, pulsed pumping was found to be less efficient than continuous pumping.

The models of Armstrong et al. (1994), Rathfelder et al. (1991), Sleep and Sykes (1989), and Brusseau et al. (1991) have incorporated macropore and micropore transport and partitioning processes between the mobile gas phase and the immobile aqueous and NAPL phases and have used first-order interphase mass transfer descriptions to determine the mass transport. Gierke et al. (1992) conducted laboratory experiments to study the vapor transport in unsaturated soil columns; their results have indicated that vapor extraction performance in moist aggregated soils will be affected by nonequilibrium transport. An experimental study by Wilkins et al. (1995) examined factors influencing rate-limited NAPL-vapor phase mass transfer in unsaturated porous media. They found that deviations of vapor phase concentrations from local equilibrium values can occur at gas-phase velocities typically encountered in soil vapor extraction systems, and more significant mass transfer rate limitations would be expected over extended periods of volatilization.

Lingineni and Dhir (1992, 1997) have developed a nonisothermal multicomponent model for soil vapor extraction. This model, and a few key results from the model, are described in the following.

A. Nonisothermal Multicomponent Model

The overall model is divided into four submodels: gaseous flow in subsurface environments; thermal regime in subsurface environments; macroscopic model for mass transfer; and model for diffusional transport in the NAPL.

1. Gaseous Flow in Subsurface Environments

Efficient design and operation of soil venting systems requires realistic predictive capability of gaseous flow created in the soil for the imposed subsurface conditions. Gas flow through soil is dependent on soil characteristics, including porosity and permeability; gas properties such as viscosity and density; and pressures and temperatures in the subsurface. The flow of gases due to pressure gradients consists of two parts—viscous flow and slip flow. Nonzero velocities at the pore walls result in slip flow. The importance of slip flow increases as the average pore radius decreases and as the pressure decreases. The effects of slip flow are found to be negligible for gas transport in most soil types including silts, sands, and gravel under the conditions typical of soil venting systems (Massmann 1989). Hence, in the following analysis, the flow of gases through the soil medium is assumed to be dominated by viscous forces and thus described by Darcy's law. It should be noted that, at high fluid velocities, microscopic drag forces can no longer be neglected and deviations from Darcy's law occur. Darcy's law is found to be in good agreement with the experimental data collected by various researchers for $Re = ud_p/\nu < 10$.

The mass continuity equation for the gas flow through porous media is given by

$$\frac{\partial(\varepsilon S_g \rho_g)}{\partial t} + \nabla \cdot (\varepsilon \rho_g \mathbf{v}_g) = Q \tag{1}$$

where $\mathbf{v}_g$ is the gas velocity, ρ_g is the gas density, ε is the porosity of the medium, and S_g is the local gas saturation. Q denotes the mass per unit volume per unit time exchanged through a source or sink. In the above formulation, it is assumed that Q is positive for a source and negative for a sink; see Eq. (1). Q is defined to have a value only at locations corresponding to either vapor extraction wells or air injection wells. Also, in this type of formulation, the model expects input in the form of air flow rates through each of the wells.

There is another equivalent formulation in which pressures imposed at wells can be provided as input. For the particular case of soil matrix being nondeformable and homogeneous, the above mass continuity equation can be written as

$$\frac{\partial(S_g \rho_g)}{\partial t} + \nabla \cdot (\rho_g \mathbf{v}_g) = Q/\varepsilon \tag{2}$$

The equation of motion for gas flow in soil, as described by Darcy's law, can be written as

$$\mathbf{v}_g = -k\frac{k_{rg}}{\mu_g}(\rho_g g \nabla z + \nabla P_g) \tag{3}$$

where k is the intrinsic permeability of the medium, which depends on the geometry of the medium solely, and k_{rg} is the gas phase relative permeability that represents the fraction of permeability available for gas flow in a multiphase system. The relative permeability k_{rg} is a function of gas phase saturation, which in turn is dependent upon the liquid contaminant (NAPL) and soil water saturations. The dependence of relative permeabilities on the fluid saturations is described by various formulations (Brooks and Corey 1964; Parker et al. 1987; Farrell and Larson 1972). One of the simplest ways of representing the increase in gas permeability with reduction in liquid saturation is to use a modified Brooks and Corey relationship, developed by Falta et al. (1989) as

$$k_{rg} = S_g^3 \tag{4}$$

Another formulation, proposed by Parker et al. (1987) to estimate three-phase relative permeabilities, can be employed to predict the gas relative permeability as

$$k_{rg} = \bar{S}_j^{1/2}\left(1 - \bar{S}_t^{1/m}\right)^{2m} \tag{5}$$

where $\bar{S}_t$ is the total pore fluid saturation, i.e., the sum of liquid contaminant saturation and water saturation, m is an empirical constant, and the function $\bar{S}$ for any phase j is defined as

$$\bar{S}_j = \frac{S_j - S_r}{1 - S_r} \tag{6}$$

In the above definition of $\bar{S}_j$, S_r is the irreducible saturation of the wetting fluid. Note that most of these types of formulation are developed for hydrologic situations at ambient temperatures, where the liquid flow ceases at residual saturations and further reduction in liquid saturation cannot be achieved by the applied pressure gradients. When advective gas flows are created through soil and soil temperatures are increased by in-situ heating methods, liquid saturations can actually go below residual saturations as a result of evaporation and the value of S_r should be modified accordingly.

Darcy's law must be substituted into the mass continuity equation to obtain a governing equation for gas pressures

$$\frac{\partial(S_g \rho_g)}{\partial t} - \nabla \cdot \left(k \frac{k_{rg}\rho_g}{\mu_g}(\rho_g g \nabla z + \nabla P_g)\right) = Q/\varepsilon \tag{7}$$

The gas density, ρ_g, is a function of gas pressure and temperature and can be obtained by using the ideal gas law

$$\rho_g = \frac{P_g}{RT} \tag{8}$$

where T is the absolute temperature and R is the ideal gas constant.

The pressure distribution within the subsurface environment can be obtained by solving the nonlinear coupled equations (7) and (8) along with the appropriate boundary conditions and source/sink conditions to represent the vapor extraction as well as the air injection wells. The gas velocities at various locations in the soil can then be calculated by using Eq. (3).

2. Thermal Region in the Subsurface Environment

Soil temperature is one of the most critical factors that influence various physical and chemical processes occurring within the soil during venting. Soil temperatures can vary during soil venting due to a variety of natural processes such as diurnal variations in soil temperature, latent heat absorption during contaminant evaporation, heat transport due to differences between ambient air temperature and soil temperature. Soil temperatures can also be increased, as discussed earlier, by employing various in-situ heating techniques such as injection or preheated air into the soil, installation of buried heat sources in the soil, radio-frequency soil heating, and steam injection in order to enhance remediation.

To account for variations in the fluid phase properties and the interphase mass transfer rates with local soil temperatures, the propagation of thermal perturbations within the subsurface needs to be modeled. Once the heat is deposited into the soil by a specific in-situ heating method, heat flow within the soil can occur mainly as a result of conduction, convection of heat by fluid flow, and latent heat transfer.

The temporal and spatial variations of temperatures in the subsurface environment can be obtained by performing an energy balance on the solid phase and the gas phase of a differential soil volume element.

Solid phase:

$$(\rho c_p)_s \frac{\partial T_s}{\partial t} = \lambda_s \nabla^2 T_s - hA_s(T_s - T_g) - \sum_{k=1}^{N} m_k'' h_{fg,k} f_1 + m_w'' h_{fg,w} f_2 \tag{9}$$

Gas phase:

$$(\rho c_p)_g \frac{\partial T_g}{\partial t} + (\rho c_p)_g \mathbf{v}_g \cdot \nabla T_g$$
$$= \lambda_g \nabla^2 T_g + hA_s(T_s - T_g) - \sum_{k=1}^{N} m_k'' h_{fg,k}(1 - f_1) + m_w'' h_{fg,w}(1 - f_2) \tag{10}$$

Due to the fact that the liquid saturations are very low in vadose zone soils, the effect of liquid phase on the overall energy balance is small. Heat flow to the solid phase can occur by conduction, interphase heat transfer from gaseous phase, and latent heat transfer from phase changes occurring within the control volume. In the case of gas phase, the main heat transfer mechanisms are convective heat transfer to and from the solid phase, and latent heat transfer. However, for the sake of completeness, a conduction term is also included in the gas phase energy conservation equation. The amount of latent heat absorbed during the contaminant evaporation and the amount of latent heat released due to condensation of water or contaminant vapor have to be partitioned between the solid and gas phases. This partitioning is represented by the fractions f_1 and f_2 in Eqs. (9) and (10). The interphase heat transfer rates from the gas phase to the solid phase are described by using an appropriate convection heat transfer coefficient for flow of gases in porous media. One such heat transfer correlation, based on experimental data for spheres and short cylinders (Whitaker 1972), is given by

$$Nu = \frac{hd_p}{\lambda_g} = \left(0.5Re^{1/2} + 0.2Re^{2/3}\right)Pr^{1/3} \tag{11}$$

where Nu is the Nusselt number, d_p is the particle diameter, and Re is the Reynolds number based on superficial gas velocity within the porous medium. For typical conditions in the soil, the transfer area available for heat transfer between the solid and gaseous phases is quite large. This would result in small temperature differences between solid and gas phases within the local soil element. Thus, it is reasonable to simplify the thermal analysis by assuming that the temperatures of all the phases in a differential soil element are equal. Then an overall energy balance can be performed on all of the phases to obtain the following energy conservation equation

$$(\rho c_p)_m \frac{\partial T}{\partial t} = \lambda_m \nabla^2 T - (\rho c_p)\mathbf{v}_g \cdot \nabla T_g - \sum_{k=1}^{N} m_k'' h_{fg,k} + m_w'' h_{fg,w} \tag{12}$$

In performing the above overall energy balance, energy storage and conduction heat transfer rates are described by defining an effective thermal conductivity and heat capacity as

$$\lambda_m = \lambda_s \varepsilon_s + \lambda_w \varepsilon_w + \lambda_n \varepsilon_n + \lambda_g \varepsilon_g \tag{13}$$

and

$$(\rho c_p)_m = \rho_g c_{pg} \varepsilon_g + \rho_s c_{ps} \varepsilon_s + \rho_w c_{pw} \varepsilon_w + \rho_n c_{pn} \varepsilon_n \tag{14}$$

The above expression for effective thermal conductivity of the soil is obtained by assuming that individual thermal resistance of each phase act in parallel. Similarly, an effective heat capacity for the soil matrix is defined as the sum of the heat capacities of various phases weighted by their volume fractions. The second term on the right-hand side of the energy equation describes the heat transfer by gaseous advection. Third and fourth terms represent the latent heat absorbed during contaminant evaporation and the latent heat released by water vapor condensing from the incoming air. If evaporation of water in the pores occurs, the fourth term will have a negative sign in front.

3. Macroscopic Mass Transport Model

The transport of contaminant species from the gas–liquid interface into the gas phase is determined by the species concentration difference that exists between the interface and the bulk gas. The macroscopic transport of the contaminant fractions in the gaseous phase is described by the following species conservation equation

$$\rho_g\left[\frac{\partial C_{k,g}}{\partial t}+\mathbf{v}_g\cdot\nabla C_{k,g}-D_{k,g}\nabla^2 C_{k,g}\right]=R_{k,s\leftrightarrow g}+R_{k,w\leftrightarrow g}+R_{k,n\leftrightarrow g} \tag{15}$$

where $C_{k,g}$ is the concentration of the kth species in the gaseous phase, $\mathbf{v}_g$ is the local velocity of the air, and $D_{k,g}$ is the diffusion coefficient of the kth species in the gaseous phase. The terms on the right-hand side of Eq. (15) represent interphase mass transfer fluxes of contaminants into the gaseous phase from solid, aqueous, and nonaqueous phases, respectively. In order to describe the evaporation of a multicomponent NAPL mixture, the interphase mass transfer rate $R_{k,n\leftrightarrow g}$ can be represented as

$$R_{k,n\leftrightarrow g}=h_m A_\ell (C_{k,\text{sat}}-C_{k,g}) \tag{16}$$

where $C_{k,\text{sat}}$ is the gas phase concentration of the kth species at the NAPL–gas interface, $C_{k,g}$ is the kth species concentration in the bulk gas phase, h_m is the mass transfer coefficient, and A_ℓ is the interfacial surface area of the evaporating liquid.

The interphase mass transfer rate of contaminant fraction from the liquid phase to the gaseous phase, as described by Eq. (16), strongly depends upon the concentration of that fraction at the gas–liquid interface. The saturation vapor concentrations of each individual species at the NAPL–gas interface, according to Raoult's law, can be expressed as

$$C_{k,\text{sat}}=x_k C_k^* \qquad k=1,2,\ldots,N \tag{17}$$

where x_k is the mole fraction of the kth species in the bulk liquid and C_k^* is the saturation vapor concentration of pure kth species at the mixture temperature.

By assuming that the gas phase behaves as an ideal gas (a good approximation at environmental temperatures and pressures), a relationship between pressure and concentration can be written as

$$C_k^* = \frac{w_k}{RT} p_k^* \qquad k = 1, 2, \ldots, N \tag{18}$$

where w_k is the molecular weight of the kth species, R is the universal gas constant, T is the temperature, and p_k^* is the saturation vapor pressure over pure liquid. The saturation vapor pressure p_k^* is a function of temperature and is calculated using the Antoine vapor pressure correlation

$$\ln p_k^* = \alpha_k - \frac{\beta_k}{T + \gamma_k} \tag{19}$$

where the constants α, β, γ depend upon the kth species and T is the absolute temperature of the interface. The above equation predicts the vapor pressure over a flat surface.

When the saturation values of the liquid contaminant are equal to or less than the residual saturation, pendular configuration of liquid is believed to exist in the soil. Rose (1958) presented a method to determine the volumes and surface areas of pendular rings that exist in the soil interstices. Assuming that the interfacial surfaces of pendular rings can be regarded as arcs of circles, the volume of the pendular rings that form at zero contact angle can be determined according to the equation

$$V_{\text{ring}} = 2\pi r_p^3 \left\{ 2 - 2\cos\theta - \tan\theta \times \left[2\sin\theta - \tan\theta + (\pi/2 - \theta)\left(\frac{\cos\theta - 1}{\cos\theta}\right)^2 \right] \right\} \tag{20}$$

where θ is the defining angle, shown in Figure 5.

The radius of curvature of the liquid contaminant can be related to the particle size and the defining angle θ through geometric relations such as

$$r_1 = r_p \left(\frac{1 - \cos\theta}{\cos\theta} \right) \tag{21}$$

Thus, at a given liquid saturation level, the volume of a single pendular ring can be calculated by assuming a regular packing for the soil. For example, consider a soil matrix with uniform size particles of radius r_p and assume that the soil particles are arranged in a cubic packing. Figure 5 shows a unit cell of such packing, which is a cube of dimension $2r_p$. Each

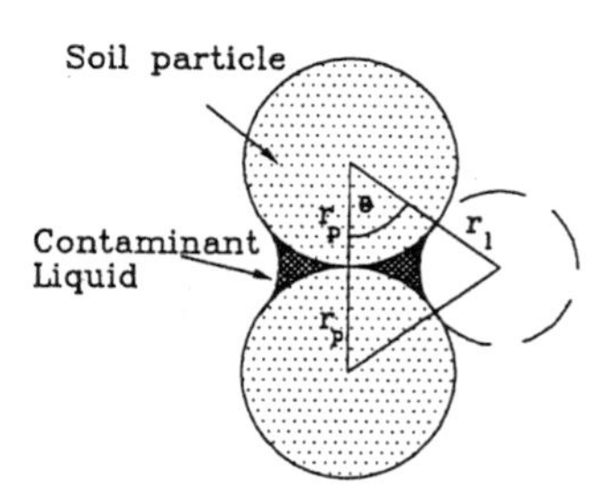

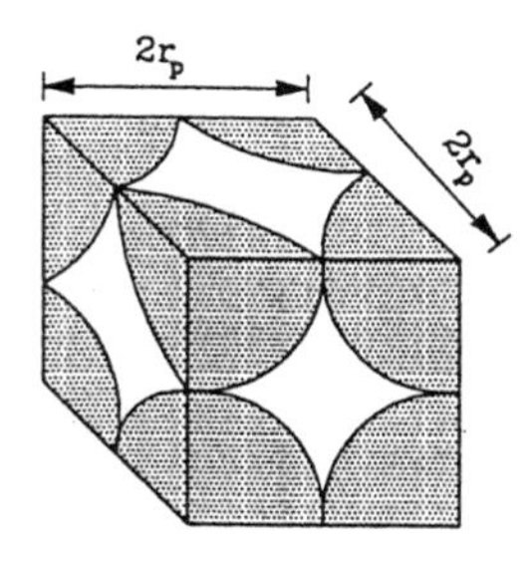

Figure 5. Model for determining the relationship between liquid–gas interfacial curvature and volume of pendular rings.

unit cell consists of twelve contact locations where liquid can exist in the form of pendular rings, and the total volume of the liquid within the cell amounts to three complete rings. Thus, the volume of each pendular ring can be written as

$$V_{\text{ring}} = \frac{8}{3} r_p^3 \varepsilon S_\ell \tag{22}$$

where ε is the porosity of the soil and S_ℓ is the liquid saturation. The defining angle θ can be determined from the volume of the pendular ring, using the relationship given by Eq. (20), which can then be used to calculate the radius of curvature of the gas–liquid interface. Using the above relations for uniform sized particles, the specific surface areas of the pendular rings can be calculated by numerical integration.

With the reduction of the contaminant in the pore space, the radius of curvature of the interface also decreases. As a result the pressures in the liquid and gas phases start to differ, with pressure in the liquid phase being smaller than that in the gas phase. This leads to reduction in vapor concentrations at the liquid–gas interface which, in turn, diminishes the evaporation rates of the contaminant species. Among other variables, this condition can define the clean-up limit that can be achieved in soil venting. Thus, any modeling effort (e.g., Lingineni 1994) should include, in the analysis, the consequences of differences in liquid and gas pressure.

The overall mass transfer coefficient, h_m, can be estimated from the relations for mass transfer in packed beds. One such relation, proposed on the basis of experimental data on mass transfer in packed beds of spheres and cylindrical pellets with single-phase fluid flow (Sherwood et al. 1975), was used to determine the mass transfer coefficients at the gas–solid interface and can be expressed as

$$\frac{h_m}{v_g Sc^{-2/3}} = 1.17 Re^{-0.415} \tag{23}$$

where Sc is the Schmidt number and v_g is the superficial velocity of the gas in the porous medium.

In the past, in models such as that by Rathfelder et al. (1991), the product of the mass transfer coefficient and the interfacial area was used and a constant value of the transfer factor was assumed during the venting process. As is clear from the above description, the transfer factors can be determined quantitatively for residual saturation under idealized conditions and large variations in their values can occur during the remediation process.

The vapor concentrations at the interface, as described by Raoult's law or Henry's law, are dependent upon the mole fractions of the individual contaminant species at the interface. Several models of venting systems determine the mole fractions at the evaporating interface by assuming that the contaminant species in the liquid phase are completely mixed at all times. However, if a particular species is being depleted from the liquid phase by evaporation, its concentration at the interface will decrease in comparison to its concentration in the interior of the liquid film (or bulk liquid phase). The concentration gradient, thus created across the liquid film, will initiate diffusional transport of that species from the bulk liquid to the interface, and the rate at which such transport can occur will determine the concentration levels at the interface. The interphase mass transfer rate terms $R_{k,s\leftrightarrow g}$ and $R_{k,w\leftrightarrow g}$ can be determined in the same manner as Eq. (16). However, the determination of interfacial area becomes much more difficult.

4. Diffusion-limited Transport in the NAPL

For a thin layer of liquid consisting of two immiscible species of volatiles, as shown in Figure 6, the initial concentrations of the species within the layer are assumed to be uniform and are specified as $C_{A,0}$ and $C_{B,0}$. The following governing equations can be written to describe the diffusional transport of each species from the bulk liquid phase to the interface:

Component A

$$\frac{\partial C_A}{\partial t} = D_A \frac{\partial^2 C_A}{\partial z^2} \tag{24}$$

Initial and boundary conditions:

$$C_A(z, 0) = C_{A,0} \qquad 0 \le z \le \delta$$

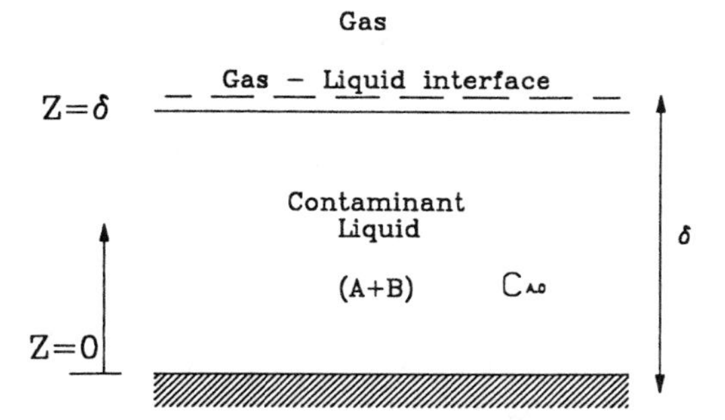

Figure 6. Evaporation of thin liquid layer consisting of two immiscible species.

$$\left(\frac{\partial C_A}{\partial z}\right)_{0,t} = 0 \qquad t > 0$$

$$-D_A\left(\frac{\partial C_A}{\partial z}\right)_{\delta,t} = h_m(C_{A,\delta} - C_{A,\infty}) \qquad t > 0$$

Component B

$$\frac{\partial C_B}{\partial t} = D_B\frac{\partial^2 C_B}{\partial z^2} \tag{25}$$

Initial and boundary conditions:

$$C_B(z, 0) = C_{B,0} \qquad 0 \leq z \leq \delta$$

$$\left(\frac{\partial C_B}{\partial z}\right)_{0,t} = 0 \qquad t > 0$$

$$-D_B\left(\frac{\partial C_B}{\partial z}\right)_{\delta,t} = h_m(C_{B,\delta} - C_{B,\infty}) \qquad t > 0$$

In writing the above boundary conditions, it is assumed that the lower boundary of the liquid layer is impermeable for species transport and the evaporation rate from the upper surface is described by the product of mass transfer coefficient h_m and the concentration difference between the interface and the bulk gas phase.

5. Comparison of Model Results with Data from Column Experiments

(a) Single Component. The governing equations described above were solved by using a finite difference numerical scheme. The input conditions to the model consisted of particle size, porosity, the thermophysical properties of the contaminant and the soil, relative saturation of the soil with contaminant, initial temperature of the soil and the contaminant, and temperature and flow rate of incoming air which was considered to be free of the contaminant. It was also assumed that there was no heat loss or gain from regions outside the region of interest. Lingineni and Dhir (1992) conducted experiments in a rectangular column (18 cm × 18 cm cross-section) filled with 360 μm glass particles. The porosity of the bed was about 0.39 and a residual saturation of 13% was established for ethyl alcohol which was used as a single-component contaminant. A vacuum pump was used to establish different flow rates through the test section. In the experiments, gas samples were taken at different locations along the test section, which was 60 cm in height. The ambient air entered the test section at a temperature of 23°C and a relative humidity of about 40%. No aqueous phase was present in the simulated soil prior to initiation of the experiment. Figure 7 shows temperature profiles obtained in the test section at different times, as predicted from the model described above, and those obtained from the

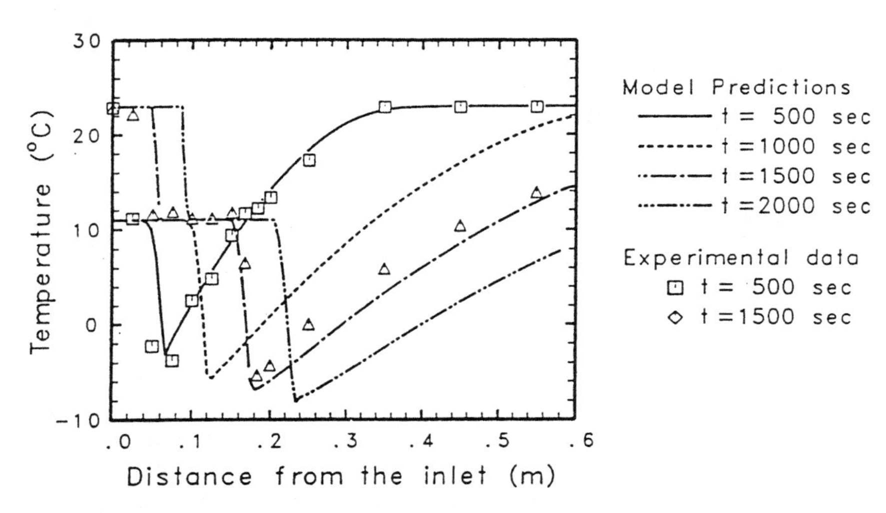

Figure 7. Temperature profiles in the test section at various times (air flow rate = 234 l/min).

experiment in which flow rate of air through the test section was 234 l/min. In exercising the model, it was assumed that, in addition to the absence of water in the pores, contaminant sorption did not occur on the particles. As inlet air passes through the test section, evaporation of liquid contaminant occurs at air–liquid interfaces. The latent heat of vaporization required to evaporate the liquid is partly provided by the incoming air, while the rest of it is obtained from the porous medium. As a result of the energy that is used to supply the latent heat of evaporation, the soil temperatures decrease with time. This reduction in local soil temperature continues until an equilibrium condition is attained between the heat transfer from the incoming air to the soil and the latent heat absorption, or until all of the contaminant at that location is completely evaporated. After the contaminant is completely evaporated at a particular location, the local soil temperature increases due to heat transfer from incoming air to the soil.

Moisture content in the air has the following effect on the soil temperature profiles and the evaporation rates. If the moisture content of air is greater than the saturation moisture content corresponding to local temperature of the soil, condensation of water vapor will occur. Latent heat released due to the vapor condensation provides additional heat, thus increasing the soil temperature at a rate higher than the value expected if heat transfer occurred from dry air only. Secondly, condensation of water in the soil changes the composition of the contaminant and hence the evaporation rate is different from that of a single-component contaminant. Since ethyl alcohol is miscible with water, the evaporation rates can be calculated using two component mixture equations. On the other hand, if the contaminant is immiscible with water, water can form a layer above the contaminant and hinder the evaporation of the contaminant. Another effect of humidity in incoming air is on the temperature to which soil is heated back once evaporation is completed. In the case of dry incoming air, the soil temperatures change from the minimum temperature to the ambient air temperature. But when moisture condensation occurs, the soil temperature increases only up to a temperature lower than ambient air temperature (close to wet bulb temperature). The soil remains at the wet bulb temperature until all of the condensed water is re-evaporated. In the temperature profiles plotted at each time level, the location of minimum soil temperature marks the upstream region where the contaminant has been completely evaporated. This location of minimum soil temperature also corresponds to the evaporation front location since all evaporation occurs downstream of that location. It can be seen from Figure 7 that, after 500 s of venting, the evaporation front has traveled a distance of 10 cm from the inlet of the test section, and the soil upstream of the location of the evaporation front is free of contaminant.

As the venting air passes through the test section, from the inlet of the test section to the location of the evaporation front, its temperature falls because of heat loss to the soil at the lower temperature. Downstream of the evaporation front, it encounters contaminated soil at higher temperatures. Due to high surface area per unit volume of the porous media, air becomes saturated with contaminant vapor within a narrow region. This saturation vapor concentration in the air depends upon the local soil temperature and the composition of the contaminant. Higher local temperatures in the soil provide higher saturation vapor concentrations at the air–liquid interfaces along the test section. The evaporative flux is proportional to the concentration difference that exists between the air–liquid interface and the incoming bulk gas. Hence, evaporation occurs in soil from the location of the evaporation front up to a point at which air becomes saturated at the soil temperature. Thus, one of the effects of lowering the soil temperature is to increase the width of the evaporation zone. At a time of 500 s, the evaporation zone extends from 10 cm to about 40 cm from the inlet. The location of minimum temperature (or the evaporation front) moves from the inlet towards the exit as venting progresses. Also, as the evaporation front moves toward the exit, vapor concentrations decrease in the outlet air. The predictions of the particle temperature are seen to be in good agreement with the data. The mass of the contaminant evaporated from the particulate layer as predicted from the model was also in good agreement with the data.

It should be noted that air flow rates in the experiments were one to two orders of magnitude larger than those normally encountered in the field tests. Because of this, in field tests the magnitude of temperature changes will be much smaller. Nevertheless, the results of experiments and the model demonstrate the need for nonisothermal models of soil venting.

Lingineni and Dhir (1992) applied the model to study the effect of preheating the incoming air on soils contaminated with other organics. In exercising the model, they considered four contaminants: benzene, trichlorethene (TCE), toluene, and *o*-oxylene. Each of these contaminants was considered as a single-component liquid present in the soil in the form of free NAPL at a given initial residual saturation. The contaminant zone was considered to be a cylindrical soil region of radius 5 m and depth 1 m. The venting well configuration consisted of a vacuum extraction well at the center of the contaminant zone and several passive air inlet wells placed along the boundary of the contaminant zone. The soil surface above the contaminant zone was assumed to be paved, and a radially symmetric air flow was assumed to be created by the wells that fully penetrated through the contamination plume in the subsurface. The initial residual saturation was assumed to be 15%, and initial temperature was taken to be 20°C. Figure 8 shows the location of the evaporation front with preheating of

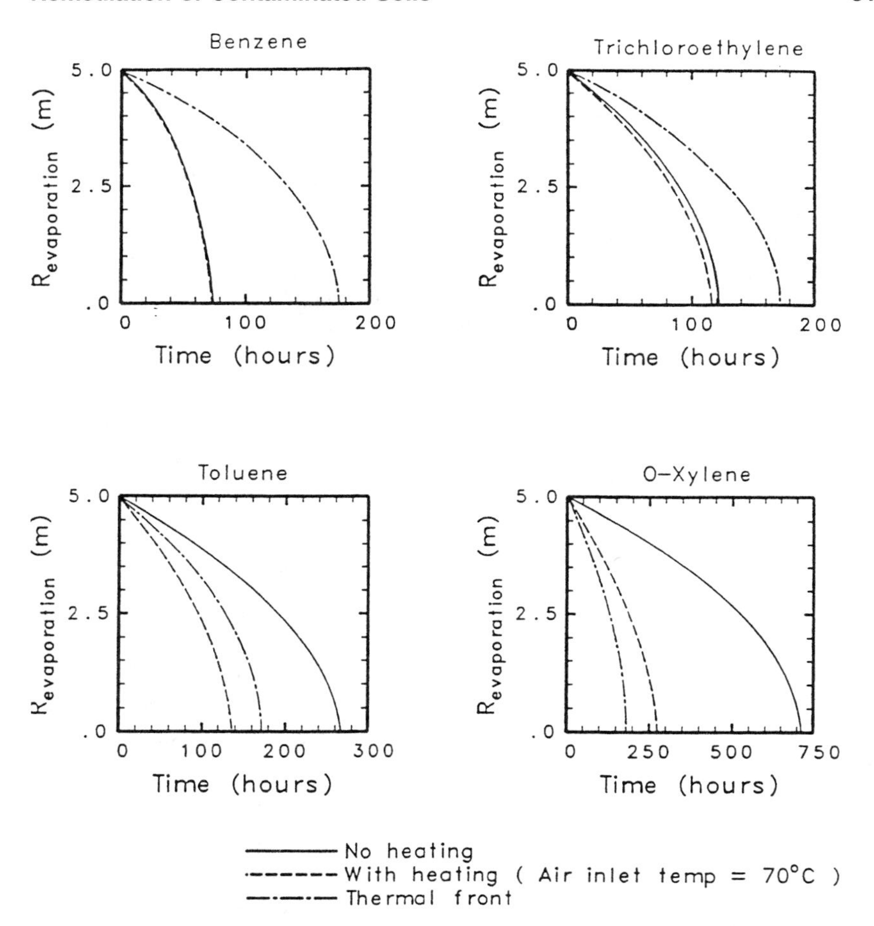

Figure 8. Effectiveness of heating on the remediation of various contaminants with different volatilities; prediction of evaporation and thermal front movements.

the inlet air to 70°C and without preheating. The location of the thermal front is also shown. As the preheated air moves through the soil, the soil temperature at a given location increases with time until an asymptotic value is reached, equal to the temperature of the inlet air. Thus, in determining the penetration of the thermal front, it was assumed that the thermal front arrived at a given location when the temperature at the location reached a mean of the initial and final asymptotic value. The velocity of the evaporation front indicates the rate at which the contaminated soil is being remediated as a function of venting time. The total time required to remediate the entire contamination zone is given by the time at which the evaporation front reaches the vacuum well. The total remediation time without preheating of the soil is seen to increase with decreasing volatility of the contaminant, it being smallest for benzene and highest for *o*-oxylene. With preheating of the incoming air to 70°C, the remediation time for less volatile contaminants such as *o*-xylene and toluene is considerably reduced whereas no such reduction in time is seen for TCE and benzene. In comparing the rate at which the remediation front moves through the soil with the propagation of a thermal front in the soil, it is found that for TCE and benzene the evaporation front moves faster than the thermal front. Thus, for these contaminants, the regions in which contaminant evaporation is occurring have not yet experienced the effect of heating, and the contaminant removal rate is not improved. By the time the thermal front reaches a particular location in the soil, the contaminant at that location has already been evaporated and the thermal energy in the incoming air is wasted in raising the temperature of the clean soil regions. On the contrary, for *o*-xylene and toluene the evaporation process is slower, and the thermal front can reach the evaporation zone prior to complete removal of the contaminant and thereby enhance the remediation process. From this observation, one can conclude that the remediation rate of soil can be improved by heating the incoming air, but only if the evaporation front for the contaminant moves slower than the rate at which the thermal front propagates in the soil. It should be noted that, in the simulation, water was not considered to be present in the soil.

(b) Multicomponent NAPL (Gasoline). Lingineni and Dhir (1997) applied the multicomponent model to study the remediation of soil contaminated with gasoline. In exercising the model, the soil was considered to be free of the aqueous phase. Gasoline has hundreds of constituents. However, while modeling, gasoline was divided into seven major components. Their mass fractions in the liquid phase, liquid density, molecular weight, and boiling temperature at one atmosphere pressure are given in Table 1. The numerical simulation was carried out with diffusion-limited transport in the liquid phase and under the assumption of complete mixing of the liquid phase. The results from the numerical simulations were compared with one-dimensional column experiments of Katsumata and Dhir (1992). These experiments were conducted in a cylindrical test section filled with 360 μm glass particles. The test section was 11.43 cm in diameter and 30 cm long. The glass particles filled the lower 20 cm of the column while leaving 10 cm open at the top. The porosity of the layer was 38% and the residual saturation of gasoline in the test section, as measured with a gamma densitometer, was found to be approximately 16%. A superficial velocity

Table 1. Properties of gasoline fractions chosen for laboratory 1-D experiment

Substance	Molecular weight (kg kmole^{-1})	Boiling point at 1 atm (°C)	Initial mass fraction in liquid	Liquid density (kg/m^3)
pentane	72.15	36.2	0.413	626
benzene	78.17	80	0.005 85	874
toluene	92.1	111	0.028 78	867
n-octane	114.2	126	0.036 65	703
p-xylene	106.2	138	0.008 69	866
n-propyl benzene	120.2	159	0.005 14	862
decane	142.3	174.3	0.502 89	703

about 2 cm/s of air at room temperature was created in the test section, using a compressed air source and a flow reservoir to reduce any fluctuations in the flow. Vapor samples of only five selected fractions of gasoline (benzene, toluene, *n*-octane, xylene, and *n*-propylbenzene) were collected from the outlet air at regular intervals and were analyzed using and HP 5890 chromatograph.

In the model, the diffusion coefficient for each of the gasoline fractions was estimated by using the Wilke–Chang technique (Reid et al. 1977) for estimating diffusion of a single species through a homogeneous solution of mixed solvents. The effective diffusion coefficient for the kth species in liquid phase was thus calculated using

$$D_{k,\ell} = \frac{(\phi\omega)^{0.5} T}{\eta_m V_k^{0.6}} \tag{26}$$

$$\phi\omega = \sum_{j=1}^{n} x_j \phi_j \omega_j \tag{27}$$

where x_j is the mole fraction, ω_j is the molecular weight, ϕ_j is the association factor of the jth species, V_k is the molar volume of solute fraction at its normal boiling temperature, and η_m is the mixture viscosity. An average association factor of 1.5 was assumed for all of the gasoline fractions.

Figures 9–13 show comparisons of vapor concentrations of selected fractions obtained from experiments with those obtained through numerical simulations. The following general trends can be noticed from these figures. The initial vapor concentrations of each species, as measured in the experiments and as observed in simulations, are very close to the values dictated by Raoult's law as combined with the saturation vapor concentrations and

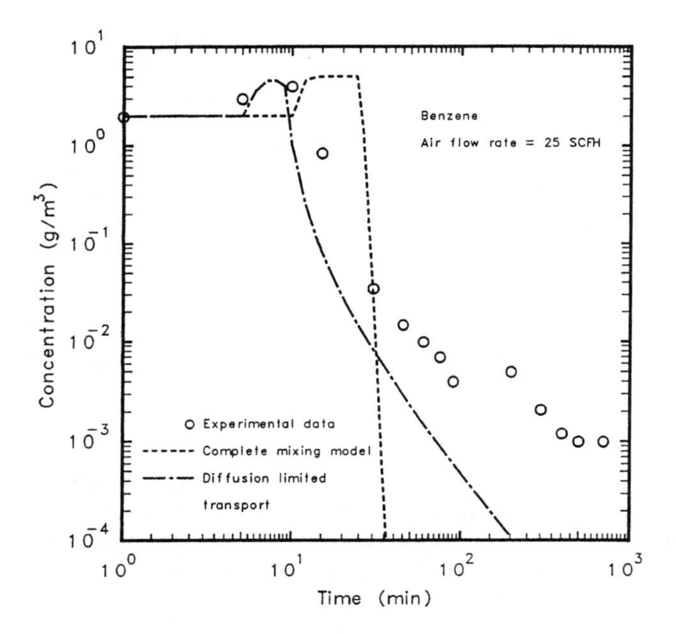

Figure 9. Vapor concentration of benzene from gasoline evaporation during venting with an air flow rate of 25 SCFH.

initial composition of the liquid contaminant. Experimental data show a gradual increase in the vapor concentrations at the initial venting periods and tailing off in the concentrations after attaining the peak value. Within the group of contaminant fractions monitored, the lighter fractions attained the peak value, followed by the heavier fractions. This can be attributed to the continuous changes in the liquid composition towards heavier fractions, thus increasing the mole fractions of the heavier fractions remaining in the liquid phase.

The results from the simulations carried out under the assumption of complete mixing of all of the species in the liquid layer show significant differences from the experimental data. Although the agreement is quite satisfactory during initial periods of evaporation, at later stages of venting the trends as well as the magnitudes observed in vapor concentrations as a function of time are considerably different. The experimental data show a gradual decrease in vapor concentrations for very long periods of the order of 10^3 min or greater. The numerical simulations show rapid reductions in the vapor concentrations and, as a result, much smaller remediation times.

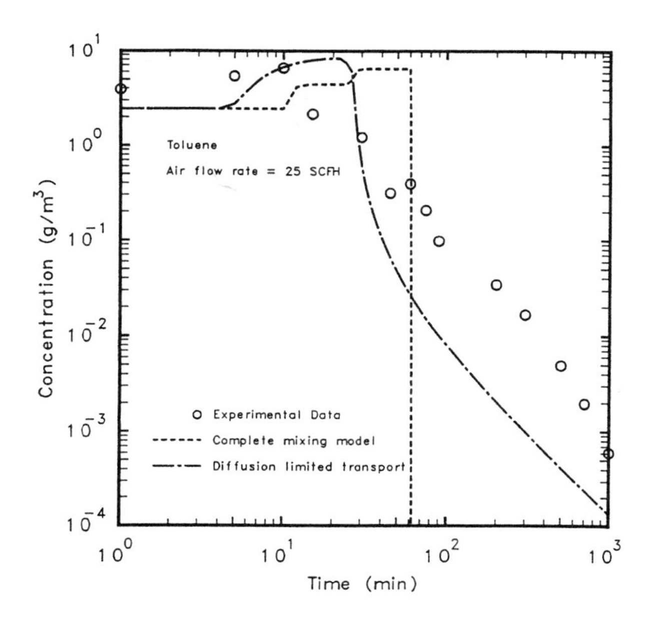

Figure 10. Vapor concentration of toluene from gasoline evaporation during venting with an air flow rate of 25 SCFH.

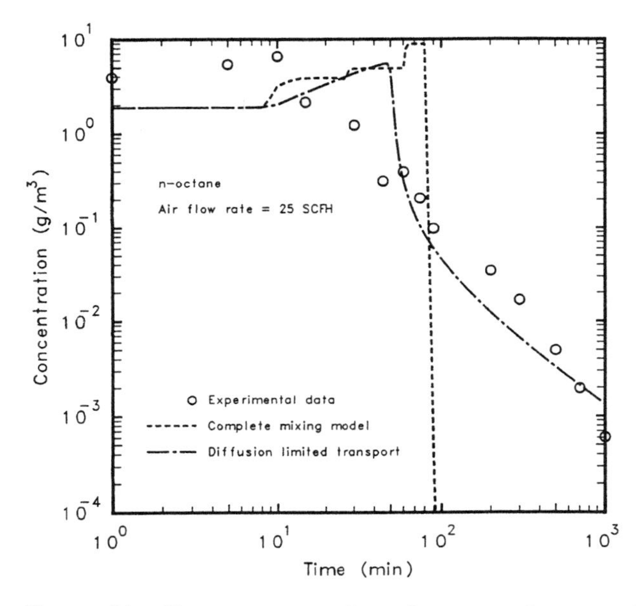

Figure 11. Vapor concentration of *n*-octane from gasoline evaporation during venting with an air flow rate of 25 SCFH.

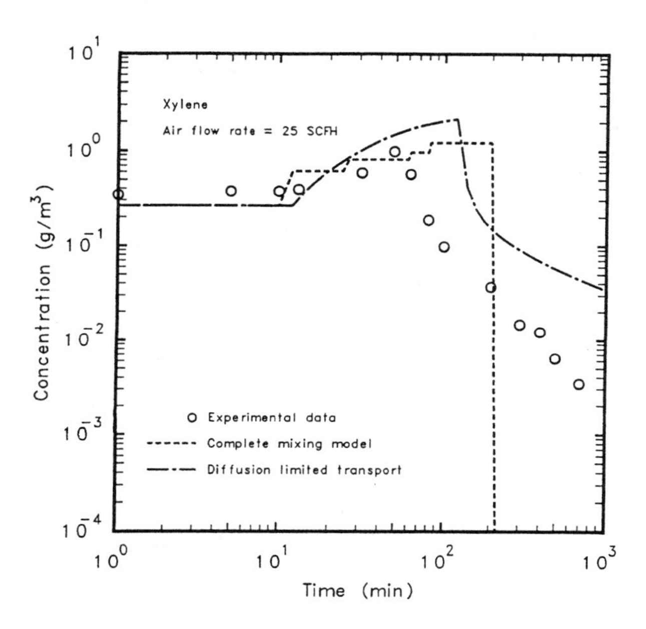

Figure 12. Vapor concentration of *p*-xylene from gasoline evaporation during venting with an air flow rate of 25 SCFH.

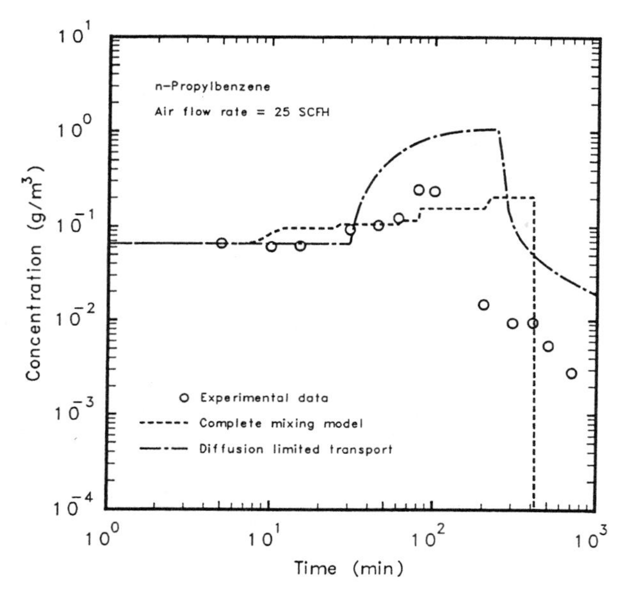

Figure 13. Vapor concentration of *n*-propylbenzene from gasoline evaporation during venting with an air flow rate of 25 SCFH.

The second set of simulations was carried out with the same input parameters and model, but using a diffusional transport model to predict the interface concentrations. The inclusion of diffusional transport was found to be an additional resistance mechanism in the removal of selected fractions from the bulk phase, thus reducing the vapor concentrations and in turn increasing the tailing-off period of the vapor removal. The results obtained with the diffusional transport model are in much closer agreement with the data from experiments.

(c) Multicomponent Contaminant (Diesel). The fraction of low volatility components in diesel fuel is much higher than that in gasoline. As such, preheating of incoming air is expected to be much more effective in enhancing the remediation process. Fontinich et al. (1999) have conducted column experiments using simulated soil contaminated with diesel fuel. The experimental apparatus was similar to that used by Katsumata and Dhir (1992). The test column was 11.4 cm in diameter and 30 cm in height, made of acrylic pipe. The test section was filled with 360 μm diameter glass particles up to a height of 22 cm with 8 cm unfilled space at the top. Four flexible heaters were attached to the outer surface of the test section to prevent heat loss to the surrounding area. Air from a pressurized source was dried and was preheated by passing it over a finned cartridge heater. The pressure of air at the inlet to the test section was near the ambient pressure. Samples of the air exiting the test section were obtained through a heated sampling port and were analyzed on an HP 6890 gas chromatograph. Temperatures in the direction of flow at the center and at the test section inner wall were recorded.

Porosity and diesel fuel residual saturation created in the test section were measured with a gamma densitometer and were found to be 0.37 and 0.12 respectively. As with gasoline, diesel fuel also contains hundreds of constituents. Because of this, numbers of hydrocarbons were grouped together and were identified by a representative carbon number. The composition of the diesel fuel used in the experiments was determined and is given in Table 2 along with that obtained by Mackay (1985). It is noted that the composition of diesel fuel from the two studies differs significantly. This suggests that the composition of diesel fuel depends on the source. Furthermore, in Table 2, the mass of a component was obtained by summation of the masses of all hydrocarbon classes included in the group. Properties of the group such as density, vapor pressure, and latent heat of vaporization were averaged on a mass basis with the use of known properties of the hydrocarbons present in the group.

In the experiments, the effect of flow rate and temperature of incoming air on the removal of various hydrocarbon classes was investigated.

Table 2. Experimentally established composition of diesel fuel used in experiment

Carbon number (1)	Experimental mass* (g) (2)	Mass*† (g) (3)
7	0.62	0.0803
8	0.23	0.0938
9	0.40	0.2627
10	3.54	2.5806
11	12.18	6.8033
12	15.33	9.1493
13	14.26	14.1697
14	11.64	13.7474
15	8.48	12.3868
16	7.06	12.3868
17	5.97	11.4484
18	4.94	9.2901
19	3.93	4.7858
20	3.44	2.8152
21	2.50	0
22	1.82	0
23	1.35	0
24	0.93	0
25	0.51	0
26	0.42	0
27	0.22	0
28	0.14	0
29	0.05	0
30	0.04	0
Totals	100	100

* Mass per 100 g of diesel.
† From Mackay (1985).

Figure 14 shows the mass fraction of three hydrocarbon classes in the exit air as a function of time for gas flow rates of 42 SCHF (superficial velocity of about 8.7 cm/s) and 14 SCHF. In both cases, the air inlet temperature was 90°C. It can be noted that, for the higher flow rate, the contaminant removal rate is faster as the peak in the mass fraction in the effluent gas occurs earlier. Each contaminant component mass fraction reaches the trace level significantly faster, and complete component removal occurs much earlier in comparison with the case of low air flow rate. It can also be noted that, as the process progresses, the time period between

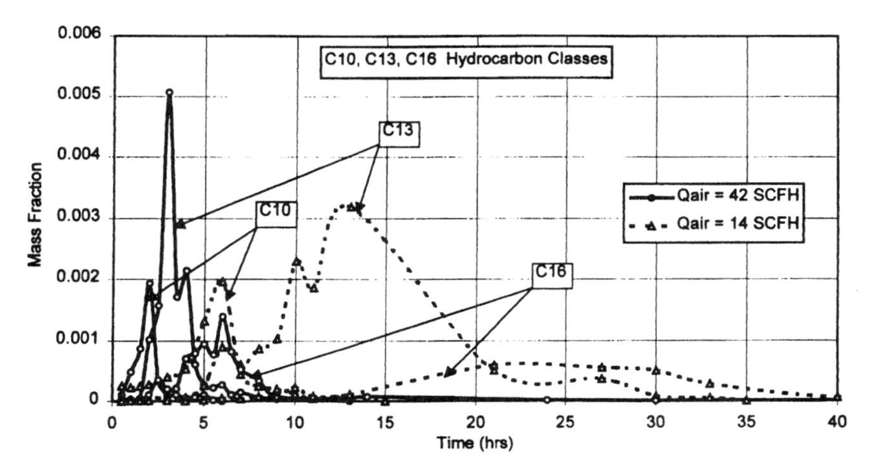

Figure 14. Mass fraction of C_{10}, C_{13}, and C_{16} hydrocarbon classes in effluent air as a function of time for experiments with different flow rates.

the peak mass fractions of various hydrocarbon classes increases. The period is ~ 4 h between the peaks for the C_{10} hydrocarbon class, whereas it is ~ 10 h for the C_{13} hydrocarbon class and ~ 17 h for the C_{16} hydrocarbon class. The increase in time periods as the volatility of the removed components decreases is an indication of the fact that, with lower flow rates, it takes longer for the temperature to increase; hence, a slowdown in the removal rate of low volatility components occurs.

Furthermore, for most of the components, the maximum mass fractions are lower with the lower air flow rate. For example, the maximum mass fraction of the C_{16} hydrocarbon class is ~ 0.0014 for the higher flow rate, whereas it is ~ 0.0006 for the lower flow rate. According to Raoult's law, the mass fraction of a component at the gas–liquid interface of a multicomponent liquid contaminant is proportional to the mass fraction of the component in the liquid phase. Since the removal rate of the diesel fuel components is slowed down in the experiment with the lower flow rate, as can be seen in Figure 14, larger amounts of components lighter than C_{16} remain in the liquid phase when the C_{16} hydrocarbon class reaches its maximum removal rate in ~ 23 h from the beginning of the experiment. Thus, the maximum mass fraction of C_{16} in the gaseous phase is considerably lower in the experiment with the lower flow rate in comparison with the experiment with the higher flow rate.

Figures 15 and 16 show the percentage of hydrocarbons C_7–C_{20} removed as a function of time in the experiments with different flow rates.

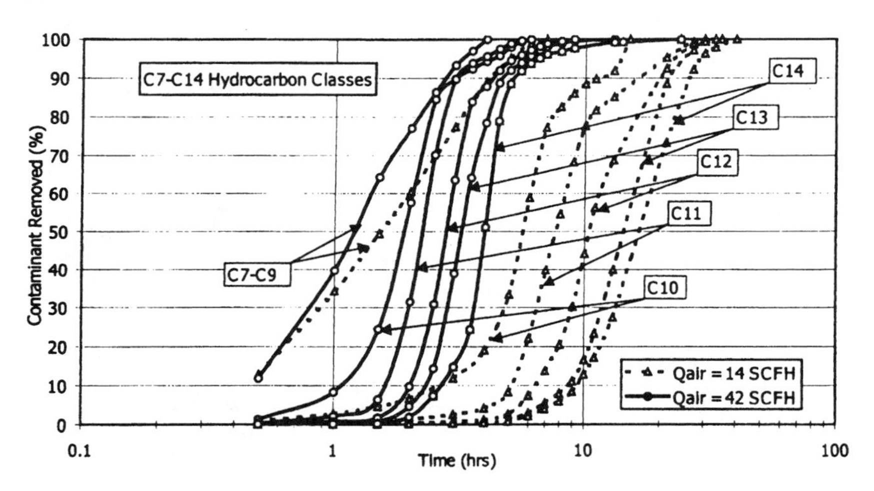

Figure 15. Percentage of C_7–C_{14} hydrocarbon classes removed as a function of time for experiments with different flow rates.

From the slope of the curves, it is possible to estimate the magnitude of removal rate of a species. Most of the experimental curves display three regimes of removal rate. Initially, the rate is found to increase until it reaches some peak value. Then, the peak removal rate is maintained until 80–90% of the component is removed. Thereafter, the removal rate gener-

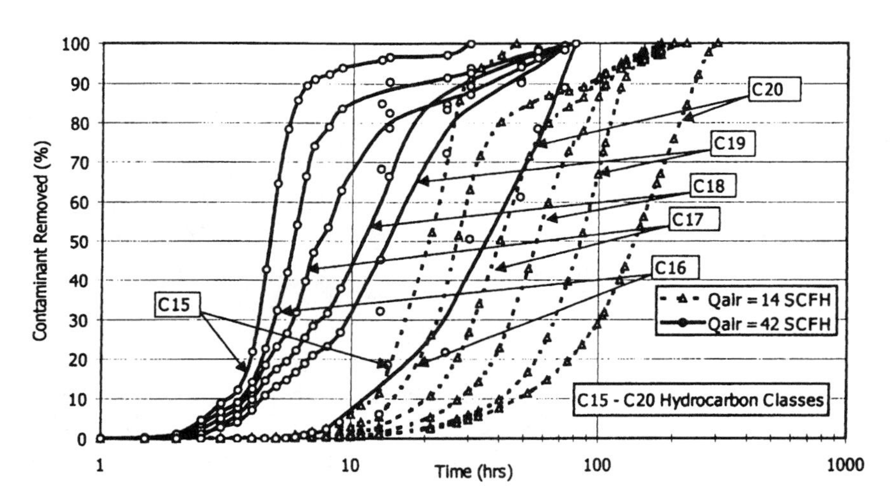

Figure 16. Percentages of C_{15}–C_{20} hydrocarbon classes removed as a function of time for experiments with different flow rates.

ally decreases until the contaminated species is completely evaporated. The group of lightest hydrocarbons, C_7–C_9, plotted in Figure 15 appear to display a slightly different removal rate pattern. These most volatile components rapidly reach their highest removal rates. Since the first data points were not taken until a few minutes after the beginning of the experiment, the slow rise period for the removal rate is not seen for the lightest hydrocarbons. However, after removal of the species from the outer layers of liquid contaminants, they must diffuse out form the interior of the liquid before they can be evaporated, consequently decreasing their removal rate. There is no significant decrease in the removal rate of the C_{20} hydrocarbon class since it is the last species remaining in the soil. For the heavier hydrocarbons, the removal rates are always slower than for the lighter species.

As can be concluded from Figures 15 and 16, the experimental curves display the time shift, which increases with the increase in carbon number. The time shift in the curves is initially ~ 3 h for the lightest hydrocarbon group and ends up to be ~ 220 h for the heaviest hydrocarbon group. For the heavy hydrocarbons, the mass concentration at the liquid–vapor interface at a given temperature is lower than that for lighter hydrocarbons. As a result, any change in concentration in the bulk gas due to flow rate affects the driving concentration difference strongly. Thus, removal of the heaviest hydrocarbons is influenced the most by the increase in flow rate.

For all hydrocarbons at any given temperature, the mass concentration difference between the evaporating interface and the bulk venting gas is larger for higher venting flow rates. The higher mass concentration difference, together with the higher energy input to the system, causes higher contaminant removal rates, which result in faster remediation of the contaminated soil.

Keeping the air flow rate constant (42 SCFH), the effect of inlet air temperature was investigated by conducting experiments with inlet air at 60°C and at 90°C. Figures 17 and 18 show the percentages of the diesel fuel components removed from the simulated soil as a function of time. Generally, the curves follow the pattern discussed earlier; that is, each curve shows three regimes: (1) removal rate increasing with time; (2) constant with time; and (3) gradually declining. The removal rate of the species is generally higher for the experiment with the higher inlet temperature. For the group of lighter components from C_7 to C_{14}, the maximum removal rates are not significantly different for the two temperatures, and the time shift of the experimental curves is not large. For the heavier components, starting with C_{15}, the removal rates at the reduced inlet air temperature are significantly lower, and the time shift increases as well. Similar to the experiments with the decreased flow rate, removal rates of the heaviest diesel components are significantly reduced as the inlet temperature is decreased.

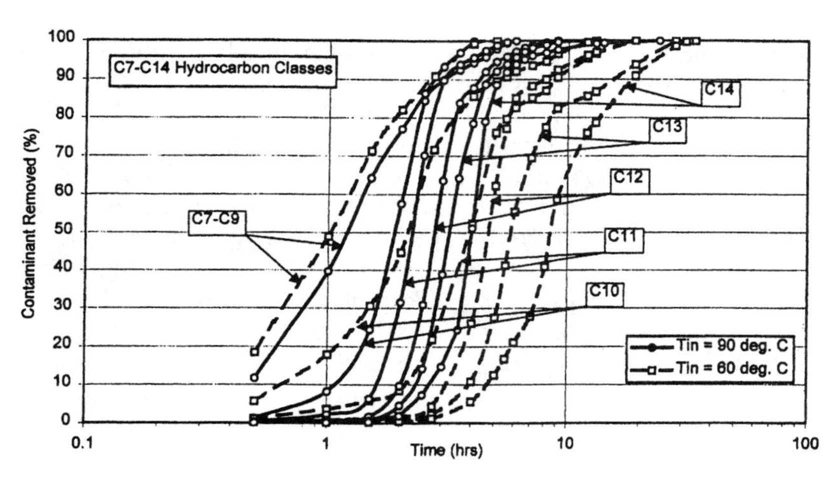

Figure 17. Percentage of C_7–C_{20} hydrocarbon classes removed as a function of time for experiments with inlet air temperatures.

For the fuel components, the effect of increased temperature results in increased vapor pressure, which causes higher component concentration at the liquid–vapor interface. The resulting increase in concentration difference between liquid–vapor and bulk gas facilitates removal of the components. Similar to the findings reported by Katsumata (1992) and Lingineni and

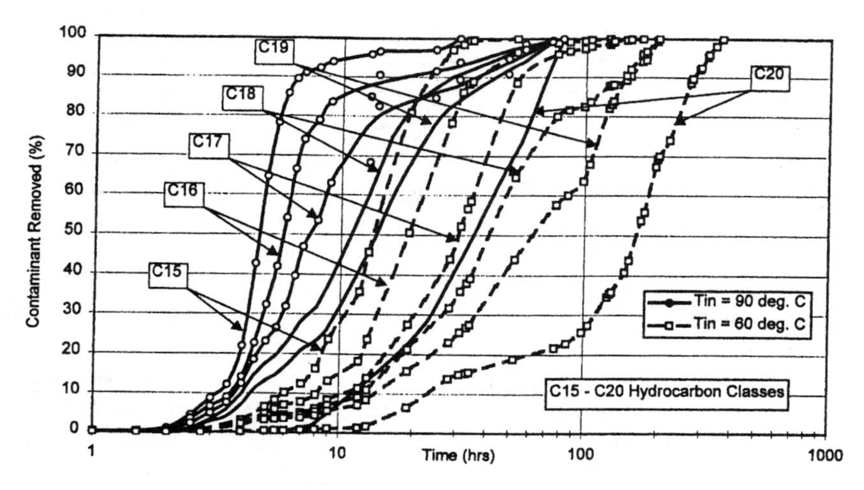

Figure 18. Percentage of C_{15}–C_{20} hydrocarbon classes removed as a function of time for experiments with inlet air temperatures.

Dhir (1992), the increased temperature plays a more significant role for removal of low volatility species.

For the venting air with the highest inlet temperature (90°C) and the highest flow rate (42 SCFH) it took about 80 h to remove residual diesel from the test section. When either inlet air temperature or the air flow rate was decreased, the time periods necessary to fully remove the contaminant increased significantly. For example, it took $\sim$ 360 h to fully remediate the soil in the experiment with a flow rate of 42 SCFH when the air inlet temperature was reduced from 90°C to 60°C. In the experiment in which air flow rate was reduced by a factor of 3 while the temperature was kept at 90°C, it took 300 h to complete the clean-up process. Thus, a 30°C increase in temperature of the incoming air caused a decrease in the clean-up time by a factor of 4.5, while the threefold increase in air flow rate reduced the clean-up time by a factor of 3.75.

The model described above was also used to predict the contaminant removal rates observed in the experiments. In exercising the model, diesel fuel was divided into five constituent groups. The group of lightest hydrocarbons, ranging from C_7 to C_{10}, was considered as the first group. The next three hydrocarbon classes, C_{11}, C_{12}, and C_{13}, were considered to form the second group, and so on. The heaviest diesel fuel component, C_{20}, was considered as the fifth and the last group. In choosing the representative vapor pressure correlation for each of the hydrocarbon groups, the following logic was used. If, in a particular group, the mass fraction of one species was much higher than that of the other species in the group, the vapor pressure of the dominant species was chosen to represent the whole group; otherwise, the vapor pressure in the middle species in each group was used. Thus, the vapor pressure correlations used in the computer model for the five groups correspond to C_{10}, C_{12}, C_{15}, C_{18}, and C_{20}, respectively.

The predicted percentage of contaminant removed versus time for the five hydrocarbon groups is compared with the experimental results in Figure 19 for the experiment in which the venting air flow rate was 42 SCFH and the inlet air temperature was 90°. The predicted removal rates of contaminant groups, as can be concluded from the slopes of the predicted curves, are mostly in good agreement with the experimental values. It can be noted in Figure 19 that the predicted curve and the experimental data for the first group, consisting of C_7 and C_{10} hydrocarbon classes, match quite well. The difference in time periods necessary to remove 95% of the hydrocarbon class, which can be considered as a criterion for comparison between the model and the experiment, is $\sim$ 0.7 h. The mass of the C_{10} hydrocarbon class contributes $\sim$ 74% to the total mass of the group, as can be seen in Table 2. Therefore, the vapor pressure of the C_{10} hydrocarbon class was used as a good representative of the group vapor pressure.

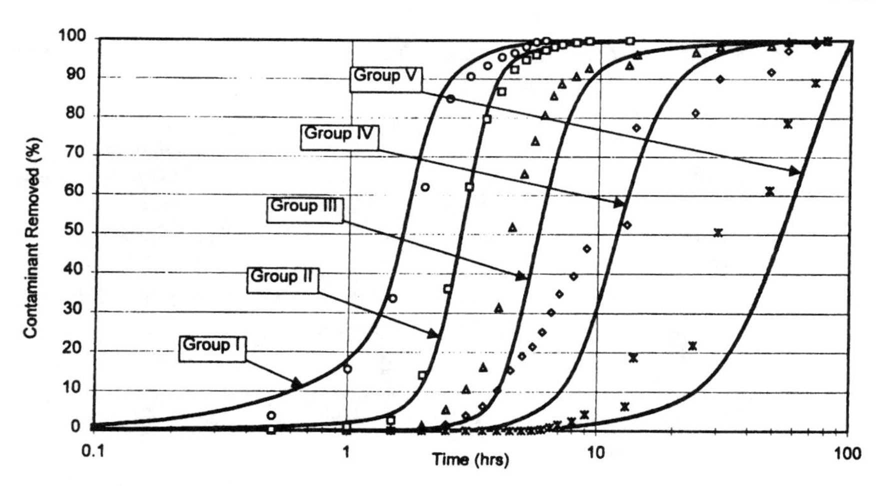

Figure 19. Experimental and predicted percentage of hydrocarbon groups removed versus time.

A comparison between experimental and predicted values for the second hydrocarbon group, which includes C_{11}–C_{13} hydrocarbon classes, is also shown in Figure 19. As can be seen in Table 2, the total mass of the hydrocarbon classes present in the group is almost equally distributed among the group constituents. Therefore, the saturation pressure of the C_{12} hydrocarbon class serves as an appropriate average pressure for the whole group. The agreement between model predictions and experimental data is excellent for the second group. The model underpredicts the time required to remove 95% of the group by only 0.5 h. For the third group, which includes C_{14}, C_{15}, and C_{16} hydrocarbon classes, the agreement between data and theory is again good except that a slight time shift can be observed in the model prediction curve and the data. The model predicts that removal of the hydrocarbon group is delayed during the first 3 h of the venting process. Judging from the experimental results, the removal begins ~ 1 h earlier than is predicted from the model. Removal of 95% of the hydrocarbon group occurs, according to the experiment, 1 h later than the model predicts. The time shift in the predicted curve and data is ~ 3 h for the fourth group of hydrocarbons, which includes C_{17}, C_{18}, and C_{19} hydrocarbon classes. The choice of the representative vapor pressure as well as the delay in removal of the third group can affect the predictions for the fourth group. It can be seen from Figure 19 that the slopes of the data points and predicted curve are very similar for the fifth hydrocarbon group, which includes the C_{20} hydrocarbon class. This suggests that the removal rate of the hydrocarbon group is predicted correctly. The most probable reason for the time shift in the data and the prediction curve in removal of the fifth group could be the delay in removal of the two previous hydrocarbon groups. The sensitivity studies performed for the model with the different representative vapor pressures for various groups showed that the proper choice of vapor pressure is vital in obtaining good agreement between model predictions and the data.

The experimental and predicted percentage of diesel fuel removed versus time is shown in Figure 20. The agreement between the data and the model predictions for the removal of total diesel fuel trapped in the test section is excellent, as can be seen from this figure. The removal rate of diesel fuel is predicted properly, and time of complete cleanup is in good agreement with the data. Recognizing that the sum of mass fractions of each group is unity, the better agreement between the data and predictions is observed because of compensation of uncertainties that exist in defining the mass fraction in each hydrocarbon group in the model versus that present in the contaminant.

B. Summary Comments

The experimental studies of venting of simulated soils were carried out in the absence of aqueous phase in the pore space prior to initiation of the reme-

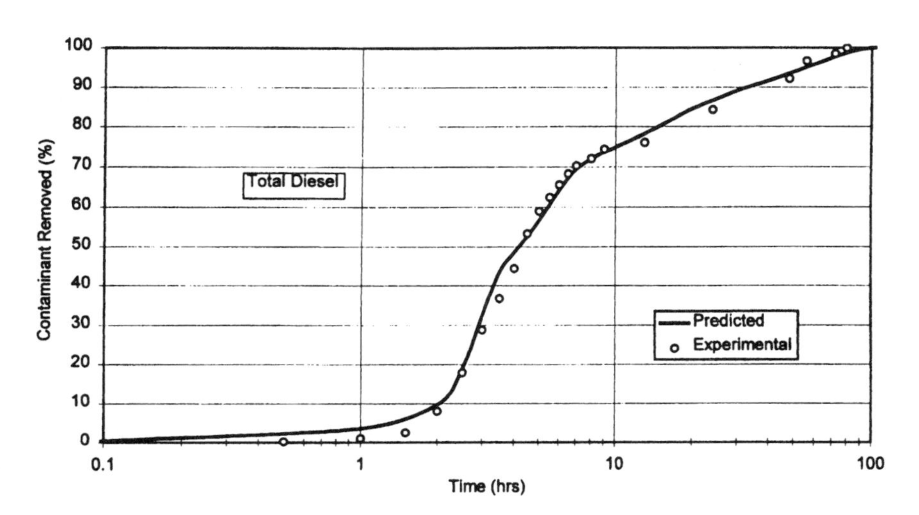

Figure 20. Experimental and predicted percentage of diesel fuel removed versus time.

diation process. The presence of water as immiscible or miscible liquid can prolong the remediation process, but can be simulated using the theoretical model described above. Also, the comparison of the model predictions with the data was carried out for simulated soils of uniform particle size and porosity. In the field, large variations in soil permeability and porosity can occur. The momentum equation described earlier in this section can appropriately be modified to include the variation of permeability and porosity with spatial coordinates. However, additional modeling for mass transfer by diffusion from less permeable regions to the more permeable region and for the temperature field in the less permeable region will have to be carried out.

The results of the above-described experimental and theoretical studies reveal that:

1. Local soil temperatures can reduce significantly during the remediation of contaminants with high latent heat of vaporization.
2. Moisture in the incoming air can condense in the soil region at lower temperatures and thereby slow down the evaporation of the contaminants.
3. Preheating of incoming air can reduce the remediation time. However, relative rates of propagation of thermal and evaporation fronts for a single-species contaminant play a significant role in determining the effectiveness of preheating the inlet air.
4. In multicomponent contaminants (NAPL), the liquid composition is found to shift towards less volatile fractions during the venting process. The preheating of inlet air is more effective in enhancing the removal rate of the less volatile contaminants.
5. The removal rates of contaminants during soil venting are found to be governed by three main limiting processes. The first is the saturation limit of the air passing through the contaminated soil. This limit is set by the air flow rates created in the soil and the local soil temperature. The second limit results from convective transport of various species from the liquid–gas interface in the pore space and the bulk gas. The mass transfer coefficient depends on the flow Reynolds number, Schmidt number, and density and superficial velocity of gas. The third limit is set by the rate at which contaminant fractions can be transported in the liquid phase to provide the required concentrations at the interface to sustain evaporation. This limit is affected by the diffusion coefficients of the contaminant fractions in the liquid phase, effective thickness of the liquid filaments in the pore space, and the composition of the liquid.

The models, as presented, have several limitations. The most important is that adsorption in the soil particles and desorption from the soil particles have not been considered. Because of the longer times that are needed for complete desorption, the clean-up times may be longer than those obtained from the above-described models. Although the models allow for the presence of the aqueous phase in the soil, no comparison of predictions with the data have been provided. Models have also not been tested when heterogeneities and three-dimensional conditions exist in the soil.

III. STEAM INJECTION

Injection of steam to enhance the recovery of oils, especially that of light oils, has been used successfully in the oil industry. According to Blevins et al. (1984), a number of mechanisms contribute to the increased recovery of light oils with steam injection. These include reduction in viscosity of oil with increase in temperature, rapid volatilization of high volatility compounds, and volatilization of low volatility organics after oil has been depleted of high volatility compounds. Following the work of Miller (1975), Hunt et al. (1988) analyzed in an approximate manner the removal of NAPLs from a water-saturated medium by injection of saturated steam. With the injection of steam, the water velocity ahead of the front increases beyond that associated with the mass flow rate of steam. This is due to the fact that equilibrium saturation of water upstream of the steam front is smaller. The increased velocity downstream of the steam front helps in mobilization of some of the trapped contaminants, but not all of them. Thus, contaminant removal mostly occurs because of the volatilization of the contaminants caused by increased temperatures of the soil, water, and the contaminants near the steam condensation front. In their one-dimensional formulation, Hunt et al. (1988) made no distinction between steam condensation (steam breakthrough) front and steam distillation front (location upstream of which soil is completely free of contaminants). An approximate expression was derived for the steady-state velocity of propagation of the steam front in a homogeneous porous medium by equating the time rate of change of enthalpy of steam as it condenses with the spatial increase in the internal energy of the soil and the water trapped in the interstitials. Subsequently, Stewart and Udell (1988) corrected this expression to account for the differences in saturations of water, steam, and contaminant far upstream and far downstream of the steam zone. From the analysis, the dimensionless steam front velocity was obtained as

$$\frac{\varepsilon \rho_s v_f}{m''_{\text{inj}} x} = \frac{\dfrac{\rho_s}{\rho_w}\left[\dfrac{h_f g}{c_p w(T_f - T_i)} + \dfrac{1}{x}\right]}{\left[\begin{array}{c} 1 - S_{\ell\infty} + \dfrac{(1-\varepsilon)\rho_p c_{pp}}{\varepsilon \rho_w c_{pw}} + \dfrac{\rho_s(1 - S_{w(-\infty)})h_{fg}}{\rho_w c_{pw}(T_f - T_i)} \\ -\dfrac{\rho_\ell - \rho_s}{\rho_w} \cdot (1 - S_{w(-\infty)} - S_{\ell\infty}) \end{array}\right]} \tag{28}$$

For a steady-state steam front propagation velocity, Hunt et al. (1988) also obtained an expression for the characteristic thickness of the heated zone ahead of the steam condensation front as

$$\delta = \frac{k_m(T_f - T_i)}{m''_{\text{inj}} x h_{fg}} \tag{29}$$

The constituents of the trapped NAPL that have boiling temperature below the steam condensation temperature evaporate with the arrival of the steam front and are carried forward in the form of vapor. Downstream of the front, where the temperature is lower than the condensation temperature, both the steam and the evaporated contaminants condense. The process repeats as the steam front advances, and the region upstream of the steam front can be considered to be free of contaminants with a boiling temperature lower than the steam condensation temperature. However, the evaporation of the low volatility components will continue to occur in a manner similar to that for soil venting.

Hunt et al. (1988) also conducted laboratory experiments to study the mobilization of NAPL by water flooding, dissolution and removal by steam injection. A horizontal porous layer of Ottawa sand with a particle diameter in the range of 160–200 μm was formed in a 5.1 cm diameter and 91 cm long Pyrex tube. The porous layer had a porosity of 0.385 and was initially completely saturated with water. The temperatures along the centerline and pressure were measured at several locations along the test section. Guard heaters were used to compensate for the radial heat loss. Experiments were conducted by injecting, at the center of the water-saturated porous layer, 18 ml of a single contaminant (TCE), a 50% by volume mixture of toluene and benzene, and gasoline.

Figure 21 shows the superficial velocity of water and concentration of total hydrocarbons at the exit of the test section when gasoline was used as the test contaminant. In the experiment, water flooding was followed by steam injection and then water flooding again. Water flooding was carried out for the time needed to replace 5.6 pore volumes. High velocity water flooding was followed by a short period of flooding at low velocity. Thereafter, saturated steam at a pressure of about 1 atm was injected for 5.6 to 6.7 pore volumes. Initially, flooding at a high velocity was not found

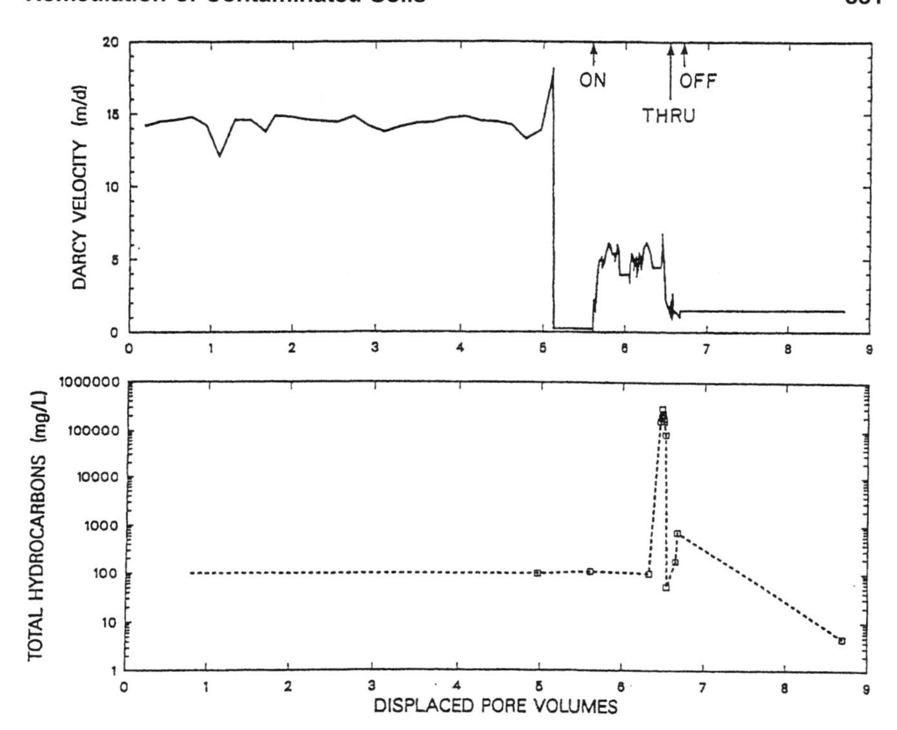

Figure 21. Experimental results for the sand column containing 18 ml of gasoline. (Top) Displaced liquid velocity during water and steam flooding. (Bottom) Total hydrocarbon concentration measured in column effluent reported per liter of displaced fluid (Hunt et al. 1988).

to lead to any displacement of gasoline as the concentration of hydrocarbons in the effluent corresponded to about the aqueous solubility limit of gasoline. Breakthrough of steam front was observed to occur at 6.544 volumes. Just prior to the steam breakthrough front, a large increase in the concentration of the hydrocarbons in the effluent was observed. In fact, about 75% of the initial amount of gasoline was recovered as a free product in the effluent just prior to or after the breakthrough of steam front. This indicated that less volatile constituents of gasoline were also evaporated by steam. At the end of the water flooding period that followed the steam injection, the concentration of total hydrocarbons in the effluent was found to be much less than that at the beginning of the experiment. The steam front velocity of 13.8 m/day was found to be consistent with that obtained from an expression similar to Eq. (28). Although no calculated value of the heated zone thickness ahead of the steam front for gasoline was provided, the calculated value of the heated zone thickness from Eq. (29) for single and two-component contaminants was found to be in agreement with the data. The measured heated zone thickness of about 1.4 cm was found to be nearly insensitive to the type of contaminant. In the experiments, the water saturation upstream of the steam zone was found to be about 0.27. The temperature gradient and heat capacities affected more of the movement of the steam front than did the pressure gradient and the permeability.

Udell and Stewart (1989) conducted field tests on a pilot scale to study the effectiveness of vacuum venting followed by steam injection for remediation of soil contaminated with organics such as xylenes, ethylbenzene, 1,2-dichlorobenzene, 1,1,1-trichloroethane, trichlorethene, tetrachloroethene, and acetone. The soil was unsaturated with contaminants and overlaid a thin unconfined aquifer located about 6–8 m below the soil surface. The soil consisted of sand silts, clay and sand, and sand regions of largely distinct permeabilities. A square area of soil measuring about 4 m in each direction and about 7 m deep was considered for remediation and six intake wells and a recovery well were installed. The recovery well was installed in the middle of the contaminated area whereas the intake wells were equally spaced on the periphery of a circle with a radius of about 1.6 m from the center of the recovery well. Vacuum extraction was carried out for 40 h with a constant flow rate of about $0.75\,m^3/min$ (25 SCFM). The exit gases were found to be saturated with contaminant vapor at a nearly constant concentration of $60\,g/m^3$. As such, about 72% of the contaminants present in the soil were recovered by vacuum extraction alone. From vacuum testing, the average permeability of the soil was found to be about $8.7 \times 10^{-12}\,m^2$.

Steam injection at a rate of about 114 kg/h was maintained for 140 h. The steam front breakthrough was observed to occur at the recovery well about 32 h after the beginning of steam injection. This compared favorably with a time of 36 h calculated from Eq. (28). The measured soil temperatures indicated the existence of a high permeability region in the lowest 1.5 m of the contaminated soil, with a region of low permeability just above it. That is, during early periods of steam injection much lower temperatures were found in the less permeable region and the region just above it. The concentration of contaminants in the noncondensible gas at the exit was found to be about the same as during vacuum venting. It was found that removal of more volatile compounds was followed by high boiling point organics after the steam had heated the soil and was flowing through it. At the retrieval well, liquid was produced at a rate of about 19 l/h for the first 90 h of operation. In addition, about 37 l of condensate was produced per hour from the effluent (gas/vapor) leaving the retrieval well. The concentrations of contaminants in the liquid retrieved with a jack pump were highest prior to steam front breakthrough and dropped off rapidly thereafter. In the condensate, high concentrations of contaminants were observed just after steam front breakoff. During the early period, the concentrations of high volatility components (boiling points lower than steam condensation temperature) dominated those of the low volatility components. However, in the later stages, the concentrations decreased and reached a steady state when mostly low volatility components were being evaporated.

Concentrations of various contaminants in the soil that remained after treatment were also determined. Large variations in the concentration of various contaminants in the vertical direction were found. High concentrations were found in low permeability regions. It was postulated that contaminant-laden condensate had imbibed into the low permeability region due to capillary action and that resistance for mass transfer from low permeability regions was high. Although a significant number of soil samples were analyzed, it was concluded that "a statistically meaningful evaluation of the effectiveness of the process was not possible." Nevertheless, it was remarked that steam injection, together with vacuum extraction, was a promising technique for rapid and effective remediation of soils contaminated with organics.

In a subsequent work, Yuan and Udell (1993) have developed a detailed one-dimensional model for removal of single-component contaminant by injection of steam in an initially water-saturated homogeneous porous medium. The model results were verified with laboratory experiments utilizing a one-dimensional horizontal column. Figure 22 shows their conceptualization of the physical regions in the vicinity of the distillation zone (the region where volatilization of the contaminant occurs). Upstream of the distillation zone, the soil is free of contaminants and steam and water enter the porous medium at a constant flow rate. Downstream of the distillation zone, an equilibrium zone exists in which the vapor mixture is saturated with the contaminant. The steam condensation front represents the downstream end of the equilibrium zone and a large temperature gradient exists ahead of the condensation front. For a quasi-static case, both the distillation front and the steam condensation front were assumed to move at constant velocities v_c and v_f, respectively.

In analyzing the physics of the evaporation process in the vicinity of the distillation zone, little or no temperature gradients were assumed to exist. As a result, diffusion of heat in the matrix was ignored and energy was assumed to be mainly transported by convection. A similar assumption was made with respect to mass transfer in the gas phase. A combined mass conservation equation for steam and water in one dimension was written as

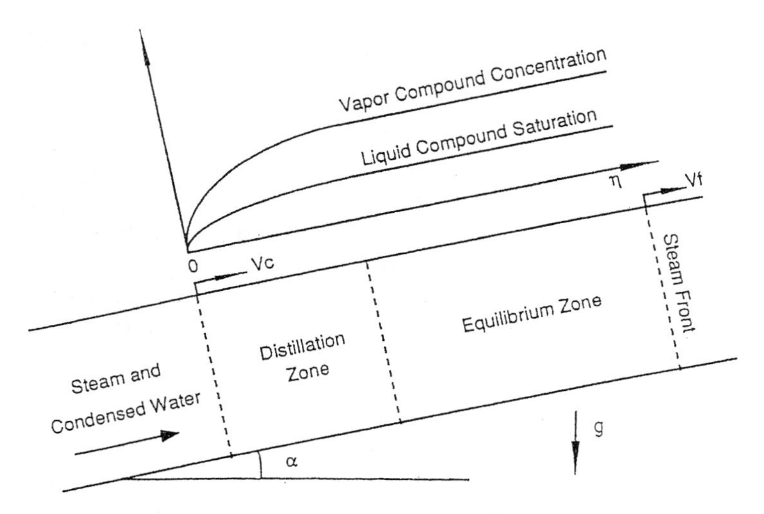

Figure 22. Steam distillation of single-component hydrocarbons in a one-dimensional porous solid (Yuan and Udell, 1993).

$$\frac{\partial}{\partial x}\left[m_v''(1 - C_v) + m_w''\right] + \varepsilon\frac{\partial}{\partial x}[\rho_v S_v(1 - C_v) + \rho_w S_w] = 0 \tag{30}$$

For a stationary contaminant, the conservation equation of mass for the liquid contaminant was obtained as

$$-\varepsilon\frac{\partial}{\partial t}(\rho_\ell S_\ell) = g_m(C_s - C_v) \tag{31}$$

In writing Eq. (31), it is assumed that the mass concentration of vapor at the liquid–vapor interface of the contaminant is always at its saturation value whereas the concentration in the gas can vary spatially and temporally.

The parameter g_m in Eq. (31) represents a product of mass transfer coefficient and the interfacial area per unit volume. A value of this constant was deduced from the experiments. A combined conservation of mass equation for the liquid and vapor phases of the contaminant was written as

$$\frac{\partial}{\partial x}[m_v'' C_v] + \varepsilon\frac{\partial}{\partial t}[\rho_v S_v C_v + \rho_\ell S_\ell] = 0 \tag{32}$$

With the pressures in the vapor and liquid differing by the capillary pressure, and assuming that the pressure gradients experienced by liquid and vapor were the same, Yuan and Udell (1993) obtained a combined momentum equation for contaminant vapor and water as

$$\frac{\partial p_c}{\partial x} - (\rho_w - \rho_v) g \sin \alpha = \frac{\mu_v m''_v}{\rho_v k k_{rv}} + \frac{\mu_w m''_w}{\rho_w k k_{rw}} \tag{33}$$

where p_c is a function of liquid saturation and is obtained through the use of the Leverett function, k is the intrinsic permeability, and k_{rv} and k_{rw} are the relative permeabilities for steam and water. The relative permeabilities are a function of liquid saturation.

In the absence of a temperature gradient, the combined energy equation for liquid and vapor phases of water and the contaminant in the porous medium was written as

$$\begin{aligned} &\frac{\partial}{\partial x}[h_s m''_v (1 - C_v) + h_v m''_v C_v + h_w m''_w] \\ &\quad + \varepsilon \frac{\partial}{\partial t}[e_v \rho_v S_v C_v + e_s \rho_v S_v (1 - C_v) + e_\ell \rho_\ell S_\ell + e_w \rho_w S_w] = 0 \end{aligned} \tag{34}$$

Under the assumption that the velocity, v_c, of the distillation front is constant and that the profiles of various variables (m''_v, m''_w, S_ℓ, S_w, and C) are frozen upstream and downstream of the distillation front, Eqs. (30)–(34) were solved by defining a similarity variable. The similarity variable combined the space coordinate and time. Under the constraint that $S_v + S_\ell + S_w = 1$, the five boundary conditions far upstream and far downstream of the distillation front, respectively, were defined as

$$C_v(-\infty, t) = 0, \qquad S_\ell(-\infty, t) = 0$$

$$m''_v(-\infty, t) = m''_{\text{inj}} x, \qquad m''_w(-\infty, t) = m''_{\text{inj}}(1 - x)$$

$$S_w(-\infty, t) = S_{w-\infty}$$

and

$$C_v(\infty, t) = C_S, \qquad S_\ell(\infty, t) = S_{\ell\infty}$$

$$m''_v(-\infty, t) = \text{constant}, \qquad m''_w(\infty, t) = \text{constant}$$

$$S_w(\infty, t) = \text{constant}$$

Hence, they were able to obtain a simple expression for the dimensionless velocity of the distillation front as

$$\frac{\varepsilon v_c \rho_s}{m''_{\text{inj}} x} = \frac{1}{\left[\left(1 - S_{w(-\infty)}\right) + \frac{\rho_\ell}{\rho_s} S_{\ell\infty}\left(\frac{1}{C_s} + \frac{h_{\ell v}}{h_{fg}} - 1\right)\right]} \tag{35}$$

Figure 23 shows a plot of dimensionless velocity of the distillation front as a function of the residual saturation for different contaminants. In this figure

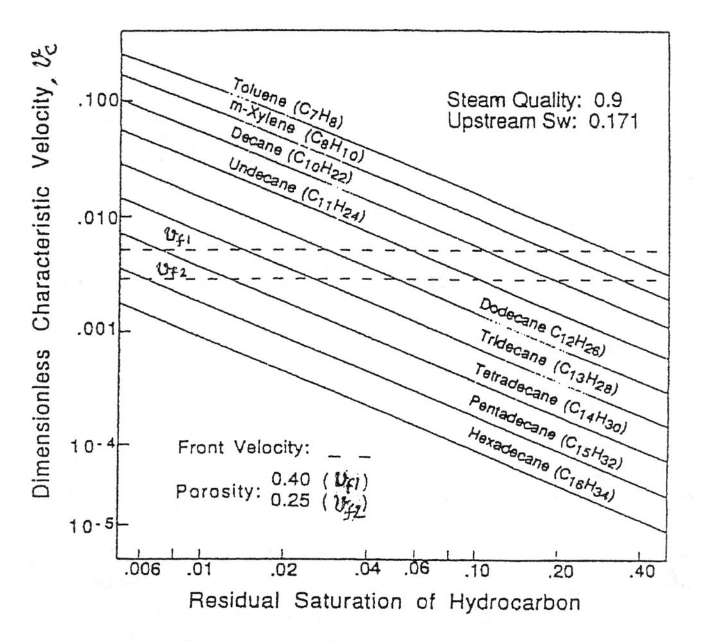

Figure 23. Dimensionless distillation front velocity versus saturation of various hydrocarbons.

the velocity of the steam condensation front for two porosities is shown as dotted lines. For the situations in which the steam condensation front velocity is smaller than the distillation front velocity, the evaporation of contaminant will occur when steam comes into contact with the contaminant. Thus the distillation zone velocity will be limited by the condensation front velocity. However, if the steam condensation front velocity is higher than the distillation zone velocity, a separate distillation zone will exist. The roles of the steam condensation front and the distillation zone front are somewhat analogous to the thermal front and the evaporation fronts discussed earlier in the context of soil venting. The concentration of a single-species contaminant (dodecane) in effluent leaving the test section was measured and found to be in good agreement with the predictions from the model. Figure 24 shows the comparison. From the results it was also deduced that the mass transfer rate g_m varied linearly with the saturation of the contaminant.

Through analysis of the region upstream and downstream of the distillation zone, Yuan and Udell (1993) were able to delineate the physics of the remediation process of a simulated soil contaminated with a single species. However, in practice, the contaminant is a mixture of several com-

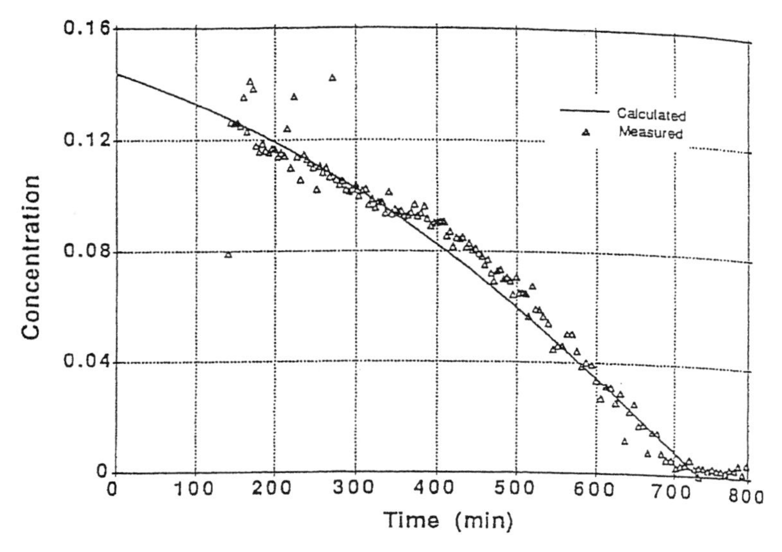

Figure 24. Dodecane concentration in the vapor phase as measured in the effluent.

ponents and simultaneous evaporation and condensation of different species can occur. As a result, the process can be quite complex. Also, depending on the degree of miscibility of the contaminant with water, large uncertainty can exist in the interfacial area of the contaminant over which evaporation occurs. In soils contaminated with a mixture of different species, the resistance to diffusion of different species within the liquid contaminant must also be considered.

A. Summary Comments

Both laboratory-scale experiments and field tests have demonstrated that steam injection can be used effectively to remediate soils contaminated with hydrocarbons. In the field tests, vacuum venting was followed by steam injection. Almost 72% of the contaminants present in the soil were recovered by vacuum extraction alone. Because of the heterogeneities in the soil and a statistically limited number of soil samples that were taken, it was noted that the effectiveness of the process could not be fully substantiated. Because of much longer times associated with diffusion of contaminant from less permeable regions of the soil and imbibition of water and contaminants from regions with large permeabilities, the remediation times can be substantially prolonged. The other results from the above experimental and theoretical studies reveal that:

1. Only a small fraction of the contaminant trapped in the interstitial has been shown to be mobilized by the inertia of water flowing through the porous medium.
2. Some dissolution of contaminants can occur in the aqueous phase.
3. Volatilization of contaminants mostly occurs because of the increased temperature of the soil matrix caused by condensation of steam.
4. A theoretical model for remediation by steam injection of water-saturated soil contaminated with a single species was developed. The model was validated with results from laboratory experiments. However, in comparing the model results with data, the overall mass transfer rates (product of mass transfer coefficient and the interfacial area) were deduced from the experiments.
5. At present, no model exists for remediation, by steam injection, of water-saturated soils containing multi-species contaminants.

IV. CONCLUDING REMARKS

In this chapter, several of the technologies proposed for remediation of soils contaminated with hydrocarbons have been briefly described, with the exception of soil venting and steam injection. For these two technologies, mathematical models for hydrodynamic, thermal, and mass transfer processes during removal of multispecies contaminants have been described in detail. The models have been shown to describe, in general, the behavior observed in the laboratory experiments. A detailed validation of the models in a typical field site is lacking. However, soil venting with preheating of incoming air and steam injection have been shown to be very effective technologies for remediation of soils contaminated with volatile and semi-volatile hydrocarbons.

In the models, large uncertainties exist with respect to desorption in the pore space and in the phenomena of simultaneous evaporation and condensation of different species. No clear-cut basis exists for the evaluation of mass transfer coefficients and interfacial areas when liquid contaminants and the aqueous phase coexist in the pore space. Adsorption and desorption of contaminants on the surface of particles and in the secondary pore space in the soil continue to be ill-understood processes. However, uncertainties in the geological properties of the soil can overwhelm any uncertainties that may exist in the modeling of the remediation process.

Research into the modeling of the transport processes that occur in the pore space as well as in the liquid contaminant retained in the pore space during remediation will be of considerable interest in the future and will pose a significant challenge. A similar statement can be made with respect to modeling of partitioning of the contaminant into different phases and the adsorption and desorption processes. The models described in this chapter will be useful in determining the optimum conditions for application of the remediation technology in the field. However, they are not a substitute for the real field experience since the geology of the soil to be remediated can significantly alter the predicted remediation times and degree of remediation. In many instances it may be helpful to combine a number of technologies (e.g., soil venting and bioremediaton). Thus, models that not only describe the physical and chemical processes but also the biological process, simultaneously, will be needed.

ACKNOWLEDGMENT

This work received support from the State of California under the University of California Toxic Substances Research and Teaching Program.

NOMENCLATURE

Roman Letters

A	interface area
C	concentration
c_p	specific heat
c_{pp}	specific heat of solid phase
$D_{k,g}$	diffusion coefficient of the kth species in the gas
$D_{k,\ell}$	diffusion coefficient of the kth species in the liquid
d_p	particle diameter
e	internal energy
f_1, f_2	energy partitioning factors in the soil matrix
g	gravitational acceleration
h	heat transfer coefficient or enthalpy
h_{fg}	latent heat of vaporization
h_m	overall mass transfer coefficient
k	soil permeability
k_r	relative permeability

m''	mass evaporation rate per unit volume
m''_{inj}	mass injection rate of steam per unit area
N	number of species in the liquid contaminant
Nu	Nusselt number
P	pressure
Pr	Prandtl number
p	partial pressure
p_k^*	saturation vapor pressure of pure kth species
Q	mass flow rate per unit volume
R	radius or gas constant
Re	Reynolds number
R_k	interphase mass transfer rate for the kth species
r_1	radius of curvature of liquid contaminant
r_p	radius of soil particle
S	saturation of any phase
S_r	residual saturation of the liquid phase
Sc	Schmidt number
T	temperature
T_f	temperature at steam condensation front
T_g	temperature of the gaseous phase
T_i	initial temperature far ahead of the steam condensation front
T_s	temperature of solid phase
t	time
u	average pore velocity of the gaseous phase
V	molar volume
V_{ring}	volume of a single liquid pendular ring
$\mathbf{v}_g$	velocity in the gas phase
v_c	velocity of distillation zone
v_f	velocity of steam condensation front
v_g	velocity in the gas phase
x	mole fraction in the liquid phase, or steam quality
z	distance in the vertical direction or normal to liquid film

Greek Letters

α	constant in Antoine equation, or angle of inclination of porous layer
β, γ	constants in Antoine correlation
ε	porosity of the soil
θ	defining angle for the liquid pendular ring
λ_g	thermal conductivity of the gas phase
μ	absolute viscosity

ν	kinematic viscosity
ρ	density
ϕ	associate factor
ω	molecular weight

Subscripts

evaporation	evaporation front
g	gaseous phase
k	kth fraction of the contaminant
ℓ	liquid phase
m	mixture
n	nonaqueous phase
p	particle material
s	solid phase, or steam
v	vapor
ω	aqueous phase
∞ or $-\infty$	far away downstream or upstream of the condensation front

REFERENCES

American Petroleum Institute. Laboratory Scale Gasoline Spill and Venting Experiment. Publication 7743-5. Washington, DC: API, 1980.

Armstrong JE, Frind FO, McCellan RD. Non-equilibrium mass transfer between the vapor, aqueous and solid phases in unsaturated soils during soil vapor extraction. Water Resour Res 30:355–368, 1994.

Baehr AL, Hoag GE, Marley MC. Removing volatile contaminants from the unsaturated zone by inducing advective air-phase transport. J Contam Hydrol 4:1–26, 1989.

Blevins TR, Duerksen JR, Ault JW. Light-oil steam flooding—an emerging technology. J Pet Technol 36:1115–1122, 1984.

Brock TD. Biology of Microorganisms. Englewood Cliffs, NJ: Prentice-Hall, 1970.

Brooks RH, Corey AT. Hydraulic properties of porous media. Hydrology Paper 3. Fort Collins: Colorado State University, 1964.

Brusseau ML, Larsen T, Christensen TH. Rate-limited sorption and nonequilibrium transport of organic chemicals in low organic carbon aquifer materials. Water Resour Res 27:1137–1145, 1991.

Chowdiah P, Misra BR, Kilbane II JJ, Srivastva VJ, Hayes, TD. Foam propagation through soils for enhanced in-situ remediation. J Haz Mater 62:263–280, 1998.

Crow WL, Anderson EP, Minugh E. Subsurface Venting of Hydrocarbon Vapors Emanating from Hydrocarbon Product on Groundwater. API Publication 4410, 1985.

Fall EW, Pickens WE. In-situ Hydrocarbon Extraction. Case study. California: Converse Environmental Consultants, 1988.
Falta RW. Javendel I, Pruess K, Witherspoon PA. Density-driven flow of gas in the unsaturated zone due to the evaporation of volatile organic compounds. Water Resour Res 25:2159–2169, 1989.
Farrell DA, Larson WE. Modeling the pore structure of porous media. Water Resour Res 8:699–706, 1972.
Ferdinandy, van Vlerken MMA. Chances for biological techniques in sediment remediation. Water Sci Technol 37:345–353, 1998.
Fotinich A, Dhir VK, Lingineni S. Remediation of simulated soils contaminated with diesel. J Environ Eng 125:36–46, 1999.
Ghassemi M. Innovative in-situ treatment technologies for cleanup of contaminated sites. J Haz Mater 17:189–206, 1988.
Gierke JS, Hutzler NJ, McKenzie DB. Vapor transport in unsaturated soil columns: Implications for vapor extraction. Water Resour Res 28:323–335, 1992.
Heath WO. In situ heating to destroy and remove organics from soils. Department of Energy/Air Force Joint Technology Review Meeting, Soil and Groundwater Remediation, Atlanta, GA, 1990.
Hunt JR, Sitar N, Udell KS. Nonaqueous phase liquid transport and cleanup—1. Analysis of mechanisms. Water Resour Res 24:1259–1269, 1988.
Johnson PC, Kemblowski MW, Colthart JD. Quantitative analysis for the cleanup of hydrocarbon contaminated soils by in-situ soil venting. Ground Water 28:413–429, 1990.
Katsumata P. An experimental study of soil venting systems with preheated air to remediate gasoline contaminated soils. MS thesis, University of California, Los Angeles, 1992.
Katsumata P, Dhir VK. An experimental study of soil venting systems with preheated air to remediate gasoline contaminated soils. Proceedings of 28th National Heat Transfer Conference and Exhibitions, San Diego, CA, 1992.
LaGrega MD, Buckingham PL, Evans JC. Hazardous Waste Management. New York: McGraw-Hill, 1994.
Lee M. The remediation of tetrachloroethylene by surfactant flushing in both the saturated an the unsaturated zone. Environ Technol 19:1073–1083, 1998.
Lingineni S. Experimental and theoretical investigation of heat and mass transfer process in soil venting systems. PhD dissertation, University of California, Los Angeles, 1994.
Lingineni S, Dhir VK. Modeling of soil venting processes to remediate unsaturated soils. J Environ Eng 118:135–152, 1992.
Lingineni S, Dhir VK. Controlling transport processes during NAPL removal by soil venting. Adv Water Resour, 20:157–169, 1997.
Mackay D. The chemistry and modeling of soil contamination with petroleum. In: Calabrese EJ, Kostecki PT, Fleischer EJ, eds. Soils Contaminated by Petroleum: Environmental and Public Health Effects. New York: John Wiley & Sons, 1985.

Marley MC. Quantitative and qualitative analysis of gasoline fractions stripped by air for the unsaturated soil zone. MS thesis, University of Connecticut, 1985.

Marley MC, Hoag GE. Induced soil venting for recovery of gasoline hydrocarbons in the vadose zone. Proceedings of NWWA/API Conference on Petroleum Hydrocarbons and Organic Chemicals in Groundwater, 1984.

Massmann JW. Applying groundwater flow models in vapor extraction system design. J Environ Eng 115:129–149, 1989.

Miller CA. Stability of moving surfaces in fluid systems with heat and mass transport, III. Stability of displacement fronts in porous media. AIChE J 21:474–479, 1975.

Niven RK, Khalili N. In situ multiphase ("upflow washing") for the remediation of hydrocarbon contaminated sands. Can Geotech J 35:938–960, 1998.

Nunno T, Hyman J, Spawn P, Healy J, Spears C, Brown M, Jonker C. In-situ steam stripping of soils. International Fact Sheet. Assessment of international technologies for superfund applications—Technology identification and selection, EPA/600/2-89-017, 1989.

Parker JC, Lenhard RJ, Kuppuswamy T. A parametric model for constitutive properties governing multi-phase flow in porous media. Water Resour Res 23:618–624, 1987.

Rathfelder K, Yeh WW-G, Mackay D. Mathematical simulation of soil vapor extraction systems: model development and numerical examples. J Contam Hydrol 8:263–297, 1991.

Reid RC, Prausnitz JM, Sherwood TK. The Properties of Gases and Liquids. 3rd ed. New York: McGraw-Hill, 1977.

Roberts DJ, Venkataraman N, Pandharkar S. The effect of metals on biological remediation of munitions-contaminated soil. Environ Eng Sci 15:265–277, 1998.

Rose W. Volumes and surface areas of pendular rings. J Appl Phys 29:687–691, 1958.

Sepehr M, Samani ZA. In-situ soil remediation using vapor extraction wells, development and testing of a three-dimensional finite-difference model. Ground Water 31:425–445, 1993.

Sherwood TK, Pigford RL, Wilke CR. Mass Transfer. New York: McGraw Hill, 1975.

Sleep BE, Sykes JF. Modeling of transport of volatile organics in variably saturated media. Water Resour Res 25:81–92, 1989.

Smith LA, Hinchee RE. In Situ Thermal Technologies for Site Remediation. Boca Raton, FL: Lewis, 1992.

Stewart LD Jr, Udell KS. Mechanisms of residual oil displacement by steam injection. SPE Reservoir Eng 1233–1242, 1988.

Texas Research Institute. Forced Venting to Remove Gasoline Vapors from a Large Scale Model Aquifer. API Publication 4431, 1984.

Thornton JS, Wootan WL. Venting for the removal of hydrocarbon vapors from gasoline contaminated soil. J Environ Sci Health A17:31–44, 1982.

Udell KS, Stewart LD Jr. Field study of in-situ steam injection and vacuum extraction for recovery of volatile organic solvents. Sanitary Engineering and Environmental Health Research Laboratory Report 89-2. Berkeley: University of California, 1989.

Whitaker S. Forced convection heat transfer correlations for flow in pipes, past flat places, single cylinders, single spheres and for flow in packed beds and tube bundles. AIChE J 18:361–371, 1972.

White MD, Oostrom M. Modeling surfactant enhanced nonaqueous-phase liquid remediation of porous media. Soil Sci 163:931–940, 1998.

Wilkins MD, Abriola LM, Pennell KD. An experimental investigation of rate-limited nonaqueous phase liquid volatilization in unsaturated porous media: steady-state mass transfer. Water Resour Res 31:2159–2172, 1995.

Yuan ZG, Udell KS. Steam distillation of a single component hydrocarbon liquid in porous media. Int J Heat Mass Transfer 36:887–897, 1993.

19

Heat Transfer During Mold Filling in Liquid Composite Manufacturing Processes

Suresh G. Advani and Kuang-Ting Hsiao
University of Delaware, Newark, Delaware

I. INTRODUCTION

To manufacture advanced polymeric composites, one has to completely impregnate the empty spaces between a stationary bed of continuous fibers with polymer resins to form an integral component. The fibers may be glass, carbon, or Kevlar, and are converted first into a fiber preform which constitutes the porous medium. Fiber preform may be a continuous fiber strand or in tows (containing from a few hundred to 48,000 individual fibers that may be either woven or stitched in various repetitive arrangements to tailor to the desired mechanical properties, as shown in Figure 1). The resin may be either a thermoplastic or a thermosett resin. Thermoplastic resins are usually very viscous, of the order of about a million times more viscous than water, and usually difficult to impregnate in the small spaces between the fibers. Thermosett resins are about 50 to 300 times more viscous than water and relatively easier to impregnate. However, thermosetts undergo an exothermic chemical reaction and cross-link and hence are difficult to recycle. There are many different ways to manufacture these composites, depending upon the type of application, the geometry of the part, and the performance desired. For an introduction to this, the reader may refer to the following composite manufacturing books (Åström 1997; Gutowski 1997; Advani 1994).

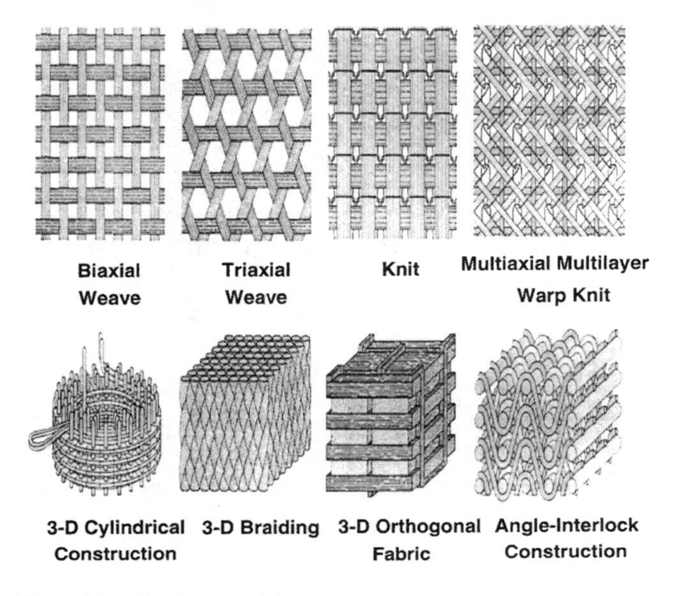

Figure 1. Preform architectures.

In this chapter we will focus on a widely used class of manufacturing processes called liquid molding. The liquid composites molding (LCM) processes include resin transfer molding (RTM), structuring reaction injection molding (SRIM), and vacuum assisted resin transfer molding (VARTM). These manufacturing techniques are used widely because they lend themselves to automation, readily reducing cost and time, and also allow one to produce nearly net shape complex-shaped composite parts. Different industries may have different expectations from the LCM process. For example, the automotive industry emphasizes the potential for high volume manufacture. On the other hand, the aerospace industry (Cochran et al. 1997; Lockheed Martin 1998; Kruckenberg and Paton 1998) is interested in the potential of producing high quality composite parts automatically. As this process is flexible enough to accommodate these needs and constraints, over the last decade research has focused on gaining a scientific understanding of the process. Many mathematical models and simulations of the process have been developed to create a virtual manufacturing environment as this would help to reduce the prototype development cost and time.

A typical LCM process can loosely be divided into five steps, as illustrated in Figure 2. The first step is to manufacture the fiber preform from either glass, carbon, or Kevlar in a form as shown in Figure 1. The second

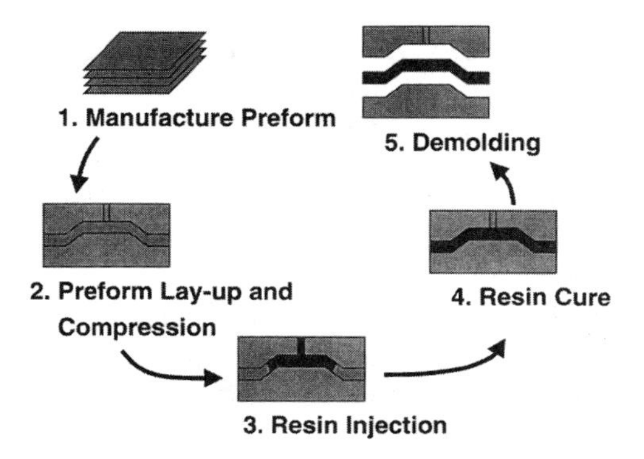

Figure 2. Manufacturing steps for a typical LCM process.

step is to put the preform in the mold cavity like a stack of pancakes. The initial thickness of the preforms is usually much higher than the final thickness. The mold is then closed and sealed, using pressure applied with a press, compacting the preforms to the desired final thickness to increase their final volume fraction in the composite. This stationary bed of compacted preforms is the porous medium. The third step is to inject a thermosett resin into the mold cavity through one or more openings, called gates, to saturate the fiber preform. One continues to inject the resin until all the empty spaces between the fibers are filled with this resin. The fourth step is the resin curing. Usually the mold is heated and a cold resin is injected inside the cavity. As the resin heats during the impregnation process, it will initiate an exothermic reaction and will cross-link, forming a solid part that is then demolded—the final step in the manufacturing cycle—as this will minimize the residual stresses caused by thermal loads.

The hot mold and preform heat the resin as it fills the mold cavity. The nonisothermal steps are the third and the fourth steps. During the third step, the resin injection, the mold and the fiber preform are usually preheated. Then the resin is injected to saturate the fiber preform and the heat diffuses from the preheated fiber preform and the mold walls into the resin. The intent is to either reduce the resin viscosity and/or trigger and expedite the curing cycle. During the fourth step, the heat is either added or removed from the mold walls to control the temperature distribution of the composite part during curing and to optimize the curing cycle.

This chapter will focus on heat transfer during the third step, the nonisothermal mold filling of the LCM processes. Calado and Advani (2000) have addressed the details of step 4. It is important to understand and predict the temperature of the resin during this filling process because, once the resin cures and cross-links, its viscosity approaches infinity and will result in an incomplete filling process and a defective part. Hence it is important to model the heat transfer phenomena during mold filling. In the book *Flow and Rheology in Composites Manufacturing* edited by S. G. Advani (1994), Tucker and Dessenberger (1994) described the governing equations one needs to solve to address the flow of heat in this composite manufacturing process.

The mold filling process is modeled as flow through fibrous porous media, using Darcy's law to predict the location of the resin front and the fluid pressure as it impregnates the fibrous preforms (Bruschke and Advani 1990, 1994; Chen et al. 1997; Lin et al. 1991; Trochu et al. 1995; Rudd and Kendall 1991; Phelan 1997). The composite parts are usually shell-like structures about 3 mm to 10 mm thick as compared to being about a meter long and wide. Hence the velocity in the thickness direction is averaged and the pressure and the flow front motion are solved only in the in-plane direction.

However, to predict the temperature of the resin and the heat flow, one needs to account for the advection of the resin locally and also for the conduction of heat in the thickness direction. Practice has converged on the use of a heat dispersion term to account for the advection of the fluid locally due to the difference in its local velocity from the average Darcy velocity. Also, due to the dominance of conduction in the thickness direction, one has to include the heat transfer in that direction to capture the physics correctly.

As this book addresses the general heat transfer phenomena and theory in porous media, we will point out the special features of the composites manufacturing process that one must translate into mathematical form if one wants to take advantage of the general theory to model the temperature and heat flow realistically in this process. The specific features and issues that we will address in this chapter that are important in composites manufacturing are:

1. The flow is usually modeled as two-dimensional but the heat transfer should be modeled as three-dimensional due to the importance of the conduction of heat in the thickness direction.
2. The local velocity profile is not calculated during the flow modeling; hence one must accurately represent the difference between the Darcy velocity and the local velocity by the use of a heat dispersion term. Hence, one needs to characterize this term as a function of preform architecture and average flow velocity.
3. When the cold resin comes in contact with the hot fibers, a question one should address is: should one invoke local equilibrium theory and assume that the fibers and the resin equilibrate to the

same temperature instantly, or should one relax this assumption and introduce more terms in the energy equation that will make the model more accurate but will require determination of more parameters? Another possibility is to treat the heat transfer separately in the solid and the liquid and have convective boundary conditions at the interface. Of course, the complication this introduces is how to handle the interface conditions correctly and represent the physics at the interface reasonably in the absence of experimental data.

4. A special case for this process is a moving flow front. How one prescribes the correct boundary condition on the front is of vital importance to what the prediction of the temperature field will be downstream.
5. The fibrous porous medium is generally a periodic porous medium due to the repetition of the fiber tow arrangements, as shown in Figure 1; hence one can take advantage of the geometry in order to predict some of the parameters such as heat dispersion and effective thermal conductivity as a function of the fiber architecture, fiber volume fraction, and the local Péclet number, which would be impossible to do in a general theory of porous media.
6. In this process, there may be a possibility that the resin will undergo an exothermic chemical reaction during the filling stage, introducing a local heat generation term that may need to be handled appropriately.
7. How does one design experiments that will reveal useful experimental data for validation of the theory? The predictions depend on reliability of the thermophysical properties of the constituents and the effective composite properties. What theories should one use to approximate the thermophysical properties of the composite?
8. One would generally like to simulate this nonisothermal process such that it can be used in a virtual environment to design the mold and the manufacturing process simultaneously in a reasonable time scale of a few seconds to a few minute. Hence, development and use of numerical and analytical methods that will address the important physics approximately is more of interest than to solve the complete problem with brute force. The key issue that the modeler of this process should remember is that the more sophisticated the model, the more parameters it needs to have to serve as input to the simulation; at present there is very scarce data and little confidence on the reliability of these parameters. Hence, a sophisticated model with incorrect parameters

can result in far less accurate predictions than a simpler approximate model with a smaller number of parameters.

In addition, there are several unresolved issues for modeling the heat transfer during the impregnation of fibrous porous media (Tucker 1996). Heat dispersion is important in such a process and must be considered (Dessenberger and Tucker 1995). Some researchers question the validity of the assumption of local thermal equilibrium in such processes (Chiu et al. 1997; Lin et al. 1991; Hsiao and Advani 1999a; Hsiao et al. 2000). Since most of the fiber preforms consist of many fiber tows, the partial saturation phenomenon in dual-scale porous media (Parseval et al. 1997) will complicate the heat transfer in such cases.

To model the heat transfer during the filling step in LCM, there are three mathematical models available. The simplest is the well-known local thermal equilibrium model (Tucker and Dessenberger 1994; Kaviany 1995; Carbonell and Whitaker 1983). This model is derived by using a volume-averaging technique and assuming that the solid phase-averaged temperature is equal to the fluid phase-averaged temperature and hence also equal to the volume-averaged temperature. The local thermal equilibrium model has one volume-averaged energy equation along with one variable, the volume-averaged temperature. In order to capture the effect of temperature difference between the solid phase and the fluid phase, the second model uses a two-phase model (two-medium treatment) (Quintard and Whitaker 1993; Zanotti and Carbonell 1984) to model the heat transfer in LCM (Chiu et al. 1997; Lin et al. 1991). The two-phase model has two phase-averaged energy equations and two coupled variables (the fluid phase-averaged temperature and the solid phase-averaged temperature). However, since the model is complicated and the macroscopic thermal coefficients are difficult to determine experimentally (Quintard et al. 1997), we do not think it is a promising approach to model the LCM processes, hence we will not discuss it in this chapter. The third, recently developed, model, the generalized model (Hsiao and Advani 1999a), which relaxes the local thermal equilibrium assumption, can be employed to model the heat transfer during the filling step in the LCM process. This model has the simplicity of the local thermal equilibrium model which has one variable, the volume-averaged temperature, and can capture the effect due to the temperature difference between the fluid phase-averaged temperature and the solid phase-averaged temperature, similarly to the two-phase model. If one assumes local thermal equilibrium, the generalized model will simplify to the local thermal equilibrium model. Hence one can use the generalized model to evaluate the local thermal equilibrium assumption in such cases. The other advantage is that the generalized model can be used in any arbitrary observation frame; hence on can formulate the energy balance boundary condition at the flow front associated with a moving observation frame. The disadvantage is that one has to determine more parameters for this model before one can predict the temperature distribution during mold filling.

Next, we will state the general mathematical model for this problem, then address the numerical approaches one has to take to simplify the solution methods. The experimental studies that have been conducted will be briefly summarized, and how their results could be used to refine the mathematical model will be illustrated. The experimental observation will show the importance of heat dispersion, the effect due to partial saturation in dual-scale porous media, and the appropriate handling of the micro-convection phenomena. A section on how one can determine some of the parameters necessary to use the generalized model by using the unit cell approach and taking advantage of periodic porous media will be introduced. Finally, in the outlook section, we will discuss what approaches will help us to resolve some of the perplexing issues in this field of heat transfer through an ordered fibrous porous medium.

II. MATHEMATICAL MODEL

A. Governing Equations for Flow of Resins

Tucker and Dessenberger (1994) and Dessenberger and Tucker (1995) have derived the governing equations for the LCM processes using the volume-averaging technique. Here we will just state them.

The local-volume-averaged continuity equation is

$$\nabla \cdot \langle \mathbf{u}_f \rangle = 0 \tag{1}$$

The operator $\langle * \rangle := \frac{1}{v}\int_V * \mathrm{d}V$ is the volume average operator. $\mathbf{u}_f$ is the velocity of the fluid phase. The momentum equation is Darcy's law

$$\langle \mathbf{u}_f \rangle = -\frac{1}{\mu}\mathbf{S} \cdot \nabla \langle P_f \rangle^f \tag{2}$$

where $\mathbf{S}$ is the permeability tensor. P_f is a modified fluid pressure, defined as

$$P_f := p_f + \rho_f g z \tag{3}$$

where p_f is the pressure in the fluid, g is the acceleration due to gravity, and z is the height above some reference point. Note that Darcy's law is an approximation, neglecting the inertial effect. Most of researchers assume that Darcy's law is valid for LCM; however, Tucker and Dessenberger (1994) performed a scaling analysis and questioned the importance of the inertial terms. For more details of the non-Darcian effects, one can refer to the work done by Vafai and Amiri (1998).

B. Governing Equations for Energy Balance

Conventionally, the local thermal equilibrium volume-averaged energy equation (Carbonell and Whitaker 1983) was widely used by modeling the heat transfer in LCM (Tucker and Dessenberger 1994; Dessenberger and Tucker 1995; Mal et al. 1998). The important characteristic of the local thermal equilibrium model is its simplicity, which comes from the local thermal equilibrium assumption. This assumption assumes that the fluid phase-averaged temperature, the solid phase-averaged temperature, and the volume-averaged temperature are equivalent. Though the local thermal equilibrium assumption does make the energy equation simple, some researchers may question its validity. Amiri and Vafai (1994, 1998) discussed the validity of the local thermal equilibrium model. The local thermal equilibrium volume-averaged energy equation is

$$\left[\varepsilon_s(\rho c_p)_s + \varepsilon_f(\rho c_p)_f\right]\frac{\partial\langle T\rangle}{\partial t} + (\rho c_p)_f\langle \mathbf{u}_f\rangle \cdot \nabla\langle T\rangle = \nabla\cdot[(\mathbf{k}_e + \mathbf{K}_D)\cdot\nabla\langle T\rangle] + \langle\dot{s}\rangle$$

where $\langle\mathbf{u}\rangle$ is the Darcy velocity and $\langle T\rangle$ is the volume-averaged temperature. The subscript and superscripts s and f represent the solid phase and fluid phase respectively. $\mathbf{k}_e$ and $\mathbf{K}_D$ are the effective thermal conductivity tensor and thermal dispersion tensor respectively. $\langle\dot{s}\rangle$ is the volume-averaged heat source term.

An advanced model, upgraded from the local thermal equilibrium model, is the two-medium treatment (Zanotti and Carbonell 1984; Quintard and Whitaker 1993; Whitaker 1999; Kuznetsov 1998). By relaxing the local thermal equilibrium assumption, they allowed the fluid phase-averaged temperature and the solid phase-averaged temperature to be the governing variables. Thus, this model requires one to solve two coupled phase-averaged energy equations (Amiri and Vafai, 1994, 1998). For example, the fluid phase-averaged energy equation is given by (Quintard et al. 1997)

$$\varepsilon_f(\rho c_p)_f\frac{\partial\langle T_f\rangle^f}{\partial t} + \varepsilon_f(\rho c_p)_f\langle\mathbf{u}_f\rangle^f\cdot\nabla\langle T_f\rangle^f - \mathbf{u}_{ff}\cdot\nabla\langle T_f\rangle^f - \mathbf{u}_{fs}\cdot\nabla\langle T_s\rangle^s = \nabla\cdot\left(\mathbf{K}_{ff}\cdot\nabla\langle T_f\rangle^f + \mathbf{K}_{fs}\cdot\nabla\langle T_s\rangle^s\right) - a_v h\left(\langle T_f\rangle^f - \langle T_s\rangle^s\right) + \langle\dot{s}_f\rangle$$

where a_v is the interfacial area per unit volume, h is the film heat transfer coefficient, and $\mathbf{u}_{ff}$, $\mathbf{u}_{fs}$ are transport coefficients in the fluid phase-averaged energy equation. $\mathbf{K}_{ff}$, $\mathbf{K}_{fs}$ are the total effective thermal conductivity tensors in the fluid phase-averaged energy equation. Note the solid phase averaged energy equation can be written in the same way.

Most researchers expected this advanced model to provide more accuracy than the local thermal equilibrium model. However, many coefficients have to be determined before we can use it to model the LCM and it is not clear how one would determine them or predict them. Moreover, the coupled two-equation system requires extensive computational effort. Though the two-medium treatment (two-phase model) is difficult to apply to LCM, some attempts have been made (Chiu et al. 1997; Lin et al. 1991). One cannot validate the models experimentally in LCM as the thermocouple can measure only one temperature and not the fluid and solid temperatures separately. Hence it makes more sense to assume some sort of average of the fluid and solid temperatures at a spatial location.

To gain in both simplicity and accuracy, one may want to have an energy balance model with only one equation but relax the local thermal equilibrium assumption. Also, note that the local thermal equilibrium model and the two-phase model were developed on the basis of the stationary observation frame, in which the solid porous medium is stationary. However, for the mold-filling step in LCM, we do have to describe the heat balance at the moving flow front and should select a moving observation which should be attached to the moving flow front. In other words, to describe the heat balance at the moving flow front for LCM processes, we need an objective volume-averaged energy balance model which can be used in any arbitrary observation frame. Next, we will introduce the generalized model that fulfills the requirements of objectivity, simplicity, accuracy for modeling heat transfer and for ease in measurement and validation in the LCM process in practice.

The generalized volume-averaged energy equation in porous media in which the thermal equilibrium between the fiber and fluid has been relaxed is given by Hsiao and Advani (1999a)

$$\begin{aligned}&\left\{\sum_{i=s,f}(\rho c_p)_i\varepsilon_i\right\}\frac{\partial\langle T\rangle}{\partial t}+\mathbf{c}_{hc}\cdot\frac{\partial(\nabla\langle T\rangle)}{\partial t}+\left\{\sum_{i=s,f}(\rho c_p)_i\langle\mathbf{u}_i\rangle\right\}\cdot\nabla\langle T\rangle\\&=\nabla\cdot[\mathbf{K}\cdot\nabla\langle T\rangle+(\mathbf{k}_{2d}\cdot\nabla)\nabla\langle T\rangle]+\sum_{i=s,f}\langle\dot{s}_i\rangle\end{aligned}\tag{4}$$

The solid forming the porous matrix (preform) and fluid (resin) values are indexed with s and f, respectively. Two volume-averaged values are related to microscopic values as follows

$$\mathbf{u}_i = \langle \mathbf{u} \rangle + \hat{\mathbf{u}}_i \tag{5}$$

where $\langle \mathbf{u} \rangle$ is the Darcy velocity and $\hat{\mathbf{u}}_i$ is the velocity deviation of phase i

$$T_i = \langle T \rangle + \mathbf{b}_i \cdot \nabla \langle T \rangle \tag{6}$$

here $\langle T \rangle$ is the volume-averaged temperature. The temperature deviation vector $\mathbf{b}_i$, which maps $T_i - \langle T \rangle$ onto $\nabla \langle T \rangle$, is periodic for a periodic porous medium and is a function of the geometry of the unit cell, the thermal properties of the materials, and the velocity field. The thermal capacity correction vector, which characterizes the sum of the difference between the volumetric heat capacity of the solid and the fluid and the difference between phase-averaged temperatures of the solid and the fluid phase, is expressed as

$$\mathbf{c}_{hc} = \sum_{i=s,f} (\rho c_p)_i \langle \mathbf{b}_i \rangle \tag{7}$$

The thermal diffusive correction vector, which characterizes the sum of the difference between the heat conductivities of the solid and the fluid and the difference between phase-averaged temperatures of the solid and the fluid phase, is expressed as

$$\mathbf{k}_{2d} = \sum_{i=s,f} k_i \langle \mathbf{b}_i \rangle \tag{8}$$

In this theory, the total effective thermal conductivity, $\mathbf{K}$, is expressed as the sum of three terms

$$\mathbf{K} = \mathbf{k}_e + \mathbf{K}_D + \mathbf{C}_{mc} \tag{9}$$

The contribution from thermal conduction (the effective thermal conductivity of the fluid-saturated porous medium) is given by

$$\mathbf{k}_e = \sum_{i=s,f} k_i \left(\varepsilon_i \mathbf{I} + \frac{1}{V} \int_{S_i} \mathbf{n} \mathbf{b}_i \mathrm{d}S \right) \tag{10}$$

here k_s and k_f refer to the thermal conductivity of the solid and the fluid, and ε_s and ε_f refer to the volume fraction of the solid and the fluid, respectively. If the porous medium is isotropic, we have $\mathbf{k}_e \equiv k_e \mathbf{I}$. Torquato (1991) suggested that the value of k_e be bounded by

$$\frac{k_f k_s}{\varepsilon_s k_f + \varepsilon_f k_s} \le k_e \le \varepsilon_s k_s + \varepsilon_f k_f \tag{11}$$

The second contribution due to heat dispersion, which characterizes the differences in the local velocity and the averaged velocity and which can be best explained by Figure 3, is given by

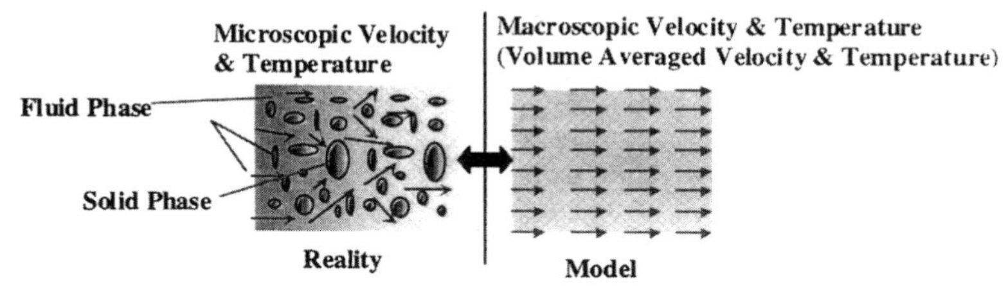

Figure 3. Definition of heat dispersion.

$$\mathbf{K}_D = \sum_{i=s,f} -\frac{(\rho c_p)_i}{V} \int_{V_i} \hat{\mathbf{u}}_i \mathbf{b}_i \mathrm{d}V \tag{12}$$

Finally, the contribution from the macro-convection along with local thermal nonequilibrium, which can be lumped into effective thermal conductivity, can be expressed as

$$\mathbf{C}_{mc} = -\langle \mathbf{u} \rangle \mathbf{c}_{hc} \tag{13}$$

Note that if local thermal equilibrium is assumed, i.e.

$$\langle T_s \rangle^s = \langle T_f \rangle^f = \langle T \rangle \Rightarrow \langle \mathbf{b}_s \rangle = \langle \mathbf{b}_f \rangle = 0 \tag{14}$$

then $\mathbf{c}_{hc}$, $\mathbf{k}_{2d}$, and $\mathbf{C}_{mc}$ will be zero. Therefore, the generalized model simplifies to the local thermal equilibrium model. Note that if the fluid and solid have $(\rho c_p)_f = (\rho c_p)_s$ and $k_f = k_s$, $\mathbf{c}_{hc}$, $\mathbf{C}_{mc}$, and $\mathbf{k}_{2d}$ will be zero even if local thermal equilibrium is not assumed (Koch et al. 1989). This is because

$$\langle \mathbf{b}_s \rangle + \langle \mathbf{b}_f \rangle = \langle \mathbf{b} \rangle = \mathbf{0} \tag{15}$$

The volume-averaged heat flow from the general theory can be derived using Fourier's law

$$\langle \mathbf{q}_{\text{total}} \rangle = \left\{ \sum_{i=s,f} (\rho c_p)_i \langle \mathbf{u}_i \rangle \right\} \langle T \rangle - \mathbf{K} \cdot \nabla \langle T \rangle - (\mathbf{k}_{2d} \cdot \nabla) \nabla \langle T \rangle \tag{16}$$

Let us consider the one-dimensional flow problem where the flow is in the x-direction. This will translate into a 2-dimensional heat transfer problem in which the x-direction is in the Darcy flow direction and the z-direction is along the thickness of the mold and perpendicular to the

Darcy flow direction. The fluid (resin) and solid (preform) are assumed to be isotropic. The geometry of the fibrous porous unit cell is assumed to be symmetric with respect to its own x-axis. Hence, the macroscopic thermal properties can be expanded in its components as

$$\left.\begin{aligned}
\mathbf{c}_{hc} &= c_{hcx}\mathbf{e}_x \\
\mathbf{k}_{2d} &= k_{2dx}\mathbf{e}_x \\
\mathbf{k}_e &= k_{exx}\mathbf{e}_x + k_{ezz}\mathbf{e}_z \approx k_e\mathbf{I} \\
\mathbf{K}_D &= K_{Dxx}\mathbf{e}_x\mathbf{e}_x + K_{Dzz}\mathbf{e}_z\mathbf{e}_z \\
\mathbf{C}_{mc} &= -\langle u\rangle_x c_{hcx}\mathbf{e}_x\mathbf{e}_x \\
\mathbf{K} &= K_{xx}\mathbf{e}_x\mathbf{e}_x + K_{zz}\mathbf{e}_z\mathbf{e}_z \\
&= (k_{exx} + K_{Dxx} - \langle u\rangle_x c_{hcx})\mathbf{e}_x\mathbf{e}_x + (k_{ezz} + K_{Dzz})\mathbf{e}_z\mathbf{e}_z
\end{aligned}\right\} \tag{17}$$

where $\mathbf{e}_x$ and $\mathbf{e}_z$ are unit vectors in the flow and transverse (thickness) direction respectively. The energy equation for this one dimensional flow and two dimensional heat transfer problem in the stationary observation frame can be rewritten as

$$\begin{aligned}
c_{hcx}\frac{\partial}{\partial t}\frac{\partial\langle T\rangle}{\partial x} + \left\{(\rho c_p)_s\varepsilon_s + (\rho c_p)_f\varepsilon_f\right\}\frac{\partial\langle T\rangle}{\partial t} + (\rho c_p)_f\langle u\rangle_x\frac{\partial\langle T\rangle}{\partial x} \\
= \left\{K_{xx}\frac{\partial^2\langle T\rangle}{\partial x^2} + K_{zz}\frac{\partial^2\langle T\rangle}{\partial z^2}\right\} \\
+ k_{2dx}\frac{\partial}{\partial x}\left(\frac{\partial^2\langle T\rangle}{\partial x^2} + \frac{\partial^2\langle T\rangle}{\partial z^2}\right) + \langle\dot{s}\rangle
\end{aligned} \tag{18}$$

The volume-averaged heat flux is

$$\begin{aligned}
\langle q_{\text{total}}\rangle = &\left\{\left[(\rho c_p)_s\langle u_s\rangle_x + (\rho c_p)_f\langle u_f\rangle_x\right]\langle T\rangle - K_{xx}\frac{\partial\langle T\rangle}{\partial x} - k_{2dx}\frac{\partial^2\langle T\rangle}{\partial x^2}\right\}\mathbf{e}_x \\
&+ \left\{\left[(\rho c_p)_s\langle u_s\rangle_z + (\rho c_p)_f\langle u_f\rangle_z\right]\langle T\rangle - K_{zz}\frac{\partial\langle T\rangle}{\partial z} - k_{2dx}\frac{\partial^2\langle T\rangle}{\partial x\partial z}\right\}\mathbf{e}_z
\end{aligned} \tag{19}$$

Usually, the mold cavity in composite processes such as RTM is very thin and long. The characteristic length along the x-direction, x_c, is very large. Hence, we assume that the effect of k_{2dx} can be neglected for such processes by comparing it with the effects of K_{xx} and K_{zz}. This saves us from the determination of an additional parameter, k_{2dx}, and use of additional boundary conditions, and we still maintain the important physics of the problem.

C. Equations for Chemical Reaction

It is possible to include the transport phenomena of the conversion of chemical species in porous media. However, to simplify the analysis, previous researchers (Tucker and Dessenberger 1994; Lin et al. 1991; Chiu et al. 1997; Tucker 1996) assumed that mass diffusion and dispersion can be neglected since the mass diffusivity is very small. Hence, the reaction equation can be expressed as

$$\varepsilon_f \frac{\partial \langle c_f \rangle^f}{\partial t} + \langle \mathbf{u}_f \rangle \cdot \nabla \langle c_f \rangle^f = \varepsilon_f R_c \left\{ \langle c_f \rangle^f, \langle T_f \rangle^f \right\} \tag{20}$$

where R_c is the reaction rate, which depends on the conversion of the chemical reaction $\langle c_f \rangle^f$ and the fluid phase-averaged temperature $\langle T_f \rangle^f$.

1. Viscosity

The viscosity of the resin is dependent on the conversion of the chemical reaction $\langle c_f \rangle^f$ and the fluid phase-averaged temperature $\langle T_f \rangle^f$. Hence, we have

$$\mu = \mu \left\{ \langle c_f \rangle^f, \langle T_f \rangle^f \right\} \tag{21}$$

Note that when the conversion approaches the gel point, the viscosity of a thermosetting resin will approach infinity.

2. Boundary Conditions

The inflow boundary conditions are

$$\langle P_f \rangle^f = P_{\text{inflow}} \qquad \text{or} \qquad \mathbf{n} \cdot \langle \mathbf{u}_f \rangle = u_{\text{inflow}} \qquad \text{at inflow} \tag{22}$$

$$\langle T \rangle = T_{\text{inflow}} \qquad \text{at inflow} \tag{23}$$

$$\langle c_f \rangle^f = \langle c_f \rangle^f_{\text{inflow}} \qquad \text{at inflow} \tag{24}$$

The wall boundary conditions are

$$\mathbf{n} \cdot \langle \mathbf{u}_f \rangle = 0 \Rightarrow \frac{\partial \langle P_f \rangle^f}{\partial_n} = 0 \qquad \frac{\text{no leakage or flow at the}}{\text{walls of the mold}} \tag{25}$$

$$\langle T \rangle = T_{\text{wall}} \qquad \text{at walls of the mold} \tag{26}$$

$$\frac{\partial \langle c_f \rangle^f}{\partial n} = 0 \qquad \text{at walls of the mold} \tag{27}$$

here T_{wall} is the temperature on the inside wall of the mold, which may vary along the wall and is different from the temperature on the outside surface of the mold. If the mold is made from a high conductivity material such as aluminum or even with hot fluid circulating through it, T_{wall} will be close to the temperature on the outside surface of the mold. On the other hand, if the mold is made of low conductivity material such as epoxy, T_{wall} will be very different from the temperature on the outside surface of the mold. The thermal effect of mold material was discussed by Young (1995). Considering there is coolant running through the mold and the temperature of the coolant is controlled, Young described the relation between the mold temperature and coolant temperature as

$$-k_{\text{wall}} \frac{\partial T_{\text{wall}}}{\partial n} = h_c (T_{\text{wall}} - T_{\text{coolant}}) \tag{28}$$

where h_c is the heat transfer coefficient of the coolant system. Another possible temperature condition is the constant heat flux boundary condition, which can be expressed as

$$\mathbf{n} \cdot [-\mathbf{K} \cdot \nabla \langle T \rangle] = q_{\text{wall}} \tag{29}$$

where q_{wall} is the heat flux from the mold wall into the porous media.

The flow front boundaries need more discussion. The flow front velocity relative to the fiber preform is

$$\mathbf{u}_{\text{front}} = \langle \mathbf{u}_f \rangle^f = \frac{\langle u \rangle_x}{\varepsilon_f} \mathbf{e}_x \tag{30}$$

The pressure at the flow front equals the atmospheric pressure or the vacuum pressure, i.e.,

$$\langle P_f \rangle^f = P_{\text{front}} = P_{\text{atm}} \text{ or } P_{\text{vacuum}} \tag{31}$$

Note that the relation between Darcy velocity and the pressure gradient can be determined from Darcy's law, i.e., Eq. (2).

An energy balance approach is used at the moving flow front. To describe the heat balance at the flow front, we have to select the moving observation frame which is attached at the flow front. Hence, the dry preform enters the flow front at a relative velocity, $-\mathbf{u}_{\text{front}}$. The dry preform is preheated to the initial temperature of the wall ($T_0 = T_{\text{wall}}$) or the temperature of the coolant ($T_0 = T_{\text{wall}} = T_{\text{coolant}}$). The volume-averaged heat flux is calculated in the flow front region where the preform is saturated

$$(\rho c_p)_s \varepsilon_s T_0 \left(-\frac{\langle u \rangle_x}{\varepsilon_f} \right) = \langle q_{\text{total}} \rangle_{x \text{ rel. to front}} \quad \text{(in the saturated region)}$$
$$= \left[(\rho c_p)_s \langle u_s \rangle_x + (\rho c_p)_f \langle u_f \rangle_x \right]_{\text{rel. to front}} \langle T \rangle - K_{xx \text{ rel. to front}} \frac{\partial \langle T \rangle}{\partial x} \tag{32}$$

where $T_0 = T_{\text{wall}}$ and $\varepsilon_s + \varepsilon_f = 1$. As $\langle u_f \rangle_{x \text{ rel. to front}} = 0$ in the same frame attached to the moving front, the energy balance can be simplified to

$$(\rho c_p)_s \varepsilon_s \left(-\frac{\langle u \rangle_x}{\varepsilon_f} \right) (T_0 - \langle T \rangle) = -K_{xx \text{ rel. to front}} \frac{\partial \langle T \rangle}{\partial x} \tag{33}$$

where

$$K_{xx \text{ rel. to front}} = k_{exx} + K_{Dxx} + \frac{\varepsilon_s}{\varepsilon_f} \langle u \rangle_x c_{hcx} \tag{34}$$

This boundary condition is usually not needed in flow through a porous medium in which one does not have a moving boundary. However, it is important to consider such moving boundaries in the manufacturing process. The experiments to validate such boundary conditions should be designed to collect temperature data during the flow of resin through heated fibrous porous media during the saturation phase.

3. Initial Conditions

For most cases, we have the following initial conditions before the resin is injected into the mold

$$\langle T \rangle = T_0 = T_{\text{wall}} = T_{\text{coolant}} \tag{35}$$

$$\langle c_f \rangle^f = 0 \tag{36}$$

It is important to conduct experiments in this field to validate models and also to find the role and value of parameters used in the generalized volume-averaged energy equation. The next section will focus on the experimental studies.

III. EXPERIMENTAL APPROACH

When modeling a nonisothermal LCM manufacturing process, it is useful to determine which terms in the general volume-averaged energy equation (Eq. (4)) play a significant role in the prediction of the temperature field. As nonisothermal conditions are more prevalent in resin transfer molding (RTM) processes in the class of LCM processes, most researchers have designed and performed experiments in RTM processes. Their intent and goal has been to understand the significance of heat dispersion as compared with conduction from the mold walls and convection in the plane of the mold during the mold filling process. The general approach has been to inject a cold resin inside a hot mold cavity containing mainly random continuous fiber preforms with thermocouples embedded inside them. The temperature history at each thermocouple is recorded during the one-dimensional filling process. The typical temperature history for a thermocouple embedded in random fiberglass preform located in the mid-plane in the mold is illustrated in Figure 4. The mold wall temperature is controlled by either holding it constant or subjecting it to a convective boundary condition by heating it with oil or water that runs through channels in the mold. The experimental setup is shown in Figure 5 (Hsiao et al. 2000).

The typical temperature history during an RTM process can be divided into a mold-filling step and a curing step after the mold filling is completed. This chapter focuses only on the mold-filling step. At the inception of the mold-filling step, the preform is dry and preheated to the same temperature as the mold wall. No sooner does the cold resin

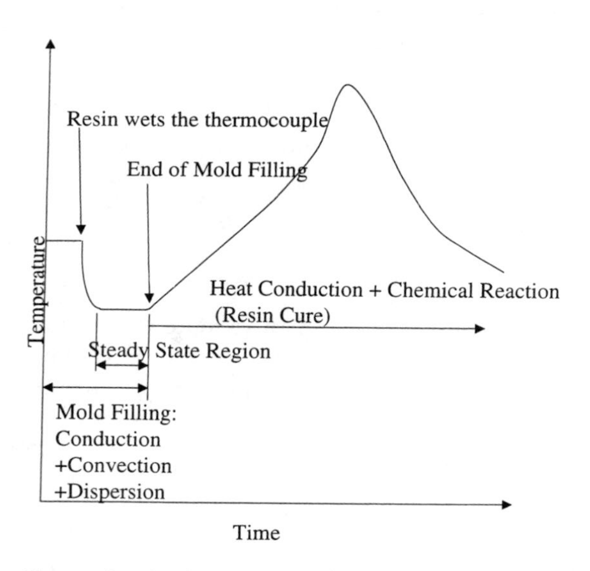

Figure 4. Typical temperature history for a thermocouple located at the mid-plane of the mold and embedded in the random fiberglass preform during the resin transfer molding process.

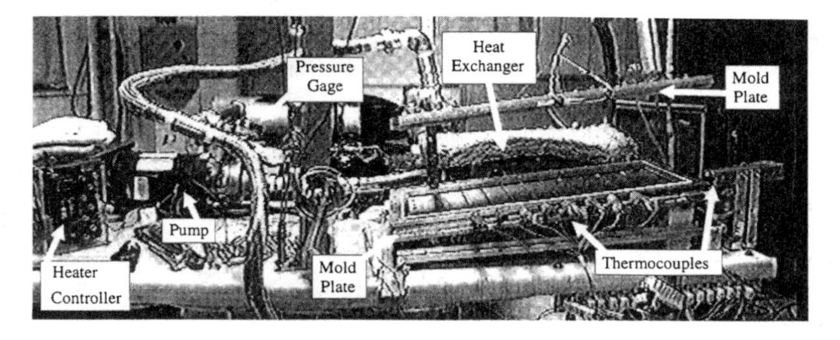

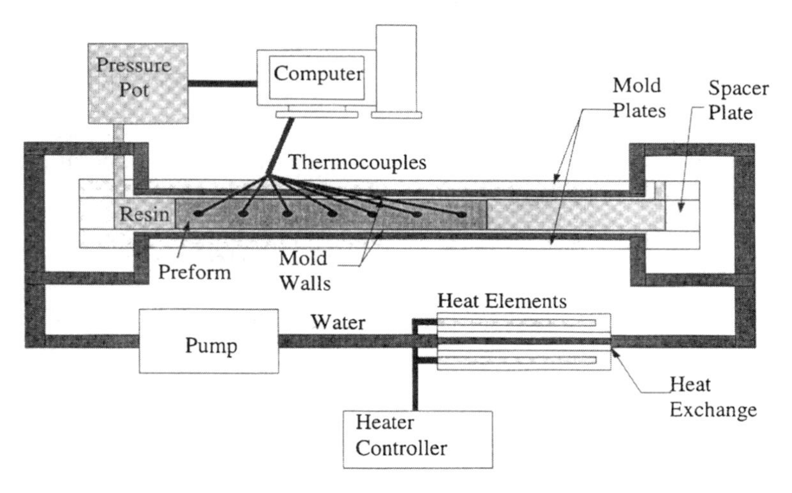

Figure 5. Experimental setup of the mold (photograph and schematic).

come in contact with the dry preform where the thermocouple is embedded, than the temperature starts to decrease. The thermocouple will register a steady state if the time to fill the mold exceeds the time for that spatial location to reach a steady state (only if the mold wall is at a constant temperature), as is the case in Figure 4. This implies that under steady-state conditions the heat conducted and dispersed in the thickness direction is gained by heat convected along the flow direction. However, in the transient region, where the temperature at the spatial location of the thermocouple drops from the initial temperature of the preform to the final steady-state value, the transient term plays an important role in addition to convection, conduction, and dispersion. It is possible that the local thermal nonequilibrium contributes during the transient time.

The larger the fraction of the time it takes from inception to steady state as compared to the mold-filling time, the more attention will the modeler have to pay to address transient issues.

To predict the mold-filling flow one needs to know the viscosity of the resin. However, the viscosity of the resin is a function of temperature, assuming that the chemical reaction is not triggered during mold filling in RTM processes. The temperature is a function of spatial location. This can cause complications for the experimentalist, who usually gets around this situation by controlling the injection flow rate of the system and hence maintaining the Darcy velocity constant.

A. Approach to Find Heat Dispersion

The temperature history data shown in Figure 4 can be collected at various embedded thermocouple junctions along the flow direction during the mold-filling process. Next, the researcher compares the steady-state temperature values with the one-dimensional flow and two-dimensional steady-state temperature solution of Eq. (4). The approach is to start with a value of zero heat dispersion in the analysis and increase the dispersion value in the mathematical model until the predicted and measured steady-state temperatures give a best fit for all the locations where the temperature was measured. Numerical or analytic solution may be used in this approach. However, the analytic solution is much more useful as the closed-form solution allows one to iterate to the optimized value in a more systematic fashion. This procedure to find the dispersion values experimentally can be repeated at different injection flow rates and hence different Darcy velocities. In this way, one can characterize the ratio of the dispersion coefficient to the thermal conductivity of the composite as a function of the Péclet number for that preform and at that particular fiber volume fraction.

The experimental studies in this field to collect such data have been basically carried out by four research groups led by Tucker (Dessenberger and Tucker 1995), Lee (Chiu et al. 1997), Gauvin (Lebrun and Gauvin 1995), and Advani (Advani et al. 1998; Hsiao et al. 2000). The effective thermal coefficients, such as heat dispersion and effective conductivity, were fitted by matching the experimental data with predicted temperatures. Most of the above researchers used numerical solution to predict the temperature history. However, as shown in Figure 4, the steady-state region existed in every RTM experimental result. Hsiao et al. (2000) developed an analytical solution to predict the temperature distribution in the steady-state region in addition to the numerical solution to predict temperature history. Below we will present the one-dimensional flow and two-dimensional steady-state heat transfer closed-form solution. Next, we will show how this solution can be used to characterize heat dispersion. After that we will focus on the transient part of the solution (here we have to use a numerical solution) and discuss the approach to predict other parameters such as K_{xx}, c_{hcx}, k_{2dx} and evaluate their importance.

B. Analytical Solution for Steady-state 1-D Flow through Porous Media

We define

$$Pe := \frac{(\rho c_p)_f \langle u \rangle d_p}{2k_f}$$

Here d_p is the diameter of the solid particle (either fiber or fiber-tow). If a low Péclet number is assumed, we can neglect the complex effects due to the inclusion of K_{xx}, c_{hcx}, k_{2dx} for a thin mold cavity. The volume-averaged energy equation (4) at steady-state simplifies to

$$(\rho c_p)_f \langle u \rangle_x \frac{\partial \langle T \rangle}{\partial x} = K_{zz} \frac{\partial^2 \langle T \rangle}{\partial z^2}, \qquad 0 < x < \infty, \qquad -z_c < z < z_c \tag{37}$$

where the thickness of the mold cavity is $2z_c$. The velocity profile is assumed uniform (one-dimensional flow) and hence $\langle u \rangle_x$ is constant. The geometry of this case is illustrated in Figure 6. For this case, we assume constant thermophysical properties. If one assumes

$$\left.\begin{aligned}
\theta &= \frac{\langle T \rangle - T_{\text{wall}}}{T_{\text{inflow}} - T_{\text{wall}}} \\
z' &= z/z_c \\
D_h &= 4z_c \\
Re &= \rho_f \langle u \rangle_x D_h / \mu \\
Pr &= \frac{\mu c_{pf}}{K_{zz}} \\
Gz &= D_h Re\, Pr/x \\
x' &= 16/Gz
\end{aligned}\right\} \tag{38}$$

for the boundary conditions of constant inflow temperature and constant wall temperature, the analytical solution is

$$\theta = \frac{4}{\pi} \sum_{n=0}^{\infty} \frac{(-1)^n}{2n+1} \cos\left[(2n+1)\frac{\pi}{2} z'\right] \exp\left[-\left(n + \frac{1}{2}\right)^2 \pi^2 x'\right] \tag{39}$$

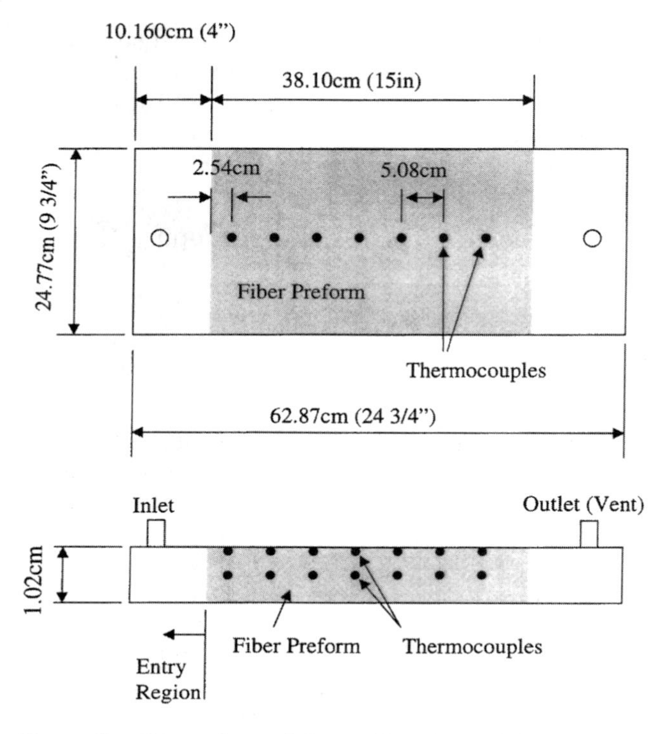

Figure 6. Dimensions of the mold cavity and the locations of thermocouples.

Note that the flow front boundary conditions do not appear in the steady-state problem. This solution provides a valuable initial guess for K_{zz} to numerically fit the experimental data for the temperature field. If we know the value of k_{ezz}, the effective thermal conductivity of the composite, then we can use this analytical approach to find the dispersion coefficient ($K_{Dzz} = K_{zz} - k_{ezz}$) by fitting the results to the experimental data. Note that these experiments have to be repeated for different flow rates to forge a relationship between dispersion coefficient and Péclet number as shown in Figure 7. Also, if the architecture of the preform changes (for example, same preform and different volume fraction, or a different preform) one would have to repeat this set of experiments. Later, we will discuss a predictive method to characterize dispersion coefficients as a function of Péclet number given the periodic structure of the porous medium. This should help in reducing the number of characterization experiments.

If the transients in this process occupy a lion's share of the filling time, it will be important to include the transients in the prediction as well. This

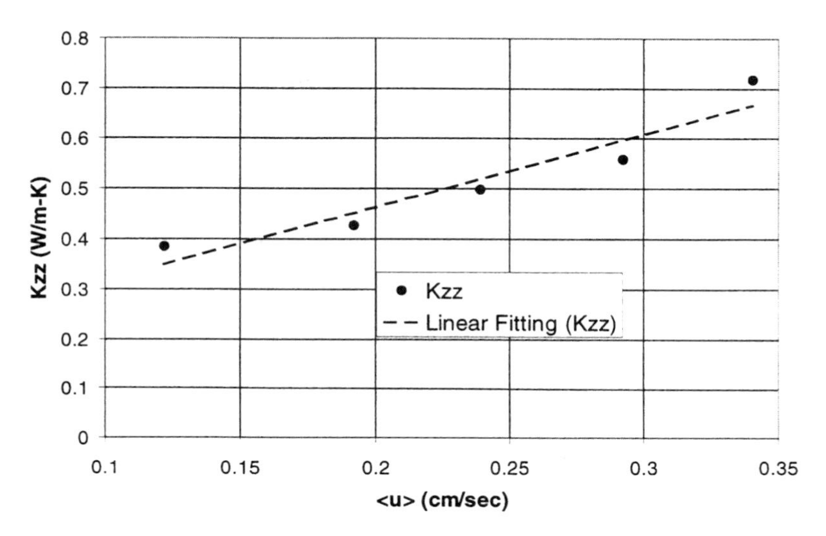

Figure 7. K_{zz} versus the Darcy velocity (or the Péclet number) for the experiments using biweave preform.

will require the researchers to characterize terms such as K_{xx} and c_{hcx} that appear during the transient phase of the heat transfer. An order-of-magnitude analysis could help in determining the significance of these terms in the process.

C. Order of Magnitude Analysis

This analytical solution can also be used to determine the characteristic length along the x direction of this problem. If we assume the characteristic length in the x direction is x_c, which corresponds to the 99% temperature drop along the centerline ($z' = 0$), we can write

$$0.01 = \frac{4}{\pi}\sum_{n=0}^{\infty}\frac{(-1)^n}{2n+1}\exp\left[-\left(n+\frac{1}{2}\right)\pi(x_c)'\right] \tag{40}$$

This allows one to solve for x_c. The characteristic thickness in the z-direction is z_c, which is the half thickness of the mold cavity. One can neglect x-direction conduction for the steady-state case if

$$K_{xx}\frac{\Delta T}{x_c^2} \ll K_{zz}\frac{\Delta T}{z_c^2} \Rightarrow \frac{K_{zz}}{K_{xx}}\left(\frac{x_c}{z_c}\right)^2 \gg 1 \tag{41}$$

However, for the transient case, we still need to find the values of K_{xx} and c_{hcx} to be used in Eq. (4) to calculate the temperature field. There is no easy way to find K_{xx} and c_{hcx}. Hence, order-of-magnitude analysis can be useful to establish the conditions under which we can neglect the effects of K_{xx} and c_{hcx} even for the transient case. Another approach is to plot the predicted temperature history using the numerical solution of Eq. (4) in which the sensitivity of K_{xx} and c_{hcx} values can be explored.

The other issue that should be explored is the flow front boundary condition. First, one should examine whether the temperature drop at the flow front is strongly influenced by x-direction conduction. In most of the experimental data, it is found that the temperature drop is insignificant, i.e., $(T_0 - \langle T\rangle)/\Delta T \ll 1$ near the resin front. Previous researchers (Tucker 1996; Mal et al. 1998) modeled such a phenomenon as a thermal wave traveling at its initial temperature T_0 right behind the flow front. From Eq. (33), we have

$$\frac{(T_0 - \langle T\rangle)}{\Delta T} = \frac{K_{xx\ \text{rel. to front}}}{(\rho c_p)_s \langle u\rangle_x x_c}\frac{\varepsilon_f}{(1-\varepsilon f)} \ll 1 \tag{42}$$

If we know the order of magnitude of both K_{xx} and $K_{xx\ \text{rel. to front}}$, we can determine the order of magnitude of c_{hcx} as

$$O\{c_{hcx}\} = O\left\{\frac{\varepsilon_f(K_{xx} - K_{xx\ \text{rel. to front}})}{\langle u\rangle_x}\right\} \tag{43}$$

By comparing the magnitude of the transient terms in the energy equation (18), we find the transient effect of c_{hcx} can be neglected if

$$|c_{hcx}| \ll \left[(\rho c_p)_s \varepsilon_s + (\rho c_p)_f \varepsilon_f\right] x_c \tag{44}$$

In summary, here is how one can determine the significance of these parameters during the transient case. First, we determine x_c from Eq. (40). A reasonable value for K_{xx} from literature (e.g., Kaviany 1995) can be used. Then we can check whether K_{xx} is negligible by using Eq. (41). If one has an estimate of the temperature drop near the flow front, one can estimate the order of $K_{xx\ \text{rel. to front}}$ from Eq. (42). Then c_{hcx} can be calculated from Eq. (43). To address the question whether the local thermal equilibrium assumption will work well for the temperature prediction for a specific case, both Eq. (44) and

$$\left|\frac{(K_{xx\ \text{rel. to front}} - K_{xx})}{(\rho c_p)_s \langle u\rangle_x x_c}\frac{\varepsilon_f}{(1-\varepsilon_f)}\right| \ll 1 \tag{45}$$

must be satisfied.

D. Experimental Setup

Most experimental setups for the collection of temperature data during the mold-filling process are similar to the experimental setup shown in Figures 5 and 6 (Advani et al. 1998). The mold wall temperature can be controlled either by electric heaters or by circulating hot water through tiny ducts inside the mold. The example presented here and seven thermocouples embedded in the preform and located along the mid-plane of the mold cavity. The injection flow rate was controlled by the computer and was held constant. Experiments were performed using a noncuring fluid to avoid exothermic influences of actual resin systems. The fluid used was a mixture of two-thirds glycerin and one-third ethylene glycol. This combination provided a viscosity similar to typical resin systems, 200–220 times more viscous than water. In addition, the thermophysical properties of the mixture were very similar to the resin systems used and are well documented. These properties are listed in Table 1. Experiments can be first completed in molds made from transparent plexiglass to ensure one-dimensional flow and guard against race-tracking (Bickerton and Advani 1999), i.e., premature exiting of the resin by flowing around instead of through the preform. During the actual experimentation, which should be carried out in a metal mold, whether race-tracking is present can be deduced by monitoring the outlet port as well as checking that the preform is fully saturated immediately after the experiment has been completed. Preforms can be weighed to calculate accurate volume fractions, and a constant flow rate should be imposed during each experiment. To determine whether dispersion is important in these types of flow, the next section details an experimental approach.

E. Significance of Heat Dispersion in the Mold Thickness Direction

To evaluate the significance of heat dispersion, an identical experiment with the porous media consisting of unidirectional fibers, first with flow in the direction of the fibers and then in the direction perpendicular to the fibers,

Table 1. Thermal material properties

Material	ρ(kg/m^2)	C_p(J/kg.°C)	k(W/m.K)
Carbon fiber	1180	712	7.8
E-glass fibers	2560	670	0.417
1/3 ethylene glycol + 2/3 glycerin	1202	2500	0.276

should be repeated. If one finds that the temperature history at the same thermocouple locations is different, then heat dispersion is playing a significant role. The larger this difference, the more significant is the dispersion coefficient. Figure 8 (Hsiao et al. 2000) presents the effect of fiber orienta-

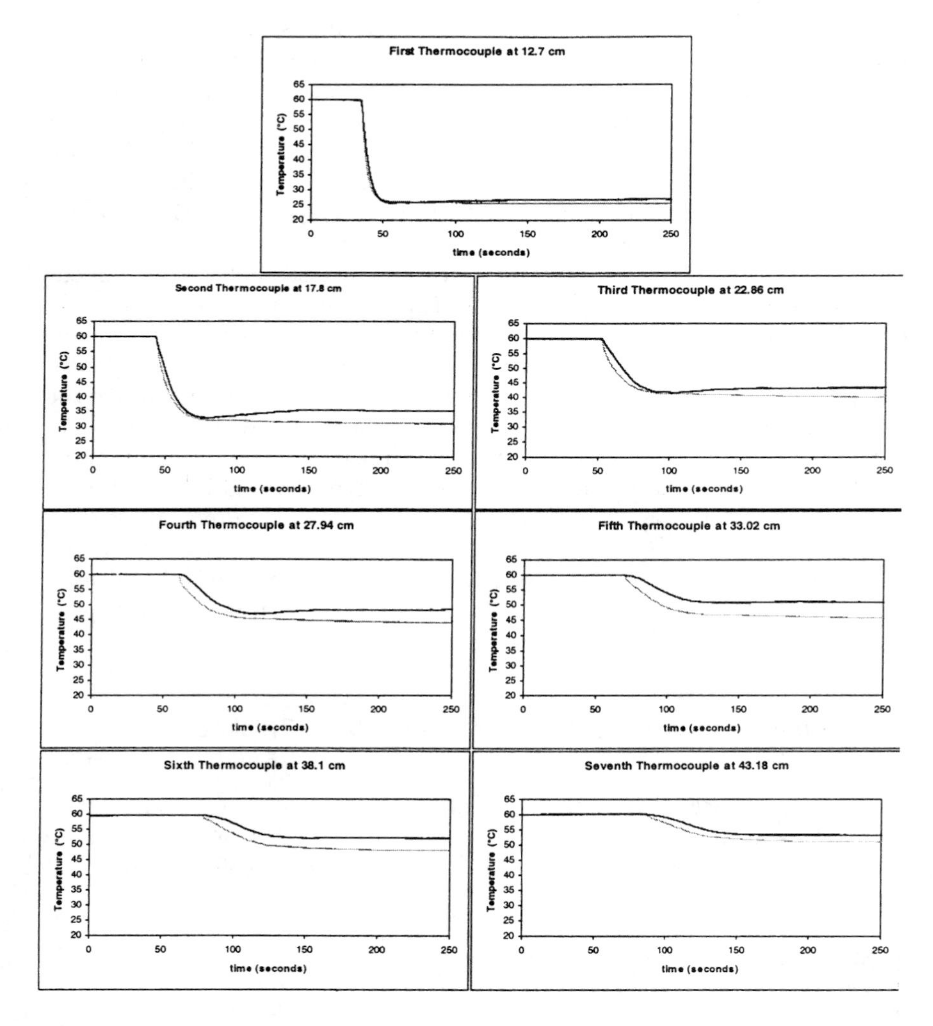

Figure 8. Comparison of the experimental temperature histories of the unidirectional fiberglass roving. Gray lines: fiber roving oriented along the flow direction. Black lines: fiber roving oriented perpendicular to the flow direction. The thermocouples are located along the mid-plane of the mold cavity.

tion on the temperature history. The temperature difference caused by the variation in fiber orientation shows that heat dispersion has a significant effect on the temperature, all other conditions being the same. By using the steady-state analytical solution with the approach stated above, one can find the K_{zz} value that best matches the steady-state temperature field along the mid-plane in the flow direction. It was found that $K_{zz} = 0.8818\,\text{W/m.K}$, which was about 3.2 times the fluid thermal conductivity, for the case when the fiber-tows were perpendicular to the flow direction and $K_{zz} = 0.7365\,\text{W/m.K}$, which was about 2.7 times the fluid thermal conductivity, for the case when the fiber-tows were oriented along the flow direction. This difference was about 20% in K_{zz} for the two cases investigated. This shows that the heat dispersion effect is not negligible and is influenced by the architecture of the porous medium.

Since $K_{zz} = K_{ezz} + K_{Dzz}$, it is important to use the correct model to predict k_{ezz}. Carbon preforms, due to their high thermal conductivity as compared with the fluid thermal conductivity, clearly show, as listed in Table 2, that the series arrangement model and the homogenization model provided by Chang (1982) predict k_{ezz} reasonably. In practice, it is interesting to examine the necessity of considering K_{Dzz} in the prediction. Figure 9 shows it is necessary to include the heat dispersion effect (K_{Dzz}) in the steady-state temperature predictions in order to match the steady-state experimental data.

Table 2. Comparison of various predictions of k_e for carbon biweave (the models were collected by Kaviany 1995)

Model	Formula	k_e(W/m.K)	k_e/k_f
parallel arrangement	$k_e = k_f\varepsilon_f + k_s\varepsilon_s$	3.59	13.01
series arrangement	$k_e = \dfrac{k_f k_s}{k_f\varepsilon_s + k_s\varepsilon_f}$	0.47	1.71
geometric mean	$k_e = (k_f)^{\varepsilon_f}(k_s)^{\varepsilon_s}$	1.17	4.24
homogenization of diffusion equation (Chang 1982) (two-dimensional periodic unit cell)	$\dfrac{k_e}{k_f} = \dfrac{(2-\varepsilon_f)k_s/k_f + 1}{2 - \varepsilon_f + k_s/k_f}$	0.39	1.40

Note: For the carbon biweave experiments, we have $\varepsilon_f = 57\%$, $0.385 \le K_{zz} \le 0.718\,\text{W/m.K}$, i.e., $1.39 \le K_{zz}/k_f \le 2.60$.

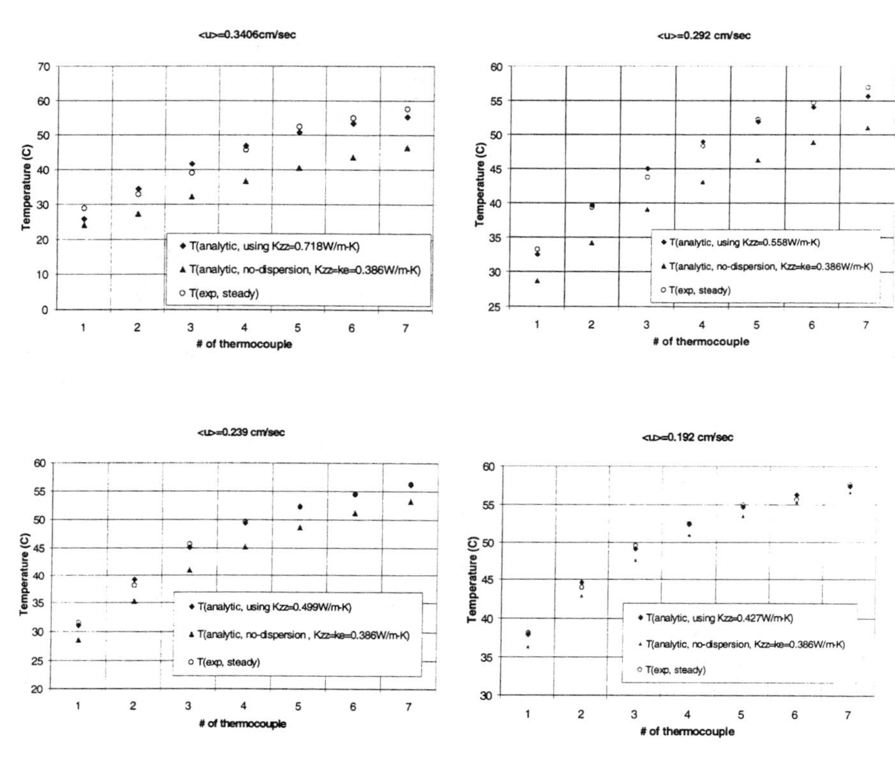

Figure 9. The significance of heat dispersion for the steady-state temperature predictions for four different Darcy velocities and hence Péclet number. The dependence of K_{zz} on the Darcy velocity must be considered to match the experimental data from the carbon biweave cases.

Note that the experiments presented above were with porous media of fiber consisting of fiber-tows. The complexity due to the dual-scale partial saturation at the flow from (Parseval et al. 1997) will mask any subtleties gained from the transient temperature analysis. The dual-scale partial saturation occurs because of the different pore sizes of the gaps between fiber-tows and the pores between single fibers. At the flow front, the resin flows through the gaps between fiber-tows but does not penetrate into the pores between single fibers due to their lower permeability. For such a partial saturation case, the volume fraction of resin continues to increase until the preform is fully saturated. However, we find no such partial saturation phenomenon if we use random fiberglass as the preform, as the length scales of all the pores are about the same. Hence, the transient temperature analysis can be best analyzed if random fiberglass is used as the preform to create the porous medium.

F. Heat Dispersion along the Flow Direction in the Random Fiberglass Preform Case

Random fiberglass is popular and commonly used in experimental investigation by almost all composites researchers (Dessenberger and Tucker 1995; Chiu et al. 1997; Lebrun and Gauvin 1995; Advani et al. 1998). The reason is that a random preform does not introduce any complications in the flow due to the dual length scale or due to race-tracking, and is also an isotropic material which will simplify the analysis considerably. Thus it is an ideal preform to analyze the transient temperature behavior. Most of the experimental results from various researchers agree qualitatively with each other. Figure 10 shows a typical temperature history from a laboratory scale experiment.

On the basis of steady-state temperature data, we can find the optimized value for K_{zz} by using Eq. (39). For example, for the case shown in Figure 10, if we assume $k_e \cong 0.29\,\mathrm{W/m.K} = 1.05k_f$, which is about the same order as the fluid conductivity by using Chang's model (1982), we find that $K_{Dzz}/k_f = 2.36$. The diameter of the fiberglass is 0.14 mm, which

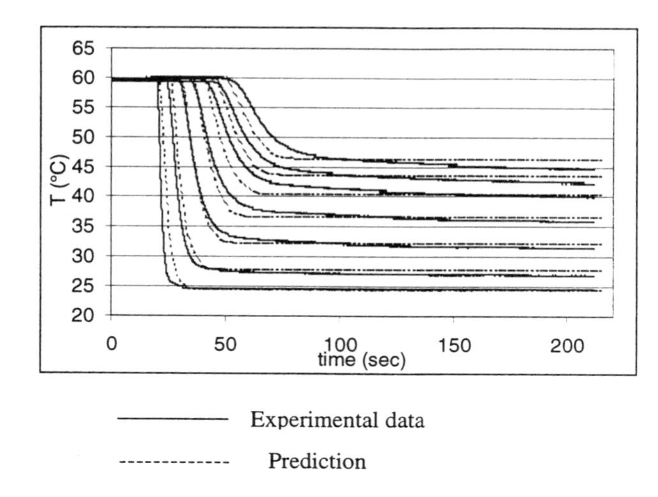

Figure 10. The centerline temperature history of all seven locations as shown in Figures 5 and 6. The material was random fiberglass with fiber volume fraction of 22% and Darcy velocity of 0.8255 cm/s.

gives $Pe := 0.5 \cdot d_p \langle u \rangle / \alpha_f = 6.28$. From the experimental investigations collected by Fried and Combarnous (1971), we get $K_{Dzz}/k_f \simeq 1.8$ and $K_{Dxx}/k_f \simeq 17.8$ for the corresponding Péclet number range. Hence the K_{Dzz} value found by this method is close to the result from Fried and Combarnous. This is encouraging, and we can proceed to perform the transient state analysis. With K_{Dzz} value known, one can now increase K_{Dxx} values within the predicted range into the finite difference code. The result was that K_{Dxx} has almost no effect on the temperature history. This is expected from the order-of-magnitude analysis applied to the energy equations. For example, when $Pe = 6.28$, i.e., $\langle u \rangle_x = 0.826\,\text{cm/s}$, we calculate $x_c = 1.429\,\text{m}$ from Eq. (40). Note that, as $z_c = 0.0102/2 = 0.005\,\text{m}$, we find from Eq. (41) that

$$\frac{K_{zz}}{K_{xx}} \left(\frac{x_c}{z_c} \right)^2 = 14{,}775 \gg 1 \tag{46}$$

This implies that K_{xx} has no significance in the energy equation, i.e., Eq. (18). If we assume local thermal equilibrium is true here, i.e., that $c_{hcx} = 0 \Rightarrow K_{xx\ \text{rel. to front}} = K_{xx}$, we can use Eq. (42) to check the significance of K_{xx} at the flow front

$$\frac{(T_0 - \langle T \rangle)}{\Delta T} = \frac{K_{xx\ \text{rel. to front}}}{(\rho c_p)_s \langle u \rangle_x x_c} \frac{\varepsilon_f}{1 - \varepsilon_f} \cong \frac{K_{xx}}{(\rho c_p)_s \langle u \rangle_x x_c} \frac{\varepsilon_f}{1 - \varepsilon_f} = 0.0033 \ll 1 \tag{47}$$

For this case, Eq. (47) shows K_{xx} has no significance at the flow front boundary either. Thus, the numerical study agrees with the order-of-magnitude analysis and shows that K_{xx} and K_{Dxx} have no significant effect on the temperature history for a typical lab-scale RTM experiment which uses random fiberglass preform as the porous medium. Note that a lab-scale RTM experiment usually has a thin mold cavity and low/mid Péclet number based on the fiber diameter. Dessenberger and Tucker (1995) also found the insignificance of K_{xx} and K_{Dxx} in RTM by performing a similar numerical study.

G. Validity of Local Thermal Equilibrium Assumption

From the experimental data shown in Figure 11, we observed no significant temperature drop at the flow front. Hence, the conditions stated in Eqs. (42), (43), and (44) are satisfied. The transient effect of the c_{hcx} term and the effects of K_{xx} can be neglected for both the steady state and the transient state analysis for this case. In other words, the local thermal assumption is

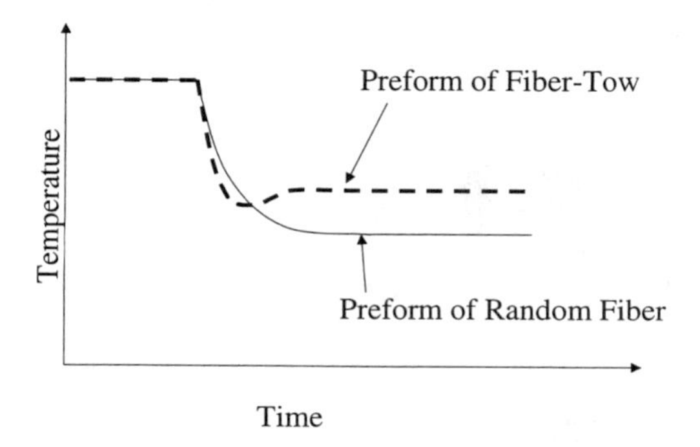

Figure 11. Different temperature histories due to different preform architectures.

valid for this case since c_{hcx} is negligible. This conclusion is based on the reasonable order of magnitude of K_{xx} and K_{zz} along with the observation of no significant temperature drop at the flow front. The most significant thermal coefficient that needs to be determined for this case is K_{zz}. Note that we can easily determine the value of K_{zz} by comparing the steady-state analytical solution and the experimental data. The finite difference method is necessary only to evaluate the effects of K_{xx} and c_{hcx}. Figure 10 shows a good agreement between the temperature predictions and the experimental data for the random fiber case at low Péclet numbers, while we consider 5–20% error may occur in the RTM experiment.

H. The Effect due to Dual-scale Partial Saturation at the Flow Front

For the fibrous preform consisting of fiber tows, because the phenomenon of dual-scale partial saturation at the flow front is not well known, it is not possible to model this problem accurately. However, the dual-scale partial saturation at the flow front does cause the pattern of temperature history to change. Figure 11 illustrates the different temperature history patterns of the porous media consisting of random fibers and of fiber-tows. We believe there are several factors that contribute to the differences. First, the fiber-tow porous media go from an air/fiber system and air/resin/fiber system to a resin/fiber system at the partial saturation region, as shown in Figure 12 (Parseval et al. 1997). In this partially saturated region, the resin flows around the fiber-tows but does not fully penetrate into the fiber-tows which consist of thousands of single fibers; thus the volume fractions of air and resin keep changing. We know that the volume fractions of air and resin will affect the volume-

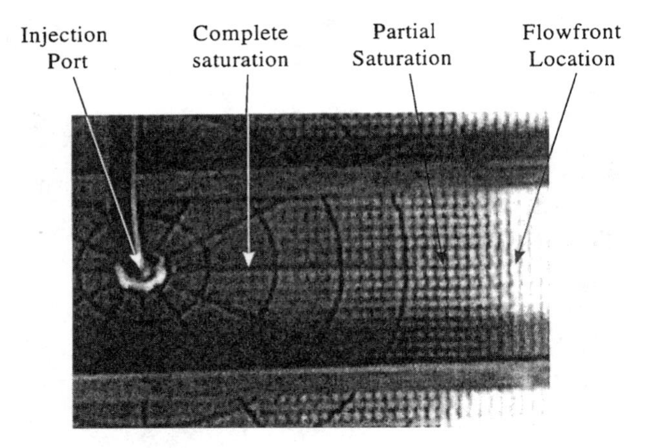

Figure 12. Partial saturation near the flow front for a preform consisting of fiberglass rovings.

averaged thermal properties. Hence, we have to be able to predict how the volume fraction changes before we can analyze the heat transfer for such processes.

Second, there are two scales in the fiber-tow porous media, i.e., the diameter of a single fiber and the diameter of a fiber-tow. We have to understand which diameter is dominant in causing the heat dispersion phenomenon in the porous media. Since the fiber-tow is commonly used as preform in industrial practice, it is of great importance to perform more in-depth studies of such porous media.

In summary, the few experimental studies performed at low Péclet number suggest that dispersion is important in such composite manufacturing methods and should be included in modeling of the process. At low Péclet number, at least for the random preform porous media, local equilibrium theory is not a bad assumption and only dispersion in the thickness direction seems to play an important role. Similar conclusions can be drawn for porous media consisting of fiber-tows after the temperature of the resin reaches a steady state. However, it is difficult to conclude what contributes to the differences in temperatures during the transient phase as compared to the random preform case. Further experimental studies can shed more light on the interaction of dispersion and other transient material parameters on the energy equation, and theoretical predictions of such material parameters can help characterize the importance of these terms.

IV. NUMERICAL METHODS TO SIMULATE THE MANUFACTURING PROCESS

For the last two decades, composite manufacturing processes in general and LCM processes in particular have seen great benefit in the use of a fundamental science base instead of a trial-and-error approach in the design and manufacture of composite structures. Mathematical models that describe the physics of the process have been developed and numerical simulations created by discretizing the governing equations using a finite element or finite volume approach over the geometry of the part to be fabricated. These equations are solved numerically with the imposed boundary conditions of pressure, flow rate, and temperature to predict the flow front location, the pressure, and the temperature field inside the mold during the filling process. If the manufacturing engineer is not satisfied with the predicted results, he or she can re-run the simulation under different conditions. This can be accomplished either by modifying the geometry of the part to be manufactured, changing the material properties of the fibers or resin, or by adjusting the processing parameters such as the locations of injection gates, the vents, the injection flow rate/pressure, the temperature of the mold, etc., by changing the boundary conditions (Liu et al. 1996). The simulation can be used not only as a design tool but also to develop a strategic controller for on-line process control (Bickerton et al. 1998). Thus the usefulness of numerical simulation is recognized widely by the manufacturing engineer and there has been a sincere effort to take advantage of the science base and reduce the trial-and-error approaches towards the manufacture of composites.

The flow simulations under isothermal conditions of the LCM processes are very well developed and widely used (Bruschke and Advani 1990, 1994; Chen et al. 1997; Lin et al. 1991; Trochu et al. 1995; Rudd and Kendall 1991; Phelan 1997). On the other hand, nonisothermal simulations are still in the early developmental stage. Although some nonisothermal simulations do exist (Bruschke and Advani 1994; Liu and Advani 1995; Lin et al. 1991; Chiu et al. 1997; Young 1995; Rudd and Kendall 1991; Mal et al. 1998), their validity is not completely established and the physics in the mathematical model is not complete. Moreover, as some resins can be processed at room temperature to fill the mold before curing, the demand and need for nonisothermal simulation is not acute for some classes of the LCM process such as vacuum assisted resins transfer molding (VARTM). However, many aerospace-grade resins require heating during mold filing, and there will be a need to address the nonisothermal issues during mold filling.

Addition of the nonisothermal model requires coupling of the energy equation with Darcy's law in addition to determination of additional parameters such as viscosity dependence on temperature and cure, parameters for cure kinetics, parameters such as effective thermal conductivity, heat capacity, and dispersion coefficient for the energy equation. Hence the numerical cost and parameter determination cost increase dramatically. In the next few sections, we will review the issues and attempts made to address nonisothermal simulation for LCM.

A. Issues

There are a few unique numerical issues that need to be addressed when developing nonisothermal simulations for LCM. First of all, in LCM the mold cavities are very thin as compared to the in-plane dimensions of the part. Hence the flow can be approximated as two-dimensional by considering the average velocity in the thickness direction. However, heat conduction and heat dispersion through the thickness are as important as convection within the plane, as revealed by the order-of-magnitude analysis in the previous section. Hence one needs to solve for the temperature field in three dimensions. Hence, the numerical coupling of two-dimensional flow and three-dimensional heat transfer requires special techniques (Liu and Advani 1995; Advani and Simacek 1998). Of course, one could solve for three-dimensional flow and heat transfer, except that the computational time will increase by an order of magnitude without much gain in accuracy.

Secondly, as most resin viscosities depend on temperature, the temperature solution and the flow solution are coupled. However, if one solves for flow in two dimensions, one also has to integrate the viscosity through the thickness before solving for the pressure field. Thirdly, to describe the physics of heat transfer in LCM, a good heat transfer model along with the correct thermal property data is necessary. To determine the thermal property data accurately for porous media is not trivial and hence raises another issue for the user, i.e., what values to use for thermal dispersion and other thermophysical properties. The numerical simulations are very useful when dealing with complex geometries. Usually the finite element model will contain several thousand elements. To conduct an isothermal simulation of the process may take the user a few minutes. However, to conduct a nonisothermal simulation make take several hours on a personal computer. If one is interested in an optimal simulation, one may have to run hundreds of these simulations and this makes nonisothermal simulation less attractive to the user. Hence, development of fast algorithms and numerical methods to speed up the simulation can prove very useful. In addition to the above issues, the moving boundary problem at the flow front introduces numerical issues such as stability and convergence (Tucker 1996).

B. Current Simulation Approaches

Lin et al. (1991) used the control volume method in their simulations. To accommodate the complex mold shape, they transferred the global coordinate system to a local coordinate system which has its z-axis always normal to the cavity surface, i.e., the walls of the mold. They implemented both the local thermal equilibrium model and the two-phase model in their code. However, they did not consider the important phenomenon of heat dispersion. Chiu et al. (1997) included the effects of heat dispersion in the same numerical code but did not validate the code. Bruschke and Advani (1994) developed their numerical code by formulating the flow simulation on the basis of the finite element/control volume method and solving the heat transfer governing equation on the basis of the finite difference/control volume method. Their method is faster than previous approaches. They proposed the convection wall boundary condition for mold walls that have channels with hot fluid circulating through them. Though they used the local thermal equilibrium model, heat dispersion was neglected in this code since it was not easy to determine. Liu and Advani (1995) developed an operator-splitting scheme for the 3-D temperature solution for RTM. The operator-splitting scheme takes advantage of the fact that advection is only important in the plane and conduction is only important in the thickness direction. They claimed that this scheme had superior stability and could use a large time step to accelerate the computation and still maintain good accuracy. Gao et al. (1995) used a finite element code to analyze heat transfer in RTM which was modeled by a local thermal equilibrium model without heat dispersion. They employed the Taylor–Galerkin method (Donea 1984) to overcome the instability associated with the classical Galerkin method when heat convection dominates in the energy equation. Aoyagi et al. (1992) used a numerical grid generation method as well as local thermal equilibrium to simulate structural reaction injection molding (SRIM). Unlike the fixed mesh used by most researchers, the numerical grid generation method generates a new finite difference mesh at each time step to solve for the moving boundary problem in an irregularly shaped domain. This method was time intensive and not very useful when the domain had multiple inserts. No dispersion effect was considered in this code. Young (1995) utilized the control volume method to study the thermal effect due to different mold materials for RTM. In this work, Young used 2-D elements to solve the flow problem and 3-D elements for heat transfer and chemical reaction. Young used the local thermal equilibrium model without heat dispersion to model the problem.

All the above simulations suffer from the fact that the computational effort is over an order of magnitude higher as compared with isothermal simulations and still there is no consensus on what values should be used for parameters such as heat dispersion coefficients. There is also a lack of experimental data to verify and validate some of the results from the simulations systematically.

Numerical simulations are most useful when they can be used to optimize the molding process design and develop strategic controllers for real-time control of the process. To achieve this goal, researchers have proposed various algorithms (Mathur et al. 1999; Youssef and Springer 1997; Yu and Young 1997). However, those optimization algorithms require more powerful numerical techniques to solve the coupled flow/heat transfer problem efficiently as the simulation may need to be run several times with varying boundary conditions.

Several heat transfer models can be implemented in the numerical simulations to various levels of sophistication. The more detailed the model, the more parameters are required to predict the temperature distribution. Those parameters are permeability, porosity, effective thermal conductivity, heat dispersion coefficient, etc. Usually, these values are measured from experiments or converted from experimental data and may differ by over 10–30% during the measurement in the same laboratory. Hence, a complex model with more uncertainties in the parameters could be less accurate than a simple model that ignores the effect of that parameter. For example, the local thermal equilibrium model may be adequate to model the RTM process. However, the accuracy of thermal properties such as effective thermal conductivity and heat dispersion still influences the predictions strongly. Hsiao et al. (2000) designed an experiment to show the significance of the heat dispersion in RTM. In their case, they also show that by varying the total effective thermal conductivity (combination of effective thermal conductivity and heat dispersion) by 30%, the predicted temperatures can change by 20%. The significance of heat dispersion was also shown by Dessenberger and Tucker (1995) and Mal et al. (1998). Note that some thermal parameters such as heat dispersion depend on the Darcy velocity. Hence this requires a built-in database or functions to determine parameters such as heat dispersion at each iteration during the computation.

In conclusion, we propose that the future of nonisothermal numerical simulation for LCM should utilize a simple heat transfer model that retains the important physics in order to enhance the speed of the simulation. The thermal properties associated with the model should be predicted from idealized studies validated with experimental measures. A stable numerical technique should be selected to solve the problem. We also need more systematic experimental data that can be used to check the nonisothermal numerical simulations.

V. PREDICTION OF MACROSCOPIC THERMAL PROPERTIES FOR PERIODIC POROUS MEDIA

One must know the values of macroscopic thermal properties such as $\mathbf{k}_e$ and $\mathbf{K}_D$ before one can predict the temperature distribution during nonisothermal impregnation in the LCM process. These macroscopic thermal properties may be determined from experiments, analytical solution, or numerical predictions (Kaviany 1995). The experimental determination is not simple or straightforward and does contain many possible sources of error. Moreover, as quantities such as $\mathbf{K}_D$ lump many microscopic phenomena, one must repeat the experiments for different fiber volume fractions, different stationary fiber bed architectures, and different injection flow velocities. The required number of experiments can amount to thousands just to characterize one quantity. The analytical approaches are usually limited to porous media with an oversimplified structure, or one that is uniformly random. The numerical approach can extend the solution to complex fibrous architectures if they are periodic in nature. Unlike the random porous media commonly used to characterize many different applications, the fibrous porous media used in the LCM can be accurately represented by periodic architectures (Figure 1). Hence, it may be beneficial to conduct numerical experiments in a periodic unit cell with various architectures to capture the effect of local convection effects in characterizing the $\mathbf{K}_D$ value. The following section will briefly disclose how this is done, with a help of a simple example.

A. Approach

To numerically predict macroscopic thermal properties such as $\mathbf{K}_D$, the first step is to discretize, using finite elements or control volume, the unit cell geometry that forms the repetitive architecture of the fibrous porous medium. A unit cell is the smallest periodic structure that can represent the important features of the structure of the porous medium. Second, the thermal properties and fluid dynamic properties of the fluid and solid(s) need to be known and are specified. Third, the governing equations for the unit cell problem should be based on or transformed from physics such as Navier–Stokes equations, continuity equation, and energy equation for each material. Fourth, the boundary condition can be approximated from the periodic assumption of the **b**-vectors. Fifth, the time dependence can be approximated from Taylor's dispersion theory (Taylor 1953, 1954). This allows most researchers (Sahraoui and Kaviany 1994; Quintard et al. 1997; Hsiao and Advani 1999b) to assume a quasi-steady state for the unit cell problem. Next, one can solve the microscopic velocity field and temperature distribution (or the related transformed field such as **b**-vectors) for the unit cell problem. Finally, one uses the average of these local quantities to compute the macroscopic thermal properties.

Based on these basic steps, there are three different methods to approach the macroscopic thermal properties for a unit cell in periodic porous media. The traditional method (Eidsath et al. 1983; Sahraoui and Kaviany 1994) is to compute the **b**-vectors for the local thermal equilibrium model, then to integrate to find the macroscopic thermal properties. In the second approach, Quintard et al. (1997) compared the **b**-fields for the two-phase model (two-medium treatment). Unlike the above two methods for solving for the transform fields, Hsiao and Advani (1999b) solved for the temperature distribution inside the unit cell and averaged the heat flux and temperature. The total effective thermal conductivity was then computed directly from the relation between average heat flux and average temperature. This method can be applied with the use of a solver for fluid flow and heat transfer. This can be a commercial package such as FIDAP, which handles complex unit cell geometries easily.

B. Direct Temperature Solution Method

Here we briefly review the direct temperature solution method proposed by Hsiao and Advani (1999b). First, we define a specific observation frame with its velocity as

$$\mathbf{u}_{pcf} = \frac{(\rho c_p)_f \varepsilon_f \langle \mathbf{u}_f \rangle^f + (\rho c_p)_s \varepsilon_s \langle \mathbf{u}_f \rangle^s}{(\rho c_p)_f \varepsilon_f + (\rho c_p)_s \varepsilon_s} \tag{48}$$

At this specific observation frame, if we further assume quasi-steady state, the volume-averaged heat flux (Eq. (16)) can be described as

$$\langle \mathbf{q}_{\text{total}} \rangle = -\mathbf{K} \cdot \nabla \langle T \rangle - (\mathbf{k}_{2d} \cdot \nabla) \nabla \langle T \rangle \tag{49}$$

If we apply a constant average temperature gradient, i.e., $\nabla \langle T \rangle \equiv \text{const}$, we have the relation between the volume-averaged heat flux and the volume-averaged temperature gradient as

$$\langle \mathbf{q}_{\text{total}} \rangle = -\mathbf{K} \cdot \nabla \langle T \rangle \tag{50}$$

The value of $\langle \mathbf{q}_{\text{total}} \rangle$ can be calculated from FIDAP for the prescribed boundary condition of constant $\nabla\langle T\rangle$ across the unit cell. The total effective thermal conductivity tensor is $\mathbf{K} = K_{xx}\mathbf{e}_x\mathbf{e}_x + K_{yy}\mathbf{e}_y\mathbf{e}_y K_{zz}\mathbf{e}_z\mathbf{e}_z$. Hence one can apply $\nabla\langle T\rangle = (\partial\langle T\rangle/\partial x)\mathbf{e}_x$ to get K_{xx}. The same treatment can be applied to solve K_{yy} and K_{zz} in the other two directions. Note that in this specific observation the macro-convection is zero. Hence this observation frame is called the "pure conduction frame". This specific velocity is also found by Zanotti and Carbonell (1984) to be associated with a pulse temperature traveling in a thick walled tube.

C. Flow Solution for the Unit Cell Problem

To compute the volume-averaged heat flux for a unit cell, we need to solve for the temperature and velocity distribution inside the unit cell. As the flow is slow, we can use the steady-state Navier–Stokes (N–S) equations along with the continuity equation to solve for the velocity field, i.e.,

$$\nabla \cdot \mathbf{u} = 0 \tag{51}$$

$$\rho\mathbf{u} \cdot \nabla\mathbf{u} = -\nabla P + \mu\nabla^2\mathbf{u} \tag{52}$$

Note, we neglect the body force in these N–S equations for the unit cell problem. For a two-dimensional unit cell, as shown in Figure 13, the velocity boundary conditions are no-flow and symmetry at both the top and the symmetry line, i.e.,

$$\frac{\partial u_x}{\partial z} = 0 \qquad \text{and} \qquad u_z = 0 \qquad \text{at both the top and the symmetry line} \tag{53}$$

and fully developed at outflow

$$\frac{\partial u_x}{\partial x} = 0 \qquad \text{and} \qquad u_z = 0 \qquad \text{at outflow} \tag{54}$$

To get the desired inflow, we can use series of precursory unit cells to observe how the flow develops. We need to input a guess for an inflow velocity profile to keep the total flow rate equal to the product of Darcy velocity and the height of the 2-D unit cell, i.e., $u_D \cdot 2h_d$. We used a fourth-order polynomial as the initial guess for the input velocity profile

$$\frac{u_{x\,\text{guessed}}(z)}{u_D} = \frac{-240}{7}\left(\frac{z}{h_d}\right)^4 + \frac{120}{7}\left(\frac{z}{h_d}\right)^2 \tag{55}$$

$$\text{and} \qquad u_z = 0 \qquad \text{for the inflow profile guess}$$

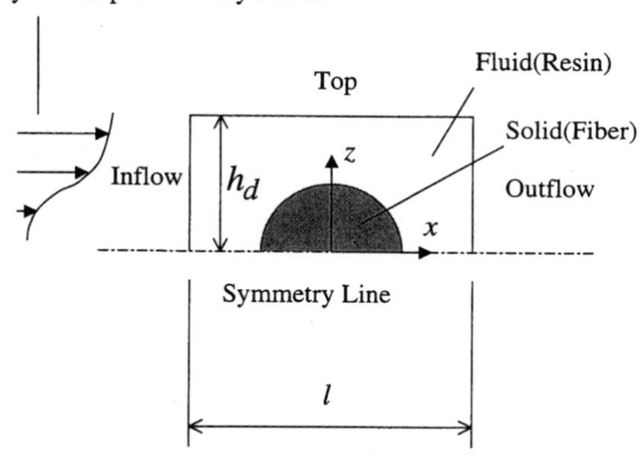

Figure 13. An in-line two-dimensional cylindrical unit cell.

and the boundary condition at the solid and fluid interface is the no-slip condition, i.e., $\mathbf{u}_s = \mathbf{u}_f$. Note that all the equations listed above are described in the stationary observation frame. One can mathematically transform them into equations for the pure conduction frame by subtracting $\mathbf{u}_{pcf}$ and using the chain rule. We found that two precursory unit cells were sufficient to make the flow periodic.

D. Total Effective Thermal Conductivity along Flow Direction (K_{xx})

A reasonable temperature distribution inside a unit cell should also satisfy the volume-averaged energy equation. This is because both the solutions of the unit cell problem and the volume-averaged energy equation are derived from the same energy balance concept. To solve for K_{xx}, we use the thermal wave-type trivial solution for the generalized volume-averaged equation (4), which is given by

$$\frac{\partial\langle T\rangle}{\partial t_{pcf}} \simeq 0 \qquad \text{and}$$
$$\nabla\langle T\rangle = \frac{\partial\langle T\rangle}{\partial x}\mathbf{e}_x = \text{const} \qquad \text{in the pure conduction frame} \tag{56}$$

where the subscript pcf means the pure conduction frame. Assuming quasi-steady state for the thermal wave, we write the microscopic energy equation for the unit cell problem

$$\rho c_p \mathbf{u} \cdot \nabla T = k\nabla^2 T \qquad \text{in the pure conduction frame} \tag{57}$$

with the semi-periodic temperature and periodic heat flux boundary conditions at the inflow and outflow

$$T(x+l, z) = T(x, z) + \frac{\partial \langle T \rangle}{\partial x} \cdot l \qquad \text{and}$$
$$\frac{\partial T}{\partial x}(x+l, z) = \frac{\partial T}{\partial x}(x, z) \qquad \text{at inflow and outflow} \tag{58}$$

At the top and symmetry line, we have the no-flux condition

$$\frac{\partial T}{\partial z} = 0 \qquad \text{at top and symmetry line} \tag{59}$$

Along the interface of fluid and solid, the no-slip temperature condition and equal heat flux are applied

$$T_s = T_F \quad \text{and} \quad \mathbf{n} \cdot k_s \nabla T_s = \mathbf{n} \cdot k_f \nabla T_f \qquad \text{along the interface} \tag{60}$$

This completes and makes the problem well defined to solve for the temperature distribution inside the unit cell. One can use flow and heat transfer solution packages such as FIDAP to find the local temperature and velocity field inside the unit cell and compute the average heat flow and determine K_{xx} using Eq. (50).

E. Total Effective Thermal Conductivity Perpendicular to Flow Direction (K_{zz})

To solve for K_{zz}, the procedure is similar and less involved. The trivial solution is given by

$$\frac{\partial \langle T \rangle}{\partial t} = 0 \qquad \text{and}$$
$$\nabla \langle T \rangle = \frac{\partial \langle T \rangle}{\partial z} \mathbf{e}_z = \text{constant} \qquad \text{in the stationary observation frame} \tag{61}$$

Hence, the energy equation for the unit cell problem reduces to

$$\rho c_p \mathbf{u} \cdot \nabla T = k\nabla^2 T \qquad \text{in the stationary observation frame} \tag{62}$$

At the inflow and outflow, we have the periodic temperature condition and equal heat flux condition.

$$T(x+l,z) = T(x,z) \qquad \text{and}$$
$$\frac{\partial T}{\partial x}(x+l,z) = \frac{\partial T}{\partial x}(x,z) \qquad \text{at inflow and outflow} \tag{63}$$

At the symmetry line, we assume

$$T(x,0) = 0 \qquad \text{along the symmetry line} \tag{64}$$

At the top, we apply a temperature difference

$$T(x,h_d) = \Delta T/2 \qquad \text{at the top} \tag{65}$$

Along the interface, we impose equal temperatures and the heat flux boundary condition similarly to the previous case. This makes the problem well posed and one can solve for the temperature distribution inside the unit cell and compute K_{zz} from Eq. (50). Figure 14 (Hsiao and Advani 1999b) shows an example of temperature distribution inside the unit cell problem found by imposing the above boundary conditions in FIDAP. This approach can be extended to three-dimensional unit cells that can account for three-dimensional woven structures of fabrics as shown in Figure 1. However, the computation effort will increase substantially.

F. Conclusions

For the in-line unit cell (Figure 15), most researchers agree that $K_{Dxx} \simeq Pe^2$ and $K_{Dzz} \simeq$ const. However, for the staggered unit cell (Figure 15) or even disordered porous medium, research shows that $K_{Dzz} \simeq Pe$. Indeed, the staggered unit cell or disordered porous medium is more realistic for the preform used in LCM.

Here we have described how to determine lumped parameters such as $\mathbf{K}_D$ by focusing on a periodic unit cell geometry. By applying the correct periodic boundary conditions, one can obtain the local velocity and temperature field. Once the local field is known, macroscopic thermal properties such as $\mathbf{K}_D$ and $\mathbf{c}_{hc}$ can be calculated as functions of Péclet number for a given geometric arrangement. One can solve this for various arrangements of unit cells based on the architecture shown in Figure 1. The values of these material parameters can then be used in nonisothermal macroscopic simulations of the LCM process, as discussed in the previous section.

VI. OUTLOOK

In this chapter, we have described how one can take the general theory of heat transfer and apply it to processes such as liquid composite molding by

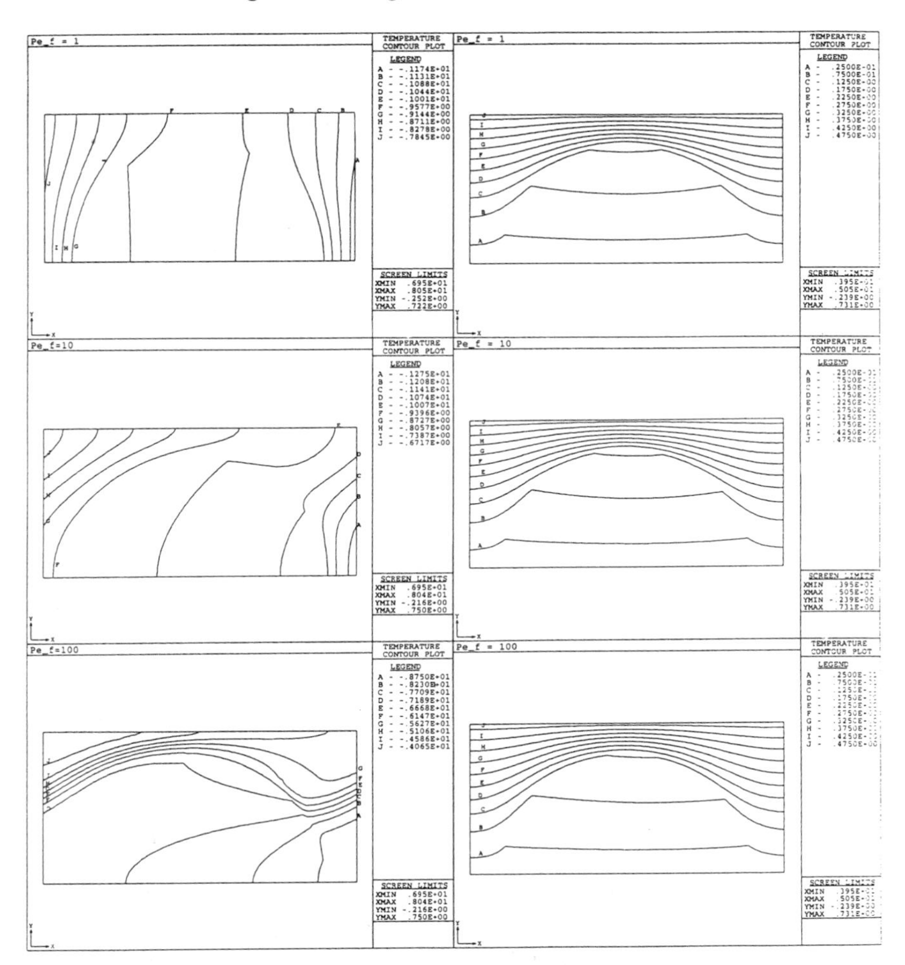

Figure 14. The microscopic temperature contours in a unit cell. Left side: the gradient of volume-averaged temperature is in the flow direction (x). Right side: the gradient of volume-averaged temperature is in the transverse direction (y). Note that $k_s/k_f = 10$, $(\rho c_p)_s/(\rho c_p)_f = 1$, $\epsilon_s = 50\%$.

taking advantage of the experiments conducted, the periodic nature of the geometry, and using order-of-magnitude analysis. The intent in composite processing is to be able to predict the temperature of the resin as it occupies the empty spaces between the porous network of hot fibers and the hot mold. These predictions are useful in design of the mold heating cycle, in

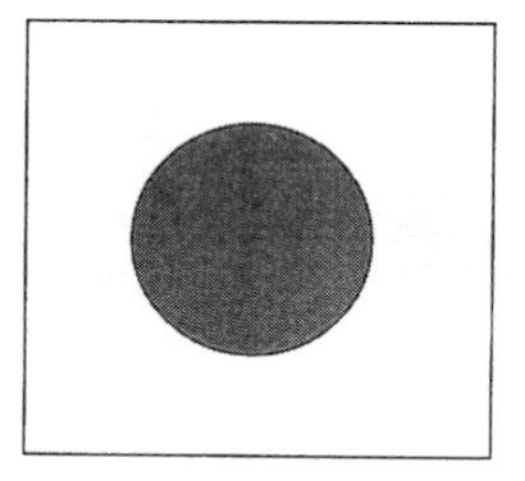

In-Line Unit Cell

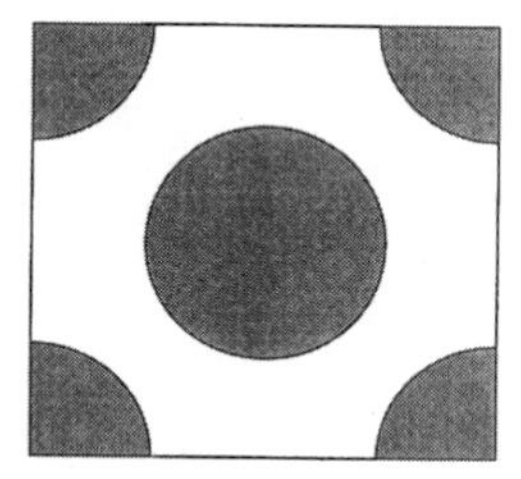

Staggered Unit Cell

Figure 15. Different configurations for unit cells.

the use of simulations in controlling the mold-filling process, and in calculation of residual stresses in the solid part.

Although reasonable progress has been made in predictions of the temperature distribution, not much has been done to validate whether these predictions are correct. The experimental results are usually matched with predictions by adjusting the heat dispersion parameter. A systematic study of how one can predict the dispersion parameter will make a big impact in this field as one can then design the network of fibers to facilitate the heat transfer uniformly once we know how the fiber architecture and fiber material properties influence heat dispersion. Carefully designed experiments that can characterize heat dispersion in various preforms should prove to be useful in validating the theory and the predictive approaches to find dispersion coefficients and other thermal parameters.

NOMENCLATURE

Roman Letters

a_v	interfacial area per unit volume (for two-phase model) (m^{-1})
$\mathbf{b}$	closure vector function (m)
c	conversion of chemical reaction
$\mathbf{c}_{hc}$	thermal capacity correction vector ($J/m^2.K$)
c_p	specific heat (J/kg.K)
$\mathbf{C}_{mc}$	correction of macro-convection (W/m.K)
d_p	particle diameter (m)
$\mathbf{e}$	unit vector
g	gravitational constant (m/s^2)
Gz	Graetz number
h	film heat transfer coefficient (for two-phase model) ($W/m^2.K$)
h_c	heat transfer coefficient ($W/m^2.K$)
h_d	half height of the 2-D unit cell (m)
$\mathbf{I}$	identity tensor
k_f	thermal conductivity of the fluid (W/m.K)
$\mathbf{k}_e$	effective thermal conductivity of the fluid saturated porous media (W/m.K)
$\mathbf{k}_{2d}$	thermal diffusive correction vector (W/K)
$\mathbf{K}$	total effective thermal conductivity tensor (W/m.K)
$\mathbf{K}_D$	effective thermal conductivity for dispersion effect (W/m.K)
$\mathbf{K}_{ff}$, $\mathbf{K}_{fs}$	total effective thermal conductivity tensors for two-phase model (W/m.K)
l	length of the 2-D unit cell (m)
$\mathbf{n}$	normal vector
n	normal direction or integer
P	pressure (Pa)
Pe	Péclet number
Pr	Prandtl number
q	heat flux (W/m^2)
$\mathbf{q}$	heat flux vector (W/m^2)
R_c	reaction rate (s^{-1})
S	surface (m^2)
$\mathbf{S}$	permeability tensor (m^2)
$\dot{s}$	heat source (W/m^3)
t	time (s)
T	temperature (K)
T_0	initial temperature (K)
$\mathbf{u}$	velocity vector (m/s)
u_D	magnitude of Darcy velocity (m/s)

u_{ff}, u_{fs} transport coefficients for the two-phase model (W/m^2.K)
$\mathbf{u}_{pcf}$ reference velocity of the pure conduction frame (m/s)
V representative volume (m^3)
x, y, z Cartesian coordinates
x_c, z_c characteristic length (m)

Greek Letters

ΔT temperature difference (K)
ε volume fraction
θ transformation of temperature
μ dynamic viscosity (Pa/s)
ρ density (kg/m^3)

Subscripts

f fluid phase
i material index
pcf pure conduction frame
s solid phase

Superscripts

f fluid phase
i material index
s sold phase
$\wedge$ deviation

Others

$\langle\ \rangle$ local volume-averaging operator

REFERENCES

Advani SG ed. Flow and Rheology in Polymer Composites Manufacturing. Amsterdam: Elsevier Science, 1994.

Advani SG, Simacek P. Modeling and simulation of flow, heat transfer and cure. In: Kruckenberg T, Paton R, eds. Resin Transfer Moulding for Aerospace Structures. Dordrecht: Kluwer Academic Publishers, 1998, Chap 8, pp. 228–281.

Advani SG, Hsiao K-T, Laudorn H. Significance of heat dispersion in the resin transfer molding process. Proceedings of the 8th Japan–US Conference on Composite Materials, Baltimore, MD, 1998, pp 87–104.

Amiri A, Vafai K. Analysis of dispersion effects and non-thermal equilibrium, non-Darcian, variable porosity, incompressible flow through porous media. Int J Heat Mass Transfer 37:939–954, 1994.

Amiri A, Vafai K. Transient analysis of incompressible flow through a packed bed. Int J Heat Mass Transfer 41:4259–4279, 1998.

Aoyagi H, Uenoyama M, Guceri SI. Analysis and simulation of structural reaction injection molding (SRIM). Int Polym Process 7:71–83, 1992.

Åström BT. Manufacturing of Polymer Composites, 1st ed. London: Chapman & Hall, 1997.

Bickerton S, Advani SG. Characterization and modeling of race tracking in liquid composite molding process. Compos Sci Technol 9:2215–2229, 1999.

Bickerton S, Stadtfeld H, Karl VS, Advani SG. Active control of resin injection for the resin transfer molding process. Proceedings of the 13th Annual Technical Conference on Composite Materials, Baltimore, 1998, pp 232–245.

Bruschke MV, Advani SG. A finite element/control volume approach to mold filling in anisotropic porous media. Polym Compos 11:398–405, 1990.

Bruschke MV, Advani SG. A numerical approach to model nonisothermal viscous flow through fibrous media with free surface. Int J Numer Methods Fluids 19:575–603, 1994.

Calado V, Advani SG. A review of cure kinetics and chemorheology for thermosett resins. To appear in: Loos A, Dave R, eds. Transport Processes in Composites, 2000 (in press).

Carbonell RG, Whitaker S. Dispersion in pulsed systems—II. Theoretical developments for passive dispersion in porous media. Chem Eng Sci 38:1795–1802, 1983.

Chang H-C. Multi-scale analysis of effective transport in periodic heterogeneous media. Chem Eng Commun 15:83–91, 1982.

Chen YF, Stelson KA, Voller VR. Prediction of filling time and vent locations for resin transfer molds. J Compos Mater 31:1141–1161, 1997.

Chiu H-T, Chen S-C, Lee LJ. Analysis of heat transfer and resin reaction in liquid composite molding. SPE ANTEC Conference, Toronto, 1997, pp 2424–2429.

Cochran R, Matson C, Thoman S, Wong D. Advanced composite processes for aerospace applications. 42nd International SAMPE Symposium, 1997, pp 635–640.

Dessenberger RB, Tucker III CL. Thermal dispersion in resin transfer molding. Polym Compos 16:495–506, 1995.

Donea J. Taylor–Galerkin method for convective transport problems. Int J Numer Methods Eng 20:101–119, 1983.

Eidsath A, Carbonell RG, Whitaker S, Herman LR. Dispersion in pulsed systems—III. Comparison between theory and experiment for packed beds. Chem Eng Sci 38: 1803–1816, 1983.

Fried JJ, Combarnous MA. Dispersion in porous media. Adv Hydrosci 7:169–282, 1971.

Gao DM, Trochu F, Gauvin R. Heat transfer analysis of non-isothermal resin transfer molding by the finite element method. Mater Manuf Process 10:57–64, 1995.

Gutowski T. Advanced Composites Manufacturing. John Wiley & Sons, 1997.

Hsiao K-T, Advani SG. A theory to describe heat transfer during laminar incompressible flow of a fluid in periodic porous media. Phys Fluids 11(7):1738–1748, 1999a.

Hsiao K-T, Advani S-G. Modified effective thermal conductivity due to heat dispersion in fibrous porous media. Int J Heat Mass Transfer 42:1237–1254, 1999b.

Hsiao K-T, Laudorn H, Advani SG. Heat dispersion due to impregnation of heated fibrous porous media with viscous fluids in composites processing. ASME J Heat Transfer (submitted), 2000.

Kaviany M. Principles of Heat Transfer in Porous Media. New York: Springer-Verlag, 1995.

Koch DL, Cox RG, Brenner H, Brady JF. The effect of order on dispersion in porous media. J Fluid Mech 200:173–188, 1989.

Kruckenberg T, Paton R, eds. Resin Transfer Moulding for Aerospace Structures. Dordrecht: Kluwer, 1998.

Kuznetsov AV. Thermal nonequilibrium forced convection in porous media. In: Ingham DB, Pop I, eds. Transport Phenomena in Porous Media. Amsterdam: Elsevier, 1998, pp 103–129.

Lebrun G, Gauvin R. Heat transfer analysis in a heated mold during the impregnation phase of the resin transfer molding process. J Mater Process Manuf Sci 4:81–104, 1995.

Lin R, Lee LJ, Liou M. Non-isothermal mold filling and curing simulation in thin cavities with preplaced fiber mats. Int Polym Process 6:356–369, 1991.

Liu B, Advani SG. Operator splitting scheme for 3-D temperature solution based on 2-D flow approximation. Comput Mech 38:74–82, 1995.

Liu D, Bickerton S, Advani SG. Modeling and simulation of RTM: Gate control, venting and dry spot prediction. Composites A 27:135–141, 1996.

Lockheed Martin. F-22 RAPTOR: Air dominance for the 21st century. Adv Mater Process 5:23–26, 1998.

Mal O, Couniot A, Dupret F. Non-isothermal simulation of the resin transfer moulding process. Composites A 29A:189–198, 1998.

Mathur R, Fink B, Advani SG. Use of genetic algorithms to optimize gate and vent locations for the resin transfer molding process. Polym Compos 20(2):167–178, 1999.

Parseval YD, Pillai KM, Advani SG. A simple model for the variation of permeability due to partial saturation in dual scale porous media. Transp Porous Media 27:243–264, 1997.

Phelan Jr FR. Simulation of the injection process in resin transfer molding. Polym Compos 18:460–476, 1997.

Quintard M, Whitaker S. One and two equation models for transient diffusion processes in two phase systems. Adv Heat Transfer 23:269–464, 1993.

Quintard M, Kaviany M, Whitaker S. Two-medium treatment of heat transfer in porous media: Numerical results for effective properties. Adv Water Resour 20:77–94, 1997.

Rudd CD, Kendall KN. Modeling non-isothermal liquid moulding processes. Proceedings of 3rd International Conference on Automated Composites, The Hague, The Netherlands, 19991, pp 30/1–30/5.

Sahraoui M, Kaviany M. Slip and no-slip temperature boundary conditions at the interface of porous, plain media: Convection. Int J Heat Mass Transfer 37:1029–1044, 1994.

Taylor GI. Dispersion of soluble matter in solvent flowing slowly through a tube. Proc R Soc (Lond) A 219:186–203, 1953.

Taylor GI. Condition under which dispersion of a solute in a stream of solvent can be used to measure molecular diffusion. Proc Soc (Lond) A 225:473–477, 1954.

Torquato S. Random heterogeneous media: Microstructure and improved bounds on effective properties. Appl Mech Rev 44:37–76, 1991.

Trochu F, Boudreault J-F, Gao DM, Gauvin R. Three-dimensional flow simulations for the resin transfer molding process. Mater Manuf Process 10:21–26, 1995.

Tucker III CL. Heat transfer and reaction issues in liquid composite molding. Polym Compos 17:60–72, 1996.

Tucker III CL, Dessenberger RB. Governing equations for flow and heat transfer in stationary fiber beds. In: Advani SG, ed. Flow and Rheology in Polymer Composites Manufacturing. Amsterdam: Elsevier Science, 1994, pp 257–323.

Vafai K, Amiri A. Non-Darcian effects in confined forced convective flows. In: Ingham DB, Pop I, eds. Transport Phenomena in Porous Media, Amsterdam: Elsevier, 1998, pp 313–329.

Whitaker S. The Method of Volume Averaging. Dordrecht: Kluwer, 1999.

Young W-B. Thermal behaviors of the resin and mold in the process of resin transfer molding. J Reinf Plast Compos 14:310–332, 1995.

Youssef MI, Springer GS. Interactive simulation of resin transfer molding. J Compos Mater 31:954–980, 1997.

Yu H-W, Young W-B. Optimal design of process parameters for resin transfer molding. J Compos Mater 31:1113–1140, 1997.

Zanotti F, Carbonell RG. Development of transport equations for multiphase systems I–III. Chem Eng Sci 39:263–311, 1984.

Index